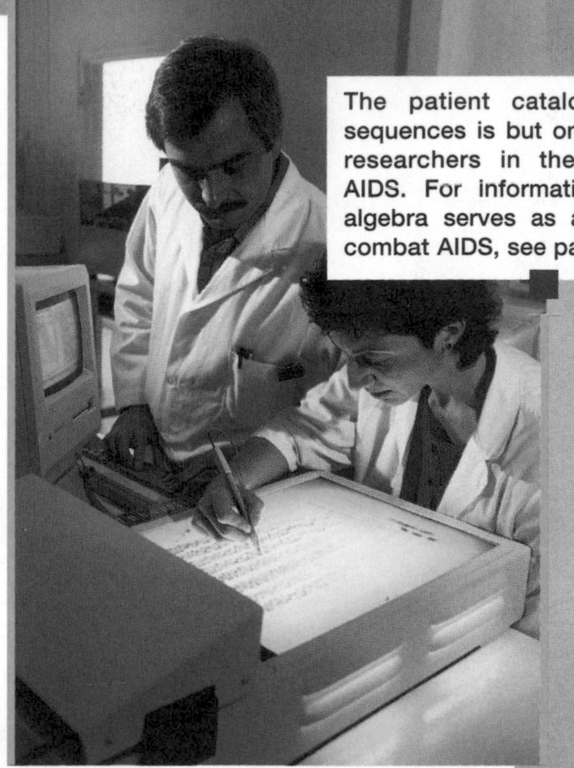

The patient cataloging of DNA sequences is but one tool used by researchers in the fight against AIDS. For information about how algebra serves as another tool to combat AIDS, see page 318.

In order to avoid excessive exposure to ultraviolet radiation, sunbathers should be aware of the intensity of the ultraviolet rays at their favorite beach. Real numbers can be used to quantify the intensity of sunlight, which varies according to the latitude of the beach. See page 254.

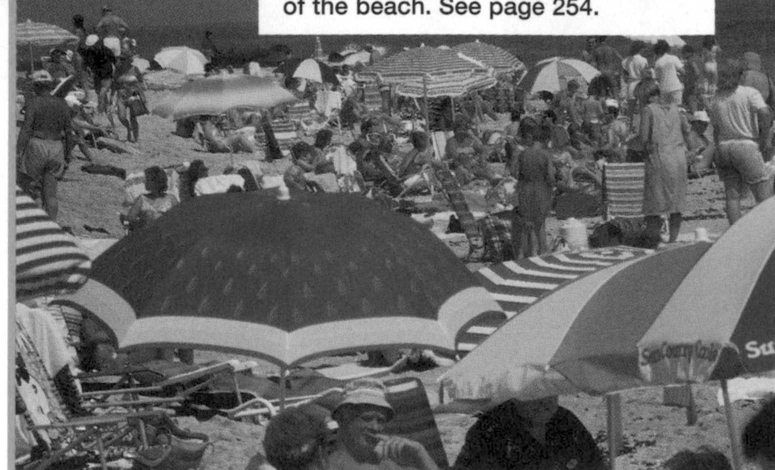

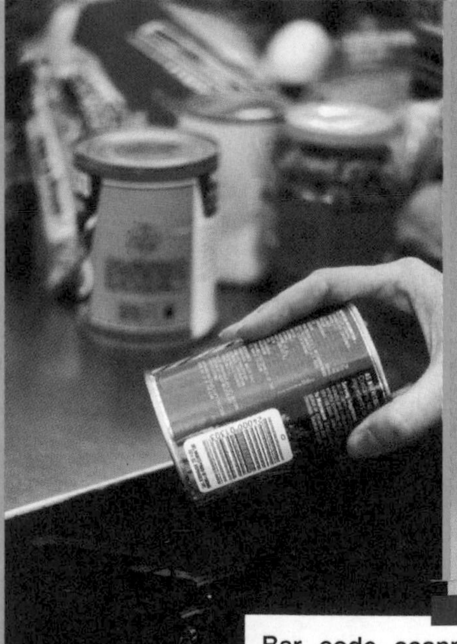

International banking relies on the dependable and secure electronic transmission of financial information. The security of these transmissions relies on the mathematics of encryption systems. To learn more about one widely used method, see page 214.

Bar code scanners not only translate the bar code to a sequence of numbers, they are also programmed to use properties of our numeration system to check that the code is valid. For another application of properties of our numeration system, see page 152.

EIGHTH EDITION

CHARLES D. MILLER

VERN E. HEEREN
American River College

E. JOHN HORNSBY, JR.
University of New Orleans

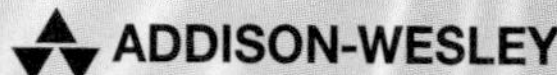

An imprint of Addison Wesley Longman, Inc.

Reading, Massachusetts • Menlo Park, California • New York • Harlow, England
Don Mills, Ontario • Sydney • Mexico City • Madrid • Amsterdam

Sponsoring Editor: Anne Kelly
Developmental Editor: Greg McRill
Project Editor: Cathy Wacaser
Design Administrator: Jess Schaal
Text and Cover Design: Lesiak/Crampton Design Inc: Lucy Lesiak
Cover Photo: Doug Armand/Tony Stone Images
Photo Researcher: Karen Koblik
Production Administrator: Randee Wire
Compositor: Interactive Composition Corporation
Printer and Binder: R.R. Donnelley & Sons Company
Cover Printer: Phoenix Color Corporation

Reprinted with corrections, April 1998

Mathematical Ideas, Eighth Edition

Library of Congress Cataloging-in-Publication Data

Miller, Charles David.
Mathematical ideas/Charles D. Miller, Vern E. Heeren, E. John Hornsby, Jr.—8th ed.
p. cm.
Includes index.
ISBN 0-673-99893-2 (Student Edition)
ISBN 0-673-98241-6 (Instructor's Annotated Exercises)
1. Mathematics. I. Heeren, Vern E. II. Hornsby, E. John. III. Title.
QA39.2.M55 1997 96-25478
510—dc20 CIP

98 99 —DOW— 9 8 7 6

Preface

Nearly thirty years after its first appearance, *Mathematical Ideas* remains one of the most popular texts in college mathematics. This eighth edition, like its predecessors, has been designed with a variety of students in mind. It is well-suited for several types of courses, including mathematics for liberal arts students, survey courses in mathematics, finite mathematics, and mathematics for prospective and in-service elementary and middle school teachers. Ample topics are included for a two-term course, yet the variety of topics and flexibility of sequence make the text suitable for shorter courses as well.

In this edition we have attempted to retain many of the features of the seventh edition, and at the same time introduce new ones that will appeal to a new generation of instructors and students. Our main objectives continue to be comprehensive coverage of topics appropriate for a mathematics survey course, logical and flexible organization, clear exposition, an abundance of examples, and well-planned exercise sets with numerous applications.

We have worked diligently to incorporate into this edition the recommendations of the *Curriculum and Evaluation Standards for School Mathematics,* prepared by the National Council of Teachers of Mathematics (NCTM), and the *Standards for Introductory College Mathematics Before Calculus,* presented by the American Mathematical Association of Two-Year Colleges (AMATYC). For example, to address the issue of cooperative learning we have introduced the feature titled "Collaborative Investigation" in each chapter. The face of mathematics education is changing drastically due to the availability of low-cost graphing technology, and we have included many screens indicating how such technology can be used to support results found analytically in the text. (It is not essential, however, that the student have a graphing calculator to study from this book; the technology component is totally optional.)

The seventh edition included an abundance of exercises new to that edition. We have retained many of them, while also adding many new applications focusing on real-life data. These include graphs, tables, and charts when appropriate. Chapters 6, 7, and 8 are especially rich in these new exercises. In addition to the usual drill exercises, we include many that test conceptual understanding. In light of the current focus on writing across the curriculum, most sets include some exercises that require the student to answer by writing a few sentences.

The popular margin notes that have appeared in previous editions have been retained and updated when necessary. New ones have been added as well. We hope that users continue to enjoy them as much as we enjoy researching and composing them.

Some topics in the first six chapters require a basic knowledge of equation solving. Depending on the background of the students, the instructor may omit topics that require this skill. On the other hand, the two algebra chapters (7 and 8) provide an excellent overview of basic algebra. Because of the flexibility of the text, they may be covered at almost any time.

Pedagogical Features

Several new or enhanced features, designed to assist students in the learning process and aid instructors in teaching, have been introduced and retained in this edition. We continue to use full color, and the art program remains one of the highlights of the book. Numerous carefully selected examples illustrate the concepts and skills being introduced. Illustrated margin notes present items of interest that explore the human dimension and historical context of mathematical ideas. "For Further Thought" boxes encourage students to share among themselves their reasoning processes to gain a deeper understanding of key mathematical concepts. Special paragraphs labeled "Problem Solving" relate the discussion of strategies to techniques that have been learned earlier or will be applied later. Most sections include a few challenging exercises that require students to extend the ideas presented in the section. These are identified in the *Instructor's Annotated Exercises* (*IAE*) by the symbol ▲. To complement the drill and application exercises, most exercise sets include a few exercises requiring a deeper understanding of the concepts introduced. In addition, many exercises require students to respond by writing a few sentences. These conceptual and writing exercises are indicated in the *IAE* by the symbols ◉ and ✍.

Overview of the Chapters

Chapter 1 (Approach to Problem Solving) introduces the student to inductive reasoning, pattern recognition, and problem-solving techniques. Chapters 2 and 3 (Sets *and* Logic) give brief overviews of set theory and elementary logic; instructors wishing to do so may cover Chapter 3 before Chapter 2. Chapter 4 (Numeration and Mathematical Systems) covers various types of numeration systems, modulo systems, and groups. In Chapter 5 (Number Theory) we present an elementary introduction to topics such as prime and composite numbers and the Fibonacci Sequence. In Chapter 6 (The Real Number System), we introduce some of the basic concepts of arithmetic of the real numbers in their various forms. Chapters 7 and 8 (The Basic Concepts of Algebra *and* Functions, Graphs, and Systems of Equations and Inequalities) form the core of the algebra component of the text, providing either an introduction to or a review of basic freshman algebra. In Chapter 9 (Geometry), the student is introduced to the basic concepts of Euclidean geometry. Several enrichment sections are included as well. Chapter 10 (Counting Methods) focuses on elementary counting techniques, in preparation for the chapter to follow. Chapter 11 (Probability) covers the basics of probability, odds, and expected value. In Chapter 12 (Statistics), we give a brief introduction to the various topics of elementary statistics. Chapter 13 (Consumer Mathematics) focuses on mathematics of interest, mortgages, and investments. In Chapter 14 (Matrices and Their Applications) we show how matrices play an important role in both theoretical and applied mathematics.

New to this Edition

In this edition we have added the following new features.

Chapter Openers A chapter opener focusing on an application of the topic to be addressed in the chapter is now included. These openers were researched by Gary Rockswold of Mankato State University. The goal of these openers is to show the student that real-world applications of mathematics do indeed exist, and hopefully they will enable the student to appreciate the relevance of mathematics in our world.

Graphing Calculator Notes While the use of graphing calculators is not a requirement for using this text, we have included many typical screens that illustrate how this modern technology can be used to supplement and support the concepts presented in the text. We often address the *limitations* of technology as well, pointing out that the calculator can only assist us if we are aware of the concepts first.

Collaborative Investigations At the end of each chapter we have provided an activity for group discussion, based on the material covered in the chapter. This is in response to instructors who like to incorporate such activities in their classes.

Increased Emphasis on Graphs, Charts, and Real Data We have included many new applications, particularly in Chapters 6, 7, 8, 12, and 13, that employ data from real-life sources. The data is often provided in the form of a graph or a chart.

A Chapter on Consumer Mathematics Chapter 13 is a newly written chapter on consumer mathematics. Such a chapter had been included for several editions but was eliminated from the seventh edition. Due to overwhelming response from users, it is back, completely rewritten for the 1990s.

A Chapter on Matrices Again, a chapter that was eliminated from the seventh edition is back. Chapter 14 is a newly written introduction to the theory and application of matrices.

Course Outline Considerations

For the most part, the chapters in the text are independent and may be covered in the order chosen by the instructor, with a few exceptions. Chapter 6 contains some material dependent on ideas found in Chapter 5. Chapter 6 should be covered before Chapter 7 if student background so dictates. Chapters 7 and 8 should be considered a "package" and covered in sequential order. A thorough coverage of Chapter 11 depends on knowledge of Chapter 10 material, although probability can be covered without learning extensive counting methods by avoiding the more difficult exercises. The latter part of Chapter 12, on inferential statistics, depends on an understanding of probability (Chapter 11).

Supplements

This edition is accompanied by an extensive supplemental package with offerings for both the instructor and the student.

For the Instructor

Instructor's Annotated Exercises This manual contains all of the exercises from the student text, with each answer printed in color next to the corresponding exercise. In addition, challenging exercises, conceptual exercises, writing exercises, and calculator exercises are identified with colored symbols so that instructors may assign these problems at their descretion. Each section of this manual is page-referenced to the corresponding exercise set in the student text.

Instructor's Test Manual This manual includes four different test forms for each chapter paralleling the chapter tests in the text, plus an additional 125–150 test items for each chapter. It also contains teaching tips and suggested course outlines.

Instructor's Solution Manual This manual contains solutions to all even-numbered exercises and a list of all conceptual, writing, and challenging exercises.

Test Generator/Editor for Mathematics with QuizMaster This is a computerized test generator that lets instructors select test questions by objective or section, or use a ready-made test for each chapter. The software is algorithm driven so that regenerated number values maintain problem types and provide a large number of test items in both multiple-choice and open-response formats for one or more test forms. The Editor lets instructors modify existing questions or create their own including graphics and accurate math symbols. Tests created with the Test Generator can be used with **QuizMaster,** which records student scores as they take tests on a single computer network, and prints reports for students, classes, or courses. CLAST and TASP versions of this package are also available (IBM, DOS, and Macintosh).

For the Student

Study Guide and Solution Manual This for-sale manual contains solutions to all odd-numbered section, extension, and appendix exercises. Also, chapter summaries review key points in the text, provide extra examples, and enumerate major topic objectives.

Guide to Florida CLAST Mathematical Competency This special for-sale study guide for use in Florida offers help in preparing for the College Level Academic Skills Test (CLAST). It includes a review of arithmetic, a CLAST pretest, supplementary exercises, and a CLAST posttest.

Computer-Assisted Tutorials The tutorials offer self-paced, interactive review in IBM and Macintosh formats. Solutions are given for all examples and exercises, as needed. Students wishing to purchase the tutorials should contact their bookstore.

Graph Explorer With this sophisticated software, available in IBM and Macintosh versions, students can graph rectangular, conic, polar, and parametric equations; zoom; transform functions; and experiment with families of equations quickly and easily. Students wishing to purchase this software should contact their bookstore.

GeoExplorer Available for IBM and Macintosh, this software package enables students to draw, measure, modify, and transform geometric shapes on the screen. Students wishing to purchase this software should contact their bookstore.

StatExplorer This software package for IBM and Macintosh computers helps students enhance their understanding of statistics by exploring a wide range of statistical representations including graphs, centers and spreads, and transformations. Students wishing to purchase this software should contact their bookstore.

Videotapes A number of videotapes relevant to the course material are available. See your local representative for details.

Acknowledgments

The seventh edition of this text was extremely successful, and we would like to once again thank the following individuals who reviewed the manuscript for that edition.

Isali Alsina, *Kean College of New Jersey*
Charles Baker, *West Liberty State College*
Dr. Carole Bauer, *Triton College*
Kathy Bavelas, *Manchester Community College*
Sally Burran, *Abraham Baldwin College*
Elizabeth Calog, *Belleville Area Community College*
David A. Capaldi, *Rhode Island College*
Beverly Connér, *Polk Community College*
Angela M. DeGroat, *SUNY–Cortland*
Sarah Donovan, *Solano Community College*
Arlene L. Donshen, Ed.D, *Widener University*
Dorris Edwards, *Motlow State Community College*
Don Eidam, *Millersville University*
Donna Ericksen, *Central Michigan University*
James Fightmaster, *Virginia Western Community College*
Anita Fleming, *Northeast Louisiana State University*
Susan Fredine, *Xavier University*
Khadiga Gamgoum, *Northern Virginia Community College*
David Gau, *California State University–Long Beach*
John Gaudio, *Waubonsee Community College*
Donald R. Goral, *Northern Virginia Community College*
Henry Gore, *Morehouse College*
Paula Gupton, *Surry Community College*
Mary Henderson, *Okaloosa–Walton Community College*
S. Paul Hess, *Prestonburg Community College*
Jeffrey C. Jones, *County College of Morris*
Vuryl J. Klassen, *California State University–Fullerton*
Donna LaLonde, *Washburn University*
Jeannie Lazaris, *East Los Angeles College*
Carole Lipkin, *Alamance Community College*
William R. Livingston, *Missouri Southern State College*
Chi-chang Lo, *Clearwater Christian College*
Philip J. Metz, *Passaic County Community College*
Diane Mitchell, *University of Northern Iowa*
Bryan Moran, *Radford University*

Darla Morgan, *Hopkinsville Community College*
David Mykita, *Morgan State University*
David Neal, *Western Kentucky University*
Virginia Powell, *Northeast Louisiana University*
Sam Robinson, *William Paterson College*
MaryKay Schippers, *Fort Hays State University*
C. Donald Smith, *Louisiana State University–Shreveport*
Carolyn Smith, *Tallahassee Community College*
Greg Stein, *College of Mt. St. Vincent*
David E. Stewart, *Dundalk Community College*
Bruce Teague, *Santa Fe Community College*
Cindie Wade, *St. Clair County Community College*
Richard Werner, *Santa Rosa Junior College*
Debra Wiens, *Rocky Mountain College*
Gail Wiltsey, *St. John's River Community College*
T. Patrick Wolff, *Montclair State College*
Susan Yellott, *Kilgore College*

We wish to also thank the following instructors who reviewed portions of the manuscript for this edition.

Cynthia Blackburn, *Hampton University*
Jonell Comerford, *Eastern Illinois University*
James Fightmaster, *Virginia Western Community College*
Richard Fritsche, *Northeast Louisiana University*
Linda L. Galloway, *Macon College*
Theresa A. Geiger, *St. Petersburg Junior College*
Donald R. Goral, *Northern Virginia Community College*
Kathleen Graziano, *St. Petersburg Junior College*
Clayton Halfyard, *Memorial University of Newfoundland*
Irene Jones, *Fullerton College*
Vijay S. Joshi, *Virginia Intermont College*
Monty Kester, *Liberty University*
Judy B. Kidd, *James Madison University*
J. Larry Martin, *Missouri Southern State College*
Maria Luisa Mendez, *Laredo Community College*
Michael A. Parker, *Norfolk State University*
Richard L. Poss, *St. Norbert College*
Rita J. Rebovich, *Edison Community College*
Susan L. Wirth, *Indian River Community College*

A project of this magnitude cannot be accomplished without the help of many other dedicated individuals. We wish to extend special thanks to our developmental editor, Greg McRill, whose mathematical and editorial expertise made this revision a pleasure to do. Thanks also to Paul Eldersveld, Ed Moura, Anne Kelly, and Cathy Wacaser for their special parts in making this edition a reality.

Vern E. Heeren
E. John Hornsby, Jr.

Contents

1

Approach to Problem Solving

Modern supercomputers can perform billions of calculations in a single second. With such computing power available, why do scientists need even faster computers to solve some problems?

Consider the question of how a steel beam supporting a bridge might fracture under extreme stress. To approach this problem, a new science called **molecular dynamics** is being used to model the motion of individual atoms in the metal. Modeling a rectangular slab of metal only 1,000 atoms wide, 2,000 atoms long, and 19 atoms thick requires over 50 hours for a supercomputer. In order to simulate the motion of 100 million atoms in a steel beam, 7.6 billion numbers must be stored in computer memory. Because an atom's movement is so rapid, scientists must calculate its velocity and location every 1 millionth of 1 billionth of a second. Each one of these steps in time takes 90 billion arithmetic calculations. Scientists have so far succeeded in modeling 600 million atoms at one time. But, a single speck of dust can contain more than a billion atoms, so only extremely small pieces of metal can currently be modeled. Modeling an entire steel beam as it fractures under stress is not within reach of today's supercomputers simply because they are too slow.

Real applications requiring high-speed computation are becoming increasingly important. From modeling the heart over long periods of time to predicting weather 24 hours in advance, real world problems are indeed very complex. Understanding them will require not only faster computers, capable of performing trillions rather than billions of calculations per second, but also new problem-solving strategies.

Source: SIAM News, Vol. 28, No. 5, 1995.

1.1 Introduction to Inductive Reasoning

The Moscow papyrus, which dates back to about 1850 B.C., provides an example of inductive reasoning employed by the early Egyptian mathematicians and has been called "the greatest Egyptian pyramid" by Eric T. Bell. Problem 14 in the document reads as follows.

> *You are given a truncated pyramid of 6 for the vertical height by 4 on the base by 2 on the top. You are to square this 4, result 16. You are to double 4, result 8. You are to square 2, result 4. You are to add the 16, the 8, and the 4, result 28. You are to take one-third of 6, result 2. You are to take 28 twice, result 56. See, it is 56. You will find it right.*

What does all this mean? A *frustum* of a pyramid is that part of the pyramid remaining after its top has been cut off by a plane parallel to the base of the pyramid. The actual formula for finding the volume of the frustum of a pyramid with a square base is

$$V = \frac{1}{3}h(b^2 + bB + B^2),$$

where b is the area of the upper base, B is the area of the lower base, and h is the height (or altitude). By following the wording of the problem, we can see that the writer is giving a method of determining the volume of the frustum of a pyramid with square bases on the

continued on page 3

While there are people who study mathematics for the sheer enjoyment of the subject, the vast majority of students of mathematics have at some point asked themselves, "When will I ever need this?" The development of mathematics can be traced to the Egyptian and Babylonian cultures as a necessity for problem solving. The application of mathematics to problem-solving situations remains the primary reason for its study.

The mathematics of the early Babylonians and Egyptians (3000 B.C.–A.D. 260) was an example of the "do thus and so" method: in order to solve a problem or perform an operation, a cookbook-like recipe was given, and it was performed over and over to solve similar problems. The classical Greek period (600 B.C.–A.D. 450) gave rise to a more formal type of mathematics, where general concepts were applied to specific problems, resulting in a structured, logical development of mathematics.

By observing that a specific method worked for a certain type of problem, the Babylonians and the Egyptians concluded that the same method would work for any similar type of problem. Such a conclusion is called a *conjecture.* A **conjecture** is an educated guess based upon repeated observations of a particular process or pattern. The method of reasoning we have just described is called *inductive reasoning.*

Inductive Reasoning

Inductive reasoning is characterized by drawing a general conclusion (making a conjecture) from repeated observations of specific examples. The conjecture may or may not be true.

In testing a conjecture obtained by inductive reasoning, it takes only one example that does not work in order to prove the conjecture false. Such an example is called a **counterexample.** Inductive reasoning provides a powerful method of drawing conclusions, but it is also important to realize that there is no assurance that the observed conjecture will always be true. For this reason, mathematicians are reluctant to accept a conjecture as an absolute truth until it is formally proved using methods of *deductive reasoning.* Deductive reasoning characterized the development and approach of Greek mathematics, as seen in the works of Euclid, Pythagoras, Archimedes, and others.

Deductive Reasoning

Deductive reasoning is characterized by applying general principles to specific examples.

Let us now look at examples of these two types of reasoning. In this chapter we will often refer to the **natural** or **counting numbers:**

$$1, 2, 3, \ldots.$$

top and bottom, with bottom base side of length 4, top base side of length 2, and height equal to 6. The very fact that these early mathematicians were able to arrive at such a method using purely inductive reasoning led Bell to his conclusion that this accomplishment outweighed that of the construction of the actual pyramids of Egypt.

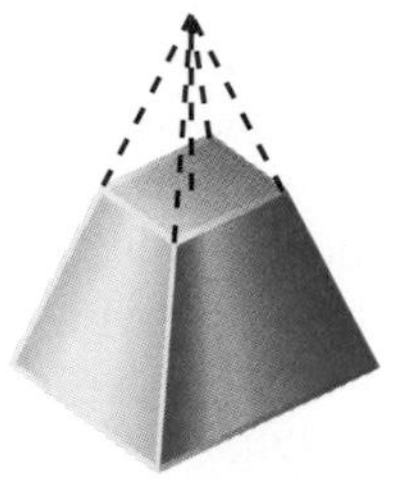

A truncated pyramid, or frustum of a pyramid

The three dots indicate that the numbers continue indefinitely in the pattern that has been established. The most probable rule for continuing this pattern is "add 1 to the previous number" and this is indeed the rule that we follow. Now consider the following list of natural numbers:

$$2, 9, 16, 23, 30.$$

What is the next number of this list? Most people would say that the next number is 37. Why? They probably reason something like this: What have 2 and 9 and 16 in common? What is the pattern?

After looking at the numbers for a while, we might see that $2 + 7 = 9$, and $9 + 7 = 16$. Is something similar true for the other numbers in this list? Do you add 7 and 16 to get 23? Do you add 7 and 23 to get 30? Yes; any number in the given list can be found by adding 7 to the preceding number. By this pattern, the next number in the list should be $30 + 7 = 37$. The number after 37 would be $37 + 7 = 44$.

You set out to find the "next number" by reasoning from your observation of the numbers in the list. You may have jumped from the facts given in the list above ($2 + 7 = 9$, $9 + 7 = 16$, and so on) to the general statement that any number in the list is 7 more than the preceding number. This is an example of inductive reasoning.

For Further Thought

The following anecdote concerning inductive reasoning appears in the first volume of the *In Mathematical Circles* series by Howard Eves (PWS-KENT Publishing Company).

> A scientist had two large jars before him on the laboratory table. The jar on his left contained a hundred fleas; the jar on his right was empty. The scientist carefully lifted a flea from the jar on the left, placed the flea on the table between the two jars, stepped back and in a loud voice said, "Jump." The flea jumped and was put in the jar on the right. A second flea was carefully lifted from the jar on the left and placed on the table between the two jars. Again the scientist stepped back and in a loud voice said, "Jump." The flea jumped and was put in the jar on the right. In the same manner, the scientist treated each of the hundred fleas in the jar on the left, and each flea jumped as ordered. The two jars were then interchanged and the experiment continued with a slight difference. This time the scientist carefully lifted a flea from the jar on the left, *yanked off its hind legs,* placed the flea on the table between the jars, stepped back and in a loud voice said, "Jump." The flea did not jump, and was put in the jar on the right. A second flea was carefully lifted from the jar on the left, its hind legs yanked off, and then placed on the table between the two jars. Again the scientist stepped back and in a loud voice said, "Jump." The flea did not jump, and was put in the jar on the right. In this manner, the scientist treated each of the hundred fleas in the jar on the left, and in no case did a flea jump when ordered. So the scientist recorded in his notebook the following induction: "A flea, if its hind legs are yanked off, cannot hear."

For Group Discussion

As a class, discuss examples from advertising on television, in newspapers, magazines, etc., that lead consumers to draw invalid conclusions.

By using inductive reasoning, we concluded that 37 was the next number in the list. But this is wrong. You were set up. You've been tricked into drawing an incorrect conclusion. Not that your logic was faulty; but the person making up the list has another answer in mind. The list of numbers

$$2, 9, 16, 23, 30$$

actually gives the dates of Mondays in June if June 1 falls on a Sunday. The next Monday after June 30 is July 7. With this pattern, the list continues as

$$2, 9, 16, 23, 30, 7, 14, 21, 28, \ldots .$$

The process you may have used to obtain the rule "add 7" in the list above reveals one main flaw of inductive reasoning. You can never be sure that what is true in a specific case will be true in general. Even a larger number of cases may not be enough. Inductive reasoning does not guarantee a true result, but it does provide a means of making a conjecture.

Deductive reasoning goes from general statements to specific situations. For example, suppose you want to show that the area of your rectangular living room is 300 square feet. You measure the living room and find it to be 15 feet by 20 feet. You then use the general formula for the area of a rectangle, area = length $\times$ width (a formula that is known to be true), and find the area of your living room: area $= 20 \times 15 = 300$ square feet. Here you reasoned from a general formula to a specific situation.

Reasoning through a problem usually requires certain *premises.* A **premise** can be an assumption, law, rule, widely held idea, or observation. Then reason inductively or deductively from the premises to obtain a **conclusion.** The premises and conclusion make up a **logical argument.**

EXAMPLE 1 Identify each premise and the conclusion in each of the following arguments. Then tell whether each argument is an example of inductive or deductive reasoning.

(a) Our house is made of red brick. Both of my next-door neighbors have red brick houses. Therefore, all houses in our neighborhood are made of red brick.

The premises are "Our house is made of red brick" and "Both of my next-door neighbors have red brick houses." The conclusion is "Therefore, all houses in our neighborhood are made of red brick." Since the reasoning goes from specific examples to a general statement, the argument is an example of inductive reasoning (although it may very well have a false conclusion).

(b) All word processors will type the letter *p*. I have a word processor. I can type the letter *p*.

Here the premises are "All word processors will type the letter *p*" and "I have a word processor." The conclusion is "I can type the letter *p*." This reasoning goes from general to specific, so deductive reasoning was used.

(c) Today is Sunday. Tomorrow will be Monday.

There is only one premise here, "Today is Sunday." The conclusion is "Tomorrow will be Monday." The fact that Monday follows Sunday is being used, even though this fact is not explicitly stated. Since the conclusion comes from general facts that apply to this special case, deductive reasoning was used. ■

The example involving dates earlier in this section illustrated how inductive reasoning may, at times, lead to false conclusions. However, in many cases inductive reasoning does provide correct results, if we look for the most *probable* answer.

EXAMPLE 2 Use inductive reasoning to determine the *probable* next number in each list below.

(a) 3, 9, 15, 21, 27

Each number in the list is obtained by adding 6 to the previous number. The probable next number is $27 + 6 = 33$.

(b) 5, 20, 80, 320

Each number after the first is obtained by multiplying the previous number by 4, so the probable next number is $320 \times 4 = 1{,}280$.

(c) 1, 1, 2, 3, 5, 8, 13, 21

Beginning with the third number in the list, each number is obtained by adding the two previous numbers in the list. That is, $1 + 1 = 2$, $1 + 2 = 3$, $2 + 3 = 5$, and so on. The probable next number in the list is $13 + 21 = 34$. (These are the first few terms of the famous *Fibonacci sequence,* covered in detail in a later chapter.)

(d) 1, 2, 4, 8, 16

It appears here that in order to obtain each number after the first, we must double the previous number. Therefore, the most probable next number is $16 \times 2 = 32$. ■

Inductive reasoning often can be used to predict an answer in a list of similarly constructed computation exercises, as shown in the next example.

EXAMPLE 3 Consider the following list of equations.

$$37 \times 3 = 111$$
$$37 \times 6 = 222$$
$$37 \times 9 = 333$$
$$37 \times 12 = 444$$

Use the list to predict the next multiplication fact in the list.

In each case, the left side of the equation has two factors, the first 37 and the second a multiple of 3, beginning with 3. The product (answer) in each case consists of three digits, all the same, beginning with 111 for 37×3. For this pattern to continue, the next multiplication fact would be $37 \times 15 = 555$, which is indeed true. ■

One of the most famous examples of possible pitfalls in inductive reasoning involves the maximum number of regions formed when chords are constructed in a circle. When two points on a circle are joined with a line segment, a *chord*

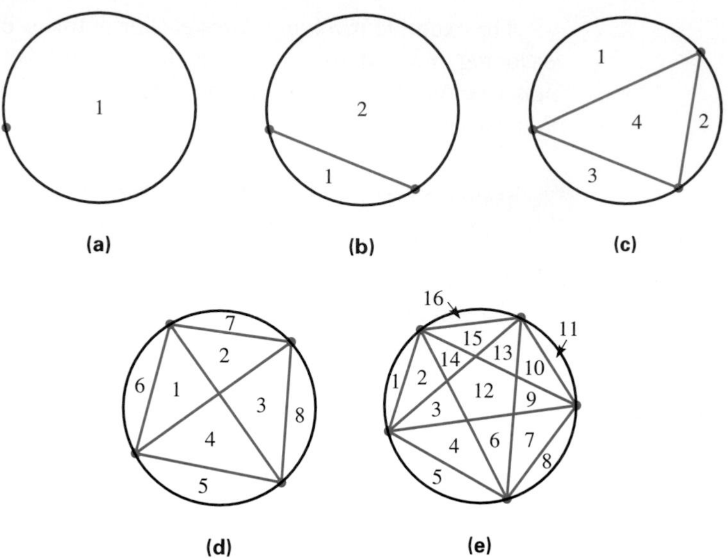

■ FIGURE 1

is formed. Locate a single point on a circle. Since no chords are formed, a single interior region is formed. See Figure 1(a).

Locate two points and draw a chord. Two interior regions are formed, as shown in Figure 1(b).

Continue this pattern. Locate three points, and draw all possible chords. Four interior regions are formed, as shown in Figure 1(c).

Four points yield 8 regions and five points yield 16 regions. See Figures 1(d) and 1(e).

The results of the preceding observations may be summarized in the following chart.

Number of Points	Number of Regions
1	1
2	2
3	4
4	8
5	16

(f)

■ FIGURE 1

The pattern formed in the column headed "Number of Regions" is the same one we saw in Example 2(d), where we predicted that the next number would be 32. It seems here that for each additional point on the circle, the number of regions doubles. A reasonable inductive conjecture would be that for six points, 32 regions would be formed. But as Figure 1(f) indicates, there are only 31 regions!

No, a region was not "missed." It happens that the pattern of doubling ends when the sixth point is considered. Adding a seventh point would yield 57 regions. The numbers obtained here are

$$1, 2, 4, 8, 16, 31, 57.$$

It has been shown that for n points on the circle, the number of regions is given by the formula

$$\frac{n^4 - 6n^3 + 23n^2 - 18n + 24}{24}.^*$$

As indicated earlier, not until a general relationship is proved can one be sure about a conjecture, since one counterexample is always sufficient to make the conjecture false.

1.1 EXERCISES

Decide whether each of the following is an example of inductive or deductive reasoning in Exercises 1–10.

1. If it is your wish, then it is my command. It is your wish. Therefore, it is my command.

2. Automobile repairs always take longer than the mechanic predicts. The mechanic told me that fixing the brakes on my car will take three hours. Therefore, my car will be in the shop for more than three hours.

3. The last four governors in our state have been Republicans. Therefore, the next governor will be Republican.

4. Joshua had 40 baseball cards. He gave 10 away. Therefore, he has 30 left.

5. If the same number is added to both sides of a true equation, the resulting equation is true. $3 + 2 = 5$. Therefore, $(3 + 2) + 9 = 5 + 9$.

6. Linda's first four children were girls. Therefore, her next child will be a girl.

7. Prudent people never buy $3,000 television sets. Lisa Wunderle is prudent. Therefore, Lisa will never buy a $3,000 television set.

8. The best player on our girls' team this year attended a basketball clinic to improve her skills. Therefore, in order to be the best player on your team, you must attend a basketball clinic.

9. For the past 20 years, a rare plant has bloomed in a South American rain forest each spring, alternating between pink and white flowers. Last spring, it bloomed with pink flowers. Therefore, this spring it will bloom with white flowers.

10. All men are mortal. Socrates is a man. Therefore, Socrates is mortal.

11. Discuss the differences between inductive and deductive reasoning. Give an example of each.

Determine the most probable next term in each list of numbers.

12. 6, 9, 12, 15, 18

13. 13, 18, 23, 28, 33

14. 3, 12, 48, 192, 768

15. 32, 16, 8, 4, 2

16. 3, 6, 9, 15, 24, 39

17. 1/3, 3/5, 5/7, 7/9, 9/11

18. 1/2, 3/4, 5/6, 7/8, 9/10

19. 1, 4, 9, 16, 25

20. 1, 8, 27, 64, 125

21. 1, 0, 1, 1, 0, 1, 1, 1, 0, 1, 1, 1, 1, 0, 1, 1

22. 1, 1, 1, 1, 0, 0, 0, 1, 1, 1, 0, 0, 0, 1, 1, 0, 0

23. Construct a list of numbers similar to those in Exercises 12–22 such that the most probable next number in the list is 60.

24. Construct a list of numbers similar to those in Exercises 12–22 such that either 20 or 25 may be a true conjecture for the next number in the list.

* For more information on this and other similar patterns, see "Counting Pizza Pieces and Other Combinatorial Problems" by Eugene Maier, in the January 1988 issue of *Mathematics Teacher,* pp. 22–26.

In Exercises 25–36, a list of equations is given. Use the list and inductive reasoning to predict the next equation, and then verify your conjecture.

25.
$$(1 \times 9) + 2 = 11$$
$$(12 \times 9) + 3 = 111$$
$$(123 \times 9) + 4 = 1{,}111$$
$$(1{,}234 \times 9) + 5 = 11{,}111$$

26.
$$(9 \times 9) + 7 = 88$$
$$(98 \times 9) + 6 = 888$$
$$(987 \times 9) + 5 = 8{,}888$$
$$(9{,}876 \times 9) + 4 = 88{,}888$$

27.
$$15{,}873 \times 7 = 111{,}111$$
$$15{,}873 \times 14 = 222{,}222$$
$$15{,}873 \times 21 = 333{,}333$$
$$15{,}873 \times 28 = 444{,}444$$

28.
$$3{,}367 \times 3 = 10{,}101$$
$$3{,}367 \times 6 = 20{,}202$$
$$3{,}367 \times 9 = 30{,}303$$
$$3{,}367 \times 12 = 40{,}404$$

29.
$$11 \times 11 = 121$$
$$111 \times 111 = 12{,}321$$
$$1{,}111 \times 1{,}111 = 1{,}234{,}321$$

30.
$$34 \times 34 = 1{,}156$$
$$334 \times 334 = 111{,}556$$
$$3{,}334 \times 3{,}334 = 11{,}115{,}556$$

31.
$$2 = 4 - 2$$
$$2 + 4 = 8 - 2$$
$$2 + 4 + 8 = 16 - 2$$
$$2 + 4 + 8 + 16 = 32 - 2$$

32.
$$3 = \frac{3(2)}{2}$$
$$3 + 6 = \frac{6(3)}{2}$$
$$3 + 6 + 9 = \frac{9(4)}{2}$$
$$3 + 6 + 9 + 12 = \frac{12(5)}{2}$$

33.
$$3 = \frac{3(3 - 1)}{2}$$
$$3 + 9 = \frac{3(9 - 1)}{2}$$
$$3 + 9 + 27 = \frac{3(27 - 1)}{2}$$
$$3 + 9 + 27 + 81 = \frac{3(81 - 1)}{2}$$

34.
$$5(6) = 6(6 - 1)$$
$$5(6) + 5(36) = 6(36 - 1)$$
$$5(6) + 5(36) + 5(216) = 6(216 - 1)$$
$$5(6) + 5(36) + 5(216) + 5(1{,}296) = 6(1{,}296 - 1)$$

35.
$$\frac{1}{1 \cdot 2} = \frac{1}{2}$$
$$\frac{1}{1 \cdot 2} + \frac{1}{2 \cdot 3} = \frac{2}{3}$$
$$\frac{1}{1 \cdot 2} + \frac{1}{2 \cdot 3} + \frac{1}{3 \cdot 4} = \frac{3}{4}$$
$$\frac{1}{1 \cdot 2} + \frac{1}{2 \cdot 3} + \frac{1}{3 \cdot 4} + \frac{1}{4 \cdot 5} = \frac{4}{5}$$

36.
$$\frac{1}{2} = 1 - \frac{1}{2}$$
$$\frac{1}{2} + \frac{1}{4} = 1 - \frac{1}{4}$$
$$\frac{1}{2} + \frac{1}{4} + \frac{1}{8} = 1 - \frac{1}{8}$$
$$\frac{1}{2} + \frac{1}{4} + \frac{1}{8} + \frac{1}{16} = 1 - \frac{1}{16}$$

A story is often told about how the great mathematician Carl Friedrich Gauss (1777–1855) at a very young age was told by his teacher to find the sum of the first 100 *counting numbers. While his classmates toiled at the problem, Carl simply wrote down a single number and handed it in to his teacher. His answer was correct. When asked how he did it, the young Carl explained that he observed that there were* 50 *pairs of numbers that each added up to* 101. *(See below.) So the sum of all the numbers must be* $50 \times 101 = 5{,}050$.

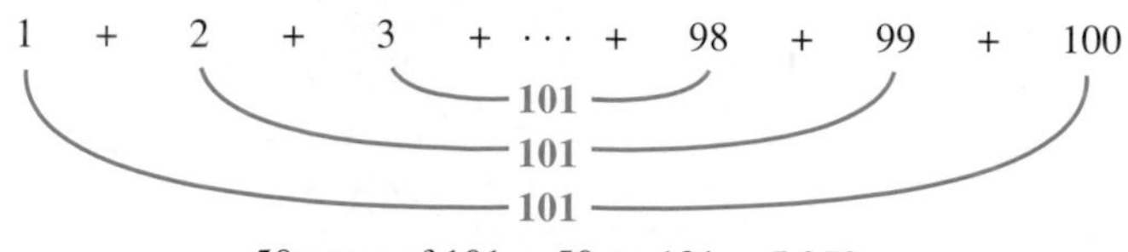

Use the method of Gauss to find each of the following sums.

37. $1 + 2 + 3 + \cdots + 150$

38. $1 + 2 + 3 + \cdots + 300$

39. $1 + 2 + 3 + \cdots + 500$

40. $1 + 2 + 3 + \cdots + 1{,}000$

41. Modify the procedure of Gauss to find the sum $1 + 2 + 3 + \cdots + 125$.

42. Explain in your own words how the procedure of Gauss can be modified to find the sum $1 + 2 + 3 + \cdots + n$, where n is an odd natural number. (An odd natural number, when divided by 2, leaves a remainder of 1.)

43. Modify the procedure of Gauss to find the sum $2 + 4 + 6 + \cdots + 100$.

44. Use the result of Exercise 43 to find the sum $4 + 8 + 12 + \cdots + 200$.

---

45. Consider the following table.

0	2	2	2	0	0	0	0	0
0	2	4	6	4	2	0	0	0
0	2	6	12	14	12	6	2	0
0	2	8	20	32	38	32	20	8

Find a pattern and predict the next row of the table.

46. Find a pattern in the following list of figures and use inductive reasoning to predict the next figure in the list.

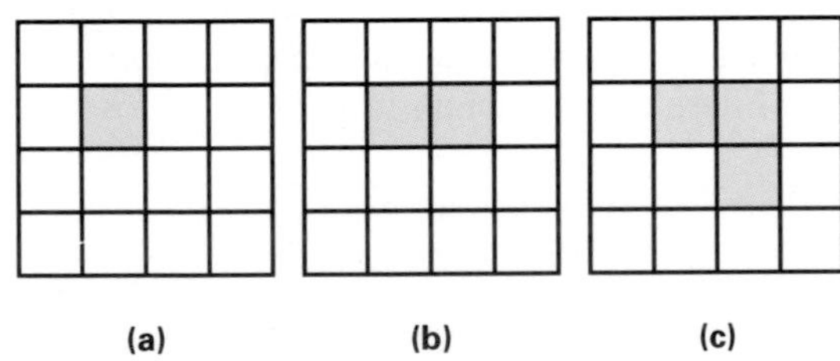

(a) (b) (c)

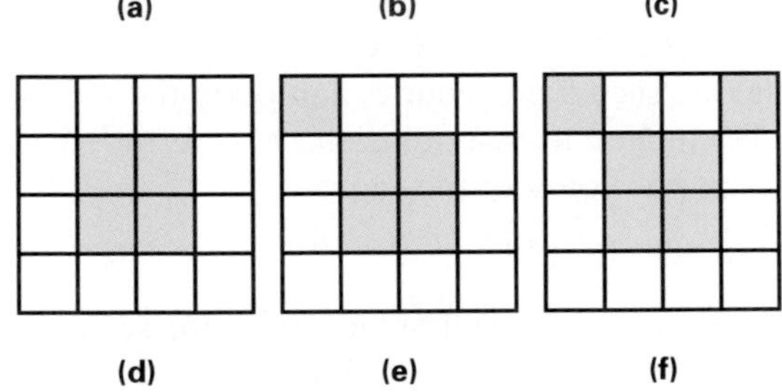

(d) (e) (f)

47. What is the next term in this list? O, T, T, F, F, S, S, E, N, T (*Hint:* Think about words and their relationship to numbers.)

48. What is the most probable next number in this list? 12, 1, 1, 1, 2, 1, 3 (*Hint:* Think about a clock.)

49. Choose any number, and follow these steps.

(a) Multiply by 2.
(b) Add 6.
(c) Divide by 2.
(d) Subtract the number you started with.
(e) Record your result.

Repeat the process, except in Step (b), add 8. Record your final result. Repeat the process once more, except in Step (b), add 10. Record your final result.

(f) Observe what you have done; use inductive reasoning to explain how to predict the final result. You may wish to use this exercise as a "number trick" to amuse your friends.

50. **(a)** Choose any three-digit number with all different digits. Now reverse the digits, and subtract the smaller from the larger. Record your result. Choose another three-digit number and repeat this process. Do this as many times as it takes for you to see a pattern in the different results you obtain. (*Hint:* What is the middle digit? What is the sum of the first and third digits?)

(b) Write an explanation of this pattern. You may wish to use this exercise as a "number trick" to amuse your friends.

51. Complete the following.

$12{,}345{,}679 \times 9 =$ _____
$12{,}345{,}679 \times 18 =$ _____
$12{,}345{,}679 \times 27 =$ _____

By what number would you have to multiply 12,345,679 in order to get an answer of 888,888,888?

52. Complete the following.

$142{,}857 \times 1 =$ _____
$142{,}857 \times 2 =$ _____
$142{,}857 \times 3 =$ _____
$142{,}857 \times 4 =$ _____
$142{,}857 \times 5 =$ _____
$142{,}857 \times 6 =$ _____

What pattern exists in the successive answers? Now multiply 142,857 by 7 to obtain an interesting result.

53. Choose two natural numbers. Add 1 to the second and divide by the first to get a third. Add 1 to the third and divide by the second to get a fourth. Add 1 to the fourth and divide by the third to get a fifth. Continue this process until you discover a pattern. What is the pattern?

54. Refer to Figures 1(b)–(f). Instead of counting interior regions of the circle, count the chords formed. Use inductive reasoning to predict the number of chords that would be formed if seven points were used.

55. Discuss one example of inductive reasoning that you have used recently in your life. Test your premises and your conjecture. Did your conclusion ultimately prove to be true or false?

56. Give an example of faulty inductive reasoning.

Investigating Number Patterns

In the previous section we were introduced to inductive reasoning, and we saw how it can be applied in predicting "what comes next" in a list of numbers or equations. In this section we will continue our investigation of number patterns.

A list of numbers such as

$$3, 9, 15, 21, 27, \ldots$$

can be considered a *sequence.* A **number sequence** is a list of numbers having a first number, a second number, a third number, and so on, called the **terms** of the sequence. To indicate that the terms of a sequence continue past the last term written, we use three dots (an ellipsis). The sequences in parts (a) and (b) of Example 2 in the previous section are called *arithmetic* and *geometric sequences,* respectively. An arithmetic sequence has a common *difference* between successive terms, while a geometric sequence has a common *ratio* between successive terms. The Fibonacci sequence in part (c) is covered in a later chapter.

Successive Differences

The sequences seen in the previous section were usually simple enough for us to make an obvious conjecture about the next term. However, some sequences may provide more difficulty in making such a conjecture, and often the **method of successive differences** may be applied to determine the next term if it is not obvious at first glance. For example, consider the sequence

$$2, 6, 22, 56, 114, \ldots.$$

Since the next term is not obvious, subtract the first term from the second term, the second from the third, the third from the fourth, and so on.

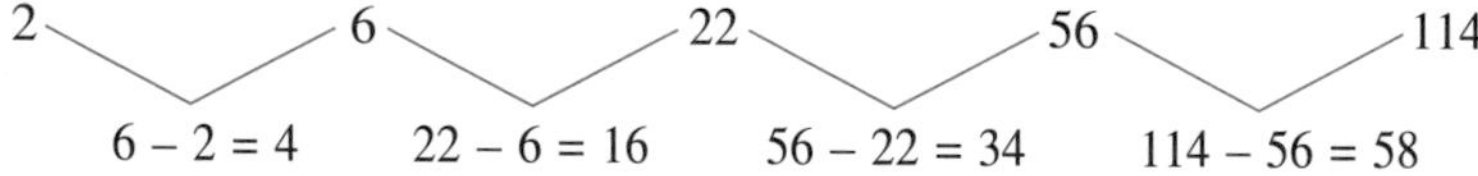

Now repeat the process with the sequence 4, 16, 34, 58 and continue repeating until the difference is a constant value, as shown in line (4) on the next page.

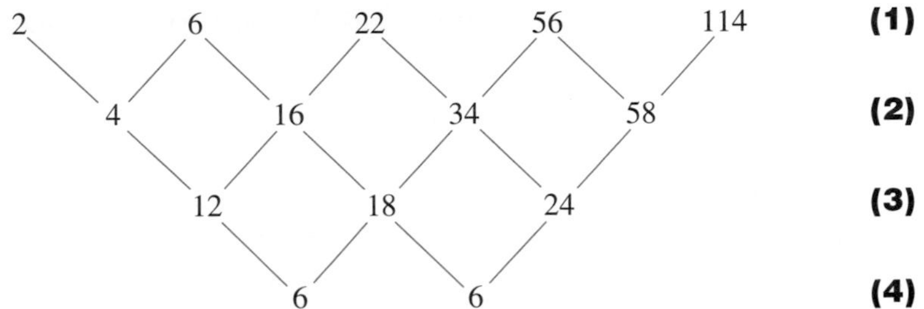

For this pattern to continue, another 6 should appear in line (4), meaning that the next term in line (3) would have to be 24 + 6 = 30. The next term in line (2) would be 58 + 30 = 88. Finally, the next term in the given sequence would be 114 + 88 = 202. The final scheme of numbers is shown below.

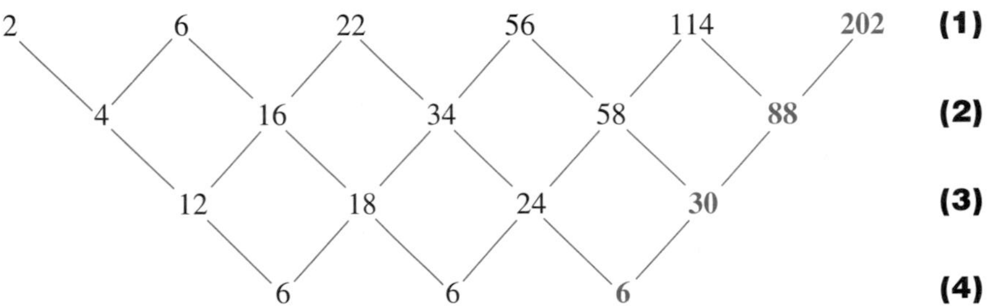

Once a line of constant values is obtained, simply work "backwards" by adding until the desired term of the given sequence is obtained.

EXAMPLE 1 Use the method of successive differences to determine the next number in each sequence.

(a) 14, 22, 32, 44, . . .
Using the scheme described above, obtain the following:

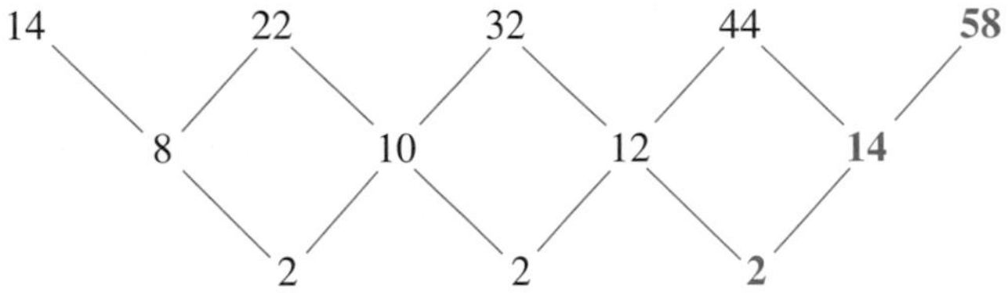

Once the row of 2s was obtained and extended, we were able to get 12 + 2 = 14, and 44 + 14 = 58, as shown above. The next number in the sequence is 58.

(b) 5, 15, 37, 77, 141, . . .
Proceeding as before, obtain the following diagram.

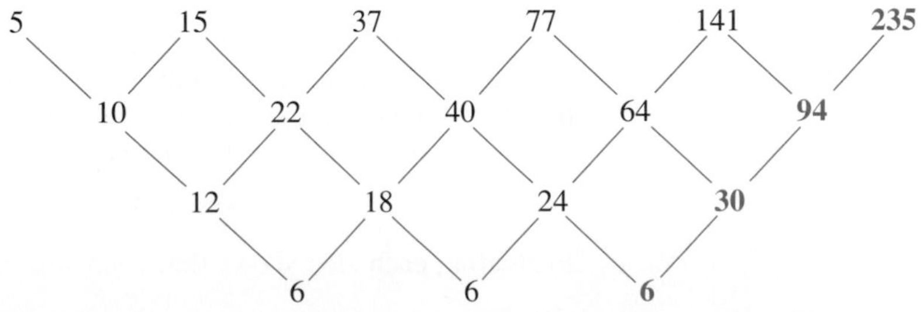

As shown above, the next number in the sequence is 235. ■

Don't get the impression that the method of successive differences will always work. For example, try it on the Fibonacci sequence in Example 2(c) of Section 1 and see what happens!

For Further Thought

Take any three-digit number whose digits are not all the same. Arrange the digits in decreasing order, and then arrange them in increasing order. Now subtract. Repeat the process, using a 0 if necessary in the event that the difference consists of only two digits. For example, suppose that we choose the digits 1, 4, and 8.

$$\begin{array}{r} 841 \\ -\ 148 \\ \hline 693 \end{array} \qquad \begin{array}{r} 963 \\ -\ 369 \\ \hline 594 \end{array} \qquad \begin{array}{r} 954 \\ -\ 459 \\ \hline 495 \end{array}$$

Notice that we have obtained the number 495, and the process will lead to 495 again. The number 495 is called a Kaprekar number, and it will eventually always be generated if this process is followed for such a three-digit number.

For Group Discussion

1. Have each student in a group of students apply the process of Kaprekar to a different two-digit number, in which the digits are not the same. (Interpret 9 as 09 if necessary.) Compare the results. What seems to be true?
2. Repeat the process for four digits, with each student in the group comparing results after several steps. What conjecture can the group make for this situation?
3. Repeat the process for a five-digit number. What conjecture can be made?

Number Patterns

One of the most beautiful aspects of mathematics is its seemingly endless variety of number patterns. Observe the following pattern:

$$\begin{aligned} 1 &= 1^2 \\ 1 + 3 &= 2^2 \\ 1 + 3 + 5 &= 3^2 \\ 1 + 3 + 5 + 7 &= 4^2 \\ 1 + 3 + 5 + 7 + 9 &= 5^2. \end{aligned}$$

In each case, the left side of the equation is the indicated sum of the consecutive odd counting numbers beginning with 1, and the right side is the square of the number of terms on the left side. You should verify this in each case. Inductive reasoning would suggest that the next line in this pattern is

$$1 + 3 + 5 + 7 + 9 + 11 = 6^2.$$

Evaluating each side shows that each side simplifies to 36.

Pythagoras (David Smith Collection) The Greek mathematician **Pythagoras** lived during the sixth century B.C. He and his fellow mathematicians formed the Pythagorean brotherhood, devoted to the study of mathematics and music. The Pythagoreans investigated the figurate numbers introduced in this section. They investigated connections between mathematics and music as well, and discovered that musical tones are related to the length of stretched strings by ratios of counting numbers. You can test this on a cello. Stop any string midway, so that the ratio of the whole string to the part is 2/1. If you pluck the free half of the string, you get the octave above the fundamental tone of the whole string. The ratio 3/2 gives you the fifth above the octave, and so on. Pythagoras noted that simple ratios of 1, 2, 3, 4 give the most harmonious musical intervals. He claimed that the intervals between planets must also be ratios of counting numbers. The idea came to be called "music of the spheres." (The planets were believed to orbit around the earth on crystal spheres.)

Can it be concluded from these observations that this pattern will continue indefinitely? The answer is no, because observation of a finite number of examples does not guarantee that the pattern will continue. However, mathematicians have proved that this pattern does indeed continue indefinitely, using a method of proof called *mathematical induction.* (See any standard college algebra text.)

Any even counting number may be written in the form $2k$, where k is a counting number. It follows that the kth odd counting number is written $2k - 1$. For example, the third odd counting number, 5, can be written $2(3) - 1$. Using these ideas, we can write the result obtained above as follows.

Sum of the First *n* Odd Counting Numbers

If n is any counting number,

$$1 + 3 + 5 + \cdots + (2n - 1) = n^2.$$

EXAMPLE 2 In each of the following, several equations are given illustrating a suspected number pattern. Determine what the next equation would be, and verify that it is indeed a true statement.

(a)

$$1^2 = 1^3$$
$$(1 + 2)^2 = 1^3 + 2^3$$
$$(1 + 2 + 3)^2 = 1^3 + 2^3 + 3^3$$
$$(1 + 2 + 3 + 4)^2 = 1^3 + 2^3 + 3^3 + 4^3$$

The left side of each equation is the square of the sum of the first n counting numbers, while the right side is the sum of their cubes. The next equation in the pattern would be

$$(1 + 2 + 3 + 4 + 5)^2 = 1^3 + 2^3 + 3^3 + 4^3 + 5^3.$$

Each side of the above equation simplifies to 225, so the pattern is true for this equation.

(b)

$$1 = 1^3$$
$$3 + 5 = 2^3$$
$$7 + 9 + 11 = 3^3$$
$$13 + 15 + 17 + 19 = 4^3$$

The left sides of the equations contain the sum of odd counting numbers, starting with the first (1) in the first equation, the second and third (3 and 5) in the second equation, the fourth, fifth, and sixth (7, 9, and 11) in the third equation, and so on. The right side contains the cube (third power) of the number of terms on the left side in each case. Following this pattern, the next equation would be

$$21 + 23 + 25 + 27 + 29 = 5^3,$$

which can be verified as true by computation.

(c)

$$1 = \frac{1 \cdot 2}{2}$$

$$1 + 2 = \frac{2 \cdot 3}{2}$$

$$1 + 2 + 3 = \frac{3 \cdot 4}{2}$$

$$1 + 2 + 3 + 4 = \frac{4 \cdot 5}{2}$$

$$1 + 2 + 3 + 4 + 5 = \frac{5 \cdot 6}{2}$$

The left side of each equation gives the indicated sum of the first n counting numbers, and the right side is always of the form

$$\frac{n(n + 1)}{2}.$$

For the pattern to continue, the next equation would be

$$1 + 2 + 3 + 4 + 5 + 6 = \frac{6 \cdot 7}{2}.$$

Since each side simplifies to 21, the pattern is true for this equation. ◆

The patterns established in Examples 2(a) and 2(c) can be written in general as follows.

Two Special Sum Formulas

For any counting number n,

$$(1 + 2 + 3 + \cdots + n)^2 = 1^3 + 2^3 + 3^3 + \cdots + n^3$$

and

$$1 + 2 + 3 + \cdots + n = \frac{n(n + 1)}{2}.$$

The second formula above is a generalization of the method first explained preceding Exercise 37 in the previous section, relating the story of young Karl Gauss. We can provide a general, deductive argument showing how this equation is obtained. Suppose that we let S represent the sum $1 + 2 + 3 + \cdots + n$. This sum can also be written as $S = n + (n - 1) + (n - 2) + \cdots + 1$. Now write these two equations as follows.

$$\begin{array}{rl} S = 1 & + 2 \quad + 3 \quad + \cdots + n \\ S = n & + (n - 1) + (n - 2) + \cdots + 1 \\ \hline 2S = (n + 1) & + (n + 1) + (n + 1) + \cdots + (n + 1) \end{array}$$

Add the corresponding sides.

Since the right side of the equation has n terms, each of them being $(n + 1)$, we can write it as n times $(n + 1)$.

$$2S = n(n + 1)$$

Divide both sides by 2 to obtain

$$S = \frac{n(n + 1)}{2}.$$

Now that this formula has been verified in a general manner, we can apply deductive reasoning to find the sum of the first n counting numbers for any given value of n. (See Exercises 21–24.)

Figurate Numbers

The Greek mathematician Pythagoras (c. 540 B.C.) was the founder of the Pythagorean brotherhood. This group studied, among other things, numbers of geometric arrangements of points, such as *triangular numbers, square numbers,* and *pentagonal numbers.* Figure 2 illustrates the first few of each of these types of numbers.

The figurate numbers possess numerous interesting patterns. Every square number greater than 1 is the sum of two consecutive triangular numbers. (For example, $9 = 3 + 6$ and $25 = 10 + 15$.) Every pentagonal number can be represented as the sum of a square number and a triangular number. (For example, $5 = 4 + 1$ and $12 = 9 + 3$.)

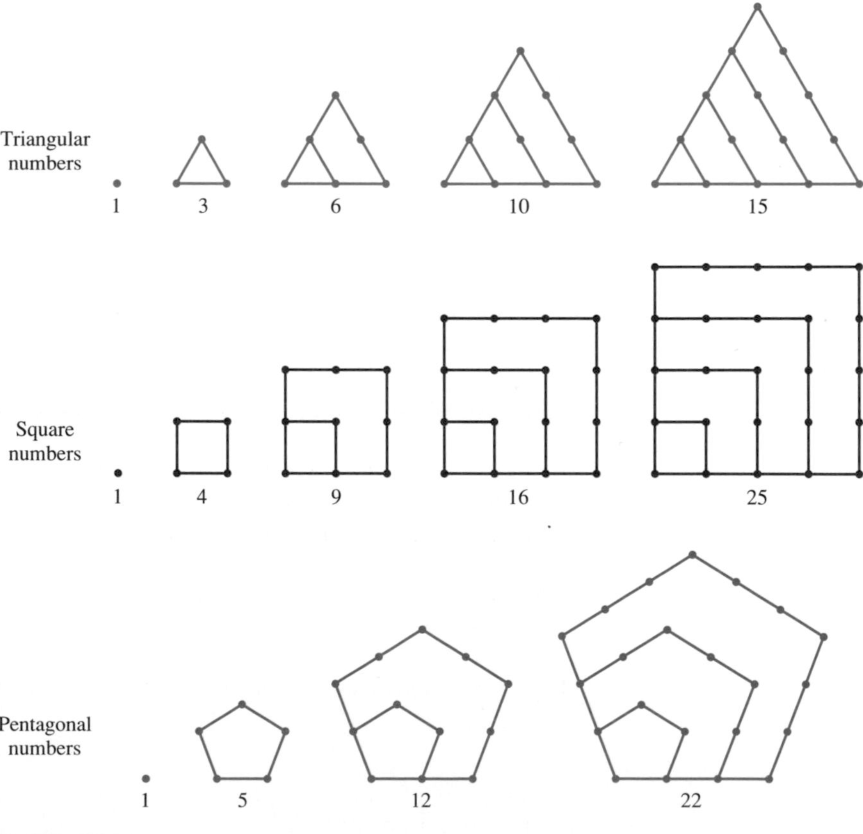

FIGURE 2

In the expression T_n, n is called a subscript. T_n is read "T sub n," and it represents the triangular number in the nth position in the sequence. For example,

$$T_1 = 1, \qquad T_2 = 3, \qquad T_3 = 6, \qquad T_4 = 10.$$

S_n and P_n represent the nth square and pentagonal numbers respectively. It can be shown that the following formulas give these figurate numbers:

$$T_n = \frac{n(n+1)}{2}, \qquad S_n = n^2, \qquad P_n = \frac{n(3n-1)}{2}.$$

EXAMPLE 3 Use the formulas above to find each of the following.

(a) the seventh triangular number

$$T_7 = \frac{7(7+1)}{2} = \frac{7(8)}{2} = \frac{56}{2} = 28$$

(b) the twelfth square number

$$S_{12} = 12^2 = 144$$

(c) the sixth pentagonal number

$$P_6 = \frac{6[3(6)-1]}{2} = \frac{6(18-1)}{2} = \frac{6(17)}{2} = 51$$

EXAMPLE 4 Show that the sixth pentagonal number is equal to 3 times the fifth triangular number, plus 6.

From Example 3(c), $P_6 = 51$. The fifth triangular number is 15. According to the problem,

$$51 = 3(15) + 6$$
$$51 = 45 + 6$$
$$51 = 51.$$

The general relationship examined in Example 4 can be written as follows:

$$P_n = 3 \cdot T_{n-1} + n.$$

Other such relationships among figurate numbers are examined in the exercises of this section.

The method of successive differences, introduced at the beginning of this section, can be used to predict the next figurate number in a sequence of figurate numbers, as shown in the final example.

EXAMPLE 5 The first five pentagonal numbers are

1, 5, 12, 22, 35.

Use the method of successive differences to predict the sixth pentagonal number.

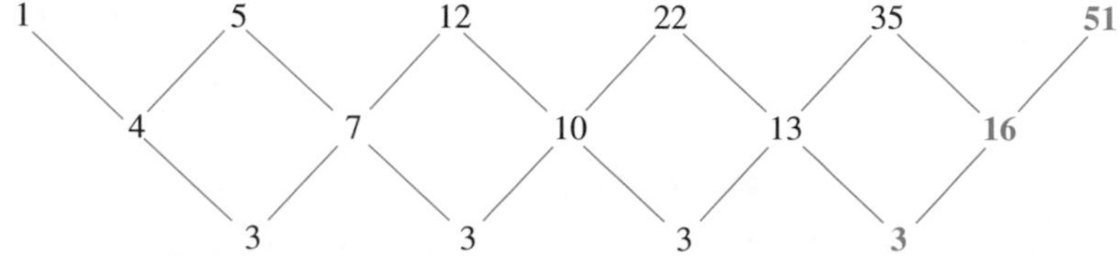

After the second line of successive differences, we can work backwards to find that the sixth pentagonal number is 51, which was also found in Example 3(c). ■

1.2 EXERCISES

Use the method of successive differences to determine the next number in each sequence.

1. 2, 8, 16, 26, . . .

2. 15, 24, 39, 60, . . .

3. 21, 34, 51, 72, 97, . . .

4. 11, 18, 29, 44, 63, . . .

5. 5, 28, 87, 200, 385, . . .

6. 11, 34, 81, 164, 295, . . .

7. 1, 15, 79, 253, 622, 1,296, . . .

8. 5, 23, 91, 269, 641, 1,315, . . .

9. Suppose that the expression $n^2 + 5$ determines the nth term in a sequence. That is, to find the first term, let $n = 1$, to find the second term, let $n = 2$, and so on.
(a) Find the first four terms of the sequence.
(b) Use the method of successive differences to predict the fifth term of the sequence.
(c) Find the fifth term of the sequence by letting $n = 5$ in the expression $n^2 + 5$. Does your result agree with the one you found in part (b)?

10. Refer to Figure 1 in the previous section. The method of successive differences can be applied to the sequence of interior regions, 1, 2, 4, 8, 16, 31, to find the number of regions determined by seven points on the circle. What is the next term in this sequence? How many regions would be determined by eight points? Verify this using the formula given at the end of that section.

In each of the following, several equations are given illustrating a suspected number pattern. Determine what the next equation would be, and verify that it is indeed a true statement.

11. $(1 \times 8) + 1 = 9$
$(12 \times 8) + 2 = 98$
$(123 \times 8) + 3 = 987$

12. $(1 \times 9) - 1 = 8$
$(21 \times 9) - 1 = 188$
$(321 \times 9) - 1 = 2{,}888$

13. $101 \times 101 = 10{,}201$
$10{,}101 \times 10{,}101 = 102{,}030{,}201$

14. $999{,}999 \times 2 = 1{,}999{,}998$
$999{,}999 \times 3 = 2{,}999{,}997$

15. $1 = 1^2$
$1 + 2 + 1 = 2^2$
$1 + 2 + 3 + 2 + 1 = 3^2$
$1 + 2 + 3 + 4 + 3 + 2 + 1 = 4^2$

16. $3^2 - 1^2 = 2^3$
$6^2 - 3^2 = 3^3$
$10^2 - 6^2 = 4^3$
$15^2 - 10^2 = 5^3$

17. $1^2 + 1 = 2^2 - 2$
$2^2 + 2 = 3^2 - 3$
$3^2 + 3 = 4^2 - 4$

18. $2^2 - 1^2 = 2 + 1$
$3^2 - 2^2 = 3 + 2$
$4^2 - 3^2 = 4 + 3$

19. $1 + 2 = 3$
$4 + 5 + 6 = 7 + 8$
$9 + 10 + 11 + 12 = 13 + 14 + 15$

20. $1 = 1 \times 1$
$1 + 5 = 2 \times 3$
$1 + 5 + 9 = 3 \times 5$

Use the formula $S = \frac{n(n + 1)}{2}$ *derived in this section to find each of the following sums.*

21. $1 + 2 + 3 + \cdots + 100$

22. $1 + 2 + 3 + \cdots + 400$

23. $1 + 2 + 3 + \cdots + 525$

24. $1 + 2 + 3 + \cdots + 600$

Use the formula $S = n^2$ *discussed in this section to find each of the following sums. (Hint: To find n, add 1 to the last term and divide by 2.)*

25. $1 + 3 + 5 + \cdots + 49$

26. $1 + 3 + 5 + \cdots + 101$

27. $1 + 3 + 5 + \cdots + 301$

28. $1 + 3 + 5 + \cdots + 999$

29. Use the formula for finding the sum $1 + 2 + 3 + \cdots + n$ to discover a formula for finding the sum $2 + 4 + 6 + \cdots + 2n$.

30. State in your own words the following formula discussed in this section: $(1 + 2 + 3 + \cdots + n)^2 = 1^3 + 2^3 + 3^3 + \cdots + n^3$.

31. Explain how the following diagram geometrically illustrates the formula $1 + 2 + 3 + 4 = \frac{4 \times 5}{2}$.

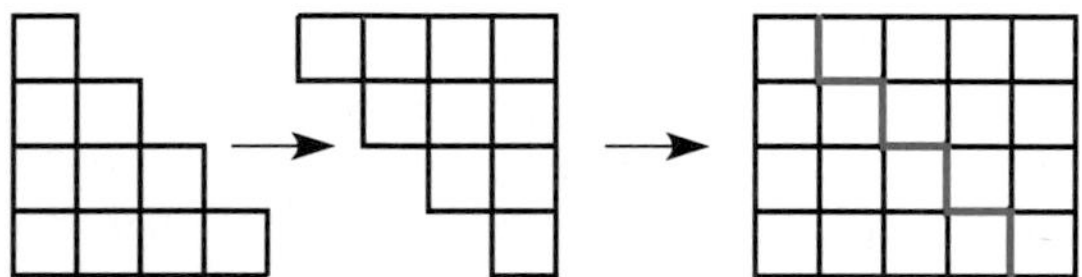

32. Explain how the following diagram geometrically illustrates the formula $1 + 3 + 5 + 7 + 9 = 5^2$.

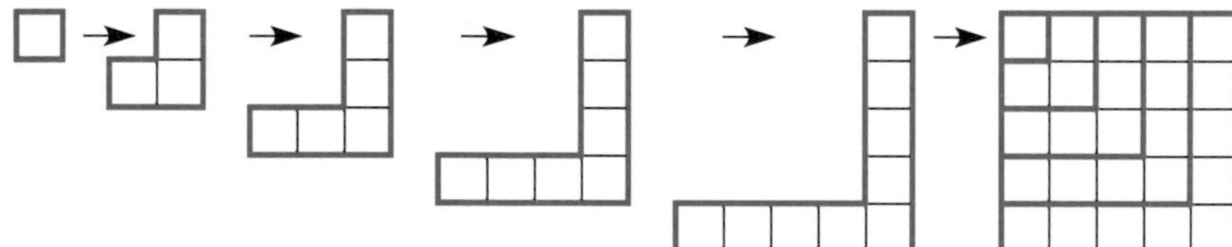

33. Use patterns to complete the table below.

Figurate Number	1st	2nd	3rd	4th	5th	6th	7th	8th
Triangular	1	3	6	10	15	21		
Square	1	4	9	16	25			
Pentagonal	1	5	12	22				
Hexagonal	1	6	15					
Heptagonal	1	7						
Octagonal	1							

34. The first five triangular, square, and pentagonal numbers may be obtained using sums of terms of sequences, as shown below.

Triangular	Square	Pentagonal
$1 = 1$	$1 = 1$	$1 = 1$
$3 = 1 + 2$	$4 = 1 + 3$	$5 = 1 + 4$
$6 = 1 + 2 + 3$	$9 = 1 + 3 + 5$	$12 = 1 + 4 + 7$
$10 = 1 + 2 + 3 + 4$	$16 = 1 + 3 + 5 + 7$	$22 = 1 + 4 + 7 + 10$
$15 = 1 + 2 + 3 + 4 + 5$	$25 = 1 + 3 + 5 + 7 + 9$	$35 = 1 + 4 + 7 + 10 + 13$

Notice the successive differences of the added terms on the right sides of the equations. The next type of figurate number is the **hexagonal** number. (A hexagon has six sides.) Use the patterns above to predict the first five hexagonal numbers.

35. Eight times any triangular number, plus 1, is a square number. Show that this is true for the first four triangular numbers.

36. Every square number can be written as the sum of two triangular numbers. For example, $16 = 6 + 10$. This can be represented geometrically by dividing a square array of dots with a line as illustrated at the right.

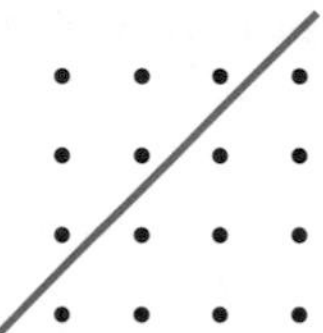

The triangular arrangement above the line represents 6, the one below the line represents 10, and the whole arrangement represents 16. Show how the square numbers 25 and 36 may likewise be geometrically represented as the sum of two triangular numbers.

37. Divide the first triangular number by 3 and record the remainder. Divide the second triangular number by 3 and record the remainder. Repeat this procedure several more times. Do you notice a pattern?

38. Repeat Exercise 37, but instead use square numbers and divide by 4. What pattern is determined?

39. Exercises 37 and 38 are specific cases of the following: In general, when the numbers in the sequence of n-agonal numbers are divided by n, the sequence of remainders obtained is a repeating sequence. Verify this for $n = 5$ and $n = 6$.

Refer to the table of figurate numbers in Exercise 33. The 2×2 *square of entries formed by the third and fourth triangular and square numbers is*

$$\begin{matrix} 6 & 10 \\ 9 & 16. \end{matrix}$$

If we add the terms in the diagonal from upper left to lower right, add the terms in the diagonal from upper right to lower left, and then subtract these sums, we obtain

$$(6 + 16) - (10 + 9) = 3.$$

Notice that this difference, 3, *is the column (the third) in which the terms on the left side of the square are located.*

Verify this property for the following 2×2 *squares from the table.*

40. $\begin{matrix} 3 & 6 \\ 4 & 9 \end{matrix}$

41. $\begin{matrix} 10 & 15 \\ 16 & 25 \end{matrix}$

42. $\begin{matrix} 9 & 16 \\ 12 & 22 \end{matrix}$

43. $\begin{matrix} 5 & 12 \\ 6 & 15 \end{matrix}$

44. The 2×2 square

$$\begin{matrix} 189 & 235 \\ 225 & 280 \end{matrix}$$

appears in an extended table of figurate numbers, similar to the one in Exercise 33. In what two columns do these numbers appear? (*Hint:* Refer to the explanation preceding Exercise 40.)

45. Complete the following table.

	n	**2**	**3**	**4**	**5**	**6**	**7**	**8**
A	Square of n							
B	(Square of n) + n							
C	One-half of Row B entry							
D	(Row A entry) − n							
E	One-half of Row D entry							

Use your results to answer the following, using inductive reasoning.

(a) What kind of figurate number is obtained when you find the average of n^2 and n? (See Row C.)

(b) If you square n and then subtract n from the result, and then divide by 2, what kind of figurate number is obtained? (See Row E.)

46. A fraction is reduced to *lowest terms* if the greatest common factor of its numerator and its denominator is 1. For example, 3/8 is reduced to lowest terms, but 4/12 is not.
(a) For $n = 2$ to $n = 8$, form the fractions

$$\frac{n\text{th square number}}{(n + 1)\text{th square number}}.$$

(b) Repeat part (a), but use triangular numbers instead.
(c) Use inductive reasoning to make a conjecture based on your results from parts (a) and (b), observing whether the fractions are reduced to lowest terms.

In addition to the formulas for T_n, S_n, *and* P_n *shown in the text, the following formulas are true for* ***hexagonal*** *numbers* (H), ***heptagonal*** *numbers* (Hp), *and* ***octagonal*** *numbers* (O):

$$H_n = \frac{n(4n - 2)}{2} \qquad Hp_n = \frac{n(5n - 3)}{2} \qquad O_n = \frac{n(6n - 4)}{2}.$$

Use these formulas to find each of the following.

47. the seventeenth square number **48.** the tenth triangular number **49.** the eighth pentagonal number

50. the sixth hexagonal number **51.** the eighth heptagonal number **52.** the sixth octagonal number

53. Observe the formulas given for H_n, Hp_n, and O_n, and use patterns and inductive reasoning to predict the formula for N_n, the nth **nonagonal** number. (A nonagon has 9 sides.) Then use the fact that the sixth nonagonal number is 111 to further confirm your conjecture.

54. Use the result of Exercise 53 to find the seventh nonagonal number.

Use inductive reasoning to answer each question in Exercises 55–58.

55. If you add two consecutive triangular numbers, what kind of figurate number do you get?

56. If you add the squares of two consecutive triangular numbers, what kind of figurate number do you get?

57. Square a triangular number. Square the next triangular number. Subtract the smaller result from the larger. What kind of number do you get?

58. Choose a value of n greater than or equal to 2. Find T_{n-1}, multiply it by 3, and add n. What kind of figurate number do you get?

1.3 Problem-Solving Strategies

In the first two sections of this chapter we stressed the importance of pattern recognition and the use of inductive reasoning, and these two approaches are important in solving problems. There are others as well, and we will investigate them in this section. These ideas will be used throughout the text, as problem solving is not a topic to be covered one day and then forgotten the next.

Probably the most famous study of problem-solving techniques was developed by George Polya (1888–1985), among whose many publications was the modern classic *How to Solve It*. In this book, Polya proposed a four-step process for problem solving.

George Polya, the author of the classic *How to Solve It,* died at the age of 97 on September 7, 1985. A native of Budapest, Hungary, he was once asked why there were so many good mathematicians to come out of Hungary at the turn of the century. He theorized that it was because mathematics is the cheapest science. It does not require any expensive equipment, only pencil and paper. He was once quoted as saying, "I was not smart enough to become a physicist and too smart to be a philosopher, so I chose mathematics." He authored or coauthored more than 250 papers in many languages, wrote a number of books, and was a brilliant lecturer and teacher. Yet, interestingly enough, he never learned to drive a car.

Polya's Four-Step Process for Problem Solving

1. **Understand the problem.** You cannot solve a problem if you do not understand what you are asked to find. The problem must be read and analyzed carefully. You will probably need to read it several times. After you have done so, ask yourself, "What must I find?"
2. **Devise a plan.** There are many ways to attack a problem and decide what plan is appropriate for the particular problem you are solving. See the list headed "Suggested Problem-Solving Strategies" for a number of possible approaches.
3. **Carry out the plan.** Once you know how to approach the problem, carry out your plan. You may run into "dead ends" and unforeseen roadblocks, but be persistent. If you are able to solve a problem without a struggle, it isn't much of a problem, is it?
4. **Look back and check.** Check your answer to see that it is reasonable. Does it satisfy the conditions of the problem? Have you answered all the questions the problem asks? Can you solve the problem a different way and come up with the same answer?

In Step 2 of Polya's problem-solving process, we are told to devise a plan. Here are some strategies that may prove useful.

Suggested Problem-Solving Strategies

Make a table or a chart.
Look for a pattern.
Solve a similar simpler problem.
Draw a sketch.
Use inductive reasoning.
Write an equation and solve it.
If a formula applies, use it.
Work backward.
Guess and check.
Use trial and error.
Use common sense.
Look for a "catch" if an answer seems too obvious, or impossible.

A particular problem solution may involve one or more of the strategies listed here, and you should try to be creative in your problem-solving techniques. The examples that follow illustrate some of these strategies. As you read through them, keep in mind that it is one thing to read a problem solution or watch a teacher solve a problem, but it is another to be able to do it yourself. Think about this: Michael Jordan makes incredible shots look easy but very few people can duplicate his feats. Even with the natural ability he has, hours of practice are necessary. Just like any skill, proficiency in problem solving requires perseverance and hard work.

Problem solving should be the central focus of the mathematics curriculum. As such, it is a primary goal of all mathematics instruction and an integral part of all mathematical activity. Problem solving is not a distinct topic but a process that should permeate the entire program and provide the context in which concepts and skills can be learned.

From *Curriculum and Evaluation Standards for School Mathematics,* 1989 (National Council of Teachers of Mathematics)

For Further Thought

Some form of the following problem has been around for many years.

In Farmer Jack's will, Jack bequeathed 1/2 of his horses to his son Johnny, 1/3 to his daughter Linda, and 1/9 to his son Jeff. Jack had 17 horses, so how were they to comply with the terms of the will? Certainly horses cannot be divided up into fractions. Their attorney, Tydings, came to their rescue, and was able to execute the will to the satisfaction of all. How did he do it?

Here is the solution:

Tydings added one of his horses to the 17, giving a total of 18. Johnny received 1/2 of 18, or 9, Linda received 1/3 of 18, or 6, and Jeff received 1/9 of 18, or 2. That accounted for a total of $9 + 6 + 2 = 17$ horses. Then Tydings took back his horse, and everyone was happy.

For Group Discussion

Discuss whether the terms of the will were actually fulfilled, according to the letter of the law.

Fibonacci (1170–1250) discovered the sequence named after him in a problem on rabbits. Fibonacci (son of Bonaccio) is one of several names for Leonardo of Pisa. His father managed a warehouse in present-day Bougie (or Bejaia), in Algeria. Thus it was that Leonardo Pisano studied with a Moorish teacher and learned the "Indian" numbers that the Moors and other Moslems brought with them in their westward drive.

Fibonacci wrote books on algebra, geometry, and trigonometry. These contain Arabian mathematics as well as his own work.

EXAMPLE 1 ***(Solving a Problem by Using a Table or a Chart)*** A man put a pair of rabbits in a cage. During the first month the rabbits produced no offspring, but each month thereafter produced one new pair of rabbits. If each new pair thus produced reproduces in the same manner, how many pairs of rabbits will there be at the end of one year?

This problem is a famous one in the history of mathematics, and first appeared in *Liber Abaci,* a book written by the Italian mathematician Leonardo Pisano (also known as Fibonacci) in the year 1202. Let us apply Polya's process to solve it.

1. **Understand the problem.** After several readings, we can reword the problem as follows. How many pairs of rabbits will the man have at the end of one year, if he starts with one pair, and they reproduce this way: During the first month of life, each pair produces no new rabbits but each month thereafter, each pair produces one new pair?
2. **Devise a plan.** Since there is a definite pattern to how the rabbits will reproduce, we can construct a table and fill in the information. The initial table will look like this:

Month	Number of Pairs at Start	Number of New Pairs Produced	Number of Pairs at End of Month
1st			
2nd			
3rd			
4th			
5th			
6th			
7th			
8th			
9th			
10th			
11th			
12th			

Once the table entries are completed, we must give the final entry in the final column as our answer.

3. **Carry out the plan.** At the start of the first month there is only one pair of rabbits, as specified in the problem. No new pairs are produced during the first month so there is $1 + 0 = 1$ pair present at the end of the first month. This pattern continues throughout the table. We add the number in the first column of numbers to the number in the second column to get the number in the third. Continue in this way through the twelfth month.

Month	Number of Pairs at Start	+	Number of New Pairs Produced	=	Number of Pairs at End of Month	
1st	1		0		1	1 + 0 = 1
2nd	1		1		2	1 + 1 = 2
3rd	2		1		3	2 + 1 = 3
4th	3		2		5	.
5th	5		3		8	.
6th	8		5		13	.
7th	13		8		21	.
8th	21		13		34	.
9th	34		21		55	.
10th	55		34		89	.
11th	89		55		144	.
12th	144		89		**233**	144 + 89 = 233

According to our table, there will be 233 pairs of rabbits at the end of one year.

4. **Look back and check.** This problem can be checked by going back and making sure that we have interpreted it correctly, which we have. Double-check the arithmetic. We have answered the question posed by the problem, so the problem is solved. ◆

The sequence shown in color in the table in Example 1 is the Fibonacci sequence, and many of its interesting properties will be investigated in a later chapter.

In the remaining examples of this section, we will use Polya's process but will not list the steps specifically as we did in Example 1.

EXAMPLE 2 ***(Solving a Problem by Working Backward)*** Kevin Connors plays poker every Friday night. One week he tripled his money, but then lost $6. He took his money back the next week, doubled it, but then lost $20. The following week he tried again, taking his money back with him. He quadrupled it, and then played well enough to take that much home with him, a total of $40. How much did he start with the first week?

This problem asks us to find Kevin's starting amount, given information about his winnings and losses. We also know his final amount. While we could write an algebraic equation to solve this problem, the method of working backward can be applied quite easily. Since his final amount was $40 and this represents four times the amount he started with on the third week, we *divide* \$40 by 4 to find that he started the third week with \$10. Before he lost \$20 the second week, he had this \$10 plus the \$20 he lost, giving him \$30. This represented double what he started with, so he started with \$30 *divided by* 2, or \$15, the second week. Repeating this process once more for the first week, before his \$6 loss he had \$15 + \$6 = \$21, which represents triple what he started with. Therefore, he started with \$21 ÷ 3, or \$7.

To check our answer, \$7, observe the following equations that depict the winnings and losses:

First week: (3 × \$7) − \$6 = \$21 − \$6 = **\$15**

Second week: (2 × **\$15**) − \$20 = \$30 − \$20 = **\$10**

Third week: (4 × **\$10**) = \$40. ■

Since he finished with \$40, the computations verify our earlier work.

Augustus De Morgan (see Example 3) was an English mathematician and philosopher, who served as professor at the University of London. He wrote numerous books, one of which was a *A Budget of Paradoxes.* His work in set theory and logic led to laws that bear his name and are covered in other chapters. He died in the same year as Charles Babbage.

EXAMPLE 3 ***(Solving a Problem by Trial and Error)*** The mathematician Augustus De Morgan lived in the nineteenth century. He once made the following statement: "I was x years old in the year x^2." In what year was De Morgan born?

We must find the year of De Morgan's birth. The problem tells us that he lived in the nineteenth century, which is another way of saying that he lived during the 1800s. According to his statement, one year of his life was a perfect square, so we must find a number between 1800 and 1900 that is a perfect square. By trial and error we find the following.

$$42^2 = 1764$$

$$43^2 = 1849$$

1849 is between 1800 and 1900.

$$44^2 = 1936$$

The only natural number whose square is between 1800 and 1900 is 43, since $43^2 = 1849$. Therefore, De Morgan was 43 years old in 1849. The final step in solving the problem is to subtract 43 from 1849 to find the year of his birth: 1849 − 43 = 1806. He was born in 1806.

While the following suggestion for a check may seem unorthodox, it does work: Look up De Morgan's birth date in a book dealing with mathematics history, such as *An Introduction to the History of Mathematics,* Sixth Edition, by Howard W. Eves. ■

The next problem dates back to Hindu mathematics, circa 850.

EXAMPLE 4 ***(Solving a Problem by Guessing and Checking)*** One-fourth of a herd of camels was seen in the forest; twice the square root of that herd had gone to the mountain slopes; and 3 times 5 camels remained on the riverbank. What is the numerical measure of that herd of camels?

While an algebraic method of solving an equation could be used to solve this problem, we will show an alternative method. We are looking for a numerical measure of a herd of camels, so the number must be a counting number. Since the problem mentions "one-fourth of a herd" and "the square root of that herd," the number of camels must be both a multiple of 4 and a perfect square, so that no fractions will be encountered. The smallest counting number that satisfies both conditions is 4. Let us write an equation where x represents the numerical measure of the herd, and then substitute 4 for x to see if it is a solution.

$\frac{1}{4}x$	$+\ 2\sqrt{x}$	$+\ 3 \cdot 5$	$= x$	
"one-fourth of the herd"	**+ "twice the square root of that herd"**	**+ "3 times 5 camels"**	**= "the numerical measure of the herd"**	
$\frac{1}{4}(4)$	$+\ 2\sqrt{4}$	$+\ 3 \cdot 5$	$= 4$	Let $x = 4$.
1	+ 4	+ 15	$\stackrel{?}{=} 4$	
		20	$\neq 4$	

Since 4 is not the solution, try 16, the next perfect square that is a multiple of 4.

$$\frac{1}{4}(16) + 2\sqrt{16} + 3 \cdot 5 = 16 \qquad \text{Let } x = 16.$$

$$4 + 8 + 15 \stackrel{?}{=} 16$$

$$27 \neq 16$$

Since 16 is not a solution, try 36.

$$\frac{1}{4}(36) + 2\sqrt{36} + 3 \cdot 5 = 36 \qquad \text{Let } x = 36.$$

$$9 + 12 + 15 \stackrel{?}{=} 36$$

$$36 = 36$$

We see from this last equation that 36 is a solution, so it is the numerical measure of the herd. Check back into the words of the problem, making this substitution: "One-fourth of 36, plus twice the square root of 36, plus 3 times 5" gives 9 plus 12 plus 15, which equals 36. ■

EXAMPLE 5 ***(Solving a Problem by Considering a Similar Simpler Problem and Looking for a Pattern)*** The digit farthest to the right in a counting number is called the *ones* or *units* digit, since it tells how many ones are contained in the number when grouping by tens is considered. What is the ones or units digit in $2^{4,000}$?

Recall that $2^{4,000}$ means that 2 is used as a factor 4,000 times:

$$2^{4,000} = \underbrace{2 \times 2 \times 2 \times \cdots \times 2.}_{4,000 \text{ factors}}$$

Certainly we are not expected to evaluate this number. In order to answer the question, we can consider examining some smaller powers of 2 and then looking for a pattern. Let us start with the exponent 1 and look at the first twelve powers of 2.

$$\begin{array}{lll} 2^1 = 2 & 2^5 = 32 & 2^9 = 512 \\ 2^2 = 4 & 2^6 = 64 & 2^{10} = 1{,}024 \\ 2^3 = 8 & 2^7 = 128 & 2^{11} = 2{,}048 \\ 2^4 = 16 & 2^8 = 256 & 2^{12} = 4{,}096 \end{array}$$

Notice that in each of the four rows above, the ones digit is the same. The final row, which contains the exponents 4, 8, and 12, has the ones digit 6. Each of these exponents is divisible by 4, and since 4,000 is divisible by 4, we can use inductive reasoning to predict that the units digit in $2^{4,000}$ is 6. (*Note:* The units digit for any other power can be found if we divide the exponent by 4 and consider the remainder. Then compare the result to the list of powers above. For example, to find the units digit of 2^{543}, divide 543 by 4 to get a quotient of 135 and a remainder of 3. The units digit is the same as that of 2^3, which is 8.)

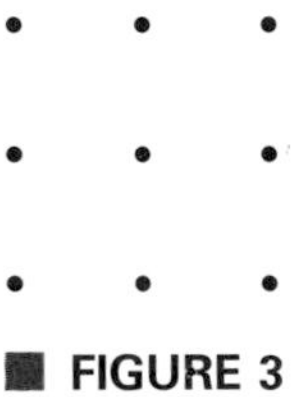

■ FIGURE 3

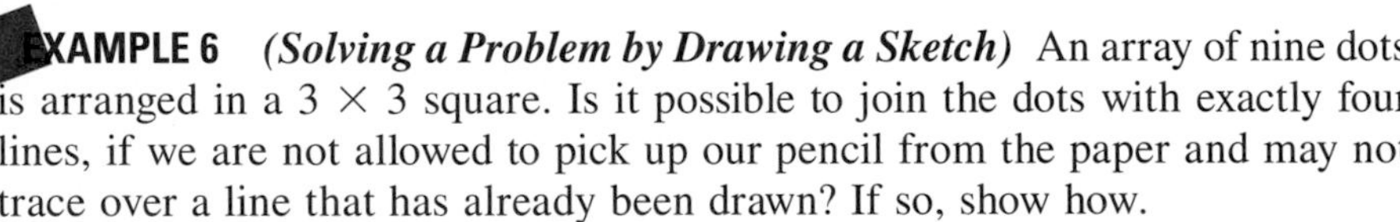

EXAMPLE 6 ***(Solving a Problem by Drawing a Sketch)*** An array of nine dots is arranged in a 3 × 3 square. Is it possible to join the dots with exactly four lines, if we are not allowed to pick up our pencil from the paper and may not trace over a line that has already been drawn? If so, show how.

We can begin by drawing the required arrangement. See Figure 3.

Now consider the following attempts at joining all nine dots with exactly four lines, without picking up our pencil from the paper. See Figure 4.

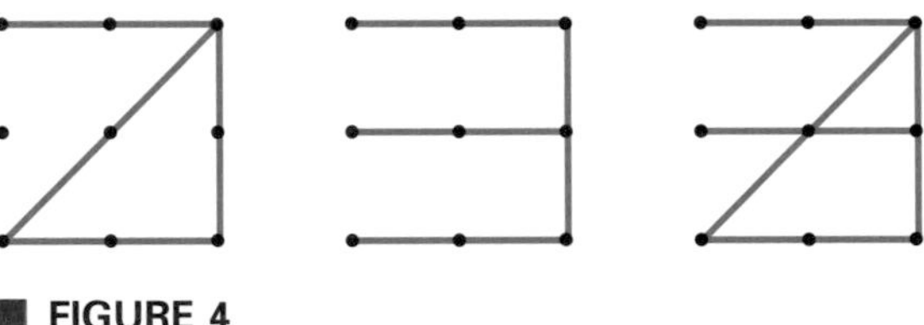

■ FIGURE 4

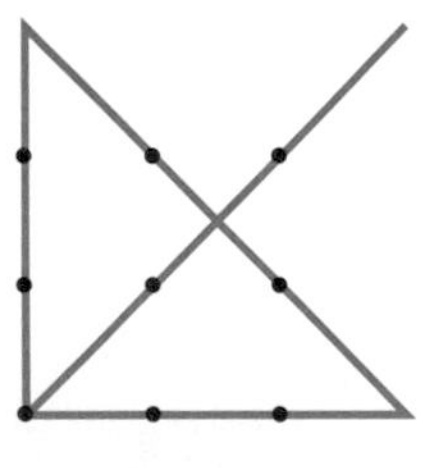

■ FIGURE 5

In each case, something is wrong. In the first sketch, one dot is not joined. In the second, we have a figure that cannot be drawn without picking up our pencil from the paper or tracing over a line that has already been drawn. In the third figure, all dots have been joined, but we have used five lines as well as retraced over the figure. Further attempts may lead to similar results. However, the conditions of the problem can be satisfied, as shown in Figure 5.

Notice that here we "went outside of the diagram," which was not prohibited by the conditions of the problem. This is an example of creative thinking—we used a strategy that is usually not considered at first, since our initial attempts involved "staying within the confines" of the figure.

Count the number of times the letter *f* (upper- or lowercase) appears in the sentence below.

Finished files are the result of years of scientific study combined with the experience of many years.

See the following page for the answer.

The final example falls into a category of problems that involve a "catch" in the wording. Such problems often seem impossible to solve at first, due to preconceived notions about the matter being discussed. Some of these problems seem too easy or perhaps impossible at first, because we tend to overlook an obvious situation. We should look carefully at the use of language in such problems. And, of course, we should never forget to use common sense.

Answer to the question in the margin on page 26. The letter *f* appears six times. Many people only count three, neglecting to count its appearance in the word *of,* which appears three times. (The authors wish to thank Tom Fandal for his contribution of this puzzle.)

EXAMPLE 7 *(Solving a Problem Using Common Sense)* Two currently minted U.S. coins together have a total value of 30¢. One is not a nickel. What are the two coins?

Our initial reaction may be "The only way to have two such coins with a total value of 30¢ is to have a quarter and a nickel, but the problem says 'one is not a nickel.'" This statement is indeed true. What we must realize here is that the one that is not a nickel is the quarter, and the *other* coin is a nickel. So the two coins are a quarter and a nickel.

1.3 EXERCISES

Use problem-solving strategies to solve each of the following problems. In many cases there is more than one possible approach, so be creative. Keep in mind that ingenuity is an important approach in problem solving.

1. If you raise 3 to the 328th power, what is the units digit of the result?

2. What is the units digit in 7^{491}?

3. Barbara Burnett bought a book for $10 and then spent half her remaining money on a train ticket. She then bought lunch for $4 and spent half her remaining money at a bazaar. She left the bazaar with $20. How much money did she start with?

4. I am thinking of a positive number. If I square it, double the result, take half of that result, and then add 12, I get 21. What is my number?

5. A frog is at the bottom of a 20-foot well. Each day it crawls up 4 feet but each night it slips back 3 feet. After how many days will the frog reach the top of the well?

6. A drawer contains 20 black socks and 20 white socks. If the light is off and you reach into the drawer to get your socks, what is the minimum number of socks you must pull out in order to be sure that you have a matching pair?

7. How many squares are in the following figure?

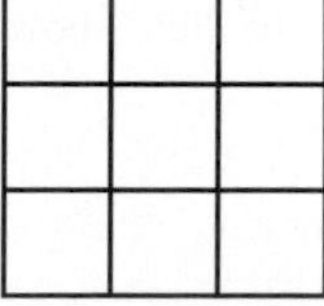

8. How many triangles are in the following figure?

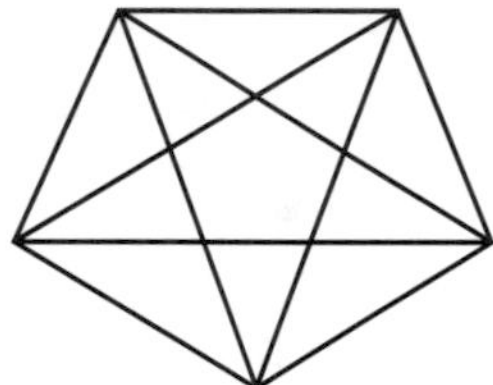

9. Some children are standing in a circular arrangement. They are evenly spaced and marked in numerical order. The fourth child is standing directly opposite the twelfth child. How many children are there in the circle?

10. A *perfect number* is a counting number that is equal to the sum of all its counting number divisors except itself. For example, 28 is a perfect number, since its divisors other than itself are 1, 2, 4, 7, and 14, and $1 + 2 + 4 + 7 + 14 = 28$. What is the smallest perfect number?

11. A *palindromic number* is a number whose digits read the same from left to right as they read from right to left. For example, 14,641 is palindromic. What three-digit palindromic numbers have the sum of their digits equal to the sum of the digits of the largest two-digit palindromic number?

12. A lily pad grows so that each day it doubles its size. On the fifteenth day of its life, it completely covers a pond. On what day was the pond half covered?

13. Comment on an interesting property of this sentence: "A man, a plan, a canal, Panama." (*Hint:* See Exercise 11.)

14. Draw a diagram that satisfies the following description, using the minimum number of birds: "Two birds above a bird, two birds below a bird, and a bird between two birds."

15. Assuming that he lives that long, one of the authors of this book will be 76 years old in the year x^2, where x is a counting number. In what year was he born?

16. The same author mentioned in Exercise 15 graduated from high school in the year that satisfies these conditions: (1) The sum of the digits is 23; (2) The hundreds digit is 3 more than the tens digit; (3) No digit is an 8. In what year did he graduate?

17. Donna is taller than David but shorter than Bill. Dan is shorter than Bob. What is the first letter in the name of the tallest person?

18. There is a two-digit number between 20 and 30 such that the sum of the cubes of its digits is equal to three times the number. What is the number?

19. Eve said to Adam, "If you give me one dollar, then we will have the same amount of money." Adam then replied, "Eve, if you give me one dollar I will have double the amount of money you are left with." How much does each have?

20. The number of hens and cows in a barnyard adds up to 11. The number of legs among them is 32. How many hens and how many cows are in the barnyard?

21. If Earl Karn weighs 170 pounds standing on one leg, how much does he weigh standing on two legs?

22. If you take 7 bowling pins from 10 bowling pins, what do you have?

23. In the addition problem below, some digits are missing as indicated by the blanks. If the problem is done correctly, what is the sum of the missing digits?

```
     _ 3 5
     8 _ 6
+    1 4 _
----------
  _, 4 0 8
```

24. Fill in the blanks so that the multiplication problem below uses all digits 0, 1, 2, 3, . . . , 9 exactly once, and is correctly worked.

```
      _ 0 2
×       3 _
-----------
  _ 5, _ _ _
```

25. A *magic square* is a square array of numbers that has the property that the sum of the numbers in each row, column, and diagonal is the same. Fill in the square below so that it becomes a magic square, and all digits 1, 2, 3, . . . , 9 are used exactly once.

6		8
	5	
		4

26. Refer to Exercise 25. Complete the magic square below so that all digits 1, 2, 3, . . . , 16 are used exactly once, and the sum in each row, column, and diagonal is 34.

6			9
	15		14
11		10	
16		13	

27. With a 5-minute sand timer and a 9-minute sand timer, what is the easiest way to time an egg to boil for 13 minutes?

28. Michael Booth has an unlimited number of cents (pennies), nickels, and dimes. In how many different ways can he pay 15¢ for a chocolate mint? (For example, one way is 1 dime and 5 pennies.)

29. What is the minimum number of pitches that a baseball pitcher who pitches a complete game can make in a regulation 9-inning baseball game?

30. What is the least natural number whose written name in the English language has its letters in alphabetical order?

31. If it takes 7 1/2 minutes to boil an egg, how long does it take to boil five eggs?

32. A hayfield has two haystacks in one corner and four haystacks in another corner. If all the hay is put into the middle of the field, how many haystacks will there be?

33. Several soldiers must cross a deep river at a point where there is no bridge. The soldiers spot two children playing in a small rowboat. The rowboat can hold only two children or one soldier. All the soldiers get across the river. How?

34. A person must take a wolf, a goat, and some cabbage across a river. The rowboat to be used has room for the person plus either the wolf, the goat, or the cabbage. If the person takes the cabbage in the boat, the wolf will eat the goat. While the wolf crosses in the boat, the cabbage will be eaten by the goat. The goat and cabbage are safe only when the person is present. Even so, the person gets everything across the river. Explain how. (This problem dates back to around the year 750.)

35. You have eight coins. Seven are genuine and one is a fake, which weighs a little less than the other seven. You have a balance scale, which you may use only three times. Tell how to locate the bad coin in three weighings. (Then show how to detect the bad coin in only *two* weighings.)

36. Three women, Ms. Thompson, Ms. Johnson, and Ms. Andersen, are sitting side by side at a meeting of the neighborhood improvement group. Ms. Thompson always tells the truth, Ms. Johnson sometimes tells the truth, and Ms. Andersen never tells the truth. The woman on the left says, "Ms. Thompson is in the middle." The woman in the middle says, "I'm Ms. Johnson," while the woman on the right says, "Ms. Andersen is in the middle." What are the correct positions of the women?

37. In the following problem, each letter stands for a specific digit. Assuming the problem is worked correctly, what do the letters stand for?

$$\begin{array}{r} PPP \\ +\quad\;\; A \\ \hline ABBB \end{array}$$

38. Repeat Exercise 37 for the following.

$$\begin{array}{r} XY \\ +\;\; Y \\ \hline YX \end{array}$$

39. Draw a square in the following figure so that no two dogs share the same region.

40. When the diagram shown is folded to form a cube, what letter is opposite the face marked Y?

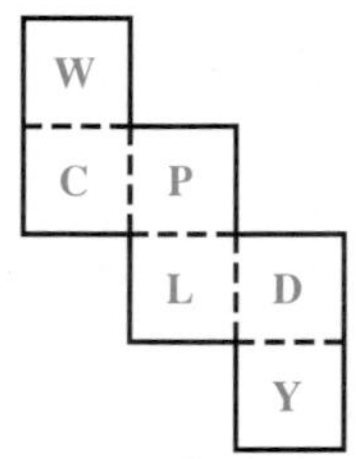

Exercises 41–42 are ancient Hindu mathematical problems.

41. Beautiful maiden with beaming eyes, tell me . . . which is the number that when multiplied by 3, then increased by 3/4 the product, then divided by 7, diminished by 1/3 of the quotient, multiplied by itself, diminished by 52, by the extraction of the square root, addition of 8, and division by 10 gives the number 2?

42. The mixed price of 9 citrons and 7 fragrant wood apples is 107; again, the mixed price of 7 citrons and 9 fragrant wood apples is 101. O you arithmetician, tell me quickly the price of a citron and of a wood apple, having distinctly separated those prices well.

43. How many times can you subtract 10 from 100?

44. At a variety store, 1 costs \$.50 while 5082 costs \$2.00. What is being sold?

45. Draw the following figure without picking up your pencil from the paper and without tracing over a line you have already drawn.

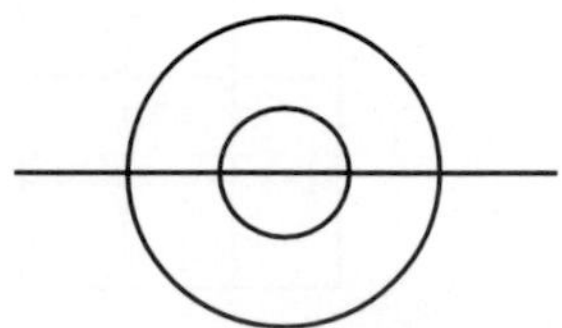

46. Repeat Exercise 45 for the figure shown here.

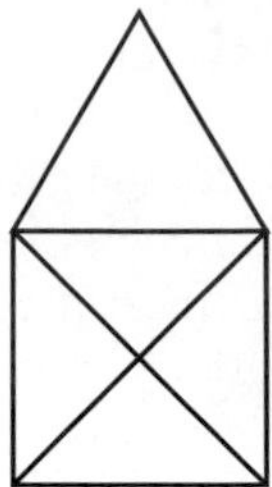

47. Volumes 1 and 2 of *Mathematics Is for Everyone* are standing in numerical order from left to right on your bookshelf. Volume 1 has 300 pages and Volume 2 has 250 pages. Excluding the covers, how many pages are between page 1 of Volume 1 and page 250 of Volume 2?

48. The brother of the chief executive officer (CEO) of a major industrial firm died. The man who died has no brother. How is this possible?

49. A teenager's age increased by 2 gives a perfect square. Her age decreased by 10 gives the square root of that perfect square. She is 5 years older than her brother. How old is her brother?

50. James, Dan, Jessica, and Cathy form a pair of married couples. Their ages are 35, 30, 29, and 28. Jessica is married to the oldest person in the group. James is older than Jessica but younger than Cathy. Who is married to whom, and what are their ages?

51. Only one of the numbers here is a perfect square. Which one is it? (No extensive calculation is necessary. *Hint:* Consider the units digit.)
(a) 23,784,855,888,784
(b) 49,552,384,781,228
(c) 33,587,988,047,112
(d) 64,987,237,001,667

52. After you have solved the problem in Exercise 51 correctly, write a few sentences explaining the procedure you used.

53. What is the maximum number of small squares in which we may place a cross (×) and not have any row, column, or diagonal completely filled with crosses? Illustrate your answer.

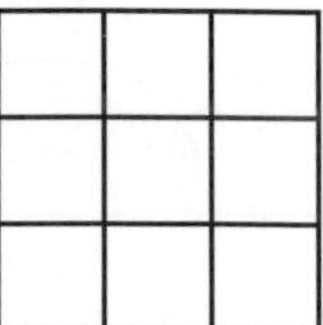

54. How much dirt is there in a cubical hole, 4 feet on each side?

55. Refer to Example 1, and observe the sequence of numbers in color, the Fibonacci sequence. Choose any four consecutive terms. Multiply the first one chosen by the fourth, and then multiply the two middle terms. Repeat this process a few more times. What do you notice when the two products are compared?

56. What is the 100th digit in the decimal representation for $1/7$?

57. In how many different ways can you make change for a half dollar using currently minted U.S. coins, if cents (pennies) are not allowed?

58. Some months have 30 days and some have 31 days. How many months have 28 days?

59. If a year has two consecutive months with Friday the thirteenth, what months must they be?

60. **(a)** Consider the following multiplication problems. They each consist of two 2-digit factors that are obtained by reversing the digits. Notice that the products are the same.

$$\begin{array}{r} 84 \\ \times\ 24 \\ \hline 2{,}016 \end{array} \qquad \begin{array}{r} 48 \\ \times\ 42 \\ \hline 2{,}016 \end{array}$$

$$\begin{array}{r} 23 \\ \times\ 64 \\ \hline 1{,}472 \end{array} \qquad \begin{array}{r} 32 \\ \times\ 46 \\ \hline 1{,}472 \end{array}$$

What pattern do you notice?

(b) Explain the necessary condition for this pattern to exist.

61. The first man introduced himself to the first woman with a brief "palindromic" greeting. What was the greeting? (*Hint:* See Exercises 11, 13, and 19.)

62. Refer to Exercises 11 and 15. Another author of this book (who expects to be somewhat older than 76 in the year x^2) has a palindromic, seven-digit phone number, say *abc-defg*, satisfying these conditions:

$b = 2 \times a$

$e = f$

a, b, and c are all multiples of 4

$a + b + c + d + e + f + g = 47.$

Determine his phone number.

1.4

Calculating, Estimating, and Reading Graphs

The photograph shows the Texas Instruments TI-1795, a typical four-function calculator.

Since the introduction of hand-held calculators in the early 1970s, the methods of everyday arithmetic have been drastically altered. At first, however, they were not inexpensive. One of the first consumer models available was the Texas Instruments SR-10, which sold for nearly $150 in 1973. It could perform the four operations of arithmetic and take square roots, but could do very little more.

Calculators are taken for granted, even by Hollywood. Alan Alda mentions that he does calculations on his Bowmar (an early manufacturer of calculators) in the 1978 movie *Same Time, Next Year.* But why were the cadets in the 1982 movie *An Officer and a Gentleman* using slide rules?

The emphasis placed on paper-and-pencil computation, which has long been a part of the school mathematics curriculum, is not as important as it once was. The search for easier ways to calculate and compute has culminated in the development of hand-held calculators and computers. Professional organizations devoted to the teaching and learning of mathematics recommend that this technology be incorporated throughout the mathematics curriculum. In light of this recommendation, this text assumes that all students have access to calculators, and the authors recommend that students become computer literate to the extent their academic resources and individual finances allow.

In this text we view calculators as a means of allowing students to spend more time on the conceptual nature of mathematics and less time on the drudgery of computation with paper and pencil. Calculators come in a large array of different types, sizes, and prices, making it difficult to decide which machine best fits your needs. For the general population, a calculator that performs the operations of arithmetic and a few other functions is sufficient. These are known as **four-function calculators.** Students who take higher mathematics courses (engineers, for example) usually need the added power of **scientific calculators. Programmable calculators,** which allow short programs to be written and performed, and **graphing calculators,** which actually plot graphs on small screens, are also available. Keep in mind that calculators differ from one manufacturer to another. For this reason, remember the following.

Always refer to your owner's manual if you need assistance in performing an operation with your calculator.

Calculating

While graphing calculators cost more than four-function, scientific, and programmable calculators, they are the wave of the future in the world of hand-held calculators. One of the main advantages of a graphing calculator at the level of mathematics studied in this course is that both the information the user inputs into the calculator and the result generated by that calculator can be viewed on the same screen. In this way, the user can verify that the information entered into the calculator is correct. While it is not necessary to have a graphing calculator to study the material presented in this text, we will occasionally include typical graphing calculator screens to support the results obtained or to provide supplemental information.

Any calculator (particularly a modern graphing calculator) consists of two components: the electronic "box" and the owner's instruction manual that explains how to use it.

Graphing calculators currently available are manufactured by several companies, most notably Texas Instruments, Casio, Sharp, and Hewlett Packard. Because it is one of the most popular models available, we have chosen to include in this text screens generated by TI-Graph Link™ software for the Texas Instruments TI-82 calculator.

The screens that follow illustrate some of the more common types of entries and operations.

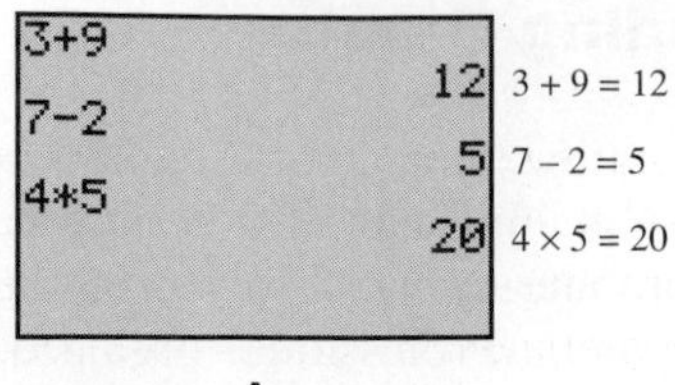

A

Screen A illustrates how two numbers can be added, subtracted, or multiplied.

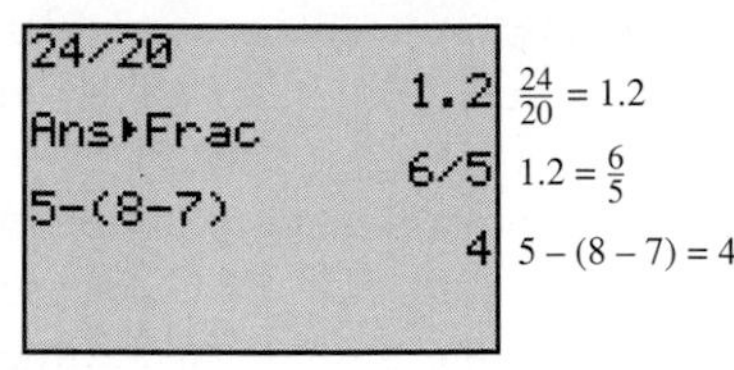

B

Screen B shows how two numbers can be divided, how the decimal quotient (stored in the memory cell ANS) can be converted into a fraction, and how parentheses can be used in a computation.

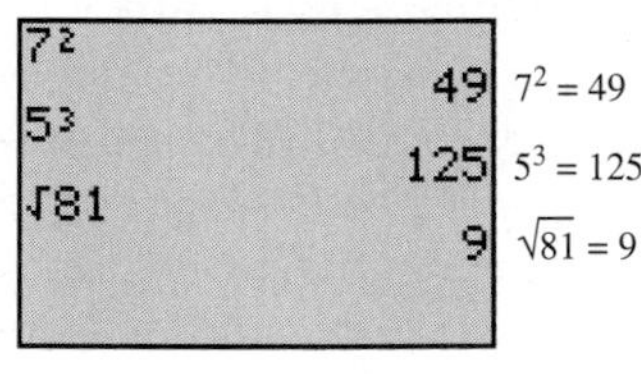

C

Screen C shows how a number can be squared, how it can be cubed, and how its square root can be taken.

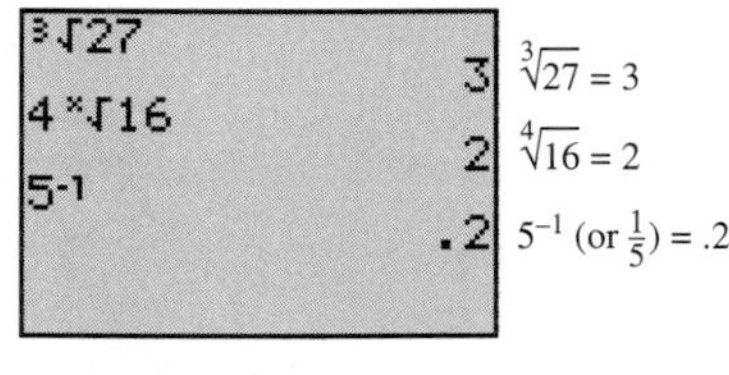

D

Screen D shows how a higher root can be found (cube root and fourth root), and how the reciprocal of a number can be found using -1 as an exponent.

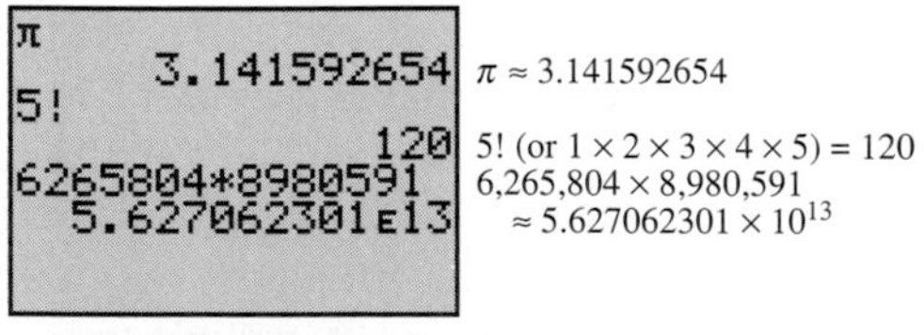

≈ indicates "is approximately equal to"

E

Screen E shows how the important number π can be accessed with its own special key, how a *factorial* (as indicated by !) can be found, and how a result might be displayed in *scientific notation.* (The "E13" following 5.627062301 means that this number is multiplied by 10^{13}. This answer still is only an approximation, because the product 6,265,804 $\times$ 8,980,591 contains more digits than the calculator is capable of displaying.)

Estimation

While calculators can make life easier when it comes to computations, we should not lose sight of the fact that many times we need only estimate an answer to a problem, and in these cases a calculator may not be necessary. In fact, calculator answers are often not appropriate for a problem, as seen in the following example.

EXAMPLE 1 A birdhouse for purple martins can accommodate up to 6 nests. How many birdhouses would be necessary to accommodate 92 nests?

If we divide 92 by 6 either by hand or with a calculator, we get an answer of 15.333333 (rounded). Can this possibly be the desired number? Of course not, since we cannot consider fractions of birdhouses. Now we must ask ourselves whether we should decide on 15 or 16 birdhouses? Because we need to provide nesting space for the nests left over after the 15 birdhouses (as indicated by the decimal fraction), we should plan to use 16 birdhouses. Notice that, in this problem, we must round our answer *up* to the next counting number. ■

The ability to estimate results from data provided in an everyday situation is important. We are bombarded by information (much of it useless to many of us) in this day and age, and much of it is in numerical form. By improving our ability to estimate, we can often obtain approximate results when precise results are not absolutely necessary.

EXAMPLE 2 Harry Kelly, who played basketball for Texas Southern during the years 1980–83, scored 3,066 points in 110 games. What was his approximate average number of points per game?

Since we are only asked to find Harry's approximate average, we can say that he scored about 3,000 points in about 100 games, for an average of about $3{,}000/100 = 30$ points per game. (A calculator shows that his average to the nearest tenth was 27.9 points. Verify this.) ■

EXAMPLE 3 In a recent year there were approximately 127,000 males working on farms in the 25–29-year age bracket. This represented part of the total of 238,000 workers in that age bracket. Of the 331,000 farm workers in the 40–44-year age bracket, 160,000 were males. Without using a calculator, determine which age bracket had a larger proportion of males.

Here it is best to think in terms of thousands, instead of dealing with all the zeros. First, let us analyze the age bracket 25–29 years. Since there were a total of 238 thousand workers, of which 127 thousand were males, there were $238 - 127 = 111$ thousand female workers. Here, more than half of the workers were males. In the 40–44-year age bracket, of the 331 thousand workers there were 160 thousand males, giving $331 - 160 = 171$ thousand females, meaning fewer than half were males. A comparison, then, shows that the 25–29-year age bracket had the larger proportion of males. ■

Reading Graphs

The use of graphs to depict information has become an efficient means of transmitting information in a concise, space-saving way. Any issue of the newspaper *USA Today* will verify this immediately. Most graphs are designed to be easily interpreted by the general public. The next example presents three types of graphs, with questions regarding interpretation.

There are many ways to represent information using graphs and charts. Bar graphs, pie charts, and line graphs are the most common. The following example illustrates one of each.

EXAMPLE 4 Use the graph to make the interpretations.

(a) The bar graph in Figure 6(a) compares the extent of personal computer ownership in the United States, Germany, and Japan. The source of this information is the February 27, 1994 issue of *The New York Times* (Section 3, page 6). Using the scale given on the left of the graph we can determine that about 35 of every 100 U.S. workers has a personal computer. *Estimate* the corresponding numbers for Germany and Japan.

Move horizontally from the top of the bar for Germany over to the scale on the left to see that about 15 of 100 German workers, or 15%, have personal computers. The top of the bar for Japan indicates about 9% of Japanese workers have personal computers.

(b) Baby boomers are expected to inherit \$10.4 trillion from their parents over the next 45 years, an average of \$50,000 each. The pie chart in Figure 6(b) shows how they plan to spend their inheritance. How much of the \$50,000 is expected to go toward the purchase of a home? How much to retirement?

The chart shows that 14% will go toward a home. This amounts to $.14(50{,}000) = \$7{,}000$. They plan to use 20% or $.20(50{,}000) = \$10{,}000$ for retirement.

(c) The line graph in Figure 6(c) shows how the Dow Jones industrial average fluctuated over a 10-day period. What was the approximate level of the Dow Jones industrial average on the first Monday shown? For the 10-day period, did it close higher or lower than the first Monday?

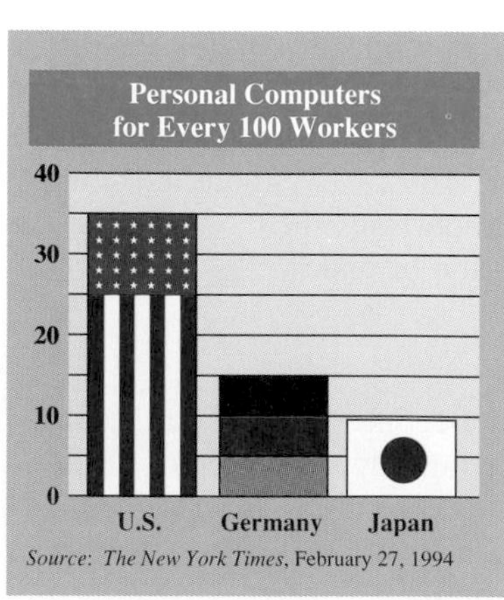

(a)

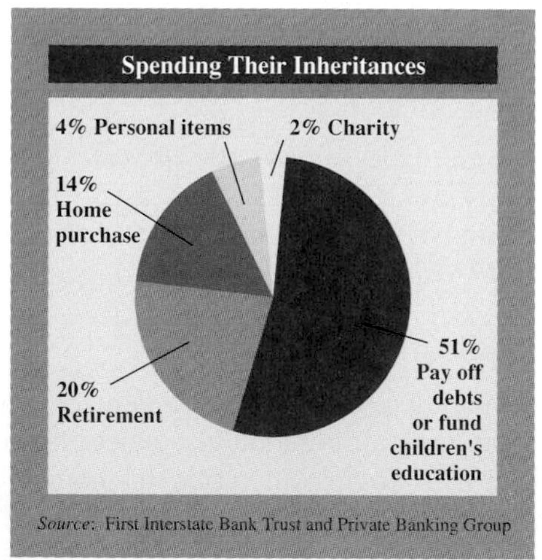

(b)

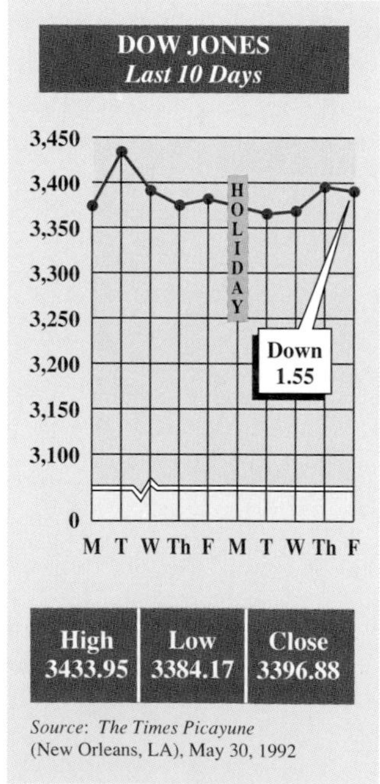

(c)

■ **FIGURE 6**

This line graph indicates that on the first Monday, the average was about halfway between 3,350 and 3,400, indicating that the average was about 3,375 (the number halfway between 3,350 and 3,400). Since the final dot (for the second Friday) lies above the dot for the first Monday, the market closed higher than at the start of the 10-day period. ◆

We will analyze graphs in more detail in the chapter on statistics.

For Further Thought
The Quest for Numeracy

Letter is to *number* as *literacy* is to *numeracy.* In recent years much has been written about how important it is that the general population be "numerate." The essay "Quantity" by James T. Fey in *On the Shoulders of Giants: New Approaches to Numeracy* contains the following description of an approach to numeracy.

> Given the fundamental role of quantitative reasoning in applications of mathematics as well as the innate human attraction to numbers, it is not surprising that number concepts and skills form the core of school mathematics. In the earliest grades all children start on a mathematical path designed to develop computational procedures of arithmetic together with corresponding conceptual understanding that is required to solve quantitative problems and make informed decisions. Children learn many ways to describe quantitative data and relationships using numerical, graphic, and symbolic representations; to plan arithmetic and algebraic operations and to execute those plans using effective procedures; and to interpret quantitative information, to draw inferences, and to test the conclusions for reasonableness.

For Group Discussion
With calculator in hand, each student in the class is to attempt to fill in the boxes with the digits 3, 4, 5, 6, 7, or 8, with each digit used at most once. Then take a poll to see who was able to come up with the closest number to the "goal number." You are allowed one minute per round. Good luck!

Round I □ × □□□□ = 30,000
Round II □ × □□□□ = 40,000
Round III □ × □□□□ = 50,000
Round IV □□ × □□□□ = 30,000
Round V □□ × □□□□ = 60,000

1.4 EXERCISES

Exercises 1–18 are designed to give you practice in learning how to do some basic operations on your calculator. Perform the indicated operations and give as many digits in your answer as shown on your calculator display. (The number of displayed digits will vary depending on the model used.)

1. $28.3 + (9.7 - 4.8)$

2. $4.2 \times (3.1 - 1.6)$

3. $-4.1 + (7.8 \times 1.4)$

4. $[1.331 \div (-11)] + 3$

5. $\sqrt{6.036849}$

6. $\sqrt{34,774,609}$

7. $\sqrt[3]{260{,}917{,}119}$ **8.** $\sqrt[3]{109.215352}$ **9.** 7.4^2

10. 8.2^2 **11.** 9.1^3 **12.** 6.55^3

13. 2.1^5 **14.** 1.8^6 **15.** $\sqrt[6]{1.29}$

16. $\sqrt[5]{9.68}$ **17.** 2π **18.** π^2

19. Use your calculator to *square* the following two-digit numbers ending in 5: 15, 25, 35, 45, 55, 65, 75, 85. Write down your results, and examine the pattern that develops. Then use inductive reasoning to predict the value of 95^2. Write an explanation of how you can mentally square a two-digit number ending in 5.

20. Choose any number consisting of five digits. Multiply it by 9 on your calculator. Now add the digits in the answer. If the sum is more than 9, add the digits again, and repeat until the sum is less than 10. Your answer will always be 9. Repeat the exercise with a number consisting of six digits. Does the same result hold?

By examining several similar computation problems and their answers obtained on a calculator, we can use inductive reasoning to make conjectures about certain rules, laws, properties, and definitions in mathematics. Perform each calculation and observe the answers. Then fill in the blank with the appropriate response. (Justification of these results will be discussed later in the book.)

21. $5 \times (-4)$; -3×8; $2.7 \times (-4.3)$
Multiplying a negative number and a positive number gives a ________ (negative/positive) product.

22. $(-3) \times (-8)$; $(-5) \times (-4)$; $(-2.7) \times (-4.3)$
Multiplying two negative numbers gives a ________ (negative/positive) product.

23. 1^2; 1^3; 1^{-3}; 1^0; 1^{13}
Raising 1 to any power gives a result of ____.

24. 5.6^0; π^0; 2^0; 120^0; $.5^0$
Raising a nonzero number to the power 0 gives a result of ____.

25. $1/7$; $1/(-9)$; $1/3$; $1/(-8)$
The sign of the reciprocal of a number is ________ (the same as/different from) the sign of the number.

26. $0/8$; $0/2$; $0/(-3)$; $0/\pi$
Zero divided by a nonzero number gives a quotient of ____.

27. $5/0$; $9/0$; $\pi/0$; $-3/0$; $0/0$
Dividing a number by 0 gives a result of ____ on a calculator. (What do you think this indicates?)

28. $\sqrt{-3}$; $\sqrt{-5}$; $\sqrt{-6}$; $\sqrt{-10}$
Taking the square root of a negative number gives a result of ____ on a calculator. (What do you think this indicates?)

29. $(-3) \times (-4) \times (-5)$; $(-3) \times (-4) \times (-5) \times (-6) \times (-7)$;
$(-3) \times (-4) \times (-5) \times (-6) \times (-7) \times (-8) \times (-9)$
Multiplying an *odd* number of negative numbers gives a ________ (positive/negative) product.

30. $(-3) \times (-4)$; $(-3) \times (-4) \times (-5) \times (-6)$;
$(-3) \times (-4) \times (-5) \times (-6) \times (-7) \times (-8)$
Multiplying an *even* number of negative numbers gives a ________ (positive/negative) product.

31. Find the decimal fraction for 2/3 on your calculator. The display will show perhaps a lead 0, a decimal point, and a string of all 6s or a string of 6s with final digit 7. Does your calculator *truncate* or *round off*?

32. From algebra we know that we are allowed to raise a negative number to a counting number power. For example,

$$(-3)^5 = (-3) \times (-3) \times (-3) \times (-3) \times (-3)$$
$$= -243.$$

However, some scientific calculators will not allow a negative number to be raised to a power, despite the fact that this is mathematically valid. Suppose your calculator gives an error message when you try to evaluate $(-3)^5$. Give an explanation of how you would use your calculator to do this computation. (*Hint:* See the result of Exercise 29.)

33. Choose any three-digit number and enter the digits into a calculator. Then enter them again to get a six-digit number. Divide this six-digit number by 7. Divide the result by 13. Divide the result by 11. What is your answer? Explain why this happens.

34. Choose any digit except 0. Multiply it by 429. Now multiply the result by 259. What is your answer? Explain why this happens.

For each of the following, perform the indicated calculation on a scientific calculator. (These exercises do not apply to typical graphing calculator display screens.) Then turn your calculator upside down to read the word that belongs in the blank in the accompanying sentence.

35. 11,669 × 3; One of teen idol Fabian's 1959 records was *Turn Me* _____.

36. 7,531,886.6 ÷ 1.4; Have you ever had a case of the _____?

37. 5 × 10,609; Imelda has an obsession for _____.

38. 128,396 − 93,016; When Mrs. Percy Pearl Washington died in 1972, she weighed about 880 pounds. She was very _____.

39. Make up your own exercise similar to Exercises 35–38. When a standard calculator is turned upside down, the digits in the display correspond to letters of the English alphabet as follows:

0 ↔ O	3 ↔ E	7 ↔ L
1 ↔ I	4 ↔ h	8 ↔ B
2 ↔ Z	5 ↔ S	9 ↔ G.

40. Displayed digits on most calculators usually show some or all of the parts in the pattern shown in the figure. For the digits 0 through 9:
(a) Which part is used most frequently?
(b) Which part is used the least?
(c) Which digit uses the most parts?
(d) Which digit uses the fewest parts?

Give the appropriate counting number answer to each question in Exercises 41–44.

41. A certain type of carrying case will hold a maximum of 48 audio cassettes. If you need to store 490 audio cassettes, how many carrying cases will you require?

42. A plastic page designed to store baseball cards will hold up to 16 cards. If you must store your collection of 484 cards, how many pages will you need?

43. Each room available for administering a placement test will hold up to 40 students. Two hundred fifty students have signed up for the test. How many rooms will be used?

44. A gardener wants to fertilize 2,000 tomato plants. Each bag of fertilizer will supply up to 150 plants. How many bags does she need to do the job?

In Exercises 45–50, use estimation to choose the letter of the choice closest to the correct answer.

45. The planet Mercury takes 88.0 Earth days to revolve around the sun. Pluto takes 90,824.2 days to do the same. When Pluto has revolved around the sun once, about how many times will Mercury have revolved around the sun?
(a) 100 **(b)** 1,000 **(c)** 10,000
(d) 100,000

46. Hale County in Texas has a population of 34,671 and covers 1,005 square miles. About how many inhabitants per square mile does the county have?
(a) 35 **(b)** 350 **(c)** 3,500 **(d)** 35,000

47. The 1990 U.S. census showed that the total population of the country was 248,709,873. Of these, 25,223,086 were in the 45–50-year age bracket. On the average, about one in every _____ citizens is in this age bracket.
(a) 5 **(b)** 10 **(c)** 15 **(d)** 20

48. The minimum start-up fee to open a Dairy Queen franchise is $375,000. Suppose that 19 people decide to put up equal amounts toward the fee. About how much would each have to contribute?
(a) $10,000 **(b)** $15,000 **(c)** $20,000
(d) $25,000

49. In voting for the portrayal of which age of Elvis that was to be depicted on a U.S. postage stamp, voters cast 851,000 votes for the younger Elvis against 277,723 votes for the more mature Elvis. A newspaper article read "Younger Elvis is Winner by _____ to 1 Margin." What number belongs in the blank?
(a) 5 **(b)** 4 **(c)** 3 **(d)** 2

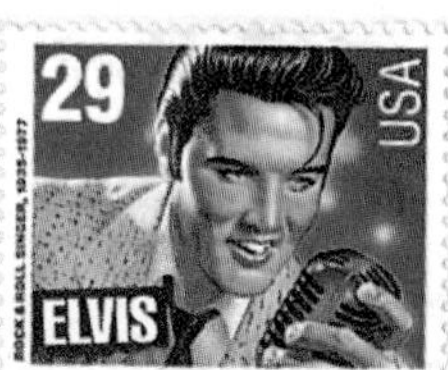

50. The distance from Springfield, MO, to Seattle, WA, is 2,009 miles. If a bus averages about 50 miles per hour, about how many hours would this trip take?
(a) 30 **(b)** 35 **(c)** 40 **(d)** 45

When the Dawkins family left Hattiesburg to visit their cousins in Columbus, their car's instrument panel looked like this.

When they arrived in Columbus, it looked like this.

51. What is the round-trip distance between Columbus and Hattiesburg?

52. If the car's tank holds 16 gallons of gasoline, how much is left?

53. How many miles per gallon did the car get?

54. How many gallons of gas will the car have left after the return trip?

Use the bar graph to answer the questions in Exercises 55–58.

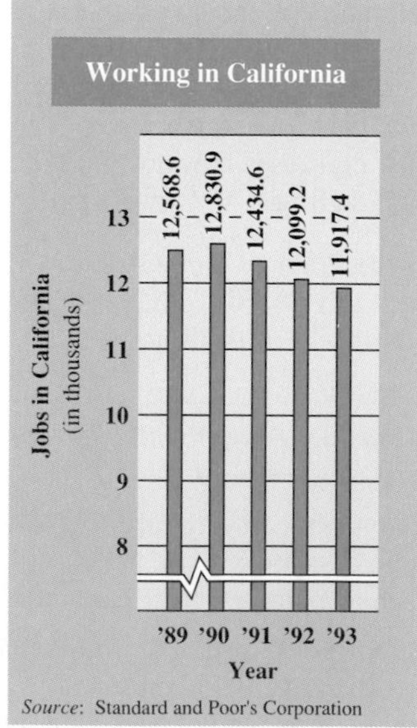

55. Between what two consecutive years shown on the graph did the number of jobs in California increase? How much was the increase?

56. How many jobs did California lose between 1990 and 1993?

57. According to Standard & Poor's Corporation, half of the losses between 1990 and 1993 were in defense and 40% were in construction. How many were lost in defense? How many were lost in construction?

58. Suppose that the same number of jobs were lost between 1993 and 1994 as were lost between 1992 and 1993. How many jobs were there in California in 1994 based on this assumption?

Use the line graph to answer the questions in Exercises 59–62.

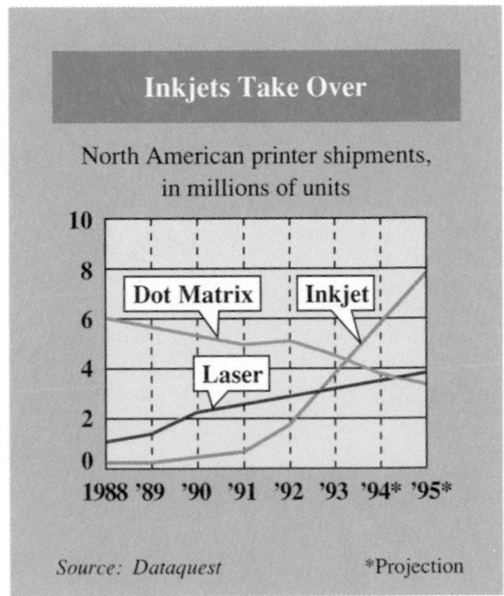

59. Which one of the following would be the best estimate for the total number of units of printer shipments in North America in 1989?
(a) 5.8 million **(b)** 1.5 million
(c) .2 million **(d)** 7.5 million

60. In what year did the number of Inkjet printer shipments first exceed the number of Laser printer shipments?

61. What type of printer has shown the most rapid increase in shipments since 1992?

62. How has the trend in shipments of Dot Matrix printers differed from the shipments of Inkjet and Laser printers since 1992?

Use the pie chart to answer the questions in Exercises 63–66.

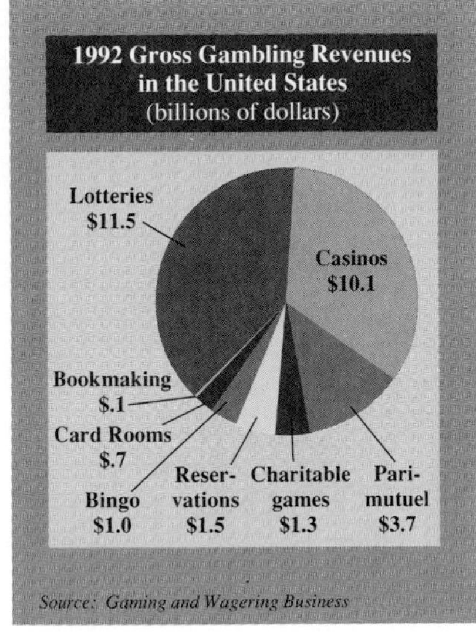

63. In 1992, the gaming industry had \$29.9 billion in gross revenues. What percent of this was from lotteries?

64. Suppose that a 4% tax had been levied on gaming revenues in 1992. How much tax would have been collected from casinos?

65. Suppose that casinos paid out 83% of their gross revenues. How much would they have paid out in 1992?

66. How much more in gross revenues did lotteries take in than bookmaking, card rooms, bingo, charitable games, reservations, and pari-mutuel combined?

67. Which type of graph studied in this section would be most suitable to illustrate the percentages of different races in a group of 1,000 people?

68. Why would a pie chart not be suitable to depict the trends in price fluctuations in gasoline over a five-year period?

The map shows the predicted weather information for a summer day in July. Use the map and the accompanying legend to answer the following questions.

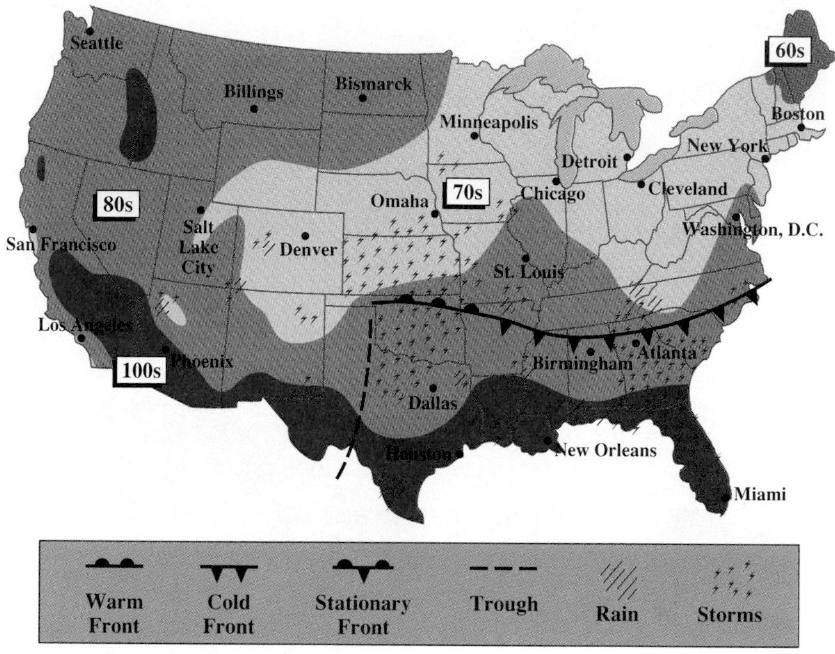

Source: Weather Data Incorporated

69. Of the cities Houston, Minneapolis, St. Louis, and Omaha, which one has a predicted temperature in the 80s?

70. Of the cities Billings, Detroit, Miami, and Washington, D.C., which one has a prediction of storms?

71. What kind of front is just north of Birmingham and Atlanta?

72. In what state is a trough located?

EXTENSION

Using Writing to Learn About Mathematics

The use of writing in the mathematics curriculum has recently received increased attention, due in large part to the recommendations of the National Council of Teachers of Mathematics (N.C.T.M.). In its 1989 publication *Curriculum and Evaluation Standards for School Mathematics,* it proposes that "the mathematics curriculum should include the continued development of language and symbolism to communicate mathematical ideas. . . ." Research has indicated that one's ability to express mathematical observations in writing can serve as a positive force in one's continued development as a mathematics student. The implementation of writing in the mathematics class can utilize several approaches.

Mathematical writing takes on many forms. One of the most famous of all author/mathematicians was Charles Dodgson (1832–1898), who began his writing by using the pen name Lewis Carroll.

continued on page 41

Dodgson was a mathematics lecturer at Oxford University in England. Queen Victoria told Dodgson how much she enjoyed *Alice's Adventures in Wonderland* and how much she wanted to read his next book; he is said to have sent her *Symbolic Logic,* his most famous mathematical work. In this book, Dodgson makes some mighty strong claims for the subject: "It [symbolic logic] will give you clearness of thought—the ability to *see your way* through a puzzle—the habit of arranging your ideas in an orderly and get-at-able form—and, more valuable than all, the power to detect *fallacies,* and to tear to pieces the flimsy illogical arguments which you will so continually encounter in books, in newspapers, in speeches, and even in sermons . . . ." The *Alice* books made Carroll famous. Late in life, however, Dodgson shunned attention and denied that he and Carroll were the same person, even though he gave away hundreds of signed copies to children and children's hospitals.

Journals

One way of using writing to help you learn about mathematics is to keep a journal in which you spend a few minutes explaining what happened in class that day. The journal entries may be general or specific, depending on the topic covered, the degree to which you understand the topic, your interest level at the time, and so on. Journal entries are usually written in informal language, and are often an effective means of communicating to yourself, your classmates, and your instructor what feelings, perceptions, and concerns you are having at the time.

Learning Logs

While journal entries are for the most part unstructured writings in which the student's thoughts are allowed to roam freely, entries in learning logs are typically more structured. An instructor may pose a specific question for a student to answer in a learning log. In this text, we intersperse writing exercises in each exercise set that are appropriate for answering in a learning log. For example, consider Exercise 11 in the exercise set for the opening section in this chapter.

> Discuss the differences between inductive and deductive reasoning. Give an example of each.

Here is a possible response to this exercise.

Deductive reasoning occurs when you go from general ideas to specific ones. For example, I know that I can multiply both sides of $\frac{1}{2}x = 6$ by 2 to get $x = 12$, because I can multiply any equation by whatever I want (except 0). Inductive reasoning goes the other way. If I make a general conclusion from specific observations, that's inductive reasoning. Example – in the numbers 4, 8, 12, 16, and so on, I can conclude the next number is 20, since to get the next number I always add 4.

Reports on Articles from Mathematics Publications

The motto "Publish or perish" has long been around, implying that a scholar in pursuit of an academic position must publish in a journal in his or her field in order to find or retain the position. Mathematics is no exception, and there are numerous journals that publish papers in mathematics research and/or mathematics education. While many such articles are beyond the level of students enrolled in a class for which this text is appropriate, there are also many that *can* be read and understood. In Activity 3 at the end of this section, we provide some suggestions of articles that have appeared within the last few years that can be understood by undergraduate nonmathematics majors. A report on such an article can help you understand what mathematicians do, and what ideas mathematics teachers use to convey the concepts of their subject to their students.

Term Papers

A term paper in a mathematics class? In increasing numbers, professors in mathematics survey courses (like the ones for which this text is intended) are requiring short term papers of their students. In this way, you can become aware of the plethora of books and articles on mathematics and mathematicians, many written specifically for the layperson. A term paper in a mathematics survey course need not delve into high-powered mathematics. For example, it may investigate the personal side of a mathematician, the problems that women have faced because of the perception of mathematics as a "man's subject," the ongoing investigation of a classic problem in mathematics, and so on. In Activities 6 and 7 at the end of this section, we provide a list of possible term paper topics.

By using writing as a tool for communicating in mathematics, you are able to make the learning process a more personal and individual endeavor. Powell and Lopez draw the following conclusion as a result of a case study of writing in mathematics.

> Reflecting critically in writing about the mathematics they are learning gives students greater potential to control their learning and to develop criteria for monitoring their progress. The development of control and monitoring capabilities engenders in students feelings of accomplishment. These feelings, in turn, have a positive effect on their affective response toward the mathematics they were learning. Finally, from acquiring greater control over their learning, developing criteria for personal standards of progress, and conceptually understanding the mathematics in which they are engaged, students derive a great deal of satisfaction with themselves as learners capable of doing *and* understanding mathematics.*

Activities

To the Student

Rather than include a typical exercise set, due to the nature of the extension we have chosen to list some suggested activities in which writing can be used to enhance your learning of mathematics and your awareness about the world of mathematics.

ACTIVITY 1 Keep a journal in this class. After each class, write for a few minutes on your feelings about the class, about the topics covered, or whatever you feel is appropriate. In your journal you may wish to use the following guidelines.

* From "Writing as a Vehicle to Learn Mathematics: A Case Study" by Arthur B. Powell and Jose A. Lopez, a research paper appearing as Chapter 13 in *Writing to Learn Mathematics and Science,* edited by Paul Connolly and Teresa Vilardi, copyright 1989 by Teachers College, Columbia University.

Journal Writing*

1. WHO should write in your journal?
   You should.
2. WHAT should you write in your journal?
   New words or new ideas or new formulas or new concepts you've learned
   Profound thoughts you've had
   Wonderings, musings, problems to solve
   Reflections on the class
   Questions—both answerable and unanswerable
   Writing ideas
3. WHEN should you write in your journal?
   After class each day
   As you are preparing, reading, or studying for class
   Anytime an insight or question hits you
4. WHERE should you write in your journal?
   Anywhere—so keep it with you when possible.
5. WHY should you write in your journal?
   It will record ideas that you might otherwise forget.
   It will be worthwhile for you to read later on so that you can note your growth.
   It will facilitate your learning, problem solving, writing, reading, and discussion in this class.
6. HOW should you write in your journal?
   In wonderful, long, flowing sentences with perfect punctuation and perfect spelling and in perfect handwriting
   Or in single words that express your ideas, in short phrases, in sketches, in numbers, in maps, in diagrams, in sentences

ACTIVITY 2 Keep a learning log, answering at least one writing exercise from each exercise set covered in your class syllabus. Ask your teacher for suggestions of other types of specific writing assignments. For example, you might wish to choose a numbered example from a section in the text and write your own solution to the problem, or comment on the method that the authors use to solve the problem. Don't be afraid to be critical of the method used in the text.

ACTIVITY 3 The National Council of Teachers of Mathematics publishes journals in mathematics education: *Teaching Children Mathematics* (formerly called *Arithmetic Teacher*) and *Mathematics Teacher* are two of them. These journals can be found in the periodicals section of most college and university libraries. We have chosen several recent articles in each of these journals that are written at a level suitable for a report by a nonmathematics major. Of course, there are thousands of other articles from which to choose. Write a short report on one of these articles according to guidelines specified by your instructor.

* Box "Journal Writing" from "No Time for Writing in Your Class?" by Margaret E. McIntosh in *Mathematics Teacher,* September 1991, p. 431. Reprinted by permission.

From *Mathematics Teacher*

1986

Sowder, Larry. "The Looking-back Step in Problem Solving." Vol. 79, No. 7, October 1986, pp. 511–513.

1987

McLure, John W. "Six-legged Math." Vol. 80, No. 7, October 1987, pp. 524–526.

1988

Ewbank, William A. "Cryptarithms: Math Made Me Daft, Momma." Vol. 81, No. 1, January 1988, pp. 54–58.

Heid, M. Kathleen. "Calculators on Tests—One Giant Step for Mathematics Education." Vol. 81, No. 9, December 1988, pp. 710–713.

1989

Garofalo, Joe. "Beliefs and Their Influence on Mathematical Performance." Vol. 82, No. 7, October 1989, pp. 502–505.

Long, Vena M., et al. "Using Calculators on Achievement Tests." Vol. 82, No. 5, May 1989, pp. 318–325.

Philipp, Randolph A. "Piano Tuners and Problem Solving." Vol. 82, No. 4, April 1989, pp. 248–249.

1990

Dehan, Harriet Stone. "A Mathematical Magic Show." Vol. 83, No. 7, October 1990, pp. 515–523.

Mitchell, Charles E. "Real-World Mathematics." Vol. 83, No. 1, January 1990, pp. 12–16.

Spieler, Robert R. "A Mathematics Magazine and Fair." Vol. 83, No. 9, December 1990, pp. 698–700.

Tuttle, Jerome E. "What I Wish I Had Learned in School." Vol. 83, No. 6, September 1990, pp. 426–429.

1992

Demana, Frank and Bert K. Waits. "A Computer for *All* Students." Vol. 85, No. 2, February 1992, pp. 94–95.

Dodd, Anne Wescott. "Insights from a Math Phobic." Vol. 85, No. 4, April 1992, pp. 296–298.

Sutton, Gail Oberholtzer. "Cooperative Learning Works in Mathematics." Vol. 85, No. 1, January 1992, pp. 63–66.

1993

Curcio, Frances R., and J. Lewis McNeese. "The Case of Video Viewing, Reading, and Writing in Mathematics Class: Solving the Mystery." Vol. 86, No. 8, November 1993, pp. 682–685.

Klein, Mary F. "Mathematics as Current Events." Vol. 86, No. 2, February 1993, pp. 114–116.

Roche, James W. "Blackstone's Mathmagic." Vol. 86, No. 9, December 1993, pp. 733–734.

1994

Callahan, Walter. "Teaching Middle School Students with Diverse Cultural Backgrounds." Vol. 87, No. 2, February 1994, pp. 122–126.

Morgan, John L., and John L. Ginther. "The Magic of Mathematics." Vol. 87, No. 3, March 1994, pp. 150–153.

Walton, Karen Doyle. "Albrecht Dürer's Renaissance Connections between Mathematics and Art." Vol. 87, No. 4, April 1994, pp. 278–282.

1995

Disher, Fan. "Graphing Art." Vol. 88, No. 2, February 1995, pp. 124–128, 134–136.

From *Arithmetic Teacher* and *Teaching Children Mathematics*

1990

Ashlock, Robert B. "Parents Can Help Children Learn Mathematics." Vol. 38, No. 3, November 1990, pp. 42–46.

Becker, Jerry P., et al. "Some Observations of Mathematics Teaching in Japanese Elementary and Junior High Schools." Vol. 38, No. 2, October 1990, pp. 12–21.

Feinberg, Miriam M. "Using Patterns to Practice Basic Facts." Vol. 37, No. 8, April 1990, pp. 38–41.

Ford, Margaret I. "The Writing Process: A Strategy for Problem Solvers." Vol. 38, No. 3, November 1990, pp. 35–38.

Lyon, Betty Clayton. "Creating Magic Squares." Vol. 38, No. 4, December 1990, pp. 48–53.

Sindar, Viji K. "Thou Shall Not Divide by Zero." Vol. 37, No. 7, March 1990, pp. 50–51.

Zaslavsky, Claudia. "Symmetry in American Folk Art." Vol. 38, No. 1, September 1990, pp. 6–12.

1991

Bobis, Janette T. "Using a Calculator to Develop Number Sense." Vol. 38, No. 5, January 1991, pp. 42–45.

Goldenberg, E. Paul. "A Mathematical Conversation with Fourth Graders." Vol. 38, No. 8, April 1991, pp. 38–43.

Harel, Guershon, and Marilyn Behr. "Ed's Strategy for Solving Division Problems." Vol. 39, No. 3, November 1991, pp. 38–40.

Krieger, Shelley. "The Tangram—It's More Than an Ancient Puzzle." Vol. 38, No. 9, May 1991, pp. 38–43.

Moniuszko, Linda K. "Reality Math." Vol. 39, No. 1, September 1991, pp. 10–16.

Norman, F. Alexander. "Figurate Numbers in the Classroom." Vol. 38, No. 7, March 1991, pp. 42–45.

Taylor, Lyn, et al. "American Indians, Mathematical Attitudes, and the *Standards.*" Vol. 38, No. 6, February 1991, pp. 14–21.

Zepp, Raymond A. "Real-Life Business Math at Enterprise Village." Vol. 39, No. 4, December 1991, pp. 10–14.

1993

Gailey, Stavroula K. "The Mathematics-Children's Literature Connection." Vol. 40, No. 5, January 1993, pp. 258–259.

Greenes, Carol, Linda Shulman, and Ripka Spungin. "Developing Sense about Numbers." Vol. 40, No. 5, January 1993, pp. 279–284.

Kliman, Marlene. "Integrating Mathematics and Literature in the Elementary Classroom." Vol. 40, No. 6, February 1993, pp. 318–321.

Lobato, Joanne E. "Making Connections with Estimation." Vol. 40, No. 6, February 1993, pp. 347–351.

1994

Goldberg, Kenneth P. "How Old Do You Want to Be?" Vol. 1, No. 3, November 1994, pp. 184–187.

Huse, Vanessa Evans, Nancy Larson Bluemel, and Rhonda Harris Taylor. "Making Connections: From Paper to Pop-Up Book." Vol. 1, No. 1, September 1994, pp. 14–17.

1995

Flores, Alfinio. "Bilingual Lessons in Early-Grade Geometry." Vol. 1, No. 7, March 1995, pp. 420–424.

Hall, Jane B. and Rita P. Acri. "A Fourth Grade Friday Math Night." Vol. 2, No. 1, September 1995, pp. 8–10.

Jaberg, Patricia. "Links to Literature: Assessment and *Geraldine's Blanket.*" Vol. 1, No. 8, April 1995, pp. 514–517.

ACTIVITY 4 The most well-known production of the Children's Television Workshop is the classic *Sesame Street.* But did you know that this group also produces a series called *Square One,* whose goal is to promote mathematics among middle-school and junior high students? In a lively, entertaining half-hour, each program teaches basic mathematical ideas using skits, music, cartoons, and magic, with an emphasis on logic and problem solving. While the intended audience is a younger age group than most students studying from this book, anyone can learn from these shows, as they interject humor and tongue-in-cheek wit while they introduce mathematical concepts.

Square One is broadcast on many PBS affiliates throughout the country. Check your local TV listings, and write a report on an episode of this series, according to the guidelines of your instructor.

ACTIVITY 5 One of the most popular mathematical films of all time is *Donald in Mathmagicland,* produced by Disney in 1959. After nearly forty years it holds up beautifully, and is considered a true classic. It is available on video, and may be rented at many video stores or purchased at outlets that sell video releases.

Spend an entertaining half hour watching this film, and write a report on it according to the guidelines of your instructor.

Activity 6 Write a report according to the guidelines of your instructor on one of the following mathematicians, philosophers, and scientists.

Abel, N.
Agnesi, M. G.
Agnesi, M. T.
Al-Khowârizmi
Apollonius
Archimedes
Aristotle
Babbage, C.
Bernoulli, Jakob
Bernoulli, Johann
Cantor, G.
Cardano, G.
Copernicus, N.
De Morgan, A.
Descartes, R.
Euler, L.
Fermat, P.
Fibonacci (Leonardo of Pisa)
Galileo (Galileo Galilei)
Galois, E.
Gauss, C.
Hilbert, D.
Kepler, J.
Kronecker, L.
Lagrange, J.
Leibniz, G.
L'Hospital, G.
Lobachevsky, N.
Mandelbrot, B.
Napier, J.
Newton, I.
Noether, E.
Pascal, B.
Plato
Pythagoras
Ramanujan, S.
Riemann, G.
Russell, B.
Somerville, M.
Tartaglia, N.
Whitehead, A.
Wiles, A.

Activity 7 Write a term paper on one of the following topics in mathematics, according to the guidelines of your instructor.

Babylonian mathematics
Egyptian mathematics
The origin of zero
Plimpton 322
The Rhind papyrus
Origins of the Pythagorean theorem
The regular (Platonic) solids
The Pythagorean brotherhood
The Golden Ratio (Golden Section)
The three famous construction problems of the Greeks
The history of the approximations of π
Euclid and his "Elements"
Early Chinese mathematics
Early Hindu mathematics
Origin of the word *algebra*
Magic squares
Figurate numbers
The Fibonacci sequence
The Cardano/Tartaglia controversy
Historical methods of computation (logarithms, the abacus, Napier's rods, the slide rule, etc.)
Pascal's triangle
The origins of probability theory
Women in mathematics
Mathematical paradoxes
Unsolved problems in mathematics
The four-color theorem
The proof of Fermat's Last Theorem
The search for large primes
Fractal geometry
The co-inventors of calculus
The role of the computer in the study of mathematics
Mathematics and music
Police mathematics
The origins of complex numbers
Goldbach's conjecture

ACTIVITY 8 Investigate a computer program that focuses on teaching children elementary mathematics and write a critical review of it, as if you were writing for a journal that contains software reviews of educational material. Be sure to address the higher-level thinking skills in addition to drill and practice.

COLLABORATIVE INVESTIGATION
Discovering Pascal's Triangle

One fascinating array of numbers, *Pascal's Triangle,* consists of rows of numbers, each of which contains one more entry than the one before. The first five rows are shown here.

```
            1
         1     1
      1     2     1
   1     3     3     1
1     4     6     4     1
```

To discover some of its patterns, divide the class into groups of four students each. Within each group designate one student as A, one as B, one as C, and one as D. Then perform the following activities in order.

1. Discuss among group members some of the properties of the triangle that are obvious from observing the first five rows shown.
2. It is fairly obvious that each row begins and ends with 1. Discover a method whereby the other entries in a row can be determined from the entries in the row immediately above it. (*Hint:* In the fifth row, $6 = 3 + 3$.) Then, as a group, find the next four rows of the triangle, and have each member prepare his or her own copy of the entire first nine rows for later reference.
3. Now each student in the group will investigate a particular property of the triangle. In some cases, a calculator will be helpful. All students should begin working at the same time. (A discussion will follow.)

   Student A: Find the sum of the entries in each row. Notice the pattern that emerges. Now write the tenth row of the triangle.

   Student B: Investigate the successive differences in the diagonals from upper left to lower right. For example, in the diagonal that begins 1, 2, 3, 4, . . . , the successive differences are all 1; in the diagonal that begins 1, 3, 6, . . . , the successive differences are 2, 3, 4, and so on. Do this up through the diagonal that begins 1, 6, 21, ...

   Student C: Find the values of the first five powers of the number 11, starting with 11^0 (recall $11^0 = 1$).

   Student D: Arrange the nine rows of the triangle with all rows "flush left," and then draw lightly dashed arrows as shown:

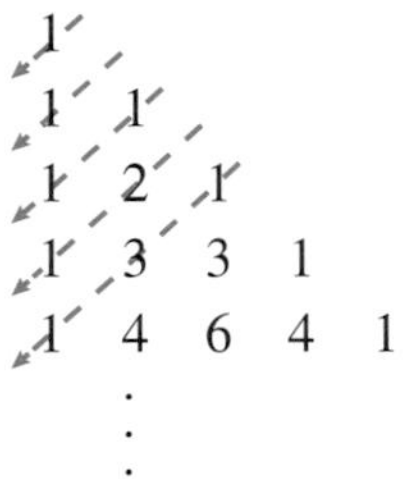

   and so on. Then add along the diagonals. Write these sums in order from left to right.
4. After all students have concluded their individual investigations in item 3, return to a group discussion.
   (a) Have student A report the result found in item 3, and then make a prediction concerning the sum of the entries in the tenth row.
   (b) Have student B report the successive differences discovered in the diagonals. Then have all students in the group in-

vestigate the successive differences in the diagonal that begins 1, 7, 28 . . . . (It may be necessary to write a few more rows of the triangle.)

(c) Have student C report the relationship between the powers of 11 found, and then determine the value of 11^5. Why does the pattern not continue here?

(d) Have student D report the sequence of numbers found. Then, as a group, predict what the next sum will be by observing the pattern in the sequence. Confirm your prediction by actual computation.

5. Choose a representative from each group to report to the entire class the observations made throughout this investigation.

Chapter 1 Test

Decide whether each of Exercises 1 and 2 is an example of inductive or deductive reasoning.

1. Angelica Canales is a salesperson for a major publishing company. She has met her annual sales goal in each of the past four years. Therefore, she will reach her goal this year.

2. If you square a positive number, you will get a positive number. The number 5 is positive. Therefore, 5^2 is a positive number.

3. What is the most probable next number in the list 1, 1/3, 1/5, 1/7, . . . ?

4. Use the list of equations and inductive reasoning to predict the next equation, and then verify your conjecture.

$$65{,}359{,}477{,}124{,}183 \times 17 = 1{,}111{,}111{,}111{,}111{,}111$$
$$65{,}359{,}477{,}124{,}183 \times 34 = 2{,}222{,}222{,}222{,}222{,}222$$
$$65{,}359{,}477{,}124{,}183 \times 51 = 3{,}333{,}333{,}333{,}333{,}333$$

5. Use the method of successive differences to find the next term in the sequence

3, 11, 31, 69, 131, 223 . . . .

6. Find the sum $1 + 2 + 3 + \cdots + 1{,}000$.

7. Consider the following equations, where the left side of each is an octagonal number.

$$1 = 1$$
$$8 = 1 + 7$$
$$21 = 1 + 7 + 13$$
$$40 = 1 + 7 + 13 + 19$$

Use the pattern established on the right sides to predict the next octagonal number. What is the next equation in the list?

8. Use the result of Exercise 7 and the method of successive differences to find the first eight octagonal numbers. Then divide each by 4 and record the remainder. What is the pattern obtained?

9. Explain the pattern used to obtain the terms of the Fibonacci sequence 1, 1, 2, 3, 5, 8, 13, 21, ...

10. What is the units digit in 2^{35}?

11. If fence posts are placed 6 feet apart, how many posts are necessary to construct a 60-foot fence (with no gates)?

12. How many rectangles are in this figure?

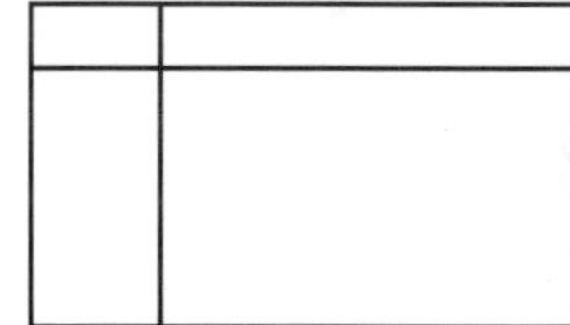

13. I am thinking of a number. If I double it, add 6 to the result, triple that result and then subtract 4, the final result is 50. What is my number?

14. Which is correct? Three cubed *is* nine or three cubed *are* nine?

15. What is the 103rd digit in the decimal representation for 1/11?

16. Based on your knowledge of elementary arithmetic, explain the pattern that can be observed when the following operations are performed: 9×1, 9×2, 9×3, . . . , 9×9. (*Hint:* Add the digits in the answers. What do you notice?)

Use your calculator to evaluate each of the following. Give as many decimal places as the calculator displays.

17. $\sqrt{17.4}$ **18.** 1.5^3

19. In 1989 Blaise Bryant, a running back for the Iowa State football team, carried the ball 299 times and scored 19 touchdowns. This means that he scored a touchdown once in about every _____ times he carried the ball.

(a) 10 **(b)** 15 **(c)** 20 **(d)** 25

20. The line graph at the right depicts how the average National Football League player's salary increased from 1987 to 1993.

(a) By how much did the salary increase from 1992 to 1993?

(b) Determine whether the following statement is true or false: The average salary in 1993 was less than three times the average salary in 1987.

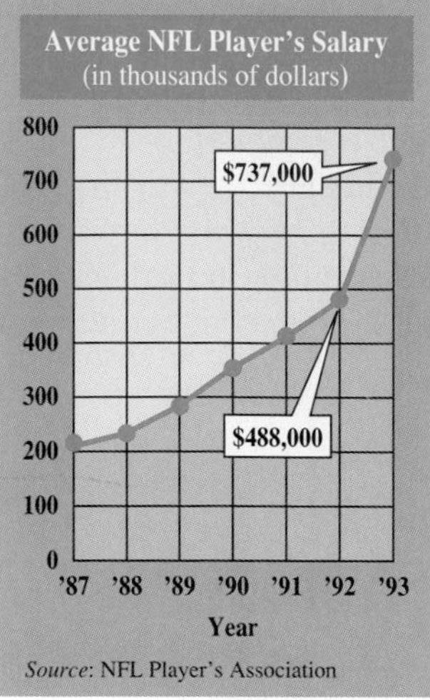

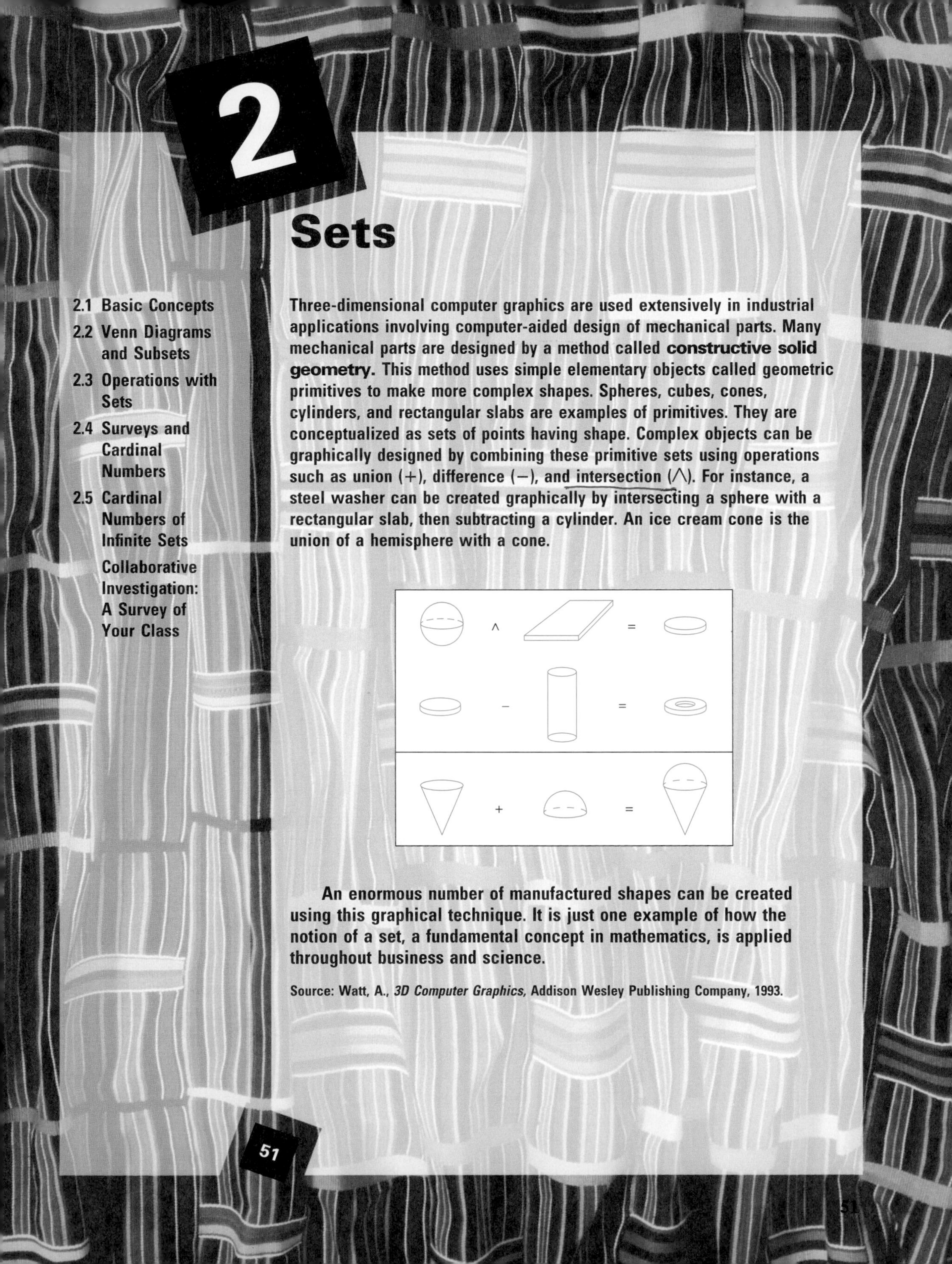

2

Sets

Three-dimensional computer graphics are used extensively in industrial applications involving computer-aided design of mechanical parts. Many mechanical parts are designed by a method called **constructive solid geometry.** This method uses simple elementary objects called geometric primitives to make more complex shapes. Spheres, cubes, cones, cylinders, and rectangular slabs are examples of primitives. They are conceptualized as sets of points having shape. Complex objects can be graphically designed by combining these primitive sets using operations such as union (+), difference (−), and intersection ($\wedge$). For instance, a steel washer can be created graphically by intersecting a sphere with a rectangular slab, then subtracting a cylinder. An ice cream cone is the union of a hemisphere with a cone.

An enormous number of manufactured shapes can be created using this graphical technique. It is just one example of how the notion of a set, a fundamental concept in mathematics, is applied throughout business and science.

Source: Watt, A., *3D Computer Graphics,* Addison Wesley Publishing Company, 1993.

The human mind likes to create collections. Instead of seeing a group of five stars as five separate items, people tend to see them as one group of stars. The mind tries to find order and patterns.

In mathematics this tendency to create collections is represented with the idea of a *set*. A set is a collection of objects. Sets occur in mathematics in many ways, for example when we group together all the solutions to a given problem, or perhaps all the numbers that have a meaning in a given situation.

Many of the topics covered in this book rely on sets. For example, sets of numbers, introduced in this chapter, will be treated in more detail and sets of outcomes for probability will appear in other chapters.

The basic ideas of set theory were developed by the German mathematician Georg Cantor (1845–1918) in about 1875. Some of the things he proved flew in the face of accepted mathematical beliefs of the times. Controversial ideas are seldom well received, and this was especially so in the 1870s. Cantor's ideas are discussed in more detail in this chapter.

2.1 Basic Concepts

A *set* was described above as a "collection of objects." The idea of set, or collection, is also conveyed by many other words, such as *group* and *assemblage*. The objects belonging to the set are called the **elements,** or **members,** of the set. Sets are designated in at least the following three ways: by (1) *word description,* (2) the *listing method,* and (3) *set-builder notation.* A given set may be more conveniently denoted by one method rather than another, but most sets can be given in any of the three ways. For example, here is a word description:

the set of even counting numbers less than ten.

This same set can be expressed by listing:

$$\{2, 4, 6, 8\},$$

or by set-builder notation:

$$\{x \mid x \text{ is an even counting number less than } 10\}.$$

In the listing and set-builder notations, the braces at the beginning and ending indicate that we are thinking of a set. Other grouping symbols, such as parentheses or square brackets, are *not* used in set notation. Also, in the listing method, the commas are essential. Other separators, such as semicolons or colons, are *not* used. The set-builder notation utilizes the algebraic idea of a variable. (Any symbol would do, but just as in other algebraic applications, the letter x is a common choice.) Before the vertical line we give the variable, which represents an element in general, and after the vertical line we state the criteria by which an element qualifies for membership in the set. By including *all* objects that meet the stated criteria, we generate the entire set. (Set-builder notation is sometimes called *set-generator* notation.)

Sets are commonly given names (usually capital letters) so that they can be easily referred to later in the discussion. If E is selected as a name for the set of all letters of the English alphabet, then we can write

$$E = \{\text{a, b, c, d, e, f, g, h, i, j, k, l, m, n, o, p, q, r, s, t, u, v, w, x, y, z}\}.$$

In many cases, the listing notation can be shortened by clearly establishing the pattern of elements included, and using an ellipsis (three dots) to indicate a continuation of the pattern. Thus, for example,

$$E = \{a, b, c, d, \ldots, x, y, z\}$$

or

$$E = \{a, b, c, d, e, \ldots, z\}.$$

Georg Cantor created a new field of theory and at the same time continued the long debate over infinity that began in ancient times. Cantor developed counting by one-to-one correspondence to determine how many objects are contained in a set. Infinite sets differ from finite sets by not obeying the familiar law that the whole is greater than any of its parts.

Cantor's teacher had been the mathematician Leopold Kronecker, who denied Cantor's "transfinite" numbers. Kronecker believed that all numbers could be expressed as combinations of integers.

EXAMPLE 1 Give a complete listing of all the elements of each of the following sets.

(a) the set of counting numbers between six and twelve

This set can be denoted {7, 8, 9, 10, 11}. (Notice that the word *between* excludes the endpoint values.)

(b) {5, 6, 7, . . . , 14}

This set begins with the element 5, then 6, then 7, and so on, with each element obtained by adding 1 to the previous element in the list. This pattern stops at 14, so a complete listing is

$$\{5, 6, 7, 8, 9, 10, 11, 12, 13, 14\}.$$

(c) $\{x \mid x$ is a counting number between 8 and 9$\}$

After a little thought, we realize that there are no counting numbers between 8 and 9, so the listing for this set will contain no elements at all. We can write the set as

$$\{\ \ \} \text{ or } \emptyset. \blacksquare$$

A set containing no elements, such as the set in Example 1(c), is called the **empty set,** or **null set.** The special symbol $\emptyset$ is often used to denote the empty set, so that $\emptyset$ and $\{\ \ \}$ have the same meaning. We do *not* denote the empty set with the symbol $\{\emptyset\}$, since this notation represents a set with one element (that element being the empty set).

Example 1 above referred to counting numbers (or natural numbers), which were introduced in Section 1.1. Other important categories of numbers, which will be used throughout the text, are summarized below. More detailed treatment of these sets can be found in a later chapter.

Important Number Sets

Natural or counting numbers {1, 2, 3, 4, . . .}
Whole numbers {0, 1, 2, 3, 4, . . .}
Integers {. . . , −3, −2, −1, 0, 1, 2, 3, . . .}
Rational numbers $\{p/q \mid p$ and q are integers, and $q \neq 0\}$
(Some examples of rational numbers are 3/5, −7/9, 5, and 0. Any rational number may be written as a terminating decimal number, like 0.25, or a repeating decimal number, like 0.666. . . .)
Real numbers $\{x \mid x$ is a real number that may be written as a decimal$\}$
Irrational numbers $\{x \mid x$ is a real number and x cannot be written as a quotient of integers$\}$
(Some examples of irrational numbers are $\sqrt{2}$, $\sqrt[3]{4}$, and π. A characteristic of irrational numbers is that their decimal representations never terminate and never repeat, that is, they never reach a point where a given pattern of digits repeats from that point on.)

The number of elements in a set is called the **cardinal number** of the set. The symbol $n(A)$, which is read "n of A," represents the cardinal number of set A.

If elements are repeated in a set listing, they should not be counted more than once when determining the cardinal number of the set. For example, the set $B = \{1, 1, 2, 2, 3\}$ has only three distinct elements, and so $n(B) = 3$.

EXAMPLE 2 Find the cardinal number of each of the following sets.

(a) $K = \{10, 12, 14, 16\}$

Set K contains four elements, so the cardinal number of set K is 4, and $n(K) = 4$.

(b) $M = \{0\}$

Set M contains only one element, zero, so $n(M) = 1$.

(c) $R = \{3, 4, 5, \ldots, 11, 12\}$

There are only five elements listed, but the ellipsis indicates that there are other elements in the set. Counting them, we find that there are ten elements, so $n(R) = 10$.

(d) The empty set, $\emptyset$, contains no elements, and $n(\emptyset) = 0$. ■

If the cardinal number of a set is a particular whole number (0 or a counting number), as in all parts of Example 2 above, we call that set a **finite set.** Given enough time, we could finish counting all the elements of any finite set and arrive at its cardinal number. Some sets, however, are so large that we could never finish the counting process. The counting numbers themselves are such a set. Whenever a set is so large that its cardinal number is not found among the whole numbers, we call that set an **infinite set.** Cardinal numbers of infinite sets will be discussed in a later section. Infinite sets can also be designated using the three methods already mentioned.

Infinity Close-up of a camera lens shows the infinity sign, ∞, defined as any distance greater than 1,000 times the focal length of a lens.

The sign was invented by the mathematician John Wallis in 1655. Wallis used $1/\infty$ to represent an infinitely small quantity. The philosopher Voltaire described the ∞ as a loveknot, and he was skeptical about the sign making the idea of infinity any clearer.

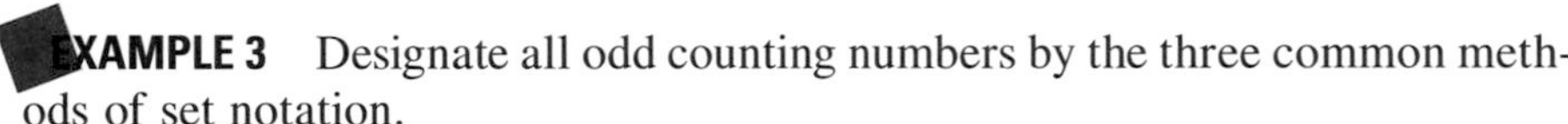

EXAMPLE 3 Designate all odd counting numbers by the three common methods of set notation.

1. word description

The set of all odd counting numbers

2. listing

$$\{1, 3, 5, 7, 9, \ldots\}$$

Notice that the ellipsis is utilized, but that there is no final element given. The listing goes on forever.

3. set-builder

$$\{x \mid x \text{ is an odd counting number}\}$$ ■

For a set to be useful, it must be **well defined.** This means that if a particular set and some particular element are given, it must be possible to tell whether or not the element belongs to the set. For example, the preceding set E of the letters of the English alphabet is well defined. If someone gives us the letter q, we know that q is an element of E. If someone gives us the Greek letter θ (theta), we know that it is not an element of set E.

However, given the set C of all fat chickens, and a particular chicken, Herman, it is not possible to say whether

Herman is an element of C

or

Herman is *not* an element of C.

The problem is the word "fat"; how fat is fat? Since we cannot necessarily decide whether or not a given chicken belongs to set C, set C is not well defined.

The letter q is an element of set E, where E is the set of all the letters of the English alphabet. To show this, $\in$ is used to replace the words "is an element of," or

$$q \in E,$$

which is read "q is an element of set E." The letter θ is not an element of E; write this with $\in$ and a slash mark:

$$\theta \notin E.$$

This is read "θ is not an element of set E."

EXAMPLE 4 Decide whether each statement is *true* or *false*.

(a) $3 \in \{-3, -1, 5, 9, 13\}$

The statement claims that the number 3 is an element of the set $\{-3, -1, 5, 9, 13\}$, which is false.

(b) $0 \in \{-3, -2, 0, 1, 2, 3\}$

Since 0 is indeed an element of the set $\{-3, -2, 0, 1, 2, 3\}$, the statement is true.

(c) $1/5 \notin \{1/3, 1/4, 1/6\}$

This statement says that $1/5$ is not an element of the set $\{1/3, 1/4, 1/6\}$, which is true. ◆

We now consider the concept of set equality.

Set Equality

Set A is **equal** to set B provided the following two conditions are met:

1. every element of A is an element of B, and
2. every element of B is an element of A.

In practice, a less formal idea may be used to determine if two sets are equal. Two sets are equal if they contain exactly the same elements, regardless of order. For example,

$$\{a, b, c, d\} = \{a, c, d, b\}$$

since both sets contain exactly the same elements.

Since repetition of elements in a set listing does not add new elements, we can say that

$$\{1, 0, 1, 2, 3, 3\} = \{0, 1, 2, 3\},$$

since these sets contain exactly the same elements.

EXAMPLE 5 Are $\{-4, 3, 2, 5\}$ and $\{-4, 0, 3, 2, 5\}$ equal sets?

Every element of the first set is an element of the second; however, 0 is an element of the second and not the first. In other words, the sets do not contain exactly the same elements, so they are not equal:

$$\{-4, 3, 2, 5\} \neq \{-4, 0, 3, 2, 5\}.$$

Although any of the three notations can designate a given set, sometimes one will be clearer than another. For example,

$$\{x \mid x \text{ is an odd counting number between 2 and 810}\}$$

may be preferable to

$$\{3, 5, 7, 9, \ldots, 809\},$$

since it clearly states the common property that identifies all elements of the set. When working with sets, it is necessary to be able to translate from one notation to another to have a clear understanding of the sets involved.

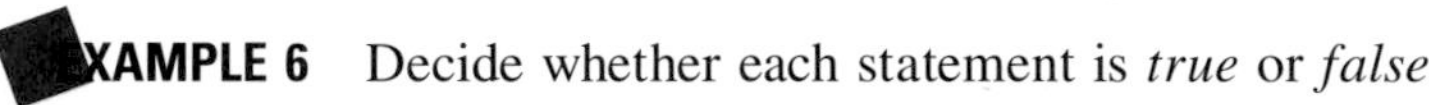

EXAMPLE 6 Decide whether each statement is *true* or *false*.

(a) $\{3\} = \{x \mid x \text{ is a counting number between 1 and 5}\}$

The set on the right contains *all* counting numbers between 1 and 5, namely 2, 3, and 4, while the set on the left contains *only* the number 3. Since the sets do not contain exactly the same elements, they are not equal. The statement is false.

(b) $\{x \mid x \text{ is a math class that requires no thinking}\} = \{y \mid y \text{ is a living tree with no roots}\}$.

Since each set is the empty set, the sets are equal. The statement is true.

2.1 EXERCISES

List all the elements of each set.

1. the set of all counting numbers less than or equal to 4

2. the set of all whole numbers greater than 8 and less than 16

3. the set of all whole numbers not greater than 6

4. the set of all counting numbers between 2 and 12

5. $\{6, 7, 8, \ldots, 14\}$

6. $\{3, 6, 9, 12, \ldots, 30\}$

7. $\{-15, -13, -11, \ldots, -1\}$

8. $\{-4, -3, -2, \ldots, 4\}$

9. $\{2, 4, 8, \ldots, 256\}$

10. $\{90, 87, 84, \ldots, 54\}$

11. $\{1, 1/3, 1/9, \ldots, 1/243\}$

12. $\{1/2, 1/4, 1/6, \ldots, 1/20\}$

13. $\{x \mid x \text{ is an even whole number less than 15}\}$

14. $\{x \mid x \text{ is an odd integer between } -8 \text{ and } 5\}$

Denote each set by the listing method. There may be more than one correct answer.

15. the set of all counting numbers greater than 20

16. the set of all integers between −200 and 500

17. the set of traditional major political parties in the United States

18. the set of all persons living on February 1, 1992 who had been President of the United States. (See the photograph.)

19. $\{x \mid x$ is a positive multiple of $4\}$

20. $\{x \mid x$ is a negative multiple of $7\}$

21. $\{x \mid x$ is the reciprocal of a natural number$\}$

22. $\{x \mid x$ is a positive integer power of $3\}$

Denote each set by set-builder notation, using x as the variable. There may be more than one correct answer.

23. the set of all rational numbers

24. the set of all even counting numbers

25. the set of all movies released this year

26. the set of all multinational corporations

27. $\{1, 3, 5, \ldots, 99\}$

28. $\{35, 40, 45, \ldots, 995\}$

Identify each set as finite *or* infinite.

29. $\{2, 4, 6, \ldots, 28\}$

30. $\{6, 12, 18, \ldots\}$

31. $\{1/2, 2/3, 3/4, \ldots, 99/100\}$

32. $\{-10, -8, -6, \ldots, 0\}$

33. $\{x \mid x$ is a counting number greater than $30\}$

34. $\{x \mid x$ is a counting number less than $30\}$

35. $\{x \mid x$ is a rational number$\}$

36. $\{x \mid x$ is a rational number between 0 and $1\}$

Find $n(A)$ for each set.

37. $A = \{0, 1, 2, 3, 4, 5, 6\}$

38. $A = \{-3, -2, -1, 0, 1, 2\}$

39. $A = \{2, 4, 6, \ldots, 1{,}000\}$

40. $A = \{0, 1, 2, 3, \ldots, 3{,}000\}$

41. $A = \{a, b, c, \ldots, z\}$

42. $A = \{x \mid x$ is a vowel in the English alphabet$\}$

43. $A =$ the set of integers between −10 and 10

44. $A =$ the set of current U.S. senators

45. $A = \{1/3, 2/4, 3/5, 4/6, \ldots, 27/29, 28/30\}$

46. $A = \{1/2, -1/2, 1/3, -1/3, \ldots, 1/10, -1/10\}$

Identify each set as well defined *or* not well defined.

47. $\{x \mid x$ is a real number$\}$

48. $\{x \mid x$ is a negative number$\}$

49. $\{x \mid x$ is a good singer$\}$

50. $\{x \mid x$ is a skillful actor$\}$

51. $\{x \mid x$ is a difficult class$\}$

52. $\{x \mid x$ is a counting number less than $1\}$

Fill each blank with either $\in$ or $\notin$ to make the following statements true.

53. 5 _____ $\{2, 4, 5, 6\}$

54. 8 _____ $\{3, -2, 5, 9, 8\}$

55. −4 _____ $\{4, 7, 9, 12\}$

56. −12 _____ $\{3, 8, 12, 16\}$

57. 0 ______ $\{-2, 0, 5, 7\}$

58. 0 ______ $\{3, 4, 7, 8, 10\}$

59. $\{3\}$ ______ $\{2, 3, 4, 5\}$

60. $\{5\}$ ______ $\{3, 4, 5, 6, 7\}$

Write true *or* false *for each of the following statements.*

61. $3 \in \{2, 5, 6, 8\}$

62. $6 \in \{-2, 5, 8, 9\}$

63. $\text{b} \in \{\text{h, c, d, a, b}\}$

64. $\text{m} \in \{1, \text{m, n, o, p}\}$

65. $9 \notin \{6, 3, 4, 8\}$

66. $2 \notin \{7, 6, 5, 4\}$

67. $\{\text{k, c, r, a}\} = \{\text{k, c, a, r}\}$

68. $\{\text{e, h, a, n}\} = \{\text{a, h, e, n}\}$

69. $\{5, 8, 9\} = \{5, 8, 9, 0\}$

70. $\{3, 7, 12, 14\} = \{3, 7, 12, 14, 0\}$

71. $\{\text{d, x, m, x, d}\} = \{\text{m, d, x}\}$

72. $\{\text{u, v, u, v}\} = \{\text{u, v}\}$

73. $\{x \mid x \text{ is a counting number less than } 3\} = \{1, 2\}$

74. $\{x \mid x \text{ is a counting number greater than } 10\} = \{11, 12, 13, \ldots\}$

Write true *or* false *for each of the following statements.*

$$\text{Let } A = \{2, 4, 6, 8, 10, 12\}$$
$$B = \{2, 4, 8, 10\}$$
$$C = \{4, 10, 12\}.$$

75. $4 \in A$

76. $8 \in B$

77. $4 \notin C$

78. $8 \notin B$

79. $10 \notin A$

80. $6 \notin A$

81. Every element of C is also an element of A.

82. Every element of C is also an element of B.

83. This chapter opened with the statement, "The human mind likes to create collections." Why do you suppose this is so? In explaining your thoughts, utilize one or more particular "collections," mathematical or otherwise.

84. Explain the difference between a well-defined set and a not well-defined set. Give examples and utilize terms introduced in this section.

Recall that two sets are called equal *if they contain identical elements. On the other hand, two sets are called* equivalent *if they contain the same number of elements (but not necessarily the same elements). For each of the following conditions, give an example or explain why it is impossible.*

85. two sets that are neither equal nor equivalent

86. two sets that are equal but not equivalent

87. two sets that are equivalent but not equal

88. two sets that are both equal and equivalent

89. Michelle Perulli is health conscious, but she does like a certain chocolate bar, each of which contains 220 calories. In order to burn off unwanted calories, Michelle participates in her favorite activities, shown below, in increments of one hour and never repeats a given activity on a given day.

Activity	Symbol	Calories Burned per Hour
Volleyball	v	160
Golf	g	260
Canoeing	c	340
Swimming	s	410
Running	r	680

(a) On Monday, Michelle has time for no more than two hours of activities. List all possible sets of activities that would burn off at least the number of calories obtained from three chocolate bars.

(b) Assume that Michelle can afford up to three hours of time for activities on Saturday. List all sets of activities that would burn off at least the number of calories in five chocolate bars.

90. The table below categorizes municipal solid waste generated in the United States.

Category	Symbol	Percentage
Food wastes	F	7
Yard wastes	Y	18
Metals	M	8
Glass	G	7
Plastics	L	8
Paper	P	40
Rubber, leather, textile, wood, other	R	12

Using the given symbols, list the elements of the following sets.
(a) $\{x \mid x$ is a category accounting for more than 15% of the wastes$\}$
(b) $\{x \mid x$ is a category accounting for less than 10% of the wastes$\}$

Venn Diagrams and Subsets

When working a problem, we can usually expect a certain type of answer. For example, a problem about money should give an answer in dollars and cents. The answer probably would not involve names, animals, or pencils.

In every problem there is either a stated or implied **universe of discourse.** The universe of discourse includes all things under discussion at a given time. For example, in studying reactions to a proposal that a certain campus raise the minimum age of individuals to whom beer may be sold, the universe of discourse might be all the students at the school, the nearby members of the public, the board of trustees of the school, or perhaps all these groups of people.

In the mathematical theory of sets, the universe of discourse is called the **universal set.** The letter U is typically used for the universal set. The universal set might well change from problem to problem. In one problem the universal set might be the set of all natural numbers, while in another problem the universal set might be the set of all females older than 25 who have two or more children.

In most areas of mathematics, our reasoning can be aided and clarified by utilizing various kinds of drawings and diagrams. In set theory, we commonly use **Venn diagrams,** developed by the logician John Venn (1834–1923). In these diagrams, the universal set is represented by a rectangle, and other sets of interest within the universal set are depicted by oval regions, or sometimes by circles or other shapes. We will use Venn diagrams to illustrate many concepts throughout the remainder of this chapter. In the Venn diagram of Figure 1, the entire region bounded by the rectangle represents the universal set U, while the portion bounded by the oval represents set A. (The size of the oval representing A is irrelevant.) Notice also in the figure that the colored region inside U and outside the oval is labeled A' (read "A prime"). This set, called the *complement* of A, contains all elements that are contained in U but not contained in A.

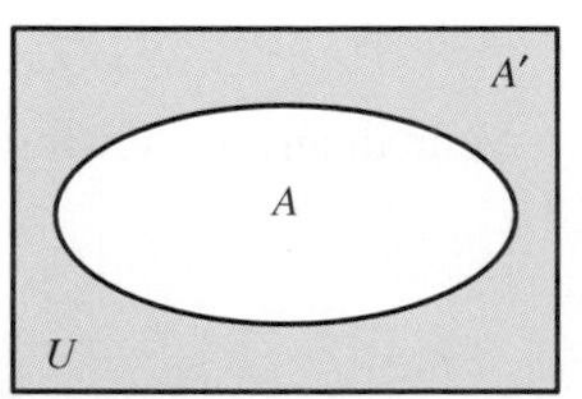

FIGURE 1

The Complement of a Set

For any set A within the universal set U, the **complement** of A, written A', is the set of elements of U that are not elements of A. That is,

$$A' = \{x \mid x \in U \text{ and } x \notin A\}.$$

EXAMPLE 1 Let
$$U = \{a, b, c, d, e, f, g, h\}$$
$$M = \{a, b, e, f\}$$
$$N = \{b, d, e, g, h\}.$$

Find each of the following sets.

(a) M'

Set M' contains all the elements of set U that are not in set M. Since set M contains the elements a, b, e, and f, these elements will be disqualified from belonging to set M', and consequently set M' will contain c, d, g, and h, or

$$M' = \{c, d, g, h\}.$$

(b) N'

Set N' contains all the elements of U that are not in set N, so $N' = \{a, c, f\}$. ■

Consider the complement of the universal set, U'. The set U' is found by selecting all the elements of U that do not belong to U. There are no such elements, so there can be no elements in set U'. This means that

$$U' = \emptyset$$

for any universal set U.

Now consider the complement of the empty set, $\emptyset'$. Since $\emptyset' = \{x \mid x \in U \text{ and } x \notin \emptyset\}$ and set $\emptyset$ contains no elements, every member of the universal set U satisfies this definition. Therefore

$$\emptyset' = U$$

for any universal set U.

Suppose that in a particular discussion, we are given that the universal set is $U = \{1, 2, 3, 4, 5\}$, while one of the sets under discussion is $A = \{1, 2, 3\}$. Every element of set A is also an element of set U. Because of this, set A is called a *subset* of set U, written

$$A \subseteq U.$$

("A is not a subset of set U" would be written $A \not\subseteq U$.) A Venn diagram showing that set M is a subset of set N is shown in Figure 2.

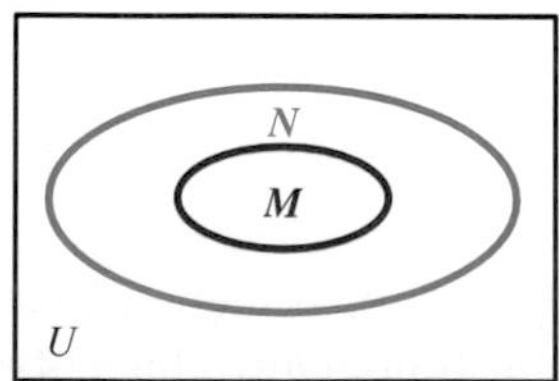

■ FIGURE 2

Subset of a Set

Set A is a **subset** of set B if every element of A is also an element of B. In symbols,

$$A \subseteq B.$$

EXAMPLE 2 Write $\subseteq$ or $\not\subseteq$ in each blank to make a true statement.

(a) {3, 4, 5, 6} ____ {3, 4, 5, 6, 8}

Since every element of {3, 4, 5, 6} is also an element of {3, 4, 5, 6, 8}, the first set is a subset of the second, so $\subseteq$ goes in the blank.

(b) {1, 2, 3} ____ {2, 4, 6, 8}

The element 1 belongs to {1, 2, 3} but not to {2, 4, 6, 8}. Place $\not\subseteq$ in the blank.

(c) {5, 6, 7, 8} ____ {5, 6, 7, 8}

Every element of {5, 6, 7, 8} is also an element of {5, 6, 7, 8}. Place $\subseteq$ in the blank. ■

As Example 2(c) suggests, every set is a subset of itself:

$$B \subseteq B \quad \text{for any set } B.$$

When studying subsets of a set B, it is common to look at subsets other than set B itself. Suppose that $B = \{5, 6, 7, 8\}$ and $A = \{6, 7\}$. A is a subset of B, but A is not all of B; there is at least one element in B that is not in A. (Actually, in this case there are two such elements, 5 and 8.) In this situation, A is called a *proper subset* of B. To indicate that A is a proper subset of B, write $A \subset B$.

Proper Subset of a Set

Set A is a **proper subset** of set B if $A \subseteq B$ and $A \neq B$. In symbols,

$$A \subset B.$$

(Notice the similarity of the subset symbols, $\subset$ and $\subseteq$, to the inequality symbols from algebra, $<$ and $\leq$.)

EXAMPLE 3 Decide whether $\subset$, $\subseteq$, or both could be placed in each blank to make a true statement.

(a) {5, 6, 7} ____ {5, 6, 7, 8}

Every element of {5, 6, 7} is contained in {5, 6, 7, 8}, so $\subseteq$ could be placed in the blank. Also, the element 8 belongs to {5, 6, 7, 8} but not to {5, 6, 7}, making {5, 6, 7} a proper subset of {5, 6, 7, 8}. This means that $\subset$ could also be placed in the blank.

(b) {a, b, c} ____ {a, b, c}

The set {a, b, c} is a subset of {a, b, c}. Since the two sets are equal, {a, b, c} is not a proper subset of {a, b, c}. Only $\subseteq$ may be placed in the blank. ■

Set A is a subset of set B if every element of set A is also an element of set B. This definition can be reworded by saying that set A is a subset of set B if there are no elements of A that are not also elements of B. This second form of the definition shows that the empty set is a subset of any set, or

$$\emptyset \subseteq B \quad \text{for any set } B.$$

This is true since it is not possible to find any elements of $\emptyset$ that are not also in B. (There are no elements in $\emptyset$.) The empty set $\emptyset$ is a proper subset of every set except itself:

$$\emptyset \subset B \text{ if } B \text{ is any set other than } \emptyset.$$

Every set (except $\emptyset$) has at least two subsets, $\emptyset$ and the set itself. Let us find a rule to tell *how many subsets* a given set has. The next example provides a starting point.

EXAMPLE 4 Find all possible subsets of each set.

(a) {7, 8}
By trial and error, the set {7, 8} has four subsets:

$$\emptyset, \quad \{7\}, \quad \{8\}, \quad \{7, 8\}.$$

(b) {a, b, c}
Here trial and error leads to 8 subsets for {a, b, c}:

$$\emptyset, \quad \{a\}, \quad \{b\}, \quad \{c\}, \quad \{a, b\}, \quad \{a, c\}, \quad \{b, c\}, \quad \{a, b, c\}.$$ ■

In Example 4, the subsets of {7, 8} and the subsets of {a, b, c} were found by trial and error. An alternative method involves drawing a **tree diagram,** a systematic way of listing all the subsets of a given set. Figures 3(a) and (b) show tree diagrams for {7, 8} and {a, b, c}.

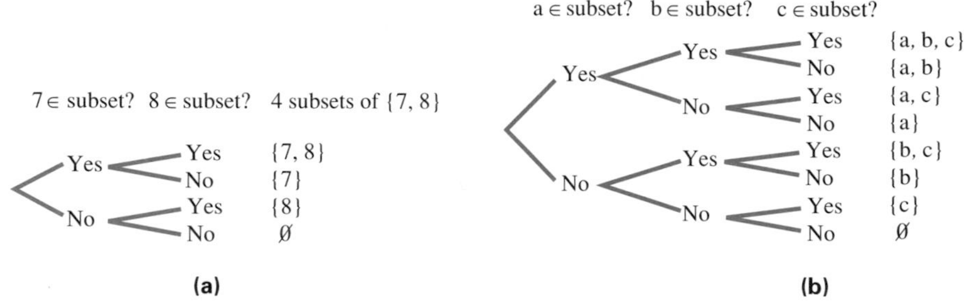

■ **FIGURE 3**

In Example 4, we determined the number of subsets of a given set by exhibiting a list of all such subsets and then counting them. The tree diagram method also produced a list of all possible subsets in each case. In many applications, we don't need to display all the subsets but simply determine how many there are. Furthermore, the trial and error method and the tree diagram method would both involve far too much work if the original set had a very large number of elements. For these reasons, it is desirable to have a formula for the number of subsets. To obtain such a formula, we can use the technique of inductive reasoning. That is, we observe particular cases to try to discover a general pattern. Begin with the set containing the least number of elements possible—the empty set. This set, $\emptyset$, has only one subset, $\emptyset$ itself. Next, a set with one element has only two subsets, itself and $\emptyset$. These facts, together with those obtained above for sets with two and three elements, are summarized here.

Number of elements	0	1	2	3
Number of subsets	1	2	4	8

Powers of 2

$2^0 = 1$
$2^1 = 2$
$2^2 = 2 \times 2 = 4$
$2^3 = 2 \times 2 \times 2 = 8$
$2^4 = 2 \times 2 \times 2 \times 2 = 16$
$2^5 = 32$
$2^6 = 64$
$2^7 = 128$
$2^8 = 256$
$2^9 = 512$
$2^{10} = 1{,}024$
$2^{11} = 2{,}048$
$2^{12} = 4{,}096$
$2^{15} = 32{,}768$
$2^{20} = 1{,}048{,}576$
$2^{25} = 33{,}554{,}432$
$2^{30} = 1{,}073{,}741{,}824$

The small numbers at the upper right are called *exponents.* Here the *base* is 2. Notice how quickly the values grow; this is the origin of the phrase *exponential growth.*

This chart suggests that as the number of elements of the set increases by one, the number of subsets doubles. This suggests that the number of subsets in each case might be a power of 2. Every number in the second row of the chart is indeed a power of 2. Add this information to the chart.

Number of elements	0	1	2	3
Number of subsets	$1 = 2^0$	$2 = 2^1$	$4 = 2^2$	$8 = 2^3$

This chart shows that the number of elements in each case is the same as the exponent on the 2. Inductive reasoning gives us the following generalization.

Number of Subsets

The number of subsets of a set with n elements is 2^n.

Since the value 2^n includes the set itself, we must subtract 1 from this value to obtain the number of proper subsets of a set containing n elements. Here is another generalization.

Number of Proper Subsets

The number of proper subsets of a set with n elements is $2^n - 1$.

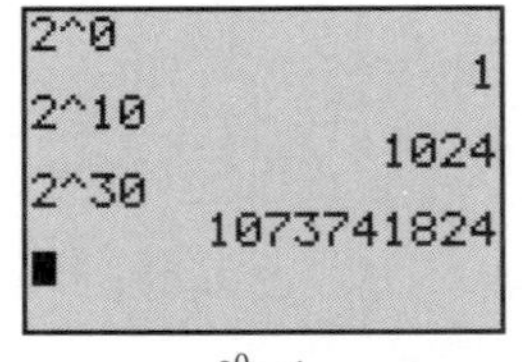

$2^0 = 1$
$2^{10} = 1{,}024$
$2^{30} = 1{,}073{,}741{,}824$

Selected Powers of 2

As shown in the chapter on problem solving, although inductive reasoning is a good way of *discovering* principles or arriving at a *conjecture,* it does not provide a proof that the conjecture is true in general. A proof must be provided by other means. The two formulas above are true, by observation, for $n = 0, 1, 2$, or 3. (For a general proof, see Exercise 66 at the end of this section.) To illustrate the use of these formulas, we consider the set {3, 8, 11, 17, 20, 25, 28}. Since this set has 7 elements (by counting), it must have $2^7 = 128$ subsets, of which $2^7 - 1 = 128 - 1 = 127$ are proper subsets.

EXAMPLE 5 Find the number of subsets and the number of proper subsets of each set.

(a) {3, 4, 5, 6, 7}

This set has 5 elements and $2^5 = 2 \times 2 \times 2 \times 2 \times 2 = 32$ subsets. Of these, 31 are proper subsets.

(b) {1, 2, 3, 4, 5, 9, 12, 14}

The set {1, 2, 3, 4, 5, 9, 12, 14} has 8 elements. There are $2^8 = 256$ subsets and 255 proper subsets. ◆

In a similar way, a set with 100 elements has 2^{100} subsets. Using more extensive mathematical tables,

$$2^{100} = 1{,}267{,}650{,}600{,}228{,}229{,}401{,}496{,}703{,}205{,}376.$$

2.2 EXERCISES

Insert $\subseteq$ *or* $\not\subseteq$ *in each blank so that the resulting statement is true.*

1. $\{-2, 0, 2\}$ ____ $\{-2, -1, 1, 2\}$
2. {Monday, Wednesday, Friday} ____ {Sunday, Monday, Tuesday, Wednesday, Thursday}
3. $\{2, 5\}$ ____ $\{0, 1, 5, 3, 4, 2\}$
4. {a, n, d} ____ {r, a, n, d, y}
5. $\emptyset$ ____ {a, b, c, d, e}
6. $\emptyset$ ____ $\emptyset$
7. $\{-7, 4, 9\}$ ____ $\{x \mid x$ is an odd integer$\}$
8. $\{2, 1/3, 5/9\}$ ____ the set of rational numbers

Decide whether $\subset$, *or* $\subseteq$, *or* both, *or* neither *can be placed in the blank to make the statement true.*

9. {B, C, D} ____ {B, C, D, F}
10. {red, green, blue, yellow} ____ {green, yellow, blue, red}
11. $\{9, 1, 7, 3, 5\}$ ____ $\{1, 3, 5, 7, 9\}$
12. {S, M, T, W, Th} ____ {M, W, Th, S}
13. $\emptyset$ ____ $\{0\}$
14. $\emptyset$ ____ $\emptyset$
15. $\{-1, 0, 1, 2, 3\}$ ____ $\{0, 1, 2, 3, 4\}$
16. $\{5/6, 9/8\}$ ____ $\{6/5, 8/9\}$

For Exercises 17–38, tell whether each statement is true *or* false.

$$\text{Let } U = \{a, b, c, d, e, f, g\}$$
$$A = \{a, e\}$$
$$B = \{a, b, e, f, g\}$$
$$C = \{b, f, g\}$$
$$D = \{d, e\}.$$

17. $A \subset U$
18. $C \subset U$
19. $D \subseteq B$
20. $D \subseteq A$
21. $A \subset B$
22. $B \subseteq C$
23. $\emptyset \subset A$
24. $\emptyset \subseteq D$
25. $\emptyset \subseteq \emptyset$
26. $D \subset B$
27. $\{g, f, b\} \subset B$
28. $\{0\} \subset D$
29. $D \not\subseteq B$
30. $A \not\subseteq B$
31. There are exactly 6 subsets of C.
32. There are exactly 31 subsets of B.
33. There are exactly 3 subsets of A.
34. There are exactly 4 subsets of D.
35. There is exactly 1 subset of $\emptyset$.
36. There are exactly 127 proper subsets of U.
37. The drawing below correctly represents the relationship among sets A, C, and U.

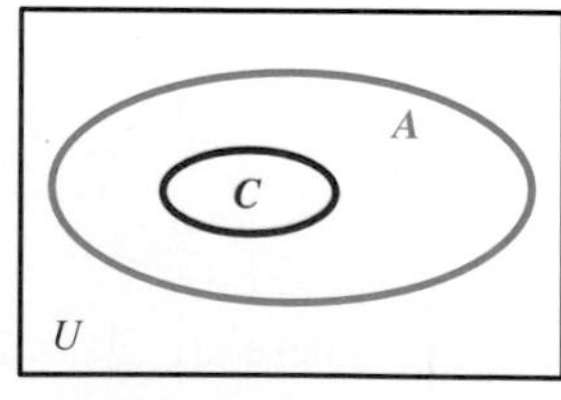

38. The drawing below correctly represents the relationship among sets B, C, and U.

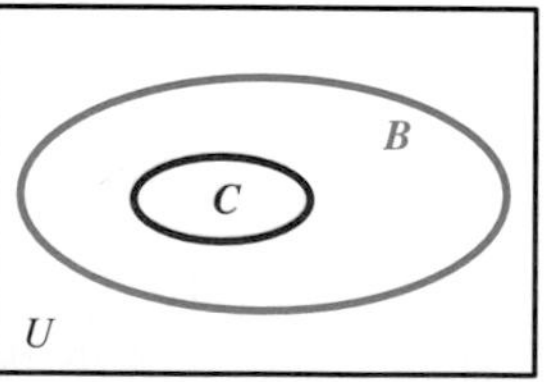

Find the number of subsets and the number of proper subsets of each of the following sets.

39. $\{1, 3, 5\}$
40. $\{-8, -6, -4, -2\}$
41. {a, b, c, d, e, f}
42. the set of days of the week
43. $\{x \mid x$ is an odd integer between -4 and $6\}$
44. $\{x \mid x$ is an even whole number less than $3\}$

Let $U = \{1, 2, 3, 4, 5, 6, 7, 8, 9, 10\}$ and find the complement of each of the following sets.

45. $\{1, 4, 6, 8\}$

46. $\{2, 5, 7, 9, 10\}$

47. $\{1, 3, 4, 5, 6, 7, 8, 9, 10\}$

48. $\{3, 5, 7, 9\}$

49. $\emptyset$

50. U

51. Greg Odjakjian and his family, wishing to see a movie this evening, have made up the following lists of characteristics for their two main options.

Go to a Movie Theater	Rent a Home Video
High cost	Low cost
Entertaining	Entertaining
Fixed schedule	Flexible schedule
Current films	Older films

Find the smallest universal set U that contains all listed characteristics of both options.

Let T represent the set of characteristics of the movie theater option, and let V represent the set of characteristics of the home video option. Using the universal set from Exercise 51, find each of the following.

52. T'

53. V'

Find the set of elements common to both of the sets in Exercises 54–57, where T and V are defined as above.

54. T and V

55. T' and V

56. T and V'

57. T' and V'

Anita Bartelson, Brett Sullivan, Curt Reynolds, Duncan MacKinnon, and Emma Price plan to meet in the hospitality suite at a sales convention to compare notes. Denoting these five people by A, B, C, D, and E, list all the possible ways that the given number of them can gather in the suite.

58. 5

59. 4

60. 3

61. 2

62. 1

63. 0

64. Find the total number of ways that members of this group can gather in the suite. (*Hint:* Add your answers to Exercises 58–63.)

65. How does your answer in Exercise 64 compare with the number of subsets of a set of 5 elements? How can you interpret the answer to Exercise 64 in terms of subsets?

66. In discovering the formula (2^n) for the number of subsets of a set with n elements, we observed that for the first few values of n, increasing the number of elements by one doubles the number of subsets. Here, you can prove the formula in general by showing that the same is true for any value of n. Assume set A has n elements and s subsets. Now add one additional element, say e, to the set A. (We now have a new set, say B, with $n + 1$ elements.) Divide the subsets of B into those that do not contain e and those that do.

(a) How many subsets of B do not contain e? (*Hint:* Each of these is a subset of the original set A.)

(b) How many subsets of B do contain e? (*Hint:* Each of these would be a subset of the original set A, with the element e thrown in.)

(c) What is the total number of subsets of B?

(d) What do you conclude?

Suppose you have available the bills shown here.

67. If you must select at least one bill, and you may select up to all of the bills, how many different sums of money could you make?

68. In Exercise 67, remove the condition "you must select at least one bill." Now, how many sums are possible?

2.3 Operations with Sets

After comparing the campaign promises of two candidates for sheriff, a voter came up with the following list of promises made by the people running for office. Each promise is given a code letter.

Honest Greg McRill	Learned Lakeisha Holiday
Spend less money, m	Spend less money, m
Emphasize traffic law enforcement, t	Crack down on crooked politicians, p
Increase service to suburban areas, s	Increase service to the city, c

The only promise common to both candidates is promise m, to spend less money. Suppose we take each candidate's promises to be a set. The promises of candidate McRill give the set $\{m, t, s\}$, while the promises of Holiday give $\{m, p, c\}$. The only element common to both sets is m; this element belongs to the *intersection* of the two sets $\{m, t, s\}$ and $\{m, p, c\}$, as shown in color in the Venn diagram in Figure 4. In symbols,

$$\{m, t, s\} \cap \{m, p, c\} = \{m\},$$

where the cap-shaped symbol $\cap$ represents intersection. Notice that the intersection of two sets is itself a set.

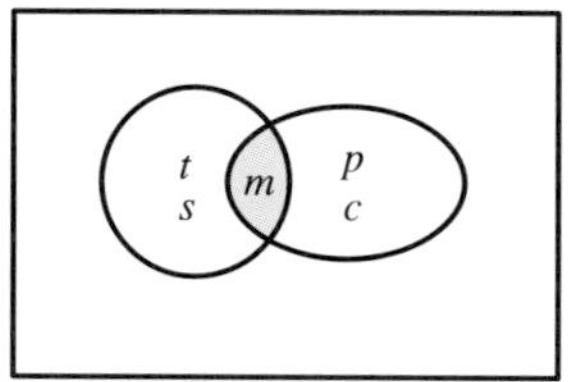

■ FIGURE 4

Intersection of Sets

The **intersection** of sets A and B, written $A \cap B$, is the set of elements common to both A and B, or

$$A \cap B = \{x \mid x \in A \text{ and } x \in B\}.$$

Form the intersection of sets A and B by taking all the elements included in both sets, as shown in color in Figure 5.

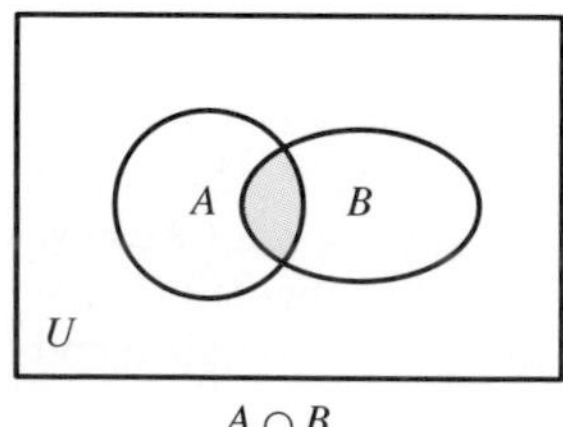

■ FIGURE 5

EXAMPLE 1 Find the intersection of the given sets.

(a) $\{3, 4, 5, 6, 7\}$ and $\{4, 6, 8, 10\}$
Since the elements common to both sets are 4 and 6,

$$\{3, 4, 5, 6, 7\} \cap \{4, 6, 8, 10\} = \{4, 6\}.$$

(b) $\{9, 14, 25, 30\}$ and $\{10, 17, 19, 38, 52\}$
These two sets have no elements in common, so

$$\{9, 14, 25, 30\} \cap \{10, 17, 19, 38, 52\} = \emptyset.$$

(c) $\{5, 9, 11\}$ and $\emptyset$
There are no elements in $\emptyset$, so there can be no elements belonging to both $\{5, 9, 11\}$ and $\emptyset$. Because of this,

$$\{5, 9, 11\} \cap \emptyset = \emptyset.$$

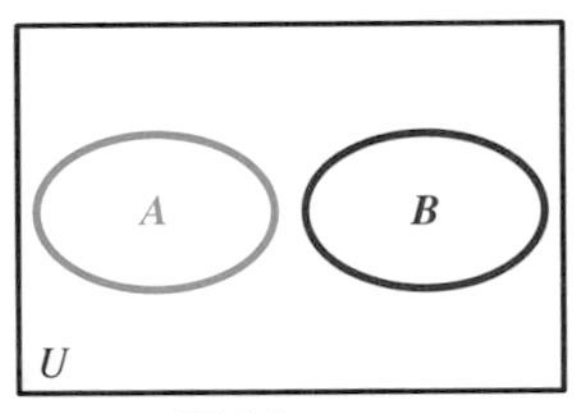

Disjoint sets

FIGURE 6

Examples 1(b) and 1(c) showed two sets that have no elements in common. Sets with no elements in common are called **disjoint sets.** A set of men and a set of women would be disjoint sets. In mathematical language, sets A and B are disjoint if $A \cap B = \emptyset$. Two disjoint sets A and B are shown in Figure 6.

For Further Thought

The arithmetic operations of addition and multiplication, when applied to numbers, have some familiar properties. If a, b, and c are real numbers, then the **commutative property of addition** says that the order of the numbers being added makes no difference: $a + b = b + a$. (Is there a commutative property of multiplication?) The **associative property of addition** says that when three numbers are added, the grouping used makes no difference: $(a + b) + c = a + (b + c)$. (Is there an associative property of multiplication?) The number 0 is called the **identity element for addition** since adding it to any number does not change that number: $a + 0 = a$. (What is the **identity element for multiplication**?) Finally, the **distributive property of multiplication over addition** says that $a(b + c) = ab + ac$. (Is there a **distributive property of addition over multiplication**?)

For Group Discussion

Now consider the operations of union and intersection, applied to sets. By recalling definitions, or by trying examples, answer the following questions.

1. Is set union commutative? How about set intersection?
2. Is set union associative? How about set intersection?
3. Is there an identity element for set union? If so, what is it? How about set intersection?
4. Is set intersection distributive over set union? Is set union distributive over set intersection?

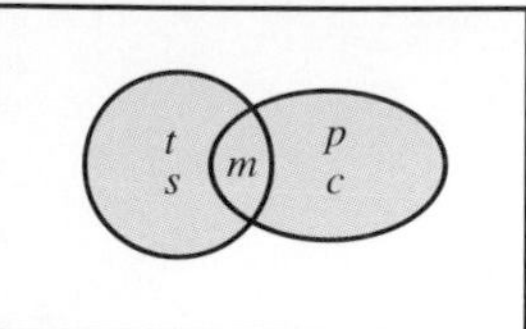

■ FIGURE 7

At the beginning of this section, we showed lists of campaign promises of two candidates running for sheriff. The intersection of those lists was found above. Suppose a student in a political science class must write a paper on the types of promises made by candidates for public office. The student would need to study *all* the promises made by *either* candidate, or the set

$$\{m, t, s, p, c\},$$

the *union* of the sets of promises made by the two candidates, as shown in color in the Venn diagram in Figure 7. In symbols,

$$\{m, t, s\} \cup \{m, p, c\} = \{m, t, s, p, c\},$$

where the cup-shaped symbol $\cup$ denotes set union. Be careful not to confuse this symbol with the universal set U. Again, the union of two sets is a set.

Union of Sets

The **union** of sets A and B, written $A \cup B$, is the set of all elements belonging to either of the sets, or

$$A \cup B = \{x \mid x \in A \text{ or } x \in B\}.$$

A B U

$A \cup B$

■ FIGURE 8

Form the union of sets A and B by taking all the elements of set A and then including the elements of set B that are not already listed, as shown in color in Figure 8.

EXAMPLE 2 Find the union of the given sets.

(a) $\{2, 4, 6\}$ and $\{4, 6, 8, 10, 12\}$

Start by listing all the elements from the first set, 2, 4, and 6. Then list all the elements from the second set that are not in the first set, or 8, 10, and 12. The union is made up of all these elements, or

$$\{2, 4, 6\} \cup \{4, 6, 8, 10, 12\} = \{2, 4, 6, 8, 10, 12\}.$$

(b) {a, b, d, f, g, h} and {c, f, g, h, k}

The union of these sets is

$$\{a, b, d, f, g, h\} \cup \{c, f, g, h, k\} = \{a, b, c, d, f, g, h, k\}.$$

(c) $\{3, 4, 5\}$ and $\emptyset$

Since there are no elements in $\emptyset$, the union of $\{3, 4, 5\}$ and $\emptyset$ contains only the elements 3, 4, and 5, or

$$\{3, 4, 5\} \cup \emptyset = \{3, 4, 5\}. \blacksquare$$

Recall from the previous section that A' represents the *complement* of set A. Set A' is formed by taking all the elements of the universal set U that are not in A.

EXAMPLE 3 Let $U = \{1, 2, 3, 4, 5, 6, 9\}$
$A = \{1, 2, 3, 4\}$
$B = \{2, 4, 6\}$
$C = \{1, 3, 6, 9\}$.

Find each set.

(a) $A' \cap B$

First identify the elements of set A', the elements of U that are not in set A:

$$A' = \{5, 6, 9\}.$$

Now find $A' \cap B$, the set of elements belonging both to A' and to B:

$$A' \cap B = \{5, 6, 9\} \cap \{2, 4, 6\} = \{6\}.$$

(b) $B' \cup C' = \{1, 3, 5, 9\} \cup \{2, 4, 5\} = \{1, 2, 3, 4, 5, 9\}$.

(c) $A \cap (B \cup C')$

First find the set inside the parentheses:

$$B \cup C' = \{2, 4, 6\} \cup \{2, 4, 5\} = \{2, 4, 5, 6\}.$$

Now finish the problem:

$$\begin{aligned} A \cap (B \cup C') &= A \cap \{2, 4, 5, 6\} \\ &= \{1, 2, 3, 4\} \cap \{2, 4, 5, 6\} \\ &= \{2, 4\}. \end{aligned}$$

(d) $(A' \cup C') \cap B'$

Set $A' = \{5, 6, 9\}$ and set $C' = \{2, 4, 5\}$, with

$$A' \cup C' = \{5, 6, 9\} \cup \{2, 4, 5\} = \{2, 4, 5, 6, 9\}.$$

Set B' is $\{1, 3, 5, 9\}$, so

$$(A' \cup C') \cap B' = \{2, 4, 5, 6, 9\} \cap \{1, 3, 5, 9\} = \{5, 9\}.$$

It is often said that mathematics is a "language." As such, it has the advantage of concise symbolism. For example, the set $(A \cap B)' \cup C$ is less easily expressed in words. One attempt is the following: "The set of all elements that are not in both A and B, or are in C." The key words here (*not, and, or*) will be treated more thoroughly in the chapter on logic.

EXAMPLE 4 Describe each of the following sets in words.

(a) $A \cap (B \cup C')$

This set might be described as "the set of all elements that are in A, and are in B or are not in C."

(b) $(A' \cup C') \cap B'$

One possibility is "the set of all elements that are not in A or not in C, and are not in B."

Another operation on sets is the *difference* of two sets. Suppose that $A = \{1, 2, 3, \ldots, 10\}$ and $B = \{2, 4, 6, 8, 10\}$. If the elements of B are excluded (or taken away) from A, the set $C = \{1, 3, 5, 7, 9\}$ is obtained. C is called the difference of sets A and B.

Difference of Sets

The **difference** of sets A and B, written $A - B$, is the set of all elements belonging to set A and not to set B, or

$$A - B = \{x \mid x \in A \text{ and } x \notin B\}.$$

Since $x \notin B$ has the same meaning as $x \in B'$, the set difference $A - B$ can also be described as $\{x \mid x \in A \text{ and } x \in B'\}$, or $A \cap B'$. Figure 9 illustrates the idea of set difference. The region in color represents $A - B$.

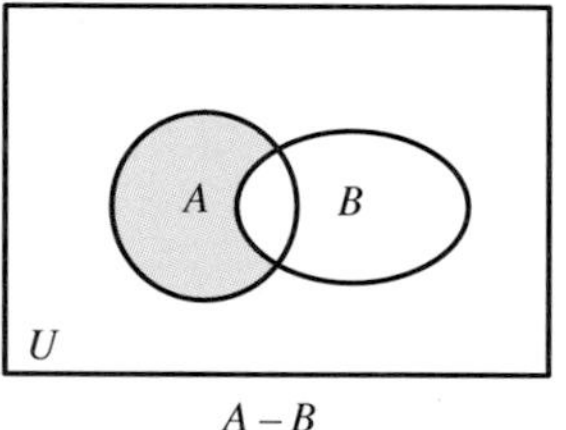

FIGURE 9

EXAMPLE 5 Let

$$\begin{aligned} U &= \{1, 2, 3, 4, 5, 6, 7\} \\ A &= \{1, 2, 3, 4, 5, 6\} \\ B &= \{2, 3, 6\} \\ C &= \{3, 5, 7\}. \end{aligned}$$

Find each set.

(a) $A - B$

Begin with set A and exclude any elements found also in set B. So, $A - B = \{1, 2, 3, 4, 5, 6\} - \{2, 3, 6\} = \{1, 4, 5\}$.

(b) $B - A$

To be in $B - A$, an element must be in set B and not in set A. But all elements of B are also in A. Thus, $B - A = \emptyset$.

(c) $(A - B) \cup C'$

From part (a), $A - B = \{1, 4, 5\}$. Also, $C' = \{1, 2, 4, 6\}$, so

$$(A - B) \cup C' = \{1, 2, 4, 5, 6\}.$$

The results in Examples 5(a) and 5(b) illustrate that, in general,

$$A - B \neq B - A.$$

When writing a set that contains several elements, the order in which the elements appear is not relevant. For example, $\{1, 5\} = \{5, 1\}$. However, there are many instances in mathematics where, when two objects are paired, the order in which the objects are written is important. This leads to the idea of the *ordered pair*. When writing ordered pairs, use parentheses (as opposed to braces, which are reserved for writing sets).

Ordered Pairs

In the **ordered pair** (a, b), a is called the **first component** and b is called the **second component.** In general, $(a, b) \neq (b, a)$.

Two ordered pairs (a, b) and (c, d) are **equal** provided that their first components are equal and their second components are equal; that is, $(a, b) = (c, d)$ if and only if $a = c$ and $b = d$.

EXAMPLE 6 Decide whether each statement is *true* or *false*.

(a) $(3, 4) = (5 - 2, 1 + 3)$

Since $3 = 5 - 2$ and $4 = 1 + 3$, the ordered pairs are equal. The statement is true.

(b) $\{3, 4\} \neq \{4, 3\}$

Since these are sets and not ordered pairs, the order in which the elements are listed is not important. Since these sets are equal, the statement is false.

(c) $(7, 4) = (4, 7)$

These ordered pairs are not equal since they do not satisfy the requirements for equality of ordered pairs. The statement is false. ■

A set may contain ordered pairs as elements. If A and B are sets, then each element of A can be paired with each element of B, and the results can be written as ordered pairs. The set of all such ordered pairs is called the *Cartesian product* of A and B, written $A \times B$ and read "A cross B." The name comes from that of the French mathematician René Descartes.

Cartesian Product of Sets

The **Cartesian product** of sets A and B, written $A \times B$, is

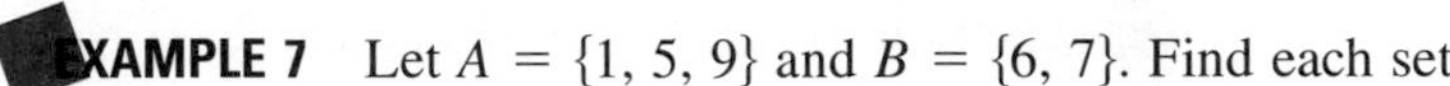

$$A \times B = \{(a, b) \mid a \in A \text{ and } b \in B\}.$$

EXAMPLE 7 Let $A = \{1, 5, 9\}$ and $B = \{6, 7\}$. Find each set.

(a) $A \times B$

Pair each element of A with each element of B. Write the results as ordered pairs, with the element of A written first and the element of B written second. Write as a set.

$$A \times B = \{(1, 6), (1, 7), (5, 6), (5, 7), (9, 6), (9, 7)\}$$

(b) $B \times A$

Since B is listed first, this set will consist of ordered pairs that have their components interchanged when compared to those in part (a).

$$B \times A = \{(6, 1), (7, 1), (6, 5), (7, 5), (6, 9), (7, 9)\}$$ ■

It should be noted that the order in which the ordered pairs themselves are listed is not important. For example, another way to write $B \times A$ in Example 7 would be

$$\{(6, 1), (6, 5), (6, 9), (7, 1), (7, 5), (7, 9)\}.$$

EXAMPLE 8 Let $A = \{1, 2, 3, 4, 5, 6\}$. Find $A \times A$.

In this example we take the Cartesian product of a set with *itself*. By pairing 1 with each element in the set, 2 with each element, and so on, we obtain the following set:

$$\begin{aligned} A \times A = \{&(1, 1), (1, 2), (1, 3), (1, 4), (1, 5), (1, 6), \\ &(2, 1), (2, 2), (2, 3), (2, 4), (2, 5), (2, 6), \\ &(3, 1), (3, 2), (3, 3), (3, 4), (3, 5), (3, 6), \\ &(4, 1), (4, 2), (4, 3), (4, 4), (4, 5), (4, 6), \\ &(5, 1), (5, 2), (5, 3), (5, 4), (5, 5), (5, 6), \\ &(6, 1), (6, 2), (6, 3), (6, 4), (6, 5), (6, 6)\}. \end{aligned}$$

It is not unusual to take the Cartesian product of a set with itself, as in Example 8. In fact, the Cartesian product in Example 8 represents all possible results that are obtained when two distinguishable dice are rolled. Determining this Cartesian product is important when studying certain problems in counting techniques and probability, as we shall see in later chapters.

From Example 7 it can be seen that, in general, $A \times B \neq B \times A$, since they do not contain exactly the same ordered pairs. However, each set contains the same number of elements, six. Furthermore, $n(A) = 3$, $n(B) = 2$, and $n(A \times B) = n(B \times A) = 6$. Since $3 \times 2 = 6$, you might conclude that the cardinal number of the Cartesian product of two sets is equal to the product of the cardinal numbers of the sets. In general, this conclusion is correct.

Cardinal Number of a Cartesian Product

If $n(A) = a$ and $n(B) = b$, then $n(A \times B) = n(B \times A) = n(A) \times n(B) = ab$.

EXAMPLE 9 Find $n(A \times B)$ and $n(B \times A)$ from the given information.

(a) $A = \{\text{a, b, c, d, e, f, g}\}$ and $B = \{2, 4, 6\}$

Since $n(A) = 7$ and $n(B) = 3$, $n(A \times B)$ and $n(B \times A)$ are both equal to 7×3, or 21.

(b) $n(A) = 24$ and $n(B) = 5$

$$n(A \times B) = n(B \times A) = 24 \times 5 = 120.$$

Finding intersections, unions, differences, Cartesian products, and complements of sets are examples of *set operations*. An **operation** is a rule or procedure by which one or more objects are used to obtain another object. The objects involved in an operation are usually sets or numbers. The most common operations on numbers are addition, subtraction, multiplication, and division. For example, starting with the numbers 5 and 7, the addition operation would produce the number $5 + 7 = 12$. With the same two numbers, 5 and 7, the multiplication operation would produce $5 \times 7 = 35$.

The most common operations on sets are summarized below, along with their Venn diagrams.

Common Set Operations

Let A and B be any sets, with U the universal set.

The **complement** of A, written A', is

$$A' = \{x \mid x \in U \text{ and } x \notin A\}.$$

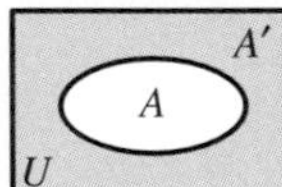

The **intersection** of A and B is

$$A \cap B = \{x \mid x \in A \text{ and } x \in B\}.$$

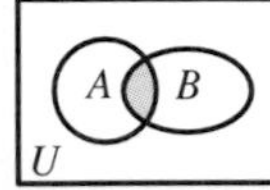

The **union** of A and B is

$$A \cup B = \{x \mid x \in A \text{ or } x \in B\}.$$

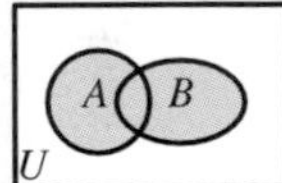

The **difference** of A and B is

$$A - B = \{x \mid x \in A \text{ and } x \notin B\}.$$

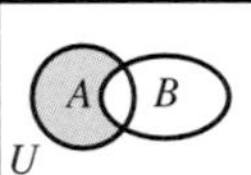

The **Cartesian product** of A and B is

$$A \times B = \{(x, y) \mid x \in A \text{ and } y \in B\}.$$

When dealing with a single set, we can use a Venn diagram as seen in Figure 10. The universal set U is divided into two regions, one representing set A and the other representing set A'.

Two sets A and B within the universal set suggest a Venn diagram as seen in Figure 11, where the four resulting regions have been numbered to provide a convenient way to refer to them. (The numbering is arbitrary.) Region 1 includes those elements outside of both set A and set B. Region 2 includes the elements belonging to A but not to B. Region 3 includes those elements belonging to both A and B. How would you describe the elements of region 4?

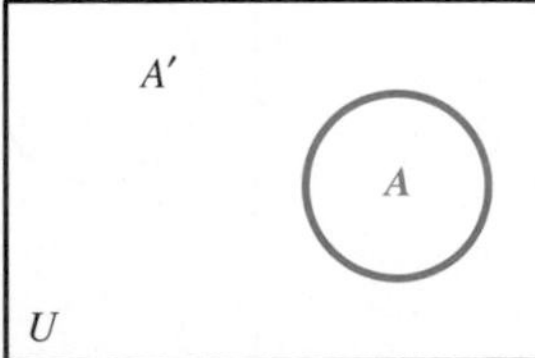

FIGURE 10

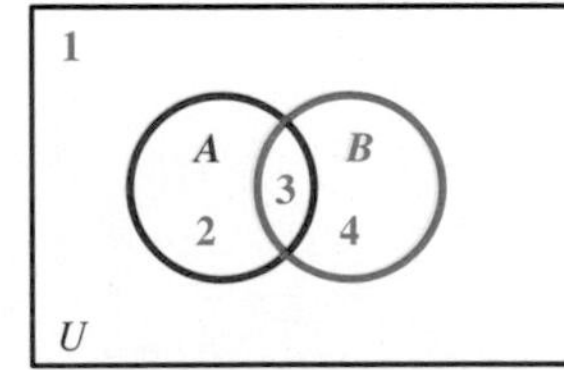

Numbering is arbitrary.

FIGURE 11

EXAMPLE 10 Draw a Venn diagram similar to Figure 11 and shade the region or regions representing the following sets.

(a) $A' \cap B$

Refer to the labeling in Figure 11. Set A' contains all the elements outside of set A, in other words, the elements in regions 1 and 4. Set B is made up of the elements in regions 3 and 4. The intersection of sets A' and B, the set $A' \cap B$, is made up of the elements in the region common to 1 and 4 and 3 and 4, that is, region 4. Thus, $A' \cap B$ is represented by region 4, which is in color in Figure 12. This region can also be described as $B - A$.

(b) $A' \cup B'$

Again, set A' is represented by regions 1 and 4, while B' is made up of regions 1 and 2. The union of A' and B', the set $A' \cup B'$, is made up of the elements belonging either to regions 1 and 4 or to regions 1 and 2. This union is composed of regions 1, 2, and 4, which are in color in Figure 13.

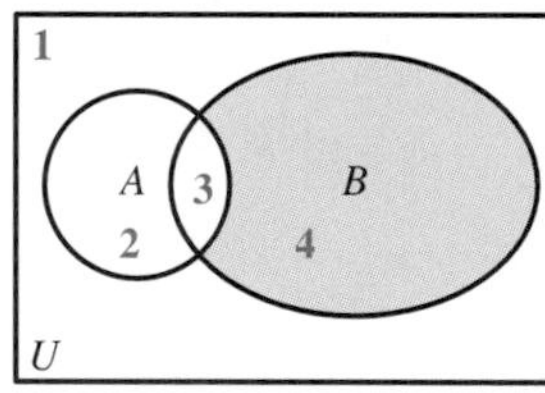

FIGURE 12

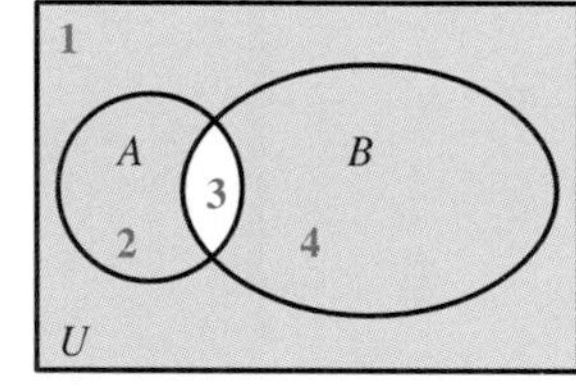

FIGURE 13

When the specific elements of sets A and B are known, it is sometimes useful to show where the various elements are located in the diagram.

EXAMPLE 11 Let

$$U = \{q, r, s, t, u, v, w, x, y, z\}$$
$$A = \{r, s, t, u, v\}$$
$$B = \{t, v, x\}.$$

Place the elements of these sets in their proper locations on a Venn diagram.

Since $A \cap B = \{t, v\}$, elements t and v are placed in region 3 in Figure 14. The remaining elements of A, that is r, s, and u, go in region 2. The figure shows the proper placement of all other elements.

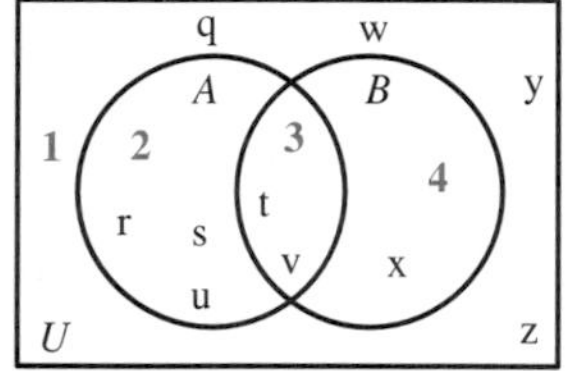

FIGURE 14

To include three sets A, B, and C in a universal set, draw a Venn diagram as in Figure 15, where again an arbitrary numbering of the regions is shown.

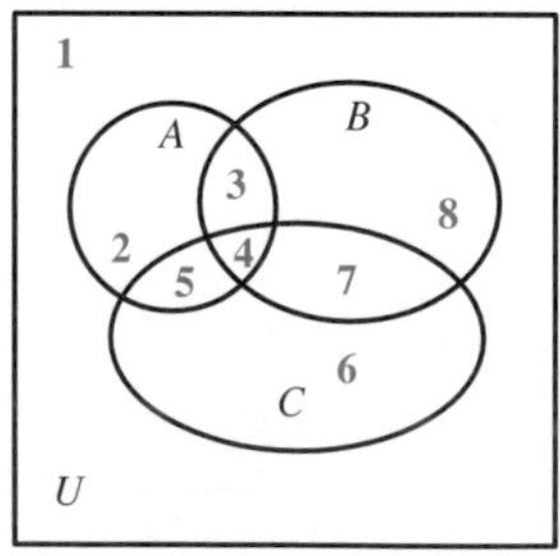

Numbering is arbitrary.

FIGURE 15

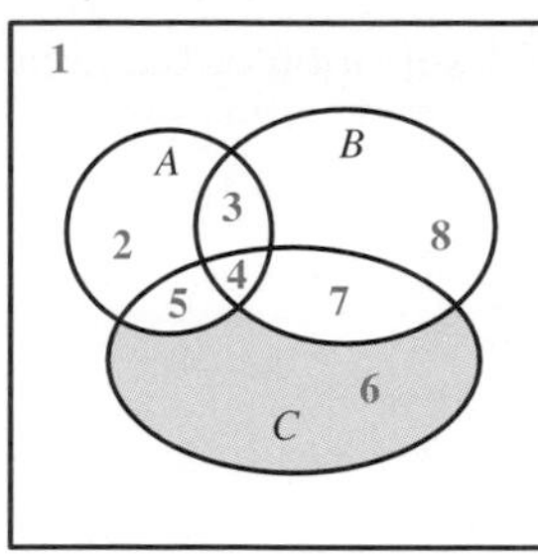

$(A' \cap B') \cap C$

FIGURE 16

EXAMPLE 12 Shade the set $(A' \cap B') \cap C$ in a Venn diagram similar to the one in Figure 15.

Work first inside the parentheses. As shown in Figure 16, set A' is made up of the regions outside set A, or regions 1, 6, 7, and 8. Set B' is made up of regions 1, 2, 5, and 6. The intersection of these sets is given by the overlap of regions 1, 6, 7, 8 and 1, 2, 5, 6, or regions 1 and 6. For the final Venn diagram, find the intersection of regions 1 and 6 with set C. As seen in Figure 16, set C is made up of regions 4, 5, 6, and 7. The overlap of regions 1, 6 and 4, 5, 6, 7 is region 6, the region in color in Figure 16.

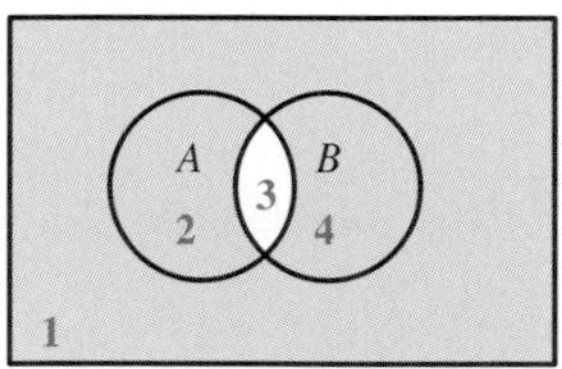

$(A \cap B)'$ is shaded.

(a)

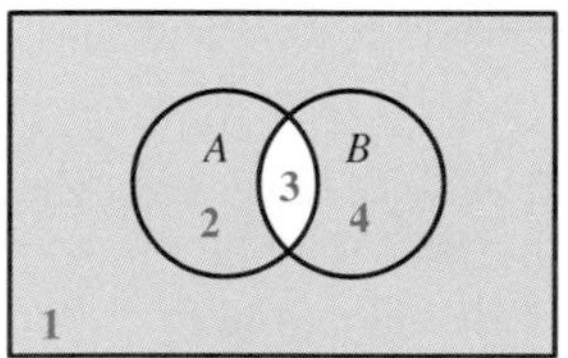

$A' \cup B'$ is shaded.

(b)

FIGURE 17

EXAMPLE 13 Is the statement

$$(A \cap B)' = A' \cup B'$$

true for every choice of sets A and B?

To help decide, use the regions labeled in Figure 11. Set $A \cap B$ is made up of region 3, so that $(A \cap B)'$ is made up of regions 1, 2, and 4. These regions are in color in Figure 17(a).

To find a Venn diagram for set $A' \cup B'$, first check that A' is made up of regions 1 and 4, while set B' includes regions 1 and 2. Finally, $A' \cup B'$ is made up of regions 1 and 4, or 1 and 2; that is, regions 1, 2, and 4. These regions are in color in Figure 17(b).

The fact that the same regions are in color in both Venn diagrams suggests that

$$(A \cap B)' = A' \cup B'.$$

This result is one of De Morgan's laws, named after the British logician Augustus De Morgan (1806–1871). (A logic version of these laws is discussed in another chapter.) De Morgan's two laws for sets follow.

De Morgan's Laws

For any sets A and B,

$$(A \cap B)' = A' \cup B'$$

and

$$(A \cup B)' = A' \cap B'.$$

The Venn diagrams in Figure 17 strongly suggest the truth of the first of De Morgan's laws. They provide a *conjecture,* as discussed in Chapter 1. Actual proofs of De Morgan's laws would require methods used in more advanced courses on set theory.

An area in a Venn diagram (perhaps set off in color) may be described using set operations. When doing this, it is a good idea to translate the region into words, remembering that intersection translates as "and," union translates as "or," and complement translates as "not." There are often several ways to describe a given region.

EXAMPLE 14 For each Venn diagram write a symbolic description of the area in color, using A, B, C, $\cap$, $\cup$, $-$, and $'$ as necessary.

(a)

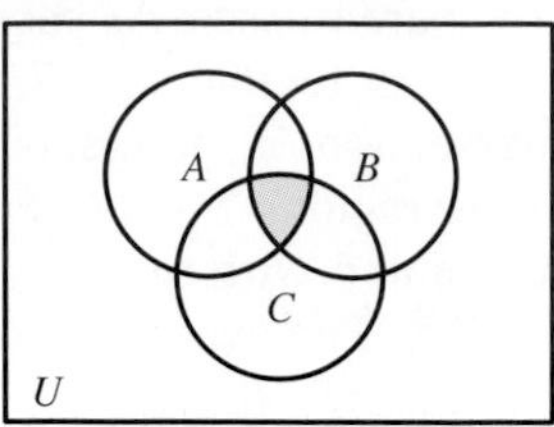

The region in color belongs to all three sets, A and B and C. Therefore, the region corresponds to $A \cap B \cap C$.

(b)

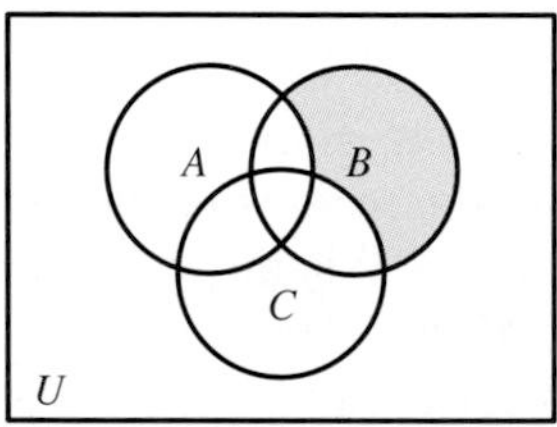

The region in color is in set B and is not in A and is not in C. Since it is not in A, it is in A', and similarly it is in C'. The region is, therefore, in B and in A' and in C', and corresponds to $B \cap A' \cap C'$.

(c) Refer to the figure in part (b) and give two additional ways of describing the region in color.

The area in color includes all of B except for the regions belonging to either A or C. This suggests the idea of set difference. The region may be described as $B - (A \cup C)$, or equivalently, $B \cap (A \cup C)'$.

EXERCISES

Perform the indicated operations.

Let $U = \{a, b, c, d, e, f, g\}$
$X = \{a, c, e, g\}$
$Y = \{a, b, c\}$
$Z = \{b, c, d, e, f\}$.

1. $X \cap Y$
2. $X \cup Y$
3. $Y \cup Z$
4. $Y \cap Z$
5. $X \cup U$
6. $Y \cap U$
7. X'
8. Y'
9. $X' \cap Y'$
10. $X' \cap Z$
11. $Z' \cap \emptyset$
12. $Y' \cup \emptyset$
13. $X \cup (Y \cap Z)$
14. $Y \cap (X \cup Z)$
15. $(Y \cap Z') \cup X$
16. $(X' \cup Y') \cup Z$
17. $(Z \cup X')' \cap Y$
18. $(Y \cap X')' \cup Z'$
19. $X - Y$
20. $Y - X$
21. $X' - Y$
22. $Y' - X$
23. $X \cap (X - Y)$
24. $Y \cup (Y - X)$

Describe each set in words.

25. $A \cup (B' \cap C')$
26. $(A \cap B') \cup (B \cap A')$
27. $(C - B) \cup A$
28. $B \cap (A' - C)$
29. $(A - C) \cup (B - C)$
30. $(A' \cap B') \cup C'$

31. The table lists some common adverse effects of prolonged tobacco and alcohol use.

Tobacco	**Alcohol**
Emphysema, e	Liver damage, l
Heart damage, h	Brain damage, b
Cancer, c	Heart damage, h

Find the smallest possible universal set U that includes all the effects listed.

Let T be the set of listed effects of tobacco and A be the set of listed effects of alcohol. (See Exercise 31.) Find each set.

32. A' **33.** T' **34.** $T \cap A$ **35.** $T \cup A$ **36.** $A \cap T'$

Describe in words each set in Exercises 37–42.

Let U = the set of all tax returns
A = the set of all tax returns with itemized deductions
B = the set of all tax returns showing business income
C = the set of all tax returns filed in 1996
D = the set of all tax returns selected for audit.

37. $B \cup C$ **38.** $A \cap D$ **39.** $C - A$

40. $D \cup A'$ **41.** $(A \cup B) - D$ **42.** $(C \cap A) \cap B'$

Assuming that A and B represent any two sets, *identify each of the following statements as either* always true *or* not always true.

43. $A \subseteq (A \cup B)$ **44.** $A \subseteq (A \cap B)$ **45.** $(A \cap B) \subseteq A$ **46.** $(A \cup B) \subseteq A$

47. $n(A \cup B) = n(A) + n(B)$ **48.** $n(A \cap B) = n(A) - n(B)$

49. $n(A \cup B) = n(A) + n(B) - n(A \cap B)$ **50.** $n(A \cap B) = n(A) + n(B) - n(A \cup B)$

51. If $B \subseteq A$, $n(A) - n(B) = n(A - B)$ **52.** $n(A - B) = n(B - A)$

For Exercises 53–60,

Let $U = \{1, 2, 3, 4, 5\}$
$X = \{1, 3, 5\}$
$Y = \{1, 2, 3\}$
$Z = \{3, 4, 5\}$.

In each case, state a general conjecture based on your observation.

53. **(a)** Find $X \cup Y$. **(b)** Find $Y \cup X$. **(c)** State a conjecture.

54. **(a)** Find $X \cap Y$. **(b)** Find $Y \cap X$. **(c)** State a conjecture.

55. **(a)** Find $X \cup (Y \cup Z)$. **(b)** Find $(X \cup Y) \cup Z$. **(c)** State a conjecture.

56. **(a)** Find $X \cap (Y \cap Z)$. **(b)** Find $(X \cap Y) \cap Z$. **(c)** State a conjecture.

57. **(a)** Find $(X \cup Y)'$. **(b)** Find $X' \cap Y'$. **(c)** State a conjecture.

58. **(a)** Find $(X \cap Y)'$. **(b)** Find $X' \cup Y'$. **(c)** State a conjecture.

59. **(a)** Find $X \cup \emptyset$. **(b)** State a conjecture.

60. **(a)** Find $X \cap \emptyset$. **(b)** State a conjecture.

Write true *or* false *for each of the following.*

61. $(3, 2) = (5 - 2, 1 + 1)$ **62.** $(10, 4) = (7 + 3, 5 - 1)$ **63.** $(4, 12) = (4, 3)$

64. $(5, 9) = (2, 9)$ **65.** $(6, 3) = (3, 6)$ **66.** $(2, 13) = (13, 2)$

67. $\{6, 3\} = \{3, 6\}$

68. $\{2, 13\} = \{13, 2\}$

69. $\{(1, 2), (3, 4)\} = \{(3, 4), (1, 2)\}$

70. $\{(5, 9), (4, 8), (4, 2)\} = \{(4, 8), (5, 9), (4, 2)\}$

Find $A \times B$ and $B \times A$, for A and B defined as follows.

71. $A = \{2, 8, 12\}, \quad B = \{4, 9\}$

72. $A = \{3, 6, 9, 12\}, \quad B = \{6, 8\}$

73. $A = \{\text{d, o, g}\}, \quad B = \{\text{p, i, g}\}$

74. $A = \{\text{b, l, u, e}\}, \quad B = \{\text{r, e, d}\}$

Use the given information to find $n(A \times B)$ and $n(B \times A)$ in Exercises 75–78.

75. the sets in Exercise 71

76. the sets in Exercise 73

77. $n(A) = 35$ and $n(B) = 6$

78. $n(A) = 13$ and $n(B) = 5$

79. If $n(A \times B) = 36$ and $n(A) = 12$, find $n(B)$.

80. If $n(A \times B) = 100$ and $n(B) = 4$, find $n(A)$.

Place the elements of these sets in the proper location on the given Venn diagram.

81. Let $U = \{\text{a, b, c, d, e, f, g}\}$
$A = \{\text{b, d, f, g}\}$
$B = \{\text{a, b, d, e, g}\}$.

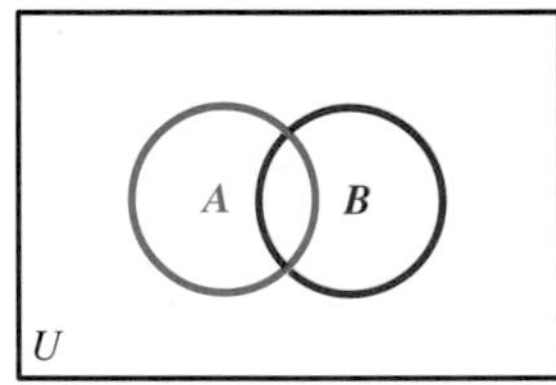

82. Let $U = \{5, 6, 7, 8, 9, 10, 11, 12, 13\}$
$M = \{5, 8, 10, 11\}$
$N = \{5, 6, 7, 9, 10\}$.

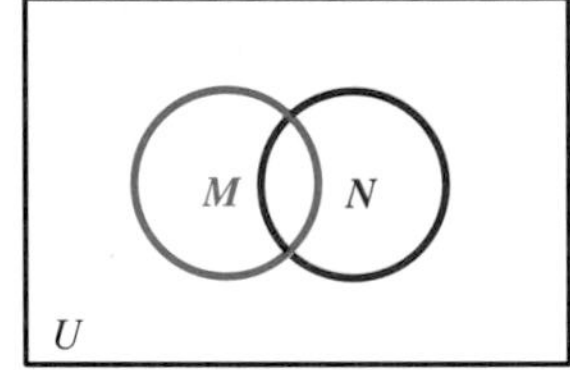

Use a Venn diagram similar to the one shown here to shade each of the following sets.

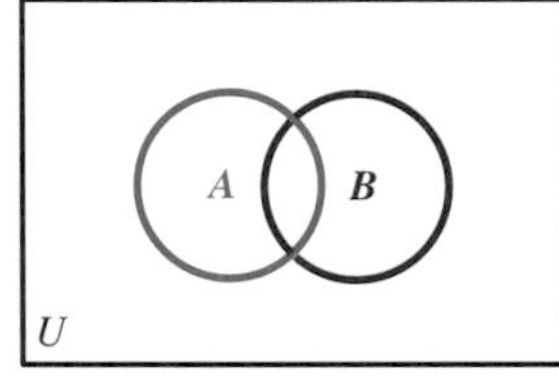

83. $B \cap A'$

84. $A \cup B$

85. $A' \cup B$

86. $A' \cap B'$

87. $B' \cup A$

88. $A' \cup A$

89. $B' \cap B$

90. $A \cap B'$

91. $B' \cup (A' \cap B')$

92. $(A \cap B) \cup B$

93. U'

94. $\emptyset'$

95. Let $U = \{\text{m, n, o, p, q, r, s, t, u, v, w}\}$
$A = \{\text{m, n, p, q, r, t}\}$
$B = \{\text{m, o, p, q, s, u}\}$
$C = \{\text{m, o, p, r, s, t, u, v}\}$.

Place the elements of these sets in the proper location on a Venn diagram similar to the one shown here.

96. Let $U = \{1, 2, 3, 4, 5, 6, 7, 8, 9\}$
$A = \{1, 3, 5, 7\}$
$B = \{1, 3, 4, 6, 8\}$
$C = \{1, 4, 5, 6, 7, 9\}$.

Place the elements of these sets in the proper location on a Venn diagram.

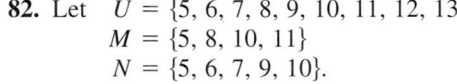

Use a Venn diagram to shade each of the following sets.

97. $(A \cap B) \cap C$

98. $(A \cap C') \cup B$

99. $(A \cap B) \cup C'$

100. $(A' \cap B) \cap C$

101. $(A' \cap B') \cap C$

102. $(A \cup B) \cup C$

103. $(A \cap B') \cup C$

104. $(A \cap C') \cap B$

105. $(A \cap B') \cap C'$

106. $(A' \cap B') \cup C$

107. $(A' \cap B') \cup C'$

108. $(A \cap B)' \cup C$

Write a description of each shaded area. Use the symbols A, B, C, ∩, ∪, −, and ′ as necessary. More than one answer may be possible.

109.

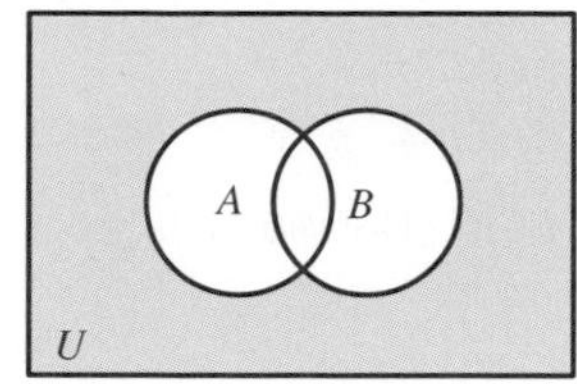

110.

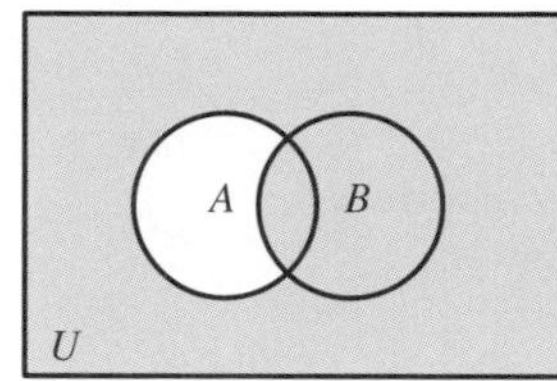

111.

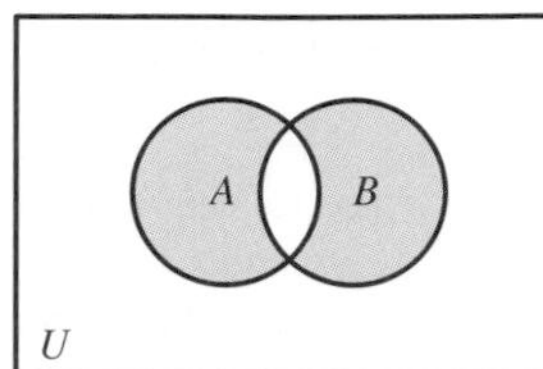

112.

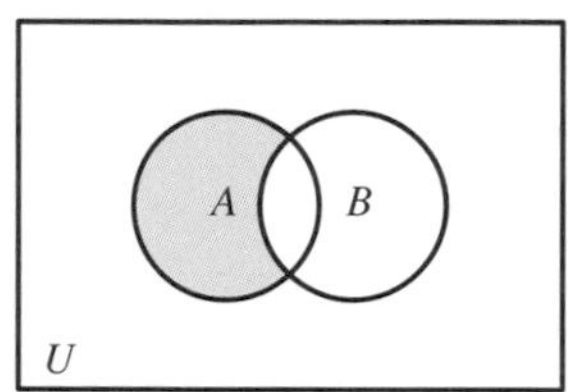

113.

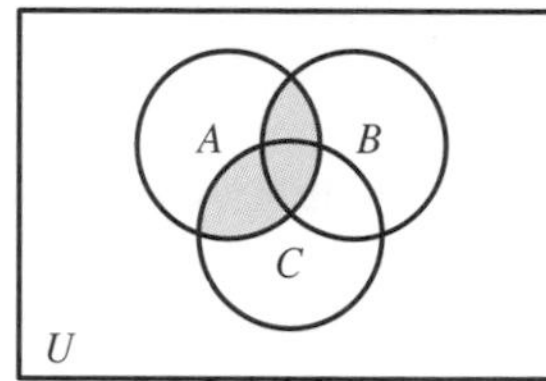

114.

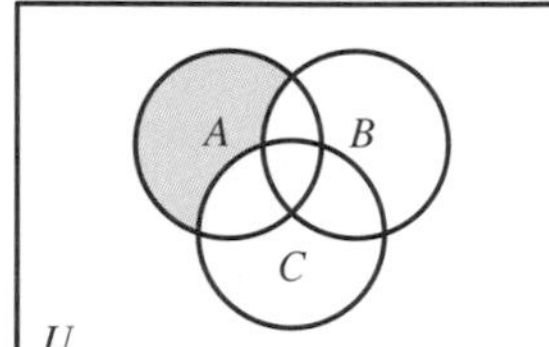

115.

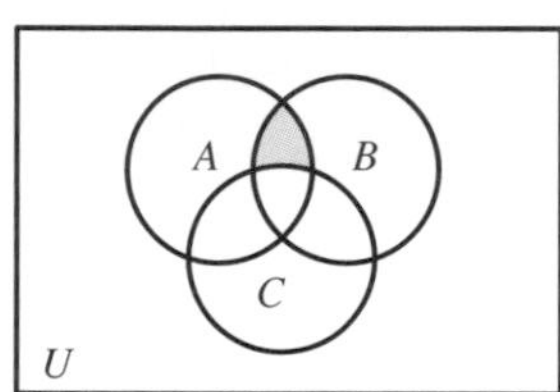

116.

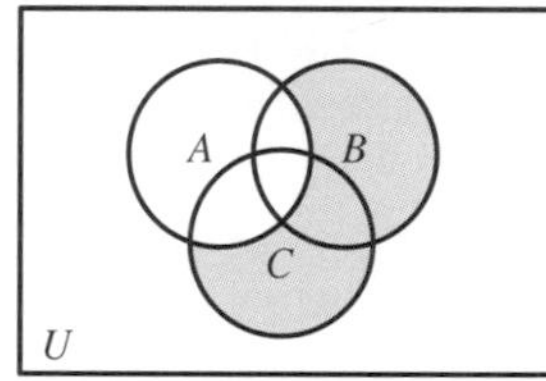

Suppose A and B are sets. Describe the conditions under which each of the following statements would be true.

117. $A = A - B$ **118.** $A = B - A$ **119.** $A = A - \emptyset$ **120.** $A = \emptyset - A$

121. $A \cup \emptyset = \emptyset$ **122.** $A \cap \emptyset = \emptyset$ **123.** $A \cap \emptyset = A$ **124.** $A \cup \emptyset = A$

125. $A \cup A = \emptyset$ **126.** $A \cap A = \emptyset$

---

127. Give examples of how the conciseness of the "language of mathematics" can be an advantage.

128. Give examples of how a language such as English, Spanish, Arabic, or Vietnamese can have an advantage over the symbolic language of mathematics.

129. If A and B are sets, is it necessarily true that $n(A - B) = n(A) - n(B)$? Explain.

130. If $Q = \{x \mid x \text{ is a rational number}\}$ and $H = \{x \mid x \text{ is an irrational number}\}$, describe each of the following sets.

(a) $Q \cup H$ **(b)** $Q \cap H$

White light can be viewed as a blending of the three primary *colors red, green, and blue. Or, we can obtain a secondary color by blending any* two *primary colors. (For example, red and blue produce magenta, blue and green produce cyan, and red and green produce yellow.) For the following exercises, refer to the photo shown here.*

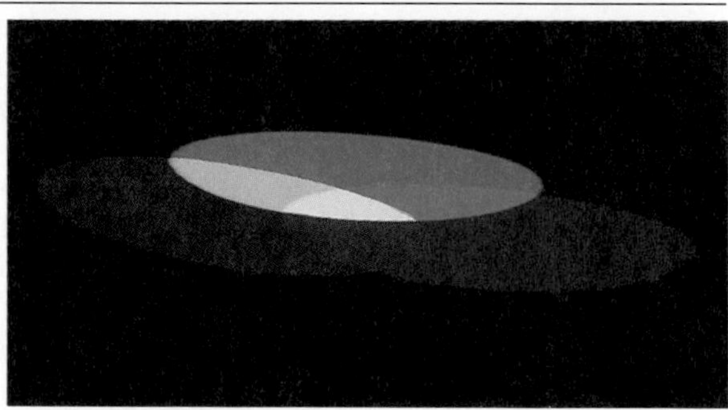

131. Name all the secondary colors of light. For each one, give its primary components.

132. In terms of set operations, white light is the three-way "intersection" of red, green, and blue. What *other* three-way intersection would also produce white light?

133. Explain why scientists sometimes refer to yellow as "minus blue."

134. What color is obtained if red is filtered out of (or subtracted from) white light?

135. What must be filtered out of white light to obtain green light?

For each of the following exercises, draw two appropriate Venn diagrams to decide whether the given statement is always true *or* not always true.

136. $A \cap A' = \emptyset$

137. $A \cup A' = U$

138. $(A \cap B) \subseteq A$

139. $(A \cup B) \subseteq A$

140. If $A \subseteq B$, then $A \cup B = A$.

141. If $A \subseteq B$, then $A \cap B = B$.

142. $(A \cup B)' = A' \cap B'$ (De Morgan's second law)

2.4 Surveys and Cardinal Numbers

Many problems involving sets of people (or other objects) require analyzing known information about certain subsets to obtain cardinal numbers of other subsets. In this section we apply three useful problem solving techniques to such problems: Venn diagrams, cardinal number formulas, and tables. The "known information" is quite often (though not always) obtained by administering a survey.

Suppose a group of college students in Arizona are questioned about their favorite musical performers, and the following information is produced.

33 like Garth Brooks	15 like Garth and Tanya
32 like Wynonna Judd	14 like Wynonna and Tanya
28 like Tanya Tucker	5 like all three
11 like Garth and Wynonna	7 like none of these performers

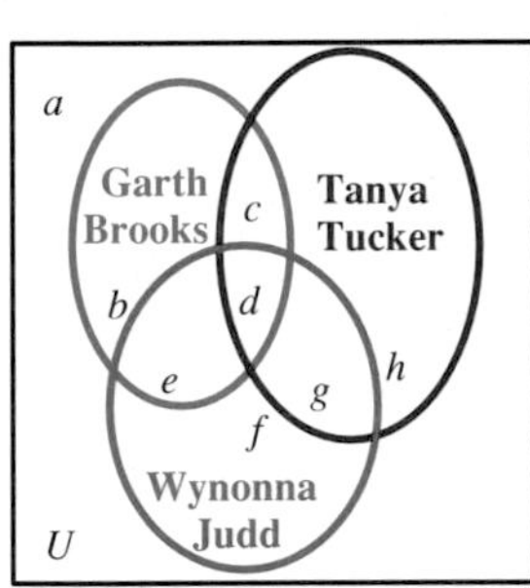

■ **FIGURE 18**

To determine the total number of students surveyed, we cannot just add the eight numbers above since there is some overlapping. For example, in Figure 18, the 33 students who like Garth Brooks should not be positioned in region b but should be distributed among regions b, c, d, and e, in a way that is consistent with all of the given data. (Region b actually contains those students who like Garth but do not like Wynonna and do not like Tanya.)

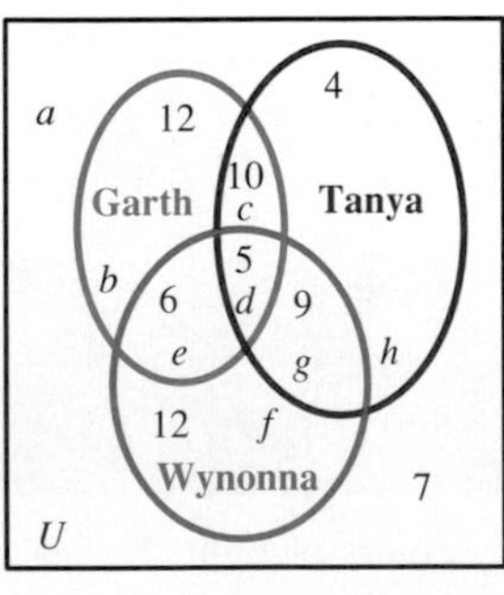

■ **FIGURE 19**

Since, at the start, we do not know how to distribute the 33 who like Garth, we look first for some more manageable data. The smallest total listed, the 5 students who like all three singers, can be placed in region d (the intersection of the three sets). And the 7 who like none of the three must go into region a. Then, the 11 who like Garth and Wynonna must go into regions d and e. Since region d already contains 5 students, we must place $11 - 5 = 6$ in region e. Since 15 like Garth and Tanya (regions c and d), we place $15 - 5 = 10$ in region c. Now that regions c, d, and e contain 10, 5, and 6 students respectively, region b receives $33 - 10 - 5 - 6 = 12$. By similar reasoning all regions are assigned their correct numbers, as shown in Figure 19.

EXAMPLE 1 Using the survey data on student preferences for performers, as summarized in Figure 19, answer the following questions.

(a) How many students like Tanya Tucker only?

A student who likes Tanya only does not like Garth and does not like Wynonna. These students are inside the regions for Tanya and outside the regions for Garth and Wynonna. Region h is the necessary region in Figure 19, and we see that four students like Tanya only.

(b) How many students like exactly two performers?

The students in regions c, e, and g like exactly two performers. The total number of such students is $10 + 6 + 9 = 25$.

(c) How many students were surveyed?

Since each student surveyed has been placed in exactly one region of Figure 19, the total number surveyed is the sum of the numbers in all eight regions:

$$7 + 12 + 10 + 5 + 6 + 12 + 9 + 4 = 65.$$

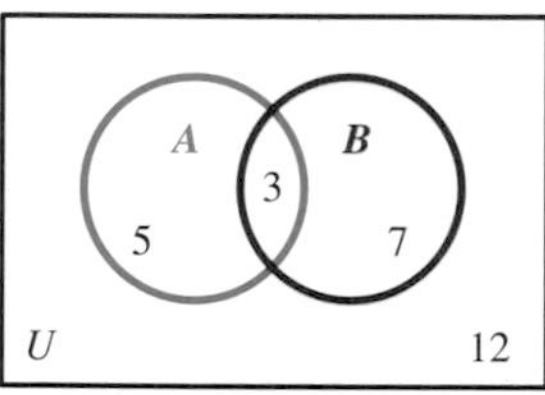

FIGURE 20

If the numbers shown in Figure 20 are the cardinal numbers of the individual regions, then $n(A) = 5 + 3 = 8$, $n(B) = 3 + 7 = 10$, $n(A \cap B) = 3$, and $n(A \cup B) = 5 + 3 + 7 = 15$. Notice that $n(A \cup B) = n(A) + n(B) - n(A \cap B)$ since $15 = 8 + 10 - 3$. This relationship is true for any two sets A and B.

Cardinal Number Formula

For any two sets A and B,

$$n(A \cup B) = n(A) + n(B) - n(A \cap B).$$

This formula can be rearranged to find any one of its four terms when the others are known.

EXAMPLE 2 Find $n(A)$ if $n(A \cup B) = 22$, $n(A \cap B) = 8$, and $n(B) = 12$.

Since the formula above can be rearranged as

$$n(A) = n(A \cup B) - n(B) + n(A \cap B),$$

we obtain $n(A) = 22 - 12 + 8 = 18$.

Sometimes, even when information is presented as in Example 2, it is more convenient to fit that information into a Venn diagram as in Example 1.

EXAMPLE 3 Bob Carlton is a section chief for an electric utility company. The employees in his section cut down tall trees, climb poles, and splice wire. Carlton recently submitted the following report to the management of the utility:

My section includes 100 employees,

T = the set of employees who can cut tall trees,

P = the set of employees who can climb poles,

W = the set of employees who can splice wire.

$n(T) = 45$	$n(P \cap W) = 20$
$n(P) = 50$	$n(T \cap W) = 25$
$n(W) = 57$	$n(T \cap P \cap W) = 11$
$n(T \cap P) = 28$	$n(T' \cap P' \cap W') = 9$

The data supplied by Carlton are reflected in Figure 21. The sum of the numbers in the diagram gives the total number of employees in the section:

$$9 + 3 + 14 + 23 + 11 + 9 + 17 + 13 = 99.$$

Carlton claimed to have 100 employees, but his data indicate only 99. The management decided that this error meant that Carlton did not qualify as section chief. He was reassigned as night-shift information operator at the North Pole. (The moral: he should have taken this course.) ■

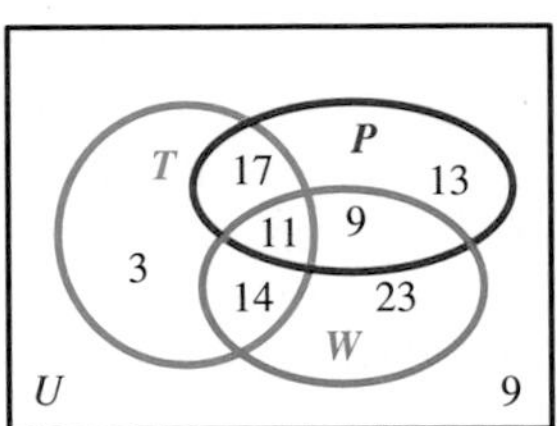

■ **FIGURE 21**

Sometimes information appears in a table rather than a Venn diagram. But the basic ideas of union and intersection still apply.

EXAMPLE 4 The officer in charge of the cafeteria on a North Carolina military base wanted to know if the beverage that enlisted men and women preferred with lunch depended on their age. So on a given day she categorized her lunch patrons according to age and according to preferred beverage, recording the resulting numbers in a table as follows.

		Beverage		
	Cola (C)	**Iced Tea (I)**	**Sweet Tea (S)**	**Totals**
Age **18–25 (Y)**	45	10	35	90
Age **26–33 (M)**	20	25	30	75
Age **Over 33 (O)**	5	30	20	55
Totals	70	65	85	220

Using the letters in the table, find the number of people in each of the following sets.

(a) $Y \cap C$

The set Y includes all personnel represented across the top row of the table (90 in all), while C includes the 70 down the left column. The intersection of these two sets is just the upper left entry: 45 people.

(b) $O' \cup I$

The set O' excludes the bottom row, so it includes the first and second rows. The set I includes the middle column only. The union of the two sets represents

$$45 + 10 + 35 + 20 + 25 + 30 + 30 = 195 \text{ people.}$$ ■

EXERCISES

Use Venn diagrams to answer each of the following.

1. Ana Pott, a psychology professor at a Southern California college, was planning a study of viewer responses to certain aspects of the movies *Casablanca, High Noon*, and *Vertigo*. Upon surveying her class of 55 students, she determined the following data:

17 had seen *Casablanca*
17 had seen *High Noon*
23 had seen *Vertigo*
6 had seen *Casablanca* and *High Noon*
8 had seen *Casablanca* and *Vertigo*
10 had seen *High Noon* and *Vertigo*
2 had seen all three of these movies.

How many students had seen
(a) exactly two of these movies?
(b) exactly one of these movies?
(c) none of these movies?
(d) *Casablanca* but neither of the others?

2. At a Florida community college, half of the 48 mathematics majors were receiving federal financial aid. Of these:

5 had Pell Grants
14 participated in the College Work Study Program
4 had Stafford Loans
2 had Stafford Loans and participated in Work Study.

Those with Pell Grants had no other federal aid.

How many of the 48 math majors had:
(a) no federal aid?
(b) more than one of these three forms of aid?
(c) federal aid other than these three forms?
(d) a Stafford Loan or Work Study?

3. The following list shows the preferences of 102 people at a wine-tasting party:

99 like Spañada
96 like Ripple
99 like Boone's Farm Apple Wine
95 like Spañada and Ripple
94 like Ripple and Boone's
96 like Spañada and Boone's
93 like all three.

How many people like:
(a) none of the three?
(b) Spañada, but not Ripple?
(c) anything but Boone's Farm?
(d) only Ripple?
(e) exactly two kinds of wine?

4. Bob Carlton (Example 3 in the text) was again reassigned, this time to the home economics department of the electric utility. He interviewed 140 people in a suburban shopping center to find out some of their cooking habits. He obtained the following results. Should he be reassigned yet one more time?

58 use microwave ovens
63 use electric ranges
58 use gas ranges
19 use microwave ovens and electric ranges
17 use microwave ovens and gas ranges
4 use both gas and electric ranges
1 uses all three
2 cook only with solar energy

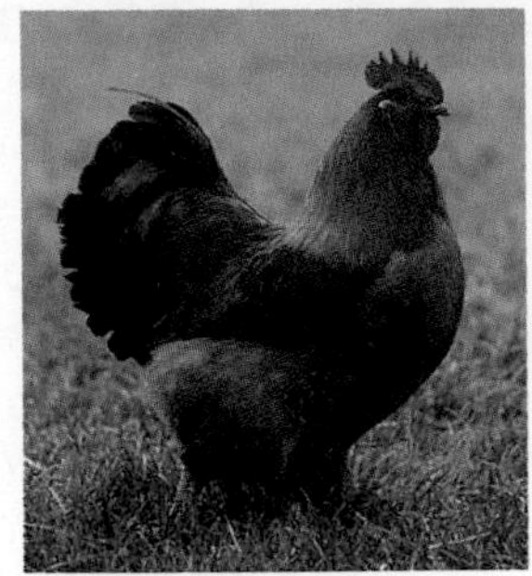

5. A chicken farmer surveyed his flock with the following results. The farmer has:

9 fat red roosters	7 thin brown hens
2 fat red hens	18 thin brown roosters
26 fat roosters	6 thin red roosters
37 fat chickens	5 thin red hens.

Answer the following questions about the flock. [*Hint:* You need a Venn diagram with circles for fat, for male (a rooster is a male, a hen is a female), and for red (assume that brown and red are opposites in the chicken world).] How many chickens are:

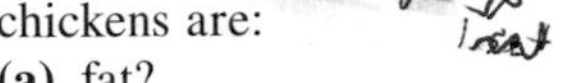

(a) fat?
(b) red?
(c) male?
(d) fat, but not male?
(e) brown, but not fat?
(f) red and fat?

6. It was once said that Country-Western songs emphasize three basic themes: love, prison, and trucks. A survey of the local Country-Western radio station produced the following data:

12 songs about a truck driver who is in love while in prison
13 about a prisoner in love
28 about a person in love
18 about a truck driver in love
3 about a truck driver in prison who is not in love
2 about people in prison who are not in love and do not drive trucks
8 about people who are out of prison, are not in love, and do not drive a truck
16 about truck drivers who are not in prison.

(a) How many songs were surveyed?
Find the number of songs about:
(b) truck drivers
(c) prisoners
(d) truck drivers in prison
(e) people not in prison
(f) people not in love.

7. Nadine Tracy conducted a survey among 75 patients admitted to the cardiac unit of a Massachusetts hospital during a two-week period.

Let B = the set of patients with high blood pressure
C = the set of patients with high cholesterol levels
S = the set of patients who smoke cigarettes.

Nadine's data are as follows:

$n(B) = 47$ $\qquad n(B \cap S) = 33$
$n(C) = 46$ $\qquad n(B \cap C) = 31$
$n(S) = 52$ $\qquad n(B \cap C \cap S) = 21.$
$n[(B \cap C) \cup (B \cap S) \cup (C \cap S)] = 51$

Find the number of these patients who:
(a) had either high blood pressure or high cholesterol levels, but not both
(b) had fewer than two of the indications listed
(c) were smokers but had neither high blood pressure nor high cholesterol levels
(d) did not have exactly two of the indications listed.

8. Stacy Thrash, who sells college textbooks, interviewed freshmen on a west coast campus to find out the main goals of today's students.

Let W = the set of those who want to become wealthy
F = the set of those who want to raise a family
E = the set of those who want to become experts in their field.

Stacy's findings are summarized here:

$n(W) = 160$ $\quad$ $n(E \cap F) = 90$
$n(F) = 140$ $\quad$ $n(W \cap F \cap E) = 80$
$n(E) = 130$ $\quad$ $n(E') = 95$
$n(W \cap F) = 95$ $\quad$ $n[(W \cup F \cup E)'] = 10.$

Find the total number of students interviewed.

9. Dwaine Tomlinson runs a basketball program in California. On the first day of the season, 60 young men showed up and were categorized by age level and by preferred basketball position, as shown in the following table.

		Position			
		Guard (G)	**Forward (F)**	**Center (N)**	**Totals**
	Junior High (J)	9	6	4	19
Age	**Senior High (S)**	12	5	9	26
	College (C)	5	8	2	15
	Totals	26	19	15	60

Using the set labels (letters) in the table, find the number of players in each of the following sets.

(a) $J \cap G$ **(b)** $S \cap N$ **(c)** $N \cup (S \cap F)$
(d) $S' \cap (G \cup N)$ **(e)** $(S \cap N') \cup (C \cap G')$ **(f)** $N' \cap (S' \cap C')$

10. A study of U.S. Army housing trends categorized personnel as commissioned officers (C), warrant officers (W), or enlisted (E), and categorized their living facilities as on-base (B), rented off-base (R), or owned off-base (O). One survey yielded the following data.

		Facilities			
		B	***R***	***O***	**Totals**
	C	12	29	54	95
Personnel	***W***	4	5	6	15
	E	374	71	285	730
	Totals	390	105	345	840

Find the number of personnel in each of the following sets.

(a) $W \cap O$ **(b)** $C \cup B$ **(c)** $R' \cup W'$
(d) $(C \cup W) \cap (B \cup R)$ **(e)** $(C \cap B) \cup (E \cap O)$ **(f)** $B \cap (W \cup R)'$

In the following exercises, make use of an appropriate formula.

11. Evaluate $n(A \cup B)$ if $n(A) = 8$, $n(B) = 14$, and $n(A \cap B) = 5$.

12. Evaluate $n(A \cap B)$ if $n(A) = 15$, $n(B) = 12$, and $n(A \cup B) = 25$.

13. Evaluate $n(A)$ if $n(B) = 20$, $n(A \cap B) = 6$, and $n(A \cup B) = 30$.

14. Evaluate $n(B)$ if $n(A) = 35$, $n(A \cap B) = 15$, and $n(A \cup B) = 55$.

Draw an appropriate Venn diagram and use the given information to fill in the number of elements in each region.

15. $n(U) = 43$, $n(A) = 25$, $n(A \cap B) = 5$, $n(B') = 30$

16. $n(A) = 19$, $n(B) = 13$, $n(A \cup B) = 25$, $n(A') = 11$

17. $n(A \cup B) = 15$, $n(A \cap B) = 8$, $n(A) = 13$, $n(A' \cup B') = 11$

18. $n(A') = 25$, $n(B) = 28$, $n(A' \cup B') = 40$, $n(A \cap B) = 10$

19. $n(A) = 24$, $n(B) = 24$, $n(C) = 26$, $n(A \cap B) = 10$, $n(B \cap C) = 8$, $n(A \cap C) = 15$, $n(A \cap B \cap C) = 6$, $n(U) = 50$

20. $n(A) = 57$, $n(A \cap B) = 35$, $n(A \cup B) = 81$, $n(A \cap B \cap C) = 15$, $n(A \cap C) = 21$, $n(B \cap C) = 25$, $n(C) = 49$, $n(B') = 52$

21. $n(A \cap B) = 21$, $n(A \cap B \cap C) = 6$, $n(A \cap C) = 26$, $n(B \cap C) = 7$, $n(A \cap C') = 20$, $n(B \cap C') = 25$, $n(C) = 40$, $n(A' \cap B' \cap C') = 2$

22. $n(A) = 15$, $n(A \cap B \cap C) = 5$, $n(A \cap C) = 13$, $n(A \cap B') = 9$, $n(B \cap C) = 8$, $n(A' \cap B' \cap C') = 21$, $n(B \cap C') = 3$, $n(B \cup C) = 32$

23. Could the information of Example 4 have been presented in a Venn diagram similar to those in Examples 1 and 3? If so, construct such a diagram. Otherwise explain the essential difference of Example 4.

24. Explain how a cardinal number formula can be derived for the case where *three* sets occur. Specifically, give a formula relating $n(A \cup B \cup C)$ to $n(A)$, $n(B)$, $n(C)$, $n(A \cap B)$, $n(A \cap C)$, $n(B \cap C)$, and $n(A \cap B \cap C)$. Illustrate with a Venn diagram.

25. In the previous section, we looked at Venn diagrams containing one, two, or three sets.* Use this information to complete the following table. (*Hint:* Make a prediction for 4 sets from the pattern of 1, 2, and 3 sets.)

Number of sets	1	2	3	4
Number of regions dividing U	2	____	____	____

The figure at the right shows U divided into 16 regions by four sets, A, B, C, and D. Find the numbers of the regions belonging to each set in Exercises 26–29.

26. $A \cap B \cap C \cap D$

27. $A \cup B \cup C \cup D$

28. $(A \cap B) \cup (C \cap D)$

29. $(A' \cap B') \cap (C \cup D)$

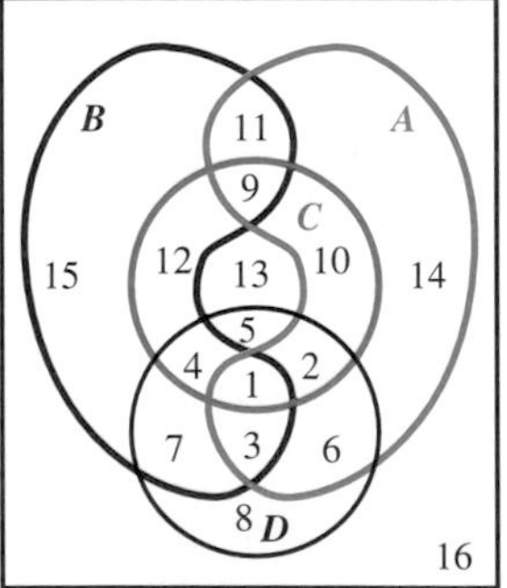

30. If we placed five most generally related sets inside U, how many regions should result? Make a general formula. If n general sets are placed inside U, how many regions would result?

* For information on using more than four sets in a Venn diagram, see "The Construction of Venn Diagrams," by Branko Grünbaum, in *The College Mathematics Journal,* June 1984.

31. A survey of 130 television viewers revealed the following facts:

52 watch football	30 watch basketball and golf
56 watch basketball	21 watch tennis and golf
62 watch tennis	3 watch football, basketball, and tennis
60 watch golf	15 watch football, basketball, and golf
21 watch football and basketball	10 watch football, tennis, and golf
19 watch football and tennis	10 watch basketball, tennis, and golf
22 watch basketball and tennis	3 watch all four of these sports
27 watch football and golf	5 don't watch any of these four sports.

Use a Venn diagram with four sets like the one in Exercises 26–29 to answer the following questions.

(a) How many of these viewers watch football, basketball, and tennis, but not golf?
(b) How many watch exactly one of these four sports?
(c) How many watch exactly two of these four sports?

Cardinal Numbers of Infinite Sets

As mentioned at the beginning of this chapter, most of the early work in set theory was done by Georg Cantor. He devoted much of his life to a study of the cardinal numbers of sets. Recall that the *cardinal number* of a finite set is the number of elements that it contains. For example, the set {5, 9, 15} contains 3 elements and has a cardinal number of 3. The cardinal number of ∅ is 0.

Cantor proved many results about the cardinal numbers of infinite sets. The proofs of Cantor are quite different from the type of proofs you may have seen in an algebra or geometry course. Because of the novelty of Cantor's methods, they were not quickly accepted by the mathematicians of his day. (In fact, some other aspects of Cantor's theory lead to paradoxes.) The results we will discuss here, however, are commonly accepted today.

The idea of the cardinal number of an infinite set depends on the idea of one-to-one correspondence. For example, each of the sets {1, 2, 3, 4} and {9, 10, 11, 12} has four elements. Corresponding elements of the two sets could be paired off in the following manner (among many other ways):

$$\begin{array}{cccc} \{1, & 2, & 3, & 4\} \\ \updownarrow & \updownarrow & \updownarrow & \updownarrow \\ \{9, & 10, & 11, & 12\}. \end{array}$$

Such a pairing is a **one-to-one correspondence** between the two sets. The "one-to-one" refers to the fact that each element of the first set is paired with exactly one element of the second set and each element of the second set is paired with exactly one element of the first set.

Two sets A and B which may be put in a one-to-one correspondence are said to be **equivalent.** Symbolically, this is written $A \sim B$. Do you see that the two sets shown above are equivalent but not equal?

The following correspondence between sets {1, 8, 12} and {6, 11},

$$\begin{array}{ccc} \{1, & 8, & 12\} \\ \updownarrow & \searrow & \swarrow \\ \{6, & & 11\}, \end{array}$$

is not one-to-one since the elements 8 and 12 from the first set are both paired with the element 11 from the second set. These sets are not equivalent.

It seems reasonable to say that if two non-empty sets have the same cardinal number, then a one-to-one correspondence can be established between the two sets. Also, if a one-to-one correspondence can be established between two sets, then the two sets must have the same cardinal number. These two facts are fundamental in discussing the cardinal numbers of infinite sets.

For Further Thought
Paradox

The word in Greek originally meant "wrong opinion" as opposed to orthodox, which meant "right opinion." Over the years, the word came to mean self-contradiction. An example is the statement "This sentence is false." By assuming it is true, we get a contradiction; likewise, by assuming it is false, we get a contradiction. Thus, it's a paradox.

Before the twentieth century it was considered a paradox that any set could be placed into one-to-one correspondence with a proper subset of itself. This paradox, called Galileo's paradox, after the sixteenth-century mathematician and scientist Galileo (see the picture), is now explained by saying that the ability to make such a correspondence is how we distinguish infinite sets from finite sets. What is true for finite sets is not necessarily true for infinite sets.

Other paradoxes include the famous paradoxes of Zeno, born about 496 B.C. in southern Italy. Two of them claim to show that a faster runner cannot overtake a slower one, and that, in fact, motion itself cannot even occur.

For Group Discussion

What is your explanation for the following two examples of Zeno's paradoxes?

1. Achilles, if he starts out behind a tortoise, can never overtake the tortoise even if he runs faster.

   Suppose Tortoise has a head start of one meter and goes one-tenth as fast as Achilles. When Achilles reaches the point where Tortoise started, Tortoise is then one-tenth meter ahead. When Achilles reaches *that* point, Tortoise is one-hundredth meter ahead. And so on. Achilles gets closer but can never catch up.

2. Motion itself cannot occur.

   You cannot travel one meter until after you have first gone a half meter. But you cannot go a half meter until after you have first gone a quarter meter. And so on. Even the tiniest motion cannot occur since a tinier motion would have to occur first.

The basic set used in discussing infinite sets is the set of counting numbers, $\{1, 2, 3, 4, 5, \ldots\}$. The set of counting numbers is said to have the infinite cardinal number $\aleph_0$ (the first Hebrew letter, aleph, with a zero subscript, read "aleph-null"). Think of $\aleph_0$ as being the "smallest" infinite cardinal number. To the question "How many counting numbers are there?" answer "There are $\aleph_0$ of them."

From the discussion above, any set that can be placed in a one-to-one correspondence with the counting numbers will have the same cardinal number as the set of counting numbers, or $\aleph_0$. It turns out that many sets of numbers have cardinal number $\aleph_0$.

Problem Solving

The next few examples show how to carry out the plan of establishing a one-to-one correspondence with the counting numbers to show that an infinite set has cardinal number $\aleph_0$.

EXAMPLE 1 Show that the set of whole numbers $\{0, 1, 2, 3, \ldots\}$ has cardinal number $\aleph_0$.

This problem is easily stated, but not quite so easily solved. All we really know about $\aleph_0$ is that it is the cardinal number of the set of counting numbers (by definition). To show that another set, such as the whole numbers, also has $\aleph_0$ as its cardinal number, we must apparently show that set to be equivalent to the set of counting numbers. And equivalence is established by a one-to-one correspondence between the two sets. This sequence of thoughts, involving just a few basic ideas, leads us to a plan: exhibit a one-to-one correspondence between the counting numbers and the whole numbers. Our strategy will be to sketch such a correspondence, showing exactly how each counting number is paired with a unique whole number. In the correspondence

$$\begin{array}{cccccccccl} \{1, & 2, & 3, & 4, & 5, & 6, & \ldots, & n, & \ldots\} & \text{Counting numbers} \\ \updownarrow & \updownarrow & \updownarrow & \updownarrow & \updownarrow & \updownarrow & & \updownarrow & & \\ \{0, & 1, & 2, & 3, & 4, & 5, & \ldots, & n-1, & \ldots\} & \text{Whole numbers} \end{array}$$

the pairing of the counting number n with the whole number $n - 1$ continues indefinitely, with neither set containing any element not used up in the pairing process. So, even though the set of whole numbers has one more element (the number 0) than the set of counting numbers, and thus should have cardinal number $\aleph_0 + 1$, the above correspondence proves that both sets have the same cardinal number. That is,

$$\aleph_0 + 1 = \aleph_0. \quad \blacksquare$$

This result shows that intuition is a poor guide for dealing with infinite sets. Intuitively, it is "obvious" that there are more whole numbers than counting numbers. However, since the sets can be placed in a one-to-one correspondence, the two sets have the same cardinal number.

The set $\{5, 6, 7\}$ is a proper subset of the set $\{5, 6, 7, 8\}$, and there is no way to place these two sets in a one-to-one correspondence. On the other hand, the set of counting numbers is a proper subset of the set of whole numbers, and

Example 1 showed that these two sets *can* be placed in a one-to-one correspondence. The only way a proper subset of a set can possibly be placed in a one-to-one correspondence with the set itself is if both sets are infinite. In fact, this important property is used as the definition of an infinite set.

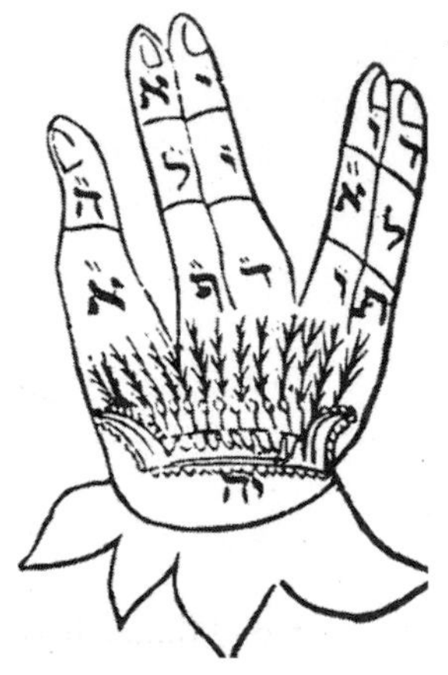

Number Lore of the Aleph-bet Aleph and other letters of the Hebrew alphabet are shown on a Kabbalistic diagram representing one of the ten emanations of God during Creation. Kabbalah, the ultra-mystical tradition within Judaism, arose in the fifth century and peaked in the sixteenth century in both Palestine and Poland.

Kabbalists believed that the Bible held mysteries that could be discovered in permutations, combinations, and anagrams of its very letters. They also "read" the numerical value of letters in a word by the technique called Gematria (from geometry?). This was possible since each letter in the aleph-bet has a numerical value (aleph = 1), and thus a numeration system exists. The letter Y stands for 10, so 15 should be YH (10 + 5). However, YH is a form of the Holy Name, so instead TW (9 + 6) is the symbol.

Infinite Set

A set is **infinite** if it can be placed in a one-to-one correspondence with a proper subset of itself.

EXAMPLE 2 Show that the set of integers $\{\ldots, -3, -2, -1, 0, 1, 2, 3, \ldots\}$ has cardinal number $\aleph_0$.

Every counting number has a corresponding negative; the negative of 8, for example, is -8. Therefore, the cardinal number of the set of integers should be $\aleph_0 + \aleph_0$, or $2\aleph_0$. However, a one-to-one correspondence can be set up between the set of integers and the set of counting numbers, as follows:

$$\begin{array}{ccccccccccc} \{1, & 2, & 3, & 4, & 5, & 6, & 7, & \ldots, & 2n, & 2n+1, & \ldots\} \\ \updownarrow & \updownarrow & \updownarrow & \updownarrow & \updownarrow & \updownarrow & \updownarrow & & \updownarrow & \updownarrow & \\ \{0, & 1, & -1, & 2, & -2, & 3, & -3, & \ldots, & n, & -n, & \ldots\}. \end{array}$$

Because of this one-to-one correspondence, the cardinal number of the set of integers is the same as the cardinal number of the set of counting numbers, or

$$2\aleph_0 = \aleph_0.$$

Notice that the one-to-one correspondence of Example 2 also proves that the set of integers is infinite. The set of integers was placed in a one-to-one correspondence with a proper subset of itself.

As shown by Example 2, there are just as many integers as there are counting numbers. This result is not at all intuitive. However, the next result is even less intuitive. We know that there is an infinite number of fractions between any two counting numbers. For example, there is an infinite set of fractions, $\{1/2, 3/4, 7/8, 15/16, 31/32, \ldots\}$ between the counting numbers 0 and 1. This should imply that there are "more" fractions than counting numbers. It turns out, however, that there are just as many fractions as counting numbers, as shown by the next example.

EXAMPLE 3 Show that the cardinal number of the set of rational numbers is $\aleph_0$.

To show that the cardinal number of the set of rational numbers is $\aleph_0$, first show that a one-to-one correspondence may be set up between the set of nonnegative rational numbers and the counting numbers. This is done by the following ingenious scheme, devised by Georg Cantor. Look at Figure 22. The nonnegative rational numbers whose denominators are 1 are written in the first row; those whose denominators are 2 are written in the second row, and so on. Every nonnegative rational number appears in this list sooner or later. For example, 327/189 is in row 189 and column 327.

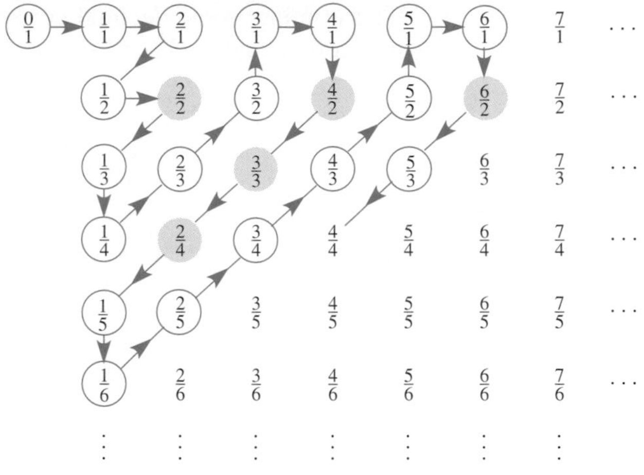

FIGURE 22

To set up a one-to-one correspondence between the set of nonnegative rationals and the set of counting numbers, follow the path drawn in Figure 22. Let 0/1 correspond to 1, let 1/1 correspond to 2, 2/1 to 3, 1/2 to 4 (skip 2/2, since $2/2 = 1/1$), 1/3 to 5, 1/4 to 6, and so on. The numbers under the colored disks are omitted, since they can be reduced to lower terms, and were thus included earlier in the listing.

This procedure sets up a one-to-one correspondence between the set of nonnegative rationals and the counting numbers, showing that both of these sets have the same cardinal number, $\aleph_0$. Now by using the method of Example 2, that is, letting each negative number follow its corresponding positive number, we can extend this correspondence to include negative rational numbers as well. Thus, the set of all rational numbers has cardinal number $\aleph_0$. ■

A set is called **countable** if it is finite, or if it has cardinal number $\aleph_0$. All the infinite sets of numbers discussed so far—the counting numbers, the whole numbers, the integers, and the rational numbers—are countable. It seems now that *every* set is countable, but this is not true. The next example shows that the set of real numbers is not countable.

EXAMPLE 4 Show that the set of all real numbers does not have cardinal number $\aleph_0$.

There are two possibilities:

1. The set of real numbers has cardinal number $\aleph_0$.
2. The set of real numbers does not have cardinal number $\aleph_0$.

Assume for the time being that the first statement is true. If the first statement is true, then a one-to-one correspondence can be set up between the set of real numbers and the set of counting numbers. We do not know what sort of correspondence this might be, but assume it can be done.

In a later chapter we show that every real number can be written as a decimal number (or simply "decimal"). Thus, in the one-to-one correspondence we are assuming, some decimal corresponds to the counting number 1, some decimal corresponds to 2, and so on. Suppose the correspondence is as follows:

$$1 \leftrightarrow .68458429006\ldots$$
$$2 \leftrightarrow .13479201038\ldots$$
$$3 \leftrightarrow .37291568341\ldots$$
$$4 \leftrightarrow .935223671611\ldots$$

and so on.

Assuming the existence of a one-to-one correspondence between the counting numbers and the real numbers means that every decimal is in the list above. Let's construct a new decimal K as follows. The first decimal in the above list has 6 as its first digit; let K start as $K = .4\ldots$. We picked 4 since $4 \neq 6$; we could have used any other digit except 6. Since the second digit of the second decimal in the list is 3, we let $K = .45\ldots$ (since $5 \neq 3$). The third digit of the third decimal is 2, so let $K = .457\ldots$ (since $7 \neq 2$). The fourth digit of the fourth decimal is 2, so let $K = .4573\ldots$ (since $3 \neq 2$). Continue K in this way.

Is K in the list that we assumed to contain all decimals? The first decimal in the list differs from K in at least the first position (K starts with 4, and the first decimal in the list starts with 6). The second decimal in the list differs from K in at least the second position, and the n-th decimal in the list differs from K in at least the n-th position. Every decimal in the list differs from K in at least one position, so that K cannot possibly be in the list. In summary:

We assume every decimal is in the list above.
The decimal K is not in the list.

Since these statements cannot both be true, the original assumption has led to a contradiction. This forces the acceptance of the only possible alternative to the original assumption: it is not possible to set up a one-to-one correspondence between the set of reals and the set of counting numbers; the cardinal number of the set of reals is not equal to $\aleph_0$. ■

The Barber Paradox This is a version of a paradox of set theory that Bertrand Russell proposed in the early twentieth century.

1. The men in a village are of two types: men who do not shave themselves and men who do.
2. The village barber shaves all men who do not shave themselves and he shaves only those men.

But who shaves the barber? The barber cannot shave himself. If he did, he would fall into the category of men who shave themselves. However, (2) above states that the barber does not shave such men.

So the barber does not shave himself. But then he falls into the category of men who do not shave themselves. According to (2), the barber shaves all of these men; hence, the barber shaves himself, too.

We find that the barber cannot shave himself, yet the barber does shave himself—a paradox.

The set of counting numbers is a proper subset of the set of real numbers. Because of this, it would seem reasonable to say that the cardinal number of the set of reals, commonly written c, is greater than $\aleph_0$. Other, even larger, infinite cardinal numbers can be constructed. For example, the set of all subsets of the set of real numbers has a cardinal number larger than c. Continuing this process of finding cardinal numbers of sets of subsets, more and more, larger and larger infinite cardinal numbers are produced.

The six most important infinite sets of numbers were listed in an earlier section. All of them have been dealt with in this section except the irrational numbers. The irrationals have decimal representations, so they are all included among the real numbers. Since the irrationals are a subset of the reals, you might guess that the irrationals have cardinal number $\aleph_0$, just like the rationals. However, since the union of the rationals and the irrationals is all the reals, that would imply that $\aleph_0 + \aleph_0 = c$. That is, $2\,\aleph_0 = c$. But in Example 2 we showed that $2\,\aleph_0 = \aleph_0$. Hence, a better guess is that the cardinal number of the irrationals is c (the same as that of the reals). This is, in fact, true. The major infinite sets of numbers, with their cardinal numbers, are summarized on the next page.

Cardinal Numbers of Infinite Number Sets

Infinite Set	Cardinal Number
Natural or counting numbers	$\aleph_0$
Whole numbers	$\aleph_0$
Integers	$\aleph_0$
Rational numbers	$\aleph_0$
Irrational numbers	c
Real numbers	c

2.5 EXERCISES

Place each pair of sets into a one-to-one correspondence, if possible.

1. {I, II, III} and {x, y, z}
2. {a, b, c, d} and {2, 4, 6}
3. {a, d, d, i, t, i, o, n} and {a, n, s, w, e, r}
4. {Reagan, Bush, Clinton} and {Nancy, Barbara, Hillary}

Give the cardinal number of each set.

5. {a, b, c, d, . . . , k}
6. {9, 12, 15, . . . , 36}
7. ∅
8. {0}
9. {300, 400, 500, . . .}
10. {−35, −28, −21, . . . , 56}
11. {−1/4, −1/8, −1/12, . . .}
12. $\{x \mid x$ is an even integer$\}$
13. $\{x \mid x$ is an odd counting number$\}$
14. {b, a, l, l, a, d}
15. {Jan, Feb, Mar, . . . , Dec}
16. {Alabama, Alaska, Arizona, . . . , Wisconsin, Wyoming}
17. Lew Lefton of the University of New Orleans has revised the old song "100 Bottles of Beer on the Wall" to illustrate a property of infinite cardinal numbers. Fill in the blank in the first line of Lefton's composition:
"$\aleph_0$ bottles of beer on the wall, $\aleph_0$ bottles of beer, take one down and pass it around, _____ bottles of beer on the wall."
18. Two one-to-one correspondences are considered "different" if some elements are paired differently in one than in the other. For example:

$$\begin{matrix}\{a, & b, & c\} \\ \updownarrow & \updownarrow & \updownarrow \\ \{a, & b, & c\}\end{matrix} \text{ and } \begin{matrix}\{a, & b, & c\} \\ \updownarrow & \updownarrow & \updownarrow \\ \{c, & b, & a\}\end{matrix} \text{ are different, while } \begin{matrix}\{a, & b, & c\} \\ \updownarrow & \updownarrow & \updownarrow \\ \{c, & a, & b\}\end{matrix} \text{ and } \begin{matrix}\{b, & c, & a\} \\ \updownarrow & \updownarrow & \updownarrow \\ \{a, & b, & c\}\end{matrix} \text{ are not.}$$

(a) How many *different* correspondences can be set up between the two sets

{George Burns, John Wayne, Chuck Norris}
and {Carlos Ray, Nathan Birnbaum, Marion Morrison}?

(b) Which one of these correspondences pairs each man with himself?

Determine whether the following pairs of sets are equal, equivalent, both, *or* neither.

19. {u, v, w}, {v, u, w}

20. {48, 6}, {4, 86}

21. {X, Y, Z}, {x, y, z}

22. {lea}, {ale}

23. $\{x \mid x$ is a positive real number$\}$, $\{x \mid x$ is a negative real number$\}$

24. $\{x \mid x$ is a positive rational number$\}$, $\{x \mid x$ is a negative real number$\}$

Show that each set has cardinal number $\aleph_0$ by setting up a one-to-one correspondence between the given set and the set of counting numbers.

25. the set of positive even numbers

26. $\{-10, -20, -30, -40, \ldots\}$

27. $\{1{,}000{,}000,\ 2{,}000{,}000,\ 3{,}000{,}000, \ldots\}$

28. the set of odd integers

29. $\{2, 4, 8, 16, 32, \ldots\}$ (*Hint:* $4 = 2^2$, $8 = 2^3$, $16 = 2^4$, and so on)

30. $\{-17, -22, -27, -32, \ldots\}$

In each of Exercises 31–34, identify the given statement as always true *or* not always true. *If not always true, give an example.*

31. If A and B are infinite sets, then $A \sim B$.

32. If set A is an infinite set and set B can be put in a one-to-one correspondence with a proper subset of A, then B must be infinite.

33. If A is an infinite set and A is not equivalent to the set of counting numbers, then $n(A) = c$.

34. If A and B are both countably infinite sets, then $n(A \cup B) = \aleph_0$.

35. The set of real numbers can be represented by an infinite line, extending indefinitely in both directions. Each point on the line corresponds to a unique real number, and each real number corresponds to a unique point on the line.

(a) Use the figure below, where the line segment between 0 and 1 has been bent into a semicircle and positioned above the line, to prove that

$$\{x \mid x \text{ is a real number between 0 and 1}\} \sim \{x \mid x \text{ is a real number}\}.$$

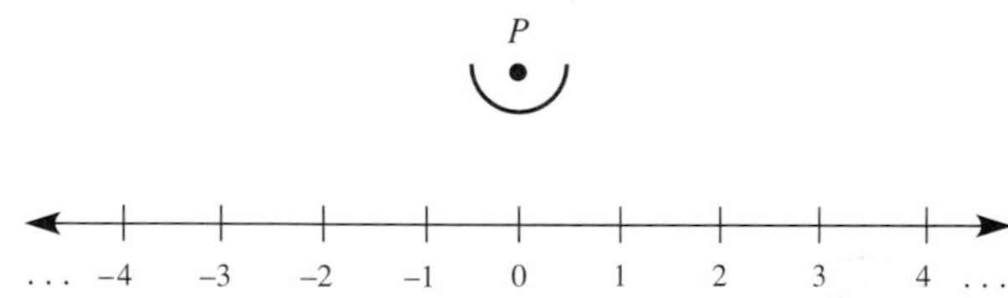

(b) What fact does part (a) establish about the set of real numbers?

36. Show that the two vertical line segments shown here both have the same number of points.

Show that each of the following sets can be placed in a one-to-one correspondence with a proper subset of itself to prove that each set is infinite.

37. {3, 6, 9, 12, . . .}
38. {4, 7, 10, 13, 16, . . .}
39. {3/4, 3/8, 3/12, 3/16, . . .}
40. {1, 4/3, 5/3, 2, . . .}
41. {1/9, 1/18, 1/27, 1/36, . . .}
42. {−3, −5, −9, −17, . . .}

43. Describe the difference between *equal* and *equivalent* sets.

44. Explain how the correspondence suggested in Example 4 shows that the set of real numbers between 0 and 1 is not countable.

COLLABORATIVE INVESTIGATION

A Survey of Your Class

This group activity is designed to determine the number of students present in your class without actually counting the members one-by-one. This will be accomplished by having each member of the class determine one particular set he or she belongs to, and then finding the sum of the cardinal numbers of the subsets.

For this activity, we will designate three sets: X, Y, and Z. Here are their descriptions:

X = {male students in the class}

Y = {students in the class 25 years old or younger}

Z = {students who have been or are currently married}.

Each student in the class will belong to one of the sets X, X', one of the sets Y, Y', and one of the sets Z, Z'. (Recall that the complement of a set consists of all elements in the universe (class) that are not in the set.)

As an example, suppose that a student is female, 23 years old, and has never been married. Then she belongs to the sets X', Y, and Z'. Joining these with intersection symbols, the set to which she belongs is

$$X' \cap Y \cap Z'.$$

Now observe the Venn diagram that follows. The eight subsets are identified by lowercase letters (*a*) through (*h*), and their symbolic descriptions. The final column will be completed when a survey is made. Each student should now determine which set he or she belongs to.

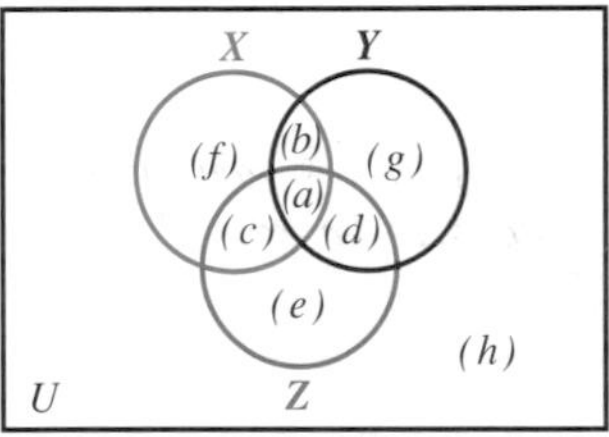

Region	Description in Terms of Set Notation	Number of Class Members in the Set
(*a*)	$X \cap Y \cap Z$	
(*b*)	$X \cap Y \cap Z'$	
(*c*)	$X \cap Y' \cap Z$	
(*d*)	$X' \cap Y \cap Z$	
(*e*)	$X' \cap Y' \cap Z$	
(*f*)	$X \cap Y' \cap Z'$	
(*g*)	$X' \cap Y \cap Z'$	
(*h*)	$X' \cap Y' \cap Z'$	

The instructor will now poll the class to see how many members are in each set. *Remember that each class member will belong to one and only one set.*

After the survey is made, find the sum of the numbers in the final column. They should add up to *exactly* the number of students present. Count the class members individually to verify this.

Topics for Discussion

1. Suppose that the final column entries do not add up to the total number of class members. What might have gone wrong?
2. Why can't a class member be a member of more than one of the eight subsets listed?

Chapter 2 Test

For Exercises 1–14,

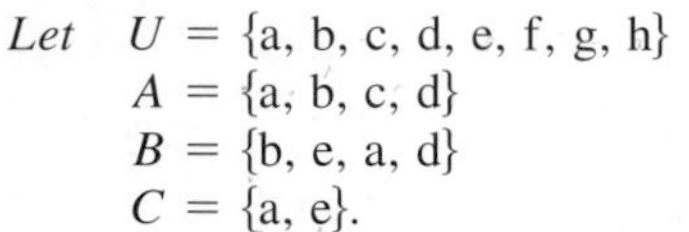

Let $U = \{a, b, c, d, e, f, g, h\}$
$A = \{a, b, c, d\}$
$B = \{b, e, a, d\}$
$C = \{a, e\}$.

Find each of the following sets.

1. $A \cup C$ **2.** $B \cap A$ **3.** B' **4.** $A - (B \cap C')$

Identify each of the following statements as true *or* false.

5. $b \in A$ **6.** $C \subseteq A$ **7.** $B \subset (A \cup C)$ **8.** $c \notin C$

9. $n[(A \cup B) - C] = 4$ **10.** $\emptyset \subset C$ **11.** $(A \cap B') \sim (B \cap A')$

12. $(A \cup B)' = A' \cap B'$

Find each of the following.

13. $n(A \times C)$ **14.** the number of proper subsets of A

Give a word description for each of the following sets.

15. $\{-3, -1, 1, 3, 5, 7, 9\}$ **16.** {January, February, March, . . . , December}

Express each of the following sets in set-builder notation.

17. $\{-1, -2, -3, -4, \ldots\}$ **18.** $\{24, 32, 40, 48, \ldots, 88\}$

Place $\subset$, $\subseteq$, both, *or* neither *in each blank to make a true statement.*

19. $\emptyset$ ______ $\{x \mid x \text{ is a counting number between 17 and 18}\}$

20. $\{4, 9, 16\}$ ______ $\{4, 5, 6, 7, 8, 9, 10\}$

Shade each of the following sets in an appropriate Venn diagram.

21. $X \cup Y'$ **22.** $X' \cap Y'$

23. $(X \cup Y) - Z$ **24.** $[(X \cap Y) \cup (Y \cap Z) \cup (X \cap Z)] - (X \cap Y \cap Z)$

The following table lists ten inventions, important directly or indirectly in our lives, together with other pertinent data.

Invention	Date	Inventor	Nation
Adding machine	1642	Pascal	France
Barometer	1643	Torricelli	Italy
Electric razor	1917	Schick	U.S.
Fiber optics	1955	Kapany	England
Geiger counter	1913	Geiger	Germany
Pendulum clock	1657	Huygens	Holland
Radar	1940	Watson-Watt	Scotland
Telegraph	1837	Morse	U.S.
Thermometer	1593	Galileo	Italy
Zipper	1891	Judson	U.S.

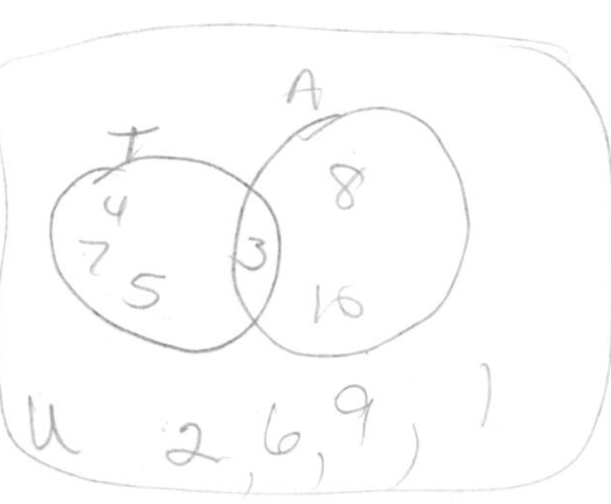

Let U = the set of all ten inventions
A = the set of items invented in the United States
T = the set of items invented in the twentieth century.

List the elements of each of the following sets.

25. $A \cap T$ **26.** $(A \cup T)'$ **27.** $A - T'$

28. State De Morgan's laws for sets in words rather than symbols.

29. Explain in your own words why, if A and B are any two nonempty sets, $A \times B \neq B \times A$ but $n(A \times B) = n(B \times A)$.

In one recent year, financial aid available to college students in the United States was nearly 30 billion dollars. (Much of it went unclaimed, mostly because qualified students were not aware of it, did not know how to obtain or fill out the required applications, or did not feel the results would be worth their effort.) The three major sources of aid are government grants, private scholarships, and the colleges themselves.

30. Jill Lynn Yuen, Financial Aid Director of a small private Midwestern college, surveyed the records of 100 sophomores and found the following:

- 49 receive government grants
- 55 receive private scholarships
- 43 receive aid from the college
- 23 receive government grants and private scholarships
- 18 receive government grants and aid from the college
- 28 receive private scholarships and aid from the college
- 8 receive help from all three sources.

How many of the students in the survey:

(a) have a government grant only?
(b) have a private scholarship but not a government grant?
(c) receive financial aid from only one of these sources?
(d) receive aid from exactly two of these sources?
(e) receive no financial aid from any of these sources?
(f) receive no aid from the college or from the government?

3

Logic

Although modern digital computers are exceedingly complex machines, their design involves only two basic components: gates and memory cells. A gate is a simple electrical switch that controls the flow of data much like a gate controls the flow of water in a canal. A memory cell can store a single bit of data. By interconnecting various gates and memory cells, a circuit can be designed to perform operations like addition and subtraction. Since the beginning of the computer industry, there has been a constant trend toward the miniaturization of electrical circuits. In 1960 only a few gates could be reliably printed onto a silicon chip. This number doubled annually throughout the 1960s and approximately quadrupled every three years thereafter. Today over a million logic gates can be printed onto a single chip that is only a few millimeters square. Some experts believe that a silicon chip containing over a billion components may be achieved within ten years. After this, fundamental physical limitations may prevent further improvements.

How is it possible to design the interconnection of so many logic gates and have them all work together properly? You will learn the answer to this question in this chapter when you study Boolean algebra and formal logic. Without Boolean algebra and formal logic, modern computers would have been impossible. An electrical circuit is either on or off, and as a result, it can be modeled by the logical states of true or false. Instead of looking at actual connections in electrical circuits, computer engineers construct truth tables to determine the output of an electrical component for certain inputs. As a result of this abstraction, design errors can be corrected prior to making the first prototype circuit. The use of logic to create complex circuitry has propelled technology forward at an incredible pace.

Source: Stallings, W., *Computer Organization and Architecture,* Macmillan Publishing Company, New York, 1993.

3.1 Statements and Quantifiers

Gottfried Leibniz (1646–1716) was a wide-ranging philosopher and a universalist who tried to patch up Catholic-Protestant conflicts. He promoted cultural exchange between Europe and the East. Chinese ideograms led him to search for a universal symbolism.

An early inventor of symbolic logic, Leibniz hoped to build a totally objective model for human reasoning and thus reduce all reasoning errors to minor calculational mistakes. His dream has so far proved futile.

This section introduces the study of *symbolic logic,* which uses letters to represent statements, and symbols for words such as *and, or, not.* One of the main applications of logic is in the study of the *truth value* (that is, the truth or falsity) of statements with many parts. The truth value of these statements depends on the components that comprise them.

Many kinds of sentences occur in ordinary language, including factual statements, opinions, commands, and questions. Symbolic logic discusses only the first type of sentence, the kind that involves facts.

A **statement** is defined as a declarative sentence that is either true or false, but not both simultaneously. For example, both of the following are statements:

Radio provides a means of communication.

$2 + 1 = 6$

Each one is either true or false. However, the following sentences are not statements based on this definition:

Paint the wall.

How do you spell relief?

Shaquille O'Neal is a better basketball player than Anfernee Hardaway.

This sentence is false.

These sentences cannot be identified as being either true or false. The first sentence is a command, and the second is a question. The third is an opinion. "This sentence is false" is a paradox; if we assume it is true, then it is false, and if we assume it is false, then it is true.

A **compound statement** may be formed by combining two or more statements. The statements making up a compound statement are called **component statements.** Various **logical connectives,** or simply **connectives,** can be used in forming compound statements. Words like *and, or, not,* and *if . . . then* are examples of connectives. (While a statement such as "Today is not Tuesday" does not consist of two component statements, for convenience it is considered compound, since its truth value is determined by noting the truth value of a different statement, "Today is Tuesday.")

EXAMPLE 1 Decide whether or not each statement is compound.

(a) Trey Yuen restaurant serves Peking duck and Pat O'Brien's serves drinks called Hurricanes.

This statement is compound, since it is made up of the component statements "Trey Yuen restaurant serves Peking duck" and "Pat O'Brien's serves drinks called Hurricanes." The connective is *and.*

(b) You can pay me now or you can pay me later.

The connective here is *or.* The statement is compound.

(c) If he said it, then it must be true.

The connective here is *if . . . then,* discussed in more detail in a later section. The statement is compound.

(d) My pistol was made by Smith and Wesson.

While the word "and" is used in this statement, it is not used as a *logical* connective, since it is part of the name of the manufacturer. The statement is not compound. ■

TEST LOGIC
1:=
2:≠
3:>
4:≥
5:<
6:≤

Negations

The sentence "Tom Jones has a red car" is a statement; the **negation** of this statement is "Tom Jones does not have a red car." The negation of a true statement is false, and the negation of a false statement is true.

EXAMPLE 2 Form the negation of each statement.

4 < 9 is true, as indicated by the 1.

(a) That state has a governor.

To negate this statement, we introduce *not* into the sentence: "That state does not have a governor."

(b) The sun is not a star.

Negation: "The sun is a star." 

4 > 9 is false, as indicated by the 0.

The TEST menu of the TI-82 calculator allows the user to test the truth or falsity of statements involving =, ≠, >, ≥, <, and ≤. If a statement is true, it returns a 1; if false, it returns a 0.

One way to detect incorrect negations is to check the truth value. A negation must have the opposite truth value from the original statement.

The next example uses some of the following inequality symbols from algebra.

Symbolism	Meaning	Examples	
$a < b$	a is less than b	$4 < 9$	$\frac{1}{2} < \frac{3}{4}$
$a > b$	a is greater than b	$6 > 2$	$-5 > -11$
$a \leq b$	a is less than or equal to b	$8 \leq 10$	$3 \leq 3$
$a \geq b$	a is greater than or equal to b	$-2 \geq -3$	$-5 \geq -5$

EXAMPLE 3 Give a negation of each inequality. Do *not* use a slash symbol.

(a) $p < 9$

The negation of "p is less than 9" is "p is *not* less than 9." Since we cannot use "not," which would require writing $p \not< 9$, phrase the negation as "p is greater than or equal to 9," or $p \geq 9$.

(b) $7x + 11y \geq 77$

The negation, with no slash, is $7x + 11y < 77$. ■

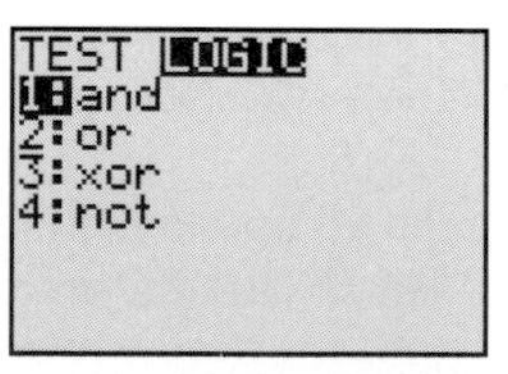

The LOGIC menu of the TI-82 calculator allows the user to test truth or falsity of statements involving *and, or, exclusive or* (see Exercise 76 in the next section), and *not.*

Symbols

To simplify work with logic, symbols are used. Statements are represented with letters, such as p, q, or r, while several symbols for connectives are shown in the following table. The table also gives the type of compound statement having the given connective.

Connective	Symbol	Type of Statement
and	$\wedge$	Conjunction
or	$\vee$	Disjunction
not	$\sim$	Negation

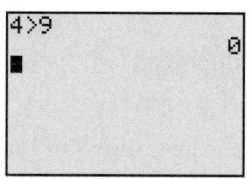

The symbol $\sim$ represents the connective *not*. If p represents the statement "Millard Fillmore was president in 1850" then $\sim p$ represents "Millard Fillmore was not president in 1850."

EXAMPLE 4 Let p represent "It is 80° today," and let q represent "It is Tuesday." Write each symbolic statement in words.

(a) $p \vee q$
From the table, $\vee$ symbolizes *or;* thus, $p \vee q$ represents

It is 80° today or it is Tuesday.

(b) $\sim p \wedge q$

It is not 80° today and it is Tuesday.

(c) $\sim(p \vee q)$

It is not the case that it is 80° today or it is Tuesday.

(d) $\sim(p \wedge q)$

It is not the case that it is 80° today and it is Tuesday. ■

Aristotle, the first to systematize the logic we use in everyday life, appears above in a detail from the painting *The School of Athens,* by Raphael. He is shown debating a point with his teacher **Plato.** Many people feel that Aristotle's logic and analysis of how things work is inseparably woven into the ways Westerners are taught to think.

The statement in part (c) of Example 4 is usually translated in English as "Neither p nor q."

Quantifiers

The words *all, each, every,* and *no(ne)* are called **universal quantifiers,** while words and phrases like *some, there exists,* and *(for) at least one* are called **existential quantifiers.** Quantifiers are used extensively in mathematics to indicate *how many* cases of a particular situation exist. Be careful when forming the negation of a statement involving quantifiers.

The negation of a statement must be false if the given statement is true and must be true if the given statement is false, in all possible cases. Consider the statement

All girls in the group are named Mary.

Many people would write the negation of this statement as "No girls in the group are named Mary" or "All girls in the group are not named Mary." But this would not be correct. To see why, look at the three groups below.

Group I: Mary Jones, Mary Smith, Mary Jackson
Group II: Mary Johnson, Betty Parker, Margaret Boyle
Group III: Shannon d'Hemecourt, Annie Ross, Patricia Gainey

These groups contain all possibilities that need to be considered. In Group I, *all* girls are named Mary; in Group II, *some* girls are named Mary (and some are not); and in Group III, *no* girls are named Mary. Look at the truth values in the chart on the next page and keep in mind that "some" means "at least one (and possibly all)."

Truth Value as Applied to:

	Group I	Group II	Group III
(1) All girls in the group are named Mary. **(Given)**	T	F	F
(2) No girls in the group are named Mary. **(Possible negation)**	F	F	T
(3) All girls in the group are not named Mary. **(Possible negation)**	F	F	T
(4) Some girls in the group are not named Mary. **(Possible negation)**	F	T	T

Negation

The negation of the given statement (1) must have opposite truth values in *all* cases. It can be seen that statements (2) and (3) do not satisfy this condition (for Group II), but statement (4) does. It may be concluded that the correct negation for "All girls in the group are named Mary" is "Some girls in the group are not named Mary." Other ways of stating the negation are:

Not all girls in the group are named Mary.

It is not the case that all girls in the group are named Mary.

At least one girl in the group is not named Mary.

The following table can be used to generalize the method of finding the negation of a statement involving quantifiers.

Negations of Quantified Statements

Statement	Negation
All do.	Some do not. (Equivalently: Not all do.)
Some do.	None do. (Equivalently: All do not.)

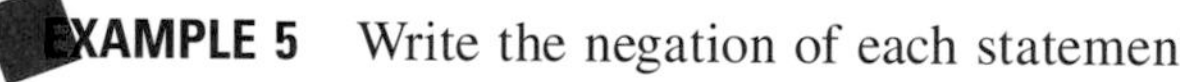

The negation of the negation of a statement is simply the statement itself. For instance, the negations of the statements in the Negation column are simply the corresponding original statements in the Statement column. As an example, the negation of "Some do not" is "All do."

EXAMPLE 5 Write the negation of each statement.

(a) Some dogs have fleas.

Since *some* means "at least one," the statement "Some dogs have fleas" is really the same as "At least one dog has fleas." The negation of this is "No dog has fleas."

(b) Some dogs do not have fleas.

This statement claims that at least one dog, somewhere, does not have fleas. The negation of this is "All dogs have fleas."

(c) No dogs have fleas.

The negation is "Some dogs have fleas." ■

The many relationships among special sets of numbers can be expressed using universal and existential quantifiers. Earlier we introduced the counting or natural numbers, which form the "building blocks" of the real number system.

Sets of real numbers are studied in algebra, and we summarize these in the box that follows.

Sets of Real Numbers

Natural or counting numbers $\{1, 2, 3, 4, \ldots\}$
Whole numbers $\{0, 1, 2, 3, 4, \ldots\}$
Integers $\{\ldots, -3, -2, -1, 0, 1, 2, 3, \ldots\}$
Rational numbers $\{p/q \mid p \text{ and } q \text{ are integers, and } q \neq 0\}$
(Some examples of rational numbers are $3/5$, $-7/9$, 5, and 0. Any rational number may be written as a terminating decimal number, like 0.25 or a repeating decimal number, like 0.666. . . .)
Real numbers $\{x \mid x \text{ is a number which may be written as a decimal}\}$
Irrational numbers $\{x \mid x \text{ is a real number and } x \text{ cannot be written as a quotient of integers}\}$
(Some examples of irrational numbers are $\sqrt{2}$, $\sqrt[3]{4}$, and π. A characteristic of irrational numbers is that their decimal representations never terminate and never repeat, that is, they never reach a point where a given pattern of digits repeats from that point on.)

EXAMPLE 6 Decide whether each of the following statements about sets of numbers involving a quantifier is *true* or *false.*

(a) There exists a whole number that is not a natural number.
Because there is such a whole number (it is 0), this statement is true.

(b) Every integer is a natural number.
This statement is false, because we can find at least one integer that is not a natural number. For example, -1 is an integer but is not a natural number. (There are infinitely many other choices we could have made.)

(c) Every natural number is a rational number.
Since every natural number can be written as a fraction with denominator 1, this statement is true.

(d) There exists an irrational number that is not real.
In order to be an irrational number, a number must first be real (see the box). Therefore, since we cannot give an irrational number that is not real, this statement is false. (Had we been able to find at least one, the statement would have then been true.) ■

EXERCISES

Decide whether or not each of the following is a statement.

1. A ZIP code for New Orleans is 70115.
2. October 12, 1949 was a Wednesday.
3. Have a nice day.
4. Stand up and be counted.
5. $8 + 15 = 23$
6. $9 - 4 = 5$ and $2 + 1 = 5$
7. Chester A. Arthur was president in 1882.
8. Not all numbers are positive.
9. Dancing is enjoyable.
10. Since 1950, more people have died in automobile accidents than of cancer.
11. Toyotas are better cars than Dodges.

12. Sit up and take notice.
13. One gallon of water weighs more than 5 pounds.
14. Kevin "Catfish" McCarthy once took a prolonged continuous shower for 340 hours, 40 minutes.

Decide whether each of the following statements is compound.

15. My sister got married in Paris.
16. I read novels and I read newspapers.
17. Denise Clark is younger than 40 years of age, and so is Rodger Klas.
18. Yesterday was Friday.
19. $4 + 2 \neq 8$
20. $5 \neq 4 + 2$
21. If Buddy is a politician, then Eddie is a crook.
22. If Earl Karn sells his quota, then Kerry Freeman will be happy.

Write a negation for each of the following statements.

23. The flowers are to be watered.
24. Her aunt's name is Lucia.
25. No rain fell in southern California today.
26. Every dog has its day.
27. All students present will get another chance.
28. Some books are longer than this book.
29. Some people have all the luck.
30. No computer repairman can play blackjack.
31. Nobody doesn't like Sara Lee.
32. Everybody loves somebody sometime.

Give a negation of each inequality. Do not use a slash symbol.

33. $x > 3$ 34. $y < -2$

35. $p \geq 4$ 36. $q \leq 12$

37. Explain why the negation of "$r > 4$" is not "$r < 4$."
38. Try to negate the sentence "The exact number of words in this sentence is ten" and see what happens. Explain the problem that arises.

Let p represent the statement "She has blue eyes" *and let q represent the statement* "He is 43 years old." *Translate each symbolic compound statement into words.*

39. $\sim p$ 40. $\sim q$

41. $p \wedge q$ 42. $p \vee q$

43. $\sim p \vee q$ 44. $p \wedge \sim q$

45. $\sim p \vee \sim q$ 46. $\sim p \wedge \sim q$

47. $\sim(\sim p \wedge q)$ 48. $\sim(p \vee \sim q)$

Let p represent the statement "Chris collects videotapes" *and let q represent the statement* "Jack is a shortstop." *Convert each of the following compound statements into symbols.*

49. Chris collects videotapes and Jack is not a shortstop.
50. Chris does not collect videotapes or Jack is not a shortstop.
51. Chris does not collect videotapes or Jack is a shortstop.
52. Jack is a shortstop and Chris does not collect videotapes.
53. Neither Chris collects videotapes nor Jack is a shortstop.
54. Either Jack is a shortstop or Chris collects videotapes, and it is not the case that both Jack is a shortstop and Chris collects videotapes.
55. Incorrect use of quantifiers is often heard in everyday language. Suppose you hear that a local electronics chain is having a 30% off sale, and the radio advertisement states "All items are not available in all stores." Do you think that, literally translated, the ad really means what it says? What do you think is really meant? Explain your answer.
56. Repeat Exercise 55 for the following: "All people don't have the time to devote to maintaining their cars properly."

Decide whether each statement involving a quantifier is true *or* false.

57. Every natural number is an integer.
58. Every whole number is an integer.
59. There exists an integer that is not a natural number.
60. There exists a rational number that is not an integer.
61. All irrational numbers are real numbers.
62. All rational numbers are real numbers.
63. Some whole numbers are not rational numbers.
64. Some rational numbers are not integers.
65. Each rational number is a positive number.
66. Each whole number is a positive number.

Refer to the sketches labeled A, B, and C, and identify the sketch (or sketches) that is (are) satisfied by the given statement involving a quantifier.

A

B

C

67. All plants have a flower.

68. At least one plant has a flower.

69. No plant has a flower.

70. All plants do not have a flower.

71. At least one plant does not have a flower.

72. No plant does not have a flower.

73. Not every plant has a flower.

74. Not every plant does not have a flower.

75. Explain the difference between the following statements:

All students did not pass the test.
Not all students passed the test.

76. Write the following statement using "every": There is no one here who has not done that at one time or another.

77. The statement "For some real number x, $x^2 \geq 0$" is true. However, your friend does not understand why, since he claims that $x^2 \geq 0$ for *all* real numbers x (and not *some*). How would you explain his misconception to him?

78. Only one of the following statements is true. Which one is it?
(a) For some real number x, $x \not< 0$.
(b) For all real numbers x, $x^3 > 0$.
(c) For all real numbers x less than 0, x^2 is also less than 0.
(d) For some real number x, $x^2 < 0$.

3.2 Truth Tables

In this section, the truth values of component statements are used to find the truth values of compound statements. To begin, let us decide on the truth values of the **conjunction** *p and q*, symbolized $p \wedge q$. In everyday language, the connective *and* implies the idea of "both." The statement

Monday immediately follows Sunday and March immediately follows February

is true, since each component statement is true. On the other hand, the statement

Monday immediately follows Sunday and March immediately follows January

is false, even though part of the statement (Monday immediately follows Sunday) is true. For the conjunction $p \wedge q$ to be true, both p and q must be true. This result is summarized by a table, called a **truth table,** which shows all four of the possible combinations of truth values for the conjunction *p and q*. The truth table for *conjunction* is shown here.

Truth Table for the Conjunction *p* and *q*

p and q

p	q	$p \wedge q$
T	T	T
T	F	F
F	T	F
F	F	F

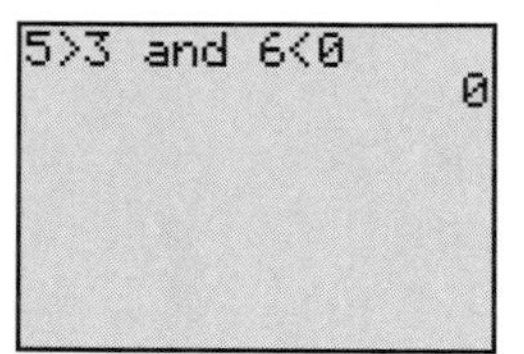

The calculator returns a "0" for $5 > 3$ *and* $6 < 0$, indicating that the statement is false.

EXAMPLE 1 Let p represent "$5 > 3$" and let q represent "$6 < 0$." Find the truth value of $p \wedge q$.

Here p is true and q is false. Looking in the second row of the conjunction truth table shows that $p \wedge q$ is false.

In some cases, the logical connective *but* is used in compound statements. For example, consider the statement

He wants to go to the mountains but she wants to go to the beach.

Here, *but* is used in place of *and* to give a different sort of emphasis to the statement. In such a case, we consider the statement as we would consider the conjunction using the word *and.* The truth table for the conjunction, given above, would apply.

In ordinary language, the word *or* can be ambiguous. The expression "this or that" can mean either "this or that or both," or "this or that but not both." For example, the statement

I will paint the wall or I will paint the ceiling

probably has the following meaning: "I will paint the wall or I will paint the ceiling or I will paint both." On the other hand, the statement

I will drive the Ford or the Datsun to the store

probably means "I will drive the Ford, or I will drive the Datsun, but I shall not drive both."

The symbol $\vee$ normally represents the first *or* described. That is, $p \vee q$ means "p or q or both." With this meaning of *or*, $p \vee q$ is called the *inclusive disjunction,* or just the **disjunction** of p and q.

In everyday language, the disjunction implies the idea of "either." For example, the disjunction

I have a quarter or I have a dime

is true whenever I have either a quarter, a dime or both. The only way this disjunction could be false would be if I had neither coin. A disjunction is false only if both component statements are false. The truth table for *disjunction* follows.

Truth Table for the Disjunction *p* or *q*

p or q

p	q	$p \vee q$
T	T	T
T	F	T
F	T	T
F	F	F

The inequality symbols $\leq$ and $\geq$ are examples of inclusive disjunction. For example, $x \leq 6$ is true if either $x < 6$ or $x = 6$.

EXAMPLE 2 The following list shows several statements and the reason that each is true.

Statement	Reason that It Is True
$8 \geq 8$	$8 = 8$
$3 \geq 1$	$3 > 1$
$-5 \leq -3$	$-5 < -3$
$-4 \leq -4$	$-4 = -4$

For Further Thought

Raymond Smullyan is one of today's foremost writers of logic puzzles. This multitalented professor of mathematics and philosophy at City University of New York has written several books on recreational logic, including *The Lady or the Tiger?, What is the Name of This Book?,* and *Alice in Puzzleland.* The title of the first of these is taken from the classic Frank Stockton short story, in which a prisoner must make a choice between two doors: behind one is a beautiful lady, and behind the other is a hungry tiger.

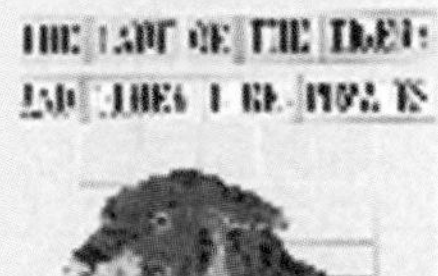

For Group Discussion

Smullyan proposes the following: What if each door has a sign, and the man knows that only one sign is true? The sign on Door 1 reads:

IN THIS ROOM THERE IS A LADY AND
IN THE OTHER ROOM THERE IS A TIGER.

The sign on Door 2 reads:

IN ONE OF THESE ROOMS THERE IS A LADY AND
IN ONE OF THESE ROOMS THERE IS A TIGER.

With this information, the man is able to choose the correct door. Can you? (The answer is on page 108.)

Answer to the Problem of *The Lady or the Tiger?*

The lady is behind Door 2. Suppose that the sign on Door 1 is true. Then the sign on Door 2 would also be true, but this is impossible. So the sign on Door 2 must be true, and the sign on Door 1 false. Since the sign on Door 1 says the lady is in Room 1, and this is false, the lady must be behind Door 2.

The **negation** of a statement p, symbolized $\sim p$, must have the opposite truth value from the statement p itself. This leads to the truth table for the negation, shown here.

Truth Table for the Negation not *p*

not p

p	$\sim p$
T	F
F	T

EXAMPLE 3 Suppose p is false, q is true, and r is false. What is the truth value of the compound statement $\sim p \wedge (q \vee \sim r)$?

Here parentheses are used to group q and $\sim r$ together. Work first inside the parentheses. Since r is false, $\sim r$ will be true. Since $\sim r$ is true and q is true, find the truth value of $q \vee \sim r$ by looking in the first row of the *or* truth table. This row gives the result T. Since p is false, $\sim p$ is true, and the final truth value of $\sim p \wedge (q \vee \sim r)$ is found in the top row of the *and* truth table. From the *and* truth table, when $\sim p$ is true, and $q \vee \sim r$ is true, the statement $\sim p \wedge (q \vee \sim r)$ is true.

The paragraph above may be interpreted using a short-cut symbolic method, letting T represent a true statement and F represent a false statement:

$$\sim p \wedge (q \vee \sim r)$$
$$\sim \text{F} \wedge (\text{T} \vee \sim \text{F})$$
$$\text{T} \wedge (\text{T} \vee \text{T}) \qquad \sim\text{F gives T.}$$
$$\text{T} \wedge \text{T} \qquad \text{T} \vee \text{T gives T.}$$
$$\text{T}. \qquad \text{T} \wedge \text{T gives T.}$$

The T in the final row indicates that the compound statement is true. ■

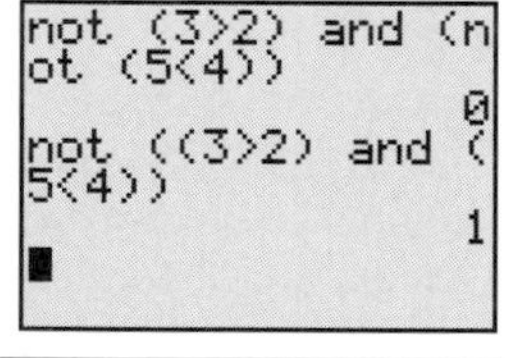

Example 4(a) explains why $\sim(3 > 2) \wedge [\sim(5 < 4)]$ is false. The calculator returns a 0. For a true statement such as $\sim[(3 > 2) \wedge (5 < 4)]$, it returns a 1.

The next two examples show the use of truth tables to determine the truth values of algebraic statements.

EXAMPLE 4 Let p represent the statement $3 > 2$, q represent $5 < 4$, and r represent $3 < 8$. Decide whether the following statements are *true* or *false.*

(a) $\sim p \wedge \sim q$

Since p is true, $\sim p$ is false. By the *and* truth table, if one part of an "and" statement is false, the entire statement is false. This makes $\sim p \wedge \sim q$ false.

(b) $\sim(p \wedge q)$

First, work within the parentheses. Since p is true and q is false, $p \wedge q$ is false by the *and* truth table. Next, apply the negation. The negation of a false statement is true, making $\sim(p \wedge q)$ a true statement.

George Boole (1815–1864) grew up in poverty. His father, a London tradesman, gave him his first math lessons and taught him to make optical instruments. Boole was largely self-educated. At 16 he worked in an elementary school and by age 20 had opened his own school. He studied mathematics in his spare time. Lung disease killed him at age 49.

Boole's ideas have been used in the design of computers and telephone systems.

(c) $(\sim p \wedge r) \vee (\sim q \wedge \sim p)$

Here p is true, q is false, and r is true. This makes $\sim p$ false and $\sim q$ true. By the *and* truth table, the statement $\sim p \wedge r$ is false, and the statement $\sim q \wedge \sim p$ is also false. Finally,

$$\underset{\text{F}}{\underset{\downarrow}{(\sim p \wedge r)}} \vee \underset{\text{F},}{\underset{\downarrow}{(\sim q \wedge \sim p)}}$$

which is false by the *or* truth table. (For an alternate solution, see Example 8 (b).) ■

When a quantifier is used with a conjunction or a disjunction, we must be careful in determining the truth value, as shown in the following example.

EXAMPLE 5 Identify each statement as *true* or *false.*

(a) For some real number x, $x < 5$ and $x > 2$.

Replacing x with 3 (as an example) gives $3 < 5$ and $3 > 2$. Since both $3 < 5$ and $3 > 2$ are true statements, the given statement is true by the *and* truth table. (Remember: *some* means "at least one.")

(b) For every real number b, $b > 0$ or $b < 1$.

No matter which real number might be tried as a replacement for b, at least one of the statements $b > 0$ and $b < 1$ will be true. Since an "or" statement is true if one or both component statements is true, the entire statement as given is true.

(c) For all real numbers, x, $x^2 > 0$.

Since the quantifier is a universal quantifier, we need only find one case in which the inequality is false to make the entire statement false. Can we find a real number whose square is not positive (that is, not greater than 0)? Yes, we can—0 itself is a real number (and the *only* real number) whose square is not positive. Therefore, this statement is false. ■

Truth Tables

In the examples above, the truth value for a given statement was found by going back to the basic truth tables. In the long run, it is easier to first create a complete truth table for the given statement itself. Then final truth values can be read directly from this table. The procedure for making new truth tables is shown in the next few examples.

In this book we will use the following standard format for listing the possible truth values in compound statements involving two statements.

p	q	**Compound Statement**
T	T	
T	F	
F	T	
F	F	

EXAMPLE 6

(a) Construct a truth table for $(\sim p \wedge q) \vee \sim q$.

Begin by listing all possible combinations of truth values for p and q, as above. Then find the truth values of $\sim p \wedge q$. Start by listing the truth values of $\sim p$, which are the opposite of those of p.

p	q	$\sim p$
T	T	F
T	F	F
F	T	T
F	F	T

Use only the "$\sim p$" column and the "q" column, along with the *and* truth table, to find the truth values of $\sim p \wedge q$. List them in a separate column.

p	q	$\sim p$	$\sim p \wedge q$
T	T	F	F
T	F	F	F
F	T	T	T
F	F	T	F

Next include a column for $\sim q$.

p	q	$\sim p$	$\sim p \wedge q$	$\sim q$
T	T	F	F	F
T	F	F	F	T
F	T	T	T	F
F	F	T	F	T

Finally, make a column for the entire compound statement. To find the truth values, use *or* to combine $\sim p \wedge q$ with $\sim q$.

p	q	$\sim p$	$\sim p \wedge q$	$\sim q$	$(\sim p \wedge q) \vee \sim q$
T	T	F	F	F	F
T	F	F	F	T	T
F	T	T	T	F	T
F	F	T	F	T	T

(b) Suppose both p and q are true. Find the truth value of $(\sim p \wedge q) \vee \sim q$.

Look in the first row of the final truth table above, where both p and q have truth value T. Read across the row to find that the compound statement is false. ■

EXAMPLE 7 Construct the truth table for $p \wedge (\sim p \vee \sim q)$.

Proceed as shown.

p	q	$\sim p$	$\sim q$	$\sim p \vee \sim q$	$p \wedge (\sim p \vee \sim q)$
T	T	F	F	F	F
T	F	F	T	T	T
F	T	T	F	T	F
F	F	T	T	T	F

■

If a compound statement involves three component statements p, q, and r, we will use the following format in setting up the truth table.

p	q	r	**Compound Statement**
T	T	T	
T	T	F	
T	F	T	
T	F	F	
F	T	T	
F	T	F	
F	F	T	
F	F	F	

EXAMPLE 8

(a) Construct a truth table for $(\sim p \wedge r) \vee (\sim q \wedge \sim p)$.

This statement has three component statements, p, q, and r. The truth table thus requires eight rows to list all possible combinations of truth values of p, q, and r. The final truth table, however, can be found in much the same way as the ones above.

p	q	r	$\sim p$	$\sim p \wedge r$	$\sim q$	$\sim q \wedge \sim p$	$(\sim p \wedge r) \vee (\sim q \wedge \sim p)$
T	T	T	F	F	F	F	F
T	T	F	F	F	F	F	F
T	F	T	F	F	T	F	F
T	F	F	F	F	T	F	F
F	T	T	T	T	F	F	T
F	T	F	T	F	F	F	F
F	F	T	T	T	T	T	T
F	F	F	T	F	T	T	T

(b) Suppose p is true, q is false, and r is true. Find the truth value of $(\sim p \wedge r) \vee (\sim q \wedge \sim p)$.

By the third row of the truth table in part (a), the compound statement is false. (This is an alternate method for working part (c) of Example 4.) ■

Problem Solving

One strategy for problem solving is noticing a pattern and using inductive reasoning. This strategy is used in the next example.

EXAMPLE 9 If n is a counting number, and a logical statement is composed of n component statements, how many rows will appear in the truth table for the compound statement?

To answer this question, let us examine some of the earlier truth tables in this section. The truth table for the negation has one statement and two rows. The truth tables for the conjunction and the disjunction have two component statements, and each has four rows. The truth table in Example 8(a) has three

Emilie, Marquise du Châtelet (1706–1749) participated in the scientific activity of the generation after Newton and Leibniz. She was educated in science, music, and literature, and she was studying mathematics at the time (1733) she began a long intellectual relationship with the philosopher François Voltaire (1694–1778). Her chateau was equipped with a physics laboratory. She and Voltaire competed independently in 1738 for a prize offered by the French Academy on the subject of fire. Although du Châtelet did not win, her dissertation was published by the academy in 1744. By that time she had published *Institutions of Physics* (expounding in part some ideas of Leibniz) and a work on Vital Forces. During the last four years of her life she translated Newton's *Principia* from Latin into French—the only French translation to date.

component statements and eight rows. Summarizing these in a table shows a pattern seen earlier.

Number of Statements	Number of Rows
1	$2 = 2^1$
2	$4 = 2^2$
3	$8 = 2^3$

Inductive reasoning leads us to the conjecture that, if a logical statement is composed of n component statements, it will have 2^n rows. This can be proved using more advanced concepts. ◆

The result of Example 9 is reminiscent of the formula for the number of subsets of a set having n elements.

A logical statement having n component statements will have 2^n rows in its truth table.

Alternate Method for Constructing Truth Tables

After making a reasonable number of truth tables, some people prefer the shortcut method shown in Example 10, which repeats Examples 6 and 8 above.

EXAMPLE 10 Construct the truth table for each statement.

(a) $(\sim p \wedge q) \vee \sim q$

Start by inserting truth values for $\sim p$ and for q.

p	q	$(\sim p$	$\wedge$	$q)$	$\vee$	$\sim q$
T	T	F		T		
T	F	F		F		
F	T	T		T		
F	F	T		F		

Next, use the *and* truth table to obtain the truth values of $\sim p \wedge q$.

p	q	$(\sim p$	$\wedge$	$q)$	$\vee$	$\sim q$
T	T	F	F	T		
T	F	F	F	F		
F	T	T	T	T		
F	F	T	F	F		

Now disregard the two preliminary columns of truth values for $\sim p$ and for q, and insert truth values for $\sim q$.

p	q	$(\sim p$	$\wedge$	$q)$	$\vee$	$\sim q$
T	T		F			F
T	F		F			T
F	T		T			F
F	F		F			T

Finally, use the *or* truth table.

p	q	$(\sim p$	$\wedge$	$q)$	$\vee$	$\sim q$
T	T		F		F	F
T	F		F		T	T
F	T		T		T	F
F	F		F		T	T

These steps can be summarized as follows.

p	q	$(\sim p$	$\wedge$	$q)$	$\vee$	$\sim q$
T	T	F	F	T	F	F
T	F	F	F	F	T	T
F	T	T	T	T	T	F
F	F	T	F	F	T	T
		①	②	①	④	③

The circled numbers indicate the order in which the various columns of the truth table were found.

(b) $(\sim p \wedge r) \vee (\sim q \wedge \sim p)$

Work as follows.

p	q	r	$(\sim p$	$\wedge$	$r)$	$\vee$	$(\sim q$	$\wedge$	$\sim p)$
T	T	T	F	F	T	F	F	F	F
T	T	F	F	F	F	F	F	F	F
T	F	T	F	F	T	F	T	F	F
T	F	F	F	F	F	F	T	F	F
F	T	T	T	T	T	T	F	F	T
F	T	F	T	F	F	F	F	F	T
F	F	T	T	T	T	T	T	T	T
F	F	F	T	F	F	T	T	T	T
			①	②	①	⑤	③	④	③

■

Equivalent Statements

One application of truth tables is illustrated by showing that two statements are equivalent; by definition, two statements are **equivalent** if they have the same truth value in *every* possible situation. The columns of each truth table that were the last to be completed will be exactly the same for equivalent statements.

EXAMPLE 11 Are the statements

$$\sim p \wedge \sim q \text{ and } \sim(p \vee q)$$

equivalent?

To find out, make a truth table for each statement, with the following results.

p	q	$\sim p \wedge \sim q$
T	T	F
T	F	F
F	T	F
F	F	T

p	q	$\sim(p \vee q)$
T	T	F
T	F	F
F	T	F
F	F	T

Since the truth values are the same in all cases, as shown in the columns in color, the statements $\sim p \wedge \sim q$ and $\sim(p \vee q)$ are equivalent. Equivalence is written with a three-bar symbol, $\equiv$. Using this symbol, $\sim p \wedge \sim q \equiv \sim(p \vee q)$. ■

In the same way, the statements $\sim p \vee \sim q$ and $\sim(p \wedge q)$ are equivalent. We call these equivalences De Morgan's laws.

De Morgan's Laws

For any statements p and q,

$$\sim(p \vee q) \equiv \sim p \wedge \sim q$$
$$\sim(p \wedge q) \equiv \sim p \vee \sim q.$$

(Compare the logic statements of De Morgan's laws with the set versions.) De Morgan's laws can be used to find the negations of certain compound statements.

EXAMPLE 12 Find a negation of each statement by applying De Morgan's laws.

(a) I got an A or I got a B.

If p represents "I got an A" and q represents "I got a B," then the compound statement is symbolized $p \vee q$. The negation of $p \vee q$ is $\sim(p \vee q)$; by one of De Morgan's laws, this is equivalent to

$$\sim p \wedge \sim q,$$

or, in words,

I didn't get an A and I didn't get a B.

This negation is reasonable—the original statement says that I got either an A or a B; the negation says that I didn't get *either* grade.

(b) She won't try and he will succeed.

From one of De Morgan's laws, $\sim(p \wedge q) \equiv \sim p \vee \sim q$, so the negation becomes

She will try or he won't succeed.

(c) $\sim p \vee (q \wedge \sim p)$

Negate both component statements and change $\vee$ to $\wedge$.

$$\sim[\sim p \vee (q \wedge \sim p)] \equiv p \wedge \sim(q \wedge \sim p)$$

Now apply De Morgan's law again.

$$\begin{aligned} p \wedge \sim(q \wedge \sim p) &\equiv p \wedge (\sim q \vee \sim(\sim p)) \\ &\equiv p \wedge (\sim q \vee p) \end{aligned}$$

A truth table will show that the statements

$$\sim p \vee (q \wedge \sim p) \qquad \text{and} \qquad p \wedge (\sim q \vee p)$$

are negations. ■

3.2 EXERCISES

Use the concepts introduced in this section to answer Exercises 1–6.

1. If we know that p is true, what do we know about the truth value of $p \vee q$ even if we are not given the truth value of q?

2. If we know that p is false, what do we know about the truth value of $p \wedge q$ even if we are not given the truth value of q?

3. If p is false, what is the truth value of $p \wedge (q \vee \sim r)$?

4. If p is true, what is the truth value of $p \vee (q \vee \sim r)$?

5. Explain in your own words the condition that must exist for a conjunction of two component statements to be true.

6. Explain in your own words the condition that must exist for a disjunction of two component statements to be false.

Let p represent a false statement and let q represent a true statement. Find the truth value of the given compound statement.

7. $\sim p$

8. $\sim q$

9. $p \vee q$

10. $p \wedge q$

11. $p \vee \sim q$

12. $\sim p \wedge q$

13. $\sim p \vee \sim q$

14. $p \wedge \sim q$

15. $\sim(p \wedge \sim q)$

16. $\sim(\sim p \vee \sim q)$

17. $\sim[\sim p \wedge (\sim q \vee p)]$

18. $\sim[(\sim p \wedge \sim q) \vee \sim q]$

19. Is the statement $3 \geq 1$ a conjunction or a disjunction?

20. Why is the statement $6 \geq 2$ true? Why is $6 \geq 6$ true?

Let p represent a true statement, and q and r represent false statements. Find the truth value of the given compound statement.

21. $(q \vee \sim r) \wedge p$

22. $(p \wedge r) \vee \sim q$

23. $(\sim p \wedge q) \vee \sim r$

24. $p \wedge (q \vee r)$

25. $(\sim r \wedge \sim q) \vee (\sim r \wedge q)$

26. $\sim(p \wedge q) \wedge (r \vee \sim q)$

27. $\sim[r \vee (\sim q \wedge \sim p)]$

28. $\sim[(\sim p \wedge q) \vee r]$

Let p represent the statement $1 > 9$, *let q represent the statement* $7 \not> 5$, *and let r represent the statement* $13 \geq 13$. *Find the truth value of the given compound statement.*

29. $p \wedge r$

30. $p \vee \sim q$

31. $\sim q \vee \sim r$

32. $\sim p \wedge \sim r$

33. $(p \wedge q) \vee r$

34. $\sim p \vee (\sim r \vee \sim q)$

35. $(\sim r \wedge q) \vee \sim p$

36. $\sim(p \vee \sim q) \vee \sim r$

Give the number of rows in the truth table for each of the following compound statements.

37. $p \vee \sim r$

38. $p \wedge (r \wedge \sim s)$

39. $(\sim p \wedge q) \vee (\sim r \vee \sim s) \wedge r$

40. $[(p \vee q) \wedge (r \wedge s)] \wedge (t \vee \sim p)$

41. $[(\sim p \wedge \sim q) \wedge (\sim r \wedge s \wedge \sim t)] \wedge (\sim u \vee \sim v)$

42. $[(\sim p \wedge \sim q) \vee (\sim r \vee \sim s)] \vee [(\sim m \wedge \sim n) \wedge (u \wedge \sim v)]$

43. If the truth table for a certain compound statement has 64 rows, how many distinct component statements does it have?

44. Is it possible for the truth table of a compound statement to have exactly 48 rows? Explain.

Construct a truth table for each compound statement.

45. $\sim p \wedge q$

46. $\sim p \vee \sim q$

47. $\sim(p \wedge q)$

48. $p \vee \sim q$

49. $(q \vee \sim p) \vee \sim q$

50. $(p \wedge \sim q) \wedge p$

51. $\sim q \wedge (\sim p \vee q)$

52. $\sim p \vee (\sim q \wedge \sim p)$

53. $(p \vee \sim q) \wedge (p \wedge q)$

54. $(\sim p \wedge \sim q) \vee (\sim p \vee q)$

55. $(\sim p \wedge q) \wedge r$

56. $r \vee (p \wedge \sim q)$

57. $(\sim p \wedge \sim q) \vee (\sim r \vee \sim p)$

58. $(\sim r \vee \sim p) \wedge (\sim p \vee \sim q)$

59. $\sim(\sim p \wedge \sim q) \vee (\sim r \vee \sim s)$

60. $(\sim r \vee s) \wedge (\sim p \wedge q)$

Use one of De Morgan's laws to write the negation of each statement.

61. You can pay me now or you can pay me later.

62. I am not going or she is going.

63. It is summer and there is no snow.

64. 1/2 is a positive number and -12 is less than zero.

65. I said yes but she said no.

66. Kelly Bell tried to sell the book, but she was unable to do so.

67. $5 - 1 = 4$ and $9 + 12 \neq 7$

68. $3 < 10$ or $7 \neq 2$

69. Dasher or Dancer will lead Santa's sleigh next Christmas.

70. The lawyer and the client appeared in court.

Identify each of the following statements as true *or* false.

71. For every real number y, $y < 12$ or $y > 4$.

72. For every real number t, $t > 3$ or $t < 3$.

73. For some integer p, $p \geq 5$ and $p \leq 5$.

74. There exists an integer n such that $n > 0$ and $n < 0$.

75. Complete the truth table for *exclusive disjunction.* The symbol $\underline{\vee}$ represents "one or the other is true, but not both."

p	q	$p \underline{\vee} q$
T	T	
T	F	
F	T	
F	F	

Exclusive disjunction

76. Lawyers sometimes use the phrase "and/or." This phrase corresponds to which usage of the word *or*: inclusive or exclusive?

Decide whether the following compound statements are true *or* false. *Here,* or *is the* exclusive disjunction; *that is, assume "either p or q is true, but not both."*

77. $3 + 1 = 4$ or $2 + 5 = 7$

78. $3 + 1 = 4$ or $2 + 5 = 9$

79. $3 + 1 = 7$ or $2 + 5 = 7$

80. $3 + 1 = 7$ or $2 + 5 = 9$

3.3 The Conditional

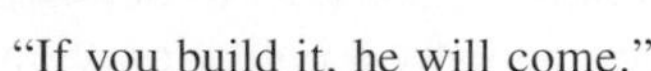

"If you build it, he will come."
—The Voice in the 1990 movie *Field of Dreams*

Ray Kinsella, an Iowa farmer in the movie *Field of Dreams,* heard a voice from the sky. Ray interpreted it as a promise that if he would build a baseball field in his cornfield, then the ghost of Shoeless Joe Jackson (a baseball star in the early days of the twentieth century) would come to play on it. The promise came in the form of a conditional statement. A **conditional** statement is a compound statement that uses the connective *if . . . then.* For example, here are a few conditional statements.

If I read for too long, *then* I get a headache.

If looks could kill, *then* I would be dead.

If he doesn't get back soon, *then* you should go look for him.

In each of these conditional statements, the component coming after the word *if* gives a condition (but not necessarily the only condition) under which the statement coming after *then* will be true. For example. "If it is over 90°, then I'll go to the mountains" tells one possible condition under which I will go to the mountains—if the temperature is over 90°.

The conditional is written with an arrow, so that "if p, then q" is symbolized as

$$p \rightarrow q.$$

We read $p \rightarrow q$ as "p implies q" or "if p, then q." In the conditional $p \rightarrow q$, the statement p is the **antecedent,** while q is the **consequent.**

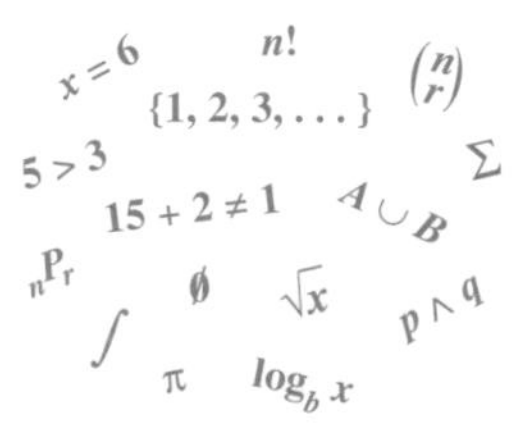

The importance of symbols was emphasized by the American philosopher-logician Charles Sanders Peirce (1839–1914), who asserted the nature of humans as symbol-using or sign-using organisms. Symbolic notation is half of mathematics, Bertrand Russell once said.

The conditional connective may not always be explicitly stated. That is, it may be "hidden" in an everyday expression. For example, the statement

Big girls don't cry

can be written in *if . . . then* form as

If you're a big girl, then you don't cry.

As another example, the statement

It is difficult to study when you are distracted

can be written

If you are distracted, then it is difficult to study.

As seen in the quote from the movie *Field of Dreams* earlier, the word "then" is sometimes not stated but understood to be there from the context of the statement. In that statement, "you build it" is the antecedent and "he will come" is the consequent.

The conditional truth table is a little harder to define than were the tables in the previous section. To see how to define the conditional truth table, let us analyze a statement made by a politician, Senator Julie Davis:

If I am elected, then taxes will go down.

As before, there are four possible combinations of truth values for the two component statements. Let p represent "I am elected," and let q represent "Taxes will go down."

As we analyze the four possibilities, it is helpful to think in terms of the following: "Did Senator Davis lie?" If she lied, then the conditional statement is considered false; if she did not lie, then the conditional statement is considered true.

Possibility	Elected?	Taxes Go Down?	
1	Yes	Yes	p is T, q is T
2	Yes	No	p is T, q is F
3	No	Yes	p is F, q is T
4	No	No	p is F, q is F

The four possibilities are as follows:

1. In the first case assume that the senator was elected and taxes did go down (p is T, q is T). The senator told the truth, so place T in the first row of the truth table. (We do not claim that taxes went down *because* she was elected; it is possible that she had nothing to do with it at all.)
2. In the second case assume that the senator was elected and taxes did not go down (p is T, q is F). Then the senator did not tell the truth (that is, she lied). So we put F in the second row of the truth table.
3. In the third case assume that the senator was defeated, but taxes went down anyway (p is F, q is T). Senator Davis did not lie; she only promised a tax reduction if she were elected. She said nothing about what would happen if she were not elected. In fact, her campaign promise gives no information about what would happen if she lost. Since we cannot say that the senator lied, place T in the third row of the truth table.
4. In the last case assume that the senator was defeated but taxes did not go down (p is F, q is F). We cannot blame her, since she only promised to reduce taxes if elected. Thus, T goes in the last row of the truth table.

The completed truth table for the conditional is defined as follows.

```
PROGRAM:SIGN
:Input A
:If A>0
:Then
:Disp "POSITIVE"
:
:
```

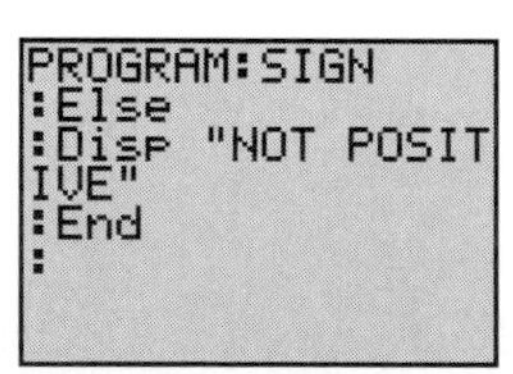

Conditional statements are useful in writing programs. This short program determines whether a number is positive. Notice the lines that begin with *If* and *Then*.

Truth Table for the Conditional If p, then q

If p, then q.

p	q	$p \rightarrow q$
T	T	T
T	F	F
F	T	T
F	F	T

It must be emphasized that the use of the conditional connective in no way implies a cause-and-effect relationship. Any two statements may have an arrow placed between them to create a compound statement. For example,

If I pass mathematics, then the sun will rise the next day

is true, since the consequent is true. (See the box after Example 1.) There is, however, no cause-and-effect connection between my passing mathematics and the sun's rising. The sun will rise no matter what grade I get in a course.

EXAMPLE 1 Given that p, q, and r are all false, find the truth value of the statement $(p \to \sim q) \to (\sim r \to q)$.

Using the short-cut method explained in Example 3 of the previous section, we can replace p, q, and r with F (since each is false) and proceed as before, using the negation and conditional truth tables as necessary.

$$(p \to \sim q) \to (\sim r \to q)$$
$$(\text{F} \to \sim\text{F}) \to (\sim\text{F} \to \text{F})$$
$$(\text{F} \to \text{T}) \to (\text{T} \to \text{F})$$
$$\text{T} \to \text{F} \qquad \text{Use the negation truth table.}$$
$$\text{F} \qquad \text{Use the conditional truth table.}$$

The statement $(p \to \sim q) \to (\sim r \to q)$ is false when p, q, and r are all false.

The following observations come from the truth table for $p \to q$.

Special Characteristics of Conditional Statements

1. $p \to q$ is false only when the antecedent is *true* and the consequent is *false.*
2. If the antecedent is *false*, then $p \to q$ is automatically *true.*
3. If the consequent is *true,* then $p \to q$ is automatically *true.*

EXAMPLE 2 Write *true* or *false* for each statement. Here T represents a true statement, and F represents a false statement.

(a) $\text{T} \to (6 = 3)$

Since the antecedent is true, while the consequent, $6 = 3$, is false, the given statement is false by the first point mentioned above.

(b) $(5 < 2) \to \text{F}$

The antecedent is false, so the given statement is true by the second observation.

(c) $(3 \neq 2 + 1) \to \text{T}$

The consequent is true, making the statement true by the third characteristic of conditional statements.

Truth tables for compound statements involving conditionals are found using the techniques described in the previous section. The next example shows how this is done.

EXAMPLE 3 Construct a truth table for each statement.

(a) $(\sim p \rightarrow \sim q) \rightarrow (\sim p \wedge q)$

First insert the truth values of $\sim p$ and of $\sim q$. Then find the truth value of $\sim p \rightarrow \sim q$.

p	q	$\sim p$	$\sim q$	$\sim p \rightarrow \sim q$
T	T	F	F	T
T	F	F	T	T
F	T	T	F	F
F	F	T	T	T

Next use $\sim p$ and q to find the truth values of $\sim p \wedge q$.

p	q	$\sim p$	$\sim q$	$\sim p \rightarrow \sim q$	$\sim p \wedge q$
T	T	F	F	T	F
T	F	F	T	T	F
F	T	T	F	F	T
F	F	T	T	T	F

Now use the conditional truth table to find the truth values of $(\sim p \rightarrow \sim q) \rightarrow (\sim p \wedge q)$.

p	q	$\sim p$	$\sim q$	$\sim p \rightarrow \sim q$	$\sim p \wedge q$	$(\sim p \rightarrow \sim q) \rightarrow (\sim p \wedge q)$
T	T	F	F	T	F	F
T	F	F	T	T	F	F
F	T	T	F	F	T	T
F	F	T	T	T	F	F

(b) $(p \rightarrow q) \rightarrow (\sim p \vee q)$

Go through steps similar to the ones above.

p	q	$p \rightarrow q$	$\sim p$	$\sim p \vee q$	$(p \rightarrow q) \rightarrow (\sim p \vee q)$
T	T	T	F	T	T
T	F	F	F	F	T
F	T	T	T	T	T
F	F	T	T	T	T

As the truth table in Example 3(b) shows, the statement $(p \rightarrow q) \rightarrow (\sim p \vee q)$ is always true, no matter what the truth values of the components. Such a statement is called a **tautology.** Other examples of tautologies (as can be checked by forming truth tables) include $p \vee \sim p$, $p \rightarrow p$, $(\sim p \vee \sim q) \rightarrow \sim(q \wedge p)$, and so on. By the way, the truth tables in Example 3 also could have been found by the alternate method shown in the previous section.

Negation of $p \rightarrow q$

Suppose that someone makes the conditional statement

"If it rains, then I take my umbrella."

When will the person have lied to you? The only case in which you would have been misled is when it rains *and* the person does *not* take the umbrella. Letting

p represent "it rains" and q represent "I take my umbrella," you might suspect that the symbolic statement

$$p \wedge \sim q$$

is a candidate for the negation of $p \rightarrow q$. That is,

$$\sim(p \rightarrow q) \equiv p \wedge \sim q.$$

It happens that this is indeed the case, as the next truth table indicates.

p	q	$p \rightarrow q$	$\sim(p \rightarrow q)$	$\sim q$	$p \wedge \sim q$
T	T	T	F	F	F
T	F	F	T	T	T
F	T	T	F	F	F
F	F	T	F	T	F

$\sim(p \rightarrow q) \equiv p \wedge \sim q$ (columns marked ≡)

Negation of $p \rightarrow q$

The negation of $p \rightarrow q$ is $p \wedge \sim q$.

Since

$$\sim(p \rightarrow q) \equiv p \wedge \sim q,$$

by negating each expression we have

$$\sim[\sim(p \rightarrow q] \equiv \sim(p \wedge \sim q).$$

The left side of the above equivalence is $p \rightarrow q$, and one of De Morgan's laws can be applied to the right side.

$$p \rightarrow q \equiv \sim p \vee \sim(\sim q)$$
$$p \rightarrow q \equiv \sim p \vee q$$

This final row indicates that a conditional may be written as a disjunction.

Writing a Conditional as an "or" Statement

$p \rightarrow q$ is equivalent to $\sim p \vee q$.

EXAMPLE 4 Write the negation of each statement.

(a) If you build it, he will come.

If b represents "you build it" and q represents "he will come," then the given statement can be symbolized by $b \rightarrow q$. The negation of $b \rightarrow q$, as shown above, is $b \wedge \sim q$, so the negation of the statement is

You build it and he will not come.

(b) All dogs have fleas.

First, we must restate the given statement in *if . . . then* form:

If it is a dog, then it has fleas.

Based on our earlier discussion, the negation is

It is a dog and it does not have fleas. ■

A common error occurs when students try to write the negation of a conditional statement as another conditional statement. As seen in Example 4, the negation of a conditional statement is written as a conjunction.

EXAMPLE 5 Write each conditional as an equivalent statement without using *if . . . then.*

(a) If the Cubs win the pennant, then Gwen will be happy.

Since the conditional $p \rightarrow q$ is equivalent to $\sim p \vee q$, let p represent "The Cubs win the pennant" and q represent "Gwen will be happy." Restate the conditional as

The Cubs do not win the pennant or Gwen will be happy.

(b) If it's Borden's, it's got to be good.

If p represents "it's Borden's" and if q represents "it's got to be good," the conditional may be restated as

It's not Borden's or it's got to be good. ■

3.3 EXERCISES

In Exercises 1–8, decide whether each statement is true *or* false.

1. If the antecedent of a conditional statement is false, the conditional statement is true.

2. If the consequent of a conditional statement is true, the conditional statement is true.

3. If q is true, then $(p \wedge q) \rightarrow q$ is true.

4. If p is true, then $\sim p \rightarrow (q \vee r)$ is true.

5. The negation of "If pigs fly, I'll believe it" is "If pigs don't fly, I won't believe it."

6. The statements "If it flies, then it's a bird" and "It does not fly or it's a bird" are logically equivalent.

7. Given that $\sim p$ is true and q is false, the conditional $p \rightarrow q$ is true.

8. Given that $\sim p$ is false and q is false, the conditional $p \rightarrow q$ is true.

9. In a few sentences, explain how we determine the truth value of a conditional statement.

10. Explain why the statement "If $3 = 5$, then $4 = 6$" is true.

Rewrite each statement using the if . . . then *connective. Rearrange the wording or add words as necessary.*

11. You can believe it if it's in *USA Today*.

12. It must be dead if it doesn't move.

13. Kathi Callahan's area code is 708.

14. Kara Gourley goes to Hawaii every summer.

15. All soldiers maintain their weapons.

16. Every dog has its day.

17. No koalas live in Mississippi.

18. No guinea pigs are scholars.

19. An alligator cannot live in these waters.

20. Romeo loves Juliet.

Tell whether each conditional is true *or* false. *Here* T *represents a true statement and* F *represents a false statement.*

21. $F \rightarrow (4 = 7)$

22. $T \rightarrow (4 < 2)$

23. $(6 = 6) \rightarrow F$

24. $F \rightarrow (3 = 3)$

25. $(4 = 11 - 7) \rightarrow (3 > 0)$

26. $(4^2 \neq 16) \rightarrow (4 + 4 = 8)$

Let s represent "I study in the library," *let p represent the statement* "I pass my psychology course," *and let m represent* "I major in mathematics." *Express each compound statement in words.*

27. $\sim m \rightarrow p$

28. $p \rightarrow \sim m$

29. $s \rightarrow (m \wedge p)$

30. $(s \wedge p) \rightarrow m$

31. $\sim p \rightarrow (\sim m \vee s)$

32. $(\sim s \vee \sim m) \rightarrow \sim p$

Let d represent "I drive my car," *let s represent* "it snows," *and let c represent* "classes are cancelled." *Write each compound statement in symbols.*

33. If it snows, then I drive my car.

34. If I drive my car, then classes are cancelled.

35. If I do not drive my car, then it does not snow.

36. If classes are cancelled, then it does not snow.

37. I drive my car, or if classes are cancelled then it snows.

38. Classes are cancelled, and if it snows then I do not drive my car.

39. I'll drive my car if it doesn't snow.

40. It snows if classes are cancelled.

Find the truth value of each statement. Assume that p and r are false, and q is true.

41. $\sim r \rightarrow q$

42. $\sim p \rightarrow \sim r$

43. $q \rightarrow p$

44. $\sim r \rightarrow p$

45. $p \rightarrow q$

46. $\sim q \rightarrow r$

47. $\sim p \rightarrow (q \wedge r)$

48. $(\sim r \vee p) \rightarrow p$

49. $\sim q \rightarrow (p \wedge r)$

50. $(\sim p \wedge \sim q) \rightarrow (p \wedge \sim r)$

51. $(p \rightarrow \sim q) \rightarrow (\sim p \wedge \sim r)$

52. $(p \rightarrow \sim q) \wedge (p \rightarrow r)$

53. Explain why, if we know that p is true, we also know that

$$[r \vee (p \vee s)] \rightarrow (p \vee q)$$

is true, even if we are not given the truth values of q, r, and s.

54. Construct a true statement involving a conditional, a conjunction, a disjunction, and a negation (not necessarily in that order), that consists of component statements p, q, and r, with all of these component statements false.

Construct a truth table for each statement. Identify any tautologies.

55. $\sim q \rightarrow p$

56. $p \rightarrow \sim q$

57. $(\sim p \rightarrow q) \rightarrow p$

58. $(\sim q \rightarrow \sim p) \rightarrow \sim q$

59. $(p \vee q) \rightarrow (q \vee p)$

60. $(p \wedge q) \rightarrow (p \vee q)$

61. $(\sim p \rightarrow \sim q) \rightarrow (p \wedge q)$

62. $r \rightarrow (p \wedge \sim q)$

63. $[(r \vee p) \wedge \sim q] \rightarrow p$

64. $(\sim r \rightarrow s) \vee (p \rightarrow \sim q)$

65. $(\sim p \wedge \sim q) \rightarrow (\sim r \rightarrow \sim s)$

66. What is the minimum number of Fs that need appear in the final column of a truth table for us to be assured that the statement is not a tautology?

Write the negation of each statement. Remember that the negation of $p \rightarrow q$ *is* $p \wedge \sim q$.

67. If you give your plants tender, loving care, they flourish.

68. If the check is in the mail, I'll be surprised.

69. If she doesn't, he will.

70. If I say yes, she says no.

71. All residents of Boise are residents of Idaho.

72. All men were once boys.

Write each statement as an equivalent statement that does not use the if . . . then *connective. Remember that $p \rightarrow q$ is equivalent to $\sim p \vee q$.*

73. If you give your plants tender, loving care, they flourish.

74. If the check is in the mail, I'll be surprised.

75. If she doesn't, he will.

76. If I say yes, she says no.

77. All residents of Boise are residents of Idaho.

78. All men were once boys.

Use truth tables to decide which of the pairs of statements are equivalent.

79. $p \rightarrow q$; $\sim p \vee q$

80. $\sim(p \rightarrow q)$; $p \wedge \sim q$

81. $p \rightarrow q$; $\sim q \rightarrow \sim p$

82. $q \rightarrow p$; $\sim p \rightarrow \sim q$

83. $\sim(\sim p)$; p

84. $p \rightarrow q$; $q \rightarrow p$

85. $p \wedge \sim q$; $\sim q \rightarrow \sim p$

86. $\sim p \wedge q$; $\sim p \rightarrow q$

EXTENSION

Circuits

One of the first nonmathematical applications of symbolic logic was seen in the master's thesis of Claude Shannon, in 1937. Shannon showed how logic could be used as an aid in designing electrical circuits. His work was immediately taken up by the designers of computers. These computers, then in the developmental stage, could be simplified and built for less money using the ideas of Shannon.

To see how Shannon's ideas work, look at the electrical switch shown in Figure 1. We assume that current will flow through this switch when it is closed and not when it is open.

FIGURE 1

FIGURE 2

Figure 2 shows two switches connected in *series;* in such a circuit, current will flow only when both switches are closed. Note how closely a series circuit corresponds to the conjunction $p \wedge q$. We know that $p \wedge q$ is true only when both p and q are true.

A circuit corresponding to the disjunction $p \vee q$ can be found by drawing a *parallel* circuit, as in Figure 3. Here, current flows if either *p or q* is closed or if both *p and q* are closed.

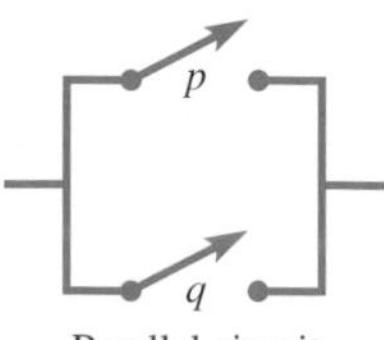

Parallel circuit

FIGURE 3

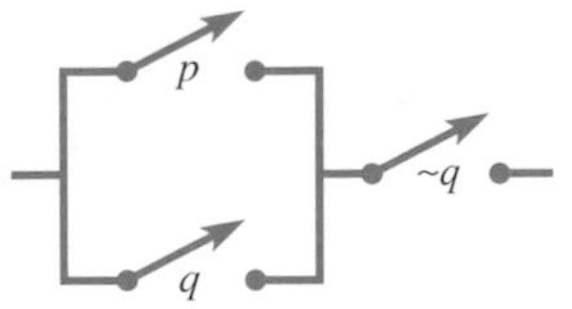

■ FIGURE 4

The circuit in Figure 4 corresponds to the statement $(p \lor q) \land \sim q$, which is a compound statement involving both a conjunction and a disjunction.

The way that logic is used to simplify an electrical circuit depends on the idea of equivalent statements, from Section 2. Recall that two statements are equivalent if they have exactly the same truth table. The symbol $\equiv$ is used to indicate that the two statements are equivalent. Some of the equivalent statements that we shall need are shown in the following box.

Equivalent Statements Used to Simplify Circuits

$$p \lor (q \land r) \equiv (p \lor q) \land (p \lor r) \qquad p \lor p \equiv p$$
$$p \land (q \lor r) \equiv (p \land q) \lor (p \land r) \qquad p \land p \equiv p$$
$$p \to q \equiv \sim q \to \sim p \qquad \sim(p \land q) \equiv \sim p \lor \sim q$$
$$p \to q \equiv \sim p \lor q \qquad \sim(p \lor q) \equiv \sim p \land \sim q$$

If T represents any true statement and F represents any false statement, then

$$p \lor \text{T} \equiv \text{T} \qquad p \lor \sim p \equiv \text{T}$$
$$p \land \text{F} \equiv \text{F} \qquad p \land \sim p \equiv \text{F}.$$

Circuits can be used as models of compound statements, with a closed switch corresponding to T, while an open switch corresponds to F. The method for simplifying circuits is explained in the following example.

EXAMPLE 1 Simplify the circuit of Figure 5.

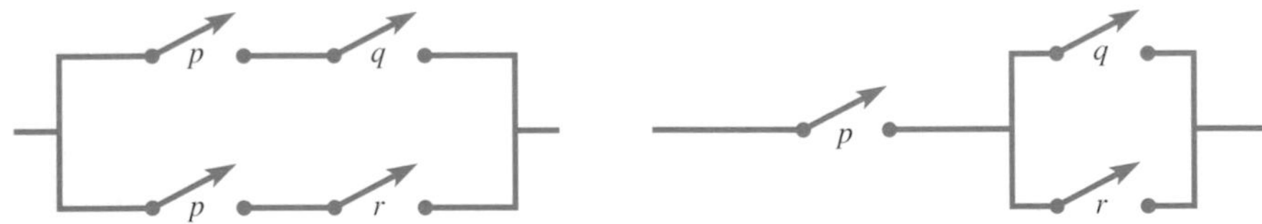

■ FIGURE 5 ■ FIGURE 6

At the top of Figure 5, p and q are connected in series, and at the bottom, p and r are connected in series. These are interpreted as the compound statements $p \land q$ and $p \land r$, respectively. These two conjunctions are connected in parallel, as indicated by the figure treated as a whole. Therefore, we write the disjunction of the two conjunctions:

$$(p \land q) \lor (p \land r).$$

(Think of the two switches labeled "p" as being controlled by the same handle.) By one of the pairs of equivalent statements above,

$$(p \land q) \lor (p \land r) \equiv p \land (q \lor r),$$

which has the circuit of Figure 6. This new circuit is logically equivalent to the one above, and yet contains only three switches instead of four—which might well lead to a large savings in manufacturing costs. ■

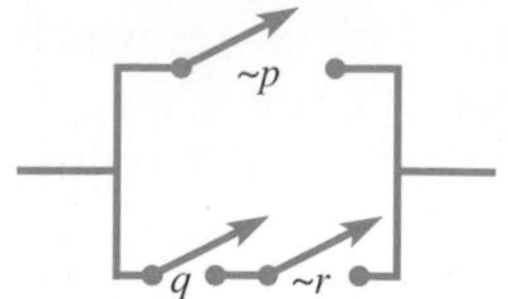

FIGURE 7

EXAMPLE 2 Draw a circuit for $p \rightarrow (q \wedge \sim r)$.

From the list of equivalent statements in the box, $p \rightarrow q$ is equivalent to $\sim p \vee q$. This equivalence gives $p \rightarrow (q \wedge \sim r) \equiv \sim p \vee (q \wedge \sim r)$, which has the circuit diagram in Figure 7.

Extension Exercises

Write a logical statement representing each of the following circuits. Simplify each circuit when possible.

1.

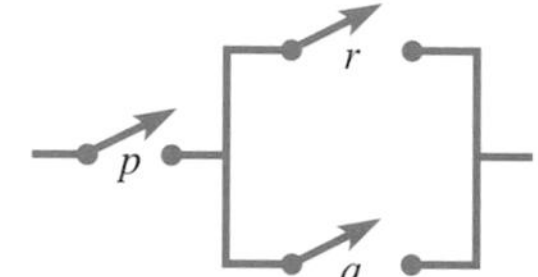

2.

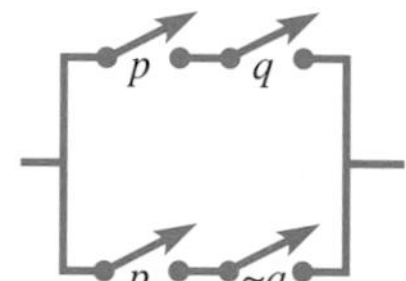

3.

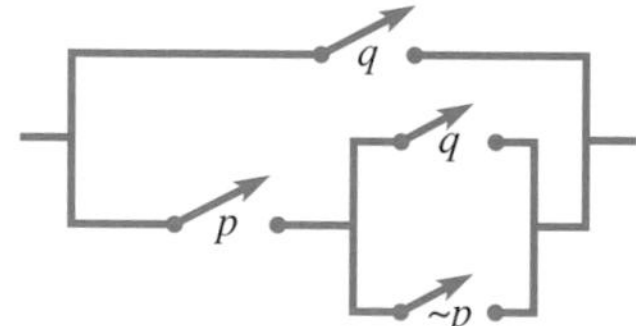

4.

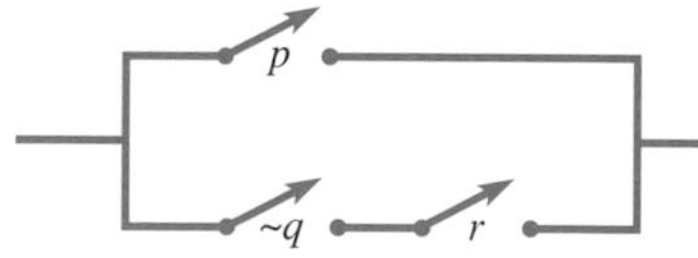

5.

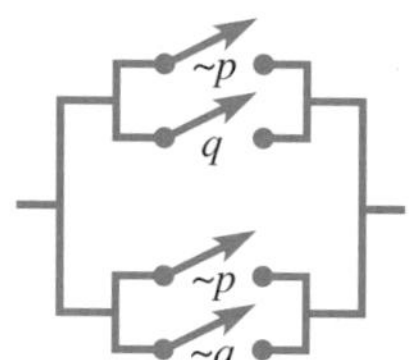

6.

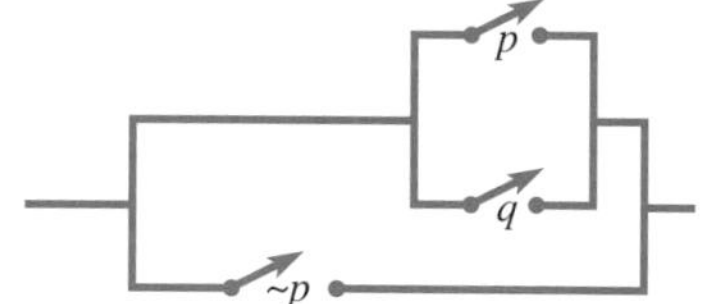

Draw circuits representing the following statements as they are given. Simplify if possible.

7. $(\sim p \wedge \sim q) \wedge \sim r$

8. $p \wedge (q \vee \sim p)$

9. $(\sim q \wedge \sim p) \vee (\sim p \vee q)$

10. $(p \vee q) \wedge (\sim p \wedge \sim q)$

11. $[(\sim p \wedge \sim r) \vee \sim q] \wedge (\sim p \wedge r)$

12. $[(p \vee q) \wedge r] \wedge \sim p$

13. $\sim p \rightarrow (\sim p \vee \sim q)$

14. $\sim q \rightarrow (\sim p \rightarrow q)$

15. Refer to Figures 5 and 6 in Example 1. Suppose the cost of the use of one switch for an hour is 3¢. By using the circuit in Figure 6 rather than the circuit in Figure 5, what is the savings for a year of 365 days, assuming that the circuit is in continuous use?

16. Explain why the circuit

will always have exactly one open switch. What does this circuit simplify to?

More on the Conditional

The conditional statement, introduced in the previous section, is one of the most important of all compound statements. Many mathematical properties and theorems are stated in *if . . . then* form. Because of their usefulness, we need to study conditional statements that are related to a statement of the form $p \rightarrow q$.

Converse, Inverse, and Contrapositive

Any conditional statement is made up of an antecedent and a consequent. If they are interchanged, negated, or both, a new conditional statement is formed. Suppose that we begin with the direct statement

If you stay, then I go,

and interchange the antecedent ("you stay") and the consequent ("I go"). We obtain the new conditional statement

If I go, then you stay.

This new conditional is called the **converse** of the given statement.

By negating both the antecedent and the consequent, we obtain the **inverse** of the given statement:

If you do not stay, then I do not go.

If the antecedent and the consequent are both interchanged *and* negated, the **contrapositive** of the given statement is formed:

If I do not go, then you do not stay.

These three related statements for the conditional $p \rightarrow q$ are summarized below. (Notice that the inverse is the contrapositive of the converse.)

Related Conditional Statements

Direct Statement	$p \rightarrow q$	(If p, then q.)
Converse	$q \rightarrow p$	(If q, then p.)
Inverse	$\sim p \rightarrow \sim q$	(If not p, then not q.)
Contrapositive	$\sim q \rightarrow \sim p$	(If not q, then not p.)

Alfred North Whitehead (1861–1947), and Bertrand Russell worked together on *Principia Mathematica.* During that time, Whitehead was teaching mathematics at Cambridge University and had written *Universal Algebra.* In 1910 he went to the University of London, exploring not only the philosophical basis of science but also the "aims of education" (as he called one of his books). It was as philosopher that he was invited to Harvard University in 1924. Whitehead died at the age of 86 in Cambridge, Massachusetts.

EXAMPLE 1 Given the direct statement

If I live in Tampa, then I live in Florida,

write each of the following.

(a) the converse

Let p represent "I live in Tampa" and q represent "I live in Florida." Then the direct statement may be written $p \rightarrow q$. The converse, $q \rightarrow p$, is

If I live in Florida, then I live in Tampa.

Notice that for this statement, the converse is not necessarily true, even though the direct statement is.

(b) the inverse
The inverse of $p \to q$ is $\sim p \to \sim q$. For the given statement, the inverse is

If I don't live in Tampa, then I don't live in Florida.

which is again not necessarily true.

(c) the contrapositive
The contrapositive, $\sim q \to \sim p$, is

If I don't live in Florida, then I don't live in Tampa.

The contrapositive, like the direct statement, is true. ■

Example 1 shows that the converse and inverse of a true statement need not be true. They *can* be true, but they need not be. The relationship between the truth values of the direct statement, converse, inverse, and contrapositive is shown in the truth table that follows.

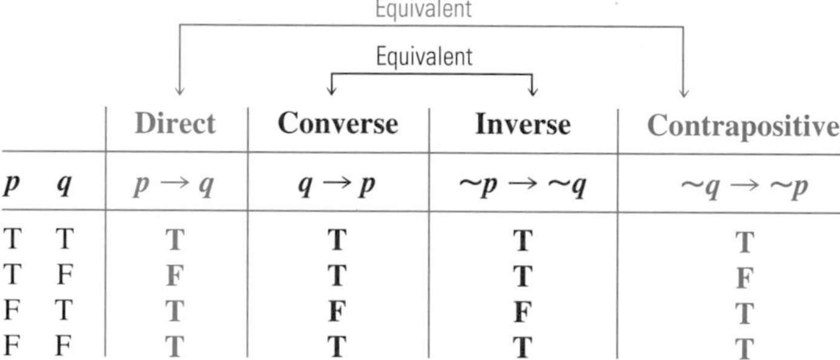

		Direct	Converse	Inverse	Contrapositive
p	q	$p \to q$	$q \to p$	$\sim p \to \sim q$	$\sim q \to \sim p$
T	T	T	T	T	T
T	F	F	T	T	F
F	T	T	F	F	T
F	F	T	T	T	T

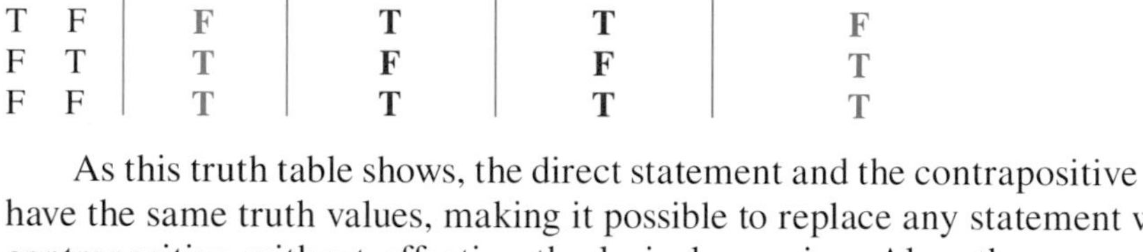

Bertrand Russell was a student of Whitehead's before they wrote the *Principia*. Like his teacher, Russell turned toward philosophy. His works include a critique of Leibniz, analyses of mind and of matter, and a history of Western thought.

Russell became a public figure because of his involvement in social issues. Deeply aware of human loneliness, he was "passionately desirous of finding ways of diminishing this tragic isolation." During World War I he was an antiwar crusader, and he was imprisoned briefly. Again in the 1960s he championed peace. He wrote many books on social issues, winning the Nobel Prize for Literature in 1950.

As this truth table shows, the direct statement and the contrapositive always have the same truth values, making it possible to replace any statement with its contrapositive without affecting the logical meaning. Also, the converse and inverse always have the same truth values.

This discussion is summarized in the following sentence.

> The direct statement and the contrapositive are equivalent, and the converse and the inverse are equivalent.

EXAMPLE 2 For the direct statement $\sim p \to q$, write each of the following.

(a) the converse
The converse of $\sim p \to q$ is $q \to \sim p$.

(b) the inverse
The inverse is $\sim(\sim p) \to \sim q$, which simplifies to $p \to \sim q$.

(c) the contrapositive
The contrapositive is $\sim q \to \sim(\sim p)$, which simplifies to $\sim q \to p$. ■

Alternate Forms of "if *p*, then *q*"

The conditional statement "if p, then q" can be stated in several other ways in English. For example,

If you go to the shopping center, then you will find a place to park can also be written

Going to the shopping center is *sufficient* for finding a place to park.

According to this statement, going to the shopping center is enough to guarantee finding a place to park. Going to other places, such as schools or office buildings, *might* also guarantee a place to park, but at least we *know* that going to the shopping center does. Thus, $p \to q$ can be written "p is sufficient for q." Knowing that p has occurred is sufficient to guarantee that q will also occur. On the other hand,

Turning on the set is necessary for watching television **(*)**

has a different meaning. Here, we are saying that one condition that is necessary for watching television is that you turn on the set. This may not be enough; the set might be broken, for example. The statement labeled (*) could be written as

If you watch television, then you turned on the set.

As this example suggests, $p \to q$ is the same as "q is necessary for p." In other words, if q doesn't happen, then neither will p. Notice how this idea is closely related to the idea of equivalence between the direct statement and its contrapositive.

Some common translations of $p \to q$ are summarized in the following box.

Common Translations of $p \to q$

The conditional $p \to q$ can be translated in any of the following ways.

If p, then q.	p is sufficient for q.
If p, q.	q is necessary for p.
p implies q.	All p's are q's.
p only if q.	q if p.

The translation of $p \to q$ into these various word forms does not in any way depend on the truth or falsity of $p \to q$.

EXAMPLE 3 The statement

If you are 18, then you can vote

can be written in any of the following ways.

You can vote if you are 18.
You are 18 only if you can vote.
Being able to vote is necessary for you to be 18.
Being 18 is sufficient for being able to vote.
All 18-year-olds can vote.
Being 18 implies that you can vote. ■

EXAMPLE 4 Write each statement in the form "if p, then q."

(a) You'll be sorry if I go.
If I go, then you'll be sorry.

(b) Today is Friday only if yesterday was Thursday.
If today is Friday, then yesterday was Thursday.

(c) All nurses wear white shoes.
If you are a nurse, then you wear white shoes. ■

For Further Thought

How many times have you heard a wise saying like "A stitch in time saves nine," "A rolling stone gathers no moss," or "Birds of a feather flock together"? In many cases, such proverbial advice can be restated as a conditional in *if . . . then* form. For example, these three statements can be restated as follows.

"If you make a stitch in time, then it will save you nine (stitches)."

"If a stone rolls, then it gathers no moss."

"If they are birds of a feather, then they flock together."

There is much wisdom in such sayings. How many times have you decided to have something repaired, even though it was working acceptably, and you could have lived with it as it was? Then it comes back from the repair shop in worse condition than before. And then, you thought of what someone once told you: "If it ain't broke, don't fix it."

For Group Discussion

1. Think of some wise sayings that have been around for a long time, and state them in *if . . . then* form.
2. You have probably heard the saying "All that glitters is not gold." Do you think that what is said here actually is what is meant? If not, restate it as you think it should be stated. (*Hint:* Write the original statement in *if . . . then* form.)

EXAMPLE 5 Let p represent "A triangle is equilateral," and let q represent "A triangle has three equal sides." Write each of the following in symbols.

(a) A triangle is equilateral if it has three equal sides.

$$q \rightarrow p$$

(b) A triangle is equilateral only if it has three equal sides.

$$p \rightarrow q$$ ■

Biconditionals

The compound statement *p if and only if q* (often abbreviated *p iff q*) is called a **biconditional.** It is symbolized $p \leftrightarrow q$, and is interpreted as the conjunction of the two conditionals $p \rightarrow q$ and $q \rightarrow p$. Using symbols, this conjunction is written

$$(q \rightarrow p) \wedge (p \rightarrow q)$$

so that, by definition,

$$p \leftrightarrow q \equiv (q \rightarrow p) \wedge (p \rightarrow q).$$

Using this definition, the truth table for the biconditional $p \leftrightarrow q$ can be determined.

Truth Table for the Biconditional *p* if and only if *q*

p if and only if *q*.

p	q	$p \leftrightarrow q$
T	T	T
T	F	F
F	T	F
F	F	T

Principia Mathematica The title chosen by Whitehead and Russell was a deliberate reference to *Philosophiae naturalis principia mathematica,* or "mathematical principles of the philosophy of nature," Isaac Newton's epochal work of 1687. Newton's Principia pictured a kind of "clockwork universe" that ran via his Law of Gravitation. Newton independently invented the calculus, unaware that Leibniz had published his own formulation of it earlier. A controversy over their priority continued into the eighteenth century.

EXAMPLE 6 Tell whether each biconditional statement is *true* or *false.*

(a) $6 + 9 = 15$ if and only if $12 + 4 = 16$

Both $6 + 9 = 15$ and $12 + 4 = 16$ are true. By the truth table for the biconditional, this biconditional is true.

(b) $5 + 2 = 10$ if and only if $17 + 19 = 36$

Since the first component $(5 + 2 = 10)$ is false, and the second is true, the entire biconditional statement is false.

(c) $6 = 5$ if and only if $12 \neq 12$

Both component statements are false, so by the last line of the truth table for the biconditional, the entire statement is true. (Understanding this might take some extra thought!)

In this section and in the previous two sections, truth tables have been derived for several important types of compound statements. The summary that follows describes how these truth tables may be remembered.

Summary of Basic Truth Tables

1. $\sim p$, the **negation** of p, has truth value opposite that of p.
2. $p \wedge q$, the **conjunction,** is true only when both p and q are true.
3. $p \vee q$, the **disjunction,** is false only when both p and q are false.
4. $p \rightarrow q$, the **conditional,** is false only when p is true and q is false.
5. $p \leftrightarrow q$, the **biconditional,** is true only when p and q have the same truth value.

3.4 EXERCISES

For each given direct statement, write **(a)** *the converse,* **(b)** *the inverse, and* **(c)** *the contrapositive in* if . . . then *form. In Exercises 3–10, it may be helpful to restate the direct statement in* if . . . then *form.*

1. If you lead, then I will follow.
2. If beauty were a minute, then you would be an hour.
3. If I had a nickel for each time that happened, I would be rich.
4. If it ain't broke, don't fix it.
5. Milk contains calcium.
6. Walking in front of a moving car is dangerous to your health.
7. A rolling stone gathers no moss.
8. Birds of a feather flock together.
9. Where there's smoke, there's fire.
10. If you build it, he will come.
11. $p \rightarrow \sim q$
12. $\sim p \rightarrow q$
13. $\sim p \rightarrow \sim q$
14. $\sim q \rightarrow \sim p$
15. $p \rightarrow (q \vee r)$ (*Hint:* Use one of De Morgan's laws as necessary.)
16. $(r \vee \sim q) \rightarrow p$ (*Hint:* Use one of De Morgan's laws as necessary.)
17. Discuss the equivalences that exist among the direct conditional statement, the converse, the inverse, and the contrapositive.
18. State the contrapositive of "If the square of a natural number is even, then the natural number is even." The two statements must have the same truth value. Use several examples and inductive reasoning to decide whether both are true or both are false.

Write each of the following statements in the form "if p, then q."

19. If I finish studying, I'll go to the party.
20. If it is muddy, I'll wear my galoshes.
21. "Today is Wednesday" implies that yesterday was Tuesday.
22. "17 is positive" implies that $17 + 1$ is positive.
23. All whole numbers are integers.
24. All integers are rational numbers.
25. Being in Fort Lauderdale is sufficient for being in Florida.
26. Doing crossword puzzles is sufficient for driving me crazy.
27. Being an environmentalist is necessary for being elected.
28. A day's growth of beard is necessary for Greg Odjakjian to shave.
29. The principal will hire more teachers only if the school board approves.
30. I can go from Park Place to Baltic Avenue only if I pass GO.
31. No integers are irrational numbers.
32. No whole numbers are not integers.
33. Newt will be a liberal when pigs fly.
34. The Indians will win the pennant when their pitching improves.
35. A parallelogram is a four-sided figure with opposite sides parallel.
36. A rectangle is a parallelogram with a right angle.
37. A square is a rectangle with two adjacent sides equal.
38. A triangle with two sides of the same length is isosceles.
39. An integer whose units digit is 0 or 5 is divisible by 5.
40. The square of a two-digit number whose units digit is 5 will end in 25.
41. One of the following statements is not equivalent to all the others. Which one is it?
    (a) r only if s.
    (b) r implies s.
    (c) If r, then s.
    (d) r is necessary for s.
42. Many students have difficulty interpreting *necessary* and *sufficient*. Use the statement "Being in Canada is sufficient for being in North America" to explain why "p is sufficient for q" translates as "if p, then q."

43. Use the statement "To be an integer, it is necessary that a number be rational" to explain why "p is necessary for q" translates as "if q, then p."

44. Explain why the statement "A week has eight days if and only if December has forty days" is true.

Identify each statement as true *or* false.

45. $5 = 9 - 4$ if and only if $8 + 2 = 10$.

46. $3 + 1 \neq 6$ if and only if $9 \neq 8$.

47. $8 + 7 \neq 15$ if and only if $3 \times 5 = 9$.

48. $6 \times 2 = 14$ if and only if $9 + 7 \neq 16$.

49. John F. Kennedy was president if and only if Ronald Reagan was not president.

50. Burger King sells Big Macs if and only if Guess sells jeans.

Two statements that can both be true about the same object are **consistent.** *For example,* "It is green" *and* "It is small" *are consistent statements. Statements that cannot both be true about the same object are called* **contrary;** "It is a Ford" *and* "It is a Chevrolet" *are contrary.*

Label the following pairs of statements as either contrary *or* consistent.

51. Elvis is alive. Elvis is dead.

52. Bill Clinton is a Democrat. Bill Clinton is a Republican.

53. That animal has four legs. That animal is a dog.

54. That book is nonfiction. That book costs more than $40.

55. This number is an integer. This number is irrational.

56. This number is positive. This number is a natural number.

57. Make up two statements that are consistent.

58. Make up two statements that are contrary.

Using Euler Diagrams to Analyze Arguments

Leonhard Euler (1707–1783) won the academy prize and edged out du Châtelet and Voltaire. That was a minor achievement, as was the invention of "Euler circles" (which antedated Venn diagrams). Euler was the most prolific mathematician of his generation despite blindness that forced him to dictate from memory.

There are two types of reasoning: inductive and deductive. With inductive reasoning we observe patterns to solve problems. Now, in this section and the next, we will study how deductive reasoning may be used to determine whether logical arguments are valid or invalid. A logical argument is made up of **premises** (assumptions, laws, rules, widely held ideas, or observations) and a **conclusion.** Together, the premises and the conclusion make up the argument. Also recall that *deductive* reasoning involves drawing specific conclusions from given general premises. When reasoning from the premises of an argument to obtain a conclusion, we want the argument to be valid.

Valid and Invalid Arguments

An argument is **valid** if the fact that all the premises are true forces the conclusion to be true. An argument that is not valid is **invalid,** or a **fallacy.**

It is very important to note that "valid" and "true" are not the same—an argument can be valid even though the conclusion is false. (See Example 4 below.)

Several techniques can be used to check whether an argument is valid. One of these is the visual technique based on **Euler diagrams,** illustrated by the following examples. (Another is the method of truth tables, shown in the next section.)

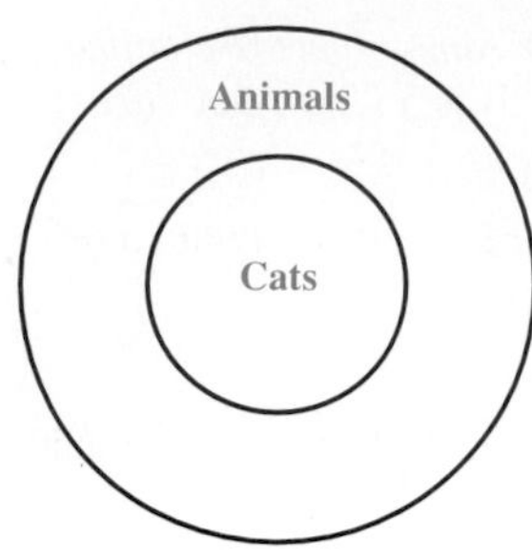

FIGURE 8

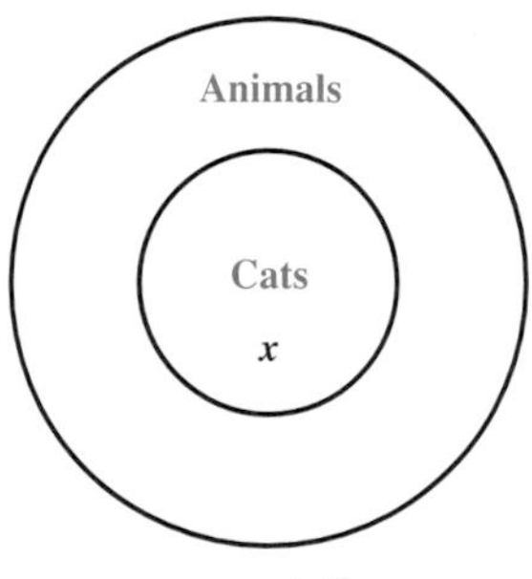

x represents Tom.

FIGURE 9

EXAMPLE 1 Is the following argument valid?

All cats are animals.
Tom is a cat.
―――――――――――
Tom is an animal.

Here we use the common method of placing one premise over another, with the conclusion below a line. To begin, draw a region to represent the first premise. This is the region for "animals." Since all cats are animals, the region for "cats" goes inside the region for "animals," as in Figure 8.

The second premise, "Tom is a cat," suggests that "Tom" would go inside the region representing "cats." Let x represent "Tom." Figure 9 shows that "Tom" is also inside the region for "animals." Therefore, if both premises are true, the conclusion that Tom is an animal must be true also. The argument is valid, as checked by Euler diagrams. ■

The method of Euler diagrams is especially useful for arguments involving the quantifiers *all, some,* or *none.*

EXAMPLE 2 Is the following argument valid?

All rainy days are cloudy.
Today is not cloudy.
―――――――――――
Today is not rainy.

In Figure 10 the region for "rainy days" is drawn entirely inside the region for "cloudy days." Since "Today is *not* cloudy," place an x for "today" *outside* the region for "cloudy days." (See Figure 11.) Placing the x outside the region for "cloudy days" forces it to be also outside the region for "rainy days." Thus, if the first two premises are true, then it is also true that today is not rainy. The argument is valid. ■

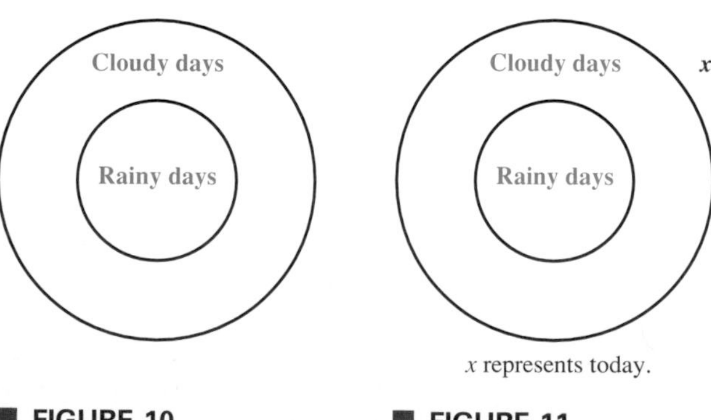

x represents today.

FIGURE 10 FIGURE 11

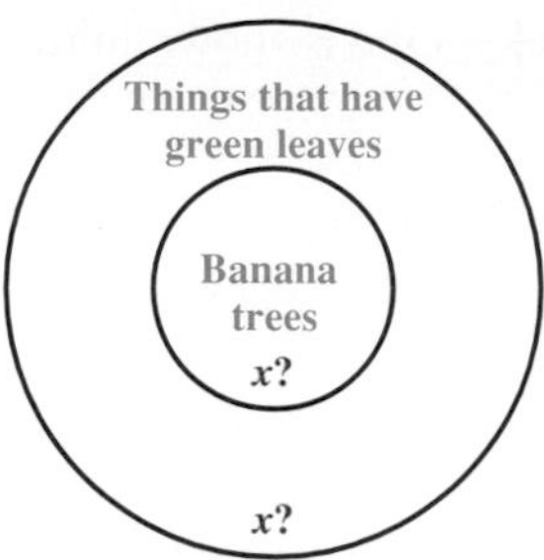

■ **FIGURE 12**

EXAMPLE 3 Is the following argument valid?

All banana trees have green leaves.
That plant has green leaves.
――――――――――――
That plant is a banana tree.

The region for "banana trees" goes entirely inside the region for "things that have green leaves." (See Figure 12.) There is a choice for locating the x that represents "that plant." The x must go inside the region for "things that have green leaves," but can go either inside or outside the region for "banana trees." Even if the premises are true, we are not forced to accept the conclusion as true. This argument is invalid; it is a fallacy. ■

As mentioned earlier, the validity of an argument is not the same as the truth of its conclusion. The argument in Example 3 was invalid, but the conclusion "That plant is a banana tree" may or may not be true. We cannot be sure.

EXAMPLE 4 Is the following argument valid?

All expensive things are desirable.
All desirable things make you feel good.
All things that make you feel good make you live longer.
――――――――――――
All expensive things make you live longer.

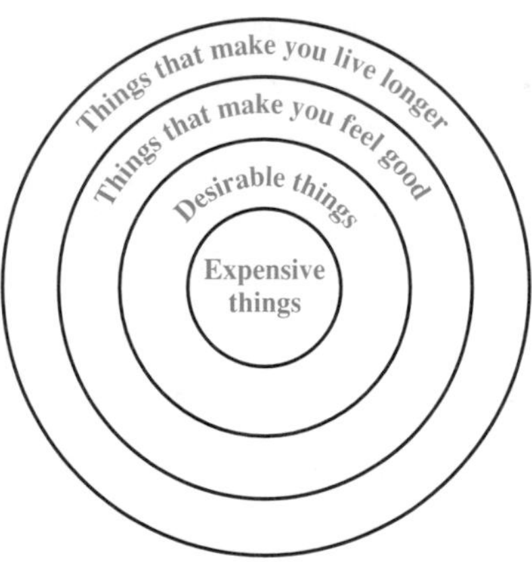

■ **FIGURE 13**

A diagram for the argument is given in Figure 13. If each premise is true, then the conclusion must be true since the region for "expensive things" lies completely within the region for "things that make you live longer." Thus, the argument is valid. (This argument is an example of the fact that a *valid* argument need *not* have a true conclusion.) ■

Arguments with the word "some" can be a little tricky. One is shown in the final example of this section.

EXAMPLE 5 Is the following argument valid?

Some students go to the beach.
I am a student.
―――――――――――
I go to the beach.

The first premise is sketched in Figure 14. As the sketch shows, some (but not necessarily *all*) students go to the beach. There are two possibilities for *I*, as shown in Figure 15.

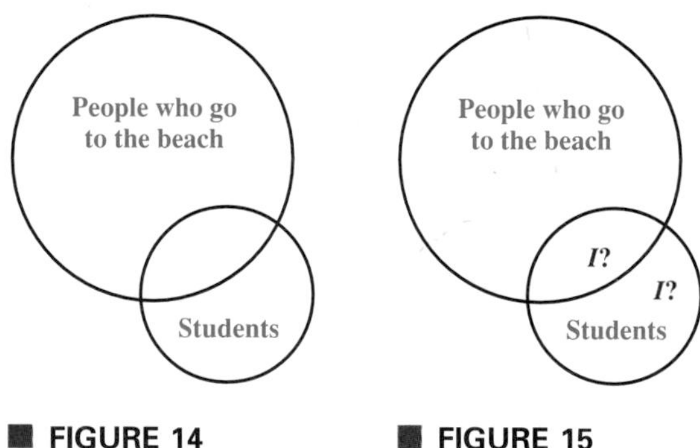

FIGURE 14 FIGURE 15

One possibility is that *I* go to the beach; the other is that *I* don't. Since the truth of the premises does not force the conclusion to be true, the argument is invalid.

3.5 EXERCISES

Decide whether each argument is valid *or* invalid.

1. All museums exhibit art.
   The Louvre is a museum.
   ―――――――――――
   The Louvre exhibits art.

2. All children love to swim.
   Troy Kroeger is a child.
   ―――――――――――
   Troy Kroeger loves to swim.

3. All homeowners have a plumber.
   Vonalaine Crowe has a plumber.
   ―――――――――――
   Vonalaine Crowe is a homeowner.

4. All politicians have questionable ethics.
   That man has questionable ethics.
   ―――――――――――
   That man is a politician.

5. All residents of St. Tammany parish live on farms.
   Jay Beckenstein lives on a farm.
   ―――――――――――
   Jay Beckenstein is a resident of St. Tammany parish.

6. All dogs love to bury bones.
   Archie is a dog.
   ―――――――――――
   Archie loves to bury bones.

7. All members of the credit union have savings accounts.
   Kristyn Wasag does not have a savings account.
   ―――――――――――
   Kristyn Wasag is not a member of the credit union.

8. All engineers need mathematics.
Shane Stagg does not need mathematics.
Shane Stagg is not an engineer.

9. All residents of New Orleans have huge utility bills in July.
Erin Kelly has a huge utility bill in July.
Erin Kelly lives in New Orleans.

10. All people applying for a home loan must provide a down payment.
Cynthia Herring provided a down payment.
Cynthia Herring applied for a home loan.

11. Some mathematicians are absent-minded.
Diane Gray is a mathematician.
Diane Gray is absent-minded.

12. Some animals are nocturnal.
Oliver Owl is an animal.
Oliver Owl is nocturnal.

13. Some cars have automatic door locks.
Some cars are red.
Some red cars have automatic door locks.

14. Some doctors appreciate classical music.
Kevin Howell is a doctor.
Kevin Howell appreciates classical music.

15. Refer to Example 4 in this section. Give a different conclusion than the one given there so that the argument is still valid.

16. Construct a valid argument based on the Euler diagram shown here.

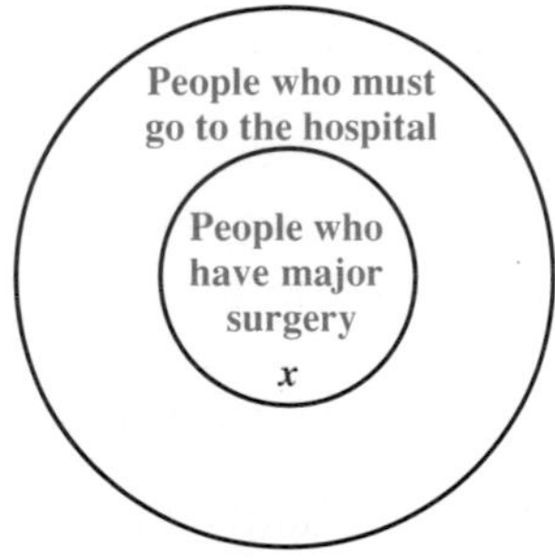

x represents Cynthia Biron.

As mentioned in the text, an argument can have a true *conclusion yet be invalid. In these exercises, each argument has a* true *conclusion. Identify each argument as* valid *or* invalid.

17. All birds fly.
All planes fly.
A bird is not a plane.

18. All cars have tires.
All tires are rubber.
All cars have rubber.

19. All chickens have a beak.
All hens are chickens.
All hens have a beak.

20. All chickens have a beak.
All birds have a beak.
All chickens are birds.

21. Quebec is northeast of Ottawa.
Quebec is northeast of Toronto.
Ottawa is northeast of Toronto.

22. Veracruz is south of Tampico.
Tampico is south of Monterrey.
Veracruz is south of Monterrey.

23. No whole numbers are negative.
-1 is negative.
-1 is not a whole number.

24. A scalene triangle has a longest side.
A scalene triangle has a largest angle.
The largest angle in a scalene triangle is opposite the longest side.

In Exercises 25–30, the premises marked A, B, and C are followed by several possible conclusions. Take each conclusion in turn, and check whether the resulting argument is valid *or* invalid.

A. *All people who drive contribute to air pollution.*
B. *All people who contribute to air pollution make life a little worse.*
C. *Some people who live in a suburb make life a little worse.*

25. Some people who live in a suburb drive.
26. Some people who live in a suburb contribute to air pollution.
27. Some people who contribute to air pollution live in a suburb.
28. Suburban residents never drive.
29. All people who drive make life a little worse.
30. Some people who make life a little worse live in a suburb.
31. Find examples of arguments in magazine ads. Check them for validity.
32. Find examples of arguments on television commercials. Check them for validity.

Logic puzzles of the type found in Exercises 33–34 can be solved by using a "grid" where information is entered in each box. If the situation is impossible, write No *in the grid. If it is possible write* Yes. *For example, consider the following puzzle.*

Norma, Harriet, Betty, and Geneva are married to Don, Bill, John, and Nathan, but not necessarily in the order given. One couple has first names that start with the same letter. Harriet is married to John. Don's wife is neither Geneva nor Norma. Pair up each husband and wife.

The grid that follows from the puzzle is shown below.

	Don	**Bill**	**John**	**Nathan**
Norma	No	No	No	Yes
Harriet	No	No	Yes	No
Betty	Yes	No	No	No
Geneva	No	Yes	No	No

From the grid, we deduce that Harriet is married to John, Norma is married to Nathan, Betty is married to Don, and Geneva is married to Bill.

Solve each of the following puzzles.

33. Juanita, Evita, Li, Fred, and Butch arrived at a party at different times. Evita arrived after Juanita but before Butch. Butch was neither the first nor the last to arrive. Juanita and Evita arrived after Fred but all three of them were present when Li got there. In what order did the five arrive?

34. There are five girls whose first names are Cherie, Leann, Chris, Mary, and Monica. (The last two are twins.) The twins have the same last name, and the four last names are Parkerson, Bingle, Plunk, and Waters. Cherie's last name begins with a "P." The twins have never met the girl whose last name is Waters, and Leann has the longest last name. Determine the girls' complete names.

3.6 Using Truth Tables to Analyze Arguments

The previous section showed how to use Euler diagrams to test the validity of arguments. While Euler diagrams often work well for simple arguments, difficulties can develop with more complex ones. These difficulties occur because Euler diagrams require a sketch showing every possible case. In complex arguments it is hard to be sure that all cases have been considered.

In deciding whether to use Euler diagrams to test the validity of an argument, look for quantifiers such as "all," "some," or "no." These words often indicate arguments best tested by Euler diagrams. If these words are absent, it may be better to use truth tables to test the validity of an argument.

As an example of this method, consider the following argument:

If the floor is dirty, then I must mop it.
The floor is dirty.
―――――――――――――
I must mop it.

In order to test the validity of this argument, we begin by identifying the *component* statements found in the argument. They are "the floor is dirty" and "I must mop it." We shall assign the letters p and q to represent these statements:

p represents "the floor is dirty."

q represents "I must mop it."

Now we write the two premises and the conclusion in symbols:

Premise 1: $p \to q$
Premise 2: p
―――――――――――――
Conclusion: q .

To decide if this argument is valid, we must determine whether the conjunction of both premises implies the conclusion for all possible cases of truth values for p and q. Therefore, write the conjunction of the premises as the antecedent of a conditional statement, and the conclusion as the consequent.

$$[(p \to q) \quad \wedge \quad p] \quad \longrightarrow \quad q$$

premise (↑ under $(p \to q)$), and (↑ under $\wedge$), premise (↑ under p), implies (↑ under $\longrightarrow$), conclusion (↑ under q)

Finally, construct the truth table for this conditional statement, as shown below.

p	q	$p \to q$	$(p \to q) \wedge p$	$[(p \to q) \wedge p] \to q$
T	T	T	T	T
T	F	F	F	T
F	T	T	F	T
F	F	T	F	T

Since the final column, shown in color, indicates that the conditional statement that represents the argument is true for all possible truth values of p and q, the statement is a tautology. Thus, the argument is valid.

The pattern of the argument in the floor-mopping example,

$$\begin{array}{l} p \to q \\ \underline{p \qquad} \\ q \end{array},$$

is a common one, and is called **modus ponens,** or the *law of detachment*.

In summary, to test the validity of an argument using truth tables, go through the steps in the box that follows.

Testing the Validity of an Argument with a Truth Table

1. Assign a letter to represent each component statement in the argument.
2. Express each premise and the conclusion symbolically.
3. Form the symbolic statement of the entire argument by writing the *conjunction* of *all* the premises as the antecedent of a conditional statement, and the conclusion of the argument as the consequent.
4. Complete the truth table for the conditional statement formed in part 3 above. If it is a tautology, then the argument is valid; otherwise, it is invalid.

EXAMPLE 1 Determine whether the argument is *valid* or *invalid.*

If my check arrives in time, I'll register for the Fall semester.
I've registered for the Fall semester.
____
My check arrived in time.

Let p represent "my check arrives (arrived) in time" and let q represent "I'll register (I've registered) for the Fall semester." Using these symbols, the argument can be written in the form

$$\begin{array}{l} p \to q \\ \underline{q \qquad} \\ p \end{array} .$$

To test for validity, construct a truth table for the statement

$$[(p \to q) \wedge q] \to p.$$

p	q	$p \to q$	$(p \to q) \wedge q$	$[(p \to q) \wedge q] \to p$
T	T	T	T	T
T	F	F	F	T
F	T	T	T	**F**
F	F	T	F	T

The third row of the final column of the truth table shows F, and this is enough to conclude that the argument is invalid. ■

If a conditional and its converse were logically equivalent, then an argument of the type found in Example 1 would be valid. Since a conditional and its converse are *not* equivalent, the argument is an example of what is sometimes called the **fallacy of the converse.**

EXAMPLE 2 Determine whether the argument is *valid* or *invalid.*

If a man could be in two places at one time, I'd be with you.
I am not with you.
____
A man can't be in two places at one time.

If p represents "a man could be in two places at one time" and q represents "I'd be with you," the argument becomes

$$\begin{array}{l} p \to q \\ \underline{\sim q} \\ \sim p \end{array}.$$

The symbolic statement of the entire argument is

$$[(p \to q) \wedge \sim q] \to \sim p.$$

The truth table for this argument, shown below, indicates a tautology, and the argument is valid.

p	q	$p \to q$	$\sim q$	$(p \to q) \wedge \sim q$	$\sim p$	$[(p \to q) \wedge \sim q] \to \sim p$
T	T	T	F	F	F	T
T	F	F	T	F	F	T
F	T	T	F	F	T	T
F	F	T	T	T	T	T

The pattern of reasoning of this example is called **modus tollens,** or the *law of contraposition,* or *indirect reasoning.* ■

With reasoning similar to that used to name the fallacy of the converse, the fallacy

$$\begin{array}{l} p \to q \\ \underline{\sim p} \\ \sim q \end{array}$$

is often called the **fallacy of the inverse.** An example of such a fallacy is "If it rains, I get wet. It doesn't rain. Therefore, I don't get wet."

EXAMPLE 3 Determine whether the argument is *valid* or *invalid.*

I'll buy a car or I'll take a vacation.
I won't buy a car.
―――――――――――――――
I'll take a vacation.

If p represents "I'll buy a car" and q represents "I'll take a vacation," the argument becomes

$$\begin{array}{l} p \vee q \\ \underline{\sim p} \\ q \end{array}.$$

We must set up a truth table for

$$[(p \vee q) \wedge \sim p] \to q.$$

p	q	$p \vee q$	$\sim p$	$(p \vee q) \wedge \sim p$	$[(p \vee q) \wedge \sim p] \to q$
T	T	T	F	F	T
T	F	T	F	F	T
F	T	T	T	T	T
F	F	F	T	F	T

The statement is a tautology and the argument is valid. Any argument of this form is valid by the law of **disjunctive syllogism.** ■

EXAMPLE 4 Determine whether the argument is *valid* or *invalid*.

If it squeaks, then I use WD-40.
If I use WD-40, then I must go to the hardware store.
―――
If it squeaks, then I must go to the hardware store.

Let p represent "it squeaks," let q represent "I use WD-40," and let r represent "I must go to the hardware store." The argument takes on the general form

$$\begin{array}{l} p \rightarrow q \\ \underline{q \rightarrow r} \\ p \rightarrow r. \end{array}$$

Make a truth table for the following statement:

$$[(p \rightarrow q) \wedge (q \rightarrow r)] \rightarrow (p \rightarrow r).$$

It will require eight rows.

p	q	r	$p \rightarrow q$	$q \rightarrow r$	$p \rightarrow r$	$(p \rightarrow q) \wedge (q \rightarrow r)$	$[(p \rightarrow q) \wedge (q \rightarrow r)] \rightarrow (p \rightarrow r)$
T	T	T	T	T	T	T	T
T	T	F	T	F	F	F	T
T	F	T	F	T	T	F	T
T	F	F	F	T	F	F	T
F	T	T	T	T	T	T	T
F	T	F	T	F	T	F	T
F	F	T	T	T	T	T	T
F	F	F	T	T	T	T	T

This argument is valid since the final statement is a tautology. The pattern of argument shown in this example is called **reasoning by transitivity,** or the *law of hypothetical syllogism.* ■

A summary of the valid and invalid forms of argument presented so far follows.

Valid Argument Forms

Modus Ponens	Modus Tollens	Disjunctive Syllogism	Reasoning by Transitivity
$p \rightarrow q$	$p \rightarrow q$	$p \vee q$	$p \rightarrow q$
$\underline{p}$	$\underline{\sim q}$	$\underline{\sim p}$	$\underline{q \rightarrow r}$
q	$\sim p$	q	$p \rightarrow r$

Invalid Argument Forms (Fallacies)

Fallacy of the Converse	Fallacy of the Inverse
$p \to q$	$p \to q$
q	$\sim p$
p	$\sim q$

When an argument contains three or more premises, it will be necessary to determine the truth value of the conjunction of all of them. Remember that if *at least one* premise in a conjunction of several premises is false, then the entire conjunction is false. This will be shown in the next example.

EXAMPLE 5 Determine whether the argument is *valid* or *invalid.*

If Eddie goes to town, then Mabel stays at home.
If Mabel does not stay at home, then Rita will cook.
Rita will not cook.
Therefore, Eddie does not go to town.

In an argument written in this manner, the premises are given first, and the conclusion is the statement which follows the word "Therefore." Let p represent "Eddie goes to town," let q represent "Mabel stays at home," and let r represent "Rita will cook." The symbolic form of the argument is

$$\begin{array}{l} p \to q \\ \sim q \to r \\ \underline{\sim r} \\ \sim p \end{array}$$

To test validity, set up a truth table for the statement

$$[(p \to q) \wedge (\sim q \to r) \wedge \sim r] \to \sim p.$$

The table is shown below.

p	q	r	$p \to q$	$\sim q$	$\sim q \to r$	$\sim r$	$[(p \to q) \wedge (\sim q \to r) \wedge \sim r]$	$\sim p$	$[(p \to q) \wedge (\sim q \to r) \wedge \sim r] \to \sim p$
T	T	T	T	F	T	F	F	F	T
T	T	F	T	F	T	T	T	F	**F**
T	F	T	F	T	T	F	F	F	T
T	F	F	F	T	F	T	F	F	T
F	T	T	T	F	T	F	F	T	T
F	T	F	T	F	T	T	T	T	T
F	F	T	T	T	T	F	F	T	T
F	F	F	T	T	F	T	F	T	T

Because the final column does not contain all Ts, the statement is not a tautology. The argument is invalid.

Consider the following poem, which has been around for many years.

For want of a nail, the shoe was lost.
For want of a shoe, the horse was lost.
For want of a horse, the rider was lost.
For want of a rider, the battle was lost.
For want of a battle, the war was lost.
Therefore, for want of a nail, the war was lost.

Each line of the poem may be written as an *if . . . then* statement. For example, the first line may be restated as "if a nail is lost, then the shoe is lost." Other statements may be worded similarly. The conclusion, "for want of a nail, the war was lost," follows from the premises since repeated use of the law of transitivity applies. Arguments used by Lewis Carroll (see the margin note in Chapter 1) often take on a similar form. The next example comes from one of his works.

Tweedlogic "I know what you're thinking about," said Tweedledum, "but it isn't so, nohow." "Contrariwise," continued Tweedledee, "if it was so, it might be; and if it were so, it would be, but as it isn't, it ain't. That's logic."

EXAMPLE 6 Supply a valid conclusion for the following premises.

Babies are illogical.
Nobody is despised who can manage a crocodile.
Illogical persons are despised.

First, write each premise in the form *if . . . then.*

If you are a baby, then you are illogical.
If you can manage a crocodile, then you are not despised.
If you are illogical, then you are despised.

Let p be "you are a baby," let q be "you are logical," let r be "you can manage a crocodile," and let s be "you are despised." With these letters, the statements can be written symbolically as

$$p \rightarrow \sim q$$
$$r \rightarrow \sim s$$
$$\sim q \rightarrow s.$$

Now begin with any letter that appears only once. Here p appears only once. Using the contrapositive of $r \rightarrow \sim s$, which is $s \rightarrow \sim r$, rearrange the three statements as follows:

$$p \rightarrow \sim q$$
$$\sim q \rightarrow s$$
$$s \rightarrow \sim r.$$

From the three statements, repeated use of reasoning by transitivity gives the valid conclusion

$$p \rightarrow \sim r.$$

In words, "If you are a baby, then you cannot manage a crocodile," or, as Lewis Carroll would have written it, "Babies cannot manage crocodiles." ■

For Further Thought

The field of advertising is notorious for logical errors. While the conclusion reached in an advertisement may be true, the logic used to reach that conclusion may be based on an invalid argument. One fallacy that is seen time and time again is the fallacy of experts. Here, an expert (or famous figure) is quoted in a field outside of that person's field of knowledge.

According to syndicated columnist Paul Harvey ("Celebrity Salesman," July 1, 1988, *Sacramento Union*), having well-known celebrities who "(seek) to transfer their own popularity to some product or service is a questionable practice." Some personalities and their endorsed products in recent years have been Andre Agassi for a camera, Ray Charles for a diet soft drink, and Ken Griffey, Jr., for a sport drink.

The list goes on and on. Yet, according to Harvey, Video Storyboard, in a survey of effectiveness of television commercials, found that the most effective commercials are those that contain no "live people" at all. Some of the best ones were those that featured the Eveready Bunny, the Pillsbury Doughboy, and the Keebler Elves.

In another illustration of the *fallacy of experts*, the following appeared as part of a question addressed to columnist Sue Rusche ("Education helps control alcoholism," July 9, 1988, *Sacramento Union*):

> The simple truth is that Congress and the press seldom listen to people who actually know something about a problem. They prefer to listen instead to those whose testimony or statements, though completely uninformed, will make a publicity splash. As a British friend of mine commented, "You Americans pick an actor who has played a drug addict in some film to testify before Congress as an expert on drugs. Lord Laurence Olivier has played Hamlet at least a thousand times, but Parliament has never called on him to testify on matters Danish."

Another common ploy in advertising is the fallacy of emotion, where an appeal is made to pity, passion, brute force, snobbishness, vanity, or some other emotion. Have you seen the tire advertisement that suggests that if you don't use a certain brand of tire, your infants will be unsafe? And have you ever seen an ad for beer that doesn't involve "beautiful people"?

For Group Discussion

1. What are some examples of the fallacy of experts that you have seen or heard?
2. What are some examples of the fallacy of emotion that you have seen or heard?

3.6 EXERCISES

Each of the following arguments is either valid by one of the forms of valid arguments discussed in this section, or a fallacy by one of the forms of invalid arguments discussed. (See the summary boxes.) Decide whether the argument is valid *or a* fallacy, *and give the form that applies.*

1. If you build it, he will come.
   If he comes, then you will see your father.

   If you build it, then you will see your father.

2. If Harry Connick, Jr. comes to town, then I will go to the concert.
   If I go to the concert, then I'll be broke until payday.

   If Harry Connick, Jr. comes to town, then I'll be broke until payday.

3. If Doug Gilbert sells his quota, he'll get a bonus.
   Doug Gilbert sold his quota.

   He got a bonus.

4. If Cyndi Keen works hard enough, she will get a raise.
   Cyndi Keen worked hard enough.

   She got a raise.

5. If she buys a dress, then she will buy shoes.
   She buys shoes.

   She buys a dress.

6. If I didn't have to write a term paper, I'd be ecstatic.
   I am ecstatic.

   I don't have to write a term paper.

7. If beggars were choosers, then I could ask for it.
   I cannot ask for it.

   Beggars aren't choosers.

8. If Roger Clemens pitches, the Red Sox will win.
   The Red Sox will not win.

   Roger Clemens will not pitch.

9. "If I have seen farther than others, it is because I have stood on the shoulders of giants." (Sir Isaac Newton)
   I have not seen farther than others.

   I have not stood on the shoulders of giants.

10. "If we evolved a race of Isaac Newtons, that would not be progress." (Aldous Huxley)
    We have not evolved a race of Isaac Newtons.

    That is progress.

11. Alice Lavin sings or Barbara Lumer dances.
    Barbara Lumer does not dance.

    Alice Lavin sings.

12. She charges it on Visa or she orders it C.O.D.
    She doesn't charge it on Visa.

    She orders it C.O.D.

Use a truth table to determine whether the argument is valid *or* invalid.

13. $p \wedge \sim q$
$\underline{p}$
$\sim q$

14. $p \vee q$
$\underline{p}$
$\sim q$

15. $p \vee \sim q$
$\underline{p}$
$\sim q$

16. $\sim p \rightarrow \sim q$
$\underline{q}$
p

17. $\sim p \rightarrow q$
$\underline{p}$
$\sim q$

18. $p \rightarrow q$
$\underline{q \rightarrow p}$
$p \wedge q$

19. $p \rightarrow \sim q$
$\underline{\sim p}$
$\sim q$

20. $p \rightarrow \sim q$
$\underline{q}$
$\sim p$

21. $(p \rightarrow q) \wedge (q \rightarrow p)$
$\underline{p}$
$p \vee q$

22. $(\sim p \vee q) \wedge (\sim p \rightarrow q)$
$\underline{p}$
$\sim q$

23. $(r \wedge p) \rightarrow (r \vee q)$
$\underline{(q \wedge p)}$
$r \vee p$

24. $(\sim p \wedge r) \rightarrow (p \vee q)$
$\underline{(\sim r \rightarrow p)}$
$q \rightarrow r$

25. Earlier we showed how to analyze arguments using Euler diagrams. Refer to Example 4 in this section, restate each premise and the conclusion using a quantifier, and then draw an Euler diagram to illustrate the relationship.

26. Explain in a few sentences how to determine the statement for which a truth table will be constructed so that the arguments in Exercises 27–36 can be analyzed for validity.

Determine whether the following arguments are valid *or* invalid.

27. Wally's hobby is amateur radio. If his wife likes to read, then Wally's hobby is not amateur radio. If his wife does not like to read, then Nikolas likes cartoons. Therefore, Nikolas likes cartoons.

28. If you are infected with a virus, then it can be transmitted. The consequences are serious and it cannot be transmitted. Therefore, if the consequences are not serious, then you are not infected with a virus.

29. Paula Abdul sings or Tom Cruise is not a hunk. If Tom Cruise is not a hunk, then Garth Brooks does not win a Grammy. Garth Brooks wins a Grammy. Therefore, Paula Abdul does not sing.

30. If Bill so desires, then Al will be the vice president. Magic is a spokesman or Al will be the vice president. Magic is not a spokesman. Therefore, Bill does not so desire.

31. The Falcons will be in the playoffs if and only if Morten is an all-pro. Janet loves the Falcons or Morten is an all-pro. Janet does not love the Falcons. Therefore, the Falcons will not be in the playoffs.

32. If you're a big girl, then you don't cry. If you don't cry, then your momma does not say, "Shame on you." You don't cry or your momma says, "Shame on you." Therefore, if you're a big girl, then your momma says, "Shame on you."

33. If I were your woman and you were my man, then I'd never stop loving you. I've stopped loving you. Therefore, I am not your woman or you are not my man.

34. If Charlie is a salesman, then he lives in Hattiesburg. Charlie lives in Hattiesburg and he loves to fish. Therefore, if Charlie does not love to fish, he is not a salesman.

35. All men are mortal. Socrates is a man. Therefore, Socrates is mortal.

36. All men are created equal. All people who are created equal are women. Therefore, all men are women.

37. Mittie Arnold made the following observation: "If I want to determine whether an argument leading to the statement

$$[(p \rightarrow q) \wedge \sim q] \rightarrow \sim p$$

is valid, I only need to consider the lines of the truth table which lead to T for the column headed $(p \rightarrow q) \wedge \sim q$." Mittie was very perceptive. Can you explain why her observation was correct?

38. Suppose that you ask someone for the time and you get the following response:

> "If I tell you the time, then we'll start chatting. If we start chatting, then you'll want to meet me at a truck stop. If we meet at a truck stop, then we'll discuss my family. If we discuss my family, then you'll find out that my daughter is available for marriage. If you find out that she is available for marriage, then you'll want to marry her. If you want to marry her, then my life will be miserable since I don't want my daughter married to some fool who can't afford a $10 watch."

Use reasoning by transitivity to draw a valid conclusion.

In the arguments used by Lewis Carroll, it is helpful to restate a premise in if . . . then *form in order to more easily lead to a valid conclusion. The following premises come from Lewis Carroll. Write each premise in* if . . . then *form.*

39. None of your sons can do logic.

40. All my poultry are ducks.

41. No teetotalers are pawnbrokers.

42. Guinea pigs are hopelessly ignorant of music.

43. Opium-eaters have no self-command.

44. No teachable kitten has green eyes.

45. All of them written on blue paper are filed.

46. I have not filed any of them that I can read.

The following exercises involve premises from Lewis Carroll. Write each premise in symbols, and then in the final part, give a valid conclusion.

47. Let p be "one is able to do logic," q be "one is fit to serve on a jury," r be "one is sane," and s be "he is your son."
   (a) Everyone who is sane can do logic.
   (b) No lunatics are fit to serve on a jury.
   (c) None of your sons can do logic.
   (d) Give a valid conclusion.

48. Let p be "it is a duck," q be "it is my poultry," r be "one is an officer," and s be "one is willing to waltz."
   (a) No ducks are willing to waltz.
   (b) No officers ever decline to waltz.
   (c) All my poultry are ducks.
   (d) Give a valid conclusion.

49. Let p be "it is a guinea pig," q be "it is hopelessly ignorant of music," r be "it keeps silent while the *Moonlight Sonata* is being played," and s be "it appreciates Beethoven."
   (a) Nobody who really appreciates Beethoven fails to keep silent while the *Moonlight Sonata* is being played.
   (b) Guinea pigs are hopelessly ignorant of music.
   (c) No one who is hopelessly ignorant of music ever keeps silent while the *Moonlight Sonata* is being played.
   (d) Give a valid conclusion.

50. Let p be "one is honest," q be "one is a pawnbroker," r be "one is a promise-breaker," s be "one is trustworthy," t be "one is very communicative," and u be "one is a wine-drinker."
   (a) Promise-breakers are untrustworthy.
   (b) Wine-drinkers are very communicative.
   (c) A person who keeps a promise is honest.
   (d) No teetotalers are pawnbrokers. (*Hint:* Assume "teetotaler" is the opposite of "wine-drinker.")
   (e) One can always trust a very communicative person.
   (f) Give a valid conclusion.

51. Let p be "he is going to a party," q be "he brushes his hair," r be "he has self-command," s be "he looks fascinating," t be "he is an opium-eater," u be "he is tidy," and v be "he wears white kid gloves."
   (a) No one who is going to a party ever fails to brush his hair.
   (b) No one looks fascinating if he is untidy.
   (c) Opium-eaters have no self-command.
   (d) Everyone who has brushed his hair looks fascinating.
   (e) No one wears white kid gloves unless he is going to a party. (*Hint:* "a unless b" $\equiv \sim b \rightarrow a$.)
   (f) A man is always untidy if he has no self-command.
   (g) Give a valid conclusion.

52. Let p be "it begins with 'Dear Sir'," q be "it is crossed," r be "it is dated," s be "it is filed," t be "it is in black ink," u be "it is in the third person," v be "I can read it," w be "it is on blue paper," x be "it is on one sheet," and y be "it is written by Brown."
   (a) All the dated letters in this room are written on blue paper.
   (b) None of them are in black ink, except those that are written in the third person.
   (c) I have not filed any of them that I can read.
   (d) None of them that are written on one sheet are undated.
   (e) All of them that are not crossed are in black ink.
   (f) All of them written by Brown begin with "Dear Sir."
   (g) All of them written on blue paper are filed.
   (h) None of them written on more than one sheet are crossed.
   (i) None of them that begin with "Dear Sir" are written in the third person.
   (j) Give a valid conclusion.

OLLABORATIVE INVESTIGATION
The Logic of a Rental Contract

The "legalese" language used in contracts can be overwhelming to the ordinary individual who enters into them. It is possible that you are currently renting the house or apartment in which you live. One place where logic can be used is in analyzing a rental contract. The many clauses, whereas's, heretofores, and furthermores make such contracts a fertile place for using symbolic logic for analysis.

The following exercises give a few clauses from a typical rental contract and then ask for the clauses to be written in symbols. In addition to the various concepts studied in this chapter, you should also be aware that "p unless q" means $\sim q \rightarrow p$. Divide up into groups, and make sure that at least one member of the group is currently renting a house or an apartment. Work these exercises in numerical order, conferring with others in the group to obtain a common answer. Then share your answers with other groups to see if they correspond.

1. This contract is noncancellable by tenants for 90 days.

Let p be "You are a tenant."
q be "It is within 90 days."
r be "The contract may be cancelled."

The contract sentence can be written.

$$(p \wedge q) \rightarrow \_\_\_\_.$$

2. Write the contrapositive of the symbolic statement from Exercise 1.

$$\_\_\_\_ \rightarrow \sim(\_\_\_\_)$$

3. By De Morgan's laws, the result of Exercise 2 becomes

$$r \rightarrow (\_\_\_\_).$$

4. Write the statement from Exercise 3 in words.

5. The tenant agrees to lease the premises for nine months, beginning on October 1 and at a monthly rental of $375.

Let p be "You are a tenant."
s be "You will lease the premises for nine months."
t be "You will begin on October 1."
u be "You will pay $375 per month."

The contract sentence becomes

$$p \rightarrow [(\_\_\_\_ \wedge t) \wedge \_\_\_\_].$$

6. Utilities, except water and garbage, are paid by the tenant.

Let p be "You are a tenant."
v be "You pay water."
w be "You pay garbage."
x be "You pay the other utilities."

The contract sentence becomes

$$p \rightarrow [\_\_\_\_ \wedge (\_\_\_\_ \vee \_\_\_\_)].$$

7. Rents are prorated to the first day of the month except for the first month.

Let p be "You are the tenant."
a be "Rents are prorated to the first."
b be "It is the first month."

The contract sentence becomes

$$p \rightarrow (a \leftrightarrow \_\_\_\_).$$

8. A late charge of $10 is added to the rent payment after the fifth of each month with the exception of the first month.

Let p be "You are the tenant."
c be "A late charge of $10 is added."
d be "It is after the fifth of the month."
b be "It is the first month."

The contract sentence becomes

$$p \rightarrow [\sim b \_\_\_\_ (d \rightarrow \_\_\_\_)].$$

Chapter 3 Test

Write a negation for each of the following statements.

1. $5 + 3 = 9$

2. Every good boy deserves favour.

3. Some people here can't play this game.

4. If it ever comes to that, I won't be here.

5. My mind is made up and you can't change it.

Let p represent "it is broken" *and let q represent* "you can fix it." *Write each of the following in symbols.*

6. If it isn't broken, then you can fix it.

7. It is broken or you can't fix it.

8. You can't fix anything that is broken.

Using the same directions as for Exercises 6–8, write each of the following in words.

9. $\sim p \wedge q$

10. $p \leftrightarrow \sim q$

In each of the following, assume that p and q are true, with r false. Find the truth value of each statement.

11. $\sim p \wedge \sim r$

12. $r \vee (p \wedge \sim q)$

13. $r \rightarrow (s \vee r)$ (The truth value of the statement s is unknown.)

14. $r \leftrightarrow (p \rightarrow \sim q)$

15. What are the necessary conditions for a conditional statement to be false? for a conjunction to be true?

16. Explain in your own words why, if p is a statement, the biconditional $p \leftrightarrow \sim p$ must be false.

Write a truth table for each of the following. Identify any tautologies.

17. $p \wedge (\sim p \vee q)$

18. $\sim(p \wedge q) \rightarrow (\sim p \vee \sim q)$

Decide whether each statement is true *or* false.

19. All positive integers are whole numbers.

20. No real numbers are integers.

Write each conditional statement in the form if . . . then.

21. All rational numbers are real numbers.

22. Being a rectangle is sufficient for a polygon to be a quadrilateral.

23. Being divisible by 2 is necessary for a number to be divisible by 6.

24. She cries only if she is hurt.

For each statement, write **(a)** *the converse,* **(b)** *the inverse, and* **(c)** *the contrapositive.*

25. If a picture paints a thousand words, the graph will help me understand it.

26. $\sim p \rightarrow (q \wedge r)$ (Use one of De Morgan's laws as necessary.)

27. Use an Euler diagram to determine whether the following argument is *valid* or *invalid.*

All members of that music club save money.
Steve Gold is a member of that music club.
―――――――――――――――――――
Steve Gold saves money.

28. Match each argument in (a)–(d) with the law that justifies its validity, or the fallacy of which it is an example.

A. Modus ponens
B. Modus tollens
C. Reasoning by transitivity
D. Disjunctive syllogism
E. Fallacy of the converse
F. Fallacy of the inverse

(a) If you like ice cream, then you'll like Blue Bell.
You don't like Blue Bell.
―――――――――――――――――――
You don't like ice cream.

(b) If I buckle up, I'll be safer.
I don't buckle up.
―――――――――――――――――――
I'm not safer.

(c) If you love me, you will let me go.
If you let me go, I'll try to forget.
―――――――――――――――――――
If you love me, I'll try to forget.

(d) It is March or April.
It's not March.
―――――――――――――――――――
It's April.

Use a truth table to determine whether each argument is valid *or* invalid.

29. If I hear that song, it reminds me of my youth. If I get sentimental, then it does not remind me of my youth. I get sentimental. Therefore, I don't hear that song.

30. $\sim p \rightarrow \sim q$
$q \rightarrow p$
―――――
$p \vee q$

4

Numeration and Mathematical Systems

In order to have a consistent numbering system for published books the International Standard Book Number (ISBN) was established. A book can be identified completely by its ISBN. It is usually composed of 10 digits with hyphens appearing at various positions. For example, the ISBN of this text is 0-673-99893-2. The 0 is for the United States and a few other English-speaking countries. The 673 is associated with the publisher, Addison-Wesley, while the next 5 digits are the book number assigned to this text. The final digit is a check digit between 0 and 10 with X representing 10. The reason for a check is to avoid errors such as transposing digits when copying an ISBN number. The check digit helps to determine if a number is a valid ISBN. To determine if it is valid one can complete the table.

0		
6	6	6
7	13	19
3	16	35
9	25	60
9	34	94
8	42	136
9	51	187
3	54	241
2	56	297

In the first column write down the 10-digit ISBN. In the second column calculate the running total of the numbers in the first column. Similarly, in the third column calculate the total of the numbers in the second column. If the final sum in the lower right corner of the table is evenly divisible by 11 ($297 \div 11 = 27$), it is considered a valid ISBN. Try changing one number in 0-673-99893-2 or transposing two adjacent digits and then test the validity of the ISBN. (See Exercises 77–82 in Section 4.4 for more on ISBN.)

Source: Hamming, R., *Coding and Information Theory,* Prentice-Hall, Englewood Cliffs, New Jersey, 1986.

Earlier we introduced and studied the concept of a *set,* a collection of elements. A set, in itself, may have no particular structure. But when we introduce *ways of combining the elements* (called *operations*) and *ways of comparing the elements* (called *relations*), we obtain a **mathematical system.**

Mathematical System

A **mathematical system** is made up of three things:

1. a set of elements;
2. one or more operations for combining the elements;
3. one or more relations for comparing the elements.

A familiar example of a mathematical system is the set of whole numbers {0, 1, 2, 3, . . .}, along with the operation of addition and the relation of equality.

Symbols designed to represent objects or ideas are among the oldest inventions of humans. These Indian symbols in Arizona are several hundred years old.

Historically, the earliest mathematical system to be developed involved the set of counting numbers, or initially a limited subset of the "smaller" counting numbers. The development of this system was perhaps the most basic, as well as one of the most useful, of all mathematical ideas.

The various ways of symbolizing and working with the counting numbers are called **numeration systems.** The symbols of a numeration system are called **numerals.** In the first half of this chapter, we relate some historical numeration systems to our own modern system, observe how operations are carried out in our system, and see how certain technical applications call for basing numeration on numbers other than ten. The second half of the chapter involves some more abstract algebraic systems and their applications.

4.1 Historical Numeration Systems

Primitive societies have little need for large numbers. Even today, the languages of some cultures contain no words for numbers beyond "one," "two," and maybe an indefinite word suggesting "many."

For example, according to UCLA physiologist Jared Diamond (*Discover,* Aug. 1987, p. 38), there are Gimi villages in New Guinea that use just two root words—*iya* for one and *rarido* for two. Slightly larger numbers are indicated using combinations of these two: for example, *rarido-rarido* is four and *rarido-rarido-iya* is five.

A practical method of keeping accounts by matching may have developed as humans established permanent settlements and began to grow crops and raise livestock. People might have kept track of the number of sheep in a flock by matching pebbles with the sheep, for example. The pebbles could then be kept as a record of the number of sheep.

A more efficient method is to keep a **tally stick.** With a tally stick, one notch or **tally** is made on a stick for each sheep. Tally sticks and tally marks have been found that appear to be many thousands of years old. Tally marks are still used today: for example, nine items are tallied by writing 𝍸 ||||.

Tally sticks like this one were used by the English in about 1400 A.D. to keep track of financial transactions. Each notch stands for one pound sterling.

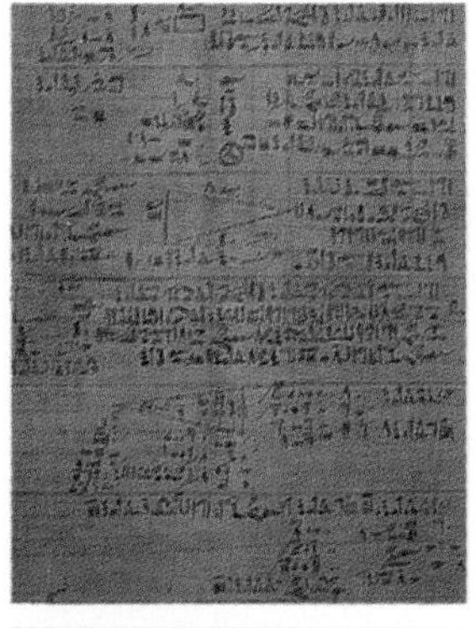

Egyptian Mathematics
Much of our knowledge of Egyptian mathematics comes from the Rhind papyrus, from about 3,800 years ago. A small portion of this papyrus, showing methods for finding the area of a triangle, is reproduced here.

Tally sticks and groups of pebbles were an important advance. By these methods, the idea of *number* began to develop. Early people began to see that a group of three chickens and a group of three dogs had something in common: the idea of *three*. Gradually, people began to think of numbers separately from the things they represented. Words and symbols were developed for various numbers.

A numeration system, like an alphabet or any other symbolic system, is a medium of "information transfer." When information is transferred from one place to another (even if only to the person standing in front of you), we can think of it as *communication*. When the transfer is from one time to another, it involves *memory*. The numerical records of ancient people give us some idea of their daily lives and create a picture of them as producers and consumers. For example, Mary and Joseph went to Bethlehem to be counted in a census—a numerical record. Even earlier than that, as long as 5,000 years ago, the Egyptian and Sumerian peoples were using large numbers in their government and business records. Ancient documents reveal some of their numerical methods, as well as those of the Greeks, Romans, Chinese, and Hindus. Numeration systems became more sophisticated as the need arose.

Ancient Egyptian Numeration—Simple Grouping

Early matching and tallying led eventually to the basic essential ingredient of all more advanced numeration systems, that of **grouping.** We will see that grouping allows for less repetition of symbols and also makes numerals easier to interpret. Most historical systems, including our own, have used groups of ten, indicating that people commonly learn to count by using the fingers (of both hands). The size of the groupings (again, usually ten) is called the **base** of the number system. Bases of five, twenty, and sixty have also been used.

The ancient Egyptian system is an example of a simple grouping system. It utilized ten as its base, and its various symbols are shown in Table 1. The symbol for 1 (|) is repeated, in a tally scheme, for 2, 3, and so on up to 9. A new symbol is introduced for 10 (∩), and that symbol is repeated for 20, 30, and so on, up to 90. This pattern enabled the Egyptians to express numbers up to 9,999,999 with just the seven symbols shown in the table.

TABLE 1 Early Egyptian Symbols

Number	Symbol	Description
1	|	Stroke
10	∩	Heel bone
100	𓍢	Scroll
1,000	𓆼	Lotus flower
10,000	𓂭	Pointing finger
100,000	𓆐	Burbot fish
1,000,000	𓁨	Astonished person

Applied Math An Egyptian tomb painting shows scribes tallying the count of a grain harvest. Egyptian mathematics was oriented more to practicality than was Greek or Babylonian mathematics, although the Egyptians did have a formula for finding the volume of a certain portion of a pyramid.

The symbols used denote the various **powers** of the base (ten):

$$10^0 = 1,\ 10^1 = 10,\ 10^2 = 100,\ 10^3 = 1{,}000,\ 10^4 = 10{,}000,$$
$$10^5 = 100{,}000, \text{ and } 10^6 = 1{,}000{,}000.$$

The smaller numerals at the right of the 10s, and slightly raised, are called **exponents.**

EXAMPLE 1 Write in our system the number below.

𓆐𓆐𓆼𓆼𓆼𓆼𓆼𓍢𓍢𓍢𓍢𓎆𓎆𓎆𓎆𓎆𓎆𓎆𓎆𓎆𓏺𓏺𓏺𓏺𓏺𓏺𓏺

Refer to Table 1 for the values of the Egyptian symbols. Each 𓆐 represents 100,000. Therefore, two 𓆐 represent $2 \times 100{,}000$, or 200,000. Proceed as follows:

two	𓆐	$2 \times 100{,}000 =$	200,000
five	𓆼	$5 \times 1{,}000 =$	5,000
four	𓍢	$4 \times 100 =$	400
nine	𓎆	$9 \times 10 =$	90
seven	𓏺	$7 \times 1 =$	7
			205,497.

The number is 205,497.

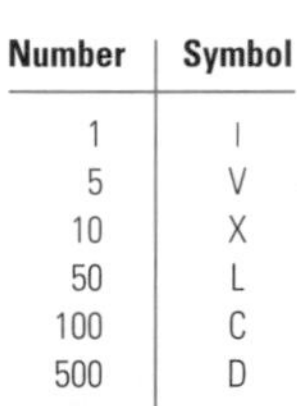

Number	Symbol
1	I
5	V
10	X
50	L
100	C
500	D
1,000	M

Roman numerals still appear today, mostly for decorative purposes: on clock faces, for chapter numbers in books, and so on. The system is essentially base ten, simple grouping, but with separate symbols for the intermediate values 5, 50, and 500, as shown above. If I is positioned left of V or X, it is subtracted rather than added. Likewise for X appearing left of L or C, and for C appearing left of D or M. Thus, for example, whereas CX denotes 110, XC denotes 90.

EXAMPLE 2 Write 376,248 in Egyptian symbols.

Writing this number requires three 𓆐s, seven 𓂭s, six 𓆼s, two 𓍢s, four 𓎆s, and eight 𓏺s, or

𓆐𓆐𓆐𓂭𓂭𓂭𓂭𓂭𓂭𓂭𓆼𓆼𓆼𓆼𓆼𓆼𓍢𓍢𓎆𓎆𓎆𓎆𓏺𓏺𓏺𓏺𓏺𓏺𓏺𓏺.

Notice that the position or order of the symbols makes no difference in a simple grouping system. Each of the numbers 𓍢𓍢𓎆𓎆𓎆𓏺𓏺𓏺𓏺, 𓏺𓏺𓏺𓏺𓎆𓎆𓎆𓍢𓍢, and 𓏺𓏺𓎆𓎆𓍢𓍢𓎆𓏺𓏺 would be interpreted as 234. The most common order, however, is that shown in Examples 1 and 2, where like symbols are grouped together and groups of higher valued symbols are positioned to the left.

A simple grouping system is well suited to addition and subtraction. For example, to add 𓆼𓆼𓍢𓍢𓎆𓎆𓎆𓏺𓏺 and 𓆼𓍢𓍢𓍢𓎆𓏺𓏺𓏺𓏺𓏺𓏺 in the early Egyptian system, work as shown. Two 𓏺s plus six 𓏺s equal eight 𓏺s, and so on.

	𓆼𓆼	𓍢𓍢	𓎆𓎆𓎆	𓏺𓏺
	+ 𓆼	𓍢𓍢𓍢	𓎆	𓏺𓏺𓏺𓏺𓏺𓏺
Sum:	𓆼𓆼𓆼	𓍢𓍢𓍢𓍢𓍢	𓎆𓎆𓎆𓎆	𓏺𓏺𓏺𓏺𓏺𓏺𓏺𓏺

While we used a + sign for convenience and drew a line under the numbers, the Egyptians did not do this.

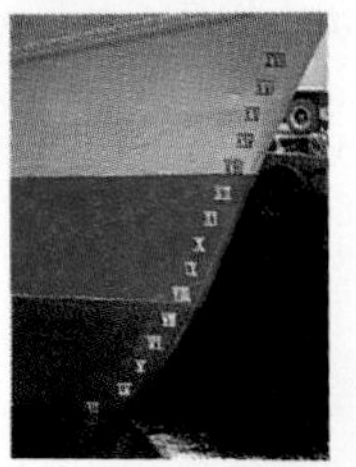

How deep is this ship in the water?

Greek Numerals

1	α	60	ξ
2	β	70	o
3	γ	80	π
4	δ	90	φ
5	ϵ	100	ρ
6	ς	200	σ
7	ζ	300	τ
8	η	400	υ
9	θ	500	ϕ
10	ι	600	χ
20	κ	700	ψ
30	λ	800	ω
40	μ	900	ϡ
50	ν		

What About the Greeks?

Classical Greeks used letters of their alphabet as numerical symbols. The base of the system was the number 10, and numbers 1 through 9 were symbolized by the first nine letters of the alphabet. Rather than using repetition or multiplication, they assigned nine more letters to multiples of 10 (through 90), and more letters to multiples of 100 (through 900). This is called a ciphered system, and it sufficed for small numbers. For example, 57 would be $\nu\zeta$; 573 would be $\phi o\gamma$; and 803 would be $\omega\gamma$. A small stroke was used with a units symbol for multiples of 1,000 (up to 9,000); thus 1,000 would be ͵α or $'\alpha$. Often M would indicate tens of thousands (M for myriad = 10,000) with the multiples written above M.

Sometimes regrouping, or "carrying," is needed, as in the example below in which the answer contains more than nine heel bones. To regroup, get rid of ten heel bones from the tens group. Compensate for this by placing an extra scroll in the hundreds group.

	𓂭 𓆼𓆼 𓍢𓍢 𓎆𓎆𓎆𓎆𓎆𓎆𓎆 𓏺𓏺
+	𓂭𓂭 𓍢𓍢𓍢 𓎆𓎆𓎆𓎆𓎆 𓏺𓏺𓏺𓏺𓏺
Sum:	𓂭𓂭𓂭 𓆼𓆼 𓍢𓍢𓍢𓍢𓍢 𓎆𓎆𓎆𓎆𓎆𓎆𓎆𓎆𓎆𓎆𓎆𓎆 𓏺𓏺𓏺𓏺𓏺𓏺𓏺
Regrouped answer:	𓂭𓂭𓂭 𓆼𓆼 𓍢𓍢𓍢𓍢𓍢𓍢 𓎆𓎆 𓏺𓏺𓏺𓏺𓏺𓏺𓏺

Subtraction is done in much the same way, as shown in the next example.

EXAMPLE 3 Subtract in each of the following.

(a)

	𓍢𓍢𓍢𓍢𓍢 𓎆𓎆𓎆𓎆 𓏺𓏺𓏺𓏺𓏺𓏺𓏺
−	𓍢𓍢𓍢 𓎆 𓏺𓏺𓏺𓏺
Difference:	𓍢𓍢 𓎆𓎆𓎆 𓏺𓏺𓏺

(b)

	𓍢𓍢 𓎆𓎆𓎆𓎆 𓏺𓏺
−	𓍢 𓎆𓎆 𓏺𓏺𓏺𓏺

In part (b), to subtract four |s from two |s, "borrow" one heel bone, which is equivalent to ten |s. Finish the problem after writing ten |s on the right.

Regrouped:	𓍢𓍢 𓎆𓎆𓎆 𓏺𓏺𓏺𓏺𓏺𓏺𓏺𓏺𓏺𓏺𓏺𓏺
−	𓍢 𓎆𓎆 𓏺𓏺𓏺𓏺
Difference:	𓍢 𓎆 𓏺𓏺𓏺𓏺𓏺𓏺𓏺𓏺

A procedure such as those described above is called an **algorithm:** a rule or method for working a problem. The Egyptians used an interesting algorithm for multiplication which requires only an ability to add and to double numbers. Example 4 illustrates the way that the Egyptians multiplied. For convenience, this example uses our symbols rather than theirs.

EXAMPLE 4 A stone used in building a pyramid has a rectangular base measuring 5 by 18 cubits. Find the area of the base.

The area of a rectangle is found by multiplying the length and the width; in this problem, 5×18. To begin, build two columns of numbers, as shown below. Start the first column with 1, and the second column with 18. Each column is built downward by doubling the number above. Keep going until the first column contains numbers that can be added to equal 5. Here $1 + 4 = 5$. To find 5×18, add only those numbers from the second column that correspond to 1 and 4. Here 18 and 72 are added to get the answer 90. The area of the base of the stone is 90 square cubits.

$1 + 4 = \mathbf{5}$	→ **1**	**18** ← Corresponds to 1	$18 + 72 = \mathbf{90}$
	2	36	
	→ **4**	**72** ← Corresponds to 4	

Finally, $5 \times 18 = 90$.

EXAMPLE 5 Use the Egyptian multiplication algorithm to find 19×70.

→ 1	70 ←
→ 2	140 ←
4	280
8	560
→ 16	1,120 ←

Form two columns, headed by 1 and by 70. Keep doubling until there are numbers in the first column that add up to 19. (Here, $1 + 2 + 16 = 19$.) Then add corresponding numbers from the second column: $70 + 140 + 1{,}120 = 1{,}330$, so that $19 \times 70 = 1{,}330$.

Traditional Chinese Numeration—Multiplicative Grouping

Examples 1 through 3 above show that simple grouping, although an improvement over tallying, still requires considerable repetition of symbols. To denote 90, for example, the ancient Egyptian system must utilize nine ∩s: ∩∩∩∩∩∩∩∩∩. If an additional symbol (a "multiplier") was introduced to represent nine, say "9," then 90 could be denoted 9 ∩. All possible numbers of repetitions of powers of the base could be handled by introducing a separate multiplier symbol for each counting number less than the base. Although the ancient Egyptian system apparently did not evolve in this direction, just such a system was developed many years ago in China. It was later adopted, for the most part, by the Japanese, with several versions occurring over the years. Here we show the predominant Chinese version, which used the symbols shown in Table 2. We call this type of system a **multiplicative grouping** system. In general, such a system would involve pairs of symbols, each pair containing a multiplier (with some counting number value less than the base) and then a power of the base. The Chinese numerals are read from top to bottom rather than from left to right.

TABLE 2

Number	Symbol
1	一
2	二
3	三
4	四
5	五
6	六
7	七
8	八
9	九
10	十
100	百
1,000	千
0	零

The figure illustrates a **quipu.** In *Ethnomathematics: A Multicultural View of Mathematical Ideas,* Marcia Ascher writes:

A quipu is an assemblage of colored knotted cotton cords. Among the Inca, cotton cloth and cordage were of great importance. Used to construct bridges, in ceremonies, for tribute, and in every phase of the life cycle from birth to death, cotton cordage and cloth were of unparalleled importance in Inca culture and, hence, not a surprising choice for its principal medium. The colors of the cords, the way the cords are connected together, the relative placement of the cords, the spaces between the cords, the types of knots on the individual cords, and the relative placement of the knots are all part of the logical-numerical recording. Just as anything from simple scratch marks to complex mathematical notation can be contained on a piece of paper or a blackboard, the logical-numerical system embedded in the cord arrays sets them quite apart from any other knotted cord usage by individuals or groups in other cultures. The quipus are distinctively Inca and unique to them.

Three features distinguish this system from a strictly pure multiplicative grouping system. First, the number of 1s is indicated using a single symbol rather than a pair. In effect, the multiplier (1, 2, 3, . . . , 9) is written but the power of the base (10^0) is not. Second, in the pair indicating 10s, if the multiplier is 1, then that multiplier is omitted. Just the symbol for 10 is written. Third, when a given power of the base is totally missing in a particular number, this omission is shown by the inclusion of the special zero symbol. (See Table 2.) If two or more consecutive powers are missing, just one zero symbol serves to note the total omission. The omission of 1s and 10s, and any other powers occurring at the extreme bottom of a numeral, need not be noted with a zero symbol. (Note that, for clarification in the examples that follow, we have emphasized the grouping into pairs by spacing and by using braces. These features are *not* part of the actual numeral.)

EXAMPLE 6 Interpret the Chinese numerals below.

(a)

三 千	$3 \times 1{,}000 = 3{,}000$
一 百	$1 \times 100 = 100$
六 十	$6 \times 10 = 60$
四	$4(\times 1) = 4$
	Total: 3,164

(b)

七 百	$7 \times 100 = 700$
零	$0(\times 10) = 00$
三	$3(\times 1) = 3$
	Total: 703

(c)

五 千	$5 \times 1{,}000 = 5{,}000$
零	$0(\times 100) = 000$ $0(\times 10) = 00$
九	$9(\times 1) = 9$
	Total: 5,009

(d)

四 千	$4 \times 1{,}000 = 4{,}000$
二 百	$2 \times 100 = 200$
	Total: 4,200

EXAMPLE 7 Write Chinese numerals for these numbers.

(a) 614

This number is made up of six 100s, one 10, and one 4, as depicted at the right.

6×100:	六 百
$(1 \times)$ 10:	十
$4(\times 1)$:	四

(b) 5,090

This number consists of five 1,000s, no 100s, and nine 10s (no 1s).

$5 \times 1{,}000$:	五 千
$0(\times 100)$:	零
9×10:	九 十

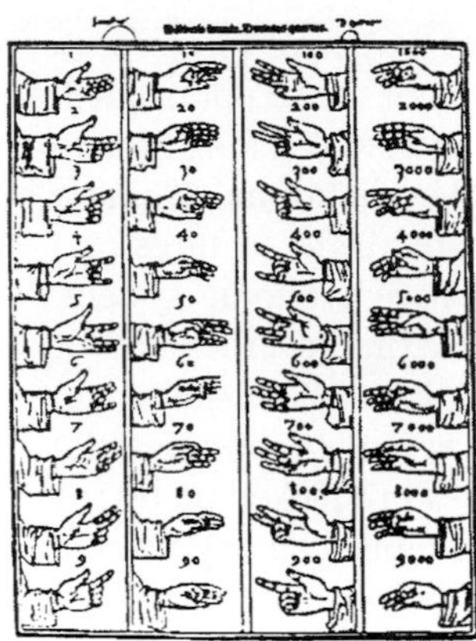

Finger Reckoning There is much evidence that early humans (in various cultures) used their fingers to represent numbers. As the various calculations of everyday life became more complicated, *finger reckoning,* as shown in this sketch, became popular. The Romans apparently became adept at this sort of calculating, carrying it to 10,000 or perhaps higher.

Hindu-Arabic Numeration—Positional System

A simple grouping system relies on repetition of symbols to denote the number of each power of the base. A multiplicative grouping system uses multipliers in place of repetition, which is more efficient. But the ultimate in efficiency is attained only when we proceed to the next step, a **positional** system, in which only the multipliers are used. The various powers of the base require no separate symbols, since the power associated with each multiplier can be understood by the position that the multiplier occupies in the numeral. If the Chinese system had evolved into a positional system, then the numeral for 7,482 could be written

七
四
八
二 rather than 七
千
四
百
八
十
二.

The lowest symbol is understood to represent two 1s (10^0), the next one up denotes eight 10s (10^1), then four 100s (10^2), and finally seven 1,000s (10^3). Each symbol in a numeral now has both a **face value,** associated with that particular symbol (the multiplier value) and a **place value** (a power of the base), associated with the place, or position, occupied by the symbol. Since these features are so important, we emphasize them as follows.

Positional Numeration

In a positional numeral, each symbol (called a **digit**) conveys two things:

1. **face value—**the inherent value of the symbol
2. **place value—**the power of the base which is associated with the position that the digit occupies in the numeral.

Number	Symbol
1	▼
10	‹

Babylonian numeration was positional, base sixty. But the face values within the positions were base ten simple grouping numerals, formed with the two symbols shown above. (These symbols resulted from the Babylonian method of writing on clay with a wedge-shaped stylus.) The numeral

‹‹▼▼▼‹‹‹‹▼

denotes 1,421 ($23 \times 60 + 41 \times 1$).

The place values in a Hindu-Arabic numeral, from right to left, are 1, 10, 100, 1,000, and so on. The three 4s in the number 46,424 all have the same face value but different place values. The first 4, on the left, denotes four 10,000s, the next one denotes four 100s, and the one on the right denotes four 1s. Place values (in base ten) are named as shown here:

This numeral is read as eight billion, three hundred twenty-one million, four hundred fifty-six thousand, seven hundred ninety-five.

To work successfully, a positional system must have a symbol for zero, to serve as a **placeholder** in case one or more powers of the base are not needed. Because of this requirement, some early numeration systems took a long time to evolve to a positional form, or never did. Although the traditional Chinese system does utilize a zero symbol, it never did incorporate all the features of a positional system, but remained essentially a multiplicative grouping system.

The one numeration system that did achieve the maximum efficiency of positional form is our own system, the one commonly known, for historical reasons, as the **Hindu-Arabic** system. It developed over many centuries. Its symbols have been traced to the Hindus of 200 B.C. They were picked up by the Arabs and eventually transmitted to Spain, where a late tenth-century version appeared like this:

1 2 3 4 5 6 7 8 9.

The earliest stages of the system evolved under the influence of navigational, trade, engineering, and military requirements. And in early modern times, the advance of astronomy and other sciences led to a structure well suited to fast and accurate computation. The purely positional form that the system finally assumed was introduced to the West by Leonardo Fibonacci of Pisa (1170–1250) early in the thirteenth century. But widespread acceptance of standardized symbols and form was not achieved until the invention of printing during the fifteenth century. Since that time, no better system of numeration has been devised, and the positional base ten Hindu-Arabic system is commonly used around the world today. (In India, where it all began, standardization is still not totally achieved, as various local systems are still used today.)

Of course, the Hindu-Arabic system that we know and use today also contains the placeholder "0" (zero). We are so accustomed to the positional structure of the system that most of us hardly give it a thought. For example, knowing that the system is based on 10, we readily interpret the numeral 5,407 as representing five 1,000s (10^3), four 100s (10^2), no 10s (10^1), and seven 1s (10^0).

In the next section we shall look in more detail at the structure of the Hindu-Arabic system and some early methods and devices for doing computation.

We have now considered the ancient Egyptian numeration system, which was simple grouping, the traditional Chinese system, which is (essentially) multiplicative grouping, and the Hindu-Arabic system, which is positional. Other well-documented numeration systems that have been used around the world at various times include the Babylonian, Greek, Roman, and Mayan systems. For our purposes here, the three systems already investigated are sufficient to illustrate the historical progression from the basic simple grouping approach to the more sophisticated and efficient positional structure.

From Tally to Tablet The clay tablet above, despite damage, shows the durability of the mud of the Babylonians. Thousands of years after the tablets were made, Babylonian algebra problems can be worked out from the original writings.

In the twentieth century, herdsmen make small "tokens" out of this clay as tallies of animals. Similar tokens have been unearthed by archaeologists in the land that was once Babylonia. Some tokens are 10,000 years old. Shaped like balls, disks, cones, and other regular forms, they rarely exceed 5 cm in diameter.

As reported in *Science News*, December 24 and 31, 1988, Denise Schmandt-Besserat, of the University of Texas at Austin, concluded after nearly two decades of study that these tokens were used 4,000 years before symbolic abstraction occurred and that they led not only to numeration systems but to writing in general.

EXERCISES

Convert each Egyptian numeral to Hindu-Arabic form.

1. 𓆼𓆼𓍢𓍢𓍢𓍢𓎆𓏺𓏺
2. 𓂭𓆼𓆼𓆼𓎆𓎆𓎆𓏺𓏺𓏺𓏺𓏺𓏺
3. 𓁨𓁨𓁨𓆼𓆼𓆼𓆼𓆼𓍢𓎆𓎆𓎆𓏺
4. 𓁨𓁨𓁨𓁨𓁨𓁨𓁨𓆐𓆐𓆐𓆐𓆐𓆐𓂭𓂭𓂭𓍢𓍢𓍢𓍢𓍢𓍢𓍢𓎆𓎆𓏺𓏺𓏺𓏺𓏺𓏺𓏺𓏺𓏺

Convert each Hindu-Arabic numeral to Egyptian form.

5. 427
6. 23,145
7. 306,090
8. 8,657,000

Chapter 1 of the book of Numbers in the Bible describes a census of the draft-eligible men of Israel after Moses led them out of Egypt into the Desert of Sinai, about 1450 B.C. Write an Egyptian numeral for the number of available men from each tribe listed.

9. 46,500 from the tribe of Reuben
10. 59,300 from the tribe of Simeon
11. 45,650 from the tribe of Gad
12. 74,600 from the tribe of Judah
13. 54,400 from the tribe of Issachar
14. 62,700 from the tribe of Dan

Convert each Chinese numeral to Hindu-Arabic form.

15. 二
百
四
十
六

16. 九
百
三
十
五

17. 四
千
九
百
零
二

18. 三
千
零
七

Convert each Hindu-Arabic numeral to Chinese.

19. 63
20. 960
21. 2,416
22. 7,012

Though Chinese art forms began before written history, their highest development was achieved during four particular dynasties. Write traditional Chinese numerals for the beginning and ending dates of each dynasty listed.

23. Han (202 B.C. to A.D. 220)
24. T'ang (618 to 907)
25. Sung (960 to 1279)
26. Ming (1368 to 1644)

Work each of the following addition or subtraction problems, using regrouping as necessary. Convert each answer to Hindu-Arabic form.

27. 𓍢𓍢𓎆𓎆𓎆𓏺𓏺𓏺𓏺𓏺𓏺𓏺
+ 𓍢𓎆𓎆𓎆𓎆𓎆𓏺𓏺𓏺𓏺𓏺

28. 𓍢𓎆𓎆𓎆𓎆𓏺𓏺
+ 𓎆𓎆𓎆𓎆𓎆𓎆𓎆𓏺𓏺𓏺𓏺

29. 𓆼𓆼𓆼𓍢𓍢𓍢𓍢𓍢𓎆𓎆𓎆𓎆𓏺𓏺
+ 𓆼𓆼𓍢𓍢𓍢𓍢𓍢𓍢𓎆𓎆𓏺𓏺𓏺𓏺𓏺𓏺

30. 𓂭𓂭𓂭𓆼𓆼𓎆𓏺𓏺𓏺
+ 𓂭𓂭𓆼𓍢𓍢𓍢𓍢𓍢𓎆𓎆𓎆𓎆𓎆𓎆𓎆𓏺𓏺𓏺𓏺𓏺𓏺𓏺𓏺𓏺

31. 𓎆𓎆𓎆𓎆𓎆𓏺𓏺𓏺𓏺𓏺𓏺
− 𓎆𓎆𓎆𓏺𓏺𓏺𓏺

32. 𓍢𓍢𓎆𓎆𓎆𓏺𓏺𓏺𓏺
− 𓍢𓎆𓎆𓏺

33. 𓆼𓍢𓍢𓍢𓍢𓍢𓎆𓎆𓎆𓏺𓏺𓏺𓏺𓏺𓏺
− 𓍢𓍢𓎆𓎆𓎆𓎆𓎆𓎆𓎆𓏺𓏺𓏺

34. 𓂭𓆼𓍢𓍢𓏺𓏺𓏺𓏺
− 𓆼𓆼𓆼𓍢𓍢𓍢𓍢𓍢𓍢𓏺𓏺𓏺𓏺𓏺𓏺

Use the Egyptian algorithm to find each product.

35. 3×19 **36.** 5×26

37. 12×93 **38.** 21×44

Convert all numbers in the following problems to Egyptian numerals. Multiply using the Egyptian algorithm, and add using the Egyptian symbols. Give the final answer using a Hindu-Arabic numeral.

39. King Solomon told the King of Tyre (now Lebanon) that Solomon needed the best cedar for his temple, and that he would "pay you for your men whatever sum you fix." Find the total bill to Solomon if the King of Tyre used the following numbers of men: 5,500 tree cutters at two shekels per week each, for a total of seven weeks; 4,600 sawers of wood at three shekels per week each, for a total of 32 weeks; and 900 sailors at one shekel per week each, for a total of 16 weeks.

40. The book of Ezra in the Bible describes the return of the exiles to Jerusalem. When they rebuilt the temple, the King of Persia gave them the following items: thirty golden basins, a thousand silver basins, four hundred ten silver bowls, and thirty golden bowls. Find the total value of this treasure, if each gold basin is worth 3,000 shekels, each silver basin is worth 500 shekels, each silver bowl is worth 50 shekels, and each golden bowl is worth 400 shekels.

Explain why each of the following steps would be an improvement in the development of numeration systems.

41. progressing from carrying groups of pebbles to making tally marks on a stick

42. progressing from tallying to simple grouping

43. progressing from simple grouping to multiplicative grouping

44. progressing from multiplicative grouping to positional numeration

Recall that the ancient Egyptian system described in this section was simple grouping, used a base of ten, and contained seven distinct symbols. The largest number expressible in that system is 9,999,999. Identify the largest number expressible in each of the following simple grouping systems. (In Exercises 49–52, d can be any counting number.)

45. base ten, five distinct symbols

46. base ten, ten distinct symbols

47. base five, five distinct symbols

48. base five, ten distinct symbols

49. base ten, d distinct symbols

50. base five, d distinct symbols

51. base seven, d distinct symbols

52. base b, d distinct symbols (where b is any counting number 2 or greater)

The Hindu-Arabic system is positional and uses ten as the base. Describe any advantages or disadvantages that may have resulted in each of the following cases.

53. Suppose the base had been larger, say twelve or twenty for example.

54. Suppose the base had been smaller, maybe eight or five.

4.2 Arithmetic in the Hindu-Arabic System

The historical development of numeration culminated in positional systems, the most successful of which is the Hindu-Arabic system. This type of system gives us the easiest way of expressing numbers. But, just as important, it gives us the easiest way of computing with numbers. This section involves some of the structure and history of the basic arithmetic operations.

As stated in Section 1, Hindu-Arabic place values are powers of the base ten. For example, 10^4 denotes the fourth power of ten. Such expressions are often called **exponential expressions.** In this case, 10 is the **base** and 4 is the **exponent.** Exponents actually indicate repeated multiplication of the base:

$$10^4 = \underbrace{10 \times 10 \times 10 \times 10}_{\text{Four factors of 10}} = 10{,}000.$$

Digits This Iranian stamp should remind us that counting on fingers (and toes) is an age-old practice. In fact, our word *digit,* referring to the numerals 0–9, comes from a Latin word for "finger" (or "toe"). It seems reasonable to connect so natural a counting method with the fact that number bases of five, ten, or twenty are the most frequent in human cultures. Aristotle first noted the relationships between fingers and base ten in Greek numeration. Anthropologists go along with the notion. Some cultures, however, have used two, three, or four as number bases, for example, counting on the joints of the fingers or the spaces between them.

In the same way, $10^2 = 10 \times 10 = 100$, $10^6 = 10 \times 10 \times 10 \times 10 \times 10 \times 10 = 1{,}000{,}000$, and so on. The base does not have to be 10; for example,

$$4^3 = 4 \times 4 \times 4 = 64, \qquad 2^2 = 2 \times 2 = 4,$$
$$3^5 = 3 \times 3 \times 3 \times 3 \times 3 = 243,$$

and so on. Expressions of this type are defined in general as follows.

Exponential Expressions

For any number a and any counting number m,

$$a^m = \underbrace{a \times a \times a \cdots \times a}_{m \text{ factors of } a}.$$

The number a is the **base,** m is the **exponent,** and a^m is read "a to the power m."

EXAMPLE 1 Find each power.

(a) $10^3 = 10 \times 10 \times 10 = 1{,}000$
(10^3 is read "10 cubed," or "10 to the third power.")

(b) $7^2 = 7 \times 7 = 49$
(7^2 is read "7 squared," or "7 to the second power.")

(c) $5^4 = 5 \times 5 \times 5 \times 5 = 625$
(5^4 is read "5 to the fourth power.")

To simplify work with exponents, it is agreed that $a^0 = 1$ for any nonzero number a. By this agreement, $7^0 = 1$, $52^0 = 1$, and so on. At the same time, $a^1 = a$ for any number a. For example, $8^1 = 8$, and $25^1 = 25$. The exponent 1 is usually omitted.

With the use of exponents, numbers can be written in **expanded form** in which the value of the digit in each position is made clear. For example, write 924 in expanded form by thinking of 924 as nine 100s plus two 10s plus four 1s, or

$$924 = 900 + 20 + 4$$
$$924 = (9 \times 100) + (2 \times 10) + (4 \times 1).$$

By the definition of exponents, $100 = 10^2$, $10 = 10^1$, and $1 = 10^0$. Use these exponents to write 924 in expanded form as follows:

$$924 = (9 \times 10^2) + (2 \times 10^1) + (4 \times 10^0).$$

EXAMPLE 2 The following are written in expanded form.

(a) $1{,}906 = (1 \times 10^3) + (9 \times 10^2) + (0 \times 10^1) + (6 \times 10^0)$

Since $0 \times 10^1 = 0$, this term could be omitted, but the form is clearer with it included.

(b) $46{,}424 = (4 \times 10^4) + (6 \times 10^3) + (4 \times 10^2) + (2 \times 10^1) + (4 \times 10^0)$

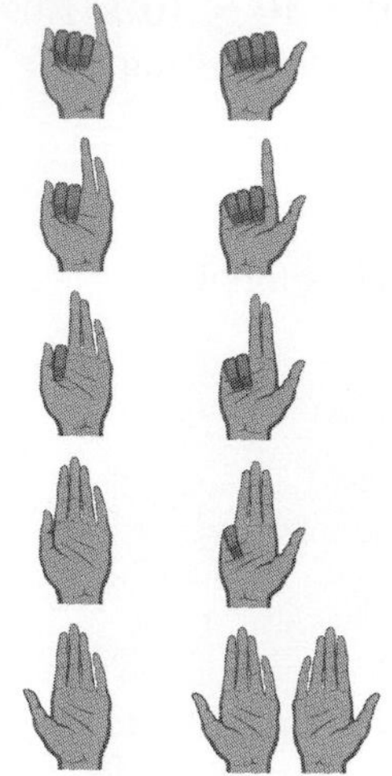

Finger Counting Many people of the world have learned to count without using the Hindu-Arabic digits 1, 2, 3, 4, 5, 6, 7, 8, 9, 0. The first digits people used were their fingers. In Africa the Zulu used the method shown here to count to ten. They started on the left hand with palm up and fist closed. The Zulu finger positions for 1–5 are shown above on the left. The Zulu finger positions for 6–10 are shown on the right.

EXAMPLE 3 Each of the following expansions is simplified.

(a) $(3 \times 10^5) + (2 \times 10^4) + (6 \times 10^3) + (8 \times 10^2) + (7 \times 10^1) + (9 \times 10^0) = 326{,}879$

(b) $(2 \times 10^1) + (8 \times 10^0) = 28$ ■

Expanded notation can be used to see why standard algorithms for addition and subtraction really work. The key idea behind these algorithms is based on the **distributive property,** which will be discussed more fully later in this chapter. It can be written in one form as follows.

Distributive Property

For all real numbers a, b, and c,

$$(b \times a) + (c \times a) = (b + c) \times a.$$

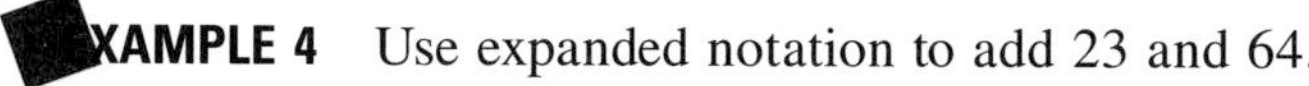

For example,

$$\begin{aligned}(3 \times 10^4) + (2 \times 10^4) &= (3 + 2) \times 10^4 \\ &= 5 \times 10^4.\end{aligned}$$

EXAMPLE 4 Use expanded notation to add 23 and 64.

$$\begin{array}{rcl} 23 &=& (2 \times 10^1) + (3 \times 10^0) \\ \underline{+\ 64} &=& \underline{(6 \times 10^1) + (4 \times 10^0)} \\ && (8 \times 10^1) + (7 \times 10^0) = 87. \end{array}$$ ■

Subtraction works in much the same way.

EXAMPLE 5 Find $695 - 254$.

$$\begin{array}{rcl} 695 &=& (6 \times 10^2) + (9 \times 10^1) + (5 \times 10^0) \\ \underline{-\ 254} &=& \underline{(2 \times 10^2) + (5 \times 10^1) + (4 \times 10^0)} \\ && (4 \times 10^2) + (4 \times 10^1) + (1 \times 10^0) = 441 \end{array}$$ ■

Expanded notation and the distributive property can also be used to show how to solve addition problems where a power of 10 ends up with a multiplier of more than one digit.

EXAMPLE 6 Use expanded notation to add 75 and 48.

$$\begin{array}{rcl} 75 &=& (7 \times 10^1) + \ (5 \times 10^0) \\ \underline{+\ 48} &=& \underline{(4 \times 10^1) + \ (8 \times 10^0)} \\ && (11 \times 10^1) + (13 \times 10^0) \end{array}$$

Since the units position (10^0) has room for only one digit, 13×10^0 must be modified:

$$\begin{aligned}13 \times 10^0 &= (10 \times 10^0) + (3 \times 10^0) \qquad \text{Distributive property} \\ &= (1 \times 10^1) + (3 \times 10^0).\end{aligned}$$

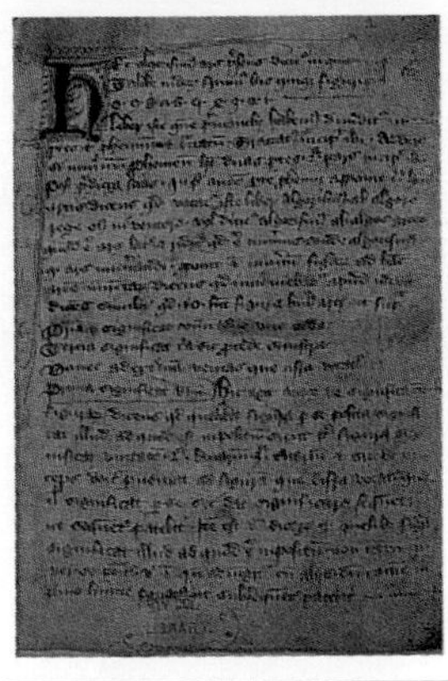

The ***Carmen de Algorismo*** (opening verses shown here) by Alexander de Villa Dei, thirteenth century, popularized the new art of "algorismus":

> *. . . from these twice five figures*
> *0 9 8 7 6 5 4 3 2 1*
> *of the Indians we benefit . . .*

The *Carmen* related that Algor, an Indian king, invented the art. But actually, "algorism" (or "algorithm") comes in a roundabout way from the name Muhammad ibn Musa al-Khorârizmi, an Arabian mathematician of the ninth century, whose arithmetic book was translated into Latin. Furthermore, this Muhammad's book on equations, *Hisab al-jabr w'almuqâbalah,* yielded the term "algebra" in a similar way.

In effect, the 1 from 13 moved to the left from the units position to the tens position. This is called "carrying." Now our sum is

$$\begin{aligned}
&\underbrace{(11 \times 10^1) + (1 \times 10^1)} + (3 \times 10^0) \\
&= (12 \times 10^1) + (3 \times 10^0) && \text{Distributive property} \\
&= (10 \times 10^1) + (2 \times 10^1) + (3 \times 10^0) \\
&= (1 \times 10^2) + (2 \times 10^1) + (3 \times 10^0) \\
&= 123.
\end{aligned}$$

Subtraction problems often require "borrowing," which can also be clarified with expanded notation.

EXAMPLE 7 Use expanded notation to subtract 186 from 364.

$$\begin{aligned}
364 &= (3 \times 10^2) + (6 \times 10^1) + (4 \times 10^0) \\
-\ 186 &= (1 \times 10^2) + (8 \times 10^1) + (6 \times 10^0)
\end{aligned}$$

Since, in the units position, we cannot subtract 6 from 4, we modify the top expansion as follows (the units position borrows from the tens position):

$$\begin{aligned}
&(3 \times 10^2) + (6 \times 10^1) + (4 \times 10^0) \\
&= (3 \times 10^2) + \underbrace{(5 \times 10^1) + (1 \times 10^1)} + (4 \times 10^0) && \text{Distributive property} \\
&= (3 \times 10^2) + (5 \times 10^1) + \underbrace{(10 \times 10^0) + (4 \times 10^0)} \\
&= (3 \times 10^2) + (5 \times 10^1) + (14 \times 10^0). && \text{Distributive property}
\end{aligned}$$

(We can now subtract 6 from 14 in the units position, but cannot take 8 from 5 in the tens position, so we continue the modification, borrowing from the hundreds to the tens position.)

$$\begin{aligned}
&(3 \times 10^2) + (5 \times 10^1) + (14 \times 10^0) \\
&= \underbrace{(2 \times 10^2) + (1 \times 10^2)} + (5 \times 10^1) + (14 \times 10^0) && \text{Distributive property} \\
&= (2 \times 10^2) + \underbrace{(10 \times 10^1) + (5 \times 10^1)} + (14 \times 10^0) \\
&= (2 \times 10^2) + (15 \times 10^1) + (14 \times 10^0) && \text{Distributive property}
\end{aligned}$$

Now we can complete the subtraction.

$$\begin{array}{rrrrr}
 & (2 \times 10^2) + & (15 \times 10^1) + & (14 \times 10^0) & \\
- & (1 \times 10^2) + & (8 \times 10^1) + & (6 \times 10^0) & \\
\hline
 & (1 \times 10^2) + & (7 \times 10^1) + & (8 \times 10^0) & = 178
\end{array}$$

Palmtop computers are the latest in the development of calculating devices.

Examples 4 through 7 used expanded notation and the distributive property to clarify our usual addition and subtraction methods. In practice, our actual work for these four problems would appear as follows:

$$\begin{array}{r} 23 \\ +\ 64 \\ \hline 87 \end{array} \qquad \begin{array}{r} 695 \\ -\ 254 \\ \hline 441 \end{array} \qquad \begin{array}{r} {}^{1}75 \\ +\ 48 \\ \hline 123 \end{array} \qquad \begin{array}{r} {}^{2}\ {}^{15}\ {}^{1} \\ 3\ \not{6}\ {}^{1}4 \\ -\ 1\ 8\ 6 \\ \hline 1\ 7\ 8. \end{array}$$

The procedures seen in this section also work for positional systems with bases other than ten.

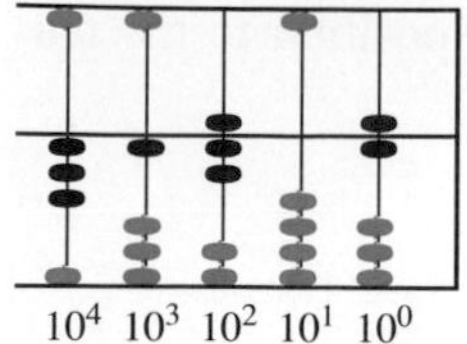

FIGURE 1

Since our numeration system is based on powers of ten, it is often called the **decimal system,** from the Latin word *decem,* meaning ten.* Over the years, many methods have been devised for speeding calculations in the decimal system. One of the oldest is the **abacus,** a device made with a series of rods with sliding beads and a dividing bar. Reading from right to left, the rods have values of 1, 10, 100, 1,000, and so on. The bead above the bar has five times the value of those below. Beads moved *toward* the bar are in the "active" position, and those toward the frame are ignored.

In our illustrations of abaci (plural form of abacus), such as in Figure 1, the activated beads are shown in black for emphasis.

EXAMPLE 8 The number on the abacus in Figure 1 is found as follows:

$$\begin{aligned}&(3 \times 10{,}000) + (1 \times 1{,}000) + [(1 \times 500) + (2 \times 100)] + 0\\&\quad + [(1 \times 5) + (1 \times 1)]\\&= 30{,}000 + 1{,}000 + 500 + 200 + 0 + 5 + 1\\&= 31{,}706\end{aligned}$$

As paper became more readily available, people gradually switched from devices like the abacus (though these are still commonly used in some areas) to paper-and-pencil methods of calculation. One early scheme, used both in India and Persia, was the **lattice method,** which arranged products of single digits into a diagonalized lattice, as shown in the following example.

Merry Math The following two rhymes come from *Marmaduke Multiply's Merry Method of Making Minor Mathematicians,* a primer published in the late 1830s in Boston:

1. Twice 2 are 4. Pray hasten on before.
2. Five times 5 are 25. I thank my stars I'm yet alive.

EXAMPLE 9 Find the product 38 × 794 by the lattice method.

Work as follows.

Step 1: Write the problem, with one number at the side and one across the top.

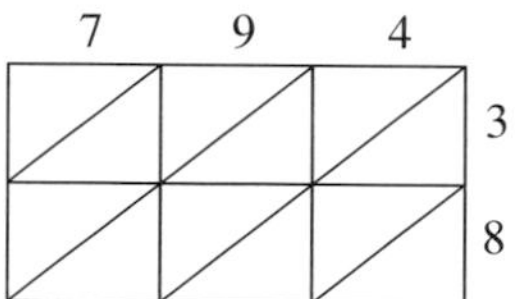

Step 2: Within the lattice, write the products of all pairs of digits from the top and side.

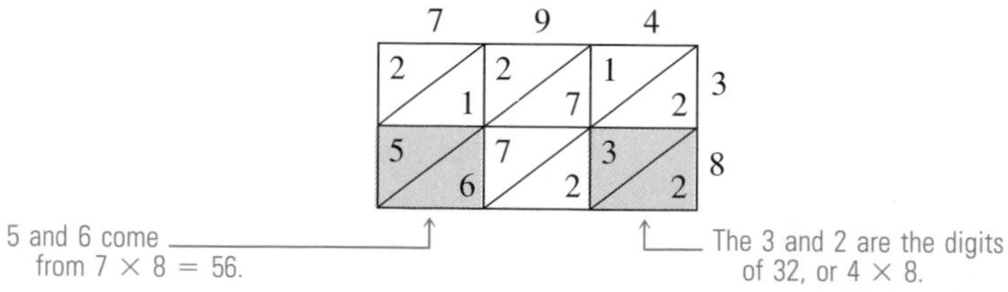

**December* was the tenth month in an old form of the calendar. It is interesting to note that *decem* became *dix* in the French language; a ten-dollar bill, called "dixie," was in use in New Orleans before the Civil War. And "Dixie Land" was a nickname for that city before Dixie came to refer to all the Southern states, as in Daniel D. Emmett's song, written in 1859.

John Napier's most significant mathematical contribution, developed over a period of at least 20 years, was the concept of *logarithms,* which, among other things, allow multiplication and division to be accomplished with addition and subtraction, a great computational advantage given the state of mathematics at the time (1614).

Napier himself regarded his interest in mathematics as a recreation, his main involvements being political and religious. A supporter of John Knox and James I, he published a widely read anti-Catholic work which analyzed the Biblical book of Revelation and concluded that the Pope was the Antichrist and that the Creator would end the world between 1688 and 1700. Napier was one of many who, over the years, have miscalculated the end of the world. One of the most recent was a former NASA rocket engineer, Edgar C. Whisenant, who wrote the book *88 Reasons Why the Rapture Will Be in 1988.* Apparently we have not yet perfected our mathematics to the point of being able to calculate all the secrets of the Creator.

Step 3: Starting at the right of the lattice add diagonally, carrying as necessary.

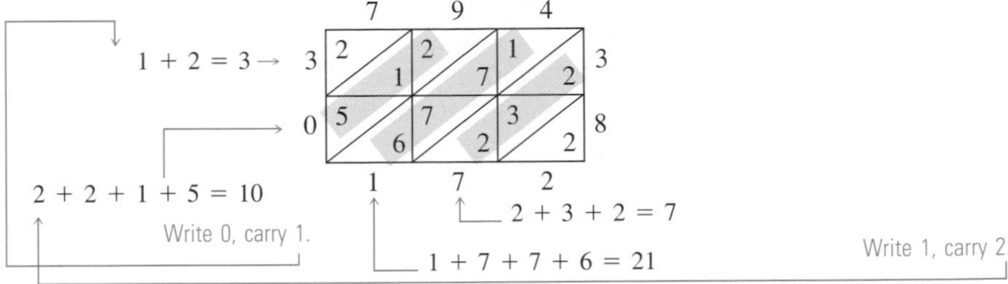

Step 4: Read the answer around the left side and bottom:

$$38 \times 794 = 30{,}172.$$

The Scottish mathematician John Napier (1550–1617) introduced a significant calculating tool called **Napier's rods,** or **Napier's bones.** Napier's invention, based on the lattice method of multiplication, is widely acknowledged as a very early forerunner of modern computers. It consisted of a set of strips, several for each digit 0 through 9, on which the multiples of each digit appeared in a sort of lattice column. See Figure 2.

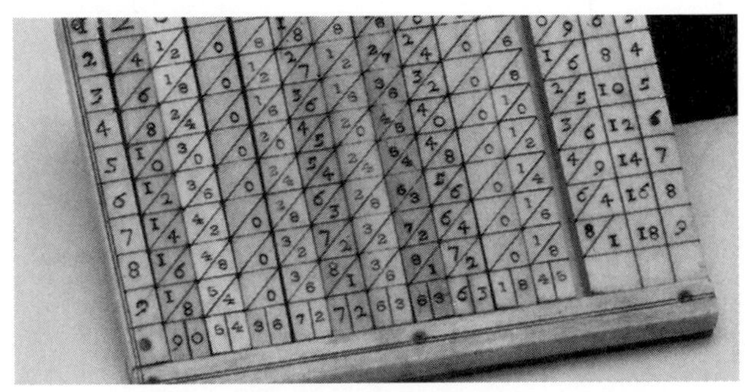

Index	0	1	2	3	4	5	6	7	8	9
1	0/0	0/1	0/2	0/3	0/4	0/5	0/6	0/7	0/8	0/9
2	0/0	0/2	0/4	0/6	0/8	1/0	1/2	1/4	1/6	1/8
3	0/0	0/3	0/6	0/9	1/2	1/5	1/8	2/1	2/4	2/7
4	0/0	0/4	0/8	1/2	1/6	2/0	2/4	2/8	3/2	3/6
5	0/0	0/5	1/0	1/5	2/0	2/5	3/0	3/5	4/0	4/5
6	0/0	0/6	1/2	1/8	2/4	3/0	3/6	4/2	4/8	5/4
7	0/0	0/7	1/4	2/1	2/8	3/5	4/2	4/9	5/6	6/3
8	0/0	0/8	1/6	2/4	3/2	4/0	4/8	5/6	6/4	7/2
9	0/0	0/9	1/8	2/7	3/6	4/5	5/4	6/3	7/2	8/1

FIGURE 2

An additional strip, called the index, could be laid beside any of the others to indicate the multiplier at each level. Napier's rods were used for mechanically multiplying, dividing, and taking square roots. Figure 3 shows how to multiply 2,806 by 7. Select the rods for 2, 8, 0, and 6, placing them side by side. Then using the index, locate the level for a multiplier of 7. The resulting lattice, shown at the bottom of the figure, gives the product 19,642.

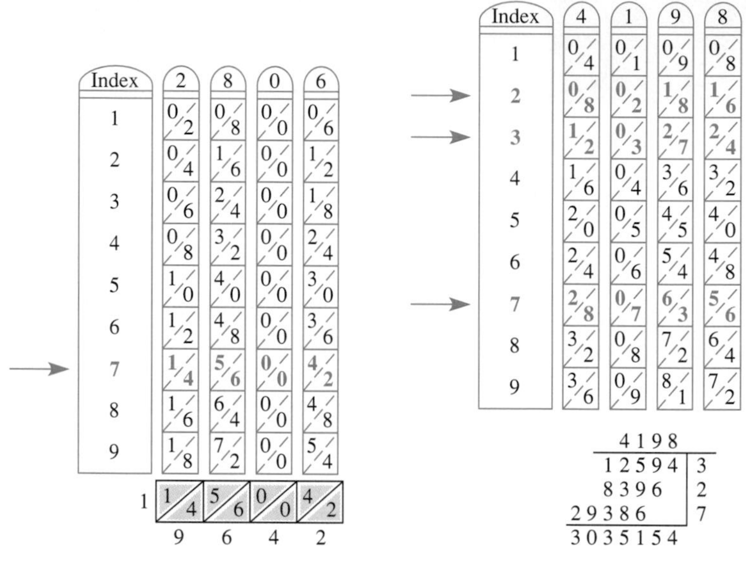

FIGURE 3

FIGURE 4

EXAMPLE 10 Use Napier's rods to find the product of 723 and 4,198.

We line up the rods for 4, 1, 9, and 8 next to the index as in Figure 4. The product $3 \times 4{,}198$ is found as described in Example 9 and written at the bottom of the figure. Then $2 \times 4{,}198$ is found similarly and written below, shifted one place to the left (why?). Finally, the product $7 \times 4{,}198$ is written shifted two places to the left. The final answer is found by addition to obtain

$$723 \times 4{,}198 = 3{,}035{,}154.$$

One other paper-and-pencil method of multiplication is the **Russian peasant method,** which is similar to the Egyptian method of doubling explained in the previous section. (In fact both of these methods work, in effect, by expanding one of the numbers to be multiplied, but in base two rather than in base ten. Base two numerals are discussed in the next section.) To multiply 37 and 42 by the Russian peasant method, make two columns headed by 37 and 42. Form the first column by dividing 37 by 2 again and again, ignoring any remainders. Stop when 1 is obtained. Form the second column by doubling each number down the column.

	37	42	
	18	84	
Divide by 2, ignoring remainders.	9	168	Double each number.
	4	336	
	2	672	
	1	1,344	

Now add up only the second column numbers that correspond to odd numbers in the first column. Omit those corresponding to even numbers in the first column.

→	**37**	**42**	←
	18	84	
Odd	**9**	**168**	← Add
numbers →	4	336	these
	2	672	numbers.
→	**1**	**1,344**	←

Finally, $37 \times 42 = 42 + 168 + 1{,}344 = 1{,}554$.

For Further Thought

The abacus has been (and still is) used to perform very rapid calculations. A simple example is adding 526 and 362. Start with 526 on the abacus:

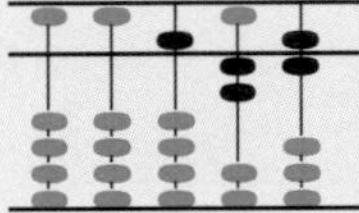

To add 362, start by "activating" an additional 2 on the 1s rod:

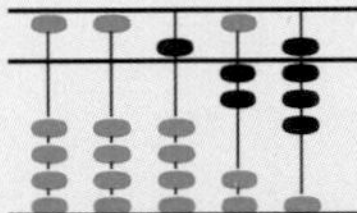

Next, activate an additional 6 on the 10s rod:

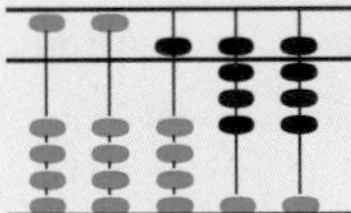

Finally, activate an additional 3 on the 100s rod:

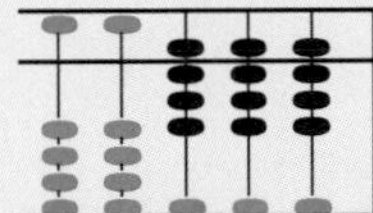

The sum, read from the abacus, is 888.

For problems where carrying or borrowing are required, it takes a little more thought and skill. Try to obtain an actual abacus (or, otherwise, make sketches) and practice some addition and subtraction problems until you can do them quickly.

For Group Discussion

1. Use an abacus to add: 13,728 + 61,455. Explain each step of your procedure.
2. Use an abacus to subtract: 6,512 − 4,816. Again, explain each step of your procedure.

4.2 EXERCISES

Write each number in expanded form.

1. 37 **2.** 814 **3.** 2,815 **4.** 15,504

5. three thousand, six hundred twenty-eight

6. fifty-three thousand, eight hundred twelve

7. thirteen million, six hundred six thousand, ninety

8. one hundred twelve million, fourteen thousand, one hundred twelve

Simplify each of the following expansions.

9. $(7 \times 10^1) + (3 \times 10^0)$

10. $(2 \times 10^2) + (6 \times 10^1) + (0 \times 10^0)$

11. $(5 \times 10^3) + (0 \times 10^2) + (7 \times 10^1) + (2 \times 10^0)$

12. $(4 \times 10^5) + (0 \times 10^4) + (7 \times 10^3) + (7 \times 10^2) + (5 \times 10^1) + (2 \times 10^0)$

13. $(5 \times 10^7) + (6 \times 10^5) + (2 \times 10^3) + (3 \times 10^0)$

14. $(6 \times 10^8) + (5 \times 10^7) + (1 \times 10^2) + (4 \times 10^0)$

In each of the following, add in expanded notation.

15. 63 + 26 **16.** 693 + 305

In each of the following, subtract in expanded notation.

17. 84 − 52 **18.** 673 − 412

Perform each addition using expanded notation.

19. 65 + 44 **20.** 536 + 279 **21.** 424 + 298 **22.** 6,755 + 4,827

Perform each subtraction using expanded notation.

23. 53 − 47 **24.** 253 − 48 **25.** 643 − 436 **26.** 826 − 345

Identify the number represented on each of these abaci.

27. **29.**

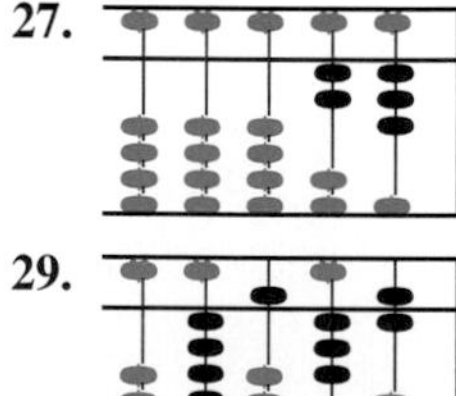

28.

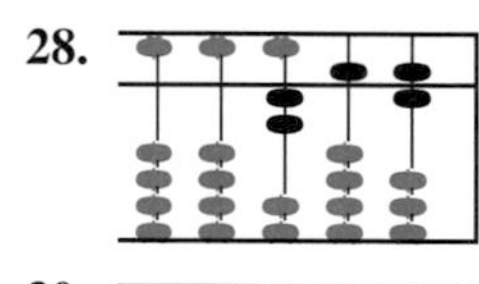

30.

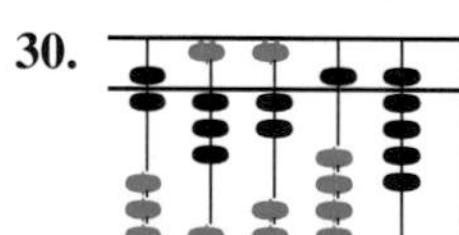

Sketch an abacus to show each number.

31. 38 **32.** 183 **33.** 2,547 **34.** 70,163

Use the lattice method to find each product.

35. 63×28 **36.** 29×635 **37.** 413×68 **38.** 845×396

Refer to Example 10, where Napier's rods were used to find the product of 723 *and* 4,198. *Then complete Exercises 39 and 40.*

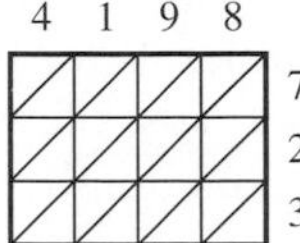

39. Find the product of 723 and 4,198 by completing the lattice process shown here.

40. Explain how Napier's rods could have been used in Example 10 to set up one complete lattice product rather than adding three individual (shifted) lattice products.

Make use of Napier's rods (Figure 2) to find each product.

41. 8×62 **42.** 32×73 **43.** $26 \times 8{,}354$ **44.** $526 \times 4{,}863$

Use the Russian peasant method to find each product.

45. 5×82 **46.** 41×33 **47.** 62×429 **48.** 135×63

4.3 Converting Between Number Bases

Most of us can work with decimal numbers effectively, having used them all our lives. But that doesn't mean we necessarily have a deep understanding of the system we use so well. You may immediately recognize that the digit 4 in the numeral 1,473 denotes 4 "hundreds," but that may not be because you know the third digit from the right represents second powers of the base. By writing numbers in unfamiliar bases, and converting from one base to another, you can come to appreciate the nature of a positional system and realize that a base of ten is a choice, not a necessity.

Although the numeration systems discussed in the opening section were all base ten, other bases have occurred historically. For example, the ancient Babylonians used 60 as their base. The Mayan Indians of Central America and Mexico used 20. In this section we consider bases other than ten, but we use the familiar Hindu-Arabic symbols. We will consistently indicate bases other than ten with a spelled-out subscript, as in the numeral 43_{five}. Whenever a number appears without a subscript, it is to be assumed that the intended base is ten. It will help to be careful how you read (or verbalize) numerals here. The numeral 43_{five} is read "four three base five." (Do *not* read it as "forty-three," as that terminology implies base ten and names a totally different number.)

For reference in doing number expansions and base conversions, Table 3 gives the first several powers of some numbers used as alternate bases in this section.

TABLE 3 Selected Powers of Some Alternate Number Bases

	Fourth Power	Third Power	Second Power	First Power	Zero Power
Base two	16	8	4	2	1
Base five	625	125	25	5	1
Base seven	2,401	343	49	7	1
Base eight	4,096	512	64	8	1
Base sixteen	65,536	4,096	256	16	1

TABLE 4

Base Ten	Base Five
0	0
1	1
2	2
3	3
4	4
5	10
6	11
7	12
8	13
9	14
10	20
11	21
12	22
13	23
14	24
15	30
16	31
17	32
18	33
19	34
20	40
21	41
22	42
23	43
24	44
25	100
26	101
27	102
28	103
29	104
30	110

We begin our illustrations with the base five system, which requires just five distinct symbols, 0, 1, 2, 3, and 4. Table 4 compares the base five and decimal (base ten) numerals for the whole numbers 0 through 30. Notice that, while the base five system uses fewer distinct symbols, it normally requires more digits to denote the same number.

EXAMPLE 1 Convert $1{,}342_{\text{five}}$ to decimal form.

Referring to the powers of five in Table 3, we see that this number has one 125, three 25s, four 5s, and two 1s, so

$$\begin{aligned} 1{,}342_{\text{five}} &= (1 \times 125) + (3 \times 25) + (4 \times 5) + (2 \times 1) \\ &= 125 + 75 + 20 + 2 \\ &= 222. \end{aligned}$$

A shortcut for converting from base five to decimal form, which is *particularly useful when you use a calculator,* can be derived as follows. (We can illustrate this by repeating the conversion of Example 1.)

$$1{,}342_{\text{five}} = (1 \times 5^3) + (3 \times 5^2) + (4 \times 5) + 2$$

Now 5 can be factored out of the three quantities in parentheses, so

$$1{,}342_{\text{five}} = ((1 \times 5^2) + (3 \times 5) + 4) \times 5 + 2.$$

Now, factoring another five out of the two "inner" quantities, we get

$$1{,}342_{\text{five}} = (((1 \times 5) + 3) \times 5 + 4) \times 5 + 2.$$

The inner parentheses around 1×5 are not needed since the product would be automatically done before the 3 is added. Therefore, we can write

$$1{,}342_{\text{five}} = ((1 \times 5 + 3) \times 5 + 4) \times 5 + 2.$$

This series of products and sums is easily done as an uninterrupted sequence of operations on a calculator, with no intermediate results written down. The same thing works for converting to base ten from any other base. The procedure is summarized as follows.

Calculator Shortcut

To convert from another base to decimal form: Start with the first digit on the left and multiply by the base. Then add the next digit, multiply again by the base, and so on. The last step is to add the last digit on the right. Do *not* multiply it by the base.

Trick or Tree? The octal number 31 is equal to the decimal number 25. This may be written as

31 OCT = 25 DEC

Does this mean that Halloween and Christmas fall on the same day of the year?

Exactly how you accomplish these steps depends on the type of calculator you use. With some, only the digits, the multiplications, and the additions need to be entered, in order. With others, you may need to press the = key following each addition of a digit. If you handle grouped expressions on your calculator by actually entering parentheses, then enter the expression just as illustrated above, and in the following example. (The number of left parentheses to start with will be two fewer than the number of digits in the original numeral.)

Yin-yang The binary (base two) symbols of the *I Ching,* a 2,000-year-old Chinese classic, permute into 8 elemental trigrams; 64 hexagrams are interpreted in casting oracles.

The basic symbol here is the ancient Chinese "yin-yang," in which the black and the white enfold each other, each containing a part of the other. A kind of duality is conveyed between destructive (yin) and beneficial (yang) aspects. Leibniz (1646–1716) studied Chinese ideograms in search of a universal symbolic language and promoted East-West cultural contact. He saw parallels between the trigrams and his binary arithmetic.

Niels Bohr (1885–1962), famous Danish Nobel laureate in physics (atomic theory), adopted the yin-yang symbol in his coat of arms to depict his principle of *complementarity,* which he believed was fundamental to reality at the deepest levels. Bohr also pushed for East-West cooperation.

(In its 1992 edition, The World Book Dictionary first judged "yin-yang" to have been used enough to become a permanent part of our ever-changing language, assigning it the definition, "made up of opposites.")

EXAMPLE 2 Use the calculator shortcut to convert $244{,}314_{\text{five}}$ to decimal form.

$$\begin{aligned} 244{,}314_{\text{five}} &= ((((2 \times 5 + 4) \times 5 + 4) \times 5 + 3) \times 5 + 1) \times 5 + 4 \\ &= 9{,}334. \end{aligned}$$

Knowledge of the base five place values (the powers of five, as in Table 3) enables us to convert from decimal form to base five as in the next example.

EXAMPLE 3 Convert 497 from decimal form to base five.

The base five place values, starting from the right, are 1, 5, 25, 125, 625, and so on. Since 497 is between 125 and 625, it will require no 625s, but some 125s, as well as possibly some 25s, 5s, and 1s. Dividing 497 by 125 determines the proper number of 125s. The quotient is 3, with remainder 122. So we need three 125s. Next, the remainder, 122, is divided by 25 (the next place value) to find the proper number of 25s. The quotient is 4, with remainder 22, so we need four 25s. Dividing 22 by 5 yields 4, with remainder 2. So we need four 5s. Dividing 2 by 1 yields 2 (with remainder 0), so we need two 1s. Finally, we see that 497 consists of three 125s, four 25s, four 5s, and two 1s, so $497 = 3{,}442_{\text{five}}$.

More concisely, this process can be written as follows.

$$\begin{aligned} 497 \div 125 &= 3 && \text{Remainder 122} \\ 122 \div 25 &= 4 && \text{Remainder 22} \\ 22 \div 5 &= 4 && \text{Remainder 2} \\ 2 \div 1 &= 2 && \text{Remainder 0} \\ 497 &= 3{,}442_{\text{five}} \end{aligned}$$

Check the answer:

$$\begin{aligned} 3{,}442_{\text{five}} &= (3 \times 125) + (4 \times 25) + (4 \times 5) + (2 \times 1) \\ &= 375 + 100 + 20 + 2 \\ &= 497. \end{aligned}$$

The calculator shortcut for converting from another base to decimal form involved repeated *multiplications* by the other base. (See Example 2.) A shortcut for converting from decimal form to another base makes use of repeated *divisions* by the other base. Just divide the original decimal numeral, and the resulting quotients in turn, by the desired base until the quotient 0 appears.

EXAMPLE 4 Repeat Example 3 using the shortcut just described.

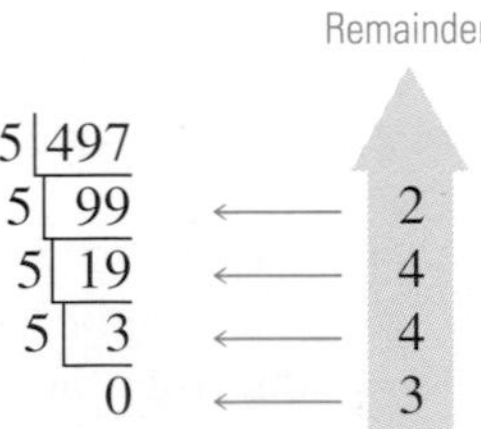

Read the answer from the remainder column, reading from the bottom up:

$$497 = 3{,}442_{\text{five}}.$$

Woven fabric is a binary system of threads going lengthwise (warp threads—white in the diagram above) and threads going crosswise (weft, or woof). At any point in a fabric, either warp or weft is on top, and the variation creates the pattern.

Weaving is done on a loom, and there must be some way to lift the warp threads wherever the pattern dictates. Nineteenth-century looms operated using punched cards, "programmed" for pattern. The looms were set up with hooked needles, the hooks holding the warp. Where there were holes in cards, the needles moved, the warp lifted, the weft passed under. Where no holes were, the warp did not lift, and the weft was on top. The system parallels the on-off system in calculators and computers. In fact, the looms described here were models in the development of modern calculating machinery.

Joseph Marie Jacquard (1752–1823) is credited with improving the mechanical loom so that mass production of fabric was feasible.

To see why this shortcut works, notice the following:

The first division shows that four hundred ninety-seven 1s are equivalent to ninety-nine 5s and two 1s. (The two 1s are set aside and account for the last digit of the answer.)

The second division shows that ninety-nine 5s are equivalent to nineteen 25s and four 5s. (The four 5s account for the next digit of the answer.)

The third division shows that nineteen 25s are equivalent to three 125s and four 25s. (The four 25s account for the next digit of the answer.)

The fourth (and final) division shows that the three 125s are equivalent to no 625s and three 125s. The remainders, as they are obtained *from top to bottom,* give the number of 1s, then 5s, then 25s, then 125s.

The methods for converting between bases ten and five, including the shortcuts, can be adapted for conversions between base ten and any other base, as illustrated in the following examples.

EXAMPLE 5 Convert $6{,}343_{\text{seven}}$ to decimal form, by expanding in powers, and by using the calculator shortcut.

$$\begin{aligned} 6{,}343_{\text{seven}} &= (6 \times 7^3) + (3 \times 7^2) + (4 \times 7^1) + (3 \times 7^0) \\ &= (6 \times 343) + (3 \times 49) + (4 \times 7) + (3 \times 1) \\ &= 2{,}058 + 147 + 28 + 3 \\ &= 2{,}236 \end{aligned}$$

Calculator shortcut:

$$6{,}343_{\text{seven}} = ((6 \times 7 + 3) \times 7 + 4) \times 7 + 3 = 2{,}236.$$

EXAMPLE 6 Convert 7,508 to base seven.

Divide 7,508 by 7, then divide the resulting quotient by 7, and so on, until a quotient of 0 results.

		Remainder
7 | 7,508		
7 | 1,072	⟵	4
7 | 153	⟵	1
7 | 21	⟵	6
7 | 3	⟵	0
0	⟵	3

From the remainders, reading bottom to top, $7{,}508 = 30{,}614_{\text{seven}}$.

Because we are accustomed to doing arithmetic in base ten, most of us would handle conversions between arbitrary bases (where neither is ten) by going from the given base to base ten and then to the desired base. This method is illustrated in the next example.

EXAMPLE 7 Convert $3{,}164_{\text{seven}}$ to base five.
First convert to decimal form.

$$\begin{aligned} 3{,}164_{\text{seven}} &= (3 \times 7^3) + (1 \times 7^2) + (6 \times 7^1) + (4 \times 7^0) \\ &= (3 \times 343) + (1 \times 49) + (6 \times 7) + (4 \times 1) \\ &= 1{,}029 + 49 + 42 + 4 \\ &= 1{,}124 \end{aligned}$$

Next convert this decimal result to base five.

		Remainder
5	1,124	
5	224	4
5	44	4
5	8	4
5	1	3
	0	1

From the remainders, $3{,}164_{\text{seven}} = 13{,}444_{\text{five}}$.

Computer Mathematics

There are three alternative base systems that are most useful in computer applications. These are the **binary** (base two), **octal** (base eight), and **hexadecimal** (base sixteen) systems. Computers and handheld calculators actually use the binary system for their internal calculations since that system consists of only two symbols, 0 and 1. All numbers can then be represented by electronic "switches," of one kind or another, where "on" indicates 1 and "off" indicates 0. The octal system is used extensively by programmers who work with internal computer codes. In a computer, the CPU (central processing unit) often uses the hexadecimal system to communicate with a printer or other output device.

The binary system is extreme in that it has only two available symbols (0 and 1); because of this, representing numbers in binary form requires more digits than in any other base. Table 5 shows the whole numbers up to 20 expressed in binary form.

Conversions between any of these three special base systems (binary, octal, and hexadecimal) and the decimal system can be done by the methods already discussed, including the shortcut methods.

TABLE 5

Base Ten (decimal)	Base Two (binary)
0	0
1	1
2	10
3	11
4	100
5	101
6	110
7	111
8	1,000
9	1,001
10	1,010
11	1,011
12	1,100
13	1,101
14	1,110
15	1,111
16	10,000
17	10,001
18	10,010
19	10,011
20	10,100

EXAMPLE 8 Convert $110{,}101_{\text{two}}$ to decimal form, by expanding in powers, and by using the calculator shortcut.

$$\begin{aligned} 110{,}101_{\text{two}} &= (1 \times 2^5) + (1 \times 2^4) + (0 \times 2^3) + (1 \times 2^2) \\ &\quad + (0 \times 2^1) + (1 \times 2^0) \\ &= (1 \times 32) + (1 \times 16) + (0 \times 8) + (1 \times 4) + (0 \times 2) \\ &\quad + (1 \times 1) \\ &= 32 + 16 + 0 + 4 + 0 + 1 \\ &= 53 \end{aligned}$$

Calculator shortcut:

$$110{,}101_{\text{two}} = ((((1 \times 2 + 1) \times 2 + 0) \times 2 + 1) \times 2 + 0) \times 2 + 1$$
$$= 53.$$

EXAMPLE 9 Convert 9,583 to octal form.

Divide repeatedly by 8, writing the remainders at the side.

		Remainder
8 | 9,583		
8 | 1,197	⟵	7
8 | 149	⟵	5
8 | 18	⟵	5
8 | 2	⟵	2
0	⟵	2

From the remainders, $9{,}583 = 22{,}557_{\text{eight}}$.

The hexadecimal system, having base 16, which is greater than 10, presents a new problem. Since distinct symbols are needed for every whole number from 0 up to one less than the base, base sixteen requires more symbols than are normally used in our decimal system. Computer programmers commonly use the letters A, B, C, D, E, and F as hexadecimal digits for the numbers ten through fifteen, respectively.

Converting Calculators A number of scientific calculators are available that will convert between decimal, binary, octal, and hexadecimal, and will also do calculations directly in all of these separate modes.

EXAMPLE 10 Convert $\text{FA5}_{\text{sixteen}}$ to decimal form.

Since the hexadecimal digits F and A represent 15 and 10, respectively,

$$\begin{aligned}\text{FA5}_{\text{sixteen}} &= (15 \times 16^2) + (10 \times 16^1) + (5 \times 16^0)\\ &= 3{,}840 + 160 + 5\\ &= 4{,}005.\end{aligned}$$

EXAMPLE 11 Convert 748 from decimal form to hexadecimal form.

Use repeated division by 16.

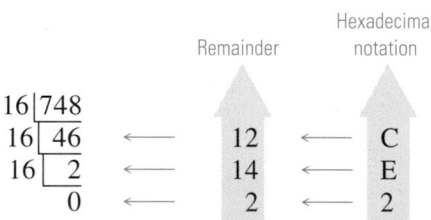

		Remainder		Hexadecimal notation
16 | 748				
16 | 46	⟵	12	⟵	C
16 | 2	⟵	14	⟵	E
0	⟵	2	⟵	2

From the remainders at the right, $748 = 2\text{EC}_{\text{sixteen}}$.

The decimal whole numbers 0 through 17 are shown in Table 6 along with their equivalents in the common computer-oriented bases (two, eight, and sixteen). Conversions among binary, octal, and hexadecimal systems can generally

TABLE 6 Some Decimal Equivalents in the Common Computer-Oriented Bases

Decimal (Base Ten)	Hexadecimal (Base Sixteen)	Octal (Base Eight)	Binary (Base Two)
0	0	0	0
1	1	1	1
2	2	2	10
3	3	3	11
4	4	4	100
5	5	5	101
6	6	6	110
7	7	7	111
8	8	10	1,000
9	9	11	1,001
10	A	12	1,010
11	B	13	1,011
12	C	14	1,100
13	D	15	1,101
14	E	16	1,110
15	F	17	1,111
16	10	20	10,000
17	11	21	10,001

be accomplished by the shortcuts explained below, and are illustrated in the next several examples.

The binary system is the natural one for internal computer workings because of its compatibility with the two-state electronic switches. It is very cumbersome, however, for human use, because so many digits occur even in the numerals for relatively small numbers. The octal and hexadecimal systems are the choices of computer programmers mainly because of their close relationship with the binary system. *Both eight and sixteen are powers of two.* And when base conversions involve one base that is a power of the other, there is a quick conversion shortcut available. For example, since $8 = 2^3$, every octal digit (0 through 7) can be expressed as a 3-digit binary numeral. See Table 7.

TABLE 7

Octal	Binary
0	000
1	001
2	010
3	011
4	100
5	101
6	110
7	111

EXAMPLE 12 Convert 473_{eight} to binary form.

Replace each octal digit with its 3-digit binary equivalent. (Leading zeros can be omitted only when they occur in the leftmost group.) Then combine all the binary equivalents into a single binary numeral.

$$\begin{array}{ccc} 4 & 7 & 3_{\text{eight}} \\ \downarrow & \downarrow & \downarrow \\ 100 & 111 & 011_{\text{two}} \end{array}$$

By this method, $473_{\text{eight}} = 100{,}111{,}011_{\text{two}}$. ■

TABLE 8

Hexadecimal	Binary
0	0000
1	0001
2	0010
3	0011
4	0100
5	0101
6	0110
7	0111
8	1000
9	1001
A	1010
B	1011
C	1100
D	1101
E	1110
F	1111

Convert from binary form to octal form in a similar way. Start at the right and break the binary numeral into groups of three digits. (Leading zeros in the leftmost group may be omitted.)

EXAMPLE 13 Convert $10{,}011{,}110_{\text{two}}$ to octal form.

Starting at the right, break the digits into groups of three. Then convert the groups to their octal equivalents.

$$\begin{array}{ccc} 10 & 011 & 110_{\text{two}} \\ \downarrow & \downarrow & \downarrow \\ 2 & 3 & 6_{\text{eight}} \end{array}$$

Finally, $10{,}011{,}110_{\text{two}} = 236_{\text{eight}}$. ■

Since $16 = 2^4$, every hexadecimal digit can be equated to a 4-digit binary numeral (see Table 8), and conversions between binary and hexadecimal forms can be done in a manner similar to that used in Examples 12 and 13.

For Further Thought

Katie Reynolds, produce buyer for a supermarket, has received five large bins filled with bags of carrots. All bins should be filled with 10-pound bags of carrots, but she has learned that, through an error, some of the bins are filled with 9-pound bags. Katie's assistant, Dusty Rainbolt, claims that he can pile some bags from all five bins together on the scale and, from just a single weighing of them, tell exactly which bins have the 9-pound bags.

For Group Discussion

1. Remembering binary notation, verify the following: Every counting number can be broken down into ones, twos, fours, eights, sixteens, and so on, with no more than one of each power of two needed.
2. Explain how Dusty can make good on his claim.

Several years ago, the Kellogg Company featured a Magic Trick Age Detector activity on specially marked packages of *Kellogg's® Rice Krispies®* cereal. The trick is simply an extension of the discussion in the text.

Kellogg's® Rice Krispies® and characters *Snap!® Crackle!®* and *Pop!®* are registered trademarks of Kellogg Company.

EXAMPLE 14 Convert $8{,}B4F_{\text{sixteen}}$ to binary form.

Each hexadecimal digit yields a 4-digit binary equivalent.

$$\begin{array}{cccc} 8 & B & 4 & F_{\text{sixteen}} \\ \downarrow & \downarrow & \downarrow & \downarrow \\ 1000 & 1011 & 0100 & 1111_{\text{two}} \end{array}$$

Combining these groups of digits, we see that

$$8{,}B4F_{\text{sixteen}} = 1{,}000{,}101{,}101{,}001{,}111_{\text{two}}.$$

Several games and tricks are based on the binary system. For example, Table 9 can be used to find the age of a person 31 years old or younger. The person need only tell you the columns that contain his or her age. For example, suppose Jack Cook says that his age appears in columns B and D only. To find his age, add the numbers from the top row of these columns:

Jack is 2 + 8 = 10 years old.

Do you see how this trick works? If not, you can get help in Exercises 70–73.

TABLE 9

A	B	C	D	E
1	2	4	8	16
3	3	5	9	17
5	6	6	10	18
7	7	7	11	19
9	10	12	12	20
11	11	13	13	21
13	14	14	14	22
15	15	15	15	23
17	18	20	24	24
19	19	21	25	25
21	22	22	26	26
23	23	23	27	27
25	26	28	28	28
27	27	29	29	29
29	30	30	30	30
31	31	31	31	31

4.3 EXERCISES

List the first twenty counting numbers in each of the following bases.

1. seven (Only digits 0 through 6 are used in base seven.)
2. eight (Only digits 0 through 7 are used.)
3. nine (Only digits 0 through 8 are used.)
4. sixteen (The digits 0, 1, 2, . . . , 9, A, B, C, D, E, F are used in base sixteen.)

For each of the following, write (in the same base) the counting numbers just before and just after the given number. (Do not convert to base ten.)

5. 14_{five} **6.** 555_{six} **7.** $\text{B6F}_{\text{sixteen}}$ **8.** $10{,}111_{\text{two}}$

Determine the number of distinct symbols needed in each of the following positional systems.

9. base three **10.** base seven **11.** base eleven **12.** base sixteen

Determine, in each of the following bases, the smallest and largest four-digit numbers and their decimal equivalents.

13. three **14.** sixteen

Convert each of the following to decimal form by expanding in powers and by using the calculator shortcut.

15. 24_{five} **16.** 62_{seven} **17.** $1{,}011_{\text{two}}$ **18.** 35_{eight}
19. $\text{3BC}_{\text{sixteen}}$ **20.** $34{,}432_{\text{five}}$ **21.** $2{,}366_{\text{seven}}$ **22.** $101{,}101{,}110_{\text{two}}$
23. $70{,}266_{\text{eight}}$ **24.** $\text{A,BCD}_{\text{sixteen}}$ **25.** $2{,}023_{\text{four}}$ **26.** $6{,}185_{\text{nine}}$
27. $41{,}533_{\text{six}}$ **28.** $88{,}703_{\text{nine}}$

Convert each of the following from decimal form to the given base.

29. 86 to base five **30.** 65 to base seven **31.** 19 to base two
32. 935 to base eight **33.** 147 to base sixteen **34.** 2,730 to base sixteen
35. 36,401 to base five **36.** 70,893 to base seven **37.** 586 to base two
38. 12,888 to base eight **39.** 8,407 to base three **40.** 11,028 to base four
41. 9,346 to base six **42.** 99,999 to base nine

Make the following conversions as indicated.

43. 43_{five} to base seven **44.** 27_{eight} to base five
45. $\text{C02}_{\text{sixteen}}$ to base seven **46.** $6{,}748_{\text{nine}}$ to base four

Convert each of the following from octal form to binary form.

47. 367_{eight} **48.** $2{,}406_{\text{eight}}$

Convert each of the following from binary form to octal form.

49. $100{,}110{,}111_{\text{two}}$ **50.** $11{,}010{,}111{,}101_{\text{two}}$

Make the following conversions as indicated.

51. $\text{DC}_{\text{sixteen}}$ to binary **52.** $\text{F,111}_{\text{sixteen}}$ to binary
53. $101{,}101_{\text{two}}$ to hexadecimal **54.** $101{,}111{,}011{,}101{,}000_{\text{two}}$ to hexadecimal

Identify the largest number from each list in Exercises 55 and 56.

55. 42_{seven}, 37_{eight}, $\text{1D}_{\text{sixteen}}$ **56.** $1{,}101{,}110_{\text{two}}$, 407_{five}, $\text{6F}_{\text{sixteen}}$

Some people think that twelve would be a better base than ten. This is mainly because twelve has more divisors (1, 2, 3, 4, 6, 12) *than ten* (1, 2, 5, 10), *which makes fractions easier in base twelve. The base twelve system is called the* duodecimal system. *Just as in the decimal system we speak of a one, a ten, and a hundred (and so on), in the duodecimal system we say a* one, *a* dozen *(twelve), and a* gross *(twelve squared, or one hundred forty-four).*

57. Altogether, Otis Taylor's clients ordered 9 gross, 10 dozen, and 11 copies of *Math for Base Runners* during the last week of January. How many copies was that in base ten?

One very common method of converting symbols into binary digits for computer processing is called ASCII (American Standard Code of Information Interchange). The upper case letters A *through* Z *are assigned the numbers* 65 *through* 90, *so* A *has binary code* 1000001 *and* Z *has code* 1011010. *Lowercase letters* a *through* z *have codes* 97 *through* 122 *(that is,* 1100001 *through* 1111010*). ASCII codes, as well as other numerical computer output, normally appear without commas.*

Write the binary code for each of the following letters.

58. C **59.** X **60.** k **61.** r

Break each of the following into groups of seven digits and write as letters.

62. 1001000100010110011001010000

63. 10000111001000101010110000111001011

Translate each word into an ASCII string of binary digits. (Be sure to distinguish upper and lowercase letters.)

64. New **65.** Orleans

66. Explain why the octal and hexadecimal systems are convenient for people who code for computers.

67. There are thirty-seven counting numbers whose base eight numerals contain two digits but whose base three numerals contain four digits. Find the smallest and largest of these numbers.

68. What is the smallest counting number (expressed in base ten) that would require six digits in its base nine representation?

69. In a pure decimal (base ten) monetary system, the first three denominations needed are pennies, dimes, and dollars. Name the first three denominations of a base five monetary system.

Refer to Table 9 for Exercises 70–73.

70. After observing the binary forms of the numbers 1–31, identify a common property of all Table 9 numbers in each of the following columns.

(a) Column A **(b)** Column B **(c)** Column C **(d)** Column D **(e)** Column E

71. Explain how the "trick" of Table 9 works.

72. How many columns would be needed for Table 9 to include all ages up to 63?

73. How many columns would be needed for Table 9 to include all numbers up to 127?

In our decimal system, we distinguish odd and even numbers by looking at their ones (or units) digit. If the ones digit is even (0, 2, 4, 6, *or* 8) *the number is even. If the ones digit is odd* (1, 3, 5, 7, *or* 9) *the number is odd. For Exercises 74–83, determine whether this same criterion works for numbers expressed in the given bases.*

74. two **75.** three **76.** four **77.** five

78. six **79.** seven **80.** eight **81.** nine

82. Consider all even bases. If it works for all, explain why. If not, find a criterion that does work for all even bases.

83. Consider all odd bases. If it works for all, explain why. If not, find a criterion that does work for all odd bases.

84. Which of the following base five numerals represent numbers that are divisible by five?

3,204, 200, 342, 2,310, 3,041

Clock Arithmetic and Modular Systems

Recall from the introduction to this chapter that a "mathematical system" consists of (1) a set of elements, along with (2) one or more operations for combining those elements, and (3) one or more relations for comparing those elements. The chapter so far has dealt with the set of counting numbers (an infinite set), and with basic questions about them: how to denote them, and how to perform the arithmetic operations on them. In the remaining sections of the chapter, some of the more familiar properties of mathematical operations will be encountered in the context of some not-so-familiar systems, built mostly upon finite sets. Some of these properties were mentioned in relation to set operations. They will appear again in relation to real number arithmetic at various other points throughout the book.

■ FIGURE 5

The first nontraditional system of this section, called the **12-hour clock system,** is based on an ordinary clock face, with the difference that 12 is replaced with 0 and the minute hand is left off. See Figure 5.

The clock face yields the finite set {0, 1, 2, 3, 4, 5, 6, 7, 8, 9, 10, 11}. As an operation for this clock system, define addition as follows: add by moving the hour hand in a clockwise direction. For example, to add 5 and 2 on a clock, first move the hour hand to 5, as in Figure 6. Then, to add 2, move the hour hand 2 more hours in a clockwise direction. The hand stops at 7, so

$$5 + 2 = 7.$$

■ FIGURE 6

This result agrees with traditional addition. However, the sum of two numbers from the 12-hour clock system is not always what might be expected, as the following example shows.

EXAMPLE 1 Find each sum in clock arithmetic.

(a) 8 + 9

Move the hour hand to 8, as in Figure 7. Then advance the hand clockwise through 9 more hours. It stops at 5, so that

$$8 + 9 = 5.$$

(b) 11 + 3

Proceed as shown in Figure 8. Check that

$$11 + 3 = 2.$$ ■

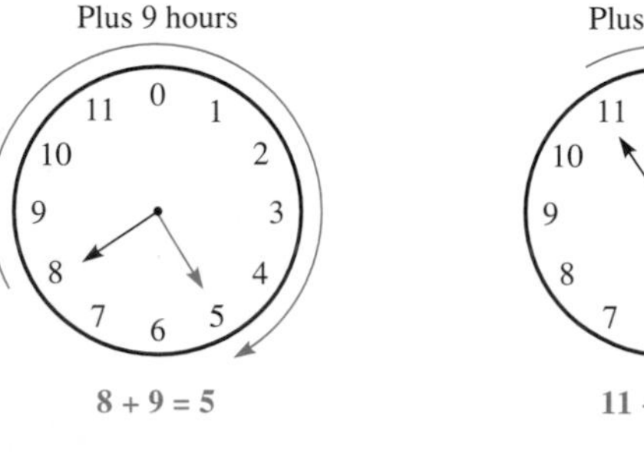

■ FIGURE 7

■ FIGURE 8

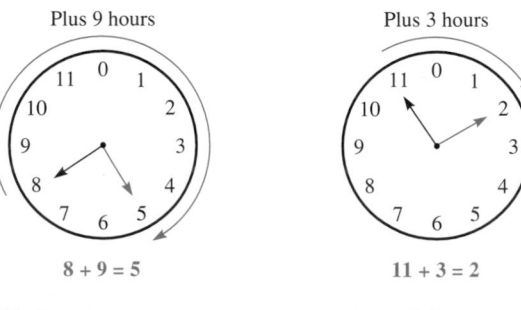

Since there are infinitely many whole numbers, it is not possible to write a complete table of addition facts for that set. Such a table, to show the sum of every possible pair of whole numbers, would have an infinite number of rows and columns, making it impossible to construct.

On the other hand, the 12-hour clock system uses only the whole numbers 0, 1, 2, 3, 4, 5, 6, 7, 8, 9, 10, and 11. A table of all possible sums for this system requires only 12 rows and 12 columns. The 12-hour clock **addition table** is shown in Table 10. (The significance of the colored diagonal line will be discussed later.) Since the 12-hour system is built upon a finite set, it is called a **finite mathematical system.**

TABLE 10 Table for 12-Hour Clock Addition

+	**0**	**1**	**2**	**3**	**4**	**5**	**6**	**7**	**8**	**9**	**10**	**11**
0	0	1	2	3	4	5	6	7	8	9	10	11
1	1	2	3	4	5	6	7	8	9	10	11	0
2	2	3	4	5	6	7	8	9	10	11	0	1
3	3	4	5	6	7	8	9	10	11	0	1	2
4	4	5	6	7	8	9	10	11	0	1	2	3
5	5	6	7	8	9	10	11	0	1	2	3	4
6	6	7	8	9	10	11	0	1	2	3	4	5
7	7	8	9	10	11	0	1	2	3	4	5	6
8	8	9	10	11	0	1	2	3	4	5	6	7
9	9	10	11	0	1	2	3	4	5	6	7	8
10	10	11	0	1	2	3	4	5	6	7	8	9
11	11	0	1	2	3	4	5	6	7	8	9	10

Until the mid-1700s, sailors had no accurate way of telling their **longitude,** their east-west position on the earth (they had fairly accurate star charts which would tell them the latitude, or north-south position). The British government offered a prize for a method of finding longitude; the winner was Harrison's Chronometer, a very accurate clock. Not surprisingly, Harrison had a difficult time collecting the prize money from the government.

EXAMPLE 2 Use the 12-hour clock addition table to find each sum.

(a) $7 + 11$

Find 7 on the left of the addition table and 11 across the top. The intersection of the row headed 7 and the column headed 11 gives the number 6. Thus, $7 + 11 = 6$.

(b) Also from the table, $11 + 1 = 0$.

So far, our 12-hour clock system consists of the set {0, 1, 2, 3, 4, 5, 6, 7, 8, 9, 10, 11}, together with the operation of clock addition. Table 10 shows that the sum of two numbers on a clock face is always a number on the clock face. In other words: if a and b are any clock numbers in the set of the system, then $a + b$ is also in the set of the system. This is called the **closure property** for clock addition. (The set of the system is said to be *closed* under addition.)

Notice also that in this system, $6 + 8$ and $8 + 6$ both yield 2. Also $7 + 11$ and $11 + 7$ both yield 6. The order of the elements being added does not seem to matter. For any clock numbers a and b, $a + b = b + a$. This is called the **commutative property** of clock addition.

A quick way to test a finite system for commutativity is to draw a diagonal line through the table of the system, from upper left to lower right (as was done in Table 10). If the part of the table *above* the diagonal is a mirror image of the part *below* the diagonal, the system has the commutative property. Notice in Table 10 that wherever a given number occurs in the body of the chart, that same number also occurs straight through the diagonal, the same distance to the other side.

A third important property that holds in many mathematical systems is the **associative property,** which says that whenever three elements are combined in a given order, it does not matter whether the initial combination involves the first and second or the second and third; symbolically (for addition):

$$(a + b) + c = a + (b + c).$$

EXAMPLE 3 Is 12-hour clock addition associative?

It would take lots of work to prove that the required relationship *always* holds. But a few examples should either disprove it (by revealing a *counterexample*—a case where it fails to hold), or should make it at least plausible. Using the clock numbers 4, 5, and 9, we see that

$$\begin{aligned}(4 + 5) + 9 &= 9 + 9 \qquad & 4 + (5 + 9) &= 4 + 2\\ &= 6 & &= 6.\end{aligned}$$

Thus, $(4 + 5) + 9 = 4 + (5 + 9)$. Try another example:

$$\begin{aligned}(7 + 6) + 3 &= 1 + 3 \qquad & 7 + (6 + 3) &= 7 + 9\\ &= 4 & &= 4.\end{aligned}$$

So $(7 + 6) + 3 = 7 + (6 + 3)$. Any other examples tried would also work. Clock addition *does* satisfy the associative property. ■

The fourth property we consider here is the **identity property.** A system satisfying the identity property must include in its set an element which, when combined with any element (in either order), produces that same element. That is, the identity element e (for addition) must satisfy $a + e = a$ and $e + a = a$ for any element a of the system. Notice in the clock addition table (Table 10) that $4 + 0 = 0 + 4 = 4$, $6 + 0 = 0 + 6 = 6$, and so on. The number 0 is the identity element (just as for ordinary addition of whole numbers), so that clock addition *does* satisfy the identity property.

Generally, if a finite system has an identity element e, it can be located easily in the operation table. Check the body of Table 10 for a column that is identical to the column at the left side of the table. Since the column under 0 meets this requirement, $a + 0 = a$ holds for all elements a in the system. Thus 0 is *possibly* the identity. Now locate 0 at the left of the table. Since the corresponding row is identical to the row at the top of the table, $0 + a = a$ also holds for all elements a, which is the other requirement of an identity element. Hence 0 is *indeed* the identity.

Subtraction can be performed on a 12-hour clock. Subtraction may be interpreted on the clock face by a movement in the counterclockwise direction. For example, to perform the subtraction $2 - 5$, begin at 2 and move 5 hours counterclockwise, ending at 9, as shown in Figure 9. Therefore, in this system,

$$2 - 5 = 9.$$

■ **FIGURE 9**

In our usual system, subtraction may be checked by addition and this is also the case in clock arithmetic. To check that $2 - 5 = 9$, simply add $9 + 5$. The result on the clock face is 2, verifying the accuracy of this subtraction.

The *additive inverse,* $-a$, of an element a in clock arithmetic, is the element that satisfies the following statement: $a + (-a) = 0$ and $(-a) + a = 0$. The next example examines this idea.

EXAMPLE 4 Determine the additive inverse of 5 in clock arithmetic.

The additive inverse for the clock number 5 is a number x such that

$$5 + x = 0.$$

Going from 5 to 0 on the clock face requires 7 more hours, so

$$5 + 7 = 0.$$

This means that 7 is the additive inverse of 5.

The method used in Example 4 may be used to verify that *every* element of the system has an additive inverse (also in the system). This is called the **inverse property** of clock addition.

Another simpler way to verify the inverse property, once you have the table, is to make sure the identity element appears exactly once in each row, and that the pair of elements that produces it also produces it in the opposite order. (This last condition is automatically true if the commutative property holds for the system.) For example, note in Table 10 that row 3 contains one 0, under the 9, so $3 + 9 = 0$, and that row 9 contains one 0, under the 3, so $9 + 3 = 0$ also. Therefore, 3 and 9 are inverses.

The following chart lists the elements and their additive inverses. Notice that one element, 6, is its own inverse for addition.

Clock value a	0	1	2	3	4	5	6	7	8	9	10	11
Additive inverse $-a$	0	11	10	9	8	7	6	5	4	3	2	1

Using the additive inverse symbol, we can say that in clock arithmetic,

$$-5 = 7, \qquad -11 = 1, \qquad -10 = 2,$$

and so on.

Now that additive inverses have been discussed, a formal definition of subtraction in clock arithmetic may be given. Notice that it is the same definition that is used in our usual system.

Subtraction on a Clock

If a and b are elements in clock arithmetic, then the **difference,** $a - b$, is defined as

$$a - b = a + (-b).$$

EXAMPLE 5 Find each of the following differences.

(a) $8 - 5 = 8 + (-5)$ Use the definition of subtraction.

$= 8 + 7$ The additive inverse of 5 is 7, from the table of inverses.

$8 - 5 = 3$

This result agrees with traditional arithmetic. Check by adding 5 and 3; the sum is 8.

(b) $6 - 11 = 6 + (-11)$

$= 6 + 1$

$6 - 11 = 7$ ■

Clock numbers can also be multiplied. For example, find the product 5×4 by adding 4 a total of 5 times:

$$5 \times 4 = 4 + 4 + 4 + 4 + 4 = 8.$$

EXAMPLE 6 Find each product, using clock arithmetic.

(a) $6 \times 9 = 9 + 9 + 9 + 9 + 9 + 9 = 6$

(b) $3 \times 4 = 4 + 4 + 4 = 0$

(c) $6 \times 0 = 0 + 0 + 0 + 0 + 0 + 0 = 0$

(d) $0 \times 8 = 0$ ■

Some properties of the system of 12-hour clock numbers with the operation of multiplication will be investigated in Exercises 6–8.

We now expand the ideas of clock arithmetic to **modulo systems** in general. Recall that 12-hour clock arithmetic was set up so that answers were always whole numbers less than 12. For example, $8 + 6 = 2$. The traditional sum, $8 + 6 = 14$, reflects the fact that moving the clock hand forward 8 hours from 0, and then forward another 6 hours, amounts to moving it forward 14 hours total. But since the final position of the clock hand is at 2, we see that 14 and 2 are, in a sense, equivalent. More formally we say that 14 and 2 are **congruent modulo** 12, which is written

$$14 \equiv 2 \pmod{12} \qquad \text{(The sign } \equiv \text{ indicates } \textit{congruence.}\text{)}$$

By observing clock hand movements, you can also see that, for example,

$$26 \equiv 2 \pmod{12}, \qquad 38 \equiv 2 \pmod{12}, \qquad \text{and so on.}$$

In each case, the congruence is true because the difference of two congruent numbers is a multiple of 12:

$$14 - 2 = 12 = 1 \times 12, \; 26 - 2 = 24 = 2 \times 12, \; 38 - 2 = 36 = 3 \times 12.$$

This suggests the following definition.

Chess Clock The double clock shown here is used to time chess, backgammon, and scrabble games. Push one button, and that clock stops—the other begins simultaneously. When a player's allotted time for the game has expired, that player will lose if he or she has not made the required number of moves.

Mathematics and chess are often thought to be closely related. Actually that is not so. Both arts demand logical thinking. Chess requires psychological acumen and knowledge of the opponent. No mathematical knowledge is needed in chess.

Emanuel Lasker was able to achieve mastery in both fields. He was best known as a World Chess Champion for 27 years, until 1921. Lasker also was famous in mathematical circles for his work concerning the theory of primary ideals, algebraic analogies of prime numbers. An important result, the Lasker-Noether theorem, bears his name along with that of Emmy Noether. Noether extended Lasker's work. Her father had been Lasker's Ph.D. advisor.

Congruence Modulo *m*

The integers a and b are **congruent modulo** $\boldsymbol{m}$ (where m is a natural number greater than 1 called the **modulus**) if and only if the difference $a - b$ is divisible by m. Symbolically, this congruence is written

$$a \equiv b \pmod{m}.$$

Since being divisible by m is the same as being a multiple of m, we can say that

$$a \equiv b \pmod{m} \text{ if and only if } a - b = km \text{ for some integer } k.$$

The basic ideas of congruence were introduced by Carl F. Gauss in 1801, when he was twenty-four years old.

EXAMPLE 7 Decide whether each statement is *true* or *false*.

(a) $16 \equiv 10 \pmod{2}$
The difference $16 - 10 = 6$ is divisible by 2, so $16 \equiv 10 \pmod{2}$ is true.

(b) $49 \equiv 32 \pmod{5}$
False, since $49 - 32 = 17$, which is not divisible by 5.

(c) $30 \equiv 345 \pmod{7}$
True, since $30 - 345 = -315$ is divisible by 7. (It doesn't matter if we find $30 - 345$ or $345 - 30$.)

There is another method of determining if two numbers, a and b, are congruent modulo m.

Criterion for Congruence

$a \equiv b \pmod{m}$ if and only if the same remainder is obtained when a and b are divided by m.

For example, we know that $27 \equiv 9 \pmod{6}$ because $27 - 9 = 18$, which is divisible by 6. Now, if 27 is divided by 6, the quotient is 4 and the remainder is 3. Also, if 9 is divided by 6, the quotient is 1 and the remainder is 3. According to the criterion above, $27 \equiv 9 \pmod{6}$ since the remainder is the same in each case.

Addition, subtraction, and multiplication can be performed in any modulo system just as with clock numbers. Since final answers should be whole numbers less than the modulus, we can first find an answer using ordinary arithmetic. Then, as long as the answer is nonnegative, simply divide it by the modulus and keep the remainder. This produces the smallest nonnegative integer that is congruent (modulo m) to the ordinary answer.

EXAMPLE 8 Find each of the following sums, differences, and products.

(a) $(9 + 14) \pmod 3$

First add 9 and 14 to get 23. Then divide 23 by 3. The remainder is 2, so $23 \equiv 2 \pmod 3$ and

$$(9 + 14) \equiv 2 \pmod 3.$$

(b) $(27 - 5) \pmod 6$

$27 - 5 = 22$. Divide 22 by 6, obtaining 4 as a remainder:

$$(27 - 5) \equiv 4 \pmod 6.$$

(c) $(50 + 34) \pmod 7$

$50 + 34 = 84$. When 84 is divided by 7, a remainder of 0 is found:

$$(50 + 34) \equiv 0 \pmod 7.$$

(d) $(8 \times 9) \pmod{10}$

Since $8 \times 9 = 72$, and 72 leaves a remainder of 2 when divided by 10,

$$(8 \times 9) \equiv 2 \pmod{10}.$$

(e) $(12 \times 10) \pmod 5$

$$(12 \times 10) = 120 \equiv 0 \pmod 5$$

Public Codes One application of modulo systems comes from the recently discovered "trapdoor codes" involving a very simple method of encoding a message and an extremely complex method of decoding. In fact, the encoding method is so simple that it can be made public without helping the adversary discover how to decode the message.

Problem Solving

Modulo systems can often be applied to questions involving cyclical changes. For example, our method of dividing time into weeks causes the days to repeatedly cycle through the same pattern of seven. Suppose today is Sunday and we want to know what day of the week it will be 45 days from now. Since we don't care how many weeks will pass between now and then, we can discard the largest whole number of weeks in 45 days and keep the remainder. (We are finding the smallest nonnegative integer that is congruent to 45 modulo 7.) Dividing 45 by 7 leaves remainder 3, so the desired day of the week is 3 days past Sunday, or *Wednesday*.

EXAMPLE 9 If today is Thursday, November 12, and *next* year is a leap year, what day of the week will it be one year from today?

A modulo 7 system applies here, but we need to know the number of days between today and one year from today. Today's date, November 12, is unimportant except that it shows we are later in the year than the end of February and therefore the next year (starting today) will contain 366 days. (This would not be so if today were, say, January 12.) Now dividing 366 by 7 produces 52 with remainder 2. Two days past Thursday is our answer. That is, one year from today will be a Saturday.

Just as in algebra, equations can be solved in modulo systems. A **modulo equation** (or just an *equation*) is a sentence such as $(3 + x) \equiv 5 \pmod 7$ that may or may not be true, depending upon the replacement value of the variable x. A method of solving these equations is illustrated in the examples that follow.

EXAMPLE 10 Solve $(3 + x) \equiv 5 \pmod 7$.

In a modulo 7 system, any integer will be congruent to one of the integers 0, 1, 2, 3, 4, 5, or 6. So, the equation $(3 + x) \equiv 5 \pmod 7$ can be solved by trying, in turn, each of these integers as a replacement for x.

$x = 0$: It it true that $(3 + 0) \equiv 5 \pmod 7$? No

$x = 1$: Is it true that $(3 + 1) \equiv 5 \pmod 7$? No

$x = 2$: Is it true that $(3 + 2) \equiv 5 \pmod 7$? Yes

Try $x = 3$, $x = 4$, $x = 5$, and $x = 6$ to see that none of them work. Of the integers from 0 through 6, only 2 is a solution of the equation $(3 + x) \equiv 5 \pmod 7$.

Since 2 is a solution, find other solutions to this mod 7 equation by repeatedly adding 7:

$$2$$
$$2 + 7 = 9$$
$$2 + 7 + 7 = 16$$
$$2 + 7 + 7 + 7 = 23,$$

and so on. The set of all positive solutions of $(3 + x) \equiv 5 \pmod 7$ is

$$\{2, 9, 16, 23, 30, 37, \ldots\}.$$ ■

EXAMPLE 11 Solve the equation $5x \equiv 4 \pmod 9$.

Because the modulus is 9, try 0, 1, 2, 3, 4, 5, 6, 7, and 8:

Is it true that $5 \times 0 \equiv 4 \pmod 9$? No

Is it true that $5 \times 1 \equiv 4 \pmod 9$? No

Continue trying numbers. You should find that none work except $x = 8$:

$$5 \times 8 = 40 \equiv 4 \pmod 9.$$

The set of all positive solutions to the equation $5x \equiv 4 \pmod 9$ is

$$\{8, 8 + 9, 8 + 9 + 9, 8 + 9 + 9 + 9, \ldots\} \quad \text{or}$$
$$\{8, 17, 26, 35, 44, 53, \ldots\}.$$ ■

EXAMPLE 12 Solve the equation $6x \equiv 3 \pmod 8$.

Try the numbers 0, 1, 2, 3, 4, 5, 6, and 7. You should find that none work. Therefore, the equation $6x \equiv 3 \pmod 8$ has no solutions at all. Write the set of all solutions as the empty set, $\emptyset$.

This result is reasonable since $6x$ will always be even, no matter which whole number is used for x. Since $6x$ is even and 3 is odd, the difference $6x - 3$ will be odd, and therefore not divisible by 8. ■

EXAMPLE 13 Solve $8x \equiv 8 \pmod 8$.

Trying the integers 0, 1, 2, 3, 4, 5, 6, and 7 shows that *any* integer can be used as a solution. ■

An equation like $8x \equiv 8 \pmod 8$ in Example 13 that is true for all values of the variable (x, y, and so on) is called an **identity.** Other examples of identities (in ordinary algebra) include $2x + 3x = 5x$ and $y^2 \cdot y^3 = y^5$.

Some problems can be solved by writing down two or more modulo equations and finding their common solutions. The next example illustrates the process.

EXAMPLE 14 Julio wants to arrange his CD collection in equal size stacks, but after trying stacks of 4, stacks of 5, and stacks of 6, he finds that there is always 1 disc left over. Assuming Julio owns more than one CD, what is the least possible number of discs in his collection?

The given information leads to three modulo equations,

$$x \equiv 1 \pmod 4, \qquad x \equiv 1 \pmod 5, \qquad x \equiv 1 \pmod 6,$$

whose sets of positive solutions are, respectively,

$$\{1, 5, 9, 13, 17, 21, 25, 29, 33, 37, 41, 45, 49, 53, 57, 61, 65, 69, \ldots\},$$
$$\{1, 6, 11, 16, 21, 26, 31, 36, 41, 46, 51, 56, 61, 66, 71, 76, \ldots\}, \quad \text{and}$$
$$\{1, 7, 13, 19, 25, 31, 37, 43, 49, 55, 61, \ldots\}.$$

The smallest common solution greater than 1 is 61, so the collection must contain 61 CDs.

For Further Thought
A Card Trick

Many card "tricks" that have been around for years are really not tricks at all, but are based on mathematical properties that allow anyone to do them with no special conjuring abilities. One of them is based on modulo 14 arithmetic.

In this trick, suits play no role. Each card has a numerical value: 1 for ace, 2 for two, . . . , 11 for jack, 12 for queen, and 13 for king. The deck is shuffled and given to a spectator who is instructed to place the deck of cards face up on a table, and is told to follow the procedure described: A card is laid down with its face up. (We shall call it the "starter" card.) The starter card will be at the bottom of a pile. In order to form a pile, note the value of the starter card, and then add cards on top of it while counting up to 13. For example, if the starter card is a six, pile up seven cards on top of it. If it is a jack, add two cards to it, and so on.

When the first pile is completed, it is picked up and placed face down. The next card becomes the starter card for the next pile, and the process is repeated. This continues until all cards are used or until there are not enough cards to complete the last pile. Any cards that are left over are put aside for later use. We shall refer to these as "leftovers."

The performer then requests that a spectator choose three piles at random. The remaining piles are added to the leftovers. The spectator is then instructed to turn over any two top cards from the piles. The performer is then able to determine the value of the third top card.

The secret to the trick is that the performer adds the values of the two top cards that were turned over, and then adds 10 to this sum. The performer then counts off this number of cards from the leftovers. The number of cards remaining in the leftovers is the value of the remaining top card!

For Group Discussion

1. Obtain a deck of playing cards and perform the "trick" as described above. (As with many activities, you'll find that doing it is simpler than describing it.) Does it work?
2. Explain why this procedure works. (If you want to see how someone else explained it, using modulo 14 arithmetic, see "An Old Card Trick Revisited," by Barry C. Felps, in the December 1976 issue of *The Mathematics Teacher.*)

EXERCISES

Find each of the following differences on the 12-hour clock.

1. 8 − 3 **2.** 4 − 9 **3.** 2 − 8 **4.** 0 − 3

5. Complete the 12-hour clock multiplication table at the right. You can use repeated addition and the addition table (for example, $3 \times 7 = 7 + 7 + 7 = 2 + 7 = 9$) or use modulo 12 multiplication techniques, as in Example 8, parts (d) and (e).

×	**0**	**1**	**2**	**3**	**4**	**5**	**6**	**7**	**8**	**9**	**10**	**11**
0	0	0	0	0	0	0	0	0	0	0	0	0
1	0	1	2	3	4	5	6	7	8	9	10	11
2	0	2	4	6	8	10		2	4		8	
3	0	3	6	9	0	3	6			3	6	
4	0	4	8			8		4			4	8
5	0	5	10	3	8		6	11	4			
6	0	6	0		0	6	0	6		6		6
7	0	7	2	9				1			10	
8	0	8	4	0				8	4		8	4
9	0	9			0		6		0			
10	0	10	8			2						2
11	0	11										1

By referring to your table in Exercise 5, determine which of the following properties hold for the system of 12-hour clock numbers with the operation of multiplication.

6. closure **7.** commutative **8.** identity

A 5-hour clock system utilizes the set {0, 1, 2, 3, 4}, *and relates to the clock face shown here.*

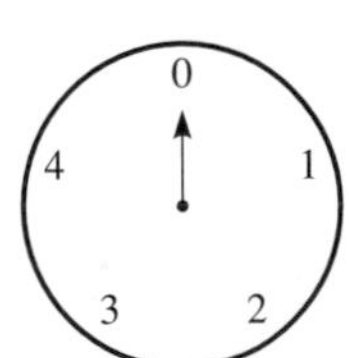

5-hour clock system

9. Complete this 5-hour clock addition table.

+	**0**	**1**	**2**	**3**	**4**
0	0	1	2	3	4
1	1	2	3	4	
2	2	3	4		1
3	3	4			
4	4				3

Which of the following properties are satisfied by the system of 5-hour clock numbers with the operation of addition?

10. closure

11. commutative

12. identity (If so, what is the identity element?)

13. inverse (If so, name the inverse of each element.)

14. Complete this 5-hour clock multiplication table.

×	**0**	**1**	**2**	**3**	**4**
0	0	0	0	0	0
1	0	1	2	3	4
2	0	2	4		
3	0	3			
4	0	4			

Which of the following properties are satisfied by the system of 5-hour clock numbers with the operation of multiplication?

15. closure

16. commutative

17. identity (If so, what is the identity element?)

In clock arithmetic, as in ordinary arithmetic, $a - b = d$ is true if and only if $b + d = a$. Similarly, $a \div b = q$ if and only if $b \times q = a$.

Use the idea above and your 5-hour clock multiplication table of Exercise 14 to find the following quotients on a 5-hour clock.

18. $1 \div 3$

19. $3 \div 1$

20. $2 \div 3$

21. $3 \div 2$

22. Is division commutative on a 5-hour clock? Explain.

23. Is there an answer for $4 \div 0$ on a 5-hour clock? Find it or explain why not.

The military uses a 24-hour clock to avoid the problems of "A.M." *and* "P.M." *For example,* 1100 hours *is* 11 A.M., *while* 2100 hours *is* 9 P.M. *(*12 *noon +* 9 *hours). In these designations, the last two digits represent minutes, and the digits before that represent hours. Find each of the following sums in the 24-hour clock system.*

24. 1400 + 500

25. 1300 + 1800

26. 0750 + 1630

27. 1545 + 0815

28. Explain how the following three statements can *all* be true. (*Hint:* Think of clocks.)

$$1145 + 1135 = 2280$$
$$1145 + 1135 = 1120$$
$$1145 + 1135 = 2320$$

Answer true *or* false *for each of the following.*

29. $5 \equiv 19 \pmod 3$

30. $35 \equiv 8 \pmod 9$

31. $5{,}445 \equiv 0 \pmod 3$

32. $7{,}021 \equiv 4{,}202 \pmod 6$

Do each of the following modular arithmetic problems.

33. $(12 + 7) \pmod 4$

34. $(62 + 95) \pmod 9$

35. $(35 - 22) \pmod 5$

36. $(82 - 45) \pmod 3$

37. $(5 \times 8) \pmod 3$

38. $(32 \times 21) \pmod 8$

39. $[4 \times (13 + 6)] \pmod{11}$

40. $[(10 + 7) \times (5 + 3)] \pmod{10}$

41. The text described how to do arithmetic modulo m when the ordinary answer comes out nonnegative. Explain what to do when the ordinary answer is negative.

Do each of the following modular arithmetic problems.

42. (3 − 27) (mod 5)

43. (16 − 60) (mod 7)

44. [(−8) × 11] (mod 3)

45. [2 × (−23)] (mod 5)

In each of Exercises 46 and 47:
(a) *Complete the given addition table.*
(b) *Decide whether the closure, commutative, identity, and inverse properties are satisfied.*
(c) *If the inverse property is satisfied, give the inverse of each number.*

46. modulo 4

+	**0**	**1**	**2**	**3**
0	0	1	2	3
1				
2				
3				

47. modulo 7

+	**0**	**1**	**2**	**3**	**4**	**5**	**6**
0	0	1	2	3	4	5	6
1	1	2	3	4	5	6	
2							
3							
4							
5							
6							

In each of Exercises 48–51:
(a) *Complete the given multiplication table.*
(b) *Decide whether the closure, commutative, identity, and inverse properties are satisfied.*
(c) *Give the inverse of each nonzero number that has an inverse.*

48. modulo 2

×	**0**	**1**
0	0	0
1	0	

49. modulo 3

×	**0**	**1**	**2**
0	0	0	0
1	0	1	2
2	0	2	

50. modulo 4

×	**0**	**1**	**2**	**3**
0	0	0	0	0
1	0	1	2	3
2	0	2		
3	0	3		

51. modulo 9

×	**0**	**1**	**2**	**3**	**4**	**5**	**6**	**7**	**8**
0	0	0	0	0	0	0	0	0	0
1	0	1	2	3	4	5	6	7	8
2	0	2	4	6	8			5	
3	0	3	6	0		6		3	6
4	0	4	8		7		6		5
5	0	5	1		2		3	8	
6	0	6	3	0	6	3	0	6	3
7	0	7	5			8			2
8	0	8	7			4	3		1

52. Complete this statement: a modulo system satisfies the inverse property for multiplication only if the modulo number is a(n) ____________ number.

Find all positive solutions for each of the following equations. Note any identities.

53. $x \equiv 3 \pmod 7$

54. $(2 + x) \equiv 7 \pmod 3$

55. $6x \equiv 2 \pmod 2$

56. $(5x - 3) \equiv 7 \pmod 4$

57. For many years automobile odometers showed five whole-number digits and a digit for tenths of a mile. For those odometers showing just five whole-number digits, totals are recorded in modulo what number?

58. If a car's five-digit whole number odometer shows a reading of 29,306, *in theory* how many miles might the car have traveled?

59. Lawrence Rosenthal finds that whether he sorts his White Sox ticket stubs into piles of 10, piles of 15, or piles of 20, there are always 2 left over. What is the least number of stubs he could have (assuming he has more than 2)?

60. Roxanna Parker has a collection of silver spoons from all over the world. She finds that she can arrange her spoons in sets of 7 with 6 left over, sets of 8 with 1 left over, or sets of 15 with 3 left over. If Roxanna has fewer than 200 spoons, how many are there?

61. Refer to Example 9 in the text. (Recall that *next* year is a leap year.) Assuming today was Thursday, January 12, answer the following questions.

(a) How many days would the next year (starting today) contain?

(b) What day of the week would occur one year from today?

62. Assume again, as in Example 9, that *next* year is a leap year. If the next year (starting today) does *not* contain 366 days, what is the range of possible dates for today?

63. Robin Strang and Kristyn Wasag, flight attendants for two different airlines, are close friends and like to get together as often as possible. Robin flies a 21-day schedule (including days off), which then repeats, while Kristyn has a repeating 30-day schedule. Both of their routines include stopovers in Chicago, New Orleans, and San Francisco. The table below shows which days of each of their individual schedules they are in these cities. (Assume the first day of a cycle is day number 1.)

	Days in Chicago	**Days in New Orleans**	**Days in San Francisco**
Robin	1, 2, 8	5, 12	6, 18, 19
Kristyn	23, 29, 30	5, 6, 17	8, 10, 15, 20, 25

If today is July 1 and both are starting their schedules today (day 1), list the days during July and August that they will be able to see each other in each of the three cities.

The basis of the complex number system is the "imaginary" number i. The powers of i cycle through a repeating pattern of just 4 distinct values as shown here:

$$i^0 = 1, \quad i^1 = i, \quad i^2 = -1, \quad i^3 = -i, \quad i^4 = 1, \quad i^5 = i, \quad \textit{and so on.}$$

Find the values of each of the following powers of i.

64. i^{16} **65.** i^{47} **66.** i^{98} **67.** i^{137}

The following formula can be used to find the day of the week on which a given year begins. Here y represents the year (after* 1582, *when our current calendar began). First calculate*

$$a = y + [\![(y-1)/4]\!] - [\![(y-1)/100]\!] + [\![(y-1)/400]\!],$$

where $[\![x]\!]$ *represents the greatest integer less than or equal to x. (For example,* $[\![9.2]\!] = 9$, and $[\![\pi]\!] = 3$.*) After finding a, find the smallest nonnegative integer b such that*

$$a \equiv b \pmod 7.$$

Then b gives the day of January 1, with $b = 0$ *representing Sunday,* $b = 1$ *Monday, and so on.*

Find the day of the week on which January 1 would occur in the following years.

68. 1812 **69.** 1865 **70.** 1997 **71.** 1999

*Given in "An Aid to the Superstitious," by G. L. Ritter, S. R. Lowry, H. B. Woodruff, and T. L. Isenhour. *The Mathematics Teacher,* May 1977, pages 456–457.

Some people believe that Friday the thirteenth is unlucky. The table below shows the months that will have a Friday the thirteenth if the first day of the year is known. A year is a leap year if it is divisible by* 4. *The only exception to this rule is that a century year* (1900, *for example*) *is a leap year only when it is divisible by* 400.

First Day of Year	Nonleap Year	Leap Year
Sunday	Jan., Oct.	Jan., April, July
Monday	April, July	Sept., Dec.
Tuesday	Sept., Dec.	June
Wednesday	June	March, Nov.
Thursday	Feb., March, Nov.	Feb., Aug.
Friday	August	May
Saturday	May	Oct.

Use the table to determine the months that have a Friday the thirteenth for the following years.

72. 1997 **73.** 1995 **74.** 1999 **75.** 2200

76. Modulo numbers can be used to create **modulo designs.** For example, to construct the design (11, 5) proceed as follows.

(a) Draw a circle and divide the circumference into 10 equal parts. Label the division points as 1, 2, 3, . . . , 10.

(b) Since $1 \times 5 \equiv 5 \pmod{11}$, connect 1 and 5. (We use 5 as a multiplier since we are making an (11, 5) design.)

(c) $2 \times 5 \equiv 10 \pmod{11}$ Therefore, connect 2 and ____.

(d) $3 \times 5 \equiv$ ____ (mod 11) Connect 3 and ____.

(e) $4 \times 5 \equiv$ ____ (mod 11) Connect 4 and ____.

(f) $5 \times 5 \equiv$ ____ (mod 11) Connect 5 and ____.

(g) $6 \times 5 \equiv$ ____ (mod 11) Connect 6 and ____.

(h) $7 \times 5 \equiv$ ____ (mod 11) Connect 7 and ____.

(i) $8 \times 5 \equiv$ ____ (mod 11) Connect 8 and ____.

(j) $9 \times 5 \equiv$ ____ (mod 11) Connect 9 and ____.

(k) $10 \times 5 \equiv$ ____ (mod 11) Connect 10 and ____.

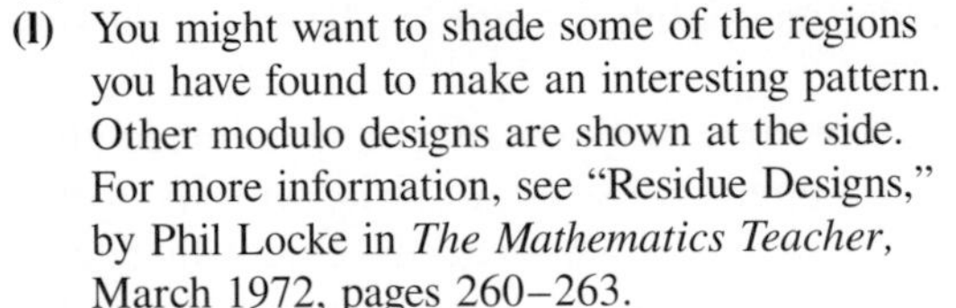

(l) You might want to shade some of the regions you have found to make an interesting pattern. Other modulo designs are shown at the side. For more information, see "Residue Designs," by Phil Locke in *The Mathematics Teacher,* March 1972, pages 260–263.

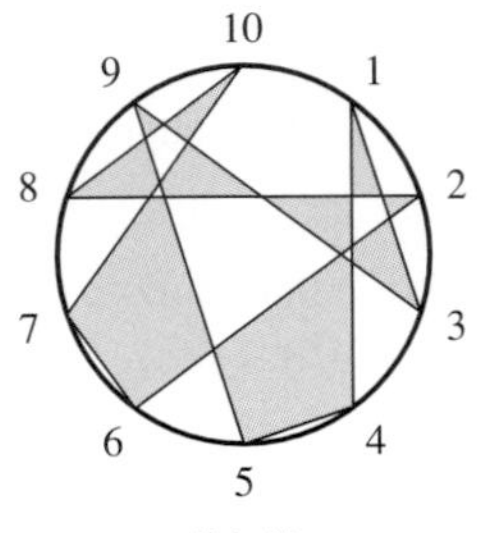

(11, 3)

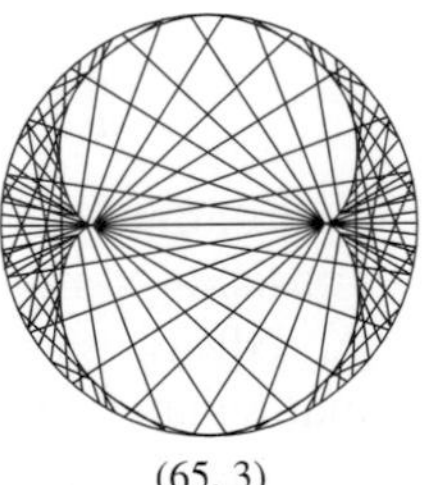

(65, 3)

Identification numbers are used in various ways for all kinds of different products.[†] *Books, for example, are assigned International Standard Book Numbers (ISBNs). Each ISBN is a ten-digit number. It includes a check digit, which is determined on the basis of modular arithmetic. The ISBN for this book is*

0-673-99893-2

*Given in "An Aid to the Superstitious," by G. L. Ritter, S. R. Lowry, H. B. Woodruff, and T. L. Isenhour. *The Mathematics Teacher,* May 1977, pages 456–457.

[†] For an interesting general discussion, see "The Mathematics of Identification Numbers," by Joseph A. Gallian in *The College Mathematics Journal,* May 1991, page 194.

The first digit, 0, *identifies the book as being published in an English-speaking country. The next digits,* 673, *identify the publisher, while* 99893 *identifies this particular book. The final digit,* 2, *is a check digit. To find this check digit, start at the left and multiply the digits of the ISBN number by* 10, 9, 8, 7, 6, 5, 4, 3, *and* 2, *respectively. Then add these products. For this book we get*

$$(10 \times 0) + (9 \times 6) + (8 \times 7) + (7 \times 3) + (6 \times 9) + (5 \times 9) + (4 \times 8) + (3 \times 9) + (2 \times 3) = 295.$$

The check digit is the smallest number that must be added to this result to get a multiple of 11. *Since* $295 + 2 = 297$, *a multiple of* 11, *the check digit is* 2. *(It is possible to have a check "digit" of* 10; *the letter X is used instead of* 10.*)*

When an order for this book is received, the ISBN is entered into a computer, and the check digit evaluated. If this result does not match the check digit on the order, the order will not be processed.

Which of the following ISBNs have correct check digits?

77. 0-399-13615-4

78. 0-691-02356-5

Find the appropriate check digit for each of the following ISBNs.

79. *The Beauty of Fractals,* by H. O. Peitgen and P. H. Richter, 3-540-15851-

80. *Women in Science,* by Vivian Gornick, 0-671-41738-

81. *Beyond Numeracy,* by John Allen Paulos, 0-394-58640-

82. *Iron John,* by Robert Bly, 0-201-51720-

NOTE Another application of modular arithmetic is found in the Collaborative Investigation at the end of this chapter.

4.5 Other Finite Mathematical Systems

We continue our study of finite mathematical systems. Some examples here will consist of numbers, but others will be made up of elements denoted by letters. And the operations are often represented by abstract symbols with no particular mathematical meaning. This is to make the point that a system is characterized by how its elements behave under its operations, not by the choice of symbols used.

To begin, let us introduce a new finite mathematical system made up of the set of elements $\{a, b, c, d\}$, and an operation we shall write with the symbol ☆. We give meaning to operation ☆ by showing an **operation table,** which tells how operation ☆ is used to find the answer for any two elements from the set $\{a, b, c, d\}$. The operation table for ☆ is shown in Table 11. To use the table to find, say, $c ☆ d$, first locate c on the left, and d across the top. This row and column give b, so that

$$c ☆ d = b.$$

TABLE 11

☆	*a*	*b*	*c*	*d*
a	*a*	*b*	*c*	*d*
b	*b*	*d*	*a*	*c*
c	*c*	*a*	*d*	*b*
d	*d*	*c*	*b*	*a*

As long as a system possesses a single operation, the properties we look for are the same as those mentioned earlier: closure, commutative, associative, identity, and inverse. The only difference now is that the operations are new and must be looked at in a new light. It is hard to have preconceived ideas about operation ☆ and the table at the left. Let us now decide which properties are satisfied by the system made up of $\{a, b, c, d\}$ and operation ☆. (All of these properties are summarized later in this section.)

Closure Property

For this system to be closed under the operation $\star$, the answer to any possible combination of elements from the system must be in the set $\{a, b, c, d\}$. A glance at Table 11 shows that the answers in the body of the table are all elements of this set. This means that the system is closed. If an element other than a, b, c, or d had appeared in the body of the table, the system would not have been closed.

Commutative Property

TABLE 12

$\star$	a	b	c	d
a	a	b	c	d
b	b	d	a	c
c	c	a	d	b
d	d	c	b	a

In order for the system to have the commutative property, it must be true that $\Gamma \star \Delta = \Delta \star \Gamma$, where Γ and Δ stand for any elements from the set $\{a, b, c, d\}$. For example,

$$c \star d = b \qquad \text{and} \qquad d \star c = b, \qquad \text{so} \qquad c \star d = d \star c.$$

To establish the commutative property for the system, apply the diagonal line test described in the previous section. Since Table 12 is symmetric about the line, $\star$ is a commutative operation for this system.

Associative Property

The system is associative if $(\Gamma \star \Delta) \star \Upsilon = \Gamma \star (\Delta \star \Upsilon)$, where Γ, Δ, and Υ represent any elements from the set $\{a, b, c, d\}$. There is no quick way to check a table for the associative property, as there is for the commutative property. All we can do is try some examples. Using the table that defines operation $\star$,

$$(a \star d) \star b = d \star b = c, \qquad \text{and} \qquad a \star (d \star b) = a \star c = c,$$

so that $\qquad (a \star d) \star b = a \star (d \star b).$

In the same way, $\qquad b \star (c \star d) = (b \star c) \star d.$

In both these examples, changing the location of parentheses did not change the answers. Since the two examples worked, we suspect that the system is associative. We cannot be sure of this, however, unless every possible choice of three letters from the set is checked. (Although we have not completely verified it here, this system does, in fact, satisfy the associative property.)

Identity Property

For the identity property to hold, there must be an element Δ from the set of the system such that $\Delta \star X = X$ and $X \star \Delta = X$, where X represents any element from the set $\{a, b, c, d\}$. We can use the same criterion here as was used for clock addition. Since the column below a (at the top) is identical to the column at the left, and the row across from a (at the left) is identical to the row at the top, a is in fact the identity element of the system. (It is shown in more advanced courses that if a system has an identity element, it has *only* one.)

Inverse Property

We found above that a is the identity element for the system using operation $\star$. Is there any inverse in this system for, say, the element b? If Δ represents the inverse of b in this system, then

$$b \star \Delta = a \qquad \text{and} \qquad \Delta \star b = a \qquad \text{(since } a \text{ is the identity element).}$$

Inspecting the table for operation $\star$ shows that Δ should be replaced with c:

$$b \star c = a \qquad \text{and} \qquad c \star b = a.$$

Just as with clock addition, we can inspect the table to see if every element of our system has an inverse in the system. We see (in Table 12) that the identity element a appears exactly once in each row, and that, in each case, the pair of elements that produces a also produces it in the opposite order. Therefore, we conclude that the system satisfies the inverse property.

In summary, the mathematical system made up of the set $\{a, b, c, d\}$ and operation $\star$ satisfies the closure, commutative, associative, identity, and inverse properties.

Let us now list the basic properties that may (or may not) be satisfied by a mathematical system involving a single operation.

Potential Properties of a Single-Operation System

Here a, b, and c represent elements from the set of the system, and $\circ$ represents the operation of the system.

Closure The system is closed if for all elements a and b,

$$a \circ b$$

is in the set of the system.

Commutative The system has the commutative property if

$$a \circ b = b \circ a$$

for all elements a and b from the system.

Associative The system has the associative property if

$$(a \circ b) \circ c = a \circ (b \circ c)$$

for every choice of three elements a, b, and c of the system.

Identity The system has an identity element e (where e is in the set of the system) if

$$a \circ e = a \qquad \text{and} \qquad e \circ a = a,$$

for every element a in the system.

Inverse The system satisfies the inverse property if, for every element a of the system, there is an element x in the system such that

$$a \circ x = e \qquad \text{and} \qquad x \circ a = e,$$

where e is the identity element of the system.

Bernard Bolzano (1781–1848) was an early exponent of rigor and precision in mathematics. Many early results in such areas as calculus were produced by the masters in the field; these masters knew what they were doing and produced accurate results. However, their sloppy arguments caused trouble in the hands of the less gifted. The work of Bolzano and others helped put mathematics on a strong footing.

×	0	1	2	3	4	5
0	0	0	0	0	0	0
1	0	1	2	3	4	5
2	0	2	4	0	2	4
3	0	3	0	3	0	3
4	0	4	2	0	4	2
5	0	5	4	3	2	1

EXAMPLE 1 The table in the margin is a multiplication table for the set {0, 1, 2, 3, 4, 5} under the operation of multiplication modulo 6. Which of the properties above are satisfied by this system?

All the numbers in the body of the table come from the set {0, 1, 2, 3, 4, 5}, so the system is closed. If we draw a line from upper left to lower right, we could fold the table along this line and have the corresponding elements match; the system has the commutative property.

To check for the associative property, try some examples:

$$2 \times (3 \times 5) = 2 \times 3 = 0 \quad \text{and} \quad (2 \times 3) \times 5 = 0 \times 5 = 0,$$

so that $\qquad 2 \times (3 \times 5) = (2 \times 3) \times 5.$

Also, $\qquad 5 \times (4 \times 2) = (5 \times 4) \times 2.$

Any other examples that we might try would also work. The system has the associative property.

Since the column at the left of the multiplication table is repeated under 1 in the body of the table, 1 is a candidate for the identity element in the system. To be sure that 1 is indeed the identity element here, check that the row corresponding to 1 at the left is identical with the row at the top of the table.

To find inverse elements, look for the identity element, 1, in the rows of the table. The identity element appears in the second row, $1 \times 1 = 1$; and in the bottom row, $5 \times 5 = 1$; so 1 and 5 are both their own inverses. There is no identity element in the rows opposite the numbers 0, 2, 3, and 4, so none of these elements has an inverse.

In summary, the system made up of the set {0, 1, 2, 3, 4, 5} under multiplication modulo 6 satisfies the closure, associative, commutative, and identity properties, but not the inverse property. ◆

×	1	2	3	4	5	6
1	1	2	3	4	5	6
2	2	4	6	1	3	5
3	3	6	2	5	1	4
4	4	1	5	2	6	3
5	5	3	1	6	4	2
6	6	5	4	3	2	1

EXAMPLE 2 The table in the margin is a multiplication table for the set of numbers {1, 2, 3, 4, 5, 6} under the operation of multiplication modulo 7. Which of the properties are satisfied by this system?

Notice here that 0 is not an element of this system. This is perfectly legitimate. Since we are defining the system, we can include (or exclude) whatever we wish. Check that the system satisfies the closure, commutative, associative, and identity properties, with identity element 1. Let us now check for inverses. The element 1 is its own inverse, since $1 \times 1 = 1$. In row 2, the identity element 1 appears under the number 4, so $2 \times 4 = 1$ (and $4 \times 2 = 1$), with 2 and 4 inverses of each other. Also, 3 and 5 are inverses of each other, and 6 is its own inverse. Since each number in the set of the system has an inverse, the system satisfies the inverse property. ◆

When a mathematical system has two operations, rather than just one, we can look for an additional, very important property, namely the **distributive property.** For example, when we studied Hindu-Arabic arithmetic of counting numbers, we saw that multiplication is distributive with respect to (or "over") addition.

$$3 \times (5 + 9) = 3 \times 5 + 3 \times 9$$

and $\qquad (5 + 9) \times 3 = 5 \times 3 + 9 \times 3.$

In each case, the factor 3 is "distributed" over the 5 and the 9.

Although the distributive property (as well as the other properties discussed here) will be applied to the real numbers in general in a later section, we state it here just for the system of integers.

Distributive Property of Multiplication with Respect to Addition

For integers a, b, and c, the **distributive property** holds for multiplication with respect to addition.

$$a \times (b + c) = a \times b + a \times c$$

and

$$(b + c) \times a = b \times a + c \times a$$

EXAMPLE 3 Is addition distributive over multiplication?

To find out, exchange $\times$ and $+$ in the first equation above:

$$a + (b \times c) = (a + b) \times (a + c).$$

We need to find out whether this statement is true for *every* choice of three numbers that we might make. Try an example. If $a = 3$, $b = 4$, and $c = 5$,

$$a + (b \times c) = 3 + (4 \times 5) = 3 + 20 = 23,$$

while

$$(a + b) \times (a + c) = (3 + 4) \times (3 + 5) = 7 \times 8 = 56.$$

Since $23 \neq 56$, we have $3 + (4 \times 5) \neq (3 + 4) \times (3 + 5)$. This false result is a *counterexample* (an example showing that a general statement is false). This counterexample shows that addition is *not* distributive over multiplication.

Because subtraction of real numbers is defined in terms of addition, the distributive property of multiplication also holds with respect to subtraction.

Distributive Property of Multiplication with Respect to Subtraction

For integers a, b, and c, the **distributive property** holds for multiplication with respect to subtraction.

$$a \times (b - c) = a \times b - a \times c$$

and

$$(b - c) \times a = b \times a - c \times a$$

The **general form of the distributive property** appears below.

General Form of the Distributive Property

Let $\star$ and $\circ$ be two operations defined for elements in the same set. Then $\star$ is distributive over $\circ$ if

$$a \star (b \circ c) = (a \star b) \circ (a \star c)$$

for every choice of elements a, b, and c from the set.

The final example illustrates how the distributive property may hold for a finite system.

EXAMPLE 4 Suppose that the set $\{a, b, c, d, e\}$ has two operations ☆ and ∘ defined by the charts below.

☆	*a*	*b*	*c*	*d*	*e*
a	*a*	*a*	*a*	*a*	*a*
b	*a*	*b*	*c*	*d*	*e*
c	*a*	*c*	*e*	*b*	*d*
d	*a*	*d*	*b*	*e*	*c*
e	*a*	*e*	*d*	*c*	*b*

∘	*a*	*b*	*c*	*d*	*e*
a	*a*	*b*	*c*	*d*	*e*
b	*b*	*c*	*d*	*e*	*a*
c	*c*	*d*	*e*	*a*	*b*
d	*d*	*e*	*a*	*b*	*c*
e	*e*	*a*	*b*	*c*	*d*

The distributive property of ☆ with respect to ∘ holds in this system. Verify for the following case: $e \star (d \circ b) = (e \star d) \circ (e \star b)$.

First evaluate the left side of the equation by using the charts.

$$e \star (d \circ b) = e \star e \qquad \text{Use the } \circ \text{ chart.}$$
$$= b \qquad \text{Use the } \star \text{ chart.}$$

Now, evaluate the right side of the equation.

$$(e \star d) \circ (e \star b) = c \circ e \qquad \text{Use the } \star \text{ chart twice.}$$
$$= b \qquad \text{Use the } \circ \text{ chart.}$$

In each case the final result is b, and the distributive property is verified for this case. ■

4.5 EXERCISES

For each system in Exercises 1–10, decide which of the properties of single-operation systems are satisfied. If the identity property is satisfied, give the identity element. If the inverse property is satisfied, give the inverse of each element. If the identity property is satisfied but the inverse property is not, name the elements that have no inverses.

1. $\{1, 2, 3, 4\}$; multiplication modulo 5

×	**1**	**2**	**3**	**4**
1	1	2	3	4
2	2	4	1	3
3	3	1	4	2
4	4	3	2	1

2. $\{1, 2\}$; multiplication modulo 3

×	**1**	**2**
1	1	2
2	2	1

3. $\{1, 2, 3, 4, 5\}$; multiplication modulo 6

×	**1**	**2**	**3**	**4**	**5**
1	1	2	3	4	5
2	2	4	0	2	4
3	3	0	3	0	3
4	4	2	0	4	2
5	5	4	3	2	1

4. $\{1, 2, 3, 4, 5, 6, 7\}$; multiplication modulo 8

×	**1**	**2**	**3**	**4**	**5**	**6**	**7**
1	1	2	3	4	5	6	7
2	2	4	6	0	2	4	6
3	3	6	1	4	7	2	5
4	4	0	4	0	4	0	4
5	5	2	7	4	1	6	3
6	6	4	2	0	6	4	2
7	7	6	5	4	3	2	1

5. {1, 3, 5, 7}; multiplication modulo 8

×	**1**	**3**	**5**	**7**
1	1	3	5	7
3	3	1	7	5
5	5	7	1	3
7	7	5	3	1

6. {1, 3, 5, 7, 9}; multiplication modulo 10

×	**1**	**3**	**5**	**7**	**9**
1	1	3	5	7	9
3	3	9	5	1	7
5	5	5	5	5	5
7	7	1	5	9	3
9	9	7	5	3	1

7. {*m*, *n*, *p*}; operation ***J***

J	***m***	***n***	***p***
m	*n*	*p*	*n*
n	*p*	*m*	*n*
p	*n*	*n*	*m*

8. {*A*, *B*, *F*}; operation ☆

☆	***A***	***B***	***F***
A	*B*	*F*	*A*
B	*F*	*A*	*B*
F	*A*	*B*	*F*

9. {*A*, *J*, *T*, *U*}; operation #

#	***A***	***J***	***T***	***U***
A	*A*	*J*	*T*	*U*
J	*J*	*T*	*U*	*A*
T	*T*	*U*	*A*	*J*
U	*U*	*A*	*J*	*T*

10. {*r*, *s*, *t*, *u*}; operation **Z**

Z	***r***	***s***	***t***	***u***
r	*u*	*t*	*r*	*s*
s	*t*	*u*	*s*	*r*
t	*r*	*s*	*t*	*u*
u	*s*	*r*	*u*	*t*

The tables in the finite mathematical systems that we developed in this section can be obtained in a variety of ways. For example, let us begin with a square, as shown in the figure. Let the symbols a, b, c, and d be defined as shown in the figure.

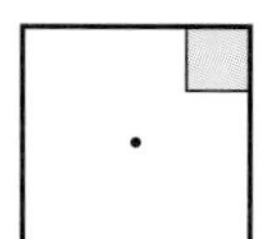

Let *a* represent zero rotation—leave the original square as is.

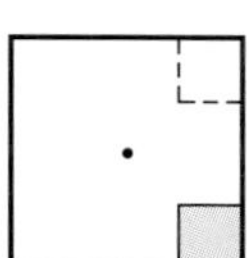

Let *b* represent rotation of 90° clockwise from original position.

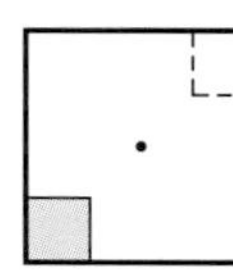

Let *c* represent rotation of 180° clockwise from original position.

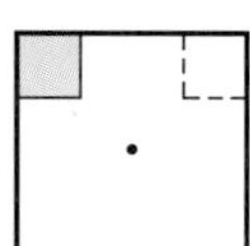

Let *d* represent rotation of 270° clockwise from original position.

Define an operation □ for these letters as follows. To evaluate b □ c, for example, first perform b by rotating the square 90°. *(See the figure.) Then perform operation c by rotating the square an additional* 180°. *The net result is the same as if we had performed d only. Thus,*

$$b \square c = d.$$

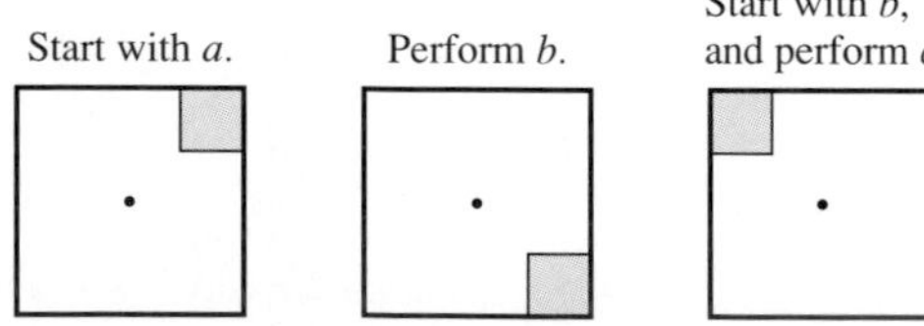
Start with *a*.
Perform *b*.
Start with *b*, and perform *c*.

Use this method to find each of the following.

11. $b \square d$

12. $b \square b$

13. $d \square b$

14. $a \square b$

15. Complete the table at the right.

□	*a*	*b*	*c*	*d*
a	*a*	*b*	*c*	*d*
b	*b*	*c*		*a*
c	*c*		*a*	
d	*d*	*a*		

16. Which of the properties from this section are satisfied by this system?

17. Define a universal set U as the set of counting numbers. Form a new set that contains all possible subsets of U. This new set of subsets together with the operation of set intersection forms a mathematical system. Which of the properties listed in this section are satisfied by this system?

18. Replace the word "intersection" with the word "union" in Exercise 17; then answer the same question.

19. Complete the table at the right so that the result is *not* the same as operation □ of Exercise 15, but so that the five properties listed in this section still hold.

	a	*b*	*c*	*d*
a				
b				
c				
d				

Try examples to help you decide whether or not the following operations, when applied to the integers, satisfy the distributive property.

20. subtraction over multiplication **21.** addition over subtraction **22.** subtraction over addition

Recall that Example 3 provided a counterexample for the general statement

$$a + (b \times c) = (a + b) \times (a + c).$$

Thus, addition is not *distributive over multiplication. Now do Exercises 23–26.*

23. Decide if the statement above is true for each of the following sets of values.

(a) $a = 2, b = -5, c = 4$ **(b)** $a = -7, b = 5, c = 3$

(c) $a = -8, b = 14, c = -5$ **(d)** $a = 1, b = 6, c = -6$

24. Find another set of a, b, and c values that make the statement true.

25. Explain what general conditions will always cause the statement above to be true.

26. Explain why, regardless of the results in Exercises 23–25, addition is still *not* distributive over multiplication.

---

27. Give the conditions under which each of the following equations would be true.

(a) $a + (b - c) = (a + b) - (a + c)$

(b) $a - (b + c) = (a - b) + (a - c)$

28. (a) Find values of a, b, and c such that

$$a - (b \times c) = (a - b) \times (a - c).$$

(b) Does this mean that subtraction is distributive over multiplication? Explain.

Verify for the mathematical system of Example 4 that the distributive property holds for the following cases.

29. $c \star (d \circ e) = (c \star d) \circ (c \star e)$

30. $a \star (a \circ b) = (a \star a) \circ (a \star b)$

31. $d \star (e \circ c) = (d \star e) \circ (d \star c)$

32. $b \star (b \circ b) = (b \star b) \circ (b \star b)$

These exercises are for students who have studied sets.

33. Use Venn diagrams to show that the distributive property for union with respect to intersection holds for sets A, B, and C. That is,

$$A \cup (B \cap C) = (A \cup B) \cap (A \cup C).$$

34. Use Venn diagrams to show that *another* distributive property holds for sets A, B, and C. It is the distributive property of intersection with respect to union.

$$A \cap (B \cup C) = (A \cap B) \cup (A \cap C)$$

These exercises are for students who have studied logic.

35. Use truth tables to show that the following distributive property holds:

$$p \vee (q \wedge r) \equiv (p \vee q) \wedge (p \vee r).$$

36. Use truth tables to show that *another* distributive property holds:

$$p \wedge (q \vee r) \equiv (p \wedge q) \vee (p \wedge r).$$

4.6 Groups

Niels Henrik Abel (1802–1829) of Norway was identified in childhood as a mathematical genius but never received in his lifetime the professional recognition his work deserved.

At 16, influenced by a perceptive teacher, he read the works of Newton, Euler, and Lagrange. In only a few years he began producing work of his own. One of Abel's achievements was the demonstration that a general formula for solving fifth-degree equations does not exist. The quadratic formula (for equations of degree 2) is well known, and formulas do exist for solving third- and fourth-degree equations. Abel's accomplishment ended a search that had lasted for years.

In the study of abstract algebra, groups which have the commutative property are referred to as **abelian** groups in honor of Abel.

When his father died, Abel assumed responsibility for his family and never escaped poverty. Even though a government grant enabled him to visit Germany and France, the leading mathematicians there failed to acknowledge his genius. He died of tuberculosis at age 27.

We have considered quite a few mathematical systems, most of which have satisfied some or all of the following properties: closure, associative, commutative, identity, inverse, and distributive. Systems are commonly classified according to which properties they satisfy. One of the most important categories ever studied is the mathematical **group,** which we define here.

Group

A mathematical system is called a **group** if, under its operation, it satisfies the closure, associative, identity, and inverse properties.

EXAMPLE 1 Does the set $\{-1, 1\}$ under the operation of multiplication form a group?

Check the necessary four properties.

Closure The given system leads to the multiplication table below. All entries in the body of the table are either -1 or 1; the system is closed.

×	**−1**	**1**
−1	1	−1
1	−1	1

Associative Both -1 and 1 are integers, and multiplication of integers is associative.

Identity The identity for multiplication is 1, an element of the set of the system $\{-1, 1\}$.

Inverse Both -1 and 1 are their own inverses for multiplication.

All four of the properties are satisfied, so the system is a group. ◆

EXAMPLE 2 Does the set $\{-1, 1\}$ under the operation of addition form a group?

The addition table below shows that closure is not satisfied, so there is no need to check further. The system is not a group.

+	**−1**	**1**
−1	−2	0
1	0	2

◆

Amalie ("Emmy") Noether (1882–1935) was an outstanding mathematician in the field of abstract algebra. She studied and worked in Germany at a time when it was very difficult for a woman to do so. At the University of Erlangen in 1900, Noether was one of only two women. Although she could sit in on classes, professors could and did deny her the right to take the exams for their courses. Not until 1904 was Noether allowed to register officially. She completed her doctorate four years later.

In 1916 Emmy Noether went to Göttingen to work with David Hilbert on the general theory of relativity. But even with Hilbert's backing and prestige, it was three years before the faculty voted to make Noether a *Privatdozent,* the lowest rank in the faculty. In 1922 Noether was made an unofficial professor (or assistant). She received no pay for this post, although she was given a small stipend to lecture in algebra.

Noether's area of interest was abstract algebra, particularly structures called rings and ideals. (Groups are structures, too, with different properties.) One special type of ring bears her name; she was the first to study its properties.

The system of Example 1 is a finite group. Let's look for an infinite group.

EXAMPLE 3 Does the set of integers $\{\ldots, -3, -2, -1, 0, 1, 2, 3, \ldots\}$ under the operation of addition form a group?

Check the required properties.

Closure The sum of any two integers is an integer; the system is closed.

Associative Try some examples:

$$2 + (5 + 8) = 2 + 13 = 15$$

and $$(2 + 5) + 8 = 7 + 8 = 15,$$

so $$2 + (5 + 8) = (2 + 5) + 8.$$

$$-4 + (7 + 14) = -4 + 21 = 17$$

and $$(-4 + 7) + 14 = 3 + 14 = 17,$$

so $$-4 + (7 + 14) = (-4 + 7) + 14.$$

Apparently, addition of integers is associative.

Identity We know that $a + 0 = a$ and $0 + a = a$ for any integer a. The identity element for addition of integers is 0.

Inverse Given any integer a, its additive inverse, $-a$, is also an integer. For example 5 and -5 are inverses. The system satisfies the inverse property.

Since all four properties are satisfied, this (infinite) system *is* a group. ■

Group structure applies not only to sets of numbers. One common group is the group of **symmetries of a square,** which we will now develop. First, cut out a small square, and label it as shown in Figure 10.

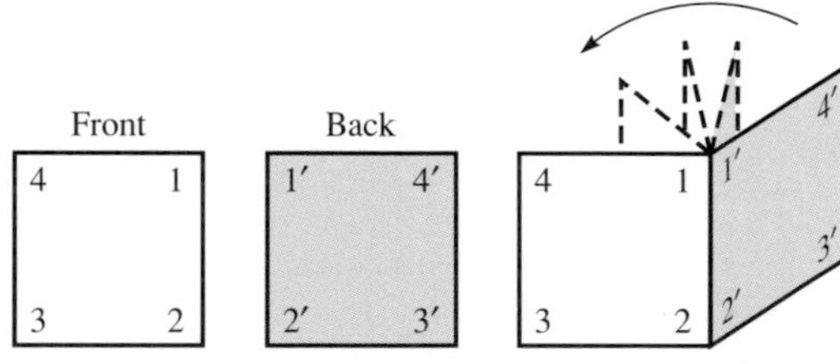

■ **FIGURE 10**

Make sure that 1 is in front of 1′, 2 is in front of 2′, 3 is in front of 3′, and 4 is in front of 4′. Let the letter M represent a clockwise rotation of 90° *about the center of the square* (marked with a dot in Figure 11). Let N represent a rotation of 180°, and so on. A complete list of the symmetries of a square is given in Figure 11.

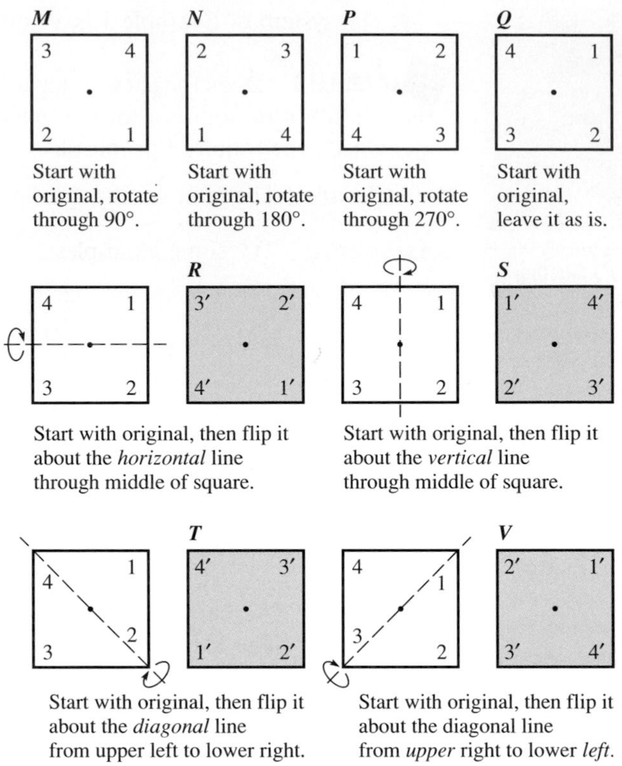

FIGURE 11

Combine symmetries as follows: Let NP represent N followed by P. Performing N and then P is the same as performing just M, so that $NP = M$. See Figure 12.

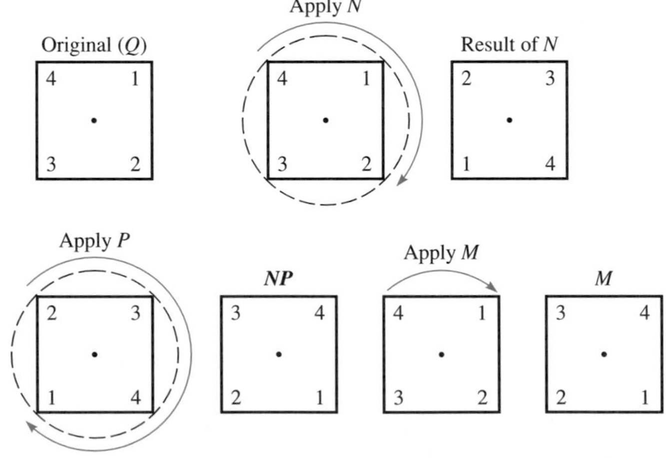

Think of N as advancing each corner two quarter turns clockwise. Thus 4 goes from upper left to lower right. To this result apply P, which advances each corner three quarter turns. Thus 2 goes from upper left to lower left. The result, NP, is the same as advancing each (original) corner one quarter turn, which M does. Thus, $NP = M$.

FIGURE 12

Évariste Galois (1811–1832) About 160 years ago, as a young Frenchman, Galois agreed to fight a duel. He had been engaged in profound mathematical research for some time. Now, anticipating the possibility of his death, he summarized the essentials of his discoveries in a letter to a friend. His letter concluded: "You will publicly ask Jacobi or Gauss to give their opinion not on the truth, but on the importance of the theorems. After this there will be, I hope, some people who will find it to their advantage to decipher all this mess." The next day Galois was killed. He was not yet 21 years old when he died.

Galois's theories were never shown to Jacobi or Gauss. It was not until 1846 that they were published. Mathematicians began to appreciate the importance of Galois's work, which centered on solving equations by using groups. Galois found a way to derive a group which corresponds to each equation. This group, called "the group of the equation," contains the important properties of the equation. So-called Galois groups form an important part of modern abstract algebra.

EXAMPLE 4 Find RT.

First, perform R by flipping the square about a horizontal line through the middle. Then, perform T by flipping the result of R about a diagonal from upper left to lower right. The result of RT is the same as performing only M, so that $RT = M$. ■

The method in Example 4 can be used to complete the following table for combining the symmetries of a square.

□	*M*	*N*	*P*	*Q*	*R*	*S*	*T*	*V*
M	*N*	*P*	*Q*	*M*	*V*	*T*	*R*	*S*
N	*P*	*Q*	*M*	*N*	*S*	*R*	*V*	*T*
P	*Q*	*M*	*N*	*P*	*T*	*V*	*S*	*R*
Q	*M*	*N*	*P*	*Q*	*R*	*S*	*T*	*V*
R	*T*	*S*	*V*	*R*	*Q*	*N*	*M*	*P*
S	*V*	*R*	*T*	*S*	*N*	*Q*	*P*	*M*
T	*S*	*V*	*R*	*T*	*P*	*M*	*Q*	*N*
V	*R*	*T*	*S*	*V*	*M*	*P*	*N*	*Q*

EXAMPLE 5 Show that the system made up of the symmetries of a square is a group.

For the system to be a group, it must satisfy the closure, associative, identity, and inverse properties.

Closure All the entries in the body of the table come from the set $\{M, N, P, Q, R, S, T, V\}$. Thus, the system is closed.

Associative Try examples:

$$P(MT) = P(R) = T.$$

Also, $$(PM)T = (Q)T = T,$$

so that $$P(MT) = (PM)T.$$

Other similar examples also work. (See Exercises 25–28.) Thus, the system has the associative property.

Identity The column at the left in the table is repeated under Q. Check that Q is indeed the identity element.

Inverse In the first row, Q appears under P. Check that M and P are inverses of each other. In fact, every element in the system has an inverse. (See Exercises 29–34.)

Since all four of the properties are satisfied, the system is a group. ■

□	*M*	*N*	*P*	*Q*
M	*N*	*P*	*Q*	*M*
N	*P*	*Q*	*M*	*N*
P	*Q*	*M*	*N*	*P*
Q	*M*	*N*	*P*	*Q*

EXAMPLE 6 Form a mathematical system by using only the set $\{M, N, P, Q\}$ from the group of symmetries of a square. Is this new system a group?

The table for the elements $\{M, N, P, Q\}$ (shown on the left) is just one corner of the table for the entire system. Verify that the system represented by this table satisfies all four properties and thus is a group. This new group is a *subgroup* of the original group of the symmetries of a square. ■

Monster Groups Much research in group theory now is devoted to "simple" groups (which are not at all simple). These groups are fundamental—other groups are built up from them. One of the largest of these simple groups has

$$2^{46} \cdot 3^{20} \cdot 5^9 \cdot 7^6 \cdot 11^2 \cdot 13^3 \cdot 17 \cdot 19 \cdot 23 \cdot 29 \cdot 31 \cdot 41 \cdot 47 \cdot 59 \cdot 71$$

different elements. This huge number of elements is why the group is called the "monster group." If this "monster group" contains too many elements, how about the "baby monster" simple group, with only

$$2^{41} \cdot 3^{13} \cdot 5^6 \cdot 7^2 \cdot 11 \cdot 13 \cdot 17 \cdot 19 \cdot 23 \cdot 31 \cdot 47$$

different elements? (For more information, see "The Classification of the Finite Simple Groups," by Michael Aschbacher in *The Mathematical Intelligencer,* Issue 2, 1981.)

Permutation Groups

A very useful example of a group comes from studying the arrangements, or permutations, of a list of numbers. Start with the symbols 1-2-3, in that order.

There are several ways in which the order could be changed—for example, 2-3-1. This rearrangement is written:

1-2-3
2-3-1.

Replace 1 with 2, replace 2 with 3, and 3 with 1. In the same way,

1-2-3
3-1-2

means replace 1 with 3, 2 with 1, and 3 with 2, while

1-2-3
3-2-1

says to replace 1 with 3, leave the 2 unchanged, and replace 3 with 1. All possible rearrangements of the symbols 1-2-3 are listed below where, for convenience, a name has been given to each rearrangement.

A^*: 1-2-3	B^*: 1-2-3	C^*: 1-2-3	D^*: 1-2-3
2-3-1	2-1-3	1-2-3	1-3-2

E^*: 1-2-3	F^*: 1-2-3
3-1-2	3-2-1

Two rearrangements can be combined as with the symmetries of a square; for example, the symbol $B^* F^*$ means to first apply B^* to 1-2-3 and then apply F^* to the result. Rearrangement B^* changes 1–2–3 into 2-1-3. Then apply F^* to this result: 1 becomes 3, 2 is unchanged, and 3 becomes 1. In summary:

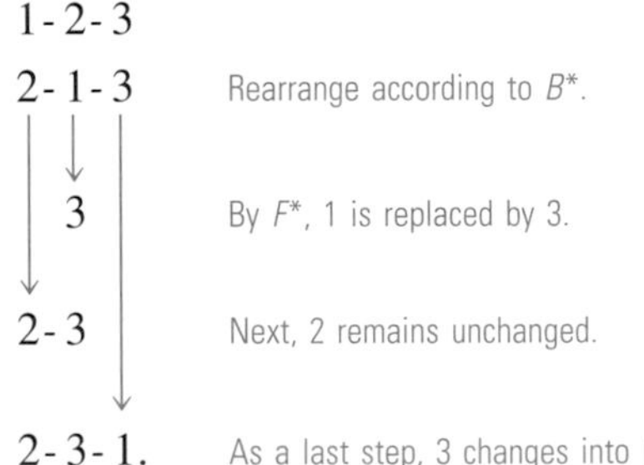

The net result of B^*F^* is to change 1-2-3 into 2-3-1, which is exactly what A^* does to 1-2-3. Therefore

$$B^*F^* = A^*.$$

EXAMPLE 7 Find D^*E^*.

Use the procedure described above.

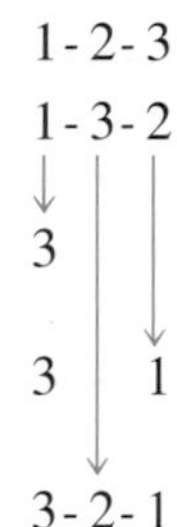

Apply D^*. To apply E^*, first replace 1 with 3. Next, replace 2 with 1. The last step is to replace 3 with 2. The result: D^*E^* converts 1-2-3 into 3-2-1, as does F^*, so

$$D^*E^* = F^*.$$

As further examples, $A^*B^* = D^*$ and $F^*E^* = B^*$.

Once again, we see that we have encountered a mathematical system: the set $\{A^*, B^*, C^*, D^*, E^*, F^*\}$ and the operation of the combination of two rearrangements. To see whether or not this system is a group, check the requirements.

Closure Combine any two rearrangements and the result is another rearrangement, so the system is closed.

Associative Try an example:

First, $(B^*D^*)A^* = E^*A^* = C^*$,

while $B^*(D^*A^*) = B^*B^* = C^*$,

so that $(B^*D^*)A^* = B^*(D^*A^*)$.

Since other examples will work out similarly, the system is associative.

Identity The identity element is C^*. If x is any rearrangement, then $xC^* = C^*x = x$.

Inverse Does each rearrangement have an inverse rearrangement? Begin with the basic order 1-2-3 and then apply, say B^*, resulting in 2-1-3. The inverse of B^* must convert this 2-1-3 back into 1-2-3, by changing 2 into 1 and 1 into 2. But B^* itself will do this. Hence $B^*B^* = C^*$ and B^* is its own inverse. By the same process, E^* and A^* are inverses of each other. Also, each of C^*, D^*, and F^* is its own inverse.

Since all four requirements are satisfied, the system is a group. Rearrangements are also referred to as *permutations,* so this group is sometimes called the **permutation group on three symbols.** The total number of different permutations of a given number of symbols can be determined by methods described in the chapter on counting.

For Further Thought

Marja, Robert, and Sonya are all members (elements?) of a small (finite?) support group that meets on campus each Wednesday. All three also happen to be in the same math class, and after studying mathematical systems they became curious as to whether their Wednesday group was actually a *group* in the mathematical sense. Marja said:

> Everyone in the group is learning to be supportive and trusting. The more the members get together the closer we become. No one has left the group all year. So I believe it must satisfy the **closure** property.

Robert responded:

> Yes, and I've noticed a couple of the members who used to be so hidden within themselves they could never talk to anyone. Now they have opened up and seem to feel comfortable associating freely with the other members. That's what makes me sure that we satisfy the **associative** property.

And Sonya added:

> I just know that before I found our group, I felt like a real nobody. But with the support of these friends, I've begun to realize that I'm somebody. I *do* have an identity of my own. And so does each of the other members. So I would say the **identity** property is surely satisfied.

For Group Discussion

1. Comment on the interpretations offered by Marja, Robert, and Sonya.
2. Do you suppose the support group satisfies the **inverse** property? Explain.

4.6 EXERCISES

What is wrong with the way in which each of the following questions is stated?

1. Do the integers form a group?
2. Does multiplication satisfy all of the group properties?

Decide whether or not each system is a group. If not a group, identify all properties that are not satisfied. (Recall that any system failing to satisfy the identity property automatically fails to satisfy the inverse property also.) For the finite systems, it may help to construct tables. For infinite systems, try some examples to help you decide.

3. $\{0\}$; addition
4. $\{0\}$; multiplication
5. $\{0\}$; subtraction
6. $\{0, 1\}$; addition
7. $\{0, 1\}$; multiplication
8. $\{-1, 1\}$; division
9. $\{-1, 0, 1\}$; addition
10. $\{-1, 0, 1\}$; multiplication
11. integers; multiplication
12. integers; subtraction
13. counting numbers; addition
14. odd integers; multiplication
15. even integers; addition
16. rational numbers; addition
17. nonzero rational numbers; multiplication
18. prime numbers; addition

19. Explain why a *finite* group based on the operation of ordinary addition of numbers cannot contain the element 1.

20. Explain why a group based on the operation of ordinary addition of numbers *must* contain the element 0.

Exercises 21–34 apply to the system of symmetries of a square presented in the text.

Find each of the following.

21. RN 22. PR 23. TV 24. VP

Verify each of the following statements.

25. $N(TR) = (NT)R$
26. $V(PS) = (VP)S$
27. $S(MR) = (SM)R$
28. $T(VN) = (TV)N$

Find the inverse of each element.

29. N 30. Q 31. R 32. S 33. T 34. V

Determine whether or not each of the following systems is a group.

35. {0, 1, 2}; addition modulo 3
36. {0, 1, 2, 3}; addition modulo 4
37. {0, 1, 2, 3, 4}; addition modulo 5
38. {0, 1, 2, 3, 4, 5}; addition modulo 6
39. Complete this statement: _____ modular addition systems are groups.

Determine whether or not each of the following systems is a group.

40. {1, 2, 3}; multiplication modulo 4
41. {1, 2, 3, 4}; multiplication modulo 5
42. {1, 2, 3, 4, 5}; multiplication modulo 6
43. {1, 2, 3, 4, 5, 6}; multiplication modulo 7
44. {1, 2, 3, 4, 5, 6, 7}; multiplication modulo 8
45. {1, 2, 3, 4, 5, 6, 7, 8}; multiplication modulo 9
46. Notice that in constructing modular multiplication systems, we excluded the element 0. Complete this statement: a modular multiplication system is a group only when the modulus is _____.

A group that also satisfies the commutative property is called a commutative group (or an abelian *group, after Niels Henrik Abel).*

Determine whether each of the following groups is commutative.

47. {0, 1, 2, 3} under addition modulo 4
48. {1, 2} under multiplication modulo 3
49. the group of symmetries of a square
50. the subgroup of Example 6
51. the integers under addition
52. the permutation group on three symbols

Give illustrations to back up your answers for Exercises 53–56.

53. Produce a mathematical system with two operations which is a group under one operation but not a group under the other operation.

54. Explain what property is gained when the system of counting numbers is extended to the system of whole numbers.

55. Explain what property is gained when the system of whole numbers is extended to the system of integers.

56. Explain what property is gained when the system of integers is extended to the system of rational numbers.

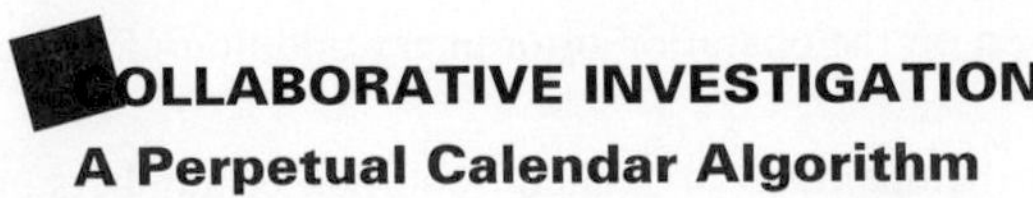

COLLABORATIVE INVESTIGATION

A Perpetual Calendar Algorithm

In the exercises for Section 4 we saw an *equation* that allows us to determine the day of the week a given day/month/year falls on. In this investigation, you will examine a four-step procedure that accomplishes the same goal.

In groups of three to five students each, read the following explanation of the procedure. (See the text preceding Exercise 72 in Section 4 describing how to determine whether a given year is a leap year.)

The Procedure

This algorithm requires several *key numbers.* Key numbers for the month, day, and century are determined by the following tables.

Month	Key
January	1 (0 if a leap year)
February	4 (3 if a leap year)
March	4
April	0
May	2
June	5
July	0
August	3
September	6
October	1
November	4
December	6

Day	Key
Saturday	0
Sunday	1
Monday	2
Tuesday	3
Wednesday	4
Thursday	5
Friday	6

Century	Key
1700s	4
1800s	2
1900s	0
2000s	6

The algorithm works as follows. (We use October 12, 1949 as an example.)

		Example
Step 1:	Obtain the following five numbers.	
	1. The number formed by the last two digits of the year	49
	2. The number in Step 1, divided by 4, with the remainder ignored	12
	3. The month key (1 for October in our example)	1
	4. The day of the month (12 for October 12)	12
	5. The century key (0 for the 1900s)	0
Step 2:	Add these five numbers.	74

Step 3: Divide the sum by 7, and retain the remainder. (74/7 = 10, with remainder 4)

Step 4: Find this remainder in the day key table. (The number 4 implies that October 12, 1949 was a Wednesday.)

Topics For Discussion

1. Have each person in the group determine the day of the week on which he or she was born.

2. Among the group members, discuss whether the following poem applies. (This is all in good fun, of course.)

> Monday's child is fair of face,
> Tuesday's child is full of grace.
> Wednesday's child is full of woe,
> Thursday's child has far to go.
> Friday's child is loving and giving,
> Saturday's child works hard for a living.
> But the child that is born on the Sabbath day
> is bonny and good, happy and gay.

3. Determine the day of the week on which the following important historical events occurred.

(a) December 7, 1941 (the bombing of Pearl Harbor)

(b) November 22, 1963 (assassination of John F. Kennedy)

(c) July 4, 1976 (bicentennial of the United States)

Chapter 4 Test

1. For the numeral 𓆼𓆼𓍢𓍢𓍢𓍢𓍢𓎆𓎆𓎆𓏺𓏺𓏺𓏺𓏺𓏺, identify the numeration system, and give the Hindu-Arabic equivalent.
2. Simplify: $(8 \times 10^3) + (3 \times 10^2) + (6 \times 10^1) + (4 \times 10^0)$.
3. Write in expanded notation: 60,923.

Convert each of the following to base ten.

4. 424_{five}
5. $100{,}110_{\text{two}}$
6. $\text{A,80C}_{\text{sixteen}}$

Convert as indicated.

7. 58 to base two
8. 1,846 to base five
9. $10{,}101{,}110_{\text{two}}$ to base eight
10. $\text{B52}_{\text{sixteen}}$ to base two

Briefly explain each of the following.

11. the advantage of multiplicative grouping over simple grouping
12. the advantage, in a positional numeration system, of a smaller base over a larger base
13. the advantage, in a positional numeration system, of a larger base over a smaller base

Do the following arithmetic problems on a 12-hour clock.

14. $11 + 9$
15. $6 - 9$
16. 8×9

Do the following modular arithmetic problems.

17. $(7 \times 18) \pmod 8$
18. $(52 + 39) \pmod 6$
19. $(23 \times 56) \pmod 2$
20. Construct an addition table for a modulo 6 system.
21. Construct a multiplication table for a modulo 6 system.

Find all positive solutions for the following.

22. $x \equiv 5 \pmod{12}$
23. $5x \equiv 3 \pmod 8$
24. Find the smallest positive integer that satisfies both $x \equiv 3 \pmod{12}$ and $x \equiv 5 \pmod 7$.
25. Describe in general what constitutes a mathematical *group*.

A mathematical system is defined by the table here.

V	***a***	***e***	***i***	***o***	***u***
a	*o*	*e*	*u*	*a*	*i*
e	*u*	*o*	*a*	*e*	*i*
i	*e*	*u*	*o*	*i*	*a*
o	*a*	*e*	*i*	*o*	*u*
u	*i*	*a*	*e*	*u*	*o*

26. **(a)** Is there an identity element in this system? **(b)** If so, what is it?
27. **(a)** Is closure satisfied by this system? **(b)** Explain.
28. **(a)** Is this system commutative? **(b)** Explain.
29. **(a)** Is the distributive property satisfied in this system? **(b)** Explain.
30. **(a)** Is the system a group? **(b)** Explain.

5

Number Theory

In today's world, information is a valuable resource. All private companies and governments must keep certain types of data secret. Codes are used to safely communicate classified information in a form which must be decoded before it can be understood. It is essential that a code be easy to use, but very difficult to break. One of the most widely used codes in industry, called the RSA code, was developed in 1977. The RSA code starts by changing the message to be coded into a large number M. After M is formed, it is then transformed into a different number C using the formula $C = M^k \pmod{N}$ where the numbers k and N are known to everyone. N usually contains more than 100 digits. The only known way to decode C into M is to first write N as a product of two unique numbers p and q called *prime numbers.* Although it is easy to factor a 2-digit number like $N = 21$ into 3×7, it has been impossible until recently to factor a 120-digit number. As a result, messages coded this way have been unintelligible to anyone who does not know the values of p and q.

In 1977 Ronald Rivest from MIT challenged researchers to decode a message that used the RSA code having an N with 129 digits. It was estimated that this message would take approximately 20,000 years to decipher without any knowledge of p and q. However, with the aid of number theory, it took 600 mathematicians in 25 different countries only 17 years to factor N into 64- and 65-digit prime numbers. The factorization is given in the margin note on page 216. The decoded message said, "The magic words are squeamish ossifrage." Since many present-day industrial codes use 130 or 140 digits for N, it may be appropriate to increase the size of N before important codes are broken.

Source: SIAM News, Vol. 27, No. 6, 1994.

The famous German mathematician Carl Friedrich Gauss once remarked, "Mathematics is the Queen of Science, and number theory is the Queen of Mathematics." This chapter is centered around the study of number theory. Number theory is the branch of mathematics devoted to the study of the properties of the natural or counting numbers. (The counting numbers are also called the natural numbers. These names will be used interchangeably in this book.)

This chapter begins with a look at *prime numbers,* those counting numbers larger than 1 that are divisible with remainder zero only by themselves and 1, and *composite numbers,* those numbers that are "composed" of prime number factors. Some interesting topics of number theory are discussed, including some famous unsolved problems in mathematics. Two important ideas of number theory are *greatest common factor* and *least common multiple,* and these are covered in the third section. We then examine one of the most famous sequences in mathematics, the *Fibonacci sequence,* and its relationship to the most aesthetically pleasing ratio in mathematics, the *golden ratio.* We conclude the chapter with an Extension on a recreational topic, *magic squares.*

5.1 Prime and Composite Numbers

In earlier chapters we discussed the set of **natural numbers,** also called the **counting numbers** or the **positive integers:**

$$\{1, 2, 3, \ldots\}.$$

Number theory deals with the study of the properties of this set of numbers, and a key concept of number theory is the idea of *divisibility*. Informally speaking, we say that one counting number is *divisible* by another if the operation of dividing the first by the second leaves a remainder 0. This can be defined formally as follows.

Divisibility

The natural number a is **divisible** by the natural number b if there exists a natural number k such that $a = bk$. If b divides a, then we write $b \mid a$ to indicate this.

Notice that if b divides a, then the quotient a/b is a natural number. For example, 4 divides 20 since there exists a natural number k such that $20 = 4k$. The value of k here is 5, since $20 = 4 \cdot 5$. (A dot is used here to denote multiplication.) The natural number 20 is not divisible by 7, for example, since there is no natural number k satisfying $20 = 7k$. Alternatively, we think "20 divided by 7 gives quotient 2 with remainder 6" and since there is a nonzero remainder, divisibility does not hold. We write $7 \nmid 20$ to indicate that 7 does not divide 20.

If the natural number a is divisible by the natural number b, then b is a **factor** (or **divisor**) of a, and a is a **multiple** of b. For example, 5 is a factor of 30, and 30 is a multiple of 5. Also, 6 is a factor of 30, and 30 is a multiple of 6. The number 30 equals $6 \cdot 5$; this product $6 \cdot 5$ is called a **factorization** of 30. Other factorizations of 30 include $3 \cdot 10$, $2 \cdot 15$, $1 \cdot 30$, and $2 \cdot 3 \cdot 5$.

Breaking Codes Finding prime factors of extremely large numbers has been considered a mere computer exercise—interesting for the improved methods of working with computers, but of no value in its own right. This has changed in recent years with new methods of *computer coding* in which very large numbers are used in an attempt to provide unbreakable codes for computer data. Just as fast as these numbers are used, other people try to find prime factors of them so that the code can be broken.

An article in *Science News,* May 7, 1994 ("Team Sieving Cracks a Huge Number," p. 292) reported that a method known as RSA was used to factor a 129-digit number. It was accomplished by more than 600 computers working together. Here is the number and its two factors:

114,381,625,757,888,867,669,235, 779,976,146,612,010,218,296,721, 242,362,562,561,842,935,706,935, 245,733,897,830,597,123,563,958, 705,058,989,075,147,599,290,026, 879,543,541 = 3,490,529,510,847, 650,949,147,849,619,903,898,133, 417,764,638,493,387,843,990,820, 577 × 32,769,132,993,266,709,549, 961,988,190,834,461,413,177,642, 967,992,942,539,798,288,533.

The ideas of even and odd natural numbers are based on the concept of divisibility. A natural number is even if it is divisible by 2, and odd if it is not.

EXAMPLE 1 Decide whether the first number in each pair is divisible by the second.

(a) 45, 9

Is there a natural number k that satisfies $45 = 9k$? The answer is yes, since $45 = 9 \cdot 5$, and 5 is a natural number. Therefore, 9 divides 45, written $9 \mid 45$.

(b) 60, 7

Since the quotient $60 \div 7$ is not a natural number, 60 is not divisible by 7, written $7 \nmid 60$.

(c) 19, 19

The quotient $19 \div 19$ is the natural number 1, so 19 is divisible by 19. (In fact, any natural number is divisible by itself and also by 1.)

(d) 26, 1

The quotient $26 \div 1$ is the natural number 26, so 26 is divisible by 1. (In fact, any natural number is divisible by 1. People often forget this.) ◆

The observations made in parts (c) and (d) of Example 1 may be generalized as follows: For any natural number a, $a \mid a$ and $1 \mid a$.

EXAMPLE 2 Find the natural number factors of each number.

(a) 36

To find the factors of 36, start by trying to divide 36 by 1, 2, 3, 4, 5, 6, and so on. Doing this gives the following list of natural number factors of 36:

1, 2, 3, 4, 6, 9, 12, 18, and 36.

(b) 50

The factors of 50 are 1, 2, 5, 10, 25, and 50.

(c) 11

The only natural number factors of 11 are 11 and 1. ◆

Like the number 19 in Example 1(c), the number 11 has only two natural number factors, itself and 1. This is a very important kind of natural number, called a *prime number*.

Prime and Composite Numbers

A natural number greater than 1 that has only itself and 1 as factors is called a **prime number.** A natural number greater than 1 that is not prime is called **composite.**

To simplify certain formulas and rules, mathematicians agree that the natural number 1 is neither prime nor composite. (People often erroneously think that 1 is prime.) An alternate definition of prime number helps to alleviate this problem.

How to Use Up Lots of Chalk In 1903, long before the age of the computer, the mathematician F. N. Cole presented before a meeting of the American Mathematical Society his discovery of a factorization of the number

$$2^{67} - 1.$$

He walked up to the chalkboard, raised 2 to the 67th power, and then subtracted 1. Then he moved over to another part of the board and multiplied out

$$193{,}707{,}721 \times 761{,}838{,}257{,}287.$$

The two calculations agreed, and Cole received a standing ovation for a presentation that did not include a single word.

A **prime number** is a natural number that has *exactly* two different natural number factors.

This alternate definition excludes the possibility of 1 being prime, since 1 has only one natural number factor (namely, 1).

There is a systematic method for identifying prime numbers in a list of numbers: 2, 3, . . . , n. The method, known as the **Sieve of Eratosthenes,** is named after the Greek geographer, poet, astronomer, and mathematician, who lived from about 276 to 192 B.C. To construct such a sieve, list all the natural numbers from 2 through some given natural number, such as 100. The number 2 is prime, but all multiples of 2 (4, 6, 8, 10, and so on) are composite. Since 2 is prime, draw a circle around 2, and cross out all the other multiples of 2. Since 3 is a prime, it should be circled, while all other multiples of 3 (6, 9, 12, 15, and so on) that are not already crossed out should be crossed out. The next prime is 5, which is circled, while all other multiples of 5 are crossed out. Circle 7, and cross out all the other multiples of 7. Continue this process for all primes less than or equal to the square root of the last number in the list. For this list, we may stop with 7, since the next prime, 11, is greater than the square root of 100, which is 10. Table 1 gives the Sieve of Eratosthenes for 2, 3, 4, . . . , 100.

Every circled number is prime. Theoretically, such a sieve can be constructed for any value of n.

TABLE 1 Sieve of Eratosthenes

	(2)	(3)	4	(5)	6	(7)	8	9	10	(11)	12	(13)	14
15	16	(17)	18	(19)	20	21	22	(23)	24	25	26	27	28
(29)	30	(31)	32	33	34	35	36	(37)	38	39	40	(41)	42
(43)	44	45	46	(47)	48	49	50	51	52	(53)	54	55	56
57	58	(59)	60	(61)	62	63	64	65	66	(67)	68	69	70
(71)	72	(73)	74	75	76	77	78	(79)	80	81	82	(83)	84
85	86	87	88	(89)	90	91	92	93	94	95	96	(97)	98
99	100												

EXAMPLE 3 Use the definitions of prime and composite numbers to decide whether each number is prime or composite.

(a) 97

Since the only factors of 97 are 97 and 1, the number 97 is prime.

(b) 59,872

The number 59,872 is even, so it is divisible by 2. It is composite. (There is only one even prime, the number 2 itself.)

(c) 697

For 697 to be composite, there must be a number other than 697 and 1 that divides into it with remainder 0. Start by trying 2, and then 3. Neither works. There is no need to try 4. (If 4 divides with remainder 0 into a number, then 2 will also.) Try 5. There is no need to try 6 or any succeeding even number. (Why?) Try 7. Try 11. (Why not try 9?) Try 13. Keep trying numbers until one works, or until a number is tried whose square exceeds the given number. Try 17:

$$697 \div 17 = 41.$$

The number 697 is composite. ◆

Writing in the October 1, 1994, issue of *Science News*, Ivars Peterson gives a fascinating account of the recent discovery of a 75-year-old factoring machine ("Cranking Out Primes: Tracking Down a Long-lost Factoring Machine"). In 1989, Jeffrey Shallit of the University of Waterloo in Ontario came across an article in an obscure 1920 French journal, in which the author, Eugene Olivier Carissan, reported his invention of the factoring apparatus. Shallit and two colleagues embarked on a search for the machine. They contacted all telephone subscribers in France named Carissan, and received a reply from Eugene Carissan's daughter. The machine was still in existence and in working condition, stored in a drawer at an astronomical observatory in Floirac, near Bordeaux.

Peterson explains in the article how the apparatus works. Using the machine, Carissan took just ten minutes to prove that 708,158,977 is a prime number, and he was able to factor a 13-digit number. While this cannot compare to what technology can accomplish today, it was a significant achievement for Carissan's day.

An aid in determining whether a natural number is divisible by another natural number is called a **divisibility test.** Some simple divisibility tests exist for small natural numbers, and they are given in Table 2. Divisibility tests for 7 and 11 are a bit involved, and they are discussed in the exercises for this section.

TABLE 2

Divisible By	Test	Example
2	Number ends in 0, 2, 4, 6, or 8. (The last digit is even.)	9,489,994 ends in 4; is divisible by 2.
3	Sum of the digits is divisible by 3.	897,432 is divisible by 3, since 8 + 9 + 7 + 4 + 3 + 2 = 33 is divisible by 3.
4	Last two digits form a number divisible by 4.	7,693,432 is divisible by 4, since 32 is divisible by 4.
5	Number ends in 0 or 5.	890 and 7,635 are divisible by 5.
6	Number is divisible by both 2 and 3.	27,342 is divisible by 6 since it is divisible by both 2 and 3.
8	Last three digits form a number divisible by 8.	1,437,816 is divisible by 8, since 816 is divisible by 8.
9	Sum of the digits is divisible by 9.	428,376,105 is divisible by 9 since sum of digits is 36, which is divisible by 9.
10	The last digit is 0.	897,463,940 is divisible by 10.
12	Number is divisible by both 4 and 3.	376,984,032 is divisible by 12.

EXAMPLE 4 Decide whether the first number listed is divisible by the second.

(a) 2,984,094; 4

The last two digits form the number 94. Since 94 is not divisible by 4, the given number is not divisible by 4.

(b) 4,119,806,514; 9

The sum of the digits is 4 + 1 + 1 + 9 + 8 + 0 + 6 + 5 + 1 + 4 = 39, which is not divisible by 9. The given number is therefore not divisible by 9. ◆

Prime Search The number $2^{64} - 1$ is one of an infinitude having the form

$$2^n - 1$$

where n is a counting number. And where n is prime, $2^n - 1$ is sometimes a prime. The first ten such primes occur for $n = 2, 3, 5, 7, 13, 17, 19, 31, 61$, and 89. A composite number, 2,047 occurs for $n = 11$, even though 11 is prime. ($2{,}047 = 23 \times 89$.) Thus not every prime n generates a prime for $2^n - 1$.

At one time, $2^{11{,}213} - 1$ was the largest known prime. To honor its discovery, the Urbana, Illinois, post office used the cancellation picture above. Since then, larger ones have been found. In 1978, two students at California State University at Hayward, Laura Nickel and Curt Noll, proved that the number $2^{21{,}701} - 1$ is prime.

As of June 1995, there were 33 such primes known.

The following program can be used on the TI-82 calculator to list all primes less than or equal to a given natural number N.

```
PROGRAM: PRIMES
: Disp "INPUT N≥2
TO GET PRIMES"
: Input N
: Disp "PRIMES≤N"
: 2→T
: Disp T
: 1→A
: Lbl 1
: A+2→A
: 3→B
```

continued on page 220

As mentioned earlier, a number such as 15, which can be written as a product of prime factors ($15 = 3 \cdot 5$), is called a *composite* number. An important theorem in mathematics states that there is only one possible way to write the prime factorization of a composite natural number. This theorem is called the *fundamental theorem of arithmetic,* a form of which was known to the ancient Greeks.*

The Fundamental Theorem of Arithmetic

Every composite natural number can be expressed in one and only one way as a product of primes (if the order of the factors is disregarded).

This theorem is sometimes called the **unique factorization theorem** to reflect the idea that there is only one (unique) prime factorization possible for any given natural number.

A composite number can be factored into primes by using a "factor tree" as illustrated in the next example.

EXAMPLE 5 Find the prime factorization of each composite number.

(a) 90

Since 2 is a factor of 90, we may begin the first branches of the factor tree by writing 90 as $2 \cdot 45$. Then continue by factoring the composite numbers that result into primes, until the end of each branch contains a prime number.

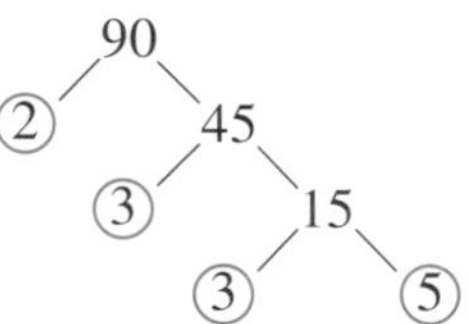

The prime factors are circled. Using exponents, the prime factorization of 90 is $2 \cdot 3^2 \cdot 5$. (**Note:** Because the fundamental theorem of arithmetic guarantees that this prime factorization is unique, we could have started the factor tree with $3 \cdot 30$, or $6 \cdot 15$, or $5 \cdot 18$, and we would have obtained the same factorization. Try it.)

(b) 504

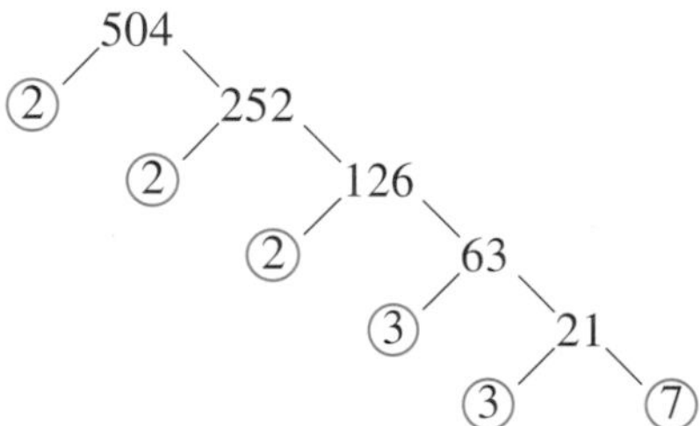

Therefore, $504 = 2^3 \cdot 3^2 \cdot 7$.

*A theorem is a statement that can be proved true from other statements. For a proof of this theorem, see *What Is Mathematics?* by Richard Courant and Herbert Robbins (Oxford University Press, 1941), page 23.

```
: If A>N
: Stop
: Lbl 2
: If B≤√A
: Goto 3
: Disp A
: Pause
: Goto 1
: Lbl 3
: If A/B≤int (A/B)
: Goto 1
: B+2→B
: Goto 2
```

This program was written by Charles W. Gantner and is provided courtesy of Texas Instruments.

The display indicates that the primes less than or equal to 6 are 2, 3, and 5.

As an alternative to the factor tree method, we may divide 504 by primes over and over, using the compact form shown below.

$$\begin{array}{r|l} 2 & 504 \\ \hline 2 & 252 \\ \hline 2 & 126 \\ \hline 3 & 63 \\ \hline 3 & 21 \\ \hline & 7 \end{array}$$

Keep going, using as necessary the primes 2, 3, 5, 7, 11, and so on, until the last number is a prime. Read the prime factors as shown in color above, to get $504 = 2^3 \cdot 3^2 \cdot 7$. ■

EXAMPLE 6 Find the smallest natural number that is divisible by all of the following: 2, 3, 4, 6, 8, 9.

To find this number, consider all prime factorizations of the composite numbers in the list, as well as the prime numbers in the list:

$$2,\ 3,\ 2^2,\ 2 \cdot 3,\ 2^3,\ 3^2$$

Choose the largest exponent on each prime that appears in the line above, and multiply these prime powers: $2^3 \cdot 3^2 = 8 \cdot 9 = 72$. The natural number 72 is the number we are seeking. ■

5.1 EXERCISES

Decide whether the following statements are true *or* false.

1. Every natural number is divisible by 1.
2. No natural number is both prime and composite.
3. There are no even prime numbers.
4. If n is a natural number and $9 \mid n$, then $3 \mid n$.
5. If n is a natural number and $5 \mid n$, then $10 \mid n$.
6. 1 is the smallest prime number.
7. Every natural number is both a factor and a multiple of itself.
8. If 16 divides a natural number, then 2, 4, and 8 must also divide that natural number.
9. The composite number 50 has exactly two prime factorizations.
10. The prime number 53 has exactly two natural number factors.

Find all natural number factors of each number.

11. 12 **12.** 18 **13.** 20 **14.** 28

15. 52 **16.** 63 **17.** 120 **18.** 172

Use divisibility tests to decide whether the given number is divisible by
(a) 2 **(b)** 3 **(c)** 4 **(d)** 5 **(e)** 6 **(f)** 8 **(g)** 9 **(h)** 10 **(i)** 12.

19. 315 **20.** 7,425 **21.** 1,092 **22.** 4,488 **23.** 630

24. 25,025 **25.** 45,815 **26.** 5,940 **27.** 123,456,789 **28.** 987,654,321

29. Continue the Sieve of Eratosthenes in Table 1 from 101 to 200. (You need only check for divisibility by primes through 13.) List the primes between 100 and 200.

30. List two primes that are consecutive natural numbers. Can there be any others?

31. Can there be three primes that are consecutive natural numbers? Explain.

32. For a natural number to be divisible by both 2 and 5, what must be true about its last digit?

33. Consider the divisibility tests for 2, 4, and 8 (all powers of 2). Use inductive reasoning to predict the divisibility test for 16. Then, use the test to show that 456,882,320 is divisible by 16.

Find the prime factorization of each composite number.

34. 240 **35.** 300 **36.** 360 **37.** 425

38. 663 **39.** 885 **40.** 1,280 **41.** 1,575

Here is a divisibility test for 7.
(a) *Double the last digit of the given number, and subtract this value from the given number with the last digit omitted.*
(b) *Repeat the process of part (a) as many times as necessary until the number obtained can easily be divided by* 7.
(c) *If the final number obtained is divisible by* 7, *then the given number is also divisible by* 7. *If the final number is not divisible by* 7, *then neither is the given number.*

Use this divisibility test to determine whether or not each of the following is divisible by 7.

42. 142,891 **43.** 409,311 **44.** 458,485 **45.** 287,824

Here is a divisibility test for 11.
(a) *Starting at the left of the given number, add together every other digit.*
(b) *Add together the remaining digits.*
(c) *Subtract the smaller of the two sums from the larger.*
(d) *If the final number obtained is divisible by* 11, *then the given number is also divisible by* 11. *If the final number is not divisible by* 11, *then neither is the given number.*

Use this divisibility test to determine whether or not each of the following is divisible by 11.

46. 8,493,969 **47.** 847,667,942 **48.** 453,896,248 **49.** 552,749,913

50. Consider the divisibility test for the composite number 6, and make a conjecture for the divisibility test for the composite number 15.

Find the smallest natural number that is divisible by all of the numbers in the group of numbers listed.

51. 2, 3, 5, 7, 8 **52.** 2, 3, 4, 9, 10 **53.** 2, 3, 4, 5, 6, 7, 8, 9 **54.** 2, 3, 4, 5, 6, 7, 8, 9, 12

55. Explain why the answers in Exercises 53 and 54 must be the same.

56. Explain why the answer in Exercise 51 would not change if 2 were omitted from the group of numbers.

Determine all possible digit replacements for x so that the first number is divisible by the second. For example, 37,58x is divisible by 2 if $x = 0, 2, 4, 6,$ *or* 8.

57. 398,87x; 2
58. 2,45x,765; 3
59. 64,537,84x; 4
60. 2,143,89x; 5
61. 985,23x; 6
62. 7,643,24x; 8
63. 4,329,7x5; 9
64. 23,x54,470; 10

There is a method to determine the **number of divisors** *of a composite number. To do this, write the composite number in its prime factored form, using exponents. Add* 1 *to each exponent and multiply these numbers. Their product gives the number of divisors of the composite number. For example,* $24 = 2^3 \cdot 3 = 2^3 \cdot 3^1$. *Now add* 1 *to each exponent:* $3 + 1 = 4$, $1 + 1 = 2$. *Multiply* 4×2 *to get* 8. *There are* 8 *divisors of* 24. *(Since* 24 *is rather small, this can be verified easily. The divisors are* 1, 2, 3, 4, 6, 8, 12, *and* 24, *a total of eight as predicted.)*

Find the number of divisors of each composite number.

65. 36
66. 48
67. 72
68. 144
69. $2^8 \cdot 3^2$
70. $2^4 \cdot 3^4 \cdot 5^2$

Leap years occur when the year number is divisible by 4. *An exception to this occurs when the year number is divisible by* 100 *(that is, it ends in two zeros). In such a case, the number must be divisible by* 400 *in order for the year to be a leap year. Determine which of the following years are leap years.*

71. 1776
72. 1948
73. 1994
74. 1894
75. 2000
76. 2400
77. 1900
78. 1800

79. Why is the following *not* a valid divisibility test for 8? "A number is divisible by 8 if it is divisible by both 4 and 2." Support your answer with an example.

80. Choose any three consecutive natural numbers, multiply them together, and divide the product by 6. Repeat this several times, using different choices of three consecutive numbers. Make a conjecture concerning the result.

81. Explain why the product of three consecutive natural numbers must be divisible by 6.

82. Choose any 6-digit number consisting of three digits followed by the same three digits in the same order (for example, 467,467). Divide by 13. Divide by 11. Divide by 7. What do you notice? Why do you think this happens?

Selected Topics from Number Theory

In an earlier chapter we introduced figurate numbers, a topic investigated by the Pythagoreans. This group of Greek mathematicians and musicians held their meetings in secret, and were led by Pythagoras. The **Pythagorean theorem,** which gives the relationship among the sides of a right triangle, is one of the most important theorems in elementary mathematics, and is discussed in another chapter. In this section we will examine some of the other special numbers that fascinated the Pythagoreans, and are still studied by mathematicians today.

Perfect Numbers

Divisors of a natural number were covered in the first section of this chapter. The **proper divisors** of a natural number include all divisors of the number except the number itself. For example, the proper divisors of 8 are 1, 2, and 4. (8 is *not* a proper divisor of 8.)

Smith Numbers The mathematician Albert Wilansky, when phoning his brother-in-law, Mr. Smith, noticed an interesting property concerning Smith's phone number (493–7775). The number 4,937,775 is composite, and its prime factorization is $3 \times 5 \times 5 \times 65{,}837$. When the digits of the phone number are added, the result, 42, is equal to the sum of the digits in the prime factors: $3 + 5 + 5 + 6 + 5 + 8 + 3 + 7 = 42$. Wilansky termed such a number a *Smith number*. In 1985 it was proved that there are infinitely many Smith numbers, but there are still many unanswered questions about them.

Perfect Numbers

A natural number is said to be **perfect** if it is equal to the sum of its proper divisors.

Is 8 perfect? No, since $1 + 2 + 4 = 7$, and $7 \neq 8$. It happens that the smallest perfect number is 6, since the proper divisors of 6 are 1, 2, and 3, and

$$1 + 2 + 3 = 6. \quad \text{6 is perfect.}$$

EXAMPLE 1 Show that 28 is a perfect number.

The proper divisors of 28 are 1, 2, 4, 7, and 14. The sum of these is 28:

$$1 + 2 + 4 + 7 + 14 = 28.$$

By the definition, 28 is perfect. ◆

The numbers 6 and 28 are the two smallest perfect numbers. The next two are 496 and 8,128. (You are asked to verify that these are perfect numbers in the exercises.)

There are still many unanswered questions about perfect numbers. It is not known if there are infinitely many perfect numbers. As of mid-1995, there were 33 known perfect numbers. All known perfect numbers are even, and it is not known whether odd perfect numbers exist. If $2^n - 1$ is prime, then $2^{n-1}(2^n - 1)$ is perfect, and conversely. Also, all even perfect numbers must be of this form. Any even perfect number will end in 6 or 28.

Deficient and Abundant Numbers

Earlier we saw that 8 is not perfect since it is greater than the sum of its proper divisors $(8 > 7)$.

Deficient and Abundant Numbers

A natural number is **deficient** if it is greater than the sum of its proper divisors. It is **abundant** if it is less than the sum of its proper divisors.

Based on this definition, a *deficient number does not have enough* proper divisors to add up to itself, while an *abundant number has more than enough* proper divisors to add up to itself.

Weird Numbers A number is said to be *weird* if it is abundant without being equal to the sum of any set of its own proper divisors. For example, 70 is weird because it is abundant $(1 + 2 + 5 + 7 + 10 + 14 + 35 = 74 > 70)$, but no set of the factors 1, 2, 5, 7, 10, 14, 35 adds up to 70.

EXAMPLE 2 Decide whether the given number is deficient or abundant.

(a) 12

The proper divisors of 12 are 1, 2, 3, 4, and 6. The sum of these is 16. Since $12 < 16$, 12 is abundant.

(b) 10

The proper divisors of 10 are 1, 2, and 5. Since $1 + 2 + 5 = 8$, and $10 > 8$, 10 is deficient. ◆

For Further Thought

One of the most remarkable books on number theory is *The Penguin Dictionary of Curious and Interesting Numbers* (1986) by David Wells. This book contains fascinating numbers and their properties, some of which are given here.

113, 199, and 337 are the only three-digit numbers that are prime and all rearrangements of the digits are prime.

Find the sum of the cubes of the digits of 136: $1^3 + 3^3 + 6^3 = 244$. Repeat the process with the digits of 244: $2^3 + 4^3 + 4^3 = 136$. We're back to where we started.

635,318,657 is the smallest number that can be expressed as the sum of two fourth powers in two ways: $635{,}318{,}657 = 59^4 + 158^4 = 133^4 + 134^4$.

The number 24,678,050 has an interesting property: $24{,}678{,}050 = 2^8 + 4^8 + 6^8 + 7^8 + 8^8 + 0^8 + 5^8 + 0^8$.

The number 54,748 has a similar interesting property: $54{,}748 = 5^5 + 4^5 + 7^5 + 4^5 + 8^5$.

The number 3,435 has this property: $3{,}435 = 3^3 + 4^4 + 3^3 + 5^5$.

For anyone whose curiosity is piqued by such facts, this book is for you!

For Group Discussion
Have each student in the class choose a three-digit number that is a multiple of 3. Add the cubes of the digits. Repeat the process until the same number is obtained over and over. Then, have the students compare their results. What is curious and interesting about this process?

Sociable Numbers An extension of the idea of amicable numbers results in sociable numbers. In a chain of sociable numbers, the sum of the proper divisors of each number is the next number in the chain, and the sum of the proper divisors of the last number in the chain is the first number. Here is a 5-link chain of sociable numbers:

12,496
14,288
15,472
14,536
14,264.

The number 14,316 starts a 28-link chain of sociable numbers. Mathematicians have termed a sociable chain with three links a "crowd" (get it?), but so far, no crowds have been found.

Amicable (Friendly) Numbers

Suppose that we add the proper divisors of 284:

$$1 + 2 + 4 + 71 + 142 = 220.$$

Their sum is 220. Now, add the proper divisors of 220:

$$1 + 2 + 4 + 5 + 10 + 11 + 20 + 22 + 44 + 55 + 110 = 284.$$

Notice that the sum of the proper divisors of 220 is 284, while the sum of the proper divisors of 284 is 220. Number pairs such as these are said to be *amicable* or *friendly*.

Amicable or Friendly Numbers

The natural numbers a and b are **amicable,** or **friendly,** if the sum of the proper divisors of a is b, and the sum of the proper divisors of b is a.

The smallest pair of amicable numbers, 220 and 284, was known to the Pythagoreans, but it was not until more than 1,000 years later that the next pair, 18,416 and 17,296, was discovered. Many more pairs were found over the next few decades, but it took a 16-year-old Italian boy named Nicolo Paganini to discover in the year 1866 that the pair of amicable numbers 1,184 and 1,210 had been overlooked for centuries!

Prime Search Goes On According to the September 16, 1989, issue of *Science News* a team of researchers at the Amdahl Corporation's Key Computer Laboratories in Fremont, California discovered that the number

$$391{,}581 \times 2^{216{,}193} - 1$$

is prime. Until the discovery that $2^{756{,}839} - 1$ is prime (see the text), it was the largest known prime, having 65,087 digits. Can you explain why it is *not* a Mersenne prime?

Mersenne Primes

Prime numbers of the form $2^n - 1$ have been studied by mathematicians for centuries. Of particular importance are prime numbers of the form $2^p - 1$, where p is a prime number. The French monk Marin Mersenne (1588–1648) studied prime numbers of this form, and they are named Mersenne primes in his honor. He acted as a sort of mathematics broadcaster, corresponding with the mathematicians of his day, spreading news about inquiries and discoveries.

Early mathematicians believed that $2^p - 1$ was prime for every prime number p. While this is true for $p = 2, 3, 5$, and 7, it was discovered in 1536 that $2^{11} - 1 = 2{,}047$ is not prime. Further investigations followed, and with the computer age upon us now, larger and larger Mersenne primes are being discovered. According to an article in the June 1992 issue of *Focus* (the newsletter of the Mathematical Association of America), a Cray-2 supercomputer discovered that the 227,832 digit number $2^{756{,}839} - 1$ is prime. This was the thirty-second known Mersenne prime, eclipsing the previous record-holder, $2^{216{,}091} - 1$. There are now thirty-three known Mersenne primes.

EXAMPLE 3 Use the Mersenne prime $2^{756{,}839} - 1$ to give an expression for a perfect number.

According to a formula discovered by Euclid, $2^n - 1$ is prime if and only if $2^{n-1}(2^n - 1)$ is perfect. Therefore, with $n = 756{,}839$, we know that

$$2^{756{,}838}(2^{756{,}839} - 1)$$

is perfect. (This perfect number has 455,663 digits in its decimal representation.) ■

Goldbach's Conjecture

Of the many unsolved problems in mathematics, one of the most famous is Goldbach's conjecture. The mathematician Christian Goldbach (1690–1764) made this conjecture (guess) about prime numbers: He said that every even number greater than 2 can be written as the sum of two prime numbers. For example, $8 = 5 + 3$ and $10 = 5 + 5$. (In many cases, there are several ways to satisfy the conjecture. For example, 10 may also be written as $7 + 3$.) For several centuries, mathematicians have tried without success to prove Goldbach's conjecture.

EXAMPLE 4 Write each even number as the sum of two primes.

(a) 18

18 = 5 + 13. Another way of writing it is 7 + 11. Notice that 1 + 17 is *not* valid, since by definition 1 is not a prime number.

(b) 60

60 = 7 + 53. Can you find other ways? Why is 3 + 57 not valid? ■

A Dull Number? The Indian mathematician Srinivasa Ramanujan (1887–1920) developed many ideas in number theory. His friend and collaborator on occasion was G. H. Hardy, also a number theorist and professor at Cambridge University in England.

The 1991 Ramanujan biography *The Man Who Knew Infinity*, by Robert Kanigel, is a compelling drama of genius in the tragic context of cultural dichotomy.

A story has been told about Ramanujan that illustrates his genius. Hardy once mentioned to Ramanujan that he had just taken a taxicab with a rather dull number: 1,729. Ramanujan countered by saying that this number isn't dull at all; it is the smallest natural number that can be expressed as the sum of two cubes in two different ways:

$$1^3 + 12^3 = 1{,}729$$

and $$9^3 + 10^3 = 1{,}729.$$

Can you show that 85 can be written as the sum of two squares in two ways?

Twin Primes

Prime numbers that differ by 2 are called twin primes. For example, some twin primes are 3 and 5, 5 and 7, 11 and 13, and so on. One of the many unproved conjectures about prime numbers deals with twin primes. It states that there are infinitely many pairs of twin primes. Like Goldbach's conjecture, this has never been proved. You may wish to verify that there are eight such pairs less than 100, using the Sieve of Eratosthenes in Table 1.

The Infinitude of Primes

While there are many unsolved problems in number theory, the theorem that there are infinitely many primes was proved by Euclid more than 2,000 years ago. It remains today as one of the most elegant proofs in all of mathematics. (An *elegant* mathematical proof is one that exhibits the desired result in a most direct, concise manner. Mathematicians strive for elegance in their proofs.) It is called a **proof by contradiction.**

A statement can be proved by contradiction as follows: We assume that the negation of the statement is true. The assumption that the negation is true is used to produce some sort of contradiction, or absurdity. The fact that the negation of the original statement leads to a contradiction means that the original statement must be true.

In order to better understand a particular part of the proof that there are infinitely many primes, it is helpful to examine the following argument.

Suppose that $M = 2 \cdot 3 \cdot 5 \cdot 7 + 1 = 211$. Now M is the product of the first four prime numbers, plus 1. If we divide 211 by each of the primes 2, 3, 5, and 7, the remainder is always 1.

$$\begin{array}{r} 105 \\ 2\overline{)211} \\ \underline{210} \\ 1 \end{array} \qquad \begin{array}{r} 70 \\ 3\overline{)211} \\ \underline{210} \\ 1 \end{array} \qquad \begin{array}{r} 42 \\ 5\overline{)211} \\ \underline{210} \\ 1 \end{array} \qquad \begin{array}{r} 30 \\ 7\overline{)211} \\ \underline{210} \\ 1 \end{array}$$

So 211 is not divisible by any of the primes 2, 3, 5, and 7.

Now we are ready to prove that there are infinitely many primes. If it can be shown that *there is no largest prime number,* then there must be infinitely many primes.

Beheading a Prime to Get a Prime The number

357,686,312,646,216,567,629,137

is, according to David Wells, the largest prime such that if you chop off the lead digit over and over, the resulting numbers are prime.

THEOREM There is no largest prime number.

Suppose that there is a largest prime number and call it P. Now form the number M such that

$$M = p_1 \cdot p_2 \cdot p_3 \cdot \cdots \cdot P + 1,$$

where $p_1, p_2, p_3, \ldots, P$ represent all the primes less than or equal to P. Now the number M must be either prime or composite.

1. Suppose that M is prime.
   M is obviously larger than P, so if M is prime, it is larger than the assumed largest prime P. We have reached a *contradiction.*
2. Suppose that M is composite.
   If M is composite, it must have a prime factor. But none of $p_1, p_2, p_3, \ldots, P$ are factors of M, since division by each will leave a remainder of 1. (Recall the above argument.) So if M has a prime factor, it must be greater than P. But this is a *contradiction,* since P is the assumed largest prime.

In either case 1 or 2, we reach a contradiction. The whole argument was based upon the assumption that a largest prime exists, but since this leads to contradictions, there must be no largest prime, or equivalently, there are infinitely many primes. ■

Fermat's Last Theorem

In any right triangle with shorter sides a and b, and longest side (hypotenuse) c, the equation $a^2 + b^2 = c^2$ will hold true. This is the famous Pythagorean theorem. For example, with $a = 3$, $b = 4$, and $c = 5$, we have

$$3^2 + 4^2 = 5^2$$
$$25 = 25.$$

It turns out that there are infinitely many such triples of numbers that satisfy this equation. But are there any triples of natural numbers that satisfy $a^n + b^n = c^n$ for natural numbers $n \geq 3$? Pierre de Fermat, profiled in a margin note in the next section thought not and, as was his custom, wrote in the margin of a book that he had "a truly wonderful proof" for this, but that the margin was "too small to contain it." Was Fermat just joking, did he indeed have a proof, or did he have an incorrect proof?

Whatever the case, Fermat's assertion was the object of more than 400 years of attempts by mathematicians to provide a suitable proof. While it was verified for many specific cases (Fermat himself proved it for $n = 3$), a proof of the general case could not be found. But after eight years of concentrated effort, Princeton mathematician Andrew Wiles presented a lecture at Cambridge based on a paper that finally proved Fermat's Last Theorem. Nigel Boston provided the following status report of the development of the proof in the March 1995 issue of the *College Mathematics Journal* ("A Taylor-made Plug for Wiles' Proof," pp. 100–105):

Andrew Wiles

> Wiles gave a course in Princeton in the spring of 1993 in which he announced his proof. Unfortunately, all but Nicholas Katz had dropped the course and so were not there to witness that historic moment! Katz and other confidants looked through Wiles' work and believed enough of it for him to go public. On June 23, 1993, at the Newton Institute in Cambridge, England, to a packed audience, most of whom had been tipped off about the amazing developments in his lectures the previous two days, as cameras flashed, Wiles announced that he had proven Fermat's Last Theorem. The next week at the Institute was far from normal. We were assailed by cameramen, requests for interviews, phones ringing off the hook, and celebrations.
>
> It appeared that Wiles hoped to have a manuscript ready for public viewing in the fall. He submitted his work to *Inventiones Mathematicae,* and a team of referees began to pick it apart. Some errors were found and taken care of. Then rumors of an unfixable gap started circulating and in November 1993 Wiles officially confirmed this.
>
> Some events have clouded the water. Marilyn vos Savant of "Ask Marilyn" fame wrote a book in three weeks, misleadingly claiming to have consulted leading mathematicians in the field and coming up with reasons why the proof is false based more on her lack of understanding of the area than on fact. Wiles was asked to model for another kind of Gap (jeans; he declined) and appeared in *People* magazine's 1993 end-of-year issue as one of the 25 most intriguing people of the year. Around April 1, 1994, hoaxes appeared saying that Elkies had found a counterexample and that Wiles' graduate students had fixed the gap.
>
> In the spring of 1994, Wiles gave a series of talks at Princeton on his work. He did not get into the details of the proof, but at least produced Chapter 1 of his paper. At the same time, his former student Taylor, now a lecturer at Cambridge University, was visiting Princeton. Thus began their joint effort, culminating in the release of the manuscripts on October 25, 1994. The manuscripts have been submitted to the *Annals of Mathematics* for publication and have been received with enthusiasm by the experts, who have yet to find fault with them. In fact, this new approach is shorter than the old and many further improvements have been suggested recently. Thus there is hope that at long last a reasonably concise, albeit advanced, proof of Fermat's Last Theorem will soon be available.

EXAMPLE 5 One of the theorems proved by Fermat is as follows: *Every odd prime can be expressed as the difference of two squares in one and only one way.* For each of the following odd primes, express as the difference of two squares.

(a) 3

$$3 = 4 - 1 = 2^2 - 1^2$$

(b) 7

$$7 = 16 - 9 = 4^2 - 3^2$$

5.2 EXERCISES

Based on your readings in this section, decide whether each of the following (Exercises 1–10) is true *or* false.

1. There are infinitely many prime numbers.
2. The prime numbers 2 and 3 are twin primes.
3. There is no perfect number between 496 and 8,128.
4. Because its last digit is 4, the number 54,034 cannot possibly be perfect.
5. Any prime number must be deficient.
6. The equation $17 + 51 = 68$ verifies Goldbach's conjecture for the number 68.
7. The number $2^5 - 1$ is an example of a Mersenne prime.
8. The number 31 cannot be represented as the difference of two squares.
9. The number $2^6(2^7 - 1)$ is perfect.
10. A natural number greater than 1 will be one and only one of the following: perfect, deficient, or abundant.
11. The proper divisors of 496 are 1, 2, 4, 8, 16, 31, 62, 124, and 248. Use this information to verify that 496 is perfect.
12. The proper divisors of 8,128 are 1, 2, 4, 8, 16, 32, 64, 127, 254, 508, 1,016, 2,032, and 4,064. Use this information to verify that 8,128 is perfect.
13. As mentioned in the text, when $2^n - 1$ is prime, $2^{n-1}(2^n - 1)$ is perfect. By letting $n = 2, 3, 5,$ and 7, we obtain the first four perfect numbers. Show that $2^n - 1$ is prime for $n = 13$, and then find the decimal digit representation for the fifth perfect number.
14. In 1988, the largest known prime number was $2^{216,091} - 1$. Use the formula in Exercise 13 to write an expression for the perfect number generated by this prime number.
15. It has been proved that the reciprocals of *all* the positive divisors of a perfect number have a sum of 2. Verify this for the perfect number 6.
16. Consider the following equations.

$$6 = 1 + 2 + 3$$
$$28 = 1 + 2 + 3 + 4 + 5 + 6 + 7$$

Show that a similar equation is valid for the third perfect number, 496.

Determine whether each number is abundant *or* deficient.

17. 36 **18.** 30 **19.** 75 **20.** 95

21. There are four abundant numbers between 1 and 25. Find them. (*Hint:* They are all even, and no prime number is abundant.)
22. Explain why a prime number must be deficient.
23. The first odd abundant number is 945. Its proper divisors are 1, 3, 5, 7, 9, 15, 21, 27, 35, 45, 63, 105, 135, 189, and 315. Use this information to verify that 945 is abundant.
24. Explain in your own words the terms perfect number, abundant number, and deficient number.
25. Nicolo Paganini's numbers 1,184 and 1,210 are amicable. The proper divisors of 1,184 are 1, 2, 4, 8, 16, 32, 37, 74, 148, 296, and 592. The proper divisors of 1,210 are 1, 2, 5, 10, 11, 22, 55, 110, 121, 242, and 605. Use the definition of amicable (friendly) numbers to show that they are indeed amicable.
26. An Arabian mathematician of the ninth century stated the following.

If the three numbers

$$x = 3 \cdot 2^{n-1} - 1,$$
$$y = 3 \cdot 2^n - 1,$$
and
$$z = 9 \cdot 2^{2n-1} - 1$$

are all prime and $n \geq 2$, then $2^n xy$ and $2^n z$ are amicable numbers.

(a) Use $n = 2$, and show that the result is the pair of amicable numbers mentioned in the discussion in the text.

(b) Use $n = 4$ to obtain another pair of amicable numbers.

Write each of the following even numbers as the sum of two primes. (There may be more than one way to do this.)

27. 14 **28.** 22 **29.** 26 **30.** 32

31. Joseph Louis Lagrange (1736–1813) conjectured that every odd natural number greater than 5 can be written as a sum $a + 2b$, where a and b are both primes. Verify this for the odd natural number 11.
32. An unproved conjecture in number theory states that every natural number multiple of 6 can be written as the difference of two primes. Verify this for 6, 12, and 18.

Find one pair of twin primes between the two numbers given.

33. 65 and 80

34. 85 and 105

35. 125 and 140

While Pierre De Fermat is probably best known for his now famous "last theorem," he did provide proofs of many other theorems in number theory. Exercises 36–40 investigate some of these theorems.

36. If p is prime and the natural numbers a and p have no common factor except 1, then $a^{p-1} - 1$ is divisible by p.
(a) Verify this for $p = 5$ and $a = 3$.
(b) Verify this for $p = 7$ and $a = 2$.

37. Every odd prime can be expressed as the difference of two squares in one and only one way.
(a) Find this one way for the prime number 5.
(b) Find this one way for the prime number 11.

38. A prime number of the form $4k + 1$ can be represented as the sum of two squares.
(a) The prime number 5 satisfies the conditions of the theorem, with $k = 1$. Verify this theorem for 5.
(b) Verify this theorem for 13 (here, $k = 3$).

39. There is only one solution in natural numbers for $a^2 + 2 = b^3$, and it is $a = 5$, $b = 3$. Verify this solution.

40. There are only two solutions in integers for $a^2 + 4 = b^3$, and they are $a = 2$, $b = 2$, and $a = 11$, $b = 5$. Verify these solutions.

While there are infinitely many primes, no formula has ever been found that will yield only prime numbers for all natural numbers n. The formula

$$p = n^2 - n + 41$$

will give prime numbers for all natural numbers $n = 1$ *through* $n = 40$. *Verify that this is true for the following values of n.*

41. $n = 1$ **42.** $n = 5$ **43.** $n = 7$ **44.** $n = 9$

45. Explain why the formula used in Exercises 41–44 fails when $n = 41$.

The formula

$$p = n^2 - 79n + 1{,}601$$

will give prime numbers for all values of n less than 80, *but fails with* $n = 80$, *since for* $n = 80$, $p = 1{,}681 = 41^2$. *Determine the primes generated by this formula for the following values of n.*

46. $n = 1$ **47.** $n = 5$ **48.** $n = 7$ **49.** $n = 9$

50. In Euclid's proof that there is no largest prime, we formed a number M by taking the product of primes and adding 1. Observe the pattern below.

$M = 2 + 1 = 3$ (3 is prime)
$M = 2 \cdot 3 + 1 = 7$ (7 is prime)
$M = 2 \cdot 3 \cdot 5 + 1 = 31$ (31 is prime)
$M = 2 \cdot 3 \cdot 5 \cdot 7 + 1 = 211$ (211 is prime)
$M = 2 \cdot 3 \cdot 5 \cdot 7 \cdot 11 + 1 = 2{,}311$ (2,311 is prime)

It seems as though this pattern will always yield a prime number. Now evaluate

$$M = 2 \cdot 3 \cdot 5 \cdot 7 \cdot 11 \cdot 13 + 1.$$

Is M prime or composite? If composite, give its prime factorization.

Greatest Common Factor and Least Common Multiple

The following program for the TI-82 calculator will give the greatest common factor (GCF) and the least common multiple (LCM) of two numbers.

```
PROGRAM: GCFLCM
: Input "1ST NUMBER
=", A
: A→C
: Input "2ND NUMBER=",
B
: B→D
: Lbl 1
: f Part (A/B) * B→R
: round (R, 0)→R
: If R=0
: Goto 2
: B→A
: R→B
: Goto 1
: Lbl 2
: Disp "GCF =", B
: CD/B→L
: Disp "LCM =", L
```

This program, written by Tim Tilton, was provided courtesy of Texas Instruments. See page 235 for an example of the display obtained after the program is run.

What is the largest natural number that is a factor (divisor) of both 36 and 54? What is the smallest natural number that is a multiple of both 72 and 150? These are the kinds of questions we will answer in this section.

The **greatest common factor** of a group of natural numbers is the largest natural number that is a factor of all the numbers in the group. For example, 18 is the greatest common factor of 36 and 54, since 18 is the largest natural number that divides both 36 and 54. Also, 1 is the greatest common factor of 7 and 18.

Greatest common factors can be found by using prime factorizations. To see how, let us find the greatest common factor of 36 and 54. First, write the prime factorization of each number:

$$36 = 2^2 \cdot 3^2 \qquad \text{and} \qquad 54 = 2^1 \cdot 3^3.$$

To find the greatest common factor, form the product of the primes common to the factorizations, with each prime raised to the power indicated by the *smallest* exponent that it has in any factorization. Here, the prime 2 has 1 as the smallest exponent (in $54 = 2^1 \cdot 3^3$), while the prime 3 has 2 as the smallest exponent (in $36 = 2^2 \cdot 3^2$). The greatest common factor of 36 and 54 is

$$2^1 \cdot 3^2 = 2 \cdot 9 = 18,$$

as stated earlier. We summarize as follows.

Finding the Greatest Common Factor (Prime Factors Method)

1. Write the prime factorization of each number.
2. Choose all primes common to *all* factorizations, with each prime raised to the *smallest* exponent that it has in any factorization.
3. Form the product of all the numbers in Step 2; this product is the greatest common factor.

EXAMPLE 1 Find the greatest common factor of 360 and 2,700.

Write the prime factorization of each number:

$$360 = 2^3 \cdot 3^2 \cdot 5 \qquad \text{and} \qquad 2{,}700 = 2^2 \cdot 3^3 \cdot 5^2.$$

Now find the primes common to both factorizations, with each prime having as exponent the *smallest* exponent from either product: 2^2, 3^2, 5. Then form the product of these numbers.

$$\text{Greatest common factor} = 2^2 \cdot 3^2 \cdot 5 = 180$$

The greatest common factor of 360 and 2,700 is 180. ■

EXAMPLE 2 Find the greatest common factor of the three numbers 720, 1,000, and 1,800.

To begin, write the prime factorization for each number:

$$720 = 2^4 \cdot 3^2 \cdot 5, \qquad 1{,}000 = 2^3 \cdot 5^3, \qquad \text{and} \qquad 1{,}800 = 2^3 \cdot 3^2 \cdot 5^2.$$

Use the smallest exponent on each prime common to the factorizations:

$$\text{Greatest common factor} = 2^3 \cdot 5 = 40.$$

(The prime 3 is not used in the greatest common factor since the prime 3 does not appear in the prime factorization of 1,000.) ■

Pierre de Fermat, a government official who did not interest himself in mathematics until he was past 30, faithfully and accurately carried out the duties of his post, devoting leisure time to the study of mathematics. He was a worthy scholar, best known for his work in number theory. His other major contributions involved certain applications in geometry and his original work in probability.

Unfortunately, much of Fermat's best work survived only on loose sheets or jotted, without proof, in the margins of works that he read. Mathematicians of subsequent generations have not always had an easy time verifying some of those results, though their truth has generally not been doubted.

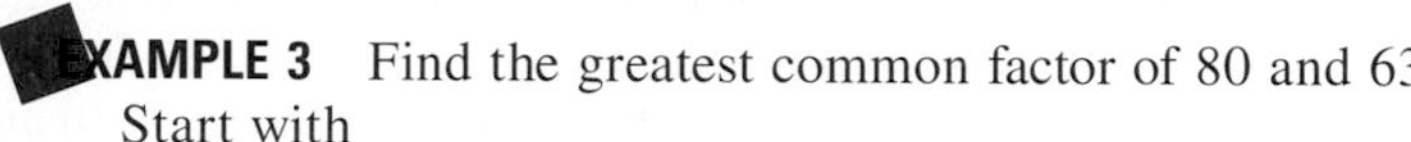

EXAMPLE 3 Find the greatest common factor of 80 and 63.

Start with

$$80 = 2^4 \cdot 5 \qquad \text{and} \qquad 63 = 3^2 \cdot 7.$$

There are no primes in common here, so

$$\text{Greatest common factor} = 1.$$

The number 1 is the largest number that will divide into both 80 and 63. ■

Two numbers, such as 80 and 63, with a greatest common factor of 1 are called **relatively prime** numbers—that is, they are prime *relative* to one another.

Another method of finding the greatest common factor involves dividing the numbers by common prime factors.

Finding the Greatest Common Factor (Dividing by Prime Factors Method)

1. Write the numbers in a row.
2. Divide each of the numbers by a common prime factor. Try 2, then try 3, and so on.
3. Divide the quotients by a common prime factor. Continue until no prime will divide into all the quotients.
4. The product of the primes in Steps 2 and 3 is the greatest common factor.

This method is illustrated in the next example.

EXAMPLE 4 Find the greatest common factor of 12, 18, and 60.

Write the numbers in a row and divide by 2.

$$\begin{array}{r|rrr} 2 & 12 & 18 & 60 \\ \hline & 6 & 9 & 30 \end{array}$$

The numbers 6, 9, and 30 are not all divisible by 2, but they are divisible by 3.

$$\begin{array}{r|rrr} 2 & 12 & 18 & 60 \\ \hline 3 & 6 & 9 & 30 \\ \hline & 2 & 3 & 10 \end{array}$$

No prime divides into 2, 3, and 10, so the greatest common factor of the numbers 12, 18, and 60 is given by the product of the primes on the left, 2 and 3.

$$\begin{array}{r|rrr} 2 & 12 & 18 & 60 \\ \hline 3 & 6 & 9 & 30 \\ \hline & 2 & 3 & 10 \end{array}$$

$$2 \cdot 3 = 6$$

The greatest common factor is 6. ■

There is yet another method of finding the greatest common factor of two numbers that does not require factoring into primes or successively dividing by primes. It is called the **Euclidean algorithm,*** and it is illustrated in the next example.

EXAMPLE 5 Use the Euclidean algorithm to find the greatest common factor of 90 and 168.

Step 1: Begin by dividing the larger, 168, by the smaller, 90. Disregard the quotient, but note the remainder.

$$90\overline{)168} = 1 \text{, } 168 - 90 = 78$$

Step 2: Divide the smaller of the two numbers by the remainder obtained in Step 1. Once again, note the remainder.

$$78\overline{)90} = 1 \text{, } 90 - 78 = 12$$

Step 3: Continue dividing the successive remainders, as many times as necessary to obtain a remainder of 0.

$$12\overline{)78} = 6 \text{, } 78 - 72 = \mathbf{6} \quad \text{Greatest common factor}$$

$$6\overline{)12} = 2 \text{, } 12 - 12 = 0$$

Step 4: The *last positive remainder* in this process is the greatest common factor of 90 and 168. It can be seen that their greatest common factor is 6. ■

The Euclidean algorithm is particularly useful if the two numbers are difficult to factor into primes. We summarize the algorithm here.

* For a proof that this process does indeed give the greatest common divisor, see *Elementary Introduction to Number Theory, Second Edition,* by Calvin T. Long, pp. 34–35.

Finding the Greatest Common Factor (Euclidean Algorithm)

To find the greatest common factor of two unequal numbers, divide the larger by the smaller. Note the remainder, and divide the previous divisor by this remainder. Continue the process until a remainder of 0 is obtained. The greatest common factor is the last positive remainder obtained in this process.

Closely related to the idea of the greatest common factor is the concept of the least common multiple. The **least common multiple** of a group of natural numbers is the smallest natural number that is a multiple of all the numbers in the group. For example, if we wish to find the least common multiple of 15 and 10, we can specify the sets of multiples of 15 and multiples of 10.

$$\text{Multiples of 15:} \quad \{15, 30, 45, 60, 75, 90, 105, \ldots\}$$
$$\text{Multiples of 10:} \quad \{10, 20, 30, 40, 50, 60, 70, \ldots\}$$

The set of natural numbers that are multiples of *both* 15 and 10 form the set of *common multiples*:

$$\{30, 60, 90, 120, \ldots\}.$$

While there are infinitely many common multiples, the *least* common multiple is observed to be 30.

A method similar to the first one given for the greatest common factor may be used to find the least common multiple of a group of numbers.

Finding the Least Common Multiple (Prime Factors Method)

1. Write the prime factorization of each number.
2. Choose all primes belonging to *any* factorization, with each prime raised to the power indicated by the *largest* exponent that it has in any factorization.
3. Form the product of all the numbers in Step 2; this product is the least common multiple.

EXAMPLE 6 Find the least common multiple of 135, 280, and 300.

Write the prime factorizations:

$$135 = 3^3 \cdot 5, \qquad 280 = 2^3 \cdot 5 \cdot 7, \qquad \text{and} \qquad 300 = 2^2 \cdot 3 \cdot 5^2.$$

Form the product of all the primes that appear in *any* of the factorizations. As exponents, use the largest exponent from any factorization.

$$\text{Least common multiple} = 2^3 \cdot 3^3 \cdot 5^2 \cdot 7 = 37{,}800$$

The smallest natural number divisible by 135, 280, and 300 is 37,800. ■

It is shown in more advanced courses that the least common multiple of two numbers m and n can be obtained by dividing their product by their greatest common factor.

Finding the Least Common Multiple (Formula)

$$\text{Least common multiple of } m \text{ and } n = \frac{m \cdot n}{\text{Greatest common factor of } m \text{ and } n}$$

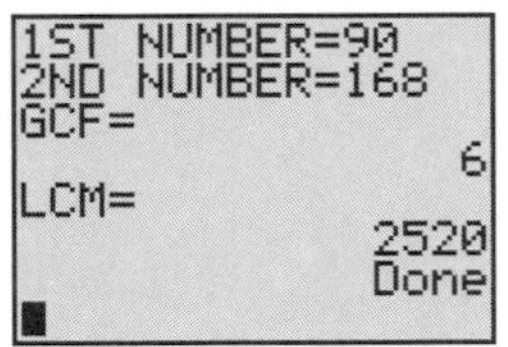

The program on page 231 was used to find the GCF and LCM of 90 and 168. Compare with Examples 5 and 7.

EXAMPLE 7 Use the formula to find the least common multiple of 90 and 168.

In Example 5 we used the Euclidean algorithm to find that the greatest common factor of 90 and 168 is 6. Therefore, the formula gives us

$$\text{Least common multiple of 90 and 168} = \frac{90 \cdot 168}{6} = 2{,}520.$$

Problem Solving

Problems that deal with questions like "How many objects will there be in each group if each group contains the same number of objects?" and "When will two events occur at the same time?" can sometimes be solved using the ideas of greatest common factor and least common multiple.

EXAMPLE 8 The King Theatre and the Star Theatre run movies continuously, and each starts its first feature at 1:00 P.M. If the movie shown at the King lasts 80 minutes and the movie shown at the Star lasts 2 hours, when will the two movies start again at the same time?

First, convert 2 hours to 120 minutes. The question can be restated as follows: "What is the smallest number of minutes it will take for the two movies to start at the same time again?" This is equivalent to saying "What is the least common multiple of 80 and 120?" Using either of the methods described in this section, it can be shown that the least common multiple of 80 and 120 is 240. Therefore, it will take 240 minutes, or $240/60 = 4$ hours for the movies to start again at the same time. By adding 4 hours to 1:00 P.M., we find that they will start together again at 5:00 P.M.

EXAMPLE 9 Joshua Hornsby has 450 football cards and 840 baseball cards. He wants to place them in stacks on a table so that each stack has the same number of cards, and no stack has different types of cards within it. What is the largest number of cards that he can have in each stack?

Here, we are looking for the largest number that will divide evenly into 450 and 840. This is, of course, the greatest common factor of 450 and 840. Using any of the methods described in this section, we find that

$$\text{Greatest common factor of 450 and 840} = 30.$$

Therefore, the largest number of cards he can have in each stack is 30.

5.3 EXERCISES

Decide whether each of the following is true *or* false.

1. Two even natural numbers cannot be relatively prime.
2. Two different prime numbers must be relatively prime.
3. If p is a prime number, then the greatest common factor of p and p^2 is p.
4. If p is a prime number, then the least common multiple of p and p^2 is p^3.
5. There is no prime number p such that the greatest common factor of p and 2 is 2.
6. The set of all common multiples of two given natural numbers is finite.
7. Two natural numbers must have at least one common factor.
8. The least common multiple of two different primes is their product.
9. Two composite numbers may be relatively prime.
10. The set of all common factors of two given natural numbers is finite.

Use the prime factors method to find the greatest common factor of each group of numbers.

11. 70 and 120
12. 180 and 300
13. 480 and 1,800
14. 168 and 504
15. 28, 35, and 56
16. 252, 308, and 504

Use the method of dividing by prime factors to find the greatest common factor of each group of numbers.

17. 60 and 84
18. 130 and 455
19. 310 and 460
20. 234 and 470
21. 12, 18, and 30
22. 450, 1,500, and 432

Use the Euclidean algorithm to find the greatest common factor of each group of numbers.

23. 36 and 60
24. 25 and 70
25. 84 and 180
26. 72 and 120
27. 210 and 560
28. 150 and 480
29. Explain in your own words how to find the greatest common factor of a group of numbers.
30. Explain in your own words how to find the least common multiple of a group of numbers.

Use the prime factors method to find the least common multiple of each group of numbers.

31. 24 and 30
32. 12 and 32
33. 56 and 96
34. 28 and 70
35. 30, 40, and 70
36. 24, 36, and 48

Use the formula given in the text and the results of Exercises 23–28 to find the least common multiple of each group of numbers.

37. 36 and 60
38. 25 and 70
39. 84 and 180
40. 72 and 120
41. 210 and 560
42. 150 and 480
43. If p, q, and r are different primes, and a, b, and c are natural numbers such that $a > b > c$,
    (a) what is the greatest common factor of $p^a q^c r^b$ and $p^b q^a r^c$?
    (b) what is the least common multiple of $p^b q^a$, $q^b r^c$, and $p^a r^b$?
44. Find **(a)** the greatest common factor and **(b)** the least common multiple of $2^{31} \cdot 5^{17} \cdot 7^{21}$ and $2^{34} \cdot 5^{22} \cdot 7^{13}$. Leave your answers in prime factored form.

It is possible to extend the Euclidean algorithm in order to find the greatest common factor of more than two numbers. For example, if we wish to find the greatest common factor of 150, 210, *and* 240, *we can first use the algorithm to find the greatest common factor of two of these (say, for example,* 150 *and* 210*). Then we find the greatest common factor of that result and the third number,* 240. *The final result is the greatest common factor of the original group of numbers.*

Use the Euclidean algorithm as described above to find the greatest common factor of each group of numbers.

45. 150, 210, and 240

46. 12, 75, and 120

47. 90, 105, and 315

48. 48, 315, and 450

49. 144, 180, and 192

50. 180, 210, and 630

If we allow repetitions of prime factors, Venn diagrams (Chapter 2) can be used to find the greatest common factor and the least common multiple of two numbers. For example, consider $36 = 2^2 \cdot 3^2$ *and* $45 = 3^2 \cdot 5$.

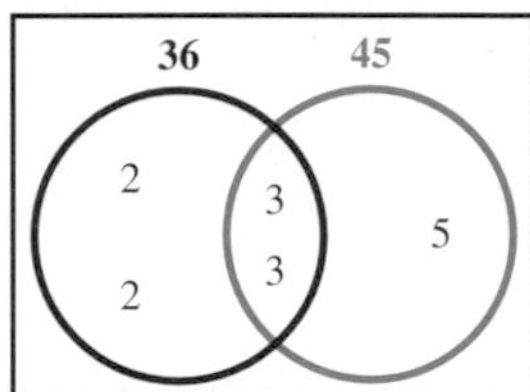

Intersection gives 3, 3.
Union gives 2, 2, 3, 3, 5.

Their greatest common factor is $3^2 = 9$, *and their least common multiple is* $2^2 \cdot 3^2 \cdot 5 = 180$.

Use this method to find **(a)** *the greatest common factor and* **(b)** *the least common multiple of the two numbers given.*

51. 12 and 18

52. 27 and 36

53. 54 and 72

---

54. Suppose that the least common multiple of p and q is q. What can we say about p and q?

55. Suppose that the least common multiple of p and q is pq. What can we say about p and q?

56. Suppose that the greatest common factor of p and q is p. What can we say about p and q?

57. Recall some of your early experiences in mathematics (for example, in the elementary grade classroom). What topic involving fractions required the use of the least common multiple? Give an example.

58. Recall some of your experiences in elementary algebra. What topics required the use of the greatest common factor? Give an example.

Refer to Examples 8 and 9 to solve the following problems.

59. Melissa Lerner and Kay LaBarre work on an assembly line, inspecting electronic calculators. Melissa inspects the electronics of every sixteenth calculator, while Kay inspects the workmanship of every thirty-sixth calculator. If they both start working at the same time, which calculator will be the first that they both inspect?

60. Daniel Lange and Donald Cole work as security guards at a publishing company. Daniel has every sixth night off, and Donald has every tenth night off. If both are off on July 1, what is the next night that they will both be off together?

61. Diane Blake has 240 pennies and 288 nickels. She wants to place the pennies and nickels in stacks so that each stack has the same number of coins, and each stack contains only one denomination of coin. What is the largest number of coins that she can place in each stack?

62. Melissa Martin and Tracy Copley are in a bicycle race, following a circular track. If they start at the same place and travel in the same direction, and Melissa completes a revolution every 40 seconds and Tracy completes a revolution every 45 seconds, how long will it take them before they reach the starting point again simultaneously?

63. John Cross sold some books at \$24 each, and had enough money to buy some concert tickets at \$50 each. He had no money left over after buying the tickets. What is the least amount of money he could have earned from selling the books? What is the least number of books he could have sold?

64. Joan Langevin has some two-by-four pieces of lumber. Some are 60 inches long, and some are 72 inches long. She wishes to cut them so as to obtain equal length pieces. What is the longest such piece she can cut so that no lumber is left over?

5.4 The Fibonacci Sequence and the Golden Ratio

One of the most famous problems in elementary mathematics comes from the book *Liber Abaci,* written in 1202 by Leonardo of Pisa, a.k.a. Fibonacci. The problem is as follows:

> A man put a pair of rabbits in a cage. During the first month the rabbits produced no offspring, but each month thereafter produced one new pair of rabbits. If each new pair thus produced reproduces in the same manner, how many pairs of rabbits will there be at the end of one year?

The solution of this problem leads to a sequence of numbers known as the **Fibonacci sequence.** Here are the first fifteen terms of the Fibonacci sequence:

$$1, 1, 2, 3, 5, 8, 13, 21, 34, 55, 89, 144, 233, 377, 610.$$

The **Fibonacci Association** is a research organization dedicated to investigation into the Fibonacci sequence and related topics. Check your library to see if it has the journal *Fibonacci Quarterly.* The first two journals of 1963 contain a basic introduction to the Fibonacci sequence.

Notice the pattern established in the sequence. After the first two terms (both 1), each term is obtained by adding the two previous terms. For example, the third term is obtained by adding $1 + 1$ to get 2, the fourth term is obtained by adding $1 + 2$ to get 3, and so on. This can be described by a mathematical formula known as a *recursion formula.* If F_n represents the Fibonacci number in the nth position in the sequence, then

$$F_1 = 1$$
$$F_2 = 1$$
$$F_n = F_{n-1} + F_{n-2} \quad \text{for } n \geq 3.$$

The Fibonacci sequence exhibits many interesting patterns, and by inductive reasoning we can make many conjectures about these patterns. However, as we have indicated many times earlier, simply observing a finite number of examples does not provide a proof of a statement. Proofs of the properties of the Fibonacci sequence often involve mathematical induction (covered in college algebra texts). Here we will simply observe the patterns, and will not attempt to provide such proofs, which may be beyond the background of the reader.

Here is an example of one of the many interesting properties of the Fibonacci sequence. Choose any term of the sequence after the first and square it. Then multiply the terms on either side of it, and subtract the smaller result from the larger. The difference is always 1. For example, choose the sixth term in the sequence, 8. The square of 8 is 64. Now multiply the terms on either side of 8: $5 \times 13 = 65$. Subtract 64 from 65 to get $65 - 64 = 1$. This pattern continues throughout the sequence.

EXAMPLE 1 Find the sum of the squares of the first n Fibonacci numbers for $n = 1, 2, 3, 4, 5$, and examine the pattern. Generalize this relationship.

$$1^2 = 1 = 1 \cdot 1 = F_1 \cdot F_2$$
$$1^2 + 1^2 = 2 = 1 \cdot 2 = F_2 \cdot F_3$$
$$1^2 + 1^2 + 2^2 = 6 = 2 \cdot 3 = F_3 \cdot F_4$$
$$1^2 + 1^2 + 2^2 + 3^2 = 15 = 3 \cdot 5 = F_4 \cdot F_5$$
$$1^2 + 1^2 + 2^2 + 3^2 + 5^2 = 40 = 5 \cdot 8 = F_5 \cdot F_6$$

The sum of the squares of the first n Fibonacci numbers seems to always be the product of F_n and F_{n+1}. This has been proven to be true, in general, using mathematical induction.

There are many other patterns similar to the one examined in Example 1, and some of them are discussed in the exercises of this section. An interesting property of the decimal value of the reciprocal of 89, the eleventh Fibonacci number, is examined in Example 2.

EXAMPLE 2 Observe the steps of the long-division process used to find the first few decimal places for 1/89.

```
      .011235. . .
  89)1.000000
     89
     110
      89
      210
      178
       320
       267
        530
        445
         850. . .
```

Notice that after the 0 in the tenths place, the next five digits are the first five terms of the Fibonacci sequence. In addition, as indicated in color in the process, the digits 1, 1, 2, 3, 5, 8 appear in the division steps. Now, look at the digits next to the ones in color, beginning with the second "1"; they, too, are 1, 1, 2, 3, 5, ...

If the division process is continued past the final step shown above, the pattern seems to stop, since to ten decimal places, $1/89 \approx .0112359551$. (The

The following program for the TI-82 utilizes the *Binet form* of the nth Fibonacci number (see Exercises 35–38) to determine its value.

```
PROGRAM: FIB
: Clr Home
: Disp "WHICH TERM?"
: Input N
: (1+√5)/2→A
: (1 − √5)/2→B
: (A ∧ N − B ∧
  N)/√5→F
: Disp F
```

The following screen indicates that the twentieth Fibonacci number is 6,765.

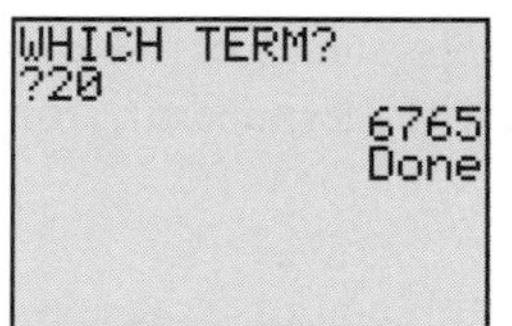

decimal representation actually begins to repeat later in the process, since 1/89 is a rational number.) However, the sum below indicates how the Fibonacci numbers are actually "hidden" in this decimal.

$$
\begin{array}{rl}
 & .01 \\
 & .001 \\
 & .0002 \\
 & .00003 \\
 & .000005 \\
 & .0000008 \\
 & .00000013 \\
 & .000000021 \\
 & .0000000034 \\
 & .00000000055 \\
 & \underline{.000000000089} \\
1/89 = & .0112359550\ldots
\end{array}
$$

■

■ For Further Thought

Or, should we say, "For Further Viewing"? The 1959 animated film *Donald in Mathmagic Land* has endured for more than 35 years as a classic. It provides a 30-minute trip with Donald Duck, led by the Spirit of Mathematics, through the world of mathematics. Several minutes of the film are devoted to the golden ratio (or, as it is termed there, the golden section). Disney provides animation to explain the golden ratio in a way that the printed word simply cannot do. The golden ratio is seen in architecture, nature, and the human body.

The film is available on video and can be purchased or rented quite easily. If you have never seen this film, or if you've seen it and want to appreciate it once again, by all means do so.

For Group Discussion

1. Verify the following Fibonacci pattern in the conifer family. Obtain a pineapple, and count spirals formed by the "scales" of the cone, first counting from lower left to upper right. Then count the spirals from lower right to upper left. What do you find? (A pattern similar to the one seen here also appears in pine cones.)
2. Two popular sizes of index cards are 3″ by 5″ and 5″ by 8″. Why do you think that these are industry-standard sizes?
3. Divide your height by the height of your navel. Find a class average. What value does this come close to?

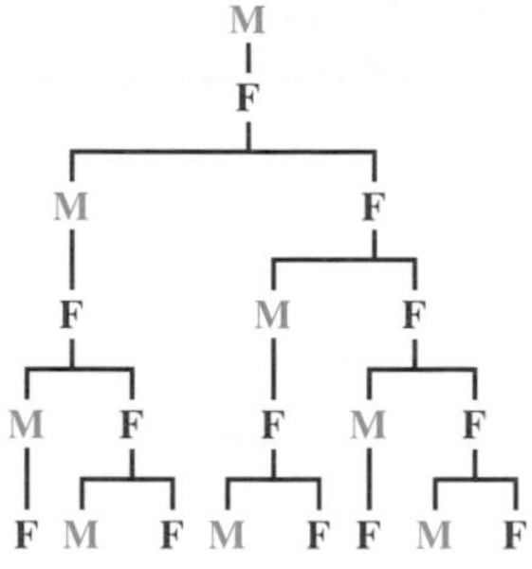

FIGURE 1

Fibonacci patterns have been found in numerous places in nature. For example, male honeybees hatch from eggs which have not been fertilized, so a male bee has only one parent, a female. On the other hand, female honeybees hatch from fertilized eggs, so a female has two parents, one male and one female. Figure 1 shows several generations of a male honeybee.

Notice that in the first generation there is 1 bee, in the second there is 1 bee, in the third there are 2 bees, and so on. These are the terms of the Fibonacci sequence. Furthermore, beginning with the second generation, the numbers of female bees form the sequence, and beginning with the third generation, the numbers of male bees also form the sequence.

Successive terms in the Fibonacci sequence also appear in plants. For example, the photo on the left below shows the double spiraling of a daisy head, with 21 clockwise spirals and 34 counterclockwise spirals. These numbers are successive terms in the sequence. Most pineapples (see the photo on the right) exhibit the Fibonacci sequence in the following way: Count the spirals formed by the "scales" of the cone, first counting from lower left to upper right. Then count the spirals from lower right to upper left. You should find that in one direction you get 8 spirals, and in the other you get 13 spirals, once again successive terms of the Fibonacci sequence. Many pine cones exhibit 5 and 8 spirals, and the cone of the giant sequoia has 3 and 5 spirals.

The Golden Ratio

If we consider the quotients of successive Fibonacci numbers, a pattern emerges.

$$\frac{1}{1} = 1 \qquad \frac{13}{8} = 1.625$$

$$\frac{2}{1} = 2 \qquad \frac{21}{13} \approx 1.615384615$$

$$\frac{3}{2} = 1.5 \qquad \frac{34}{21} \approx 1.619047619$$

$$\frac{5}{3} = 1.666\ldots \qquad \frac{55}{34} \approx 1.617647059$$

$$\frac{8}{5} = 1.6 \qquad \frac{89}{55} = 1.618181818\ldots$$

Continuing the Discussion of the Golden Ratio A fraction such as

$$1+\cfrac{1}{1+\cfrac{1}{1+\cfrac{1}{1+\cdots}}}$$

is called a *continued fraction.* This continued fraction can be evaluated as follows.

Let $x = 1+\cfrac{1}{1+\cfrac{1}{1+\cdots}}$

Then
$$x = 1 + \frac{1}{x}$$
$$x^2 = x + 1$$
$$x^2 - x - 1 = 0.$$

By the quadratic formula from algebra,

$$x = \frac{1 \pm \sqrt{1 - 4(1)(-1)}}{2(1)}$$
$$x = \frac{1 \pm \sqrt{5}}{2}.$$

Notice that the positive solution

$$\frac{1 + \sqrt{5}}{2}$$

is the golden ratio.

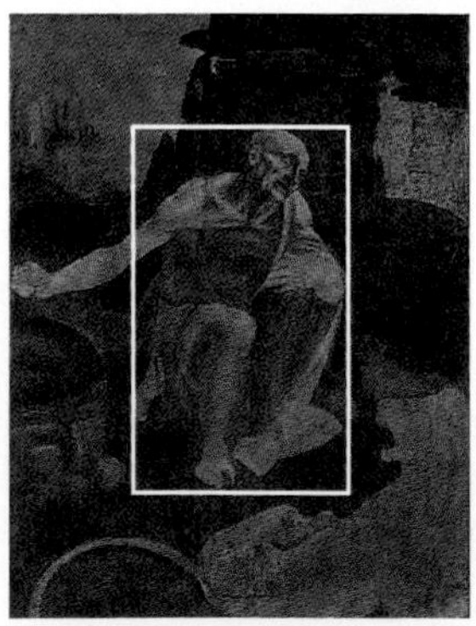

A Golden Rectangle in Art The rectangle outlining the figure in *St. Jerome* by Leonardo da Vinci is an example of a golden rectangle.

It seems as if these quotients are approaching some "limiting value" close to 1.618. This is indeed the case. As we go farther and farther out into the sequence, these quotients approach the number

$$\frac{1 + \sqrt{5}}{2},$$

known as the **golden ratio**, and often symbolized by ϕ, the Greek letter phi.

The golden ratio appears over and over in art, architecture, music, and nature. Its origins go back to the days of the ancient Greeks, who thought that a golden rectangle exhibited the most aesthetically pleasing proportion. A golden rectangle is defined as a rectangle whose dimensions satisfy the equation

$$\frac{\text{Length}}{\text{Width}} = \frac{\text{Length} + \text{Width}}{\text{Length}}.$$

If we let L represent length and W represent width, we have

$$\frac{L}{W} = \frac{L + W}{L}$$
$$\frac{L}{W} = \frac{L}{L} + \frac{W}{L}.$$

Since $L/W = \phi$, $L/L = 1$, and $W/L = 1/\phi$, this equation can be written

$$\phi = 1 + \frac{1}{\phi}.$$

Multiply both sides by ϕ to get

$$\phi^2 = \phi + 1.$$

Subtract ϕ and subtract 1 from both sides to get the quadratic equation

$$\phi^2 - \phi - 1 = 0.$$

Using the quadratic formula from algebra, the positive solution of this equation is found to be $(1 + \sqrt{5})/2 \approx 1.618033989$, the golden ratio.

An example of a golden rectangle is shown in Figure 2. The Parthenon (see the photo), built on the Acropolis in ancient Athens during the fifth century B.C., is an example of architecture exhibiting many examples of golden rectangles.

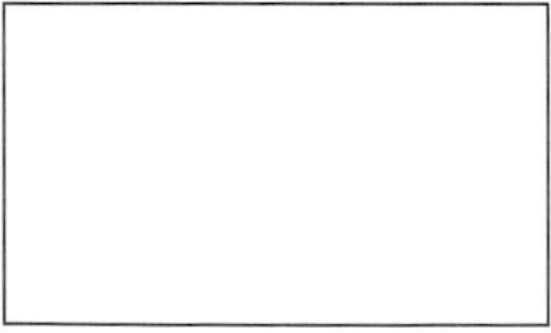

FIGURE 2

To see how the terms of the Fibonacci sequence relate geometrically to the golden ratio, a rectangle that measures 89 by 55 units is constructed. (See Figure 3.) This is a very close approximation to a golden rectangle. Within this

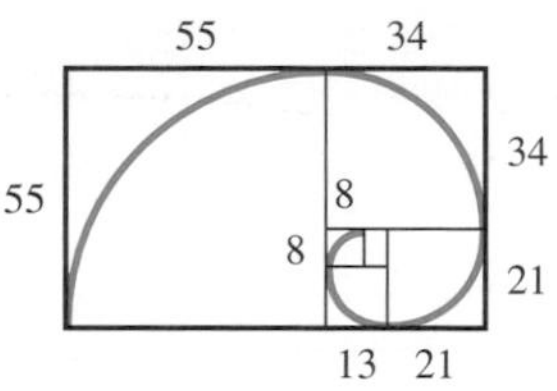

■ **FIGURE 3**

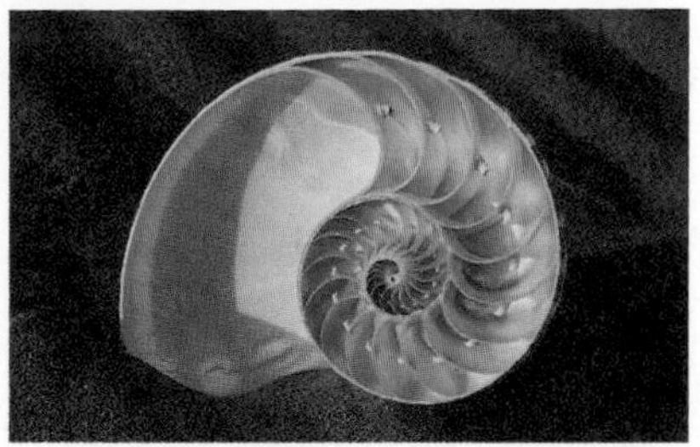

rectangle a square is then constructed, 55 units on a side. The remaining rectangle is also approximately a golden rectangle, measuring 55 units by 34 units. Each time this process is repeated, a square and an approximate golden rectangle are formed. As indicated in the figure, vertices of the square may be joined by a smooth curve known as a *spiral.* This spiral resembles the outline of a cross section of the shell of the chambered nautilus, as shown in the photograph.

EXERCISES

Answer each of the following questions concerning the Fibonacci sequence or the golden ratio.

1. The sixteenth Fibonacci number is 987 and the seventeenth Fibonacci number is 1,597. What is the eighteenth Fibonacci number?

2. Recall that F_n represents the Fibonacci number in the nth position in the sequence. What are the only two values of n such that $F_n = n$?

3. $F_{23} = 28{,}657$ and $F_{25} = 75{,}025$. What is the value of F_{24}?

4. If two successive terms of the Fibonacci sequence are both odd, is the next term even or odd?

5. What is the exact value of the golden ratio?

6. What is the approximate value of the golden ratio to the nearest thousandth?

In Exercises 7–14, a pattern is established involving terms of the Fibonacci sequence. Use inductive reasoning to make a conjecture concerning the next equation in the pattern, and verify it. You may wish to refer to the first few terms of the sequence given in the text.

7. $1 = 2 - 1$
$1 + 1 = 3 - 1$
$1 + 1 + 2 = 5 - 1$
$1 + 1 + 2 + 3 = 8 - 1$
$1 + 1 + 2 + 3 + 5 = 13 - 1$

8. $1 = 2 - 1$
$1 + 3 = 5 - 1$
$1 + 3 + 8 = 13 - 1$
$1 + 3 + 8 + 21 = 34 - 1$
$1 + 3 + 8 + 21 + 55 = 89 - 1$

9. $1 = 1$
$1 + 2 = 3$
$1 + 2 + 5 = 8$
$1 + 2 + 5 + 13 = 21$
$1 + 2 + 5 + 13 + 34 = 55$

10. $1^2 + 1^2 = 2$
$1^2 + 2^2 = 5$
$2^2 + 3^2 = 13$
$3^2 + 5^2 = 34$
$5^2 + 8^2 = 89$

11. $2^2 - 1^2 = 3$
$3^2 - 1^2 = 8$
$5^2 - 2^2 = 21$
$8^2 - 3^2 = 55$

12. $2^3 + 1^3 - 1^3 = 8$
$3^3 + 2^3 - 1^3 = 34$
$5^3 + 3^3 - 2^3 = 144$
$8^3 + 5^3 - 3^3 = 610$

13. $1 = 1^2$
$1 - 2 = -1^2$
$1 - 2 + 5 = 2^2$
$1 - 2 + 5 - 13 = -3^2$
$1 - 2 + 5 - 13 + 34 = 5^2$

14. $1 - 1 = -1 + 1$
$1 - 1 + 2 = 1 + 1$
$1 - 1 + 2 - 3 = -2 + 1$
$1 - 1 + 2 - 3 + 5 = 3 + 1$
$1 - 1 + 2 - 3 + 5 - 8 = -5 + 1$

---

15. Every natural number can be expressed as a sum of Fibonacci numbers, where no number is used more than once. For example, $25 = 21 + 3 + 1$. Express each of the following in this way.
(a) 37 **(b)** 40 **(c)** 52

16. It has been shown that if m divides n, then F_m is a factor of F_n. Show that this is true for the following values of m and n.
(a) $m = 2, n = 6$ **(b)** $m = 3, n = 9$
(c) $m = 4, n = 8$

17. It has been shown that if the greatest common factor of m and n is r, then the greatest common factor of F_m and F_n is F_r. Show that this is true for the following values of m and n.
(a) $m = 10, n = 4$ **(b)** $m = 12, n = 6$
(c) $m = 14, n = 6$

18. For any prime number p except 2 or 5, either F_{p+1} or F_{p-1} is divisible by p. Show that this is true for the following values of p.
(a) $p = 3$ **(b)** $p = 7$ **(c)** $p = 11$

19. Earlier we saw that if a term of the Fibonacci sequence is squared and then the product of the terms on each side of the term is found, there will always be a difference of 1. Follow the steps below, choosing the seventh Fibonacci number, 13.
(a) Square 13. Multiply the terms of the sequence two positions away from 13 (i.e., 5 and 34). Subtract the smaller result from the larger, and record your answer.
(b) Square 13. Multiply the terms of the sequence three positions away from 13. Once again, subtract the smaller result from the larger, and record your answer.
(c) Repeat the process, moving four terms away from 13.
(d) Make a conjecture about what will happen when you repeat the process, moving five terms away. Verify your answer.

20. Here is a number trick that you can perform. Ask someone to pick any two numbers at random and to write them down. Ask the person to determine a third number by adding the first and second, a fourth number by adding the second and third, and so on, until ten numbers are determined. Then ask the person to add these ten numbers. You will be able to give the sum before the person even completes the list, because the sum will always be 11 times the seventh number in the list. Verify that this is true, by using x and y as the first two numbers arbitrarily chosen. (*Hint:* Remember the distributive property from algebra.)

Another Fibonacci-type *sequence that has been studied by mathematicians is the* **Lucas sequence,** *named after a French mathematician of the nineteenth century. The first ten terms of the Lucas sequence are*

1, 3, 4, 7, 11, 18, 29, 47, 76, 123.

Exercises 21–26 pertain to the Lucas sequence.

21. What is the eleventh term of the Lucas sequence?

22. Choose any term of the Lucas sequence and square it. Then multiply the terms on either side of the one you chose. Subtract the smaller result from the larger. Repeat this for a different term of the sequence. Do you get the same result? Make a conjecture about this pattern.

23. The first term of the Lucas sequence is 1. Add the first and third terms. Record your answer. Now add the first, third, and fifth terms and record your answer. Continue this pattern, each time adding another term that is in an *odd* position in the sequence. What do you notice about all of your results?

24. The second term of the Lucas sequence is 3. Add the second and fourth terms. Record your answer. Now add the second, fourth, and sixth terms and record your answer. Continue this pattern, each time adding another term that is in an *even* position of the sequence. What do you notice about all of your sums?

25. Many interesting patterns exist between the terms of the Fibonacci sequence and the Lucas sequence. Make a conjecture about the next equation that would appear in each of the lists and then verify it.

(a)
$1 \cdot 1 = 1$
$1 \cdot 3 = 3$
$2 \cdot 4 = 8$
$3 \cdot 7 = 21$
$5 \cdot 11 = 55$

(b)
$1 + 2 = 3$
$1 + 3 = 4$
$2 + 5 = 7$
$3 + 8 = 11$
$5 + 13 = 18$

(c)
$1 + 1 = 2 \cdot 1$
$1 + 3 = 2 \cdot 2$
$2 + 4 = 2 \cdot 3$
$3 + 7 = 2 \cdot 5$
$5 + 11 = 2 \cdot 8$

26. In the text we illustrate that the quotients of successive terms of the Fibonacci sequence approach the golden ratio. Make a similar observation for the terms of the Lucas sequence; that is, find the decimal approximations for the quotients

$$\frac{3}{1}, \frac{4}{3}, \frac{7}{4}, \frac{11}{7}, \frac{18}{11}, \frac{29}{18},$$

and so on, using a calculator. Then make a conjecture about what seems to be happening.

Recall the **Pythagorean theorem** *from geometry: If a right triangle has legs of lengths a and b and hypotenuse of length c, then $a^2 + b^2 = c^2$. Suppose that we choose any four successive terms of the Fibonacci sequence. Multiply the first and fourth. Double the product of the second and third. Add the squares of the second and third. The three results obtained form a Pythagorean triple (three numbers that satisfy the equation $a^2 + b^2 = c^2$). Find the Pythagorean triple obtained this way using the four given successive terms of the Fibonacci sequence.*

27. 1, 1, 2, 3 **28.** 1, 2, 3, 5 **29.** 2, 3, 5, 8

30. Look at the values of the hypotenuse (c) in the answers to Exercises 27–29. What do you notice about each of them?

31. The following array of numbers is called **Pascal's triangle.**

1
1 1
1 2 1
1 3 3 1
1 4 6 4 1
1 5 10 10 5 1
1 6 15 20 15 6 1

This array is important in the study of counting techniques and probability (see later chapters) and appears in algebra in the binomial theorem. If the triangular array is written in a different form, as follows, and the sums along the diagonals as indicated by the dashed lines are found, an interesting thing occurs. What do you find when the numbers are added?

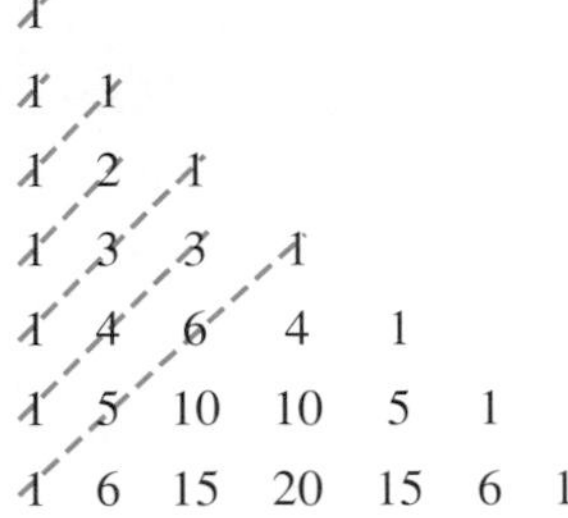

32. Write a paragraph explaining some of the occurrences of the Fibonacci sequence and the golden ratio in your everyday surroundings.

Exercises 33–38 require the use of a scientific calculator.

33. The positive solution of the equation $x^2 - x - 1 = 0$ is $(1 + \sqrt{5})/2$, as indicated in the text. The negative solution is $(1 - \sqrt{5})/2$. Find decimal approximations for both. What similarity do you notice between the two decimals?

34. In some cases, writers define the golden ratio to be the *reciprocal* of $(1 + \sqrt{5})/2$. Find a decimal approximation for the reciprocal of $(1 + \sqrt{5})/2$. What similarity do you notice between the decimals for $(1 + \sqrt{5})/2$ and its reciprocal?

A remarkable relationship exists between the two solutions of $x^2 - x - 1 = 0$,

$$\phi = \frac{1 + \sqrt{5}}{2} \quad \text{and} \quad \overline{\phi} = \frac{1 - \sqrt{5}}{2},$$

and the Fibonacci numbers. To find the nth Fibonacci number without using the recursion formula, evaluate

$$\frac{\phi^n - \overline{\phi}^n}{\sqrt{5}}$$

using a calculator. For example, to find the thirteenth Fibonacci number, evaluate

$$\frac{\left(\frac{1 + \sqrt{5}}{2}\right)^{13} - \left(\frac{1 - \sqrt{5}}{2}\right)^{13}}{\sqrt{5}}.$$

This form is known as the **Binet form** *of the nth Fibonacci number. Use the Binet form and a calculator to find the nth Fibonacci number for each of the following values of n.*

35. $n = 14$ **36.** $n = 20$

37. $n = 22$ **38.** $n = 25$

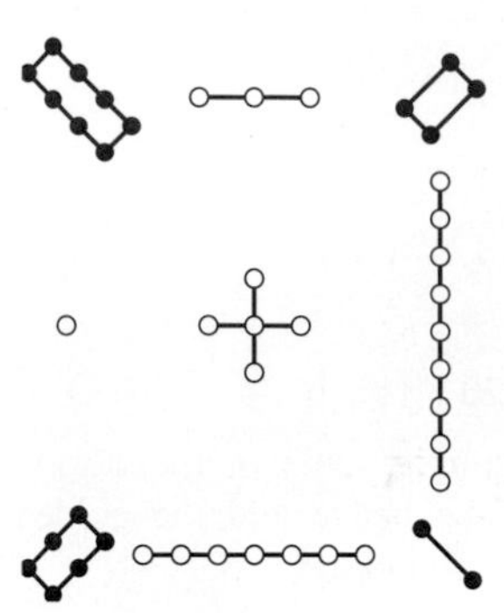

FIGURE 4

8	3	4
1	5	9
6	7	2

FIGURE 5

EXTENSION

Magic Squares

Legend has it that in about 2200 B.C. the Chinese Emperor Yu discovered on the bank of the Yellow River a tortoise whose shell bore the diagram in Figure 4. This so-called *lo-shu* is an early example of a **magic square.** If the numbers of dots are counted and arranged in a square fashion, the array in Figure 5 is obtained. A magic square is a square array of numbers with the property that the sum along each row, column, and diagonal is the same. This common value is called the "magic sum." The **order** of a magic square is simply the number of rows (and columns) in the square. The magic square of Figure 5 is an order 3 magic square.

By using the formula for the sum of the first n terms of an arithmetic sequence, it can be shown that if a magic square of order n has entries 1, 2, 3, . . . , n^2, then the sum of *all entries* in the square is

$$\frac{n^2(n^2 + 1)}{2}.$$

Since there are n rows (and columns), the magic sum of the square may be found by dividing the above expression by n. This results in the following rule for finding the magic sum.

Magic Sum Formula

If a magic square of order n has entries 1, 2, 3, . . . , n^2, then the magic sum MS is given by the formula

$$\text{MS} = \frac{n(n^2 + 1)}{2}.$$

With $n = 3$ in the previous formula we find that the magic sum of the square in Figure 5 is

$$\text{MS} = \frac{3(3^2 + 1)}{2} = 15,$$

which may be verified by direct addition.

There is a method of constructing an odd-order magic square which is attributed to an early French envoy, *de la Loubere,* that is sometimes referred to as the "staircase method." The method is described below for an order 5 square, with entries 1, 2, 3, . . . , 25.

Benjamin Franklin admitted that he would amuse himself while in the Pennsylvania Assembly with magic squares or circles "or any thing to avoid Weariness." He wrote about the usefulness of mathematics in the *Gazette* in 1735 saying that no employment can be managed without arithmetic, no mechanical invention without geometry. He also thought that mathematical demonstrations are better than academic logic for training the mind to reason with exactness and distinguish truth from falsity even outside of mathematics.

The square shown here is one developed by Franklin. It has a sum of 2,056 in each row and diagonal, and, according to Franklin, has the additional property "that a four-square hole being cut in a piece of paper of such size as to take in and show through it just 16 of the little squares, when laid on the greater square, the sum of the 16 numbers so appearing through the hole, wherever it was placed on the greater square should likewise make 2,056." He claimed that it was "the most magically magic square ever made by any magician."

200	217	232	249	8	25	40	57	72	89	104	121	136	153	168	185
58	39	26	7	250	231	218	199	186	167	154	135	122	103	90	71
198	219	230	251	6	27	38	59	70	91	102	123	134	155	166	187
60	37	28	5	252	229	220	197	188	165	156	133	124	101	92	69
201	216	233	248	9	24	41	56	73	88	105	120	137	152	169	184
55	42	23	10	247	234	215	202	183	170	151	138	119	106	87	74
203	214	235	246	11	22	43	54	75	86	107	118	139	150	171	182
53	44	21	12	245	236	213	204	181	172	149	140	117	108	85	76
205	212	237	244	13	20	45	52	77	84	109	116	141	148	173	180
51	46	19	14	243	238	211	206	179	174	147	142	115	110	83	78
207	210	239	242	15	18	47	50	79	82	111	114	143	146	175	178
49	48	17	16	241	240	209	208	177	176	145	144	113	112	81	80
196	221	228	253	4	29	36	61	68	93	100	125	132	157	164	189
62	35	30	3	254	227	222	195	190	163	158	131	126	99	94	67
194	223	226	255	2	31	34	63	66	95	98	127	130	159	162	191
64	33	32	1	256	225	224	193	192	161	160	129	128	97	96	65

B. Franklin inv. J. Ferguson delin. J. Mynde sc.

You might wish to verify the following property of this magic square: The sum of any four numbers that are opposite each other and at equal distances from the center is 514 (which is one-fourth of the magic sum).

Begin by sketching a square as shown in Figure 6. Entries are entered in each cell (small square) of the magic square as follows:

1. Write 1 in the middle cell of the top row.
2. Always try to enter numbers in sequence in the cells by moving diagonally from lower left to upper right. There are two exceptions to this:
   (a) If you go outside of the magic square, move all the way across the row or column to enter the number. Then proceed to move diagonally.
   (b) If you run into a cell which is already occupied (that is, you are "blocked"), drop down one cell from the last entry written and enter the next number there. Then proceed to move diagonally.
3. Your last entry, 25, will be in the middle cell of the bottom row.

Figure 7 shows the completed magic square. Its magic sum is 65.

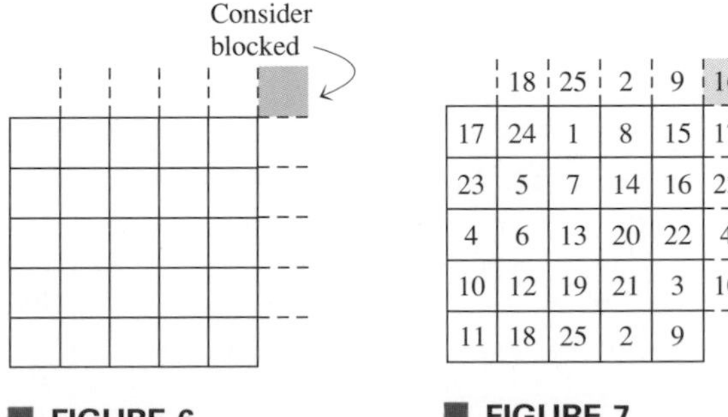

■ **FIGURE 6** ■ **FIGURE 7**

Extension Exercises

Given a magic square, other magic squares may be obtained by rotating the given one. For example, starting with the magic square in Figure 5, a 90° rotation in a clockwise direction gives the magic square shown at the right.

6	1	8
7	5	3
2	9	4

Start with Figure 5 and give the magic square obtained by each rotation described.

1. 180° in a clockwise direction

2. 90° in a counterclockwise direction

Start with Figure 7 and give the magic square obtained by each rotation described.

3. 90° in a clockwise direction

4. 180° in a clockwise direction

5. 90° in a counterclockwise direction

6. Try to construct an order 2 magic square containing the entries 1, 2, 3, 4. What happens?

Given a magic square, other magic squares may be obtained by adding or subtracting a constant value to each entry, multiplying each entry by a constant value, or dividing each entry by a nonzero constant value. In Exercises 7–10, start with the magic square whose figure number is indicated, and perform the operation described to find a new magic square. Give the new magic sum.

7. Figure 5, multiply by 3

8. Figure 5, add 7

9. Figure 7, divide by 2

10. Figure 7, subtract 10

According to a fanciful story by Charles Trigg in Mathematics Magazine *(September 1976, page 212), the Emperor Charlemagne (742–814) ordered a five-sided fort to be built at an important point in his kingdom. As good-luck charms, he had magic squares placed on all five sides of the fort. He had one restriction for these magic squares: all the numbers in them must be prime.*

The magic squares are given in Exercises 11–15, with one missing entry. Find the missing entry in each square.

11.

	71	257
47	269	491
281	467	59

12.

389		227
107	269	431
311	347	149

13.

389	227	191
71	269	
347	311	149

14.

401	227	179
47	269	491
359		137

15.

401	257	149
17		521
389	281	137

16. Compare the magic sums in Exercises 11–15. Charlemagne had stipulated that each magic sum should be the year in which the fort was built. What was that year?

Find the missing entries in each magic square.

17.

75	68	**(a)**
(b)	72	**(c)**
71	76	**(d)**

18.

1	8	13	**(a)**
(b)	14	7	2
16	9	4	**(c)**
(d)	**(e)**	**(f)**	15

19.

3	20	**(a)**	24	11
(b)	14	1	18	10
9	21	13	**(c)**	17
16	8	25	12	**(d)**
(e)	2	**(f)**	**(g)**	**(h)**

20.

3	36	2	35	31	4
10	12	**(a)**	26	7	27
21	13	17	14	**(b)**	22
16	**(c)**	23	**(d)**	18	15
28	30	8	**(e)**	25	9
(f)	1	32	5	6	34

21. Use the "staircase method" to construct a magic square of order 7, containing the entries 1, 2, 3, . . . , 49.

The magic square shown in the photograph is from a woodcut by Albrecht Dürer entitled Melancholia.

The two numbers in the center of the bottom row give 1514, *the year the woodcut was created. Refer to this magic square to answer Exercises 22–30.*

16	3	2	13
5	10	11	8
9	6	7	12
4	15	14	1

Dürer's Magic Square

22. What is the magic sum?

23. Verify: The sum of the entries in the four corners is equal to the magic sum.

24. Verify: The sum of the entries in any 2 by 2 square at a corner of the given magic square is equal to the magic sum.

25. Verify: The sum of the entries in the diagonals is equal to the sum of the entries not in the diagonals.

26. Verify: The sum of the squares of the entries in the diagonals is equal to the sum of the squares of the entries not in the diagonals.

27. Verify: The sum of the cubes of the entries in the diagonals is equal to the sum of the cubes of the entries not in the diagonals.

28. Verify: The sum of the squares of the entries in the top two rows is equal to the sum of the squares of the entries in the bottom two rows.

29. Verify: The sum of the squares of the entries in the first and third rows is equal to the sum of the squares of the entries in the second and fourth rows.

30. Find another interesting property of Dürer's magic square and state it.

31. A magic square of order 4 may be constructed as follows. Lightly sketch in the diagonals of the blank magic square. Beginning at the upper left, move across each row from left to right, counting the cells as you go along. If the cell is on a diagonal, count it but do not enter its number. If it is not on a diagonal, enter its number. When this is completed, reverse the procedure, beginning at the bottom right and moving across from right to left. As you count the cells, enter the number if the cell is not occupied. If it is already occupied, count it but do not enter its number. You should obtain a magic square similar to the one given for Exercises 22–30. How do they differ?

With chosen values for a, b, and c, an order 3 magic square can be constructed by substituting these values in the generalized form shown here.

$a + b$	$a - b - c$	$a + c$
$a - b + c$	a	$a + b - c$
$a - c$	$a + b + c$	$a - b$

Use the given values of a, b, and c to construct an order 3 magic square, using this generalized form.

32. $a = 5, \quad b = 1, \quad c = -3$

33. $a = 16, \quad b = 2, \quad c = -6$

34. $a = 5, \quad b = 4, \quad c = -8$

35. It can be shown that if an order n magic square has least entry k, and its entries are consecutive counting numbers, then its magic sum is given by the formula

$$\text{MS} = \frac{n(2k + n^2 - 1)}{2}.$$

Construct an order 7 magic square whose least entry is 10 using the staircase method. What is its magic sum?

36. Use the formula of Exercise 35 to find the missing entries in the following order 4 magic square whose least entry is 24.

(a)	38	37	27
35	**(b)**	30	32
31	33	**(c)**	28
(d)	26	25	**(e)**

In a 1769 letter from Benjamin Franklin to a Mr. Peter Collinson, Franklin exhibited the following magic square of order 8.

52	61	4	13	20	29	36	45
14	3	62	51	46	35	30	19
53	60	5	12	21	28	37	44
11	6	59	54	43	38	27	22
55	58	7	10	23	26	39	42
9	8	57	56	41	40	25	24
50	63	2	15	18	31	34	47
16	1	64	49	48	33	32	17

37. What is the magic sum?

Verify the following properties of this magic square.

38. The first half of each row and the second half of each row are each equal to half the magic sum.

39. The four corner entries added to the four center entries is equal to the magic sum.

40. The "bent diagonals" consisting of eight entries, going up four entries from left to right and down four entries from left to right, give the magic sum. (For example, starting with 16, one bent diagonal sum is $16 + 63 + 57 + 10 + 23 + 40 + 34 + 17$.)

If we use a "knight's move" (up two, right one) from chess, a variation of the staircase method gives rise to the magic square shown here. (When blocked, we move to the cell just below the previous entry.)

10	18	1	14	22
11	24	7	20	3
17	5	13	21	9
23	6	19	2	15
4	12	25	8	16

Use a similar process to construct an order 5 *magic square, starting with* 1 *in the cell described.*

41. fourth row, second column (up two, right one; when blocked, move to the cell just below the entry)

42. third row, third column (up one, right two; when blocked, move to the cell just to the left of the previous entry)

COLLABORATIVE INVESTIGATION

Investigating an Interesting Property of Number Squares

In the Extension at the end of this chapter, we looked at magic squares. Now in this group activity we will investigate another property of squares of numbers. Begin by dividing up into groups of three or four students. Each student in the group should prepare a square of numbers like the one that follows:

1	2	3	4	5
6	7	8	9	10
11	12	13	14	15
16	17	18	19	20
21	22	23	24	25

Topics for Discussion

1. Each student should do the following individually:

Choose any number in the first row. Circle it, and cross out all entries in the column below it. (For example, if you circle 4, cross out 9, 14, 19, and 24.) Now circle any remaining number in the second row, and cross out all entries in the column below it.

Repeat this procedure for the third and fourth rows, and then circle the final remaining number in the fifth row.

Now each student in the group should add the circled numbers and compare his or her sum with all others in the group. What do you notice?

2. How does the sum obtained in Exercise 1 compare with the magic sum for an order 5 magic square?
3. Prepare another square like the one above and repeat Exercise 1, except this time start with a number in the first *column* and cross out remaining numbers in *rows*. Discuss the result with other members in your group.
4. Prepare a square of the natural numbers from 1 through 36. Then repeat Exercise 1. Discuss your results. How does the sum compare with the magic sum for an order 6 magic square?
5. Prepare another square as you did in Exercise 4. This time, start with a number in the first column and cross out remaining numbers in rows. Discuss the result with other members of your group.

Chapter 5 Test

Decide whether each statement is true *or* false *(Exercises 1–5).*

1. No two prime numbers differ by 1.
2. There are infinitely many prime numbers.
3. If a natural number is divisible by 9, then it must also be divisible by 3.
4. If p and q are different primes, 1 is their greatest common factor and pq is their least common multiple.
5. For all natural numbers n, 1 is a factor of n and n is a multiple of n.
6. Use divisibility tests to determine whether the number

   331,153,470

   is divisible by each of the following.

   (a) 2 **(b)** 3 **(c)** 4
   (d) 5 **(e)** 6 **(f)** 8
   (g) 9 **(h)** 10 **(i)** 12
7. Decide whether each number is prime, composite, or neither.

   (a) 93 **(b)** 1 **(c)** 59
8. Give the prime factorization of 1,440.
9. In your own words state the Fundamental Theorem of Arithmetic.
10. Decide whether each number is perfect, deficient, or abundant.

    (a) 17 **(b)** 6 **(c)** 24
11. Which of the following statements is false?
    (a) There are no known odd perfect numbers.
    (b) Every even perfect number must end in 6 or 28.
    (c) Goldbach's Conjecture for the number 8 is verified by the equation $8 = 7 + 1$.
12. Give a pair of twin primes between 40 and 50.
13. Find the greatest common factor of 270 and 450.
14. Find the least common multiple of 24, 36, and 60.
15. Both Sherrie Firavich and Della Daniel work at a fast-food outlet. Sherrie has every sixth day off and Della has every fourth day off. If they are both off on Wednesday of this week, what will be the day of the week that they are next off together?
16. The twenty-second Fibonacci number is 17,711 and the twenty-third Fibonacci number is 28,657. What is the twenty-fourth Fibonacci number?
17. Make a conjecture about the next equation in the following list, and verify it.

$$8 - (1 + 1 + 2 + 3) = 1$$
$$13 - (1 + 2 + 3 + 5) = 2$$
$$21 - (2 + 3 + 5 + 8) = 3$$
$$34 - (3 + 5 + 8 + 13) = 5$$
$$55 - (5 + 8 + 13 + 21) = 8$$

18. **(a)** Give the first eight terms of a Fibonacci-type sequence with first term 1 and second term 5.

(b) Choose any term after the first in the sequence just formed. Square it. Multiply the two terms on either side of it. Subtract the smaller result from the larger. Now repeat the process with a different term. Make a conjecture about what this process will yield for any term of the sequence.

19. Which one of the following is the *exact* value of the golden ratio?

(a) $\frac{1 + \sqrt{5}}{2}$ **(b)** $\frac{1 - \sqrt{5}}{2}$ **(c)** 1.6

(d) 1.618

20. Write a short paragraph explaining what Fermat's Last Theorem says and giving an account of the recent developments concerning it.

6

The Real Number System

Sunbathing is a popular pastime. However, excessive exposure to ultraviolet light can be responsible for both skin and eye damage. Ultraviolet light from the sun is responsible for both the tanning and burning of exposed skin. Only about 6 percent of solar radiation reaching the earth is in the form of ultraviolet light. The intensity of ultraviolet light is affected by seasonal changes in the ozone layer, cloud cover, and time of day. The table shows maximum ultraviolet intensity, measured in milliwatts per square meter for various latitudes and dates.

Latitude	Mar. 21	June 21	Sept. 21	Dec. 21
0°	325	254	325	272
10°	311	275	280	220
20°	249	292	256	143
30°	179	248	182	80
40°	99	199	127	34
50°	57	143	75	13

If a student from Seattle, located at a latitude of 47.5°, spends spring break in Hawaii with a latitude of 20°, the sun's ultraviolet rays in Hawaii will be approximately $249/57 \approx 4.37$ times more intense than in Seattle. Hawaii's spring sun is approximately $249/143 \approx 1.74$ times stronger than Seattle's most intense sun in the summer. Because ultraviolet light is scattered through reflection, clouds and shade do not stop sunburn completely. A person sitting in the shade can receive 40 percent of the ultraviolet light occurring in sunlit areas. Thin clouds transmit 80 percent of ultraviolet light while a person swimming 1 1/2 feet below the surface of the water also receives 80 percent of the sun's ultraviolet radiation. So, even sitting in the shade our Seattle student will still feel a more intense sun in Hawaii.

Real numbers are needed to quantify the effect of ultraviolet light on humans. Quantities are often described using different types of numbers like whole numbers, decimals, fractions, and percentages. This chapter introduces the numbers which are essential in the description of real-world phenomena.

Source: Williams, J., *The USA Today Weather Almanac 1995,* Vintage Books, New York, 1994.

As mathematics developed, it was discovered that the *counting,* or *natural, numbers* did not satisfy all requirements of mathematicians. Consequently, new, expanded number systems were created. The mathematician Leopold Kronecker (1823–1891) once made the statement, "God made the integers, all the rest is the work of man." In this chapter we look at those sets that, according to Kronecker, are the work of mankind.

The *whole numbers* consist of the counting numbers with the important number 0 included. By taking the negatives of the counting numbers along with the whole numbers, the set of *integers* is formed. The concept of fractional quantities is introduced when quotients of integers are formed. Counting numbers, whole numbers, integers, and quotients of integers all are examples of *rational numbers.* There are other useful numbers, such as $\sqrt{2}$ and π, that are not rational numbers. They cannot be represented exactly as quotients of integers, although they may be approximated that way. The very large set containing all of these numbers is called the set of *irrational numbers.* The union of the set of rational numbers and the set of irrational numbers is called the set of *real numbers.*

6.1 Introduction to Sets of Real Numbers

The Origins of Zero The Mayan Indians of Mexico and Central America had one of the earliest numeration systems that included a symbol for zero. The very early Babylonians had a positional system, but they placed only a space between "digits" to indicate a missing power. When the Greeks absorbed Babylonian astronomy, they quickly saw the need for a symbol to represent "no powers." As early as 150 A.D. they used o or $\bar{o}$ to represent "no power," or "zero." By the year 500 A.D., the Greeks were using the letter omicron, o, of their alphabet for "zero." The Greek numeration system was gradually replaced by the Roman numeration system throughout Western Europe.

The Roman system was the one most commonly used in Europe from the time of Christ until perhaps 1400 A.D., when the Hindu-Arabic system began to

continued on page 256

The natural numbers are those numbers with which we count discrete objects. By including 0 in the set, we obtain the set of whole numbers.

Natural Numbers

{1, 2, 3, 4, . . .} is the set of **natural numbers.**

Whole Numbers

{0, 1, 2, 3, . . .} is the set of **whole numbers.**

These numbers, along with many others, can be represented on **number lines** like the one pictured in Figure 1. We draw a number line by locating any point on the line and calling it 0. Choose any point to the right of 0 and call it 1. The distance between 0 and 1 gives a unit of measure used to locate other points, as shown in Figure 1. The points labeled in Figure 1 and those continuing in the same way to the right correspond to the set of whole numbers.

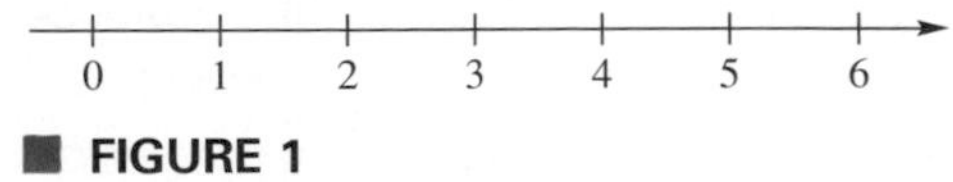

FIGURE 1

All the whole numbers starting with 1 are located to the right of 0 on the number line. But numbers may also be placed to the left of 0. These numbers, written −1, −2, −3, and so on, are shown in Figure 2. (The negative sign is used to show that the numbers are located to the *left* of 0.)

take over. Zero was fundamental to the Hindu-Arabic system. The original Hindu word for zero was *sunya,* meaning "void." The Arabs adopted this word as *sifr,* or "vacant." There was a considerable battle over the new system in Europe, with most people sticking with the Roman system. Gradually, however, the advantages of the new Hindu-Arabic system became clear, and it replaced the cumbersome Roman system. The word *sifr* passed into Latin as *zephirum,* which over the years became *zevero, zepiro,* and finally, *zero.*

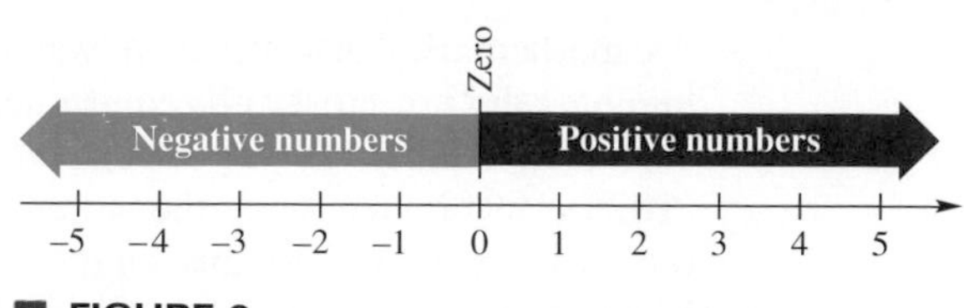

FIGURE 2

The numbers to the *left* of 0 are **negative numbers.** The numbers to the *right* of 0 are **positive numbers.** The number 0 itself is neither positive nor negative. Positive numbers and negative numbers are called **signed numbers.**

There are many practical applications of negative numbers. For example, temperatures sometimes fall below zero. The lowest temperature ever recorded in meteorological records was -128.6°F at Vostok, Antarctica, on July 21, 1983. A business that spends more than it takes in has a negative "profit." Altitudes below sea level can be represented by negative numbers. The shore surrounding the Dead Sea is 1,312 feet below sea level; this can be represented as $-1{,}312$ feet.

The set of numbers marked on the number line in Figure 2, including positive and negative numbers and zero, is part of the set of *integers.*

Integers

$\{\ldots, -3, -2, -1, 0, 1, 2, 3, \ldots\}$ is the set of **integers.**

The Origins of Negative Numbers Negative numbers can be traced back to the Chinese between 200 B.C. and 220 A.D. Mathematicians at first found negative numbers ugly and unpleasant, even though they kept cropping up in the solutions of problems. For example, an Indian text of about 1150 A.D. gives the solution of an equation as -5 and then makes fun of anything so useless.

Leonardo of Pisa (Fibonacci), while working on a financial problem, was forced to conclude that the solution must be a negative number, that is, a financial loss. In 1545 A.D., the rules governing operations with negative numbers were published by Girolamo Cardano in his *Ars Magna* (Great Art).

Not all numbers are integers. For example, $1/2$ is not; it is a number halfway between the integers 0 and 1. Also, $3\ 1/4$ is not an integer. Several numbers that are not integers are *graphed* in Figure 3. The **graph** of a number is a point on the number line. Think of the graph of a set of numbers as a picture of the set. All the numbers in Figure 3 can be written as quotients of integers. These numbers are examples of *rational numbers.*

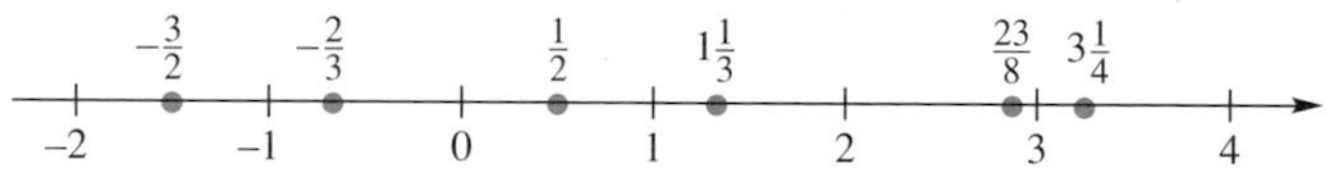

FIGURE 3

Rational Numbers

$\{x \mid x$ is a quotient of two integers, with denominator not equal to $0\}$ is the set of **rational numbers.**
(Read the part in the braces as "the set of all numbers x such that x is a quotient of two integers, with denominator not equal to 0.")

The set symbolism used in the definition of rational numbers,

$$\{x \mid x \text{ has a certain property}\},$$

is called **set-builder notation.** This notation is convenient to use when it is not possible to list all the elements of the set.

Since any integer can be written as the quotient of itself and 1, all integers also are rational numbers.

All numbers that can be represented by points on the number line are called *real numbers.*

Real Numbers

$\{x \mid x$ is a number that can be represented by a point on the number line$\}$ is the set of **real numbers.**

Although a great many numbers are rational, not all are. For example, a floor tile one foot on a side has a diagonal whose length is the square root of 2 (written $\sqrt{2}$). It will be shown later that $\sqrt{2}$ cannot be written as a quotient of integers. Because of this, $\sqrt{2}$ is not rational; it is *irrational.*

Irrational Numbers

$\{x \mid x$ is a real number that is not rational$\}$ is the set of **irrational numbers.**

Examples of irrational numbers include $\sqrt{3}$, $\sqrt{7}$, $-\sqrt{10}$, and π, which is the ratio of the distance around a circle to the distance across it.

Real numbers can be written as decimal numbers. Any rational number can be written as a decimal that will come to an end (terminate), or repeat in a fixed "block" of digits. For example, $2/5 = .4$ and $27/100 = .27$ are rational numbers with terminating decimals; $1/3 = .3333\ldots$ and $3/11 = .27272727\ldots$ are repeating decimals. The decimal representation of an irrational number will neither terminate nor repeat. Decimal representations of rational and irrational numbers will be discussed further later in this chapter.

Figure 4 on the next page illustrates two ways to represent the relationships among the various sets of real numbers.

EXAMPLE 1 List the numbers in the set

$$\left\{-5,\ -\frac{2}{3},\ 0,\ \sqrt{2},\ \frac{13}{4},\ 5,\ 5.8\right\}$$

that belong to each of the following sets of numbers.

(a) natural numbers

The only natural number in the set is 5.

(b) whole numbers

The whole numbers consist of the natural numbers and 0. So the elements of the set that are whole numbers are 0 and 5.

(c) integers

The integers in the set are -5, 0, and 5.

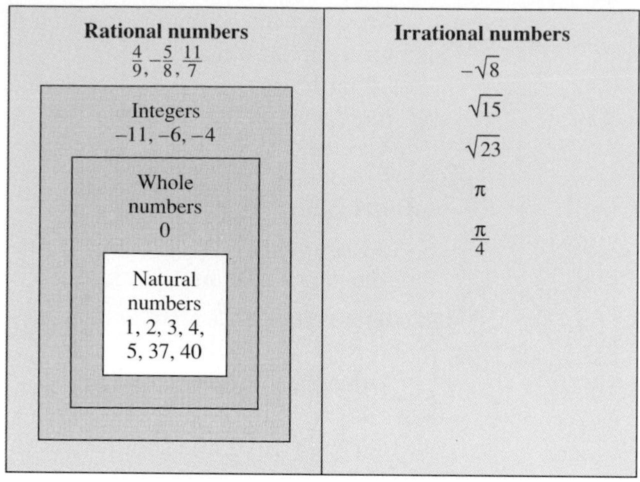

All numbers shown are real numbers.

(a)

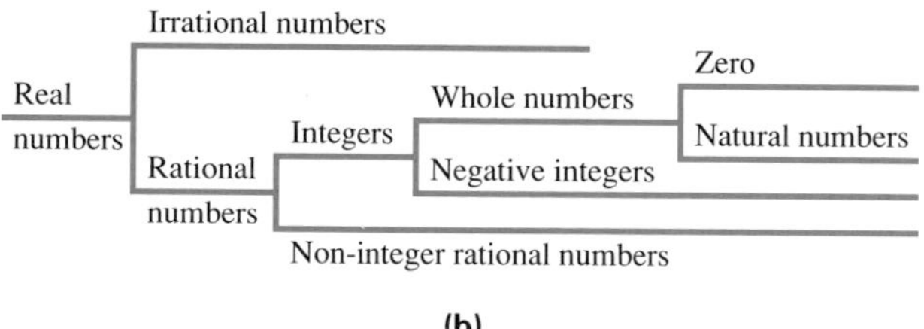

(b)

FIGURE 4

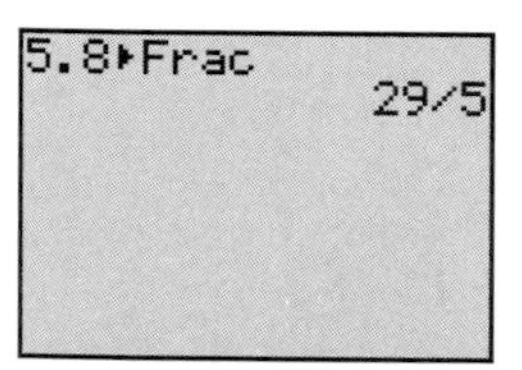

The TI-82 calculator will convert a decimal to a fraction. See Example 1(d).

(d) rational numbers
The rational numbers are -5, $-2/3$, 0, $13/4$, 5, and 5.8, since each of these numbers *can* be written as the quotient of two integers. For example, $5.8 = 58/10 = 29/5$.

(e) irrational numbers
The only irrational number in the set is $\sqrt{2}$.

(f) real numbers
All the numbers in the set are real numbers. ■

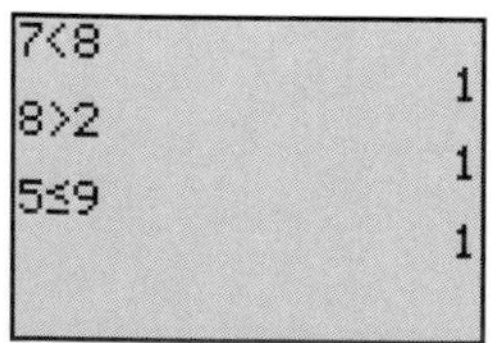

The calculator returns a 1 for these statements of inequality, signifying that each is true.

Two real numbers may be compared using the ideas of equality and inequality. Suppose that a and b represent two real numbers. If their graphs on the number line are the same point, they are **equal.** If the graph of a lies to the left of b, a **is less than** b, and if the graph of a lies to the right of b, a **is greater than** b. We use symbols to represent these ideas.

When read from left to right, the symbol $<$ represents "is less than," so that "7 is less than 8" is written

$$7 < 8.$$

Also, we write "6 is less than 9" as $6 < 9$.

The symbol for equality, $=$, was first introduced by the Englishman Robert Recorde in his 1557 algebra text *The Whetstone of Witte.* He used two parallel line segments because, he claimed, no two things can be more equal.

The symbols for order relationships, $<$ and $>$, were first used by Thomas Harriot (1560–1621), another Englishman. These symbols were not immediately adopted by other mathematicians.

The symbol $>$ means "is greater than." We write "8 is greater than 2" as

$$8 > 2.$$

The statement "17 is greater than 11" becomes $17 > 11$.

We can keep the meanings of the symbols $<$ and $>$ clear by remembering that the symbol always points to the smaller number. For example, write "8 is less than 15" by pointing the symbol toward the 8:

$$8 < 15.$$

Two other symbols, $\leq$ and $\geq$, also represent the idea of inequality. The symbol $\leq$ means "is less than or equal to," so that

$$5 \leq 9$$

means "5 is less than or equal to 9." This statement is true, since $5 < 9$ is true. If either the $<$ part or the $=$ part is true, then the inequality $\leq$ is true.

The symbol $\geq$ means "is greater than or equal to." Again,

$$9 \geq 5$$

is true because $9 > 5$ is true. Also, $8 \leq 8$ is true since $8 = 8$ is true. But it is not true that $13 \leq 9$ because neither $13 < 9$ nor $13 = 9$ is true.

EXAMPLE 2 Determine whether each statement is *true* or *false*.

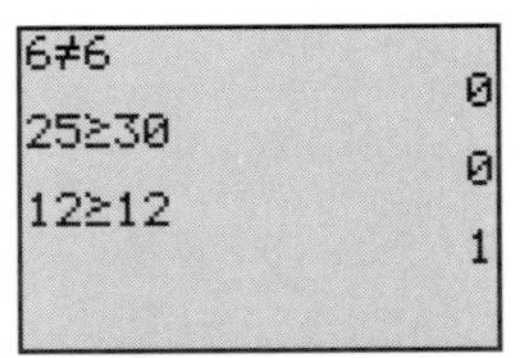

The inequalities in Example 2(a) and (d) are false, as signified by the 0. The statement in Example 2(e) is true.

(a) $6 \neq 6$
The statement is false because 6 *is equal to* 6.

(b) $5 < 19$
Since 5 represents a number that is indeed less than 19, this statement is true.

(c) $15 \leq 20$
The statement $15 \leq 20$ is true, since $15 < 20$.

(d) $25 \geq 30$
Both $25 > 30$ and $25 = 30$ are false. Because of this, $25 \geq 30$ is false.

(e) $12 \geq 12$
Since $12 = 12$, this statement is true. ■

For any real number x (except 0), there is exactly one number on the number line the same distance from 0 as x but on the opposite side of 0. For example, Figure 5 shows that the numbers 3 and -3 are both the same distance from 0 but are on opposite sides of 0. The numbers 3 and -3 are called **additive inverses,** or **opposites,** of each other.

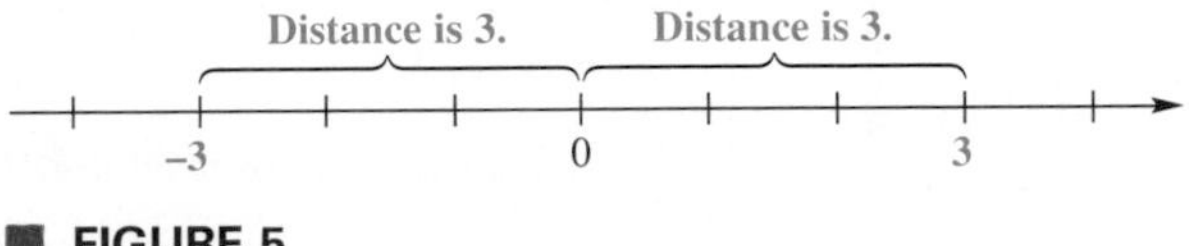

FIGURE 5

The additive inverse of the number 0 is 0 itself. This makes 0 the only real number that is its own additive inverse. Other additive inverses occur in pairs.

For example, 4 and -4, and 5 and -5, are additive inverses of each other. Several pairs of additive inverses are shown in Figure 6.

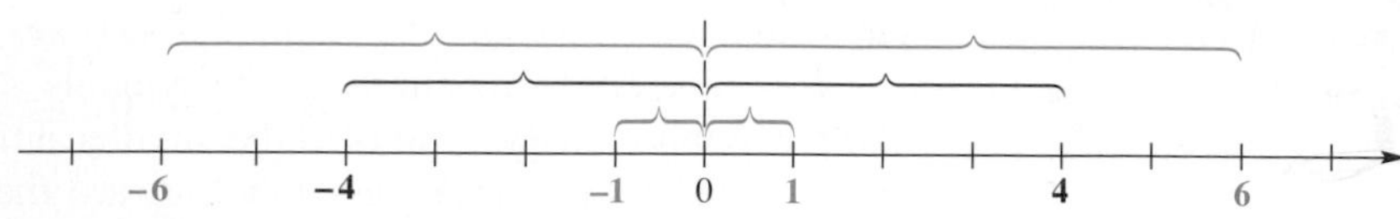

FIGURE 6

EXAMPLE 3 The following chart shows several numbers and their additive inverses.

Number	Additive Inverse
-4	$-(-4)$, or 4
0	0
-2	2
19	-19
3	-3

An important property of additive inverses will be studied in more detail in a later section of this chapter: $a + (-a) = (-a) + a = 0$ for all real numbers a.

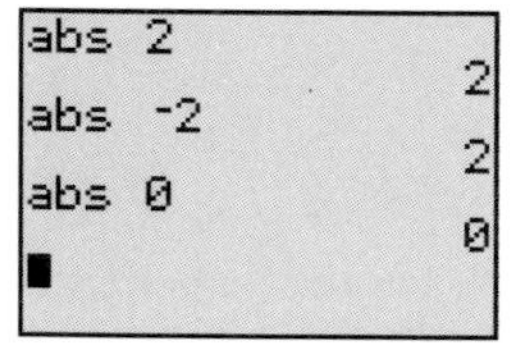

The command abs returns the absolute value of the number following it.

As mentioned above, additive inverses are numbers that are the same distance from 0 on the number line. This idea can also be expressed by saying that a number and its additive inverse have the same absolute value. The **absolute value** of a real number can be defined as the distance between 0 and the number on the number line. The symbol for the absolute value of the number x is $|x|$, read "the absolute value of x." For example, the distance between 2 and 0 on the number line is 2 units, so that

$$|2| = 2.$$

Because the distance between -2 and 0 on the number line is also 2 units,

$$|-2| = 2.$$

Since distance is never negative, the absolute value of a number is never negative. For example, $|12| = 12$ and $|-12| = 12$, since both 12 and -12 lie at a distance of 12 units from 0 on the number line. Also, since 0 is a distance 0 units from 0, $|0| = 0$.

In symbols, the absolute value of x is defined as follows.

Formal Definition of Absolute Value

$$|x| = \begin{cases} x \text{ if } x \geq 0 \\ -x \text{ if } x < 0 \end{cases}$$

By this definition, if x is a positive number or 0, then its absolute value is x itself. For example, since 8 is a positive number, $|8| = 8$. However, if x is a negative number, then its absolute value is the additive inverse of x. For example, since

-9 is a negative number, $|-9| = -(-9) = 9$, since the additive inverse of -9 is 9.

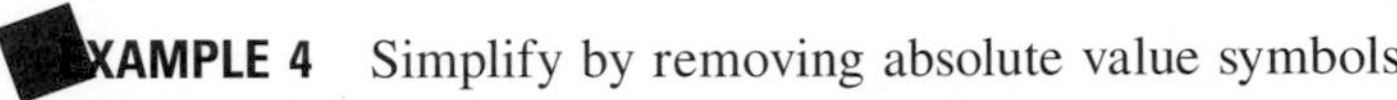

EXAMPLE 4 Simplify by removing absolute value symbols.

(a) $|5| = 5$

(b) $|-5| = -(-5) = 5$

(c) $-|5| = -(5) = -5$

(d) $-|-14| = -(14) = -14$

(e) $|8 - 2| = |6| = 6$ ■

6.1 EXERCISES

Graph each set on a number line.

1. $\{2, 3, 4, 5\}$

2. $\{0, 2, 4, 6, 8\}$

3. $\{-4, -3, -2, -1, 0, 1\}$

4. $\{-6, -5, -4, -3, -2\}$

5. $\left\{-\frac{1}{2}, \frac{3}{4}, \frac{5}{3}, \frac{7}{2}\right\}$

6. $\left\{-\frac{3}{5}, -\frac{1}{10}, \frac{9}{8}, \frac{12}{5}, \frac{13}{4}\right\}$

List the numbers in the given set that belong to **(a)** *the natural numbers,* **(b)** *the whole numbers,* **(c)** *the integers,* **(d)** *the rational numbers,* **(e)** *the irrational numbers, and* **(f)** *the real numbers.*

7. $\left\{-9, -\sqrt{7}, -\frac{5}{4}, -\frac{3}{5}, 0, \sqrt{5}, 3, 5.9, 7\right\}$

8. $\left\{-5.3, -5, -\sqrt{3}, -1, -\frac{1}{9}, 0, 1.2, 1.8, 3, \sqrt{11}\right\}$

Decide whether each of the following statements is true *or* false.

9. $-2 < -1$

10. $-8 < -4$

11. $-3 \geq -7$

12. $-9 \geq -12$

13. $-15 \leq -20$

14. $-21 \leq -27$

15. $-8 \leq -(-4)$

16. $-9 \leq -(-6)$

17. $0 \leq -(-4)$

18. $0 \geq -(-6)$

19. $6 > -(-2)$

20. $-8 > -(-2)$

Give **(a)** *the additive inverse and* **(b)** *the absolute value of each of the following.*

21. 5

22. 9

23. -6

24. -8

---

25. A statement commonly heard is "Absolute value is always positive." Is this true? If not, explain.

26. If a is a negative number, then is $-|-a|$ positive or negative?

27. Fill in the blanks with the correct values: The opposite of -3 is _____, while the absolute value of -3 is _____. The additive inverse of -3 is _____, while the additive inverse of the absolute value of -3 is _____.

28. True or false: For all real numbers a and b,

$$|a - b| = |b - a|.$$

Specify each set by listing its elements. If there are no elements, write $\emptyset$.

29. $\{x \mid x \text{ is a natural number less than } 7\}$

30. $\{m \mid m \text{ is a whole number less than } 9\}$

31. $\{a \mid a \text{ is an even integer greater than } 10\}$

32. $\{k \mid k \text{ is a natural number less than } 1\}$

33. $\{x \mid x \text{ is an irrational number that is also rational}\}$

34. $\{r \mid r \text{ is a negative integer greater than } -1\}$

35. $\{p \mid p \text{ is a number whose absolute value is } 3\}$

36. $\{w \mid w \text{ is a number whose absolute value is } 12\}$

37. $\{z \mid z \text{ is a whole number multiple of } 5\}$

38. $\{n \mid n \text{ is a natural number multiple of } 3\}$

Give three examples of numbers that satisfy the given condition.

39. positive real numbers but not integers

40. real numbers but not positive numbers

41. real numbers but not whole numbers

42. rational numbers but not integers

43. real numbers but not rational numbers

44. rational numbers but not negative numbers

Tell whether each statement is true *or* false.

45. Every rational number is an integer.

46. Every natural number is an integer.

47. Every integer is a rational number.

48. Every whole number is a real number.

49. Some rational numbers are irrational.

50. Some natural numbers are whole numbers.

51. Some rational numbers are integers.

52. Some real numbers are integers.

53. Every rational number is a real number.

54. Some integers are not real numbers.

55. Every integer is positive.

56. Every whole number is positive.

57. Some irrational numbers are negative.

58. Some real numbers are not rational.

59. Not every rational number is positive.

60. Some whole numbers are not integers.

Simplify each expression by removing absolute value symbols.

61. $|3|$ **62.** $|-8|$

63. $-|7|$ **64.** $-|-9|$

65. $|7 - 4|$ **66.** $-|12 - 3|$

67. $-|-(5 - 1)|$ **68.** $|-(3 - 1)|$

Decide which of the following sets the given number belongs to: **(a)** *whole numbers,* **(b)** *integers,* **(c)** *rational numbers,* **(d)** *irrational numbers,* **(e)** *real numbers.*

69. 12 **70.** -4

71. $\frac{2}{3}$ **72.** $\frac{4}{7}$

73. 9.25 **74.** -7.335

75. 0 **76.** $.888\ldots$

77. $\sqrt{2}$ **78.** $\sqrt{-1}$

*A **bar graph,** as shown in the figure, is a convenient method for depicting data. Here we see that in 1989, $28 million was the figure for direct impact of moviemakers' spending in the state of Louisiana. In 1990, it was $23 million, and so on. Use the bar graph to answer the questions in Exercises 79–84.*

Graph, "It's Showtime: Moviemakers' spending a hit in Louisiana," from *The Times-Picayune*, August 17, 1994. Copyright © 1994 by *The Times-Picayune*. Reprinted by permission.

79. In what years was the direct impact greater than $23 million?

80. In what years was the direct impact less than $29 million?

81. In what years was the direct impact less than or equal to $29 million?

82. Explain why the answers to Exercises 80 and 81 are the same.

83. The total economic impact is usually determined by multiplying the direct impact by 3. What was the total economic impact from 1989 through 1992?

84. (See Exercise 83.) What was the total economic impact in 1993?

6.2 Operations, Properties, and Applications of Real Numbers

The answer to an addition problem is called the **sum.** The rules for addition of real numbers are given on the next page.

Practical Arithmetic From the time of Egyptian and Babylonian merchants, practical aspects of arithmetic complemented mystical (or "Pythagorean") tendencies. This was certainly true in the time of Adam Riese (1489–1559), a "reckon master" influential when commerce was growing in Northern Europe. Riese's likeness on the stamp above comes from the title page of one of his popular books on *Rechnung* (or "reckoning"). He championed new methods of reckoning using Hindu-Arabic numerals and quill pens. (The Roman methods then in common use moved counters on a ruled board.) Riese thus fulfilled Fibonacci's efforts 300 years earlier to supplant Roman numerals and methods.

Adding Real Numbers

Like Signs Add two numbers with the *same* sign by adding their absolute values. The sign of the sum (either + or −) is the same as the sign of the two numbers.

Unlike Signs Add two numbers with *different* signs by subtracting the smaller absolute value from the larger. The sum is positive if the positive number has the larger absolute value. The sum is negative if the negative number has the larger absolute value.

For example, to add −12 and −8, first find their absolute values:

$$|-12| = 12 \qquad \text{and} \qquad |-8| = 8.$$

Since these numbers have the *same* sign, add their absolute values: 12 + 8 = 20. Give the sum the sign of the two numbers. Since both numbers are negative, the sum is negative and

$$-12 + (-8) = -20.$$

Find −17 + 11 by subtracting the absolute values, since these numbers have different signs.

$$|-17| = 17 \qquad \text{and} \qquad |11| = 11$$
$$17 - 11 = 6$$

Give the result the sign of the number with the larger absolute value.

$$-17 + 11 = -6$$

Negative since $|-17| > |11|$

EXAMPLE 1 Find each of the following sums.

(a) $(-6) + (-3) = -(6 + 3) = -9$

(b) $(-12) + (-4) = -(12 + 4) = -16$

(c) $4 + (-1) = 3$

(d) $-9 + 16 = 7$

(e) $-16 + 12 = -4$

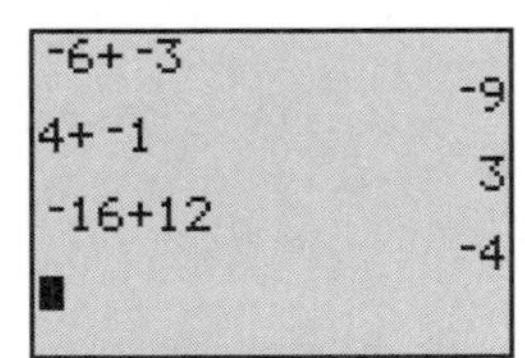

The calculator supports the results in Example 1(a), (c), and (e).

We now turn our attention to subtraction of real numbers. The result of subtraction is called the **difference.** Thus, the difference between 7 and 5 is 2. To see how subtraction should be defined, compare the two statements below.

$$7 - 5 = 2$$
$$7 + (-5) = 2$$

In a similar way,

$$9 - 3 = 9 + (-3).$$

That is, to subtract 3 from 9, add the additive inverse of 3 to 9. These examples suggest the following rule for subtraction.

Definition of Subtraction

For all real numbers a and b,

$$a - b = a + (-b).$$

(Change the sign of the second number and add.)

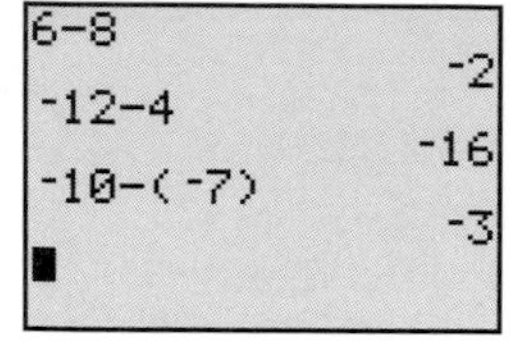

The calculator supports the results of Example 2(a), (b), and (c). Notice how the *negative* (negation) sign differs from the *minus* (subtraction) sign. There are different keys on the calculator for these purposes.

EXAMPLE 2 Perform the indicated operations.

Change to addition.
Change sign of second number.

(a) $6 - 8 = 6 + (-8) = -2$

Change to addition.
Sign changed.

(b) $-12 - 4 = -12 + (-4) = -16$

(c) $-10 - (-7) = -10 + [-(-7)]$ This step can be omitted.
$= -10 + 7$
$= -3$

(d) $15 - (-3) - 5 - 12$

When a problem with both addition and subtraction is being worked, perform the additions and subtractions in order from left to right.

$$\begin{aligned} 15 - (-3) - 5 - 12 &= (15 + 3) - 5 - 12 \\ &= 18 - 5 - 12 \\ &= 13 - 12 \\ &= 1 \end{aligned}$$

We now turn our attention to the operations of multiplication and division of real numbers. Any rules for multiplication with negative real numbers should be consistent with the usual rules for multiplication of positive real numbers and zero. To inductively obtain a rule for multiplying a positive real number and a negative real number, observe the pattern of products below. (The **product** is the result of a multiplication problem.)

$$\begin{aligned} 4 \cdot 5 &= 20 \\ 4 \cdot 4 &= 16 \\ 4 \cdot 3 &= 12 \\ 4 \cdot 2 &= 8 \\ 4 \cdot 1 &= 4 \\ 4 \cdot 0 &= 0 \\ 4 \cdot (-1) &= ? \end{aligned}$$

What number must be assigned as the product $4 \cdot (-1)$ so that the pattern is maintained? The numbers just to the left of the equals signs decrease by 1 each time, and the products to the right decrease by 4 each time. To maintain the

Early ways of writing the four basic operation symbols were quite different from those used today. The addition symbol shown below was derived from the Italian word *piú* (plus) in the sixteenth century. The + sign used today is shorthand for the Latin *et* (and).

The subtraction symbol shown below was used by Diophantus in Greece sometime during the second or third century A.D. Our subtraction bar may be derived from a bar used by medieval traders to mark differences in weights of products.

In the seventeenth century, Leibniz used the symbol below for multiplication to avoid × as too similar to the "unknown" x. The multiplication symbol × is based on St. Andrew's Cross.

The division symbol shown below was used by Gallimard in the eighteenth century. The familiar ÷ symbol may come from the fraction bar, embellished with the dots above and below.

pattern, the number to the right in the bottom equation must be 4 less than 0, which is -4, or

$$4 \cdot (-1) = -4.$$

The pattern continues with

$$\begin{aligned} 4 \cdot (-2) &= -8 \\ 4 \cdot (-3) &= -12 \\ 4 \cdot (-4) &= -16, \end{aligned}$$

and so on. In the same way,

$$\begin{aligned} -4 \cdot 2 &= -8 \\ -4 \cdot 3 &= -12 \\ -4 \cdot 4 &= -16, \end{aligned}$$

and so on.

A similar observation can be made about the product of two negative real numbers. Look at the pattern that follows.

$$\begin{aligned} -5 \cdot 4 &= -20 \\ -5 \cdot 3 &= -15 \\ -5 \cdot 2 &= -10 \\ -5 \cdot 1 &= -5 \\ -5 \cdot 0 &= 0 \\ -5 \cdot (-1) &= ? \end{aligned}$$

The numbers just to the left of the equals signs decrease by 1 each time. The products on the right increase by 5 each time. To maintain the pattern, the product $-5 \cdot (-1)$ must be 5 more than 0, or

$$-5 \cdot (-1) = 5.$$

Continuing this pattern gives

$$\begin{aligned} -5 \cdot (-2) &= 10 \\ -5 \cdot (-3) &= 15 \\ -5 \cdot (-4) &= 20, \end{aligned}$$

and so on.

These observations lead to the following rules for multiplication.

Multiplying Real Numbers

Like Signs Multiply two numbers with the *same* sign by multiplying their absolute values. The product is positive.

Unlike Signs Multiply two numbers with *different* signs by multiplying their absolute values. The product is negative.

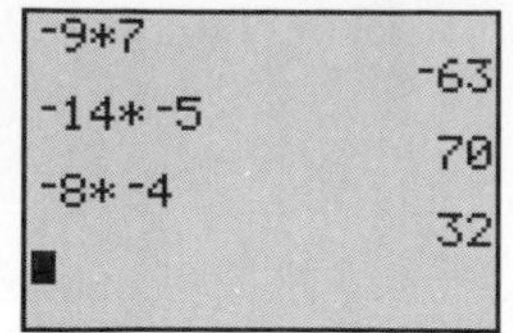

An asterisk represents multiplication on this screen. The display supports the results of Example 3.

EXAMPLE 3 Find each of the following products.

(a) $-9 \cdot 7 = -63$

(b) $-14 \cdot (-5) = 70$

(c) $-8 \cdot (-4) = 32$

Let us now consider the operation of division. The result obtained by dividing real numbers is called the **quotient.** For real numbers a, b, and c, where $b \neq 0$, $a/b = c$ means that $a = b \cdot c$. To illustrate this, consider the division problem

$$\frac{10}{-2}.$$

The value of this quotient is obtained by asking "What number multiplied by -2 gives 10?" From our discussion of multiplication, the answer to this question must be "-5." Therefore,

$$\frac{10}{-2} = -5,$$

because $(-2) \cdot (-5) = 10$. Similar reasoning leads to the following results.

$$\frac{-10}{2} = -5 \qquad \frac{-10}{-2} = 5$$

These facts, along with the fact that the quotient of two positive numbers is positive, lead to the following rules for division.

Dividing Real Numbers

Like Signs Divide two numbers with the *same* sign by dividing their absolute values. The quotient is positive.

Unlike Signs Divide two numbers with *different* signs by dividing their absolute values. The quotient is negative.

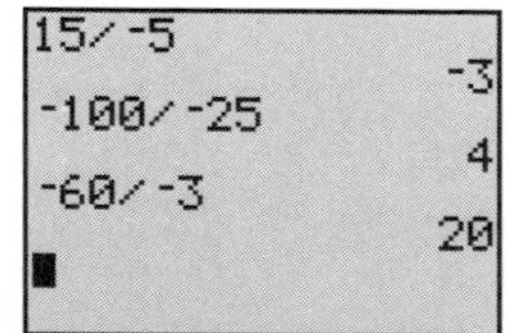

The division operation is represented by a slash (/). This screen supports the results of Example 4.

EXAMPLE 4 Find each of the following quotients.

(a) $\frac{15}{-5} = -3$ This is true because $-5 \cdot (-3) = 15$.

(b) $\frac{-100}{-25} = 4$

(c) $\frac{-60}{-3} = 20$

If 0 is divided by a nonzero number, the quotient is 0. That is,

$$\frac{0}{a} = 0 \quad \text{for } a \neq 0.$$

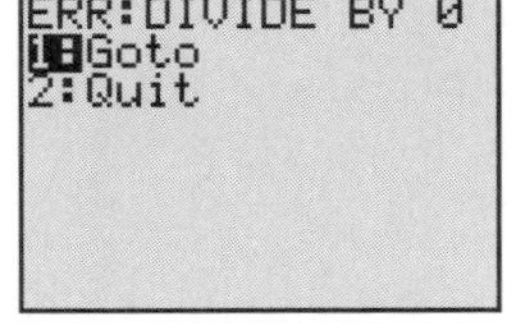

Dividing by zero leads to this message on the TI-82.

This is true because $a \cdot 0 = 0$. However, we cannot divide by 0. There is a good reason for this. Whenever a division is performed, we want to obtain one and only one quotient. Now consider the division problem

$$\frac{7}{0}.$$

We must ask ourselves "What number multiplied by 0 gives 7?" There is no such number, since the product of 0 and any number is zero. On the other hand, if we consider the quotient

$$\frac{0}{0},$$

there are infinitely many answers to the question, "What number multiplied by 0 gives 0?" Since division by 0 does not yield a *unique* quotient, it is not permitted. To summarize these two situations, we make the following statement.

Division by 0 is undefined.

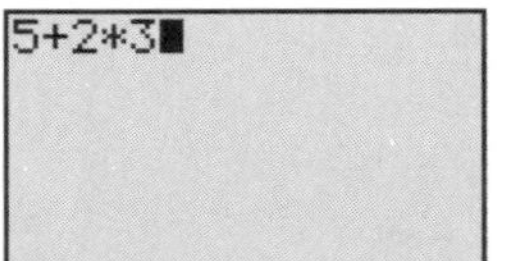

What result does the calculator give? The order of operations determines the answer. (See Example 5(a).)

Order of Operations

Given a problem such as $5 + 2 \cdot 3$, should 5 and 2 be added first or should 2 and 3 be multiplied first? When a problem involves more than one operation, we use the following **order of operations.** (This is the order used by computers and many calculators.)

The sentence "Please excuse my dear Aunt Sally" is often used to help remember the rule for order of operations. The letters P, E, M, D, A, S are the first letters of the words of the sentence, and they stand for *parentheses, exponents, multiply, divide, add, subtract.*

Order of Operations

If parentheses or square brackets are present:

Step 1: Work separately above and below any fraction bar.

Step 2: Use the rules below within each set of parentheses or square brackets. Start with the innermost set and work outward.

If no parentheses or brackets are present:

Step 1: Apply all exponents.

Step 2: Do any multiplications or divisions in the order in which they occur, working from left to right.

Step 3: Do any additions or subtractions in the order in which they occur, working from left to right.

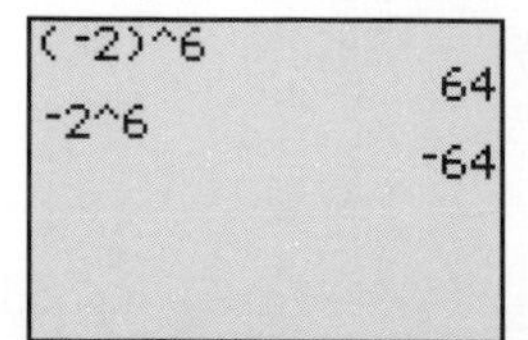

Notice the difference in the two expressions. This supports $(-2)^6 \neq -2^6$.

It is important to be careful when evaluating an exponential expression that involves a negative sign. In particular, we should be aware that $(-a)^n$ and $-a^n$ do not necessarily represent the same quantity. For example, if $a = 2$ and $n = 6$,

$$(-2)^6 = (-2)(-2)(-2)(-2)(-2)(-2) = 64$$

while

$$-2^6 = -(2 \cdot 2 \cdot 2 \cdot 2 \cdot 2 \cdot 2) = -64.$$

Notice that for $(-2)^6$, the $-$ sign is used in each factor of the expanded form, while for -2^6, the $-$ sign is treated as a factor of -1, and the base of the exponential expression is 2 (not -2).

EXAMPLE 5 Use the order of operations to simplify each of the following.

(a) $5 + 2 \cdot 3$
First multiply, and then add.

$$\begin{aligned} 5 + 2 \cdot 3 &= 5 + 6 && \text{Multiply.} \\ &= 11 && \text{Add.} \end{aligned}$$

(b) $4 \cdot 3^2 + 7 - (2 + 8)$
Work inside the parentheses.

$$4 \cdot 3^2 + 7 - (2 + 8) = 4 \cdot 3^2 + 7 - 10$$

Apply all exponents. Since $3^2 = 3 \cdot 3 = 9$,

$$4 \cdot 3^2 + 7 - 10 = 4 \cdot 9 + 7 - 10.$$

Do all multiplications or divisions, working from left to right.

$$4 \cdot 9 + 7 - 10 = 36 + 7 - 10$$

Finally, do all additions or subtractions, working from left to right.

$$\begin{aligned} 36 + 7 - 10 &= 43 - 10 \\ &= 33 \end{aligned}$$

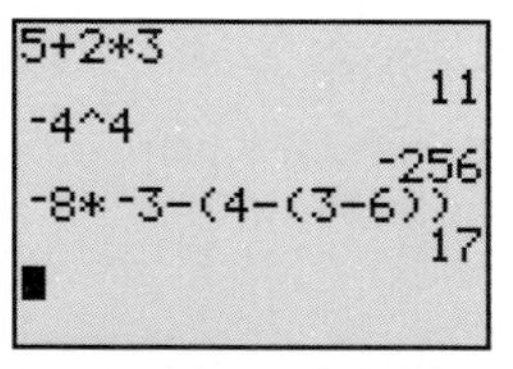

The calculator supports the results in Example 5(a), (d), and (f).

(c) $\dfrac{2(8 - 12) - 11(4)}{5(-2) - 3} = \dfrac{2(-4) - 11(4)}{5(-2) - 3} = \dfrac{-8 - 44}{-10 - 3} = \dfrac{-52}{-13} = 4$

(d) $-4^4 = -(4 \cdot 4 \cdot 4 \cdot 4) = -256$

(e) $(-4)^4 = (-4)(-4)(-4)(-4) = 256$

(f) $$\begin{aligned} (-8)(-3) - [4 - (3 - 6)] &= (-8)(-3) - [4 - (-3)] \\ &= (-8)(-3) - [4 + 3] \\ &= (-8)(-3) - 7 \\ &= 24 - 7 \\ &= 17 \end{aligned}$$

Properties of Addition and Multiplication of Real Numbers

Several properties of addition and multiplication of real numbers that are essential to our study in this chapter are summarized in the following box.

Properties of Addition and Multiplication

For real numbers a, b, and c, the following properties hold.

Closure Properties If a and b are real numbers, then $a + b$ and ab are real numbers.

Commutative Properties $a + b = b + a \qquad ab = ba$

Associative Properties $(a + b) + c = a + (b + c)$

$(ab)c = a(bc)$

Identity Properties There is a real number 0 such that

$a + 0 = a$ and $0 + a = a$.

There is a real number 1 such that

$a \cdot 1 = a$ and $1 \cdot a = a$.

Inverse Properties For each real number a, there is a single real number $-a$ such that

$a + (-a) = 0$ and $(-a) + a = 0$.

For each nonzero real number a, there is a single real number $1/a$ such that

$a \cdot \frac{1}{a} = 1$ and $\frac{1}{a} \cdot a = 1$.

Distributive Property of Multiplication with Respect to Addition $a(b + c) = ab + ac$

$(b + c)a = ba + ca$

The set of real numbers is said to be closed with respect to the operations of addition and multiplication. This means that the sum of two real numbers and the product of two real numbers are themselves real numbers. The commutative properties state that two real numbers may be added or multiplied in any order without affecting the result. The associative properties allow us to group terms or factors in any manner we wish without affecting the result. The number 0 is called the **identity element for addition,** and it may be added to any real number to obtain that real number as a sum. Similarly, 1 is called the **identity element for multiplication,** and multiplying a real number by 1 will always yield that real number. Each real number a has an **additive inverse,** $-a$, such that their sum is the additive identity element 0. Each nonzero real number a has a **multiplicative inverse,** or **reciprocal,** $1/a$, such that their product is the multiplicative identity element 1. The distributive property allows us to change a product to a sum or a sum to a product.

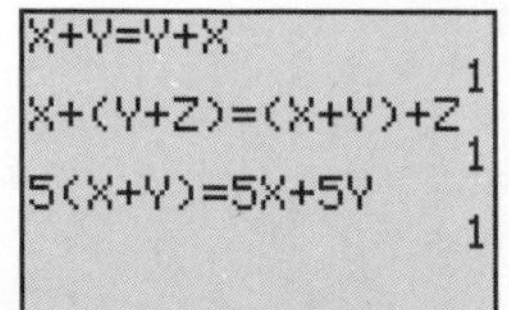

No matter what values are stored in *X*, *Y*, and *Z*, the commutative, associative, and distributive properties assure us that these statements are true.

EXAMPLE 6 Some specific examples of the properties of addition and multiplication of real numbers are given here.

(a) $5 + 7$ is a real number. Closure property for addition

(b) $5 + (6 + 8) = (5 + 6) + 8$ Associative property of addition

(c) $8 + 0 = 8$ Identity property of addition

(d) $-4(-1/4) = 1$ Inverse property of multiplication

(e) $4 + (3 + 9) = 4 + (9 + 3)$ Commutative property of addition

(f) $5(x + y) = 5x + 5y$ Distributive property

Applications of Real Numbers

The usefulness of negative numbers can be seen by considering situations that arise in everyday life. For example, we need negative numbers to express the temperatures on January days in Anchorage, Alaska, where they often drop below zero. If a company loses money, its "profits" are negative. Such money-losing companies are said to be "in the red" (an expression from the Renaissance, when losses were written in red ink and profits in black ink). And, of course, haven't we all experienced a checking account balance below zero, with hopes that our deposit will make it to the bank before our outstanding checks "bounce"?

Problem Solving

When problems deal with gains and losses, the gains may be interpreted as positive numbers and the losses as negative numbers. Temperatures below 0° are negative, and those above 0° are positive. Altitudes above sea level are considered positive and those below sea level are considered negative. The next examples show applications of these ideas.

EXAMPLE 7 Paul Van Erden gained 3 yards on the first play from scrimmage, lost 12 yards on the second play, and then gained 13 yards on the third play. How many yards did Paul gain or lose altogether?

The gains are represented by positive numbers and the loss by a negative number.

$$3 + (-12) + 13$$

Add from left to right.

$$3 + (-12) + 13 = [3 + (-12)] + 13 = (-9) + 13 = 4$$

Paul gained 4 yards altogether.

EXAMPLE 8 The record high temperature of 134°F in the United States was recorded in Death Valley, California, in 1913. The record low was −80°F at Prospect Creek, Alaska, in 1971. What is the difference between the highest and lowest temperatures?

We must find the value of the highest temperature minus the lowest temperature.

$$134 - (-80) = 134 + 80 \quad \text{Use the definition of subtraction.}$$
$$= 214 \quad \text{Add.}$$

The difference between the highest and the lowest temperatures is 214°F. ■

6.2 EXERCISES

Decide whether each statement is true *or* false.

1. The sum of two negative numbers must be negative.
2. The difference between two negative numbers must be negative.
3. The product of two negative numbers must be negative.
4. The quotient of two negative numbers must be negative.
5. If $a > 0$ and $b < 0$, then $\frac{a}{b} < 0$.
6. The product of 648 and -927 is a real number.

Perform the indicated operations, using the order of operations as necessary.

7. $-12 + (-8)$
8. $-5 + (-2)$
9. $12 + (-16)$
10. $-6 + 17$
11. $-12 - (-1)$
12. $-3 - (-8)$
13. $-5 + 11 + 3$
14. $-9 + 16 + 5$
15. $12 - (-3) - (-5)$
16. $15 - (-6) - (-8)$
17. $-9 - (-11) - (4 - 6)$
18. $-4 - (-13) + (-5 + 10)$
19. $(-12)(-2)$
20. $(-3)(-5)$
21. $9(-12)(-4)(-1)3$
22. $-5(-17)(2)(-2)4$
23. $\frac{-18}{-3}$
24. $\frac{-100}{-50}$
25. $\frac{36}{-6}$
26. $\frac{52}{-13}$
27. $\frac{0}{12}$
28. $\frac{0}{-7}$
29. $-6 + [5 - (3 + 2)]$
30. $-8[4 + (7 - 8)]$
31. $-8(-2) - [(4^2) + (7 - 3)]$
32. $-7(-3) - [(2^3) - (3 - 4)]$
33. $-4 - 3(-2) + 5^2$
34. $-6 - 5(-8) + 3^2$
35. $(-6 - 3)(-2 - 3)$
36. $(-8 - 5)(-2 - 1)$
37. $\frac{(-10 + 4) \cdot (-3)}{-7 - 2}$
38. $\frac{(-6 + 3) \cdot (-4)}{-5 - 1}$
39. Which of the following expressions are undefined?
    (a) $\frac{8}{0}$ (b) $\frac{9}{6 - 6}$ (c) $\frac{4 - 4}{5 - 5}$ (d) $\frac{0}{-1}$
40. If you have no money in your pocket and you divide it equally among your three siblings, how much does each get? Use this situation to explain division of zero by a nonzero number.

Identify the property illustrated by each of the following statements.

41. $6 + 9 = 9 + 6$
42. $8 \cdot 4 = 4 \cdot 8$
43. $7 + (2 + 5) = (7 + 2) + 5$
44. $(3 \cdot 5) \cdot 4 = 4 \cdot (3 \cdot 5)$
45. $9 + (-9) = 0$
46. $12 + 0 = 12$

47. $9 \cdot 1 = 9$

48. $(-1/3) \cdot (-3) = 1$

49. $0 + 283 = 283$

50. $6 \cdot (4 \cdot 2) = (6 \cdot 4) \cdot 2$

51. $2 \cdot (4 + 3) = 2 \cdot 4 + 2 \cdot 3$

52. $9 \cdot 6 + 9 \cdot 8 = 9 \cdot (6 + 8)$

53. $19 + 12$ is a real number.

54. $19 \cdot 12$ is a real number.

---

55. **(a)** Evaluate $6 - 8$ and $8 - 6$.
(b) By the results of part (a), we may conclude that subtraction is not a(n) ______ operation.
(c) Are there *any* real numbers a and b for which $a - b = b - a$? If so, give an example.

56. **(a)** Evaluate $4 \div 8$ and $8 \div 4$.
(b) By the results of part (a), we may conclude that division is not a(n) ______ operation.
(c) Are there *any* real numbers a and b for which $a \div b = b \div a$? If so, give an example.

57. Many everyday occurrences can be thought of as operations that have opposites or inverses. For example, the inverse operation for "going to sleep" is "waking up." For each of the given activities, specify its inverse activity.
(a) cleaning up your room
(b) earning money
(c) increasing the volume on your portable radio

58. Many everyday activities are commutative; that is, the order in which they occur does not affect the outcome. For example, "putting on your shirt" and "putting on your pants" are commutative operations. Decide whether the given activities are commutative.
(a) putting on your shoes; putting on your socks
(b) getting dressed; taking a shower
(c) combing your hair; brushing your teeth

59. The following conversation actually took place between one of the authors of this text and his son, Jack, when Jack was four years old.

Daddy: "Jack, what is 3 + 0?"
Jack: "3"
Daddy: "Jack, what is 4 + 0?"
Jack: "4 . . . and Daddy, *string* plus zero equals *string*!"

What property of addition of real numbers did Jack recognize?

60. The phrase *defective merchandise counter* is an example of a phrase that can have different meanings depending upon how the words are grouped (think of the associative properties). For example, (*defective merchandise*) *counter* is a location at which we would return an item that does not work, while *defective* (*merchandise counter*) is a broken place where items are bought and sold. For each of the following phrases, explain why the associative property does not hold.
(a) difficult test question
(b) woman fearing husband
(c) man biting dog

61. The distributive property holds for multiplication with respect to addition. Does the distributive property hold for addition with respect to multiplication? That is, is $a + (b \cdot c) = (a + b) \cdot (a + c)$ true for all values of a, b, and c? (*Hint:* Let $a = 2$, $b = 3$, and $c = 4$.)

62. Suppose someone makes the following claim: The distributive property for addition with respect to multiplication (from Exercise 61) is valid, and here's why: Let $a = 2$, $b = -4$, and $c = 3$. Then $a + (b \cdot c) = 2 + (-4 \cdot 3) = 2 + (-12) = -10$, and $(a + b) \cdot (a + c) = [2 + (-4)] \cdot [2 + 3] = -2 \cdot 5 = -10$. Since both expressions are equal, the property must be valid. How would you respond to this reasoning?

63. Suppose that a student shows you the following work.

$$-3(4 - 6) = -3(4) - 3(6) = -12 - 18 = -30$$

The student has made a very common error in applying the distributive property. Write a short paragraph explaining the student's mistake, and work the problem correctly.

64. Work the following problem in two ways, first using the order of operations, and then using the distributive property: Evaluate $9(11 + 15)$.

Recall from the text that an expression such as -2^4 *is evaluated as follows:*

$$-2^4 = -(2 \cdot 2 \cdot 2 \cdot 2) = -16.$$

The expression $(-2)^4$ *is evaluated as follows:*

$$(-2)^4 = (-2)(-2)(-2)(-2) = 16.$$

Each of the expressions in Exercises 65–72 is equal to either 81 *or* −81. *Decide which of these is the correct value.*

65. -3^4 **66.** $-(3^4)$ **67.** $(-3)^4$ **68.** $-(-3^4)$

69. $-(-3)^4$ **70.** $[-(-3)]^4$ **71.** $-[-(-3)]^4$ **72.** $-[-(-3^4)]$

Solve the problem by evaluating a sum or difference of real numbers. No variables are needed.

73. In 1990, the net incomes of savings institutions (in millions of dollars) for states in the northeast United States were as follows.

State	Net Income (in millions of dollars)
Maine	0
New Hampshire	−24
Vermont	2
Massachusetts	−212
Rhode Island	−13
Connecticut	−149

What was the total of these net incomes? (*Source:* U.S. Office of Thrift Supervision)

74. The 1991 state general fund balances (in millions of dollars) for states in the Pacific region of the United States were as follows.

State	General Fund Balance (in millions of dollars)
Washington	468
Oregon	380
California	−1,259
Alaska	791
Hawaii	347

What was the total of these balances? (*Source:* National Association of State Budget Officers)

75. The two charts show the heights of some selected mountains and the depths of some selected trenches. Use the information given to find the answers in parts (a) through (d). (*Source: The World Almanac and Book of Facts,* 1994)

Mountain	Height (in feet)
Foraker	17,400
Wilson	14,246
Pikes Peak	14,110

Trench	Depth in Feet (as a negative number)
Philippine	−32,995
Cayman	−24,721
Java	−23,376

(a) What is the difference between the height of Mt. Foraker and the depth of the Philippine Trench?

(b) What is the difference between the height of Pikes Peak and the depth of the Java Trench?

(c) How much deeper is the Cayman Trench than the Java Trench?

(d) How much deeper is the Philippine Trench than the Cayman Trench?

Solve each problem.

76. Shalita's checking account balance is $54.00. She then takes a gamble by writing a check for $89.00. What is her new balance? (Write the balance as a signed number.)

77. The surface, or rim, of a canyon is at altitude 0. On a hike down into the canyon, a party of hikers stops for a rest at 130 meters below the surface. They then descend another 54 meters. What is their new altitude? (Write the altitude as a signed number.)

78. A pilot announces to his passengers that the current altitude of their plane is 34,000 feet. Because of some unexpected turbulence, he is forced to descend 2,100 feet. What is the new altitude of the plane? (Write the altitude as a signed number.)

79. The lowest temperature ever recorded in Little Rock, Arkansas, was $-5°$F. The highest temperature ever recorded there was 117°F more than the lowest. What was this highest temperature?

80. On January 23, 1943, the temperature rose 49°F in two minutes in Spearfish, South Dakota. If the starting temperature was $-4°$, what was the temperature two minutes later?

81. On a series of three consecutive running plays, Herschel Walker gained 4 yards, lost 3 yards, and lost 2 yards. What positive or negative number represents his total net yardage for the series of plays?

82. On three consecutive passing plays, Troy Aikman passed for a gain of 6 yards, was sacked for a loss of 12 yards, and passed for a gain of 43 yards. What positive or negative number represents the total net yardage for the plays?

83. The top of Mount Whitney, visible from Death Valley, has an altitude of 14,494 feet above sea level. The bottom of Death Valley is 282 feet below sea level. Using zero as sea level, find the difference between these two elevations.

84. The highest point in Louisiana is Driskill Mountain, at an altitude of 535 feet. The lowest point is at Spanish Fort, 8 feet below sea level. Using zero as sea level, find the difference between these two elevations.

85. A certain Greek mathematician was born in 426 B.C. His father was born 43 years earlier. In what year was his father born?

86. Kevin Carlson enjoys playing Triominoes every Thursday night. Last Thursday, on four successive turns, his scores were -19, 28, -5, and 13. What was his total score for the four turns?

87. David Fleming enjoys scuba diving. He dives to 34 feet below the surface of a lake. His partner, Kim Walrath, dives to 40 feet below the surface, but then ascends 20 feet. What is the vertical distance between Kim and David?

88. Fontaine Evaldo, a pilot for a major airline, announced to her passengers that their plane, currently at 34,000 feet, would descend 2,500 feet to avoid turbulence, and then ascend 3,000 feet once they were out of danger from the turbulence. What would their final altitude be?

89. During the years 1980–1986, Rickey Henderson was the American League leader in stolen bases each year. The *bar graph* gives a representation of the number of steals each year. Use a signed number to represent the change in the number of steals from one year to the next. For example, from 1980 to 1981 the change was $56 - 100 = -44$.

(a) from 1981 to 1982
(b) from 1982 to 1983
(c) from 1983 to 1984
(d) from 1984 to 1985
(e) from 1985 to 1986

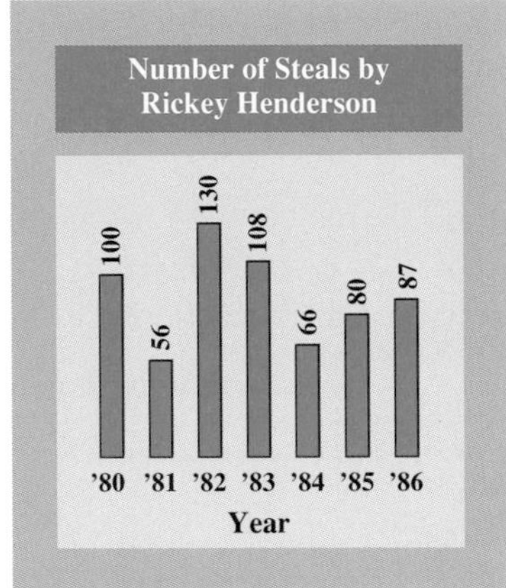

90. During the years 1985–1990, Vince Coleman was the National League leader in stolen bases each year. The *line graph* gives a representation of the number of steals each year. Use a signed number to represent the change in the number of steals from one year to the next. For example, from 1985 to 1986 the change was $107 - 110 = -3$.

(a) from 1986 to 1987
(b) from 1987 to 1988
(c) from 1988 to 1989
(d) from 1989 to 1990

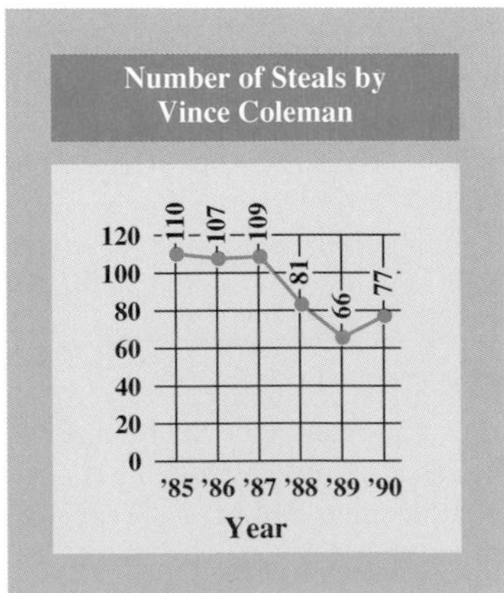

6.3 Rational Numbers and Decimals

Benjamin Banneker (1731–1806) spent the first half of his life tending a farm in Maryland. He gained a reputation locally for his mechanical skills and abilities in mathematical problem solving. In 1772 he acquired astronomy books from a neighbor and devoted himself to learning astronomy, observing the skies, and making calculations. In 1789 Banneker joined the team that surveyed what is now the District of Columbia.

Banneker published almanacs yearly from 1792 to 1802. He sent a copy of his first almanac to Thomas Jefferson along with an impassioned letter against slavery. Jefferson subsequently championed the cause of this early African-American mathematician.

The set of real numbers is composed of two important mutually exclusive subsets: the rational numbers and the irrational numbers. (Two sets are *mutually exclusive* if they contain no elements in common.) In this section and the next, we will look at these two sets in some detail. Let us begin by examining the set of rational numbers.

Quotients of integers are called **rational numbers.** Think of the rational numbers as being made up of all the fractions (quotients of integers with denominator not equal to zero) and all the integers. Any integer can be written as the quotient of two integers. For example, the integer 9 can be written as the quotient $9/1$, or $18/2$, or $27/3$, and so on. Also, -5 can be expressed as a quotient of integers as $-5/1$ or $-10/2$, and so on. (How can the integer 0 be written as a quotient of integers?) Since both fractions and integers can be written as quotients of integers, the set of rational numbers is defined as follows.

Rational Numbers

Rational numbers $= \{x \mid x$ is a quotient of two integers, with denominator not $0\}$

A rational number is said to be in **lowest terms** if the greatest common factor of the numerator (top number) and the denominator (bottom number) is 1. Rational numbers are written in lowest terms by using the *fundamental property of rational numbers.*

Fundamental Property of Rational Numbers

If a, b, and k are integers with $b \neq 0$ and $k \neq 0$, then

$$\frac{a \cdot k}{b \cdot k} = \frac{a}{b}.$$

The following example illustrates the fundamental property of rational numbers. We find the greatest common factor of the numerator and denominator and use it for k in the statement of the property.

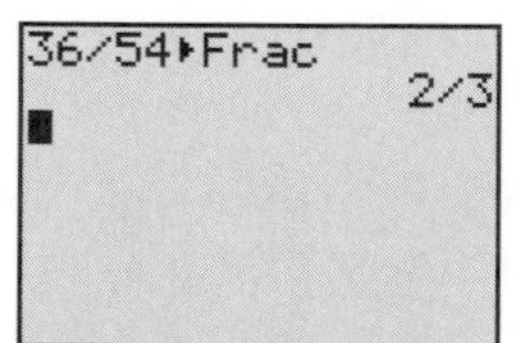

The calculator reduces 36/54 to lowest terms, as illustrated in Example 1.

EXAMPLE 1 Reduce $36/54$ to lowest terms.

Since the greatest common factor of 36 and 54 is 18,

$$\frac{36}{54} = \frac{2 \cdot 18}{3 \cdot 18} = \frac{2}{3}.$$

In the above example it was shown that $36/54 = 2/3$. If we multiply the numerator of the fraction on the left by the denominator of the fraction on the right, we obtain $36 \cdot 3 = 108$. Now if we multiply the denominator of the fraction on the left by the numerator of the fraction on the right, we obtain

$54 \cdot 2 = 108$. It is not just coincidence that the result is the same in both cases. In fact, one way of determining whether two fractions are equal is to perform this test. If the product of the "extremes" (36 and 3 in this case) equals the product of the "means" (54 and 2), the fractions are equal. This method of checking for equality of rational numbers is called the **cross-product method.** It is given in the following box.

Cross-Product Test for Equality of Rational Numbers

For rational numbers a/b and c/d, $b \neq 0, d \neq 0$,

$$\frac{a}{b} = \frac{c}{d} \quad \textbf{if and only if} \quad a \cdot d = b \cdot c.$$

The operation of addition of rational numbers can be illustrated by the sketches in Figure 7. The rectangle on the left is divided into three equal portions, with one of the portions in color. The rectangle on the right is divided into five equal parts, with two of them in color.

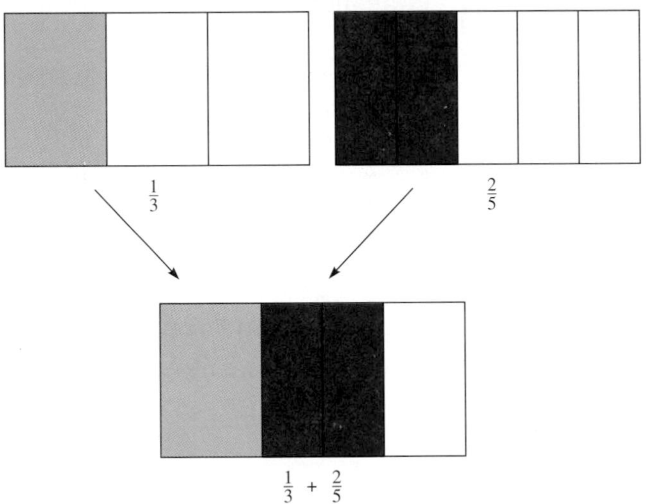

FIGURE 7

The total of the areas in color is represented by the sum

$$\frac{1}{3} + \frac{2}{5}.$$

To evaluate this sum, the areas in color must be redrawn in terms of a common unit. Since the least common multiple of 3 and 5 is 15, redraw both rectangles with 15 parts. See Figure 8. In the figure, 11 of the small rectangles are in color, so

$$\frac{1}{3} + \frac{2}{5} = \frac{5}{15} + \frac{6}{15} = \frac{11}{15}.$$

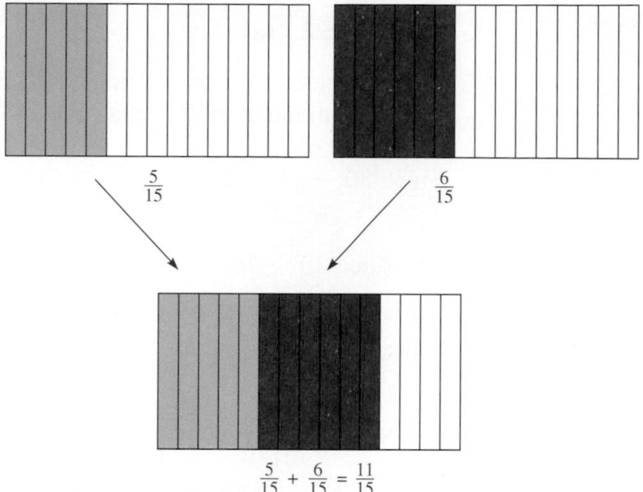

■ **FIGURE 8**

In general, the sum

$$\frac{a}{b} + \frac{c}{d}$$

may be found by writing a/b and c/d with the common denominator bd, retaining this denominator in the sum, and adding the numerators:

$$\frac{a}{b} + \frac{c}{d} = \frac{ad}{bd} + \frac{bc}{bd} = \frac{ad + bc}{bd}.$$

A similar case can be made for the difference between rational numbers. A formal definition of addition and subtraction of rational numbers follows.

Adding and Subtracting Rational Numbers

If a/b and c/d are rational numbers, then

$$\frac{a}{b} + \frac{c}{d} = \frac{ad + bc}{bd} \quad \text{and} \quad \frac{a}{b} - \frac{c}{d} = \frac{ad - bc}{bd}.$$

This formal definition is seldom used in practice. In practical problems involving addition and subtraction of rational numbers, we usually rewrite the fractions with the least common multiple of their denominators. This is called the **least common denominator,** and finding the least common denominator may be done by inspection, or by the method of prime factorization. Rewrite the fractions so that the least common multiple of their denominators becomes the denominator of each one.

The results of Example 2 are illustrated in this screen.

EXAMPLE 2 **(a)** Add: $\dfrac{2}{15} + \dfrac{1}{10}$.

The least common multiple of 15 and 10 is 30. Now write $2/15$ and $1/10$ with denominators of 30, and then add the numerators. Proceed as follows:

Since $30 \div 15 = 2$, $\qquad \dfrac{2}{15} = \dfrac{2 \cdot 2}{15 \cdot 2} = \dfrac{4}{30}$,

and since $30 \div 10 = 3$, $\qquad \dfrac{1}{10} = \dfrac{1 \cdot 3}{10 \cdot 3} = \dfrac{3}{30}$.

Thus, $\qquad \dfrac{2}{15} + \dfrac{1}{10} = \dfrac{4}{30} + \dfrac{3}{30} = \dfrac{7}{30}$.

(b) Subtract: $\dfrac{173}{180} - \dfrac{69}{1{,}200}$.

The least common multiple of 180 and 1,200 is 3,600.

$$\frac{173}{180} - \frac{69}{1{,}200} = \frac{3{,}460}{3{,}600} - \frac{207}{3{,}600} = \frac{3{,}460 - 207}{3{,}600} = \frac{3{,}253}{3{,}600}$$

The product of two rational numbers is defined as follows.

Multiplying Rational Numbers

If a/b and c/d are rational numbers, then

$$\frac{a}{b} \cdot \frac{c}{d} = \frac{ac}{bd}.$$

Example 3 illustrates how this definition is used.

(3/4)*(7/10)▸Frac
21/40
(5/18)*(3/10)▸Frac
1/12

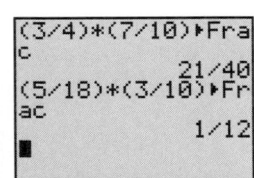

To illustrate the results of Example 3, we use parentheses around the fraction factors.

EXAMPLE 3 Find each of the following products.

(a) $\dfrac{3}{4} \cdot \dfrac{7}{10} = \dfrac{3 \cdot 7}{4 \cdot 10} = \dfrac{21}{40}$

(b) $\dfrac{5}{18} \cdot \dfrac{3}{10} = \dfrac{5 \cdot 3}{18 \cdot 10} = \dfrac{15}{180} = \dfrac{1 \cdot 15}{12 \cdot 15} = \dfrac{1}{12}$

In practice, a multiplication problem such as this is often solved by using slash marks to indicate that common factors have been divided out of the numerator and denominator.

$$\frac{\overset{1}{\cancel{5}}}{\underset{6}{\cancel{18}}} \cdot \frac{\overset{1}{\cancel{3}}}{\underset{2}{\cancel{10}}}$$

3 is divided out of the terms 3 and 18; 5 is divided out of 5 and 10.

$$= \frac{1}{6} \cdot \frac{1}{2}$$

$$= \frac{1}{12}$$

In a fraction, the fraction bar indicates the operation of division. Recall that, in the previous section, we defined the multiplicative inverse, or reciprocal, of the nonzero number b. The multiplicative inverse of b is $1/b$. We can now define division using multiplicative inverses.

Definition of Division

If a and b are real numbers, $b \neq 0$, then

$$\frac{a}{b} = a \cdot \frac{1}{b}.$$

You have probably heard the rule, "To divide fractions, invert the divisor and multiply." But have you ever wondered why this rule works? To illustrate it, suppose that you have 7/8 of a gallon of milk and you wish to find how many quarts you have. Since a quart is 1/4 of a gallon, you must ask yourself, "How many 1/4s are there in 7/8?" This would be interpreted as

$$\frac{7}{8} \div \frac{1}{4} \quad \text{or} \quad \frac{\frac{7}{8}}{\frac{1}{4}}.$$

The fundamental property of rational numbers discussed earlier can be extended to rational number values of a, b, and k. With $a = 7/8$, $b = 1/4$, and $k = 4$,

$$\frac{a}{b} = \frac{a \cdot k}{b \cdot k} = \frac{\frac{7}{8} \cdot 4}{\frac{1}{4} \cdot 4} = \frac{\frac{7}{8} \cdot 4}{1} = \frac{7}{8} \cdot \frac{4}{1}.$$

Now notice that we began with the division problem $7/8 \div 1/4$ which, through a series of equivalent expressions, led to the multiplication problem $(7/8) \cdot (4/1)$. So dividing by 1/4 is equivalent to multiplying by its reciprocal, 4/1. By the definition of multiplication of fractions,

$$\frac{7}{8} \cdot \frac{4}{1} = \frac{28}{8} = \frac{7}{2},$$

and thus there are 7/2 or 3 1/2 quarts in 7/8 gallon.*

We now state the rule for dividing a/b by c/d.

Dividing Rational Numbers

If a/b and c/d are rational numbers, where $c/d \neq 0$, then

$$\frac{a}{b} \div \frac{c}{d} = \frac{a}{b} \cdot \frac{d}{c} = \frac{ad}{bc}.$$

*3 1/2 is a **mixed number.** Mixed numbers are covered in the exercises for this section.

```
(-4/7)/(3/14)▸Fr
ac
            -8/3
(2/9)/4▸Frac
            1/18
```

This screen supports the results of Example 4(b) and (c).

EXAMPLE 4 Find each of the following quotients.

(a) $\dfrac{3}{5} \div \dfrac{7}{15} = \dfrac{3}{5} \cdot \dfrac{15}{7} = \dfrac{45}{35} = \dfrac{9 \cdot 5}{7 \cdot 5} = \dfrac{9}{7}$

(b) $\dfrac{-4}{7} \div \dfrac{3}{14} = \dfrac{-4}{7} \cdot \dfrac{14}{3} = \dfrac{-56}{21} = \dfrac{-8 \cdot 7}{3 \cdot 7} = \dfrac{-8}{3} = -\dfrac{8}{3}$

(c) $\dfrac{2}{9} \div 4 = \dfrac{2}{9} \div \dfrac{4}{1} = \dfrac{2}{9} \cdot \dfrac{1}{4} = \dfrac{\overset{1}{\cancel{2}}}{9} \cdot \dfrac{1}{\underset{2}{\cancel{4}}} = \dfrac{1}{18}$ ■

There is no integer between two consecutive integers, such as 3 and 4. However, a rational number can always be found between any two distinct rational numbers. For this reason, the set of rational numbers is said to be *dense*.

Density Property of the Rational Numbers

If r and t are distinct rational numbers, with $r < t$, then there exists a rational number s such that

$$r < s < t.$$

Simon Stevin (1548–1620) worked as a bookkeeper in Belgium and became an engineer in the Netherlands army. He is usually given credit for the development of decimals.

Example 5 shows how to find the rational number that is halfway between two given rational numbers—the average of the numbers.

EXAMPLE 5 Find the rational number halfway between 2/3 and 5/6.

Add the numbers:

$$\frac{2}{3} + \frac{5}{6} = \frac{4}{6} + \frac{5}{6} = \frac{9}{6} = \frac{3}{2}.$$

Take half this sum:

$$\frac{1}{2} \cdot \frac{3}{2} = \frac{3}{4}.$$

The number 3/4 is halfway between 2/3 and 5/6. ■

This screen shows how the problem in Example 5 can be worked with a calculator.

Repeated application of the density property implies that between two given rational numbers are *infinitely many* rational numbers. It is also true that between any two *real* numbers there are infinitely many *real* numbers. Thus, we may say that the set of real numbers is dense.

Decimal Form of Rational Numbers

Up to now in this section, we have discussed rational numbers in the form of quotients of integers. Rational numbers can also be expressed as decimals. Decimal numerals have place values that are powers of 10. For example, the decimal numeral 483.039475 is read "four hundred eighty-three and thirty-nine thousand, four hundred seventy-five millionths." The place values are as shown here.

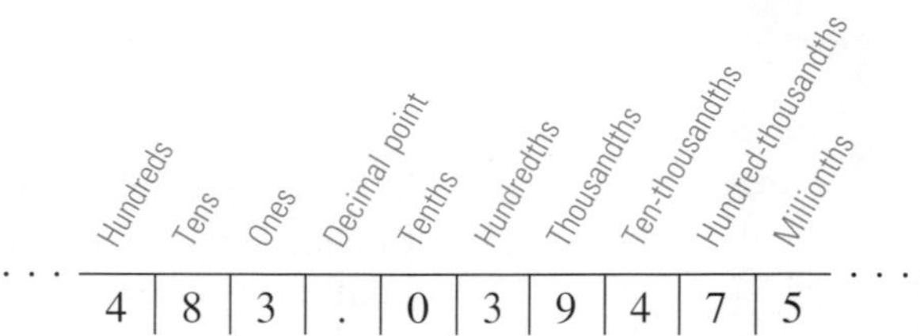

Given a rational number in the form a/b, it can be expressed as a decimal most easily by entering it into a calculator. For example, to write $3/8$ as a decimal, enter 3, then enter the operation of division, then enter 8. Press the equals key to find the following equivalence.

$$\frac{3}{8} = .375$$

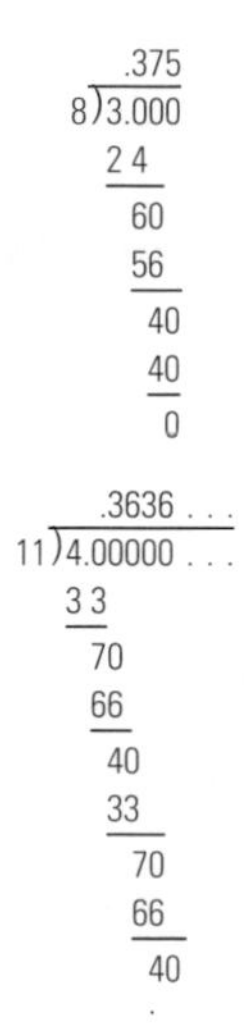

Of course, this same result may be obtained by long division, as shown in the margin. By this result, the rational number $3/8$ is the same as the decimal .375. A decimal such as .375, which stops, is called a **terminating decimal.** Other examples of terminating decimals are

$$\frac{1}{4} = .25, \qquad \frac{7}{10} = .7, \qquad \text{and} \qquad \frac{89}{1{,}000} = .089.$$

Not all rational numbers can be represented by terminating decimals. For example, convert $4/11$ into a decimal by dividing 11 into 4 using a calculator. The display shows

.36363636363, or perhaps .363636364.

However, we see that the long division process, shown in the margin, indicates that we will actually get .3636. . . , with the digits 36 repeating over and over indefinitely. To indicate this, we write a bar (called a *vinculum*) over the "block" of digits that repeats. Therefore, we can write

$$\frac{4}{11} = .\overline{36}.$$

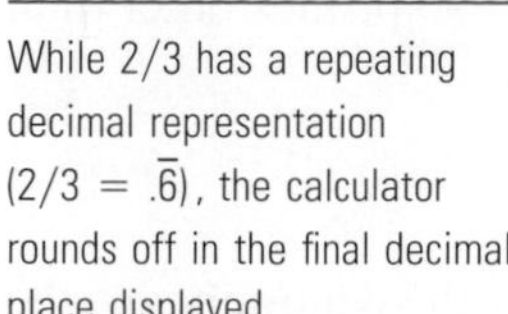

While 2/3 has a repeating decimal representation ($2/3 = .\overline{6}$), the calculator rounds off in the final decimal place displayed.

Because of the limitations of the display of a calculator, and because some rational numbers have repeating decimals, it is important to be able to interpret calculator results accordingly when obtaining repeating decimals.

While we shall distinguish between *terminating* and *repeating* decimals in this book, some mathematicians prefer to consider all rational numbers as repeating decimals. This can be justified by thinking this way: if the division process leads to a remainder of 0, then zeros repeat indefinitely in the decimal form. For example, we can consider the decimal form of $3/4$ as follows.

$$\frac{3}{4} = .75\overline{0}$$

In order to find a baseball player's batting average, we divide the number of hits by the number of at-bats. A surprising paradox exists concerning averages; it is possible for Player *A* to have a higher batting average than Player *B* in each of two successive years, yet for the two-year period, Player *B* can have a higher total batting average. Look at the chart.

Year	Player *A*	Player *B*
1991	$\frac{20}{40} = .500$	$\frac{90}{200} = .450$
1992	$\frac{60}{200} = .300$	$\frac{10}{40} = .250$
Two-year total	$\frac{80}{240} = .333$	$\frac{100}{240} = .417$

In both individual years, Player *A* had a higher average, but for the two-year period, Player *B* had the higher average. This is an example of *Simpson's paradox* from statistics.

EXAMPLE 6 A calculator or long division shows the following decimals for the given quotients of integers.

(a) $\frac{5}{11} = .\overline{45}$ **(b)** $\frac{1}{3} = .\overline{3}$ **(c)** $\frac{5}{6} = .8\overline{3}$

By considering the possible remainders that may be obtained when converting a quotient of integers to a decimal, we can draw an important conclusion about the decimal form of rational numbers. If the remainder is never zero, the division will produce a repeating decimal. This happens because each step of the division process must produce a remainder that is less than the divisor. Since the number of different possible remainders is less than the divisor, the remainders must eventually begin to repeat. This makes the digits of the quotient repeat, producing a repeating decimal.

Any rational number can be expressed as either a terminating decimal or a repeating decimal.

To determine whether the decimal form of a quotient of integers will terminate or repeat, we use the following rule.

Criteria for Terminating and Repeating Decimals

A rational number a/b in lowest terms results in a **terminating decimal** if the only prime factors of the denominator are 2 or 5 (or both).

A rational number a/b in lowest terms results in a **repeating decimal** if a prime other than 2 or 5 appears in the prime factorization of the denominator.

The justification of this rule is based on the fact that the prime factors of 10 are 2 and 5, and the decimal system uses ten as its base.

EXAMPLE 7 Without actually dividing, determine whether the decimal form of the given rational number terminates or repeats.

(a) $\frac{7}{8}$

Since 8 factors as 2^3, the decimal form will terminate. No primes other than 2 or 5 divide the denominator.

(b) $\frac{13}{150}$

$150 = 2 \cdot 3 \cdot 5^2$. Since 3 appears as a prime factor of the denominator, the decimal form will repeat.

(c) $\frac{6}{75}$

Before performing the test it is necessary to reduce the rational number to lowest terms.

$$\frac{6}{75} = \frac{2}{25}$$

Since $25 = 5^2$, the decimal form will terminate. ■

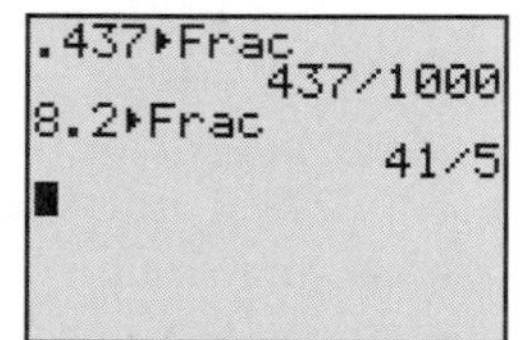

The results of Example 8 are supported in this screen.

We have seen that a rational number will be represented by either a terminating or a repeating decimal. What about the reverse process? That is, must a terminating decimal or a repeating decimal represent a rational number? The answer is *yes*. For example, the terminating decimal .6 represents a rational number.

$$.6 = \frac{6}{10} = \frac{3}{5}.$$

EXAMPLE 8 Write each terminating decimal as a quotient of integers.

(a) .437

Think of .437 in words as "four hundred thirty-seven thousandths," or

$$.437 = \frac{437}{1{,}000}.$$

(b) $8.2 = 8 + \frac{2}{10} = \frac{82}{10} = \frac{41}{5}$ ■

Repeating decimals cannot be converted into quotients of integers quite so quickly. The steps for making this conversion are given in the next example. (This example uses basic algebra.)

EXAMPLE 9 Find a quotient of two integers equal to $.\overline{85}$.

Step 1: Let $x = .\overline{85}$, so that $x = .858585\ldots$.

Step 2: Multiply both sides of the equation $x = .858585\ldots$ by 100. (Use 100 since there are two digits in the part that repeats.)

$$x = .858585\ldots$$
$$100x = 100(.858585\ldots)$$
$$100x = 85.858585\ldots$$

Step 3: Subtract the expressions in Step 1 from the final expressions in Step 2.

$$\begin{array}{r} 100x = 85.858585\ldots \\ x = .858585\ldots \\ \hline 99x = 85 \end{array}$$

(Recall that $x = 1x$ and $100x - x = 99x$.)

$1 = .999999999999\ldots$

Terminating or Repeating? One of the most baffling truths of elementary mathematics is the following:

$$1 = .9999\ldots$$

Most people believe that $.\overline{9}$ has to be less than 1, but this is not the case. The following argument shows otherwise, Let $x = .9999\ldots$ Then

$$\begin{array}{rl} 10x = 9.9999\ldots & \\ x = .9999\ldots & \\ \hline 9x = 9 & \text{Subtract.} \\ x = 1. & \end{array}$$

Therefore, $1 = .9999\ldots$. Similarly, it can be shown that any terminating decimal can be represented as a repeating decimal with an endless string of 9s. For example, $.5 = .49999\ldots$ and $2.6 = 2.59999\ldots$. This is another

continued on page 284

way of justifying what was stated in the text: any rational number may be represented as a repeating decimal.

For more on the fact that $1 = .\overline{9}$, see the article "Persuasive Arguments: .9999. . . = 1" by Lucien T. Hall, Jr., in the December 1971 issue of *The Mathematics Teacher.* See also Exercises 105 and 106 at the end of this section.

Step 4: Solve the equation $99x = 85$ by dividing both sides by 99:

$$99x = 85$$

$$\frac{99x}{99} = \frac{85}{99}$$

$$x = \frac{85}{99}.$$

Since x equals $.\overline{85}$,

$$.\overline{85} = \frac{85}{99}.$$

This result may be checked with a calculator. Remember, however, that the calculator will only show a finite number of decimal places, and may round off in the final decimal place shown. ■

6.3 EXERCISES

Decide whether each statement is true *or* false.

1. If p and q are different prime numbers, the rational number p/q is reduced to lowest terms.

2. The same number may be added to both the numerator and the denominator of a fraction without changing the value of the fraction.

3. A nonzero fraction and its reciprocal will always have the same sign.

4. The set of integers has the property of density.

Use the fundamental property of rational numbers to write each of the following in lowest terms.

5. $\frac{16}{48}$ **6.** $\frac{21}{28}$ **7.** $-\frac{15}{35}$ **8.** $-\frac{8}{48}$

Use the fundamental property to write each of the following in three other ways.

9. $\frac{3}{8}$ **10.** $\frac{9}{10}$ **11.** $-\frac{5}{7}$ **12.** $-\frac{7}{12}$

13. For each of the following, write a fraction in lowest terms that represents the portion of the figure that is in color.

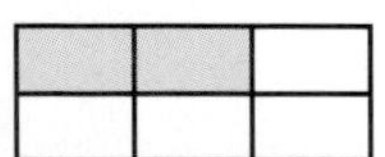

(b)

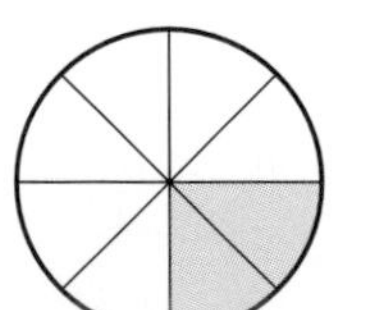

(c)

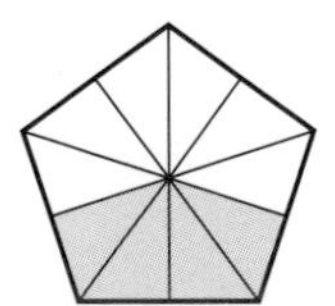

(d)

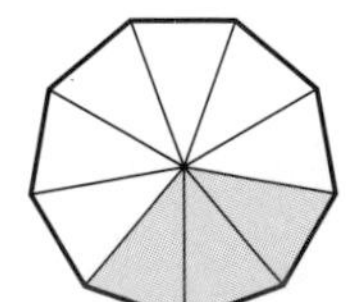

14. For each of the following, write a fraction in lowest terms that represents the region described.
(a) the dots in the rectangle as a part of the dots in the entire figure
(b) the dots in the triangle as a part of the dots in the entire figure
(c) the dots in the rectangle as a part of the dots in the union of the triangle and the rectangle
(d) the dots in the intersection of the triangle and the rectangle as a part of the dots in the union of the triangle and the rectangle

15. Refer to the figure for Exercise 14 and write a description of the region that is represented by the fraction 1/12.

16. In the local softball league, the first five games produced the following results: David Glenn got 8 hits in 20 at-bats, and Chalon Bridges got 12 hits in 30 at-bats. David claims that he and Chalon did equally well. Is he correct? Why or why not?

17. After ten games in the local softball league, the following batting statistics were obtained.

Player	At-bats	Hits	Home Runs
Bishop, Kelley	40	9	2
Carlton, Robert	36	12	3
Dykler, Francoise	11	5	1
Crowe, Vonalaine	16	8	0
Marshall, James	20	10	2

Answer each of the following, using estimation skills as necessary.
(a) Which player got a hit in exactly 1/3 of his or her at-bats?
(b) Which player got a hit in just less than 1/2 of his or her at-bats?
(c) Which player got a home run in just less than 1/10 of his or her at-bats?
(d) Which player got a hit in just less than 1/4 of his or her at-bats?
(e) Which two players got hits in exactly the same fractional parts of their at-bats? What was the fractional part, reduced to lowest terms?

18. Use estimation skills to determine the best approximation for the following sum:

$$\frac{14}{26} + \frac{98}{99} + \frac{100}{51} + \frac{90}{31} + \frac{13}{27}.$$

(a) 6 **(b)** 7 **(c)** 5 **(d)** 8

Perform the indicated operations and express answers in lowest terms. Use the order of operations as necessary.

19. $\frac{3}{8} + \frac{1}{8}$ **20.** $\frac{7}{9} + \frac{1}{9}$ **21.** $\frac{5}{16} + \frac{7}{12}$ **22.** $\frac{1}{15} + \frac{7}{18}$

23. $\frac{2}{3} - \frac{7}{8}$ **24.** $\frac{13}{20} - \frac{5}{12}$ **25.** $\frac{5}{8} - \frac{3}{14}$ **26.** $\frac{19}{15} - \frac{7}{12}$

27. $\frac{3}{4} \cdot \frac{9}{5}$ **28.** $\frac{3}{8} \cdot \frac{2}{7}$ **29.** $-\frac{2}{3} \cdot -\frac{5}{8}$ **30.** $-\frac{2}{4} \cdot \frac{3}{9}$

31. $\frac{5}{12} \div \frac{15}{4}$ **32.** $\frac{15}{16} \div \frac{30}{8}$ **33.** $-\frac{9}{16} \div -\frac{3}{8}$ **34.** $-\frac{3}{8} \div \frac{5}{4}$

35. $\left(\frac{1}{3} \div \frac{1}{2}\right) + \frac{5}{6}$ **36.** $\frac{2}{5} \div \left(-\frac{4}{5} \div \frac{3}{10}\right)$

37. On Thursday, April 7, 1994, Hewlett-Packard stock closed at 83 3/4 dollars (per share). This was 2 1/4 dollars above the price at the start of the day. How much did one share of this stock cost at the start of the day?

38. On Friday, April 8, 1994, the New York Stock Exchange reported that the 52-week high for IBM was 60 (dollars per share) while the low was 40 5/8 (dollars per share). What was the difference between these prices?

39. The following chart appears on a package of Quaker Quick Grits.

Microwave		Stove Top		
Servings	**1**	**1**	**4**	**6**
Water	$\frac{3}{4}$ cup	1 cup	3 cups	4 cups
Grits	3 Tbsp	3 Tbsp	$\frac{3}{4}$ cup	1 cup
Salt (optional)	dash	dash	$\frac{1}{4}$ tsp	$\frac{1}{2}$ tsp

(a) How many cups of water would be needed for 6 microwave servings?
(b) How many cups of grits would be needed for 5 stove top servings? (*Hint:* 5 is halfway between 4 and 6.)

40. The pie chart shown depicts the makeup of the population of Mexico. Use the chart to answer the following.
(a) What fractional part of the population is Caucasian?
(b) What fractional part of the population is composed of either Mestizo or Indian?

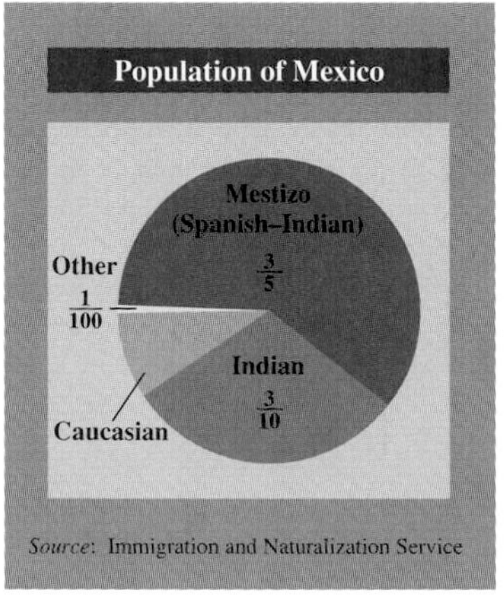

The **mixed number** 2 5/8 *represents the sum* 2 + 5/8. *We can convert* 2 5/8 *to a fraction as follows:*

$$2\frac{5}{8} = 2 + \frac{5}{8} = \frac{2}{1} + \frac{5}{8} = \frac{16}{8} + \frac{5}{8} = \frac{21}{8}.$$

The fraction 21/8 *can be converted back to a mixed number by dividing* 8 *into* 21. *The quotient is* 2 *and the remainder is* 5.

Convert each mixed number in the following exercises to a fraction, and convert each fraction to a mixed number.

41. $4\frac{1}{3}$ **42.** $3\frac{7}{8}$ **43.** $2\frac{9}{10}$ **44.** $\frac{18}{5}$ **45.** $\frac{27}{4}$ **46.** $\frac{19}{3}$

It is possible to add mixed numbers by first converting them to fractions, adding, and then converting the sum back to a mixed number. For example,

$$2\frac{1}{3} + 3\frac{1}{2} = \frac{7}{3} + \frac{7}{2} = \frac{14}{6} + \frac{21}{6} = \frac{35}{6} = 5\frac{5}{6}.$$

The other operations with mixed numbers may be performed in a similar manner.

Perform each operation and express your answer as a mixed number.

47. $3\frac{1}{4} + 2\frac{7}{8}$ **48.** $6\frac{1}{5} - 2\frac{7}{15}$ **49.** $-4\frac{7}{8} \cdot 3\frac{2}{3}$ **50.** $-4\frac{1}{6} \div 1\frac{2}{3}$

A quotient of fractions (with denominator not zero) is called a **complex fraction.** *There are two methods that are used to simplify a complex fraction.*

Method 1: *Simplify the numerator and denominator separately. Then rewrite as a division problem, and proceed as you would when dividing fractions.*

Method 2: *Multiply both the numerator and denominator by the least common denominator of all the fractions found within the complex fraction. (This is, in effect, multiplying the fraction by 1, which does not change its value.) Apply the distributive property, if necessary, and simplify.*

Use one of the methods above to simplify each of the following complex fractions.

51. $\dfrac{\frac{1}{2}+\frac{1}{4}}{\frac{1}{2}-\frac{1}{4}}$ **52.** $\dfrac{\frac{2}{3}+\frac{1}{6}}{\frac{2}{3}-\frac{1}{6}}$ **53.** $\dfrac{\frac{5}{8}-\frac{1}{4}}{\frac{1}{8}+\frac{3}{4}}$ **54.** $\dfrac{\frac{3}{16}+\frac{1}{2}}{\frac{5}{16}-\frac{1}{8}}$ **55.** $\dfrac{\frac{7}{11}+\frac{3}{10}}{\frac{1}{11}-\frac{9}{10}}$ **56.** $\dfrac{\frac{11}{15}+\frac{1}{9}}{\frac{13}{15}-\frac{2}{3}}$

The **continued fraction** *corresponding to the rational number p/q is an expression of the form*

$$a_1 + \cfrac{1}{a_2 + \cfrac{1}{a_3 + \cfrac{1}{a_4 + \ddots}}}$$

where each of the a's is an integer. For example, the continued fraction for 29/8 may be found using the following procedure.

$$\frac{29}{8} = 3 + \frac{5}{8} = 3 + \frac{1}{\frac{8}{5}} = 3 + \frac{1}{1 + \frac{3}{5}} = 3 + \cfrac{1}{1 + \cfrac{1}{\frac{5}{3}}}$$

$$= 3 + \cfrac{1}{1 + \cfrac{1}{1 + \frac{2}{3}}} = 3 + \cfrac{1}{1 + \cfrac{1}{1 + \cfrac{1}{\frac{3}{2}}}} = 3 + \cfrac{1}{1 + \cfrac{1}{1 + \cfrac{1}{1 + \frac{1}{2}}}}$$

Use this procedure to find the continued fraction representation for each of the following rational numbers.

57. $\frac{28}{13}$ **58.** $\frac{73}{31}$ **59.** $\frac{52}{11}$ **60.** $\frac{29}{13}$

Write each of the following continued fractions in the form p/q, reduced to lowest terms. (Hint: Start at the bottom, and work upward.)

61. $2 + \cfrac{1}{1 + \cfrac{1}{3 + \cfrac{1}{2}}}$

62. $4 + \cfrac{1}{2 + \cfrac{1}{1 + \cfrac{1}{3}}}$

Find the rational number halfway between the two given rational numbers.

63. $\frac{1}{2}, \frac{3}{4}$

64. $\frac{1}{3}, \frac{5}{12}$

65. $\frac{3}{5}, \frac{2}{3}$

66. $\frac{7}{12}, \frac{5}{8}$

67. $-\frac{2}{3}, -\frac{5}{6}$

68. $-3, -\frac{5}{2}$

In the March 1973 issue of The Mathematics Teacher *there appeared an article by Laurence Sherzer, an eighth-grade mathematics teacher, that immortalized one of his students, Robert McKay. The class was studying the density property and Sherzer was explaining how to find a rational number between two given positive rational numbers by finding the average. McKay pointed out that there was no need to go to all that trouble. To find a number (not necessarily their average) between two positive rational numbers a/b and c/d, he claimed, simply "add the tops and add the bottoms." Much to Sherzer's surprise, this method really does work. For example, to find a rational number between* 1/3 *and* 1/4, *add* 1 + 1 = 2 *to get the numerator and* 3 + 4 = 7 *to get the denominator. Therefore, by* **McKay's theorem,** 2/7 *is between* 1/3 *and* 1/4. *Sherzer provided a proof of this method in the article.*

Use McKay's theorem *to find a rational number between the two given rational numbers.*

69. $\frac{5}{6}$ and $\frac{9}{13}$

70. $\frac{10}{11}$ and $\frac{13}{19}$

71. $\frac{4}{13}$ and $\frac{9}{16}$

72. $\frac{6}{11}$ and $\frac{8}{9}$

73. 2 and 3

74. 3 and 4

75. Apply McKay's theorem to any pair of consecutive integers, and make a conjecture about what happens in this case.

76. Explain in your own words how to find the rational number that is one-fourth of the way between two different rational numbers.

Convert each rational number into either a repeating or a terminating decimal. Use a calculator if your instructor so allows.

77. $\frac{3}{4}$

78. $\frac{7}{8}$

79. $\frac{3}{16}$

80. $\frac{9}{32}$

81. $\frac{3}{11}$

82. $\frac{9}{11}$

83. $\frac{2}{7}$

84. $\frac{11}{15}$

Convert each terminating decimal into a quotient of integers. Write each in lowest terms.

85. .4

86. .9

87. .85

88. .105

89. .934

90. .7984

Convert each repeating decimal into a quotient of integers. Write each in lowest terms.

91. $.\overline{8}$

92. $.\overline{1}$

93. $.\overline{54}$

94. $.\overline{36}$

95. $.4\overline{3}$

96. $.2\overline{6}$

97. $1.\overline{9}$

98. $3.0\overline{9}$

Use the method of Example 7 to decide whether each of the following rational numbers would yield a repeating or a terminating decimal. (Hint: Write in lowest terms before trying to decide.)

99. $\frac{8}{15}$ **100.** $\frac{8}{35}$ **101.** $\frac{13}{125}$ **102.** $\frac{3}{24}$ **103.** $\frac{22}{55}$ **104.** $\frac{24}{75}$

105. Follow through on each part of this exercise in order.
- **(a)** Find the decimal for $1/3$.
- **(b)** Find the decimal for $2/3$.
- **(c)** By adding the decimal expressions obtained in parts (a) and (b), obtain a decimal expression for $1/3 + 2/3 = 3/3 = 1$.
- **(d)** Does your result seem bothersome? Read the margin note on terminating and repeating decimals in this section, which refers to this idea.

106. It is a fact that $1/3 = .333\ldots$ Multiply both sides of this equation by 3. Does your answer bother you? See the margin note on terminating and repeating decimals in this section.

6.4 Irrational Numbers and Decimals

In the previous section we saw that any rational number has a decimal form that terminates or repeats. Also, every repeating or terminating decimal represents a rational number. Some decimals, however, neither repeat nor terminate. For example, the decimal

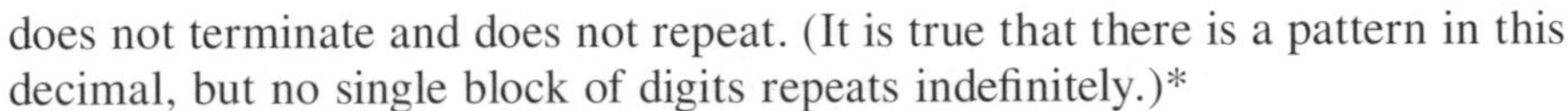

$$.102001000200001000002\ldots$$

does not terminate and does not repeat. (It is true that there is a pattern in this decimal, but no single block of digits repeats indefinitely.)*

A number represented by a nonrepeating, nonterminating decimal is called an **irrational number.** As the name implies, it cannot be represented as a quotient of integers. The decimal number above is an irrational number. Other irrational numbers include $\sqrt{2}$, $\sqrt{7}$, and π. The number π represents the ratio of the circumference of a circle to its diameter, and is one of the most important irrational numbers. (See "For Further Thought" in this section.) There are infinitely many irrational numbers.

Tsu Ch'ung-chih (about 500 A.D.), the Chinese mathematician honored in the above stamp, investigated the digits of π.
Aryabhata, his Indian contemporary, gave 3.1416 as the value.

The first number determined to be irrational was $\sqrt{2}$, discovered by the Pythagoreans in about 500 B.C. This discovery was a great setback to their philosophy that everything is based upon the whole numbers. The Pythagoreans kept their findings secret, and legend has it that members of the group who divulged this discovery were sent out to sea, and, according to Proclus (410–485), "perished in a shipwreck, to a man."

It is not difficult to construct a line segment whose length is $\sqrt{2}$. We begin with a square, one unit on a side. A diagonal of this square cuts the square into

* In this section we will assume that the digits of a number such as this continue indefinitely in the pattern established. The next few digits would be 000000100000002, and so on.

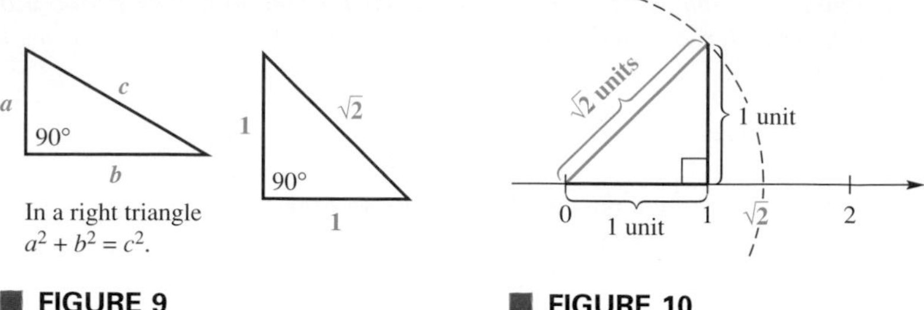

FIGURE 9 **FIGURE 10**

two right triangles. Consider one of the triangles, as in Figure 9. By the Pythagorean theorem, the length of the diagonal (the *hypotenuse* of the right triangle) is given by

$$\begin{aligned} c^2 &= 1^2 + 1^2 \\ &= 1 + 1 \\ c^2 &= 2, \quad \text{or} \quad c = \sqrt{2}. \end{aligned}$$

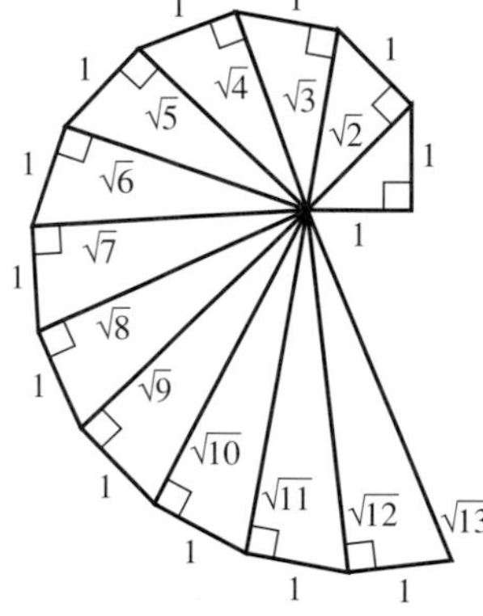

An interesting way to represent the lengths corresponding to $\sqrt{2}$, $\sqrt{3}$, $\sqrt{4}$, $\sqrt{5}$, and so on, is shown in the figure. Use the Pythagorean theorem to verify the lengths in the figure.

By this result, the diagonal of a square one unit on a side is given by an irrational number. To find $\sqrt{2}$ on the number line, first construct a right triangle with each short side 1 unit in length. The hypotenuse must then be $\sqrt{2}$ units. Draw a circle with the point 0 as center and the hypotenuse of the triangle as radius. The dashed line in Figure 10 indicates an arc of the circle. The arc intersects the number line at the point $\sqrt{2}$.

We will now prove that $\sqrt{2}$ is irrational. This proof is a classic example of a **proof by contradiction.** We begin by assuming that $\sqrt{2}$ is rational, and lead to a contradiction, or absurdity. The method is also called **reductio ad absurdum** (Latin for "reduce to the absurd").

In order to understand the proof, we consider three preliminary facts:

1. When a rational number is reduced to lowest terms, the greatest common factor of the numerator and denominator is 1.
2. If an integer is even, then it has 2 as a factor and may be written in the form $2k$, where k is an integer.
3. If a perfect square is even, then its square root is even.

THEOREM $\sqrt{2}$ is an irrational number.

Proof: Assume that $\sqrt{2}$ is a rational number. Then by definition,

$$\sqrt{2} = \frac{p}{q}, \quad \text{for some integers } p \text{ and } q.$$

Furthermore, assume that p/q is the form of $\sqrt{2}$ that is reduced to lowest terms, so that the greatest common factor of p and q is 1. Squaring both sides of the equation gives

$$2 = \frac{p^2}{q^2}$$

and multiplying through by q^2 gives

$$2q^2 = p^2.$$

This last equation indicates that 2 is a factor of p^2. So p^2 is even, and thus p is even. Since p is even, it may be written in the form $2k$, where k is an integer.

Now, substitute $2k$ for p in the last equation and simplify:

$$2q^2 = (2k)^2$$
$$2q^2 = 4k^2$$
$$q^2 = 2k^2.$$

Since 2 is a factor of q^2, q^2 must be even, and thus, q must be even. However, this leads to a contradiction: p and q cannot both be even because they would then have a common factor of 2, although it was assumed that their greatest common factor is 1.

Therefore, since the original assumption that $\sqrt{2}$ is rational has led to a contradiction, it must follow that $\sqrt{2}$ is irrational. ■

The number *e* is a fundamental number in our universe. For this reason, e, like π, is called a *universal constant.* If there are intelligent beings elsewhere, they too will have to use e to do higher mathematics.

The letter e is used to honor Leonhard Euler, who published extensive results on the number in 1748. The first few digits of the decimal value of e are 2.7182818. Since it is an irrational number, its decimal expansion never terminates and never repeats.

The properties of e are used in calculus and in higher mathematics extensively. The formula

$$e^{i\pi} + 1 = 0$$

relates five of the most important constants in mathematics (see the discussion of i in the Extension on complex numbers at the end of the chapter), and it has been given a mystical significance by some mathematicians. It is a special case of what is known as an "Euler identity."

A calculator with a square root key can give approximations of square roots of numbers that are not perfect squares. To show that they are approximations, we use the $\approx$ symbol to indicate "is approximately equal to." Some such calculator approximations are as follows:

$$\sqrt{2} \approx 1.414213562$$
$$\sqrt{7} \approx 2.645751311$$
$$\sqrt{1{,}949} \approx 44.14748011.$$

Not all square roots are irrational. For example, $\sqrt{4} = 2$, $\sqrt{36} = 6$, and $\sqrt{100} = 10$ are all rational numbers. However, if n is a positive integer that is not the square of an integer, then $\sqrt{n}$ is an irrational number.

EXAMPLE 1 The chart below shows some examples of rational numbers and irrational numbers.

Rational Numbers	Irrational Numbers	
$\frac{3}{4}$	$\sqrt{2}$	
.64	.23233233323333 . . .	
$.\overline{74}$	$\sqrt{5}$	
$\sqrt{16}$	π	
1.618	$\frac{1+\sqrt{5}}{2}$	The exact value of the golden ratio
2.718	e	An important number in higher mathematics—see the margin note. ■

One of the most useful irrational numbers is π, the ratio of the circumference to the diameter of a circle. Many formulas from geometry involve π, such as the formulas for area of a circle ($A = \pi r^2$) and volume of a sphere $\left(V = \frac{4}{3}\pi r^3\right)$. For some 4,000 years mathematicians have been finding better and better approximations for π. The ancient Egyptians used a method for finding the area of a circle that is equivalent to a value of 3.1605 for π. The Babylonians used numbers that give 3 1/8 for π. In the Bible (I Kings 7:23), a verse describes a circular pool at King Solomon's temple, about 1,000 B.C. The pool is said to be ten cubits across, "and a line of 30 cubits did compass it round about." This implies a value of 3 for π.

The symbol for π in use today is a Greek letter. It was first used in England in the 1700s. In 1859 the symbol for π shown above was proposed by Professor Benjamin Peirce of Harvard.

This poem, dedicated to Archimedes ("the immortal Syracusan"), allows us to learn the first 30 digits of the decimal representation of π. By replacing each word with the number of letters it contains, with a decimal point following the initial 3, the decimal is found. The poem was written by one A. C. Orr, and appeared in the *Literary Digest* in 1906.

Now I, even I, would celebrate
In rhymes unapt, the great
Immortal Syracusan, rivaled
nevermore,
Who in his wondrous lore
Passed on before,
Left men his guidance
How to circles mensurate.

From this poem, we can determine these digits of π: 3.141592653589793238462643383279.

Archimedes was able to use circles inscribed and circumscribed by polygons to find that the value of π is somewhere

continued on page 293

For Further Thought

The March 22, 1992, issue of *The New Yorker* contains the article "The Mountains of Pi," profiling Gregory and David Chudnovsky. With their own computer, housed in their Manhattan apartment, the Chudnovskys have become two of today's foremost researchers on the investigation of the decimal digits of π. Their research deals with observing the digits for possible patterns in the digits. For example, one conjecture about π deals with what mathematicians term "normality." The normality conjecture says that all digits appear with the same average frequency. According to Gregory "There is absolutely no doubt that π is a 'normal' number. Yet we can't prove it. We don't even know how to *try* to prove it."

In mid-1991, the Chudnovskys reached 2,260,321,336 digits of π. If printed in ordinary type, they would stretch from New York to Southern California. Their work continues today.

The computation of π has fascinated mathematicians and laymen for centuries. In the nineteenth century the British mathematician William Shanks spent many years of his life calculating π to 707 decimal places. It turned out that only the first 527 were correct. The advent of the computer greatly revolutionized the quest in calculating π. In fact, the accuracy of computers and computer programs is sometimes tested by performing the computation of π.

Despite the fact that, in 1767, J. H. Lambert proved that π is irrational (and thus its decimal will never terminate and never repeat), the 1897 Indiana state legislature considered a bill that would have *legislated* the value of π. In one part of the bill, the value was stated to be 4, and in another part, 3.2. Amazingly, the bill passed the House, but the Senate postponed action on the bill indefinitely!

The following expressions are some which may be used to compute π to more and more decimal places.

$$\frac{\pi}{2} = \frac{2 \cdot 2 \cdot 4 \cdot 4 \cdot 6 \cdot 6 \cdot 8 \ldots}{1 \cdot 3 \cdot 3 \cdot 5 \cdot 5 \cdot 7 \cdot 7 \ldots}$$

$$\frac{\pi}{4} = 1 - \frac{1}{3} + \frac{1}{5} - \frac{1}{7} + \cdots$$

$$\frac{2}{\pi} = \frac{\sqrt{2}}{2} \cdot \frac{\sqrt{2 + \sqrt{2}}}{2} \cdot \frac{\sqrt{2 + \sqrt{2 + \sqrt{2}}}}{2} \cdots$$

The fascinating history of π has been chronicled by Petr Beckman in the book *A History of Pi.*

For Group Discussion

1. Have each class member ask someone outside of class "What is π?" Then as a class, discuss the various responses obtained.
2. As with Mount Everest, some people enjoy climbing the mountain of π simply because it is there. Have you ever tackled a project for no reason other than to simply say "I did it"? Share any such experiences with the class.
3. Divide the class into three groups, and, armed with calculators, spend a few minutes calculating the expressions above to approximate π. Compare your results to see which one of the expressions converges toward π the fastest.

between 223/71 and 22/7. Because it is not difficult to find the perimeter of the polygons inscribed and circumscribed around the circle, he concluded that the circumference of the circle must be between these two perimeters. By choosing polygons of larger and larger numbers of sides, he was able to approximate the circumference of the circle (and thus the value of π) with greater and greater accuracy.

Through the centuries mathematicians have calculated π in various ways. For example, in 1579 François Viète used polygons having 393,216 sides to find π to nine decimal places. A computer was programmed by Shanks and Wrench in 1961 to compute 100,265 decimal places of π. In recent years, the Japanese mathematician Yasumasa Kanada and the brothers Gregory and David Chudnovsky, associated with Columbia University, have been at the forefront of the research being done in the evaluation of decimal digits of π. (See "For Further Thought" in this section.) Several years ago, the Chudnovskys evaluated the first 2,260,321,336 digits of π.

Mathematicians have done research in observing patterns in the decimal for π. For example, six 9s in a row appear relatively early in the decimal, within the first 800 decimal places. And past the half-billion mark appears the sequence 123456789.

Square Roots

In everyday mathematical work, nearly all of our calculations deal with rational numbers, usually in decimal form. In our *study* of the various branches of mathematics, however, we are sometimes required to perform operations with irrational numbers, and in many instances the irrational numbers are square roots. Recall that $\sqrt{a}$, for $a \geq 0$, is the nonnegative number whose square is a; that is, $(\sqrt{a})^2 = a$. We will now look at some simple operations with square roots.

Notice that

$$\sqrt{4} \cdot \sqrt{9} = 2 \cdot 3 = 6$$

and

$$\sqrt{4 \cdot 9} = \sqrt{36} = 6.$$

It is no coincidence that $\sqrt{4} \cdot \sqrt{9}$ is equal to $\sqrt{4 \cdot 9}$. This is a particular case of the following product rule.

Product Rule for Square Roots

For nonnegative real numbers a and b,

$$\sqrt{a} \cdot \sqrt{b} = \sqrt{a \cdot b}.$$

The symbol above comes from the Latin word for root, *radix.* It was first used by Leonardo da Pisa (Fibonacci) in 1220. The sixteenth-century German symbol we use today probably is also derived from the letter r.

Just as every rational number a/b can be written in simplest (lowest) terms (by using the fundamental property of rational numbers), every square root radical has a simplest form. A square root radical is in **simplified form** if the three following conditions are met.

Simplified Form of a Square Root Radical

1. The number under the radical (**radicand**) has no factor (except 1) that is a perfect square.
2. The radicand has no fractions.
3. No denominator contains a radical.

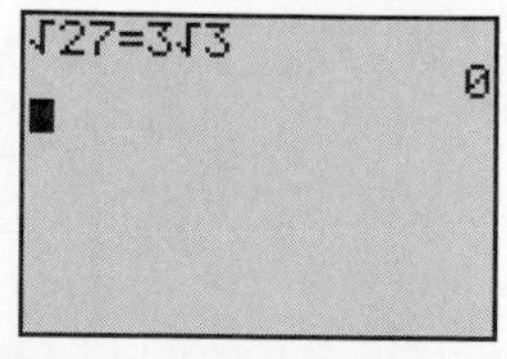

What's Going on Here?
Despite the fact that $\sqrt{27} = 3\sqrt{3}$ is true, the calculator returns a 0, indicating that it is false. Here is an excellent example of why we must be aware of the limitations of technology. The TI-82 will not always return a correct truth value when the numbers involved are irrational. Let the user beware!

EXAMPLE 2 Simplify $\sqrt{27}$.

Since 9 is a factor of 27 and 9 is a perfect square, $\sqrt{27}$ is not in simplified form. The first condition in the box above is not met. We use the product rule to simplify as follows.

$$\begin{aligned} \sqrt{27} &= \sqrt{9 \cdot 3} \\ &= \sqrt{9} \cdot \sqrt{3} && \text{Use the product rule.} \\ &= 3\sqrt{3} && \sqrt{9} = 3, \text{ since } 3^2 = 9. \end{aligned}$$

The simplified form of $\sqrt{27}$ is $3\sqrt{3}$.

Expressions like $\sqrt{27}$ and $3\sqrt{3}$ are called *exact values* of the square root of 27. We can use a calculator to strengthen our understanding that $\sqrt{27}$ and $3\sqrt{3}$ are equal. If we use the square root key of a calculator, we find

$$\sqrt{27} \approx 5.196152423.$$

If we find $\sqrt{3}$ and then multiply the result by 3, we get

$$3\sqrt{3} \approx 3(1.732050808) \approx 5.196152423.$$

Notice that these approximations are the same, as we would expect. (Due to various methods of calculating, there may be a discrepancy in the final digit of the calculation.) Understand, however, that the calculator approximations do not actually *prove* that the two numbers are equal, but only strongly suggest the equality. The work done in Example 2 actually provides the mathematical justification that they are indeed equal.

A rule similar to the product rule exists for quotients.

Quotient Rule for Square Roots

For nonnegative numbers a and positive numbers b,

$$\frac{\sqrt{a}}{\sqrt{b}} = \sqrt{\frac{a}{b}}.$$

EXAMPLE 3 Simplify each radical.

(a) $\sqrt{\frac{25}{9}}$

Because the radicand contains a fraction, the radical expression is not simplified. (See condition 2 in the box preceding Example 2.) Use the quotient rule as follows.

$$\sqrt{\frac{25}{9}} = \frac{\sqrt{25}}{\sqrt{9}} = \frac{5}{3}$$

(b) $\sqrt{\frac{3}{4}} = \frac{\sqrt{3}}{\sqrt{4}} = \frac{\sqrt{3}}{2}$

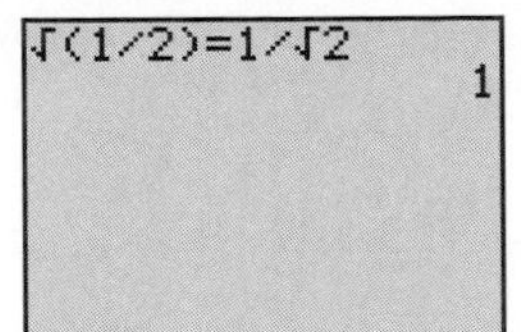

In this case, the TI-82 is able to return the correct response. See Example 3(c).

(c) $\sqrt{\frac{1}{2}}$

$$\sqrt{\frac{1}{2}} = \frac{\sqrt{1}}{\sqrt{2}}$$
$$= \frac{1}{\sqrt{2}}$$

This expression is not in simplified form, since condition 3 is not met. In order to give an equivalent expression with no radical in the denominator, we use a procedure called **rationalizing the denominator.** Multiply $1/\sqrt{2}$ by $\sqrt{2}/\sqrt{2}$, which is a form of 1, the identity element for multiplication.

$$\frac{1}{\sqrt{2}} = \frac{1}{\sqrt{2}} \cdot \frac{\sqrt{2}}{\sqrt{2}}$$
$$= \frac{\sqrt{2}}{2} \qquad \sqrt{2} \cdot \sqrt{2} = 2$$

The simplified form of $\sqrt{1/2}$ is $\sqrt{2}/2$.

(The results in each part of this example may be illustrated by calculator approximations. Verify in (a) that each expression is equal to $1.\overline{6}$, in (b) that each is approximately equal to .866025404, and in (c) that each is approximately equal to .707106781.) ■

The ancient Greeks found approximate values of π expressing the ratio of the circumference of a circle to its diameter d as a sum of fractions. One sum of this sort gave

$$\pi \approx 3 + \frac{8}{60} + \frac{30}{(60)^2}$$
$$= 3.141\overline{6}$$

which is correct to four decimal places. A later mathematician, Nehemiah, approximated the value of π by writing the area of a circle as

$$A = d^2 - \frac{d^2}{7} - \frac{d^2}{14}$$
$$= \left(1 - \frac{1}{7} - \frac{1}{14}\right)d^2$$
$$= \frac{11}{14}d^2.$$

The actual area of a circle is

$$A = \pi r^2 = \pi\left(\frac{1}{2}d\right)^2 = \frac{\pi}{4}d^2.$$

Thus, Nehemiah found

$$\frac{\pi}{4} \approx \frac{11}{14}$$
$$\pi \approx \frac{44}{14} \approx 3.1429.$$

Is it true that $\sqrt{4} + \sqrt{9}$ is equal to $\sqrt{4 + 9}$? Simple computation here shows that the answer is "no," since $\sqrt{4} + \sqrt{9} = 2 + 3 = 5$, while $\sqrt{4 + 9} = \sqrt{13}$, and $5 \neq \sqrt{13}$. Square root radicals may be added, however, if they have the same radicand. Such radicals are **like** or **similar radicals.** We add (and subtract) similar radicals with the distributive property, as shown in the next example.

EXAMPLE 4 **(a)** Add: $3\sqrt{6} + 4\sqrt{6}$.

Since both terms contain $\sqrt{6}$, they are like radicals, and may be combined.

$$3\sqrt{6} + 4\sqrt{6} = (3 + 4)\sqrt{6} \qquad \text{Distributive property}$$
$$= 7\sqrt{6} \qquad 3 + 4 = 7$$

(b) Subtract: $\sqrt{18} - \sqrt{32}$.

At first glance it seems that we cannot combine these terms. However, if we first simplify $\sqrt{18}$ and $\sqrt{32}$, then it can be done.

$$\sqrt{18} - \sqrt{32} = \sqrt{9 \cdot 2} - \sqrt{16 \cdot 2}$$
$$= \sqrt{9} \cdot \sqrt{2} - \sqrt{16} \cdot \sqrt{2} \qquad \text{Product rule}$$
$$= 3\sqrt{2} - 4\sqrt{2} \qquad \text{Take square roots.}$$
$$= (3 - 4)\sqrt{2} \qquad \text{Distributive property}$$
$$= -1\sqrt{2} \qquad 3 - 4 = -1$$
$$= -\sqrt{2} \qquad -1 \cdot a = -a$$

Use a calculator to verify that in part (a), both expressions are approximately equal to 17.1464282, and in part (b) both are approximately equal to −1.414213562. ■

From Example 4, we see that like radicals may be added or subtracted by adding or subtracting their coefficients (the numbers by which they are multiplied) and keeping the same radical. For example,

$$9\sqrt{7} + 8\sqrt{7} = 17\sqrt{7} \quad (\text{since } 9 + 8 = 17)$$
$$4\sqrt{3} - 12\sqrt{3} = -8\sqrt{3}, \quad (\text{since } 4 - 12 = -8)$$

and so on.

In the statements of the product and quotient rules for square roots, the radicands could not be negative. While $-\sqrt{2}$ is a real number, for example, $\sqrt{-2}$ is not: there is no real number whose square is -2. The same may be said for any negative radicand. In order to handle this situation, mathematicians have extended our number system to include *complex numbers,* discussed in the Extension at the end of this chapter.

6.4 EXERCISES

Identify each of the following as rational *or* irrational.

1. $\frac{4}{7}$ **2.** $\frac{5}{8}$ **3.** $\sqrt{6}$ **4.** $\sqrt{13}$

5. .89 **6.** .76 **7.** $.\overline{89}$ **8.** $.\overline{76}$

9. .878778777877778 . . . **10.** .434334333433334 . . . **11.** 3.14159

12. $\frac{22}{7}$ **13.** π **14.** 0

15. **(a)** Find the following sum.

$$\begin{array}{r} .272772777277772\ldots \\ +\ .616116111611116\ldots \\ \hline \end{array}$$

(b) Based on the result of part (a), we can conclude that the sum of two _____ numbers may be a(n) _____ number.

16. **(a)** Find the following sum.

$$\begin{array}{r} .010110111011110\ldots \\ +\ .252552555255552\ldots \\ \hline \end{array}$$

(b) Based on the result of part (a), we can conclude that the sum of two _____ numbers may be a(n) _____ number.

Use a calculator to find a rational decimal approximation for each of the following irrational numbers. Give as many places as your calculator shows.

17. $\sqrt{39}$ **18.** $\sqrt{44}$ **19.** $\sqrt{15.1}$ **20.** $\sqrt{33.6}$

21. $\sqrt{884}$ **22.** $\sqrt{643}$ **23.** $\sqrt{\frac{9}{8}}$ **24.** $\sqrt{\frac{6}{5}}$

Use the methods of Examples 2 and 3 to simplify each of the following expressions. Then, use a calculator to approximate both the given expression and the simplified expression. (Both should be the same.)

25. $\sqrt{50}$ **26.** $\sqrt{32}$ **27.** $\sqrt{75}$ **28.** $\sqrt{150}$ **29.** $\sqrt{288}$ **30.** $\sqrt{200}$

31. $\frac{5}{\sqrt{6}}$ **32.** $\frac{3}{\sqrt{2}}$ **33.** $\sqrt{\frac{7}{4}}$ **34.** $\sqrt{\frac{8}{9}}$ **35.** $\sqrt{\frac{7}{3}}$ **36.** $\sqrt{\frac{14}{5}}$

Use the method of Example 4 to perform the indicated operations.

37. $\sqrt{17} + 2\sqrt{17}$ **38.** $3\sqrt{19} + \sqrt{19}$ **39.** $5\sqrt{7} - \sqrt{7}$ **40.** $3\sqrt{27} - \sqrt{27}$

41. $3\sqrt{18} + \sqrt{2}$ **42.** $2\sqrt{48} - \sqrt{3}$ **43.** $-\sqrt{12} + \sqrt{75}$ **44.** $2\sqrt{27} - \sqrt{300}$

Solve the problems in Exercises 45–50. Use a calculator as necessary.

45. *Heron's formula* gives a method of finding the area of a triangle if the lengths of its sides are known. Suppose that a, b, and c are the lengths of the sides. Let s denote one-half of the perimeter of the triangle (called the *semiperimeter*); that is,

$$s = \frac{1}{2}(a + b + c).$$

Then the area of the triangle is

$$A = \sqrt{s(s - a)(s - b)(s - c)}.$$

Find the area of the Bermuda Triangle, if the "sides" of this triangle measure approximately 850 miles, 925 miles, and 1,300 miles. Round your answer to the nearest thousand.

46. According to an article in *The World Scanner Report* (August 1991), the distance, D, in miles, to the horizon from an observer's point of view over water or "flat" earth is given by

$$D = \sqrt{2H},$$

where H is the height of the point of view, in feet. If a person whose eyes are 6 feet above ground level is standing at the top of a hill 28 feet above the "flat" earth, approximately how far to the horizon will she be able to see? Round your answer to the nearest tenth of a mile.

47. The illumination I, in footcandles, produced by a light source is related to the distance d, in feet, from the light source by the equation

$$d = \sqrt{\frac{k}{I}},$$

where k is a constant. If $k = 640$, how far from the light source will the illumination be 2 footcandles? Give the exact value, and then round to the nearest tenth of a foot.

48. The length of the diagonal of a box is given by

$$D = \sqrt{L^2 + W^2 + H^2},$$

where L, W, and H are the length, the width, and the height of the box. Find the length of the diagonal, D, of a box that is 4 feet long, 3 feet high, and 2 feet wide. Give the exact value, then round to the nearest tenth of a foot.

49. The period of a pendulum in seconds depends on its length, L, in feet, and is given by the formula

$$P = 2\pi\sqrt{\frac{L}{32}}.$$

If a pendulum is 5.1 feet long, what is its period? Use 3.14 for π.

50. The length, L, of an animal, in centimeters, is related to its surface area, S, by the formula

$$L = \sqrt{\frac{S}{a}},$$

where a is a constant that depends on the type of animal. If an animal has $a = .51$ and has a surface area of 1,379 square centimeters, what is the length of the animal?

51. Find the first eight digits in the decimal for 355/113. Compare the result to the decimal for π given in the margin note. What do you find?

52. Using a calculator with a square root key, divide 2,143 by 22, and then press the square root key twice. Compare your result to the decimal for π given in the margin note.

53. A **mnemonic** device is a scheme whereby one is able to recall facts by memorizing something completely unrelated to the facts. One way of learning the first few digits of the decimal for π is to memorize a sentence (or several sentences) and count the letters in each word of the sentence. For example, "See, I know a digit," will give the first 5 digits of π: "See" has 3 letters, "I" has 1 letter, "know" has 4 letters, "a" has 1 letter, and "digit" has 5 letters. So the first five digits are 3.1415.

Verify that the following mnemonic devices work. Use the decimal for π given in the margin note.

(a) "May I have a large container of coffee?"

(b) "See, I have a rhyme assisting my feeble brain, its tasks ofttimes resisting."

(c) "How I want a drink, alcoholic of course, after the heavy lectures involving quantum mechanics."

54. Make up your own mnemonic device to obtain the first eight digits of π.

55. You may have seen the statements "use 22/7 for π" and "use 3.14 for π." Since 22/7 is the quotient of two integers, and 3.14 is a terminating decimal, do these statements suggest that π is rational?

56. In algebra the expression $a^{1/2}$ is defined to be $\sqrt{a}$ for nonnegative values of a. Use a calculator with an exponential key to evaluate each of the following, and compare with the value obtained with the square root key. (Both should be the same.)
(a) $2^{1/2}$ **(b)** $7^{1/2}$
(c) $13.2^{1/2}$ **(d)** $25^{1/2}$

57. The method for simplifying square root radicals, as explained in Examples 2 and 3, can be generalized to cube roots, using the perfect cubes $8 = 2^3$, $27 = 3^3$, $64 = 4^3$, $125 = 5^3$, $216 = 6^3$, and so on. Simplify each of the following cube roots.
(a) $\sqrt[3]{16}$ **(b)** $\sqrt[3]{54}$
(c) $\sqrt[3]{24}$ **(d)** $\sqrt[3]{250}$

58. Based on Exercises 56 and 57, answer the following.
(a) How do you think that the expression $a^{1/3}$ is defined as a radical?
(b) Use a calculator with a cube root key to approximate $\sqrt[3]{16}$.
(c) Use a calculator with an exponential key to approximate $16^{1/3}$.
(d) Compare your results in parts (b) and (c). What do you find?
(e) Compare the calculator approximation for the answer in Exercise 57(a) to the results in parts (b) and (c). What do you find?

59. The threshold weight, T, for a person is the weight above which the risk of death increases greatly. One researcher found that the threshold weight in pounds for men aged 40–49 is related to height, h, in inches by the equation

$$h = 12.3\sqrt[3]{T}.$$

What height corresponds to a threshold of 180 pounds for a 43-year-old man? Round your answer to the nearest tenth of a foot, and then to the nearest inch.

60. Use a calculator with an exponential key to find values for the following: $(1.1)^{10}$, $(1.01)^{100}$, $(1.001)^{1,000}$, $(1.0001)^{10,000}$, and $(1.00001)^{100,000}$. Compare your results to the approximation given for the irrational number e in the margin note in this section. What do you find?

6.5 Applications of Decimals and Percents

Perhaps the most frequent use of mathematics in everyday life concerns operations with decimal numbers and the concept of percent. When we use dollars and cents, we are dealing with decimal numbers. Sales tax on purchases made at the grocery store is computed using percent. The educated consumer must have a working knowledge of decimals and percent in financial matters. Look at any newspaper and you will see countless references to percent and percentages. In this section we will study the basic ideas of operations with decimals, the concept of percent, and applications of decimals and percents.

Because calculators have, for the most part, replaced paper-and-pencil methods for operations with decimals and percent, we will only briefly mention these latter methods. *We strongly suggest that the work in this section be done with a calculator at hand.*

Addition and Subtraction of Decimals

To add or subtract decimal numbers, line up the decimal points in a column and perform the operation.

```
.46+3.9+12.58
        16.94
12.1-8.723
        3.377
```

This screen supports the results in Example 1.

EXAMPLE 1 **(a)** To compute the sum .46 + 3.9 + 12.58, use the following method.

$$\begin{array}{r} .46 \\ 3.9 \\ \underline{12.58} \\ 16.94 \end{array} \leftarrow \text{Sum}$$

Contrary to the fears of many, the availability of calculators and computers has expanded students' capability of performing calculations. There is no evidence to suggest that the availability of calculators makes students dependent on them for simple calculations. Students should be able to decide when they need to calculate and whether they require an exact or approximate answer. They should be able to select and use the most appropriate tool. Students should have a balanced approach to calculation, be able to choose appropriate procedures, find answers, and judge the validity of those answers.

From *Curriculum and Evaluation Standards for School Mathematics* (NCTM), p. 9

(b) To compute the difference 12.1 − 8.723, use this method.

$$\begin{array}{r} 12.100 \\ -\ \ 8.723 \\ \hline 3.377 \end{array}$$

Attach zeros. ← Difference

Recall that when two numbers are multiplied, the numbers are called *factors* and the answer is called their *product*. When two numbers are divided, the number being divided is called the *dividend*, the number doing the dividing is called the *divisor*, and the answer is called the *quotient*. The rules for paper-and-pencil multiplication and division of decimals follow.

Multiplication and Division of Decimals

1. To multiply decimals, perform the multiplication in the same manner as integers are multiplied. The number of decimal places to the right of the decimal point in the product is the *sum* of the numbers of places to the right of the decimal points in the factors.
2. To divide decimals, move the decimal point to the right the same number of places in the divisor and the dividend so as to obtain a whole number in the divisor. Perform the division in the same manner as integers are divided. The number of decimal places to the right of the decimal point in the quotient is the same as the number of places to the right in the dividend.

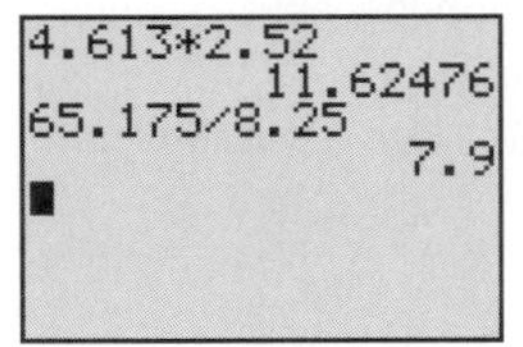

This screen supports the results in Example 2.

EXAMPLE 2

(a) To find the product 4.613 × 2.52, use the following method.

$$\begin{array}{r} 4.6\,13 \\ \times\ \ \ 2.52 \\ \hline 9\,2\,26 \\ 2\,30\,6\,5\ \ \\ 9\,22\,6\ \ \ \ \\ \hline 11.62\,4\,76 \end{array}$$

← 3 decimal places
← 2 decimal places
← 3 + 2 = 5 decimal places

(b) To find the quotient 65.175 ÷ 8.25, follow these steps.

$$8.25\overline{)65.175} \longrightarrow \begin{array}{r} 7.9 \\ 825\overline{)6517.5} \\ 5775\ \ \\ \hline 742\,5 \\ 742\,5 \\ \hline 0 \end{array}$$

Rounding Decimals

Operations with decimals often result in long strings of digits in the decimal places. Since all these digits may not be needed in a practical problem, it is common to *round* a decimal to the necessary number of decimal places. For

The TI-82 calculator is set to round the answer to two decimal places.

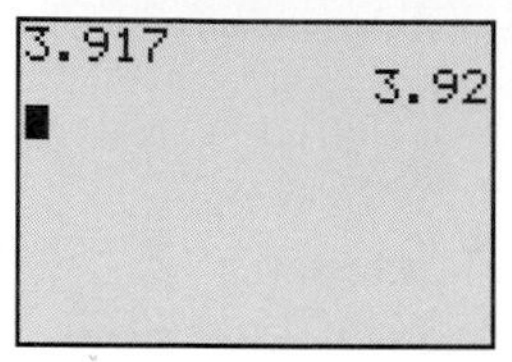

The calculator rounds 3.917 to the nearest hundredth.

example, in preparing federal income tax, money amounts are rounded to the nearest dollar. Round as shown in the next example.

EXAMPLE 3 Round 3.917 to the nearest hundredth.

The hundredths place in 3.917 contains the digit 1.

3.917
↑ Hundredths place

To round this decimal, think of a number line. Locate 3.91 and 3.92 on the number line as in Figure 11.

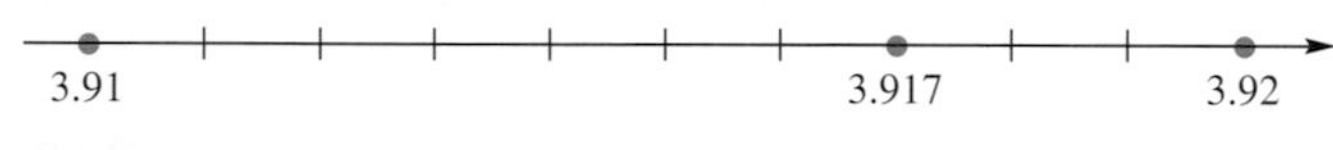

FIGURE 11

The distance from 3.91 to 3.92 is divided into ten equal parts. The seventh of these ten parts locates the number 3.917. As the number line shows, 3.917 is closer to 3.92 than it is to 3.91, so

3.917 rounded to the nearest hundredth is 3.92.

For Further Thought

Do you have the "knack"? We are all born with different abilities. That's what makes us all unique individuals. Some people have musical talents, while others can't carry a tune. Some people can repair automobiles, sew, or build things while others can't do any of these. And some people have a knack for percents, while others are completely befuddled by them. Do you have the knack?

For example, suppose that you need to compute 20% of 50. You have the knack if you use one of these methods:

1. You think "Well, 20% means 1/5, and to find 1/5 of something I divide by 5, so 50 divided by 5 is 10. The answer is 10."
2. You think "20% is twice 10%, and to find 10% of something I move the decimal point one place to the left. So, 10% of 50 is 5, and 20% is twice 5, or 10. The answer is 10."

If you don't have the knack, you probably search for a calculator whenever you need to compute a percent, and hope like crazy that you'll work it correctly. Keep in mind one thing, however: there's nothing to be ashamed of if you don't have the knack. The methods explained in this section allow you to learn how to compute percents using tried-and-true mathematical methods. And just because you don't have the knack, it doesn't mean that you can't succeed in mathematics or can't learn other concepts. One ability does not necessarily assure success in another seemingly similar area. Case in point:

The nineteenth century German Zacharias Dase was a lightning-swift calculator, and could do things like multiply 79,532,853 by 93,758,479 in his head in less than one minute, but had no concept of theoretical mathematics!

HERMAN®

"I got 6 percent in math. Is that good or bad?"

For Group Discussion

Have class members share their experiences on how they compute percents.

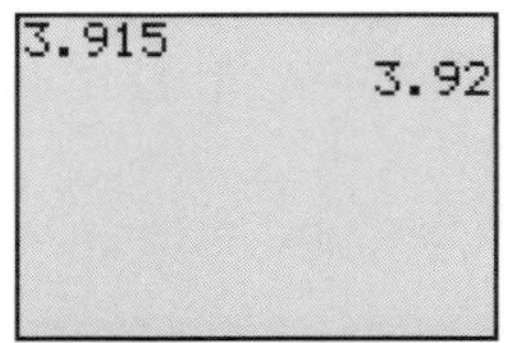

The calculator rounds 3.915 *up* to 3.92.

If the number line method of Example 3 were used to round 3.915 to the nearest hundredth, a problem would develop—the number 3.915 is exactly halfway between 3.91 and 3.92. An arbitrary decision is then made to round *up:* 3.915 rounded to the nearest hundredth is 3.92.

The number line method in Example 3 suggests the following rules for rounding decimals.

Rules for Rounding Decimals

Step 1: Locate the **place** to which the number is being rounded.

Step 2: Look at the next **digit to the right** of the place to which the number is being rounded.

Step 3A: If this digit is **less than 5,** drop all digits to the right of the place to which the number is being rounded. Do *not change* the digit in the place to which the number is being rounded.

Step 3B: If this digit is **5 or greater,** drop all digits to the right of the place to which the number is being rounded. *Add one* to the digit in the place to which the number is being rounded.

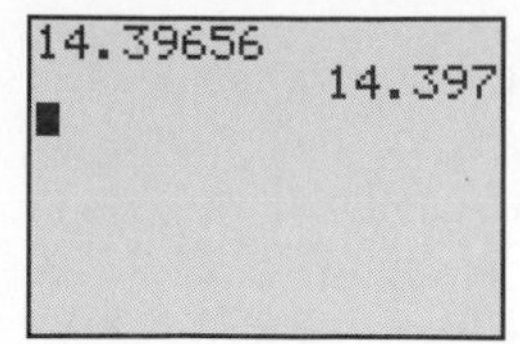

By directing the calculator to round to three decimal places, the result of Example 4 is supported.

**EXAMPLE 4** Round 14.39656 to the nearest thousandth.

Step 1: Use an arrow to locate the place to which the number is being rounded.

$$14.39\underset{\uparrow}{6}56 \quad \text{Thousandths place}$$

Step 2: Check to see if the first digit to the right of the arrow is 5 or greater.

$$14.39\underset{\uparrow}{6}\ \textcircled{5}\ 6 \quad \text{Digit to the right of the arrow is 5.}$$

Step 3: If the digit to the right of the arrow is 5 or greater, increase by 1 the digit to which the arrow is pointing. Drop all digits to the right of the arrow.

$$14.39\underset{\uparrow}{6}56 \quad \text{Drop.}$$

$$14.397 \quad \text{Increase by 1.}$$

Finally, 14.39656 rounded to the nearest thousandth is 14.397.

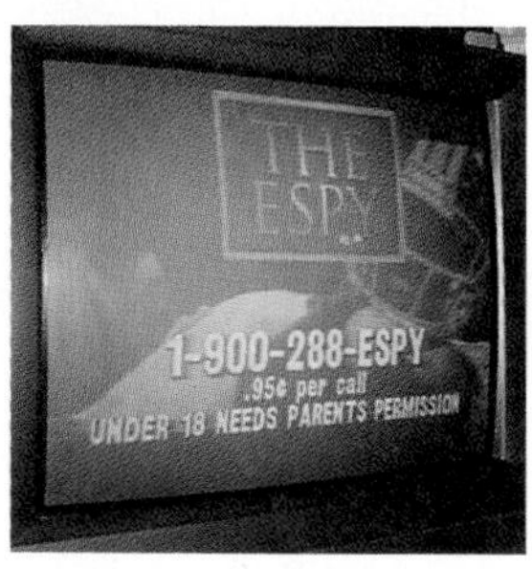

Using a decimal point erroneously with a ¢ symbol is seen almost on a daily basis. This ad indicates that a call costs .95¢. What do you think the call *really* costs?

Percent

One of the main applications of decimals comes from problems involving **percents.** Percents are widely used. In consumer mathematics, interest rates and discounts are often given as percents.

The word *percent* means "per hundred." This idea is used in the basic definition of percents. The symbol % represents "percent."

$$1\% = \frac{1}{100} = .01$$

According to information in the thirty-first yearbook of the National Council of Teachers of Mathematics, the percent sign, %, probably evolved from a symbol introduced in an Italian manuscript of 1425. Instead of "per 100," "P 100" or "P cento," which were common at that time, the author used "P$\overset{\circ}{\text{c}}$." By about 1650 the $\overset{\circ}{\text{c}}$ had become $\frac{0}{0}$, so "per $\frac{0}{0}$" was often used. Finally the "per" was dropped, leaving $\frac{0}{0}$ or %.

For example, 45% = 45(1%) = 45(.01) = .45.

EXAMPLE 5 Convert each percent to a decimal.

(a) 98% = 98(1%) = 98(.01) = .98

(b) 3.4% = .034

(c) .2% = .002

A decimal can be converted to a percent in much the same way.

EXAMPLE 6 Convert each decimal to a percent.

(a) .13 = 13(.01) = 13(1%) = 13%

(b) .532 = 53.2(.01) = 53.2(1%) = 53.2%

(c) 2.3 = 230%

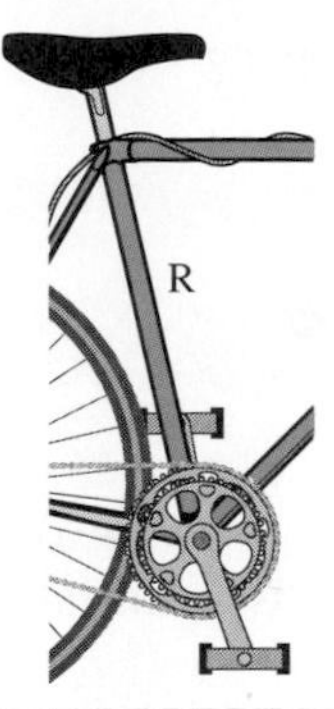

Bicycle Mathematics
Experiments done in England, using racing cyclists on stationary bicycles, show that the most efficient saddle height is 109% of a cyclist's inside-leg measurement. You can get the most mileage out of your leg work by following these directions:

1. Stand up straight, without shoes. Have someone measure your leg on the inside (from floor to crotch bone).
2. Multiply this length by 109% (that is, by 1.09) to get a measure R.
3. Adjust your saddle so that the measure R equals the distance between the top of the saddle and the lower pedal spindle when the pedals are positioned as in the diagram above.

From Examples 5 and 6, we see that the following procedures can be used when converting between percents and decimals.

Converting Between Decimals and Percents

1. *To convert a percent to a decimal,* drop the % sign and move the decimal point two places to the left, inserting zeros as placeholders if necessary.
2. *To convert a decimal to a percent,* move the decimal point two places to the right, inserting zeros as placeholders if necessary, and attach a % sign.

EXAMPLE 7 Convert each fraction to a percent.

(a) $\frac{3}{5}$

First write 3/5 as a decimal. Dividing 5 into 3 gives $3/5 = .6 = 60\%$.

(b) $\frac{14}{25} = .56 = 56\%$ ■

The procedure of Example 7 is summarized as follows.

Converting a Fraction to a Percent

To convert a fraction to a percent, convert the fraction to a decimal, and then convert the decimal to a percent.

The following examples show how to work the various types of problems involving percents. In each example, three methods are shown. The second method in each case involves the cross-product method from Section 3, and some basic algebra. The third method involves the percent key of a calculator.

EXAMPLE 8 Find 18% of 250.

Method 1: The key word here is "of." The word "of" translates as "times," with 18% of 250 given by

$$(18\%)(250) = (.18)(250) = 45.$$

Method 2: Think "18 is to 100 as what (x) is to 250?" This translates into the equation

$$\frac{18}{100} = \frac{x}{250}$$

$$100x = 18 \cdot 250 \qquad a/b = c/d \text{ if and only if } ad = bc.$$

$$x = \frac{18 \cdot 250}{100} = 45. \qquad \text{Divide by 100 and simplify.}$$

Method 3: Use the percent key on a calculator with the following keystrokes:

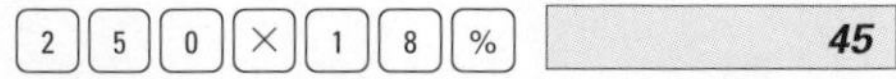

With any of these methods, we find that 18% of 250 is 45. ■

EXAMPLE 9 What percent of 500 is 75?

Method 1: Let the phrase "what percent" be represented by $x \cdot 1\%$ or $.01x$. Again the word "of" translates as "times," while "is" translates as "equals." Thus,

$$.01x \cdot (500) = 75$$

$$5x = 75 \quad \text{Multiply on the left side.}$$

$$x = 15. \quad \text{Divide by 5 on both sides.}$$

Method 2: Think "What (x) is to 100 as 75 is to 500?" This translates as

$$\frac{x}{100} = \frac{75}{500}$$

$$500x = 7{,}500$$

$$x = 15.$$

Method 3: 7 5 ÷ 5 0 0 % → 15

In each case, 15 is the percent, so we may conclude that 75 is 15% of 500. ■

EXAMPLE 10 38 is 5% of what number?

Method 1:

$$38 = .05x$$

$$x = \frac{38}{.05}$$

$$x = 760$$

Method 2: Think "38 is to what number (x) as 5 is to 100?"

$$\frac{38}{x} = \frac{5}{100}$$

$$5x = 3{,}800$$

$$x = 760$$

Method 3: 3 8 ÷ 5 % → 760

Each method shows us that 38 is 5% of 760. ■

Problem Solving

The methods of working with percents described in Examples 8, 9, and 10 can be extended to include more meaningful work. When applying percent, it is a good idea to restate the given problem as a question similar to those found in the preceding examples, and then proceed to answer that question. One method of problem solving deals with solving a simpler, similar problem. This is an excellent chance for us to utilize this problem-solving method. And because approximations are often sufficient for everyday use, we also can apply estimation techniques.

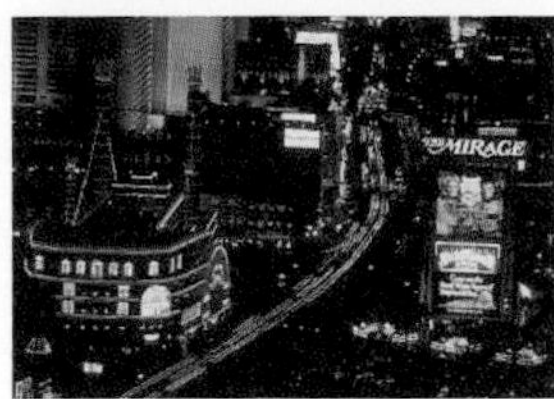

EXAMPLE 11 Las Vegas, Nevada, has the fastest growth rate of any city in the United States. In 1980, its population was approximately 463,000, and by 1990 it had grown to approximately 741,500.

(a) Use estimation skills to approximate the percent increase over these ten years.

If we round the figures to 460,000 and 740,000, respectively, we can easily determine that the increase in population is approximately 740,000 − 460,000 = 280,000. Then we must answer the question "What percent of 460,000 (the *original* population) is 280,000?" Since 280 is a little less than $2/3$ of 460 (thinking in terms of thousands), and $2/3 \approx 67\%$, the percent increase is a bit less than 67%.

(b) Use mathematics to find the percent increase over these ten years.

We must find the difference between the two populations, and then determine what percent of 463,000 comprises this difference.

$$\underbrace{741{,}500}_{\text{Population in 1990}} - \underbrace{463{,}000}_{\text{Population in 1980}} = \underbrace{278{,}500}_{\text{Increase in population}}$$

Now solve the problem "What percent of 463,000 is 278,500?" This is similar to the problem in Example 9. Any of the methods explained there will show that the answer is approximately 60%.

Problem Solving

Much information can be depicted by using a graph, and in many instances, information in the graph is given with percents. We should be able to use this information and our general knowledge of percent to be able to answer questions like those found in the next example. Again, the simple problems found in Examples 8, 9, and 10 provide techniques to solve more realistic ones as in the next example.

EXAMPLE 12 The graphs shown in Figure 12 accompanied a newspaper article. According to the article, 17,252 tornadoes were reported sighted in the past twenty years. Use the graphs to answer the following questions.

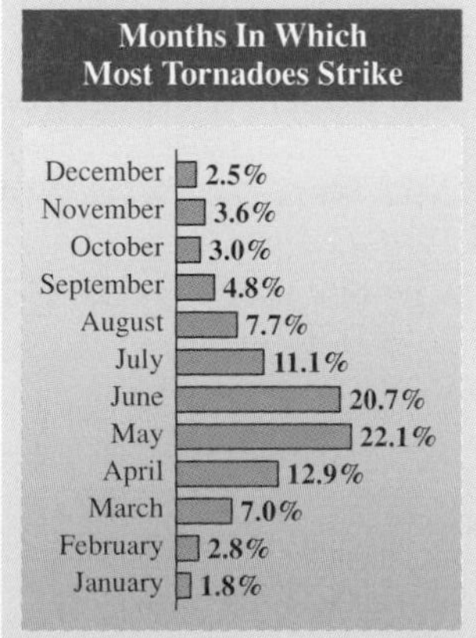

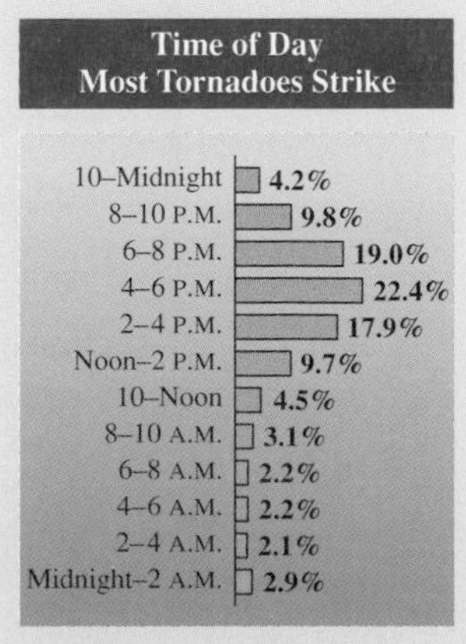

FIGURE 12

(a) How many of the reported tornadoes occurred during the hours of 6–8 P.M.?

According to the graph on the right, 19% of the tornadoes, occurred during this time, so we must ask the question "What is 19% of 17,252?" This is similar to the problem posed in Example 8. Using any of the methods described there, we find that the answer is approximately 3,278. (**Note:** If we only need a "ballpark figure" we can use the fact that 19% is about 20%, and 20% is twice 10%. Ten percent of about 17,000 is 17,000/10 = 1,700, and twice 1,700 is 3,400. Compare this result to the answer found above.)

(b) How many more tornadoes occurred during March than October?

Method 1: Seven percent occurred in March and 3% occurred in October, so 7% − 3% = 4% of the total occurrences will give us the answer. Again, as in Example 8, we must find a percent of a number: "Find 4% of 17,252." Any of the described methods gives approximately 690.

Method 2: We can find 7% of 17,252 and then 3% of 17,252, and subtract the results.

Number of tornadoes in March: 7% of 17,252 is $.07(17{,}252) \approx 1{,}208$

Number of tornadoes in October: 3% of 17,252 is $.03(17{,}252) \approx 518$

Therefore, there were approximately 1,208 − 518 = 690 more tornadoes in March than October. ■

Problem Solving

In many applications we are asked to find the percent increase or percent decrease from one quantity to another. The following guidelines summarize how to do this.

Finding Percent Increase or Decrease

1. To find the percent increase from a to b, where $b > a$, subtract a from b, and divide this result by a. Convert to a percent.

Example: The percent increase from 4 to 7 is $\frac{7-4}{4} = \frac{3}{4} = 75\%$.

2. To find the percent decrease from a to b, where $b < a$, subtract b from a, and divide this result by a. Convert to a percent.

Example: The percent decrease from 8 to 6 is $\frac{8-6}{8} = \frac{2}{8} = \frac{1}{4} = 25\%$.

These ideas will be used in some of the applications in the exercises for this section.

EXERCISES

Decide whether each of the following is true *or* false.

1. 50% of a quantity is the same as 1/2 of the quantity.
2. 200% of 8 is 16.
3. When 435.67 is rounded to the nearest ten, the answer is 435.7.
4. When 668.342 is rounded to the nearest hundredth, the answer is 668.34.
5. A football team that wins 10 games and loses 6 games has a winning percentage of 60%.
6. To find 25% of a quantity, we may simply divide the quantity by 4.
7. If 60% is a passing grade and a test has 40 items, then answering more than 22 items correctly will assure you of a passing grade.
8. To find 40% of a quantity, we may find 10% of that quantity and multiply our answer by 4.
9. 15 is less than 30% of 45.
10. If an item usually costs $50.00 and it is discounted 10%, the sale price is $5.00.

Work each of the following using either a calculator or paper-and-pencil methods, as directed by your instructor.

11. $8.53 + 2.785$ **12.** $9.358 + 7.2137$ **13.** $8.74 - 12.955$ **14.** $2.41 - 3.997$

15. $25.7 \times .032$ **16.** 45.1×8.344 **17.** $1{,}019.825 \div 21.47$

18. $-262.563 \div 125.03$ **19.** $\dfrac{118.5}{1.45 + 2.3}$ **20.** $2.45(1.2 + 3.4 - 5.6)$

Solve each of the following problems. Use a calculator as necessary.

The table shows the annual percent change in the Consumer Price Index for the years 1986 and 1987. Use the information provided to answer Exercises 21–24.

	Percent Change	
Category	**1986**	**1987**
Food	3.2	4.1
Shelter	5.5	4.7
Rent, residential	5.8	4.1
Fuel and other utilities	−2.3	−1.1
Apparel and upkeep	.9	4.4
Private transportation	−4.7	−3.0
New cars	4.2	3.6
Gasoline	−21.9	−4.0
Public transportation	5.9	3.5
Medical care	7.5	6.6
Entertainment	3.4	3.3
Commodities	−.9	3.2

Source: Bureau of Labor Statistics, U.S. Dept. of Labor

21. What category of what year represents the greatest drop?
22. Which percent change is represented by a number with a larger absolute value: 1986 fuel and other utilities or 1987 gasoline?

23. True or false? The absolute value of the change in apparel and upkeep in 1986 is less than the absolute value of the change in commodities during the same year.

24. True or false? The absolute value of the change in private transportation in 1986 was less than the absolute value of the change in private transportation in 1987.

---

25. The bank balance of Tammy's Hobby Shop was \$1,856.12 on March 1. During March, Tammy deposited \$1,742.18 received from the sale of goods, \$9,271.94 paid by customers on their accounts, and a \$28.37 tax refund. She paid out \$7,195.14 for merchandise, \$511.09 for salaries, and \$1,291.03 for other expenses.

(a) How much did Tammy deposit during March?
(b) How much did she pay out?
(c) What was her bank balance at the end of March?

26. On a recent trip to Target, David Horwitz bought three curtain rods at \$4.57 apiece, five picture frames at \$2.99 each, and twelve packs of gum at \$.39 per pack. If 6% sales tax was added to his purchase, what was his total bill?

Exercises 27–30 are based on formulas found in Auto Math Handbook: Mathematical Calculations, Theory, and Formulas for Automotive Enthusiasts, *by John Lawlor (1991, HP Books).*

27. The Blood Alcohol Concentration (BAC) of a person who has been drinking is given by the formula

$$\text{BAC} = \frac{(\text{ounces} \times \text{percent alcohol} \times .075)}{\text{body weight in lb}} - (\text{hours of drinking} \times .015).$$

Suppose a policeman stops a 190-pound man who, in two hours, has ingested four 12-ounce beers with each beer having a 3.2 percent alcohol content. The formula would then read

$$\text{BAC} = \frac{[(4 \times 12) \times 3.2 \times .075]}{190} - (2 \times .015).$$

(a) Find this BAC.
(b) Find the BAC for a 135-pound woman who, in three hours, has drunk three 12-ounce beers with each beer having a 4.0 percent alcohol content.

28. The approximate rate of an automobile in miles per hour (MPH) can be found in terms of the engine's revolutions per minute (rpm), the tire diameter in inches, and the overall gear ratio by the formula

$$\text{MPH} = \frac{\text{rpm} \times \text{tire diameter}}{\text{gear ratio} \times 336}.$$

If a certain automobile has an rpm of 5,600, a tire diameter of 26 inches, and a gear ratio of 3.12, what is its approximate rate (MPH)?

29. Horsepower can be found from indicated mean effective pressure (mep) in pounds per square inch, engine displacement in cubic inches, and revolutions per minute (rpm) using the formula

$$\text{Horsepower} = \frac{\text{mep} \times \text{displacement} \times \text{rpm}}{792{,}000}.$$

Suppose that an engine has displacement of 302 cubic inches, and indicated mep of 195 pounds per square inch at 4,000 rpm. What is its approximate horsepower?

30. To determine the torque at a given value of rpm, the formula below applies:

$$\text{Torque} = \frac{5{,}252 \times \text{horsepower}}{\text{rpm}}.$$

If the horsepower of a certain vehicle is 400 at 4,500 rpm, what is the approximate torque?

Round each of the following numbers to the nearest **(a)** *tenth;* **(b)** *hundredth. Always round from the original number.*

31. 78.414 **32.** 3,689.537
33. .0837 **34.** .0658
35. 12.68925 **36.** 43.99613

Convert each decimal to a percent.

37. .42 **38.** .87
39. .365 **40.** .792
41. .008 **42.** .0093
43. 2.1 **44.** 8.9

Convert each fraction to a percent.

45. $\frac{1}{5}$ **46.** $\frac{2}{5}$
47. $\frac{1}{100}$ **48.** $\frac{1}{50}$
49. $\frac{3}{8}$ **50.** $\frac{5}{6}$
51. $\frac{3}{2}$ **52.** $\frac{7}{4}$

53. Explain the difference between $1/2$ of a quantity and $1/2\%$ of the quantity.

54. In the left column of the chart below there are some common percents, found in many everyday situations. In the right column are fractional equivalents of these percents. Match the fractions in the right column with their equivalent percents in the left column.

(a) 25%	**A.** $\frac{1}{3}$
(b) 10%	**B.** $\frac{1}{50}$
(c) 2%	**C.** $\frac{3}{4}$
(d) 20%	**D.** $\frac{1}{10}$
(e) 75%	**E.** $\frac{1}{4}$
(f) $33\frac{1}{3}\%$	**F.** $\frac{1}{5}$

55. Fill in each blank with the appropriate numerical response.
(a) 5% means ____ in every 100.
(b) 25% means 6 in every ____ .
(c) 200% means ____ for every 4.
(d) .5% means ____ in every 100.
(e) ____% means 12 for every 2.

56. The following Venn diagram shows the number of elements in the four regions formed.

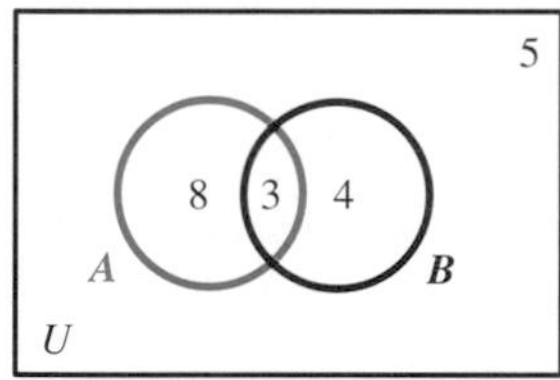

(a) What percent of the elements in the universe are in $A \cap B$?
(b) What percent of the elements in the universe are in A but not in B?
(c) What percent of the elements in $A \cup B$ are in $A \cap B$?
(d) What percent of the elements in the universe are in neither A nor B?

57. Suppose that an item regularly costs \$50.00 and it is discounted 10%. If it is then marked up 10%, is the resulting price \$50.00? If not, what is it?

58. At the start of play on July 24, 1995, the standings of the teams in the Central Division of the American League were as follows:

	Wins	**Losses**
Cleveland	56	22
Milwaukee	40	39
Kansas City	37	40
Chicago	33	45
Minnesota	28	51

"Winning percentage" is commonly expressed as a decimal rounded to the nearest thousandth. To find the winning percentage of a team, divide the number of wins by the total number of games played. Find the winning percentage of each of the teams in the division.

59. Refer to the figures in Exercise 13 of Section 3 of this chapter, and express each of the fractional parts represented by the shaded areas as percents.

60. If there is a 40% chance of rain tomorrow, what *fraction* in lowest terms represents the chance that it will not rain?

Work each of the following problems involving percent.

61. What is 26% of 480?

62. Find 38% of 12.

63. Find 10.5% of 28.

64. What is 48.6% of 19?

65. What percent of 30 is 45?

66. What percent of 48 is 20?

67. 25% of what number is 150?

68. 12% of what number is 3,600?

69. .392 is what percent of 28?

70. 78.84 is what percent of 292?

Use mental techniques to answer the questions in Exercises 71–74. Try to avoid using paper and pencil or a calculator.

71. Johnny Cross's allowance was raised from \$4.00 per week to \$5.00 per week. What was the percent of the increase?
(a) 25% **(b)** 20% **(c)** 50% **(d)** 30%

72. Jennifer Twomey bought a boat five years ago for \$5,000 and sold it this year for \$2,000. What percent of her original purchase did she lose on the sale?
(a) 40% **(b)** 50% **(c)** 20% **(d)** 60%

73. The 1990 U.S. census showed that the population of Alabama was 4,040,587, with 25.3% representing African-Americans. What is the best estimate of the African-American population of Alabama?
(a) 500,000 **(b)** 750,000
(c) 1,000,000 **(d)** 1,500,000

74. The 1990 U.S. census showed that the population of New Mexico was 1,515,069, with 38.2% being Hispanic. What is the best estimate of the Hispanic population of New Mexico?
(a) 600,000 **(b)** 60,000 **(c)** 750,000
(d) 38,000

The total revenue of all securities firms in the United States between 1987 and 1990 is illustrated in the accompanying bar graph. Use the graph to answer the questions in Exercises 75 and 76.

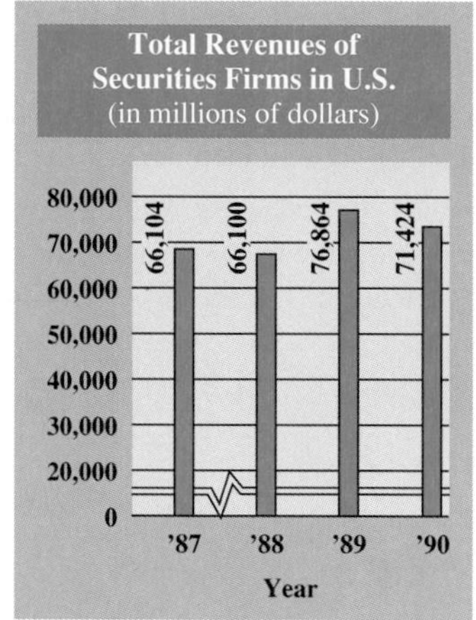

75. What was the percent increase from 1988 to 1989?

76. What was the percent decrease from 1989 to 1990?

Work each of the following problems. Round all money amounts to the nearest dollar.

77. According to a Knight-Ridder Newspapers report, as of May 31, 1992, the nation's "consumer-debt burden" was 16.4%. This means that the average American had consumer debts, such as credit card bills, auto loans, and so on, totaling 16.4% of his or her take-home pay. Suppose that George Duda has a take-home pay of $3,250 per month. What is 16.4% of his monthly take-home pay?

78. In 1992 General Motors announced that it would raise prices on its 1993 vehicles by an average of 1.6%. If a certain vehicle had a 1992 price of $10,526 and this price was raised 1.6%, what would the 1993 price be?

79. The 1916 dime minted in Denver is quite rare. The 1979 edition of *A Guide Book of United States Coins* listed its value in Extremely Fine condition as $625.00. The 1995 value had increased to $2,400.00. What was the percent increase in the value of this coin?

80. In 1963, the value of a 1903 Morgan dollar minted in New Orleans in Uncirculated condition was $1,500. Due to a discovery of a large hoard of these dollars late that year, the value plummeted. Its value as listed in the 1995 edition of *A Guide Book of United States Coins* was $175. What percent of its 1963 value was its 1995 value?

81. The manufacturer's suggested retail price of a Mazda MX 3 was $14,295. A dealer advertised a $2,300 discount. What percent discount was this?

82. According to a report by Freeport-McMoRan Copper Co., Inc., its Grasberg prospect in Indonesia has a large copper find. It has at least 50 million metric tons of ore at an average grade of 1.4% copper and 1.3 grams of gold per ton. Assuming that there actually are 50 million metric tons of ore,
(a) find the number of metric tons of pure copper;
(b) find the number of grams of gold.

It is customary in our society to "tip" waiters and waitresses when eating in restaurants. The usual rate of tipping is 15%. A quick way of figuring a tip that will give a close approximation of 15% is as follows:

1. *Round off the bill to the nearest dollar.*
2. *Find 10% of this amount by moving the decimal point one place to the left.*
3. *Take half of the amount obtained in Step 2 and add it to the result of Step 2.*

This will give you approximately 15% of the bill. The amount obtained in Step 3 is 5%, and 10% + 5% = 15%.

Use the method above to find an approximation of 15% of each of the following restaurant bills.

83. $29.57 **84.** $38.32 **85.** $5.15 **86.** $7.89

87. Example 2(a) shows a paper-and-pencil method of multiplying 4.613×2.52. The following discussion gives a mathematical justification of this method. Fill in the blanks with the appropriate responses.

$$4.613 = 4\frac{613}{1{,}000} = \frac{4{,}613}{1{,}000} = \frac{4{,}613}{10^3}$$

$$2.52 = 2\frac{52}{100} = \frac{252}{100} = \frac{252}{10^2}$$

$$4.613 \times 2.52 = \frac{4{,}613}{10^3} \cdot \frac{252}{10^2} \qquad (*)$$

(a) In algebra, we learn that multiplying powers of the same number is accomplished by *adding* the exponents. Thus,

$$10^3 \cdot 10^2 = 10^{\_\_+\_\_} = 10^{\_\_}.$$

(b) The product in the line indicated by (*) is obtained by multiplying the fractions.

$$\frac{4{,}613 \cdot 252}{10^5} = \frac{1{,}162{,}476}{10^5} = 11.62476$$

The _____ places to the right of the decimal point in the product are the result of division by $10^{\_\_}$.

88. Develop an argument justifying the paper-and-pencil method of dividing decimal numbers, as shown in Example 2(b).

89. A television reporter once asked a professional wrist-wrestler what percent of his sport was physical and what percent was mental. The athlete responded "I would say it's 50% physical and 90% mental." Comment on this response.

90. We often hear the claim "(S)he gave 110%." Comment on this claim. Do you think that this is actually possible?

EXTENSION

Complex Numbers

Early mathematicians would often come up with a solution to a problem such as $-2 + \sqrt{-16}$. If negative numbers and square roots were bad enough, what sense could be made from the square root of a negative number? These numbers were called *imaginary* by the early mathematicians, who would not permit these numbers to be used as solutions to problems.

Gradually, however, applications were found that required the use of these numbers, making it necessary to enlarge the set of real numbers to form the set of **complex numbers.** By doing this, an end is reached: the set of complex numbers provides a solution for just about any equation that can be written.

To develop the basic ideas of the complex number system, consider the equation $x^2 + 1 = 0$. It has no real number solution, since any solution must be a number whose square is -1. In the set of real numbers all squares are nonnegative numbers, because the product of either two positive numbers or two negative numbers is positive. To provide a solution for the equation $x^2 + 1 = 0$, a new number i is defined so that

$$i^2 = -1.$$

That is, i is a number whose square is -1. This definition of i makes it possible to define the square root of any negative number as follows.

For any positive real number b, $\sqrt{-b} = i\sqrt{b}$.

EXAMPLE 1 Write each number as a product of a real number and i.

(a) $\sqrt{-100} = i\sqrt{100} = 10i$

(b) $\sqrt{-2} = \sqrt{2}i = i\sqrt{2}$

It is easy to mistake $\sqrt{2}i$ for $\sqrt{2i}$, with the i under the radical. For this reason, it is common to write $\sqrt{2}i$ as $i\sqrt{2}$. ■

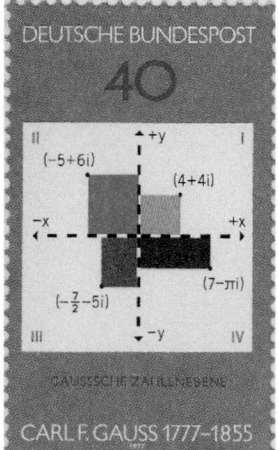

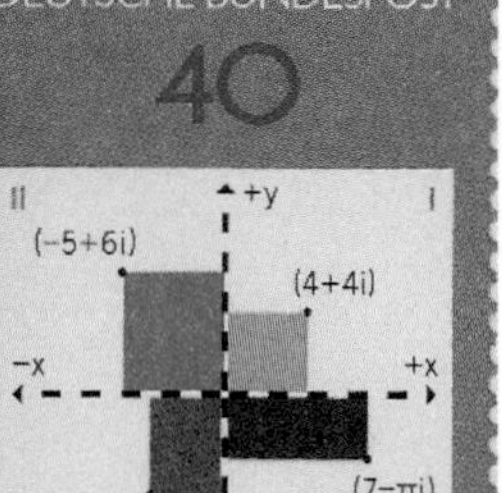

Gauss and the Complex Numbers The above stamp honors the many contributions made by Gauss to our understanding of complex numbers. In about 1831 he was able to show that numbers of the form $a + bi$ can be represented as points on the plane (as the stamp shows) just as real numbers are. He shares this contribution with Robert Argand, a bookkeeper in Paris, who wrote an essay on the geometry of the complex numbers in 1806. This went unnoticed at the time.

When finding a product such as $\sqrt{-4} \cdot \sqrt{-9}$, the product rule for radicals cannot be used, since that rule applies only when both radicals represent real numbers. For this reason, always change $\sqrt{-b}$ $(b > 0)$ to the form $i\sqrt{b}$ before performing any multiplications or divisions. For example,

$$\sqrt{-4} \cdot \sqrt{-9} = i\sqrt{4} \cdot i\sqrt{9} = i \cdot 2 \cdot i \cdot 3 = 6i^2.$$

Since $i^2 = -1$,

$$6i^2 = 6(-1) = -6.$$

An *incorrect* use of the product rule for radicals would give a wrong answer.

$$\sqrt{-4} \cdot \sqrt{-9} = \sqrt{(-4)(-9)} = \sqrt{36} = 6 \quad \text{Incorrect}$$

EXAMPLE 2 Multiply.

(a) $\sqrt{-3} \cdot \sqrt{-7} = i\sqrt{3} \cdot i\sqrt{7} = i^2\sqrt{3 \cdot 7} = (-1)\sqrt{21} = -\sqrt{21}$

(b) $\sqrt{-2} \cdot \sqrt{-8} = i\sqrt{2} \cdot i\sqrt{8} = i^2\sqrt{2 \cdot 8} = (-1)\sqrt{16} = (-1)4 = -4$

(c) $\sqrt{-5} \cdot \sqrt{6} = i\sqrt{5} \cdot \sqrt{6} = i\sqrt{30}$ ■

The methods used to find products also apply to quotients, as the next example shows.

EXAMPLE 3 Divide.

(a) $\dfrac{\sqrt{-75}}{\sqrt{-3}} = \dfrac{i\sqrt{75}}{i\sqrt{3}} = \sqrt{\dfrac{75}{3}} = \sqrt{25} = 5$

(b) $\dfrac{\sqrt{-32}}{\sqrt{8}} = \dfrac{i\sqrt{32}}{\sqrt{8}} = i\sqrt{\dfrac{32}{8}} = i\sqrt{4} = 2i$ ■

With the new number i and the real numbers, a new set of numbers can be formed that includes the real numbers as a subset. The *complex numbers* are defined as follows.

Complex Numbers

If a and b are real numbers, then any number of the form $a + bi$ is called a **complex number.**

In the complex number $a + bi$, the number a is called the **real part** and b is called the **imaginary part.*** When $b = 0$, $a + bi$ is a real number, so the real numbers are a subset of the complex numbers. Complex numbers with $b \neq 0$ are called **imaginary numbers.** In spite of their name, imaginary numbers are very useful in applications, particularly in work with electricity.

* In some texts, bi is called the imaginary part.

The relationships among the various sets of numbers discussed in this chapter are shown in Figure 13.

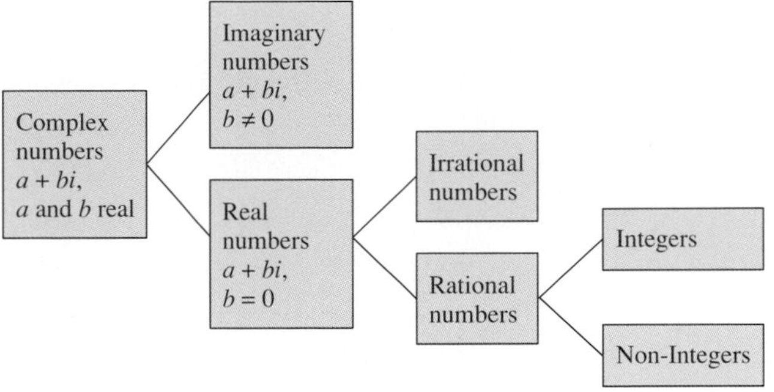

■ **FIGURE 13**

An interesting pattern emerges when we consider various powers of i. By definition, $i^0 = 1$, and $i^1 = i$. We have seen that $i^2 = -1$, and higher powers of i can be found as shown in the following list.

$$i^3 = i \cdot i^2 = i(-1) = -i \qquad i^6 = i^2 \cdot i^4 = (-1) \cdot 1 = -1$$
$$i^4 = i^2 \cdot i^2 = (-1)(-1) = 1 \qquad i^7 = i^3 \cdot i^4 = (-i) \cdot 1 = -i$$
$$i^5 = i \cdot i^4 = i \cdot 1 - i \qquad i^8 = i^4 \cdot i^4 = 1 \cdot 1 = 1$$

A few powers of i are listed here.

Powers of *i*

$i^1 = i$	$i^5 = i$	$i^9 = i$	$i^{13} = i$
$i^2 = -1$	$i^6 = -1$	$i^{10} = -1$	$i^{14} = -1$
$i^3 = -i$	$i^7 = -i$	$i^{11} = -i$	$i^{15} = -i$
$i^4 = 1$	$i^8 = 1$	$i^{12} = 1$	$i^{16} = 1$

As these examples suggest, the powers of i rotate through the four numbers i, -1, $-i$, and 1. Larger powers of i can be simplified by using the fact that $i^4 = 1$. For example, $i^{75} = (i^4)^{18} \cdot i^3 = 1^{18} \cdot i^3 = 1 \cdot i^3 = -i$. This example suggests a quick method for simplifying large powers of i.

Simplifying Large Powers of *i*

Step 1: Divide the exponent by 4.

Step 2: Observe the remainder obtained in Step 1. The large power of i is the same as i raised to the power determined by this remainder. Refer to the chart above to complete the simplification. (If the remainder is 0, the power simplifies to $i^0 = 1$.)

EXAMPLE 4 Find each power of i.

(a) $i^{12} = (i^4)^3 = 1^3 = 1$

(b) i^{39}

We can use the guidelines shown above. Start by dividing 39 by 4 (Step 1).

$$\begin{array}{r} 9 \\ 4\overline{)39} \\ \underline{36} \\ 3 \leftarrow \text{Remainder} \end{array}$$

The remainder is 3. So $i^{39} = i^3 = -i$ (Step 2).

Another way of simplifying i^{39} is as follows.

$$i^{39} = i^{36} \cdot i^3 = (i^4)^9 \cdot i^3 = 1^9 \cdot (-i) = -i$$

Extension Exercises

Use the method of Examples 1–3 to write each as a real number or a product of a real number and i.

1. $\sqrt{-144}$ **2.** $\sqrt{-196}$ **3.** $-\sqrt{-225}$ **4.** $-\sqrt{-400}$

5. $\sqrt{-3}$ **6.** $\sqrt{-19}$ **7.** $\sqrt{-75}$ **8.** $\sqrt{-125}$

9. $\sqrt{-5} \cdot \sqrt{-5}$ **10.** $\sqrt{-3} \cdot \sqrt{-3}$ **11.** $\sqrt{-9} \cdot \sqrt{-36}$

12. $\sqrt{-4} \cdot \sqrt{-81}$ **13.** $\sqrt{-16} \cdot \sqrt{-100}$ **14.** $\sqrt{-81} \cdot \sqrt{-121}$

15. $\dfrac{\sqrt{-200}}{\sqrt{-100}}$ **16.** $\dfrac{\sqrt{-50}}{\sqrt{-2}}$ **17.** $\dfrac{\sqrt{-54}}{\sqrt{6}}$

18. $\dfrac{\sqrt{-90}}{\sqrt{10}}$ **19.** $\dfrac{\sqrt{-288}}{\sqrt{-8}}$ **20.** $\dfrac{\sqrt{-48} \cdot \sqrt{-3}}{\sqrt{-2}}$

21. Why is it incorrect to use the product rule for radicals to multiply $\sqrt{-3} \cdot \sqrt{-12}$?

22. In your own words describe the relationship between complex numbers and real numbers.

Use the method of Example 4 to find each power of i.

23. i^8 **24.** i^{16} **25.** i^{42} **26.** i^{86}

27. i^{47} **28.** i^{63} **29.** i^{101} **30.** i^{141}

31. Explain the difference between $\sqrt{-1}$ and $-\sqrt{1}$. Which one of these is defined as i?

32. Is it possible to give an example of a real number that is not a complex number? Why or why not?

COLLABORATIVE INVESTIGATION
Babylonian Square Roots

With our modern conveniences like calculators and computers, we sometimes forget that our predecessors in the development of mathematics had it a lot tougher than we have it. To find a square root, all we do is enter a number and press a square root key. The Babylonians did not have such a luxury and as a result developed a method of finding a square root using simple arithmetic operations. Historians have determined that they used the following formula to approximate square roots:

$$\text{If } n = a^2 + b, \text{ then } \sqrt{n} \approx a + \frac{b}{2a},$$

for $0 < |b| < a^2$.

Topics for Discussion

Work in groups of four students each, and investigate the process of approximating the square root of 20. Designate the four group members as A, B, C, and D, and then have the members follow these directions.

Member A: Since $20 = 4^2 + 4$, approximate $\sqrt{20}$ with $a = 4$ and $b = 4$.

Member B: Since $20 = 5^2 + (-5)$, approximate $\sqrt{20}$ with $a = 5$ and $b = -5$.

Member C: Since $20 = 6^2 + (-16)$, approximate $\sqrt{20}$ with $a = 6$ and $b = -16$.

Member D: Since $20 = (9/2)^2 + (-1/4)$, approximate $\sqrt{20}$ with $a = 9/2$ and $b = -1/4$.

After each member has completed his or her task, compare the results. Then make a conjecture, as a group, about the closest approximation to $\sqrt{20}$.

Now use the modern convenience of a calculator to obtain a decimal approximation for $\sqrt{20}$. How close was the group conjecture? How does your group conjecture compare with those of other groups? (Aren't you glad you live in the days of technology?)

Chapter 6 Test

1. List the numbers in the set $\{-8, -\sqrt{6}, -4/3, -.6, 0, \sqrt{2}, 3.9, 10\}$ that are **(a)** natural numbers, **(b)** whole numbers, **(c)** integers, **(d)** rational numbers, **(e)** irrational numbers, **(f)** real numbers.
2. Explain what is wrong with the following statement: "The absolute value of a number is always positive."
3. Specify the set $\{x \mid x \text{ is a positive integer less than } 4\}$ by listing its elements.
4. Give three examples of a number that is a positive rational number, but not an integer.
5. True or false: The absolute value of -5 is $-(-5)$.
6. Perform the indicated operations, using the order of operations as necessary.
   (a) $5^2 - 3(2 + 6)$
   (b) $(-3)(-2) - [5 + (8 - 10)]$
   (c) $\dfrac{(-8 + 3) - (5 + 10)}{7 - 9}$
7. The U.S. official reserve assets for the years 1990–1992 are shown in the accompanying bar graph. What is the sum of the assets for these years? (*Source:* Bureau of Economic Analysis)

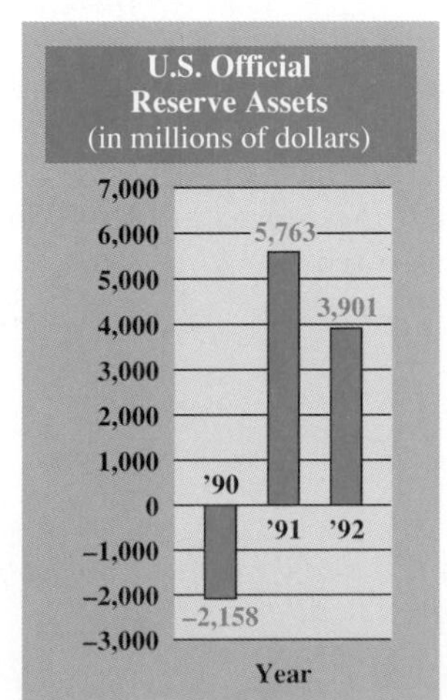

8. Match each of the statements on the left with the property that justifies it.

(a) $7 \cdot (8 \cdot 5) = (7 \cdot 8) \cdot 5$	**A.** Distributive property
(b) $3x + 3y = 3(x + y)$	**B.** Identity property
(c) $8 \cdot 1 = 1 \cdot 8 = 8$	**C.** Closure property
(d) $7 + (6 + 9) = (6 + 9) + 7$	**D.** Commutative property
(e) $9 + (-9) = -9 + 9 = 0$	**E.** Associative property
(f) $5 \cdot 8$ is a real number.	**F.** Inverse property

9. The highest temperature ever recorded in Wyoming was 114°F, while the lowest was −63°F. What is the difference between the highest and lowest temperatures?

10. The funds available from the National Endowment for the Arts for the years 1986 through 1990 are shown in the chart. Determine the change from one year to the next by subtraction for each year indicated with a blank. For example, from 1986 to 1987, the change was $170.9 - 167.1 = 3.8$ million dollars. (*Source:* U.S. National Endowment for the Arts)

Year	Funds Available (in millions of dollars)	Change
1986	167.1	
1987	170.9	3.8
1988	171.1	______
1989	166.7	______
1990	170.8	______

11. The five starters on the local high school basketball team had the following shooting statistics after the first three games.

Player	Field Goal Attempts	Field Goals Made
Camp, Jim	40	13
Cooper, Daniel	10	4
Cornett, Bill	20	8
Hickman, Chuck	6	4
Levinson, Harold	7	2

Answer each of the following, using estimation skills as necessary.

(a) Which player made more than half of his attempts?
(b) Which players made just less than 1/3 of their attempts?
(c) Which player made exactly 2/3 of his attempts?
(d) Which two players made the same fractional parts of their attempts? What was the fractional part, reduced to lowest terms?

Perform each operation. Reduce your answer to lowest terms.

12. $\frac{3}{16} + \frac{1}{2}$

13. $\frac{9}{20} - \frac{3}{32}$

14. $\frac{3}{8} \cdot \left(-\frac{16}{15}\right)$

15. $\frac{7}{9} \div \frac{14}{27}$

16. Based on the pie chart shown, what fractional part of total legal gaming revenue in 1992 came from sources other than lotteries, casinos, and pari-mutuel wagering? (*Source:* Gaming and Wagering Business)

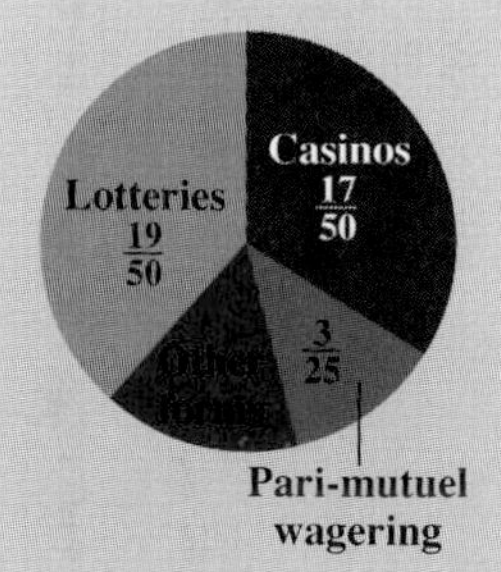

Convert each rational number into a repeating or terminating decimal. Use a calculator if your instructor so allows.

17. $\frac{9}{20}$

18. $\frac{5}{12}$

Convert each decimal into a quotient of integers, reduced to lowest terms.

19. .72

20. $.\overline{58}$

21. Identify each number as rational or irrational.

(a) $\sqrt{10}$ **(b)** $\sqrt{16}$ **(c)** .01 **(d)** $.\overline{01}$ **(e)** .0101101110 . . .

For each of the following **(a)** *use a calculator to find a decimal approximation and* **(b)** *simplify the radical according to the guidelines in this chapter.*

22. $\sqrt{150}$

23. $\dfrac{13}{\sqrt{7}}$

24. $2\sqrt{32} - 5\sqrt{128}$

25. A student using her powerful new calculator states that the *exact* value of $\sqrt{65}$ is 8.062257748. Is she correct? If not, explain.

Work each of the following using either a calculator or paper-and-pencil methods, as directed by your instructor.

26. 4.6 + 9.21

27. 12 − 3.725 − 8.59

28. 86(.45)

29. 236.439 ÷ (−9.73)

30. Round 9.0449 to the following place values: **(a)** hundredths **(b)** thousandths.

31. Find 18.5% of 90.

32. What number is 145% of 70?

33. Use mental techniques to answer the following: In 1995, James Ertl sold \$150,000 worth of books. In 1996, he sold \$450,000 worth of books. His 1996 sales were _____ of his 1995 sales.

(a) 200% **(b)** 33 1/3% **(c)** 300% **(d)** 30%

34. The bar graph shows the number of units of elementary/high school texts, in millions, sold during the years 1987 through 1990. What was the percent increase from 1988 to 1989?

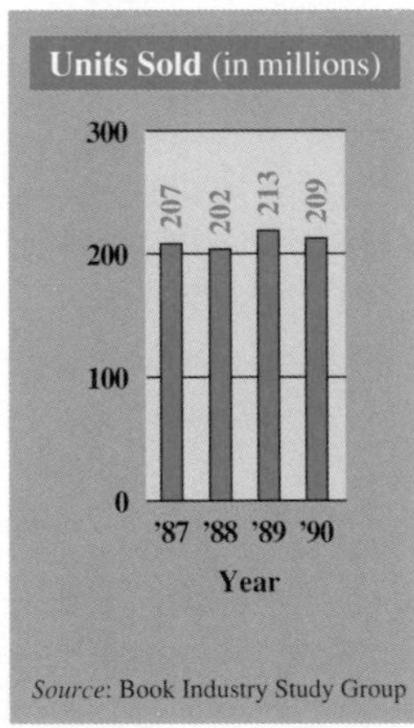

35. Consider the figure.

(a) What percent of the total number of shapes are circles?

(b) What percent of the total number of shapes are not stars?

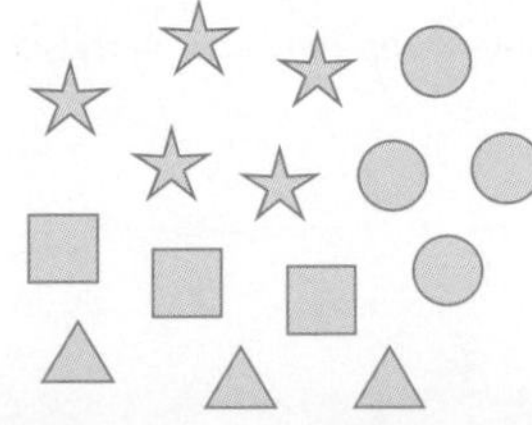

7

The Basic Concepts of Algebra

In 1981 the first case of AIDS (acquired immune deficiency syndrome) was reported in the United States. According to the Centers of Disease Control and Prevention over 360,000 individuals have been diagnosed with AIDS and of them over 220,000 have died. AIDS is one of the most devastating diseases of our time. The first two columns of the table list the total (cumulative) number of AIDS cases diagnosed in the United States to date.

Year	AIDS Cases Reported	Predicted Number of Cases
1982	1,563	1,563
1983	4,647	2,876
1984	10,845	10,589
1985	22,620	24,702
1986	41,662	45,215
1987	70,222	72,128
1988	105,489	105,441
1989	147,170	145,154
1990	193,245	191,267
1991	248,023	243,780
1992	315,329	302,693
1993	361,509	368,006

Mathematics can be used to make predictions about the number of future AIDS cases. A relatively simple algebraic equation that models this data is

$$N = 3{,}200(Y - 1982)^2 - 1{,}887(Y - 1982) + 1{,}563,$$

where N is the number of AIDS cases predicted in year Y. The number of AIDS cases predicted by the model is given in the third column of the table. Although this equation does not model the data exactly, it is capable of giving an estimate of future AIDS cases. To predict the number of cases in the year 2000, let $Y = 2000$ in this equation. This chapter will discuss other algebraic equations that are often used to model real situations.

Source: Wright, J. (editor), *The Universal Almanac,* Universal Press Syndicate Company, 1994.

In earlier chapters we have occasionally used some basic ideas of equation solving in our approaches to the solution of certain types of problems. As earlier civilizations learned rather quickly, the ability to solve equations is essential in applications of mathematics. Algebra dates back to the Babylonians of 2000 B.C. They developed methods of solving quadratic equations (those containing x^2). The Egyptians also worked problems in algebra, but the problems were not as complex as those of the Babylonians. Further advances in algebra had to wait until the time of the Hindus, in about the sixth century. The Hindus developed methods for solving problems involving interest, discounts, and partnerships.

Many Hindu and Greek works on mathematics were preserved only because Moslem scholars made translations of them. These translations, made generally from 750 to 1250, were done mostly in Baghdad. The Arabs took the work of the Greeks and Hindus and greatly expanded it. For example, Mohammed ibn Mûsâ al-Khowârizmî wrote books on algebra and on the Hindu numeration system (the one we use). His books had a tremendous influence in Western Europe; his name is remembered today in the word *algorithm.* Perhaps al-Khowârizmî's most famous book was *Hisâb al-jabr w'al muquâbalah,* from whose title we get the very word "algebra."

Throughout the fifteenth and sixteenth centuries, a main interest of mathematicians was in solving more and more complicated equations. Gradually, algebra found much use in the development of other branches of mathematics, such as calculus.

In this chapter we give a brief introduction to the basic concepts of algebra.

7.1 Linear Equations

An **algebraic expression** is a representation of the basic operations of addition, subtraction, multiplication, or division (except by 0), or extraction of roots on any collection of variables and numbers. Some examples of algebraic expressions include

$$8x + 9, \qquad \sqrt{y} + 4, \qquad \text{and} \qquad \frac{x^3y^8}{z}.$$

Applications of mathematics often lead to **equations,** statements that two algebraic expressions are equal. A *linear equation* in one variable involves only real numbers and one variable. Examples include

$$x + 1 = -2, \qquad y - 3 = 5, \qquad \text{and} \qquad 2k + 5 = 10.$$

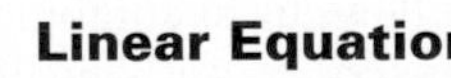

Linear Equation

An equation in the variable x is **linear** if it can be written in the form

$$ax + b = c,$$

where a, b, and c are real numbers, with $a \neq 0$.

al-jabr, algebrista, algebra The word *algebra* comes from the title of the work *Hisâb al-jabr w'al muquâbalah,* a ninth century treatise by the Arab Muhammed ibn Mûsâ al-Khowârizmî. The title translates as "the science of reunion and reduction," or more generally, "the science of transposition and cancellation."

In the title of Khowârizmî's book, *jabr* ("restoration") refers to transposing negative quantities across the equals sign in solving equations. From Latin versions of Khowârizmî's text, "al-jabr" became the broad term covering the arts of equation solving. (The prefix *al* means "the.")

In Spain under Moslem rule, the word *algebrista* referred to the person who restores (resets) broken bones. You would have seen signs outside barber shops saying *Algebrista y Sangrador* (bonesetter and bloodletter). Such services were part of the barber's trade. The traditional red-and-white striped barber pole symbolizes blood and bandages.

A linear equation in one variable is also called a **first-degree** equation, since the highest power on the variable is one.

If the variable in an equation is replaced by a real number that makes the statement true, then that number is a **solution** of the equation. For example, 8 is a solution of the equation $y - 3 = 5$, since replacing y with 8 gives a true statement. An equation is **solved** by finding its **solution set,** the set of all solutions. The solution set of the equation $y - 3 = 5$ is $\{8\}$.

Equivalent equations are equations with the same solution set. Equations are generally solved by starting with a given equation and producing a series of simpler equivalent equations. For example,

$$8x + 1 = 17, \qquad 8x = 16, \qquad \text{and} \qquad x = 2$$

are all equivalent equations since each has the same solution set, $\{2\}$. We use the addition and multiplication properties of equality to produce equivalent equations.

Addition and Multiplication Properties of Equality

Addition Property of Equality

For all real numbers a, b, and c, the equations

$$a = b \qquad \text{and} \qquad a + c = b + c$$

are equivalent. (The same number may be added to both sides of an equation without changing the solution set.)

Multiplication Property of Equality

For all real numbers a, b, and c, where $c \neq 0$, the equations

$$a = b \qquad \text{and} \qquad ac = bc$$

are equivalent. (Both sides of an equation may be multiplied by the same nonzero number without changing the solution set.)

By the addition property, the same number may be added to both sides of an equation without affecting the solution set. By the multiplication property, both sides of an equation may be multiplied by the same nonzero number to produce an equivalent equation. Because subtraction and division are defined in terms of addition and multiplication, respectively, these properties can be extended: The same number may be subtracted from both sides of an equation, and both sides may be divided by the same nonzero number without affecting the solution set.

The distributive property allows us to combine *like terms,* such as $4y$ and $2y$. For example, $4y - 2y = (4 - 2)y = 2y$. This procedure will be used often in the examples that follow.

EXAMPLE 1 Solve $4x - 2x - 5 = 4 + 6x + 3$.

First, combine terms separately on both sides of the equation to get

$$2x - 5 = 7 + 6x.$$

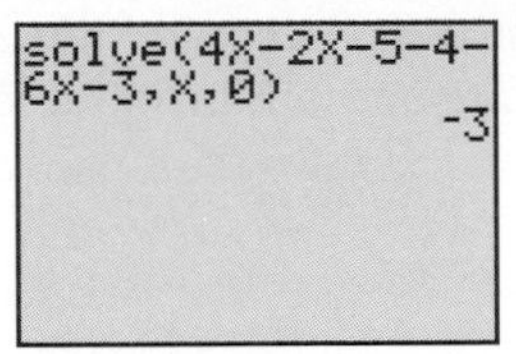

Modern graphing calculators can solve equations of the form

expression in $x = 0$.

The TI-82 requires that you input the expression, the variable for which you are solving, and a "guess," separated by commas. The screen here shows how to solve the equation in Example 1.

The problem-solving method of guessing and testing, discussed in Chapter 1, was actually used by the early Egyptians in equation solving. The method, called the **Rule of False Position,** involved making an initial guess at the solution of an equation, and then following up with an adjustment in the likely event that the guess was incorrect. For example (using our modern notation), if the equation

$$6x + 2x = 32$$

was to be solved, an initial guess was made. Suppose the guess was $x = 3$. Substituting 3 for x on the left side gives

$$6(3) + 2(3) = 32 \quad ?$$
$$18 + 6 = 32 \quad ?$$
$$24 = 32. \quad \text{False}$$

The guess, 3, gives a value (24) which is smaller than the desired value (32). Since 24 is 3/4 of 32, the guess, 3, is 3/4 of the actual solution. The actual solution, therefore, must be 4, since 3 is 3/4 of 4.

Use the methods explained in this section to verify this result.

Next, use the addition property to get the terms with x on the same side of the equation and the remaining terms (the numbers) on the other side. One way to do this is first to add 5 to both sides.

$$2x - 5 + 5 = 7 + 6x + 5 \qquad \text{Add 5.}$$
$$2x = 12 + 6x$$

Now subtract $6x$ from both sides.

$$2x - 6x = 12 + 6x - 6x \qquad \text{Subtract } 6x.$$
$$-4x = 12$$

Finally, divide both sides by -4 to get just x on the left.

$$\frac{-4x}{-4} = \frac{12}{-4} \qquad \text{Divide by } -4.$$
$$x = -3$$

To be sure that -3 is the solution, check by substituting back into the *original* equation (not an intermediate one).

$$4x - 2x - 5 = 4 + 6x + 3 \qquad \text{Given equation}$$
$$4(-3) - 2(-3) - 5 = 4 + 6(-3) + 3 \quad ? \qquad \text{Let } x = -3.$$
$$-12 + 6 - 5 = 4 - 18 + 3 \quad ? \qquad \text{Multiply.}$$
$$-11 = -11 \qquad \text{True}$$

Since a true statement is obtained, -3 is the solution. The solution set is $\{-3\}$. ■

The steps used to solve a linear equation in one variable are as follows. (Not all equations require all of these steps.)

Solving a Linear Equation in One Variable

Step 1: **Clear fractions.** Eliminate any fractions by multiplying both sides of the equation by a common denominator.

Step 2: **Simplify each side separately.** Simplify each side of the equation as much as possible by using the distributive property to clear parentheses and by combining like terms as needed.

Step 3: **Isolate the variable terms on one side.** Use the addition property of equality to get all terms with variables on one side of the equation and all numbers on the other.

Step 4: **Transform so that the coefficient of the variable is 1.** Use the multiplication property of equality to get an equation with just the variable (with coefficient 1) on one side.

Step 5: **Check.** Check by substituting back into the original equation.

In Example 1 we did not use Step 1 and the distributive property in Step 2 as given above. Many other equations, however, will require one or both of these steps, as shown in the next examples.

EXAMPLE 2 Solve $2(k - 5) + 3k = k + 6$.

Since there are no fractions in this equation, Step 1 does not apply. Begin by using the distributive property to simplify and combine terms on the left side of the equation (Step 2).

$$2(k - 5) + 3k = k + 6$$

$$2k - 10 + 3k = k + 6 \quad \text{Distributive property}$$

$$5k - 10 = k + 6 \quad \text{Combine like terms.}$$

Next, add 10 to both sides (Step 3).

$$5k - 10 + 10 = k + 6 + 10 \quad \text{Add 10.}$$

$$5k = k + 16$$

Now subtract k from both sides.

$$5k - k = k + 16 - k \quad \text{Subtract } k.$$

$$4k = 16 \quad \text{Combine like terms.}$$

An extension of the multiplication property of equality is used to get just k on the left. Divide both sides by 4 (Step 4).

$$\frac{4k}{4} = \frac{16}{4} \quad \text{Divide by 4.}$$

$$k = 4$$

Check that the solution set is $\{4\}$ by substituting 4 for k in the original equation (Step 5).

François Viète (1540–1603) was a lawyer at the court of Henry IV of France and studied equations. Viète simplified the notation of algebra and was among the first to use letters to represent numbers. For centuries, algebra and arithmetic were expressed in a cumbersome way with words and occasional symbols. Since the time of Viète, algebra has gone beyond equation solving; the abstract nature of higher algebra depends on its symbolic language.

In the rest of the examples in this section, we will not identify the steps by number.

When fractions or decimals appear as coefficients in equations, our work can be made easier if we multiply both sides of the equation by the least common denominator of all the fractions. This is an application of the multiplication property of equality, and it produces an equivalent equation with integer coefficients. The next examples illustrate this idea.

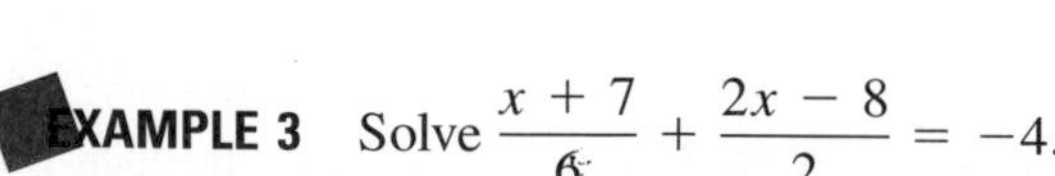

EXAMPLE 3 Solve $\frac{x + 7}{6} + \frac{2x - 8}{2} = -4$.

Start by eliminating the fractions. Multiply both sides by 6.

$$6\left[\frac{x+7}{6}+\frac{2x-8}{2}\right]=6\cdot(-4)$$

$$6\left(\frac{x+7}{6}\right)+6\left(\frac{2x-8}{2}\right)=6(-4) \qquad \text{Distributive property}$$

$$x+7+3(2x-8)=-24$$

$$x+7+6x-24=-24 \qquad \text{Distributive property}$$

$$7x-17=-24 \qquad \text{Combine terms.}$$

$$7x-17+17=-24+17 \qquad \text{Add 17.}$$

$$7x=-7$$

$$\frac{7x}{7}=\frac{-7}{7} \qquad \text{Divide by 7.}$$

$$x=-1$$

Check to see that $\{-1\}$ is the solution set. ■

For Further Thought
The Axioms of Equality

When we solve an equation, we must make sure that it remains "balanced"—that is, any operation that is performed on one side of an equation must also be performed on the other side in order to assure that the set of solutions remains the same.

Underlying the rules for solving equations are four axioms of equality, listed below. For all real numbers *a*, *b*, and *c*,

1. **Reflexive axiom** $a = a$
2. **Symmetric axiom** If $a = b$, then $b = a$.
3. **Transitive axiom** If $a = b$ and $b = c$, then $a = c$.
4. **Substitution axiom** If $a = b$, then a may replace b in any statement without affecting the truth or falsity of the statement.

A relation, such as equality, which satisfies the first three of these axioms (reflexive, symmetric, and transitive), is called an **equivalence relation.**

For Group Discussion

1. Give an example of an everyday relation that does not satisfy the symmetric axiom.
2. Does the transitive axiom hold in sports competition, with the relation "defeats"?
3. Give an example of a relation that does not satisfy the transitive axiom.

Problem Solving

In the next section we will solve problems involving interest rates and concentrations of solutions. These problems involve percents that are converted to decimal numbers. The equations that are used to solve such problems involve decimal coefficients. We can clear these decimals by multiplying by the largest power of 10 necessary to obtain integer coefficients. The next example shows how this is done.

EXAMPLE 4 Solve $.06x + .09(15 - x) = .07(15)$.

Since each decimal number is given in hundredths, multiply both sides of the equation by 100. (This is done by moving the decimal points two places to the right.)

$$
\begin{aligned}
.06x + .09(15 - x) &= .07(15) \\
6x + 9(15 - x) &= 7(15) && \text{Multiply by 100.} \\
6x + 9(15) - 9x &= 105 && \text{Distributive property} \\
-3x + 135 &= 105 && \text{Combine like terms.} \\
-3x + 135 - \mathbf{135} &= 105 - \mathbf{135} && \text{Subtract 135.} \\
-3x &= -30 \\
\frac{-3x}{\mathbf{-3}} &= \frac{-30}{\mathbf{-3}} && \text{Divide by } -3. \\
x &= 10
\end{aligned}
$$

Check to verify that the solution set is $\{10\}$.

When multiplying the term $.09(15 - x)$ by 100 in Example 4, do not multiply both .09 and $15 - x$ by 100. This step is not an application of the distributive property, but of the associative property. The correct procedure is

$$
\begin{aligned}
100[.09(15 - x)] &= [100(.09)](15 - x) && \text{Associative property} \\
&= 9(15 - x). && \text{Multiply.}
\end{aligned}
$$

Each of the equations above had a solution set containing one element; for example, $2x + 1 = 13$ has solution set $\{6\}$, containing only the single number 6. An equation that has a finite (but nonzero) number of elements in its solution set is a **conditional equation.** Sometimes an equation has no solutions. Such an equation is a **contradiction** and its solution set is $\emptyset$. It is also possible for an equation to have an infinite number of solutions. An equation that is satisfied by every number for which both sides are defined is called an **identity.** The next example shows how to recognize these types of equations.

Sofia Kovalevskaya (1850–1891) was the most widely known Russian mathematician in the late nineteenth century. She did most of her work in the theory of differential equations—equations invaluable for expressing rates of change. For example, in biology, the rate of growth of a population, say of microbes, can be precisely stated by differential equations.

Kovalevskaya studied privately because public lectures were not open to women. She eventually received a degree (1874) from the University of Göttingen, Germany. In 1884 she became a lecturer at the University of Stockholm and later was appointed professor of higher mathematics.

Kovalevskaya was well known as a writer. Besides novels about Russian life, notably *The Sisters Rajevski* and *Vera Vorontzoff,* she wrote her *Recollections of Childhood,* which has been translated into English.

EXAMPLE 5 Solve each equation. Decide whether it is a conditional equation, an identity, or a contradiction.

(a) $5x - 9 = 4(x - 3)$
Work as in the previous examples.

$$\begin{aligned} 5x - 9 &= 4(x - 3) && \\ 5x - 9 &= 4x - 12 && \text{Distributive property} \\ 5x - 9 - 4x &= 4x - 12 - 4x && \text{Subtract } 4x. \\ x - 9 &= -12 && \text{Combine like terms.} \\ x - 9 + 9 &= -12 + 9 && \text{Add 9.} \\ x &= -3 && \end{aligned}$$

The solution set, $\{-3\}$, has one element, so $5x - 9 = 4(x - 3)$ is a *conditional equation.*

(b) $5x - 15 = 5(x - 3)$
Use the distributive property on the right side.

$$5x - 15 = 5x - 15$$

Both sides of the equation are *exactly the same,* so any real number would make the equation true. For this reason, the solution set is the set of all real numbers, and the equation $5x - 15 = 5(x - 3)$ is an *identity.*

(c) $5x - 15 = 5(x - 4)$
Use the distributive property.

$$\begin{aligned} 5x - 15 &= 5x - 20 && \text{Distributive property} \\ 5x - 15 - 5x &= 5x - 20 - 5x && \text{Subtract } 5x. \\ -15 &= -20 && \text{False} \end{aligned}$$

Since the result, $-15 = -20$, is *false,* the equation has no solution. The solution set is $\emptyset$. The equation $5x - 15 = 5(x - 4)$ is a *contradiction.* ■

The solution of a problem in algebra often depends on the use of a mathematical statement or **formula** in which more than one letter is used to express a relationship. Examples of formulas are

$$d = rt, \qquad I = prt, \qquad \text{and} \qquad P = 2L + 2W.$$

In some applications, the necessary formula is solved for one of its variables, which may not be the unknown number that must be found. The following examples show how to solve a formula for any one of its variables. This process is called **solving for a specified variable.** Notice how the steps used in these examples are very similar to those used in solving a linear equation. Keep in mind that when you are solving for a specified variable, treat that variable as if it were the only one, and treat all other variables as if they were numbers.

While the process of solving for a specified variable uses the same steps used in solving a linear equation, the following additional suggestions may be helpful.

Solving for a Specified Variable

Step 1: Use the addition or multiplication properties as necessary to get all terms containing the specified variable on one side of the equation.

Step 2: All terms not containing the specified variable should be on the other side of the equation.

Step 3: If necessary, use the distributive property to write the side with the specified variable as the product of that variable and a sum of terms.

In general, follow the steps given earlier for solving linear equations.

EXAMPLE 6 Solve the formula $P = 2L + 2W$ for W.

This formula gives the relationship between the perimeter of a rectangle, P, the length of the rectangle, L, and the width of the rectangle, W. See Figure 1.

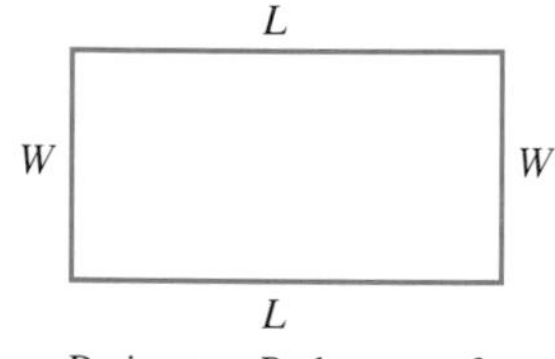

Perimeter, P, the sum of the lengths of the sides of a rectangle, is given by $P = 2L + 2W$.

FIGURE 1

Solve the formula for W by getting W alone on one side of the equals sign. To begin, subtract $2L$ from both sides.

$$P = 2L + 2W$$

$$P - 2L = 2L + 2W - 2L \quad \text{Subtract } 2L.$$

$$P - 2L = 2W$$

Divide both sides by 2.

$$\frac{P - 2L}{2} = \frac{2W}{2}$$

$$\frac{P - 2L}{2} = W$$

7.1 EXERCISES

Decide whether the given number is a solution of the equation.

1. $-6x = -24;\ 4$

2. $8r = 56;\ 7$

3. $5x + 2 = 3;\ \frac{2}{5}$

4. $6y - 4 = 4;\ \frac{1}{2}$

5. $9x + 2x = 6x;\ 0$

6. $-2p + 10p = 7p;\ 0$

7. Which one of the following equations is not a linear equation in x?

(a) $3x + x - 1 = 0$ **(b)** $8 = x^2$ **(c)** $6x + 2 = 9$ **(d)** $\frac{1}{2}x - \frac{1}{4}x = 0$

8. If two equations are equivalent, they have the same ____________ ______ .

Solve the equation.

9. $7k + 8 = 1$

10. $5m - 4 = 21$

11. $8 - 8x = -16$

12. $9 - 2r = 15$

13. $7y - 5y + 15 = y + 8$

14. $2x + 4 - x = 4x - 5$

15. $12w + 15w - 9 + 5 = -3w + 5 - 9$

16. $-4t + 5t - 8 + 4 = 6t - 4$

17. $2(x + 3) = -4(x + 1)$

18. $4(y - 9) = 8(y + 3)$

19. $3(2w + 1) - 2(w - 2) = 5$

20. $4(x - 2) + 2(x + 3) = 6$

21. $2x + 3(x - 4) = 2(x - 3)$

22. $6y - 3(5y + 2) = 4(1 - y)$

23. $6p - 4(3 - 2p) = 5(p - 4) - 10$

24. $-2k - 3(4 - 2k) = 2(k - 3) + 2$

25. $-[2z - (5z + 2)] = 2 + (2z + 7)$

26. $-[6x - (4x + 8)] = 9 + (6x + 3)$

27. $-(9 - 3a) - (4 + 2a) - 3 = -(2 - 5a) + (-a) + 1$

28. $-(-2 + 4x) - (3 - 4x) + 5 = -(-3 + 6x) + x + 1$

29. Explain in your own words the steps used to solve a linear equation.

30. In order to solve the linear equation $\frac{8y}{3} - \frac{2y}{4} = 13$, we may multiply both sides by the least common denominator of all the fractions in the equation. What is this least common denominator?

31. In order to solve the linear equation $.05y + .12(y + 5{,}000) = 940$, we may multiply both sides by a power of 10 so that all coefficients are integers. What is the smallest power of 10 that will accomplish this goal?

32. Suppose that in solving the equation

$$\frac{1}{3}y + \frac{1}{2}y = \frac{1}{6}y,$$

you begin by multiplying both sides by 12, rather than the *least* common denominator, 6. Should you get the correct solution anyway?

Solve the equation.

33. $\frac{3x}{4} + \frac{5x}{2} = 13$

34. $\frac{8y}{3} - \frac{2y}{4} = -13$

35. $\frac{x - 8}{5} + \frac{8}{5} = -\frac{x}{3}$

36. $\frac{2r - 3}{7} + \frac{3}{7} = -\frac{r}{3}$

37. $\frac{4t + 1}{3} = \frac{t + 5}{6} + \frac{t - 3}{6}$

38. $\frac{2x + 5}{5} = \frac{3x + 1}{2} + \frac{-x + 7}{2}$

39. $.05y + .12(y + 5{,}000) = 940$

40. $.09k + .13(k + 300) = 61$

41. $.02(50) + .08r = .04(50 + r)$

42. $.20(14{,}000) + .14t = .18(14{,}000 + t)$

43. $.05x + .10(200 - x) = .45x$

44. $.08x + .12(260 - x) = .48x$

45. The equation $x^2 + 2 = x^2 + 2$ is called a(n) ____________, because its solution set is {all real numbers}. The equation $x + 1 = x + 2$ is called a(n) ____________, because it has no solutions.

46. Which one of the following is a conditional equation?
(a) $2x + 1 = 3$ **(b)** $x = 3x - 2x$ **(c)** $2(x + 2) = 2x + 2$
(d) $5x - 3 = 4x + x - 5 + 2$

Decide whether the equation is conditional, *an* identity, *or a* contradiction. *Give the solution set.*

47. $-2p + 5p - 9 = 3(p - 4) - 5$

48. $-6k + 2k - 11 = -2(2k - 3) + 4$

49. $-11m + 4(m - 3) + 6m = 4m - 12$

50. $3p - 5(p + 4) + 9 = -11 + 15p$

51. $7[2 - (3 + 4r)] - 2r = -9 + 2(1 - 15r)$

52. $4[6 - (1 + 2m)] + 10m = 2(10 - 3m) + 8m$

Solve the problem.

53. The mathematical model $y = 420x + 720$ approximates the worldwide credit card fraud losses between the years 1989 and 1993, where $x = 0$ corresponds to 1989, $x = 1$ corresponds to 1990, and so on, and y is in millions of dollars. Based on this model, what would be the approximate amount of credit card fraud losses in 1994? In what year would losses reach 3,660 million dollars (that is, \$3,660,000,000)?

54. According to research done by the Beverage Marketing Corporation, ready-to-drink iced tea sales have boomed during the past few years. The model $y = 310x + 260$ approximates the revenue generated, where $x = 0$ corresponds to 1991 and y is in millions of dollars. Based on this model, what would be the revenue generated in 1992? In what year would revenue reach 2,430 million dollars?

55. The accompanying bar graph gives a pictorial representation of bank credit card charges in each year from 1989 to 1992. Use the graph to estimate the charges in each of the following years:
(a) 1989 **(b)** 1990
(c) 1991 **(d)** 1992.

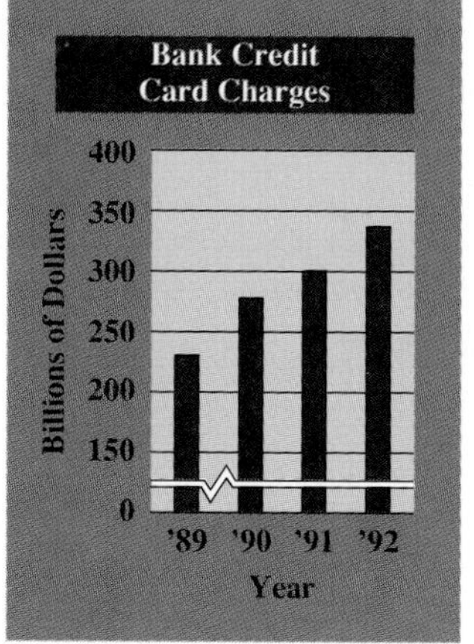

56. Based on information from RAM Research, the amount of bank credit card charges in each year between 1989 and 1992 can be approximated by the linear model $y = 37x + 230$, where $x = 0$ corresponds to 1989, and y is in billions of dollars.
(a) According to this model, what would be the charges in 1993?
(b) In 1993, the actual charges were approximately 422 billion dollars. How might you explain the discrepancy between the predicted amount from part (a) and the actual amount charged?

57. Suppose the formula $A = 2HW + 2LW + 2LH$ is "solved for L" as follows:

$$A = 2HW + 2LW + 2LH$$
$$A - 2LW - 2HW = 2LH$$
$$\frac{A - 2LW - 2HW}{2H} = L.$$

While there are no algebraic errors here, what is wrong with the final line, if we are interested in solving for L?

58. When a formula is solved for a particular variable, several different equivalent forms may be possible. If we solve $A = \frac{1}{2}bh$ for h, one possible correct answer is

$$h = \frac{2A}{b}.$$

Which one of the following is *not* equivalent to this?

(a) $h = 2\left(\frac{A}{b}\right)$ **(b)** $h = 2A\left(\frac{1}{b}\right)$ **(c)** $h = \frac{A}{\frac{1}{2}b}$ **(d)** $h = \frac{\frac{1}{2}A}{b}$

Solve the formula for the specified variable.

59. $d = rt$; for r (distance)

60. $I = prt$; for r (simple interest)

61. $A = bh$; for b (area of a parallelogram)

62. $P = 2L + 2W$; for L (perimeter of a rectangle)

63. $P = a + b + c$; for a (perimeter of a triangle)

64. $V = LWH$; for W (volume of a rectangular solid)

65. $A = \frac{1}{2}bh$; for h (area of a triangle)

66. $C = 2\pi r$; for r (circumference of a circle)

67. $S = 2\pi rh + 2\pi r^2$; for h (surface area of a right circular cylinder)

68. $A = \frac{1}{2}(B + b)h$; for B (area of a trapezoid)

69. $\mathrm{C} = \frac{5}{9}(\mathrm{F} - 32)$; for F (Fahrenheit to Celsius)

70. $\mathrm{F} = \frac{9}{5}\mathrm{C} + 32$; for C (Celsius to Fahrenheit)

71. $A = 2HW + 2LW + 2LH$; for H (surface area of a rectangular solid)

72. $V = \frac{1}{3}Bh$; for h (volume of a right pyramid)

7.2 Applications of Linear Equations

One of the most important uses of algebra is its application to solving a large variety of types of problems. In an earlier chapter we were first introduced to Polya's problem-solving procedure. In this section we will apply this procedure to problems that will be solved algebraically.

When algebra is used to solve practical applications, it is necessary to translate the verbal statements of the problems into mathematical statements.

Problem Solving

Usually there are key words and phrases in the verbal problem that translate into mathematical expressions involving the operations of addition, subtraction, multiplication, and division. Translations of some commonly used expressions are listed next.

Translation from Words to Mathematical Expressions

Verbal Expression	Mathematical Expression
Addition	
The sum of a number and 7	$x + 7$
6 more than a number	$x + 6$
3 plus 8	$3 + 8$
24 added to a number	$x + 24$
A number increased by 5	$x + 5$
The sum of two numbers	$x + y$
Subtraction	
2 less than a number	$x - 2$
12 minus a number	$12 - x$
A number decreased by 12	$x - 12$
The difference between two numbers	$x - y$
A number subtracted from 10	$10 - x$
Multiplication	
16 times a number	$16x$
Some number multiplied by 6	$6x$
2/3 of some number (used only with fractions and percent)	$\frac{2}{3}x$
Twice (2 times) a number	$2x$
The product of two numbers	xy
Division	
The quotient of 8 and some number	$\frac{8}{x}$
A number divided by 13	$\frac{x}{13}$
The ratio of two numbers or the quotient of two numbers	$\frac{x}{y}$

The symbol of equality, =, is often indicated by the word "is." In fact, since equal mathematical expressions represent different names for the same number, words that indicate the idea of "sameness" indicate translation to =. For example the verbal sentence

If the product of a number and 12 is decreased by 7, the result is 105

translates to the mathematical equation

$$12x - 7 = 105$$

where x represents the unknown number. (Why would $7 - 12x = 105$ be incorrect?)

Problem Solving

George Polya's problem-solving procedure can be adapted to applications of algebra as seen in the steps below. Steps 1 and 2 make up the first stage of Polya's procedure *(Understand the Problem),* Step 3 forms the second stage *(Devise a Plan),* Step 4 comprises the third stage *(Carry Out the Plan),* and Steps 5 and 6 form the last stage *(Look Back).*

Solving an Applied Problem Using Algebra

Step 1: **Determine what you are asked to find.** Read the problem carefully. Decide what is given and what must be found. Choose a variable and write down exactly what it represents.

Step 2: **Write down any other pertinent information.** If there are other unknown quantities, express them using the variable. Draw figures or diagrams if they apply.

Step 3: **Write an equation.** Write an equation expressing the relationships among the quantities given in the problem.

Step 4: **Solve the equation.** Use the methods of the previous section to solve the equation.

Step 5: **Answer the question(s) of the problem.** Reread the problem and make sure that you answer the question or questions posed. In some cases, you will need to give more than just the solution of the equation.

Step 6: **Check.** Check your solution by using the original words of the problem. Be sure that your answer makes sense.

Here is an application of linear equations, taken from the **Greek Anthology** (about 500 A.D.), a group of 46 number problems.

Demochares has lived a fourth of his life as a boy, a fifth as a youth, a third as a man, and has spent 13 years in his dotage. How old is he?

(Answer: 60 years old)

The next example illustrates the use of these steps.

EXAMPLE 1 The Perry brothers, Jim and Gaylord, were two outstanding pitchers in the major leagues during the past few decades. Together they won 529 games. Gaylord won 99 more games than Jim. How many games did each brother win?

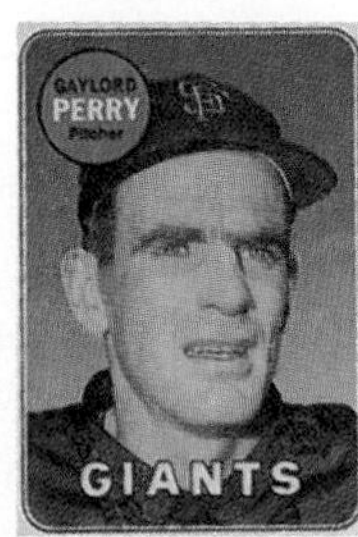

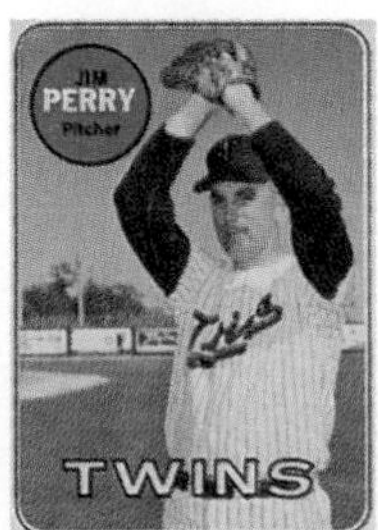

Step 1: We are asked to find the number of games each brother won. We must choose a variable to represent the number of wins of one of the men.

Let j = the number of wins for Jim.

Step 2: We must also find the number of wins for Gaylord. Since he won 99 more games than Jim,

$$j + 99 = \text{Gaylord's number of wins.}$$

Step 3: The sum of the numbers of wins is 529, so we can now write an equation.

$$\underbrace{j}_{\text{Jim's wins}} + \underbrace{(j + 99)}_{\text{Gaylord's wins}} = \underbrace{529}_{529}$$

Step 4: Solve the equation.

$$\begin{aligned} j + (j + 99) &= 529 \\ 2j + 99 &= 529 && \text{Combine like terms.} \\ 2j &= 430 && \text{Subtract 99.} \\ j &= 215 && \text{Divide by 2.} \end{aligned}$$

Step 5: Since j represents the number of Jim's wins, Jim won 215 games. Gaylord won $j + 99 = \mathbf{215} + 99 = 314$ games.

Step 6: 314 is 99 more than 215, and the sum of 314 and 215 is 529.

The words of the problem are satisfied, and our solution checks. ■

A common error in solving applied problems is forgetting to answer all the questions asked in the problem. In Example 1, we were asked for the number of wins for *each* brother, so there was an extra step at the end in order to find Gaylord's number.

Problem Solving

Sometimes it is necessary to find three unknown quantities in an applied problem. Frequently the three unknowns are compared in *pairs*. When this happens, it is usually easiest to let the variable represent the unknown found in both pairs.

The next example illustrates how we can find more than two unknown quantities in an application.

EXAMPLE 2 Meg Holden has a piece of board 70 inches long. She cuts it into three pieces. The longest piece is twice the length of the middle-sized piece, and the shortest piece is 10 inches shorter than the middle-sized piece. How long are the three pieces?

Steps 1 and 2: Since the middle-sized piece appears in both pairs of comparisons, let x represent the length of the middle-sized piece. We have

$$\begin{aligned} x &= \text{the length of the middle-sized piece} \\ 2x &= \text{the length of the longest piece} \\ x - 10 &= \text{the length of the shortest piece.} \end{aligned}$$

A sketch is helpful here. (See Figure 2.)

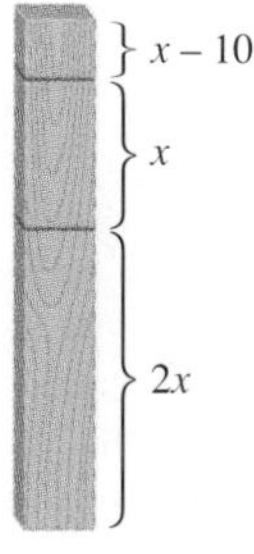

FIGURE 2

$$\underset{\text{Longest}}{2x} + \underset{\text{Middle-sized}}{x} + \underset{\text{Shortest}}{(x - 10)} = \underset{\text{Total length}}{70}$$

Step 3: (above)

Step 4:

$$4x - 10 = 70 \quad \text{Combine terms.}$$
$$4x - 10 + \mathbf{10} = 70 + \mathbf{10} \quad \text{Add 10 to each side.}$$
$$4x = 80 \quad \text{Combine terms.}$$
$$\frac{4x}{\mathbf{4}} = \frac{80}{\mathbf{4}} \quad \text{Divide by 4 on each side.}$$
$$x = 20$$

Step 5: The middle-sized piece is 20 inches long, the longest piece is 2(20) = 40 inches long, and the shortest piece is 20 − 10 = 10 inches long.

Step 6: Check to see that the sum of the lengths is 70 inches, and that all conditions of the problem are satisfied. ■

Percents are often used in problems that involve concentrations or rates. In general, we multiply the rate by the total amount to get the percentage. (The percentage may be an amount of pure substance, or an amount of money, as seen in the examples in this section.) In order to prepare to solve mixture, investment, and money problems, Example 3 illustrates this basic idea.

EXAMPLE 3 **(a)** If a chemist has 40 liters of a 35% acid solution, then the amount of pure acid in the solution is

$$\underset{\text{Amount of solution}}{40} \times \underset{\text{Rate of concentration}}{.35} = \underset{\text{Amount of pure acid}}{14 \text{ liters.}}$$

(b) If $1,300 is invested for one year at 7% simple interest, the amount of interest earned in the year is

$$\underset{\text{Principal}}{\$1{,}300} \times \underset{\text{Interest rate}}{.07} = \underset{\text{Interest earned}}{\$91.}$$

(c) If a jar contains 37 quarters, the monetary amount of the coins is

$$\underset{\text{Number of coins}}{37} \times \underset{\text{Denomination}}{\$.25} = \underset{\text{Monetary value}}{\$9.25.}$$ ■

Problem Solving

In the examples that follow, we will use *box diagrams* to organize the information in the problems. (Some students may prefer to use charts.) Either method enables us to more easily set up the equation for the problem, which is usually the most difficult part of the problem-solving process. The six steps as described in this section will be used but will not specifically be numbered.

In the next example, we will use percent to solve a mixture problem.

EXAMPLE 4 A chemist needs to mix 20 liters of 40% acid solution with some 70% solution to get a mixture that is 50% acid. How many liters of the 70% solution should be used?

Let x = the number of liters of 70% solution that are needed.

Recall from part (a) of Example 3 that the amount of pure acid in this solution will be given by the product of the percent of strength and the number of liters of solution, or

$$\text{liters of pure acid in } x \text{ liters of 70\% solution} = .70x.$$

The amount of pure acid in the 20 liters of 40% solution is

$$\text{liters of pure acid in the 40\% solution} = .40(20) = 8.$$

The new solution will contain $20 + x$ liters of 50% solution. The amount of pure acid in this solution is

$$\text{liters of pure acid in the 50\% solution} = .50(20 + x).$$

The given information can be summarized in the box diagram of Figure 3.

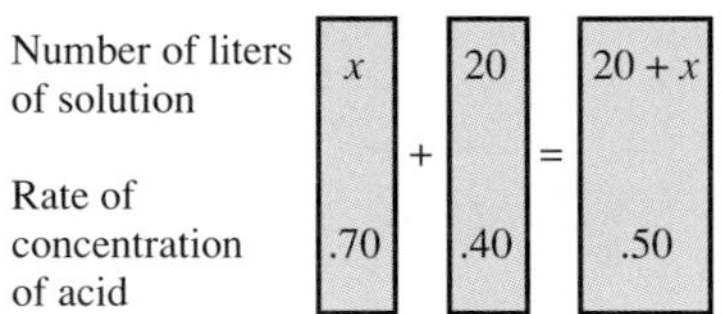

FIGURE 3

As I was going to St. Ives
I met a man with seven wives.
Every wife had seven sacks,
Every sack had seven cats,
Every cat had seven kits.
Kits, cats, sacks, and wives,
How many were going to
St. Ives?

Do you know the "trick" answer to this nursery rhyme? The rhyme is a derivation of an old application found in the Rhind papyrus, an Egyptian manuscript that dates back to about 1650 B.C. Leonardo of Pisa (Fibonacci) also included a similar problem in *Liber Abaci* in 1202.

The answer to the question is 1. Only "I" was *going* to St. Ives.

The number of liters of pure acid in the 70% solution added to the number of liters of pure acid in the 40% solution will equal the number of liters of pure acid in the final mixture, so the equation is

Pure acid in 70%	plus	pure acid in 40%	is	pure acid in 50%.
↓	↓	↓	↓	↓
$.70x$	$+$	$.40(20)$	$=$	$.50(20 + x)$.

Multiply by 100 to clear decimals.

$$70x + 40(20) = 50(20 + x)$$

Solve for x.

$$70x + 800 = 1{,}000 + 50x \quad \text{Distributive property}$$
$$20x + 800 = 1{,}000 \quad \text{Subtract } 50x.$$
$$20x = 200 \quad \text{Subtract 800.}$$
$$x = 10 \quad \text{Divide by 20.}$$

Check this solution to see that the chemist needs to use 10 liters of 70% solution.

The next example uses the formula for simple interest, $I = prt$. Note that when $t = 1$, the formula becomes $I = pr$, and once again the idea of multiplying the total amount (principal) by the rate (rate of interest) gives the percentage (amount of interest).

EXAMPLE 5 Elizabeth Thornton receives an inheritance. She invests part of it at 9% and \$2,000 more than this amount at 10%. Altogether, she makes \$1,150 per year in interest. How much does she have invested at each rate?

Let $x =$ the amount invested at 9% (in dollars);

$x + 2,000 =$ the amount invested at 10% (in dollars).

Use box diagrams to arrange the information given in the problem. See Figure 4.

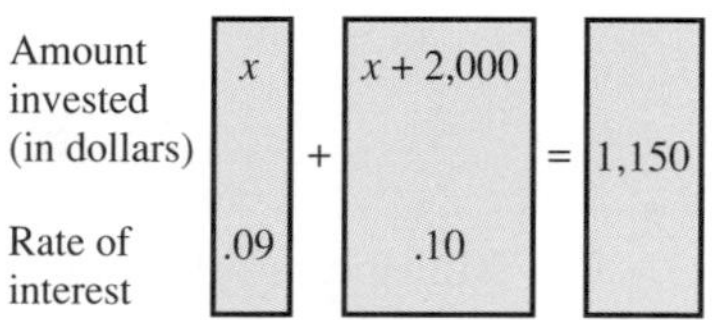

■ **FIGURE 4**

In each box on the left side, multiply amount by rate to get the interest earned. Since the sum of the interest amounts is \$1,150, the equation is

Interest at 9%	plus	interest at 10%	is	total interest.
↓	↓	↓	↓	↓
$.09x$	$+$	$.10(x + 2,000)$	$=$	$1,150.$

Multiply by 100 to clear decimals.

$$9x + 10(x + 2,000) = 115,000$$

Now solve for x.

$$9x + 10x + 20,000 = 115,000 \qquad \text{Distributive property}$$
$$19x + 20,000 = 115,000 \qquad \text{Combine terms.}$$
$$19x = 95,000 \qquad \text{Subtract 20,000.}$$
$$x = 5,000 \qquad \text{Divide by 19.}$$

She has \$5,000 invested at 9% and \$5,000 + \$2,000 = \$7,000 invested at 10%. ■

The next example illustrates a problem that can be solved using the same ideas as those in Examples 4 and 5. It deals with different denominations of money.

EXAMPLE 6 A bank teller has 25 more five-dollar bills than ten-dollar bills. The total value of the money is \$200. How many of each denomination of bill does he have?

We must find the number of each denomination of bill that the teller has.

Let $\quad x =$ the number of ten-dollar bills;

then $\quad x + 25 =$ the number of five-dollar bills.

The information given in the problem can once again be organized in a box diagram. See Figure 5.

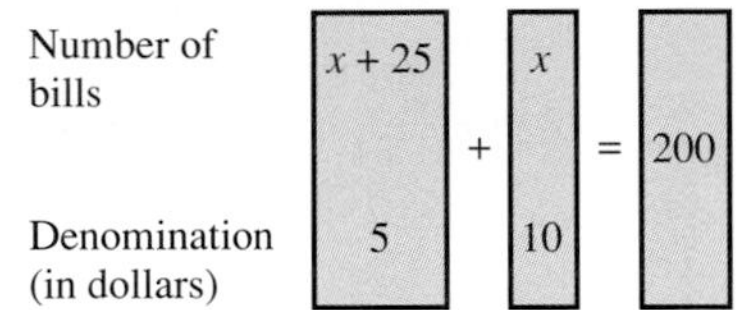

FIGURE 5

Multiplying the number of bills by the denomination gives the monetary value. The value of the tens added to the value of the fives must be \$200:

Value of fives	plus	value of tens	is	\$200.
↓	↓	↓	↓	↓
$5(x + 25)$	$+$	$10x$	$=$	$200.$

Solve this equation.

$$5x + 125 + 10x = 200 \qquad \text{Distributive property}$$
$$15x + 125 = 200 \qquad \text{Combine terms.}$$
$$15x = 75 \qquad \text{Subtract 125.}$$
$$x = 5 \qquad \text{Divide by 15.}$$

Since x represents the number of tens, the teller has 5 tens and $5 + 25 = 30$ fives. Check that the value of this money is $5(\$10) + 30(\$5) = \$200$.

In Examples 4, 5, and 6 we saw how *rates* are applied as percent concentration in a mixture, in computing interest, and as monetary denominations. Problems involving motion also use the idea of rate. For example, if an automobile travels at an average rate of 50 miles per hour for two hours, then it travels $50 \times 2 = 100$ miles. This is an example of the basic relationship between distance, rate, and time:

$$\text{distance} = \text{rate} \times \text{time}.$$

This relationship is given by the formula $d = rt$. By solving, in turn, for r and t in the formula, we obtain two other equivalent forms of the formula. The three forms are given below.

Distance, Rate, Time Relationship

$$d = rt \qquad r = \frac{d}{t} \qquad t = \frac{d}{r}$$

The following example illustrates the uses of these formulas.

EXAMPLE 7 **(a)** The speed of sound is 1,088 feet per second at sea level at 32°F. In 5 seconds under these conditions, sound travels

$$\underset{\text{Rate}}{1{,}088} \times \underset{\text{Time}}{5} = \underset{\text{Distance}}{5{,}440 \text{ feet.}}$$

Here, we found distance given rate and time, using $d = rt$.

(b) Over a short distance, an elephant can travel at a rate of 25 miles per hour. In order to travel 1/4 mile, it would take an elephant

$$\frac{\frac{1}{4}\ (\text{Distance})}{25\ (\text{Rate})} = \frac{1}{4} \times \frac{1}{25} = \frac{1}{100} \text{ hour.} \leftarrow \text{Time}$$

Here, we find time given rate and distance, using $t = d/r$. To convert 1/100 hour to minutes, multiply 1/100 by 60 to get 60/100 or 3/5 minute. To convert 3/5 minute to seconds, multiply 3/5 by 60 to get 36 seconds.

(c) In the 1988 Olympic Games, the USSR won the 400-meter relay with a time of 38.19 seconds. The rate of the team was

$$\frac{400\ (\text{Distance})}{38.19\ (\text{Time})} = 10.47 \text{ (rounded) meters per second.} \leftarrow \text{Rate}$$

This answer was obtained using a calculator. Here, we found rate given distance and time, using $r = d/t$. ■

Can we average averages? A car travels from A to B at 40 miles per hour and returns at 60 miles per hour. What is its rate for the entire trip?

The correct answer is not 50 miles per hour! Remembering the distance-rate-time relationship and letting $x =$ the distance between A and B, we can simplify a complex fraction to find the correct answer.

$$\begin{aligned}\text{Rate for entire trip} &= \frac{\text{Total distance}}{\text{Total time}} \\ &= \frac{x + x}{\frac{x}{40} + \frac{x}{60}} \\ &= \frac{2x}{\frac{3x}{120} + \frac{2x}{120}} \\ &= \frac{2x}{\frac{5x}{120}} \\ &= 2x \cdot \frac{120}{5x} \\ &= 48\end{aligned}$$

The rate for the entire trip is 48 miles per hour.

Problem Solving

The next example shows how to solve a typical application of the formula $d = rt$. A strategy for solving such problems involves two major steps:

Solving Motion Problems

Step 1: Set up a sketch showing what is happening in the problem.

Step 2: Make a chart using the information given in the problem, along with the unknown quantities.

The chart will help you organize the information, and the sketch will help you set up the equation.

EXAMPLE 8 Two cars leave Baton Rouge, Louisiana, at the same time and travel east on Interstate 12. One travels at a constant speed of 55 miles per hour and the other travels at a constant speed of 63 miles per hour. In how many hours will the distance between them be 24 miles?

Since we are looking for time,

let $t =$ the number of hours until the distance between them is 24 miles.

The sketch in Figure 6 shows what is happening in the problem. Now, construct a chart like the one below.

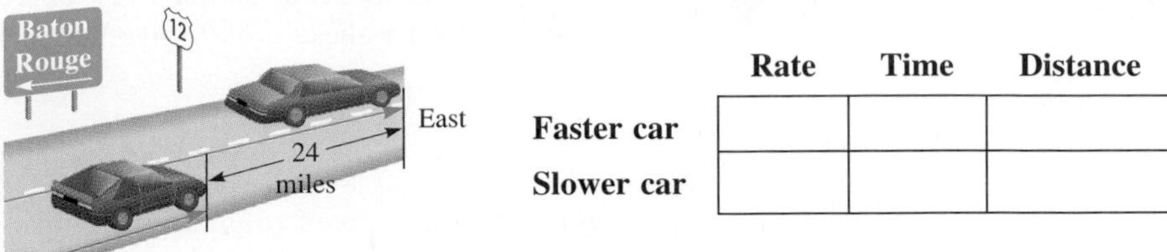

	Rate	Time	Distance
Faster car			
Slower car			

FIGURE 6

Fill in the information given in the problem, and use t for the time traveled by each car. Multiply rate by time to get the expressions of distances traveled.

	Rate ×	Time =	Distance
Faster car	63	t	$63t$
Slower car	55	t	$55t$

Difference is 24 miles.

The quantities $63t$ and $55t$ represent the different distances. Refer to Figure 6 and notice that the *difference* between the larger distance and the smaller distance is 24 miles. Now write the equation and solve it.

$$63t - 55t = 24$$

$$8t = 24 \quad \text{Combine terms.}$$

$$t = 3 \quad \text{Divide by 8.}$$

After 3 hours the faster car will have traveled $63 \times 3 = 189$ miles, and the slower car will have traveled $55 \times 3 = 165$ miles. Since $189 - 165 = 24$, the conditions of the problem are satisfied. It will take 3 hours for the distance between them to be 24 miles. ■

In motion problems like the one in Example 8, once you have filled in two pieces of information in each row of the chart, you should automatically fill in the third piece of information, using the appropriate form of the formula relating distance, rate, and time. Set up the equation based upon your sketch and the information in the chart.

7.2 EXERCISES

1. In your own words, list the six steps suggested in the box marked "Solving an Applied Problem" in this section.

2. If x represents the number, express the following using algebraic symbols:
(a) 9 less than a number
(b) 9 is less than a number.
Which one of these is an *expression*?

Translate the verbal phrase into a mathematical expression. Use x to represent the unknown number.

3. a number decreased by 13

4. a number decreased by 12

5. 7 increased by a number

6. 12 more than a number

7. the product of 8 and 12 more than a number

8. the product of 9 less than a number and 6 more than the number

9. the quotient of a number and 6

10. the quotient of 6 and a nonzero number

11. the ratio of a number and 12

12. the ratio of 12 and a nonzero number

13. $\frac{6}{7}$ of a number

14. 19% of a number

15. Which one of the following is *not* a valid translation of "30% of a number"?
(a) $.30x$
(b) $.3x$
(c) $\frac{3x}{10}$
(d) $30x$

16. Explain the difference between the answers to Exercises 11 and 12.

Use the methods of Examples 1 and 2 or your own method to solve the problem.

17. The U.S. Senate has 100 members. In 1995, there were 6 fewer Democrats than Republicans, with no other parties represented. How many members of each party were there in the Senate?

18. The total number of Democrats and Republicans in the U.S. House of Representatives in 1995 was 434. There were 26 more Republicans than Democrats. How many members of each party were there in the House of Representatives?

19. In his coaching career with the Boston Celtics, Red Auerbach had 558 more wins than losses. His total number of games coached was 1,516. How many wins did Auerbach have?

20. In the first Super Bowl, played in 1966, Green Bay and Kansas City scored a total of 45 points. Green Bay won by 25 points. What was the score of the first Super Bowl?

21. On Professor Brandsma's algebra test, the highest grade was 34 points higher than the lowest grade. The sum of the two grades was 160. What were the highest and lowest grades?

22. In one day, Gwen Boyle received 13 packages. Federal Express delivered three times as many as Airborne Express, while Airborne Express delivered two more than United Parcel Service. How many packages did each service deliver to Gwen?

23. In her job at the post office, Janie Quintana works a $6\frac{1}{2}$-hour day. She sorts mail, sells stamps, and does supervisory work. On one day, she sold stamps twice as long as she sorted mail, and sold stamps 1 hour longer than the time she spent doing supervisory work. How many hours did she spend at each task?

24. Venus is 31.2 million miles farther from the sun than Mercury, while Earth is 25.7 million miles farther from the sun than Venus. If the total of the distances for these three planets from the sun is 196.1 million miles, how far away from the sun is Mercury? (All distances given here are *mean (average)* distances.)

25. It is believed that Saturn has 5 more satellites (moons) than the known number of satellites for Jupiter, and 20 more satellites than the known number for Mars. If the total of these numbers is 41, how many satellites does Mars have?

26. On a recent model of the Eagle Premier, the suggested list price for the antilock brake system is $\frac{10}{3}$ the suggested list price of power door locks. Together, these two options cost $1,040. What is the suggested list price for each of these options?

27. Labrador retrievers and Rottweilers rank as the top two dog breeds using figures provided by the American Kennel Club. In a recent year, there were 25,434 more Labrador retrievers registered than Rottweilers, and together there were 216,324 dogs of these two breeds registered. How many of each breed were registered?

28. According to figures provided by the Air Transport Association of America, the Boeing B747–400 and the McDonnell Douglas L1011–100/200 are among the air carriers with maximum passenger seating. The Boeing seats 110 more passengers than the McDonnell Douglas, and together the two models seat 696 passengers. What is the seating capacity of each model?

Refer to the accompanying graph to help you answer the questions in Exercises 29 and 30.

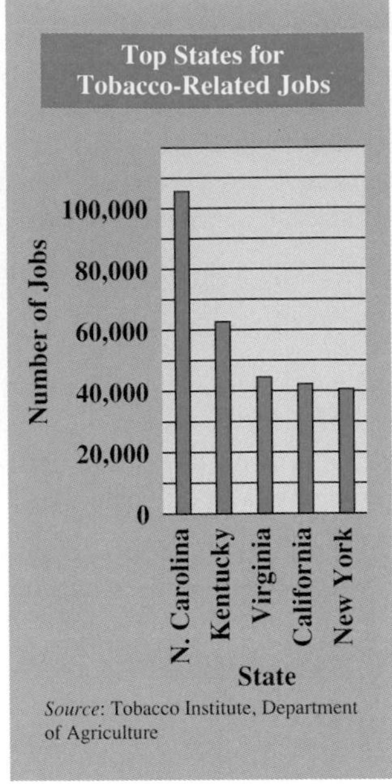

Source: Tobacco Institute, Department of Agriculture

29. According to figures provided by the Tobacco Institute, Department of Agriculture, the top two states in the tobacco industry together employ 167,281 people. The leading state employs 17,663 fewer than twice as many as the second leading state. What are these states and how many people does each employ?

30. New York employs 1,446 fewer people than California, and California employs 2,511 fewer people than Virginia. The three states together employ 127,143 people. How many people work in the tobacco industry in each of these states?

Use the method of Example 3 or your own method to solve the problem.

31. How much pure acid is in 250 milliliters of a 14% acid solution?

32. How much pure alcohol is in 150 liters of a 30% alcohol solution?

33. If \$10,000 is invested for one year at 3.5% simple interest, how much interest is earned?

34. If \$25,000 is invested at 3% simple interest for 2 years, how much interest is earned?

35. What is the monetary amount of 283 nickels?

36. What is the monetary amount of 35 half-dollars?

Use the method of Example 4 or your own method to solve the problem.

37. How many gallons of 50% antifreeze must be mixed with 80 gallons of 20% antifreeze to get a mixture that is 40% antifreeze?

38. How many liters of 25% acid solution must be added to 80 liters of 40% solution to get a solution that is 30% acid?

39. How many gallons of milk that is 2% butterfat must be mixed with milk that is 3.5% butterfat to get 10 gallons of milk that is 3% butterfat? (*Hint:* Let x represent the number of gallons that are 2% butterfat. Then $10 - x$ represents the number of gallons that are 3.5% butterfat.)

40. A pharmacist has 20 liters of a 10% drug solution. How many liters of 5% solution must be added to get a mixture that is 8%?

41. How many liters of a 60% acid solution must be mixed with a 75% acid solution to get 20 liters of a 72% solution?

42. How many gallons of a 12% indicator solution must be mixed with a 20% indicator solution to get 10 gallons of a 14% solution?

43. Minoxidil is a drug that has recently proven to be effective in treating male pattern baldness. A pharmacist wishes to mix a solution that is 2% minoxidil. She has on hand 50 milliliters of a 1% solution, and she wishes to add some 4% solution to it to obtain the desired 2% solution. How much 4% solution should she add?

44. Water must be added to 20 milliliters of a 4% minoxidil solution to dilute it to a 2% solution. How many milliliters of water should be used?

Use the concepts of this section to solve the problem.

45. Suppose that a chemist is mixing two acid solutions, one of 20% concentration and the other of 30% concentration. Which one of the following concentrations could *not* be obtained?
(a) 22% **(b)** 24% **(c)** 28% **(d)** 32%

46. Suppose that pure alcohol is added to a 24% alcohol mixture. Which one of the following concentrations could *not* be obtained?
(a) 22% **(b)** 26% **(c)** 28% **(d)** 30%

Use the method of Example 5 or your own method to solve the problem.

47. Li Nguyen invested some money at 3% and \$4,000 less than that amount at 5%. The two investments produced a total of \$200 interest in one year. How much was invested at each rate?

48. LaShondra Williams inherited some money from her uncle. She deposited part of the money in a savings account paying 2%, and \$3,000 more than that amount in a different account paying 3%. Her annual interest income was \$690. How much did she deposit at each rate?

49. Fran Liberto sold a painting that she found at a garage sale to an art dealer. With the \$12,000 she got, she invested part at 4% in a certificate of deposit and the rest at 5% in a municipal bond. Her total annual income from the two investments was \$515. How much did she invest at each rate? (*Hint:* Let x represent the amount invested at 4%. Then $12{,}000 - x$ represents the amount invested at 5%.)

50. Two investments produce an annual interest income of \$114. The total amount of money invested is \$4,000, and the two interest rates paid are 2.5% and 3.5%. How much is invested at each rate?

51. With income earned by selling the rights to his life story, an actor invests some of the money at 3% and \$30,000 more than twice as much at 4%. The total annual interest earned from the investments is \$5,600. How much is invested at each rate?

52. An artist invests her earnings in two ways. Some goes into a tax-free bond paying 6%, and \$6,000 more than three times as much goes into mutual funds paying 5%. Her total annual interest income from the investments is \$825. How much does she invest at each rate?

The problems in Exercises 53–56 involve percent. Use your own method to solve the problem.

53. The pie chart shows the approximate percents for different sizes of Health Maintenance Organizations (HMOs) for the year 1991. During that year there were approximately 560 such organizations. Find the approximate number for each size group.

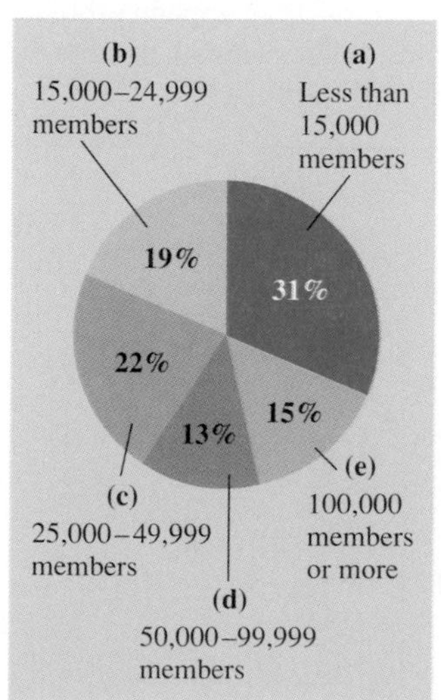

54. At the start of play on September 18, 1995, the standings of the Central Division of the American League were as shown. "Winning percentage" is commonly expressed as a decimal rounded to the nearest thousandth. To find the winning percentage of a team, divide the number of wins by the total number of games played. Find the winning percentage of each of the following teams.
(a) Cleveland
(b) Chicago
(c) Milwaukee

	Won	Lost
Cleveland	91	41
Kansas City	67	63
Milwaukee	62	69
Chicago	60	70
Minnesota	48	81

55. The pie chart shows the breakdown, by approximate percents, of age groups buying aerobic shoes in 1990. According to figures of the National Sporting Goods Association, \$582,000,000 was spent on aerobic shoes that year. How much was spent by each age group?
(a) younger than 14
(b) 18 to 24
(c) 45 to 64

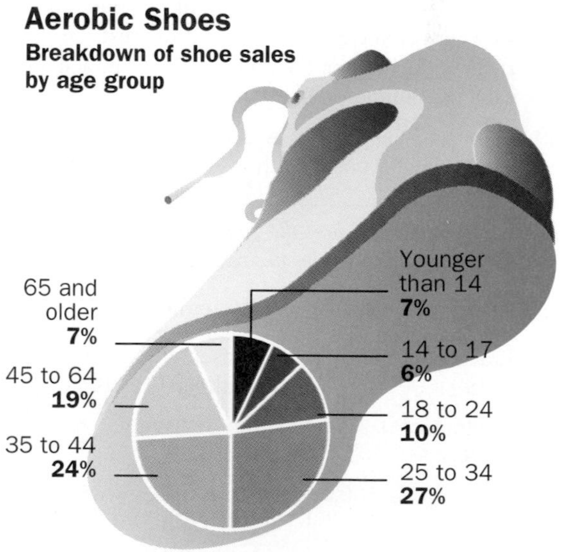

56. The pie chart shows the breakdown for the year 1990 regarding the sales and purchases of exercise equipment. If $2,295,000,000 was spent that year, determine the amount spent by each age group.
(a) 24 to 34 **(b)** 35 to 44 **(c)** 65 and older

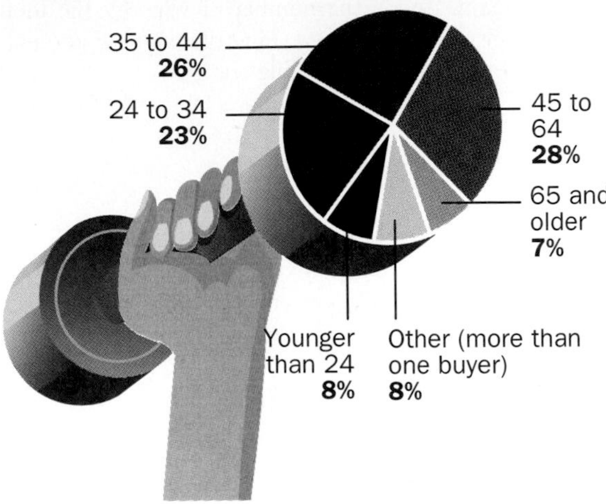

Use the method of Example 6 or your own method to solve the problem.

57. Leslie Cobar's piggy bank has 36 coins. Some are quarters and the rest are half-dollars. If the total value of the coins is $14.75, how many of each denomination does Leslie have?

58. Liz Harold has a jar in her office that contains 47 coins. Some are pennies and the rest are dimes. If the total value of the coins is $2.18, how many of each denomination does she have?

59. Sam Abo-zahrah has a box of coins that he uses when playing poker with his friends. The box currently contains 44 coins, consisting of pennies, dimes, and quarters. The number of pennies is equal to the number of dimes, and the total value is $4.37. How many of each denomination of coin does he have in the box?

60. Roma Sherry found some coins while looking under her sofa pillows. There were equal numbers of nickels and quarters, and twice as many half-dollars as quarters. If she found $2.60 in all, how many of each denomination of coin did she find?

61. In the nineteenth century, the United States minted two-cent and three-cent pieces. Toni Spahn has three times as many three-cent pieces as two-cent pieces, and the face value of these coins is $1.21. How many of each denomination does she have?

62. Kathy Jordan collects U.S. gold coins. She has a collection of 80 coins. Some are $10 coins and the rest are $20 coins. If the face value of the coins is $1,060, how many of each denomination does she have?

63. The school production of *Our Town* was a big success. For opening night, 410 tickets were sold. Students paid $1.50 each, while nonstudents paid $3.50 each. If a total of $825 was collected, how many students and how many nonstudents attended?

64. A total of 550 people attended a Frankie Valli concert. Floor tickets cost $20 each, while balcony tickets cost $14 each. If a total of $10,400 was collected, how many of each type of ticket were sold?

65. In your own words, explain how the problems in Exercises 57–64 are similar to the investment problem explained in Example 5.

66. In the nineteenth century, the United States minted half-cent coins. If an applied problem involved half-cent coins, what decimal number would represent this denomination?

In Exercises 67–70, find the time based on the information provided. Use a calculator and round your answer to the nearest thousandth.

Event and Year	Participant	Distance	Rate
67. Indianapolis 500, 1992	Al Unser, Jr. (Galmer-Chevrolet)	500 miles	134.479 mph
68. Daytona 500, 1992	Davey Allison (Ford)	500 miles	160.256 mph
69. Indianapolis 500, 1980	Johnny Rutherford (Hy-Gain McLaren/ Goodyear)	255 miles (rain-shortened)	148.725 mph
70. Indianapolis 500, 1975	Bobby Unser (Jorgensen Eagle)	435 miles (rain-shortened)	149.213 mph

In Exercises 71–74, find the rate based on the information provided. Use a calculator and round your answer to the nearest hundredth.

	Event and Year	Participant	Distance	Time
71.	Summer Olympics 400-meter Hurdles, Women, 1992	Sally Gunnell (Great Britain)	400 meters	53.23 seconds
72.	Summer Olympics, 100-meter Dash, Women, 1992	Gail Devers (United States)	100 meters	10.82 seconds
73.	Summer Olympics, 100-meter Dash, Men, 1988	Carl Lewis (United States)	100 meters	9.92 seconds
74.	Winter Olympics, 500-meter Speed Skating, Women, 1992	Bonnie Blair (United States)	500 meters	40.33 seconds

75. A driver averaged 53 miles per hour and took 10 hours to travel from Memphis to Chicago. What is the distance between Memphis and Chicago?

76. A small plane traveled from Warsaw to Rome, averaging 164 miles per hour. The trip took two hours. What is the distance from Warsaw to Rome?

77. Suppose that an automobile averages 45 miles per hour, and travels for 30 minutes. Is the distance traveled $45 \times 30 = 1{,}350$ miles? If not, explain why not, and give the correct distance.

78. Which of the following choices is the best *estimate* for the average speed of a trip of 405 miles that lasted 8.2 hours?
(a) 50 miles per hour **(b)** 30 miles per hour
(c) 60 miles per hour **(d)** 40 miles per hour

Use the method of Example 8 or your own method to solve the problem.

79. St. Louis and Portland are 2,060 miles apart. A small plane leaves Portland, traveling toward St. Louis at an average speed of 90 miles per hour. Another plane leaves St. Louis at the same time, traveling toward Portland, averaging 116 miles per hour. How long will it take them to meet?

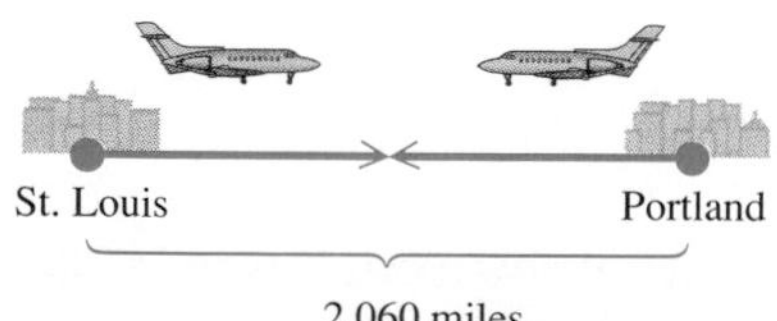

80. Atlanta and Cincinnati are 440 miles apart. John leaves Cincinnati, driving toward Atlanta at an average speed of 60 miles per hour. Pat leaves Atlanta at the same time, driving toward Cincinnati in her antique auto, averaging 28 miles per hour. How long will it take them to meet?

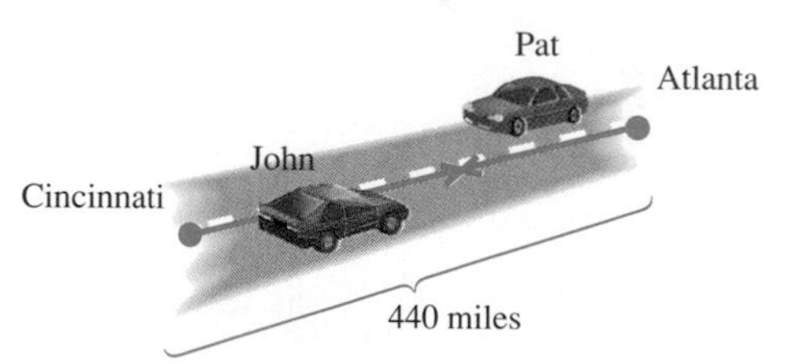

81. A train leaves Little Rock, Arkansas, and travels north at 85 kilometers per hour. Another train leaves at the same time and travels south at 95 kilometers per hour. How long will it take before they are 315 kilometers apart?

82. Two steamers leave a port on a river at the same time, traveling in opposite directions. Each is traveling 22 miles per hour. How long will it take for them to be 110 miles apart?

83. Joey and Mark commute to work, traveling in opposite directions. Joey leaves the house at 8:00 A.M. and averages 35 miles per hour. Mark leaves at 8:15 A.M. and averages 40 miles per hour. At what time will they be 140 miles apart?

84. Jeff leaves his house on his bicycle at 8:30 A.M. and averages 5 miles per hour. His wife, Joan, leaves at 9:00 A.M., following the same path and averaging 8 miles per hour. At what time will Joan catch up with Jeff?

85. When Tri drives his car to work, the trip takes 30 minutes. When he rides the bus, it takes 45 minutes. The average speed of the bus is 12 miles per hour less than his speed when driving. Find the distance he travels to work.

86. Latoya can get to school in 15 minutes if she rides her bike. It takes her 45 minutes if she walks. Her speed when walking is 10 miles per hour slower than her speed when riding. What is her speed when she rides?

7.3 Ratio, Proportion, and Variation

One of the most frequently used mathematical concepts in everyday life is ratio. A baseball player's batting average is actually a ratio. The slope, or pitch, of a roof on a building may be expressed as a ratio.

Problem Solving

Here is an example of a type of problem that is typical of many problems we are faced with on a routine basis.

> A carpet cleaning service charges $45.00 to clean 2 similarly sized rooms of carpet. How much would it cost to clean 5 rooms of carpet?

Assuming that the cleaning service does not discount its prices for cleaning additional rooms after the first two, the reasoning for solving this problem might be as follows: if it costs $45.00 to clean 2 rooms, then it would cost $\$45.00/2 = \$22.50$ per room. So, the total cost for cleaning 5 rooms would be $5 \times \$22.50 = \$112.50$.

The quotient $\$45.00/2$ is an example of a ratio of price to number of rooms. Ratios provide a way of comparing two numbers or quantities. A **ratio** is a quotient of two quantities. The ratio of the number a to the number b is written as follows.

Ratio

$$a \text{ to } b, \qquad \frac{a}{b}, \qquad \text{or} \qquad a:b$$

When ratios are used in comparing units of measure, the units should be the same. This is shown in Example 1.

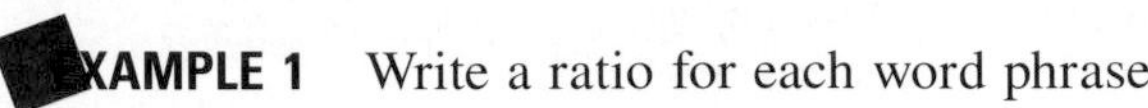

EXAMPLE 1 Write a ratio for each word phrase.

(a) the ratio of 5 hours to 3 hours
This ratio can be written as $5/3$.

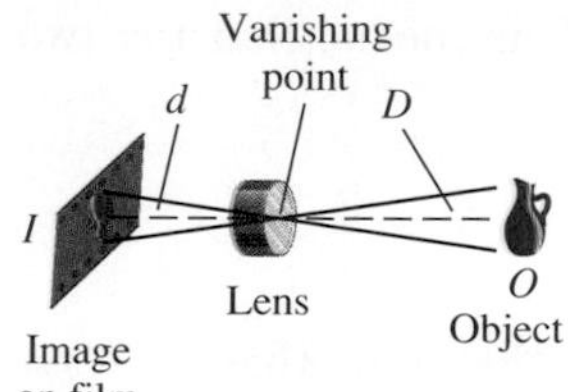

When you look a long way down a straight road or railroad track, it seems to narrow as it vanishes in the distance. The point where the sides seem to touch is called the **vanishing point.** The same thing occurs in the lens of a camera, as shown in the figure. Suppose I represents the length of the image, O the length of the object, d the distance from the lens to the film, and D the distance from the lens to the object. Then

$$\frac{\text{Image length}}{\text{Object length}} = \frac{\text{Image distance}}{\text{Object distance}}$$

or

$$\frac{I}{O} = \frac{d}{D}.$$

Given the length of the image on the film and its distance from the lens, then the length of the object determines how far away the lens must be from the object to fit on the film.

(b) the ratio of 5 hours to 3 days

First convert 3 days to hours: 3 days $= 3 \cdot 24 = 72$ hours. The ratio of 5 hours to 3 days is thus $5/72$. ◆

We now define *proportion.*

Proportion

A **proportion** is a statement that says that two ratios are equal.

For example,

$$\frac{3}{4} = \frac{15}{20}$$

is a proportion that says that the ratios $3/4$ and $15/20$ are equal. In the proportion

$$\frac{a}{b} = \frac{c}{d},$$

a, b, c, and d are the **terms** of the proportion. Beginning with the proportion

$$\frac{a}{b} = \frac{c}{d}$$

and multiplying both sides by the common denominator, bd, gives

$$bd \cdot \frac{a}{b} = bd \cdot \frac{c}{d}$$

$$ad = bc.$$

The products ad and bc can be found by multiplying diagonally.

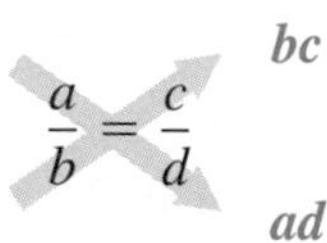

This is called **cross multiplication** and ad and bc are called **cross products.**

Cross Products

If $\frac{a}{b} = \frac{c}{d}$, then the cross products ad and bc are equal.

Also, if $ad = bc$, then $\frac{a}{b} = \frac{c}{d}$ (as long as $b \neq 0$, $d \neq 0$).

From the rule given above,

$$\text{if} \quad \frac{a}{b} = \frac{c}{d} \quad \text{then} \quad ad = bc.$$

However, if $a/c = b/d$, then $ad = cb$, or $ad = bc$. This means that the two proportions are equivalent and

$$\text{the proportion } \frac{a}{b} = \frac{c}{d} \text{ can also be written as } \frac{a}{c} = \frac{b}{d}.$$

Sometimes one form is more convenient to work with than the other.

Four numbers are used in a proportion. If any three of these numbers are known, the fourth can be found.

EXAMPLE 2 **(a)** Solve for x in the proportion

$$\frac{63}{x} = \frac{9}{5}.$$

The cross products must be equal, so

$$63 \cdot 5 = 9x$$
$$315 = 9x.$$

Divide both sides by 9 to get

$$35 = x.$$

The solution set is $\{35\}$.

(b) Solve for r in the proportion $\dfrac{8}{5} = \dfrac{12}{r}$.

Set the cross products equal to each other.

$$8r = 5 \cdot 12$$
$$8r = 60$$
$$r = \frac{60}{8} = \frac{15}{2}$$

The solution set is $\left\{\dfrac{15}{2}\right\}$. ■

EXAMPLE 3 Solve the equation

$$\frac{m-2}{5} = \frac{m+1}{3}.$$

Find the cross products, and set them equal to each other.

$$3(m-2) = 5(m+1) \quad \text{Be sure to use parentheses.}$$
$$3m - 6 = 5m + 5 \quad \text{Distributive property}$$
$$3m = 5m + 11 \quad \text{Add 6.}$$
$$-2m = 11 \quad \text{Subtract } 5m.$$
$$m = -\frac{11}{2} \quad \text{Divide by } -2.$$

Solution set: $\left\{-\dfrac{11}{2}\right\}$ ■

While the cross product method is useful in solving equations of the types found in Examples 2 and 3, it cannot be used directly if there is more than one term on either side. For example, you cannot use the method directly to solve the equation

$$\frac{4}{x} + 3 = \frac{1}{9},$$

because there are two terms on the left side.

EXAMPLE 4 Biologists can use algebra to estimate the number of fish in a lake. They first catch a sample of fish and mark each specimen with a harmless tag. Some weeks later, they catch a similar sample of fish from the same areas of the lake and determine the proportion of previously tagged fish in the new sample. The total fish population is estimated by assuming that the proportion of tagged fish in the new sample is the same as the proportion of tagged fish in the entire lake.

For example, suppose the biologists tag 300 fish on May 1. When they return and take a new sample of 400 fish on June 1, 5 of the 400 were previously tagged. Estimate the number of fish in the lake.

Let x represent the number of fish in the lake. The following proportion can be set up.

$$\frac{\text{Tagged fish on May 1}}{\text{Total fish in the lake}} = \frac{\text{Tagged fish in June 1 sample}}{\text{Total number of June 1 sample}}$$

$$\frac{300}{x} = \frac{5}{400}$$

$$5x = 120{,}000$$

$$x = 24{,}000$$

There are approximately 24,000 fish in the lake. ◆

Problem Solving

In supermarkets we sometimes see shoppers carrying hand-held calculators to assist them in their job of budgeting. While a common use is to make sure that the shopper does not go over budget, another use is to see which size of an item offered in different sizes produces the best price per unit. In order to do this, simply divide the price of the item by the unit of measure in which the item is labeled. The next example illustrates this idea.

EXAMPLE 5 The local supermarket charges the following prices for a popular brand of pancake syrup.

Size	Price
36-ounce	\$3.89
24-ounce	\$2.79
12-ounce	\$1.89

Which size is the best buy? That is, which size has the lowest unit price?

To find the best buy, divide the price by the number of ounces to get the price per ounce. Each result in the following table was found by using a calculator and rounding the answer to three decimal places.

Size	Unit Cost (dollars per ounce)
36-ounce	$\frac{\$3.89}{36} = \$.108$ ← The best buy
24-ounce	$\frac{\$2.79}{24} = \$.116$
12-ounce	$\frac{\$1.89}{12} = \$.158$

Since the 36-ounce size produces the lowest price per unit, it would be the best buy. (Be careful: Sometimes the largest container *does not* produce the lowest price per unit.) ■

Variation

Refer to the carpet-cleaning problem at the beginning of this section. The following chart shows a relationship between the number of rooms cleaned and the cost of the total job for 1 through 5 rooms.

Number of Rooms	Cost of the Job
1	\$ 22.50
2	\$ 45.00
3	\$ 67.50
4	\$ 90.00
5	\$112.50

If we divide the cost of the job by the number of rooms, in each case we obtain the quotient, or ratio, 22.50 (dollars per room). Suppose that we let x represent the number of rooms and let y represent the cost for cleaning that number of rooms. Then we find the relationship between x and y is given by the equation

$$\frac{y}{x} = 22.50$$

or

$$y = 22.50x.$$

This relationship between x and y is an example of *direct variation.*

Direct Variation

y **varies directly** as x, or y is **directly proportional** to x, if there exists a nonzero constant k such that

$$y = kx$$

or, equivalently,

$$\frac{y}{x} = k.$$

The constant k is called the **constant of variation.**

EXAMPLE 6 Suppose y varies directly as x, and $y = 50$ when $x = 20$. Find y when $x = 14$.

Since y varies directly as x, there exists a constant k such that $y = kx$. Find k by replacing y with 50 and x with 20.

$$y = kx$$
$$50 = k \cdot 20$$
$$\frac{5}{2} = k$$

Since $y = kx$ and $k = 5/2$,

$$y = \frac{5}{2}x.$$

Now find y when $x = 14$.

$$y = \frac{5}{2} \cdot 14 = 35$$

The value of y is 35 when $x = 14$. ■

EXAMPLE 7 Hooke's law for an elastic spring states that the distance a spring stretches is directly proportional to the force applied. If a force of 150 pounds stretches a certain spring 8 centimeters, how much will a force of 400 pounds stretch the spring?

If d is the distance the spring stretches and f is the force applied, then $d = kf$ for some constant k. Since a force of 150 pounds stretches the spring 8 centimeters,

$$d = kf \qquad \text{Formula}$$
$$8 = k \cdot 150 \qquad d = 8,\ f = 150$$
$$k = \frac{8}{150} = \frac{4}{75}, \qquad \text{Find } k.$$

and $d = \frac{4}{75}f$.

For a force of 400 pounds

$$d = \frac{4}{75}(400) \qquad \text{Let } f = 400.$$
$$= \frac{64}{3}.$$

The spring will stretch 64/3 centimeters if a force of 400 pounds is applied. ■

In some cases one quantity will vary directly as a *power* of another.

Direct Variation as a Power

y** varies directly as the* ***n*th power of *x if there exists a real number k such that

$$y = kx^n.$$

An example of direct variation as a power involves the area of a circle. The formula for the area of a circle is

$$A = \pi r^2.$$

Here, π is the constant of variation, and the area varies directly as the square of the radius.

EXAMPLE 8 The distance a body falls from rest varies directly as the square of the time it falls (here we disregard air resistance). If an object falls 64 feet in 2 seconds, how far will it fall in 8 seconds?

If d represents the distance the object falls and t the time it takes to fall,

$$d = kt^2$$

for some constant k. To find the value of k, use the fact that the object falls 64 feet in 2 seconds.

$$\begin{aligned} d &= kt^2 && \text{Formula} \\ 64 &= k(2)^2 && \text{Let } d = 64 \text{ and } t = 2. \\ k &= 16 && \text{Find } k. \end{aligned}$$

With this result, the variation equation becomes

$$d = 16t^2.$$

Now let $t = 8$ to find the number of feet the object will fall in 8 seconds.

$$\begin{aligned} d &= 16(8)^2 && \text{Let } t = 8. \\ &= 1{,}024 \end{aligned}$$

The object will fall 1,024 feet in 8 seconds.

Einstein's famous equation showing the connection between energy and mass, $E = mc^2$ (where c is the speed of light), is perhaps the most famous equation in science.

In direct variation where $k > 0$, as x increases, y increases, and similarly as x decreases, y decreases. Another type of variation is *inverse variation.*

Inverse Variation

y* varies inversely as *x if there exists a real number k such that

$$y = \frac{k}{x}$$

or, equivalently,

$$xy = k.$$

Also, ***y* varies inversely as the *n*th power of *x*** if there exists a real number k such that

$$y = \frac{k}{x^n}.$$

EXAMPLE 9 The weight of an object above the earth varies inversely as the square of its distance from the center of the earth. A space vehicle in an elliptical orbit has a maximum distance from the center of the earth (apogee) of 6,700 miles. Its minimum distance from the center of the earth (perigee) is 4,090 miles. See Figure 7. If an astronaut in the vehicle weighs 57 pounds at its apogee, what does the astronaut weigh at the perigee?

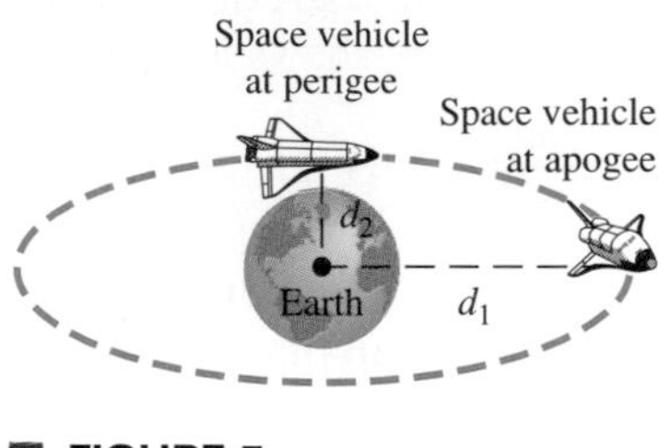

FIGURE 7

If w is the weight and d is the distance from the center of the earth, then

$$w = \frac{k}{d^2}$$

for some constant k. At the apogee the astronaut weighs 57 pounds and the distance from the center of the earth is 6,700 miles. Use these values to find k.

$$57 = \frac{k}{(6{,}700)^2} \quad \text{Let } w = 57 \text{ and } d = 6{,}700.$$

$$k = 57(6{,}700)^2$$

Then the weight at the perigee with $d = 4{,}090$ miles is

$$w = \frac{57(6{,}700)^2}{(4{,}090)^2} \approx 153 \text{ pounds.} \quad \text{Use a calculator.}$$

Johann Kepler (1571–1630) established the importance of the ellipse in 1609, when he discovered that the orbits of the planets around the sun were elliptical, not circular. The orbits of the planets are nearly circular. However, Halley's comet, which has been studied since 467 B.C., has an elliptical orbit which is long and narrow, with one axis much longer than the other. This comet was named for the British astronomer and mathematician Edmund Halley (1656–1742), who predicted its return after observing it in 1682. The comet appears regularly every 76 years.

It is common for one variable to depend on several others. For example, if one variable varies as the product of several other variables (perhaps raised to powers), the first variable is said to **vary jointly** as the others.

EXAMPLE 10 The strength of a rectangular beam varies jointly as its width and the square of its depth. If the strength of a beam 2 inches wide by 10 inches deep is 1,000 pounds per square inch, what is the strength of a beam 4 inches wide and 8 inches deep?

If S represents the strength, w the width, and d the depth, then

$$S = kwd^2$$

for some constant k. Since $S = 1{,}000$ if $w = 2$ and $d = 10$,

$$1{,}000 = k(2)(10)^2. \quad \text{Let } S = 1{,}000,\ w = 2, \text{ and } d = 10.$$

Solving this equation for k gives

$$1{,}000 = k \cdot 2 \cdot 100$$

$$1{,}000 = 200k$$

$$k = 5,$$

so

$$S = 5wd^2.$$

Find S when $w = 4$ and $d = 8$ by substitution in $S = 5wd^2$.

$$S = 5(4)(8)^2 \qquad \text{Let } w = 4 \text{ and } d = 8.$$
$$= 1{,}280$$

The strength of the beam is 1,280 pounds per square inch. ■

There are many combinations of direct and inverse variation. The final example shows a typical **combined variation** problem.

EXAMPLE 11 The maximum load that a cylindrical column with a circular cross section can hold varies directly as the fourth power of the diameter of the cross section and inversely as the square of the height. A 9-meter column 1 meter in diameter will support 8 metric tons. How many metric tons can be supported by a column 12 meters high and 2/3 meter in diameter?

Let L represent the load, d the diameter, and h the height. Then

$$L = \frac{kd^4}{h^2}. \qquad \begin{array}{l} \leftarrow \text{Load varies directly as the 4th power of the diameter.} \\ \leftarrow \text{Load varies inversely as the square of the height.} \end{array}$$

Now find k. Let $h = 9$, $d = 1$, and $L = 8$.

$$8 = \frac{k(1)^4}{9^2} \qquad h = 9,\ d = 1,\ L = 8$$

Solve for k.

$$8 = \frac{k}{81}$$
$$k = 648$$

Substitute 648 for k in the first equation.

$$L = \frac{648d^4}{h^2}$$

Now find L when $h = 12$ and $d = 2/3$ by substituting the values into the last equation.

$$L = \frac{648\left(\frac{2}{3}\right)^4}{12^2} \qquad h = 12,\ d = \frac{2}{3}$$

$$L = \frac{648\left(\frac{16}{81}\right)}{144}$$
$$= 648 \cdot \frac{16}{81} \cdot \frac{1}{144} = \frac{8}{9}$$

The maximum load is 8/9 metric ton. ■

7.3 EXERCISES

1. Which one of the following ratios is not the same as 3 to 4?
   (a) .75 **(b)** 6 to 8 **(c)** 4 to 3 **(d)** 30 to 40

2. Give three ratios that are equivalent to 3 to 1.

Determine the ratio and write it in lowest terms.

3. 40 miles to 30 miles
4. 60 feet to 70 feet
5. 120 people to 90 people
6. 72 dollars to 220 dollars
7. 20 yards to 8 feet
8. 30 inches to 8 feet
9. 24 minutes to 2 hours
10. 16 minutes to 1 hour
11. 8 days to 40 hours
12. 50 hours to 5 days
13. Explain the distinction between *ratio* and *proportion.*
14. Suppose that someone told you to use cross products in order to multiply fractions. How would you explain to the person what is wrong with his or her thinking?

Decide whether the proportion is true *or* false.

15. $\frac{5}{35} = \frac{8}{56}$
16. $\frac{4}{12} = \frac{7}{21}$
17. $\frac{120}{82} = \frac{7}{10}$
18. $\frac{27}{160} = \frac{18}{110}$
19. $\frac{\frac{1}{2}}{5} = \frac{1}{10}$
20. $\frac{\frac{1}{3}}{6} = \frac{1}{18}$

Solve the equation.

21. $\frac{k}{4} = \frac{175}{20}$
22. $\frac{49}{56} = \frac{z}{8}$
23. $\frac{x}{6} = \frac{18}{4}$
24. $\frac{z}{80} = \frac{20}{100}$
25. $\frac{3y - 2}{5} = \frac{6y - 5}{11}$
26. $\frac{2p + 7}{3} = \frac{p - 1}{4}$
27. $\frac{2r + 8}{4} = \frac{3r - 9}{3}$
28. $\frac{5k + 1}{6} = \frac{3k - 2}{3}$

Solve the problem by setting up and solving a proportion.

29. According to the Home and Garden Bulletin No. 72, four spears of asparagus contain 15 calories. How many spears of asparagus would contain 50 calories?
30. According to the source indicated in Exercise 29, three ounces of bluefish baked with butter or margarine provide 22 grams of protein. How many ounces would provide 242 grams of protein?
31. A chain saw requires a mixture of 2-cycle engine oil and gasoline. According to the directions on a bottle of Oregon 2-cycle Engine Oil, for a 50 to 1 ratio requirement, 2.5 fluid ounces of oil are required for 1 gallon of gasoline. If the tank of the chain saw holds 2.75 gallons, how many fluid ounces of oil are required?
32. The directions on the bottle mentioned in Exercise 31 indicate that if the ratio requirement is 24 to 1, 5.5 ounces of oil are required for 1 gallon of gasoline. If gasoline is to be mixed with 22 ounces of oil, how much gasoline is to be used?
33. In 1992, the average exchange rate between U.S. dollars and United Kingdom pounds was 1 pound to \$1.7655. Margaret went to London and exchanged her U.S. currency for U.K. pounds, and received 400 pounds. How much in U.S. money did Margaret exchange?
34. If 3 U.S. dollars can be exchanged for 4.2186 Swiss francs, how many Swiss francs can be obtained for \$49.20?

35. If 6 gallons of premium unleaded gasoline cost $9.36, how much would it cost to completely fill a 15-gallon tank?

36. If sales tax on a $16.00 compact disc is $1.32, how much would the sales tax be on a $120.00 compact disc player?

37. The distance between Kansas City, Missouri, and Denver is 600 miles. On a certain wall map, this is represented by a length of 2.4 feet. On the map, how many feet would there be between Memphis and Philadelphia, two cities that are actually 1,000 miles apart?

38. The distance between Singapore and Tokyo is 3,300 miles. On a certain wall map, this distance is represented by 11 inches. The actual distance between Mexico City and Cairo is 7,700 miles. How far apart are they on the same map?

39. A recipe for green salad for 70 people calls for 18 heads of lettuce. How many heads of lettuce would be needed if 175 people were to be served?

40. A recipe for oatmeal macaroons calls for 1 2/3 cups of flour to make four dozen cookies. How many cups of flour would be needed for six dozen cookies?

41. A piece of property assessed at $42,000 requires an annual property tax of $273. How much property tax would be charged for a similar piece of property assessed at $52,000?

42. If 4 pounds of fertilizer will cover 50 square feet of garden, how many pounds would be needed for 225 square feet?

43. Biologists tagged 250 fish in Willow Lake on October 5. On a later date they found 7 tagged fish in a sample of 350. Estimate the total number of fish in Willow Lake to the nearest hundred.

44. On May 13 researchers at Argyle Lake tagged 420 fish. When they returned a few weeks later, their sample of 500 fish contained 9 that were tagged. Give an approximation of the fish population in Argyle Lake to the nearest hundred.

45. The pie chart indicates the number of people who responded in various ways to a survey. Suppose that 1,680 people participated. How many would we expect to answer in each way if 4,200 people were surveyed?

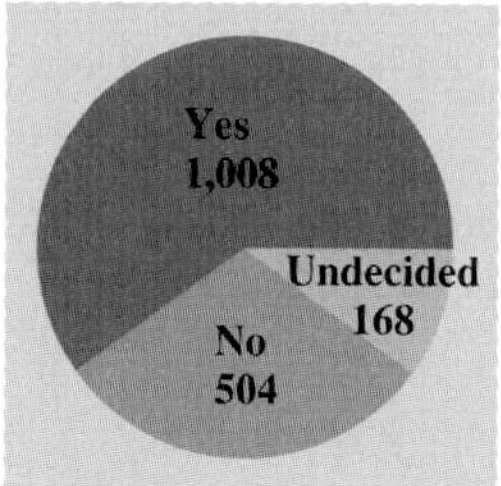

46. According to McDonald's Corporation, Canada has 40,113 people for every McDonald's restaurant. The 1992 estimated population of Canada was 27,351,000. According to these figures, about how many McDonald's restaurants would we have expected Canada to have in 1992?

A supermarket was surveyed to find the prices charged for items in various sizes. Find the best buy (based on price per unit) for the particular item.

47. Trash bags
20-count: $3.09
30-count: $4.59

48. Black pepper
1-ounce size: $ .99
2-ounce size: $1.65
4-ounce size: $4.39

49. Breakfast cereal
15-ounce size: $2.99
25-ounce size: $4.49
31-ounce size: $5.49

50. Cocoa mix
8-ounce size: $1.39
16-ounce size: $2.19
32-ounce size: $2.99

51. Tomato ketchup
14-ounce size: $ .89
32-ounce size: $1.19
64-ounce size: $2.95

52. Cut green beans
8-ounce size: $ .45
16-ounce size: $ .49
50-ounce size: $1.59

Two triangles are said to be ***similar*** *if they have the same shape (but not necessarily the same size). Similar triangles have sides that are proportional. For example, the figure shows two similar triangles. Notice that the ratios of the corresponding sides are all equal to* $\frac{3}{2}$:

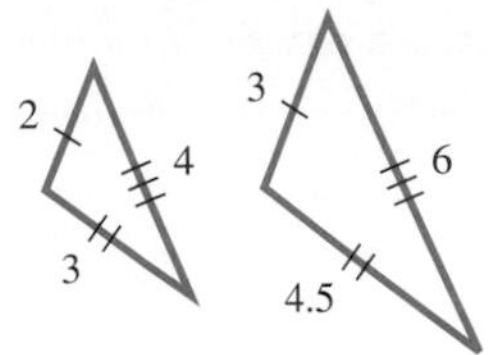

$$\frac{3}{2} = \frac{3}{2} \qquad \frac{4.5}{3} = \frac{3}{2} \qquad \frac{6}{4} = \frac{3}{2}.$$

If we know that two triangles are similar, we can set up a proportion to solve for the length of an unknown side. Use a proportion to find the length x, given that the pair of triangles are similar.

53.

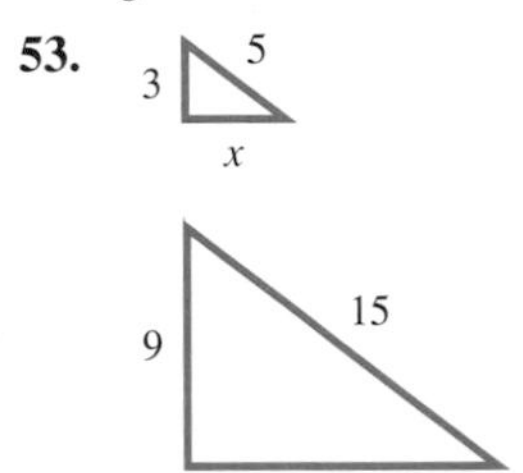

54.

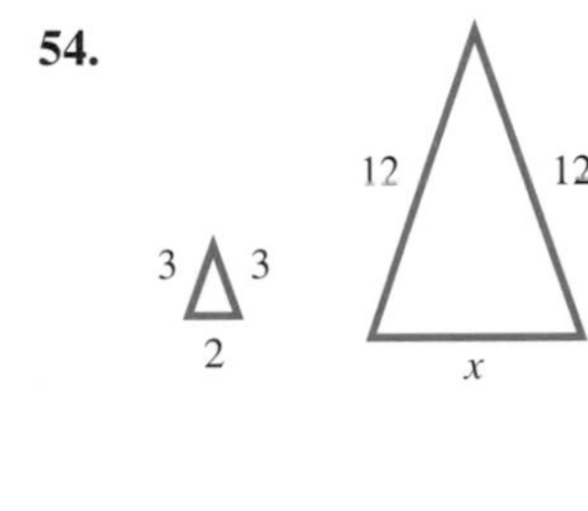

55.

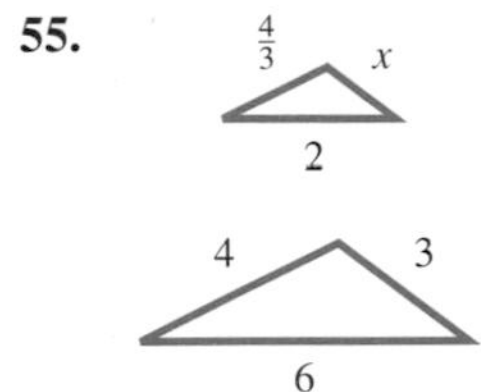

56.

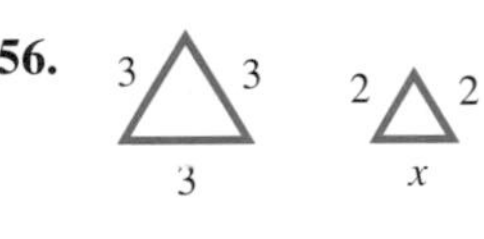

For the problems in Exercises 57 and 58, **(a)** *draw a sketch consisting of two right triangles, depicting the situation described, and* **(b)** *solve the problem.*

57. An enlarged version of the chair used by George Washington at the Constitutional Convention casts a shadow 18 feet long at the same time a vertical pole 12 feet high casts a shadow 4 feet long. How tall is the chair?

58. One of the tallest candles ever constructed was exhibited at the 1897 Stockholm Exhibition. If it cast a shadow 5 feet long at the same time a vertical pole 32 feet high cast a shadow 2 feet long, how tall was the candle?

The Consumer Price Index, issued by the U.S. Bureau of Labor Statistics, provides a means of determining the purchasing power of the U.S. dollar from one year to the next. Using the period from 1982 to 1984 as a measure of 100.0, *the Consumer Price Index for electricity from 1975 to 1991 is shown here.*

Year	Consumer Price Index	Year	Consumer Price Index
1975	50.0	1983	98.9
1976	53.1	1984	105.3
1977	56.6	1985	108.9
1978	60.9	1986	110.4
1979	65.6	1987	110.0
1980	75.8	1988	111.5
1981	87.2	1989	114.7
1982	95.8	1990	117.4
		1991	121.8

To use the Consumer Price Index, we can set up a proportion as follows:

$$\frac{\text{price of electricity in year } a}{\text{Consumer Price Index in year } a} = \frac{\text{price of electricity in year } b}{\text{Consumer Price Index in year } b}.$$

For example, if an electricity bill in 1985 was \$150, *we could predict the amount of the bill in 1989 by using the following proportion:*

$$\begin{matrix}\text{1985 price} \rightarrow \\ \text{1985 index} \rightarrow\end{matrix} \frac{150}{108.9} = \frac{x}{114.7}. \begin{matrix}\leftarrow \text{1989 price} \\ \leftarrow \text{1989 index}\end{matrix}$$

By using cross products and a calculator, we can determine that the 1989 price would be approximately \$158.

Use the Consumer Price Index figures given to find the amount that would be charged for the use of the same amount of electricity that cost \$225 *in 1980. Give your answer to the nearest dollar.*

59. in 1983 **60.** in 1986 **61.** in 1988 **62.** in 1991

63. The Consumer Price Index figures for shelter for the years A and B are 90.5 and 146.3. If shelter for a particular family cost \$3,000 in year A, what would be the comparable cost in year B? Give your answer to the nearest dollar.

64. Due to a volatile fuel oil market in the early 1980s, the price of fuel decreased during the first three quarters of the decade. The Consumer Price Index figures for 1982 and 1986 were 105.0 and 74.1. If it cost you \$21.50 to fill your tank with fuel oil in 1982, how much would it have cost to fill the same tank in 1986? Give your answer to the nearest cent.

Solve the problem involving variation.

65. If x varies directly as y, and $x = 27$ when $y = 6$, find x when $y = 2$.

66. If z varies directly as x, and $z = 30$ when $x = 8$, find z when $x = 4$.

67. If m varies directly as p^2, and $m = 20$ when $p = 2$, find m when p is 5.

68. If a varies directly as b^2, and $a = 48$ when $b = 4$, find a when $b = 7$.

69. If p varies inversely as q^2, and $p = 4$ when $q = \frac{1}{2}$, find p when $q = \frac{3}{2}$.

70. If z varies inversely as x^2, and $z = 9$ when $x = \frac{2}{3}$, find z when $x = \frac{5}{4}$.

71. Assume that the constant of variation, k, is positive.

(a) If y varies directly as x, then as y increases, x ____?____ (decreases/increases).

(b) If y varies inversely as x, then as y increases, x ____?____ (decreases/increases).

72. (a) The more gasoline you pump, the more you will have to pay. Is this an example of direct or inverse variation?

(b) The longer the term of your subscription to *Monitoring Times,* the less you will have to pay per year. Is this an example of direct or inverse variation?

Solve the problem.

73. The interest on an investment varies directly as the rate of interest. If the interest is \$48 when the interest rate is 5%, find the interest when the rate is 4.2%.

74. For a given base, the area of a triangle varies directly as its height. Find the area of a triangle with a height of 6 inches, if the area is 10 square inches when the height is 4 inches.

75. Over a specified distance, speed varies inversely with time. If a car goes a certain distance in one-half hour at 30 miles per hour, what speed is needed to go the same distance in three-fourths of an hour?

76. For a constant area, the length of a rectangle varies inversely as the width. The length of a rectangle is 27 feet when the width is 10 feet. Find the length of a rectangle with the same area if the width is 18 feet.

77. The weight of an object on the moon varies directly as the weight of the object on Earth. According to *The Guiness Book of World Records,* "Shad," a goat owned by a couple in California, is the largest known goat, weighing 352 pounds. Shad would weigh about 59 pounds on the moon. A bull moose weighing 1,800 pounds was shot in Canada and is the largest confirmed moose. How much would the moose have weighed on the moon?

78. According to *The Guiness Book of World Records,* the longest recorded voyage in a paddle boat is 2,226 miles in 103 days by the foot power of two boaters down the Mississippi River. Assuming a constant rate, how far would they have gone if they had traveled 120 days? (Distance varies directly as time.)

79. The pressure exerted by a certain liquid at a given point varies directly as the depth of the point beneath the surface of the liquid. The pressure at a depth of 10 feet is 50 pounds per square inch. What is the pressure at a depth of 20 feet?

80. If the volume is constant, the pressure of gas in a container varies directly as the temperature. If the pressure is 5 pounds per square inch at a temperature of 200 Kelvin, what is the pressure at a temperature of 300 Kelvin?

81. If the temperature is constant, the pressure of a gas in a container varies inversely as the volume of the container. If the pressure is 10 pounds per square foot in a container with 3 cubic feet, what is the pressure in a container with 1.5 cubic feet?

82. The force required to compress a spring varies directly as the change in the length of the spring. If a force of 12 pounds is required to compress a certain spring 3 inches, how much force is required to compress the spring 5 inches?

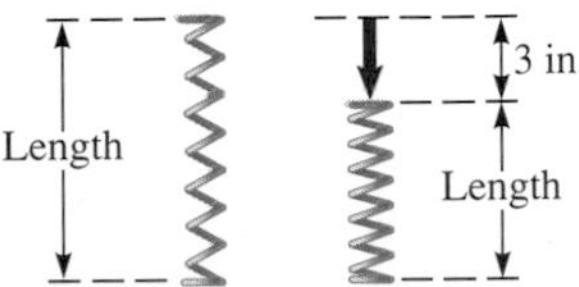

83. For a body falling freely from rest (disregarding air resistance), the distance the body falls varies directly as the square of the time. If an object is dropped from the top of a tower 400 feet high and hits the ground in 5 seconds, how far did it fall in the first 3 seconds?

84. The illumination produced by a light source varies inversely as the square of the distance from the source. If the illumination produced 4 feet from a light source is 75 foot-candles, find the illumination produced 9 feet from the same source.

7.4 Linear Inequalities

An **inequality** says that two expressions are *not* equal. A **linear inequality in one variable** is an inequality such as

$$x + 5 < 2, \qquad y - 3 \geq 5, \qquad \text{or} \qquad 2k + 5 \leq 10.$$

Linear Inequality

A linear inequality in one variable can be written in the form

$$ax + b < c,$$

where a, b, and c are real numbers, with $a \neq 0$.

(Throughout this section we give the definitions and rules only for $<$, but they are also valid for $>$, $\leq$, and $\geq$.)

An inequality is solved by finding all numbers that make the inequality true. Usually, an inequality has an infinite number of solutions. These solutions, like the solutions of equations, are found by producing a series of simpler equivalent inequalities. **Equivalent inequalities** are inequalities with the same solution set. The inequalities in this chain of equivalent inequalities are found with the addition and multiplication properties of inequality.

Archimedes, one of the greatest mathematicians of antiquity, is shown on this Italian stamp. He was born in the Greek city of Syracuse about 287 B.C.

There are numerous colorful stories about the life of Archimedes. The best known of these relates his reaction to one of his discoveries. While taking a bath, he noticed that the water level rose as he lay in the water and realized that an immersed object, if heavier than a fluid, "will, if placed in it, descend to the bottom of the fluid, and the solid will, when weighed in the fluid, be lighter than its true weight by the weight of the fluid displaced." This discovery so excited him that he ran through the streets shouting "Eureka" ("I have found it") without bothering to clothe himself!

Archimedes met his death during the pillage of Syracuse at the hand of a Roman soldier. As was his custom, he was using a sand tray to draw his geometric figures when the soldier came upon him. He ordered the soldier to move clear of his "circles," and the soldier obliged by killing him. Archimedes was 75 years old when he died.

Addition Property of Inequality

For all real numbers a, b, and c, the inequalities

$$a < b \qquad \text{and} \qquad a + c < b + c$$

are equivalent. (The same number may be added to both sides of an inequality without changing the solution set.)

As with equations, the addition property can be used to *subtract* the same number from both sides of an inequality.

EXAMPLE 1 Solve $x - 7 < -12$.

Add 7 to both sides.

$$x - 7 + 7 < -12 + 7$$
$$x < -5$$

Using set-builder notation, the solution set of this inequality is written $\{x \mid x < -5\}$. ■

It is customary to express solution sets of inequalities with number line graphs or with **interval notation.** The solution set in Example 1 can be pictured on a number line as shown in Figure 8.

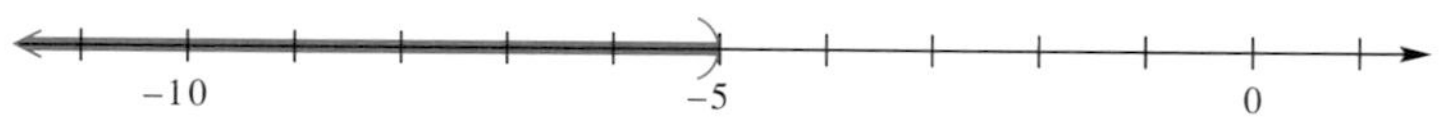

■ **FIGURE 8**

Notice that a parenthesis is used for the graph in Figure 8 to indicate that the number -5 is not included in the solution set. Had the solution set been $\{x \mid x \leq -5\}$, a square bracket would have been used to indicate the inclusion of -5. The interval notation for the set of numbers indicated in Figure 8 is $(-\infty, -5)$. The symbol $-\infty$ does not indicate a number; it is used to show that the interval includes all real numbers less than -5. The parenthesis indicates that -5 is not included. The interval $(-\infty, -5)$ is an example of an **open interval** since the endpoints are not included. Intervals that include the endpoints, as in the example $\{x \mid 0 \leq x \leq 5\}$, are **closed intervals.** Closed intervals are indicated with square brackets. The interval $\{x \mid 0 \leq x \leq 5\}$ is written as $[0, 5]$. An interval like $(2, 5]$, that is open on one end and closed on the other, is a **half-open interval.** Examples of other sets written in interval notation are shown in the following chart. In these intervals, assume that $a < b$. Note that a parenthesis is always used with the symbols $-\infty$ or ∞.

Type of Interval	Set	Interval Notation	Graph
Open interval	$\{x \mid a < x\}$	(a, ∞)	
	$\{x \mid a < x < b\}$	(a, b)	
	$\{x \mid x < b\}$	$(-\infty, b)$	
Half-open interval	$\{x \mid a \leq x\}$	$[a, \infty)$	
	$\{x \mid a < x \leq b\}$	$(a, b]$	
	$\{x \mid a \leq x < b\}$	$[a, b)$	
	$\{x \mid x \leq b\}$	$(-\infty, b]$	
Closed interval	$\{x \mid a \leq x \leq b\}$	$[a, b]$	

We sometimes use $(-\infty, \infty)$ to represent the set of all real numbers.

EXAMPLE 2 Solve the inequality $14 + 2m \leq 3m$. Give the solution set in interval form, and graph the solution set as well.

First, subtract $2m$ from both sides.

$$14 + 2m - 2m \leq 3m - 2m \qquad \text{Subtract } 2m.$$

$$14 \leq m \qquad \text{Combine like terms.}$$

The inequality $14 \leq m$ (14 is less than or equal to m) can also be written $m \geq 14$ (m is greater than or equal to 14). Notice that in each case, the inequality symbol points to the smaller expression, 14. The solution set in interval notation is $[14, \infty)$. The graph is shown in Figure 9. ■

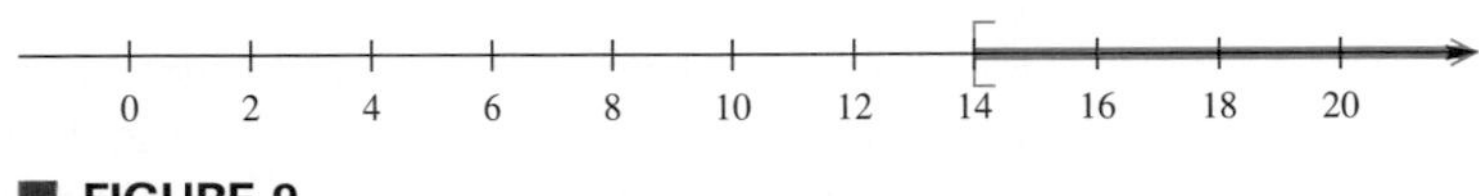

FIGURE 9

Errors often occur in graphing inequalities where the variable term is on the right side. (This is probably due to the fact that we read from left to right.) To guard against such errors, it is a good idea to rewrite these inequalities so that the variable is on the left, as shown in Example 2.

An inequality such as $3x \leq 15$ can be solved by dividing both sides by 3. This is done with the multiplication property of inequality, which is a little more involved than the corresponding property for equations. To see how this property works, start with the true statement

$$-2 < 5.$$

Multiply both sides by, say, 8.

$$-2(8) < 5(8)$$
$$-16 < 40 \quad \text{True}$$

This gives a true statement. Start again with $-2 < 5$, and this time multiply both sides by -8.

$$-2(-8) < 5(-8)$$
$$16 < -40 \quad \text{False}$$

The result, $16 < -40$, is false. To make it true, change the direction of the inequality symbol to get

$$16 > -40.$$

As these examples suggest, multiplying both sides of an inequality by a *negative* number forces the direction of the inequality symbol to be reversed. The same is true for dividing by a negative number, since division is defined in terms of multiplication.

It is a common error to forget to reverse the direction of the inequality symbol when multiplying or dividing by a negative number.

Multiplication Property of Inequality

For all real numbers a, b, and c, with $c \neq 0$,

(a) the inequalities

$$a < b \quad \text{and} \quad ac < bc$$

are equivalent if $c > 0$;

(b) the inequalities

$$a < b \quad \text{and} \quad ac > bc$$

are equivalent if $c < 0$. (Both sides of an inequality may be multiplied or divided by a *positive* number without changing the direction of the inequality symbol. Remember, multiplying or dividing by a *negative* number requires that the inequality symbol be reversed.)

EXAMPLE 3 Solve $-3(x + 4) + 2 \geq 8 - x$. Give the solution set in both interval and graphical forms.

Begin by using the distributive property on the left.

$$-3x - 12 + 2 \geq 8 - x \quad \text{Distributive property}$$
$$-3x - 10 \geq 8 - x$$

Next, the addition property is used. First add x to both sides, and then add 10 to both sides.

$$-3x - 10 + x \geq 8 - x + x \qquad \text{Add } x.$$
$$-2x - 10 \geq 8$$
$$-2x - 10 + \mathbf{10} \geq 8 + \mathbf{10} \qquad \text{Add 10.}$$
$$-2x \geq 18$$

Finally, use the multiplication property and divide both sides of the inequality by -2. Dividing by a negative number requires changing $\geq$ to $\leq$.

$$\frac{-2x}{\mathbf{-2}} \leq \frac{18}{\mathbf{-2}} \qquad \text{Divide by } -2\text{; reverse inequality symbol.}$$
$$x \leq -9$$

Figure 10 shows the graph of the solution set, $(-\infty, -9]$. ■

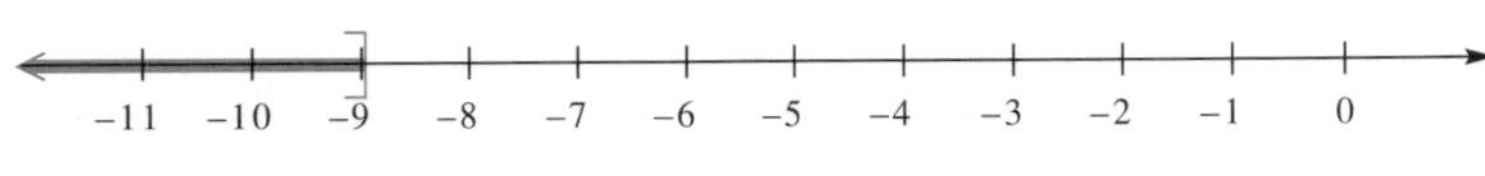

■ **FIGURE 10**

Most linear inequalities, like the one in Example 3, require the use of both the addition and multiplication properties. The steps used in solving a linear inequality are summarized below.

Solving a Linear Inequality

Step 1: Simplify each side of the inequality as much as possible by using the distributive property to clear parentheses and by combining like terms as needed.

Step 2: Use the addition property of inequality to change the inequality so that all terms with variables are on one side and all terms without variables are on the other side.

Step 3: Use the multiplication property to change the inequality to the form $x < k$ or $x > k$.

Remember: Reverse the direction of the inequality symbol **only** when multiplying or dividing both sides of an inequality by a **negative** number, and never otherwise.

Inequalities can be used to express the fact that a quantity lies between two other quantities. For example,

$$2 < 4 < 7$$

says that $2 < 4$ *and* $4 < 7$. This statement is a *conjunction* and is true because both parts of the statement are true. The three-part inequality

$$3 < x + 2 < 8$$

is true when the expression $x + 2$ is between 3 and 8. To solve this inequality, subtract 2 from each of the three parts of the inequality, giving

$$3 - 2 < x + 2 - 2 < 8 - 2$$
$$1 < x < 6.$$

The solution set, (1, 6), is graphed in Figure 11.

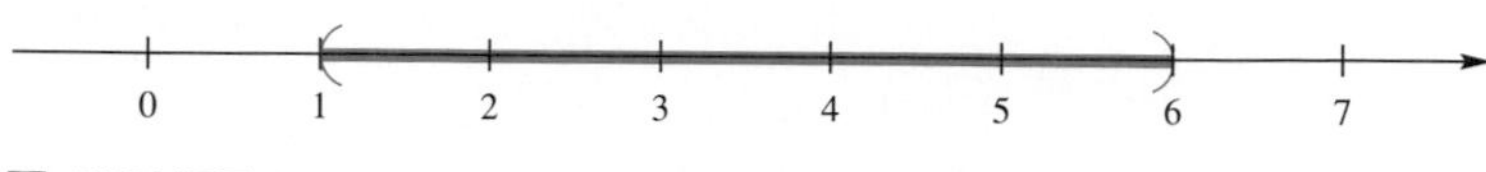

FIGURE 11

When using inequalities with three parts like the one above, it is important to have the numbers in the correct positions. If would be *wrong* to write the inequality as $8 < x + 2 < 3$, since this would imply that $8 < 3$, a false statement. In general, three-part inequalities are written so that the symbols point in the same direction, and they both point toward the smaller number.

EXAMPLE 4 Solve the inequality $-2 \leq 3k - 1 \leq 5$ and graph the solution set.

To begin, add 1 to each of the three parts.

$$-2 + 1 \leq 3k - 1 + 1 \leq 5 + 1 \qquad \text{Add 1.}$$
$$-1 \leq 3k \leq 6$$

Now divide each of the three parts by the positive number 3.

$$\frac{-1}{3} \leq \frac{3k}{3} \leq \frac{6}{3} \qquad \text{Divide by 3.}$$
$$-\frac{1}{3} \leq k \leq 2$$

A graph of the solution set, $[-1/3, 2]$, is shown in Figure 12. ■

FIGURE 12

There are several phrases that denote inequality. In addition to the familiar "is less than" and "is greater than" (which are examples of **strict** inequalities), the expressions "is no more than," "is at least," and others also denote inequalities. (These are called **nonstrict** inequalities.)

Problem Solving

Expressions for nonstrict inequalities sometimes appear in applied problems that are solved using inequalities. The chart below shows how these expressions are interpreted.

Word Expression	Interpretation	Word Expression	Interpretation
a is at least b	$a \geq b$	a is at most b	$a \leq b$
a is no less than b	$a \geq b$	a is no more than b	$a \leq b$

Do not confuse statements like "5 is more than a number" with the phrase "5 more than a number." The first of these is expressed as "$5 > x$" while the second is expressed with addition, as "$x + 5$."

Problem Solving

The next example shows an application of algebra that is important to anyone who has ever asked himself or herself "What score can I make on my next test and have a (particular grade) in this course?" It uses the idea of finding the average of a number of grades. In general, to find the average of n numbers, add the numbers, and divide by n.

EXAMPLE 5 Bill Bryden has test grades of 86, 88, and 78 on his first three tests in geometry. If he wants an average of at least 80 after his fourth test, what are the possible scores he can make on his fourth test?

Let x = Bill's score on his fourth test. To find his average after 4 tests, add the test scores and divide by 4.

Average — is at least — 80.

$$\frac{86 + 88 + 78 + x}{4} \geq 80$$

$$\frac{252 + x}{4} \geq 80 \qquad \text{Add the known scores.}$$

$$4\left(\frac{252 + x}{4}\right) \geq 4(80) \qquad \text{Multiply by 4.}$$

$$252 + x \geq 320$$

$$252 - \mathbf{252} + x \geq 320 - \mathbf{252} \qquad \text{Subtract 252.}$$

$$x \geq 68 \qquad \text{Combine terms.}$$

He must score 68 or more on the fourth test to have an average of *at least* 80. ■

EXAMPLE 6 A rental company charges \$15.00 to rent a chain saw, plus \$2.00 per hour. Maureen Kavanagh can spend no more than \$35.00 to clear some logs from her yard. What is the maximum amount of time she can use the rented saw?

Let h = the number of hours she can rent the saw. She must pay \$15.00, plus \$2.00h, to rent the saw for h hours, and this amount must be *no more than* \$35.00.

Cost of renting — is no more than — 35 dollars.

$$15 + 2h \leq 35$$

$$15 + 2h - \mathbf{15} \leq 35 - \mathbf{15} \qquad \text{Subtract 15.}$$

$$2h \leq 20$$

$$h \leq 10 \qquad \text{Divide by 2.}$$

She can use the saw for a maximum of 10 hours. (Of course, she may use it for less time, as indicated by the inequality $h \leq 10$.) ■

7.4 EXERCISES

Write each of the following sets using interval notation.

1. $\{x \mid x < 3\}$
2. $\{x \mid x < -2\}$
3. $\{y \mid y \leq 8\}$
4. $\{t \mid t \leq 1\}$
5. $\{r \mid r > 8\}$
6. $\{s \mid s > 4\}$
7. $\{w \mid w \geq 2\}$
8. $\{m \mid m \geq 0\}$
9. $\{x \mid -3 \leq x \leq 4\}$
10. $\{x \mid -2 < x \leq 0\}$
11. $\{x \mid 2 < x \leq 9\}$
12. $\{t \mid -1 < t < 1\}$

Using the variable x, write each of the following intervals as an inequality with set-builder notation.

13. $(-\infty, 4)$
14. $(-\infty, 0)$
15. $[1.5, \infty)$
16. $[3.2, \infty)$
17. $[-3, 10)$
18. $(-2, 9]$

19.

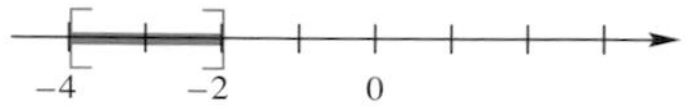

20.

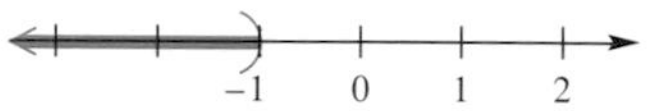

21. Match each set given in interval notation with its description.

(a) $(0, \infty)$ **A.** positive real numbers
(b) $[0, \infty)$ **B.** negative real numbers
(c) $(-\infty, 0]$ **C.** nonpositive real numbers
(d) $(-\infty, 0)$ **D.** nonnegative real numbers

22. Explain how to determine whether a parenthesis or a square bracket is used when graphing the solution set of a linear inequality.

Solve each inequality. Give the solution set in both interval and graph forms.

23. $4x > 8$
24. $6y > 18$
25. $2m \leq -6$
26. $5k \leq -10$
27. $3r + 1 \geq 16$
28. $2m - 5 \geq 15$
29. $-r \leq -7$
30. $-m > -12$
31. $-4x + 3 < 15$
32. $-6p - 2 \geq 16$
33. $-3(z - 6) > 2z - 5$
34. $-2(y + 4) \leq 6y + 8$
35. $-2(m - 4) \leq -3(m + 1)$
36. $-(9 + k) - 5 + 4k \geq 1$
37. $-3 < x - 5 < 6$
38. $-6 < x + 1 < 8$
39. $-19 \leq 3x - 5 \leq -9$
40. $-16 < 3t + 2 < -11$
41. $-4 \leq \dfrac{2x - 5}{6} \leq 5$
42. $-8 \leq \dfrac{3m + 1}{4} \leq 3$

Answer the questions in Exercises 43–46 based on the given graph.

43. In which months did the percent of tornadoes exceed 7.7%?

44. In which months was the percent of tornadoes at least 12.9%?

45. The data used to determine the graph was based on the number of tornadoes sighted in the United States during the last twenty years. A total of 17,252 tornadoes were reported. In which months were fewer than 1,500 reported?

46. How many more tornadoes occurred during March than October? (Use the total given in Exercise 45.)

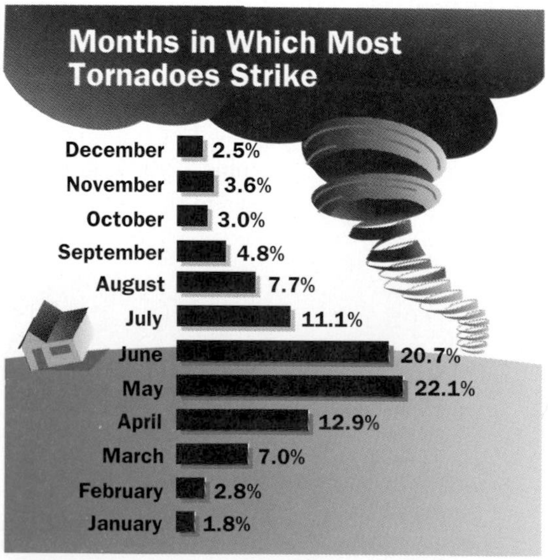

Source: The USA Today Weather Book

Solve the problem.

47. According to information provided by the Taxi and Limousine Commission, in January 1990, taxicabs in New York charged \$1.50 for the first 1/5 mile and \$.25 for each additional 1/5 mile. Amos has only \$3.75 in his pocket. What is the maximum distance he can travel (not including a tip for the cabbie)?

48. In July 1979, New York taxicab fares were \$.90 for the first 1/7 mile and \$.10 for each additional 1/7 mile. Based on the information given in Exercise 47 and the answer you found, how much farther could Amos have traveled in 1979 than in 1990?

49. Clare Lynch earned scores of 90 and 82 on her first two tests in English Literature. What score must she make on her third test to keep an average of 84 or greater?

50. Jay O'Callaghan scored 92 and 96 on his first two tests in Methods in Teaching Mathematics. What score must he make on his third test to keep an average of 90 or greater?

51. A couple wishes to rent a car for one day while on vacation. Ford Automobile Rental wants \$15.00 per day and 14¢ per mile, while Chevrolet-For-A-Day wants \$14.00 per day and 16¢ per mile. After how many miles would the price to rent the Chevrolet exceed the price to rent a Ford?

52. Jane and Terry Brandsma went to Long Island for a week. They needed to rent a car, so they checked out two rental firms. Avis wanted \$28 per day, with no mileage fee. Downtown Toyota wanted \$108 per week and 14¢ per mile. How many miles would they have to drive before the Avis price is less than the Toyota price?

A product will produce a profit only when the revenue R from selling the product exceeds the cost C of producing it. In Exercises 53 and 54 find the smallest whole number of units x that must be sold for the business to show a profit for the item described.

53. Peripheral Visions, Inc. finds that the cost to produce x studio quality videotapes is $C = 20x + 100$, while the revenue produced from them is $R = 24x$ (C and R in dollars).

54. Speedy Delivery finds that the cost to make x deliveries is $C = 3x + 2{,}300$, while the revenue produced from them is $R = 5.50x$ (C and R in dollars).

55. Write an explanation of what is wrong with the following argument.

Let a and b be numbers, with $a > b$. Certainly, $2 > 1$. Multiply both sides of the inequality by $b - a$.

$$2(b - a) > 1(b - a)$$
$$2b - 2a > b - a$$
$$2b - b > 2a - a$$
$$b > a$$

But the final inequality is impossible, since we know that $a > b$ from the given information.

56. Assume that $0 < a < b$, and go through the following steps.

$$a < b$$
$$ab < b^2$$
$$ab - b^2 < 0$$
$$b(a - b) < 0$$

Divide both sides by $a - b$ to get $b < 0$. This implies that b is negative. We originally assumed that b is positive. What is wrong with this argument?

Properties of Exponents and Scientific Notation

Exponents are used to write products of repeated factors. For example, the product $3 \cdot 3 \cdot 3 \cdot 3$ is written

$$3 \cdot 3 \cdot 3 \cdot 3 = 3^4.$$

The number 4 shows that 3 appears as a factor four times. The number 4 is the *exponent* and 3 is the *base*. The quantity 3^4 is called an *exponential expression.* Read 3^4 as "3 to the fourth power," or "3 to the fourth." Multiplying out the four 3s gives

$$3^4 = 3 \cdot 3 \cdot 3 \cdot 3 = 81.$$

$10^{100000^{0^{\cdot^{\cdot^{\cdot^{0}}}}}}$

The term **googol,** meaning 10^{100}, was coined by Professor Edward Kasner of Columbia University. A googol is made up of a 1 with one hundred zeros following it. This number exceeds the estimated number of electrons in the universe, which is 10^{79}.

If a googol isn't big enough for you, try a **googolplex:**

$$\text{googolplex} = 10^{\text{googol}}.$$

Exponents

If a is a real number and n is a natural number,

$$a^n = \underbrace{a \cdot a \cdot a \ldots a}_{n \text{ factors of } a}.$$

EXAMPLE 1 Evaluate each exponential expression.

(a) $7^2 = 7 \cdot 7 = 49$ Read 7^2 as "7 squared."

(b) $5^3 = 5 \cdot 5 \cdot 5 = 125$ Read 5^3 as "5 cubed."

(c) $(-2)^4 = (-2)(-2)(-2)(-2) = 16$

(d) $(-2)^5 = (-2)(-2)(-2)(-2)(-2) = -32$

(e) $5^1 = 5$

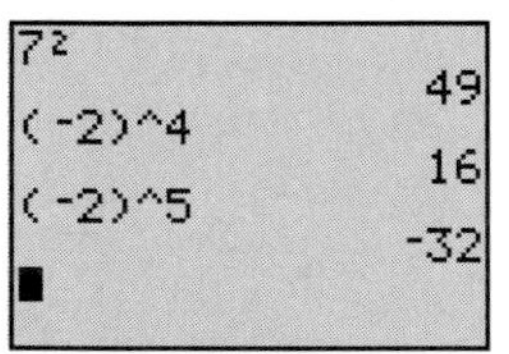

This screen supports the results in parts (a), (c), and (d) of Example 1.

In the exponential expression $3z^7$, the base of the exponent 7 is z, and *not* $3z$. That is,

$$3z^7 = 3 \cdot z \cdot z \cdot z \cdot z \cdot z \cdot z \cdot z$$

while

$$(3z)^7 = (3z)(3z)(3z)(3z)(3z)(3z)(3z).$$

Suppose that we wish to evaluate $(-2)^6$. The parentheses around -2 indicate that the base is -2, and so

$$(-2)^6 = (-2)(-2)(-2)(-2)(-2)(-2) = 64.$$

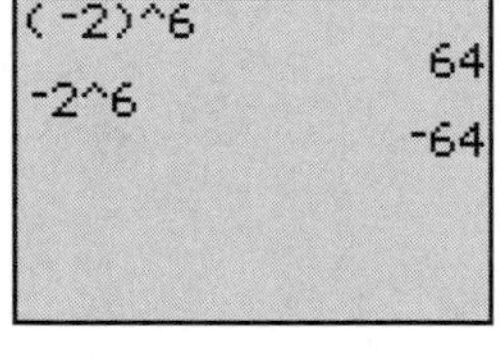

Now, let us consider the exponential expression -2^6. In this expression, the base is 2, and *not* -2. The $-$ sign tells us to find the negative, or additive inverse, of 2^6. It acts as a symbol for the factor -1. We evaluate -2^6 as follows.

$$-2^6 = -(2 \cdot 2 \cdot 2 \cdot 2 \cdot 2 \cdot 2) = -64$$

Therefore,

$$(-2)^6 \neq -2^6,$$

since $64 \neq -64$. (The calculator screen supports this).

-4²
-16
-8^4
-4096
-2^4
-16

This screen supports the results in Example 2.

EXAMPLE 2 Evaluate each exponential expression.

(a) $-4^2 = -(4 \cdot 4) = -16$

(b) $-8^4 = -(8 \cdot 8 \cdot 8 \cdot 8) = -4{,}096$

(c) $-2^4 = -(2 \cdot 2 \cdot 2 \cdot 2) = -16$

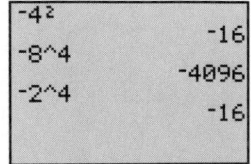

There are several useful rules that simplify work with exponents. For example, the product $2^5 \cdot 2^3$ can be simplified as follows.

$5 + 3 = 8$

$\cdot 2 \cdot 2 \cdot 2)(2 \cdot 2 \cdot 2) = 2^8$

itial expressions with the same base are found lized as the **product rule for exponents.**

ents

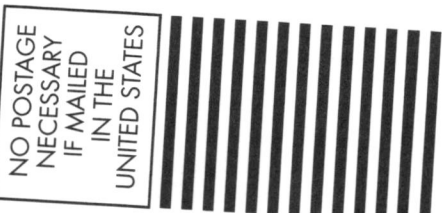

rs and a is any real number, then

$^m \cdot a^n = a^{m+n}.$

ct rule for exponents in each case.

5^4

3

mmutative properties as necessary to multiply the ables.

$)(-3y^4) = 5(-3)y^2y^4$

$= -15y^{2+4}$

$= -15y^6$

$^5qq^2 = 14p^8q^3$ ■

^{3}, can be simplified in much the same way as a his type, assume that the denominator is not zero.) ponent,

$$= \frac{a \cdot a \cdot a \cdot a \cdot a \cdot a \cdot a \cdot a}{a \cdot a \cdot a}$$

$$= a \cdot a \cdot a \cdot a \cdot a = a^5.$$

the same way,

$$= \frac{a \cdot a \cdot a}{a \cdot a \cdot a \cdot a \cdot a \cdot a \cdot a \cdot a}$$

$$= \frac{1}{a^5}.$$

Here again, $8 - 3 = 5$. In the first example, a^8/a^3, the exponent in the denominator was subtracted from the one in the numerator; the reverse was true in the second example, a^3/a^8. The order of subtracting the exponents depends on the larger exponent. In the second example, subtracting $3 - 8$ gives an exponent of -5. To simplify the rule for quotients, we first define a **negative exponent.**

Negative Exponent

For any natural number n and any nonzero real number a,

$$a^{-n} = \frac{1}{a^n}.$$

With this definition, the expression a^n is meaningful for any nonzero integer exponent n and any nonzero real number a.

```
2^-3▸Frac
            1/8
3^-2▸Frac
            1/9
6-1▸Frac
            1/6
```

This screen supports the results in parts (a)–(c) of Example 4.

EXAMPLE 4 Write the following expressions with only positive exponents.

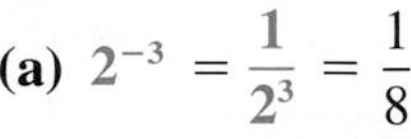

(a) $2^{-3} = \frac{1}{2^3} = \frac{1}{8}$ **(b)** $3^{-2} = \frac{1}{3^2} = \frac{1}{9}$

(c) $6^{-1} = \frac{1}{6^1} = \frac{1}{6}$ **(d)** $(5z)^{-3} = \frac{1}{(5z)^3}, \quad z \neq 0$

(e) $5z^{-3} = 5\left(\frac{1}{z^3}\right) = \frac{5}{z^3}, \quad z \neq 0$ **(f)** $(5z^2)^{-3} = \frac{1}{(5z^2)^3}, \quad z \neq 0$

(g) $-m^{-2} = -\frac{1}{m^2}, \quad m \neq 0$ **(h)** $(-m)^{-4} = \frac{1}{(-m)^4}, \quad m \neq 0$

```
3-1+4-1▸Frac
           7/12
5-1-2-1▸Frac
          -3/10
(2^-3)/(3^-2)▸Fr
ac
            9/8
```

This screen supports the results in parts (a), (b), and (e) of Example 5.

EXAMPLE 5 Evaluate each of the following expressions.

(a) $3^{-1} + 4^{-1}$

Since $3^{-1} = \frac{1}{3}$ and $4^{-1} = \frac{1}{4}$,

$$3^{-1} + 4^{-1} = \frac{1}{3} + \frac{1}{4} = \frac{4}{12} + \frac{3}{12} = \frac{7}{12}.$$

(b) $5^{-1} - 2^{-1} = \frac{1}{5} - \frac{1}{2} = \frac{2}{10} - \frac{5}{10} = -\frac{3}{10}$

(c) $\frac{1}{2^{-3}} = \frac{1}{\frac{1}{2^3}} = 1 \div \frac{1}{2^3} = 1 \cdot \frac{2^3}{1} = 2^3 = 8$

(d) $\frac{1}{3^{-2}} = \frac{1}{\frac{1}{3^2}} = 3^2 = 9$

(e) $\frac{2^{-3}}{3^{-2}} = \frac{\frac{1}{2^3}}{\frac{1}{3^2}} = \frac{1}{2^3} \cdot \frac{3^2}{1} = \frac{3^2}{2^3} = \frac{9}{8}$

Parts (c), (d), and (e) of Example 5 suggest the following generalizations.

Special Rules for Negative Exponents

If $a \neq 0$ and $b \neq 0$, $\dfrac{1}{a^{-n}} = a^n$ and $\dfrac{a^{-n}}{b^{-m}} = \dfrac{b^m}{a^n}$.

When multiplying expressions such as a^m and a^n where the base is the same, *add* the exponents; when dividing, *subtract* the exponents, as stated in the following **quotient rule for exponents.**

Quotient Rule for Exponents

If a is any nonzero real number and m and n are nonzero integers, then

$$\frac{a^m}{a^n} = a^{m-n}.$$

EXAMPLE 6 Apply the quotient rule for exponents in each case.

(a) $\dfrac{3^7}{3^2} = 3^{7-2} = 3^5$

Numerator exponent; Denominator exponent; Minus sign

(b) $\dfrac{p^6}{p^2} = p^{6-2} = p^4 \quad (p \neq 0)$

(c) $\dfrac{12^{10}}{12^9} = 12^{10-9} = 12^1 = 12$

(d) $\dfrac{7^4}{7^6} = 7^{4-6} = 7^{-2} = \dfrac{1}{7^2}$

(e) $\dfrac{k^7}{k^{12}} = k^{7-12} = k^{-5} = \dfrac{1}{k^5} \quad (k \neq 0)$

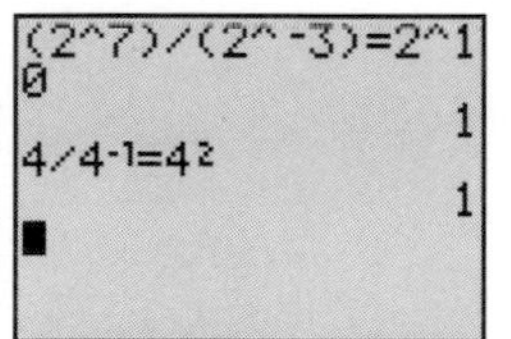

This screen supports the results in parts (a) and (d) of Example 7.

EXAMPLE 7 Write each quotient using only positive exponents.

(a) $\dfrac{2^7}{2^{-3}} = 2^{7-(-3)}$

Since $7 - (-3) = 10$, $\dfrac{2^7}{2^{-3}} = 2^{10}$.

(b) $\dfrac{8^{-2}}{8^5} = 8^{-2-5} = 8^{-7} = \dfrac{1}{8^7}$

(c) $\dfrac{6^{-5}}{6^{-2}} = 6^{-5-(-2)} = 6^{-3} = \dfrac{1}{6^3}$

(d) $\dfrac{4}{4^{-1}} = \dfrac{4^1}{4^{-1}} = 4^{1-(-1)} = 4^2$

(e) $\dfrac{z^{-5}}{z^{-8}} = z^{-5-(-8)} = z^3 \quad (z \neq 0)$

By the quotient rule, $a^m/a^n = a^{m-n}$. Suppose that the exponents in the numerator and the denominator are the same. Then, for example, $8^4/8^4 = 8^{4-4} = 8^0$. Since any nonzero number divided by itself is 1, $8^4/8^4 = 1$. Therefore

$$1 = \frac{8^4}{8^4} = 8^0.$$

Based on this argument, for consistency with past rules of exponents, the expression a^0 is defined to equal 1 for any nonzero number a.

Zero Exponent

If a is any nonzero real number, then

$$a^0 = 1.$$

The symbol 0^0 is undefined.

```
(-6)^0
                1
-6^0
               -1
5^0+12^0
                2
```

This screen supports the results in parts (b), (c), and (d) of Example 8.

EXAMPLE 8 Evaluate each expression.

(a) $12^0 = 1$ **(b)** $(-6)^0 = 1$

(c) $-6^0 = -(6^0) = -1$ **(d)** $5^0 + 12^0 = 1 + 1 = 2$

(e) $(8k)^0 = 1 \quad (k \neq 0)$

The expression $(3^4)^2$ can be simplified as $(3^4)^2 = 3^4 \cdot 3^4 = 3^{4+4} = 3^8$, where $4 \cdot 2 = 8$. This example suggests the first of the **power rules for exponents;** the other two parts can be demonstrated with similar examples.

Power Rules for Exponents

If a and b are real numbers, and m and n are integers, then

$$(a^m)^n = a^{mn}$$
$$(ab)^m = a^m b^m$$
$$\left(\frac{a}{b}\right)^m = \frac{a^m}{b^m} \quad (b \neq 0).$$

In the statements of rules for exponents, we always assume that zero never appears to a negative power.

EXAMPLE 9 Use a power rule in each case.

(a) $(p^8)^3 = p^{8 \cdot 3} = p^{24}$ **(b)** $\left(\frac{2}{3}\right)^4 = \frac{2^4}{3^4} = \frac{16}{81}$

(c) $(3y)^4 = 3^4y^4 = 81y^4$ **(d)** $(6p^7)^2 = 6^2p^{7 \cdot 2} = 6^2p^{14} = 36p^{14}$

(e) $\left(\frac{-2m^5}{z}\right)^3 = \frac{(-2)^3m^{5 \cdot 3}}{z^3} = \frac{(-2)^3m^{15}}{z^3} = \frac{-8m^{15}}{z^3} \quad (z \neq 0)$

Notice that

$$6^{-3} = \left(\frac{1}{6}\right)^3 = \frac{1}{216}$$

and

$$\left(\frac{2}{3}\right)^{-2} = \left(\frac{3}{2}\right)^2 = \frac{9}{4}.$$

These are examples of two special rules for negative exponents that may be applied when working with fractions.

Special Rules for Negative Exponents

Any nonzero number raised to the negative nth power is equal to the reciprocal of that number raised to the nth power. That is, if $a \neq 0$ and $b \neq 0$,

$$a^{-n} = \left(\frac{1}{a}\right)^n \quad \text{and} \quad \left(\frac{a}{b}\right)^{-n} = \left(\frac{b}{a}\right)^n.$$

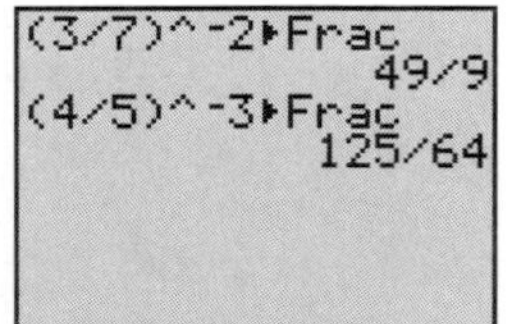

This screen supports the results of Example 10.

EXAMPLE 10 Write the following expressions with only positive exponents and then evaluate.

(a) $\left(\frac{3}{7}\right)^{-2} = \left(\frac{7}{3}\right)^2 = \frac{49}{9}$ **(b)** $\left(\frac{4}{5}\right)^{-3} = \left(\frac{5}{4}\right)^3 = \frac{125}{64}$

The next example shows how the definitions and rules for exponents are used to write expressions in equivalent forms.

EXAMPLE 11 Simplify each expression so that no negative exponents appear in the final result.

(a) $3^2 \cdot 3^{-5} = 3^{2+(-5)} = 3^{-3} = \frac{1}{3^3}$ or $\frac{1}{27}$

(b) $x^{-3} \cdot x^{-4} \cdot x^2 = x^{-3+(-4)+2} = x^{-5} = \frac{1}{x^5}$ $(x \neq 0)$

(c) $(2^5)^{-3} = 2^{5(-3)} = 2^{-15} = \frac{1}{2^{15}}$

(d) $(4^{-2})^{-5} = 4^{(-2)(-5)} = 4^{10}$

(e) $(x^{-4})^6 = x^{(-4)6} = x^{-24} = \frac{1}{x^{24}}$ $(x \neq 0)$

$$\textbf{(f)}\quad \frac{x^{-4}y^2}{x^2y^{-5}} = \frac{x^{-4}}{x^2} \cdot \frac{y^2}{y^{-5}}$$
$$= x^{-4-2} \cdot y^{2-(-5)}$$
$$= x^{-6}y^7$$
$$= \frac{y^7}{x^6} \quad (x, y \neq 0)$$

$$\textbf{(g)}\quad (2^3x^{-2})^{-2} = (2^3)^{-2} \cdot (x^{-2})^{-2}$$
$$= 2^{-6}x^4$$
$$= \frac{x^4}{2^6} \quad \text{or} \quad \frac{x^4}{64}$$ ■

The idea of exponent has now been expanded to include *all* integers, positive, negative, or zero. This has been done in such a way that all past rules for exponents are still valid. These definitions and rules are summarized below.

Definitions and Rules for Exponents

For all integers m and n and all real numbers a and b:

Product Rule	$a^m \cdot a^n = a^{m+n}$
Quotient Rule	$\frac{a^m}{a^n} = a^{m-n} \quad (a \neq 0)$
Zero Exponent	$a^0 = 1 \quad (a \neq 0)$
Negative Exponent	$a^{-n} = \frac{1}{a^n} \quad (a \neq 0)$
Power Rules	$(a^m)^n = a^{mn}$
	$(ab)^m = a^m b^m$
	$\left(\frac{a}{b}\right)^m = \frac{a^m}{b^m} \quad (b \neq 0)$
	$a^{-n} = \left(\frac{1}{a}\right)^n \quad (a \neq 0)$
	$\left(\frac{a}{b}\right)^{-n} = \left(\frac{b}{a}\right)^n \quad (a, b \neq 0).$

Scientific Notation

Many of the numbers that occur in science are very large, such as the number of one-celled organisms that will sustain a whale for a few hours: 400,000,000,000,000. Other numbers are very small, such as the shortest wavelength of visible light, about .0000004 meter. Writing these numbers is simplified by using *scientific notation.*

Scientific Notation

A number is written in **scientific notation** when it is expressed in the form

$$a \times 10^n,$$

where $1 \leq |a| < 10$, and n is an integer.

As stated in the definition, scientific notation requires that the number be written as a product of a number between 1 and 10 (or -1 and -10) and some integer power of 10. (1 and -1 are allowed as values of a.) For example, since

$$8{,}000 = 8 \cdot 1{,}000 = 8 \cdot 10^3,$$

the number 8,000 is written in scientific notation as

$$8{,}000 = 8 \times 10^3.$$

(When using scientific notation, it is customary to use × instead of a dot to show multiplication.)

The steps involved in writing a number in scientific notation are given below. (If the number is negative, ignore the minus sign, go through these steps, and then attach a minus sign to the result.)

Converting to Scientific Notation

Step 1: Place a caret, ‸, to the right of the first nonzero digit.

Step 2: Count the number of digits from the caret to the decimal point. This number gives the absolute value of the exponent on ten.

Step 3: Decide whether multiplying by 10^n should make the number larger or smaller. The exponent should be positive to make the product larger; it should be negative to make the product smaller.

```
820000
            8.2E5
.000072
           7.2E-5
```

If a graphing calculator is set in scientific notation mode, it will give results as shown here. E5 means "times 10^5" and E-5 means "times 10^{-5}". Compare to the results of Example 12.

EXAMPLE 12 Write each number in scientific notation.

(a) 820,000

Place a caret to the right of the 8 (the first nonzero digit).

8‸20,000

Count from the caret to the decimal point, which is understood to be after the last 0.

8‸20,000. ← Decimal point

Count 5 places.

Since the number 8.2 is to be made larger, the exponent on 10 is positive.

$$820{,}000 = 8.2 \times 10^5$$

(b) .000072

Count from right to left.

$$.00007\underbrace{}_{}2$$

.00007 2
5 places

Since the number 7.2 is to be made smaller, the exponent on 10 is negative.

$$.000072 = 7.2 \times 10^{-5}$$

To convert a number written in scientific notation to standard notation, just work in reverse.

Converting from Scientific Notation

Multiplying a number by a positive power of 10 makes the number larger, so move the decimal point to the right if n is positive in 10^n. Multiplying by a negative power of 10 makes a number smaller, so move the decimal point to the left if n is negative. If n is zero, leave the decimal point where it is.

EXAMPLE 13 Write the following numbers without scientific notation.

(a) 6.93×10^5

6.93000
5 places

The decimal point was moved 5 places to the right. (It was necessary to attach 3 zeros.)

$$6.93 \times 10^5 = 693{,}000$$

(b) $3.52 \times 10^7 = 35{,}200{,}000$

(c) 4.7×10^{-6}

000004.7
6 places

The decimal point was moved 6 places to the left.

$$4.7 \times 10^{-6} = .0000047$$

(d) $1.083 \times 10^0 = 1.083$

The next example shows how scientific notation and the rules for exponents can be used to simplify calculations.

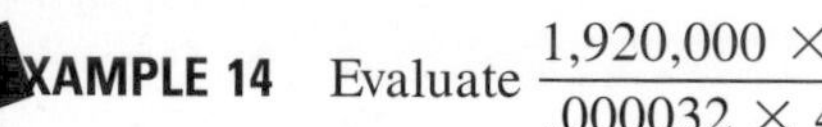

EXAMPLE 14 Evaluate $\dfrac{1{,}920{,}000 \times .0015}{.000032 \times 45{,}000}$ by using scientific notation.

First, express all numbers in scientific notation.

$$\frac{1{,}920{,}000 \times .0015}{.000032 \times 45{,}000} = \frac{1.92 \times 10^6 \times 1.5 \times 10^{-3}}{3.2 \times 10^{-5} \times 4.5 \times 10^4}$$

Next, use the commutative and associative properties and the rules for exponents to simplify the expression.

$$\begin{aligned}\frac{1{,}920{,}000 \times .0015}{.000032 \times 45{,}000} &= \frac{1.92 \times 1.5 \times 10^6 \times 10^{-3}}{3.2 \times 4.5 \times 10^{-5} \times 10^4}\\ &= \frac{1.92 \times 1.5}{3.2 \times 4.5} \times 10^4\\ &= .2 \times 10^4\\ &= (2 \times 10^{-1}) \times 10^4\\ &= 2 \times 10^3\end{aligned}$$

The expression is equal to 2×10^3, or 2,000. ■

Problem Solving

When problems require operations with numbers that are very large and/or very small, it is often advantageous to write the numbers in scientific notation first, and then perform the calculations using the rules for exponents. The next example illustrates this.

EXAMPLE 15 A certain computer can execute an algorithm in .00000000036 second. How long would it take the computer to execute one billion of these algorithms? (One billion = 1,000,000,000)

In order to solve this problem, we must multiply the time per algorithm by the number of algorithms.

$$.00000000036 \times 1{,}000{,}000{,}000 = \text{total time in seconds}$$

Write each number in scientific notation, and then use the rules for exponents.

$$\begin{aligned}&.00000000036 \times 1{,}000{,}000{,}000\\ &= (3.6 \times 10^{-10}) \times (1 \times 10^9)\\ &= (3.6 \times 1)(10^{-10} \times 10^9) && \text{Commutative and associative properties}\\ &= 3.6 \times 10^{-1} && \text{Product rule}\\ &= .36 && \text{Convert to standard notation.}\end{aligned}$$

It would take the computer .36 second. ■

7.5 EXERCISES

Evaluate each exponential expression.

1. 5^4 **2.** 10^3 **3.** $\left(\frac{5}{3}\right)^2$ **4.** $\left(\frac{3}{5}\right)^3$ **5.** $(-2)^5$

6. $(-5)^4$ **7.** -2^3 **8.** -3^2 **9.** $-(-3)^4$ **10.** $-(-5)^2$

11. Do $(-a)^n$ and $-a^n$ mean the same thing? Explain.

12. In *some* cases, $-a^n$ and $(-a)^n$ do give the same result. Using $a = 2$, and $n = 2, 3, 4$, and 5, draw a conclusion as to when they are equal and when they are opposites.

13. Which one of the following is equal to 1 ($a \neq 0$)?
(a) $3a^0$ **(b)** $-3a^0$ **(c)** $(3a)^0$ **(d)** $3(-a)^0$

14. Which one of the following represents a negative number?
(a) $(-3)^{-2}$ **(b)** $(-1{,}000)^0$ **(c)** $(-4)^0 - (-3)^0$ **(d)** $(-5)^{-3}$

Evaluate each exponential expression.

15. 7^{-2} **16.** 4^{-1} **17.** -7^{-2} **18.** -4^{-1} **19.** $\frac{2}{(-4)^{-3}}$

20. $\frac{2^{-3}}{3^{-2}}$ **21.** $\frac{5^{-1}}{4^{-2}}$ **22.** $\left(\frac{1}{2}\right)^{-3}$ **23.** $\left(\frac{1}{5}\right)^{-3}$ **24.** $\left(\frac{2}{3}\right)^{-2}$

25. $\left(\frac{4}{5}\right)^{-2}$ **26.** $3^{-1} + 2^{-1}$ **27.** $4^{-1} + 5^{-1}$ **28.** 8^0 **29.** 12^0

30. $(-23)^0$ **31.** $(-4)^0$ **32.** -2^0 **33.** $3^0 - 4^0$ **34.** $-8^0 - 7^0$

35. In order to raise a fraction to a negative power, we may change the fraction to its _____ and change the exponent to the _____ _____ of the original exponent.

36. Explain in your own words how we raise a power to a power.

37. Which one of the following is correct?
(a) $-\frac{3}{4} = \left(\frac{3}{4}\right)^{-1}$ **(b)** $\frac{3^{-1}}{4^{-1}} = \left(\frac{4}{3}\right)^{-1}$ **(c)** $\frac{3^{-1}}{4} = \frac{3}{4^{-1}}$ **(d)** $\frac{3^{-1}}{4^{-1}} = \left(\frac{3}{4}\right)^{-1}$

38. Which one of the following is incorrect?
(a) $(3r)^{-2} = 3^{-2}r^{-2}$ **(b)** $3r^{-2} = (3r)^{-2}$ **(c)** $(3r)^{-2} = \frac{1}{(3r)^2}$ **(d)** $(3r)^{-2} = \frac{r^{-2}}{9}$

Use the product, quotient, and power rules to simplify each expression. Write answers with only positive exponents. Assume that all variables represent nonzero real numbers.

39. $x^{12} \cdot x^4$ **40.** $\frac{x^{12}}{x^4}$ **41.** $\frac{5^{17}}{5^{16}}$ **42.** $\frac{3^{12}}{3^{13}}$

43. $\frac{3^{-5}}{3^{-2}}$ **44.** $\frac{2^{-4}}{2^{-3}}$ **45.** $\frac{9^{-1}}{9}$ **46.** $\frac{12}{12^{-1}}$

47. t^5t^{-12} **48.** p^5p^{-6} **49.** $(3x)^2$ **50.** $(-2x^{-2})^2$

51. $a^{-3}a^2a^{-4}$ **52.** $k^{-5}k^{-3}k^4$ **53.** $\frac{x^7}{x^{-4}}$ **54.** $\frac{p^{-3}}{p^5}$

55. $\frac{r^3r^{-4}}{r^{-2}r^{-5}}$ **56.** $\frac{z^{-4}z^{-2}}{z^3z^{-1}}$ **57.** $7k^2(-2k)(4k^{-5})$ **58.** $3a^2(-5a^{-6})(-2a)$

59. $(z^3)^{-2}z^2$ **60.** $(p^{-1})^3p^{-4}$ **61.** $-3r^{-1}(r^{-3})^2$ **62.** $2(y^{-3})^4(y^6)$

63. $(3a^{-2})^3(a^3)^{-4}$ **64.** $(m^5)^{-2}(3m^{-2})^3$ **65.** $(x^{-5}y^2)^{-1}$ **66.** $(a^{-3}b^{-5})^2$

67. Which one of the following does *not* represent the reciprocal of x ($x \neq 0$)?
(a) x^{-1} **(b)** $\frac{1}{x}$ **(c)** $\left(\frac{1}{x^{-1}}\right)^{-1}$ **(d)** $-x$

68. Which one of the following is *not* in scientific notation?
(a) 6.02×10^{23} **(b)** 14×10^{-6} **(c)** 1.4×10^{-5} **(d)** 3.8×10^3

Write each number in scientific notation.

69. 230 **70.** 46,500 **71.** .02 **72.** .0051

Write each number without scientific notation.

73. 6.5×10^3 **74.** 2.317×10^5 **75.** 1.52×10^{-2} **76.** 1.63×10^{-4}

Use scientific notation to perform each of the following computations. Leave the answers in scientific notation.

77. $\frac{.002 \times 3{,}900}{.000013}$ **78.** $\frac{.009 \times 600}{.02}$ **79.** $\frac{.0004 \times 56{,}000}{.000112}$

80. $\frac{.018 \times 20{,}000}{300 \times .0004}$ **81.** $\frac{840{,}000 \times .03}{.00021 \times 600}$ **82.** $\frac{28 \times .0045}{140 \times 1{,}500}$

If the number in the statement is written in scientific notation, write it without exponents. If it is written without exponents, write it in scientific notation.

83. The number of possible hands in contract bridge is about 6.35×10^{11}.

84. If there are 40 numbers to choose from in a lottery, and a player must choose 6 different ones, the player has about 3.84×10^6 ways to make a choice.

85. In a recent year, ESPN's regular season baseball telecasts averaged slightly fewer than 1,150,000 viewers.

86. In a recent year, ESPN's estimated losses were at least 36,000,000 dollars.

87. The body of a 150-pound person contains about 2.3×10^{-4} pound of copper and about 6×10^{-3} pound of iron.

88. The mean distance from Venus to the sun is about 6.7×10^6 miles.

The quote is taken from the source cited. Write the number given in scientific notation in the quote without exponents.

89. The muon, a close relative of the electron produced by the bombardment of cosmic rays against the upper atmosphere, has a half-life of 2 millionths of a second (2×10^{-6}s). (Excerpt from *Conceptual Physics,* 6th Edition by Paul G. Hewitt. Copyright © by Paul G. Hewitt. Published by HarperCollins College Publishers.)

90. There are 13 red balls and 39 black balls in a box. Mix them up and draw 13 out one at a time without returning any ball . . . the probability that the 13 drawings each will produce a red ball is . . . 1.6×10^{-12}. (Warren Weaver, *Lady Luck,* New York: Doubleday & Company, Inc., 1963, pp. 298–299.)

The quote is taken from the source cited. Write the given number(s) in scientific notation.

91. An electron and a positron attract each other in two ways: the electromagnetic attraction of their opposite electric charges, and the gravitational attraction of their two masses. The electromagnetic attraction is 4,200,000,000,000,000,000,000,000,000,000,000,000,000,000 times as strong as the gravitational. (Isaac Asimov, *Isaac Asimov's Book of Facts,* New York: Bell Publishing Company, 1981, p. 106.)

92. How is it that the average CEO in Japan receives an income of \$300,000, while the average CEO in the United States earns \$2.8 million? (Andrew Zimbalist, *Baseball and Billions.* New York: BasicBooks, 1992, p. 78.)

Solve the problem.

93. The distance to Earth from the planet Pluto is 4.58×10^9 kilometers. In April 1983, Pioneer 10 transmitted radio signals from Pluto to Earth at the speed of light, 3.00×10^5 kilometers per second. How long (in seconds) did it take for the signals to reach Earth?

94. In Exercise 93, how many hours did it take for the signals to reach Earth?

95. A *light-year* is the distance that light travels in one year. Find the number of miles in a light-year if light travels 1.86×10^5 miles per second.

96. The planet Mercury has a mean distance from the sun of 3.6×10^7 miles, while the mean distance of Venus from the sun is 6.7×10^7 miles. How long would it take a spacecraft traveling at 1.55×10^3 miles per hour to travel the distance represented by the difference of these two planets' mean distances from the sun?

97. When the distance between the centers of the moon and the earth is 4.60×10^8 meters, an object on the line joining the centers of the moon and the earth exerts the same gravitational force on each when it is 4.14×10^8 meters from the center of the earth. How far is the object from the center of the moon at that point?

98. Assume that the volume of the earth is 5×10^{14} cubic meters and that the volume of a bacterium is 2.5×10^{-16} cubic meter. If the earth could be packed full of bacteria, how many would it contain?

99. The graph shows the number of engineers employed in the United States during the years 1984–1993. (*Source:* Engineering Workforce Commission)

(a) Use the graph to approximate the number of engineers employed during 1992 and write the number in scientific notation.

(b) Suppose the average annual salary for an engineer in 1992 was \$$5.5 \times 10^4$. What would be the total of all salaries of engineers in that year? (This is hypothetical and not based on actual data.)

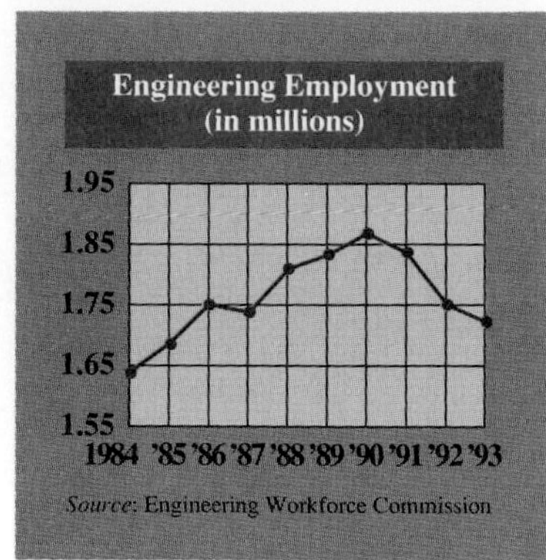

100. In 1992, the state of Texas had about 1.3×10^6 farms with an average of 7.1×10^2 acres per farm. What was the total number of acres devoted to farmland in Texas that year? (*Source:* National Agricultural Statistics Service, U.S. Dept. of Agriculture)

7.6 Polynomials and Factoring

A **term,** or **monomial,** is defined to be a number, a variable, or a product of numbers and variables. A **polynomial** is a term or a finite sum of terms, with only nonnegative integer exponents permitted on the variables. If the terms of a polynomial contain only the variable x, then the polynomial is called a **polynomial in x.** (Polynomials in other variables are defined similarly.) Examples of polynomials include

$$5x^3 - 8x^2 + 7x - 4, \qquad 9p^5 - 3, \qquad 8r^2, \qquad \text{and} \qquad 6.$$

The expression $9x^2 - 4x - 6/x$ is not a polynomial because of the presence of $-6/x$. The terms of a polynomial cannot have variables in a denominator.

The greatest exponent in a polynomial in one variable is the **degree** of the polynomial. A nonzero constant is said to have degree 0. (The polynomial 0 has no degree.) For example, $3x^6 - 5x^2 + 2x + 3$ is a polynomial of degree 6.

A polynomial can have more than one variable. A term containing more than one variable has degree equal to the sum of all the exponents appearing on the variables in the term. For example, $-3x^4y^3z^5$ is of degree $4 + 3 + 5 = 12$. The degree of a polynomial in more than one variable is equal to the greatest degree of any term appearing in the polynomial. By this definition, the polynomial

$$2x^4y^3 - 3x^5y + x^6y^2$$

is of degree 8 because of the x^6y^2 term.

A polynomial containing exactly three terms is called a **trinomial** and one containing exactly two terms is a **binomial.** For example, $7x^9 - 8x^4 + 1$ is a trinomial of degree 9.

EXAMPLE 1 The list below shows several polynomials, gives the degree of each, and identifies each as a monomial, binomial, trinomial, or none of these.

Polynomial	Degree	Type
$9p^7 - 4p^3 + 8p^2$	7	Trinomial
$29x^{11} + 8x^{15}$	15	Binomial
$-10r^6s^8$	14	Monomial
$5a^3b^7 - 3a^5b^5 + 4a^2b^9 - a^{10}$	11	None of these

Addition and Subtraction

Since the variables used in polynomials represent real numbers, a polynomial represents a real number. This means that all the properties of the real numbers mentioned in this chapter hold for polynomials. In particular, the distributive property holds, so

$$3m^5 - 7m^5 = (3 - 7)m^5 = -4m^5.$$

Like terms are terms that have the exact same variable factors. Thus, polynomials are added by adding coefficients of like terms; polynomials are subtracted by subtracting coefficients of like terms.

EXAMPLE 2 Add or subtract, as indicated.

(a) $(2y^4 - 3y^2 + y) + (4y^4 + 7y^2 + 6y)$
$= (2 + 4)y^4 + (-3 + 7)y^2 + (1 + 6)y$
$= 6y^4 + 4y^2 + 7y$

(b) $(-3m^3 - 8m^2 + 4) - (m^3 + 7m^2 - 3)$
$= (-3 - 1)m^3 + (-8 - 7)m^2 + [4 - (-3)]$
$= -4m^3 - 15m^2 + 7$

(c) $8m^4p^5 - 9m^3p^5 + (11m^4p^5 + 15m^3p^5) = 19m^4p^5 + 6m^3p^5$

(d) $4(x^2 - 3x + 7) - 5(2x^2 - 8x - 4)$
$= 4x^2 - 4(3x) + 4(7) - 5(2x^2) - 5(-8x) - 5(-4)$ Distributive property
$= 4x^2 - 12x + 28 - 10x^2 + 40x + 20$ Associative property
$= -6x^2 + 28x + 48$ Add like terms.

As shown in parts (a), (b), and (d) of Example 2, polynomials in one variable are often written with their terms in *descending powers;* so the term of greatest degree is first, the one with the next greatest degree is second, and so on.

Multiplication

The associative and distributive properties, together with the properties of exponents, can also be used to find the product of two polynomials. For example, to find the product of $3x - 4$ and $2x^2 - 3x + 5$, treat $3x - 4$ as a single expression and use the distributive property as follows.

$$(3x - 4)(2x^2 - 3x + 5) = (3x - 4)(2x^2) - (3x - 4)(3x) + (3x - 4)(5)$$

Now use the distributive property three separate times on the right of the equals sign to get

$$\begin{aligned}(3x - 4)(2x^2 - 3x + 5)& \\ &= (3x)(2x^2) - 4(2x^2) - (3x)(3x) - (-4)(3x) + (3x)5 - 4(5) \\ &= 6x^3 - 8x^2 - 9x^2 + 12x + 15x - 20 \\ &= 6x^3 - 17x^2 + 27x - 20.\end{aligned}$$

It is sometimes more convenient to write such a product vertically, as follows.

$$\begin{array}{rrrr} & 2x^2 & -\ 3x & +\ 5 \\ & & 3x & -\ 4 \\ \hline & -\ 8x^2 & +\ 12x & -\ 20 \\ 6x^3 & -\ 9x^2 & +\ 15x & \\ \hline 6x^3 & -\ 17x^2 & +\ 27x & -\ 20 \end{array}$$

Add in columns.

EXAMPLE 3 Multiply $(3p^2 - 4p + 1)(p^3 + 2p - 8)$.

Multiply each term of the second polynomial by each term of the first and add these products. It is most efficient to work vertically with polynomials of more than two terms, so that like terms can be placed in columns.

$$\begin{array}{rrrrrr} & & & 3p^2 & -\ 4p & +\ 1 \\ & & p^3 & + & 2p & -\ 8 \\ \hline & & & -\ 24p^2 & +\ 32p & -\ 8 \\ & & 6p^3 & -\ 8p^2 & +\ 2p & \\ 3p^5 & -\ 4p^4 & +\ p^3 & & & \\ \hline 3p^5 & -\ 4p^4 & +\ 7p^3 & -\ 32p^2 & +\ 34p & -\ 8 \end{array}$$

Multiply $3p^2 - 4p + 1$ by -8.
Multiply $3p^2 - 4p + 1$ by $2p$.
Multiply $3p^2 - 4p + 1$ by p^3.
Add in columns.

The FOIL method is a convenient way to find the product of two binomials. The memory aid FOIL (for First, Outside, Inside, Last) gives the pairs of terms to be multiplied to get the product, as shown in the next examples.

EXAMPLE 4 Find each product.

(a) $(6m + 1)(4m - 3) = \overset{F}{(6m)(4m)} + \overset{O}{(6m)(-3)} + \overset{I}{1(4m)} + \overset{L}{1(-3)}$
$= 24m^2 - 14m - 3$

(b) $(2x + 7)(2x - 7) = 4x^2 - 14x + 14x - 49$
$= 4x^2 - 49$

In part (a) of Example 4, the product of two binomials was a trinomial, while in part (b) the product of two binomials was a binomial. The product of two binomials of the forms $x + y$ and $x - y$ is always a binomial. Check by multiplying that the following is true.

Product of the Sum and Difference of Two Terms

$$(x + y)(x - y) = x^2 - y^2$$

This product is called the **difference of two squares.** Since products of this type occur frequently, it is important to be able to recognize when this pattern should be used.

EXAMPLE 5 Find each product.

(a) $(3p + 11)(3p - 11)$

Using the pattern discussed above, replace x with $3p$ and y with 11.

$$(3p + 11)(3p - 11) = (3p)^2 - 11^2 = 9p^2 - 121$$

(b) $(5m^3 - 3)(5m^3 + 3) = (5m^3)^2 - 3^2 = 25m^6 - 9$

(c) $(9k - 11r^3)(9k + 11r^3) = (9k)^2 - (11r^3)^2 = 81k^2 - 121r^6$ ■

The **squares of binomials** are also special products. The products $(x + y)^2$ and $(x - y)^2$ are shown below.

Squares of Binomials

$$(x + y)^2 = x^2 + 2xy + y^2$$
$$(x - y)^2 = x^2 - 2xy + y^2$$

	a	b
a	Area: a^2	Area: ab
b	Area: ab	Area: b^2

The special product

$$(a + b)^2 = a^2 + 2ab + b^2$$

can be illustrated geometrically using the diagram shown here. The side of the large square has length $a + b$, so the area of the square is

$$(a + b)^2.$$

The large square is made up of two smaller squares and two congruent rectangles. The sum of the areas of these figures is

$$a^2 + 2ab + b^2.$$

Since these expressions represent the same quantity, they must be equal, thus giving us the pattern for squaring a binomial.

These patterns also occur frequently.

EXAMPLE 6 Find each product.

(a) $(2m + 5)^2 = (2m)^2 + 2(2m)(5) + (5)^2$
$= 4m^2 + 20m + 25$

(b) $(3x - 7y^4)^2 = (3x)^2 - 2(3x)(7y^4) + (7y^4)^2$
$= 9x^2 - 42xy^4 + 49y^8$ ■

As shown in Example 6, the square of a binomial has three terms. Students often mistakenly give $a^2 + b^2$ as equivalent to the product $(a + b)^2$. Be careful to avoid that error.

The process of finding polynomials whose product equals a given polynomial is called **factoring.** For example, since $4x + 12 = 4(x + 3)$, both 4 and $x + 3$ are called **factors** of $4x + 12$. Also, $4(x + 3)$ is called the **factored form** of $4x + 12$. A polynomial that cannot be written as a product of two polynomials with integer coefficients is a **prime** or **irreducible polynomial.** A polynomial is **factored completely** when it is written as a product of prime polynomials with integer coefficients.

Factoring Out the Greatest Common Factor

Polynomials are factored by using the distributive property. For example, to factor $6x^2y^3 + 9xy^4 + 18y^5$, we look for a monomial that is the greatest common factor of all the terms of the polynomial. For this polynomial, $3y^3$ is the greatest common factor. By the distributive property,

$$6x^2y^3 + 9xy^4 + 18y^5 = (3y^3)(2x^2) + (3y^3)(3xy) + (3y^3)(6y^2)$$
$$= 3y^3(2x^2 + 3xy + 6y^2).$$

EXAMPLE 7 Factor out the greatest common factor from each polynomial.

(a) $9y^5 + y^2$

The greatest common factor is y^2.

$$9y^5 + y^2 = y^2 \cdot 9y^3 + y^2 \cdot 1$$
$$= y^2(9y^3 + 1)$$

(b) $6x^2t + 8xt + 12t = 2t(3x^2 + 4x + 6)$

(c) $14m^4(m + 1) - 28m^3(m + 1) - 7m^2(m + 1)$

The greatest common factor is $7m^2(m + 1)$. Use the distributive property as follows.

$$14m^4(m + 1) - 28m^3(m + 1) - 7m^2(m + 1)$$
$$= [7m^2(m + 1)](2m^2 - 4m - 1)$$
$$= 7m^2(m + 1)(2m^2 - 4m - 1)$$

Factoring by Grouping

When a polynomial has more than three terms, it can sometimes be factored by a method called **factoring by grouping.** For example, to factor

$$ax + ay + 6x + 6y,$$

collect the terms into two groups so that each group has a common factor.

$$ax + ay + 6x + 6y = (ax + ay) + (6x + 6y)$$

Factor each group, getting

$$ax + ay + 6x + 6y = a(x + y) + 6(x + y).$$

The quantity $(x + y)$ is now a common factor, which can be factored out, producing

$$ax + ay + 6x + 6y = (x + y)(a + 6).$$

It is not always obvious which terms should be grouped. Experience and repeated trials are the most reliable tools for factoring by grouping.

EXAMPLE 8 Factor by grouping.

(a) $mp^2 + 7m + 3p^2 + 21$

Group the terms as follows.

$$mp^2 + 7m + 3p^2 + 21 = (mp^2 + 7m) + (3p^2 + 21)$$

Factor out the greatest common factor from each group.

$$(mp^2 + 7m) + (3p^2 + 21) = m(p^2 + 7) + 3(p^2 + 7)$$
$$= (p^2 + 7)(m + 3)$$

$p^2 + 7$ is a common factor.

(b) $2y^2 - 2z - ay^2 + az$

Grouping terms as above gives

$$2y^2 - 2z - ay^2 + az = (2y^2 - 2z) + (-ay^2 + az)$$
$$= 2(y^2 - z) + a(-y^2 + z).$$

The expression $-y^2 + z$ is the negative of $y^2 - z$, so the terms should be grouped as follows.

$$\begin{aligned} 2y^2 - 2z - ay^2 + az &= (2y^2 - 2z) - (ay^2 - az) \\ &= 2(y^2 - z) - a(y^2 - z) && \text{Factor each group.} \\ &= (y^2 - z)(2 - a) && \text{Factor out } y^2 - z. \end{aligned}$$

Factoring Trinomials

Factoring is the opposite of multiplication. Since the product of two binomials is usually a trinomial, we can expect factorable trinomials (that have terms with no common factor) to have two binomial factors. Thus, factoring trinomials requires using FOIL backwards.

EXAMPLE 9 Factor each trinomial.

(a) $4y^2 - 11y + 6$

To factor this polynomial, we must find integers a, b, c, and d such that

$$4y^2 - 11y + 6 = (ay + b)(cy + d).$$

By using FOIL, we see that $ac = 4$ and $bd = 6$. The positive factors of 4 are 4 and 1 or 2 and 2. Since the middle term is negative, we consider only negative factors of 6. The possibilities are -2 and -3 or -1 and -6. Now we try various arrangements of these factors until we find one that gives the correct coefficient of y.

$$\begin{aligned} (2y - 1)(2y - 6) &= 4y^2 - \mathbf{14y} + 6 && \text{Incorrect} \\ (2y - 2)(2y - 3) &= 4y^2 - \mathbf{10y} + 6 && \text{Incorrect} \\ (y - 2)(4y - 3) &= 4y^2 - \mathbf{11y} + 6 && \text{Correct} \end{aligned}$$

The last trial gives the correct factorization.

(b) $6p^2 - 7p - 5$

Again, we try various possibilities. The positive factors of 6 could be 2 and 3 or 1 and 6. As factors of -5 we have only -1 and 5 or -5 and 1. Try different combinations of these factors until the correct one is found.

$$\begin{aligned} (2p - 5)(3p + 1) &= 6p^2 - \mathbf{13p} - 5 && \text{Incorrect} \\ (3p - 5)(2p + 1) &= 6p^2 - \mathbf{7p} - 5 && \text{Correct} \end{aligned}$$

Thus, $6p^2 - 7p - 5$ factors as $(3p - 5)(2p + 1)$.

Each of the special patterns of multiplication given earlier can be used in reverse to get a pattern for factoring. Perfect square trinomials can be factored as follows.

Perfect Square Trinomials

$$x^2 + 2xy + y^2 = (x + y)^2$$
$$x^2 - 2xy + y^2 = (x - y)^2$$

The special product

$$(a + b)(a - b) = a^2 - b^2$$

can be used to perform some multiplication problems. For example,

$$51 \times 49 = (50 + 1)(50 - 1)$$
$$= 50^2 - 1^2 = 2{,}500 - 1$$
$$= 2{,}499$$
$$102 \times 98 = (100 + 2)(100 - 2)$$
$$= 100^2 - 2^2$$
$$= 10{,}000 - 4$$
$$= 9{,}996.$$

Once these patterns are recognized, multiplications of this type can be done mentally.

EXAMPLE 10 Factor each polynomial.

(a) $16p^2 - 40pq + 25q^2$

Since $16p^2 = (4p)^2$ and $25q^2 = (5q)^2$, use the second pattern shown above with $4p$ replacing x and $5q$ replacing y to get

$$16p^2 - 40pq + 25q^2 = (4p)^2 - 2(4p)(5q) + (5q)^2$$
$$= (4p - 5q)^2.$$

Make sure that the middle term of the trinomial being factored, $-40pq$ here, is twice the product of the two terms in the binomial $4p - 5q$.

$$-40pq = 2(4p)(-5q)$$

(b) $169x^2 + 104xy^2 + 16y^4 = (13x + 4y^2)^2$, since $2(13x)(4y^2) = 104xy^2$.

Factoring Binomials

The pattern for the product of the sum and difference of two terms gives the following factorization.

Difference of Two Squares

$$x^2 - y^2 = (x + y)(x - y)$$

EXAMPLE 11 Factor each of the following polynomials.

(a) $4m^2 - 9$

First, recognize that $4m^2 - 9$ is the difference of two squares, since $4m^2 = (2m)^2$ and $9 = 3^2$. Use the pattern for the difference of two squares with $2m$ replacing x and 3 replacing y. Doing this gives

$$4m^2 - 9 = (2m)^2 - 3^2$$
$$= (2m + 3)(2m - 3).$$

(b) $256k^4 - 625m^4$

Use the difference of two squares pattern twice, as follows:

$$256k^4 - 625m^4 = (16k^2)^2 - (25m^2)^2$$
$$= (16k^2 + 25m^2)(16k^2 - 25m^2)$$
$$= (16k^2 + 25m^2)(4k + 5m)(4k - 5m).$$

(c) $x^2 - 6x + 9 - y^4$

Group the first three terms to obtain a perfect square trinomial. Then use the difference of squares pattern.

$$x^2 - 6x + 9 - y^4 = (x^2 - 6x + 9) - y^4$$
$$= (x - 3)^2 - (y^2)^2$$
$$= [(x - 3) + y^2][(x - 3) - y^2]$$
$$= (x - 3 + y^2)(x - 3 - y^2)$$

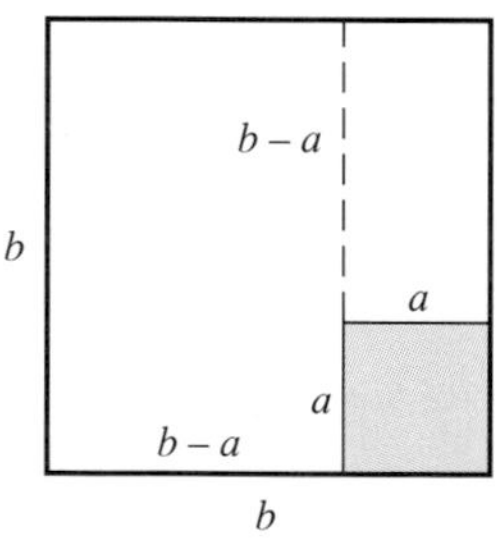

A geometric proof for the difference of squares property is shown above. (The proof is only valid for $b > a > 0$.)

$$b^2 - a^2 = b(b - a) + a(b - a)$$
$$= (b + a)(b - a)$$

Factor out $b - a$ in the second step.

Two other special results of factoring are listed below. Each can be verified by multiplying on the right side of the equation.

Difference and Sum of Two Cubes

Difference of Two Cubes $x^3 - y^3 = (x - y)(x^2 + xy + y^2)$

Sum of Two Cubes $x^3 + y^3 = (x + y)(x^2 - xy + y^2)$

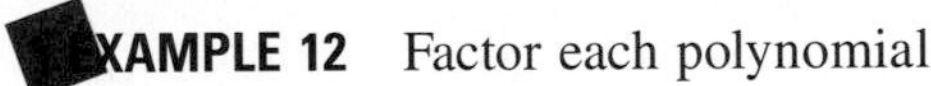

EXAMPLE 12 Factor each polynomial.

(a) $x^3 + 27$

Notice that $27 = 3^3$, so the expression is a sum of two cubes. Use the second pattern given above.

$$x^3 + 27 = x^3 + 3^3 = (x + 3)(x^2 - 3x + 9)$$

(b) $m^3 - 64n^3$

Since $64n^3 = (4n)^3$, the given polynomial is a difference of two cubes. To factor, use the first pattern in the box above, replacing x with m and y with $4n$.

$$\begin{aligned} m^3 - 64n^3 &= m^3 - (4n)^3 \\ &= (m - 4n)[m^2 + m(4n) + (4n)^2] \\ &= (m - 4n)(m^2 + 4mn + 16n^2) \end{aligned}$$

(c) $8q^6 + 125p^9$

Write $8q^6$ as $(2q^2)^3$ and $125p^9$ as $(5p^3)^3$, so that the given polynomial is a sum of two cubes.

$$\begin{aligned} 8q^6 + 125p^9 &= (2q^2)^3 + (5p^3)^3 \\ &= (2q^2 + 5p^3)[(2q^2)^2 - (2q^2)(5p^3) + (5p^3)^2] \\ &= (2q^2 + 5p^3)(4q^4 - 10q^2p^3 + 25p^6) \end{aligned}$$ ■

7.6 EXERCISES

Find each of the following sums and differences.

1. $(3x^2 - 4x + 5) + (-2x^2 + 3x - 2)$

2. $(4m^3 - 3m^2 + 5) + (-3m^3 - m^2 + 5)$

3. $(12y^2 - 8y + 6) - (3y^2 - 4y + 2)$

4. $(8p^2 - 5p) - (3p^2 - 2p + 4)$

5. $(6m^4 - 3m^2 + m) - (2m^3 + 5m^2 + 4m) + (m^2 - m)$

6. $-(8x^3 + x - 3) + (2x^3 + x^2) - (4x^2 + 3x - 1)$

7. $5(2x^2 - 3x + 7) - 2(6x^2 - x + 12)$

8. $8x^2y - 3xy^2 + 2x^2y - 9xy^2$

Find each of the following products.

9. $(x + 3)(x - 8)$

10. $(y - 3)(y - 9)$

11. $(4r - 1)(7r + 2)$

12. $(5m - 6)(3m + 4)$

13. $4x^2(3x^3 + 2x^2 - 5x + 1)$

14. $2b^3(b^2 - 4b + 3)$

15. $(2m + 3)(2m - 3)$

16. $(8s - 3t)(8s + 3t)$

17. $(4m + 2n)^2$

18. $(a - 6b)^2$

19. $(5r + 3t^2)^2$

20. $(2z^4 - 3y)^2$

21. $(2z - 1)(-z^2 + 3z - 4)$

22. $(k + 2)(12k^3 - 3k^2 + k + 1)$

23. $(m - n + k)(m + 2n - 3k)$

24. $(r - 3s + t)(2r - s + t)$

25. $(a - b + 2c)^2$

26. $(k - y + 3m)^2$

27. Which one of the following is a trinomial in descending powers, having degree 6?
(a) $5x^6 - 4x^5 + 12$ **(b)** $6x^5 - x^6 + 4$ **(c)** $2x + 4x^2 - x^6$
(d) $4x^6 - 6x^4 + 9x + 1$

28. Give an example of a polynomial of four terms in the variable x, having degree 5, written in descending powers, lacking a fourth degree term.

29. The exponent in the expression 6^3 is 3. Explain why the degree of 6^3 is not 3. What is its degree?

30. Explain in your own words how to square a binomial.

Factor the greatest common factor from each polynomial.

31. $8m^4 + 6m^3 - 12m^2$
32. $2p^5 - 10p^4 + 16p^3$
33. $4k^2m^3 + 8k^4m^3 - 12k^2m^4$
34. $28r^4s^2 + 7r^3s - 35r^4s^3$
35. $2(a + b) + 4m(a + b)$
36. $4(y - 2)^2 + 3(y - 2)$
37. $2(m - 1) - 3(m - 1)^2 + 2(m - 1)^3$
38. $5(a + 3)^3 - 2(a + 3) + (a + 3)^2$

Factor each of the following polynomials by grouping.

39. $6st + 9t - 10s - 15$
40. $10ab - 6b + 35a - 21$
41. $rt^3 + rs^2 - pt^3 - ps^2$
42. $2m^4 + 6 - am^4 - 3a$
43. $16a^2 + 10ab - 24ab - 15b^2$
44. $15 - 5m^2 - 3r^2 + m^2r^2$
45. $20z^2 - 8zx - 45zx + 18x^2$

46. Consider the polynomial $1 - a + ab - b$. One acceptable factored form is $(1 - a)(1 - b)$. However there are other acceptable factored forms. Which one is *not* a factored form of this polynomial?
(a) $(a - 1)(b - 1)$ **(b)** $(-a + 1)(-b + 1)$ **(c)** $(-1 + a)(-1 + b)$
(d) $(1 - a)(b + 1)$

Factor each trinomial.

47. $x^2 - 2x - 15$
48. $r^2 + 8r + 12$
49. $y^2 + 2y - 35$
50. $x^2 - 7x + 6$
51. $6a^2 - 48a - 120$
52. $8h^2 - 24h - 320$
53. $3m^3 + 12m^2 + 9m$
54. $9y^4 - 54y^3 + 45y^2$
55. $6k^2 + 5kp - 6p^2$
56. $14m^2 + 11mr - 15r^2$
57. $5a^2 - 7ab - 6b^2$
58. $12s^2 + 11st - 5t^2$
59. $9x^2 - 6x^3 + x^4$
60. $30a^2 + am - m^2$
61. $24a^4 + 10a^3b - 4a^2b^2$
62. $18x^5 + 15x^4z - 75x^3z^2$

63. When a student was given the polynomial $4x^2 + 2x - 20$ to factor completely on a test, she lost some credit by giving the answer $(4x + 10)(x - 2)$. She complained to her teacher that the product $(4x + 10)(x - 2)$ is indeed $4x^2 + 2x - 20$. Do you think that the teacher was justified in not giving her full credit? Explain.

64. Write an explanation as to why most people would find it more difficult to factor $36x^2 - 44x - 15$ than $37x^2 - 183x - 10$.

Factor each, using the method for factoring a perfect square trinomial. It may be necessary to factor out a common factor first.

65. $9m^2 - 12m + 4$
66. $16p^2 - 40p + 25$
67. $32a^2 - 48ab + 18b^2$
68. $20p^2 - 100pq + 125q^2$
69. $4x^2y^2 + 28xy + 49$
70. $9m^2n^2 - 12mn + 4$

Factor each difference of two squares.

71. $x^2 - 36$
72. $t^2 - 64$
73. $y^2 - w^2$
74. $25 - w^2$
75. $9a^2 - 16$
76. $16q^2 - 25$
77. $25s^4 - 9t^2$
78. $36z^2 - 81y^4$
79. $p^4 - 625$
80. $m^4 - 81$

Factor each sum or difference of cubes.

81. $8 - a^3$
82. $r^3 + 27$
83. $125x^3 - 27$
84. $8m^3 - 27n^3$
85. $27y^9 + 125z^6$
86. $27z^3 + 729y^3$

Each of the following may be factored using one of the methods described in this section. Decide on the method, and then factor the polynomial completely.

87. $x^2 + xy - 5x - 5y$
88. $8r^2 - 10rs - 3s^2$
89. $p^4(m - 2n) + q(m - 2n)$
90. $36a^2 + 60a + 25$
91. $4z^2 + 28z + 49$
92. $6p^4 + 7p^2 - 3$
93. $1{,}000x^3 + 343y^3$
94. $b^2 + 8b + 16 - a^2$
95. $125m^6 - 216$
96. $q^2 + 6q + 9 - p^2$
97. $12m^2 + 16mn - 35n^2$
98. $216p^3 + 125q^3$

99. The sum of two squares usually cannot be factored. For example, $x^2 + y^2$ is prime. Notice that

$$x^2 + y^2 \neq (x + y)(x + y).$$

By choosing $x = 4$ and $y = 2$, show that the above inequality is true.

100. The binomial $9x^2 + 36$ is a sum of two squares. Can it be factored? If so, factor it.

7.7 Quadratic Equations and Applications

In Section 1 we learned how to solve linear equations. Recall that a linear equation is one that can be written in the form $ax + b = c$, where a, b, and c are real numbers, and $a \neq 0$. We will now examine methods of solving quadratic equations.

Quadratic Equation

An equation that can be written in the form

$$ax^2 + bx + c = 0$$

where a, b, and c are real numbers, with $a \neq 0$, is a **quadratic equation.**

(Why is the restriction $a \neq 0$ necessary?) A quadratic equation written in the form $ax^2 + bx + c = 0$ is in *standard form.*

The simplest method of solving a quadratic equation, but one that is not always easily applied, is by factoring. This method depends on the following property.

Zero-Factor Property

If $ab = 0$, then $a = 0$ or $b = 0$ or both.

The next example shows how the zero-factor property is used to solve a quadratic equation.

The *solve* feature gives the two solutions of the equation in Example 1. Notice that the guess 0 yields the solution 1/3, while the guess −5 yields the solution −3/2. Compare to Example 1.

EXAMPLE 1 Solve $6x^2 + 7x = 3$.

First write the equation in standard form as

$$6x^2 + 7x - 3 = 0.$$

Now factor $6x^2 + 7x - 3$ to get

$$(3x - 1)(2x + 3) = 0.$$

By the zero-factor property, the product $(3x - 1)(2x + 3)$ can equal 0 only if

$$3x - 1 = 0 \quad \text{or} \quad 2x + 3 = 0$$
$$3x = 1 \qquad\qquad 2x = -3$$
$$x = \frac{1}{3} \qquad\qquad x = -\frac{3}{2}.$$

The solution set is $\left\{\frac{1}{3}, -\frac{3}{2}\right\}$.

A quadratic equation of the form $x^2 = k$, where $k \geq 0$, can be solved by factoring using the following sequence of equivalent equations.

$$x^2 = k$$
$$x^2 - k = 0$$
$$(x + \sqrt{k})(x - \sqrt{k}) = 0$$
$$x + \sqrt{k} = 0 \quad \text{or} \quad x - \sqrt{k} = 0$$
$$x = -\sqrt{k} \qquad\qquad x = \sqrt{k}$$

This proves the square root property for solving equations.

Square Root Property

If $k \geq 0$, the solutions of $x^2 = k$ are $x = \pm\sqrt{k}$.

If $k > 0$, the equation $x^2 = k$ has two real solutions. If $k = 0$, there is only one solution, 0. If $k < 0$, there are no real solutions. (However, in this case, there *are* imaginary solutions. See a standard algebra text for more about imaginary numbers.)

EXAMPLE 2 Use the square root property to solve each quadratic equation for real solutions.

(a) $x^2 = 25$

Since $\sqrt{25} = 5$, the solution set is $\{5, -5\}$. (This may be abbreviated $\{\pm 5\}$.)

(b) $r^2 = 18$

Since $18 = 9 \cdot 2$,

$$\sqrt{18} = \sqrt{9 \cdot 2} \qquad \text{Product rule for radicals}$$
$$18 = 3\sqrt{2}. \qquad \sqrt{9} = 3$$

The solution set is $\{\pm\ 3\sqrt{2}\}$.

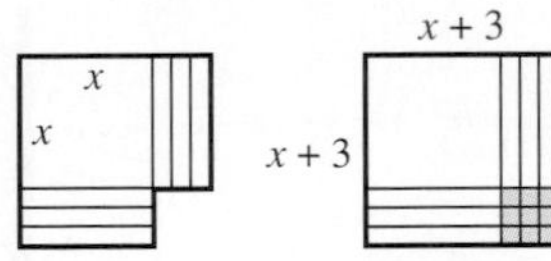

The procedure of **completing the square,** used in deriving the quadratic formula, has a number of important applications in algebra. In order to make the expression $x^2 + kx$ the square of a binomial, we must add to it the square of half the coefficient of x; that is, $[(1/2)k]^2 = k^2/4$. We then get

$$x^2 + kx + \frac{k^2}{4} = \left(x + \frac{k}{2}\right)^2.$$

For example, to make $x^2 + 6x$ the square of a binomial, we add 9, since $9 = [(1/2)6]^2$. This results in the trinomial $x^2 + 6x + 9$, which is equal to $(x + 3)^2$.

The Greeks had a method of completing the square geometrically. For example, to complete the square for $x^2 + 6x$, begin with a square of side x. Add three rectangles of width 1 and length x to the right side and the bottom. Each rectangle has area $1x$ or x, so the total area of the figure is now $x^2 + 6x$. To fill in the corner (that is, "complete the square"), we must add 9 1-by-1 squares as shown. The new larger square has sides of length $x + 3$ and area $(x + 3)^2 = x^2 + 6x + 9$.

(c) $z^2 = -3$

Since $-3 < 0$, there are no real roots, and the solution set is $\emptyset$.

(d) $(y - 4)^2 = 12$

Use a generalization of the square root property, working as follows.

$$\begin{aligned} (y - 4)^2 &= 12 \\ y - 4 &= \pm\sqrt{12} \\ y &= 4 \pm \sqrt{12} \\ y &= 4 \pm \sqrt{4 \cdot 3} \\ y &= 4 \pm 2\sqrt{3} \end{aligned}$$

The solution set is $\{4 \pm 2\sqrt{3}\}$. ■

The Quadratic Formula

By using a procedure called *completing the square* we can derive one of the most important formulas in algebra, the quadratic formula. We begin with the standard form

$$ax^2 + bx + c = 0, \quad a \neq 0,$$

and divide both sides by a to get

$$x^2 + \frac{b}{a}x + \frac{c}{a} = 0.$$

Add $-c/a$ to both sides.

$$x^2 + \frac{b}{a}x = -\frac{c}{a}$$

The polynomial on the left will be the square of a binomial if we add $b^2/(4a^2)$ to both sides of the equation. (This procedure, called *completing* the square, is discussed further in the adjoining margin note.)

$$x^2 + \frac{b}{a}x + \frac{b^2}{4a^2} = \frac{b^2}{4a^2} - \frac{c}{a}$$

Factor the left side and combine terms on the right to get

$$\left(x + \frac{b}{2a}\right)^2 = \frac{b^2 - 4ac}{4a^2}.$$

Applying the square root property gives

$$x + \frac{b}{2a} = \pm\sqrt{\frac{b^2 - 4ac}{4a^2}}$$

$$x + \frac{b}{2a} = \pm\frac{\sqrt{b^2 - 4ac}}{\sqrt{4a^2}} \qquad \text{Quotient rule for radicals}$$

$$x = -\frac{b}{2a} \pm \frac{\sqrt{b^2 - 4ac}}{2a}$$

$$x = \frac{-b \pm \sqrt{b^2 - 4ac}}{2a}. \qquad \text{Combine terms.}$$

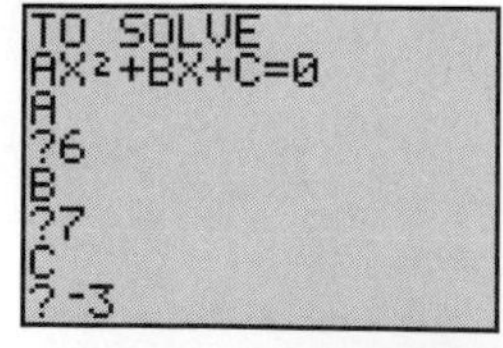

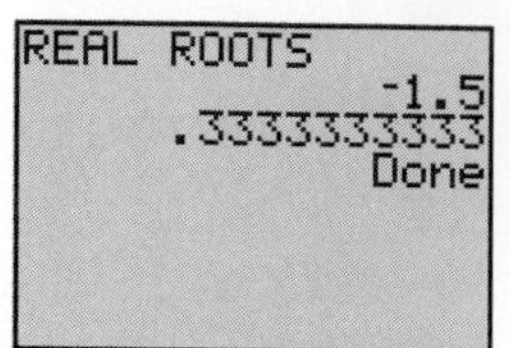

A quadratic formula program yields the solutions found in Example 1. Programs such as this one are available from calculator user-groups or can be written by the user.*

Methods of solving linear and quadratic equations have been known since the time of the Babylonians. However, for centuries mathematicians wrestled with finding a formula that could solve cubic (third-degree) equations. A story from sixteenth-century Italy concerns two main characters, **Girolamo Cardano** and **Niccolo Tartaglia.** In those days, mathematicians often kept secret their methods and participated in contests. Tartaglia had developed a method of solving a cubic equation of the form $x^3 + mx = n$ and had used it in one of these contests. Cardano begged to know Tartaglia's method and after he was told was sworn to secrecy. Now Cardano was not the kindest of men; he supposedly cut off the ears of

continued on page 391

Quadratic Formula

The solutions of $ax^2 + bx + c = 0$, $a \neq 0$, are

$$x = \frac{-b \pm \sqrt{b^2 - 4ac}}{2a}$$

Notice that the fraction bar in the quadratic formula extends under the $-b$ term in the numerator.

EXAMPLE 3 Solve $x^2 - 4x + 2 = 0$.

Here $a = 1$, $b = -4$, and $c = 2$. Substitute these values into the quadratic formula to get

$$\begin{aligned} x &= \frac{-b \pm \sqrt{b^2 - 4ac}}{2a} \\ &= \frac{-(-4) \pm \sqrt{(-4)^2 - 4(1)2}}{2(1)} && a = 1, b = -4, c = 2 \\ &= \frac{4 \pm \sqrt{16 - 8}}{2} \\ &= \frac{4 \pm 2\sqrt{2}}{2} && \sqrt{16 - 8} = \sqrt{8} = 2\sqrt{2} \\ &= \frac{2(2 \pm \sqrt{2})}{2} && \text{Factor out a 2 in the numerator.} \\ &= 2 \pm \sqrt{2}. && \text{Lowest terms} \end{aligned}$$

The solution set is $\{2 + \sqrt{2}, 2 - \sqrt{2}\}$, abbreviated as $\{2 \pm \sqrt{2}\}$. ■

EXAMPLE 4 Solve $2y^2 = y + 4$.

To find the values of a, b, and c, first rewrite the equation in standard form as $2y^2 - y - 4 = 0$. Then $a = 2$, $b = -1$, and $c = -4$. By the quadratic formula,

$$\begin{aligned} y &= \frac{-(-1) \pm \sqrt{(-1)^2 - 4(2)(-4)}}{2(2)} \\ &= \frac{1 \pm \sqrt{1 + 32}}{4} \\ &= \frac{1 \pm \sqrt{33}}{4}. \end{aligned}$$

The solution set is $\left\{\frac{1 \pm \sqrt{33}}{4}\right\}$. ■

* The program can be found in Section 2.3 of *College Algebra,* 7e by Margaret L. Lial, E. John Hornsby, Jr., and David I. Schneider, Addison-Wesley Educational Publishers Inc., 1997.

one of his sons. True to form, Cardano published Tartaglia's method in his 1545 work *Ars Magna,* despite the vow of secrecy (although he did give Tartaglia credit).

The formula for finding one real solution of the above equation is

$$x = \sqrt[3]{\frac{n}{2} + \sqrt{\left(\frac{n}{2}\right)^2 + \left(\frac{m}{3}\right)^3}} - \sqrt[3]{-\left(\frac{n}{2}\right) + \sqrt{\left(\frac{n}{2}\right)^2 + \left(\frac{m}{3}\right)^3}}$$

Try solving for one solution of the equation $x^3 + 9x = 26$ using this formula. (The solution given by the formula is 2.)

Applications

Applied problems often lead to quadratic equations. We will now look at several of these.

Problem Solving

When solving problems that lead to quadratic equations, we may get a solution that does not satisfy the physical constraints of the problem. For example, if x represents a width and the two solutions of the quadratic equation are -9 and 1, the value -9 must be rejected, since a width must be a positive number.

The first example involves the use of the Pythagorean theorem.

EXAMPLE 5 A lot is in the shape of a right triangle. The longer leg of the triangle is 20 meters longer than twice the length of the shorter leg. The hypotenuse is 10 meters longer than the longer leg. Find the lengths of the three sides of the lot.

Let s = length of the shorter leg (in meters). Then $2s + 20$ represents the length of the longer leg, and $(2s + 20) + 10 = 2s + 30$ represents the length of the hypotenuse. See Figure 13. By the Pythagorean theorem,

$$(\text{leg})^2 + (\text{other leg})^2 = (\text{hypotenuse})^2$$

which leads to

$$\begin{aligned} s^2 + (2s + 20)^2 &= (2s + 30)^2 \\ s^2 + 4s^2 + 80s + 400 &= 4s^2 + 120s + 900 \\ s^2 - 40s - 500 &= 0 \\ (s - 50)(s + 10) &= 0 \\ s = 50 \quad \text{or} \quad s &= -10. \end{aligned}$$

Since s represents a length, the value -10 is not reasonable. The shorter leg is 50 meters long, the longer leg 120 meters long, and the hypotenuse 130 meters long. ■

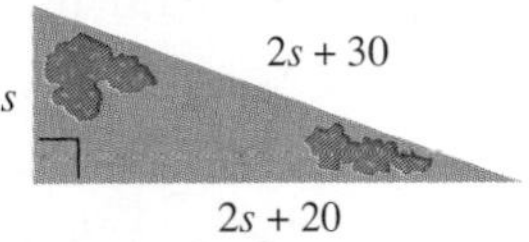

■ FIGURE 13

EXAMPLE 6 If a projectile is shot vertically upward with an initial velocity of 100 feet per second, neglecting air resistance, its height s (in feet) above the ground t seconds after projection is given by

$$s = -16t^2 + 100t.$$

(a) After how many seconds will it be 50 feet above the ground?

We must find the value of t so that $s = 50$. Let $s = 50$ in the equation, and use the quadratic formula.

$$
\begin{aligned}
50 &= -16t^2 + 100t \\
16t^2 - 100t + 50 &= 0 \quad \text{Standard form} \\
8t^2 - 50t + 25 &= 0 \quad \text{Divide by 2.} \\
t &= \frac{-(-50) \pm \sqrt{(-50)^2 - 4(8)(25)}}{2(8)} \\
t &= \frac{50 \pm \sqrt{1{,}700}}{16} \\
t \approx .55 \quad &\text{or} \quad t \approx 5.70 \quad \text{Use a calculator.}
\end{aligned}
$$

Here, both solutions are acceptable, since the projectile reaches 50 feet twice: once on its way up (after .55 second) and once on its way down (after 5.70 seconds).

(b) How long will it take for the projectile to return to the ground?

When it returns to the ground, its height s will be 0 feet, so let $s = 0$ in the equation.

$$0 = -16t^2 + 100t$$

This can be solved by factoring.

$$
\begin{aligned}
0 &= -4t(4t - 25) \\
-4t = 0 \quad &\text{or} \quad 4t - 25 = 0 \\
t = 0 \quad & \quad 4t = 25 \\
& \quad t = 6.25
\end{aligned}
$$

The first solution, 0, represents the time at which the projectile was on the ground at the instant of launch, so it does not answer the question. The projectile will return to the ground 6.25 seconds after it is launched. ◆

EXAMPLE 7 To determine the appropriate landing speed of an airplane, the formula $.1s^2 - 3s + 22 = D$ is used, where s is the initial landing speed in feet per second and D is the distance needed in feet. If the landing speed is too fast, the pilot may run out of runway; if the speed is too slow, the plane may stall. What is the appropriate landing speed if the runway is 800 feet long?

Let $D = 800$ in the given formula, and solve by using the quadratic formula.

$$
\begin{aligned}
.1s^2 - 3s + 22 &= 800 \\
.1s^2 - 3s - 778 &= 0 \\
s &= \frac{-(-3) \pm \sqrt{(-3)^2 - 4(.1)(-778)}}{2(.1)} \\
&= \frac{3 \pm \sqrt{9 + 311.2}}{.2} \\
&\approx \frac{3 \pm 17.9}{.2} \quad \text{Use a calculator.} \\
&= 104.5, -74.5
\end{aligned}
$$

Of the two solutions, we must reject -74.5. Therefore, the landing speed must be approximately 104.5 feet per second. ◆

A Radical Departure from the Other Methods of Evaluating the Golden Ratio

Recall from a previous chapter that the golden ratio is found in numerous places in mathematics, art, and nature. In a margin note there, we showed that

$$1 + \cfrac{1}{1 + \cfrac{1}{1 + \cfrac{1}{1 + \cdots}}}$$

is equal to the golden ratio, $(1 + \sqrt{5})/2$. Now consider this "nested" radical:

$$\sqrt{1 + \sqrt{1 + \sqrt{1 + \ldots}}}$$

Let x represent this radical. Because it appears "within itself", we can write

$$
\begin{aligned}
x &= \sqrt{1 + x} \\
x^2 &= 1 + x \quad \text{Square both sides.} \\
x^2 - x - 1 &= 0. \quad \text{Write in standard form.}
\end{aligned}
$$

Using the quadratic formula, with $a = 1$, $b = -1$, and $c = -1$, it can be shown that the positive solution of this equation, and thus the value of the nested radical is . . . (you guessed it!) the golden ratio.

7.7 EXERCISES

Solve each of the following equations by the zero-factor property.

1. $(x + 3)(x - 9) = 0$
2. $(m + 6)(m + 4) = 0$
3. $(2t - 7)(5t + 1) = 0$
4. $(7y - 3)(6y + 4) = 0$
5. $x^2 - x - 12 = 0$
6. $m^2 + 4m - 5 = 0$
7. $y^2 + 9y + 14 = 0$
8. $15r^2 + 7r = 2$
9. $12x^2 + 4x = 1$
10. $x(x + 3) = 4$
11. $(x + 4)(x - 6) = -16$
12. $(w - 1)(3w + 2) = 4w$
13. $(r - 5)(r - 3) = 3r(r - 3)$

14. In trying to solve $(x + 4)(x - 1) = 1$, a student reasons that since $1 \cdot 1 = 1$, the equation is solved by solving

$$x + 4 = 1 \quad \text{or} \quad x - 1 = 1.$$

Explain the error in this reasoning. What is the correct way to solve this equation?

Solve each of the following by using the square root property. Give only real number solutions.

15. $x^2 = 64$
16. $w^2 = 16$
17. $t^2 = 7$
18. $p^2 = 13$
19. $x^2 = 24$
20. $x^2 = 48$
21. $r^2 = -5$
22. $y^2 = -10$
23. $(x - 4)^2 = 3$
24. $(x + 3)^2 = 11$
25. $(2x - 5)^2 = 13$
26. $(4x + 1)^2 = 19$

Solve each of the following by the quadratic formula. Give only real number solutions.

27. $4x^2 - 8x + 1 = 0$
28. $m^2 + 2m - 5 = 0$
29. $2y^2 = 2y + 1$
30. $9r^2 + 6r = 1$
31. $q^2 - 1 = q$
32. $2p^2 - 4p = 5$
33. $4k(k + 1) = 1$
34. $4r(r - 1) = 19$
35. $(g + 2)(g - 3) = 1$
36. $(y - 5)(y + 2) = 6$
37. $m^2 - 6m = -14$
38. $y^2 = 2y - 2$

39. Can the quadratic formula be used to solve the equation $2x^2 - 5 = 0$? Explain, and solve it if the answer is yes.

40. Can the quadratic formula be used to solve the equation $4y^2 + 3y = 0$? Explain, and solve it if the answer is yes.

41. Explain why the quadratic formula cannot be used to solve the equation $2x^3 + 3x - 4 = 0$.

42. A student gave the quadratic formula incorrectly as follows: $x = -b \pm \dfrac{\sqrt{b^2 - 4ac}}{2a}$.

What is wrong with this?

The expression $b^2 - 4ac$, the radicand in the quadratic formula, is called the ***discriminant*** *of the quadratic equation $ax^2 + bx + c = 0$, $a \neq 0$. By evaluating it we can determine, without actually solving the equation, the number and nature of the solutions of the equation. Suppose that a, b, and c are integers. Then the following chart shows how the discriminant can be used to analyze the solutions.*

Discriminant	Solutions
Positive, and the square of an integer	Two different rational solutions
Positive, but not the square of an integer	Two different irrational solutions
Zero	One rational solution (a double solution)
Negative	No real solutions

In Exercises 43–48, evaluate the discriminant, and then determine whether the equation has **(a)** *two different rational solutions,* **(b)** *two different irrational solutions,* **(c)** *one rational solution (a double solution), or* **(d)** *no real solutions.*

43. $x^2 + 6x + 9 = 0$

44. $4x^2 + 20x + 25 = 0$

45. $6m^2 + 7m - 3 = 0$

46. $2x^2 + x - 3 = 0$

47. $9x^2 - 30x + 15 = 0$

48. $2x^2 - x + 1 = 0$

Solve the problem by using a quadratic equation. Use a calculator as necessary, and round the answer to the nearest tenth.

49. The Mart Hotel in Dallas, Texas, is 400 feet high. Suppose that a ball is projected upward from the top of the Mart, and its position s in feet above the ground is given by the equation $s = -16t^2 + 45t + 400$, where t is the number of seconds elapsed. How long will it take for the ball to reach a height of 200 feet above the ground?

50. The Toronto Dominion Center in Winnipeg, Manitoba, is 407 feet high. Suppose that a ball is projected upward from the top of the Center, and its position s in feet above the ground is given by the equation $s = -16t^2 + 75t + 407$, where t is the number of seconds elapsed. How long will it take for the ball to reach a height of 450 feet above the ground?

51. Refer to the equations in Exercises 49 and 50. Suppose that the first sentence in each problem did not give the height of the building. How could you use the equation to determine the height of the building?

52. In Exercises 49 and 50, one problem has only one solution while the other has two solutions. Explain why this is so.

53. A search light moves horizontally back and forth along a wall with the distance of the light from a starting point at t minutes given by $s = 100t^2 - 300t$. How long will it take before the light returns to the starting point?

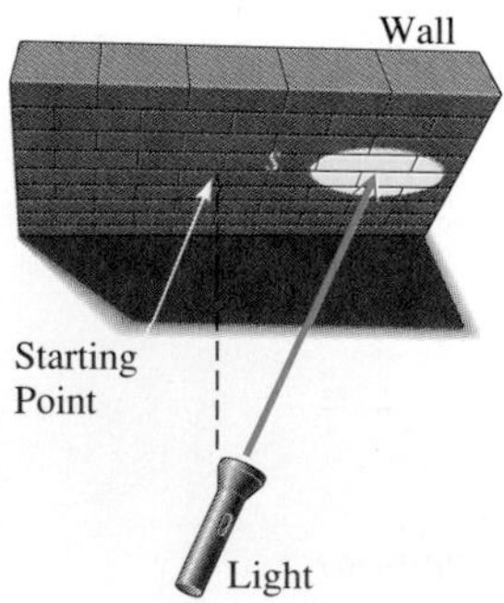

54. An object is projected directly upward from the ground. After t seconds its distance in feet above the ground is $s = 144t - 16t^2$.

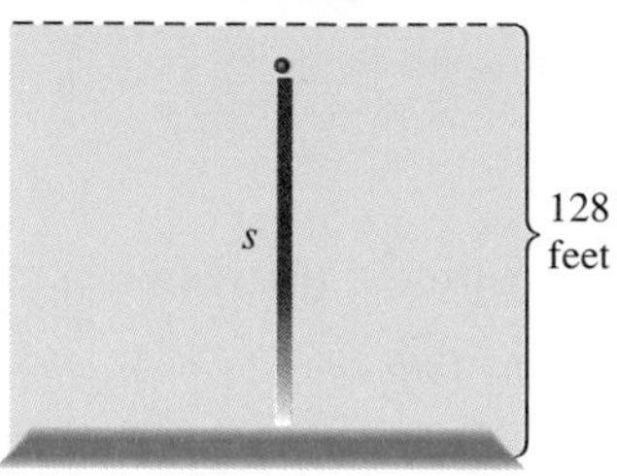

(a) After how many seconds will the object be 128 feet above the ground? (*Hint:* Look for a common factor before solving the equation.)

(b) When does the object strike the ground?

55. The formula $D = 100t - 13t^2$ gives the distance in feet a car going approximately 68 miles per hour (initially) will skid in t seconds. Find the time it would take for the car to skid 190 feet. (*Hint:* Your answer must be less than the time it takes the car to stop, which is 3.8 seconds.)

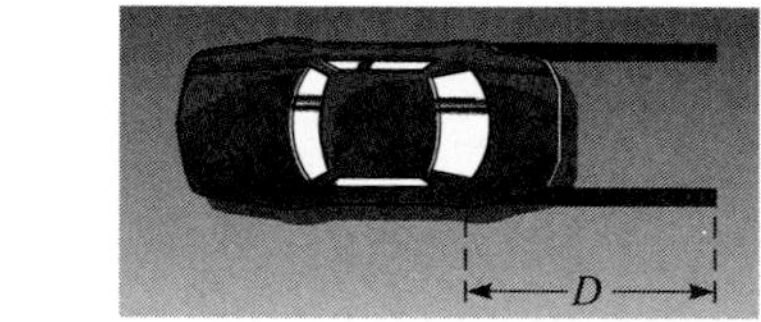

56. The formula in Exercise 55 becomes $D = 73t - 13t^2$ for a car going 50 miles per hour initially. Find the time for such a car to skid 100 feet. (*Hint:* The car will stop at about 2.8 seconds.)

57. Find the lengths of the sides of the triangle.

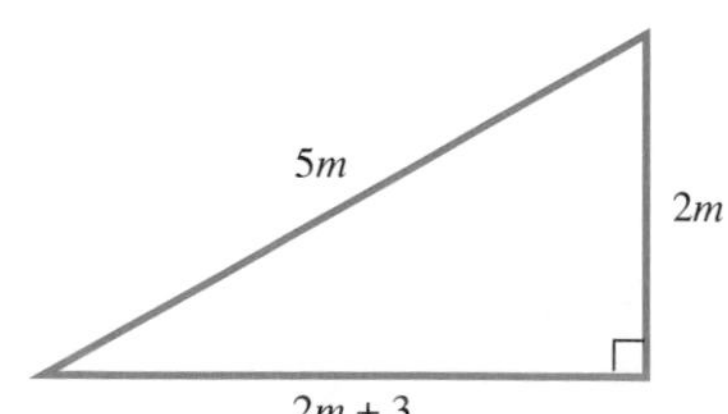

58. Find the lengths of the sides of the triangle.

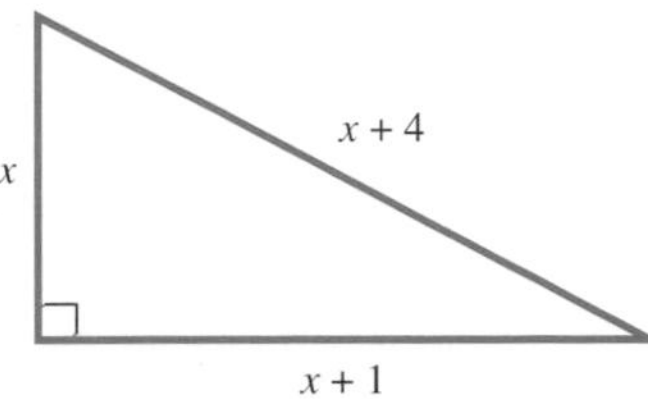

59. Refer to Exercise 49. Suppose that a wire is attached to the top of the Mart, and pulled tight. It is attached to the ground 100 feet from the base of the building, as shown in the figure. How long is the wire?

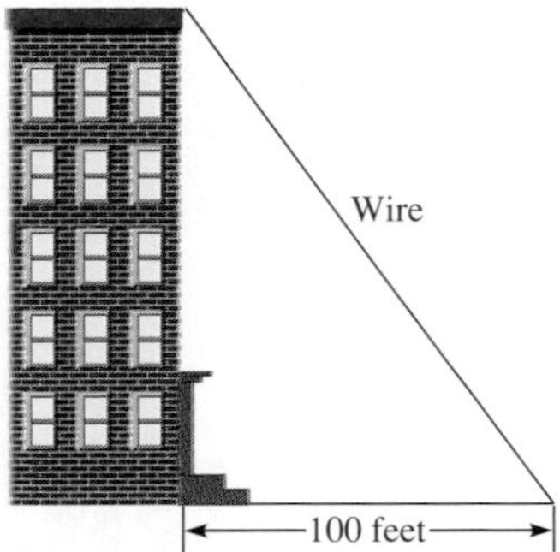

60. Refer to Exercise 50. Suppose that a wire is attached to the top of the Center, and pulled tight. The length of the wire is twice the distance between the base of the Center and the point on the ground where the wire is attached. How long is the wire?

61. Two ships leave port at the same time, one heading due south and the other heading due east. Several hours later, they are 170 miles apart. If the ship traveling south has traveled 70 miles farther than the other, how many miles has each traveled?

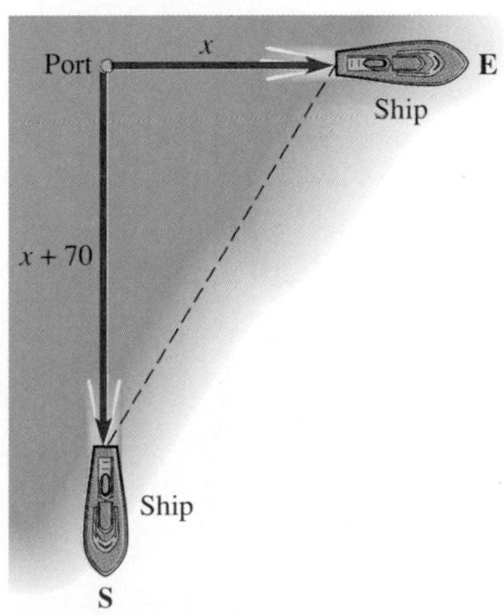

62. Joyce Smith is flying a kite that is 30 feet farther above her hand than its horizontal distance from her. The string from her hand to the kite is 150 feet long. How high is the kite?

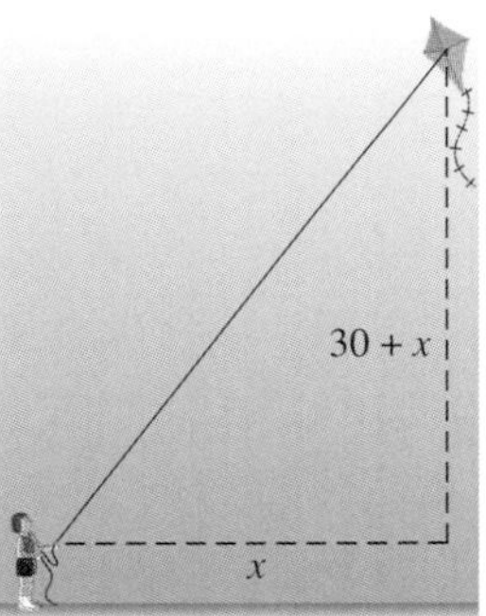

63. A toy manufacturer needs a piece of plastic in the shape of a right triangle with the longer leg 2 centimeters more than twice as long as the shorter leg, and the hypotenuse 1 centimeter more than the longer leg. How long should the three sides of the triangular piece be?

64. Adair Milmoe, a developer, owns a piece of land enclosed on three sides by streets, giving it the shape of a right triangle. The hypotenuse is 8 meters longer than the longer leg, and the shorter leg is 9 meters shorter than the hypotenuse. Find the lengths of the three sides of the property.

65. Two pieces of a large wooden puzzle fit together to form a rectangle with a length 1 centimeter less than twice the width. The diagonal, where the two pieces meet, is 2.5 centimeters in length. Find the length and width of the rectangle.

66. A 13-foot ladder is leaning against a house. The distance from the bottom of the ladder to the house is 7 feet less than the distance from the top of the ladder to the ground. How far is the bottom of the ladder from the house?

67. Catarina and José want to buy a rug for a room that is 15 by 20 feet. They want to leave an even strip of flooring uncovered around the edges of the room. How wide a strip will they have if they buy a rug with an area of 234 square feet?

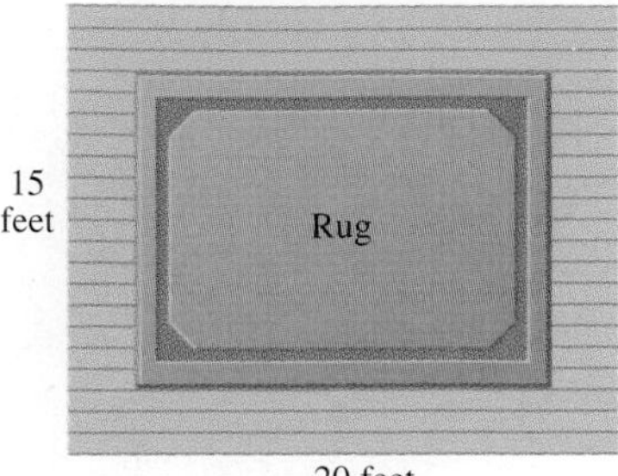

68. A club swimming pool is 30 feet wide and 40 feet long. The club members want an exposed aggregate border in a strip of uniform width around the pool. They have enough material for 296 square feet. How wide can the strip be?

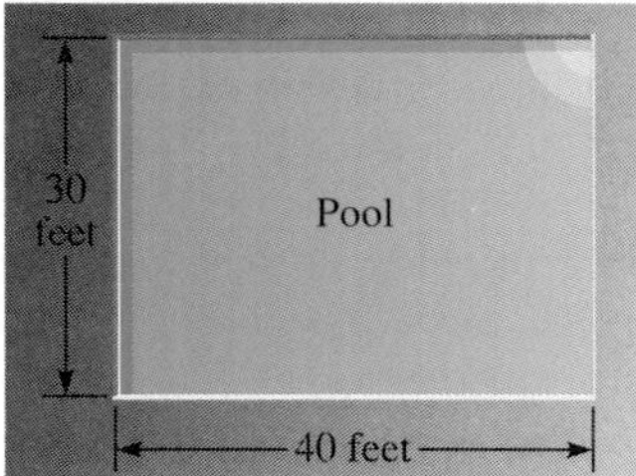

69. Arif's backyard is 20 by 30 meters. He wants to put a flower garden in the middle of the backyard, leaving a strip of grass of uniform width around the flower garden. To be happy, Arif must have 184 square meters of grass. Under these conditions what will the length and width of the garden be?

70. A rectangular piece of sheet metal has a length that is 4 inches less than twice the width. A square piece 2 inches on a side is cut from each corner. The sides are then turned up to form an uncovered box of volume 256 cubic inches. Find the length and width of the original piece of metal.

71. If a square piece of cardboard has 3-inch squares cut from its corners and then has the flaps folded up to form an open-top box, the volume of the box is given by the formula $V = 3(x - 6)^2$, where x is the length of each side of the original piece of cardboard in inches. What original length would yield a box with a volume of 432 cubic inches?

72. The adjusted poverty threshold for a single person from the year 1984 to the year 1990 is approximated by the quadratic model $T = 18.7x^2 + 105.3x + 4814.1$, where $x = 0$ corresponds to 1984, and T is in dollars. In what year during this period was the threshold T approximately \$5,300? (*Source:* Congressional Budget Office)

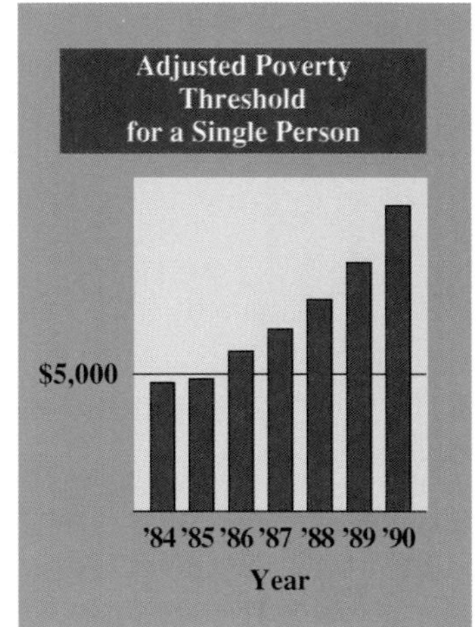

Recall from the exercises in Section 3 of this chapter that the corresponding sides of similar triangles are proportional. Use this fact to find the lengths of the indicated sides of the pair of similar triangles. Check all possible solutions in both triangles. Sides of a triangle cannot be negative.

73. side AC

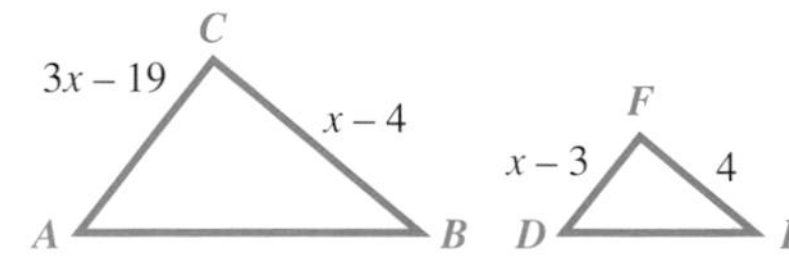

74. side RQ

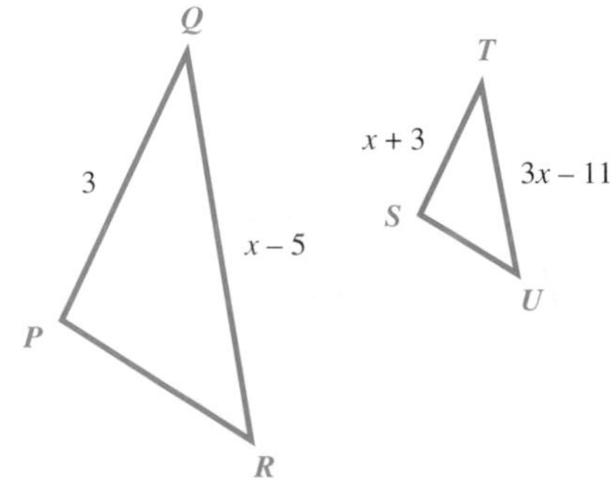

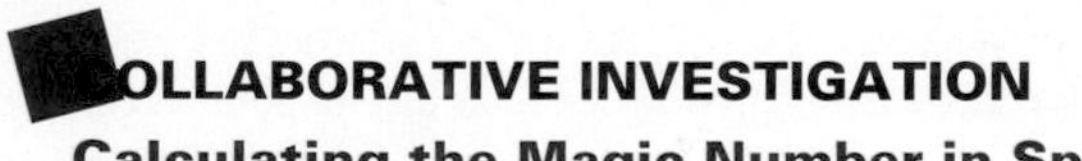

COLLABORATIVE INVESTIGATION
Calculating the Magic Number in Sports

Near the end of a major league baseball season, fans are often interested in the current first-place team's "magic number." The magic number is the sum of the required number of wins of the first-place team and the number of losses of the second-place team (for the remaining games) necessary to clinch the pennant. (In a regulation major league season, each team plays 162 games.)

To calculate the magic number M for a first-place team prior to the end of a season, we can use the formula

$$M = W_2 + N_2 - W_1 + 1, \quad \text{where}$$

W_2 = the current number of wins of the second-place team;

N_2 = the number of remaining games of the second-place team; and

W_1 = the current number of wins of the first-place team.

On Wednesday, September 6, 1995, baseball fans woke up to the following standings in their local newspapers.

National League

East Division	W	L	Pct.	GB
Atlanta	76	44	.633	—
Phila.	61	60	.504	14½
Montreal	58	62	.483	18
Florida	54	64	.458	21
New York	52	67	.437	23½

Central Division	W	L	Pct.	GB
Cincinnati	74	45	.622	—
Chicago	61	59	.508	13½
Houston	61	59	.508	13½
Pittsburgh	51	69	.425	23½
St. Louis	50	71	.413	25

West Division	W	L	Pct.	GB
Los Angeles	63	58	.521	—
Colorado	62	58	.517	½
San Diego	59	60	.496	2½
San Fran.	57	63	.475	5

American League

East Division	W	L	Pct.	GB
Boston	74	45	.622	—
New York	60	60	.500	14½
Baltimore	56	65	.463	19
Toronto	50	70	.417	24½
Detroit	50	70	.417	24½

Central Division	W	L	Pct.	GB
Cleveland	82	37	.689	—
Kansas City	61	58	.513	21
Milwaukee	59	61	.492	23½
Chicago	54	64	.458	27½
Minnesota	45	74	.378	37

West Division	W	L	Pct.	GB
California	68	54	.557	—
Seattle	61	59	.508	6
Texas	60	60	.500	7
Oakland	59	62	.488	8½

In 1995, due to the strike that shortened the season, each team played 144 games rather than the usual 162 games. To calculate Atlanta's magic number, we note that $W_2 = 61$ (the number of wins for Philadelphia), $N_2 = 144 - (61 + 60) = 23$ (the number of games Philadelphia had remaining), and $W_1 = 76$ (the number of wins Atlanta had). Therefore, the magic number for Atlanta was

$$M = 61 + 23 - 76 + 1 = 9.$$

Later in the season, when the total of Atlanta wins and Philadelphia losses became 9, Atlanta clinched the pennant.

Topics for Discussion

Divide the class into groups, and have each group choose one of the following teams: Boston, Cleveland, California, or Cincinnati. Determine the magic number for the team chosen. (*Note:* California "folded" and Seattle won the West Division of the American League!)

Chapter 7 Test

Solve each equation.

1. $5x - 3 + 2x = 3(x - 2) + 11$

2. $\dfrac{2p - 1}{3} + \dfrac{p + 1}{4} = \dfrac{43}{12}$

3. Decide whether the equation

$$3x - (2 - x) + 4x = 7x - 2 - (-x)$$

is conditional, an identity, or a contradiction. Solve and give its solution set.

4. Solve for v: $S = vt - 16t^2$.

Solve each of the following applications.

5. The three largest islands in the Hawaiian island chain are Hawaii (the Big Island), Maui, and Kauai. Together, their areas total 5,300 square miles. The island of Hawaii is 3,293 square miles larger than the island of Maui, and Maui is 177 square miles larger than Kauai. What is the area of each island?

6. How many liters of a 20% solution of a chemical should Michelle Jennings mix with 10 liters of a 50% solution to obtain a mixture that is 40% chemical?

7. A passenger train and a freight train leave a town at the same time and travel in opposite directions. Their speeds are 60 mph and 75 mph, respectively. How long will it take for them to be 297 miles apart?

8. Which is the better buy for processed cheese slices: 8 slices for $2.19 or 12 slices for $3.30?

9. The distance between Milwaukee and Boston is 1,050 miles. On a certain map this distance is represented by 21 inches. On the same map Seattle and Cincinnati are 46 inches apart. What is the actual distance between Seattle and Cincinnati?

10. The current in a simple electrical circuit is inversely proportional to the resistance. If the current is 80 amps when the resistance is 30 ohms, find the current when the resistance is 12 ohms.

Solve the inequality. Give the solution set in both interval and graph forms.

11. $-4x + 2(x - 3) \geq 4x - (3 + 5x) - 7$

12. $-10 < 3k - 4 \leq 14$

13. Which one of the following inequalities is equivalent to $x < -3$?
(a) $-3x < 9$ **(b)** $-3x > -9$ **(c)** $-3x > 9$ **(d)** $-3x < -9$

14. Edison Diest has grades of 83, 76, and 79 on his first three tests in Math 1031 (Survey of Mathematics). If he wants an average of at least 80 after his fourth test, what are the possible scores he can make on his fourth test?

Evaluate each exponential expression.

15. $\left(\dfrac{4}{3}\right)^2$

16. $-(-2)^6$

17. $\left(\dfrac{3}{4}\right)^{-3}$

18. $-5^0 + (-5)^0$

Use the properties of exponents to simplify each expression. Write answers with positive exponents only. Assume that all variables represent nonzero real numbers.

19. $9(4p^3)(6p^{-7})$

20. $\dfrac{m^{-2}(m^3)^{-3}}{m^{-4}m^7}$

21. Use scientific notation to evaluate $\dfrac{(2{,}500{,}000)(.00003)}{(.05)(5{,}000{,}000)}$. Leave the answer in scientific notation.

22. In the food chain that links the largest sea creature, the whale, to the smallest, the diatom, 4×10^{14} diatoms sustain a medium-sized whale for only a few hours. Write this number without using exponents.

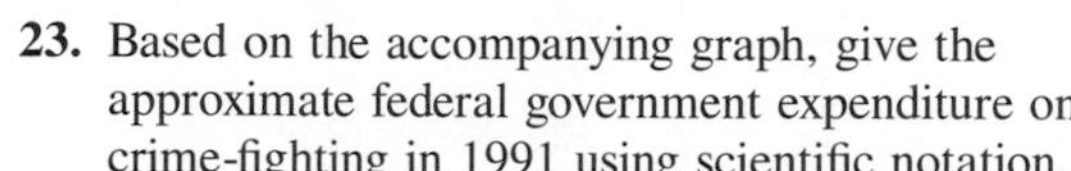

23. Based on the accompanying graph, give the approximate federal government expenditure on crime-fighting in 1991 using scientific notation.

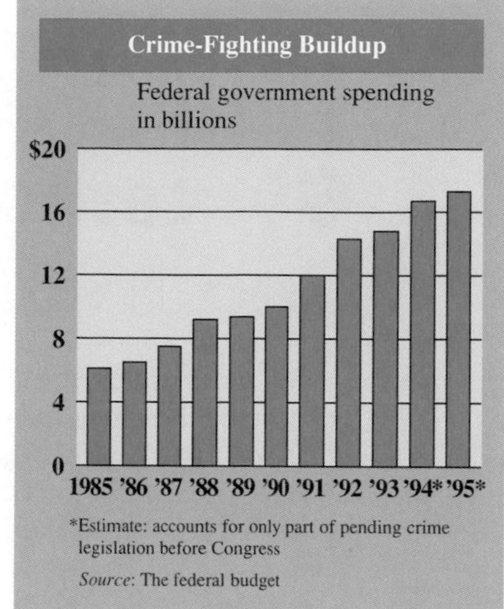

Perform the indicated operations.

24. $(3k^3 - 5k^2 + 8k - 2) - (3k^3 - 9k^2 + 2k - 12)$

25. $(5x + 2)(3x - 4)$

26. $(4x^2 - 3)(4x^2 + 3)$

27. $(x + 4)(3x^2 + 8x - 9)$

28. Give an example of a polynomial in the variable t, such that it is fifth degree, in descending powers of the variable, with exactly six terms, and having a negative coefficient for its quadratic term.

Factor each polynomial completely.

29. $2p^2 - 5pq + 3q^2$

30. $100x^2 - 49y^2$

31. $27y^3 - 125x^3$

32. $4x + 4y - mx - my$

Solve each quadratic equation.

33. $6x^2 + 7x - 3 = 0$

34. $x^2 - x = 7$

Solve the problem. Use a calculator and round the answer to the nearest hundredth.

35. The equation $s = 16t^2 + 15t$ gives the distance s in feet an object thrown off a building has descended in t seconds. Find the time t when the object has descended 25 feet.

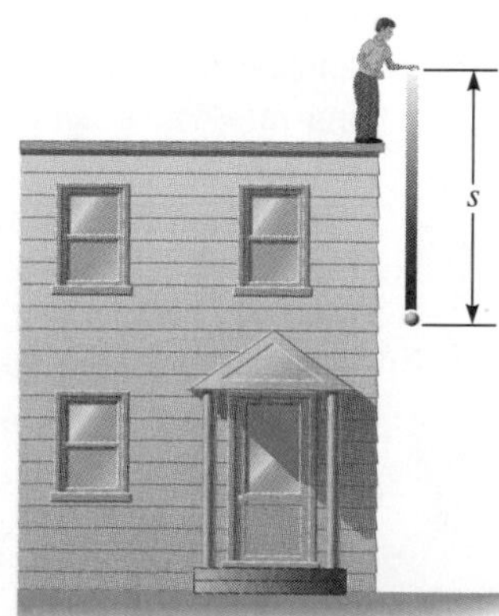

8

Functions, Graphs, and Systems of Equations and Inequalities

In 1896 the Swedish scientist Svante Arrhenius first predicted the greenhouse effect resulting from emissions of carbon dioxide into the air by the industrialized countries. The burning of fossil fuels, deforestation, and changes in land use from 1850 to 1986 input approximately 312 billion tons of carbon into the atmosphere, mostly in the form of carbon dioxide. Burning of fossil fuels continues to produce 5.4 billion tons of carbon each year, which is absorbed by both the atmosphere and the oceans. A critical aspect of the accumulation of carbon dioxide in the atmosphere is that it is essentially irreversible and requires hundreds of years for its effect to disappear. In 1990 the International Panel on Climate Change (IPCC) predicted that if current trends continue, the earth's average global temperature would increase. The table shows future amounts of atmospheric carbon dioxide in parts per million (ppm) and projected global temperature increases in degrees Celsius.

Year	Carbon Dioxide	Temp Increase
1980	330	0
2000	372	.42
2020	419	.84
2040	472	1.26
2060	532	1.68
2080	600	2.10
2100	676	2.52

During each twenty-year period, the temperature increases by .42° Celsius. However, the amount of carbon dioxide does not increase by a constant amount every twenty years, but rather continues to grow at an ever increasing rate. For example, during the twenty years from 1980 to 2000 carbon dioxide increases 42 ppm, whereas from 2080 to 2100 it increases 76 ppm. In this model, IPCC predicts that temperature will increase *linearly* while carbon dioxide levels will increase *exponentially*. The graph accompanying Exercises 47–50 in Section 8.6 illustrates past changes in global temperature together with two predictions for future temperature increases which are represented by the dashed line and curve. Compare these two predictions with the IPCC model for the year 2040.

Source: Clime W., *The Economics of Global Warming,* Institute for International Economics, Washington, DC, 1992.

The Rectangular Coordinate System and Circles

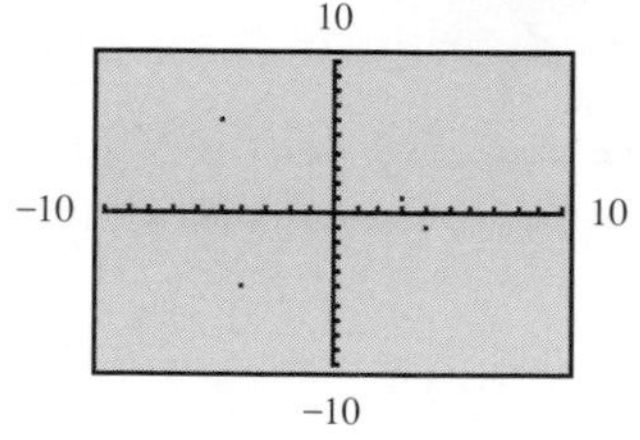

The points in Figure 2 are plotted on this calculator screen. Why is $E(-5, 0)$ not visible?

Each of the pairs of numbers $(1, 2)$, $(-1, 5)$ and $(3, 7)$ is an example of an **ordered pair;** that is, a pair of numbers written within parentheses in which the order of the numbers is important. The two numbers are the **components** of the ordered pair. An ordered pair is graphed using two number lines that intersect at right angles at the zero points, as shown in Figure 1. The common zero point is called the origin. The horizontal line, the ***x*-axis,** represents the first number in an ordered pair, and the vertical line, the ***y*-axis,** represents the second. The x-axis and the y-axis make up a **rectangular** (or Cartesian) **coordinate system.** The axes form four **quadrants,** numbered I, II, III, and IV as shown in Figure 2.

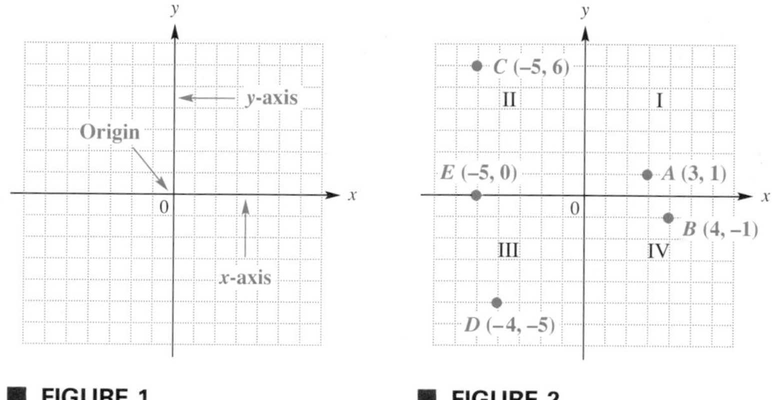

■ FIGURE 1 ■ FIGURE 2

We locate, or **plot,** the point on the graph that corresponds to the ordered pair $(3, 1)$ by going three units from zero to the right along the x-axis, and then one unit up parallel to the y-axis. The point corresponding to the ordered pair $(3, 1)$ is labeled A in Figure 2. The point $(4, -1)$ is labeled B, $(-5, 6)$ is labeled C, and $(-4, -5)$ is labeled D. Point E corresponds to $(-5, 0)$. The phrase "the point corresponding to the ordered pair $(3, 1)$" is often abbreviated "the point $(3, 1)$." The numbers in an ordered pair are called the **coordinates** of the corresponding point.

The parentheses used to represent an ordered pair are also used to represent an open interval (introduced in an earlier chapter). In general, there is no confusion between these symbols because the context of the discussion tells us whether we are discussing ordered pairs or open intervals.

Distance Formula

Suppose that we wish to find the distance between two points, say $(3, -4)$ and $(-5, 3)$. The Pythagorean theorem allows us to do this. In Figure 3, we see that the vertical line through $(-5, 3)$ and the horizontal line through $(3, -4)$ intersect at the point $(-5, -4)$. Thus, the point $(-5, -4)$ becomes the vertex of the right angle in a right triangle. By the Pythagorean theorem, the square of the

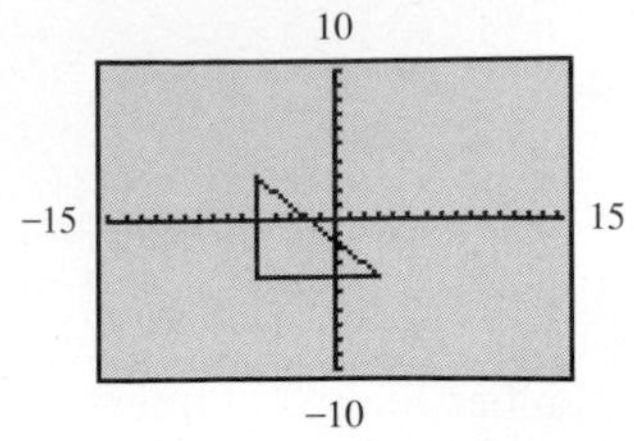

The right triangle in Figure 3 is "drawn" in this screen. A square window is used to obtain the correct perspective.

Double Descartes After the French postal service issued the above stamp in honor of René Descartes, sharp eyes noticed that the title of Descartes' most famous book was wrong. Thus a second stamp (see facing page) was issued with the correct title. The book in question, *Discourse on Method,* appeared in 1637. In it Descartes rejected traditional Aristotelian philosophy, outlining a universal system of knowledge that was to have the certainty of mathematics. He first adopted a skeptical view of everything, seeking "clear and distinct" ideas that any rational person could not doubt. One of these is his famous statement, "I think, therefore I am." For Descartes, method was *analysis,* going from self-evident truths step-by-step to more distant and more general truths. (Thomas Jefferson, also a rationalist, began the *Declaration* with the words, "We hold these truths to be self-evident.")

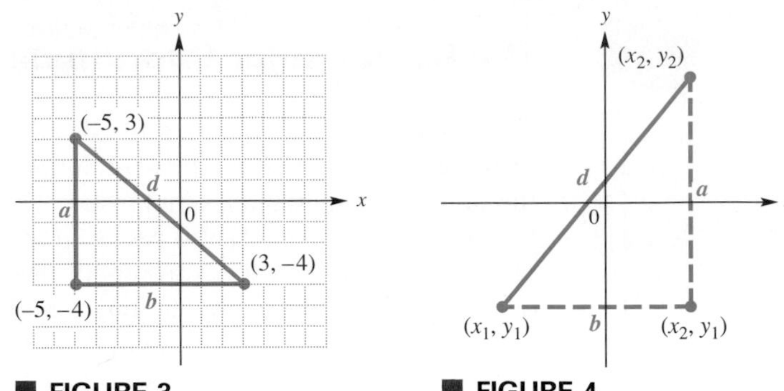

FIGURE 3 **FIGURE 4**

length of the hypotenuse, d, of the right triangle in Figure 3, is equal to the sum of the squares of the lengths of the two legs a and b:

$$d^2 = a^2 + b^2.$$

The length a is the difference between the coordinates of the endpoints. Since the x-coordinate of both points is -5, the side is vertical, and we can find a by finding the difference between the y-coordinates. Subtract -4 from 3 to get a positive value for a.

$$a = 3 - (-4) = 7$$

Similarly, find b by subtracting -5 from 3.

$$b = 3 - (-5) = 8$$

Substituting these values into the formula, we have

$$d^2 = a^2 + b^2$$

$$d^2 = 7^2 + 8^2 \qquad \text{Let } a = 7 \text{ and } b = 8.$$

$$d^2 = 49 + 64$$

$$d^2 = 113$$

$$d = \sqrt{113}. \qquad \text{Use the square root property.}$$

Therefore, the distance between $(-5, 3)$ and $(3, -4)$ is $\sqrt{113}$.

This result can be generalized. Figure 4 shows the two different points (x_1, y_1) and (x_2, y_2). To find a formula for the distance d between these two points, notice that the distance between (x_2, y_2) and (x_2, y_1) is given by $a = y_2 - y_1$, and the distance between (x_1, y_1) and (x_2, y_1) is given by $b = x_2 - x_1$. From the Pythagorean theorem,

$$d^2 = (x_2 - x_1)^2 + (y_2 - y_1)^2,$$

and by using the square root property, the distance formula is obtained.

Distance Formula

The distance between the points (x_1, y_1) and (x_2, y_2) is

$$d = \sqrt{(x_2 - x_1)^2 + (y_2 - y_1)^2}.$$

This result is called the **distance formula.**

The small numbers 1 and 2 in the ordered pairs (x_1, y_1) and (x_2, y_2) are called *subscripts.* We read x_1 as "x sub 1." Subscripts are used to distinguish between different values of a variable that have a common property. For example, in the ordered pairs $(-3, 5)$ and $(6, 4)$, -3 can be designated as x_1 and 6 as x_2. Their common property is that they are both x components of ordered pairs. This idea is used in the following example.

```
D=SQUARE ROOT OF
              82
OR
     9.055385138
            Done
```

A program can be written for the distance formula. This one supports the result in Example 1, since $\sqrt{82} \approx 9.055385138$.

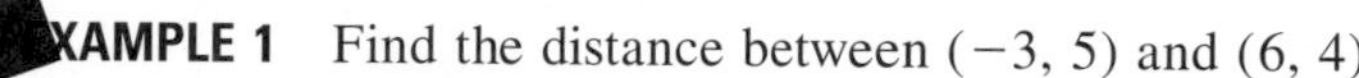

EXAMPLE 1 Find the distance between $(-3, 5)$ and $(6, 4)$.

When using the distance formula to find the distance between two points, designating the points as (x_1, y_1) and (x_2, y_2) is arbitrary. Let us choose $(x_1, y_1) = (-3, 5)$ and $(x_2, y_2) = (6, 4)$.

$$\begin{aligned} d &= \sqrt{(x_2 - x_1)^2 + (y_2 - y_1)^2} \\ &= \sqrt{(6 - (-3))^2 + (4 - 5)^2} \qquad x_2 = 6,\ y_2 = 4,\ x_1 = -3,\ y_1 = 5 \\ &= \sqrt{9^2 + (-1)^2} \\ &= \sqrt{82} \end{aligned}$$

Midpoint Formula

Given the coordinates of the two endpoints of a line segment, it is not difficult to find the coordinates of the midpoint of the segment. The midpoint of a line segment is the point on the segment that is equidistant from both endpoints. In Exercise 61, we outline a proof of the validity of the following formula.

Descartes wrote his *Geometry* as an application of his method; it was published as an appendix to the *Discourse.* His attempts to unify algebra and geometry influenced the creation of what became coordinate geometry and influenced the development of calculus by Newton and Leibniz in the next generation. In 1649 he went to Sweden to tutor Queen Christina. She preferred working in the unheated castle in the early morning; Descartes was used to staying in bed until noon. The rigors of the Swedish winter proved too much for him, and he died less than a year later.

Midpoint Formula

The coordinates of the midpoint of the segment with endpoints (x_1, y_1) and (x_2, y_2) are

$$\left(\frac{x_1 + x_2}{2}, \frac{y_1 + y_2}{2}\right).$$

In words, the midpoint formula says that the coordinates of the midpoint of a line segment are found by calculating the averages of the x- and y-coordinates of the endpoints.

EXAMPLE 2 Find the coordinates of the midpoint of the line segment with endpoints $(8, -4)$ and $(-9, 6)$.

Using the midpoint formula, we find that the coordinates of the midpoint are

$$\left(\frac{8 + (-9)}{2}, \frac{-4 + 6}{2}\right) = \left(-\frac{1}{2}, 1\right).$$

Circles

An application of the distance formula gives rise to one of the most familiar shapes in geometry, the circle. A **circle** is the set of all points in a plane that lie a fixed distance from a fixed point. The fixed point is called the **center** and the fixed distance is called the **radius.** The next example shows how the distance formula can be used to find an equation of a circle.

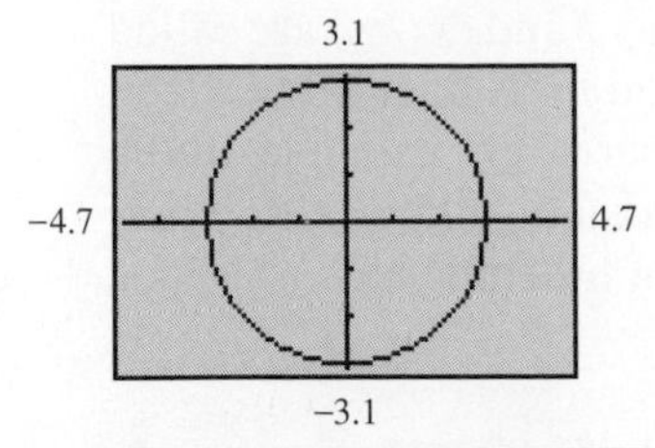

To graph $x^2 + y^2 = 9$, we solve for y to get $y_1 = \sqrt{9 - x^2}$ and $y_2 = -\sqrt{9 - x^2}$. Then we graph in a square window.

EXAMPLE 3 Find an equation of the circle with radius 3 and center at (0, 0), and graph the circle.

If the point (x, y) is on the circle, the distance from (x, y) to the center (0, 0) is 3, as shown in Figure 5. By the distance formula,

$$\sqrt{(x_2 - x_1)^2 + (y_2 - y_1)^2} = d$$

$$\sqrt{(x - 0)^2 + (y - 0)^2} = 3 \qquad x_1 = 0,\ y_1 = 0,\ x_2 = x,\ y_2 = y$$

$$x^2 + y^2 = 9. \qquad \text{Square both sides.}$$

An equation of this circle is $x^2 + y^2 = 9$. It may be graphed by using compasses, and locating all points three units from the origin.

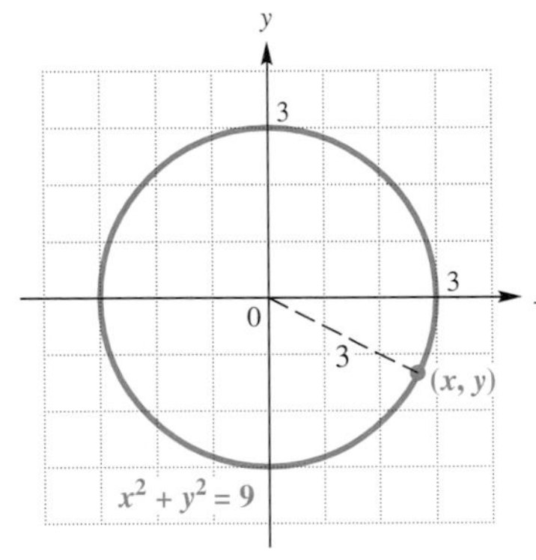

FIGURE 5

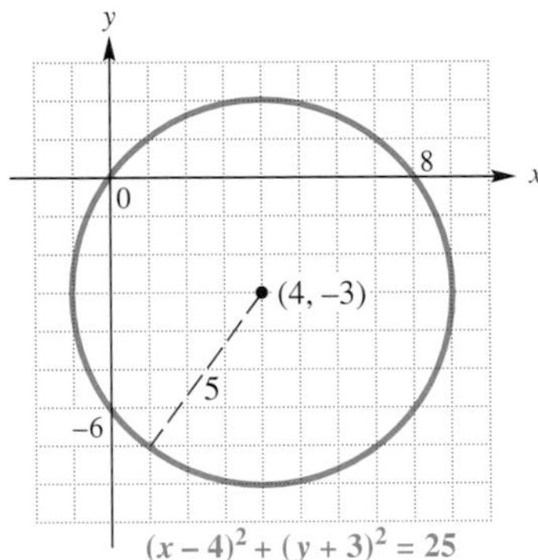

FIGURE 6

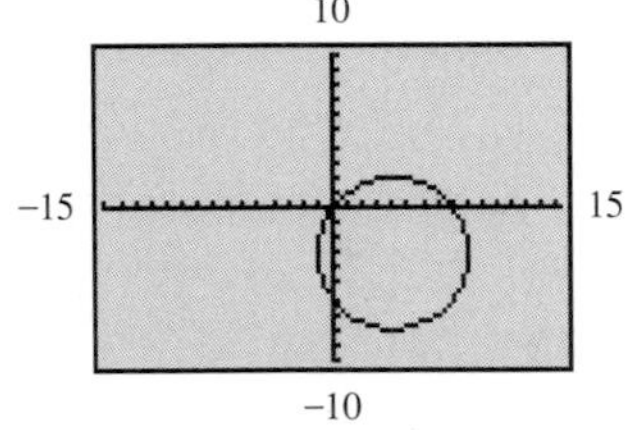

A circle can be "drawn" in a square window by using the appropriate command, entering the coordinates of the center and the radius. Compare with Figure 6.

EXAMPLE 4 Find an equation for the circle that has its center at $(4, -3)$ and radius 5, and graph the circle.

Again use the distance formula.

$$\sqrt{(x - 4)^2 + (y + 3)^2} = 5$$

$$(x - 4)^2 + (y + 3)^2 = 25 \qquad \text{Square both sides.}$$

The graph of this circle is shown in Figure 6.

Examples 3 and 4 can be generalized to get an equation of a circle with radius r and center at (h, k). If (x, y) is a point on the circle, the distance from the center (h, k) to the point (x, y) is r. Then by the distance formula,

$$\sqrt{(x - h)^2 + (y - k)^2} = r.$$

Squaring both sides gives the following equation of a circle.

Equation of a Circle

$$(x - h)^2 + (y - k)^2 = r^2$$

is an equation of a circle of radius r with center at (h, k). In particular, a circle of radius r with center at the origin has an equation

$$x^2 + y^2 = r^2.$$

EXAMPLE 5 Find an equation of the circle with center at $(-1, 2)$ and radius 4.

Let $h = -1$, $k = 2$, and $r = 4$ in the general equation above to get

$$(x - h)^2 + (y - k)^2 = r^2$$
$$[x - (-1)]^2 + (y - 2)^2 = (4)^2$$
$$(x + 1)^2 + (y - 2)^2 = 16.$$

In the equation found in Example 4, multiplying out $(x - 4)^2$ and $(y + 3)^2$ and then combining like terms gives

$$(x - 4)^2 + (y + 3)^2 = 25$$
$$x^2 - 8x + 16 + y^2 + 6y + 9 = 25$$
$$x^2 + y^2 - 8x + 6y = 0.$$

This result suggests that an equation that has both x^2 and y^2 terms may represent a circle. The next example shows how to tell, using the method of completing the square.

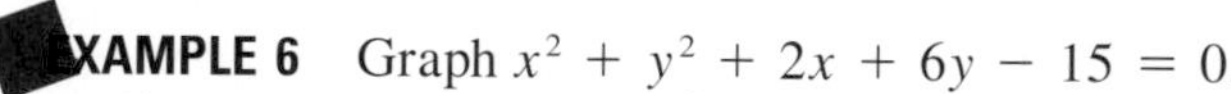

EXAMPLE 6 Graph $x^2 + y^2 + 2x + 6y - 15 = 0$.

Since the equation has x^2 and y^2 terms with equal coefficients, its graph might be that of a circle. To find the center and radius, complete the square on x and y as follows. (See the previous chapter, where completing the square is introduced.)

$$x^2 + y^2 + 2x + 6y = 15 \qquad \text{Move the constant to the right.}$$
$$(x^2 + 2x \qquad) + (y^2 + 6y \qquad) = 15 \qquad \text{Rewrite in anticipation of completing the square.}$$
$$(x^2 + 2x + 1) + (y^2 + 6y + 9) = 15 + 1 + 9 \qquad \text{Complete the square in both } x \text{ and } y.$$
$$(x + 1)^2 + (y + 3)^2 = 25 \qquad \text{Factor on the left and add on the right.}$$

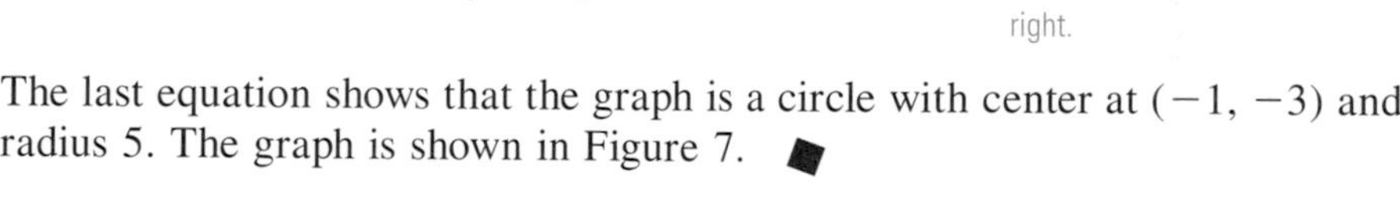

The last equation shows that the graph is a circle with center at $(-1, -3)$ and radius 5. The graph is shown in Figure 7.

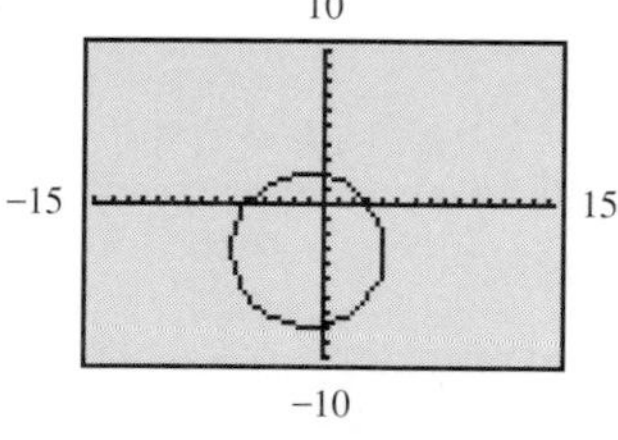

This is a calculator-generated picture of the circle in Figure 7.

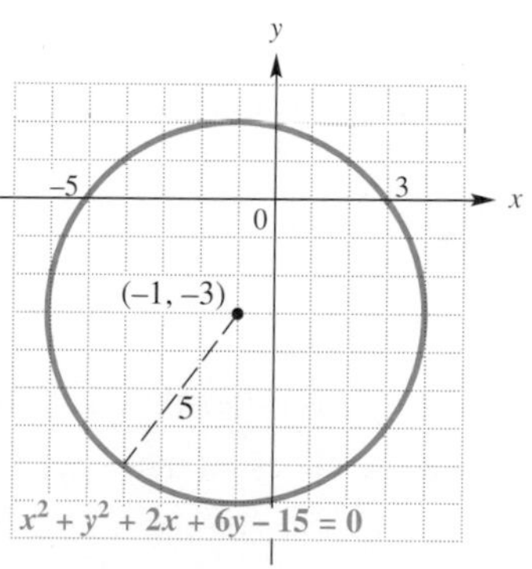

FIGURE 7

The final example in this section shows how equations of circles can be used in locating the epicenter of an earthquake.

EXAMPLE 7 Seismologists can locate the epicenter of an earthquake by determining the intersection of three circles. The radii of these circles represent the distances from the epicenter to each of three receiving stations. The centers of the circles represent the receiving stations.

Suppose receiving stations A, B, and C are located on a coordinate plane at the points $(1, 4)$, $(-3, -1)$, and $(5, 2)$. Let the distances from the earthquake epicenter to the stations be 2 units, 5 units, and 4 units, respectively. See Figure 8. Where on the coordinate plane is the epicenter located?

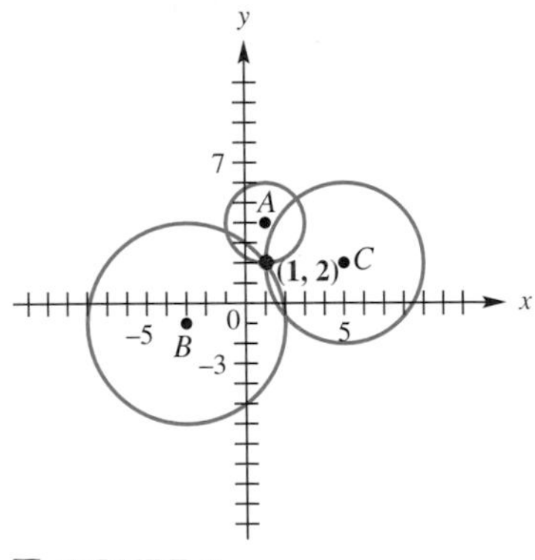

FIGURE 8

Graphically, it appears that the epicenter is located at $(1, 2)$. To check this algebraically, determine the equation for each circle and substitute $x = 1$ and $y = 2$.

Station A:
$$(x - 1)^2 + (y - 4)^2 = 4$$
$$(1 - 1)^2 + (2 - 4)^2 = 4$$
$$0 + 4 = 4$$
$$4 = 4$$

Station B:
$$(x + 3)^2 + (y + 1)^2 = 25$$
$$(1 + 3)^2 + (2 + 1)^2 = 25$$
$$16 + 9 = 25$$
$$25 = 25$$

Station C:
$$(x - 5)^2 + (y - 2)^2 = 16$$
$$(1 - 5)^2 + (2 - 2)^2 = 16$$
$$16 + 0 = 16$$
$$16 = 16$$

Thus, we can be sure that the epicenter lies at $(1, 2)$.

8.1 EXERCISES

Fill in the blank with the correct response.

1. The point with coordinates $(0, 0)$ is called the ______ of a rectangular coordinate system.
2. For any value of x, the point $(x, 0)$ lies on the ______ -axis.
3. For any value of y, the point $(0, y)$ lies on the ______ -axis.
4. The circle $x^2 + y^2 = 9$ has the point ______ as its center.
5. The point (______, 0) is the center of the circle $(x - 2)^2 + y^2 = 16$.
6. What is the radius of the circle $(x + 1)^2 + (y - 12)^2 = 36$?

Name the quadrant, if any, in which the point is located.

7. (a) (1, 6)
(b) (−4, −2)
(c) (−3, 6)
(d) (7, −5)
(e) (−3, 0)

8. (a) (−2, −10)
(b) (4, 8)
(c) (−9, 12)
(d) (3, −9)
(e) (0, −8)

9. Use the given information to determine the possible quadrants in which the point (x, y) must lie.

(a) $xy > 0$ **(b)** $xy < 0$ **(c)** $\frac{x}{y} < 0$ **(d)** $\frac{x}{y} > 0$

10. What must be true about the coordinates of any point that lies along an axis?

Locate the following points on the rectangular coordinate system.

11. (2, 3) **12.** (−1, 2) **13.** (−3, −2) **14.** (1, −4) **15.** (0, 5)
16. (−2, −4) **17.** (−2, 4) **18.** (3, 0) **19.** (−2, 0) **20.** (3, −3)

Find **(a)** *the distance between the pair of points, and* **(b)** *the coordinates of the midpoint of the segment having the points as endpoints.*

21. (3, 4) and (−2, 1) **22.** (−2, 1) and (3, −2) **23.** (−2, 4) and (3, −2)
24. (1, −5) and (6, 3) **25.** (−3, 7) and (2, −4) **26.** (0, 5) and (−3, 12)

Match the equation with the correct graph.

27. $(x - 3)^2 + (y - 2)^2 = 25$
28. $(x - 3)^2 + (y + 2)^2 = 25$
29. $(x + 3)^2 + (y - 2)^2 = 25$
30. $(x + 3)^2 + (y + 2)^2 = 25$

A.

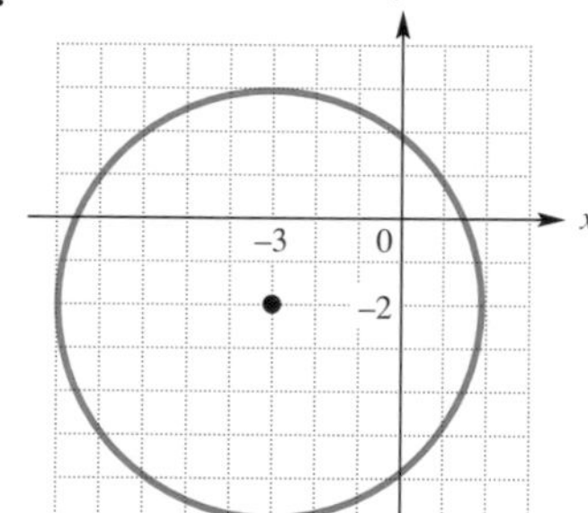

B.

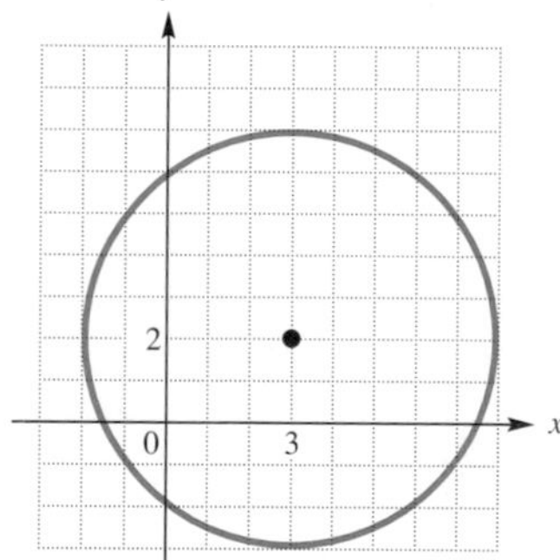

C.

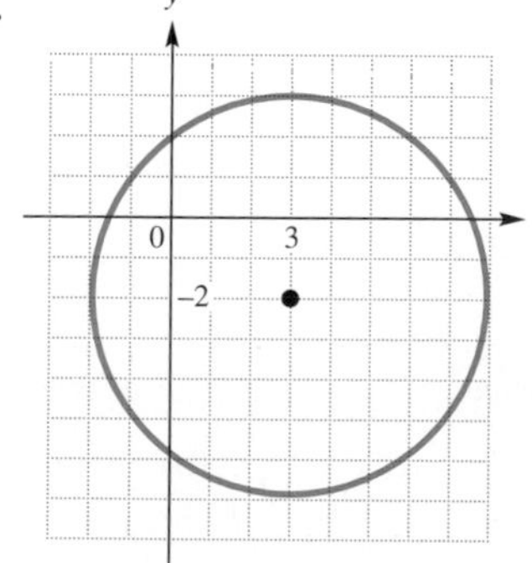

D.

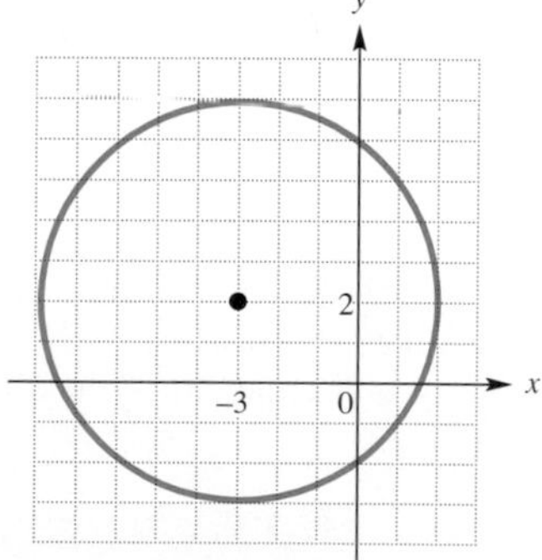

Write an equation of the circle with the given center and radius.

31. (0, 0); $r = 6$ **32.** (0, 0); $r = 5$ **33.** (−1, 3); $r = 4$
34. (2, −2); $r = 3$ **35.** (0, 4); $r = \sqrt{3}$ **36.** (−2, 0); $r = \sqrt{5}$

37. Suppose that a circle has an equation of the form $x^2 + y^2 = r^2$, $r > 0$. What is the center of the circle? What is the radius of the circle?

38. **(a)** How many points are there on the graph of $(x - 4)^2 + (y - 1)^2 = 0$? Explain your answer.
(b) How many points are there on the graph of $(x - 4)^2 + (y - 1)^2 = -1$? Explain your answer.

Find the center and the radius of the circle. (Hint: In Exercises 43 and 44 divide both sides by the greatest common factor.)

39. $x^2 + y^2 + 4x + 6y + 9 = 0$

40. $x^2 + y^2 - 8x - 12y + 3 = 0$

41. $x^2 + y^2 + 10x - 14y - 7 = 0$

42. $x^2 + y^2 - 2x + 4y - 4 = 0$

43. $3x^2 + 3y^2 - 12x - 24y + 12 = 0$

44. $2x^2 + 2y^2 + 20x + 16y + 10 = 0$

Graph the circle.

45. $x^2 + y^2 = 36$

46. $x^2 + y^2 = 81$

47. $(x - 2)^2 + y^2 = 36$

48. $x^2 + (y + 3)^2 = 49$

49. $(x + 2)^2 + (y - 5)^2 = 16$

50. $(x - 4)^2 + (y - 3)^2 = 25$

51. $(x + 3)^2 + (y + 2)^2 = 36$

52. $(x - 5)^2 + (y + 4)^2 = 49$

---

53. An alternate form of the distance formula is

$$d = \sqrt{(x_1 - x_2)^2 + (y_1 - y_2)^2}.$$

Compare this to the form given in this section, and explain why the two forms are equivalent.

54. A student was asked to find the distance between the points (5, 8) and (2, 14), and wrote the following:

$$d = \sqrt{(5 - 8)^2 + (2 - 14)^2}.$$

Explain why this is incorrect.

55. A circle can be drawn on a piece of posterboard by fastening one end of a string, pulling the string taut with a pencil, and tracing a curve as shown in the figure. Explain why this method works.

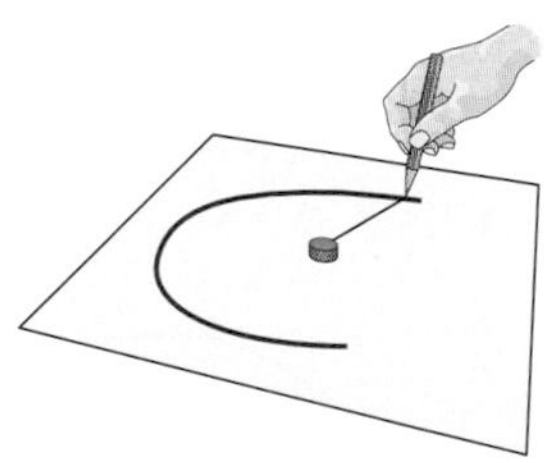

56. This figure shows how the crawfish race is held at the Crawfish Festival in Breaux Bridge, Louisiana. Explain why a circular "racetrack" is appropriate for such a race.

57. Show algebraically that if three receiving stations at (1, 4), (−6, 0), and (5, −2) record distances to an earthquake epicenter of 4 units, 5 units, and 10 units, respectively, the epicenter would lie at (−3, 4).

58. Three receiving stations record the presence of an earthquake. The location of the receiving center and the distance to the epicenter are contained in the following three equations: $(x - 2)^2 + (y - 1)^2 = 25$, $(x + 2)^2 + (y - 2)^2 = 16$, and $(x - 1)^2 + (y + 2)^2 = 9$. Graph the circles and determine the location of the earthquake epicenter.

59. Without actually graphing, state whether or not the graphs of $x^2 + y^2 = 4$ and $x^2 + y^2 = 25$ will intersect. Explain your answer.

60. Can a circle have its center at (2, 4) and be tangent to both axes? (*Tangent to* means touching in one point.) Explain.

61. Suppose that the endpoints of a line segment have coordinates (x_1, y_1) and (x_2, y_2).
(a) Show that the distance between (x_1, y_1) and $\left(\frac{x_1 + x_2}{2}, \frac{y_1 + y_2}{2}\right)$ is the same as the distance between (x_2, y_2) and $\left(\frac{x_1 + x_2}{2}, \frac{y_1 + y_2}{2}\right)$.
(b) Show that the sum of the distances between (x_1, y_1) and $\left(\frac{x_1 + x_2}{2}, \frac{y_1 + y_2}{2}\right)$, and (x_2, y_2) and $\left(\frac{x_1 + x_2}{2}, \frac{y_1 + y_2}{2}\right)$ is equal to the distance between (x_1, y_1) and (x_2, y_2).

(c) From the results of parts (a) and (b), what conclusion can be made?

62. If the coordinates of one endpoint of a line segment are $(3, -8)$ and the coordinates of the midpoint of the segment are $(6, 5)$, what are the coordinates of the other endpoint?

63. Which one of the following has a circle as its graph?
(a) $x^2 - y^2 = 9$ **(b)** $x^2 = 9 - y^2$
(c) $y^2 - x^2 = 9$ **(d)** $-x^2 - y^2 = 9$

64. For the three choices that are not circles in Exercise 63, explain why their equations are not those of circles.

8.2 Lines and Their Slopes

In the previous chapter we studied linear equations in a single variable. The solution of such an equation is a real number. We will now examine linear equations in *two* variables. An equation with two variables will have solutions written as ordered pairs. Unlike linear equations in a single variable, equations with two variables will, in general, have an infinite number of solutions.

To find the ordered pairs that satisfy the equation, select any number for one of the variables, substitute it into the equation for that variable, and then solve for the other variable. For example, suppose $x = 0$ in the equation $2x + 3y = 6$. Then, by substitution,

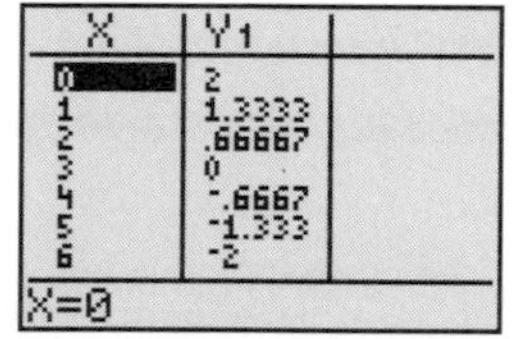

Modern graphing calculators can generate tables of ordered pairs. Here is an example of one for $2x + 3y = 6$. We must solve for y to get $y_1 = (6 - 2x)/3$ before generating the table.

$$2x + 3y = 6$$

becomes

$$2(\mathbf{0}) + 3y = 6 \qquad \text{Let } x = 0.$$
$$0 + 3y = 6$$
$$3y = 6$$
$$y = 2,$$

giving the ordered pair $(0, 2)$. Other ordered pairs satisfying $2x + 3y = 6$ include $(6, -2)$, $(3, 0)$, $(-3, 4)$, and $(9, -4)$, for example.

The equation $2x + 3y = 6$ is graphed by first plotting all the ordered pairs mentioned above. These are shown in Figure 9(a). The resulting points appear to lie on a straight line. If all the ordered pairs that satisfy the equation $2x + 3y = 6$ were graphed, they would form a straight line. In fact, the graph of any first-degree equation in two variables is a straight line. The graph of $2x + 3y = 6$ is the line shown in Figure 9(b).

All first-degree equations with two variables have straight-line graphs. Since a straight line is determined if any two different points on the line are known, finding two different points is enough to graph the line.

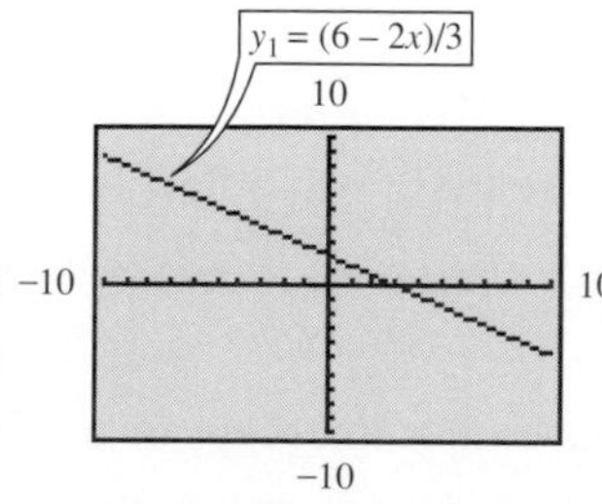

This is a calculator graph of the line shown in Figure 9(b).

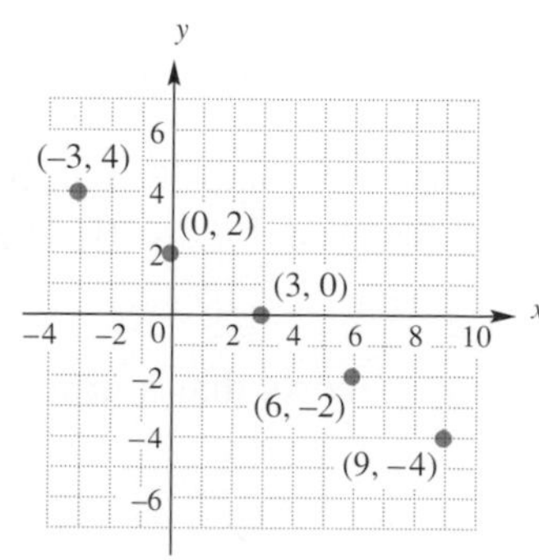

(a)

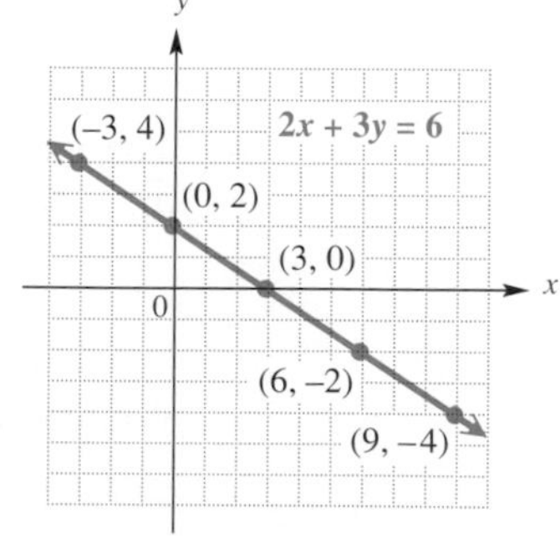

(b)

FIGURE 9

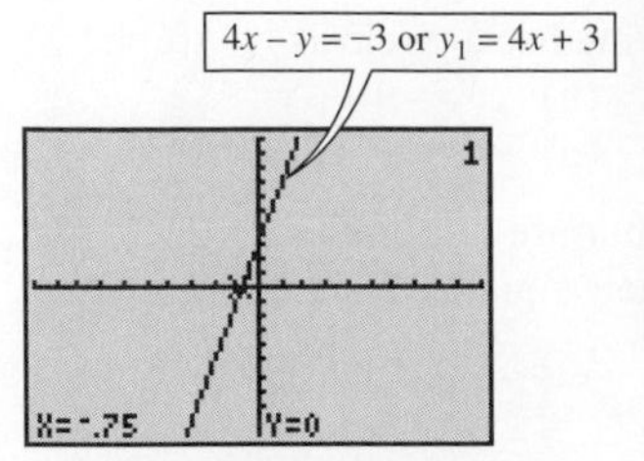

The display at the bottom of the screen supports the fact that $\left(-\frac{3}{4}, 0\right)$ is the x-intercept of the line in Figure 10.

Standard Form of a Linear Equation in Two Variables

An equation that can be written in the form

$$Ax + By = C \quad (A \text{ and } B \text{ not both } 0)$$

is a linear equation. This form is called the **standard form.**

Two points that are useful for graphing lines are the x- and y-intercepts. The ***x*-intercept** is the point (if any) where the line crosses the x-axis, and the ***y*-intercept** is the point (if any) where the line crosses the y-axis. (*Note:* In many texts, the intercepts are defined as numbers, and not points. However, in this book we will refer to intercepts as points.)

Intercepts can be found as follows.

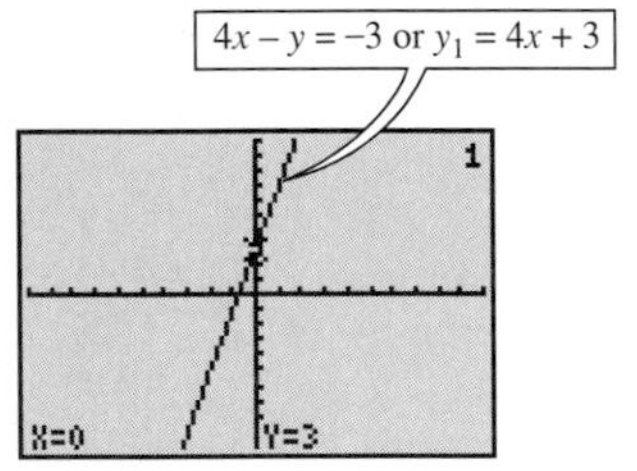

The display at the bottom of the screen supports the fact that (0, 3) is the y-intercept of the line in Figure 10.

Intercepts

Let $y = 0$ to find the x-intercept; let $x = 0$ to find the y-intercept.

EXAMPLE 1 Find the x- and y-intercepts of $4x - y = -3$, and graph the equation.

Find the x-intercept by letting $y = 0$.

$$\begin{aligned} 4x - \mathbf{0} &= -3 && \text{Let } y = 0. \\ 4x &= -3 \\ x &= -\frac{3}{4} && x\text{-intercept is } \left(-\frac{3}{4}, 0\right). \end{aligned}$$

For the y-intercept, let $x = 0$.

$$\begin{aligned} 4(\mathbf{0}) - y &= -3 && \text{Let } x = 0. \\ -y &= -3 \\ y &= 3 && y\text{-intercept is } (0, 3). \end{aligned}$$

The intercepts are the two points $(-3/4, 0)$ and $(0, 3)$. Use these two points to draw the graph, as shown in Figure 10.

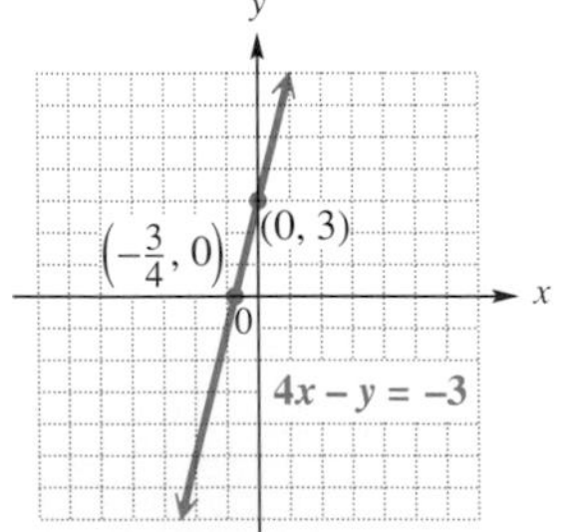

■ FIGURE 10

The next example shows that a graph can fail to have an x-intercept or a y-intercept.

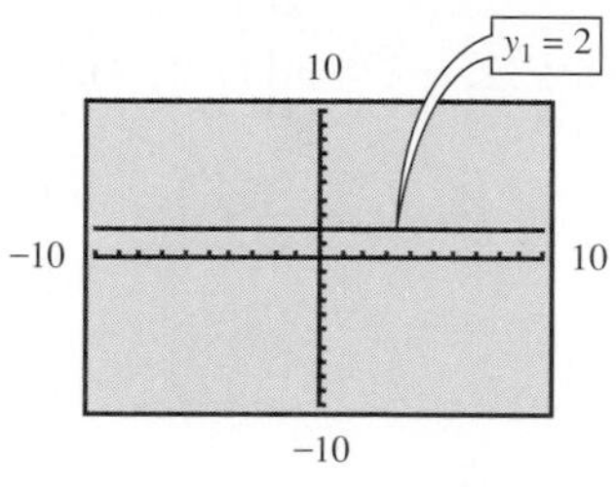

Compare this graph with the one in Figure 11.

EXAMPLE 2 **(a)** Graph $y = 2$.

Writing $y = 2$ as $0x + 1y = 2$ shows that any value of x, including $x = 0$, gives $y = 2$, making the y-intercept $(0, 2)$. Since y is always 2, there is no value of x corresponding to $y = 0$, and so the graph has no x-intercept. The graph, shown in Figure 11, is a horizontal line.

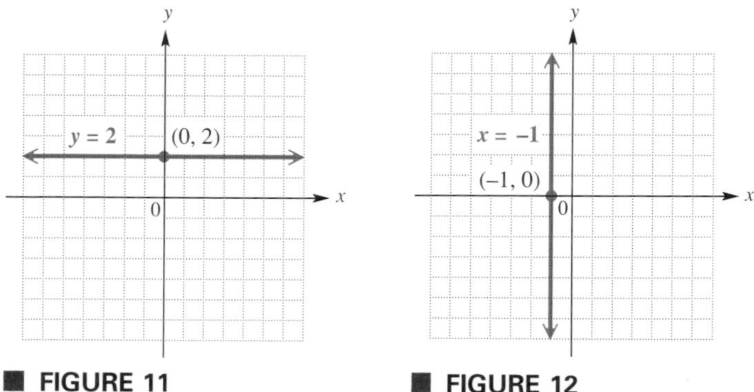

■ **FIGURE 11**

■ **FIGURE 12**

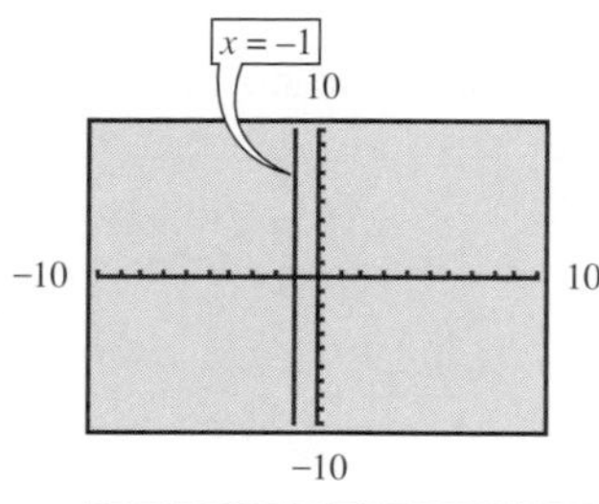

This vertical line is not an example of a *function* (see Section 4 in this chapter), so we must use a *draw* command to obtain it. Compare with Figure 12.

(b) Graph $x = -1$.

The form $1x + 0y = -1$ shows that every value of y leads to $x = -1$, and so no value of y makes $x = 0$. The graph, therefore, has no y-intercept. The only way a straight line can have no y-intercept is to be vertical, as shown in Figure 12. ■

Slope

Two different points determine a line. A line also can be determined by a point on the line and some measure of the "steepness" of the line. The measure of the steepness of a line is called the *slope* of the line. One way to get a measure of the steepness of a line is to compare the vertical change in the line (the *rise*) to the horizontal change (the *run*) while moving along the line from one fixed point to another.

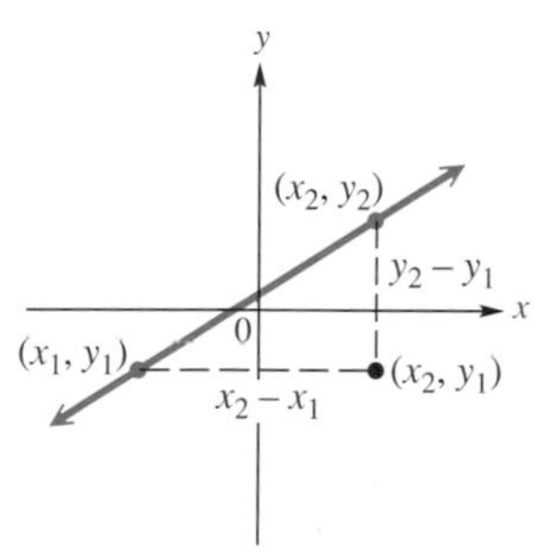

■ **FIGURE 13**

Suppose that (x_1, y_1) and (x_2, y_2) are two different points on a line. Then, going along the line from (x_1, y_1) to (x_2, y_2), the y-value changes from y_1 to y_2, an amount equal to $y_2 - y_1$. As y changes from y_1 to y_2, the value of x changes from x_1 to x_2 by the amount $x_2 - x_1$. See Figure 13. The ratio of the change in y to the change in x is called the **slope** of the line, with the letter m used for the slope.

Slope

If $x_1 \neq x_2$, the slope of the line through the distinct points (x_1, y_1) and (x_2, y_2) is

$$m = \frac{\textbf{rise}}{\textbf{run}} = \frac{\textbf{change in } y}{\textbf{change in } x} = \frac{y_2 - y_1}{x_2 - x_1}.$$

EXAMPLE 3 Find the slope of the line through the points $(2, -1)$ and $(-5, 3)$.

If $(2, -1) = (x_1, y_1)$ and $(-5, 3) = (x_2, y_2)$, then

$$m = \frac{y_2 - y_1}{x_2 - x_1}$$
$$= \frac{3 - (-1)}{-5 - 2} = \frac{4}{-7} = -\frac{4}{7}.$$

See Figure 14. On the other hand, if $(2, -1) = (x_2, y_2)$ and $(-5, 3) = (x_1, y_1)$, the slope would be

$$m = \frac{-1 - 3}{2 - (-5)} = \frac{-4}{7} = -\frac{4}{7},$$

the same answer. This example suggests that the slope is the same no matter which point is considered first. Also, using similar triangles from geometry, it can be shown that the slope is the same no matter which two different points on the line are chosen.

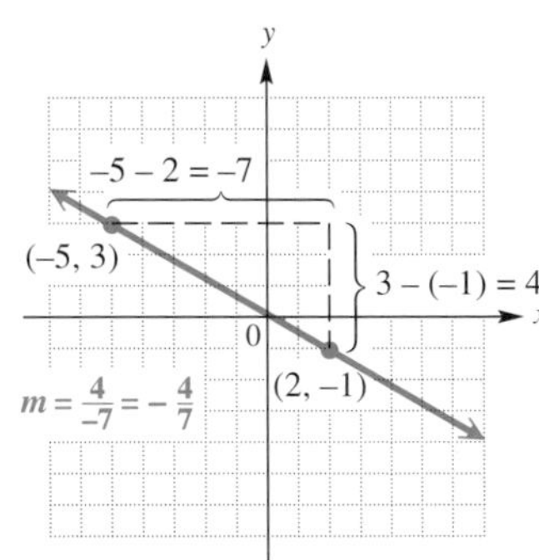

FIGURE 14

If we apply the slope formula to a vertical or a horizontal line, we find that one of the terms in the fraction is 0. The next example illustrates this.

EXAMPLE 4 Find the slope, if possible, of each of the following lines.

(a) $x = -3$

By inspection, $(-3, 5)$ and $(-3, -4)$ are two points that satisfy the equation $x = -3$. Use these two points to find the slope.

$$m = \frac{-4 - 5}{-3 - (-3)} = \frac{-9}{0}$$

Since division by zero is undefined, the slope is undefined. This is why the definition of slope includes the restriction that $x_1 \neq x_2$.

(b) $y = 5$

Find the slope by selecting two different points on the line, such as $(3, 5)$ and $(-1, 5)$, and by using the definition of slope.

$$m = \frac{5 - 5}{3 - (-1)} = \frac{0}{4} = 0$$

Hill

Slope Warning Take special care when you see this traffic symbol. You're headed for a downgrade that may be long, steep, or sharply curved.

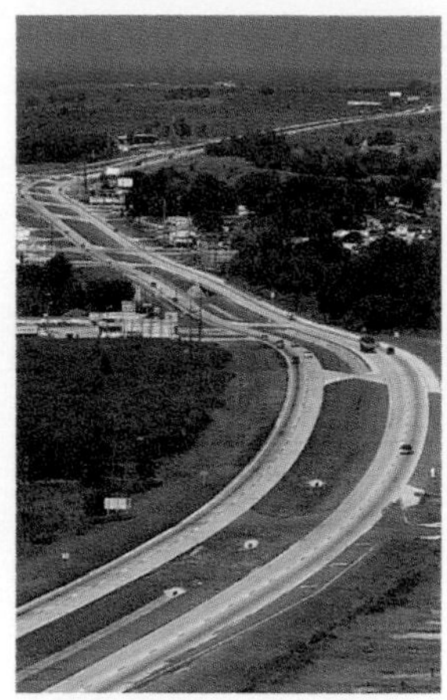

Highway slopes are measured in percent. For example, a slope of 3% means that the road gains 3 feet in altitude for each 100 feet that the road travels horizontally. Interstate highways cannot exceed a slope of 6%. While this may not seem like much of a slope, there are probably stretches of interstate highways that would be hard work for a distance runner.

As we observed in Example 2, $x = -1$ has a graph that is a vertical line, and $y = 2$ has a graph that is a horizontal line. Generalizing from those results and the results of Example 4, we can make the following statements about vertical and horizontal lines.

Vertical and Horizontal Lines

A vertical line has an equation of the form $x = k$, where k is a real number, and its slope is undefined. A horizontal line has an equation of the form $y = k$, and its slope is 0.

If we know the slope of a line and a point contained on the line, then we can graph the line using the method shown in the next example.

EXAMPLE 5 Graph the line that has slope 2/3 and goes through the point $(-1, 4)$.

First locate the point $(-1, 4)$ on a graph as shown in Figure 15. Then, from the definition of slope,

$$m = \frac{\text{change in } y}{\text{change in } x} = \frac{2}{3}.$$

Move *up* 2 units in the y-direction and then 3 units to the *right* in the x-direction to locate another point on the graph (labeled P). The line through $(-1, 4)$ and P is the required graph. ■

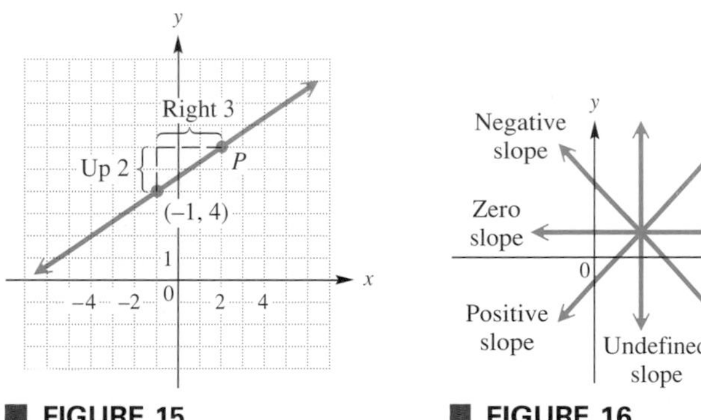

■ FIGURE 15 ■ FIGURE 16

The line graphed in Figure 14 has a negative slope, $-4/7$, and the line goes down from left to right. On the other hand, the line graphed in Figure 15 has a positive slope, 2/3, and it goes up from left to right. These are particular cases of a general statement that can be made about slopes. (Figure 16 shows lines of positive, zero, negative, and undefined slopes.)

Positive and Negative Slopes

A line with a positive slope goes up (rises) from left to right, while a line with a negative slope goes down (falls) from left to right.

The slopes of a pair of parallel or perpendicular lines are related in a special way. The slope of a line measures the steepness of a line. Since parallel lines have equal steepness, their slopes also must be equal. Also, lines with the same slope are parallel.

Slopes of Parallel Lines

Two nonvertical lines with the same slope are parallel; two nonvertical parallel lines have the same slope. Furthermore, all vertical lines are parallel.

EXAMPLE 6 Are the lines L_1, through $(-2, 1)$ and $(4, 5)$, and L_2, through $(3, 0)$ and $(0, -2)$, parallel?

The slope of L_1 is

$$m_1 = \frac{5 - 1}{4 - (-2)} = \frac{4}{6} = \frac{2}{3}.$$

The slope of L_2 is

$$m_2 = \frac{-2 - 0}{0 - 3} = \frac{-2}{-3} = \frac{2}{3}.$$

Since the slopes are equal, the lines are parallel. ■

Perpendicular lines are lines that meet at right angles. It can be shown that the slopes of perpendicular lines have a product of -1, provided that neither line is vertical. For example, if the slope of a line is $3/4$, then any line perpendicular to it has slope $-4/3$, because $(3/4)(-4/3) = -1$.

Slopes of Perpendicular Lines

If neither is vertical, perpendicular lines have slopes that are negative reciprocals; that is, their product is -1. Also, lines with slopes that are negative reciprocals are perpendicular. Every vertical line is perpendicular to every horizontal line.

EXAMPLE 7 Are the lines L_1, through $(0, -3)$ and $(2, 0)$, and L_2, through $(-3, 0)$ and $(0, -2)$, perpendicular?

The slope of L_1 is

$$m_1 = \frac{0 - (-3)}{2 - 0} = \frac{3}{2}.$$

The slope of L_2 is

$$m_2 = \frac{-2 - 0}{0 - (-3)} = -\frac{2}{3}.$$

Since the product of the slopes of the two lines is $(3/2)(-2/3) = -1$, the lines are perpendicular. ■

Problem Solving

We have seen how the slope of a line is the ratio of the change in y (vertical change) to the change in x (horizontal change). This idea can be extended to real-life situations as follows: the slope gives the average rate of change of y per unit of change in x, where the value of *y is dependent upon the value of x*. The next example illustrates this idea of average rate of change. We assume a linear relationship between x and y.

EXAMPLE 8 An environmental researcher finds that when a certain chemical pollutant is introduced into a large lake, the reproduction of redfish declines. In a given period of time, dumping 5 tons of the chemical results in a redfish population of 24,000. Also, dumping 15 tons of the chemical results in a redfish population of 10,000. Let y be the redfish population when x tons of pollutant are introduced into the lake, and find the average rate of change of y per unit change in x.

Since 5 tons of chemical yields a population of 24,000, $(x_1, y_1) = (5, 24{,}000)$. Similarly, since 15 tons yields a population of 10,000, $(x_2, y_2) = (15, 10{,}000)$. The average rate of change of y per unit change in x is found by using the slope formula.

$$\begin{aligned}\begin{array}{r}\text{Average rate of change}\\ \text{of } y \text{ per unit change in } x\end{array} &= \frac{y_2 - y_1}{x_2 - x_1} = \frac{10{,}000 - 24{,}000}{15 - 5}\\ &= \frac{-14{,}000}{10} = -1{,}400\end{aligned}$$

The result, $-1{,}400$, indicates that there is a decrease of 1,400 fish for every ton of pollutant introduced into the lake. Geometrically, this would indicate that the line joining the points (5, 24,000) and (15, 10,000) has slope $-1{,}400$. ◆

8.2 EXERCISES

Complete the given ordered pairs for the equation. Then graph the equation.

1. $2x + y = 5$; $(0, \ )$, $(\ , 0)$, $(1, \ )$, $(\ , 1)$

2. $3x - 4y = 24$; $(0, \ )$, $(\ , 0)$, $(6, \ )$, $(\ , -3)$

3. $x - y = 4$; $(0, \ )$, $(\ , 0)$, $(2, \ )$, $(\ , -1)$

4. $x + 3y = 12$; $(0, \ )$, $(\ , 0)$, $(3, \ )$, $(\ , 6)$

5. $4x + 5y = 20$; $(0, \ )$, $(\ , 0)$, $(3, \ )$, $(\ , 2)$

6. $2x - 5y = 12$; $(0, \ )$, $(\ , 0)$, $(\ , -2)$, $(-2, \ )$

7. $3x + 2y = 8$

x	y
0	
	0
2	
	−2

8. $5x + y = 12$

x	y
0	
	0
	−3
2	

9. Explain how to find the x-intercept of a linear equation in two variables.

10. Explain how to find the y-intercept of a linear equation in two variables.

11. Which one of the following has as its graph a horizontal line?
(a) $2y = 6$ **(b)** $2x = 6$ **(c)** $x - 4 = 0$ **(d)** $x + y = 0$

12. What is the minimum number of points that must be determined in order to graph a linear equation in two variables?

For each equation, give the x-intercept and the y-intercept. Then graph the equation.

13. $3x + 2y = 12$ **14.** $2x + 5y = 10$ **15.** $5x + 6y = 10$ **16.** $3y + x = 6$

17. $2x - y = 5$ **18.** $3x - 2y = 4$ **19.** $x - 3y = 2$ **20.** $y - 4x = 3$

21. $y + x = 0$ **22.** $2x - y = 0$ **23.** $3x = y$ **24.** $x = -4y$

25. $x = 2$ **26.** $y = -3$ **27.** $y = 4$ **28.** $x = -2$

In Exercises 29–36, match the equation with the figure that most closely resembles its graph.

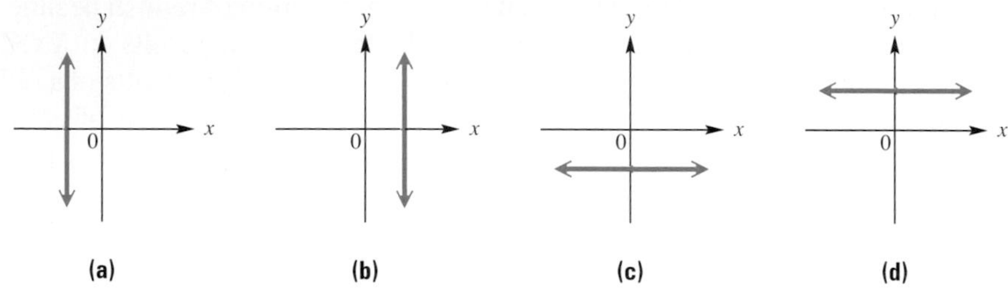

29. $y + 2 = 0$ **30.** $y + 4 = 0$ **31.** $x + 3 = 0$ **32.** $x + 7 = 0$

33. $y - 2 = 0$ **34.** $y - 4 = 0$ **35.** $x - 3 = 0$ **36.** $x - 7 = 0$

37. If a walkway rises 2 feet for every 10 feet on the horizontal, which of the following express its slope (or grade)? (There are several correct choices.)

(a) .2 **(b)** $\frac{2}{10}$ **(c)** $\frac{1}{5}$ **(d)** 20%

(e) 5 **(f)** $\frac{20}{100}$ **(g)** 500% **(h)** $\frac{10}{2}$

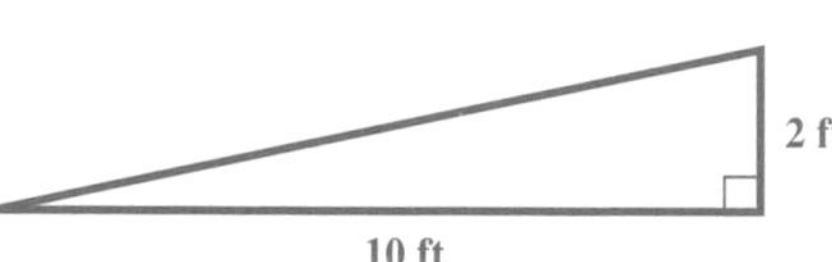

38. If the pitch of a roof is $\frac{1}{4}$, how many feet in the horizontal direction corresponds to a rise of 3 feet?

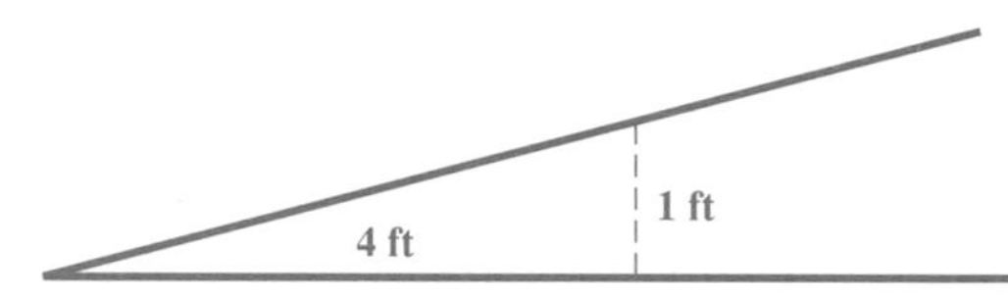

Find the slope of the line through the pair of points by using the slope formula.

39. $(-2, -3)$ and $(-1, 5)$ **40.** $(-4, 3)$ and $(-3, 4)$ **41.** $(-4, 1)$ and $(2, 6)$

42. $(-3, -3)$ and $(5, 6)$ **43.** $(2, 4)$ and $(-4, 4)$ **44.** $(-6, 3)$ and $(2, 3)$

Tell whether the slope of the given line is positive, negative, zero, *or* undefined.

45.

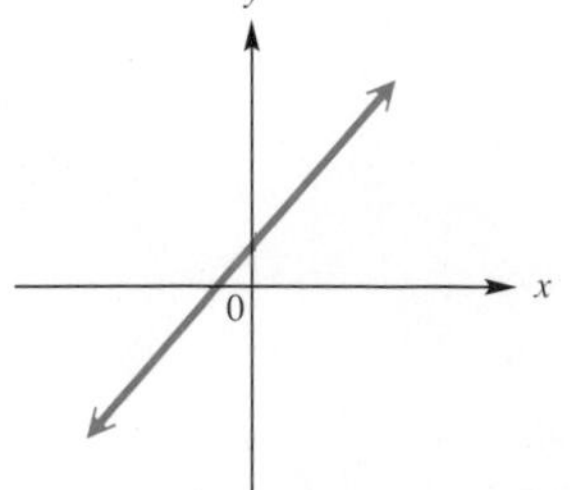

46.

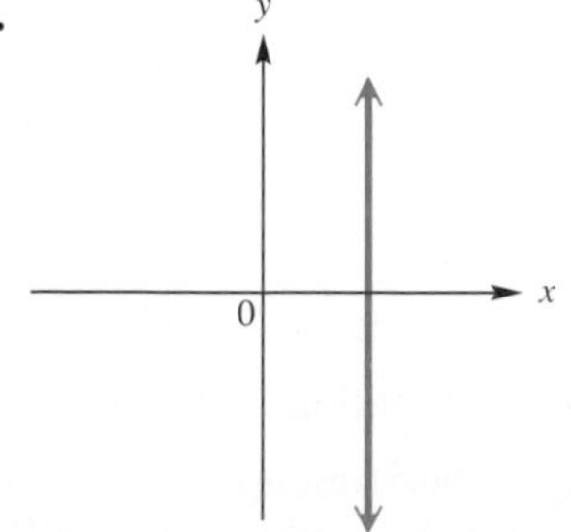

47.

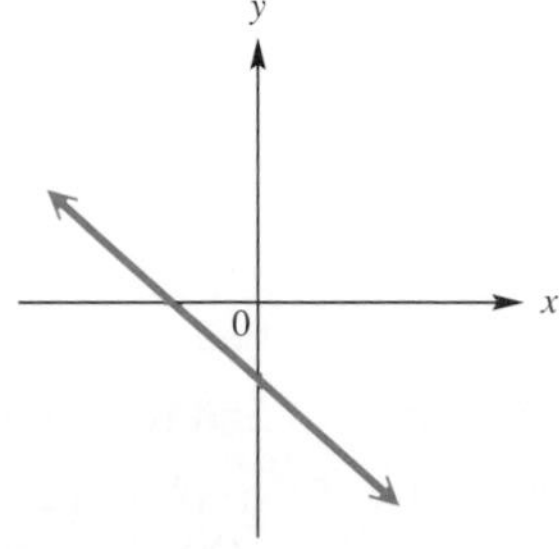

48.

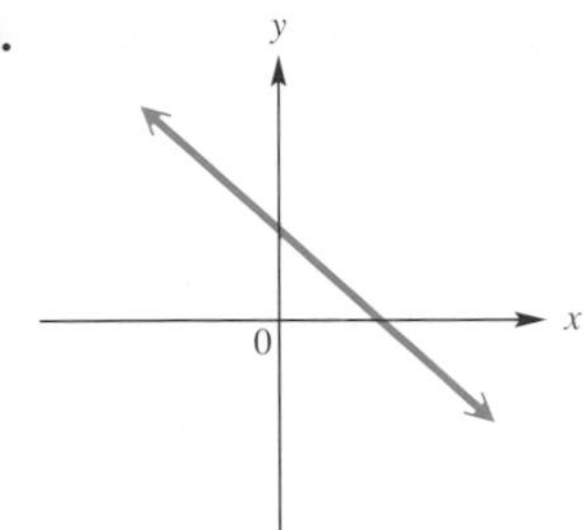

49.

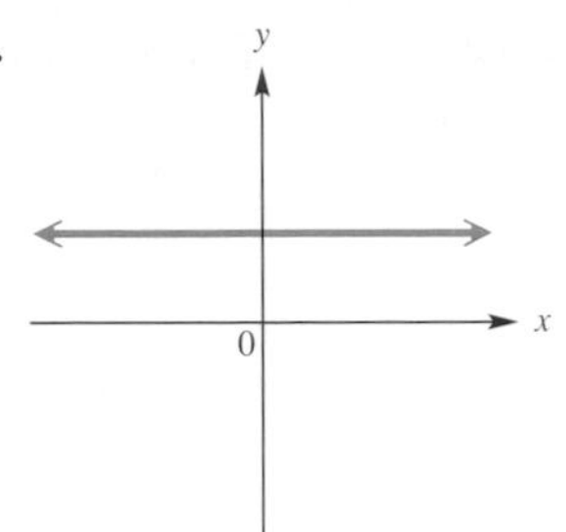

50.

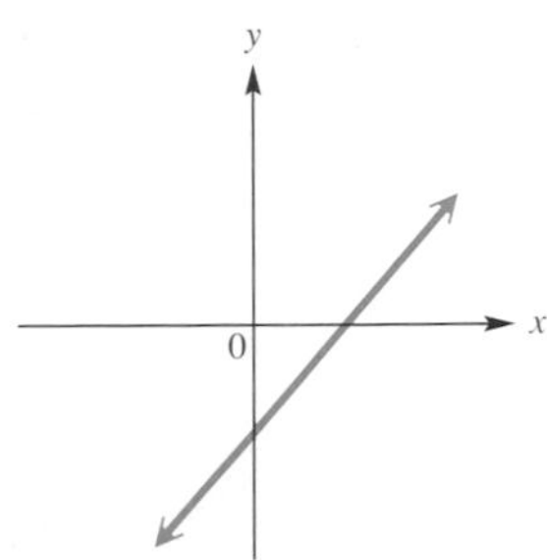

Use the method of Example 5 to graph each of the following lines.

51. $m = \frac{1}{2}$, through $(-3, 2)$

52. $m = \frac{2}{3}$, through $(0, 1)$

53. $m = -\frac{5}{4}$, through $(-2, -1)$

54. $m = -\frac{3}{2}$, through $(-1, -2)$

55. $m = -2$, through $(-1, -4)$

56. $m = 3$, through $(1, 2)$

57. $m = 0$, through $(2, -5)$

58. undefined slope, through $(-3, 1)$

Determine whether the lines described are parallel, perpendicular, *or* neither parallel nor perpendicular.

59. L_1 through $(4, 6)$ and $(-8, 7)$, and L_2 through $(7, 4)$ and $(-5, 5)$

60. L_1 through $(9, 15)$ and $(-7, 12)$, and L_2 through $(-4, 8)$ and $(-20, 5)$

61. L_1 through $(2, 0)$ and $(5, 4)$, and L_2 through $(6, 1)$ and $(2, 4)$

62. L_1 through $(0, -7)$ and $(2, 3)$, and L_2 through $(0, -3)$ and $(1, -2)$

63. L_1 through $(0, 1)$ and $(2, -3)$, and L_2 through $(10, 8)$ and $(5, 3)$

64. L_1 through $(1, 2)$ and $(-7, -2)$, and L_2 through $(1, -1)$ and $(5, -9)$

Use the concept of slope or average rate of change to solve the problem.

65. The upper deck at the new Comiskey Park in Chicago has produced, among other complaints, displeasure with its steepness. It's been compared to a ski jump. It is 160 feet from home plate to the front of the upper deck and 250 feet from home plate to the back. The top of the upper deck is 63 feet above the bottom. What is its slope?

66. When designing the new arena in Boston to replace the old Boston Garden, architects were careful to design the ramps leading up to the entrances so that circus elephants would be able to march up the ramps. The maximum grade (or slope) that an elephant will walk on is 13%. Suppose that such a ramp was constructed with a horizontal run of 150 feet. What would be the maximum vertical rise the architects could use?

67. The graph shows how average monthly rates for cable television increased from 1980 to 1992. The graph can be approximated by a straight line.

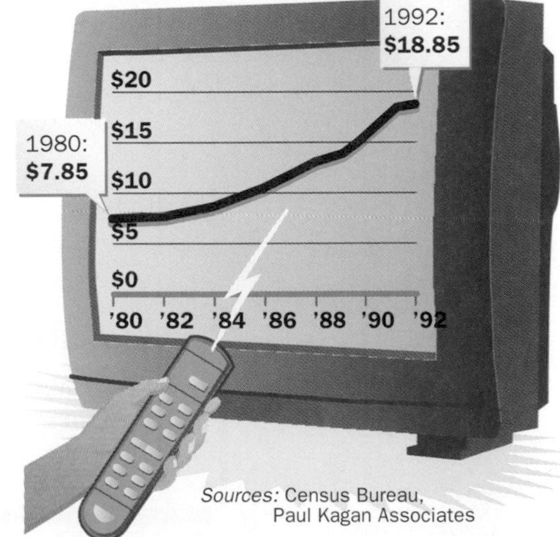

(a) Use the information provided for 1980 and 1992 to determine the average rate of change in price per year.

(b) How does a *positive* rate of change affect the consumer in a situation such as the one illustrated by this graph?

68. Assuming a linear relationship, what is the average rate of change for cable industry revenues over the period from 1990 to 1992?

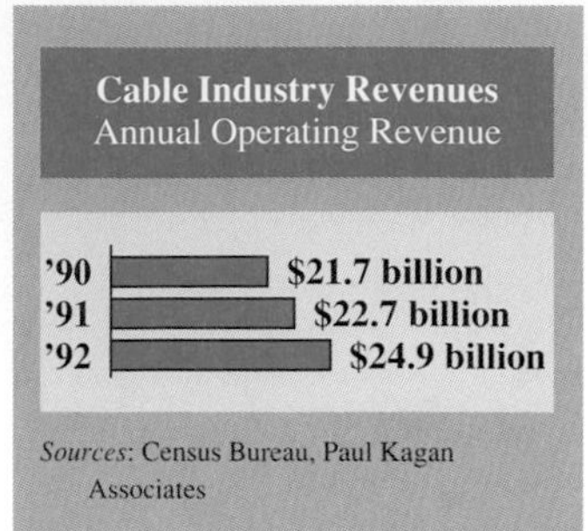

69. The 1993 Annual Report of AT&T cites the following figures concerning the global growth of international telephone calls: From 47.5 billion minutes in 1993, traffic is expected to rise to 60 billion minutes in 1995. Assuming a linear relationship, what is the average rate of change for this time period? (*Source:* TeleGeography, 1993, Washington, D.C.)

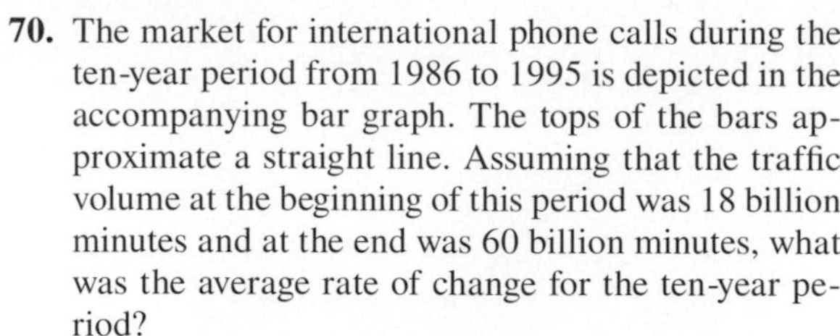

70. The market for international phone calls during the ten-year period from 1986 to 1995 is depicted in the accompanying bar graph. The tops of the bars approximate a straight line. Assuming that the traffic volume at the beginning of this period was 18 billion minutes and at the end was 60 billion minutes, what was the average rate of change for the ten-year period?

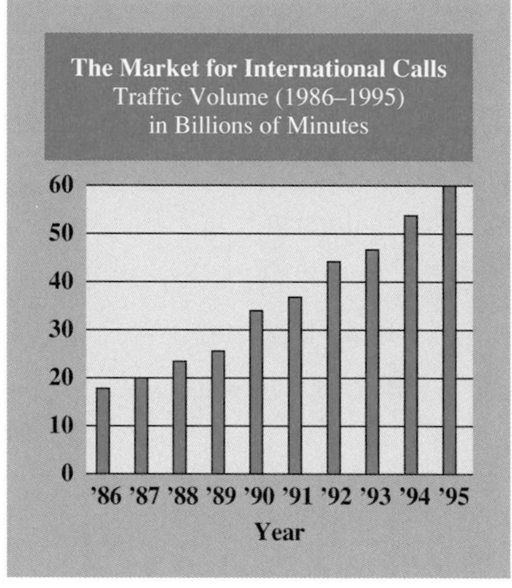

8.3 Equations of Lines

If the slope of a line and a particular point on the line are known, it is possible to find an equation of the line. Suppose that the slope of a line is m and (x_1, y_1) is a particular point on the line. Let (x, y) be any other point on the line. Then, by the definition of slope,

$$m = \frac{y - y_1}{x - x_1}.$$

Multiplying both sides by $x - x_1$ gives the following result, called the *point-slope form* of the equation of the line.

Point-Slope Form

The equation of the line through (x_1, y_1) with slope m is written in **point-slope form** as

$$y - y_1 = m(x - x_1).$$

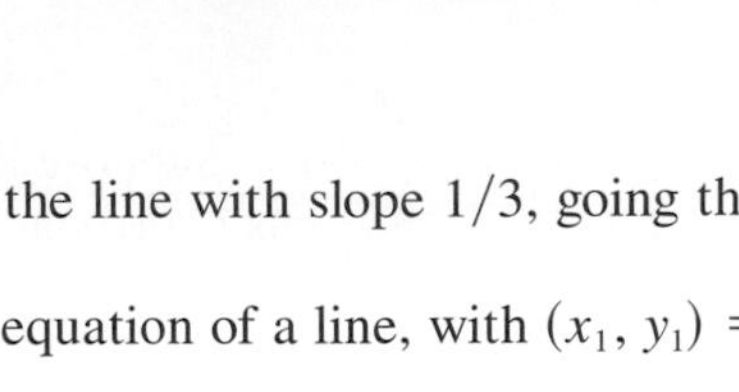

EXAMPLE 1 Find the equation of the line with slope 1/3, going through the point $(-2, 5)$.

Use the point-slope form of the equation of a line, with $(x_1, y_1) = (-2, 5)$ and $m = 1/3$.

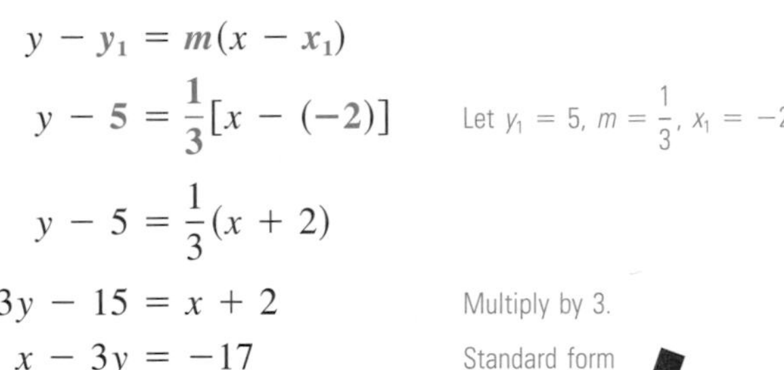

$$y - y_1 = m(x - x_1)$$

$$y - 5 = \frac{1}{3}[x - (-2)] \qquad \text{Let } y_1 = 5,\ m = \frac{1}{3},\ x_1 = -2.$$

$$y - 5 = \frac{1}{3}(x + 2)$$

$$3y - 15 = x + 2 \qquad \text{Multiply by 3.}$$

or

$$x - 3y = -17 \qquad \text{Standard form}$$

If two points on a line are known, it is possible to find an equation of the line. First, find the slope using the slope formula, and then use the slope with one of the given points in the point-slope form. This is illustrated in the next example.

EXAMPLE 2 Find an equation of the line through the points $(-4, 3)$ and $(5, -7)$.

First find the slope, using the definition.

$$m = \frac{-7 - 3}{5 - (-4)} = -\frac{10}{9}$$

Either $(-4, 3)$ or $(5, -7)$ may be used as (x_1, y_1) in the point-slope form of the equation of the line. If $(-4, 3)$ is used, then $-4 = x_1$ and $3 = y_1$.

$$y - y_1 = m(x - x_1)$$

$$y - 3 = -\frac{10}{9}[x - (-4)] \qquad \text{Let } y_1 = 3,\ m = -\frac{10}{9},\ x_1 = -4.$$

$$y - 3 = -\frac{10}{9}(x + 4)$$

$$9(y - 3) = -10(x + 4) \qquad \text{Multiply by 9.}$$

$$9y - 27 = -10x - 40 \qquad \text{Distributive property}$$

$$10x + 9y = -13 \qquad \text{Standard form}$$

Maria Gaetana Agnesi (1719–1799) did much of her mathematical work in coordinate geometry. She grew up in a scholarly atmosphere; her father was a mathematician on the faculty at the University of Bologna. In a larger sense she was an heir to the long tradition of Italian mathematicians.

Maria Agnesi was fluent in several languages by the age of 13, but she chose mathematics over literature. The curve shown below is studied in analytic geometry courses, and it is called the *witch of Agnesi.*

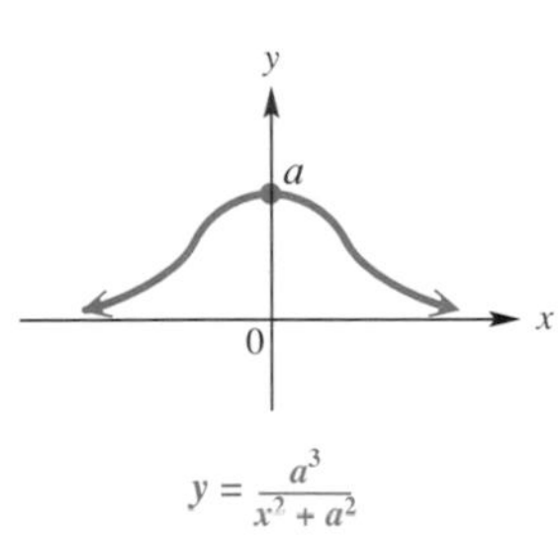

In the previous section, we saw that vertical and horizontal lines have special equations. We can analyze this further. Notice that the point-slope form does not apply to a vertical line, since the slope of a vertical line is undefined. A vertical line through the point (k, y) where k is a constant and y represents any real number, has equation $x = k$.

A horizontal line has slope 0. From the point-slope form, the equation of a horizontal line through the point (x, k), where x is any real number and k is a constant, is

$$y - y_1 = m(x - x_1)$$

$$y - k = 0(x - x) \qquad y_1 = k,\ x_1 = x$$

$$y - k = 0$$

$$y = k.$$

Suppose that the slope m of a line is known, and the y-intercept of the line has coordinates $(0, b)$. Then substituting into the point-slope form gives

$$\begin{aligned} y - y_1 &= m(x - x_1) \\ y - b &= m(x - 0) && x_1 = 0,\ y_1 = b \\ y - b &= mx \\ y &= mx + b && \text{Add } b \text{ to both sides.} \end{aligned}$$

This last result is known as the *slope-intercept form* of the equation of the line.

Slope-Intercept Form

The equation of a line with slope m and y-intercept $(0, b)$ is written in **slope-intercept form** as

$$y = \underset{\substack{\uparrow \\ \text{Slope}}}{m}x + \underset{\substack{\uparrow \\ y\text{-intercept is } (0,\ b).}}{b}.$$

The importance of the slope-intercept form of a linear equation cannot be overemphasized. First, every linear equation (of a nonvertical line) has a *unique* (one and only one) slope-intercept form. Second, in the next section, we will study *linear functions,* where the slope-intercept form is necessary in specifying such functions.

EXAMPLE 3 **(a)** Write the equation of the line described in Example 1 in slope-intercept form.

We determined the standard form of the equation of the line to be

$$x - 3y = -17.$$

Solve for y to obtain the slope-intercept form.

$$\begin{aligned} -3y &= -x - 17 \\ y &= \frac{1}{3}x + \frac{17}{3} && \text{Multiply by } -\frac{1}{3}. \end{aligned}$$

The slope is $1/3$ and the y-intercept is $(0, 17/3)$.

(b) Write the equation of the line described in Example 2 in slope-intercept form.

$$\begin{aligned} 10x + 9y &= -13 && \text{Standard form} \\ 9y &= -10x - 13 \\ y &= -\frac{10}{9}x - \frac{13}{9} \end{aligned}$$

The slope is $-10/9$ and the y-intercept is $(0, -13/9)$. ■

If the slope-intercept form of the equation of a line is known, the method of graphing described in Example 5 of the previous section can be used to graph the line, as shown in the next example.

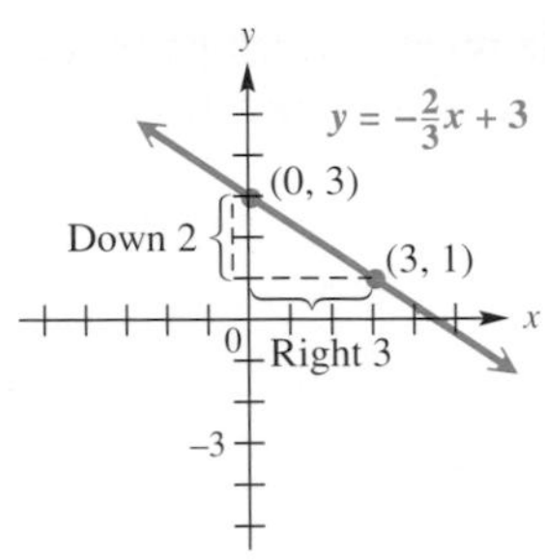

FIGURE 17

EXAMPLE 4 Graph the line with the equation $y = -\dfrac{2}{3}x + 3$.

Since the equation is given in slope-intercept form, we can easily see that the slope is $-2/3$ and the y-intercept is $(0, 3)$. Plot the point $(0, 3)$, and then, using the "rise over run" interpretation of slope, move *down* 2 units (because of the -2 in the numerator of the slope) and to the *right* 3 units (because of the 3 in the denominator). We arrive at the point $(3, 1)$. Plot the point $(3, 1)$, and join the two points with a line, as shown in Figure 17. (We could also have interpreted $-2/3$ as $2/(-3)$ and obtained a different second point; however, the line would be the same.) ■

As mentioned in the last section, parallel lines have the same slope and perpendicular lines have slopes with a product of -1. These results are used in the next example.

EXAMPLE 5 Find the slope-intercept form of the equation of the line passing through the point $(-4, 5)$, and **(a)** parallel to the line $2x + 3y = 6$; **(b)** perpendicular to the line $2x + 3y = 6$.

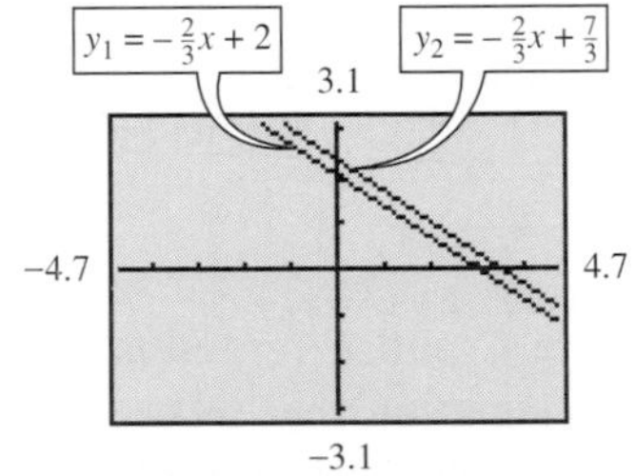

This screen gives support to the result in Example 5(a).

(a) The slope of the graph of $2x + 3y = 6$ can be found by solving for y.

$$2x + 3y = 6$$

$$3y = -2x + 6 \qquad \text{Subtract } 2x \text{ on both sides.}$$

$$y = -\frac{2}{3}x + 2 \qquad \text{Divide both sides by 3.}$$

The slope is given by the coefficient of x, so $m = -2/3$.

$$y = \underbrace{-\frac{2}{3}}_{\text{Slope}}x + 2$$

This means that the required equation of the line through $(-4, 5)$ and parallel to $2x + 3y = 6$ also has slope $-2/3$. Therefore, its equation will be of the form $y = -\frac{2}{3}x + b$. To find b, substitute -4 for x and 5 for y, and solve.

$$y = -\frac{2}{3}x + b$$

$$\mathbf{5} = -\frac{2}{3}(\mathbf{-4}) + b$$

$$5 = \frac{8}{3} + b$$

$$5 - \frac{8}{3} = b$$

$$\frac{7}{3} = b$$

The equation of the described line is $y = -\frac{2}{3}x + \frac{7}{3}$.

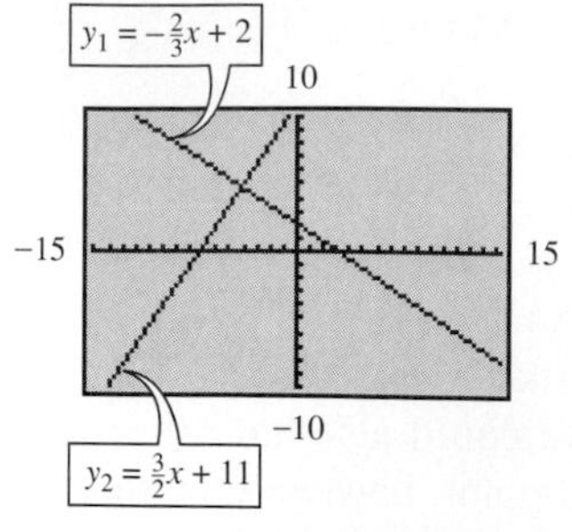

This square window gives support to the result in Example 5(b).

(b) To be perpendicular to the line $2x + 3y = 6$, a line must have slope that is the negative reciprocal of $-2/3$, which is $3/2$. Use $m = 3/2$ and the point $(-4, 5)$ in the equation $y = mx + b$ to find b.

$$5 = \frac{3}{2}(-4) + b$$

$$5 = -6 + b$$

$$b = 11$$

The equation of the line is $y = \frac{3}{2}x + 11$. ■

Problem Solving

Many natural phenomena can be described by linear equations. One common example is the relationship between Celsius and Fahrenheit temperatures. If we know two ordered pairs relating Celsius and Fahrenheit, then we can derive the equation that relates them. This application is shown in the next example.

EXAMPLE 6 There is a linear relationship between Celsius and Fahrenheit temperatures. It is common knowledge that when $C = 0°$, $F = 32°$, and when $C = 100°$, $F = 212°$.

(a) Use this information to express F in terms of C.

Think of ordered pairs of temperatures (C, F), where C and F represent corresponding Celsius and Fahrenheit temperatures. The equation that relates the two scales has a straight-line graph that contains the points $(0, 32)$ and $(100, 212)$. The slope of this line can be found by using the slope formula.

$$m = \frac{212 - 32}{100 - 0} = \frac{180}{100} = \frac{9}{5}$$

Now, think of the point-slope form of the equation in terms of C and F, where C replaces x and F replaces y. Use $m = 9/5$, and $(C_1, F_1) = (0, 32)$.

$$F - F_1 = m(C - C_1)$$

$$F - 32 = \frac{9}{5}(C - 0) \qquad F_1 = 32,\ m = \frac{9}{5},\ C_1 = 0$$

$$F - 32 = \frac{9}{5}C$$

$$F = \frac{9}{5}C + 32 \qquad \text{Solve for } F.$$

This last result gives F in terms of C. (This formula can be solved for C in terms of F to get the alternate form, $C = \frac{5}{9}(F - 32)$.)

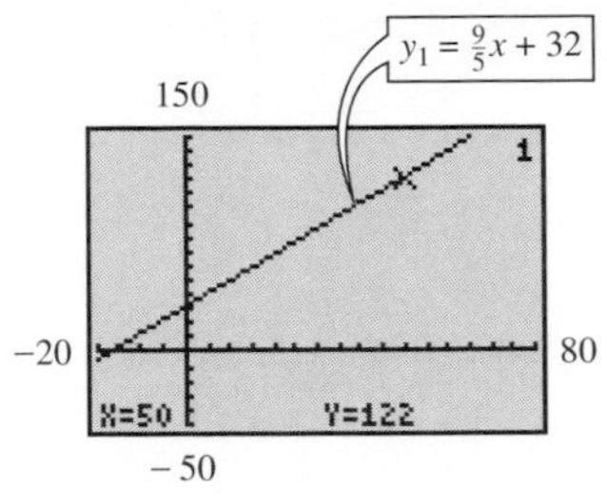

The display at the bottom of this screen indicates that when $C = x = 50$, $F = y = 122$, supporting the result in Example 6(b).

(b) Find the Fahrenheit temperature when Celsius temperature is 50°.

Let $C = 50$ in the formula found in part (a) to find the corresponding Fahrenheit temperature.

$$F = \frac{9}{5}C + 32$$

$$F = \frac{9}{5}(\mathbf{50}) + 32 \qquad \text{Let } C = 50.$$

$$F = 90 + 32$$

$$F = 122$$

When the Celsius temperature is 50°, the Fahrenheit temperature is 122°. ■

A summary of the various forms of linear equations concludes this section.

Summary of Forms of Linear Equations

$Ax + By = C$	**Standard form** (Neither A nor B is 0.)
$x = k$	**Vertical line** Undefined slope x-intercept is $(k, 0)$.
$y = k$	**Horizontal line** Slope is 0. y-intercept is $(0, k)$.
$y = mx + b$	**Slope-intercept form** Slope is m. y-intercept is $(0, b)$.
$y - y_1 = m(x - x_1)$	**Point-slope form** Slope is m. Line passes through (x_1, y_1).

EXERCISES

In Exercises 1–8, choose the one of the four graphs given here which most closely resembles the graph of the given equation. Each equation is given in the form $y = mx + b$, so consider the signs of m and b in making your choice.

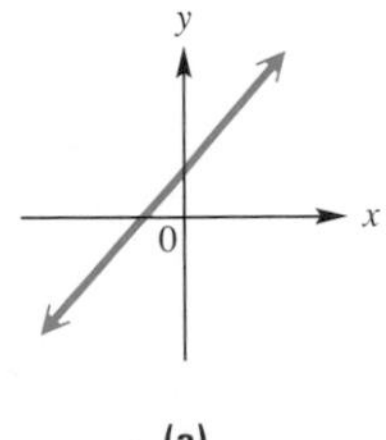

(a) **(b)** **(c)** **(d)**

1. $y = 3x + 6$

2. $y = 4x + 5$

3. $y = -3x + 6$

4. $y = -4x + 5$

5. $y = 3x - 6$

6. $y = 4x - 5$

7. $y = -3x - 6$

8. $y = -4x - 5$

Write the equation of the line satisfying the given conditions.

9. through $(-2, 4)$; $m = -\frac{3}{4}$

10. through $(-1, 6)$; $m = -\frac{5}{6}$

11. through $(5, 8)$; $m = -2$

12. through $(12, 10)$; $m = 1$

13. through $(-5, 4)$; $m = \frac{1}{2}$

14. through $(7, -2)$; $m = \frac{1}{4}$

15. through $(-4, 12)$; horizontal

16. through $(1, 5)$; horizontal

17. x-intercept $(3, 0)$; $m = 4$

18. x-intercept $(-2, 0)$; $m = -5$

19. Explain why the point-slope form of an equation cannot be used to find the equation of a vertical line.

20. Which one of the following equations is in standard form, according to the definition of standard form given in this text?
(a) $3x + 2y - 6 = 0$ **(b)** $y = 5x - 12$ **(c)** $2y = 3x + 4$ **(d)** $6x - 5y = 12$

Write an equation in the form $x = k$ or $y = k$ for some constant k for the line described.

21. through $(9, 10)$; undefined slope

22. through $(-2, 8)$; undefined slope

23. through $(.5, .2)$; vertical

24. through $\left(\frac{5}{8}, \frac{2}{9}\right)$; vertical

25. through $(-7, 8)$; horizontal

26. through $(\sqrt{2}, \sqrt{7})$; horizontal

Write the standard form of the equation of the line passing through the two points.

27. $(3, 4)$ and $(5, 8)$

28. $(5, -2)$ and $(-3, 14)$

29. $(6, 1)$ and $(-2, 5)$

30. $(-2, 5)$ and $(-8, 1)$

31. $(2, 5)$ and $(1, 5)$

32. $(-2, 2)$ and $(4, 2)$

33. $(7, 6)$ and $(7, -8)$

34. $(13, 5)$ and $(13, -1)$

Find the equation of the line satisfying the given conditions. Write it in slope-intercept form.

35. $m = 5$; $b = 15$

36. $m = -2$; $b = 12$

37. $m = -\frac{2}{3}$; $b = \frac{4}{5}$

38. $m = -\frac{5}{8}$; $b = -\frac{1}{3}$

39. slope $\frac{2}{5}$; y-intercept $(0, 5)$

40. slope $-\frac{3}{4}$; y-intercept $(0, 7)$

For the given equation **(a)** *write in slope-intercept form,* **(b)** *give the slope of the line, and* **(c)** *give the y-intercept.*

41. $x + y = 12$

42. $x - y = 14$

43. $5x + 2y = 20$

44. $6x + 5y = 40$

45. $2x - 3y = 10$

46. $4x - 3y = 10$

Use the method of Example 4 to graph the line whose equation is given in slope-intercept form.

47. $y = \frac{1}{3}x + 2$

48. $y = \frac{1}{4}x + 3$

49. $y = -\frac{3}{2}x - 4$

50. $y = -\frac{3}{5}x + 2$

51. $y = -4x - 7$

52. $y = -3x + 8$

Write an equation in standard form of the line satisfying the given conditions.

53. through $(7, 2)$; parallel to $3x - y = 8$

54. through $(4, 1)$; parallel to $2x + 5y = 10$

55. through $(-2, -2)$; parallel to $-x + 2y = 10$

56. through $(-1, 3)$; parallel to $-x + 3y = 12$

57. through (8, 5); perpendicular to $2x - y = 7$

58. through (2, −7); perpendicular to $5x + 2y = 18$

59. through (−2, 7); perpendicular to $x = 9$

60. through (8, 4); perpendicular to $x = -3$

Many real-world situations can be modeled approximately by straight-line graphs. One way to find the equation of such a line is to use two typical data points from the information provided, and then apply the point-slope form of the equation of a line. Because of the usefulness of the slope-intercept form, such equations are often given in the form $y = mx + b$. Assume that the situation described can be modeled by a straight-line graph, and use the information to find the $y = mx + b$ form of the equation of the line.

61. Let $x = 0$ represent the year 1985. In 1985, U.S. sales of Volkswagen vehicles were about 220,000. In 1993, sales had declined to 43,000. Let y represent sales in thousands. (Thus, the data points for this model are (0, 220) and (8, 43).) (*Source:* Autodata)

62. Let $x = 0$ represent the year 1989. In 1989, the number of auto accidents resulting in "catastrophic claims"—those amounting to \$100,000 or more—led to an average expected payment of \$575,000. In 1991, the average expected payment had risen to \$892,000. Let y represent the average expected payment in thousands of dollars. (*Source:* Insurance Research Council)

63. The number of post offices in the United States has been declining in recent years. Use the information given on the bar graph for the years 1984 and 1990, letting $x = 0$ represent the year 1984 and letting y represent the number of post offices. (*Source:* U.S. Postal Service, Annual Report of the Postmaster General)

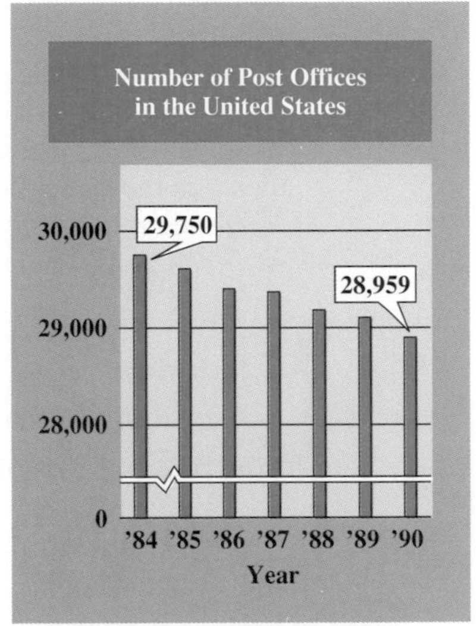

64. The number of motor vehicle deaths in the United States declined from 1988 to 1990 as seen in the accompanying bar graph. Use the information given, with $x = 0$ representing 1988 and y representing the number of deaths. (*Source:* National Safety Council)

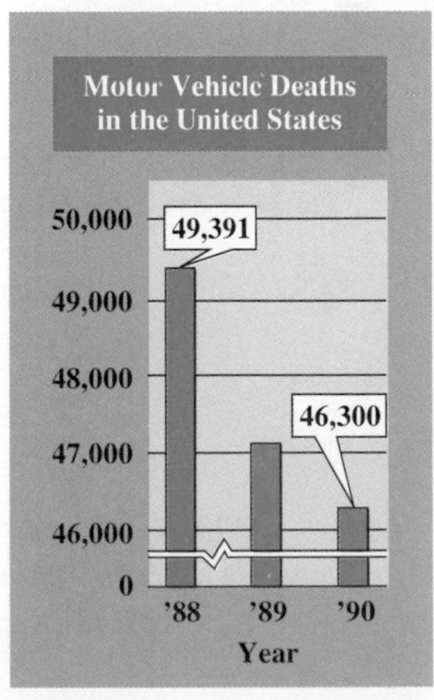

8.4 An Introduction to Functions: Linear Functions and Applications

It is often useful to describe one quantity in terms of another. For example, the growth of a plant is related to the amount of light it receives, the demand for a product is related to the price of the product, the cost of a trip is related to the distance traveled, and so on. To represent these corresponding quantities, it is helpful to use ordered pairs.

For example, we can indicate the relationship between the demand for a product and its price by writing ordered pairs in which the first number represents the price and the second number represents the demand. The ordered pair (5, 1,000) then could indicate a demand for 1,000 items when the price of the item is \$5. If it is assumed that the demand depends on the price charged, we place the price first and the demand second. The ordered pair is an abbreviation for the sentence "If the price is 5 (dollars), then the demand is for 1,000 (items)." Similarly, the ordered pairs (3, 5,000) and (10, 250) show that a price of \$3 produces a demand for 5,000 items, and a price of \$10 produces a demand for 250 items.

In this example, the demand depends on the price of the item. For this reason, demand is called the *dependent variable,* and price the *independent variable.* Generalizing, if the value of the variable y depends on the variable x, then y is the **dependent variable** and x is the **independent variable.**

Since related quantities can be written using ordered pairs, the concept of *relation* can be defined as follows.

Relation

A **relation** is a set of ordered pairs.

A special kind of relation, called a *function,* is very important in mathematics and its applications.

Function

A **function** is a relation in which, for each value of the first component of the ordered pairs, there is exactly one value of the second component.

EXAMPLE 1 Determine whether each of the following relations is a function.

$$F = \{(1, 2), (-2, 5), (3, -1)\}$$
$$G = \{(-2, 1), (-1, 0), (0, 1), (1, 2), (2, 2)\}$$
$$H = \{(-4, 1), (-2, 1), (-2, 0)\}$$

Relations F and G are functions, because for each x-value, there is only one y-value. Notice that in G, the last two ordered pairs have the same y-value. This does not violate the definition of function, since each first component (x-value) has only one second component (y-value).

Relation H is not a function, because the last two ordered pairs have the same x-value, but different y-values. ■

EXAMPLE 2 A function can also be expressed as a correspondence or *mapping* from one set to another. The mapping in Figure 18 is a function that assigns to a state its population (in millions) expected by the year 2000. ■

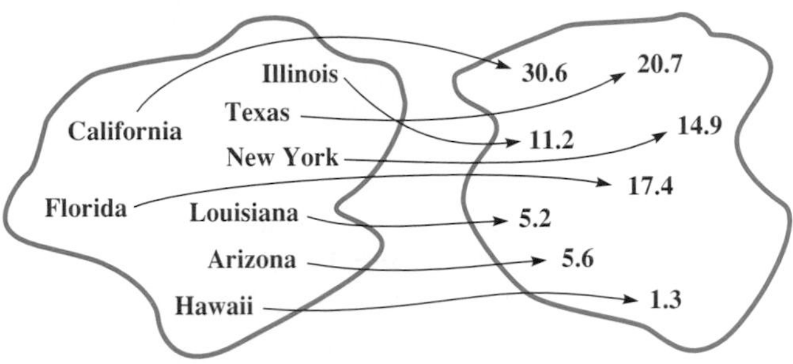

■ **FIGURE 18**

The set of all first components (x-values) of the ordered pairs of a relation is called the **domain** of the relation, and the set of all second components (y-values) is called the **range.** For example, the domain of function F in Example 1 is $\{1, -2, 3\}$; the range is $\{2, 5, -1\}$. Also, the domain of function G is $\{-2, -1, 0, 1, 2\}$ and the range is $\{0, 1, 2\}$. Domains and ranges can also be defined in terms of independent and dependent variables.

Domain and Range

In a relation, the set of all values of the independent variable (x) is the **domain;** the set of all values of the dependent variable (y) is the **range.**

EXAMPLE 3 Give the domain and range of each function.

(a) $\{(3, -1), (4, 2), (0, 5)\}$

The domain, the set of x-values, is $\{3, 4, 0\}$; the range is the set of y-values, $\{-1, 2, 5\}$.

(b) the function depicted in Figure 18

The domain is {Illinois, Texas, California, New York, Florida, Louisiana, Arizona, Hawaii} and the range is $\{1.3, 5.2, 5.6, 11.2, 14.9, 17.4, 20.7, 30.6\}$. ■

The **graph** of a relation is the graph of its ordered pairs. The graph gives a picture of the relation.

EXAMPLE 4 Three relations are graphed in Figure 19. Give the domain and range of each.

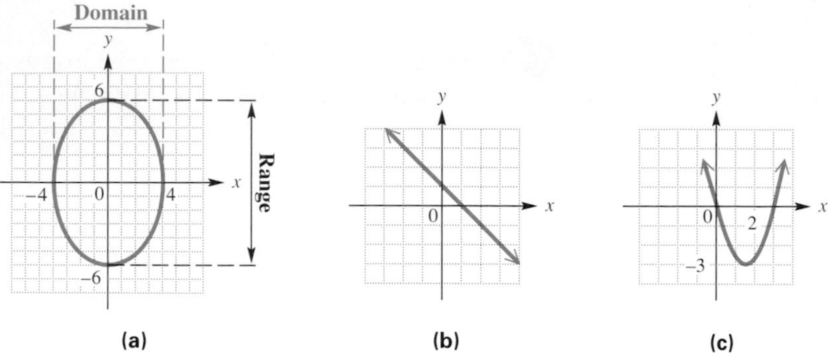

FIGURE 19

(a) In Figure 19(a), the x-values of the points on the graph include all numbers between -4 and 4, inclusive. The y-values include all numbers between -6 and 6, inclusive. Using interval notation from the previous chapter,

the domain is $[-4, 4]$;

the range is $[-6, 6]$.

(b) In Figure 19(b), the arrowheads indicate that the line extends indefinitely left and right, as well as up and down. Therefore, both the domain and the range are the set of all real numbers, written $(-\infty, \infty)$.

(c) In Figure 19(c), the arrowheads indicate that the graph extends indefinitely left and right, as well as upward. The domain is $(-\infty, \infty)$. Because there is a least y-value, -3, the range includes all numbers greater than or equal to -3, written $[-3, \infty)$.

Relations are often defined by equations such as $y = 2x + 3$ and $y^2 = x$. It is sometimes necessary to determine the domain of a relation when given the equation that defines the relation. The following agreement on the domain of a relation is assumed.

Agreement on Domain

Unless otherwise specified, the domain of a relation is assumed to be all real numbers that produce real numbers when substituted for the independent variable.

To illustrate this agreement, suppose that we consider the function $y = 2x + 3$. Since any real number can be used as a replacement for x in $y = 2x + 3$, the domain of this function is the set of real numbers. As another example, the function defined by $y = 1/x$ has all real numbers except 0 as its domain, since the denominator cannot be 0.

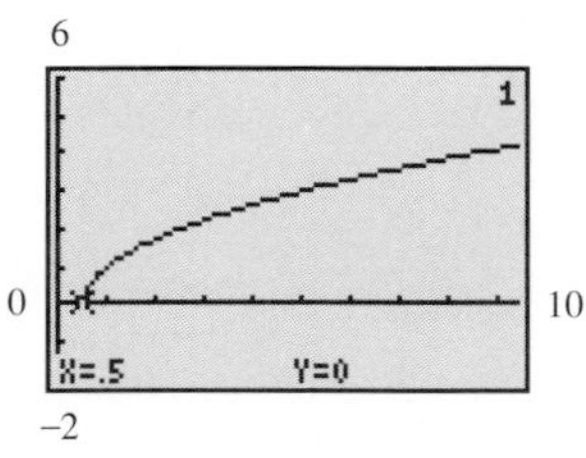

The graph of $y = \sqrt{2x - 1}$ supports the result in Example 5(a). The domain is $\left[\frac{1}{2}, \infty\right)$.

EXAMPLE 5 For each of the following, decide whether y is a function of x, and give the domain.

(a) $y = \sqrt{2x - 1}$

Here, for any choice of x in the domain, there is exactly one corresponding value for y (the radical is a nonnegative number), so this equation defines a function. Since the radicand cannot be negative,

$$2x - 1 \geq 0$$
$$2x \geq 1$$
$$x \geq \frac{1}{2}.$$

The domain is $\left[\frac{1}{2}, \infty\right)$.

(b) $y^2 = x$

The ordered pairs (16, 4) and (16, −4) both satisfy this equation. Since one value of x, 16, corresponds to two values of y, 4 and −4, this equation does not define a function. Solving $y^2 = x$ for y gives $y = \sqrt{x}$ or $y = -\sqrt{x}$, which shows that two values of y correspond to each positive value of x. Because x is equal to the square of y, the values of x must always be nonnegative. The domain is $[0, \infty)$.

(c) $y = \dfrac{5}{x^2 - 1}$

Given any value of x in the domain, we find y by squaring x, subtracting 1, then dividing the result into 5. This process produces exactly one value of y for each x-value, so this equation defines a function. The domain includes all real numbers except those that make the denominator zero. We find these numbers by setting the denominator equal to zero and solving for x.

$$x^2 - 1 = 0$$
$$x^2 = 1$$
$$x = 1 \quad \text{or} \quad x = -1 \qquad \text{Square root property}$$

Thus, the domain includes all real numbers except 1 and −1. In interval notation this is written as

$$(-\infty, -1) \cup (-1, 1) \cup (1, \infty).$$

In a function each value of x leads to only one value of y. Therefore, any vertical line drawn through the graph of a function would have to intersect the graph in at most one point. This leads to the **vertical line test for a function.**

Vertical Line Test

A vertical line will intersect the graph of a function in at most one point.

As an illustration, the graph shown in Figure 20(a) is not the graph of a function, since a vertical line can intersect the graph in more than one point, while the graph in Figure 20(b) does represent a function.

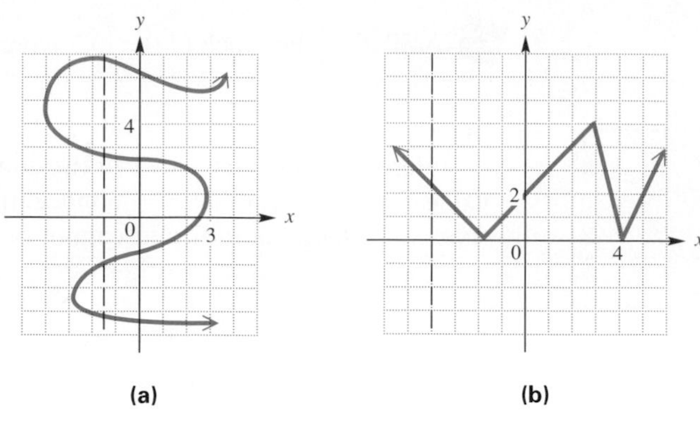

■ **FIGURE 20**

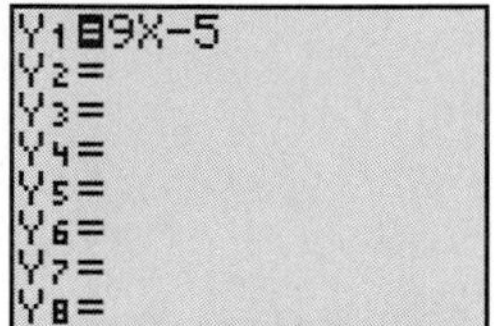

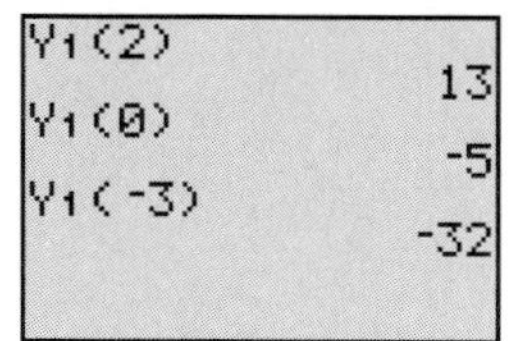

For $Y_1 = 9x - 5$, the TI-82 function notation capability supports the discussion here.

To say that y is a function of x means that for each value of x from the domain of the function, there is exactly one value of y. To emphasize that *y is a function of x*, or that y depends on x, it is common to write

$$y = f(x),$$

with $f(x)$ read "f of x." (In this special notation, the parentheses do *not* indicate multiplication.) For example, if $y = 9x - 5$, we emphasize that y is a function of x by writing $y = 9x - 5$ as

$$f(x) = 9x - 5.$$

Letters other than f, such as g and h, are often used to name functions. Since the choice of the letter is arbitrary, the function $g(x) = 9x - 5$ is the same as the function $f(x) = 9x - 5$.

This **function notation** can be used to simplify certain statements. For example, if $y = 9x - 5$, then replacing x with 2 gives

$$\begin{aligned} y &= 9 \cdot \mathbf{2} - 5 \\ &= 18 - 5 \\ &= 13. \end{aligned}$$

The statement "if $x = 2$, then $y = 13$" is abbreviated with function notation as

$$f(\mathbf{2}) = 13,$$

where f is the name of the function. Also, $f(0) = 9 \cdot 0 - 5 = -5$, and $f(-3) = -32$.

These ideas and the symbols used to represent them can be illustrated as follows.

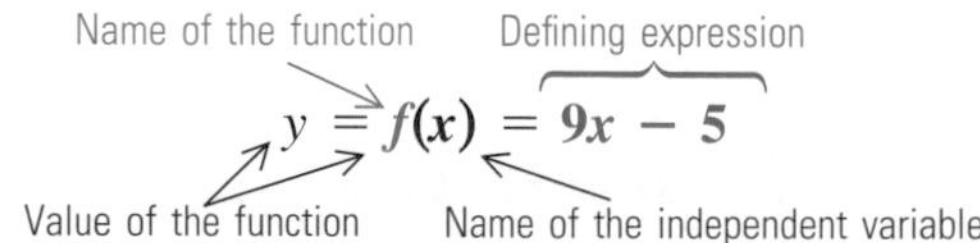

EXAMPLE 6 Let $f(x) = -x^2 + 5x - 3$. Find the following.

(a) $f(2)$

Replace x with 2.

$$f(\mathbf{2}) = -\mathbf{2}^2 + 5 \cdot \mathbf{2} - 3 = -4 + 10 - 3 = 3$$

(b) $f(-1)$

$$f(-1) = -(-1)^2 + 5(-1) - 3 = -1 - 5 - 3 = -9$$

(c) $f(2x)$

Replace x with $2x$.

$$\begin{aligned} f(2x) &= -(2x)^2 + 5(2x) - 3 \\ &= -4x^2 + 10x - 3 \end{aligned}$$ ■

Linear Functions

An important type of elementary function is the *linear function*.

> **Linear Function**
>
> A function that can be written in the form
>
> $$f(x) = mx + b$$
>
> for real numbers m and b is a **linear function.**

Notice that the form of the equation defining a linear function is the same as that of the slope-intercept form of the equation of a line, first seen in the previous section. We know that the graph of $f(x) = mx + b$ will be a line with slope m and y-intercept $(0, b)$.

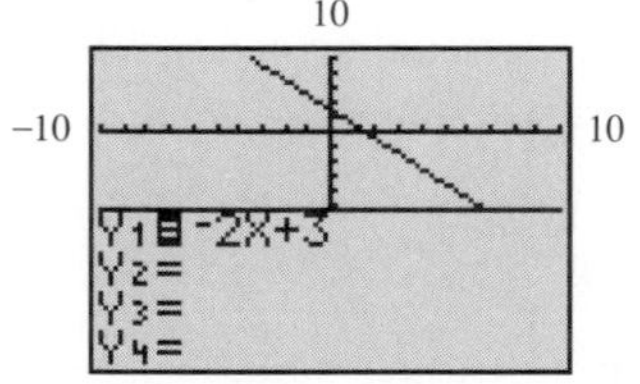

Compare with Example 7(a) and Figure 21(a).

EXAMPLE 7 Graph each linear function.

(a) $f(x) = -2x + 3$

To graph the function, locate the y-intercept, $(0, 3)$. From this point, use the slope $-2 = -2/1$ to go down 2 and right 1. This second point is used to obtain the graph in Figure 21(a).

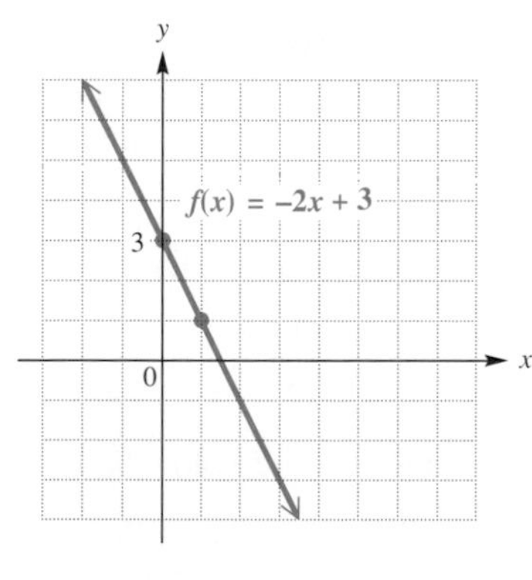

(a)

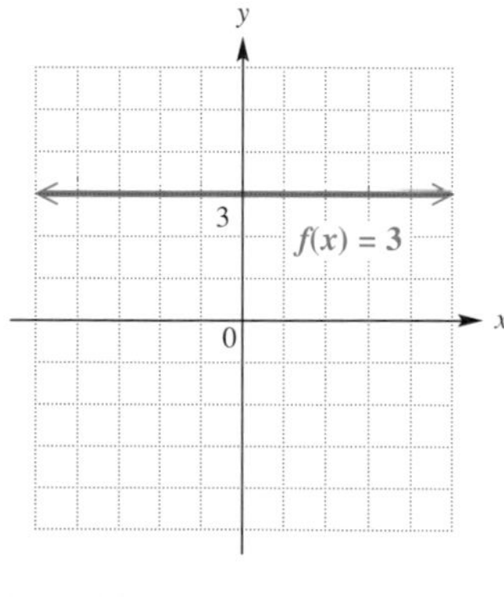

(b)

■ **FIGURE 21**

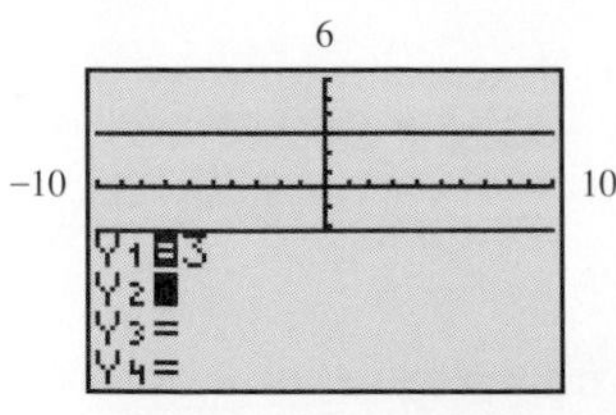

Compare with Example 7(b) and Figure 21(b).

(b) $f(x) = 3$

From the previous section, we know that the graph of $y = 3$ is a horizontal line. Therefore, the graph of $f(x) = 3$ is a horizontal line with y-intercept $(0, 3)$ as shown in Figure 21(b). ■

The function defined and graphed in Example 7(b) and Figure 21(b) is an example of a constant function. A **constant function** is a linear function of the form $f(x) = k$, where k is a real number. The domain of any linear function is $(-\infty, \infty)$. The range of a nonconstant linear function (like in Example 7(a)) is also $(-\infty, \infty)$, while the range of the constant function $f(x) = k$ is $\{k\}$.

Problem Solving

By expressing a company's cost of producing a product and the revenue from selling the product as linear functions, the idea of **break-even analysis** can be explained using the graphs of these functions. When cost is greater than revenue earned, the company loses money; when cost is less than revenue, the company makes money; and when cost equals revenue, the company breaks even.

EXAMPLE 8 Peripheral Visions, Inc., produces studio quality audiotapes of live concerts. The company places an ad in a trade newsletter. The cost of the ad is \$100. Each tape costs \$20 to produce, and the company charges \$24 per tape.

(a) Express the cost C as a function of x, the number of tapes produced.

The *fixed cost* is \$100, and for each tape produced, the *variable cost* is \$20. Therefore, the cost C can be expressed as a function of x, the number of tapes produced:

$$C(x) = 20x + 100 \quad (C \text{ in dollars}).$$

(b) Express the revenue R as a function of x, the number of tapes sold.

Since each tape sells for \$24, the revenue R is given by $R(x) = 24x$ (R in dollars).

(c) For what value of x does revenue equal cost?

The company will just break even (no profit and no loss), as long as revenue just equals cost, or $R(x) = C(x)$. This is true whenever

$$R(x) = C(x)$$
$$24x = 20x + 100 \quad \text{Substitute for } R(x) \text{ and } C(x).$$
$$4x = 100$$
$$x = 25.$$

If 25 tapes are produced and sold, the company will break even.

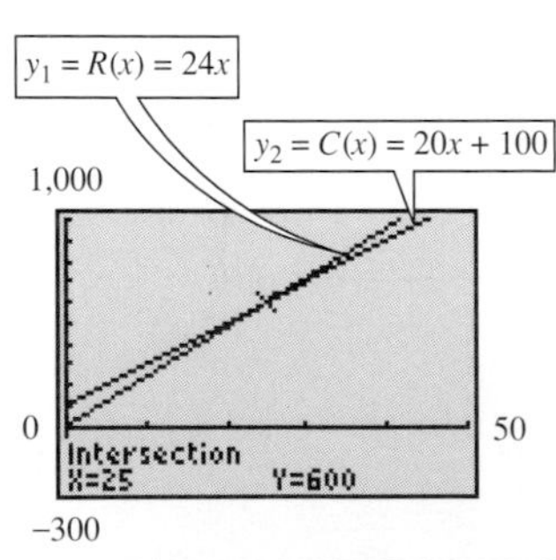

The break-even point is (25, 600), as indicated at the bottom of the screen. The calculator can find the point of intersection of the graphs.

(d) Graph $C(x) = 20x + 100$ and $R(x) = 24x$ on the same coordinate system, and interpret the graph.

Figure 22 shows the graphs of the two functions. At the break-even point, we see that when 25 tapes are produced and sold, both the cost and the revenue are \$600. If fewer than 25 tapes are produced and sold (that is, when $x < 25$), the company loses money. When more than 25 tapes are produced and sold (that is, when $x > 25$), there is a profit. ■

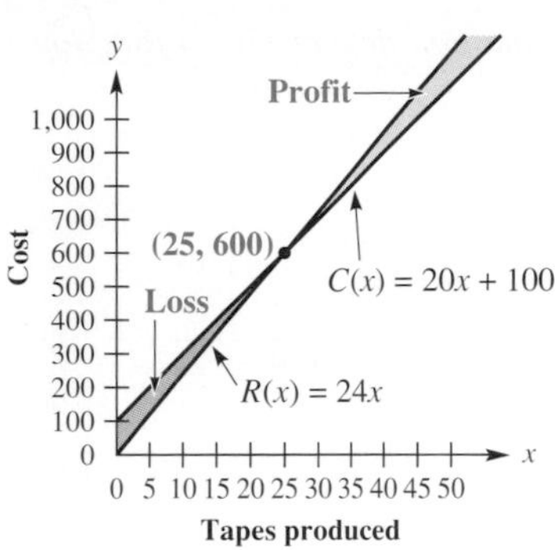

FIGURE 22

8.4 EXERCISES

1. Explain the meaning of the term.
   (a) relation (b) domain of a relation
   (c) range of a relation (d) function

2. Give an example of a relation that is not a function having domain $\{-3, 2, 6\}$ and range $\{4, 6\}$. (There are many possible correct answers.)

Decide whether the relation is a function and give the domain and the range of the relation. Use the vertical line test in Exercises 9–12.

3. $\{(5, 1), (3, 2), (4, 9), (7, 3)\}$

4. $\{(8, 0), (5, 4), (9, 3), (3, 9)\}$

5. $\{(2, 4), (0, 2), (2, 6)\}$

6. $\{(9, -2), (-3, 5), (9, 1)\}$

7.

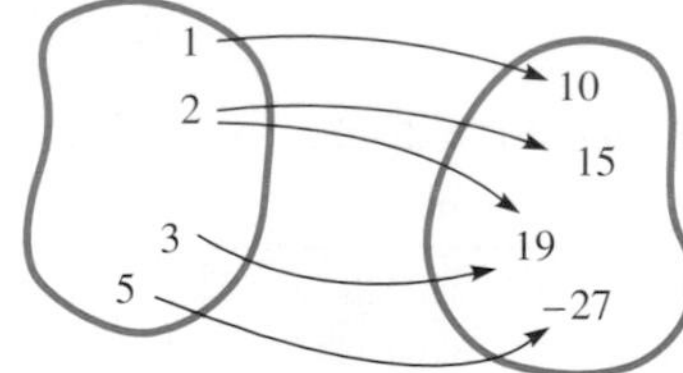

8.

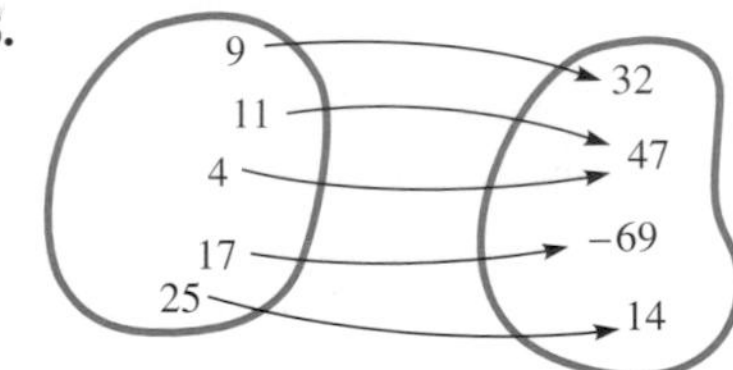

9.

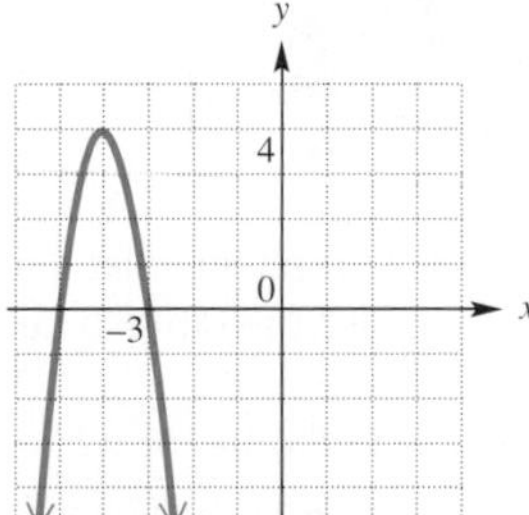

10.

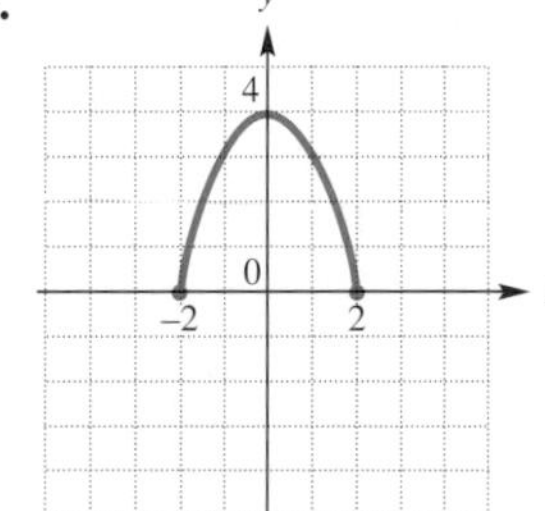

11.

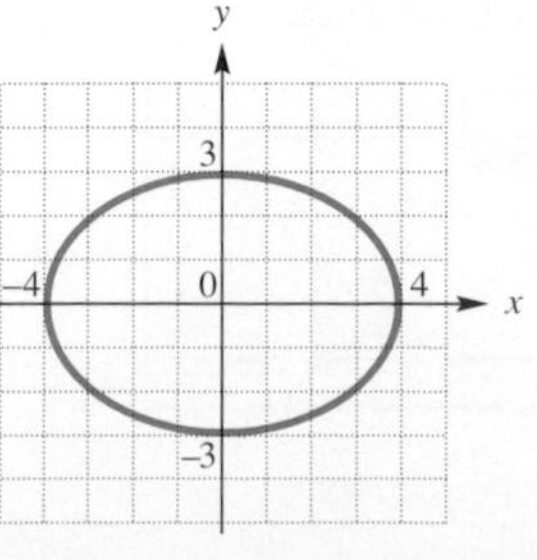

12.

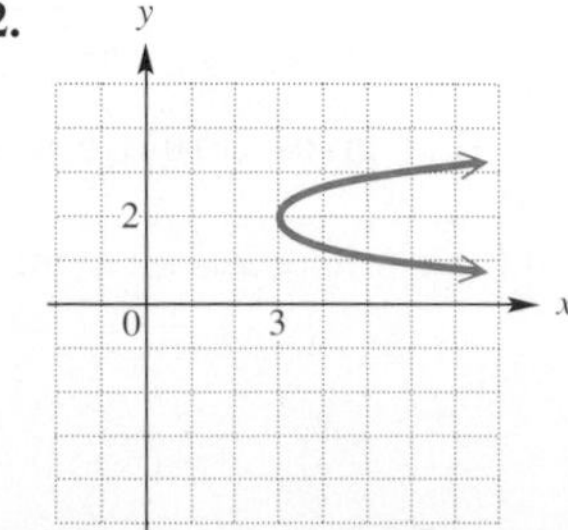

Decide whether the given equation defines y as a function of x. Give the domain. Identify any linear functions.

13. $y = x^2$ **14.** $y = x^3$ **15.** $x = y^2$ **16.** $x = y^4$

17. $x + y < 4$ **18.** $x - y < 3$ **19.** $y = \sqrt{x}$ **20.** $y = -\sqrt{x}$

21. $xy = 1$ **22.** $xy = -3$ **23.** $y = 2x - 6$ **24.** $y = -6x + 8$

25. $y = \sqrt{4x + 2}$ **26.** $y = \sqrt{9 - 2x}$ **27.** $y = \dfrac{2}{x^2 - 9}$ **28.** $y = \dfrac{-7}{x^2 - 16}$

Let $f(x) = 3 + 2x$ and $g(x) = x^2 - 2$. *Find the function value.*

29. $f(1)$ **30.** $f(4)$ **31.** $g(2)$ **32.** $g(0)$

33. $g(-1)$ **34.** $g(-3)$ **35.** $f(-8)$ **36.** $f(-5)$

Sketch the graph of each linear function. Give the domain and range.

37. $f(x) = -2x + 5$ **38.** $g(x) = 4x - 1$ **39.** $h(x) = \dfrac{1}{2}x + 2$ **40.** $F(x) = -\dfrac{1}{4}x + 1$

41. $G(x) = 2x$ **42.** $H(x) = -3x$ **43.** $f(x) = 5$ **44.** $g(x) = -4$

---

45. Fill in the blanks with the correct responses.

The equation $2x + y = 4$ has a straight ______ as its graph. One point that lies on the line is (3, ______). If we solve the equation for y and use function notation, we have a linear function $f(x) =$ ____________. For this function, $f(3) =$ ______, meaning that the point (______, ______) lies on the graph of the function.

46. Which one of the following defines a linear function?

(a) $y = \dfrac{x - 5}{4}$ **(b)** $y = \sqrt[3]{x}$

(c) $y = x^2$ **(d)** $y = x^{1/2}$

47. Which one of the functions in Exercise 46 has domain $[0, \infty)$?

48. Which one of the functions in Exercise 46 does not have negative numbers in its domain?

49. (a) Suppose that a taxicab driver charges \$.50 per mile. Fill in the chart with the correct response for the price $R(x)$ she charges for a trip of x miles.

x	$R(x)$
0	
1	
2	
3	

(b) The linear function that gives a rule for the amount charged is $R(x) =$ ______.

(c) Graph this function for x, where x is an element of {0, 1, 2, 3}.

50. Suppose that a package weighing x pounds costs $C(x)$ dollars to mail to a given location, where

$$C(x) = 2.75x.$$

(a) What is the value of $C(3)$?

(b) In your own words, describe what 3 and the value $C(3)$ mean in part (a), using the terminology *domain* and *range*.

51. The graphing calculator screen shows the graph of a linear function $y = f(x)$, along with the display of coordinates of a point on the graph. Use function notation to write what the display indicates.

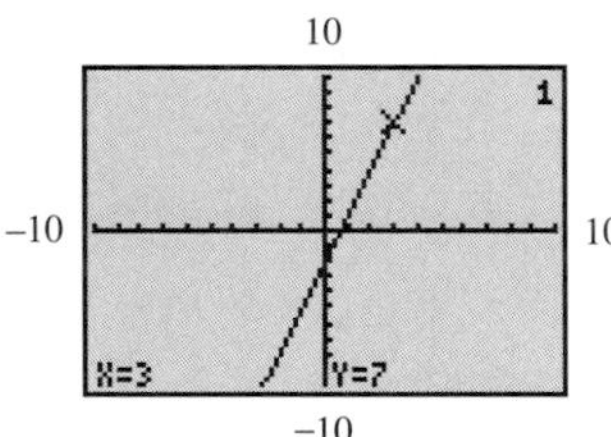

52. The table was generated by a graphing calculator for a linear function $y_1 = f(x)$. Use the table to answer.

(a) What is $f(2)$?

(b) If $f(x) = -3.7$, what is the value of x?

(c) What is the slope of the line?

(d) What is the y-intercept of the line?

(e) Find the expression for $f(x)$.

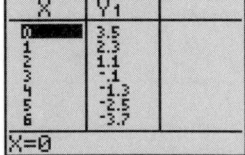

X	Y_1
0	3.5
1	2.3
2	1.1
3	-.1
4	-1.3
5	-2.5
6	-3.7

X=0

53. The two screens show the graph of the same linear function $y = f(x)$. Find the expression for $f(x)$.

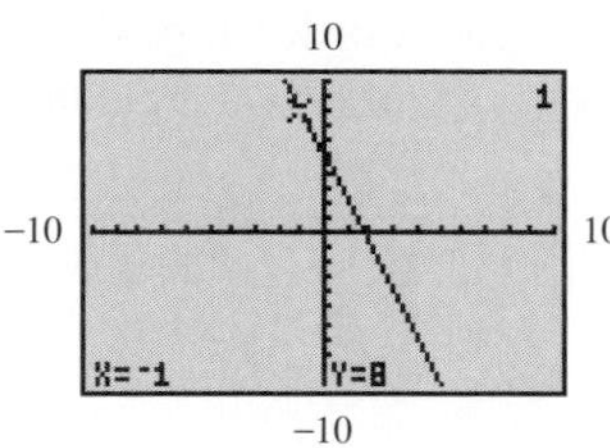

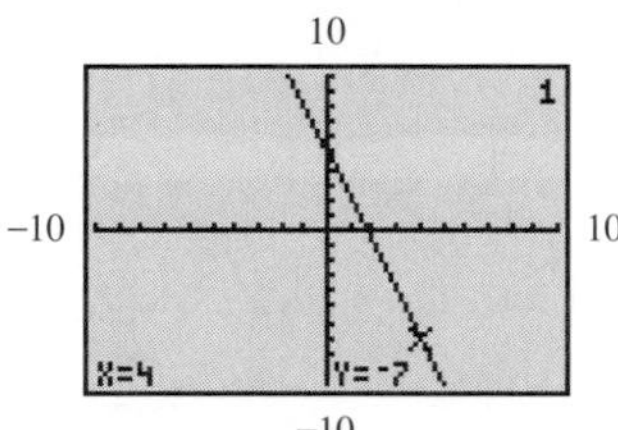

54. As determined in the previous section, the formula for converting Celsius to Fahrenheit is $F = 1.8C + 32$. If we graph $y = f(x) = 1.8x + 32$ on a graphing calculator screen, we obtain the accompanying picture. (We used the interval $[-50, 50]$ on the x-axis and $[-75, 50]$ on the y-axis.) The point $(-40, -40)$ lies on the graph, as indicated by the display. Interpret the meaning of this in the context of this exercise.

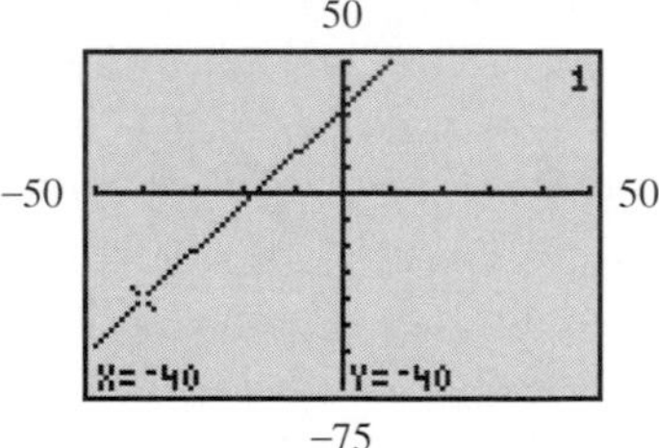

55. The linear function $f(x) = -123x + 29{,}685$ provides a model for the number of post offices in the United States from 1984 to 1990, where $x = 0$ corresponds to 1984, $x = 1$ corresponds to 1985, and so on. Use this model to give the approximate number of post offices during the given year. (*Source:* U.S. Postal Service, Annual Report of the Postmaster General)

(a) 1985 **(b)** 1987 **(c)** 1990

(d) The graphing calculator screen shows a portion of the graph of $y = f(x)$ with the coordinates of a point on the graph displayed at the bottom of the screen. Interpret the meaning of the display in the context of this exercise.

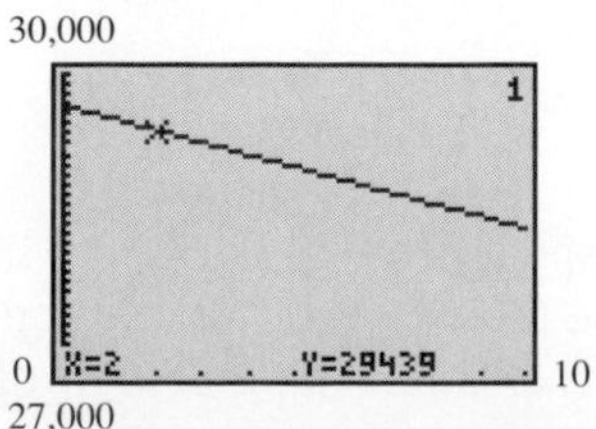

56. The linear function $f(x) = 1{,}650x + 3{,}817$ provides a model for the United States defense budget for the decade of the 1980s, where $x = 0$ corresponds to 1980, $x = 1$ corresponds to 1981, and so on, with $f(x)$ representing the budget in millions of dollars. Use this model to approximate the defense budget during the given year. (*Source:* U.S. Office of Management and Budget)

(a) 1983 **(b)** 1985 **(c)** 1988

(d) The graphing calculator screen shows a portion of the graph of $y = f(x)$ with the coordinates of a point on the graph displayed at the bottom of the screen. Interpret the meaning of the display in the context of this exercise.

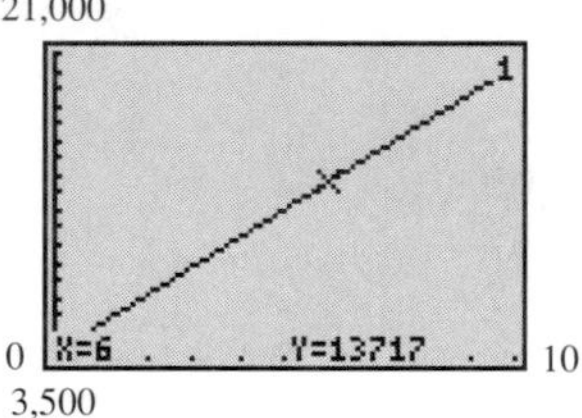

In each of the following, **(a)** *express the cost C as a function of x, where x represents the quantity of items as given;* **(b)** *express the revenue R as a function of x;* **(c)** *determine the value of x for which revenue equals cost;* **(d)** *graph $y = C(x)$ and $y = R(x)$ on the same axes, and interpret the graph.*

57. Christy Lawrence stuffs envelopes for extra income during her spare time. Her initial cost to obtain the necessary information for the job was $200.00. Each envelope costs $.02 and she gets paid $.04 per envelope stuffed. Let x represent the number of envelopes stuffed.

58. Evan Turner runs a copying service in his home. He paid $3,500 for the copier and a lifetime service contract. Each sheet of paper he uses costs $.01, and he gets paid $.05 per copy he makes. Let x represent the number of copies he makes.

59. Ray Kelley operates a delivery service in a southern city. His start-up costs amounted to $2,300. He estimates that it costs him (in terms of gasoline, wear and tear on his car, etc.) $3.00 per delivery. He charges $5.50 per delivery. Let x represent the number of deliveries he makes.

60. Julie Campbell bakes cakes and sells them at county fairs. Her initial cost for the Washington Parish fair in 1996 was $40.00. She figures that each cake costs $2.50 to make, and she charges $6.50 per cake. Let x represent the number of cakes sold. (Assume that there were no cakes left over.)

Quadratic Functions and Applications

In the previous section we studied linear functions, those that are defined by first-degree polynomials. In this section we will look at *quadratic functions,* those defined by second-degree polynomials.

> **Quadratic Function**
>
> A function f is a **quadratic function** if
>
> $$f(x) = ax^2 + bx + c,$$
>
> where a, b, and c are real numbers, with $a \neq 0$.

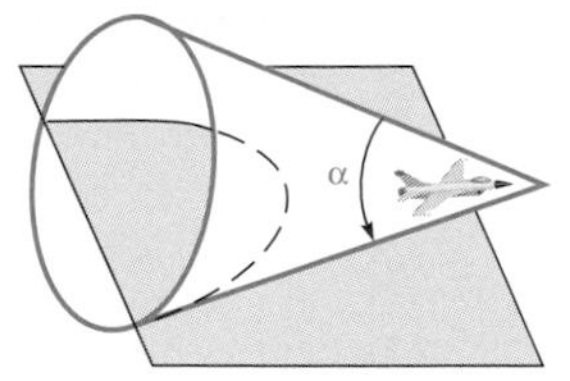

A sonic boom is a loud explosive sound caused by the shock wave that accompanies an aircraft traveling at supersonic speed. The sonic boom shock wave has the shape of a cone, and it intersects the ground in one branch of a curve known as a **hyperbola.** Everyone located along the hyperbolic curve on the ground hears the sound at the same time.

The simplest quadratic function is defined by $f(x) = x^2$. This function can be graphed by finding several ordered pairs that satisfy the equation: for example $(0, 0)$, $(1, 1)$, $(-1, 1)$, $(2, 4)$, $(-2, 4)$, $(1/2, 1/4)$, $(-1/2, 1/4)$, $(3/2, 9/4)$, and $(-3/2, 9/4)$. Plotting these points and drawing a smooth curve through them gives the graph shown in Figure 23. This graph is called a **parabola.** Every quadratic function has a graph that is a parabola.

Parabolas are symmetric about a line (the y-axis in Figure 23). Intuitively, this means that if the graph were folded along the line of symmetry, the two sides would coincide. The line of symmetry for a parabola is called the **axis** of the parabola. The point where the axis intersects the parabola is the **vertex** of the parabola. The vertex is the lowest (or highest) point of a vertical parabola.

Parabolas have many practical applications. For example, the reflectors of solar ovens and flashlights are made by revolving a parabola about its axis. The **focus** of a parabola is a point on its axis that determines the curvature. See Figure 24. When the parabolic reflector of a solar oven is aimed at the sun, the light rays bounce off the reflector and collect at the focus, creating intense heat at that point. On the other hand, when a lightbulb is placed at the focus of a parabolic reflector, light rays reflect out parallel to the axis.

The first example shows how the constant a affects the graph of a function of the form $g(x) = ax^2$.

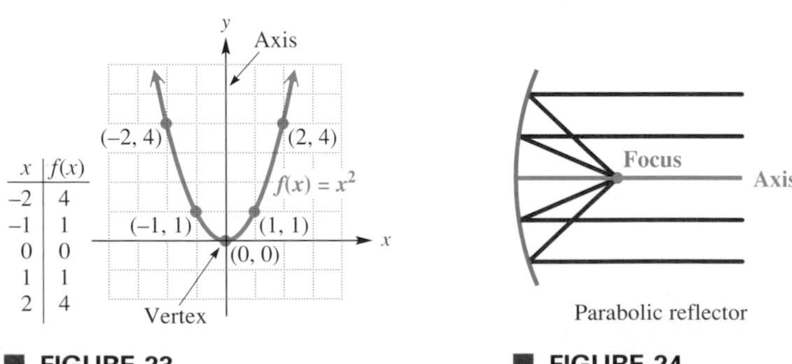

■ FIGURE 23

■ FIGURE 24

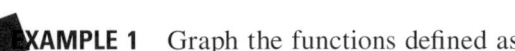

EXAMPLE 1 Graph the functions defined as follows.

(a) $g(x) = -x^2$

For a given value of x, the corresponding value of $g(x)$ will be the negative of what it was for $f(x) = x^2$. (See the table of values with Figure 25.) Because of this, the graph of $g(x) = -x^2$ is the same shape as that of $f(x)$, but opens

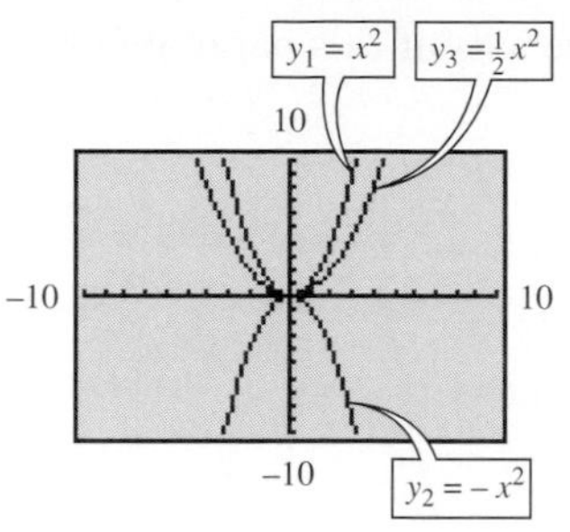

The screen illustrates the three graphs considered in Example 1 and Figures 25 and 26.

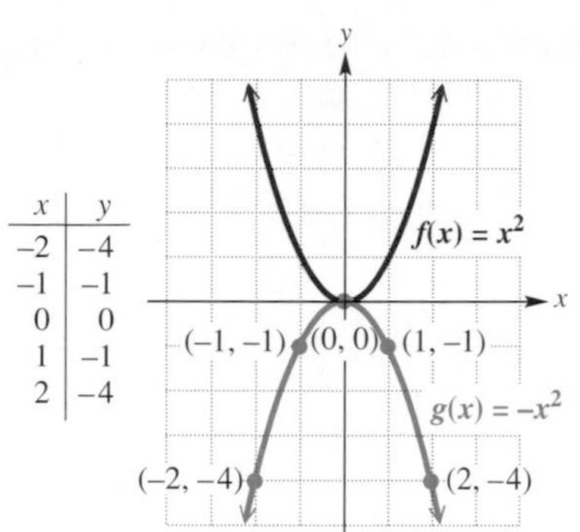

x	y
−2	−4
−1	−1
0	0
1	−1
2	−4

■ **FIGURE 25**

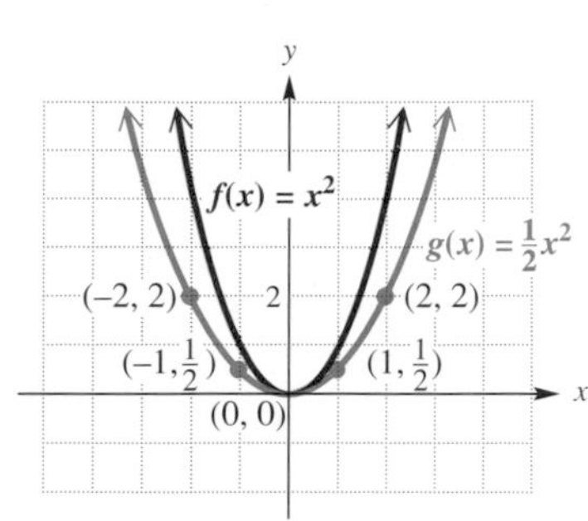

■ **FIGURE 26**

downward. See Figure 25. This is generally true; the graph of $f(x) = ax^2 + bx + c$ opens downward whenever $a < 0$.

(b) $g(x) = \frac{1}{2}x^2$

Choose a value of x, and then find $g(x)$. The coefficient $1/2$ will cause the resulting value of $g(x)$ to be smaller than for $f(x) = x^2$, making the parabola wider than the graph of $f(x) = x^2$. See Figure 26. In both parabolas of this example, the axis is the vertical line $x = 0$ and the vertex is the origin $(0, 0)$. ■

The next few examples show the results of horizontal and vertical shifts, called **translations,** of the graph of $f(x) = x^2$.

EXAMPLE 2 Graph $g(x) = x^2 - 4$.

By comparing the tables of values for $g(x) = x^2 - 4$ and $f(x) = x^2$ shown with Figure 27, we can see that for corresponding x-values, the y-values of g are each 4 less than those for f. Thus, the graph of $g(x) = x^2 - 4$ is the same as that of $f(x) = x^2$, but translated 4 units down. See Figure 27. The vertex of this parabola (here the lowest point) is at $(0, -4)$. The axis of the parabola is the vertical line $x = 0$. ■

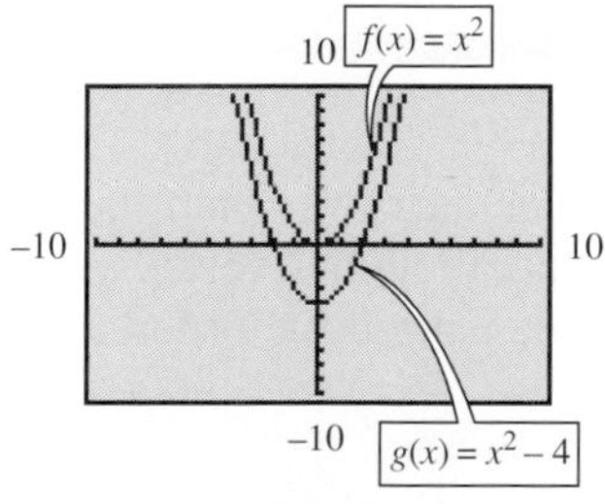

Compare with Figure 27.

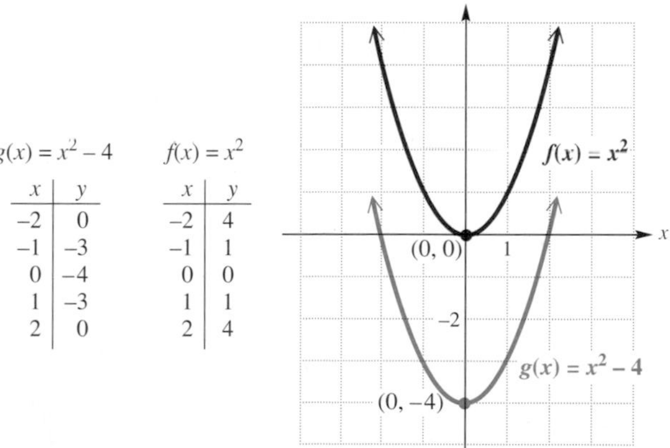

$g(x) = x^2 - 4$

x	y
−2	0
−1	−3
0	−4
1	−3
2	0

$f(x) = x^2$

x	y
−2	4
−1	1
0	0
1	1
2	4

■ **FIGURE 27**

EXAMPLE 3 Graph $g(x) = (x - 4)^2$.

Comparing the tables of values shown with Figure 28 shows that the graph of $g(x) = (x - 4)^2$ is the same as that of $f(x) = x^2$, but translated 4 units to the right. The vertex is at (4, 0). As shown in Figure 28, the axis of this parabola is the vertical line $x = 4$. ■

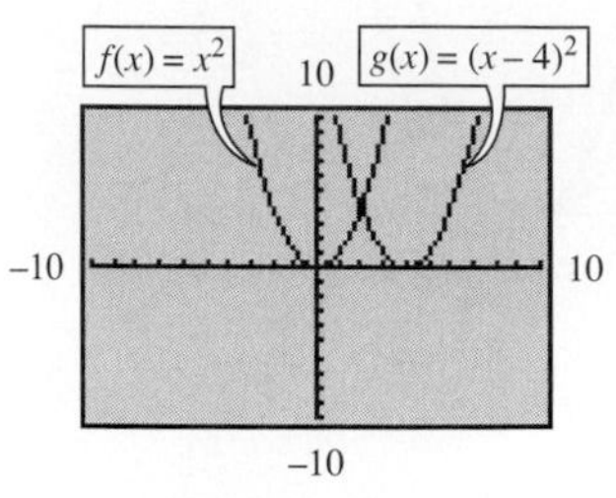

Compare with Figure 28.

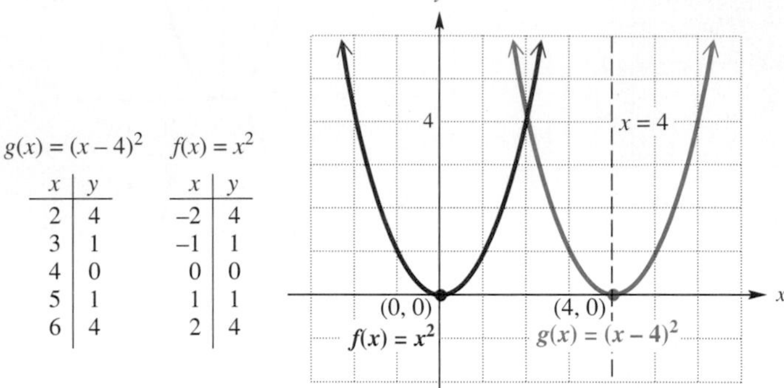

$g(x) = (x-4)^2$

x	y
2	4
3	1
4	0
5	1
6	4

$f(x) = x^2$

x	y
–2	4
–1	1
0	0
1	1
2	4

■ **FIGURE 28**

Be aware that errors frequently occur when horizontal shifts are involved. In order to determine the direction and magnitude of horizontal shifts, find the value that would cause the expression $x - h$ to equal 0. For example, the graph of $f(x) = (x - 5)^2$ would be shifted 5 units to the *right,* because $+5$ would cause $x - 5$ to equal 0. On the other hand, the graph of $f(x) = (x + 4)^2$ would be shifted 4 units to the *left*, because -4 would cause $x + 4$ to equal 0.

The following general principles apply for graphing functions of the form $f(x) = a(x - h)^2 + k$.

General Principles

1. The graph of $f(x) = a(x - h)^2 + k$, where $a \neq 0$, is a parabola with vertex (h, k), and the vertical line $x = h$ as axis.
2. The graph opens upward if a is positive and downward if a is negative.
3. The graph is wider than that of $f(x) = x^2$ if $0 < |a| < 1$. The graph is narrower than that of $f(x) = x^2$ if $|a| > 1$.

EXAMPLE 4 Graph $f(x) = -2(x + 3)^2 + 4$.

The parabola opens downward (because $a < 0$), and is narrower than the graph of $f(x) = x^2$, since $a = -2$, and $|-2| > 1$. This parabola has vertex at $(-3, 4)$, as shown in Figure 29. To complete the graph, we plotted the ordered pairs $(-4, 2)$ and $(-2, 2)$. ■

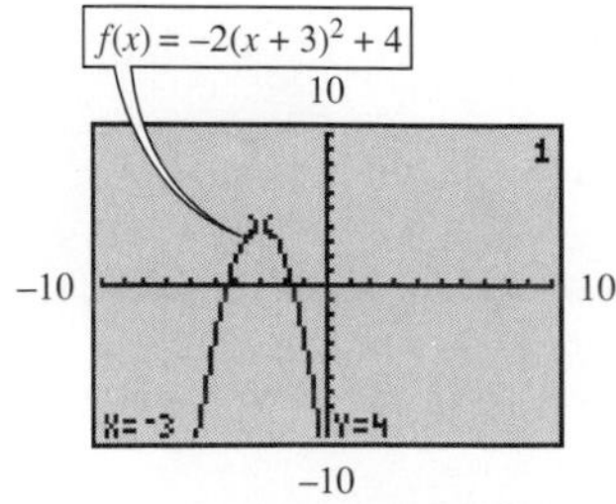

Compare with Figure 29. The vertex is $(-3, 4)$.

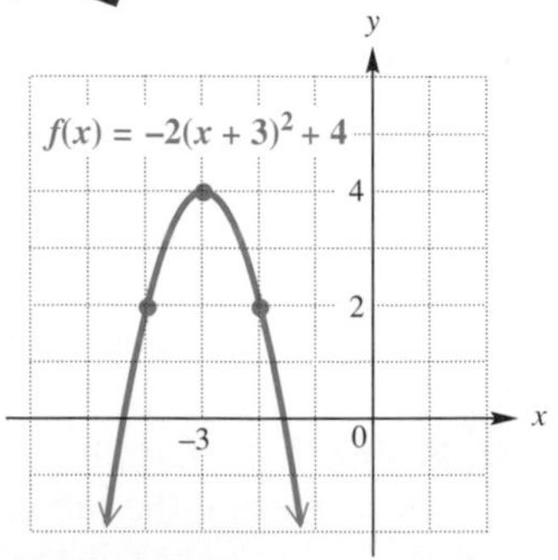

■ **FIGURE 29**

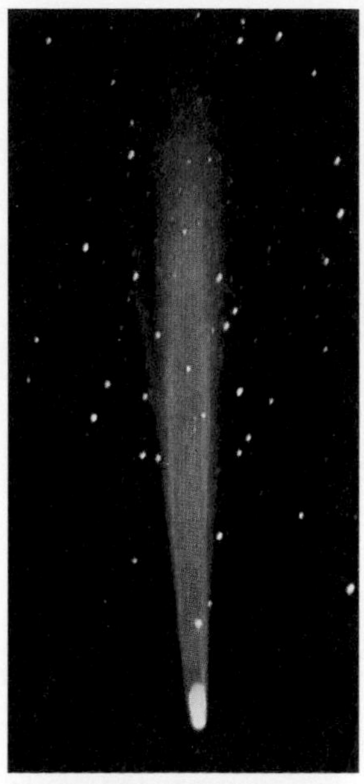

Johann Kepler (1571–1630) established the importance of a curve called an **ellipse** in 1609, when he discovered that the orbits of the planets around the sun were elliptical, not circular. The orbit of Halley's comet, shown here, is also elliptical.

When the equation of a parabola is given in the form $f(x) = ax^2 + bx + c$, it is necessary to locate the vertex in order to sketch an accurate graph. This can be done in two ways. The first is by completing the square, as shown in Example 5. The second is by using a formula which can be derived by completing the square.

EXAMPLE 5 Find the vertex of the graph of $f(x) = x^2 - 4x + 5$.

To find the vertex, we need to express $x^2 - 4x + 5$ in the form $(x - h)^2 + k$. This is done by completing the square. (See the previous chapter.) To simplify the notation, replace $f(x)$ by y.

$$y = x^2 - 4x + 5$$

$$y - 5 = x^2 - 4x \qquad \text{Transform so that the constant term is on the left.}$$

$$y - 5 + \mathbf{4} = x^2 - 4x + \mathbf{4} \qquad \text{Half of } -4 \text{ is } -2;\ (-2)^2 = 4. \text{ Add 4 to both sides.}$$

$$y - 1 = (x - 2)^2 \qquad \text{Combine terms on the left and factor on the right.}$$

$$y = (x - 2)^2 + \mathbf{1} \qquad \text{Add 1 to both sides.}$$

Now write the original equation as $f(x) = (x - 2)^2 + 1$. As shown earlier, the vertex of this parabola is $(2, 1)$. ■

A formula for the vertex of the graph of the quadratic function $y = ax^2 + bx + c$ can be found by completing the square for the general form of the equation. In doing so, we begin by dividing by a, since the coefficient of x^2 must be 1.

$$y = ax^2 + bx + c \quad (a \neq 0)$$

$$\frac{y}{a} = x^2 + \frac{b}{a}x + \frac{c}{a} \qquad \text{Divide by } a.$$

$$\frac{y}{a} - \frac{c}{a} = x^2 + \frac{b}{a}x \qquad \text{Subtract } \frac{c}{a}.$$

$$\frac{y}{a} - \frac{c}{a} + \frac{b^2}{4a^2} = x^2 + \frac{b}{a}x + \frac{b^2}{4a^2} \qquad \text{Add } \frac{b^2}{4a^2}.$$

$$\frac{y}{a} + \frac{b^2 - 4ac}{4a^2} = \left(x + \frac{b}{2a}\right)^2 \qquad \text{Combine terms on left and factor on right.}$$

$$\frac{y}{a} = \left(x + \frac{b}{2a}\right)^2 - \frac{b^2 - 4ac}{4a^2} \qquad \text{Transform so that the } y \text{ term is alone on the left.}$$

$$y = a\left(x + \frac{b}{2a}\right)^2 + \frac{4ac - b^2}{4a} \qquad \text{Multiply by } a.$$

$$y = a\left[x - \underbrace{\left(\frac{-b}{2a}\right)}_{h}\right]^2 + \underbrace{\frac{4ac - b^2}{4a}}_{k}$$

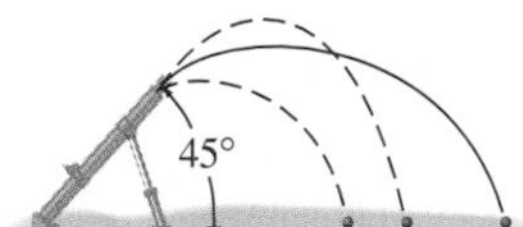

The trajectory of a shell fired from a cannon is a **parabola.** To reach the maximum range with a cannon, it is shown in calculus that the muzzle must be set at 45°. If the muzzle is elevated above 45°, the shell goes too high and falls too soon. If the muzzle is set below 45°, the shell is rapidly pulled to Earth by gravity.

The final equation shows that the vertex (h, k) can be expressed in terms of a, b, and c. However, it is not necessary to memorize the expression for k, since it can be obtained by replacing x by $\frac{-b}{2a}$. Using function notation, the y-value of the vertex is $f\left(\frac{-b}{2a}\right)$.

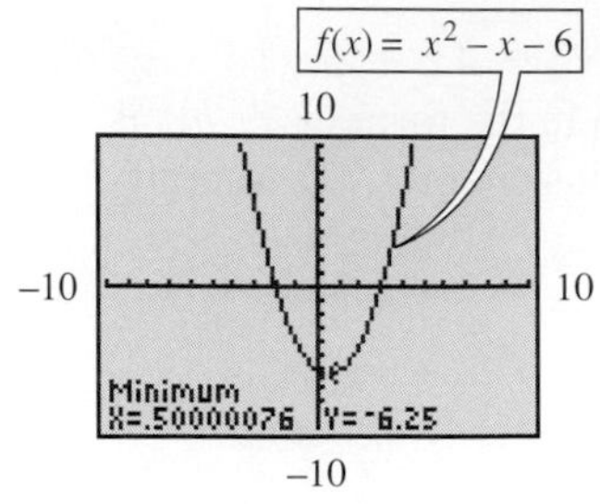

Notice the slight discrepancy when we instruct the calculator to find the vertex (a *minimum* here). This reinforces the fact that *we must understand the concepts and not totally rely on technology!*

EXAMPLE 6 Use the vertex formula to find the vertex of the graph of the function

$$f(x) = x^2 - x - 6.$$

For this function, $a = 1$, $b = -1$, and $c = -6$. The x-coordinate of the vertex of the parabola is given by

$$\frac{-b}{2a} = \frac{-(-1)}{2(\mathbf{1})} = \frac{1}{2}.$$

The y-coordinate is $f\left(\frac{-b}{2a}\right) = f\left(\frac{1}{2}\right)$.

$$\begin{aligned} f\left(\frac{1}{2}\right) &= \left(\frac{1}{2}\right)^2 - \frac{1}{2} - 6 \\ &= \frac{1}{4} - \frac{1}{2} - 6 \\ &= -\frac{25}{4} \end{aligned}$$

Finally, the vertex is $\left(\frac{1}{2}, -\frac{25}{4}\right)$. ■

A general approach to graphing quadratic functions using intercepts and the vertex is now given.

Graphing a Quadratic Function $f(x) = ax^2 + bx + c$

Step 1: Determine whether the graph opens upward (if $a > 0$) or opens downward (if $a < 0$) to aid in the graphing process.

Step 2: **Find the y-intercept.** Find the y-intercept by evaluating $f(0)$.

Step 3: **Find the x-intercepts.** Find the x-intercepts, if any, by solving $f(x) = 0$.

Step 4: **Find the vertex.** Find the vertex either by using the formula or by completing the square.

Step 5: **Complete the graph.** Find and plot additional points as needed, using the symmetry about the axis.

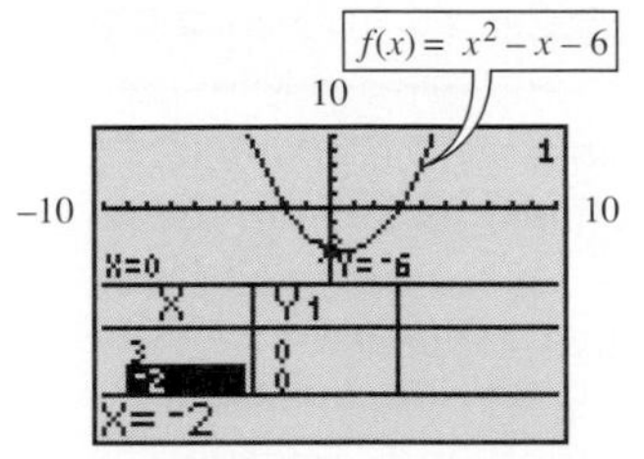

This split screen illustrates that the points $(0, -6)$, $(3, 0)$, and $(-2, 0)$ lie on the graph of $f(x) = x^2 - x - 6$. Compare with Figure 30.

X	Y1
-2	0
-1	-4
0	-6
1	-6
2	-4
3	0
4	6

X=-2

This table provides other points on the graph of $Y_1 = f(x) = x^2 - x - 6$.

EXAMPLE 7 Graph the quadratic function $f(x) = x^2 - x - 6$.

Because $a > 0$, the parabola will open upward. Now find the y-intercept.

$$f(x) = x^2 - x - 6$$
$$f(\mathbf{0}) = \mathbf{0}^2 - \mathbf{0} - 6 \qquad \text{Find } f(0).$$
$$f(0) = -6$$

The y-intercept is $(0, -6)$. Now find any x-intercepts.

$$f(x) = x^2 - x - 6$$
$$\mathbf{0} = x^2 - x - 6 \qquad \text{Let } f(x) = 0.$$
$$0 = (x - 3)(x + 2) \qquad \text{Factor.}$$
$$x - 3 = 0 \quad \text{or} \quad x + 2 = 0 \qquad \text{Set each factor equal to 0 and solve.}$$
$$x = 3 \quad \text{or} \quad x = -2$$

The x-intercepts are $(3, 0)$ and $(-2, 0)$. The vertex, found in Example 6, is $(1/2, -25/4)$. Plot the points found so far, and plot any additional points as needed. The symmetry of the graph is helpful here. The graph is shown in Figure 30.

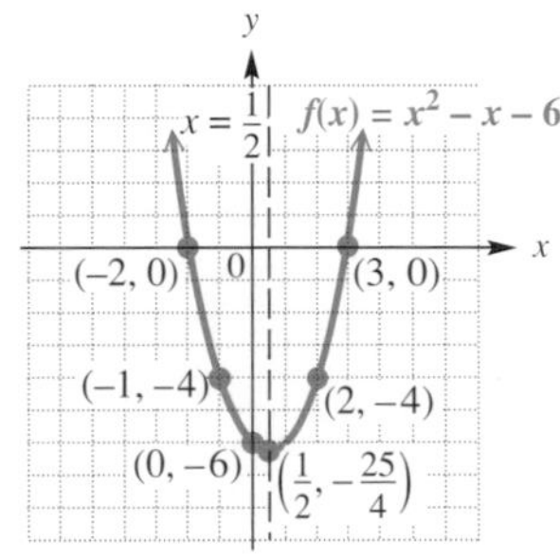

FIGURE 30

As we have seen, the vertex of a vertical parabola is either the highest or the lowest point on the parabola. The y-value of the vertex gives the maximum or minimum value of y, while the x-value tells where that maximum or minimum occurs.

Problem Solving

In many practical problems we want to know the largest or smallest value of some quantity. When that quantity can be expressed using a quadratic function $f(x) = ax^2 + bx + c$, as in the next example, the vertex can be used to find the desired value.

Galileo Galilei (1564–1642) died in the year Newton was born; his work was important in Newton's development of calculus. The idea of *function* is implicit in Galileo's analysis of the parabolic path of a projectile, where height and range are functions (in our terms) of the angle of elevation and the initial velocity.

Galileo did more than construct theories to explain physical phenomena—he set up experiments to test his ideas. According to legend, Galileo dropped objects of different weights from the tower of Pisa to disprove the Aristotelian view that heavier objects fall faster than lighter objects. He developed a formula for freely falling objects that is described by

$$d = 16t^2,$$

where d is the distance in feet that a given object falls (discounting air resistance) in a given time t, in seconds, regardless of weight.

For Further Thought

The circle, introduced in the first section of this chapter, the parabola, the ellipse, and the hyperbola are known as **conic sections.** As seen in the accompanying figure, each of these geometric shapes can be obtained by intersecting a plane and an infinite cone (made up of two *nappes*).

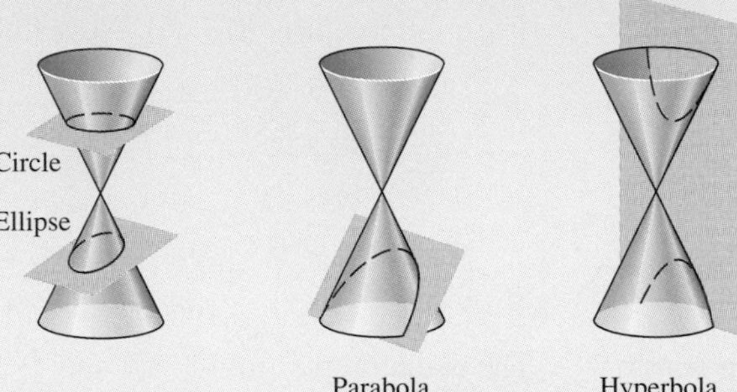

The Greek geometer Apollonius (c. 225 B.C.) was also an astronomer, and his classic work *Conic Sections* thoroughly investigated these figures. Apollonius is responsible for the names "ellipse," "parabola," and "hyperbola." The margin notes in this section show how these figures appear in the world around us.

For Group Discussion

1. The terms *ellipse, parabola,* and *hyperbola* are similar to the terms *ellipsis, parable,* and *hyperbole.* What do these latter three terms mean? You might want to do some investigation as to the similarities between the mathematical terminology and these language-related terms.
2. Name some places in the world around you where conic sections are encountered.
3. The accompanying figure shows how an ellipse can be drawn using tacks and string. Have a class member volunteer to go to the board and using string and chalk, modify the method to draw a circle. Then have two class members work together to draw an ellipse. (*Hint:* Press hard!)

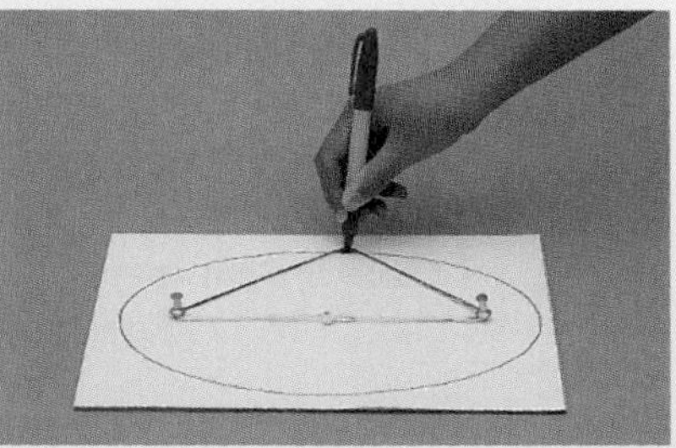

EXAMPLE 8 A farmer has 120 feet of fencing. He wants to put a fence around three sides of a rectangular plot of land next to a river. Find the maximum area he can enclose. What dimensions give this area?

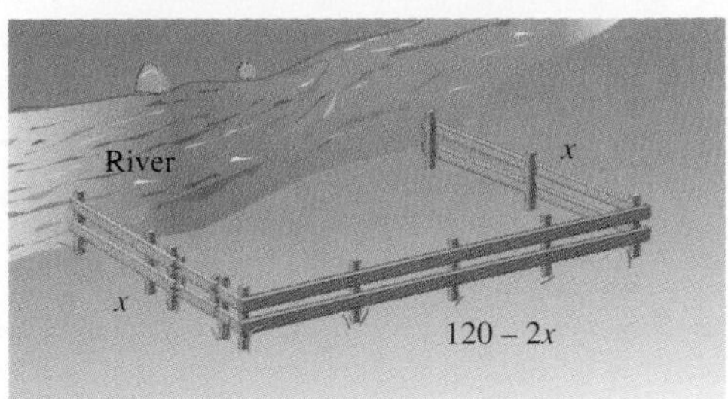

FIGURE 31

Figure 31 shows the plot. Let x represent its width. Then, since there are 120 feet of fencing,

$$x + x + \text{length} = 120 \quad \text{Sum of the sides is 120 feet.}$$
$$2x + \text{length} = 120 \quad \text{Combine terms.}$$
$$\text{length} = 120 - 2x. \quad \text{Subtract } 2x.$$

The area is given by the product of the length and width, or

$$A(x) = (120 - 2x)x = 120x - 2x^2.$$

To make the area (and thus $120x - 2x^2$) as large as possible, first find the vertex of the graph of the function $A(x) = 120x - 2x^2$.

$$A(x) = -2x^2 + 120x$$

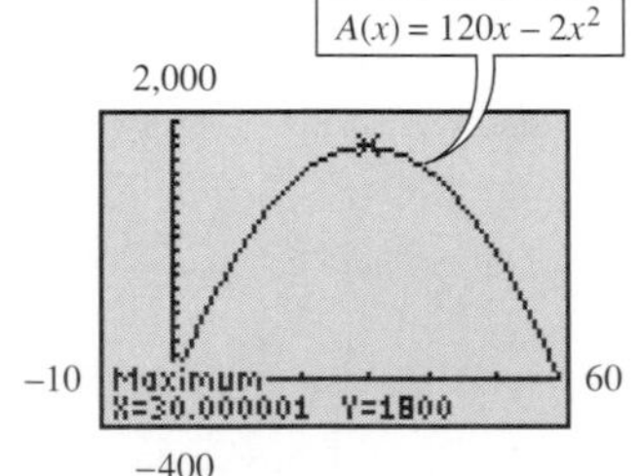

The vertex is (30, 1,800), supporting the analytic result in Example 8.

Here we have $a = -2$ and $b = 120$. The x-coordinate of the vertex is

$$\frac{-b}{2a} = \frac{-120}{2(-2)} = 30.$$

The vertex is a maximum point (since $a < 0$), so the maximum area that the farmer can enclose is

$$A(30) = -2(30)^2 + 120(30) = 1{,}800 \text{ square feet.}$$

The farmer can enclose a maximum area of 1,800 square feet, when the width of the plot is 30 feet and the length is $120 - 2(30) = 60$ feet. ◆

As seen in Example 8, it is important to be careful when interpreting the meanings of the coordinates of the vertex in problems involving maximum or minimum values. The first coordinate, x, gives the value for which the *function value* is a maximum or a minimum. It is always necessary to read the problem carefully to determine whether you are asked to find the value of the independent variable, the function value, or both.

8.5 EXERCISES

1. Explain in your own words the meaning of the term.
(a) vertex of a parabola **(b)** axis of a parabola

2. Explain why the axis of the graph of a quadratic function cannot be a horizontal line.

Identify the vertex of the graph of the quadratic function.

3. $f(x) = -3x^2$ **4.** $f(x) = -.5x^2$ **5.** $f(x) = x^2 + 4$ **6.** $f(x) = x^2 - 4$

7. $f(x) = (x - 1)^2$ **8.** $f(x) = (x + 3)^2$ **9.** $f(x) = (x + 3)^2 - 4$ **10.** $f(x) = (x - 5)^2 - 8$

For the quadratic function, tell whether the graph opens upward or downward, and tell whether the graph is wider, narrower, or the same as the graph of $f(x) = x^2$.

11. $f(x) = -2x^2$ **12.** $f(x) = -3x^2 + 1$ **13.** $f(x) = .5x^2$ **14.** $f(x) = \frac{2}{3}x^2 - 4$

15. Describe how the graph of each parabola in Exercises 9 and 10 is shifted compared to the graph of $y = x^2$.

16. Describe how the sign and the absolute value of a in $f(x) = a(x - h)^2 + k$ affects the graph of the function when compared to the graph of $y = x^2$.

17. For $f(x) = a(x - h)^2 + k$, in what quadrant is the vertex if:
(a) $h > 0, k > 0$; **(b)** $h > 0, k < 0$; **(c)** $h < 0, k > 0$; **(d)** $h < 0, k < 0$?

18. **(a)** What is the value of h if the graph of $f(x) = a(x - h)^2 + k$ has vertex on the y-axis?
(b) What is the value of k if the graph of $f(x) = a(x - h)^2 + k$ has vertex on the x-axis?

Match the graph of each equation with the graph at the right that it most closely resembles.

19. $f(x) = x^2 + 2$

20. $g(x) = x^2 - 5$

21. $h(x) = -x^2 + 4$

22. $k(x) = -x^2 - 4$

23. $F(x) = (x - 1)^2$

24. $G(x) = (x + 1)^2$

25. $H(x) = (x - 1)^2 + 1$

26. $K(x) = (x + 1)^2 + 1$

A. **B.** **C.** **D.**

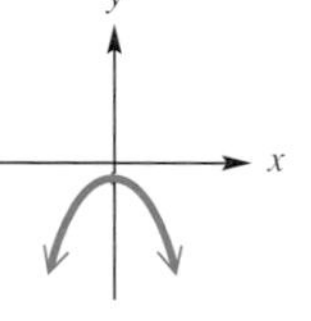
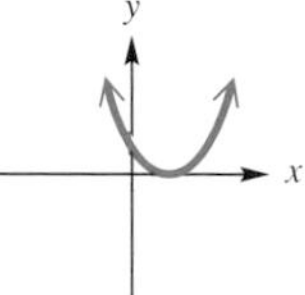

E. **F.** **G.** **H.**

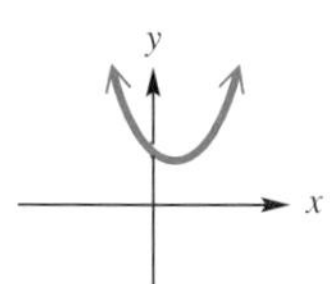
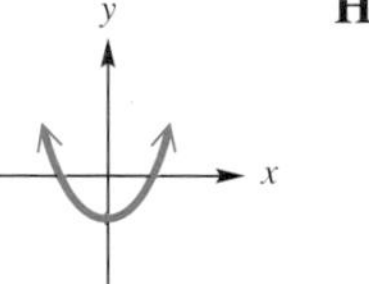
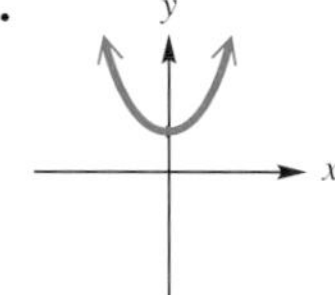

Sketch the graph of each quadratic function using the methods described in this section. Indicate two points on each graph.

27. $f(x) = 3x^2$ **28.** $f(x) = -2x^2$ **29.** $f(x) = -\frac{1}{4}x^2$ **30.** $f(x) = \frac{1}{3}x^2$

31. $f(x) = x^2 - 1$ **32.** $f(x) = x^2 + 3$ **33.** $f(x) = -x^2 + 2$ **34.** $f(x) = -x^2 - 4$

35. $f(x) = 2x^2 - 2$ **36.** $f(x) = -3x^2 + 1$ **37.** $f(x) = (x - 4)^2$ **38.** $f(x) = (x - 3)^2$

39. $f(x) = 3(x + 1)^2$ **40.** $f(x) = -2(x + 1)^2$ **41.** $f(x) = (x + 1)^2 - 2$ **42.** $f(x) = (x - 2)^2 + 3$

Sketch the graph of each quadratic function. Indicate the coordinates of the vertex of the graph.

43. $f(x) = x^2 + 8x + 14$ **44.** $f(x) = x^2 + 10x + 23$ **45.** $f(x) = x^2 + 2x - 4$

46. $f(x) = 3x^2 - 9x + 8$ **47.** $f(x) = -2x^2 + 4x + 5$ **48.** $f(x) = -5x^2 - 10x + 2$

The accompanying bar graph shows the annual average number of nonfarm payroll jobs in California for the years 1988 through 1992. If the tops of the bars were joined by a smooth curve, the curve would resemble the graph of a quadratic function (that is, a parabola). Using a technique from statistics it can be determined that this function can be described approximately as

$$f(x) = -.10x^2 + .42x + 11.90,$$

where $x = 0$ corresponds to 1988, $x = 1$ corresponds to 1989, and so on, and $f(x)$ represents the number of payroll jobs in millions. (Source: California Employment Development Department)

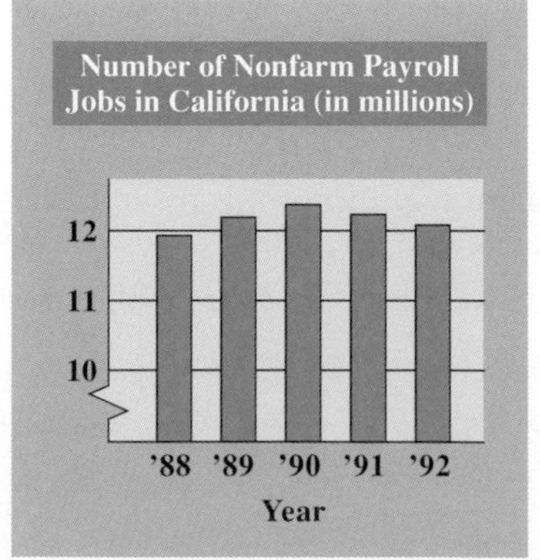

49. Explain why the coefficient of x^2 in the function is negative, based on the graph formed by joining the tops of the bars.

50. Determine the coordinates of the vertex of the graph using algebraic methods.

51. How does the x-coordinate of the vertex of the parabola indicate that during the time period under consideration, the maximum number of payroll jobs was in 1990?

52. What does the y-coordinate of the vertex of the parabola indicate?

Solve the problem.

53. Keisha Hughes has 100 meters of fencing material to enclose a rectangular exercise run for her dog. What width will give the enclosure the maximum area?

54. Morgan's Department Store wants to construct a rectangular parking lot on land bordered on one side by a highway. It has 280 feet of fencing that is to be used to fence off the other three sides. What should be the dimensions of the lot if the enclosed area is to be a maximum? What is the maximum area?

55. The world consumption of oil in millions of barrels daily is given by

$$f(x) = -.1x^2 + 2x + 58,$$

where x is the number of years since 1985 (1985 corresponds to 0). According to this equation, when will oil consumption reach a maximum? What will the maximum consumption be? Discuss how realistically this equation describes oil consumption.

56. Recent annual inflation rates (in percent) in Mexico are given by the function

$$f(x) = 4x^2 - 48x + 154,$$

where x represents the number of years since 1987. In what year does this equation indicate that the inflation rate will be a minimum? What is the minimum rate? Discuss how realistic this equation may or may not be.

57. If an object is thrown upward with an initial velocity of 32 feet per second, then its height after t seconds is given by

$$h(t) = 32t - 16t^2.$$

Find the maximum height attained by the object and the number of seconds it takes to hit the ground.

58. A projectile is fired straight upward so that its distance (in feet) above the ground t seconds after firing is

$$s(t) = -16t^2 + 400t.$$

Find the maximum height it reaches and the number of seconds it takes to reach that height.

59. If air resistance is neglected, a projectile shot straight upward with an initial velocity of 40 meters per second will be at a height s in meters given by the function

$$s(t) = -4.9t^2 + 40t,$$

where t is the number of seconds elapsed after projection. After how many seconds will it reach its maximum height, and what is this maximum height? Round your answers to the nearest tenth.

60. For a trip to a resort, a charter bus company charges a fare of \$48 per person, plus \$2 per person for each unsold seat on the bus. If the bus has 42 seats and x represents the number of unsold seats, obtain the following:

(a) a function that defines the total revenue, R, from the trip (*Hint:* Multiply the total number riding, $42 - x$, by the price per ticket, $48 + 2x$);
(b) the graph of the function from part (a);
(c) the number of unsold seats that produces the maximum revenue;
(d) the maximum revenue.

8.6 Exponential and Logarithmic Functions and Applications

In this section we introduce two new types of functions. The first of these is the exponential function.

Exponential Function

An exponential function with base b, where $b > 0$ and $b \neq 1$, is a function of the form

$$f(x) = b^x,$$

where x is any real number.

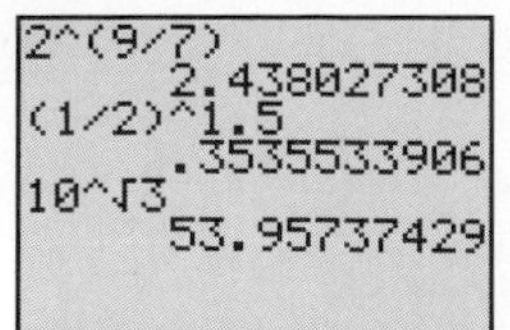

Compare with the discussion in the text.

Thus far, we have only defined integer exponents. In the definition of exponential function, we allow x to take on any real number value. By using methods not discussed in this book, expressions such as

$$2^{9/7}, \qquad \left(\frac{1}{2}\right)^{1.5}, \qquad \text{and} \qquad 10^{\sqrt{3}}$$

can be approximated. A scientific or graphing calculator is capable of determining approximations for these numbers.

$$2^{9/7} \approx 2.438027308$$
$$\left(\frac{1}{2}\right)^{1.5} \approx .3535533906$$
$$10^{\sqrt{3}} \approx 53.95737429$$

Notice that in the definition of exponential function, the base b is restricted to positive numbers, with $b \neq 1$. The first example shows the graphs of three exponential functions.

EXAMPLE 1 The graphs of $f(x) = 2^x$, $g(x) = (1/2)^x$, and $h(x) = 10^x$ are shown in Figure 32(a), (b), and (c). In each case, a table of selected points is shown. The points are joined with a smooth curve, typical of the graphs of exponential functions. Notice that for each graph, the curve approaches but does not intersect the x-axis. For this reason, the x-axis is called the **horizontal asymptote** of the graph.

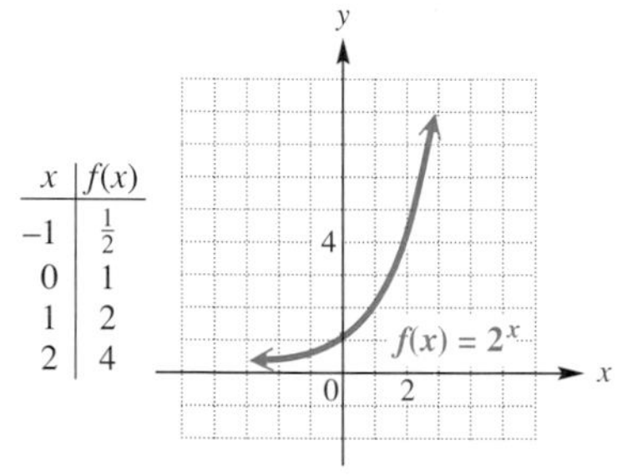

x	$f(x)$
-1	$\frac{1}{2}$
0	1
1	2
2	4

(a)

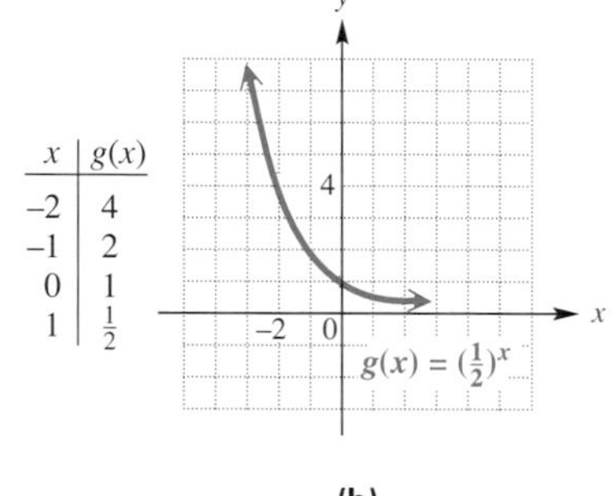

x	$g(x)$
-2	4
-1	2
0	1
1	$\frac{1}{2}$

(b)

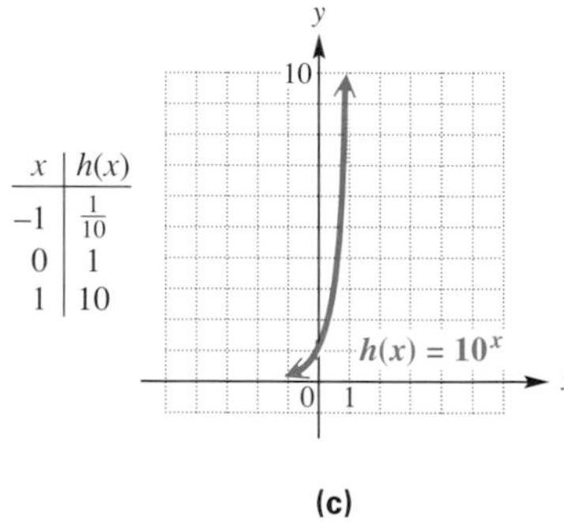

x	$h(x)$
-1	$\frac{1}{10}$
0	1
1	10

(c)

FIGURE 32

$g(x) = \left(\frac{1}{2}\right)^x$ $h(x) = 10^x$ $f(x) = 2^x$

20

−10 10

0

Compare with Figure 32.

Example 1 illustrates the following facts about the graph of an exponential function.

Graph of $f(x) = b^x$

1. The graph will always contain the point $(0, 1)$, since $b^0 = 1$.
2. When $b > 1$, the graph will rise from left to right (as in the example for $b = 2$ and $b = 10$). When $0 < b < 1$, the graph will fall from left to right (as in the example for $b = 1/2$).
3. The x-axis is the horizontal asymptote.
4. The domain is $(-\infty, \infty)$ and the range is $(0, \infty)$.

Probably the most important exponential function has the base e. The number e is named after Leonhard Euler (1707–1783), and is approximately 2.718281828. It is an irrational number, and its value is approached by the expression

$$\left(1 + \frac{1}{n}\right)^n$$

as n takes on larger and larger values. We write

$$\text{as } n \to \infty, \left(1 + \frac{1}{n}\right)^n \to e \approx 2.718281828.$$

See the chart that follows.

X	Y1
1	2
10	2.5937
100	2.7048
1000	2.7169
10000	2.7181
100000	2.7183
1E6	2.7183

Y1=2.71828046932

This table shows selected values for $Y_1 = \left(1 + \frac{1}{X}\right)^X$.

n	**Approximate Value of** $\left(1 + \frac{1}{n}\right)^n$
1	2
2	2.25
5	2.48832
10	2.59374
25	2.66584
50	2.69159
100	2.70481
500	2.71557
1,000	2.71692
10,000	2.71815
1,000,000	2.71828

Powers of e can be approximated on a scientific or graphing calculator. Some powers of e obtained on a calculator are

$$e^{-2} \approx .1353352832$$
$$e^{1.3} \approx 3.669296668$$
$$e^{4} \approx 54.59815003.$$

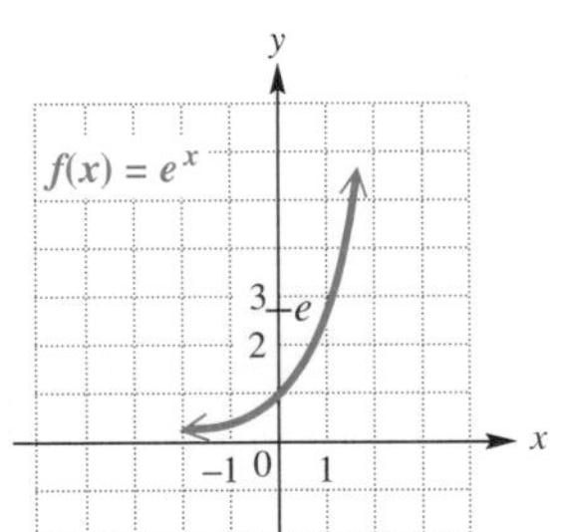

■ **FIGURE 33**

The graph of the function $f(x) = e^x$ is shown in Figure 33.

Applications of Exponential Functions

A real-life application of exponential functions occurs in the computation of compound interest.

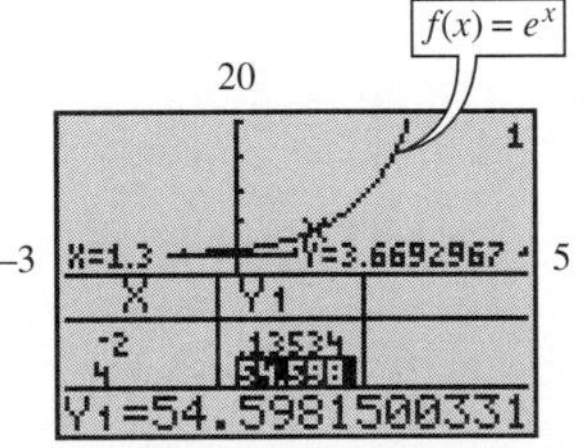

The values for e^{-2}, $e^{1.3}$, and e^4 are approximated in this split-screen graph of $f(x) = e^x$.

Compound Interest Formula

Suppose that a principal of P dollars is invested at an annual interest rate r (in percent), compounded n times per year. Then the amount A accumulated after t years is given by the formula

$$A = P\left(1 + \frac{r}{n}\right)^{nt}.$$

EXAMPLE 2 Suppose that \$1,000 is invested at an annual rate of 8%, compounded quarterly (four times per year). Find the total amount in the account after ten years if no withdrawals are made.

Use the compound interest formula above, with $P = 1{,}000$, $r = .08$, $n = 4$, and $t = 10$.

$$A = P\left(1 + \frac{r}{n}\right)^{nt}$$

$$A = 1{,}000\left(1 + \frac{.08}{4}\right)^{4 \cdot 10}$$

$$A = 1{,}000(1.02)^{40}$$

Using a calculator with an exponential key, we find that $1.02^{40} \approx 2.20804$. Multiply by 1,000 to get

$$A = 2{,}208.04.$$

There would be \$2,208.04 in the account at the end of ten years.

The compounding formula given earlier applies if the financial institution compounds interest for a finite number of compounding periods annually. Theoretically, the number of compounding periods per year can get larger and larger (quarterly, monthly, daily, etc.), and if n is allowed to approach infinity, we say that interest is compounded *continuously*. It can be shown that the formula for **continuous compounding** involves the number e.

Continuous Compounding Formula

Suppose that a principal of P dollars is invested at an annual interest rate r (in percent), compounded continuously. Then the amount A accumulated after t years is given by the formula

$$A = Pe^{rt}.$$

EXAMPLE 3 Suppose that \$5,000 is invested at an annual rate of 6.5%, compounded continuously. Find the total amount in the account after four years if no withdrawals are made.

The continuous compounding formula applies here, with $P = 5{,}000$, $r = .065$ and $t = 4$.

$$A = Pe^{rt}$$

$$A = 5{,}000e^{.065(4)}$$

$$A = 5{,}000e^{.26}$$

Use the e^x key on a calculator to find that $e^{.26} \approx 1.29693$, so $A \approx 5{,}000(1.29693) = 6{,}484.65$. There will be \$6,484.65 in the account after four years.

The continuous compounding formula is an example of an **exponential growth function.** It can be shown that in situations involving growth or decay of a quantity, the amount or number present at time t can often be closely approximated by a function of the form

$$A(t) = A_0 e^{kt},$$

where A_0 represents the amount or number present at time $t = 0$, and k is a constant. If $k > 0$, there is exponential growth; if $k < 0$, there is exponential decay.

For Further Thought

In the February 5, 1989 issue of *Parade,* Carl Sagan related an oft-told legend involving **exponential growth** in the article "The Secret of the Persian Chessboard." Once upon a time, a Persian king wanted to please his executive officer, the Grand Vizier, with a gift of his choice. The Grand Vizier explained that he would like to be able to use his chessboard to accumulate wheat. A single grain of wheat would be received for the first square on the board, two grains would be received for the second square, four grains for the third, and so on, doubling the number of grains for each of the 64 squares on the board. This doubling procedure is an example of exponential growth, and is defined by the exponential function

$$f(x) = 2^x,$$

where x corresponds to the number of the chessboard square, and $f(x)$ represents the number of grains of wheat corresponding to that square.

How many grains of wheat would be accumulated? As unlikely as it may seem, the number of grains would total 18.5 quintillion! Even with today's methods of production, this amount would take 150 years to produce. The Grand Vizier evidently knew his mathematics.

For Group Discussion

1. If a lily pad doubles in size each day, and it covers a pond after eight days of growth, when is the pond half covered? One-fourth covered?
2. Have each member of the class estimate the answer to this problem.

   If you earn 1¢ on January 1, 2¢ on January 2, 4¢ on January 3, 8¢ on January 4, and so on, doubling your salary each day, how much will you earn during the month?

   As a calculator exercise, determine the answer and see who was able to estimate the answer most accurately.
3. In a chain letter you receive a list with six names on it. You are to send one dollar to the name at the top of the list, cross off that name, add your name at the bottom, and mail the list to six people, each of whom repeats all these steps. The first person sends six letters, and each of the six recipients sends six letters. The process continues in the same manner over and over. The population of the United States is more than 250,000,000. By the eleventh mailing, more letters will have been sent than there are residents of the country, and the people who started the chain will have made a tidy sum. Chain letters are illegal, however, and postal inspectors go after the originators with a zeal not always shown by mail handlers. Relate any stories you might have heard about chain letters.

Logarithmic Functions

Consider the exponential equation

$$2^3 = 8.$$

Here we see that 3 is the exponent (or power) to which 2 must be raised in order to obtain 8. The exponent 3 is called the *logarithm* to the base 2 of 8, and this is written

$$3 = \log_2 8.$$

In general, we have the following relationship.

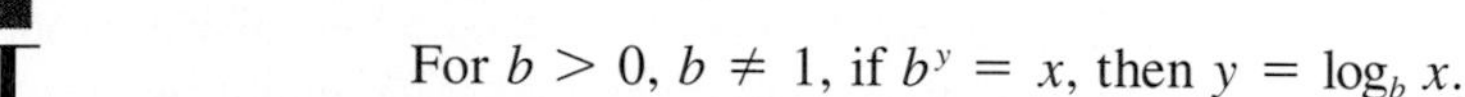

For $b > 0$, $b \neq 1$, if $b^y = x$, then $y = \log_b x$.

The following example illustrates the relationship between exponential equations and logarithmic equations.

EXAMPLE 4 The table below shows equivalent exponential and logarithmic equations.

Exponential Equation	Logarithmic Equation
$3^4 = 81$	$4 = \log_3 81$
$(1/2)^{-4} = 16$	$-4 = \log_{1/2} 16$
$10^0 = 1$	$0 = \log_{10} 1$

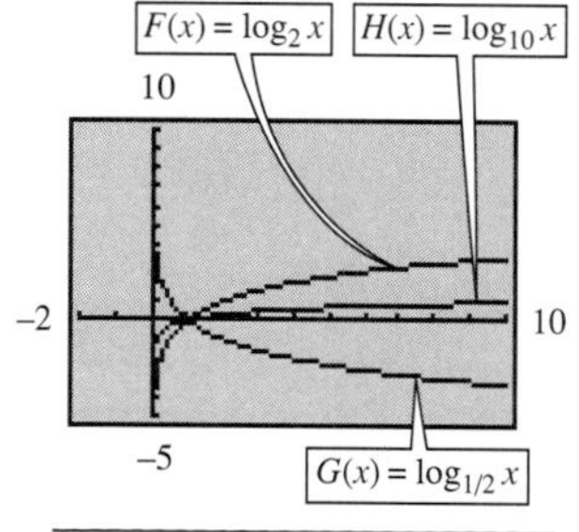

Compare with Figure 34. Graphs of logarithmic functions with bases other than 10 and *e* are accomplished with the use of the change-of-base rule from algebra:

$$\log_a x = \frac{\log x}{\log a} = \frac{\ln x}{\ln a}.$$

Alternatively, they can be "drawn" by using the capability that allows the user to obtain the graph of the inverse.

The concept of inverse functions (studied in more advanced algebra courses) leads us to the definition of the logarithmic function with base b.

Logarithmic Function

A logarithmic function with base b, where $b > 0$ and $b \neq 1$, is a function of the form

$$g(x) = \log_b x,$$

where $x > 0$.

The graph of the function $g(x) = \log_b x$ can be found by interchanging the roles of x and y in the function $f(x) = b^x$. Geometrically, this is accomplished by reflecting the graph of $f(x) = b^x$ about the line $y = x$.

EXAMPLE 5 The graphs of $F(x) = \log_2 x$, $G(x) = \log_{1/2} x$, and $H(x) = \log_{10} x$ are shown in Figure 34(a), (b), and (c). In each case, a table of selected points is shown. These points were obtained by interchanging the roles of x and y in the tables of points given in Figure 32. The points are joined with a smooth curve, typical of the graphs of logarithmic functions. Notice that for each graph, the

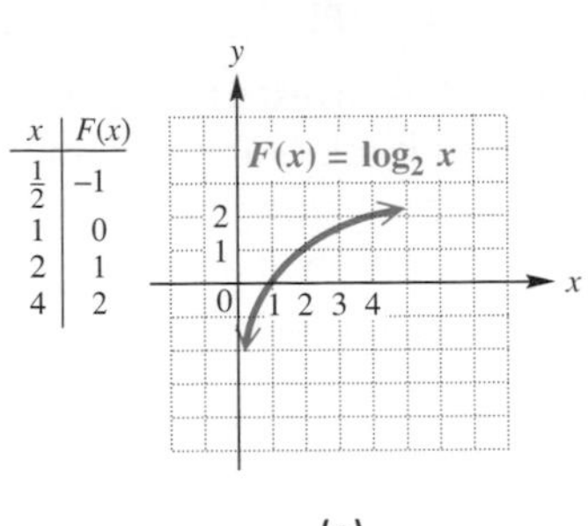

(a)

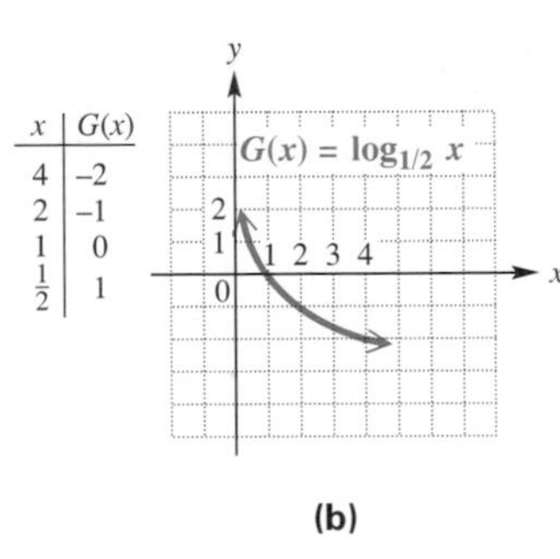

(b)

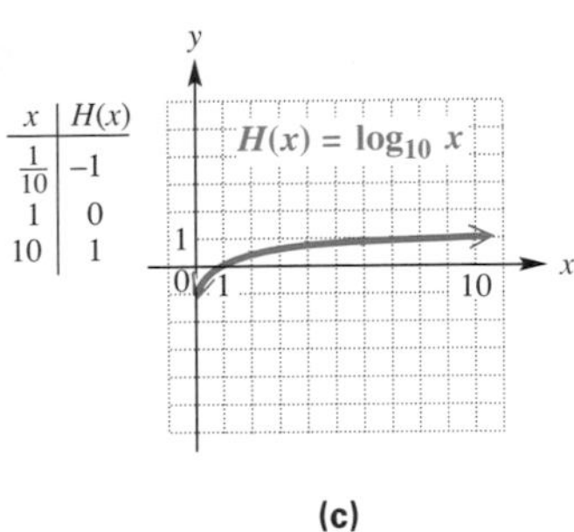

(c)

■ **FIGURE 34**

curve approaches but does not intersect the y-axis. For this reason, the y-axis is called the **vertical asymptote** of the graph. ◆

Example 5 illustrates the following facts about the graph of a logarithmic function.

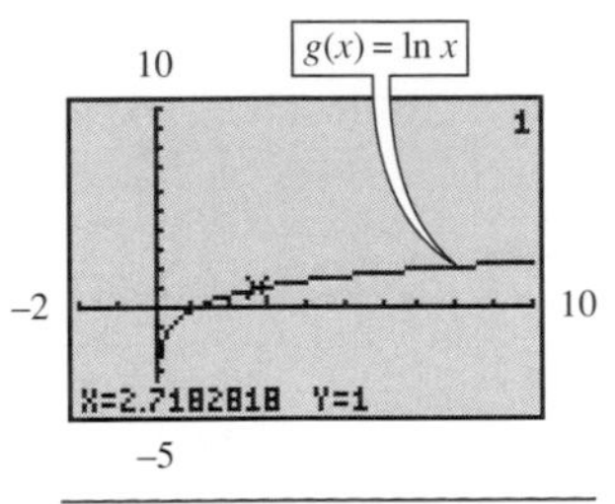

Notice that when $x = e \approx 2.7182818$, $y = \ln e = 1$.

Graph of $g(x) = \log_b x$

1. The graph will always contain the point $(1, 0)$, since $\log_b 1 = 0$.
2. When $b > 1$, the graph will rise from left to right, from the fourth quadrant to the first (as in the example for $b = 2$ and $b = 10$). When $0 < b < 1$, the graph will fall from left to right, from the first quadrant to the fourth (as in the example for $b = 1/2$).
3. The y-axis is the vertical asymptote.
4. The domain is $(0, \infty)$ and the range is $(-\infty, \infty)$.

One of the most important logarithmic functions is the function with base e. If we interchange the roles of x and y in the graph of $f(x) = e^x$ (Figure 33), we obtain the graph of $g(x) = \log_e x$. There is a special symbol for $\log_e x$: it is $\ln x$. That is,

$$\ln x = \log_e x.$$

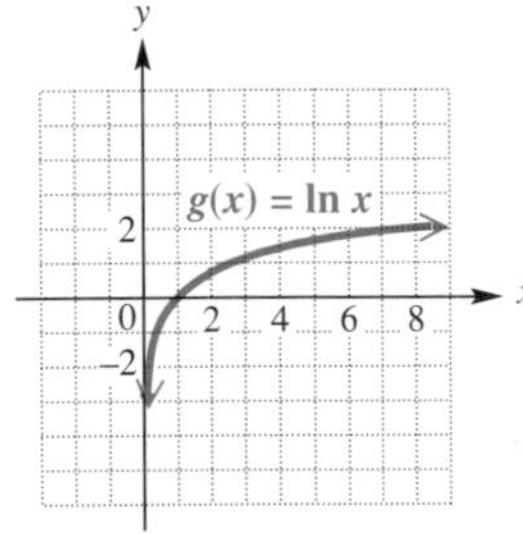

■ **FIGURE 35**

Figure 35 shows the graph of $g(x) = \ln x$, called the **natural logarithmic function.**

The expression

$$\ln e^k$$

is the exponent to which the base e must be raised in order to obtain e^k. There is only one such number that will do this, and it is k itself. Therefore, we make the following important statement.

For all real numbers k, $\ln e^k = k$.

This fact will be used in the examples that follow.

EXAMPLE 6 Suppose that a certain amount P is invested at an annual rate of 6.5%, compounded continuously. How long will it take for the amount to triple?

We wish to find the value of t in the continuous compounding formula that will make the amount A equal to $3P$ (since we want the initial investment, P, to triple). We use the formula $A = Pe^{rt}$, with $3P$ substituted for A and .065 substituted for r.

$$A = Pe^{rt}$$
$$3P = Pe^{.065t}$$
$$3 = e^{.065t} \qquad \text{Divide both sides by } P.$$

Now take the natural logarithm of both sides.

$$\ln 3 = \ln e^{.065t}$$
$$\ln 3 = .065t \qquad \text{Use the fact that } \ln e^k = k.$$

Solve for t by dividing both sides by .065.

$$t = \frac{\ln 3}{.065}$$

A scientific calculator can be used to find that $\ln 3 \approx 1.098612289$. Dividing this by .065 gives

$$t \approx 16.9$$

to the nearest tenth. Therefore, it would take about 16.9 years for any initial investment P to triple under the given conditions. ■

The number e is named in honor of Leonhard Euler (1707–83), the prolific Swiss mathematician. The value of e can be expressed as an infinite series.

$$e = 1 + \frac{1}{1} + \frac{1}{1 \cdot 2} + \frac{1}{1 \cdot 2 \cdot 3} + \frac{1}{1 \cdot 2 \cdot 3 \cdot 4} + \cdots$$

It can also be expressed in two ways using continued fractions.

$$e = 2 + \cfrac{1}{1 + \cfrac{1}{2 + \cfrac{1}{1 + \cfrac{1}{4 + \cdots}}}}$$

$$e = 2 + \cfrac{1}{1 + \cfrac{1}{2 + \cfrac{2}{3 + \cfrac{3}{4 + \cfrac{4}{\cdots}}}}}$$

EXAMPLE 7 Nuclear energy derived from radioactive isotopes can be used to supply power to space vehicles. Suppose that the output of the radioactive power supply for a certain satellite is given by the function

$$y = 40e^{-.004t},$$

where y is measured in watts and t is the time in days.

(a) What is the initial output of the power supply?

Let $t = 0$ in the equation.

$$y = 40e^{-.004t} \qquad \text{Given function}$$
$$y = 40e^{-.004(0)} \qquad \text{Let } t = 0.$$
$$y = 40e^0$$
$$y = 40 \qquad e^0 = 1$$

The initial output is 40 watts.

(b) After how many days will the output be reduced to 35 watts?
Let $y = 35$, and solve for t.

$$35 = 40e^{-.004t}$$

$$\frac{35}{40} = e^{-.004t} \qquad \text{Divide by 40.}$$

$$\ln\left(\frac{35}{40}\right) = \ln e^{-.004t} \qquad \text{Take the natural logarithm.}$$

$$\ln\left(\frac{35}{40}\right) = -.004t \qquad \ln e^k = k$$

$$t = \frac{\ln\left(\frac{35}{40}\right)}{-.004} \qquad \text{Divide by } -.004.$$

Using a calculator, we find that $t \approx 33.4$. It will take about 33.4 days for the output to be reduced to 35 watts. ■

Radioactive materials, such as the one discussed in Example 7, disintegrate according to exponential decay functions. The **half-life** of a quantity that decays exponentially is the amount of time that it takes for any initial amount to decay to half its initial value.

EXAMPLE 8 Carbon 14 is a radioactive form of carbon that is found in all living plants and animals. After a plant or animal dies, the radiocarbon disintegrates. Scientists determine the age of the remains by comparing the amount of carbon 14 present with the amount found in living plants and animals. The amount of carbon 14 present after t years is approximated by the exponential equation

$$y = y_0e^{-.0001216t},$$

where y_0 represents the initial amount. What is the half-life of carbon 14?
To find the half-life, let $y = \frac{1}{2}y_0$ in the equation.

$$\frac{1}{2}y_0 = y_0e^{-.0001216t}$$

$$\frac{1}{2} = e^{-.0001216t}$$

$$\ln\left(\frac{1}{2}\right) = -.0001216t$$

$$t \approx 5{,}700 \qquad \text{Use a calculator.}$$

The half-life of carbon 14 is about 5,700 years. ■

For Further Thought

So what does that "log" button on your scientific calculator mean? We will answer that question shortly. But let's back up to 1624 when the British mathematician Henry Briggs published what was, at that time, the greatest breakthrough ever for aiding computation: tables of common logarithms. A common, or base ten logarithm, of a positive number x is the exponent to which 10 must be raised in order to obtain x. Symbolically, we simply write $\log x$ to denote $\log_{10} x$.

Since logarithms are exponents, their properties are the same as those of exponents. These properties allowed users of tables of common logarithms to multiply by adding, divide by subtracting, raise to powers by multiplying, and take roots by dividing. For example, to multiply 534.4 by 296.7, the user would find log 534.4 in the table, then find log 296.7 in the table, and add these values. Then, by reading the table in reverse, the user would find the "antilog" of the sum in order to find the product of the original two numbers. More complicated operations were possible as well.

Reproduced here is a homework problem that someone was asked to do on May 15, 1930. It was found on a well-preserved sheet of paper stuck inside an old mathematics text purchased in a used-book store in New Orleans. Aren't you glad that we now use calculators? (Incidentally, the answer given is incorrect.)

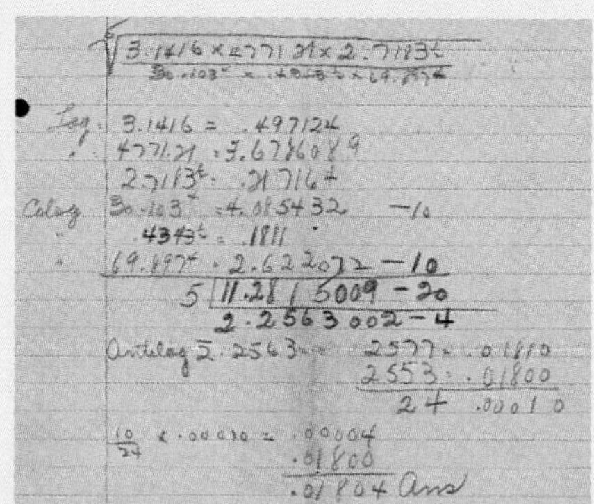

So why include a base ten logarithm key on a scientific calculator? Common logarithms are still used in certain applications in science. For example, the pH of a substance is the negative of the common logarithm of the hydronium ion concentration in moles per liter. The pH value is a measure of the acidity or alkalinity of a solution. If pH > 7.0, the solution is alkaline; if pH < 7.0, it is acidic. The Richter scale, used to measure the intensity of earthquakes, is based on logarithms, and sound intensities, measured in decibels, also have a logarithmic basis.

For Group Discussion

1. Divide the class into small groups, and use the log key of a scientific calculator to multiply 458.3 by 294.6. Remember that once the logarithms have been added, it is necessary then to raise 10 to that power to obtain the product. The 10^x key will be required for this.

2. Ask someone who studied computation with logarithms in an algebra or trigonometry course before the advent of scientific calculators what it was like. Then report their recollections to the class.
3. What is a slide rule? Have someone in the class find one in an attic or closet, and bring it to class. The slide rule was the accepted method of quick calculation through the 1960s. Anyone coming across a large stash of unused slide rules could imbed them in plastic and sell them as nostalgia items!

8.6 EXERCISES

Fill in each blank with the correct response.

1. For an exponential function $f(x) = a^x$, if $a > 1$, the graph ________ (rises/falls) from left to right. If $0 < a < 1$, the graph ________ (rises/falls) from left to right.
2. The y-intercept of the graph of $y = a^x$ is ______.
3. The graph of the exponential function $f(x) = a^x$ ________ (does/does not) have an x-intercept.
4. The point (2, ______) is on the graph of $f(x) = 3^{4x-3}$.
5. For a logarithmic function $g(x) = \log_a x$, if $a > 1$, the graph ________ (rises/falls) from left to right. If $0 < a < 1$, the graph ________ (rises/falls) from left to right.
6. The x-intercept of the graph of $y = \log_a x$ is ______.
7. The graph of the exponential function $g(x) = \log_a x$ ________ (does/does not) have a y-intercept.
8. The point (98, ______) lies on the graph of $g(x) = \log_{10}(x + 2)$.

Use a calculator to find an approximation for each of the following numbers. Give as many digits as the calculator displays.

9. $9^{3/7}$ **10.** $14^{2/7}$ **11.** $(.83)^{-1.2}$ **12.** $(.97)^{3.4}$

13. $(\sqrt{6})^{\sqrt{5}}$ **14.** $(\sqrt{7})^{\sqrt{3}}$ **15.** $\left(\frac{1}{3}\right)^{9.8}$ **16.** $\left(\frac{2}{5}\right)^{8.1}$

Sketch the graph of each of the following functions.

17. $f(x) = 3^x$ **18.** $f(x) = 5^x$ **19.** $f(x) = \left(\frac{1}{4}\right)^x$ **20.** $f(x) = \left(\frac{1}{3}\right)^x$

Use a calculator to find each of the following. Give as many digits as the calculator displays.

21. e^3 **22.** e^4 **23.** e^{-4} **24.** e^{-3}

In Exercises 25–28, rewrite the exponential equation as a logarithmic equation. In Exercises 29–32, rewrite the logarithmic equation as an exponential equation.

25. $4^2 = 16$ **26.** $5^3 = 125$ **27.** $\left(\frac{2}{3}\right)^{-3} = \frac{27}{8}$ **28.** $\left(\frac{1}{10}\right)^{-4} = 10{,}000$

29. $5 = \log_2 32$ **30.** $3 = \log_4 64$ **31.** $1 = \log_3 3$ **32.** $0 = \log_{12} 1$

Use a calculator to find each of the following. Give as many digits as the calculator displays.

33. ln 4 **34.** ln 6 **35.** ln .35 **36.** ln 2.45

Sketch the graph of each of the following functions. (Hint: Use the graphs of the exponential functions in Exercises 17–20 to help.)

37. $g(x) = \log_3 x$ **38.** $g(x) = \log_5 x$ **39.** $g(x) = \log_{1/4} x$ **40.** $g(x) = \log_{1/3} x$

Determine the amount of money that will be accumulated in an account that pays compound interest, given the initial principal in each of the following.

41. \$4,292 at 6% compounded annually for 10 years

42. \$8,906.54 at 5% compounded semiannually for 9 years

43. \$56,780 at 5.3% compounded quarterly for 23 quarters

44. \$45,788 at 6% compounded daily (ignoring leap years) for 11 years of 365 days

45. Cathy Wacaser invests a \$25,000 inheritance in a fund paying 5% per year compounded continuously. What will be the amount on deposit after each of the following time periods?
(a) 1 year **(b)** 5 years **(c)** 10 years

46. Linda Youngman, who is self-employed, wants to invest \$60,000 in a pension plan. One investment offers 7% compounded quarterly. Another offers 6.75% compounded continuously. Which investment will earn more interest in 5 years? How much more will the better plan earn?

The figure shown here accompanied the article "Is Our World Warming?" which appeared in the October 1990 issue of National Geographic. *It shows projected temperature increases using two graphs: one an exponential-type curve, and the other linear. From the figure, approximate the increase* **(a)** *for the exponential curve, and* **(b)** *for the linear graph for each of the following years.*

47. 2000

48. 2010

49. 2020

50. 2040

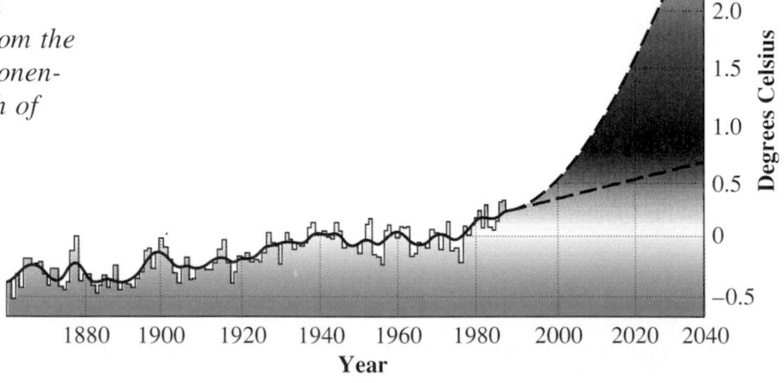

Graph, "Zero Equals Average Global Temperature for the Period 1950–1979" by Dale D. Glasgow from *National Geographic*, October 1990. Reprinted by permission of the National Geographic Society.

Solve the problem. Use a calculator as needed. The equations in Exercises 51 and 52 are models based on the assumption that current growth rates remain the same.

51. The population of Brazil, in millions, is approximated by the function

$$f(x) = 155.3(2)^{.025x},$$

where $x = 0$ corresponds to 1994, $x = 1$ corresponds to 1995, and so on.

(a) What will be the population of Brazil in the year 2000 according to this model?

(b) What will be the population in 2034?

(c) How will the population in 2034 compare to the population in 1994?

52. The population of Pakistan, in millions, is approximated by the function

$$f(x) = 126.4(2)^{.04x},$$

where $x = 0$ corresponds to 1994, $x = 1$ corresponds to 1995, and so on.

(a) What was the population of Pakistan in the year 1994?

(b) What will the population be in 2019?

(c) How will the population in 2019 compare to the population in 1994?

53. The amount of radioactive material in an ore sample is given by the function

$$A(t) = 100(3.2)^{-.5t},$$

where $A(t)$ is the amount present, in grams, of the sample t months after the initial measurement.

(a) How much was present at the initial measurement? (*Hint:* $t = 0$.)

(b) How much was present 2 months later?

(c) How much was present 10 months later?

54. A small business estimates that the value $V(t)$ of a copy machine is decreasing according to the function

$$V(t) = 5{,}000(2)^{-.15t},$$

where t is the number of years that have elapsed since the machine was purchased, and $V(t)$ is in dollars.

(a) What was the original value of the machine?

(b) What is the value of the machine 5 years after purchase? Give your answer to the nearest dollar.

(c) What is the value of the machine 10 years after purchase? Give your answer to the nearest dollar.

55. Refer to the function in Exercise 54. When will the value of the machine be \$2,500? (*Hint:* Let $V(t) =$ 2,500, divide both sides by 5,000, and use the method of Example 7(b).)

56. Refer to the function in Exercise 54. When will the value of the machine be \$1,250?

The bar graph shows the average annual major league baseball player's salary for each year since free agency began. Using a technique from statistics, it was determined that the function

$$S(x) = 74{,}741(1.17)^x$$

approximates the salary, where $x = 0$ corresponds to 1976, and so on, up to $x = 18$ representing 1994. (Salary is in dollars.)

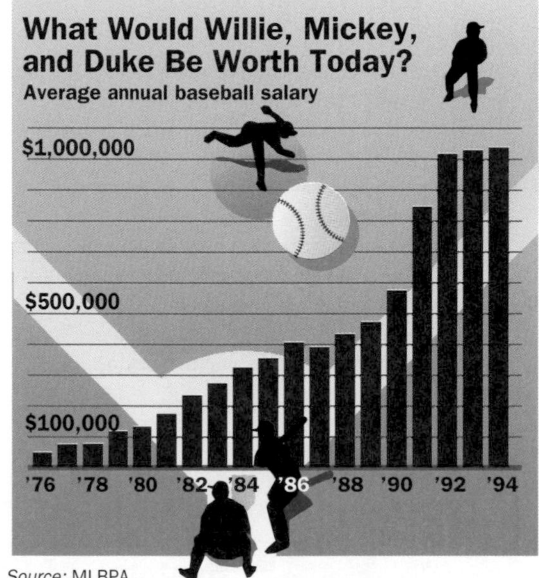

57. Based on this model, what was the average salary in 1986?

58. Based on the graph, in what year did the average salary first exceed \$1,000,000?

59. The accompanying graphing calculator screen shows this function graphed from $x = 0$ to $x = 20$. Interpret the display at the bottom of the screen.

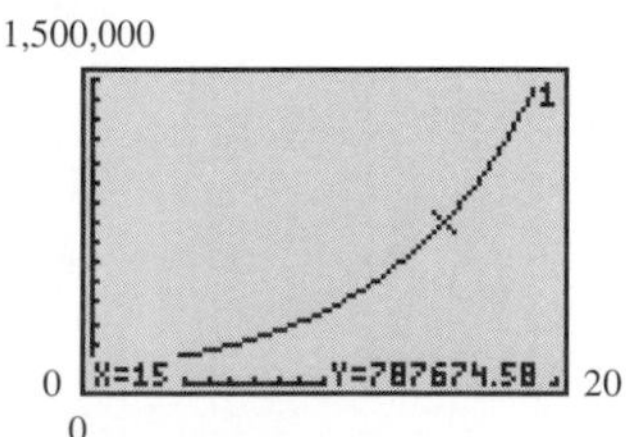

60. In the bar graph, we see that the tops of the bars rise from left to right, and in the calculator-generated graph, the curve rises from left to right. What in the model equation $S(x) = 74{,}741(1.17)^x$ indicates that during the time period, baseball salaries were *rising*?

61. Radioactive strontium decays according to the function

$$y = y_0 e^{-.0239t},$$

where t is time in years.

(a) If an initial sample contains $y_0 = 5$ grams of radioactive strontium, how many grams will be present after 20 years?

(b) How many grams of the initial 5-gram sample will be present after 60 years?

(c) What is the half-life of radioactive strontium?

62. In the United States, the intensity of an earthquake is rated using the *Richter scale.* The Richter scale rating of an earthquake of intensity x is given by

$$R = \log_{10} \frac{x}{x_0},$$

where x_0 is the intensity of an earthquake of a certain (small) size. The figure shows Richter scale ratings for major Southern California earthquakes since 1920. As the figure indicates, earthquakes "come in bunches" and the 1990s have been an especially busy time.

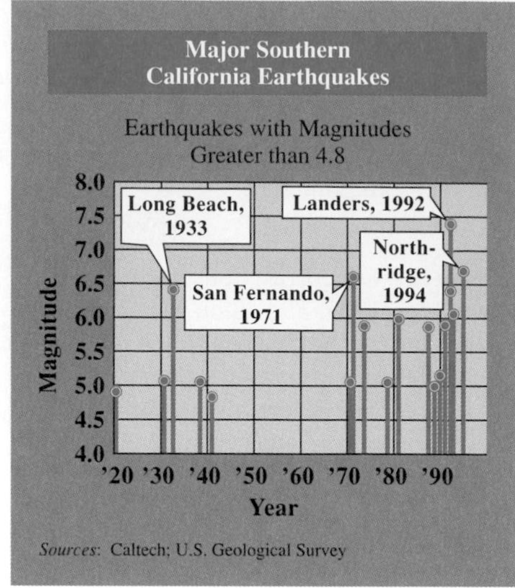

Writing the logarithmic equation just given in exponential form, we get

$$10^R = \frac{x}{x_0} \quad \text{or} \quad x = 10^R x_0.$$

The 1994 Northridge earthquake had a Richter scale rating of 6.7; the Landers earthquake had a rating of 7.3.

(a) How many times as powerful was the Landers earthquake compared to the Northridge earthquake?

(b) Compare the smallest rated earthquake in the figure (at 4.8) with the Landers quake. How many times as powerful was the Landers quake?

8.7 Systems of Equations and Applications

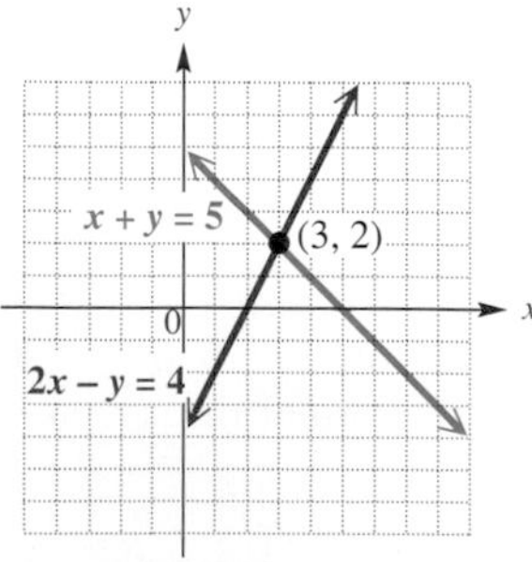

FIGURE 36

Many applications of mathematics require the simultaneous solution of a large number of equations having many variables. Such a set of equations is called a **system of equations.** The definition of a linear equation given earlier can be extended to more variables. Any equation of the form

$$a_1x_1 + a_2x_2 + \cdots + a_nx_n = b$$

for real numbers $a_1, a_2, \ldots, a_n$ (not all of which are 0), and b, is a **linear equation.** If all the equations in a system are linear, the system is a **system of linear equations,** or a **linear system.**

In Figure 36, the two linear equations $x + y = 5$ and $2x - y = 4$ are graphed on the same axes. Notice that they intersect at the point (3, 2). Because (3, 2) is the only ordered pair that satisfies both equations at the same time, we say that $\{(3, 2)\}$ is the solution set of the system

$$\begin{aligned} x + y &= 5 \\ 2x - y &= 4. \end{aligned}$$

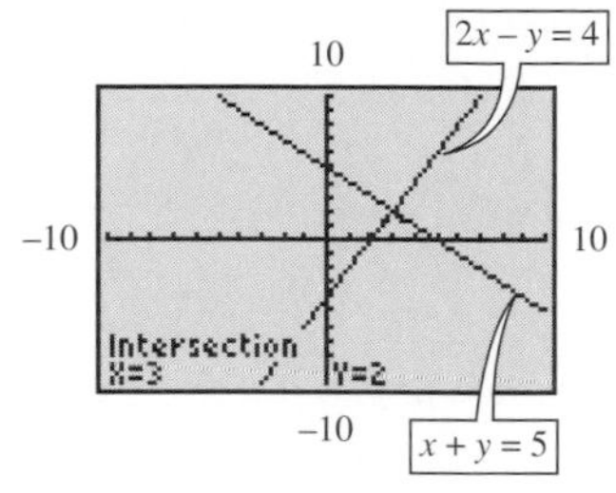

A graphing calculator supports our statement that (3, 2) is the solution of the system

$$x + y = 5$$
$$2x - y = 4.$$

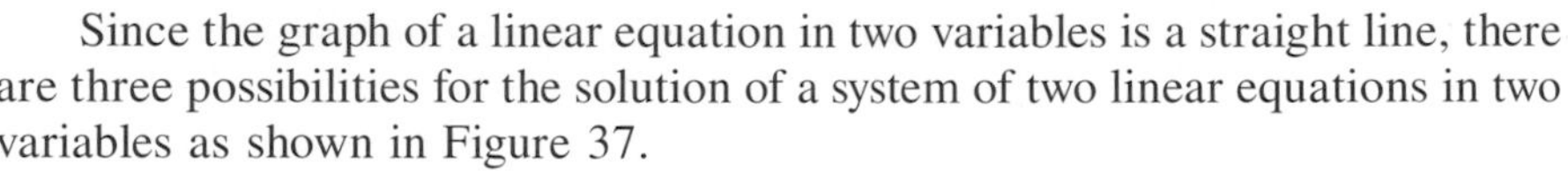

Since the graph of a linear equation in two variables is a straight line, there are three possibilities for the solution of a system of two linear equations in two variables as shown in Figure 37.

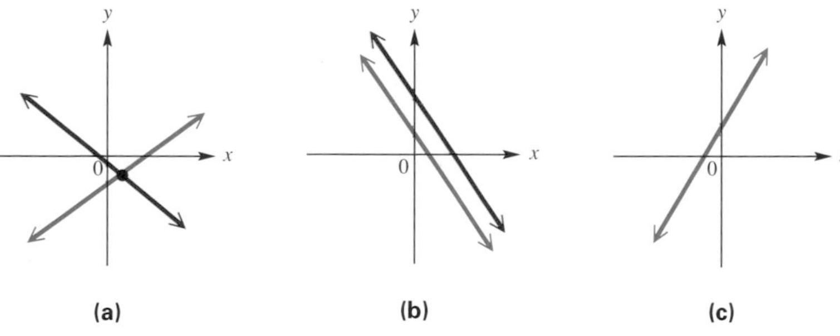

FIGURE 37

The solution of systems of equations of graphs more complicated than straight lines is the principle behind the **mattang** (shown on this stamp), a stick chart used by the people of the Marshall Islands in the Pacific. A mattang is made of roots tied together with coconut fibers, and it shows the wave patterns found when approaching an island.

Graphs of a Linear System in Two Variables

1. The two graphs intersect in a single point. The coordinates of this point give the solution of the system. This is the most common case. See Figure 37(a).
2. The graphs are parallel lines. In this case the system is **inconsistent;** that is, there is no solution common to both equations of the system, and the solution set is $\emptyset$. See Figure 37(b).
3. The graphs are the same line. In this case the equations are **dependent,** since any solution of one equation of the system is also a solution of the other. The solution set is an infinite set of ordered pairs representing the points on the line. See Figure 37(c).

It is possible to find the solution of a system of equations by graphing. However, since it can be hard to read exact coordinates from a graph, an algebraic method is usually used to solve a system. One such algebraic method, called the **elimination method** (or the **addition method**), is explained in the following examples.

EXAMPLE 1 Solve the system

$$\begin{aligned} 2x + 3y &= 2 \\ x - 3y &= 10. \end{aligned}$$

The elimination method involves combining the two equations so that one variable is eliminated. This is done using the following fact.

If $a = b$ and $c = d$, then $a + c = b + d$.

Adding corresponding sides of the two equations gives

$$\begin{aligned} 2x + 3y &= 2 \\ x - 3y &= 10 \\ \hline 3x \quad\quad &= 12. \end{aligned}$$

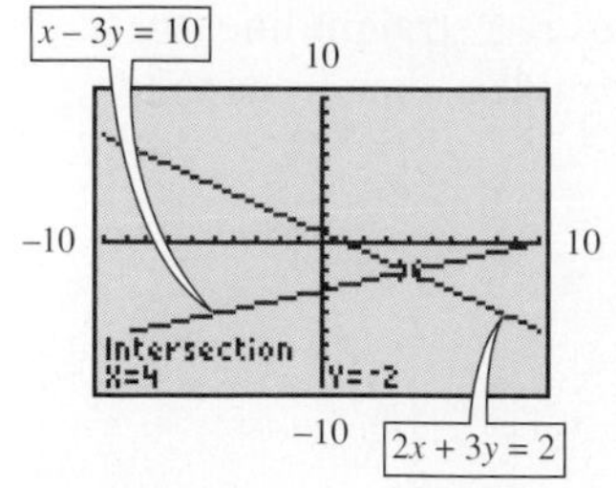

This screen supports the result in Example 1.

Dividing both sides of the equation $3x = 12$ by 3, we get

$$x = 4.$$

To find y, replace x with 4 in either one of the original equations. Using the first one gives

$$\begin{aligned} 2x + 3y &= 2 \\ 2(4) + 3y &= 2 && \text{Substitute 4 for } x. \\ 8 + 3y &= 2 && \text{Multiply.} \\ 3y &= -6 && \text{Subtract 8.} \\ y &= -2. && \text{Divide by 3.} \end{aligned}$$

The solution of the system is $x = 4$ and $y = -2$, written as the ordered pair $(4, -2)$. The solution set is $\{(4, -2)\}$. Check this solution by substituting 4 for x and -2 for y in both equations of the original system. ◆

By adding the equations in Example 1, the variable y was eliminated because the coefficients of y were opposites. In many cases the coefficients will *not* be opposites. In these cases it is necessary to transform one or both equations so that the coefficients of one of the variables are opposites. The general method of solving a system by the elimination method is summarized as follows.

Solving Linear Systems in Two Variables by Elimination

Step 1: Write both equations in the form $Ax + By = C$.

Step 2: Multiply one or both equations by appropriate numbers so that the sum of the coefficients of either x or y is zero.

Step 3: Add the new equations. The sum should be an equation with just one variable.

Step 4: Solve the equation from Step 3.

Step 5: Substitute the result of Step 4 into either of the given equations and solve for the other variable.

Step 6: Check the solution in both of the given equations.

EXAMPLE 2 Solve the system

$$\begin{aligned} 5x - 2y &= 4 \\ 2x + 3y &= 13. \end{aligned}$$

Both equations are in the form $Ax + By = C$. Suppose that we wish to eliminate the variable x. In order to do this, we can multiply the first equation by 2 and the second equation by -5.

$$\begin{aligned} 10x - 4y &= 8 && \text{2 times each side of the first equation} \\ -10x - 15y &= -65 && \text{−5 times each side of the second equation} \end{aligned}$$

Now add.

$$\begin{array}{r} 10x - 4y = 8 \\ -10x - 15y = -65 \\ \hline -19y = -57 \\ y = 3 \end{array}$$

To find x, substitute 3 for y in either of the original equations. Substituting into the second equation gives

$$\begin{aligned} 2x + 3y &= 13 \\ 2x + 3(3) &= 13 && \text{Substitute 3 for } y. \\ 2x + 9 &= 13 \\ 2x &= 4 && \text{Subtract 9.} \\ x &= 2. && \text{Divide by 2.} \end{aligned}$$

The solution set of the system is $\{(2, 3)\}$. Check this solution in both equations of the given system. ■

Another way to solve linear systems is by the substitution method. This method is most easily applied when one equation is solved for one of the variables; however, it can be applied to any system. The method of solving a system by substitution is summarized as follows.

Solving Linear Systems in Two Variables by Substitution

Step 1: Solve one of the equations for either variable. (If one of the variables has coefficient 1 or -1, choose it, since the substitution method is usually easier this way.)

Step 2: Substitute for that variable in the other equation. The result should be an equation with just one variable.

Step 3: Solve the equation from Step 2.

Step 4: Substitute the result from Step 3 into the equation from Step 1 to find the value of the other variable.

Step 5: Check the solution in both of the given equations.

The next example illustrates this method.

EXAMPLE 3 Solve the system

$$\begin{aligned} 3x + 2y &= 13 \\ 4x - y &= -1. \end{aligned}$$

To use the substitution method, first solve one of the equations for either x or y. Since the coefficient of y in the second equation is -1, it is easiest to solve for y in the second equation.

$$\begin{aligned} -y &= -1 - 4x \\ y &= 1 + 4x \end{aligned}$$

Substitute $1 + 4x$ for y in the first equation, and solve for x.

$$
\begin{aligned}
3x + 2(\mathbf{1 + 4x}) &= 13 && \text{Substitute } 1 + 4x \text{ for } y. \\
3x + 2 + 8x &= 13 && \text{Distributive property} \\
11x &= 11 && \text{Combine terms; subtract 2.} \\
x &= 1 && \text{Divide by 11.}
\end{aligned}
$$

Since $y = 1 + 4x$, let $x = 1$ to get

$$y = 1 + 4(\mathbf{1}) = 5.$$

Check that the solution set is $\{(1, 5)\}$. ■

The next example illustrates special cases that may result when systems are solved. (We will use the elimination method, but the same conclusions will follow when the substitution method is used.)

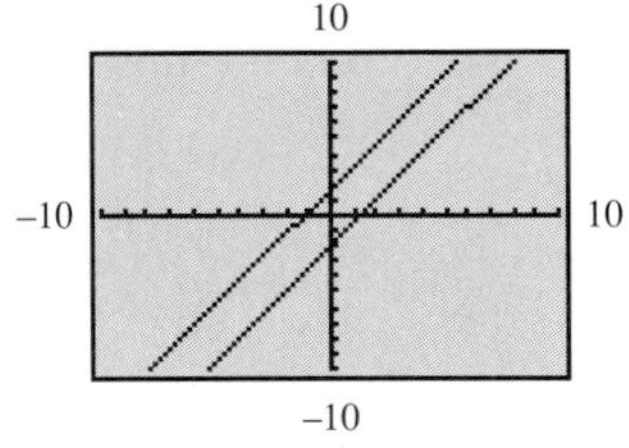

The graphs of the equations in Example 4(a) are parallel. There are no solutions.

EXAMPLE 4 **(a)** Solve the system

$$
\begin{aligned}
3x - 2y &= 4 \\
-6x + 4y &= 7.
\end{aligned}
$$

The variable x can be eliminated by multiplying both sides of the first equation by 2 and then adding.

$$
\begin{aligned}
6x - 4y &= 8 && \text{2 times the first equation} \\
-6x + 4y &= 7 && \\
\hline
\mathbf{0} &= \mathbf{15} && \text{False}
\end{aligned}
$$

Both variables were eliminated here, leaving the false statement $0 = 15$, a signal that these two equations have no solutions in common. The system is inconsistent, with the empty set $\emptyset$ as the solution set.

10

−10 10

−10

The graphs of the equations in Example 4(b) coincide. We see only one line. There are infinitely many solutions.

(b) Solve the system

$$
\begin{aligned}
-4x + y &= 2 \\
8x - 2y &= -4.
\end{aligned}
$$

Eliminate x by multiplying both sides of the first equation by 2 and then adding the result to the second equation.

$$
\begin{aligned}
-8x + 2y &= 4 && \text{2 times the first equation} \\
8x + 2y &= -4 && \\
\hline
\mathbf{0} &= \mathbf{0} && \text{True}
\end{aligned}
$$

This true statement, $0 = 0$, indicates that a solution of one equation is also a solution of the other, so the solution set is an infinite set of ordered pairs. The two equations are dependent.

We will write the solution set of a system of dependent equations as a set of ordered pairs by expressing x in terms of y as follows. Choose either equation and solve for x. Choosing the first equation gives

$$
\begin{aligned}
-4x + y &= 2 \\
x &= \frac{2 - y}{-4} = \frac{y - 2}{4}.
\end{aligned}
$$

The solution set is written as

$$\left\{\left(\frac{y-2}{4}, y\right)\right\}.$$

By selecting values for y and calculating the corresponding values for x, individual ordered pairs of the solution set can be found. For example, if $y = -2$, $x = (-2 - 2)/4 = -1$ and the ordered pair $(-1, -2)$ is a solution. ■

A solution of an equation in three variables, such as $2x + 3y - z = 4$, is called an **ordered triple** and is written (x, y, z). For example, the ordered triples $(1, 1, 1)$ and $(10, -3, 7)$ are both solutions of $2x + 3y - z = 4$, since the numbers in these ordered triples satisfy the equation when used as replacements for x, y, and z, respectively. The methods of solving systems of two equations in two variables can be extended to solving systems of equations in three variables, such as

$$\begin{aligned} 4x + 8y + z &= 2 \\ x + 7y - 3z &= -14 \\ 2x - 3y + 2z &= 3. \end{aligned}$$

Theoretically, a system of this type can be solved by graphing. However, the graph of a linear equation with three variables is a *plane* and not a line. Since the graph of each equation of the system is a plane, which requires three-dimensional graphing, this method is not practical. However, it does serve to illustrate the number of solutions possible for such systems, as Figure 38 shows. (Although Figure 38 shows one example of each possible type of solution, other examples could be given.)

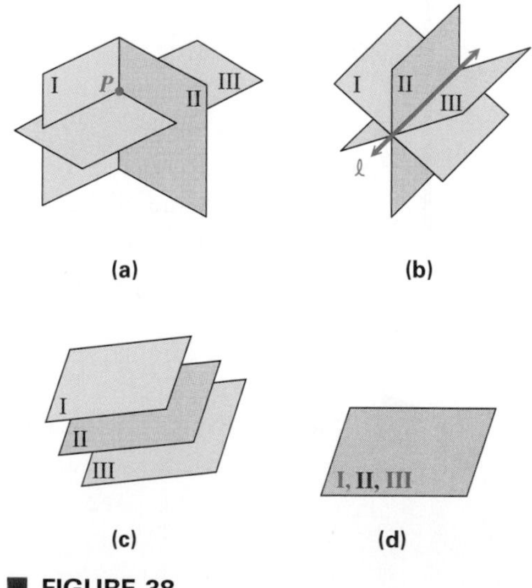

■ **FIGURE 38**

Graphs of Linear Systems in Three Variables

1. The three planes may meet at a single, common point that is the solution of the system. See Figure 38(a).
2. The three planes may have the points of a line in common so that the set of points that satisfy the equation of the line is the solution set of the system. See Figure 38(b).
3. The planes may have no points common to all three so that there is no solution for the system. See Figure 38(c).
4. The three planes may coincide so that the solution set of the system is the set of all points on a plane. See Figure 38(d).

Since a graphical solution of a system of equations of three variables is impractical, these systems may be solved with an extension of the elimination method, summarized as follows.

Solving Linear Systems in Three Variables by Elimination

Step 1: Use the elimination method to eliminate any variable from any two of the given equations. The result is an equation in two variables.

Step 2: Eliminate the *same* variable from a *different* pair of equations. The result is an equation in the same two variables as in Step 1.

Step 3: Use the elimination method to eliminate a second variable from the two equations in two variables that result from Steps 1 and 2. The result is an equation in one variable that gives the value of that variable.

Step 4: Substitute the value of the variable found in Step 3 into either of the equations in two variables to find the value of the second variable.

Step 5: Use the values of the two variables from Steps 3 and 4 to find the value of the third variable by substituting into any of the original equations.

In the next example, we will identify equations by number.

EXAMPLE 5 Solve the system

$$4x + 8y + z = 2 \quad (1)$$

$$x + 7y - 3z = -14 \quad (2)$$

$$2x - 3y + 2z = 3. \quad (3)$$

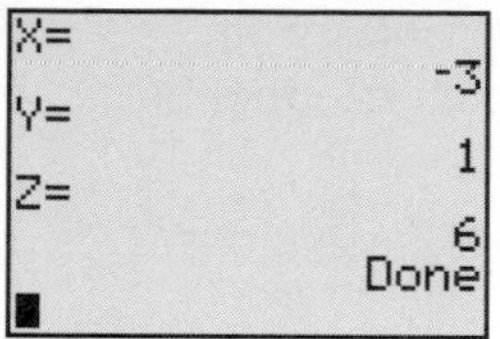

A graphing calculator can be *programmed* to solve a system such as the one in Example 5. Compare the result here to the solution in the text.

To begin we must choose a pair of equations and eliminate a single variable from them. Let us eliminate x by multiplying equation (2) by -4 and adding the result to equation (1).

$$\begin{array}{rl} 4x + 8y + z = 2 & \text{(1)} \\ -4x - 28y + 12z = 56 & -4 \text{ times equation (2)} \\ \hline -20y + 13z = 58 & \textbf{(4)} \end{array}$$

Next, choose a different pair of equations and eliminate x from them. Multiply equation (2) by -2 and add to equation (3).

$$\begin{array}{rl} -2x - 14y + 6z = 28 & -2 \text{ times equation (2)} \\ 2x - 3y + 2z = 3 & \text{(3)} \\ \hline -17y + 8z = 31 & \textbf{(5)} \end{array}$$

Now eliminate y from equations (4) and (5). Multiply equation (4) by 17 and equation (5) by -20 and add.

$$\begin{array}{rl} -340y + 221z = 986 & 17 \text{ times equation (4)} \\ 340y - 160z = -620 & -20 \text{ times equation (5)} \\ \hline 61z = 366 & \\ z = 6 & \textbf{(6)} \end{array}$$

From equation (6) we see that the value of z in the solution is 6. To find the value of y, we may go back to either equation (4) or equation (5). Let us go back to equation (4).

$$\begin{array}{rl} -20y + 13z = 58 & \text{(4)} \\ -20y + 13(6) = 58 & \text{Let } z = 6. \\ -20y + 78 = 58 & \\ -20y = -20 & \\ y = 1 & \end{array}$$

Now go back to any of the original equations to find x. Choosing equation (1) gives

$$\begin{array}{rl} 4x + 8(1) + 6 = 2 & \text{Let } y = 1 \text{ and } z = 6 \text{ in equation (1).} \\ 4x + 14 = 2 & \\ 4x = -12 & \\ x = -3. & \end{array}$$

The ordered triple $(-3, 1, 6)$ is the only solution of the system. Check that the solution satisfies all three equations in the system, so that the solution set is $\{(-3, 1, 6)\}$. ■

Problem Solving

Many problems involve more than one unknown quantity. Although some problems with two unknowns can be solved using just one variable, many times it is easier to use two variables. To solve a problem with two unknowns, we may write two equations that relate the unknown quantities; the system formed by the pair of equations then can be solved using the methods of this section.

The following steps give a strategy for solving problems using more than one variable.

Solving an Applied Problem by Writing a System of Equations

Step 1: **Determine what you are to find.** Assign a variable for each unknown and *write down* what it represents.

Step 2: **Write down other information.** If appropriate, draw a figure or a diagram and label it using the variables from Step 1. Use a chart or a box diagram to summarize the information.

Step 3: **Write a system of equations.** Write as many equations as there are unknowns.

Step 4: **Solve the system.** Use any method.

Step 5: **Answer the question(s).** Be sure you have answered all questions posed.

Step 6: **Check.** Check your solution(s) in the original problem. Be sure your answer makes sense.

The following example shows how to use two variables to solve a problem about two unknown numbers.

EXAMPLE 6 The sum of two numbers is 63. Their difference is 19. Find the two numbers.

Step 1: Let x = one number;
y = other number.

Step 2: A figure or diagram will not help in this problem, so go on to Step 3.

Step 3: From the information in the problem, set up a system of equations.

$$x + y = 63 \quad \text{The sum is 63.}$$
$$x - y = 19 \quad \text{The difference is 19.}$$

Step 4: Solve the system from Step 3. Here the elimination method is used. Adding gives

$$\begin{array}{r} x + y = 63 \\ x - y = 19 \\ \hline 2x \quad = 82. \end{array}$$

From this last equation, $x = 41$. Substitute 41 for x in either equation to find $y = 22$.

Step 5: The numbers required in the problem are 41 and 22.

Step 6: The sum of 41 and 22 is 63, and their difference is $41 - 22 = 19$. The solution satisfies the conditions of the problem. ■

If an applied problem asks for *two* values (as in Example 6), be sure to give both of them in your answer. Avoid the common error of giving only one of the values.

Here is a way to find a person's age and the date of his or her birth. Ask the person to do the following:

1. Multiply the number of the month of birth by *100*.
2. Add the date of the month.
3. Multiply the sum by *2*.
4. Add *8*.
5. Multiply this result by *5*.
6. Add *4*.
7. Multiply by *10*.
8. Add *4*.
9. Add the person's age to this.

Ask the person the number obtained. You should then subtract *444* from the number you are given. The final two figures give the age, the next figure (or figures) will identify the date of the month, and the first figure (or figures) will give the number of the month.

The next example shows how to solve a common type of applied problem involving two quantities and their costs.

Problem Solving

We can use a table or a box diagram to organize the information in order to solve an applied problem with two unknowns. A table is used in the next example.

EXAMPLE 7 Admission prices at a football game were \$6 for adults and \$2 for children. The total value of the tickets sold was \$2,528, and 454 tickets were sold. How many adults and how many children attended the game?

Step 1: Let $a =$ the number of adults' tickets sold;
$c =$ the number of childrens' tickets sold.

Step 2: The information given in the problem is summarized in the table. The entries in the "total value" column were found by multiplying the number of tickets sold by the price per ticket.

Kind of Ticket	Number Sold	Cost of Each (in dollars)	Total Value (in dollars)
Adult	a	6	$6a$
Child	c	2	$2c$
Total	454	———	2,528

The total number of tickets sold was 454, so

$$a + c = 454.$$

Since the total value was \$2,528, the final column leads to

$$6a + 2c = 2{,}528.$$

Step 3: These two equations give the following system.

$$\begin{aligned} a + c &= 454 \\ 6a + 2c &= 2{,}528 \end{aligned}$$

Step 4: We solve the system of equations with the elimination method. First, multiply both sides of the first equation by -2 to get

$$-2a - 2c = -908.$$

Then add this result to the second equation.

$$\begin{aligned} -2a - 2c &= -908 && \text{Multiply the first equation by } -2. \\ 6a + 2c &= 2{,}528 \\ \hline 4a \qquad &= 1{,}620 && \text{Add.} \\ a &= 405 && \text{Divide by 4.} \end{aligned}$$

Substitute 405 for a in the first equation to get

$$\begin{aligned} \mathbf{405} + c &= 454 && \text{Let } a = 405. \\ c &= 49. && \text{Subtract 405.} \end{aligned}$$

Step 5: There were 405 adults and 49 children at the game.

Step 6: Since 405 adults paid \$6 each and 49 children paid \$2 each, the value of the tickets sold should be $405(6) + 49(2) = 2{,}528$, or \$2,528. This result agrees with the given information. ■

Problem Solving

Many mixture problems can also be solved using two variables. In the next example, we show how to solve a mixture problem using a system of equations. The "box diagram" method introduced earlier is used here once again.

EXAMPLE 8 A pharmacist needs 100 liters of 50% alcohol solution. She has on hand 30% alcohol solution and 80% alcohol solution, which she can mix. How many liters of each will be required to make the 100 liters of 50% alcohol solution?

Step 1: Let x = the number of liters of 30% alcohol needed;
y = the number of liters of 80% alcohol needed.

Step 2: Summarize the information using a box diagram. See Figure 39.

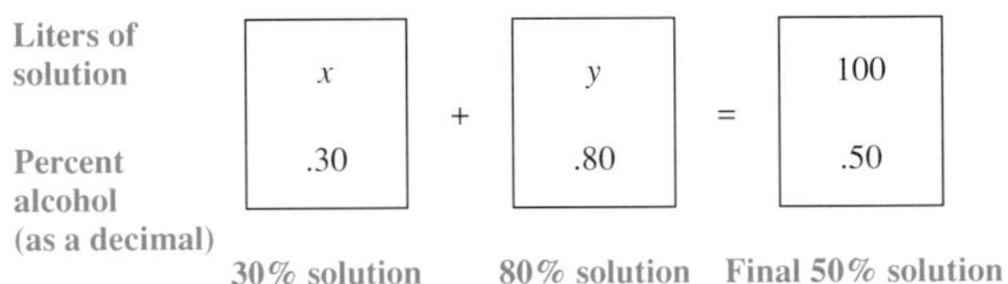

■ **FIGURE 39**

Step 3: We must write two equations. Since the total number of liters in the final mixture should be 100, the first equation is

$$x + y = 100.$$

To find the amount of pure alcohol in each mixture, multiply the number of liters by the concentration. The amount of pure alcohol in the 30% solution added to the amount of pure alcohol in the 80% solution will equal the amount of pure alcohol in the final 50% solution. This gives the second equation,

$$.30x + .80y = .50(100).$$

These two equations give the system

$$x + y = 100$$
$$.30x + .80y = 50. \qquad .50(100) = 50$$

Step 4: Solve this system by the substitution method. Solving the first equation of the system for x gives $x = 100 - y$. Substitute $100 - y$ for x in the second equation to get

$$
\begin{aligned}
.30(\mathbf{100} - y) + .80y &= 50 && \text{Let } x = 100 - y.\\
30 - .30y + .80y &= 50 && \text{Distributive property}\\
.50y &= 20 && \text{Combine terms; subtract 30.}\\
y &= 40. && \text{Divide by .50.}
\end{aligned}
$$

Then $x = 100 - y = 100 - 40 = 60$.

Step 5: The pharmacist should use 60 liters of the 30% solution and 40 liters of the 80% solution.

Step 6: Since $60 + 40 = 100$ and $.30(60) + .80(40) = 50$, this mixture will give the 100 liters of 50% solution, as required in the original problem.

The system in this problem could have been solved by the addition method. Also, we could have cleared decimals by multiplying both sides of the second equation by 100. ◆

François Viète, a mathematician of sixteenth-century France, did much for the symbolism of mathematics. Before his time, different symbols were often used for different powers of a quantity. Viète used the same letter with a description of the power and the coefficient. According to Howard Eves in *An Introduction to the History of Mathematics*, Viète would have written

$$5BA^2 - 2CA + A^3 = D$$

as

*B*5 in *A* quad − *C* plano 2 in
A + *A* cub aequatur *D* solido.

Problem Solving

Problems that use the formula relating distance, rate, and time were first introduced in an earlier chapter. In some cases, these problems can be solved by using a system of two linear equations. Keep in mind that setting up a chart and drawing a sketch will help in solving such problems.

EXAMPLE 9 Two executives in cities 400 miles apart drive to a business meeting at a location on the line between their cities. They meet after 4 hours. Find the speed of each car if one car travels 20 miles per hour faster than the other.

Step 1: Let x = the speed of the faster car;
y = the speed of the slower car.

Step 2: Use the formula that relates distance, rate, and time, $d = rt$. Since each car travels for 4 hours, the time, t, for each car is 4. This information is shown in the chart. The distance is found by using the formula $d = rt$ and the expressions already entered in the chart.

	r	t	d
Faster car	x	4	$4x$
Slower car	y	4	$4y$

Find d from $d = rt$.

Draw a sketch showing what is happening in the problem. See Figure 40.

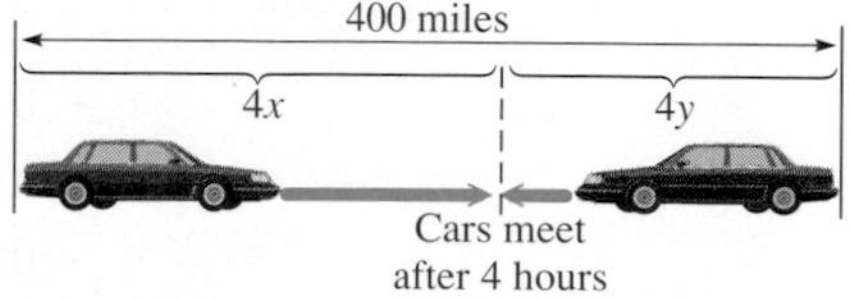

■ **FIGURE 40**

Step 3: As shown in the figure, since the total distance traveled by both cars is 400 miles, one equation is

$$4x + 4y = 400.$$

Because the faster car goes 20 miles per hour faster than the slower car, the second equation is

$$x = 20 + y.$$

Step 4: This system of equations,

$$\begin{aligned} 4x + 4y &= 400 \\ x &= 20 + y, \end{aligned}$$

can be solved by substitution. Replace x with $20 + y$ in the first equation of the system and solve for y.

$$\begin{aligned} 4(20 + y) + 4y &= 400 && \text{Let } x = 20 + y. \\ 80 + 4y + 4y &= 400 && \text{Distributive property} \\ 80 + 8y &= 400 && \text{Combine like terms.} \\ 8y &= 320 && \text{Subtract 80.} \\ y &= 40 && \text{Divide by 8.} \end{aligned}$$

Step 5: Since $x = 20 + y$, and $y = 40$,

$$x = 20 + 40 = 60.$$

The speeds of the two cars are 40 miles per hour and 60 miles per hour.

Step 6: Check the answers. Since each car travels for 4 hours, the total distance traveled is

$$4(40) + 4(60) = 160 + 240 = 400$$

miles, as required. ■

Some applied problems may be solved by using systems of three equations in three unknowns. The final example illustrates this. In this example, we do not specifically number the problem-solving steps as we did in the earlier examples. However, the same basic procedures are used.

EXAMPLE 10 Maureen Kavanagh bought apples, hamburger, and milk at the grocery store. Apples cost \$.70 a pound, hamburger was \$1.50 a pound, and milk was \$.80 a quart. She bought twice as many pounds of apples as hamburger. The number of quarts of milk was one more than the number of pounds of hamburger. If her total bill was \$8.20, how much of each item did she buy?

First choose variables to represent the three unknowns.

Let x = the number of pounds of apples;

y = the number of pounds of hamburger;

z = the number of quarts of milk.

Next, use the information in the problem to write three equations. Since Maureen bought twice as many pounds of apples as hamburger,

$$x = 2y$$

or

$$x - 2y = 0.$$

The number of quarts of milk amounted to one more than the number of pounds of hamburger, so

$$z = 1 + y$$

or

$$-y + z = 1.$$

Multiply the cost of each item by the amount of that item; adding gives the total bill.

$$.70x + 1.50y + .80z = 8.20$$

Multiply both sides of this equation by 10 to clear it of decimals.

$$7z + 15y + 8z = 82$$

Use the method shown earlier in this section to solve the system

$$\begin{aligned} x - 2y &= 0 \\ -y + z &= 1 \\ 7x + 15y + 8z &= 82. \end{aligned}$$

Verify that the solution is (4, 2, 3). Now go back to the statements defining the variables to decide what the numbers of the solution represent. Doing this shows that Maureen bought 4 pounds of apples, 2 pounds of hamburger, and 3 quarts of milk. ◆

8.7 EXERCISES

1. What is meant by a system of equations?
2. A solution of a system of linear equations in two variables is a(n) _____ ____________ of real numbers.

Decide whether the ordered pair is a solution of the given system.

3. $x + y = 6$ (5, 1)
   $x - y = 4$
4. $x - y = 17$ (8, −9)
   $x + y = -1$
5. $2x - y = 8$ (5, 2)
   $3x + 2y = 20$
6. $3x - 5y = -12$ (−1, 2)
   $x - y = 1$

Solve the system by graphing.

7. $x + y = 4$
   $2x - y = 2$
8. $x + y = -5$
   $-2x + y = 1$

Solve the system by elimination.

9. $2x - 5y = 11$
   $3x + y = 8$
10. $-2x + 3y = 1$
    $-4x + y = -3$
11. $3x + 4y = -6$
    $5x + 3y = 1$
12. $4x + 3y = 1$
    $3x + 2y = 2$
13. $3x + 3y = 0$
    $4x + 2y = 3$
14. $8x + 4y = 0$
    $4x - 2y = 2$

15. $7x + 2y = 6$
$-14x - 4y = -12$

16. $x - 4y = 2$
$4x - 16y = 8$

17. $\frac{x}{2} + \frac{y}{3} = -\frac{1}{3}$
$\frac{x}{2} + 2y = -7$

18. $\frac{x}{5} + y = \frac{6}{5}$
$\frac{x}{10} + \frac{y}{3} = \frac{5}{6}$

19. $5x - 5y = 3$
$x - y = 12$

20. $2x - 3y = 7$
$-4x + 6y = 14$

Solve the system by substitution.

21. $4x + y = 6$
$y = 2x$

22. $2x - y = 6$
$y = 5x$

23. $3x - 4y = -22$
$-3x + y = 0$

24. $-3x + y = -5$
$x + 2y = 0$

25. $-x - 4y = -14$
$2x = y + 1$

26. $-3x - 5y = -17$
$4x = y - 8$

27. $5x - 4y = 9$
$3 - 2y = -x$

28. $6x - y = -9$
$4 + 7x = -y$

29. $x = 3y + 5$
$x = \frac{3}{2}y$

30. $x = 6y - 2$
$x = \frac{3}{4}y$

31. $\frac{1}{2}x + \frac{1}{3}y = 3$
$y = 3x$

32. $\frac{1}{4}x - \frac{1}{5}y = 9$
$y = 5x$

33. Explain what the following statement means: The solution set of the system
$2x + y + z = 3$
$3x - y + z = -2$ is $\{(-1, 2, 3)\}$.
$4x - y + 2z = 0$

34. Write a system of three linear equations in three variables that has solution set $\{(3, 1, 2)\}$. Then solve the system. (*Hint:* Start with the solution and make up three equations that are satisfied by the solution. There are many ways to do this.)

Solve the system of equations in three variables.

35. $3x + 2y + z = 8$
$2x - 3y + 2z = -16$
$x + 4y - z = 20$

36. $-3x + y - z = -10$
$-4x + 2y + 3z = -1$
$2x + 3y - 2z = -5$

37. $2x + 5y + 2z = 0$
$4x - 7y - 3z = 1$
$3x - 8y - 2z = -6$

38. $5x - 2y + 3z = -9$
$4x + 3y + 5z = 4$
$2x + 4y - 2z = 14$

39. $x + y - z = -2$
$2x - y + z = -5$
$-x + 2y - 3z = -4$

40. $x + 2y + 3z = 1$
$-x - y + 3z = 2$
$-6x + y + z = -2$

41. $2x - 3y + 2z = -1$
$x + 2y + z = 17$
$2y - z = 7$

42. $2x - y + 3z = 6$
$x + 2y - z = 8$
$2y + z = 1$

43. $4x + 2y - 3z = 6$
$x - 4y + z = -4$
$-x + 2z = 2$

44. $2x + 3y - 4z = 4$
$x - 6y + z = -16$
$-x + 3z = 8$

45. $2x + y = 6$
$3y - 2z = -4$
$3x - 5z = -7$

46. $4x - 8y = -7$
$4y + z = 7$
$-8x + z = -4$

47. The accompanying graph shows the trends during the years 1966–1990 relating to bachelor's degrees awarded in the United States. (*Source:* National Science Foundation)

(a) Between what years shown on the horizontal axis did the number of degrees in all fields for men and women reach equal numbers?

(b) When the number of degrees for men and women reached equal numbers, what was that number (approximately)?

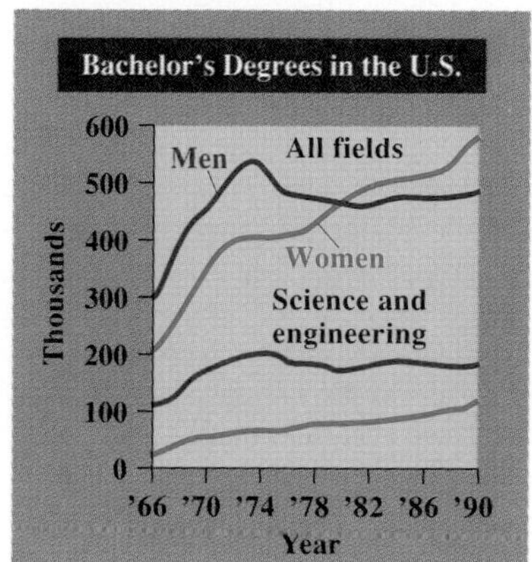

48. The accompanying graph shows how the production of vinyl LPs, audiocassettes, and compact discs (CDs) changed over the years from 1983 to 1993. (*Source:* Recording Industry of America)

(a) In what year did cassette production and CD production reach equal levels? What was that level?

(b) Express as an ordered pair of the form (year, production level) the point of intersection of the graphs of LP production and CD production.

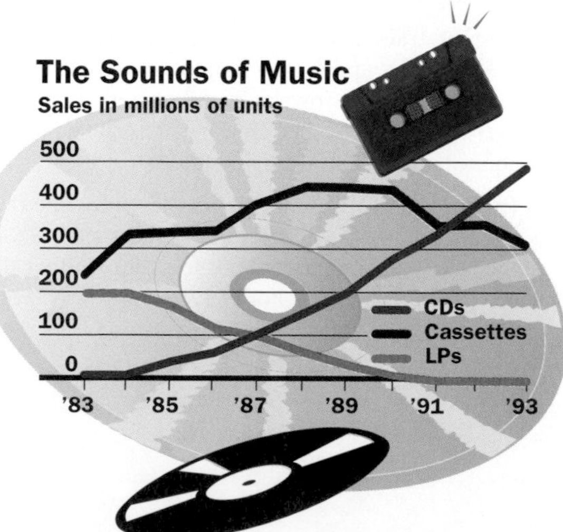

Solve the problem by using a system of two equations in two variables.

49. Andre and Monica measured the perimeter of a tennis court and found that it was 42 feet longer than it was wide, and had a perimeter of 228 feet. What were the length and the width of the tennis court?

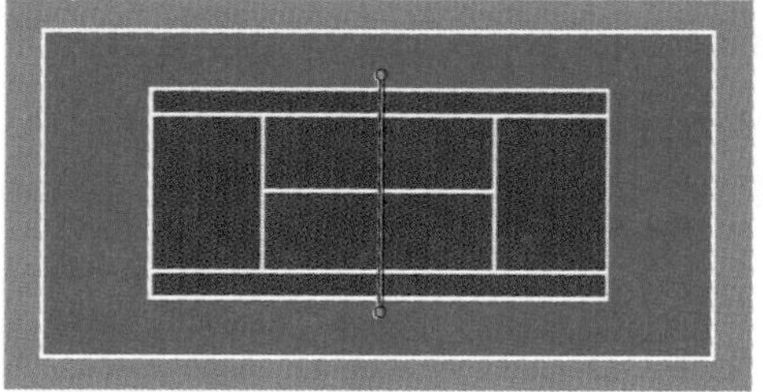

50. Kareem and Manute found that the width of their basketball court was 44 feet less than the length. If the perimeter was 288 feet, what were the length and the width of their court?

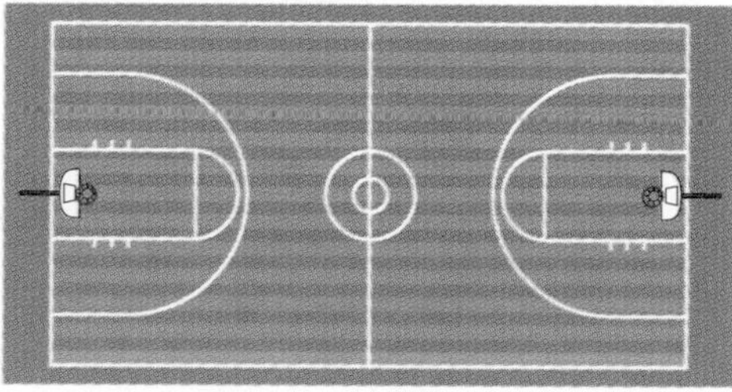

51. The length of a rectangle is 7 feet more than the width. If the length were decreased by 3 feet and the width were increased by 2 feet, the perimeter would be 32 feet. Find the length and width of the original rectangle.

52. The side of a square is 4 centimeters longer than the side of an equilateral triangle. The perimeter of the square is 24 centimeters more than the perimeter of the triangle. Find the lengths of a side of the square and a side of the triangle.

53. How many gallons each of 25% alcohol and 35% alcohol should be mixed to get 20 gallons of 32% alcohol?

54. How many liters each of 15% acid and 33% acid should be mixed to get 40 liters of 21% acid?

55. Pure acid is to be added to a 10% acid solution to obtain 27 liters of a 20% acid solution. What amounts of each should be used?

56. A truck radiator holds 18 liters of fluid. How much pure antifreeze must be added to a mixture that is 4% antifreeze in order to fill the radiator with a mixture that is 20% antifreeze?

57. A party mix is made by adding nuts that sell for \$2.50 a kilogram to a cereal mixture that sells for \$1 a kilogram. How much of each should be added to get 30 kilograms of a mix that will sell for \$1.70 a kilogram?

58. A popular fruit drink is made by mixing fruit juices. Such a mixture with 50% juice is to be mixed with another mixture that is 30% juice to get 200 liters of a mixture that is 45% juice. How much of each should be used?

59. Tickets to a production of *Othello* at Nicholls State University cost \$2.50 for general admission or \$2.00 with student identification. If 184 people paid to see a performance and \$406 was collected, how many of each type of admission were sold?

60. A grocer plans to mix candy that sells for \$1.20 a pound with candy that sells for \$2.40 a pound to get a mixture that he plans to sell for \$1.65 a pound. How much of the \$1.20 and \$2.40 candy should he use if he wants 80 pounds of the mix?

61. Angelica Canales has been saving dimes and quarters. She has 94 coins in all. If the total value is \$19.30, how many dimes and how many quarters does she have?

62. Gayle Sloan is a teller at the Seagull branch of Parish National Bank. She received a checking account deposit in \$20 bills and \$50 bills. There was a total of 70 bills, and the amount of the deposit was \$3,200. How many of each denomination were deposited?

63. A total of \$3,000 is invested, part at 2% simple interest and part at 4%. If the total annual return from the two investments is \$100, how much is invested at each rate?

64. A man must invest a total of \$15,000 in two accounts, one paying 4% annual simple interest, and the other 3%. If he wants to earn \$550 annual interest, how much should he invest at each rate?

65. San Jacinto College has decided to supply its mathematics labs with color monitors. A trip to the local electronics outlet leads to the following information: 4 CGA monitors and 6 VGA monitors can be purchased for \$4,600, while 6 CGA monitors and 4 VGA monitors will cost \$4,400. What are the prices of a single CGA monitor and a single VGA monitor?

66. The Cleveland Indians gift shop will sell you 5 New Era baseball caps and 3 Diamond Collection jerseys for \$430, or 3 New Era caps and 4 Diamond Collection jerseys for \$500. What are the prices of a single cap and a single jersey?

67. A factory makes use of two basic machines, A and B, which turn out two different products, yarn and thread. Each unit of yarn requires 1 hour on machine A and 2 hours on machine B, while each unit of thread requires 1 hour on A and 1 hour on B. Machine A runs 8 hours per day, while machine B runs 14 hours per day. How many units each of yarn and thread should the factory make each day to keep its machines running at capacity?

68. A company that makes personal computers has found that each standard model requires 4 hours to manufacture electronics and 2 hours for the case. The top-of-the-line model requires 5 hours for the electronics and 1.5 hours for the case. On a particular production run, the company has available 200 hours in the electronics department and 76 hours in the cabinet department. How many of each model can be made?

69. Theodis bought 2 kilograms of dark clay and 3 kilograms of light clay, paying \$22 for the clay. He later needed 1 kilogram of dark clay and 2 kilograms of light clay, costing \$13 altogether. How much did he pay per kilogram for each type of clay?

70. A biologist wants to grow two types of algae, green and brown. She has 15 kilograms of nutrient X and 26 kilograms of nutrient Y. A vat of green algae needs 2 kilograms of nutrient X and 3 kilograms of nutrient Y, while a vat of brown algae needs 1 kilogram of nutrient X and 2 kilograms of nutrient Y. How many vats of each type of algae should the biologist grow in order to use all the nutrients?

71. A freight train and an express train leave towns 390 kilometers apart, traveling toward one another. The freight train travels 30 kilometers per hour slower than the express train. They pass one another 3 hours later. What are their speeds?

72. A train travels 150 kilometers in the same time that a plane covers 400 kilometers. If the speed of the plane is 20 kilometers per hour less than 3 times the speed of the train, find both speeds.

73. In his motorboat, Phan Nguyen travels upstream at top speed to his favorite fishing spot, a distance of 36 miles, in two hours. Returning, he finds that the trip downstream, still at top speed, takes only 1.5 hours. Find the speed of Phan's boat and the speed of the current.

74. Traveling for three hours into a steady headwind, a plane makes a trip of 1,650 miles. The pilot determines that flying with the same wind for two hours, he could make a trip of 1,300 miles. What is the speed of the plane and the speed of the wind?

Solve the problem by using a system of three equations in three variables. (In Exercises 75–78 use the fact that the sum of the measures of the angles of a triangle is 180°.)

75. In the figure shown, $z = x + 10$ and $x + y = 100$. Determine a third equation involving x, y, and z, and then find the measures of the three angles.

76. In the figure shown, x is 10 less than y and 20 less than z. Write a system of equations and find the measures of the three angles.

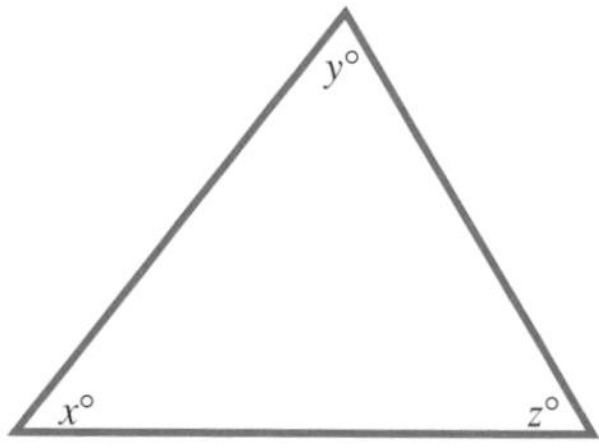

77. In a certain triangle, the measure of the second angle is 10° more than three times the first. The third angle measure is equal to the sum of the measures of the other two. Find the measures of the three angles.

78. The measure of the largest angle of a triangle is 12° less than the sum of the measures of the other two. The smallest angle measures 58° less than the largest. Find the measures of the angles.

79. The perimeter of a triangle is 70 centimeters. The longest side is 4 centimeters less than the sum of the other two sides. Twice the shortest side is 9 centimeters less than the longest side. Find the length of each side of the triangle.

80. The perimeter of a triangle is 56 inches. The longest side measures 4 inches less than the sum of the other two sides. Three times the shortest side is 4 inches more than the longest side. Find the lengths of the three sides.

81. Sharon Dempsey, a Mardi Gras trinket manufacturer, supplies three wholesalers, A, B, and C. The output from a day's production is 320 cases of trinkets. She must send wholesaler A three times as many cases as she sends B, and she must send wholesaler C 160 cases less than she provides A and B together. How many cases should she send to each wholesaler to distribute the entire day's production to them?

82. A hardware supplier manufactures three kinds of clamps, types A, B, and C. Production restrictions require them to make 10 units more type C clamps than the total of the other types and twice as many type B clamps as type A. The shop must produce a total of 490 units of clamps per day. How many units of each type can be made per day?

83. The manager of a candy store wants to feature a special Easter candy mixture of jelly beans, small chocolate eggs, and marshmallow chicks. She plans to make 15 pounds of mix to sell at \$1 a pound. Jelly beans sell for \$.80 a pound, chocolate eggs for \$2 a pound, and marshmallow chicks for \$1 a pound. She will use twice as many pounds of jelly beans as eggs and chicks combined and five times as many pounds of jelly beans as chocolate eggs. How many pounds of each candy should she use?

84. Three kinds of tickets are available for a Rhonda Rock concert: "up close," "in the middle," and "far out." "Up close" tickets cost \$2 more than "in the middle" tickets, while "in the middle" tickets cost \$1 more than "far out" tickets. Twice the cost of an "up close" ticket is \$1 less than 3 times the cost of a "far out" ticket. Find the price of each kind of ticket.

8.8 Linear Inequalities and Systems of Inequalities

Linear inequalities with one variable were graphed on the number line in an earlier chapter. In this section linear inequalities in two variables are graphed in a rectangular coordinate system.

Linear Inequality

An inequality that can be written as

$$Ax + By < C \qquad \text{or} \qquad Ax + By > C,$$

where A, B, and C are real numbers and A and B are not both 0, is a **linear inequality in two variables.**

Also, $\leq$ and $\geq$ may replace $<$ and $>$ in the definition.

A line divides the plane into three regions: the line itself and the two half-planes on either side of the line. Recall that the graphs of linear inequalities in one variable are intervals on the number line that sometimes include an endpoint. The graphs of linear inequalities in two variables are *regions* in the real number plane and may include a *boundary line.* The **boundary line** for the inequality $Ax + By < C$ or $Ax + By > C$ is the graph of the *equation* $Ax + By = C$. To graph a linear inequality, we go through the following steps.

Graphing a Linear Inequality

Step 1: **Draw the boundary.** Draw the graph of the straight line that is the boundary. Make the line solid if the inequality involves $\leq$ or $\geq$; make the line dashed if the inequality involves $<$ or $>$.

Step 2: **Choose a test point.** Choose any point not on the line as a test point.

Step 3: **Shade the appropriate region.** Shade the region that includes the test point if it satisfies the original inequality; otherwise, shade the region on the other side of the boundary line.

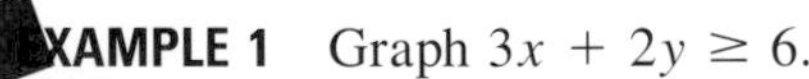

EXAMPLE 1 Graph $3x + 2y \geq 6$.

First graph the straight line $3x + 2y = 6$. The graph of this line, the boundary of the graph of the inequality, is shown in Figure 41. The graph of the inequality $3x + 2y \geq 6$ includes the points of the line $3x + 2y = 6$, and either the points *above* the line $3x + 2y = 6$ or the points *below* that line. To decide which, select any point not on the line $3x + 2y = 6$ as a test point. The origin,

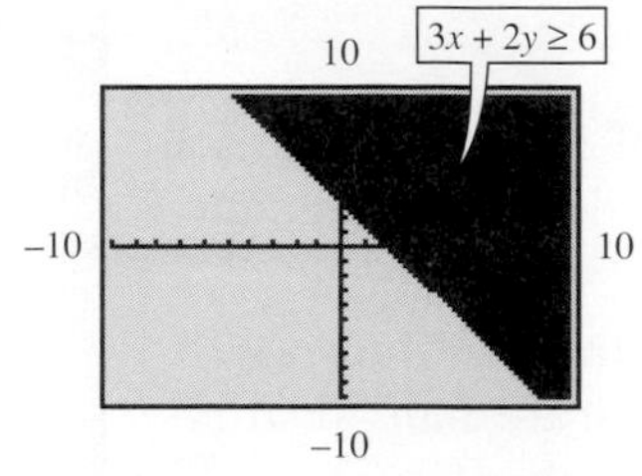

The TI-82 allows us to shade the appropriate region for an inequality. Compare with Figure 41.

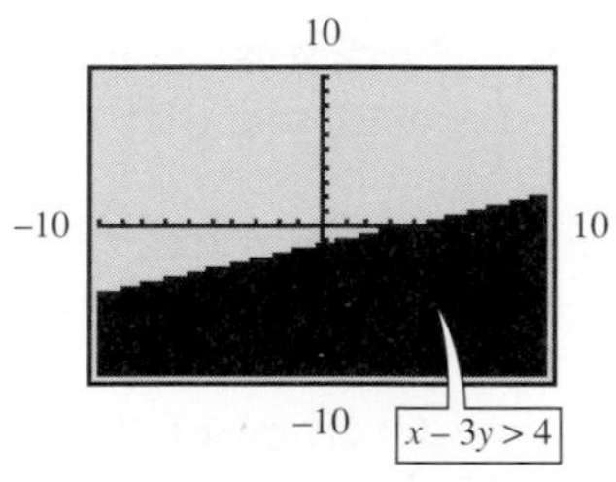

Compare with Figure 42.

(0, 0), is often a good choice. Substitute the values from the test point (0, 0) for x and y in the inequality $3x + 2y > 6$.

$$3(0) + 2(0) > 6 \quad ?$$
$$0 > 6 \quad \text{False}$$

Since the result is false, (0, 0) does not satisfy the inequality, and so the solution includes all points on the other side of the line. This region is shaded in Figure 41. ■

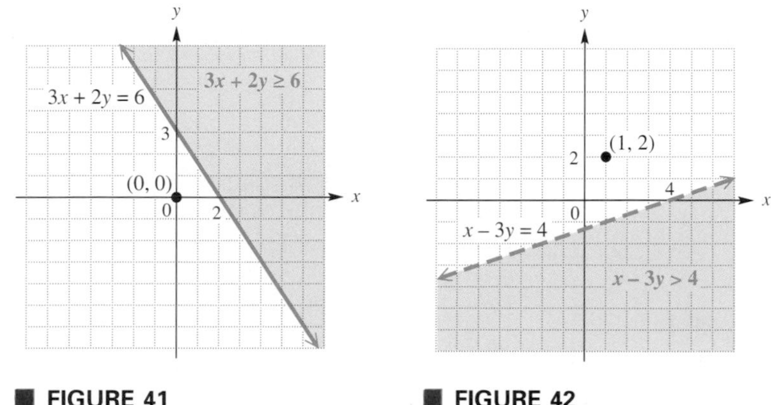

■ FIGURE 41 ■ FIGURE 42

EXAMPLE 2 Graph $x - 3y > 4$.

First graph the boundary line, $x - 3y = 4$. The graph is shown in Figure 42. The points of the boundary line do not belong to the inequality $x - 3y > 4$ (since the inequality symbol is $>$ and not $\geq$). For this reason, the line is dashed. To decide which side of the line is the graph of the solution, choose any point that is not on the line, say (1, 2). Substitute 1 for x and 2 for y in the original inequality.

$$1 - 3(2) > 4 \quad ?$$
$$-5 > 4 \quad \text{False}$$

Because of this false result, the solution lies on the side of the boundary line that does *not* contain the test point (1, 2). The solution, graphed in Figure 42, includes only those points in the shaded region (not those on the line). ■

Systems of Inequalities

Methods of solving systems of *equations* were discussed in the previous section. Systems of inequalities with two variables may be solved by graphing their solutions. A system of linear inequalities consists of two or more such inequalities, and a solution of such a system consists of all points that make all the inequalities true at the same time.

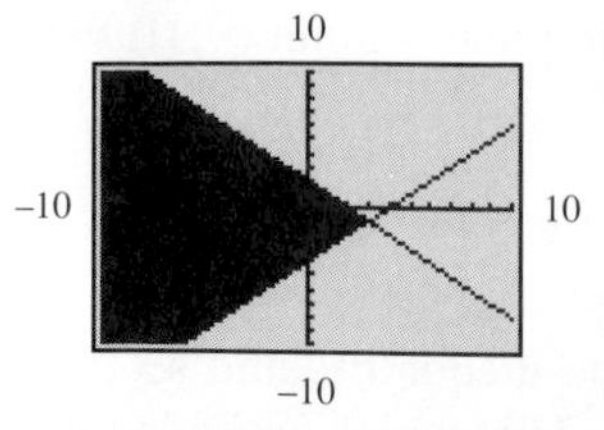

Here is the solution of the system

$$y > x - 4$$
$$y < -x + 2.$$

Graphing a System of Linear Inequalities

Step 1: **Graph each inequality on the same set of axes.** Graph each inequality in the system, using the method described in Examples 1 and 2.

Step 2: **Find the intersection of the regions of solution.** Indicate the intersection of the regions of solutions of the individual inequalities. This is the solution of the system.

EXAMPLE 3 Graph the solution of the linear system

$$3x + 2y \leq 6$$
$$2x - 5y \geq 10.$$

Begin by graphing $3x + 2y \leq 6$. To do this, graph $3x + 2y = 6$ as a solid line. Since (0, 0) makes the inequality true, shade the region containing (0, 0), as shown in Figure 43.

Now graph $2x - 5y \geq 10$. The solid line boundary is the graph of $2x - 5y = 10$. Since (0, 0) makes the inequality false, shade the region that does not contain (0, 0), as shown in Figure 44.

The solution of the system is given by the intersection (overlap) of the regions of the graphs in Figures 43 and 44. The solution is the shaded region in Figure 45, and includes portions of the two boundary lines. ■

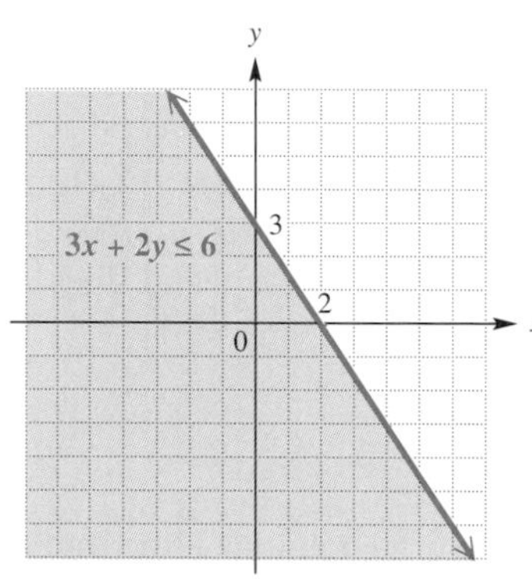

■ **FIGURE 43**

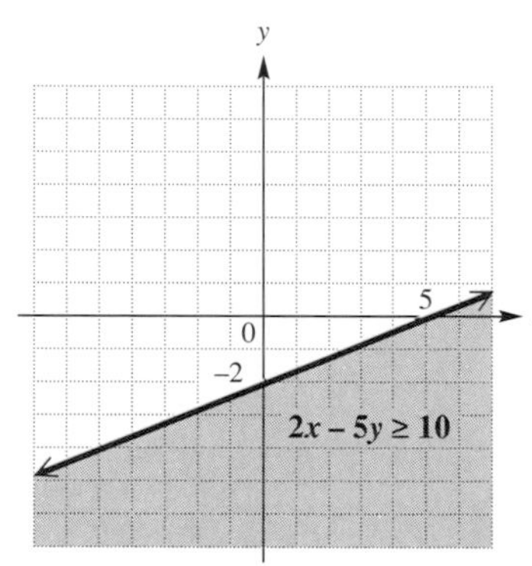

■ **FIGURE 44**

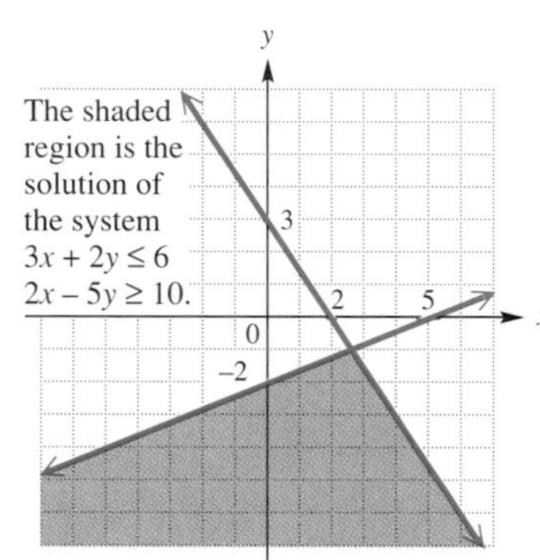

■ **FIGURE 45**

In practice, we usually do all the work on one set of axes at the same time. In the following example, only one graph is shown.

EXAMPLE 4 Graph the solution of the linear system

$$2x + 3y \geq 12$$
$$7x + 4y \geq 28$$
$$y \leq 6$$
$$x \leq 5.$$

The graph is obtained by graphing the four inequalities on the same axes and shading the region common to all four as shown in Figure 46. As the graph shows, the boundary lines are all solid. ■

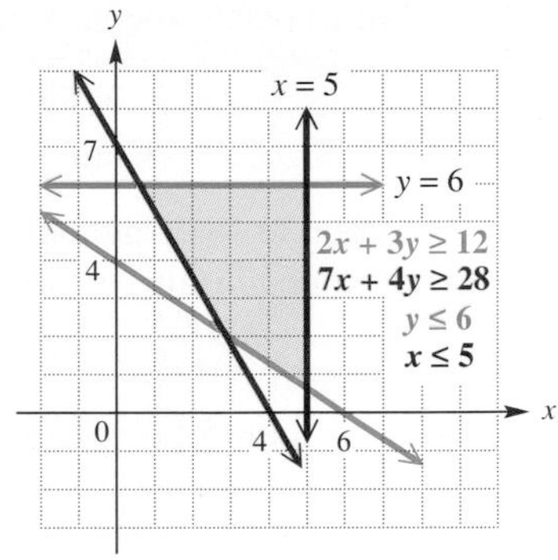

■ **FIGURE 46**

8.8 EXERCISES

Graph the linear inequality.

1. $x + y \leq 2$ **2.** $x - y \geq -3$ **3.** $4x - y \leq 5$ **4.** $3x + y \geq 6$

5. $x + 3y \geq -2$ **6.** $4x + 6y \leq -3$ **7.** $x + 2y \leq -5$ **8.** $2x - 4y \leq 3$

9. $4x - 3y < 12$ **10.** $5x + 3y > 15$ **11.** $y > -x$ **12.** $y < x$

Graph the system of inequalities.

13. $x + y \leq 1$, $x \geq 0$

14. $3x - 4y \leq 6$, $y \geq 1$

15. $2x - y \geq 1$, $3x + 2y \geq 6$

16. $x + 3y \geq 6$, $3x - 4y \leq 12$

17. $-x - y < 5$, $x - y \leq 3$

18. $6x - 4y < 8$, $x + 2y \geq 4$

19. Explain how to determine whether a dashed line or a solid line is used for the boundary when graphing a linear inequality in two variables.

20. The graph of $y > -3x + 2$ consists of all points that are ____________ (above/below) the line with the equation ____________.

EXTENSION

Linear Programming

A very important application of mathematics to business and social science is called **linear programming.** Linear programming is used to find minimum cost, maximum profit, the maximum amount of earning that can take place under given conditions, and so on. The procedures for solving linear programming problems were developed in 1947 by George Dantzig, while he was working on a problem of allocating supplies for the Air Force in a way that minimized total cost.

EXAMPLE The Kosch-Granger Company makes two products, tape decks and amplifiers. Each tape deck gives a profit of \$3, while each amplifier gives a profit of \$7. The company must manufacture at least 1 tape deck per day to satisfy one of its customers, but no more than 5 because of production problems. Also, the number of amplifiers produced cannot exceed 6 per day. As a further requirement, the number of tape decks cannot exceed the number of amplifiers. How many of each should the company manufacture in order to obtain the maximum profit?

George B. Dantzig of Stanford University has been one of the key people behind **operations research** (OR), which is making significant contributions to business and industry. As a management science, OR is not a single discipline, but draws from mathematics, probability theory, statistics, and economics. The name given to this "multiplex" shows its historical origins in World War II, when operations of a military nature called forth the efforts of many scientists to research their fields for applications to the war effort and to solve tactical problems. Applications of mathematical and statistical techniques were found to be feasible by Dantzig and other scientists.

Operations research is an approach to problem solving and decision making. First of all, the problem has to be clarified. Quantities involved have to be designated as variables, and the objectives as functions. Use of **models** is an important aspect of OR.

We translate the statements of the problem into symbols by letting

x = number of tape decks to be produced daily

y = number of amplifiers to be produced daily.

According to the statement of the problem, the company must produce at least one tape deck (one or more), so that

$$x \geq 1.$$

No more than 5 tape decks may be produced:

$$x \leq 5.$$

Not more than 6 amplifiers may be made in one day:

$$y \leq 6.$$

The number of tape decks may not exceed the number of amplifiers:

$$x \leq y.$$

The number of tape decks and of amplifiers cannot be negative:

$$x \geq 0 \quad \text{and} \quad y \geq 0.$$

All restrictions, or **constraints,** that are placed on production can now be summarized:

$$x \geq 1, \quad x \leq 5, \quad y \leq 6, \quad x \leq y, \quad x \geq 0, \quad y \geq 0.$$

The maximum possible profit that the company can make, subject to these constraints, is found by sketching the graph of the solution of the system. See Figure 47. The only feasible values of x and y are those that satisfy all constraints. These values correspond to points that lie on the boundary or in the shaded region, called the **region of feasible solutions.**

Since each tape deck gives a profit of \$3, the daily profit from the production of x tape decks is $3x$ dollars. Also, the profit from the production of y amplifiers will be $7y$ dollars per day. The total daily profit is thus given by the following **objective function:**

$$\text{Profit} = 3x + 7y.$$

The problem of the Kosch-Granger Company may now be stated as follows: find values of x and y in the region of feasible solutions as shown in Figure 47 that will produce the maximum possible value of $3x + 7y$.

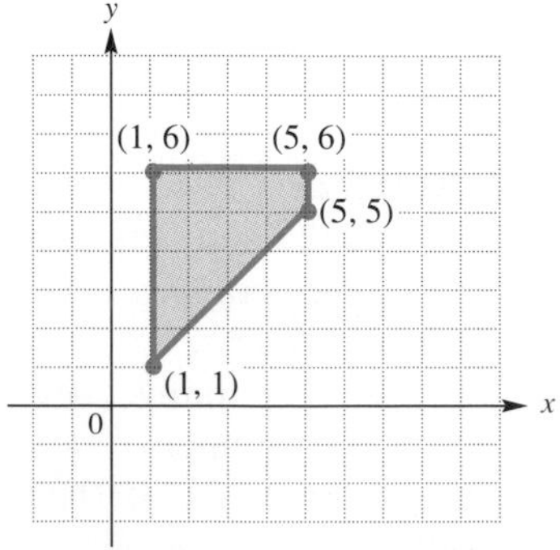

■ **FIGURE 47**

It Pays to Do Your Homework George Dantzig, profiled on the opposite page, was interviewed in the September 1986, issue of the *College Mathematics Journal.* In the interview he relates the story of how he obtained his degree without actually writing a thesis.

It seems that one day Dantzig arrived late to one of his classes and on the board were two problems. Assuming that they were homework problems, he worked on them and handed them in a few days later, apologizing to his professor for taking so long to do them. They had seemed to be a bit harder than usual. Several weeks later he received an early morning visit from his professor, who had just gotten around to looking at Dantzig's correct solutions. The problems had not been intended as homework problems; they were actually two famous *unsolved* problems in statistics! Later, when Dantzig began to worry about a thesis topic, his professor told him not to worry and that the two solutions would serve as his thesis.

It can be shown that any optimum value (maximum or minimum) will always occur at a **vertex** (or **corner point**) of the region of feasible solutions. Locate the point (x, y) that gives the maximum profit by checking the coordinates of the vertices, shown in Figure 47 and listed below. Find the profit that corresponds to each coordinate pair and choose the one that gives the maximum profit.

Point	Profit $= 3x + 7y$
(1, 1)	$3(1) + 7(1) = 10$
(1, 6)	$3(1) + 7(6) = 45$
(5, 6)	$3(5) + 7(6) = 57$ ← Maximum
(5, 5)	$3(5) + 7(5) = 50$

The maximum profit of \$57 is obtained when 5 tape decks and 6 amplifiers are produced each day. ■

Problem Solving

To solve a linear programming problem in general, use the following steps.

Solving a Linear Programming Problem

1. Write all necessary constraints and the objective function.
2. Graph the feasible region.
3. Identify all vertices.
4. Find the value of the objective function at each vertex.
5. The solution is given by the vertex producing the optimum value of the objective function.

Extension Exercises

Exercises 1 and 2 show regions of feasible solutions. Find the maximum and minimum values of the given expressions.

1. $3x + 5y$

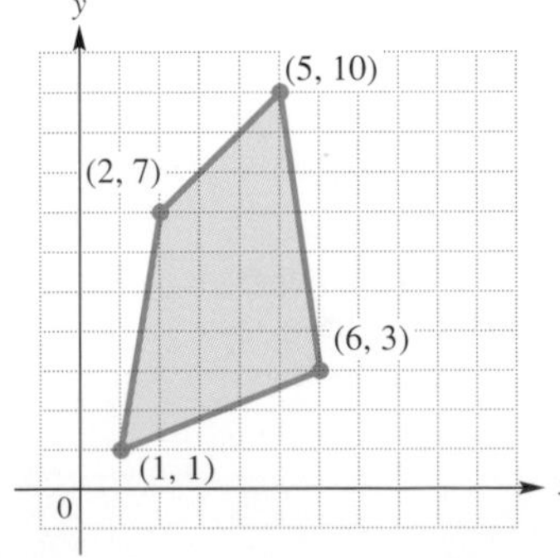

2. $40x + 75y$

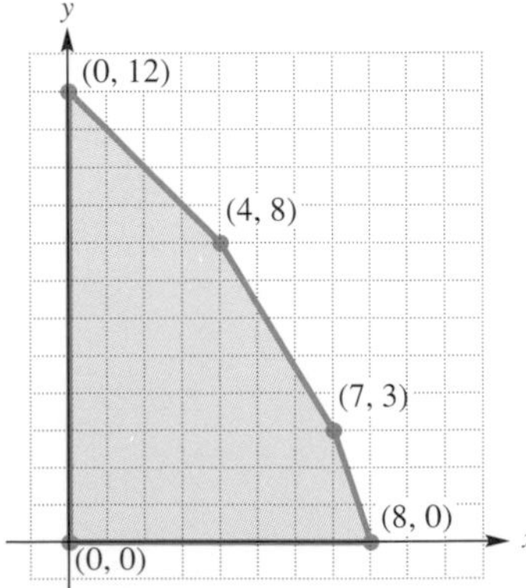

In Exercises 3–6, use graphical methods to find values of x and y satisfying the given conditions. (It may be necessary to solve a system of equations in order to find vertices.) Find the value of the maximum or minimum.

3. Find $x \geq 0$ and $y \geq 0$ such that

$$2x + 3y \leq 6$$
$$4x + y \leq 6$$

and $5x + 2y$ is maximized.

4. Find $x \geq 0$ and $y \geq 0$ such that

$$x + y \leq 10$$
$$5x + 2y \geq 20$$
$$2y \geq x$$

and $x + 3y$ is minimized.

5. Find $x \geq 2$ and $y \geq 5$ such that

$$3x - y \geq 12$$
$$x + y \leq 15$$

and $2x + y$ is minimized.

6. Find $x \geq 10$ and $y \geq 20$ such that

$$2x + 3y \leq 100$$
$$5x + 4y \leq 200$$

and $x + 3y$ is maximized.

Solve each of the following linear programming problems.

7. Gwen, who is dieting, requires two food supplements, I and II. She can get these supplements from two different products, A and B. Product A provides 3 g per serving of supplement I and 2 g per serving of supplement II. Product B provides 2 g per serving of supplement I and 4 g per serving of supplement II. Her dietician, Dr. Shoemake, has recommended that she include at least 15 g of each supplement in her daily diet. If product A costs 25¢ per serving and product B costs 40¢ per serving, how can she satisfy her requirements most economically?

8. A manufacturer of refrigerators must ship at least 100 refrigerators to its two West coast warehouses. Each warehouse holds a maximum of 100 refrigerators. Warehouse A holds 25 refrigerators already, while warehouse B has 20 on hand. It costs $12 to ship a refrigerator to warehouse A and $10 to ship one to warehouse B. How many refrigerators should be shipped to each warehouse to minimize cost? What is the minimum cost?

9. A machine shop manufactures two types of bolts. Each can be made on any of three groups of machines, but the time required on each group differs, as shown in the table below.

		Machine Groups		
		I	**II**	**III**
Bolts	**Type 1**	.1 min	.1min	.1 min
	Type 2	.1 min	.4 min	.5 min

Production schedules are made up one day at a time. In a day there are 240, 720, and 160 minutes available, respectively, on these machines. Type 1 bolts sell for 10¢ and type 2 bolts for 12¢. How many of each type of bolt should be manufactured per day to maximize revenue? What is the maximum revenue?

10. Karin Wagner takes vitamin pills. Each day, she must have at least 16 units of Vitamin A, at least 5 units of Vitamin B_1, and at least 20 units of Vitamin C. She can choose between red pills costing 10¢ each that contain 8 units of A, 1 of B_1, and 2 of C; and blue pills that cost 20¢ each and contain 2 units of A, 1 of B_1, and 7 of C. How many of each pill should she take in order to minimize her cost and yet fulfill her daily requirements?

11. A bakery makes both cakes and cookies. Each batch of cakes requires two hours in the oven and three hours in the decorating room. Each batch of cookies needs one and a half hours in the oven and two thirds of an hour in the decorating room. The oven is available no more than 15 hours a day, while the decorating room can be used no more than 13 hours a day. How many batches of cakes and cookies should the bakery make in order to maximize profits if cookies produce a profit of \$20 per batch and cakes produce a profit of \$30 per batch?

12. A manufacturing process requires that oil refineries manufacture at least 2 gallons of gasoline for each gallon of fuel oil. To meet the winter demand for fuel oil, at least 3 million gallons a day must be produced. The demand for gasoline is no more than 6.4 million gallons per day. If the price of gasoline is \$1.90 per gallon and the price of fuel oil is \$1.50 per gallon, how much of each should be produced to maximize revenue?

13. Earthquake victims in China need medical supplies and bottled water. Each medical kit measures 1 cubic foot and weighs 10 pounds. Each container of water is also 1 cubic foot but weighs 20 pounds. The plane can only carry 80,000 pounds with a total volume of 6,000 cubic feet. Each medical kit will aid 4 people, while each container of water will serve 10 people. How many of each should be sent in order to maximize the number of people aided?

14. If each medical kit could aid 6 people instead of 4, how would the results in Exercise 13 change?

COLLABORATIVE INVESTIGATION
Modeling Data Using a Linear Function

In this chapter we learned how to find the equation of a line joining two points. In statistics, a technique known as *regression* allows us to find the equation of the line that provides the best possible "fit" of points that lie approximately in a line. This technique is discussed in the chapter on statistics that follows. However, we can use the concepts of this chapter to illustrate this idea at a simpler level.

Consider the following: According to an article in *USA TODAY*, the total amount (in millions of dollars) of government guaranteed student loans during the years 1986 through 1994 is given in this table.

Year	Amount
1986	8.6
1987	9.8
1988	11.8
1989	12.5
1990	12.3
1991	13.5
1992	14.7
1993	16.5
1994	18.2

This data can be graphed as points in the plane. One way to approach this is to let $x = 0$ represent 1986, $x = 1$ represent 1987, and so on, up to $x = 8$ representing 1994. Then we graph the points (0, 8.6), (1, 9.8), and so on, up to (8, 18.2). The figure here shows these points. Notice that they lie approximately in a line.

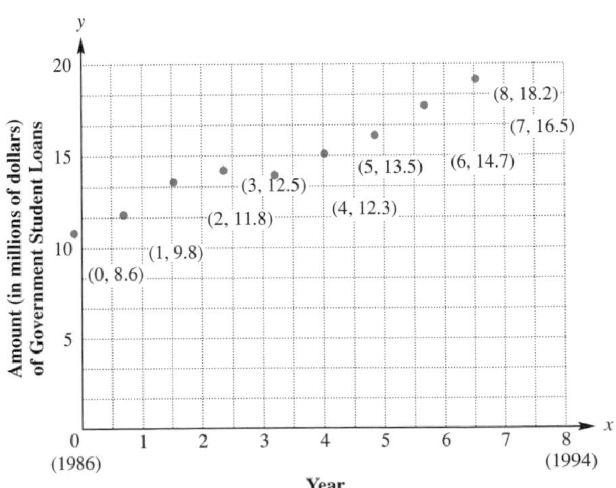

Topics for Discussion

With the class divided into groups of several students each, go through the following steps.

1. Use a visual observation to determine a pair of points that lie on or close to the line that best fits these data points. Record the coordinates of the points.
2. Find the slope of the line joining these two points, and then use the point-slope form to find an equation of the line.
3. Write the line in the form $f(x) = mx + b$ by solving the equation for y and replacing y with $f(x)$.
4. The function found in Item 3 is an example of a linear model that describes the data. It approximates the amount in millions of dollars as a function of the year. Evaluate it for a particular x-value, such as $x = 5$ for the year 1991, and compare the result to the actual dollar amount in the table. How much of a discrepancy is there between your approximation and the actual amount?
5. The method of *regression* gives the best-fitting linear function as $f(x) = 1.088x + 8.747$. Compare the various results obtained in the different groups, and see which group came closest to this.

A calculator-generated graph of the data points and the regression line is shown here.

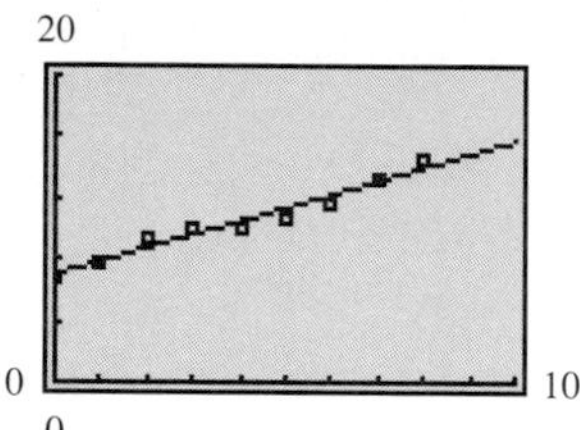

The function $f(x) = 1.088x + 8.747$ provides the best fit for the data.

Chapter 8 Test

1. Find the distance between the points $(-3, 5)$ and $(2, 1)$.
2. Find an equation of the circle whose center has coordinates $(-1, 2)$, with radius 3. Sketch its graph.
3. Find the x- and y-intercepts of the graph of $3x - 2y = 8$, and graph the equation.
4. Find the slope of the line passing through the points $(6, 4)$ and $(-1, 2)$.
5. Find the slope-intercept form of the equation of the line described.
   (a) passing through the point $(-1, 3)$, with slope $-2/5$
   (b) passing through $(-7, 2)$ and perpendicular to $y = 2x$
   (c) the line shown in the figures (Look at the displays at the bottom.)

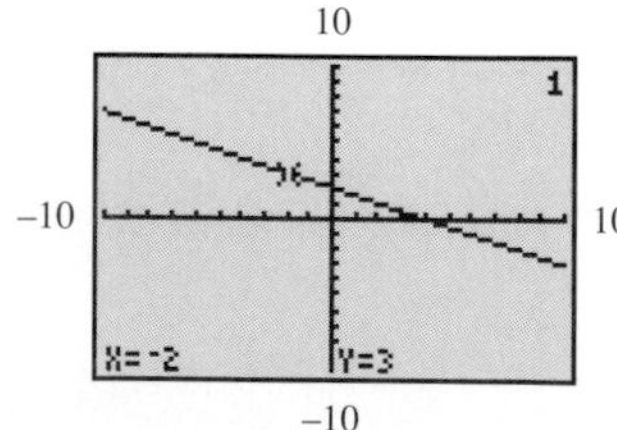

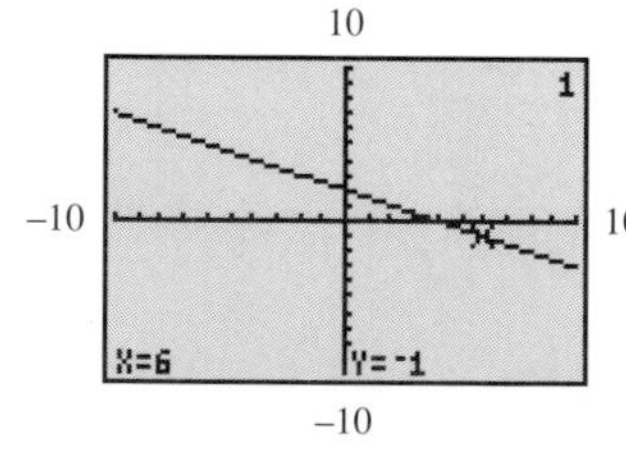

6. Which one of the following has positive slope and negative y-coordinate for its y-intercept?

(a) (b) (c) (d)

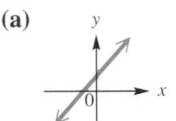

7. The linear equation $y = 825x + 8{,}689$ provides a model for the number of cases served by the Child Support Enforcement program from 1985 to 1990, where $x = 0$ corresponds to 1985, $x = 1$ corresponds to 1986, and so on. Use this model to approximate the number of cases served during 1989. (*Source:* Office of Child Support Enforcement)

8. What does the number 825 in the equation in Exercise 7 refer to
   (a) with respect to the graph?
   (b) in the context of the problem?

9. The graph indicates the percentage of all U.S. workers without any private or public health insurance. (*Source:* Census Bureau, Employee Benefit Research Institute)

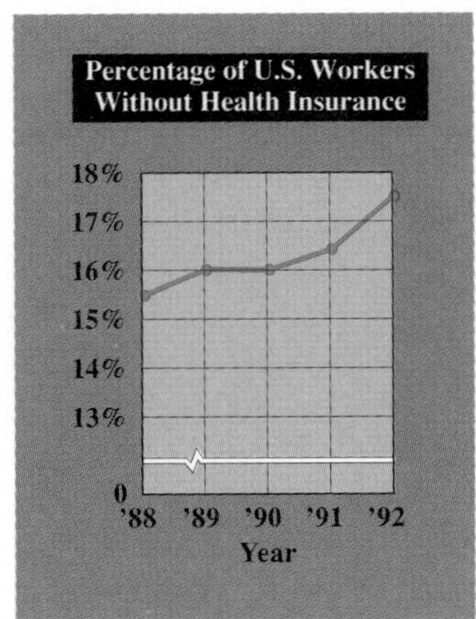

   (a) Between which two years was the percentage approximately the same?
   (b) Between which two years was the increase the greatest?
   (c) In what year was the percent about 16.5%?

10. For the function $f(x) = x^2 - 3x + 12$,
   (a) give its domain; (b) find $f(-2)$.

11. What is the domain of the function

$$g(x) = \sqrt{9 - 4x}\ ?$$

12. If the cost to produce x units of calculators is $C(x) = 50x + 5{,}000$ dollars, while the revenue is $R(x) = 60x$ dollars, find the number of units of calculators that must be produced in order to break even. What is the revenue at the break-even point?

13. Graph the quadratic function $f(x) = -(x + 3)^2 + 4$. Give the axis, the vertex, the domain, and the range.

14. Miami-Dade Community College wants to construct a rectangular parking lot on land bordered on one side by a highway. It has 320 ft of fencing with which to fence off the other three sides. What should be the dimensions of the lot if the enclosed area is to be a maximum?

15. Use a scientific calculator to find an approximation of each of the following. Give as many digits as the calculator displays.
   (a) $5.1^{4.7}$ (b) $e^{-1.85}$ (c) $\ln 23.56$

16. Which one of the following is a false statement?
   (a) The domain of the function $f(x) = \log_2 x$ is $(-\infty, \infty)$.
   (b) The graph of $F(x) = 3^x$ intersects the y-axis.
   (c) The graph of $G(x) = \log_3 x$ intersects the x-axis.
   (d) The expression $\ln x$ represents the exponent to which e must be raised in order to obtain x.

17. Suppose that \$12,000 is invested in an account that pays 4% annual interest, and is left untouched for 3 years. How much will be in the account if
   (a) interest is compounded quarterly (four times per year);
   (b) interest is compounded continuously?

18. Since 1950, the growth in the world population in millions closely fits the exponential function defined by

$$A(t) = 2{,}600e^{.018t},$$

where t is the number of years since 1950.
   (a) The world population was about 3,700 million in 1970. How closely does the function approximate this value?

(b) Use the function to approximate the population in 1990. (The actual 1990 population was about 5,320 million.)

(c) Use the function to estimate the population in the year 2000.

Solve each system by using elimination, substitution, or a combination of the two methods.

19. $2x + 3y = 2$
$3x - 4y = 20$

20. $2x + y + z = 3$
$x + 2y - z = 3$
$3x - y + z = 5$

21. $2x + 3y - 6z = 11$
$x - y + 2z = -2$
$4x + y - 2z = 7$

Solve the problem by using a system of equations.

22. A manufacturer of portable compact disc players shipped 200 of the players to its two Quebec warehouses. It costs \$3 per unit to ship to Warehouse A, and \$2.50 per unit to ship to Warehouse B. If the total shipping cost was \$537.50, how many were shipped to each warehouse?

23. Natalie Jackson sells undeveloped land. On three recent sales, she made a 10% commission, a 6% commission, and a 5% commission. Her total commissions on these sales were \$8,500, and she sold property worth \$140,000. If the 5% sale amounted to the sum of the other two, what were the three sales prices?

24. The accompanying screen was generated by a graphing calculator. Which one of the following linear inequalities in two variables does it illustrate?

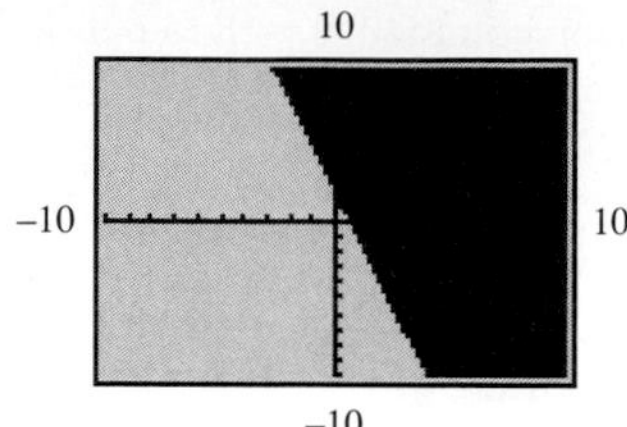

(a) $y > -3x + 2$
(b) $y < -3x + 2$
(c) $y > 3x + 2$
(d) $y < 3x + 2$

25. Graph the solution of the system of inequalities

$$x + y \leq 6$$
$$2x - y \geq 3.$$

9

Geometry

A new imaging technique called *electrical impedance tomography* (EIT) is being applied to many situations where visualization improves diagnostic capabilities. It currently is used to measure the amount of contaminants in the soil, detect corrosive damage in aircraft, and locate valuable mineral deposits in the ground. In medicine, EIT is being researched for displaying images of internal organs. To produce these images, a set of 32 electrodes is attached around the patient's torso. Then an imperceptible electrical current, oscillating 30,000 times per second, is sent through the electrodes. A TV monitor displays a cross section of the chest of the patient. Lungs appear as two lobes—blue when inhaling and orange when exhaling. The image throbbing at a regular rate is the heart.

Because various tissues and fluids in the body conduct electricity at different rates, scientists and mathematicians are able to make extensive use of both geometry and linear equations to construct an image of the body. For example, blood is a good conductor while air is not. This makes it possible to distinguish between lungs inflated with air and arteries carrying blood. Cancer cells have a much higher conductivity than normal cells, and as a result, can be detected in the image. EIT is different from X rays and CAT scans. Although both X rays and CAT scans give clearer images, EIT can display dynamic images of physiological processes like breathing as they occur. In the near future researchers hope to use EIT to measure cardiac output and avoid the invasive procedure of inserting a catheter into the heart.

Source: SIAM News, Vol. 27, No. 6, 1994.

Euclid's *Elements* as translated by Billingsley appeared in 1570 and was the first English language translation of the text—the most influential geometry text ever written.

Unfortunately, no copy of *Elements* exists that dates back to the time of Euclid (circa 300 B.C.), and most current translations are based upon a revision of the work prepared by Theon of Alexandria.

Although *Elements* was only one of several works of Euclid, it is, by far, the most important. It ranks second only to the Bible as the most published book in history.

Let no one unversed in geometry enter here.
—Motto over the door of Plato's Academy

To the ancient Greeks, mathematics meant geometry above all—a rigid kind of geometry from a twentieth-century point of view. They assumed that geometric figures could only be rotated or moved about without change—no deformations in figures were even considered. It may have seemed perfectly natural to sculptors and architects working in wood or stone that shape and size were constant properties of objects. The Greeks studied the properties of figures identical in shape and size (congruent figures) as well as figures identical in shape but not necessarily in size (similar figures). They absorbed ideas about area and volume from the Egyptians and Babylonians and established general formulas. The Greeks were the first to insist that statements in geometry be given rigorous proof.

The Greek view of geometry (and other mathematical ideas) was summarized in *Elements,* written by Euclid about 300 B.C. The power this book exerted is extraordinary; it has been studied virtually unchanged to this day as a geometry textbook and as *the* model of deductive logic.

Euclid's *Elements* begins with definitions of basic ideas such as point, line, and plane. Euclid then gives five postulates providing the foundation of all that follows.

Next, Euclid lists five axioms that he views as general truths and not just facts about geometry. (To some of the Greek writers, postulates were truths about a particular field, while axioms were general truths. Today, "axiom" is used in either case.)

Euclid's Postulates	**Euclid's Axioms**
1. Two points determine one and only one straight line.	**6.** Things equal to the same thing are equal to each other.
2. A straight line extends indefinitely far in either direction.	**7.** If equals are added to equals, the sums are equal.
3. A circle may be drawn with any given center and any given radius.	**8.** If equals are subtracted from equals, the remainders are equal.
4. All right angles are equal.	**9.** Figures that can be made to coincide are equal.
5. Given a line k and a point P not on the line, there exists one and only one line m through P that is parallel to k.*	**10.** The whole is greater than any of its parts.

Using only these ten statements and the basic rules of logic, Euclid was able to prove a large number of "propositions" about geometric figures.

The most basic ideas in geometry are **point, line,** and **plane.** In fact, it is not really possible to define them with other words. Euclid defined a point as "that which has no part," but this definition is so vague as to be meaningless. Do you think you could decide what a point is from this definition? But from your experience in saying "this point in time" or in sharpening a pencil, you have an idea of what he was getting at. Even though we don't try to define *point,* we do agree that, intuitively, a point has no magnitude and no size.

*See the discussion of John Playfair and Playfair's Axiom in Section 6 of this chapter.

Euclid defined a line as "that which has breadthless length." Again, this definition is vague. Based on our experience, however, we know what Euclid meant. The drawings that we use for lines have properties of no thickness and no width, and they extend indefinitely in two directions.

What do you visualize when you read Euclid's definition of a plane: "a surface which lies evenly with the straight lines on itself"? Do you think of a flat surface, such as a tabletop or a page in a book? That is what Euclid intended.

The geometry of Euclid is a model of deductive reasoning. In this chapter, we will present geometry from an inductive viewpoint, using objects and situations found in the world around us as models for study.

9.1 Points, Lines, Planes, and Angles

Moiré Patterns A set of parallel lines with equidistant spacing intersects an identical set, but at a small angle. The result is a moiré pattern, named after the fabric *moiré* ("watered") *silk*. You often see similar effects looking through window screens with bulges. Moiré patterns are related to periodic functions, which describe regular recurring phenomena (wave patterns such as heartbeats or business cycles). Moirés thus apply to the study of electromagnetic, sound, and water waves, to crystal structure, and to other wave phenomena.

There are certain universally accepted conventions and symbols used to represent points, lines, planes, and angles. A capital letter usually represents a point. A line may be named by two capital letters representing points that lie on the line, or by a single (usually lower-case) letter, such as ℓ. Subscripts are sometimes used to distinguish one line from another when a lower-case letter is used. For example, ℓ_1 and ℓ_2 would represent two distinct lines. A plane may be named by three capital letters representing points that lie in the plane, or by a letter of the Greek alphabet, such as α (alpha), β (beta), or γ (gamma).

Figure 1 depicts a plane that may be represented either as α or as plane *ADE*. Contained in the plane is the line *DE* (or, equivalently, line *ED*), which is also labeled ℓ in the figure.

Selecting any point on a line divides the line into three parts: the point itself, and two **half-lines,** one on each side of the point. For example, in Figure 2, point *A* divides the line into three parts, *A* itself and two half-lines. Point *A* belongs to neither half-line. As the figure suggests, each half-line extends indefinitely in the direction opposite to the other half-line.

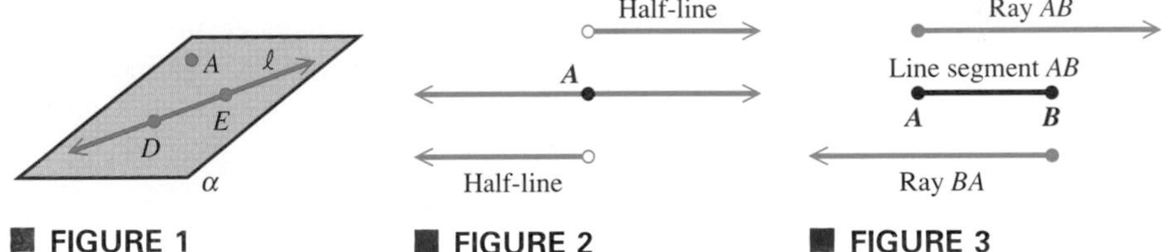

FIGURE 1 **FIGURE 2** **FIGURE 3**

Including an initial point with a half-line gives a **ray.** A ray is named with two letters, one for the initial point of the ray, and one for another point contained in the half-line. For example, in Figure 3 ray *AB* has initial point *A* and extends in the direction of *B*. On the other hand, ray *BA* has *B* as its initial point and extends in the direction of *A*.

A **line segment** includes both endpoints and is named by its endpoints. Figure 3 shows line segment *AB*, which may also be designated as line segment *BA*.

Special symbols are used to distinguish between lines, half-lines, rays, and segments. The following chart shows the figures along with the symbols used to represent them.

Name	Figure	Symbol
Line *AB* or line *BA*	A B	$\overleftrightarrow{AB}$ or $\overleftrightarrow{BA}$
Half-line *AB*	A B	$\overset{\circ\!\rightarrow}{AB}$
Half-line *BA*	A B	$\overset{\circ\!\rightarrow}{BA}$
Ray *AB*	A B	$\overrightarrow{AB}$
Ray *BA*	A B	$\overrightarrow{BA}$
Segment *AB* or segment *BA*	A B	$\overline{AB}$ or $\overline{BA}$

Today no attempt would be made to define the basic idea of a plane, even though properties of planes could be studied. For example, given any three points that are not in a straight line, a plane can be passed through the points. That is why camera tripods have three legs—no matter how irregular the surface, the tips of the three legs determine a plane. On the other hand, a camera support with four legs would wobble unless each leg were carefully extended just the right amount.

Notice that for a line, the symbol above the two letters shows two arrowheads, indicating that the line extends indefinitely in both directions. For half-lines and rays, only one arrowhead is used, since these extend in only one direction. An open circle is used for a half-line to show that the endpoint is not included, while a solid circle is used for a ray to indicate the inclusion of the endpoint. Since a segment includes both endpoints and does not extend in either direction, solid circles are used to indicate endpoints of line segments.

The geometric definitions of "parallel" and "intersecting" apply to two or more lines or planes. (See Figure 4.) **Parallel lines** lie in the same plane and never meet, no matter how far they are extended. However, **intersecting lines** do meet. If two distinct lines intersect, they intersect in one and only one point.

We use the symbol $\|$ to denote parallelism. If ℓ_1 and ℓ_2 are parallel lines, as in Figure 4, then this may be indicated as $\ell_1 \| \ell_2$.

Parallel planes also never meet, no matter how far they are extended. Two distinct **intersecting planes** form a straight line, the one and only line they have in common. **Skew lines** do not lie in the same plane, and they never meet, no matter how far they are extended.

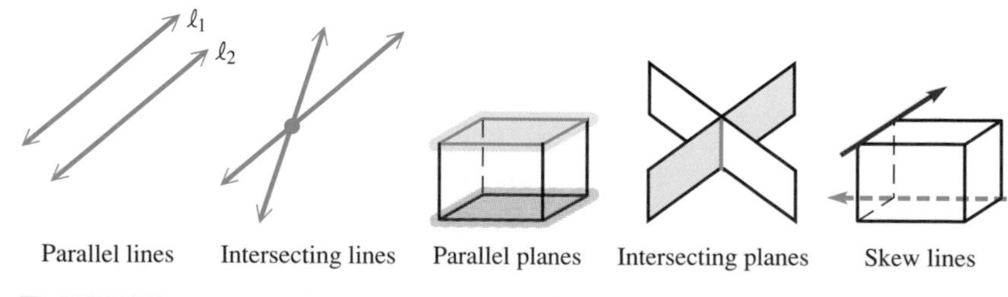

FIGURE 4

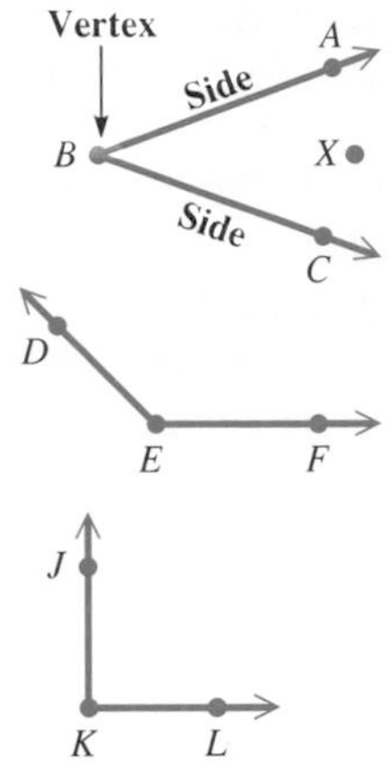

■ **FIGURE 5**

An **angle** is the union of two rays that have a common endpoint, as shown in Figure 5. It is important to remember that the angle is formed by points on the rays themselves, and no other points. In Figure 5, point X is *not* a point on the angle. (It is said to be in the *interior* of the angle.) Notice that "angle" is the first basic term in this section that is actually defined, using the undefined terms *ray* and *endpoint*.

The rays forming an angle are called its **sides.** The common endpoint of the rays is the **vertex** of the angle. There are two standard ways of naming angles. If no confusion will result, an angle can be named with the letter marking its vertex. Using this method, the angles in Figure 5 can be named, respectively, angle B, angle E, and angle K. Angles also can be named with three letters: the first letter names a point on one side of the angle; the middle letter names the vertex; the third names a point on the other side of the angle. In this system, the angles in the figure can be named angle ABC, angle DEF, and angle JKL.

Just as there are symbols to represent lines, half-lines, rays, and segments, there is a symbol for representing an angle. It is $\measuredangle$. Rather than writing "angle ABC", we may write "$\measuredangle ABC$."

An angle can be associated with an amount of rotation. For example, in Figure 6(a), we let $\overleftrightarrow{BA}$ first coincide with $\overleftrightarrow{BC}$—as though they were the same ray. We then rotate $\overleftrightarrow{BA}$ (the endpoint remains fixed) in a counterclockwise direction to form $\measuredangle ABC$.

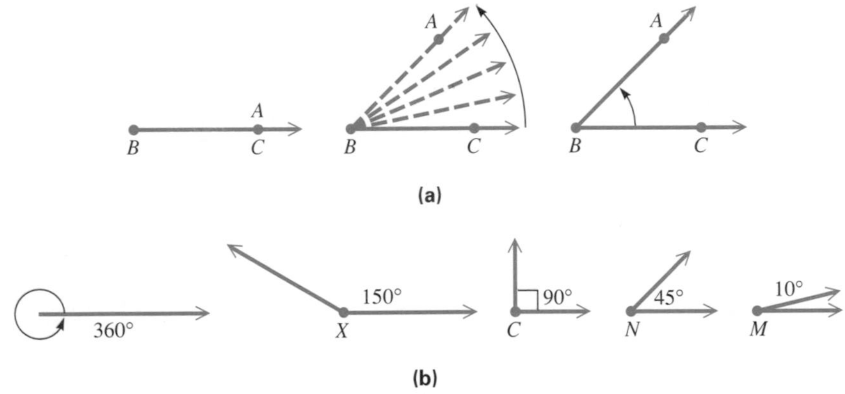

■ **FIGURE 6**

Angles are key to the study of **geodesy,** the measurement of distances on the earth's surface.

Angles are measured by the amount of rotation, using a system that dates back to the Babylonians some two centuries before Christ. Babylonian astronomers chose the number 360 to represent the amount of rotation of a ray back onto itself. Using 360 as the amount of rotation of a ray back onto itself, **one degree,** written 1°, is defined to be 1/360 of a complete rotation. Figure 6(b) shows angles of various degree measures.

Angles are classified and named with reference to their degree measures. An angle whose measure is between 0° and 90° is called an **acute angle.** Angles M and N in Figure 6(b) are acute. An angle that measures 90° is called a **right angle.** Angle C in the figure is a right angle. The squared symbol at the vertex denotes a right angle. Angles that measure more than 90° but less than 180° are said to be **obtuse angles** (angle X, for example). An angle that measures 180° is a **straight angle.** Its sides form a straight line.

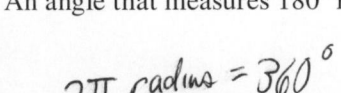

Our work in this section will be devoted primarily to angles whose measures are less than or equal to 180°. Angles whose measures are greater than 180° are studied in more detail in trigonometry courses.

A tool called a **protractor** can be used to measure angles. Figure 7 shows a protractor measuring an angle. To use a protractor, position the hole (or dot) of the protractor on the vertex of the angle. The 0-degree measure on the protractor should be placed on one side of the angle, while the other side should extend to the degree measure of the angle. The figure indicates an angle whose measure is 135°.

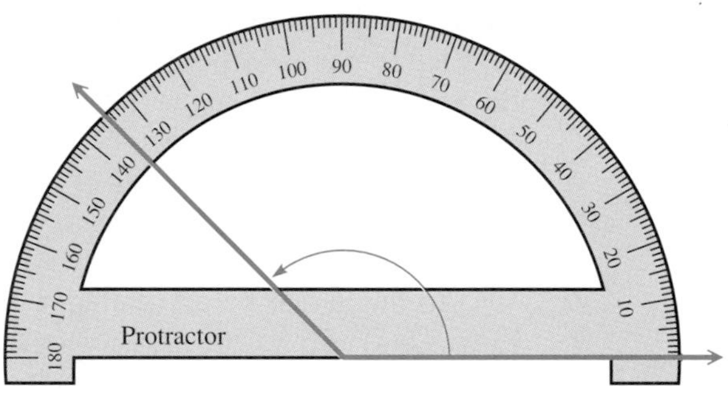

■ FIGURE 7

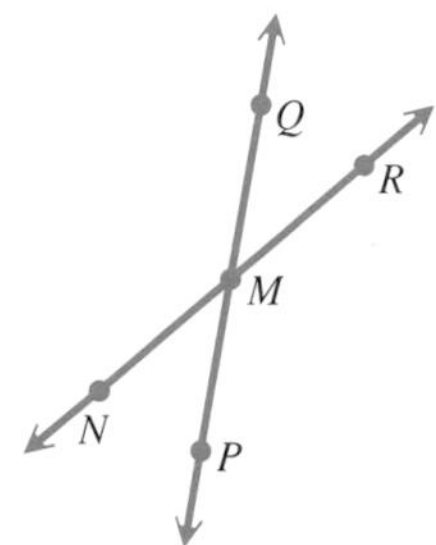

■ FIGURE 8

When two lines intersect to form right angles they are called **perpendicular lines.** Our sense of *vertical* and *horizontal* depends on perpendicularity. Carpenters and stonemasons through the ages have built amazing structures with as few tools as a plumb line (a weight at the end of a string) and a right angle (a T-square).

In Figure 8, the sides of $\measuredangle NMP$ have been extended to form another angle, $\measuredangle RMQ$. The pair $\measuredangle NMP$ and $\measuredangle RMQ$ are called **vertical angles.** Another pair of vertical angles have been formed at the same time. They are $\measuredangle NMQ$ and $\measuredangle PMR$.

An important property of vertical angles follows.

Property of Vertical Angles

Vertical angles have equal measures.

This property is verified in Exercise 78 of this section. As an example, $\measuredangle NMP$ and $\measuredangle RMQ$ in Figure 8 have equal measures. Can you name another pair of angles in the figure that have equal measures?

EXAMPLE 1 Refer to the appropriate figure to solve each problem.

(a) Find the measure of each marked angle in Figure 9.

Since the marked angles are vertical angles, they have the same measures. Set $4x + 19$ equal to $6x - 5$ and solve.

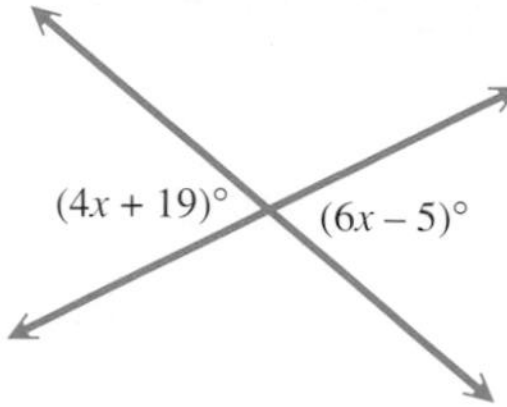

■ **FIGURE 9**

$$
\begin{aligned}
4x + 19 &= 6x - 5 \\
-4x + 4x + 19 &= -4x + 6x - 5 && \text{Add } -4x. \\
19 &= 2x - 5 \\
19 + 5 &= 2x - 5 + 5 && \text{Add 5.} \\
24 &= 2x \\
12 &= x && \text{Divide by 2.}
\end{aligned}
$$

Since $x = 12$, one angle has measure $4(12) + 19 = 67$ degrees. The other has the same measure, since $6(12) - 5 = 67$ as well. Each angle measures 67°.

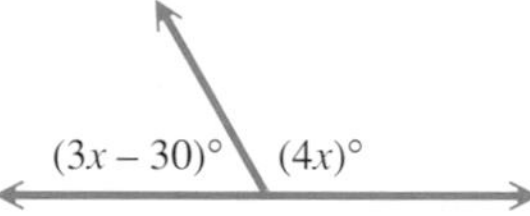

■ **FIGURE 10**

(b) Find the measure of each marked angle in Figure 10.

The measures of the marked angles must add to 180° since together they form a straight angle. The equation to solve is

$$
\begin{aligned}
(3x - 30) + 4x &= 180. \\
7x - 30 &= 180 && \text{Combine like terms.} \\
7x - 30 + 30 &= 180 + 30 && \text{Add 30.} \\
7x &= 210 \\
x &= 30 && \text{Divide by 7.}
\end{aligned}
$$

To find the measures of the angles, replace x with 30 in the two expressions.

$$
\begin{aligned}
3x - 30 &= 3(30) - 30 = 90 - 30 = 60 \\
4x &= 4(30) = 120
\end{aligned}
$$

The two angle measures are 60° and 120°. ■

Why 360? The use of the number 360 goes back to the Babylonian culture. There are several theories regarding why 360 was chosen for the number of degrees in a complete rotation around a circle. One says that 360 was chosen because it is close to the number of days in a year, and is conveniently divisible by 2, 3, 4, 5, 6, 8, 9, 10, 12, and other numbers.

If the sum of the measures of two acute angles is 90°, the angles are said to be **complementary,** and each angle is called the *complement* of the other. For example, angles measuring 40° and 50° are complementary angles, because $40° + 50° = 90°$. If two angles have a sum of 180°, they are **supplementary.** The *supplement* of an angle whose measure is 40° is an angle whose measure is 140°, because $40° + 140° = 180°$. If x represents the degree measure of an angle, $90 - x$ represents the measure of its complement, and $180 - x$ represents the measure of its supplement.

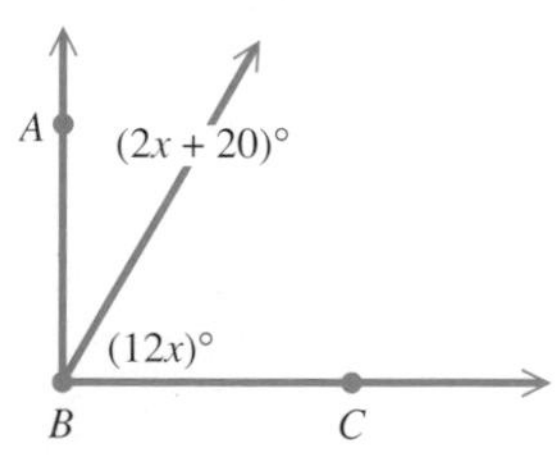

■ **FIGURE 11**

EXAMPLE 2 Find the measures of the angles in Figure 11, given that $\measuredangle ABC$ is a right angle.

The sum of the measures of the two acute angles is 90° (that is, they are complementary), since they form a right angle. We add their measures to get 90 and solve the resulting equation.

$$
\begin{aligned}
(2x + 20) + 12x &= 90 \\
14x + 20 &= 90 \\
14x &= 90 - 20 \\
14x &= 70 \\
x &= 5
\end{aligned}
$$

The value of x is 5. Therefore, the measures of the angles are

$$2x + 20 = 2(5) + 20 = 30 \text{ degrees}$$

and

$$12x = 12(5) = 60 \text{ degrees.}$$

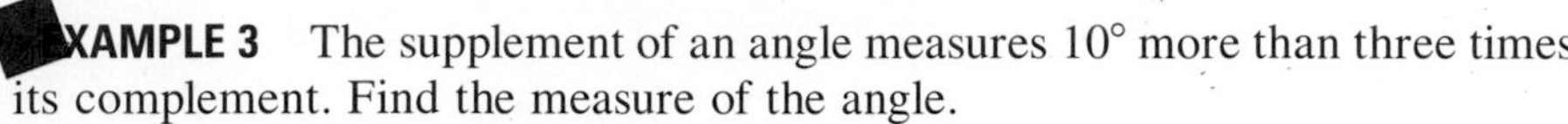

EXAMPLE 3 The supplement of an angle measures 10° more than three times its complement. Find the measure of the angle.

Let x = the degree measure of the angle. Then,
$180 - x$ = the degree measure of its supplement, and
$90 - x$ = the degree measure of its complement.

Now use the words of the problem to write the equation.

$$\underbrace{180 - x}_{\text{Supplement}} \quad \underbrace{=}_{\text{measures}} \quad \underbrace{10 +}_{\text{10 more than}} \quad \underbrace{3(90 - x)}_{\text{three times its complement.}}$$

Solve the equation.

$$\begin{aligned} 180 - x &= 10 + 270 - 3x && \text{Distributive property} \\ 180 - x &= 280 - 3x && \\ 2x &= 100 && \text{Add } 3x\text{; subtract 180.} \\ x &= 50 && \text{Divide by 2.} \end{aligned}$$

The angle measures 50°. Since its supplement (130°) is 10° more than three times its complement (40°), that is,

$$130 = 10 + 3(40) \quad \text{is true,}$$

the answer checks.

Parallel lines are lines that lie in the same plane and do not intersect. Figure 12 shows parallel lines m and n. When a line q intersects two parallel lines, q is called a **transversal.** In Figure 12, the transversal intersecting the parallel lines forms eight angles, indicated by numbers. Angles 1 through 8 in the figure possess some special properties regarding their degree measures. The following chart gives their names with respect to each other, and rules regarding their measures.

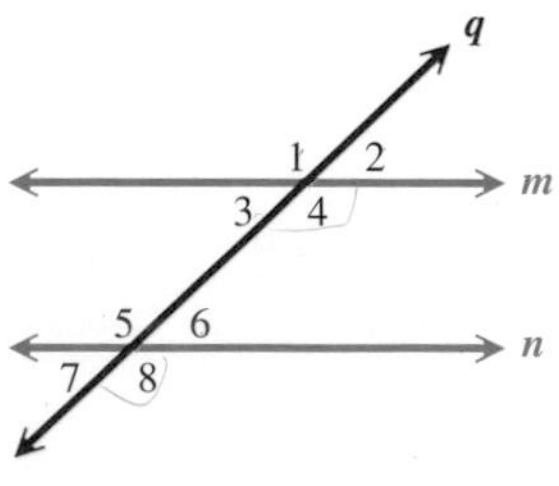

FIGURE 12

Name	Sketch	Rule
Alternate interior angles	(also 3 and 6)	Angle measures are equal.
Alternate exterior angles	(also 2 and 7)	Angle measures are equal.
Interior angles on same side of transversal	(also 3 and 5)	Angle measures add to 180°.
Corresponding angles	(also 1 and 5, 3 and 7, 4 and 8)	Angle measures are equal.

The converses of the above also are true. That is, if alternate interior angles are equal, then the lines are parallel, with similar results valid for alternate exterior angles, interior angles on the same side of a transversal, and corresponding angles.

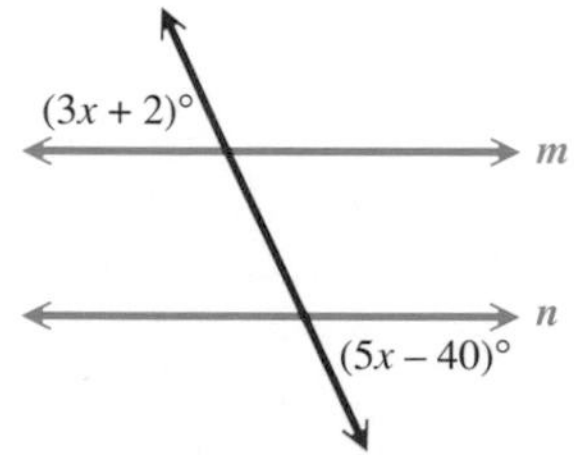

FIGURE 13

EXAMPLE 4 Find the measure of each marked angle in Figure 13, given that lines m and n are parallel.

The marked angles are alternate exterior angles, which are equal. This gives

$$3x + 2 = 5x - 40$$
$$42 = 2x \qquad \text{Subtract } 3x \text{ and add } 40.$$
$$21 = x. \qquad \text{Divide by } 2.$$

One angle has a measure of $3x + 2 = 3 \cdot 21 + 2 = 65$ degrees, and the other has a measure of $5x - 40 = 5 \cdot 21 - 40 = 65$ degrees. ■

9.1 EXERCISES

Decide whether each statement is true *or* false.

1. A line segment has two endpoints.
2. A ray has one endpoint.
3. A half-line has one endpoint.
4. If *A* and *B* are distinct points on a line, then ray *AB* and ray *BA* represent the same set of points.
5. If two lines intersect, they lie in the same plane.
6. If two lines are parallel, they lie in the same plane.
7. If two lines do not intersect, they must be parallel.
8. The sum of the measures of two right angles is the measure of a straight angle.
9. The supplement of an acute angle must be an obtuse angle.
10. Segment *AB* and segment *BA* represent the same set of points.
11. The sum of the measures of two obtuse angles cannot equal the measure of a straight angle.
12. There is no angle that is its own complement.
13. There is no angle that is its own supplement.
14. The origin of the use of the degree as a unit of measure of an angle goes back to the Egyptians.

Exercises 15–24 name portions of the line shown. For each exercise, **(a)** *give the symbol that represents the portion of the line named, and* **(b)** *draw a figure showing just the portion named, including all labeled points.*

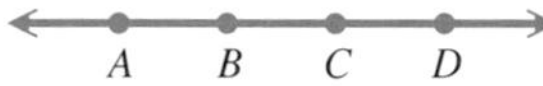

15. line segment *AB* **16.** ray *BC* **17.** ray *CB* **18.** line segment *AD*

19. half-line *BC* **20.** half-line *AD* **21.** ray *BA* **22.** ray *DA*

23. line segment *CA* **24.** line segment *DA*

In Exercises 25–32, match the figure in column I with the figure in column II that names the same set of points, based on the given figure.

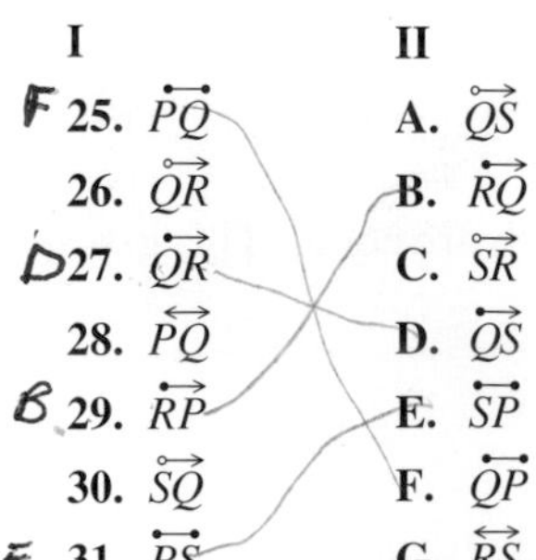

I	II
25. $\overset{\bullet\!-\!\bullet}{PQ}$	**A.** $\overset{\circ\!\!\longrightarrow}{QS}$
26. $\overset{\circ\!\!\longrightarrow}{QR}$	**B.** $\overset{\bullet\!\!\longrightarrow}{RQ}$
27. $\overset{\bullet\!\!\longrightarrow}{QR}$	**C.** $\overset{\circ\!\!\longrightarrow}{SR}$
28. $\overleftrightarrow{PQ}$	**D.** $\overset{\bullet\!\!\longrightarrow}{QS}$
29. $\overset{\bullet\!\!\longrightarrow}{RP}$	**E.** $\overset{\bullet\!-\!\bullet}{SP}$
30. $\overset{\circ\!\!\longrightarrow}{SQ}$	**F.** $\overset{\bullet\!-\!\bullet}{QP}$
31. $\overset{\bullet\!-\!\bullet}{PS}$	**G.** $\overleftrightarrow{RS}$
32. $\overset{\circ\!\!\longrightarrow}{PS}$	**H.** none of these

Lines, rays, half-lines, and segments may be considered sets of points. The **intersection** *(symbolized* $\cap$*) of two sets is composed of all elements common to both sets, while the* **union** *(symbolized* $\cup$*) of two sets is composed of all elements found in at least one of the two sets. Based on the figure below, specify each of the sets given in Exercises 33–40 in a simpler way.*

33. $\overline{MN} \cup \overline{NO}$ **34.** $\overline{MN} \cap \overline{NO}$ **35.** $\overrightarrow{MO} \cap \overrightarrow{OM}$ **36.** $\overrightarrow{MO} \cup \overrightarrow{OM}$

37. $\overset{\circ\!\rightarrow}{OP} \cap O$ **38.** $\overset{\circ\!\rightarrow}{OP} \cup O$ **39.** $\overline{NP} \cap \overline{OP}$ **40.** $\overline{NP} \cup \overline{OP}$

Give the measure of the complement of each angle.

41. 28° **42.** 32° **43.** 89° **44.** 45° **45.** $x°$ **46.** $(90 - x)°$

Give the measure of the supplement of each angle.

47. 132° **48.** 105° **49.** 26° **50.** 90° **51.** $y°$ **52.** $(180 - y)°$

Name all pairs of vertical angles in each figure.

53.

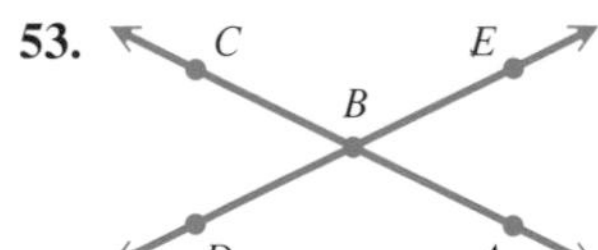

54.

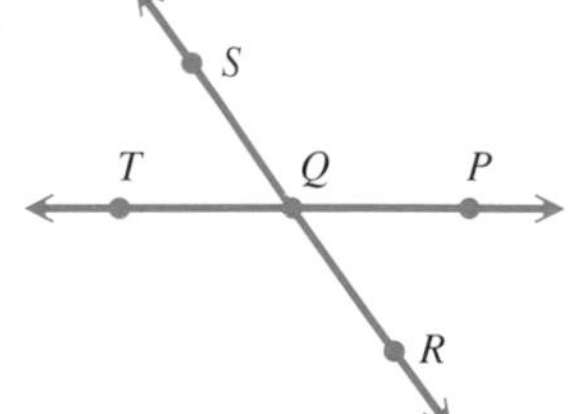

55. In Exercise 53, if ∡ABE has a measure of 52°, find the measures of the angles.
(a) ∡CBD **(b)** ∡CBE

56. In Exercise 54, if ∡SQP has a measure of 126°, find the measures of the angles.
(a) ∡TQR **(b)** ∡PQR

Find the measure of each marked angle.

57.

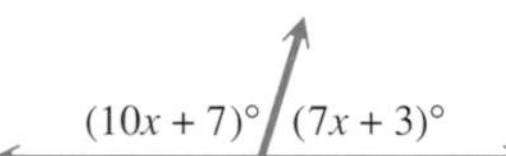

58.

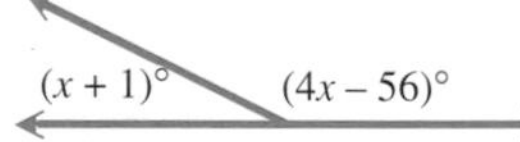

59.

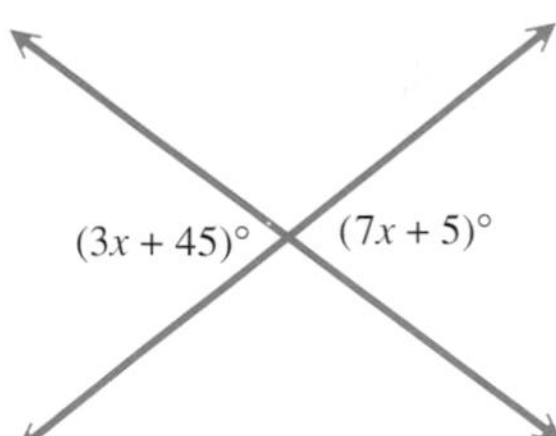

60.

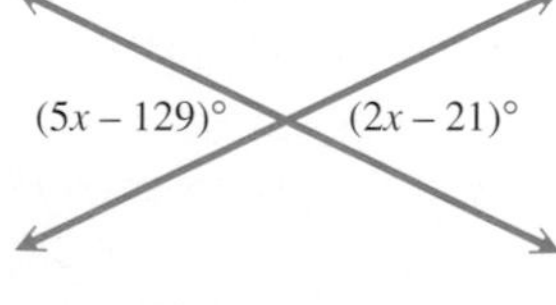

61.

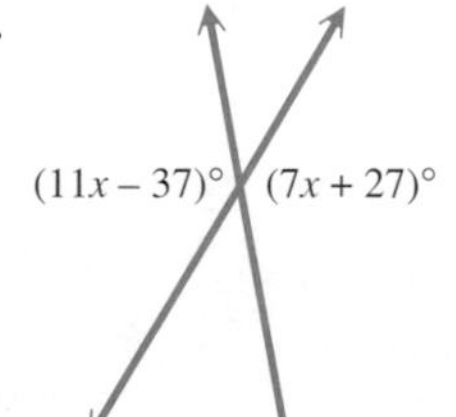

62.

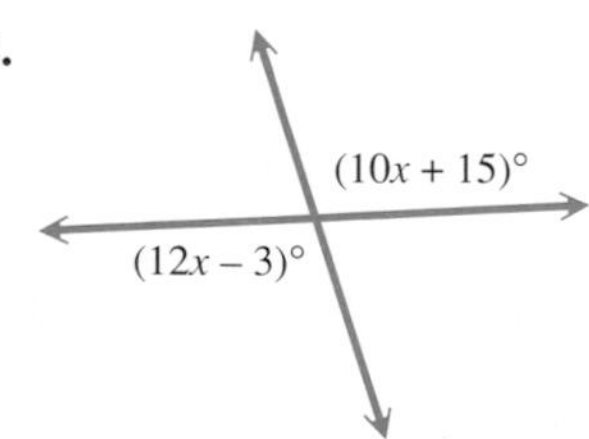

63.

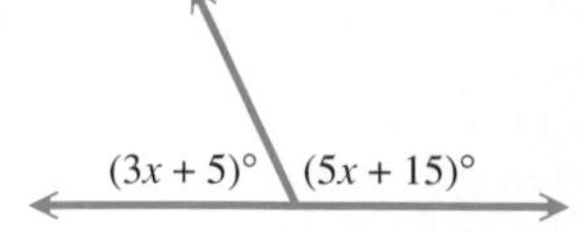

64.

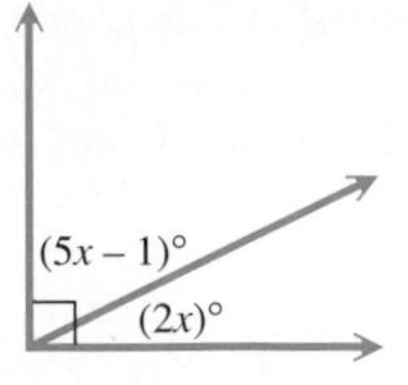

65.

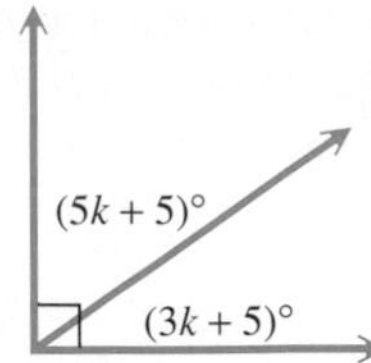

66. The sketch shows parallel lines m and n cut by a transversal q. (Recall that alternate interior angles have the same measure.) Using the figure, go through the following steps to prove that alternate exterior angles have the same measure.

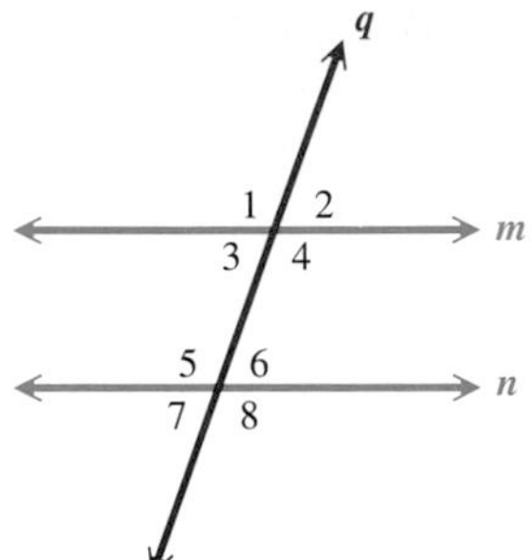

(a) Measure of ∡2 = measure of ∡ ______, since they are vertical angles.

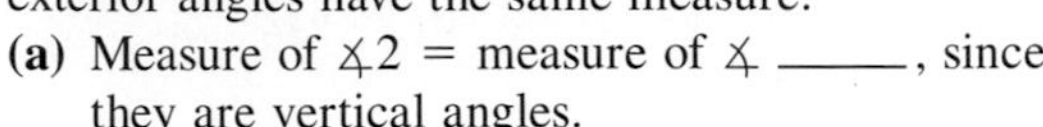

(b) Measure of ∡3 = measure of ∡ ______, since they are alternate interior angles.

(c) Measure of ∡6 = measure ∡ ______, since they are vertical angles.

(d) By the results of parts (a), (b), and (c), the measure of ∡2 must equal the measure of ∡ ______, showing that alternate ______ angles have equal measures.

In Exercises 67–70, assume that lines m and n are parallel, and find the measure of each marked angle.

67.

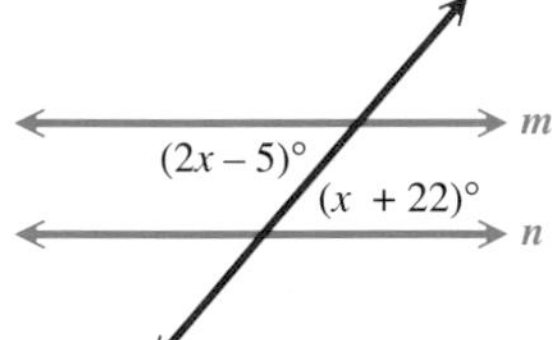

68.

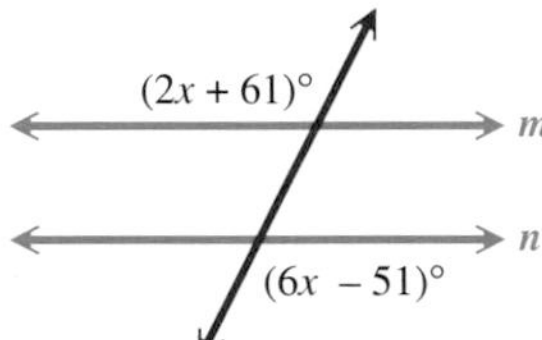

69.

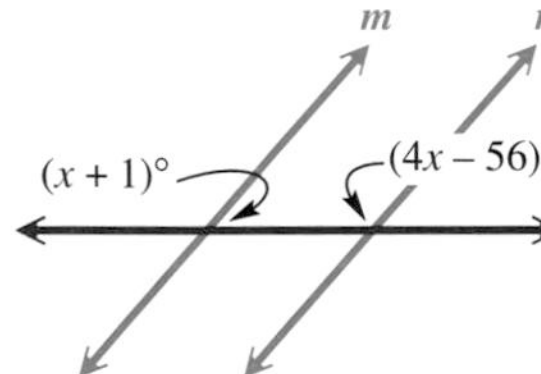

70.

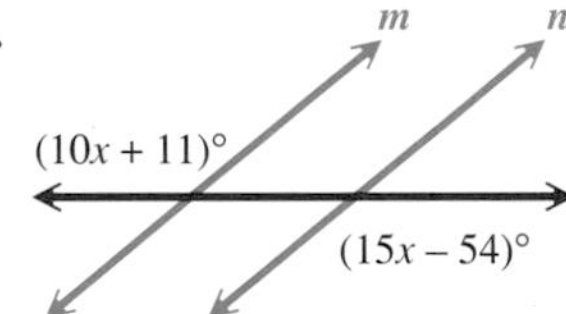

Solve each problem in Exercises 71–74.

71. The supplement of an angle measures 25° more than twice its complement. Find the measure of the angle.

72. The complement of an angle measures 10° less than one-fifth of its supplement. Find the measure of the angle.

73. The supplement of an angle added to the complement of the angle gives 210°. What is the measure of the angle?

74. Half the supplement of an angle is 12° less than twice the complement of the angle. Find the measure of the angle.

75. Write a problem similar to the one in Exercise 71, and solve it.

76. Write a problem similar to the one in Exercise 72, and solve it.

77. Use the sketch to find the measure of each numbered angle. Assume that $m \parallel n$.

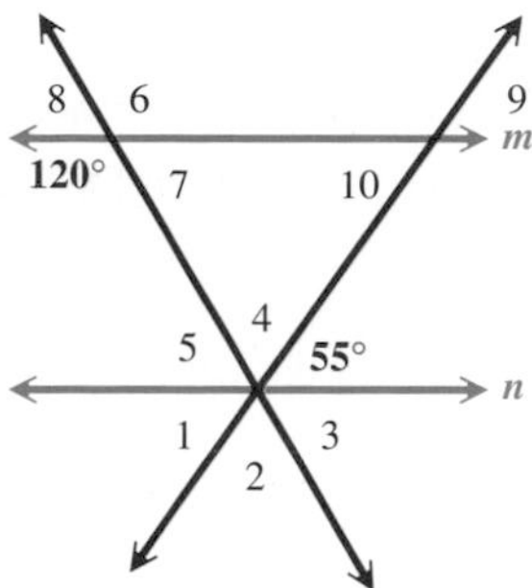

78. Complete these steps in the proof that vertical angles have equal measure. In this exercise, m(∡ ____) means "the measure of angle ____." Use the figure.

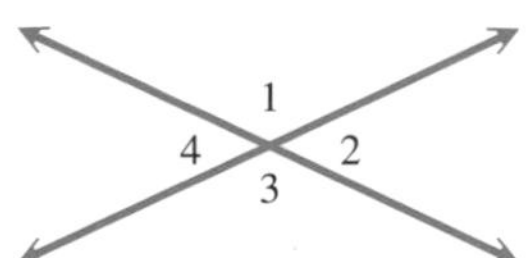

(a) m(∡1) + m(∡2) = ____ °
(b) m(∡2) + m(∡3) = ____ °
(c) Subtract the equation in part (b) from the equation in part (a) to get [m(∡1) + m(∡2)] − [m(∡2) + m(∡3)] = ____ ° − ____ °.
(d) m(∡1) + m(∡2) − m(∡2) − m(∡3) = ____ °
(e) m(∡1) − m(∡3) = ____ °
(f) m(∡1) = m(____)

79. Use the approach of Exercise 66 to prove that interior angles on the same side of a transversal are supplementary.

80. Find the values of x and y in the figure, given that $x + y = 40$.

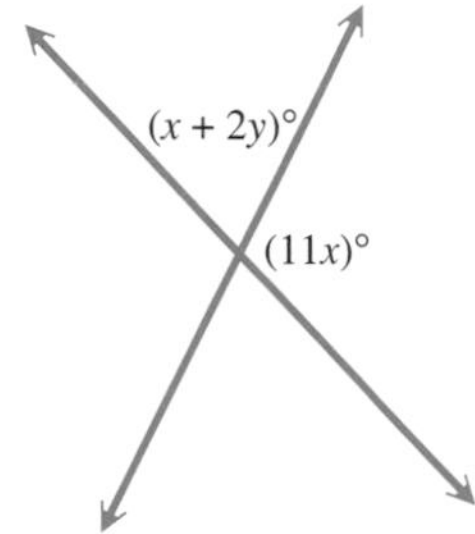

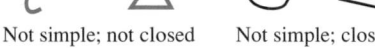

9.2 Curves, Polygons, and Circles

The basic undefined term *curve* is used for describing figures in the plane. While this term is used without any attempt to define it, common types of curves can be defined. (See the examples in Figure 14.)

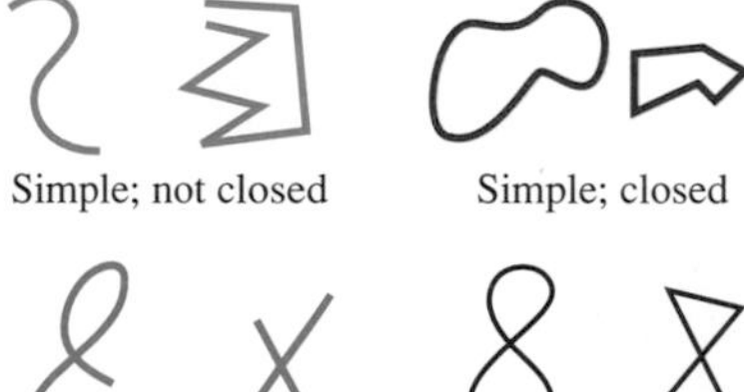

FIGURE 14

Simple Curve; Closed Curve

A **simple curve** can be drawn without lifting the pencil from the paper, and without passing through any point twice.

A **closed curve** has its starting and ending points the same, and is also drawn without lifting the pencil from the paper.

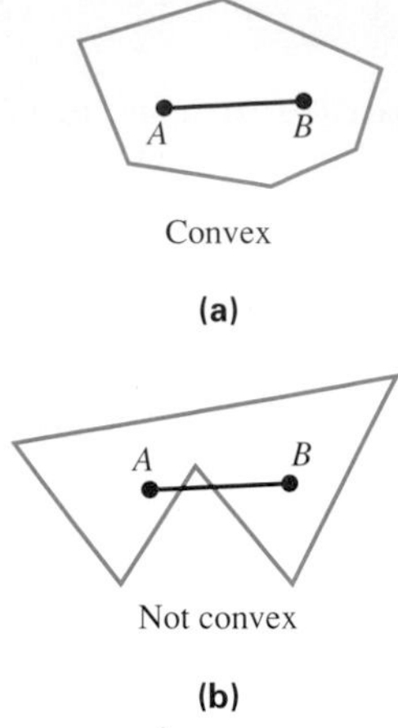

■ FIGURE 15

A figure is said to be **convex** if, for any two points A and B inside the figure, the line segment AB (that is, $\overleftrightarrow{AB}$) is always completely inside the figure. In Figure 15, part (a) shows a convex figure while (b) shows one that is not convex.

Among the most common types of curves in mathematics are those that are both simple and closed, and perhaps the most important of these are *polygons.* A **polygon** is a simple closed curve made up of only straight line segments. The line segments are called the *sides,* and the points at which the sides meet are called *vertices* (singular: *vertex*). Polygons are classified according to the number of line segments used as sides. The chart below gives the special names. In general, if a polygon has n sides, and no particular value of n is specified, it is called an n-*gon.*

Classification of Polygons According to Number of Sides

Number of Sides	Name
3	triangle
4	quadrilateral
5	pentagon
6	hexagon
7	heptagon
8	octagon
9	nonagon
10	decagon

Some examples of polygons are shown in Figure 16. A polygon may or may not be convex. Polygons with all sides equal and all angles equal are **regular polygons.**

Polygons are simple closed curves made up of straight line segments.

(a)

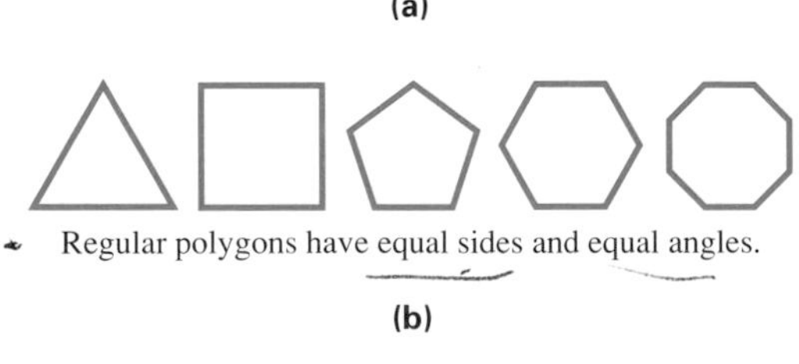

Regular polygons have equal sides and equal angles.

(b)

■ FIGURE 16

Triangles and Quadrilaterals

Two of the most common types of polygons are triangles and quadrilaterals. Triangles are classified by measures of angles as well as by number of equal sides, as shown in the following box. (Notice that tick marks are used in the bottom three figures to show how side lengths are related.)

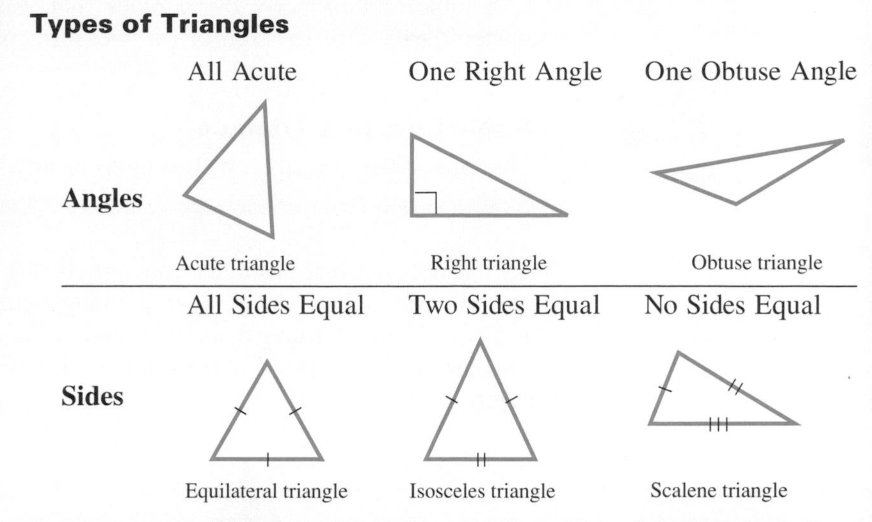

Ptolemy's geometric model of the solar system was fairly complex, with the earth at the center and the planets and sun moving in circular orbits and on epicycles. Ptolemy meant it to be a means of predicting accurately the motions of the planets, not a true picture of reality. However, Ptolemy's system was accepted on authority during the Middle Ages up until the time of the Copernican Revolution, named after **Nicolas Copernicus.**

Copernicus worked out his own system with the sun at center (heliocentric) instead of Earth at center (geocentric). A small change, you think—but it marked a revolution in thinking and the beginning of the modern world.

Quadrilaterals are classified by sides and angles, as described in the box. It can be seen below that an important distinction involving quadrilaterals is whether one or more pairs of sides are parallel. The most familiar quadrilaterals are parallelograms, and the most familiar of these are rectangles and squares. Both the square and the rhombus have all sides equal, but only the square is a regular quadrilateral. (Why?)

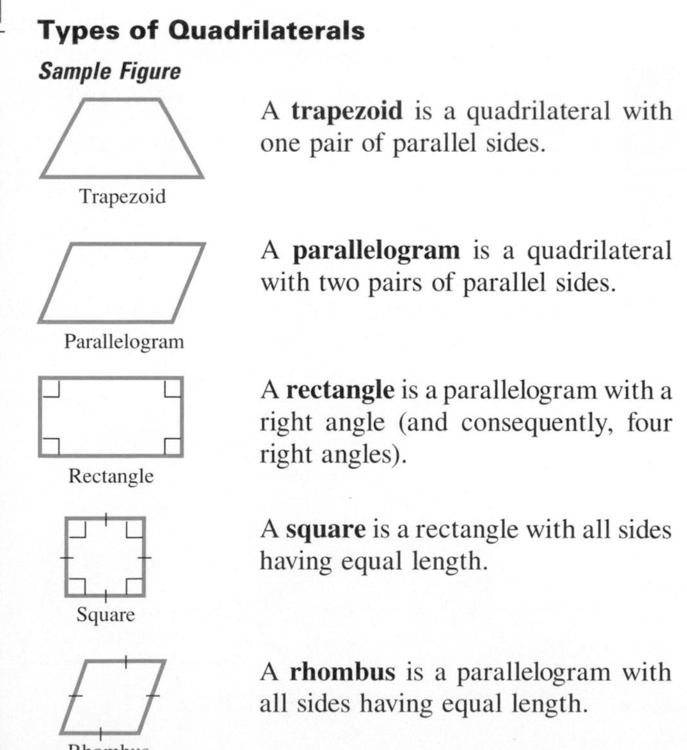

Types of Quadrilaterals
Sample Figure
Trapezoid
A trapezoid is a quadrilateral with one pair of parallel sides.
Parallelogram
A parallelogram is a quadrilateral with two pairs of parallel sides.
Rectangle
A rectangle is a parallelogram with a right angle (and consequently, four right angles).
Square
A square is a rectangle with all sides having equal length.
Rhombus
A rhombus is a parallelogram with all sides having equal length.

Tangrams The puzzle-game above comes from China, where it has been a popular amusement for centuries. The figure on the left is a tangram. Any tangram is composed of the same set of seven tans (the pieces making up the square are shown on the right).

Mathematicians have described various properties of tangrams. While each tan is convex, only 13 convex tangrams are possible. All others, like the figure on the left, are not convex.

An important property of triangles that was first proved by the Greek geometers deals with the sum of the measures of the angles of any triangle.

Angle Sum of a Triangle

The sum of the measures of the angles of any triangle is 180°.

While it is not an actual proof, a rather convincing argument for the truth of this statement can be given using any size triangle cut from a piece of paper. Tear each corner from the triangle, as suggested in Figure 17(a). You should be able to rearrange the pieces so that the three angles form a straight angle, as shown in Figure 17(b).

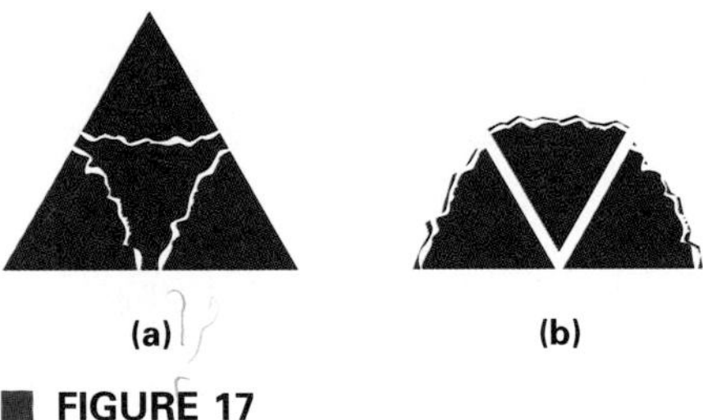

FIGURE 17

EXAMPLE 1 Find the measure of each angle in the triangle of Figure 18.

By the angle sum relationship, the three angle measures must add up to 180°. Write the equation indicating this, and then solve.

$$
\begin{aligned}
x + (x + 20) + (210 - 3x) &= 180 \\
-x + 230 &= 180 && \text{Combine like terms.} \\
-x &= 180 - 230 && \text{Subtract 230.} \\
-x &= -50 \\
x &= 50 && \text{Divide by } -1.
\end{aligned}
$$

One angle measures 50°, another measures $x + 20 = 50 + 20 = 70°$, and the third measures $210 - 3x = 210 - 3(50) = 60°$. Since $50° + 70° + 60° = 180°$, the answers are correct.

Geometria Renaissance sculptor Antonio de Pollaiolo (1433–1498) depicted Geometry in this bronze relief, one of ten reliefs on the sloping base of the bronze tomb of Sixtus IV, pope from 1471 to 1484. The important feature in this photo is the inclusion of compasses, one of two instruments allowed in classical geometry for constructions.

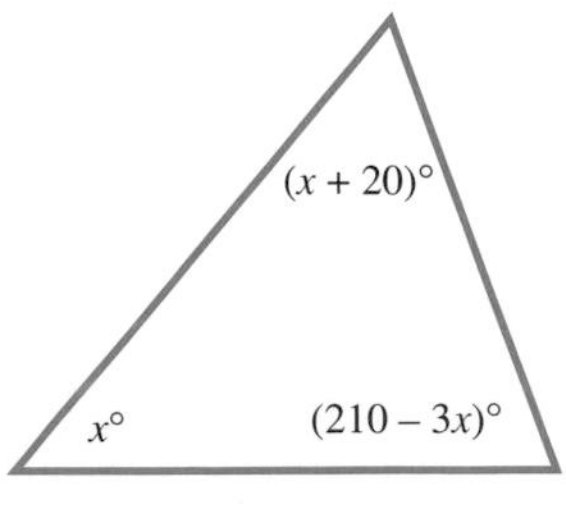

FIGURE 18

Euclidean tools, the compasses and unmarked straightedge, proved to be sufficient for Greek geometers to accomplish a great number of geometric constructions. Basic constructions such as copying an angle, constructing the perpendicular bisector of a segment, and bisecting an angle are easily performed and verified.

There were, however, three constructions that the Greeks were not able to accomplish with these tools. Now known as the *three famous problems of antiquity,* they are:

1. To trisect an arbitrary angle;
2. To construct the length of the edge of a cube having twice the volume of a given cube;
3. To construct a square having the same area as that of a given circle.

In the nineteenth century it was learned that these constructions are, in fact, impossible to accomplish with Euclidean tools. Over the years other methods have been devised to accomplish them. For example, trisecting an arbitrary angle can be accomplished if one allows the luxury of marking on the straightedge! But this violates the rules followed by the Greeks.

Despite the fact that these problems have been *proven* to be impossible to solve with compasses and a straightedge, there are people who claim to have found solutions. It seems that mathematics will always be plagued with these "angle trisectors" and "circle squarers."

In the triangle shown in Figure 19, angles 1, 2, and 3 are called interior angles, while angles 4, 5, and 6 are called exterior angles of the triangle. Using the fact that the sum of the angle measures of any triangle is 180°, and a straight angle also measures 180°, the following property may be deduced.

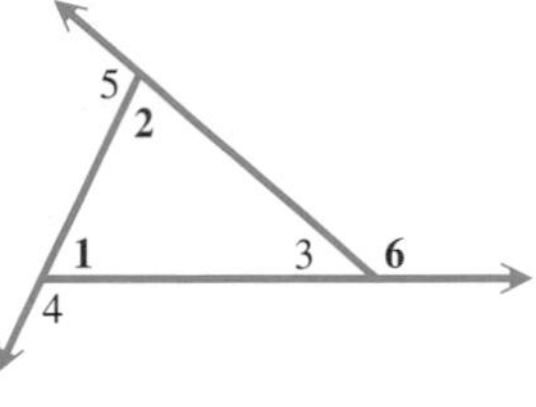

FIGURE 19

> The measure of an exterior angle of a triangle is equal to the sum of the measures of the two opposite interior angles.

Again referring to Figure 19, we can state that the measure of angle 6 is equal to the sum of the measures of angles 1 and 2. Two other such statements can be made. What are they?

A proof of the property above is required in Exercise 55 of this section.

EXAMPLE 2 Find the measures of interior angles A, B, and C of the triangle in Figure 20, and the measure of exterior angle BCD.

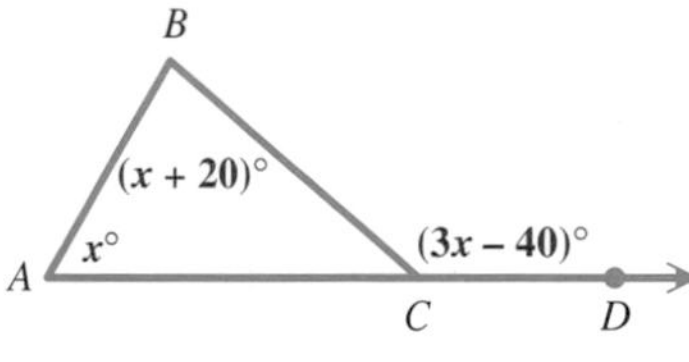

FIGURE 20

By the property concerning exterior angles, the sum of the measures of interior angles A and B must equal the measure of angle BCD. Thus,

$$\begin{aligned} x + (x + 20) &= 3x - 40 \\ 2x + 20 &= 3x - 40 \\ 2x - 3x &= -40 - 20 \\ -x &= -60 \\ x &= 60. \end{aligned}$$

Since the value of x is 60, we refer back to the figure to find the required angle measures:

$$\text{interior angle } A = 60^\circ$$
$$\text{interior angle } B = (60 + 20)^\circ = 80^\circ$$
$$\text{interior angle } C = 180^\circ - (60^\circ + 80^\circ) = 40^\circ$$
$$\text{exterior angle } BCD = [3(60) - 40]^\circ = 140^\circ.$$

Circles

One of the most important plane curves is the circle. It is a simple closed curve defined as follows.

Circle

A **circle** is a set of points in a plane, each of which is the same distance from a fixed point.

A circle may be physically constructed with a pair of compasses, where the spike leg remains fixed and the other leg swings around to construct the circle. A string may also be used to draw a circle. For example, loop a piece of chalk on one end of a piece of string. Hold the other end in a fixed position on a chalkboard, and pull the string taut. Then swing the chalk end around to draw the circle.

A circle, along with several lines and segments, is shown in Figure 21. The points P, Q, and R lie on the circle itself. Each lies the same distance from point O, which is called the **center** of the circle. (It is the "fixed point" referred to in the definition.) $\overline{OP}$, $\overline{OQ}$, and $\overline{OR}$ are segments whose endpoints are the center and a point on the circle. Each is called a **radius** of the circle (plural: **radii**). $\overline{PQ}$ is a segment whose endpoints both lie on the circle and is an example of a **chord.** $\overline{PR}$ is a chord that passes through the center and is called a **diameter** of the circle. Notice that the measure of a diameter is twice that of a radius.

$\overleftrightarrow{RT}$ is a line that touches (intersects) the circle in only one point, R, and is called a **tangent** to the circle. R is the point of tangency. Line $\overleftrightarrow{PQ}$, which intersects the circle in two points, is called a **secant** line. (What is the distinction between a chord and a secant?)

The portion of the circle shown in red is an example of an **arc** of the circle. It consists of two endpoints (P and Q) and all points on the circle "between"

Thales made his fortune merely to prove how easy it is to become wealthy; he cornered all the oil presses during a year of an exceptionally large olive crop. Legend records that Thales studied for a time in Egypt and then introduced geometry to Greece, where he attempted to apply the principles of Greek logic to his newly learned subject.

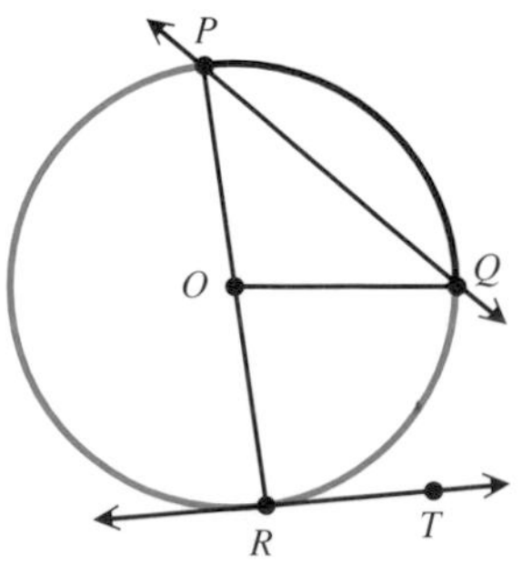

FIGURE 21

these endpoints. The colored portion is called arc *PQ* (or *QP*), denoted in symbols as $\overset{\frown}{PQ}$ (or $\overset{\frown}{QP}$).

A diameter such as $\overleftrightarrow{PR}$ in Figure 21 divides a circle into two parts of equal size, each of which is called a **semicircle.** As mentioned earlier, the Greeks were the first to insist that all propositions, or **theorems,** about geometry be given a rigorous proof before being accepted. According to tradition, the first theorem to receive such a proof was the following.

Any angle inscribed in a semicircle must be a right angle.

To be **inscribed** in a semicircle, the vertex of the angle must be on the circle with the sides of the angle going through the endpoints of the diameter at the base of the semicircle. (See Figure 22.) This first proof was said to have been given by the Greek philosopher Thales.

FIGURE 22

9.2 EXERCISES

Decide whether each statement in Exercises 1–10 is true *or* false.

1. A rhombus is an example of a regular polygon.

2. If a triangle is isosceles, then it is not scalene.

3. A triangle can have more than one obtuse angle.

4. A square is both a rectangle and a parallelogram.

5. A square must be a rhombus.

6. A rhombus must be a square.

7. A diameter of a circle is a chord of the circle.

8. The length of a diameter of a circle is twice the length of a radius of the same circle.

9. A triangle can have at most one right angle.

10. A rectangle must be a parallelogram, but a parallelogram might not be a rectangle.

11. In your own words, explain the distinction between a square and a rhombus.

12. What common traffic sign in the U.S. is in the shape of an octagon?

In Exercises 13–20, identify each curve as simple, closed, both, *or* neither.

13.

14.

15.

16.

17.

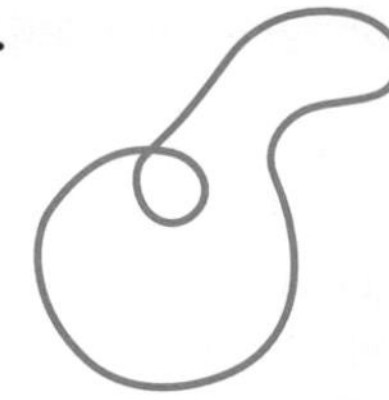

18.

19.

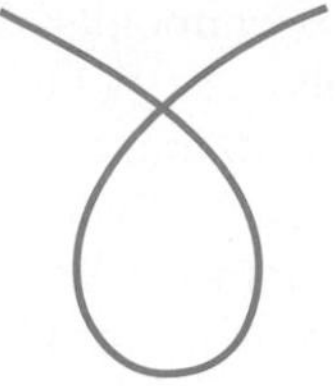

20.

In Exercises 21–26, decide whether each figure is convex *or* not convex.

21.

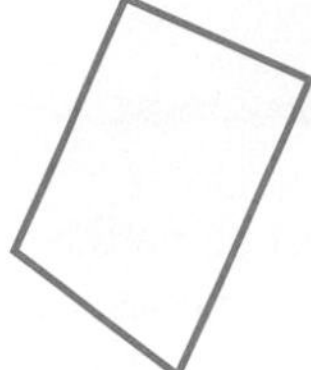

22.

23.

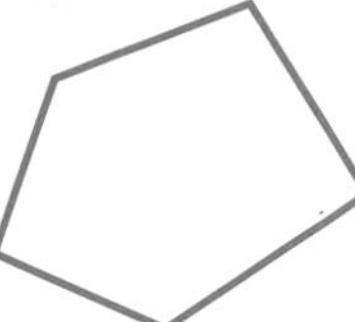

24.

25.

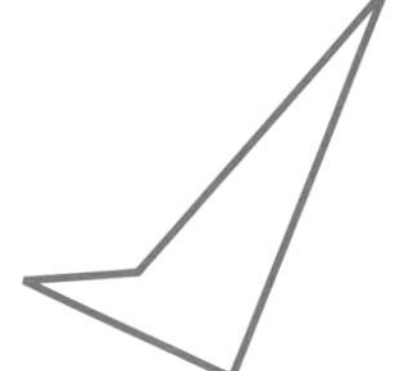

26.

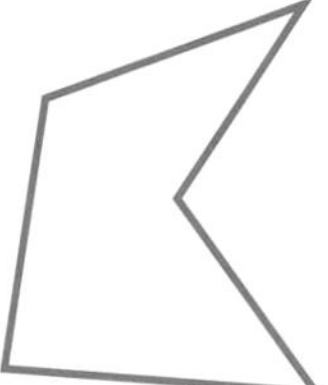

Classify each triangle in Exercises 27–38 as either acute, right, *or* obtuse. *Also classify each as either* equilateral, isosceles, *or* scalene.

27.

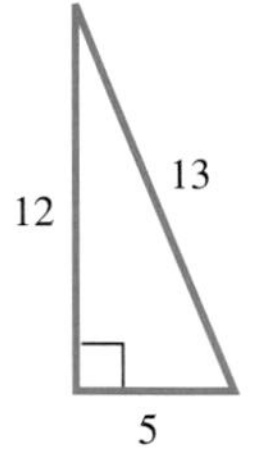

28.

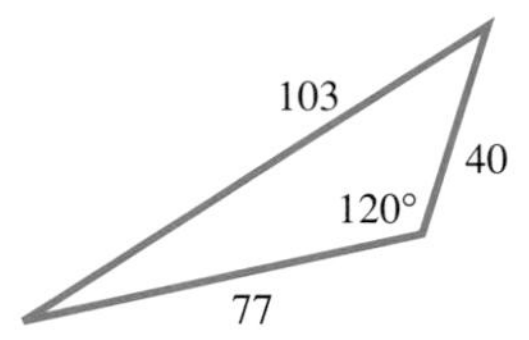

29.

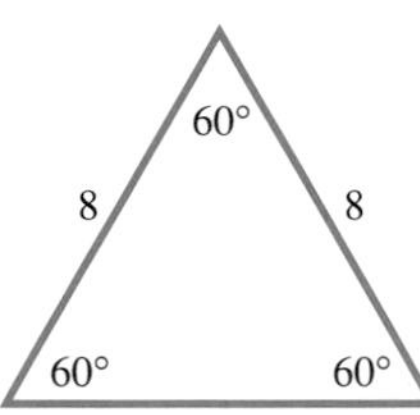

30.

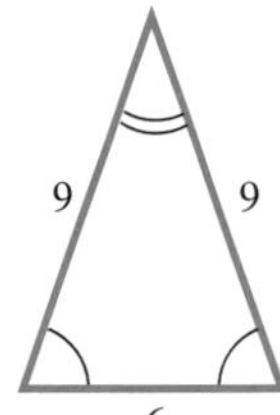

31.

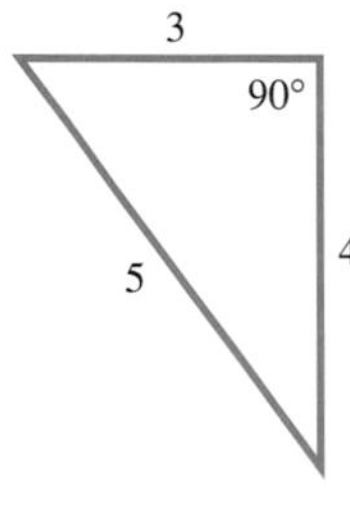

32.

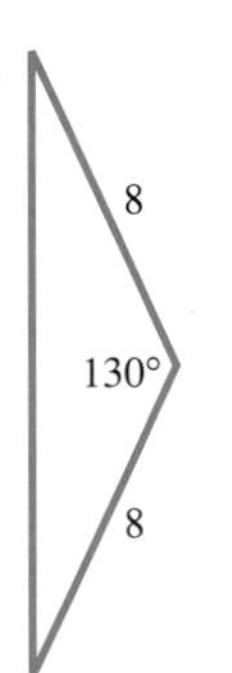

33.

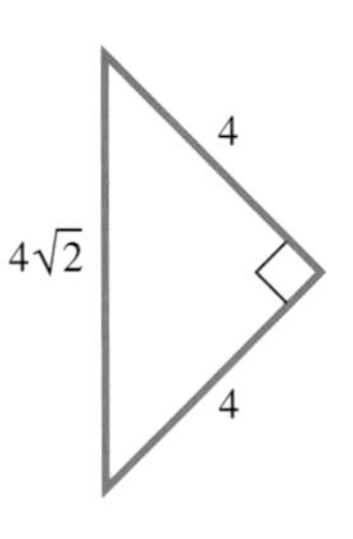

34.

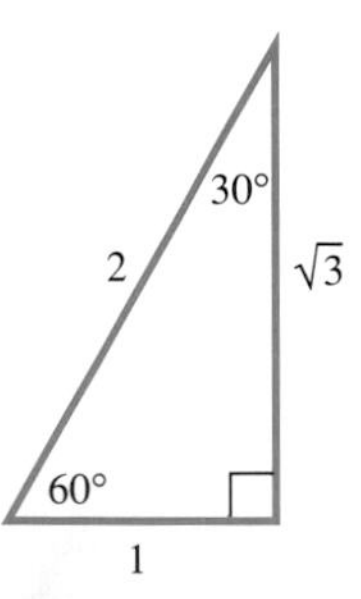

35.

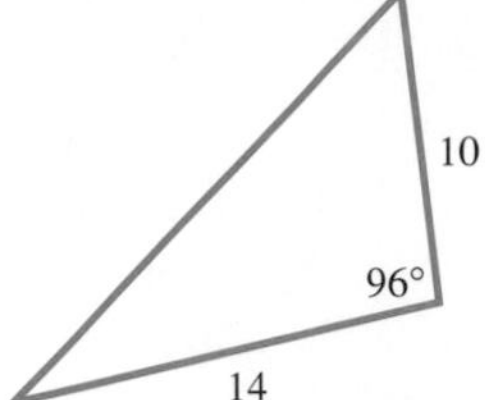

36.

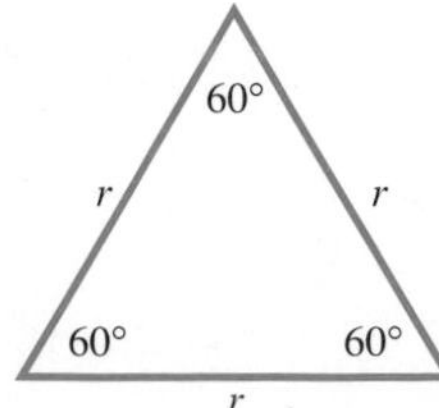

37.

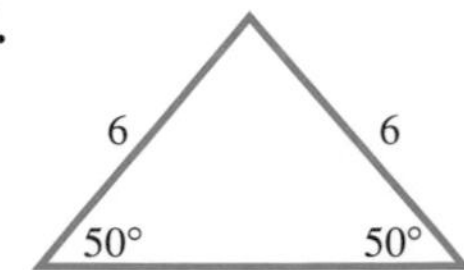

38.

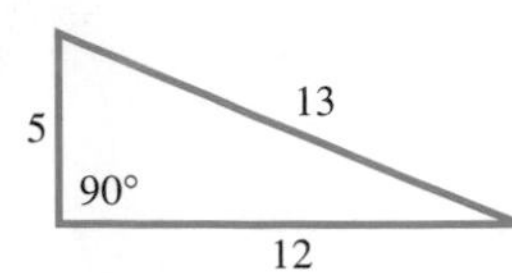

39. Write a definition of *isosceles right triangle.*

40. Explain why the sum of the lengths of any two sides of a triangle must be greater than the length of the third side.

41. Can a triangle be both right and obtuse? Explain.

42. In the classic 1939 movie *The Wizard of Oz,* the scarecrow, upon getting a brain, says the following: "The sum of the square roots of any two sides of an isosceles triangle is equal to the square root of the remaining side." Give an example to show that his statement is incorrect.

Find the measure of each angle in triangle ABC.

43.

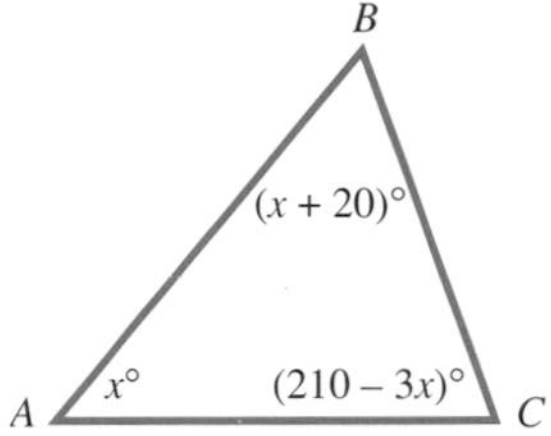

44.

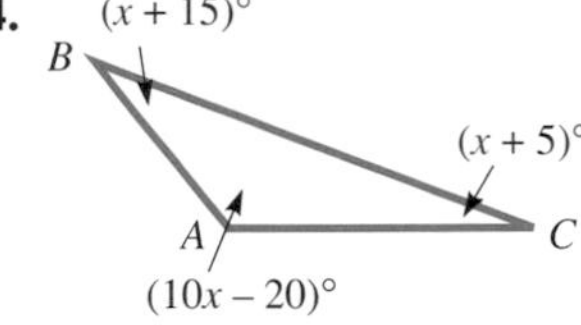

45.

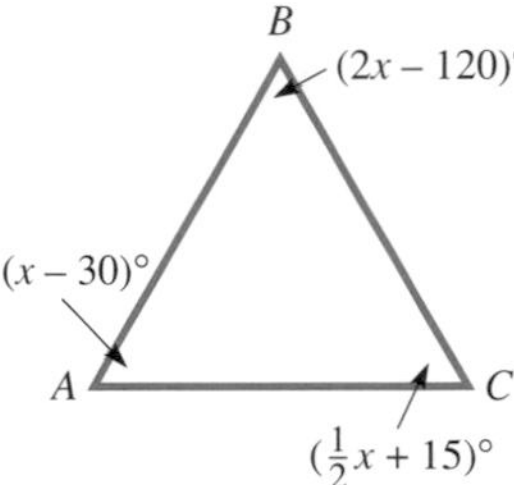

46.

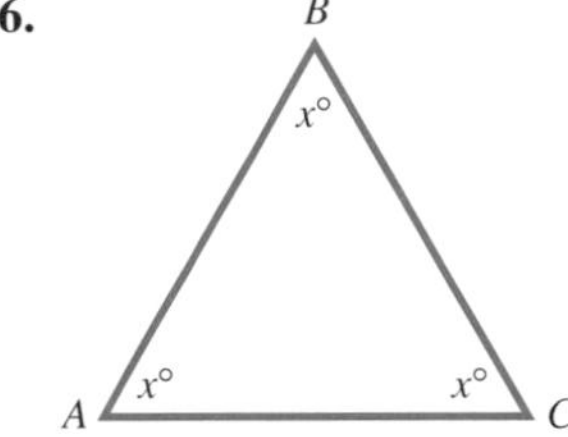

47. In triangle ABC, angles A and B have the same measure, while the measure of angle C is 24 degrees larger than the measure of each of A and B. What are the measures of the three angles?

48. In triangle ABC, the measure of angle A is 30 degrees more than the measure of angle B. The measure of angle B is the same as the measure of angle C. Find the measure of each angle.

In each triangle, find the measure of exterior angle BCD.

49.

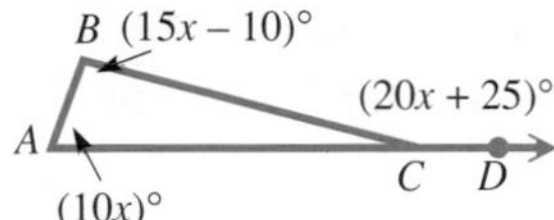

50.

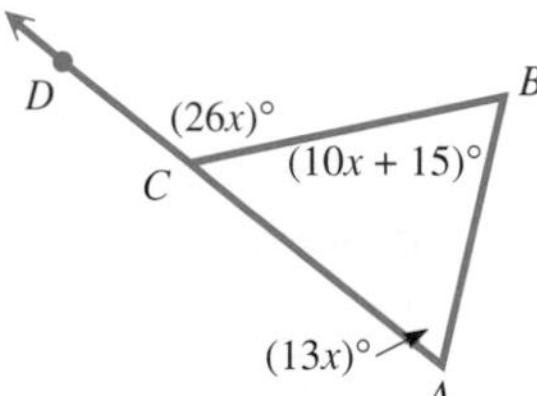

51.

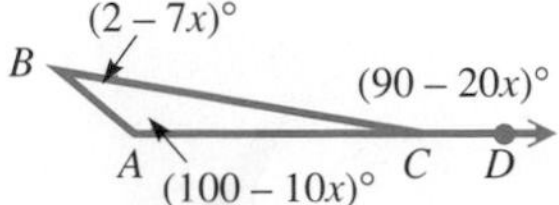

52.

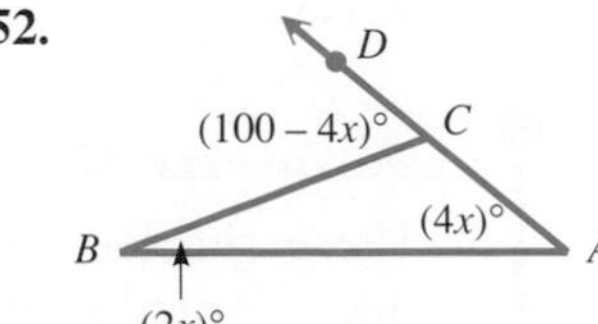

53. Go through the following argument provided by Richard Crouse in a letter to the editor of *Mathematics Teacher* in the February 1988 issue.

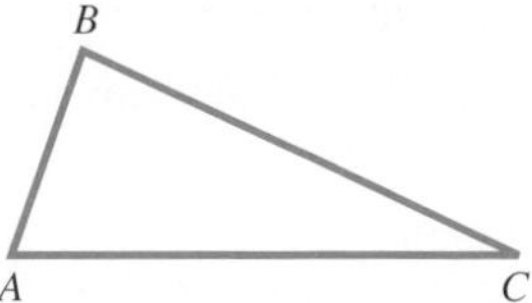

(a) Place the eraser end of a pencil on vertex A of the triangle and let the pencil coincide with side AC of the triangle.
(b) With the eraser fixed at A, rotate the pencil counterclockwise until it coincides with side AB.
(c) With the pencil fixed at point B, rotate the eraser end counterclockwise until the pencil coincides with side BC.
(d) With the pencil fixed at point C, rotate the point end of the pencil counterclockwise until the pencil coincides with side AC.
(e) Notice that the pencil is now pointing in the opposite direction. What concept from this section does this exercise reinforce?

54. Using the points, segments, and lines in the figure, list all parts of the circle.

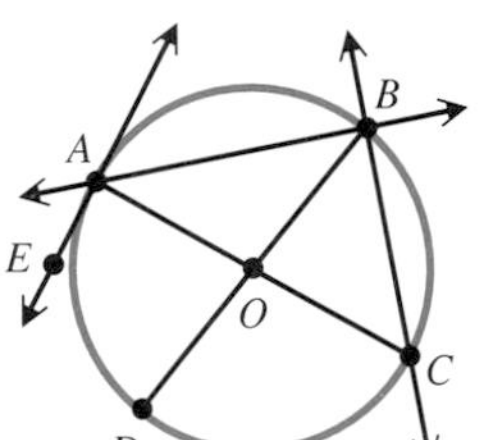

(a) center
(b) radii
(c) diameters
(d) chords
(e) secants
(f) tangents

55. Refer to angles 1, 2, and 6 in Figure 19. Prove that the sum of the measures of angles 1 and 2 is equal to the measure of angle 6.

9.3 Perimeter, Area, and Circumference

One of the most important uses of geometry in the world around us involves the measurement of objects of various shapes. In dealing with plane figures, we are often required to find the "distance around" the figure, or the amount of surface covered by the figure. These ideas are expressed as *perimeter* and *area.*

Perimeter and Area

The **perimeter** of a plane figure composed of line segments is the sum of the measures of the line segments. The **area** of a plane figure is the measure of the surface covered by the figure. Perimeter is measured in *linear units,* while area is measured in *square units.*

The simplest polygon is a triangle. If a triangle has sides of lengths a, b, and c, then to find its perimeter we simply find the sum of a, b, and c. This can be expressed as a formula, as shown below.

Perimeter of a Triangle

The perimeter P of a triangle with sides of lengths a, b, and c is given by the formula

$$P = a + b + c.$$

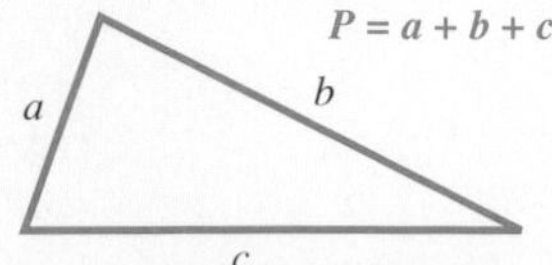

Since a rectangle is made up of two pairs of sides with each side in each pair equal in length, the formula for the perimeter of a rectangle may be stated as follows.

Perimeter of a Rectangle

The perimeter P of a rectangle with length ℓ and width w is given by the formula

$$P = 2\ell + 2w,$$

or equivalently,

$$P = 2(\ell + w).$$

$P = 2\ell + 2w$
or
$P = 2(\ell + w)$

EXAMPLE 1 A plot of land is in the shape of a rectangle. If it has length 50 feet and width 26 feet, how much fencing would be needed to completely enclose the plot?

Since we must find the distance around the plot of land, the formula for the perimeter of a rectangle is needed. Substitute 50 for ℓ and 26 for w to find P.

$$P = 2\ell + 2w$$
$$P = 2(\mathbf{50}) + 2(\mathbf{26}) \qquad \ell = 50,\ w = 26$$
$$P = 100 + 52$$
$$P = 152$$

The perimeter is 152 feet, so 152 feet of fencing is required. ■

A square is a rectangle with four sides of equal length. The formula for the perimeter of a square is a special case of the formula for the perimeter of a rectangle.

Perimeter of a Square

The perimeter P of a square with all sides of length s is given by the formula

$$P = 4s.$$

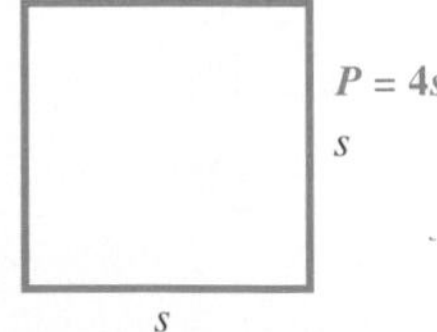

EXAMPLE 2 A square has perimeter 54 inches. What is the length of each side?

Use the formula $P = 4s$ with $P = 54$, and solve for s.

$$P = 4s$$
$$\mathbf{54} = 4s \qquad P = 54$$
$$s = 13.5 \qquad \text{Divide by 4.}$$

Each side has a measure of 13.5 inches. 

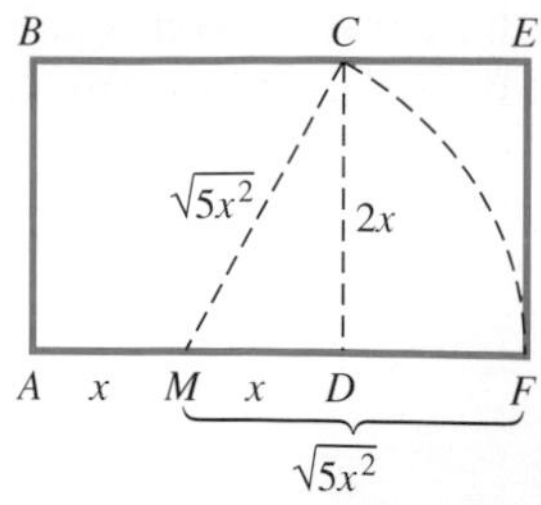

To construct a **golden rectangle,** one in which the ratio of the length to the width is equal to the ratio of the length plus the width to the length, begin with a square *ABCD*. With the spike of the compass at *M*, the midpoint of $\overleftrightarrow{AD}$, swing an arc of radius *MC* to intersect the extension of $\overleftrightarrow{AD}$ at *F*. Construct a perpendicular at *F*, and have it intersect the extension of $\overleftrightarrow{BC}$ at *E*. Then *ABEF* is a golden rectangle with golden ratio $(1 + \sqrt{5})/2$. (See Section 5.4 for more on the golden ratio.)

To verify this construction, let $AM = x$, so that $AD = 2x$. Then, by the Pythagorean theorem,

$$MC = \sqrt{x^2 + (2x)^2}$$
$$= \sqrt{x^2 + 4x^2} = \sqrt{5x^2}.$$

Since $\overset{\frown}{CF}$ is an arc of the circle with radius *MC*, $MF = MC = \sqrt{5x^2}$. Then the ratio of length *BE* to width *EF* is

$$\frac{BE}{EF} = \frac{x + \sqrt{5x^2}}{2x}$$
$$= \frac{x + x\sqrt{5}}{2x}$$
$$= \frac{x(1 + \sqrt{5})}{2x}$$
$$= \frac{1 + \sqrt{5}}{2}.$$

Problem Solving

The six-step method of solving an applied problem using algebra can be used to solve problems involving geometric figures. We will use these steps to solve the problem in the next example.

EXAMPLE 3 The length of a rectangular-shaped label is 1 centimeter more than twice the width. The perimeter is 110 centimeters. Find the length and the width.

Step 1: **Determine what you are asked to find.** We must find the length and the width. Let W represent the width. Then $1 + 2W$ can represent the length, since the length is 1 centimeter more than twice the width.

Step 2: **Write down any other pertinent information.** Figure 23 shows a diagram of the label. The perimeter formula that we will use is $P = 2\ell + 2w$.

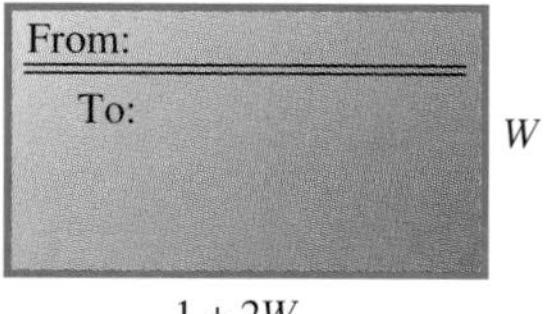

FIGURE 23

Step 3: **Write an equation.** In the formula $P = 2\ell + 2w$, replace w with W, ℓ with $1 + 2W$, and P with 110, since the perimeter is 110 centimeters.

$$110 = 2(1 + 2W) + 2W$$

Step 4: **Solve the equation.**

$110 = 2 + 4W + 2W$	Distributive property
$110 = 2 + 6W$	Combine terms.
$108 = 6W$	Subtract 2.
$18 = W$	Divide by 6.

Step 5: **Answer the question(s) of the problem.** Since $W = 18$, the width is 18 centimeters and the length is $1 + 2W = 1 + 2(18) = 37$ centimeters.

Step 6: **Check.** Because 37 is 1 more than twice 18, and because $2(37) + 2(18) = 110$, the answer is correct.

Defining the **area** of a figure requires a basic *unit of area.* One that is commonly used is the *square centimeter,* abbreviated cm^2. One square centimeter, or 1 cm^2, is the area of a square one centimeter on a side. In place of 1 cm^2, the basic unit of area could have been 1 in.^{2}, 1 ft^2, 1 m^2, or any appropriate unit.

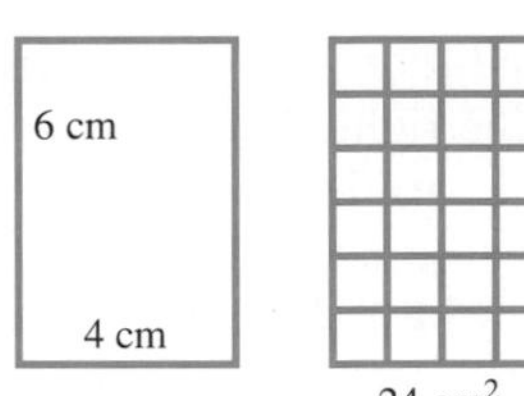

■ **FIGURE 24**

As an example, let us calculate the area of the rectangle shown in Figure 24. Using the basic 1 cm^2 unit, Figure 24 shows that 4 squares, each 1 cm on a side, can be laid off horizontally while 6 such squares can be laid off vertically. A total of $24 = 4 \cdot 6$ of the small squares are needed to cover the large rectangle. Thus, the area of the large rectangle is 24 cm^2.

We can generalize the above illustration to obtain a formula for the area of a rectangle.

Area of a Rectangle

The area A of a rectangle with length ℓ and width w is given by the formula

$$A = \ell w.$$

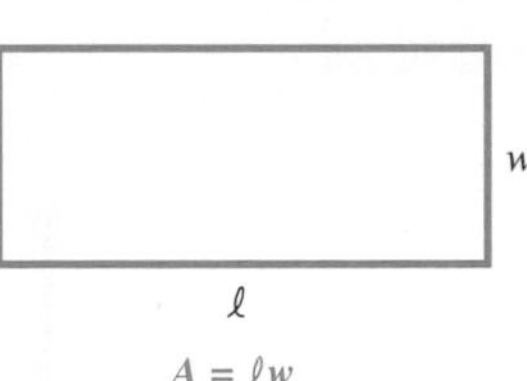

The formula for the area of a rectangle can be used to find formulas for the areas of other figures. For example, a square is a rectangle having all sides equal. The area of a square can be found by using the formula for the area of a rectangle, $A = \ell w$. If the letter s represents the equal lengths of the sides of the square, then $A = s \cdot s = s^2$.

Area of a Square

The area A of a square with all sides of length s is given by the formula

$$A = s^2.$$

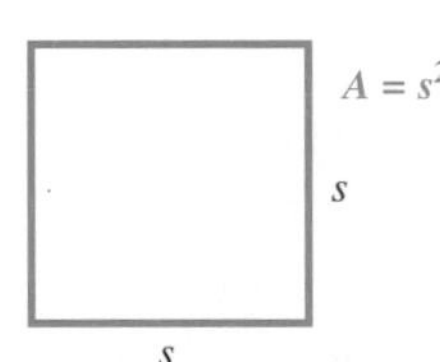

■ **FIGURE 25**

EXAMPLE 4 Figure 25 shows the floor plan of a building, made up of various rectangles. If each length given is in meters, how many square meters of carpet would be required to supply the building?

The dotted lines in the figure break up the floor area into rectangles. The areas of the various rectangles that result are

$$10 \cdot 12 = 120 \text{ m}^2$$
$$3 \cdot 10 = 30 \text{ m}^2$$
$$3 \cdot 7 = 21 \text{ m}^2$$
$$15 \cdot 25 = 375 \text{ m}^2.$$

The total area is $120 + 30 + 21 + 375 = 546 \text{ m}^2$, so 546 square meters of carpet would be needed. ◆

As mentioned earlier in this chapter, a **parallelogram** is a four-sided figure having both pairs of opposite sides parallel. Since a parallelogram need not be a rectangle, the formula for the area of a rectangle cannot be used directly for

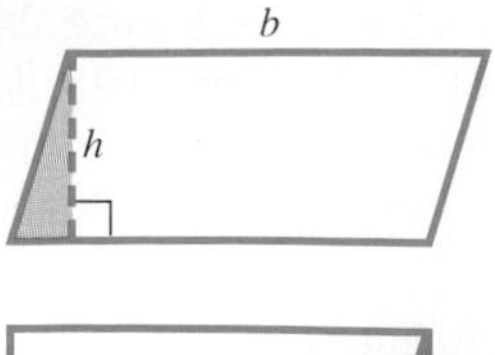

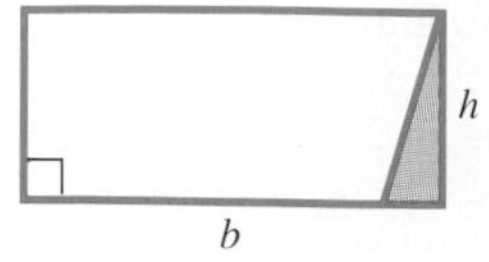

■ **FIGURE 26**

a parallelogram. However, this formula can be used indirectly, as shown in Figure 26. Cut off the triangle in color, and attach it at the right. The resulting figure is a rectangle having the same area as the original parallelogram.

The *height* of the parallelogram is the perpendicular distance between two of the parallel sides, and is denoted by h in the figure. The width of the rectangle equals the height of the parallelogram, and the length of the rectangle is the base b of the parallelogram, so that

$$A = \text{length} \cdot \text{width}$$

$$A = \text{base} \cdot \text{height}.$$

This leads to the following formula.

Area of a Parallelogram

The area A of a parallelogram with height h and base b is given by the formula

$$A = bh.$$

Note: h is not the length of a side.

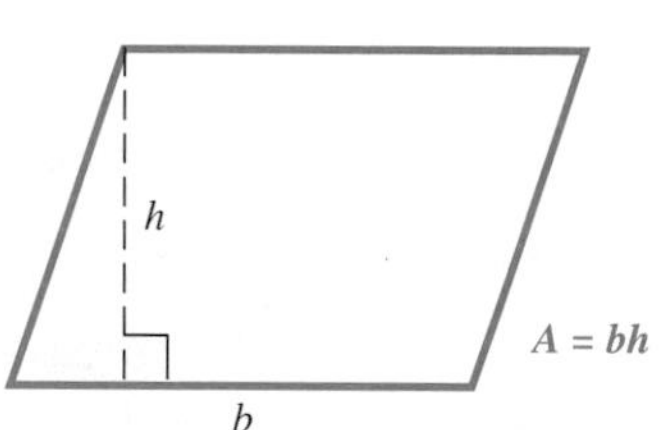

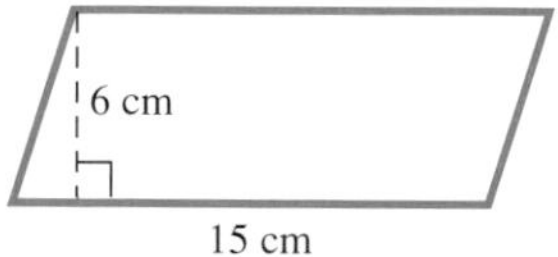

■ **FIGURE 27**

EXAMPLE 5 Find the area of the parallelogram in Figure 27.

Here the base has a length of 15 centimeters, while the height is 6 centimeters. Thus, $b = 15$ and $h = 6$.

$$A = bh$$

$$A = 15 \cdot 6$$

$$A = 90$$

The area is 90 cm^2. ■

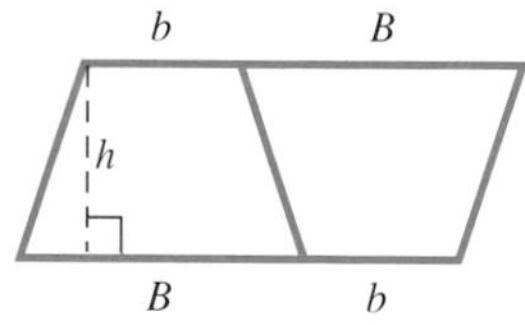

■ **FIGURE 28**

Figure 28 shows how we can find a formula for the area of a trapezoid. Notice that the figure as a whole is a parallelogram. It is made up of two trapezoids, each of which has height h, shorter base b, and longer base B. The area of the parallelogram is found by multiplying the height h by the base of the parallelogram, $b + B$.

$$\text{area of the parallelogram} = h(b + B)$$

Since the area of the parallelogram is twice the area of each trapezoid, the area of each trapezoid is half the area of the parallelogram, Thus

$$\text{area of each trapezoid} = \frac{1}{2}h(b + B).$$

Area of a Trapezoid

The area A of a trapezoid with parallel bases b and B and height h is given by the formula

$$A = \frac{1}{2}h(b + B).$$

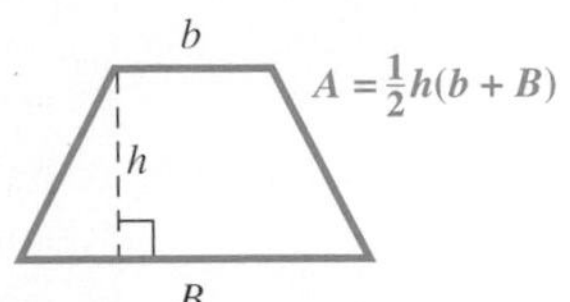

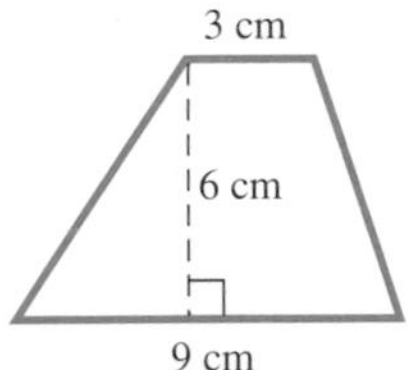

FIGURE 29

EXAMPLE 6 Find the area of the trapezoid in Figure 29.

Here $h = 6$, $b = 3$, and $B = 9$.

Thus,

$$A = \frac{1}{2}h(B + b)$$

$$A = \frac{1}{2}(6)(9 + 3)$$

$$A = \frac{1}{2}(6)(12)$$

$$A = 36.$$

The area of the trapezoid is 36 cm^2.

The formula for the area of a triangle can be found from the formula for the area of a parallelogram. In Figure 30 the triangle whose vertices are at A, B, and C has been broken into two parts, one shown in color and one shown in gray. Repeating the part shown in color and the part in gray gives a parallelogram. The area of this parallelogram is A = base · height, or $A = bh$. However, the parallelogram has *twice* the area of the triangle; in other words, the area of the triangle is *half* the area of the parallelogram.

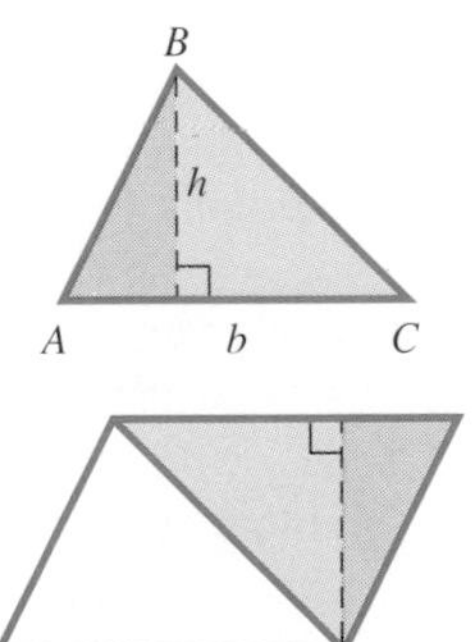

FIGURE 30

Area of a Triangle

The area A of a triangle with height h and base b is given by the formula

$$A = \frac{1}{2}bh.$$

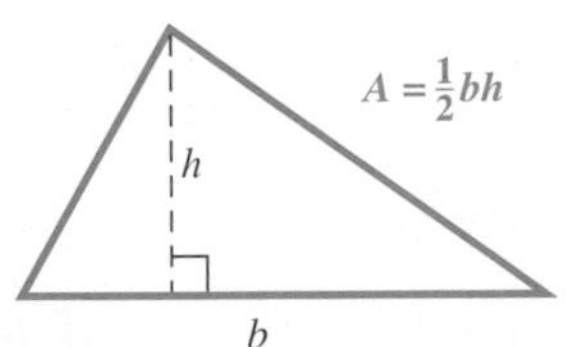

When applying the formula for the area of a triangle, remember that the height is the perpendicular distance between a vertex and the opposite side (or the extension of that side). See Figure 31.

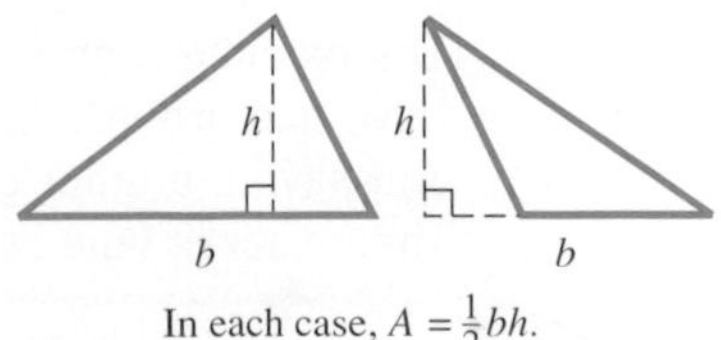

In each case, $A = \frac{1}{2}bh$.

FIGURE 31

For Further Thought

The most commonly used formula for the area of a triangle, $A = (1/2)bh$, is given in this section. In order to use it, we must know the height to one of the sides of the triangle. Suppose, however, that we know only the lengths of the three sides. Is there a way to determine the area from only this information?

The answer is yes, and it leads us to the formula known as Heron's formula. Heron of Alexandria lived during the second half of the first century A.D., and although the formula is named after him, there is evidence that it was known to Archimedes several centuries earlier.

Heron's Area Formula

Let a, b, and c be lengths of the sides of any triangle. Let $s = (1/2)(a + b + c)$ represent the semiperimeter. Then the area A of the triangle is given by the formula

$$A = \sqrt{s(s - a)(s - b)(s - c)}.$$

For example, if the sides of a triangle measure 6, 8, and 12 feet, we have $s = (1/2)(6 + 8 + 12) = 13$, and so the area of the triangle is A, where

$$A = \sqrt{13(13 - 6)(13 - 8)(13 - 12)} = \sqrt{455} \approx 21.33 \text{ square feet.}$$

For Group Discussion

1. If your class is held in a rectangular-shaped room, measure the length and the width, and then multiply them to find the area. Now, measure a diagonal of the room, and use Heron's formula with the length, the width, and the diagonal to find the area of half the room. Double this result. Do your area calculations agree?
2. For a triangle to exist, the sum of the lengths of any two sides must exceed the length of the remaining side. Have half of the class try to calculate the area of a "triangle" with $a = 4$, $b = 8$, and $c = 12$, while the other half tries to calculate the area of the "triangle" with $a = 10$, $b = 20$, and $c = 34$. In both cases, use Heron's formula. Then, discuss the results, drawing diagrams on the chalkboard to support the results obtained.
3. A popular textbook for mathematics survey courses contains the following problem and diagram.

   Find the perimeter and area of the shaded region.

   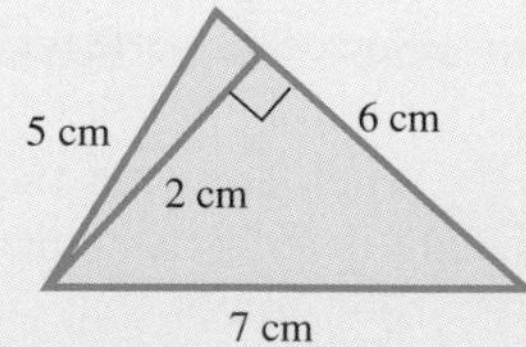

   The perimeter, obviously, is 18 cm. Divide the class into two groups. Have one group determine the area using area = (1/2)(base)(height), and have the other group determine the area using Heron's formula. Then discuss your results. (Does something seem wrong here?)

The distance around a circle is called its **circumference** (rather than "perimeter"). A simple experiment can aid in understanding the formula for the circumference of a circle. Use a piece of string to measure the distance around a circle. Measure its diameter and then divide the circumference by the diameter. It turns out that this quotient is the same, no matter what the size of the circle. The result of this measurement is an approximation for the number π. We have

$$\pi = \frac{\text{circumference}}{\text{diameter}} = \frac{C}{d}.$$

An alternate version of this formula is

$$C = \pi d.$$

Since the diameter of a circle measures twice the radius, we have $d = 2r$. These relationships allow us to state the following formulas for the circumference of a circle.

Circumference of a Circle

The circumference C of a circle of diameter d is given by the formula

$$C = \pi d.$$

Also, the circumference C of a circle of radius r is given by the formula

$$C = 2\pi r.$$

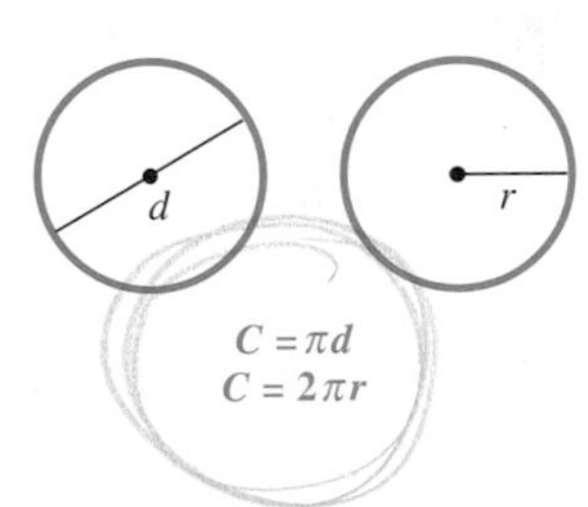

Recall that π is not a rational number. In this chapter we will use 3.14 as an approximation for π when one is required.

EXAMPLE 7 **(a)** A circle has a diameter of 12.6 cm. Find its circumference. Use $\pi \approx 3.14$.

Since $C = \pi d$,

$$C \approx 3.14(12.6) = 39.564$$

or 39.6 centimeters, rounded to the nearest tenth.

(b) The radius of a circle is 1.7 m. Find its circumference, using $\pi \approx 3.14$.

$$C = 2\pi r$$
$$C \approx 2(3.14)(1.7)$$
$$C \approx 10.7$$

The circumference is approximately 10.7 meters. ■

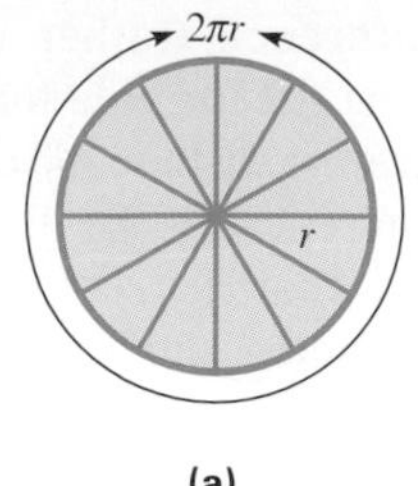

(a)

πr

r

$A \approx \pi r^2$

(b)

FIGURE 32

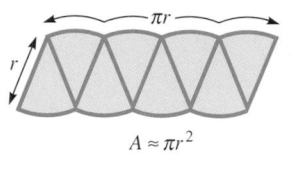

The formula for the area of a circle can be justified as follows. Start with a circle as shown in Figure 32(a), divided into many equal pie-shaped pieces **(sectors).** Rearrange the pieces into an approximate rectangle as shown in Figure 32(b). The circle has circumference $2\pi r$, so the "length" of the approximate rectangle is one-half of the circumference, or $(1/2)(2\pi r) = \pi r$, while its "width" is r. The area of the approximate rectangle is length times width, or $(\pi r)r = \pi r^2$. By choosing smaller and smaller sectors, the figure becomes closer and closer to a rectangle, so its area becomes closer and closer to πr^2. This "limiting" procedure leads to the following formula.

Area of a Circle

The area A of a circle with radius r is given by the formula

$$A = \pi r^2.$$

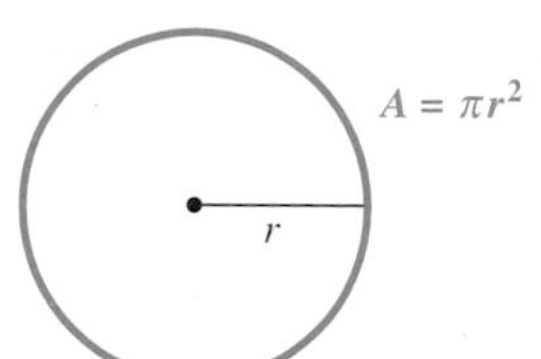

Problem Solving

The formula for the area of a circle can be used to determine the best value for the money the next time you purchase a pizza. The next example uses the idea of unit pricing.

EXAMPLE 8 Checkers delivers pizza. The price of an 8-inch diameter pepperoni pizza is \$6.99, while the price of a 16-inch diameter pizza is \$13.98. Which is the better buy?

To determine which pizza is the better value for the money, we must first find the area of each, and divide the price by the area to determine the price per square inch.

$$\text{8-inch diameter pizza area} = \pi(4)^2 \approx 50.24 \text{ in.}^2$$

↑ Radius is (1/2)(8) = 4 inches.

$$\text{16-inch diameter pizza area} = \pi(8)^2 \approx 200.96 \text{ in.}^2$$

↑ Radius is (1/2)(16) = 8 inches.

The price per square inch for the 8-inch pizza is $\$6.99/50.24 \approx 13.9¢$, while the price per square inch for the 16-inch pizza is $\$13.98/200.96 \approx 7.0¢$. Therefore, the 16-inch pizza is the better buy, since it costs approximately half as much per square inch. ■

9.3 EXERCISES

Decide whether each statement is true *or* false.

1. The perimeter of an equilateral triangle with side equal to 6 in. is the same as the perimeter of a rectangle with length 7 in. and width 2 in.
2. The area of a circle with radius r inches is greater than the area of a square with side r inches.

3. A square with area 16 cm^2 has a perimeter of 16 cm.

4. If the area of a certain triangle is 24 square units, and the base is 8 units, then the height must be 3 units.

5. The area of a circle with radius r inches is doubled if the radius is doubled.

6. The perimeter of a rectangle is doubled if both the length and the width are doubled.

Use the formulas of this section to find the area of each figure. In Exercises 19–22, use 3.14 as an approximation for π.

7.

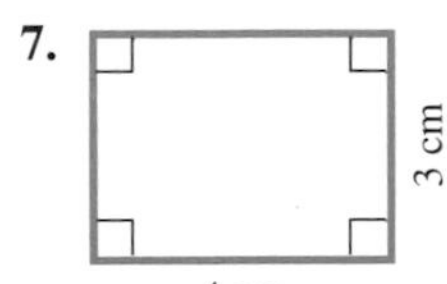

8.

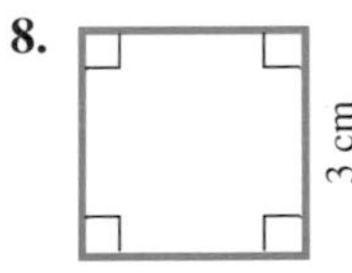

9.

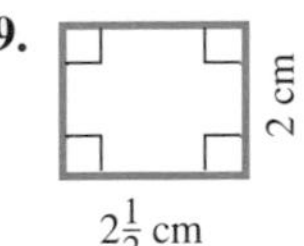

10.

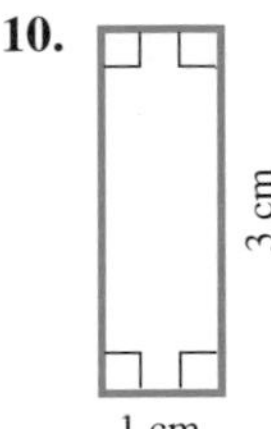

11.

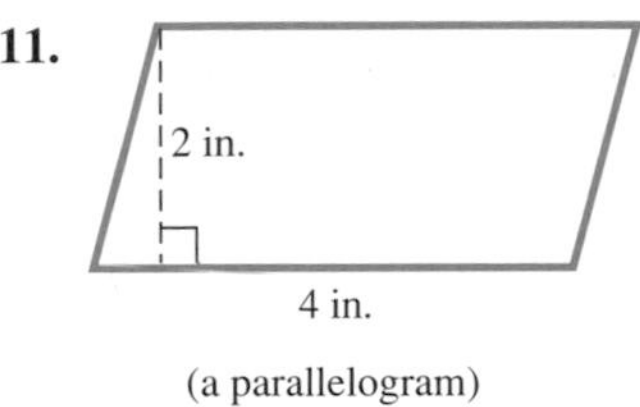

(a parallelogram)

12.

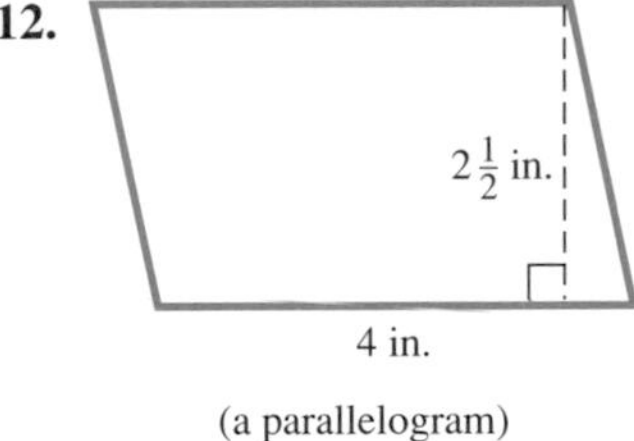

(a parallelogram)

13.

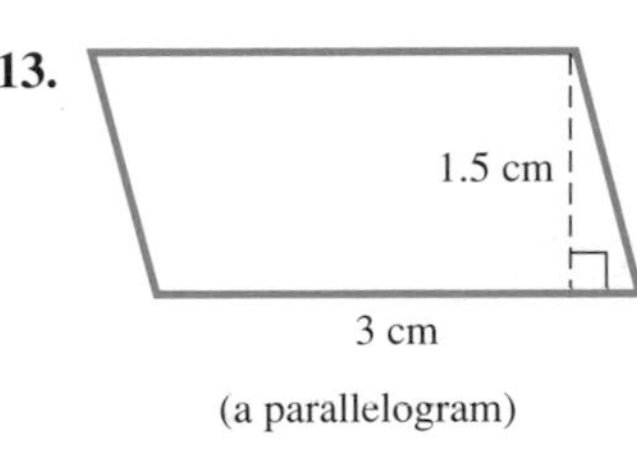

(a parallelogram)

14.

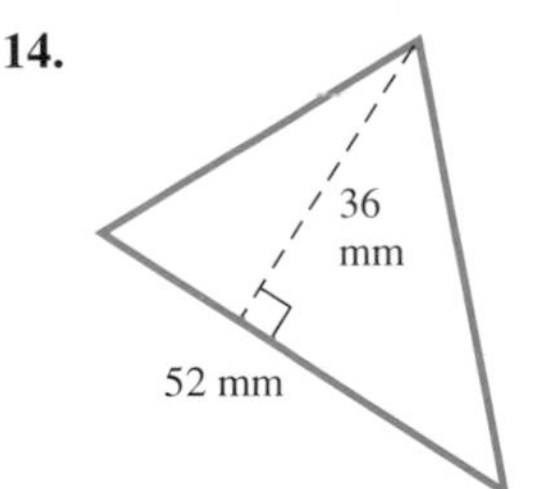

15.

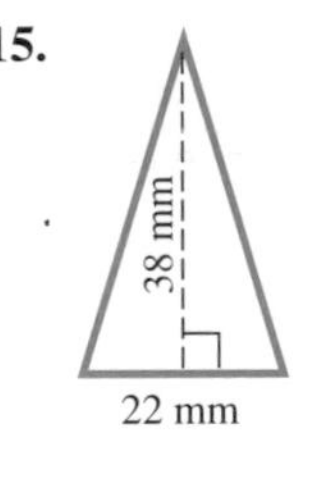

16.

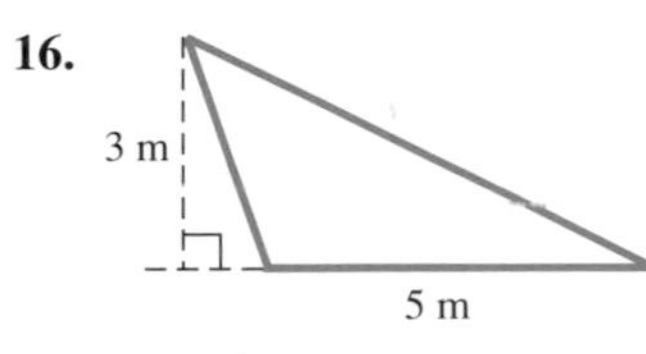

17.

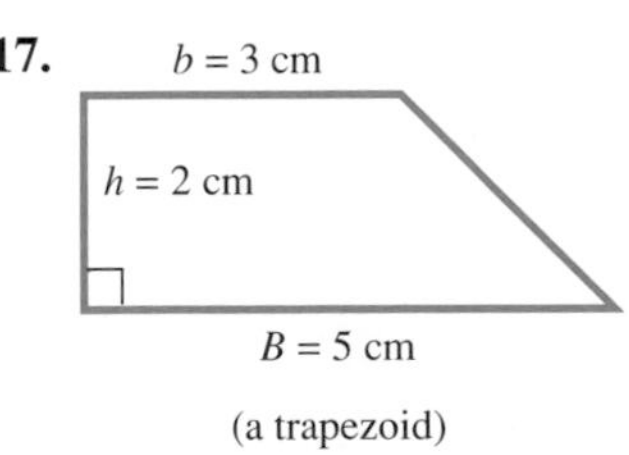

(a trapezoid)

18.

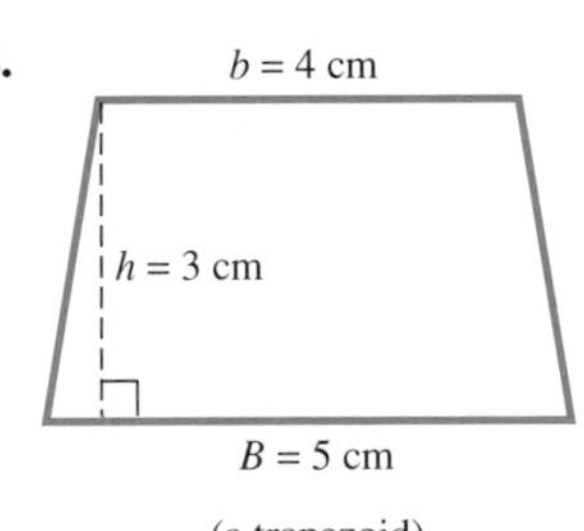

(a trapezoid)

19.

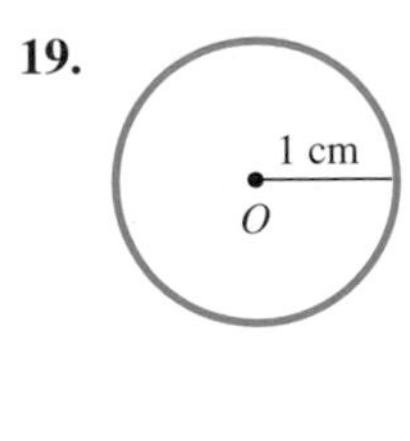

20.

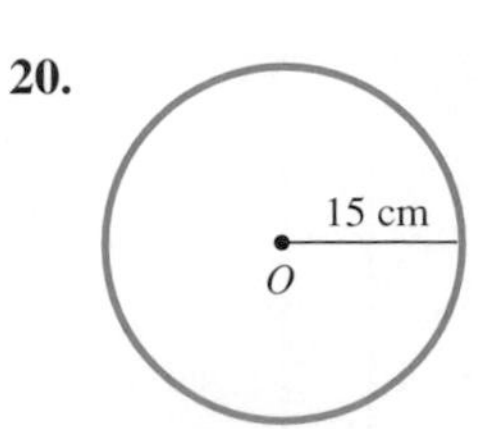

21.

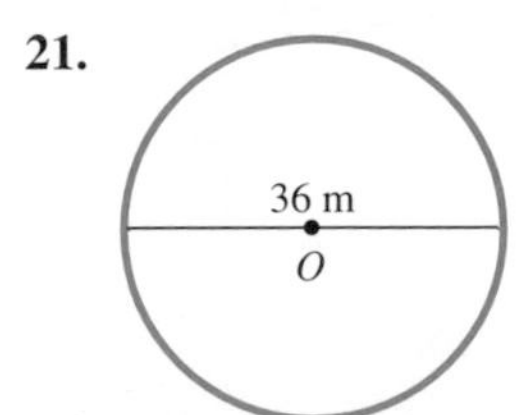

22.

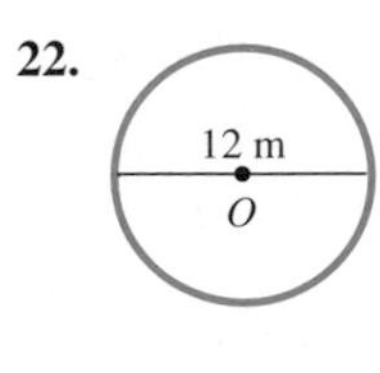

Solve each of the following problems.

23. A stained-glass window in a church is in the shape of a square. The perimeter of the square is 7 times the length of a side in meters, decreased by 12. Find the length of a side of the window.

24. A video rental establishment displayed a rectangular cardboard standup advertisement for the movie *Forrest Gump*. The length was 20 in. more than the width, and the perimeter was 176 in. What were the dimensions of the rectangle?

25. A lot is in the shape of a triangle. One side is 100 ft longer than the shortest side, while the third side is 200 ft longer than the shortest side. The perimeter of the lot is 1,200 ft. Find the lengths of the sides of the lot.

26. A wall pennant is in the shape of an isosceles triangle. (Two sides have the same length.) Each of the two equal sides measures 18 in. more than the third side, and the perimeter of the triangle is 54 in. What are the lengths of the sides of the pennant?

27. The Peachtree Plaza Hotel in Atlanta is in the shape of a cylinder, with a circular foundation. The circumference of the foundation is 6 times the radius, increased by 12.88 ft. Find the radius of the circular foundation. (Use 3.14 as an approximation for π.)

28. If the radius of a certain circle is tripled, with 8.2 cm then added, the result is the circumference of the circle. Find the radius of the circle. (Use 3.14 as an approximation for π.)

29. The survey plat in the figure shows two lots that form a trapezoid. The measures of the parallel sides are 115.80 ft and 171.00 ft. The height of the trapezoid is 165.97 ft. Find the combined area of the two lots. Round your answer to the nearest hundredth of a square foot.

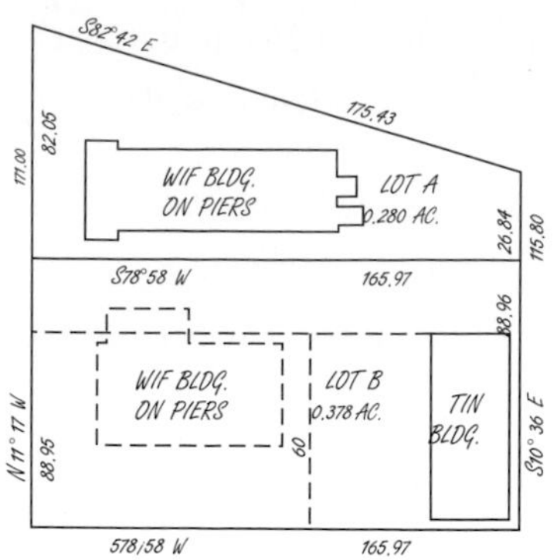

30. Lot A in the figure is in the shape of a trapezoid. The parallel sides measure 26.84 ft and 82.05 ft. The height of the trapezoid is 165.97 ft. Find the area of Lot A. Round your answer to the nearest hundredth of a square foot.

31. In order to purchase fencing to go around a rectangular yard, would you need to use perimeter or area to decide how much to buy?

32. In order to purchase fertilizer for the lawn of a yard, would you need to use perimeter or area to decide how much to buy?

In the chart below, one of the values r (radius), d (diameter), C (circumference), or A (area) is given for a particular circle. Find the remaining three values. Leave π in your answers.

	r	d	C	A
33.	6 in.			
34.	9 in.			
35.		10 ft		
36.		40 ft		
37.			12π cm	
38.			18π cm	
39.				100π in.^{2}
40.				256π in.^{2}

Each of the following figures has perimeter as indicated. Find the value of x.

41. $P = 58$

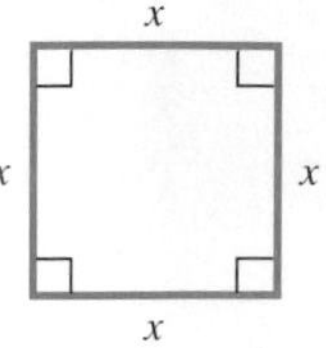

42. $P = 42$

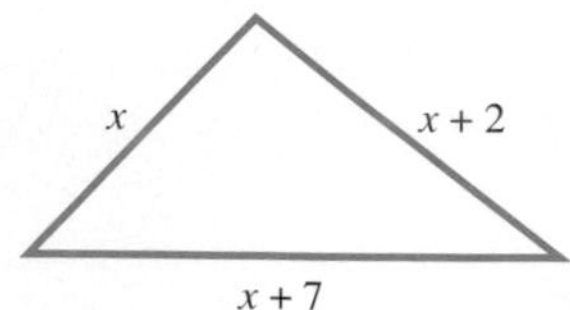

43. $P = 38$

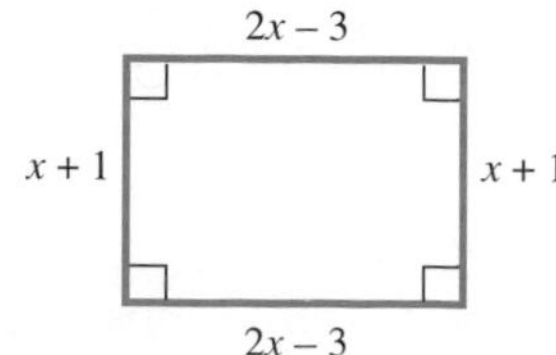

44. $P = 278$

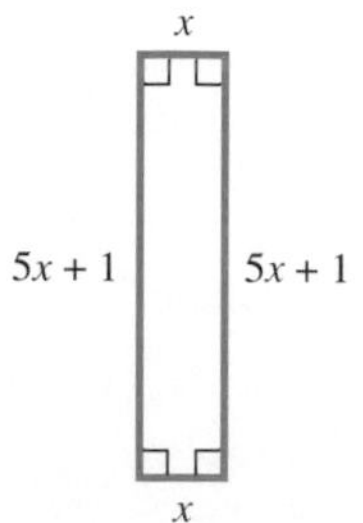

Each of the following figures has area as indicated. Find the value of x.

45. $A = 26.01$

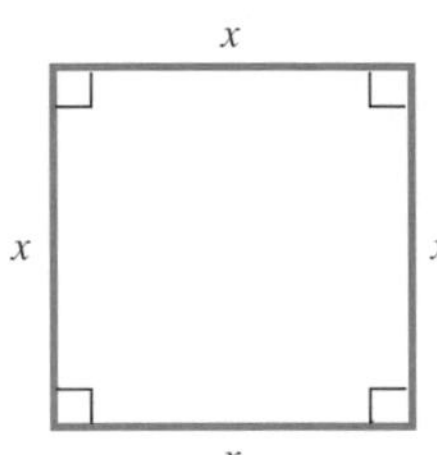

46. $A = 28$

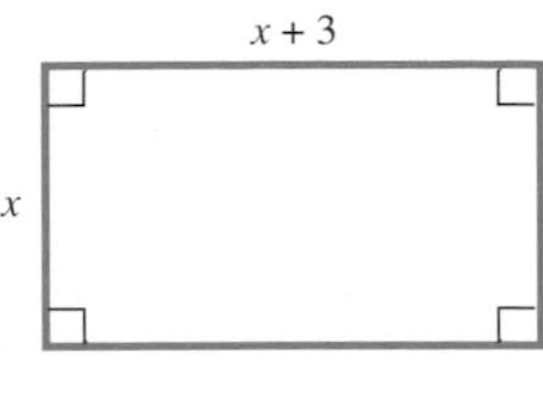

47. $A = 15$

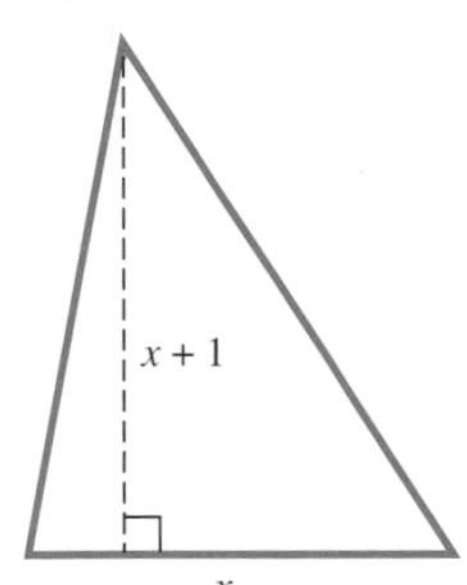

48. $A = 30$

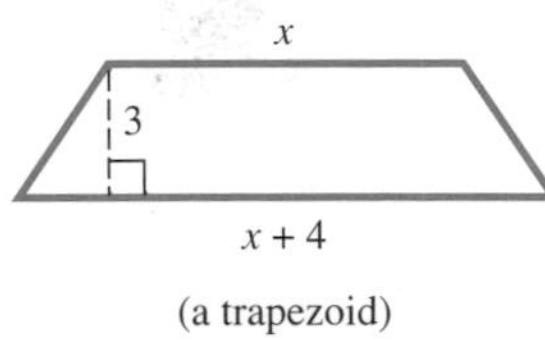

(a trapezoid)

Each of the following circles has circumference or area as indicated. Find the value of x. Use 3.14 as an approximation for π.

49. $C = 37.68$

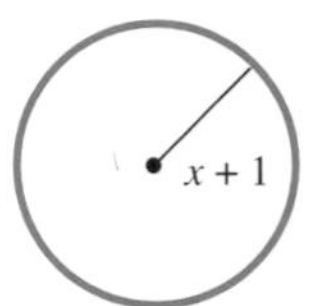

50. $C = 54.95$

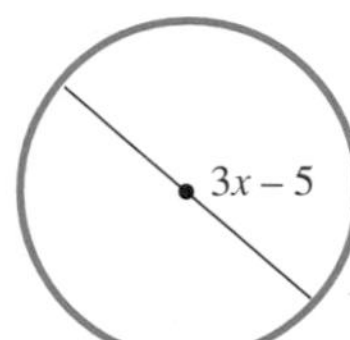

51. $A = 18.0864$

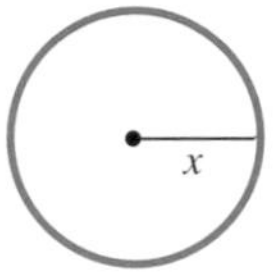

52. $A = 28.26$

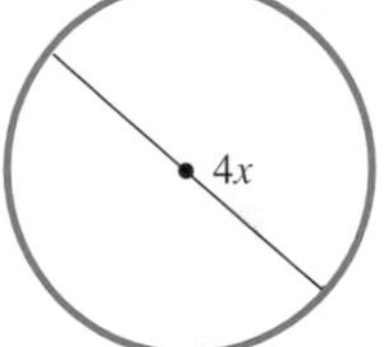

53. Work through the parts of this exercise in order, and use it to make a generalization concerning areas of rectangles.

(a) Find the area of a rectangle 4 cm by 5 cm.
(b) Find the area of a rectangle 8 cm by 10 cm.
(c) Find the area of a rectangle 12 cm by 15 cm.
(d) Find the area of a rectangle 16 cm by 20 cm.
(e) The rectangle in part (b) had sides twice as long as the sides of the rectangle in part (a). Divide the larger area by the smaller. By doubling the sides, the area increased _____ times.
(f) To get the rectangle in part (c) each side of the rectangle of part (a) was multiplied by _____. This made the larger area _____ times the size of the smaller area.
(g) To get the rectangle of part (d) each side of the rectangle of part (a) was multiplied by _____. This made the area increase to _____ times what it was originally.
(h) In general, if the length of each side of a rectangle is multiplied by n, the area is multiplied by _____.

Use the results of Exercise 53 to solve each of the following.

54. A ceiling measuring 9 ft by 15 ft can be painted for $60. How much would it cost to paint a ceiling 18 ft by 30 ft?

55. Suppose carpet for a 10 ft by 12 ft room costs $200. Find the cost to carpet a room 20 ft by 24 ft.

56. A carpet cleaner charges $80 to do an area 31 ft by 31 ft. What would be the charge for an area 93 ft by 93 ft?

57. Use the logic of Exercise 53 to answer the following: If the radius of a circle is multiplied by n, then the area of the circle is multiplied by _____.

58. Use the logic of Exercise 53 to answer the following: If the height of a triangle is multiplied by n and the base length remains the same, then the area of the triangle is multiplied by _____.

By considering total area as the sum of the areas of all of its parts, areas of figures such as those in Exercises 59–62 may be determined. Find the total area of each figure. Use 3.14 as an approximation for π in Exercises 61 and 62.

59.

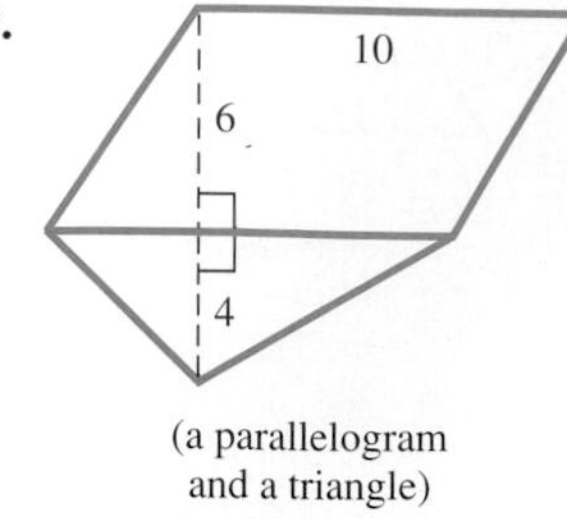

(a parallelogram and a triangle)

60.

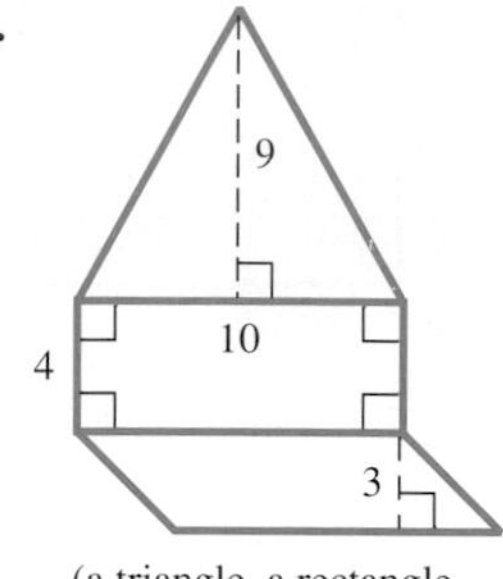

(a triangle, a rectangle, and a parallelogram)

61.

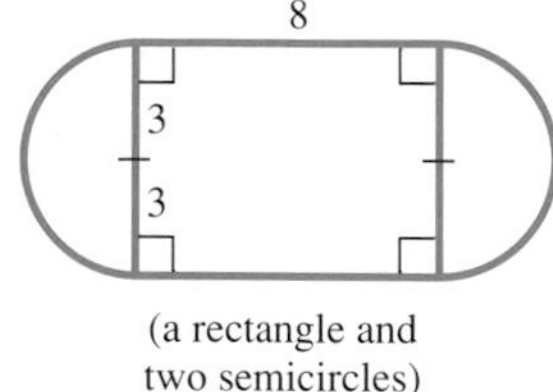

(a rectangle and two semicircles)

62.

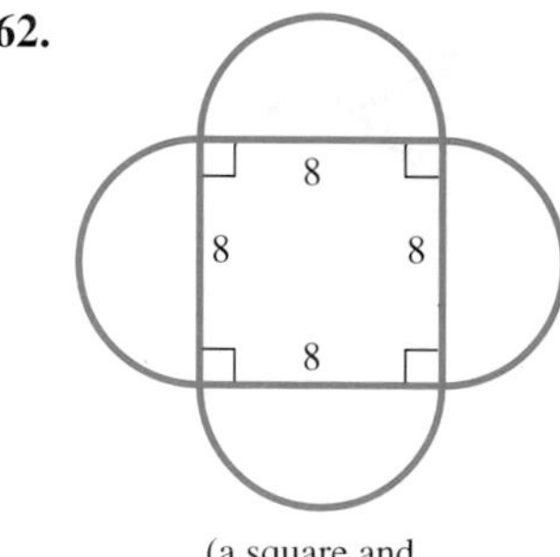

(a square and four semicircles)

The shaded areas of the figures in Exercises 63–68 may be found by subtracting the area of the unshaded portion from the total area of the figure. Use this approach to find the area of the shaded portion. Use 3.14 as an approximation for π in Exercises 66–68, and round to the nearest hundredth.

63.

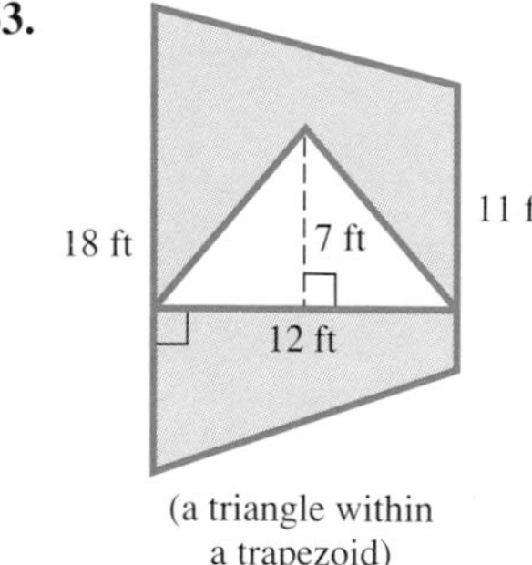

(a triangle within a trapezoid)

64.

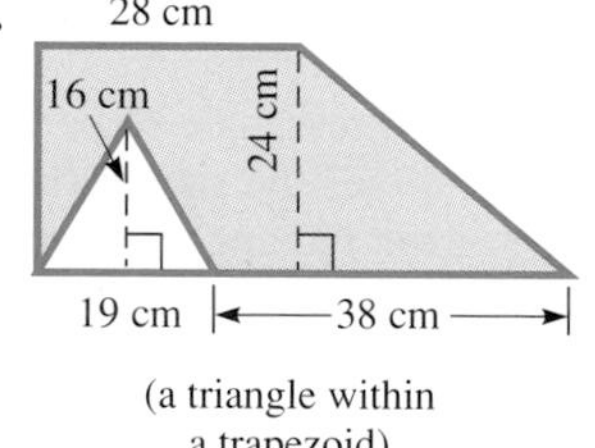

(a triangle within a trapezoid)

65.

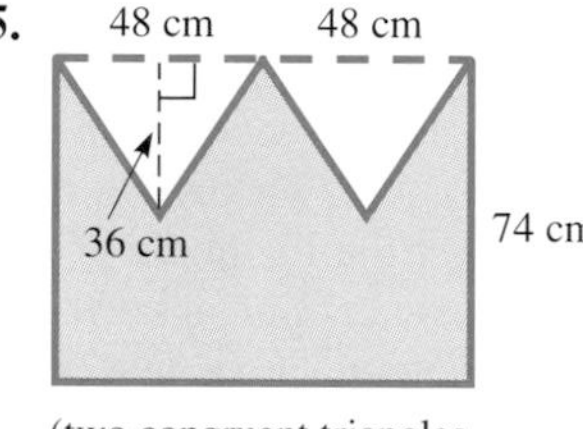

(two congruent triangles within a rectangle)

66.

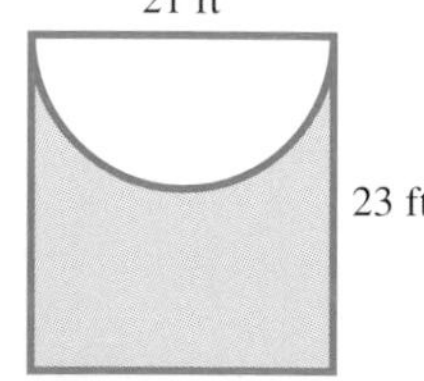

(a semicircle within a rectangle)

67.

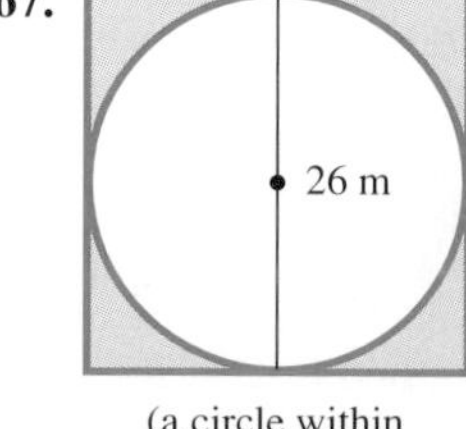

(a circle within a square)

68.

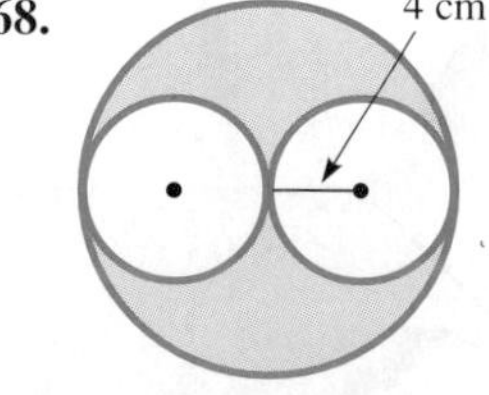

(two circles within a circle)

The following exercises show prices actually charged by a local pizzeria. In each case, the dimension is the diameter of the pizza. Find the best buy.

69. Cheese pizza: 10-in. pizza sells for \$5.99, 12-in. pizza sells for \$7.99, 14-in. pizza sells for \$8.99.

70. Cheese pizza with two toppings: 10-in. pizza sells for \$7.99, 12-in. pizza sells for \$9.99, 14-in. pizza sells for \$10.99.

71. All Feasts pizza: 10-in. pizza sells for \$9.99, 12-in. pizza sells for \$11.99, 14-in. pizza sells for \$12.99.

72. Extravaganza pizza: 10-in. pizza sells for \$11.99, 12-in. pizza sells for \$13.99, 14-in. pizza sells for \$14.99.

A polygon may be inscribed within a circle or circumscribed about a circle. In the figure, triangle ABC is inscribed within the circle, while square WXYZ is circumscribed about it. These ideas will be used in some of the remaining exercises in this section and later in this chapter.

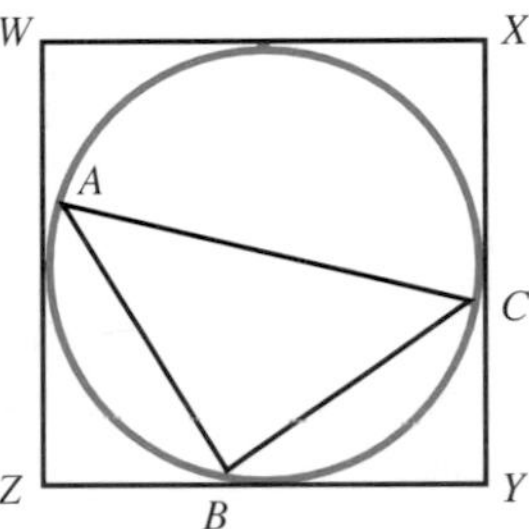

Exercises 73–80 require some ingenuity, but all may be solved using the concepts presented so far in this chapter.

73. Given the circle with center O and rectangle $ABCO$, find the diameter of the circle.

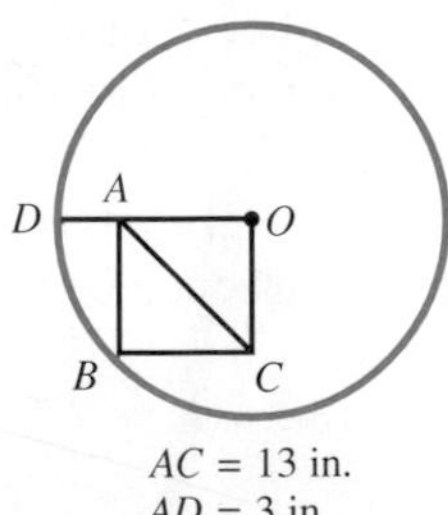

$AC = 13$ in.
$AD = 3$ in.

74. What is the perimeter of $\triangle AEB$, if $AD = 20$ in., $DC = 30$ in., and $AC = 34$ in.?

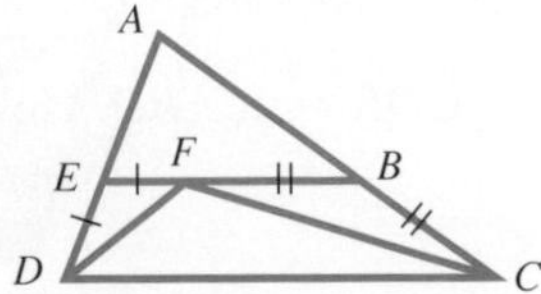

75. The area of square $PQRS$ is 1,250 square feet. T, U, V, and W are the midpoints of $\overleftrightarrow{PQ}$, $\overleftrightarrow{QR}$, $\overleftrightarrow{RS}$, and $\overleftrightarrow{SP}$, respectively. What is the area of square $TUVW$?

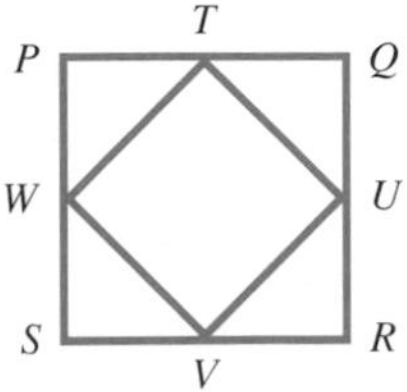

76. The rectangle $ABCD$ has length twice the width. If P, Q, R, and S are the midpoints of the sides, and the perimeter of $ABCD$ is 96 in., what is the area of quadrilateral $PQRS$?

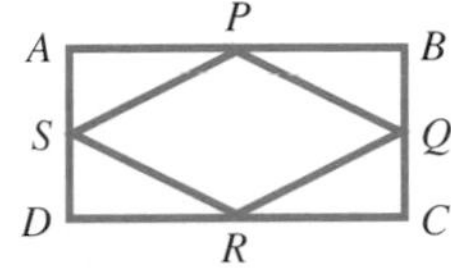

77. If $ABCD$ is a square with each side measuring 36 in., what is the area of the shaded region?

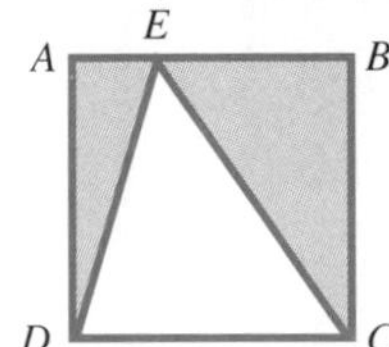

78. Can the perimeter of the polygon shown be determined from the given information? If so, what is the perimeter?

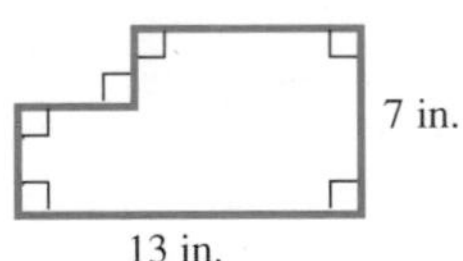

79. Express the area of the shaded region in terms of r, given that the circle is inscribed in the square.

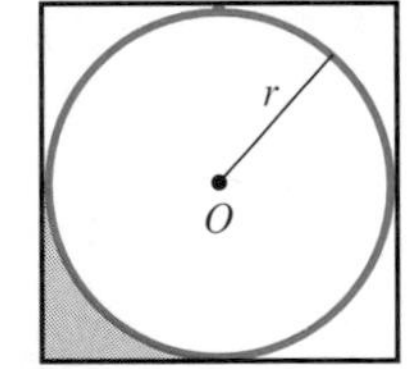

80. Find the area of trapezoid $ABCD$, given that the area of right triangle ABE is 30 in.^{2}.

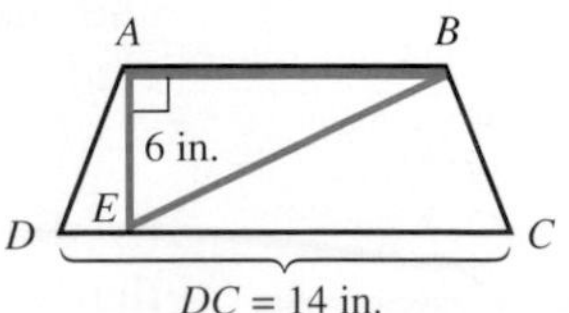

9.4 The Geometry of Triangles: Congruence, Similarity, and the Pythagorean Theorem

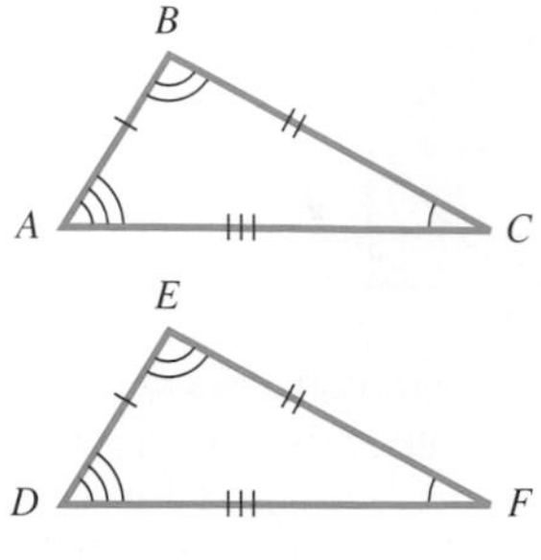

FIGURE 33

In this section we investigate special properties of triangles. Triangles that are both the same size and the same shape are called **congruent triangles.** Informally speaking, if two triangles are congruent, then it is possible to pick up one of them and place it on top of the other so that they coincide exactly. An everyday example of congruent triangles would be the triangular supports for a child's swing set, machine-produced with exactly the same dimensions each time. Figure 33 illustrates two congruent triangles, $\triangle ABC$ and $\triangle DEF$. Notice how the angles and sides are marked to indicate which angles are congruent and which sides are congruent. (Using precise terminology, we refer to angles or sides as being *congruent,* while the *measures* of congruent angles or congruent sides are *equal.* We will often use the terms "equal angles" or "equal sides" to describe angles of equal measure or sides of equal measure.)

In geometry, the following properties are used to prove that two triangles are congruent.

Plimpton 322 Our knowledge of the mathematics of the Babylonians of Mesopotamia is based largely on archaeological discoveries of thousands of clay tablets, some of which are strictly mathematical in nature. Probably the most remarkable of all the tablets is one labeled Plimpton 322. There are several columns of inscriptions that represent numbers. Research has determined that the far right column is simply one that serves to number the lines, but two other columns represent values of hypotenuses and legs of right triangles with integer-valued sides. Thus, it seems that while the famous theorem relating right-triangle side lengths is named for the Greek Pythagoras, the relationship was known more than 1,000 years prior to the time of Pythagoras.

Congruence Properties

Side-Angle-Side (SAS) If two sides and the included angle of one triangle are equal, respectively, to two sides and the included angle of a second triangle, then the triangles are congruent.

Angle-Side-Angle (ASA) If two angles and the included side of one triangle are equal, respectively, to two angles and the included side of a second triangle, then the triangles are congruent.

Side-Side-Side (SSS) If three sides of one triangle are equal, respectively, to three sides of a second triangle, then the triangles are congruent.

Geometry courses, such as those taught in high school and at the undergraduate college level, often investigate these congruence properties in a rigorous manner and require proofs based on them and other geometric properties. In this book we will not investigate this type of proof, but in Example 1 we show how triangles can be verified as congruent by the three properties above. (We may also conclude that all corresponding parts of congruent triangles have equal measures.) The symbol $\cong$ is used to indicate congruence of triangles.

EXAMPLE 1 **(a)** In Figure 34, we are given that $AE = EB$ and $CE = ED$. How can we verify that $\triangle AEC \cong \triangle BED$?

$\measuredangle AEC = \measuredangle BED$ because they are vertical angles, so we have two sides and the included angle of $\triangle AEC$ equal, respectively, to two sides and the included angle of $\triangle BED$. Thus, the Side-Angle-Side (SAS) property holds, and $\triangle AEC \cong \triangle BED$.

(b) In Figure 35, $\measuredangle ADB = \measuredangle CBD$ and $\measuredangle CDB = \measuredangle ABD$. How can we verify that $\triangle ADB \cong \triangle CBD$?

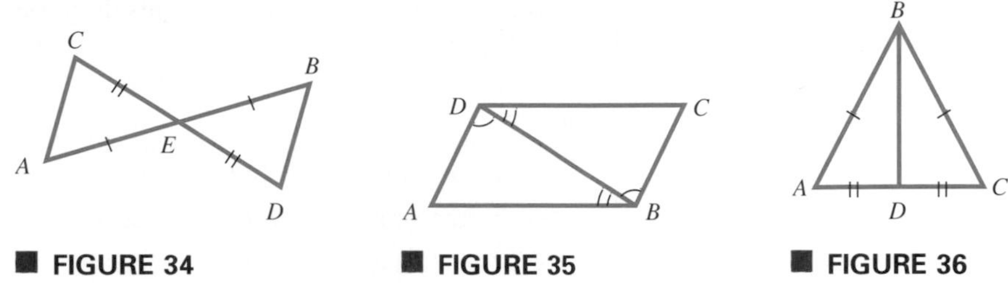

FIGURE 34 FIGURE 35 FIGURE 36

Side DB is equal to itself; therefore, by the Angle-Side-Angle (ASA) property, $\triangle ADB \cong \triangle CBD$.

(c) In Figure 36, $\triangle ABC$ is isosceles, with $AB = CB$. D is the midpoint of $\overleftrightarrow{AC}$. How can we verify that $\triangle ABD \cong \triangle CBD$?

Because D is the midpoint of $\overleftrightarrow{AC}$, we have $AD = CD$. Also, $BD = BD$, so by the Side-Side-Side (SSS) property, $\triangle ABD \cong \triangle CBD$. ■

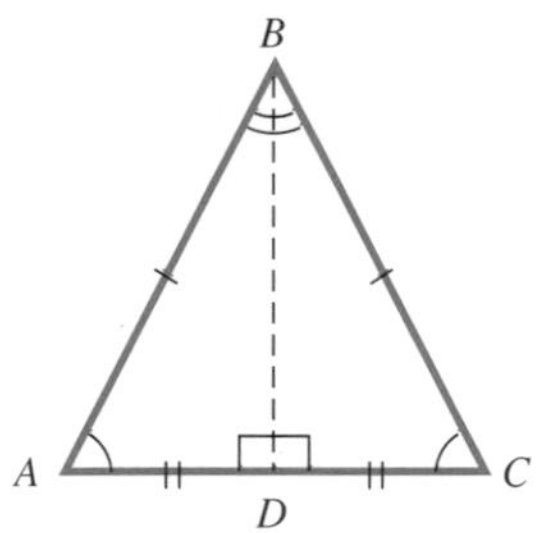

FIGURE 37

The result of Example 1(c) allows us to make several important statements about an isosceles triangle. They are indicated symbolically in Figure 37, and stated in the following box.

Important Statements About Isosceles Triangles

If $\triangle ABC$ is an isosceles triangle with $AB = CB$, and if D is the midpoint of the base AC, then the following properties hold.

1. The base angles A and C are equal.
2. Angles ABD and CBD are equal.
3. Angles ADB and CDB are both right angles.

Similar Triangles

We now turn our attention to similar triangles. Many of the key ideas of geometry depend on **similar triangles,** pairs of triangles that are exactly the same shape but not necessarily the same size. Figure 38 shows three pairs of similar triangles.

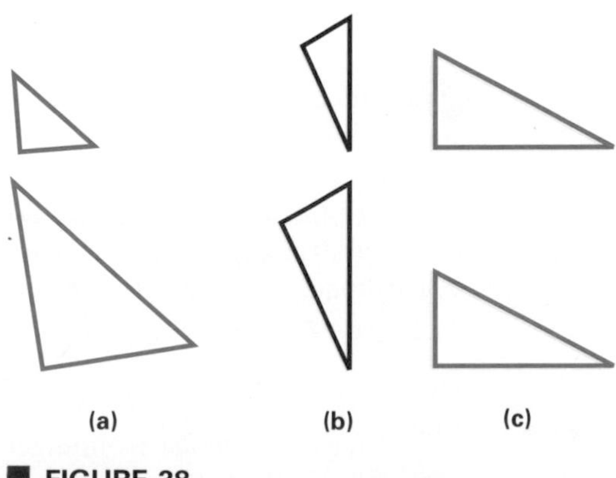

FIGURE 38

As seen in Figure 38(a), the triangles do not need to be oriented in the same fashion in order to be similar.

Suppose that a correspondence between two triangles *ABC* and *DEF* is set up as follows.

$\measuredangle A$ corresponds to $\measuredangle D$	side *AB* corresponds to side *DE*
$\measuredangle B$ corresponds to $\measuredangle E$	side *BC* corresponds to side *EF*
$\measuredangle C$ corresponds to $\measuredangle F$	side *AC* corresponds to side *DF*

For triangle *ABC* to be similar to triangle *DEF*, the following conditions must hold.

1. Corresponding angles must have the same measure.
2. The ratios of the corresponding sides must be constant; that is, the corresponding sides are proportional.

By showing that either of these conditions holds in a pair of triangles, we may conclude that the triangles are similar.

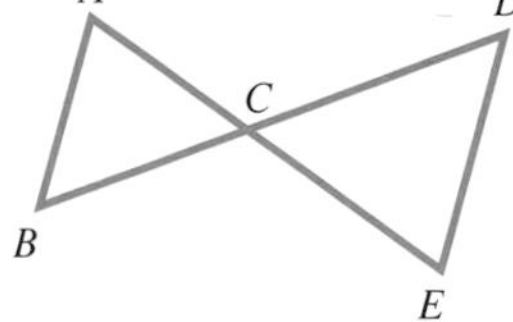

FIGURE 39

EXAMPLE 2 In Figure 39, $\overleftrightarrow{AB}$ is parallel to $\overleftrightarrow{ED}$. How can we verify that $\triangle ABC$ is similar to $\triangle EDC$?

Because $\overleftrightarrow{AB}$ is parallel to $\overleftrightarrow{ED}$, the transversal $\overleftrightarrow{BD}$ forms equal alternate interior angles *ABC* and *EDC*. Also, transversal $\overleftrightarrow{AE}$ forms equal alternate interior angles *BAC* and *DEC*. We know that $\measuredangle ACB = \measuredangle ECD$, because they are vertical angles. Because the corresponding angles have the same measures in triangles *ABC* and *EDC*, the triangles are similar. ■

Once we have shown that two angles of one triangle are equal to the two corresponding angles of a second triangle, it is not necessary to show the same for the third angle. Since, in any triangle, the sum of the angles equals 180°, we may conclude that the measures of the remaining angles *must* be equal. This leads to the following Angle-Angle Similarity property.

Angle-Angle (AA) Similarity Property

If the measures of two angles of one triangle are equal to those of two corresponding angles of a second triangle, then the two triangles are similar.

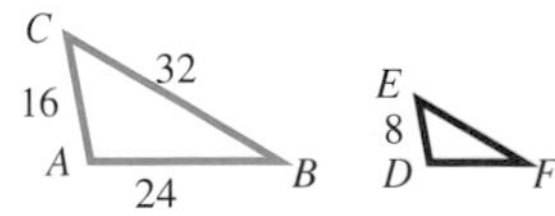

FIGURE 40

EXAMPLE 3 In Figure 40, $\triangle EDF$ is similar to $\triangle CAB$. Find the unknown side lengths in $\triangle EDF$.

As mentioned above, similar triangles have corresponding sides in proportion. Use this fact to find the unknown sides in the smaller triangle. Side *DF* of the small triangle corresponds to side *AB* of the larger one, and sides *DE* and *AC* correspond. This leads to the proportion

$$\frac{8}{16} = \frac{DF}{24}.$$

Using a technique from algebra, set the two cross-products equal. (As an alternate method of solution, multiply both sides by 48, the least common multiple of 16 and 24.) Setting cross-products equal gives

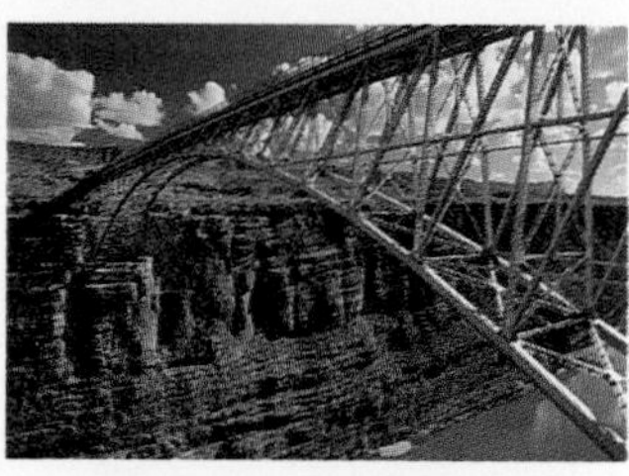

Arizona's Navajo Bridge over the Marble Canyon of the Colorado River shows **congruent and similar triangles**. Corresponding triangles on the left and the right sides of the bridge are congruent, while the triangles are similar as they get smaller toward the center of the bridge.

$$8(24) = 16(DF)$$
$$192 = 16(DF)$$
$$12 = DF.$$

Side DF has a length of 12.

Side EF corresponds to side CB. This leads to another proportion.

$$\frac{8}{16} = \frac{EF}{32}$$
$$\frac{1}{2} = \frac{EF}{32}$$
$$2EF = 32$$
$$EF = 16$$

Side EF has a length of 16. ■

EXAMPLE 4 Find the measures of the unknown parts of the similar triangles STU and ZXY in Figure 41.

Here angles X and T correspond, as do angles Y and U, and angles Z and S. Since angles Z and S correspond and since angle S is 52°, angle Z also must be 52°. The sum of the angles of any triangle is 180°. In the larger triangle $X = 71°$ and $Z = 52°$. To find Y, set up an equation and solve for Y.

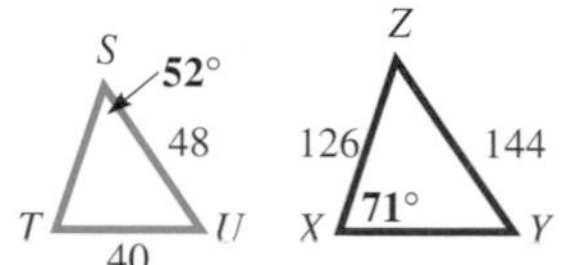

■ **FIGURE 41**

$$X + Y + Z = 180$$
$$71 + Y + 52 = 180$$
$$123 + Y = 180$$
$$Y = 57$$

Angle Y is 57°. Since angles Y and U correspond, $U = 57°$ also.

Now find the unknown sides. Sides SU and ZY correspond, as do XZ and TS, leading to the proportion

$$\frac{48}{144} = \frac{ST}{126}$$
$$\frac{1}{3} = \frac{ST}{126}$$
$$3ST = 126$$
$$ST = 42.$$

Also,

$$\frac{XY}{TU} = \frac{ZY}{SU}$$
$$\frac{XY}{40} = \frac{144}{48}$$
$$\frac{XY}{40} = \frac{3}{1}$$
$$XY = 120.$$

Side ST has a length of 42, and side XY has a length of 120. ■

EXAMPLE 5 The workers at the Abita Springs Post Office want to measure the height of the office flagpole. They notice that at the instant when the shadow of the station is 18 feet long, the shadow of the flagpole is 99 feet long. The building is 10 feet high. Find the height of the flagpole.

Figure 42 shows the information given in the problem. The two triangles shown there are similar, so that corresponding sides are in proportion, with

$$\frac{MN}{10} = \frac{99}{18}$$

or

$$\frac{MN}{10} = \frac{11}{2}$$

$$2MN = 110$$

$$MN = 55.$$

The flagpole is 55 feet high.

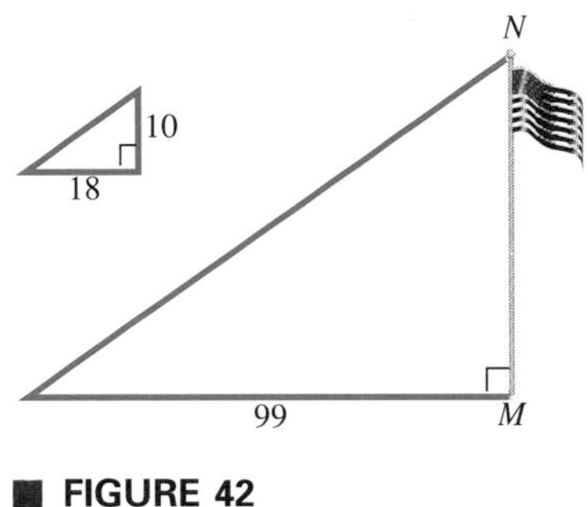

FIGURE 42

The Pythagorean Theorem

We have used the Pythagorean theorem earlier in this book, and because of its importance in mathematics, we will investigate it further in this section on the geometry of triangles. Recall that in a right triangle, the side opposite the right angle (and consequently, the longest side) is called the **hypotenuse.** The other two sides, which are perpendicular, are called the **legs.**

Pythagorean Theorem

If the two legs of a right triangle have lengths a and b, and the hypotenuse has length c, then

$$a^2 + b^2 = c^2.$$

That is, the sum of the squares of the lengths of the legs is equal to the square of the hypotenuse.

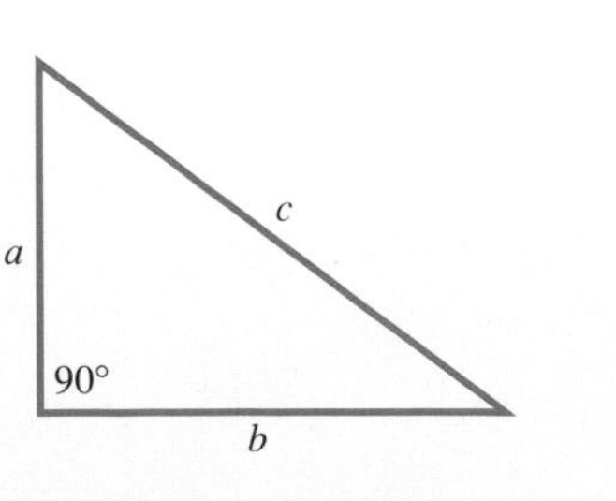

A proof of the Pythagorean theorem is outlined in For Further Thought in this section.

For Further Thought
Proving the Pythagorean Theorem

The Pythagorean theorem has probably been proved in more different ways than any theorem in mathematics. A book titled *The Pythagorean Proposition,* by Elisha Scott Loomis, was first published in 1927. It contained more than 250 different proofs of the theorem. It was reissued in 1968 by the National Council of Teachers of Mathematics as the first title in a series of "Classics in Mathematics Education."

One of the most popular proofs of the theorem is outlined in the group discussion exercise that follows.

For Group Discussion

Copy the accompanying figure on the board in order to prove the Pythagorean theorem. Keep in mind that the area of the large square must be the same, no matter how it is determined. It is made up of four right triangles and a smaller square.

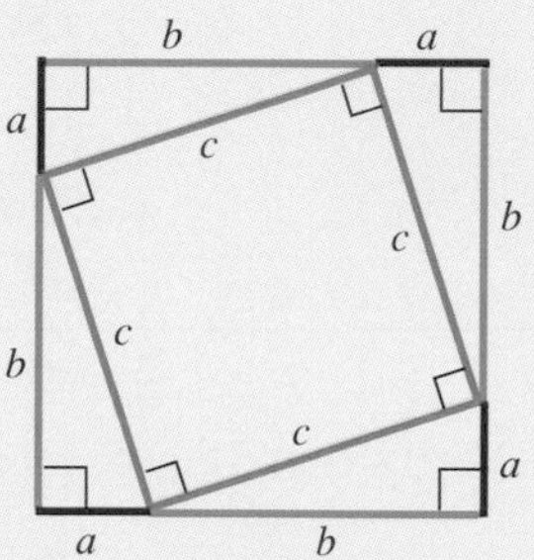

(a) The length of a side of the large square is _____, so its area is (_____$)^2$ or _____.

(b) The area of the large square may also be found by obtaining the sum of the areas of the four right triangles and the smaller square. The area of each right triangle is _____, so the sum of the areas of the four right triangles is _____. The area of the smaller square is _____.

(c) The sum of the areas of the four right triangles and the smaller square is _____.

(d) Since the areas in (a) and (c) represent the area of the same figure, the expressions there must be equal. Setting them equal to each other we obtain _____ = _____.

(e) Subtract $2ab$ from each side of the equation in (d) to obtain the desired result _____ = _____.

FIGURE 43

Pythagoras did not actually discover the theorem that was named after him, although legend tells that he sacrificed 100 oxen to the gods in gratitude for the discovery. There is evidence that the Babylonians knew the concept quite well. The first proof, however, may have come from Pythagoras.

Figure 43 illustrates the theorem in a simple way, by using a sort of tile pattern. In the figure, the side of the square along the hypotenuse measures 5 units,

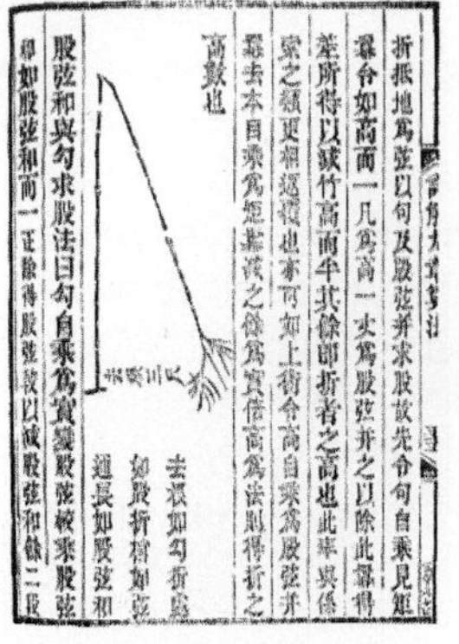

Problem of the broken bamboo (See Exercise 75.)

while those along the legs measure 3 and 4 units. If we let $a = 3$, $b = 4$, and $c = 5$, we see that the equation of the Pythagorean theorem is satisfied.

$$\begin{aligned} a^2 + b^2 &= c^2 \\ 3^2 + 4^2 &= 5^2 \\ 9 + 16 &= 25 \\ 25 &= 25 \end{aligned}$$

The natural numbers 3, 4, 5 form a **Pythagorean triple** since they satisfy the equation of the Pythagorean theorem. There are infinitely many such triples, as implied in Exercises 51–56 of this section.

If two sides of a right triangle are known, the third may always be found. This is illustrated in the next example.

EXAMPLE 6 Find the length a in the right triangle shown in Figure 44.

Let $b = 36$ and $c = 39$. Substituting these values into the equation $a^2 + b^2 = c^2$ allows us to solve for a.

$$\begin{aligned} a^2 + b^2 &= c^2 \\ a^2 + 36^2 &= 39^2 \\ a^2 + 1{,}296 &= 1{,}521 \\ a^2 &= 225 && \text{Subtract 1,296 from both sides.} \\ a &= 15 && \text{Choose the positive square root, since } a > 0. \end{aligned}$$

Verify that 15, 36, 39 is a Pythagorean triple as a check. ■

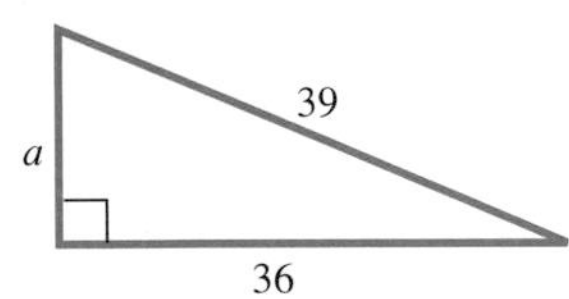

■ **FIGURE 44**

Problem Solving

The Cairo Mathematical Papyrus is an Egyptian document that dates back to about 300 B.C. It was discovered in 1938 and examined in 1962. It contains forty problems, and nine of these deal with the Pythagorean relationship. One of these problems is solved in Example 7.

EXAMPLE 7 A ladder of length 10 cubits has its foot 6 cubits from a wall. To what height does the ladder reach?

As suggested by Figure 45, the ladder forms the hypotenuse of a right triangle, and the ground and wall form the legs. Let x represent the distance from the base of the wall to the top of the ladder. Then, by the Pythagorean theorem,

$$\begin{aligned} x^2 + 6^2 &= 10^2 \\ x^2 + 36 &= 100 \\ x^2 &= 100 - 36 \\ x^2 &= 64 \\ x &= 8. && \text{We choose the positive square root of 64, since } x \text{ represents a length.} \end{aligned}$$

The ladder reaches a height of 8 cubits. ■

■ **FIGURE 45**

The 10 Mathematical Formulas that Changed the Face of the Earth Nicaragua issued a series of ten stamps commemorating mathematical formulas, including this one featuring the Pythagorean theorem, $a^2 + b^2 = c^2$.

Compasses dominate this stamp as they dominate geometry. In fact, the ordinary constructions of elementary geometry can be done with compasses alone. This was demonstrated in the late eighteenth century by Lorenzo Mascheroni.

The application of geometry to architecture is epitomized in the pictures of the Parthenon and its floor plan. The basic idea is the right angle—the angle of pillar to ground, of pillar to roof; the angle where the walls meet; the angle of the triangle where $a^2 + b^2 = c^2$ is the law, binding on all right triangles within the Euclidean system.

The statement of the Pythagorean theorem is an *if . . . then* statement. If the antecedent (the statement following the word "if") and the consequent (the statement following the word "then") are interchanged, the new statement is called the *converse* of the original one. Although the converse of a true statement may not be true, the *converse* of the Pythagorean theorem *is* also a true statement and can be used to determine if a triangle is a right triangle, given the lengths of the three sides.

Converse of the Pythagorean Theorem

If a triangle has sides of lengths a, b, and c, where c is the length of the longest side, and if $a^2 + b^2 = c^2$, then the triangle is a right triangle.

Problem Solving

If a carpenter is building a floor for a rectangular room, it is essential that the corners of the floor represent right angles; otherwise, problems will occur when the walls are constructed, when flooring is laid, and so on. To check that the floor is "squared off," the carpenter can use the converse of the Pythagorean theorem.

EXAMPLE 8 Jack of All Trades, Inc. has been contracted to add an 8-foot by 12-foot laundry room onto an existing house. After the floor has been built, Jack finds that the length of the diagonal of the floor is 14 feet, 8 inches. Is the floor "squared off" properly?

Since 14 feet, 8 inches = 14 2/3 feet, Jack must check to see if

$$8^2 + 12^2 = \left(14\frac{2}{3}\right)^2$$

is a true statement.

$$8^2 + 12^2 \stackrel{?}{=} \left(\frac{44}{3}\right)^2$$

$$208 \stackrel{?}{=} \frac{1{,}936}{9}$$

$$208 \neq 215\frac{1}{9}$$

Jack has a real problem now, since his diagonal, which measures 14 feet, 8 inches, should actually measure $\sqrt{208} \approx 14.4 \approx$ 14 feet, 5 inches. Jack must correct his error to avoid major problems later. ■

9.4 EXERCISES

In each figure, you are given certain information. Tell which congruence property you would use to verify the congruence required, and give the additional information needed to apply the property.

1. Given: $AC = BD$; $AD = BC$. Show that $\triangle ABD \cong \triangle BAC$.

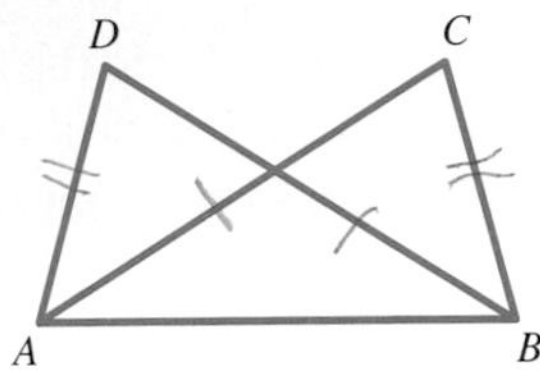

2. Given: $AC = BC$; $AD = DB$. Show that $\triangle ADC \cong \triangle BDC$.

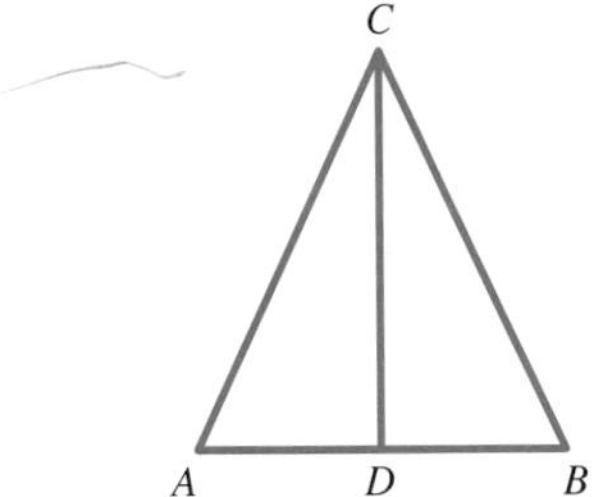

3. Given: $\overleftrightarrow{DB}$ is perpendicular to $\overleftrightarrow{AC}$; $AB = BC$. Show that $\triangle ABD \cong \triangle CBD$.

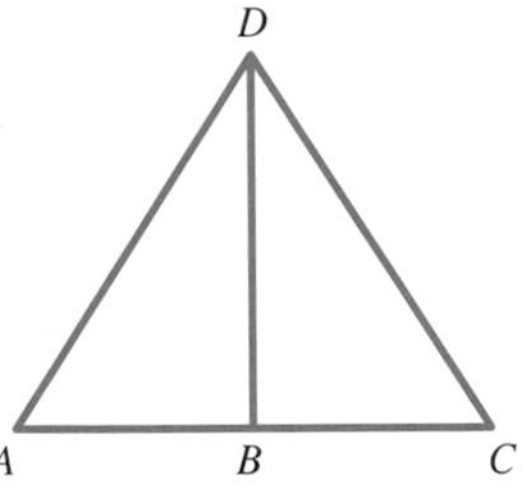

4. Given: $BC = BA$; $\measuredangle 1 = \measuredangle 2$. Show that $\triangle DBC \cong \triangle DBA$.

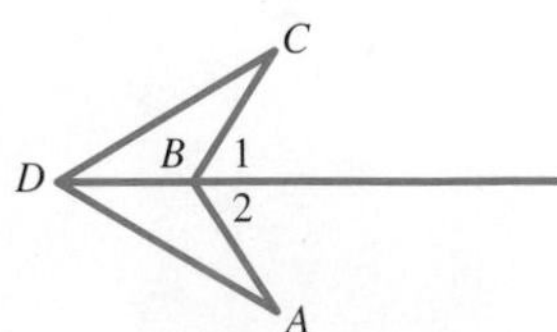

5. Given: $\measuredangle BAC = \measuredangle DAC$; $\measuredangle BCA = \measuredangle DCA$. Show that $\triangle ABC \cong \triangle ADC$.

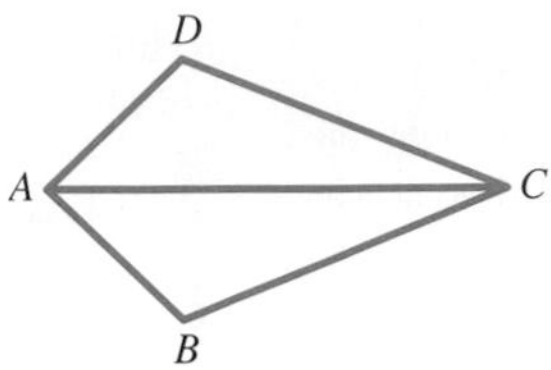

6. Given: $BO = OE$; $\overleftrightarrow{OB}$ is perpendicular to $\overleftrightarrow{AC}$; $\overleftrightarrow{OE}$ is perpendicular to $\overleftrightarrow{DF}$. Show that $\triangle AOB \cong \triangle FOE$.

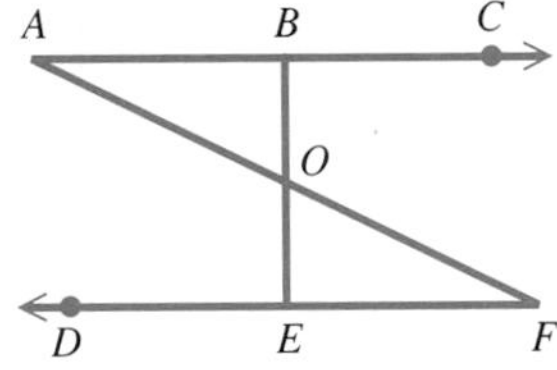

Exercises 7–10 refer to the given figure, an isosceles triangle with $AB = BC$.

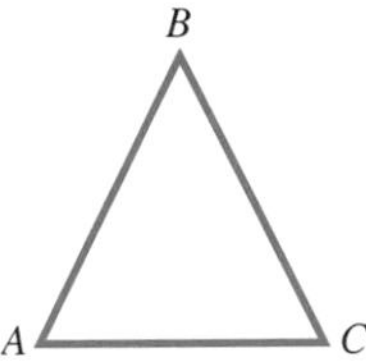

7. If $\measuredangle B$ measures 46°, then $\measuredangle A$ measures ______ and $\measuredangle C$ measures ______.
8. If $\measuredangle C$ measures 52°, what is the measure of $\measuredangle B$?
9. If $BC = 12$ in., and the perimeter of $\triangle ABC$ is 30 in., what is the length AC?
10. If the perimeter of $\triangle ABC = 40$ in., and $AC = 10$ in., what is the length AB?
11. Explain why all equilateral triangles must be similar.
12. Explain why two congruent triangles must be similar, but two similar triangles might not be congruent.

Name the corresponding angles and the corresponding sides for each of the following pairs of similar triangles.

13.

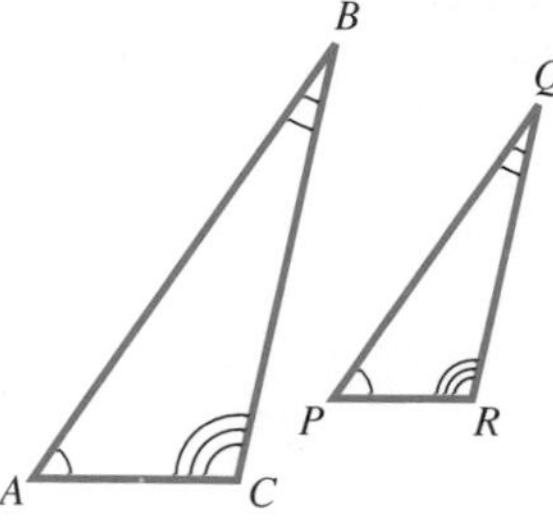

14.

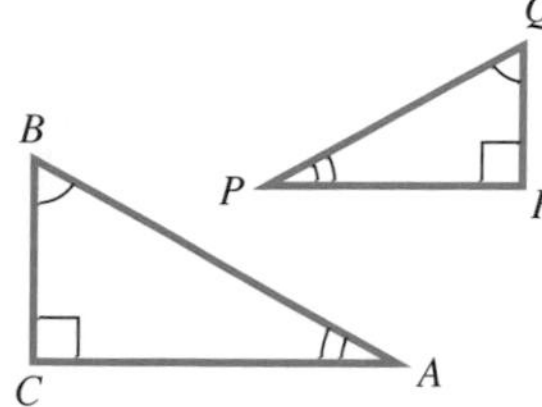

15.

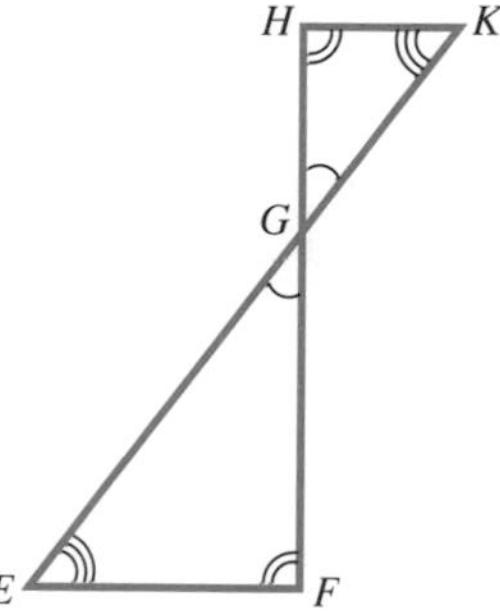

16.

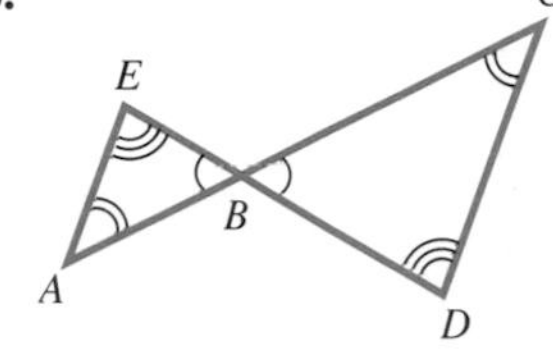

Find all unknown angle measures in each pair of similar triangles.

17.

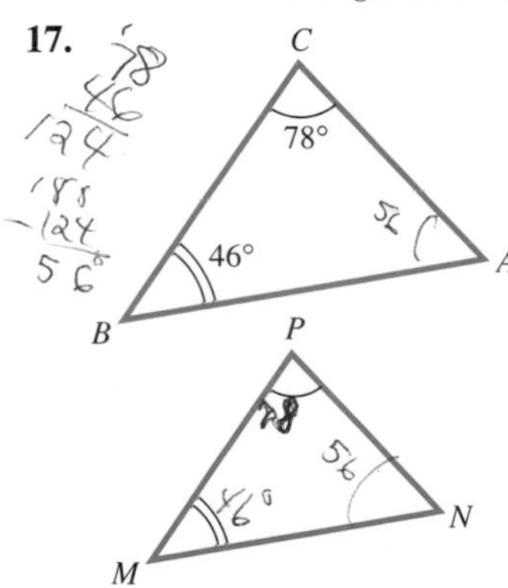

18.

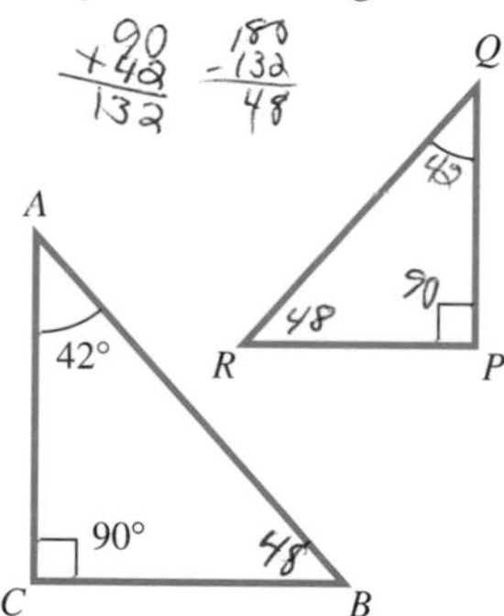

19.

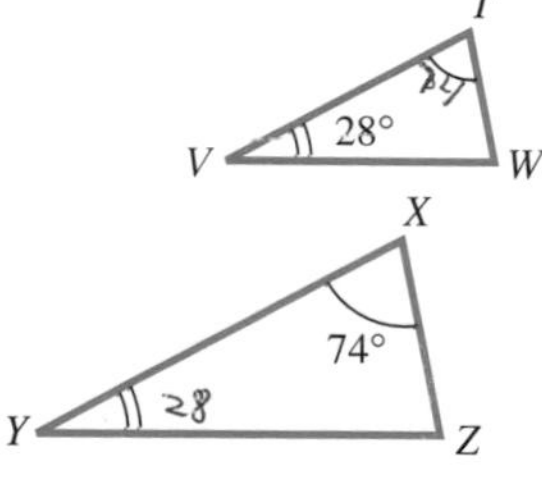

20.

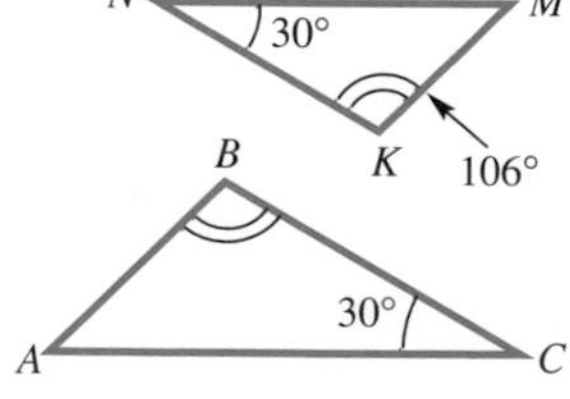

21.

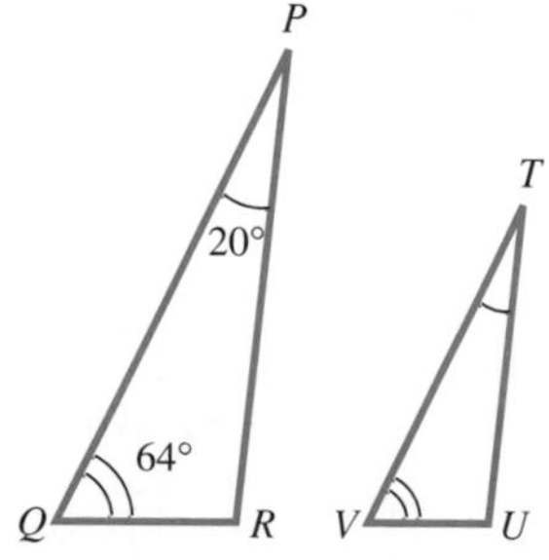

22.

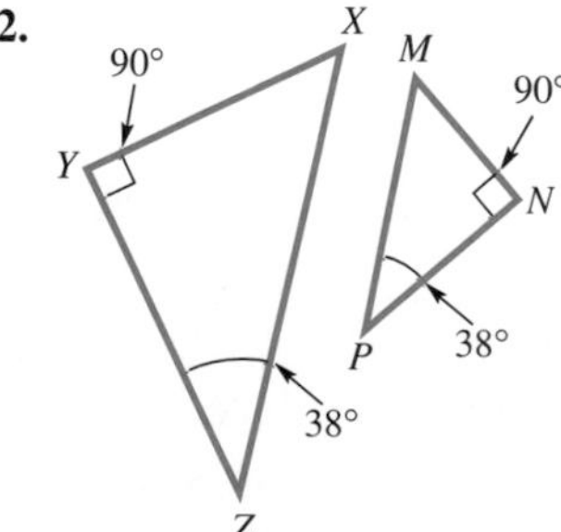

Find the unknown side lengths in each pair of similar triangles.

23.

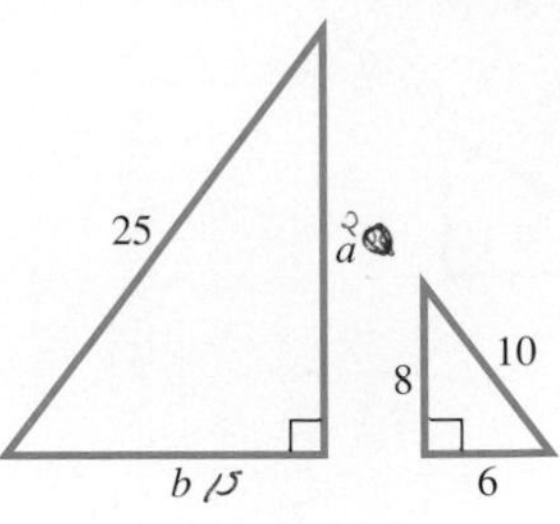

24.

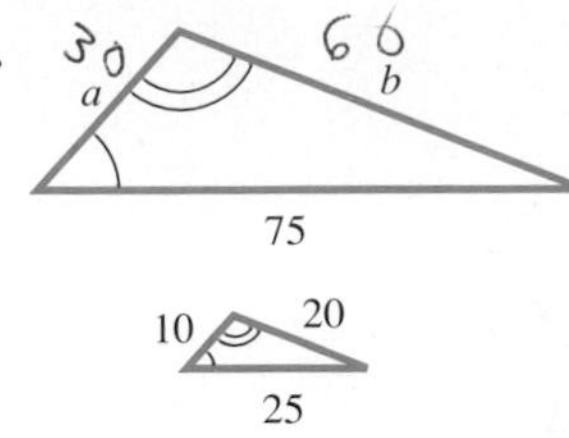

25.

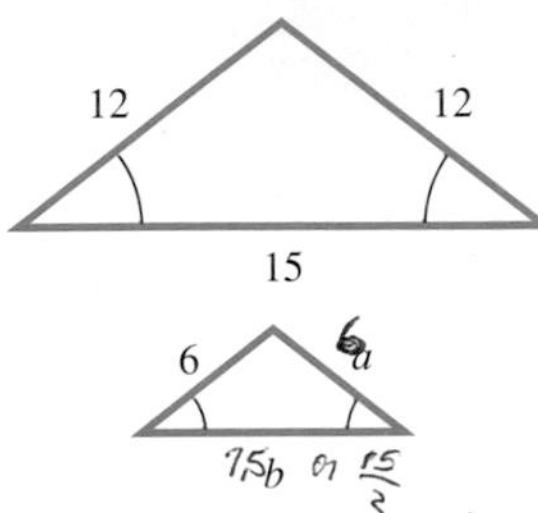

26.

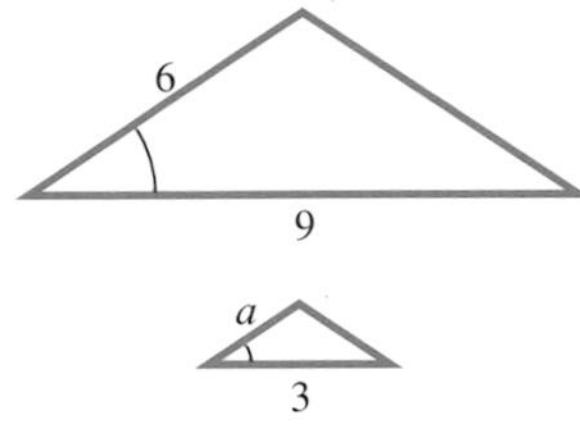

27.

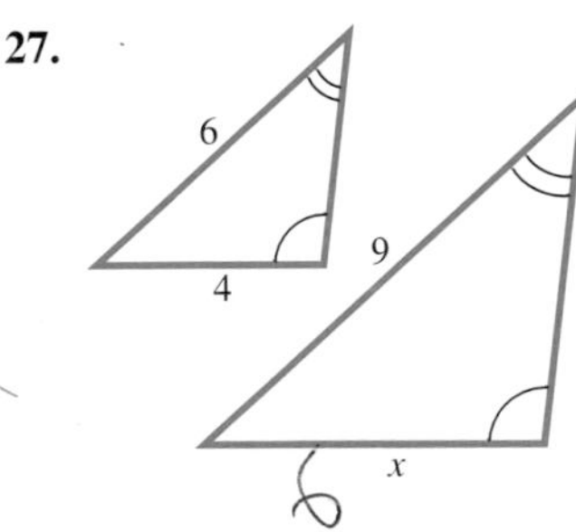

28.

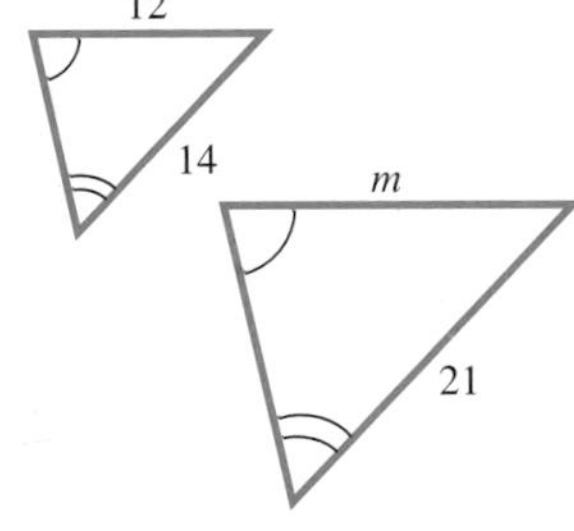

In each diagram, there are two similar triangles. Find the unknown measurement in each. (Hint: In the figure for Exercise 29, the side of length 100 in the smaller triangle corresponds to a side of length 100 + 120 = 220 in the larger triangle.)

29.

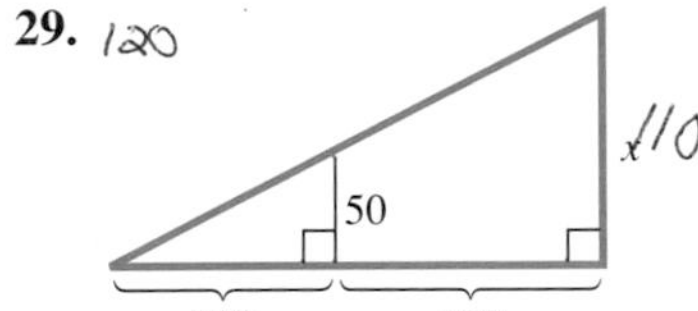

30.

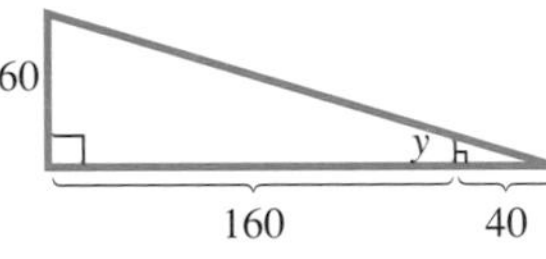

31.

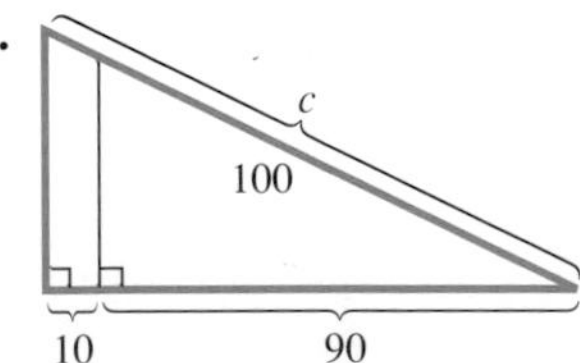

32.

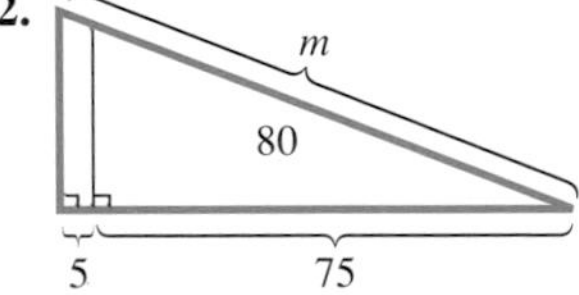

Solve the following problems.

33. A tree casts a shadow 45 m long. At the same time, the shadow cast by a vertical 2-m stick is 3 m long. Find the height of the tree.

34. A forest fire lookout tower casts a shadow 180 ft long at the same time that the shadow of a 9-ft truck is 15 ft long. Find the height of the tower.

35. On a photograph of a triangular piece of land, the lengths of three sides are 4 cm, 5 cm, and 7 cm, respectively. The shortest side of the actual piece of land is 400 m long. Find the lengths of the other two sides.

36. The Santa Cruz lighthouse is 14 m tall and casts a shadow 28 m long at 7 P.M. At the same time, the shadow of the lighthouse keeper is 3.5 m long. How tall is she?

37. A house is 15 ft tall. Its shadow is 40 ft long at the same time the shadow of a nearby building is 300 ft long. Find the height of the building.

38. By drawing lines on a map, a triangle can be formed by the cities of Phoenix, Tucson, and Yuma. On the map, the distance between Phoenix and Tucson is 8 cm, the distance between Phoenix and Yuma is 12 cm, and the distance between Tucson and Yuma is 17 cm. The actual straight-line distance from Phoenix to Yuma is 230 km. Find the distances between the other pairs of cities.

39. The photograph shows the maintenance of the Mount Rushmore head of Lincoln. Assume that Lincoln was 6 1/3 ft tall and his head 3/4 ft long. Knowing that the carved head of Lincoln is 60 ft tall, find out how tall his entire body would be if it were carved into the mountain.

40. Robert Wadlow was the tallest human being ever recorded. When a 6-ft stick cast a shadow 24 in., Robert would cast a shadow 35.7 in. How tall was he?

In Exercises 41–46, a and b represent the two legs of a right triangle, while c represents the hypotenuse. Find the lengths of the unknown sides.

41.

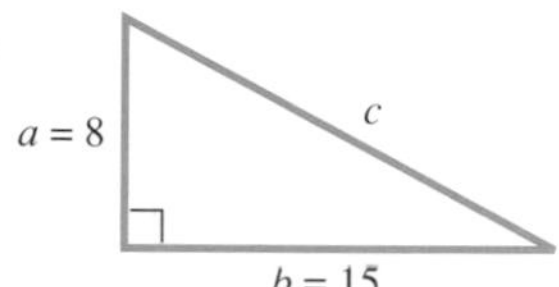

42.

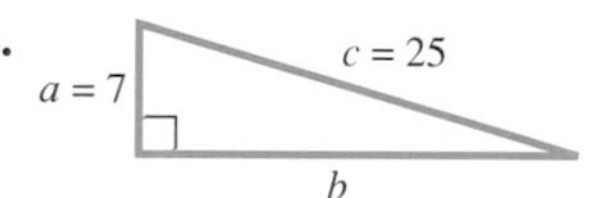

43.

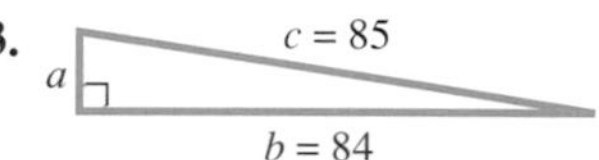

44. $a = 24$ cm; $c = 25$ cm

45. $a = 14$ m; $b = 48$ m

46. $a = 28$ km; $c = 100$ km

Find the areas of the squares on the sides of the triangles in Exercises 47 and 48 to decide whether the triangle formed is a right triangle.

47.

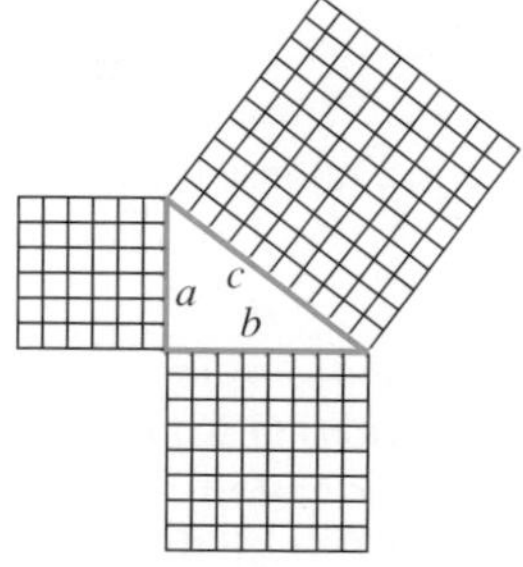

48.

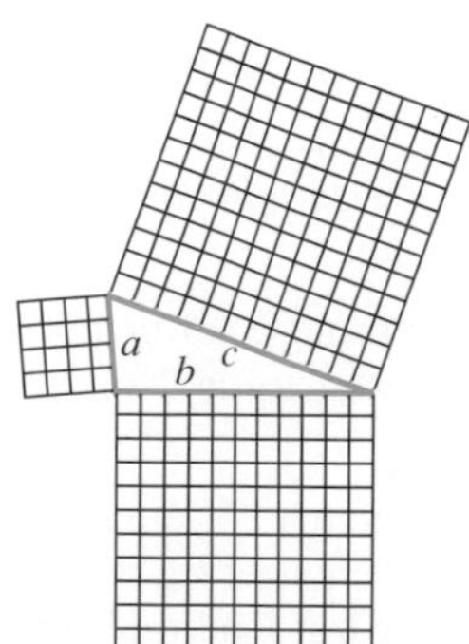

49. Refer to Exercise 42 in Section 9.2. Correct the scarecrow's statement.

50. Show that if $a^2 + b^2 = c^2$, then it is not necessarily true that $a + b = c$.

There are various formulas that will generate Pythagorean triples. For example, if we choose positive integers r and s, with $r > s$, then the set of equations

$$a = r^2 - s^2, \qquad b = 2rs, \qquad c = r^2 + s^2$$

generates a Pythagorean triple (a, b, c). Use the values of r and s given in Exercises 51–56 to generate a Pythagorean triple using this method.

51. $r = 2, s = 1$ **52.** $r = 3, s = 2$ **53.** $r = 4, s = 3$

54. $r = 3, s = 1$ **55.** $r = 4, s = 2$ **56.** $r = 4, s = 1$

57. Show that the formula given for Exercises 51–56 actually satisfies $a^2 + b^2 = c^2$.

58. It can be shown that if $(x, x + 1, y)$ is a Pythagorean triple, then so is

$$(3x + 2y + 1, 3x + 2y + 2, 4x + 3y + 2).$$

Use this idea to find three more Pythagorean triples, starting with 3, 4, 5. (*Hint:* Here, $x = 3$ and $y = 5$.)

If m is an odd positive integer greater than 1, then

$$\left(m, \frac{m^2 - 1}{2}, \frac{m^2 + 1}{2}\right)$$

is a Pythagorean triple. Use this to find the Pythagorean triple generated by each value of m in Exercises 59–62.

59. $m = 3$ **60.** $m = 5$ **61.** $m = 7$ **62.** $m = 9$

63. Show that the expressions in the directions for Exercises 59–62 actually satisfy $a^2 + b^2 = c^2$.

64. Show why (6, 8, 10) is the only Pythagorean triple consisting of consecutive even numbers.

For any integer n greater than 1,

$$(2n, n^2 - 1, n^2 + 1)$$

is a Pythagorean triple. Use this to find the Pythagorean triple generated by each value of n in Exercises 65–68.

65. $n = 2$ **66.** $n = 3$ **67.** $n = 4$ **68.** $n = 5$

69. Show that the expressions in the directions for Exercises 65–68 actually satisfy $a^2 + b^2 = c^2$.

70. Can an isosceles right triangle have sides with integer lengths? Why or why not?

Solve each problem. (You may wish to review quadratic equations from algebra.)

71. If the hypotenuse of a right triangle is 1 m more than the longer leg, and the shorter leg is 7 m, find the length of the longer leg.

72. The hypotenuse of a right triangle is 1 cm more than twice the shorter leg, and the longer leg is 9 cm less than three times the shorter leg. Find the lengths of the three sides of the triangle.

73. At a point on the ground 30 ft from the base of a tower, the distance to the top of the tower is 2 ft more than twice the height of the tower. Find the height of the tower.

74. The length of a rectangle is 2 in. less than twice the width. The diagonal is 5 in. Find the length and width of the rectangle.

75. (Problem of the broken bamboo, from the Chinese work *Arithmetic in Nine Sections* (1261)) There is a bamboo 10 ft high, the upper end of which, being broken, reaches the ground 3 ft from the stem. Find the height of the break.

76. (Adapted from *Arithmetic in Nine Sections*) There grows in the middle of a circular pond 10 ft in diameter a reed which projects 1 ft out of the water. When it is drawn down it just reaches the edge of the pond. How deep is the water?

Imagine that you are a carpenter building the floor of a rectangular room. What must the diagonal of the room measure if your floor is to be squared off properly, given the dimensions in Exercises 77–80? Give your answer to the nearest inch.

77. 12 ft by 15 ft **78.** 14 ft by 20 ft

79. 16 ft by 24 ft **80.** 20 ft by 32 ft

81. James A. Garfield, the twentieth president of the United States, provided a proof of the Pythagorean theorem using the given figure. Supply the required information in each part below in order to follow his proof.

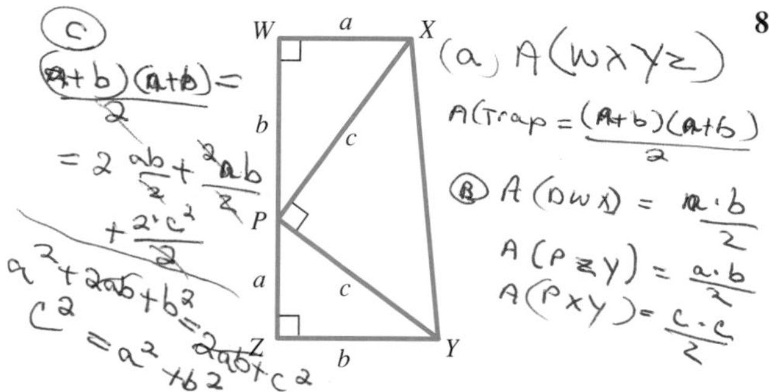

(a) Find the area of the trapezoid $WXYZ$ using the formula for the area of a trapezoid.

(b) Find the area of each of the right triangles PWX, PZY, and PXY.

(c) Since the sum of the areas of the three right triangles must equal the area of the trapezoid, set the expression from part (a) equal to the sum of the three expressions from part (b). Simplify the equation as much as possible.

82. In the figure, right triangles ABC, CBD, and ACD are similar. This may be used to prove the Pythagorean theorem. Fill in the blanks with the appropriate responses.

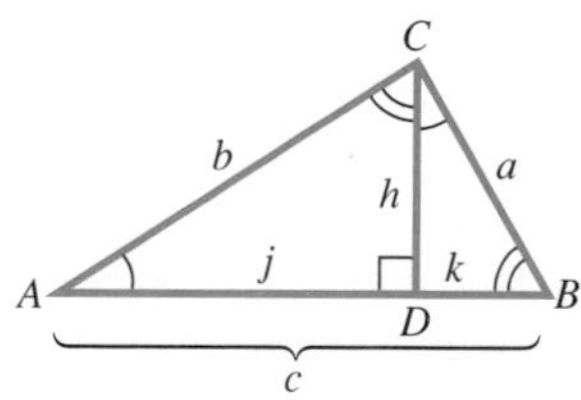

(a) By proportion, we have $c/b =$ _____ $/j$.

(b) By proportion, we also have $c/a = a/$_____.

(c) From part (a), $b^2 =$ _____.

(d) From part (b), $a^2 =$ _____.

(e) From the results of parts (c) and (d) and factoring, $a^2 + b^2 = c($_____$)$. But since _____ $= c$, it follows that _____.

Exercises 83–90 require some ingenuity, but all may be solved using the concepts presented so far in this chapter.

83. Find the area of quadrilateral $ABCD$, if angles A and C are right angles.

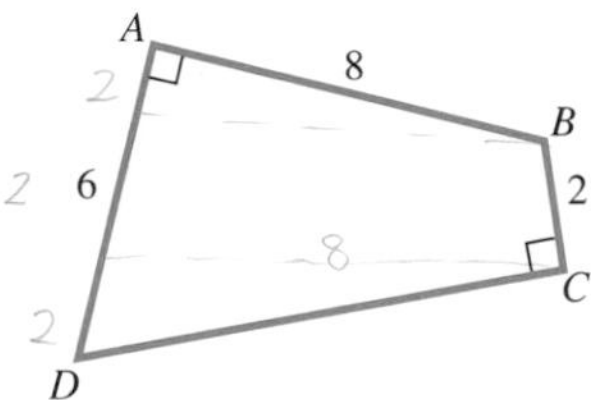

84. The perimeter of the isosceles triangle ABC (with $AB = BC$) is 128 in. The altitude BD is 48 in. What is the area of triangle ABC?

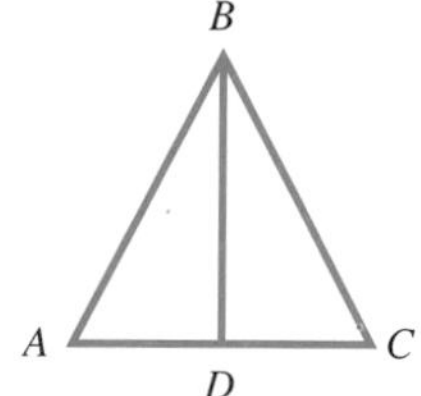

85. An isosceles triangle has a base of 24 and two sides of 13. What other base measure can an isosceles triangle with equal sides of 13 have and still have the same area as the given triangle?

86. In right triangle ABC, if $AD = DB + 8$, what is the value of CD?

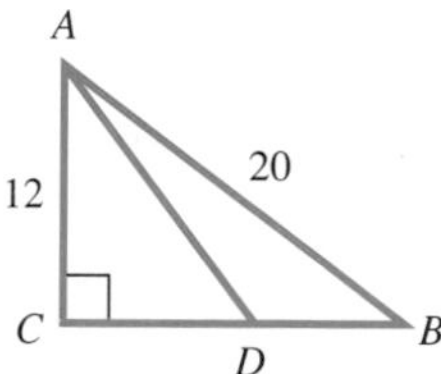

87. In the figure, pentagon $PQRST$ is formed by a square and an equilateral triangle such that $PQ = QR = RS = ST = PT$. The perimeter of the pentagon is 80. Find the area of the pentagon.

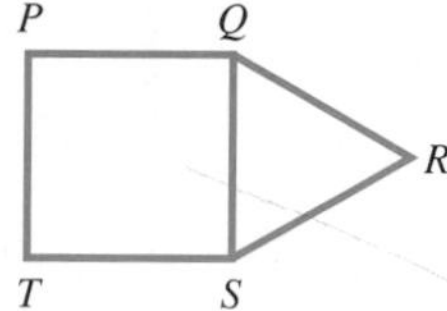

88. (A segment that *bisects* an angle divides the angle into two equal angles.) In the figure, angle A measures 50°. $\overline{OB}$ bisects angle ABC, and $\overline{OC}$ bisects angle ACB. What is the measure of angle BOC?

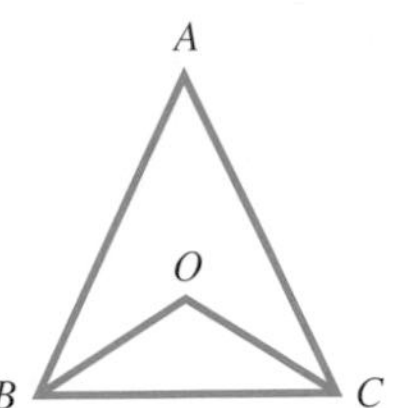

Exercises 89 and 90 refer to the given figure. The center of the circle is O.

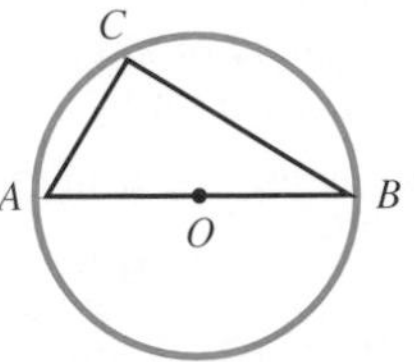

89. If $\overline{AC}$ measures 6 in. and $\overline{BC}$ measures 8 in., what is the radius of the circle?

90. If $\overline{AB}$ measures 13 cm, and the length of $\overline{BC}$ is 7 cm more than the length of $\overline{AC}$, what are the lengths of $\overline{BC}$ and $\overline{AC}$?

EXTENSION

Right Triangle Trigonometry

The foundations of trigonometry go back at least 3,000 years. The ancient Egyptians, Babylonians, and Greeks developed trigonometry to find the lengths of the sides of triangles and the measures of their angles. In Egypt, trigonometry was used to reestablish land boundaries after the annual flood of the Nile River. In Babylonia it was used in astronomy. The very word *trigonometry* comes from the Greek words for triangle *(trigon)* and measurement *(metry)*. Today trigonometry is used in electronics, surveying, and other engineering areas, and is necessary for further courses in mathematics, such as calculus.

In this brief treatment of trigonometry, we will investigate only one aspect of the subject: finding unknown parts of a right triangle, given information about the triangle. If we are given the lengths of two sides of a right triangle, or the length of one side and the measure of an acute angle of the right triangle, then we can use trigonometry to solve for the measures of the other parts. To do this, we define the trigonometric ratios, *sine, cosine,* and *tangent.*

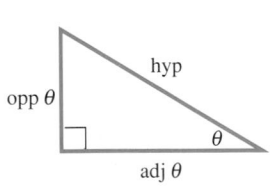

FIGURE 46

Trigonometric Ratios

Let θ (the Greek letter theta) represent an acute angle of a right triangle. Then the legs of the right triangle may be designated as sides **opposite** θ and **adjacent to** θ, abbreviated **opp** θ and **adj** θ, as shown in the figure. The **hypotenuse,** abbreviated **hyp,** is the side opposite the right angle. (See Figure 46.) Then the sine, cosine, and tangent ratios for θ, abbreviated **sin** θ, **cos** θ, and **tan** θ, are defined as follows.

$$\sin \theta = \frac{\text{opp } \theta}{\text{hyp}} \qquad \cos \theta = \frac{\text{adj } \theta}{\text{hyp}} \qquad \tan \theta = \frac{\text{opp } \theta}{\text{adj } \theta}$$

The size of the triangle does not affect the ratio for a given value of θ, since two right triangles that each have an acute angle θ must be similar, and in similar triangles, sides are proportional.

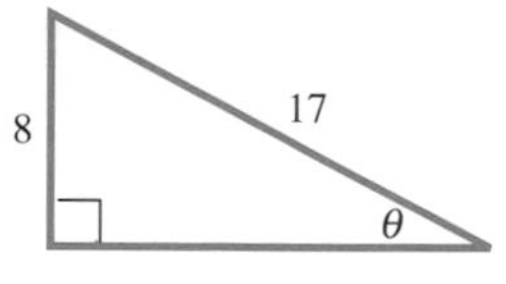

FIGURE 47

EXAMPLE 1 Find the sine, cosine, and tangent of θ in Figure 47.

From the figure, we see that the length of the side opposite θ is 8 and the length of the hypotenuse is 17. To find the length of the side adjacent to θ, we use the Pythagorean theorem. If x represents this length, then

$$\begin{aligned} x^2 + 8^2 &= 17^2 \\ x^2 + 64 &= 289 \\ x^2 &= 225 \\ x &= 15. \end{aligned}$$

Choose the positive square root, since x represents a length.

We now have opp $\theta = 8$, adj $\theta = 15$, and hyp $= 17$. By the definitions of the trigonometric ratios,

$$\sin\theta = \frac{\text{opp }\theta}{\text{hyp}} = \frac{8}{17}, \quad \cos\theta = \frac{\text{adj }\theta}{\text{hyp}} = \frac{15}{17}, \quad \text{and} \quad \tan\theta = \frac{\text{opp }\theta}{\text{adj }\theta} = \frac{8}{15}.$$

The trigonometric ratios for several special acute angles, such as 30°, 45°, and 60°, can be evaluated exactly using elementary geometry as suggested in Exercises 9 and 10. For example, $\sin 30° = 1/2$, $\cos 30° = \sqrt{3}/2$, $\tan 45° = 1$, and $\tan 60° = \sqrt{3}$. In this extension we will concentrate on approximations of trigonometric ratios, which can be obtained using the appropriate functions of a scientific or graphing calculator.

You should refer to your owner's manual to see how to find these values. Remember that for the applications discussed here, the calculator should be in *degree mode.* The following example shows how trigonometry can be applied to a problem situation.

EXAMPLE 2 A 12-foot ladder is leaning against a wall, as shown in Figure 48. Find the distance from the base of the wall to the top of the ladder, if the ladder makes an angle of 44° with the ground.

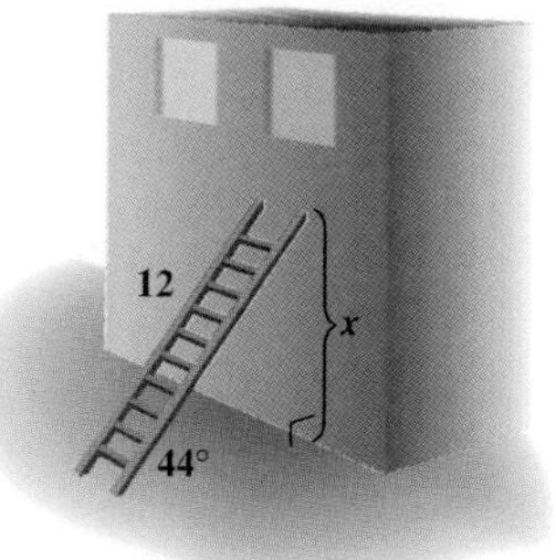

FIGURE 48

Since we want to find the height of the top of the ladder, let x represent this height. In the right triangle formed, x represents the length of the side opposite the 44° angle, and the length of the hypotenuse is 12. Since the opposite side and the hypotenuse are involved in the sine ratio, we can say that

$$\sin 44° = \frac{x}{12}.$$

By using a calculator, we know that $\sin 44° \approx .69465837$, so

$$.69465837 = \frac{x}{12}$$
$$x = 12(.69465837)$$
$$x \approx 8.3.$$

The ladder goes up the wall about 8.3 feet. ■

Many applications of right triangles involve the angle of elevation or the angle of depression. The **angle of elevation** from point X to point Y (above X) is the angle made by ray XY and a horizontal ray with endpoint at X. The angle of elevation is always measured from the horizontal. See Figure 49. The **angle of depression** from point X to point Y (below X) is the angle made by ray XY and a horizontal ray with endpoint X. Again, see Figure 49.

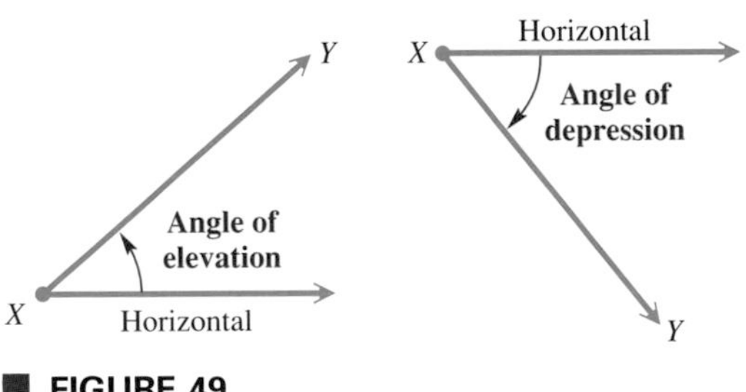

■ **FIGURE 49**

The concept of angle of elevation is used in the next example.

EXAMPLE 3 Donna Garbarino knows that when she stands 123 feet from the base of a flagpole, the angle of elevation to the top is 26°. If her eyes are 5.3 feet above the ground, find the height of the flagpole.

The length of the side adjacent to Donna is known and the length of the side opposite her is to be found. (See Figure 50.) The ratio that involves these two values is the tangent.

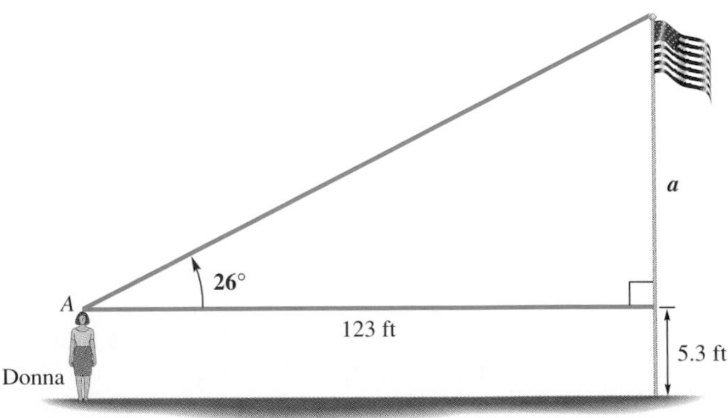

■ **FIGURE 50**

$$\tan A = \frac{\text{side opposite}}{\text{side adjacent}}$$

$$\tan 26° = \frac{a}{123}$$

$$a = 123 \tan 26°$$

$$a \approx 60.0 \text{ feet}$$

Since Donna's eyes are 5.3 feet above the ground, the height of the flagpole is about

$$60.0 + 5.3 = 65.3 \text{ feet.}$$ ■

If the lengths of two sides of a right triangle are known, then it is possible to find the measures of the acute angles of the triangle. Look back to Figure 47 in Example 1. We know that $\sin \theta = 8/17$. To find the measure of θ, we must use the *inverse sine* key of a calculator. This is most often labeled $\boxed{\sin^{-1}}$. Since we must find the measure of the angle whose sine is $8/17$, we enter $8 \div 17$ and then take inverse sine:

$$\sin^{-1} \frac{8}{17} \approx 28.072487°.$$

Therefore, the measure of θ in Figure 47 is about 28.1°. The other acute angle measures approximately 61.9°. Why is this so?

EXAMPLE 4 The length of the shadow of a building 34.1 meters tall is 37.6 meters. Find the angle of elevation of the sun.

As shown in Figure 51, the angle of elevation of the sun is angle B. Since the side opposite B and the side adjacent to B are known, the tangent ratio applies.

$$\tan B = \frac{34.1}{37.6}$$

Since we must find the *measure* of the angle B, we use the inverse tangent key on a calculator.

$$B = \tan^{-1} \frac{34.1}{37.6}$$

$$B \approx 42° \quad \text{(to the nearest degree)}$$

The angle of elevation of the sun is approximately 42°. ■

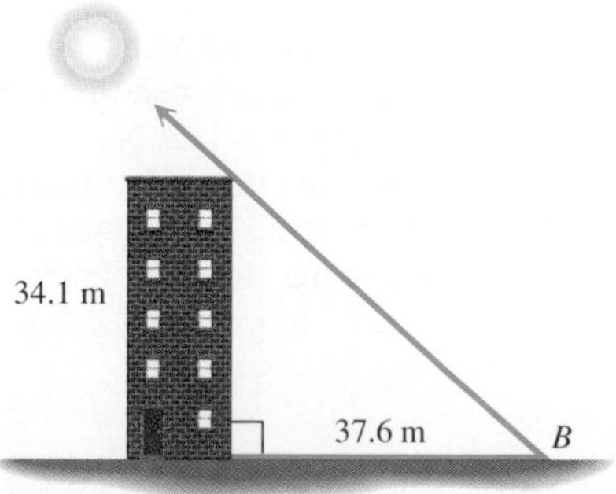

■ **FIGURE 51**

Extension Exercises

Find sin θ, cos θ, *and* tan θ *for each angle* θ. *(Do not use a calculator.)*

1.

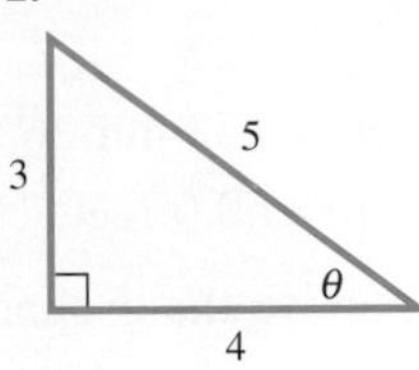

2.

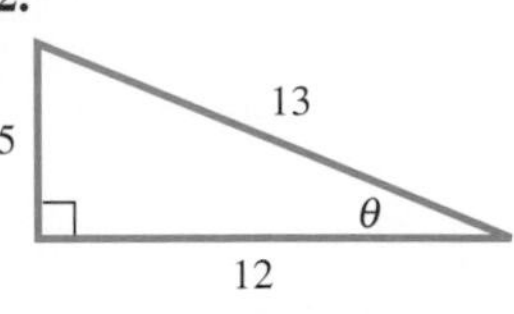

3.

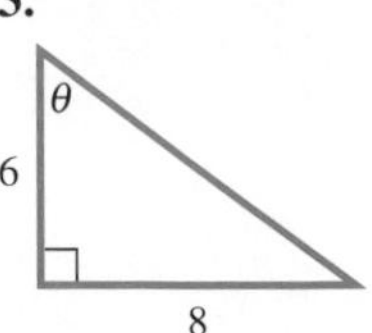

4.

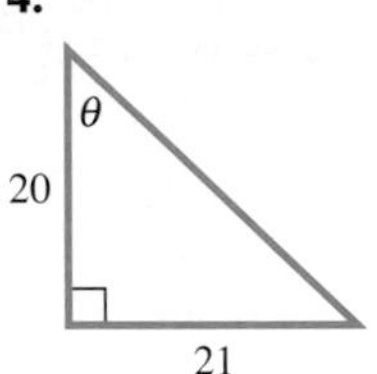

5.

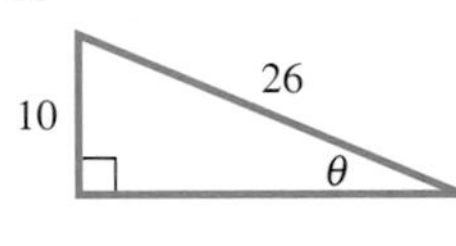

6.

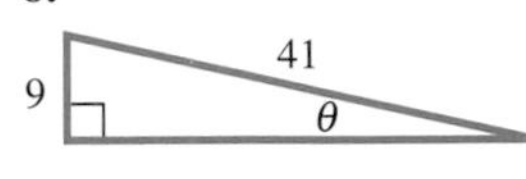

7. Refer to the figure to respond to each part of this exercise.

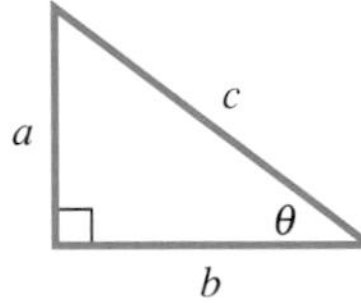

(a) Since the sum of the acute angles of a right triangle is 90°, the two acute angles of a right triangle are _____ of each other.

(b) Express the measure of the unmarked acute angle in terms of θ.

(c) Explain why the sine of angle θ is equal to the cosine of the unmarked acute angle of the triangle.

(d) The sine and cosine ratios are called **cofunctions** (the prefix "co" representing "complementary"). Based on your answers above, what can be concluded about sin θ and cos $(90° - \theta)$?

8. Using the accompanying figure, it can be determined that the sine of a 30° angle is $1/2$. Now complete the following statement: In a right triangle with an acute angle 30°, the side opposite the 30° angle measures _____ the hypotenuse of the triangle.

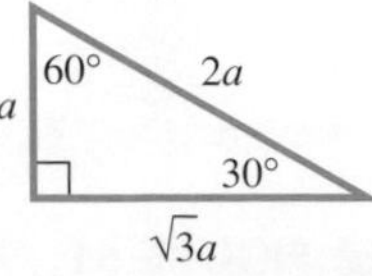

9. Use the figure in Exercise 8 to find the *exact* values of the sine, cosine, and tangent of a 30° angle and of a 60° angle. (That is, use the definitions, not a calculator.)

10. Use the figure to find the *exact* values of sin 45°, cos 45°, and tan 45°.

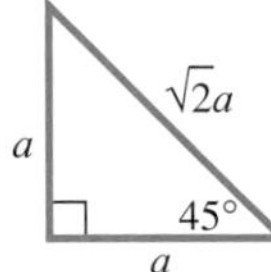

You will need a scientific calculator for Exercises 11–24.

Use an inverse trigonometric function key to find the measure of each angle θ *as described. Express the measure to the nearest tenth of a degree.*

11. the angle θ in the figure of Exercise 1

12. the angle θ in the figure of Exercise 2

13. the angle θ in the figure of Exercise 4

14. the angle θ in the figure of Exercise 6

Solve each problem. Express angle measures to the nearest degree, and lengths to the nearest tenth.

15. The angle of the sun above the horizon is 28°. Find the length of the shadow of a person 4.8 ft tall. (Find x in the figure.)

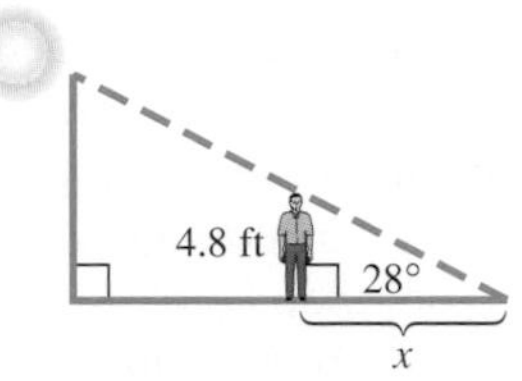

16. Find the angle of the sun above the horizon when a person 5.3 ft tall casts a shadow 8.7 ft long. (Find θ in the figure.)

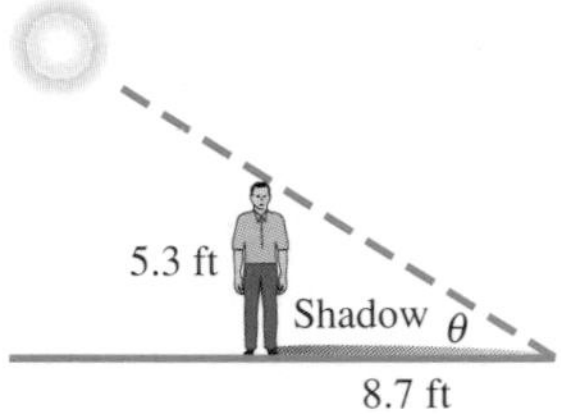

17. In one area, the lowest angle of elevation of the sun in winter is 23°. Find the minimum distance x that a plant needing full sun can be placed from a fence 4.7 ft high. (See the figure.)

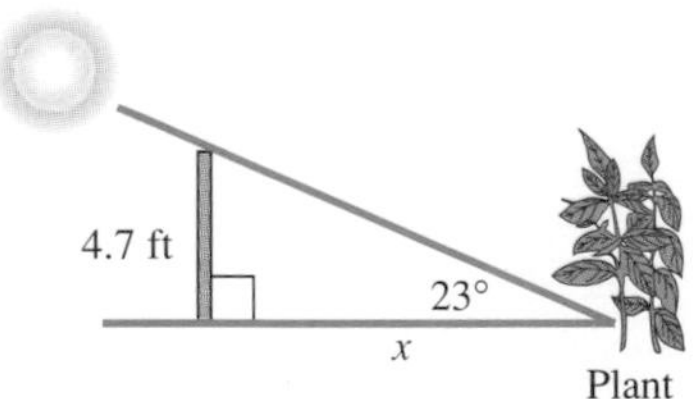

18. The shadow of a vertical tower is 40.6 m long when the angle of elevation of the sun is 35°. Find the height of the tower.

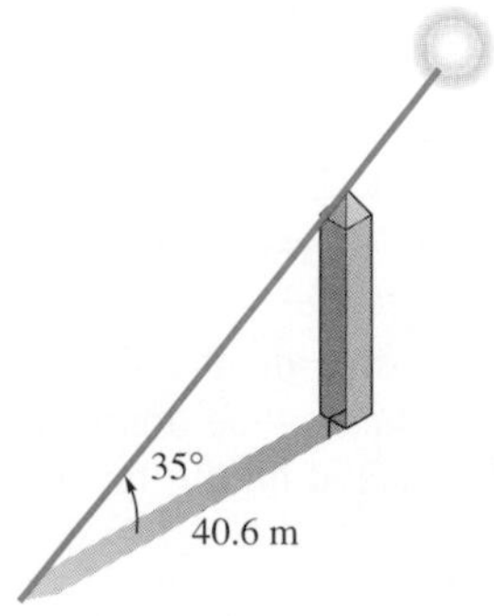

19. To measure the height of a flagpole, Laura Vanek, whose eyes are 6.0 ft above the ground, finds that the angle of elevation measured from eye level to the top of the flagpole is 38°. If Laura is standing at a point 24.8 ft from the base of the flagpole, how tall is the flagpole?

20. The angle of depression from the top of a building to a point on the ground is 32°. How far is the point on the ground from the top of the building if the building is 252.0 m high?

21. A guy wire 77.4 m long is attached to the top of an antenna mast that is 71.3 m high. Find the angle that the wire makes with the ground.

22. A rectangular piece of land is 528.2 ft by 630.7 ft. Find the acute angles made by the diagonal of the rectangle.

23. To find the distance RS across a lake, a surveyor lays off RT = 53.1 m, with angle $T = 32°$, and angle $S = 58°$. Find length RS.

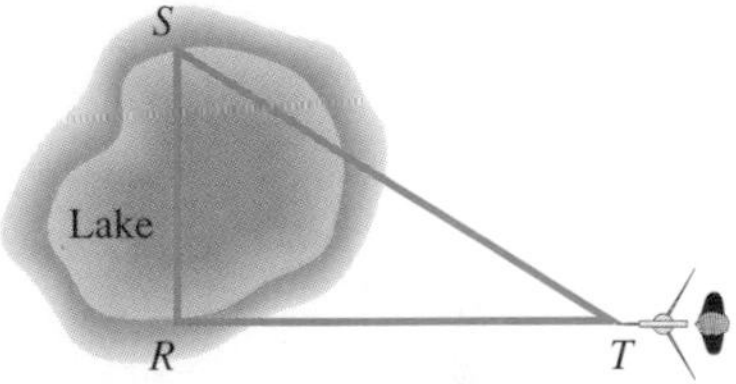

24. A surveyor must find the distance QM across a depressed freeway. She lays off QN = 769.0 ft along one side of the freeway, with angle $N = 22°$, and with angle $M = 68°$. Find QM.

25. The sine and the cosine of an acute angle cannot be greater than or equal to 1. Explain why this is so.

26. Try to calculate $\sin^{-1}(3/2)$ on a calculator. What happens? Why?

Space Figures, Volume, and Surface Area

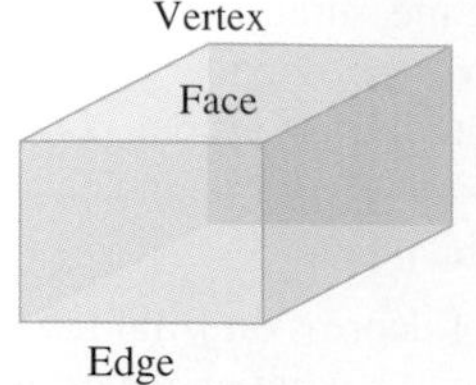

■ **FIGURE 52**

Thus far, this chapter has discussed only **plane figures**—figures that can be drawn completely in the plane of a piece of paper. However, it takes the three dimensions of space to represent the solid world around us. For example, Figure 52 shows a "box" (a **rectangular parallelepiped** in mathematical terminology). The *faces* of a box are rectangles. The faces meet at *edges*; the "corners" are called *vertices* (plural of vertex—the same word as for the "corner" of an angle).

Boxes are one kind of space figure belonging to an important group called **polyhedrons,** the faces of which are made only of polygons.

Perhaps the most interesting polyhedrons are the *regular polyhedrons.* Recall that a *regular polygon* is a polygon with all sides equal and all angles equal. A regular polyhedron is a space figure, the faces of which are only one kind of regular polygon. It turns out that there are only five different regular polyhedrons. They are shown in Figure 53. A **tetrahedron** is composed of four equilateral triangles, each three of which meet in a point. Use the figure to verify that there are four faces, four vertices, and six edges.

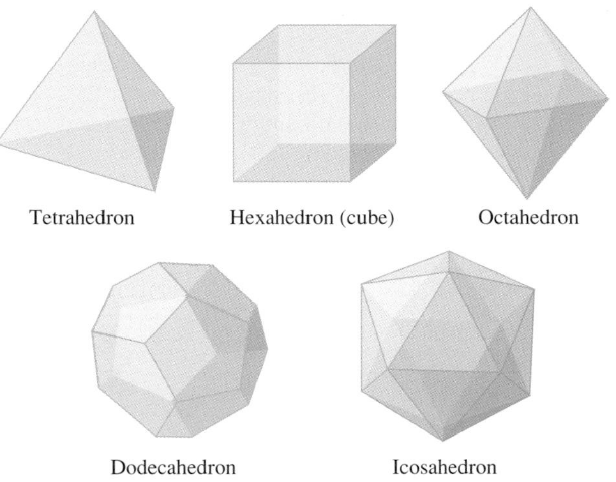

■ **FIGURE 53**

■ For Further Thought

Many crystals and some viruses are constructed in the shapes of regular polyhedrons. Many viruses that infect animals were once thought to have spherical shapes, until X-ray analysis revealed that their shells are icosahedral in nature. The sketch at the top left on the next page shows a polio virus.

Radiolaria, as shown on the right, is a group of microorganisms. The skeletons of these single-celled animals take on beautiful geometric shapes. The one pictured is based on the tetrahedron.

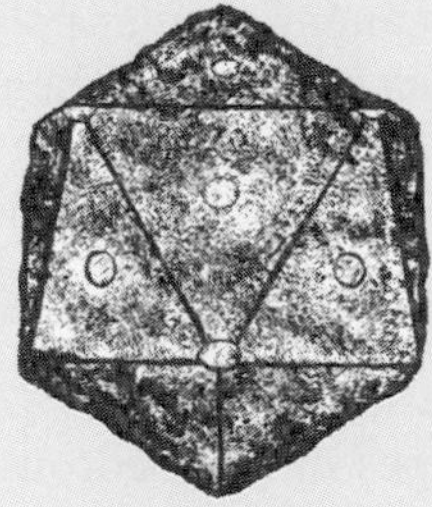

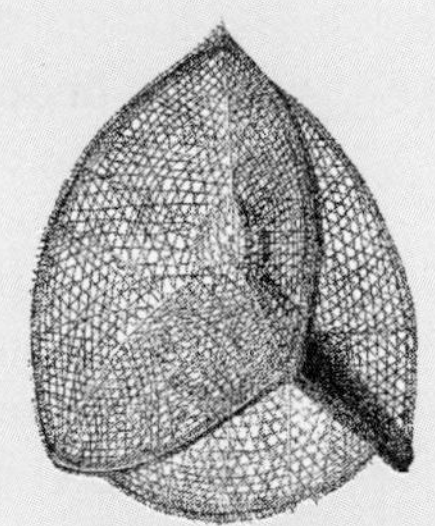

The salt crystal depicted on the stamp at the left is revealed under X-ray photography to have a cubic structure, with alternating ions of sodium and chlorine. The stamp was issued in honor of Sir Lawrence Bragg and his father, Sir William Henry Bragg, who jointly received a Nobel prize in 1915 for their early work in the study of X-ray diffraction and solid state physics.

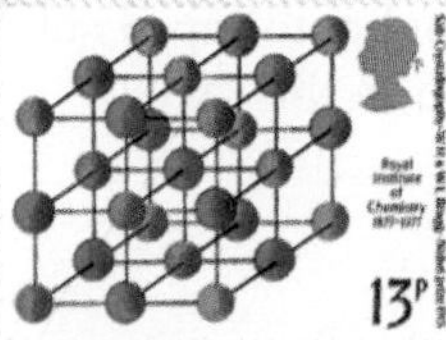

In the text we mention that a cube has 6 faces, 8 vertices, and 12 edges, while a tetrahedron has 4 faces, 4 vertices, and 6 edges. Leonhard Euler (1707–1783) investigated a remarkable relationship among these numbers. The group discussion exercise below will allow you to discover it as well.

For Group Discussion

As a class, use the figures in the text, or models brought to class by your instructor or class members, to complete the following table.

Polyhedron	**Faces (F)**	**Vertices (V)**	**Edges (E)**	**Value of $F + V - E$**
Tetrahedron				
Hexahedron (Cube)				
Octahedron				
Dodecahedron				
Icosahedron				

Now, state a conjecture concerning the value of $F + V - E$ for every polyhedron. This result, called **Euler's formula,** is given on page 552 at the end of the exercise set for this section.

A regular quadrilateral is called a square. Six squares, each three of which meet at a point, form a **hexahedron,** or **cube.** Again, use Figure 53 to verify that a cube has 6 faces, 8 vertices, and 12 edges.

Tetrahedron

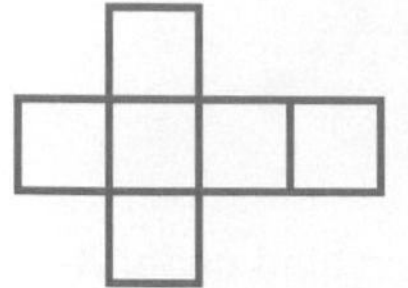

Hexahedron (cube)

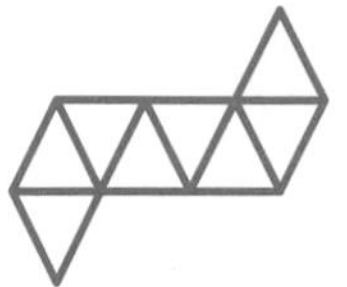

Octahedron

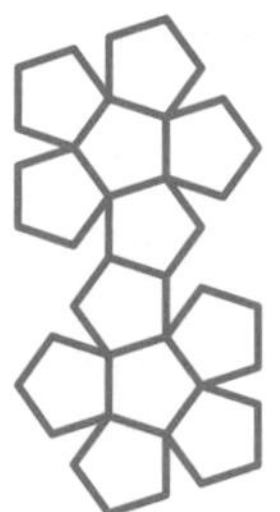

Dodecahedron

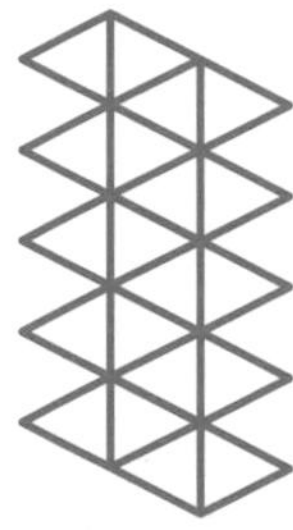

Icosahedron

Patterns such as these may be used to actually construct three-dimensional models of the regular polyhedrons.

The three remaining regular polyhedrons are the **octahedron,** the **dodecahedron,** and the **icosahedron.** The octahedron is composed of groups of four regular triangles. The dodecahedron is formed by groups of three regular pentagons, while the icosahedron is made up of groups of five regular triangles. The relationships among the number of faces, vertices, and edges in each case are discussed in For Further Thought in this section, where a formula named after Euler is investigated.

Two other types of polyhedrons are familiar space figures: pyramids and prisms. **Pyramids** are made of triangular sides and a polygonal base. **Prisms** have two faces in parallel planes; they are congruent polygons. The remaining faces of a prism are all parallelograms. (See Figure 54.) By this definition, a box is also a prism.

The circle, although a plane figure, is not a polygon. (Why?) Figure 54 also shows space figures made up in part of circles, including right circular cones and right circular cylinders. The figure also shows how circles can generate a torus. (The torus is a doughnut-shaped solid that has interesting topological properties—see Section 6 in this chapter.)

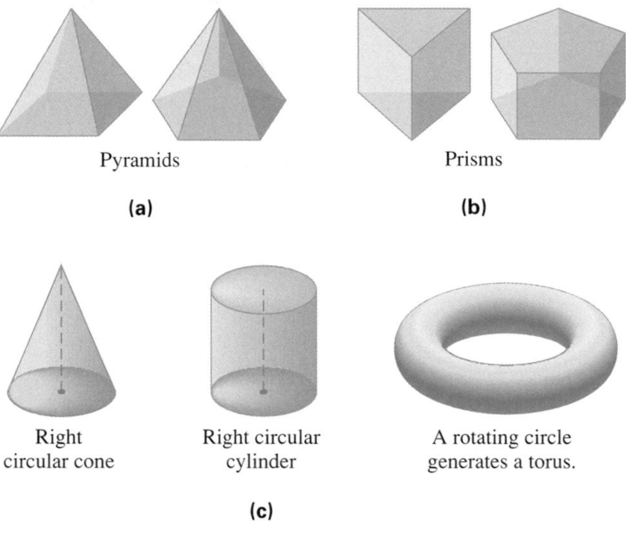

FIGURE 54

Volume and Surface Area

While area is a measure of surface covered by a plane figure, **volume** is a measure of capacity of a space figure. Volume is measured in *cubic* units. For example a cube with edge measuring 1 cm has volume 1 cubic cm, which is also written as 1 cm^3, or 1 cc. The **surface area** is the total area that would be covered if the space figure were "peeled" and the peel laid flat. Surface area is measured in square units.

The formulas for volume and surface area of a rectangular parallelepiped, or "box", are now given.

Volume and Surface Area of a Box

Suppose that a box has length ℓ, width w, and height h. Then the volume V and the surface area S are given by the formulas

$$V = \ell wh \qquad \text{and} \qquad S = 2\ell w + 2\ell h + 2hw.$$

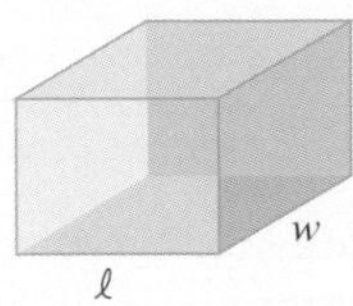

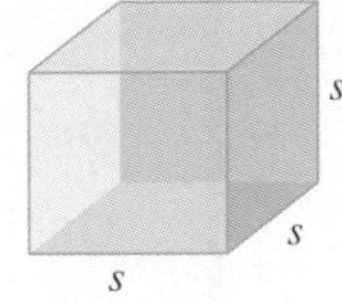

In particular, if the box is a cube with edge of length s,

$$V = s^3 \qquad \text{and} \qquad S = 6s^2.$$

EXAMPLE 1 Find the volume V and the surface area S of the box shown in Figure 55.

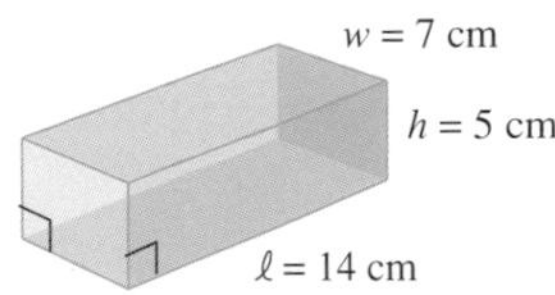

FIGURE 55

First, to find the volume, use the formula $V = \ell wh$ with $\ell = 14$, $w = 7$, and $h = 5$.

$$V = \ell wh$$
$$V = 14 \cdot 7 \cdot 5$$
$$V = 490$$

Volume is measured in cubic units, so the volume of the box is 490 cubic centimeters, or 490 cm^3.

Next, to find the surface area, use the formula $S = 2\ell w + 2\ell h + 2hw$.

$$S = 2(14)(7) + 2(14)(5) + 2(5)(7)$$
$$S = 196 + 140 + 70$$
$$S = 406$$

Like areas of plane figures, surface areas of space figures are given in square measure, so the surface area of the box is 406 square centimeters, or 406 cm^2. ◆

A typical tin can is an example of a **right circular cylinder.** The formulas for the volume and surface area of a right circular cylinder are now given.

Volume and Surface Area of a Right Circular Cylinder

If a right circular cylinder has height h and radius of its base equal to r, then the volume V and the surface area S are given by the formulas

$$V = \pi r^2 h$$

and $$S = 2\pi rh + 2\pi r^2.$$

(In the formula for S, the areas of the top and bottom are included.)

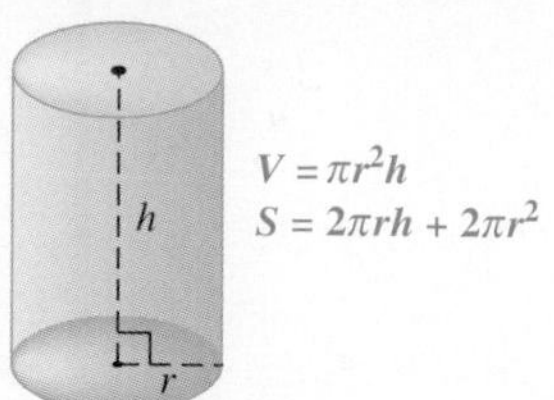

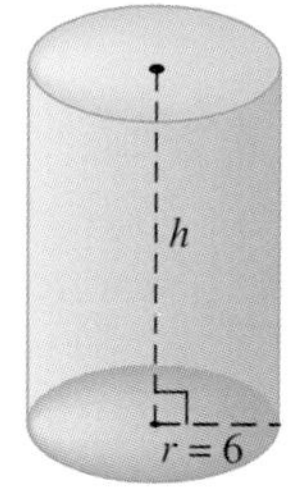

FIGURE 56

EXAMPLE 2 In Figure 56, the right circular cylinder has surface area 288π square inches, and the radius of its base is 6 inches.

(a) Find the height of the cylinder.

Since we know that $S = 288\pi$ and $r = 6$, substitute into the formula for surface area to find h.

$$\begin{aligned} S &= 2\pi rh + 2\pi r^2 \\ 288\pi &= 2\pi(6)h + 2\pi(6)^2 \\ 288\pi &= 12\pi h + 72\pi \\ 288\pi - 72\pi &= 12\pi h \\ 216\pi &= 12\pi h \\ h &= 18 \qquad \text{Divide by } 12\pi. \end{aligned}$$

The height is 18 inches.

(b) Find the volume of the cylinder.

Use the formula for volume, with $r = 6$ and $h = 18$.

$$\begin{aligned} V &= \pi r^2 h \\ V &= \pi(6)^2(18) \\ V &= 648\pi \end{aligned}$$

The exact volume is 648π cubic inches, or approximately 2,034.72 cubic inches, using $\pi \approx 3.14$.

The three-dimensional analogue of a circle is a sphere. It is defined by replacing the word "plane" with "space" in the definition of a circle (Section 2 of this chapter). The formulas for the volume and the surface area of a sphere are as follows.

Volume and Surface Area of a Sphere

If a sphere has radius r, then the volume V and the surface area S are given by the formulas

$$V = \frac{4}{3}\pi r^3 \quad \text{and} \quad S = 4\pi r^2.$$

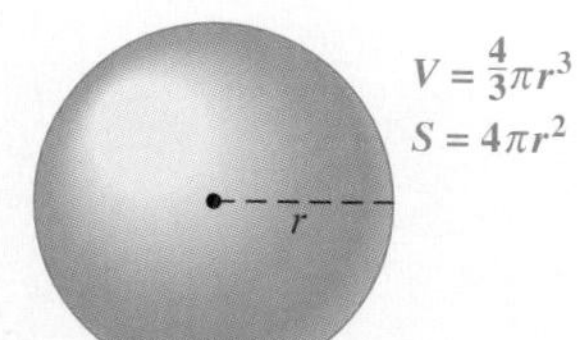

EXAMPLE 3 Suppose that a spherical tank having radius 3 meters can be filled with liquid fuel for $200. How much will it cost to fill a spherical tank of radius 6 meters with the same fuel?

We must first find the volume of the tank with radius 3 meters. Call it V_1.

$$V = \frac{4}{3}\pi r^3 \qquad \text{Formula for volume of a sphere}$$

$$V_1 = \frac{4}{3}\pi(3)^3$$

$$V_1 = \frac{4}{3}\pi(27) = 36\pi$$

Now find V_2, the volume of the tank having radius 6 meters.

$$V_2 = \frac{4}{3}\pi(6)^3$$

$$V_2 = \frac{4}{3}\pi(216) = 288\pi$$

■ FIGURE 57

Notice that by doubling the radius of the sphere from 3 meters to 6 meters, the volume has increased 8 times, since

$$V_2 = 288\pi = 8V_1 = 8(36\pi).$$

Therefore, the cost to fill the larger tank is eight times the cost to fill the smaller one: 8($200) = $1,600. ■

The space figure shown in Figure 57 is a right circular cone. The formulas for the volume and the surface area of a right circular cone are now given.

Volume and Surface Area of a Right Circular Cone

If a right circular cone has height h and the radius of its circular base is r, then the volume V and the surface area S are given by the formulas

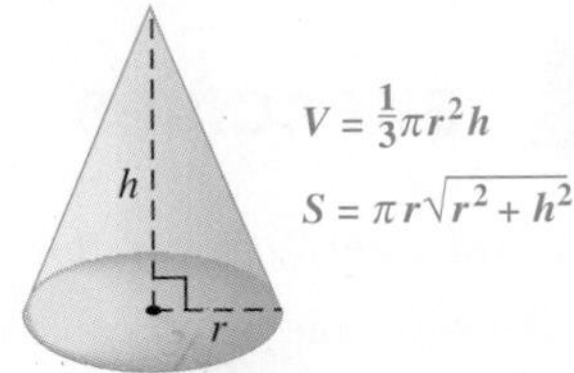

$$V = \frac{1}{3}\pi r^2 h$$

and $\quad S = \pi r\sqrt{r^2 + h^2}.$

A pyramid is a space figure having a polygonal base and triangular sides. Figure 58 shows a pyramid with a square base. The formula for the volume of a pyramid follows.

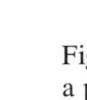

■ FIGURE 58

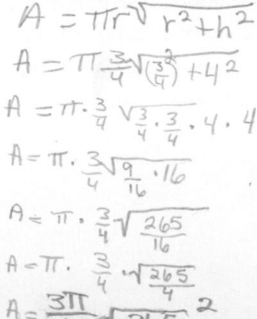

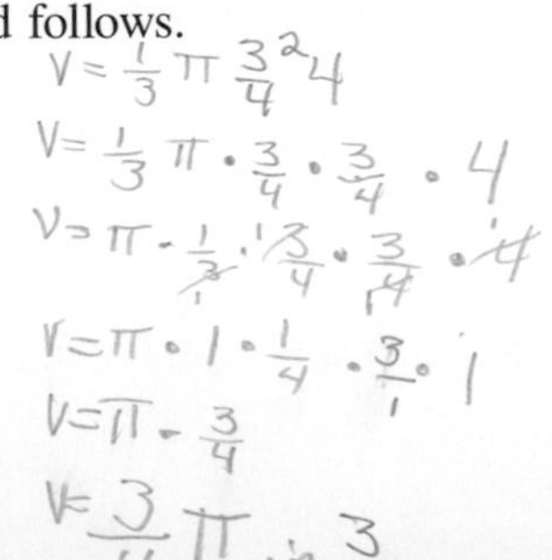

The Transamerica Tower in San Francisco is a pyramid with a square base. Each side of the base has a length of 52 meters, while the height of the building is 260 meters. The formula for the volume of a pyramid indicates that the volume of the building is about 234,000 cubic meters.

Volume of a Pyramid

If B represents the area of the base of a pyramid, and h represents the height (that is, the perpendicular distance from the top, or apex, to the base), then the volume V is given by the formula

$$V = \frac{1}{3}Bh.$$

$V = \frac{1}{3}Bh$ where B is the area of the base

EXAMPLE 4 What is the ratio of the volume of a right circular cone with radius of base r and height h to the volume of a pyramid having a square base, with each side of length r, and height h?

Using the formula for the volume of a cone, we have

$$V_1 = \text{Volume of the cone} = \frac{1}{3}\pi r^2 h.$$

Since the pyramid has a square base, the area B of its base is r^2. Using the formula for the volume of a pyramid, we get

$$V_2 = \text{Volume of the pyramid} = \frac{1}{3}Bh = \frac{1}{3}(r^2)h.$$

The ratio of the first volume to the second is

$$\frac{V_1}{V_2} = \frac{\frac{1}{3}\pi r^2 h}{\frac{1}{3}r^2 h} = \pi.$$

9.5 EXERCISES

Decide whether each of the following statements is true *or* false.

1. A cube with volume 64 cubic inches has surface area 96 square inches.
2. A tetrahedron has the same number of faces as vertices.
3. A dodecahedron can be used as a model for a calendar for a given year, where each face of the dodecahedron contains a calendar for a single month, and there are no faces left over.
4. Each face of an octahedron is an octagon.
5. If you double the length of the edge of a cube, the new cube will have a volume that is twice the volume of the original cube.
6. The numerical value of the volume of a sphere is $r/3$ times the numerical value of its surface area, where r is the measure of the radius.

Find **(a)** *the volume and* **(b)** *the surface area of each of the following space figures. When necessary, use 3.14 as an approximation for π, and round answers to the nearest hundredth.*

7. 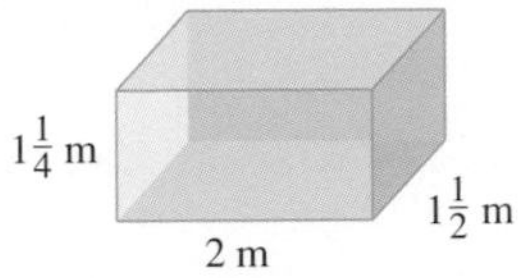 $1\frac{1}{4}$ m, 2 m, $1\frac{1}{2}$ m
(a box)

8. 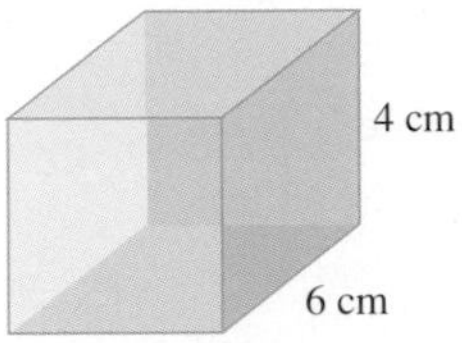 4 cm, 6 cm, 4 cm
(a box)

9. 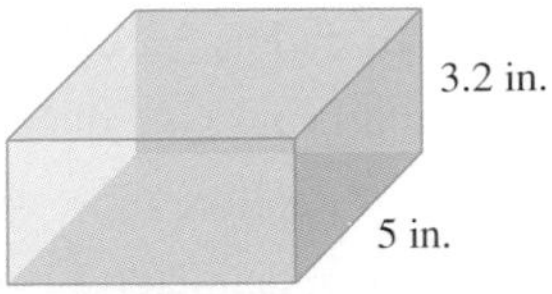 3.2 in., 5 in., 6 in.
(a box)

10. 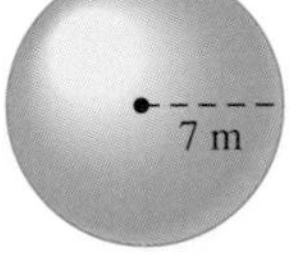 7 m
(a sphere)

11. 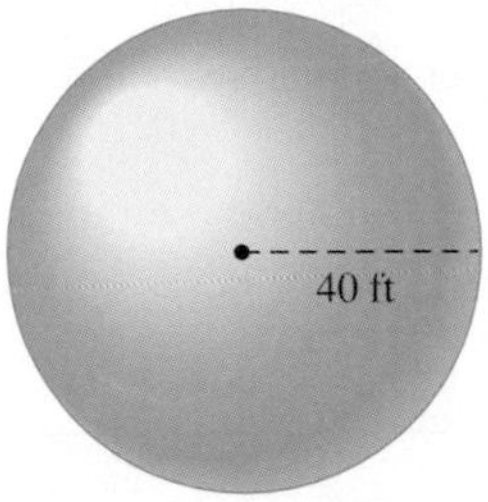 40 ft
(a sphere)

12. 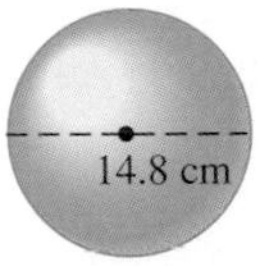 14.8 cm
(a sphere)

13. 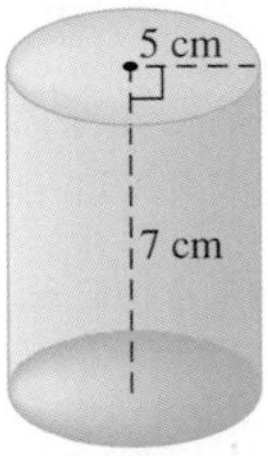 5 cm, 7 cm
(a right circular cylinder)

14. 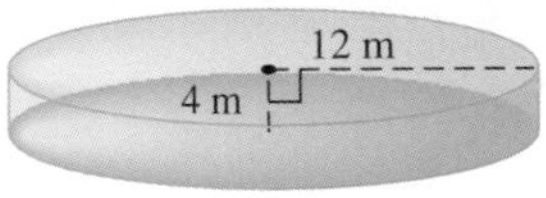 12 m, 4 m
(a right circular cylinder)

15. 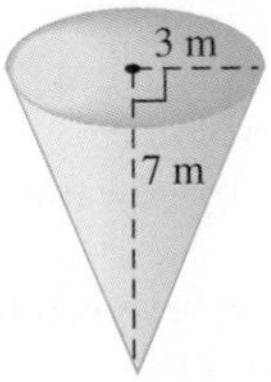 3 m, 7 m
(a right circular cone)

16. 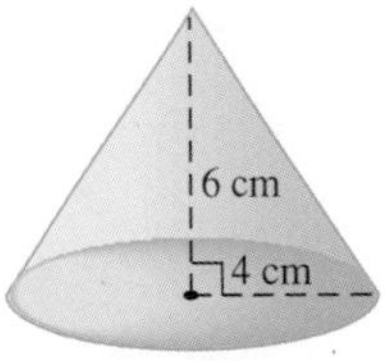 6 cm, 4 cm
(a right circular cone)

Find the volume of each pyramid. In each case, the base is a rectangle.

17. **18.**

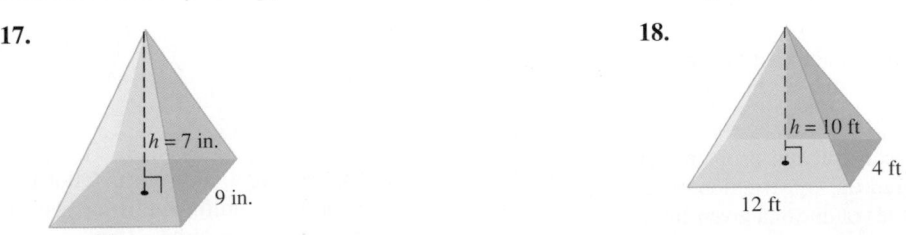

Find the volume of each of the following objects. Use 3.14 as an approximation for π when necessary.

19. a coffee can, radius 6.3 cm and height 15.8 cm

20. a soup can, radius 3.2 cm and height 9.5 cm

21. a pork-and-beans can, diameter 7.2 cm and height 10.5 cm

22. a cardboard mailing tube, diameter 2 in. and height 40 in.

23. a coffee mug, diameter 9 cm and height 8 cm

24. a bottle of typewriter correction fluid, diameter 3 cm and height 4.3 cm

25. the Great Pyramid of Cheops, near Cairo—its base is a square 230 m on a side, while the height is 137 m

26. the Peachtree Plaza Hotel in Atlanta, a cylinder with a base radius of 46 m and a height of 220 m

27. a road construction marker, a cone with height 2 m and base radius 1/2 m

28. the conical portion of a witch's hat for a Halloween costume, with height 12 in. and base radius 4 in.

In the chart below, one of the values r (radius), d (diameter), V (volume), or S (surface area) is given for a particular sphere. Find the remaining three values. Leave π in your answers.

	r	d	V	S
29.	6 in.			
30.	9 in.			
31.		10 ft		
32.		40 ft		
33.			$\frac{32}{3}\pi$ cm^3	
34.			$\frac{256}{3}\pi$ cm^3	
35.				4π m^2
36.				144π m^2

37. In order to determine the amount of liquid a spherical tank will hold, would you need to use volume or surface area?

38. In order to determine the amount of leather it would take to manufacture a basketball, would you need to use volume or surface area?

39. One of the three famous construction problems of Greek mathematics required the construction of an edge of a cube with twice the volume of a given cube. If the length of each side of the given cube is x, what would be the length of each side of a cube with twice the original volume?

40. Work through the parts of this exercise in order, and use it to make a generalization concerning volumes of spheres. Leave answers in terms of π.

(a) Find the volume of a sphere having a radius of 1 m.

(b) Suppose the radius is doubled to 2 m. What is the volume?

(c) When the radius was doubled, by how many times did the volume increase? (To find out, divide the answer for part (b) by the answer for part (a).)

(d) Suppose the radius of the sphere from part (a) is tripled to 3 m. What is the volume?

(e) When the radius was tripled, by how many times did the volume increase?

(f) In general, if the radius of a sphere is multiplied by n, the volume is multiplied by _____.

If a spherical tank 2 m *in diameter can be filled with a liquid for* \$300, *find the cost to fill tanks of the following diameters.*

41. 6 m

42. 8 m

43. 10 m

44. Use the logic of Exercise 40 to answer the following: If the radius of a sphere is multiplied by n, then the surface area of the sphere is multiplied by _____.

Each of the following figures has volume as indicated. Find the value of x.

45. $V = 60$

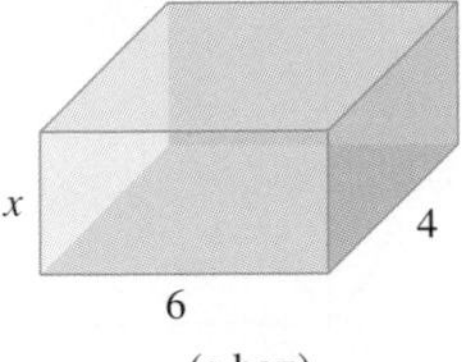

(a box)

46. $V = 450$

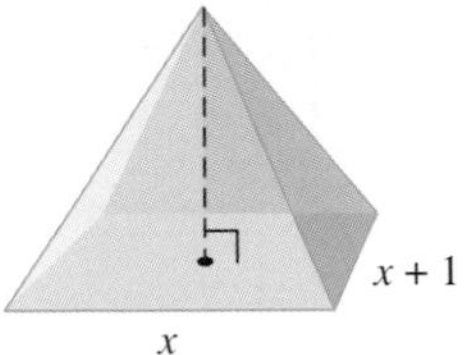

$h = 15$ cm
Base is a rectangle.

(a pyramid)

47. $V = 36\pi$

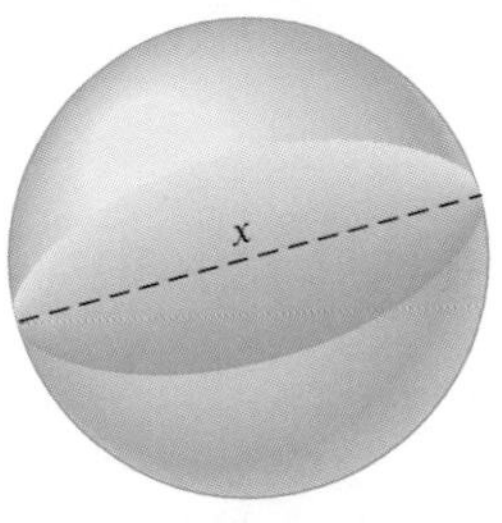

(a sphere)

48. $V = 245\pi$

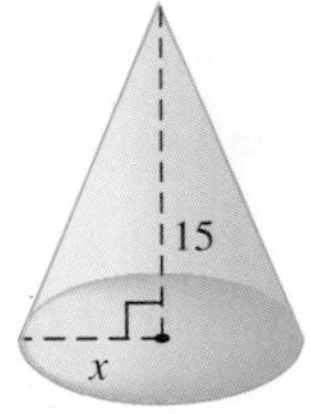

(a right circular cone)

Exercises 49–56 require some ingenuity, but all can be solved using the concepts presented so far in this chapter.

49. The areas of the sides of a rectangular box are 30 in.^{2}, 35 in.^{2}, and 42 in.^{2}. What is the volume of the box?

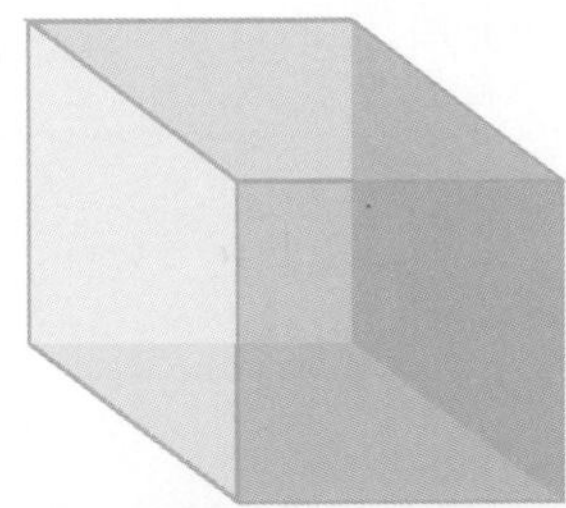

50. In the figure, a right circular cone is inscribed in a hemisphere. What is the ratio of the volume of the cone to the volume of the hemisphere?

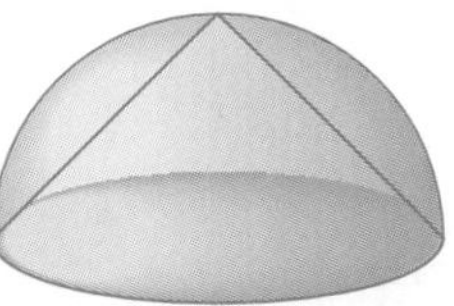

51. A plane intersects a sphere to form a circle as shown in the figure. If the area of the circle formed by the intersection is 576π in.^{2}, what is the volume of the sphere?

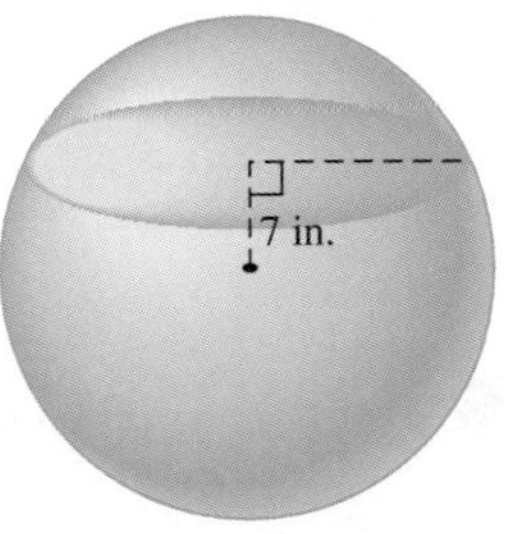

52. If the height of a right circular cylinder is halved and the diameter is tripled, how is the volume changed?

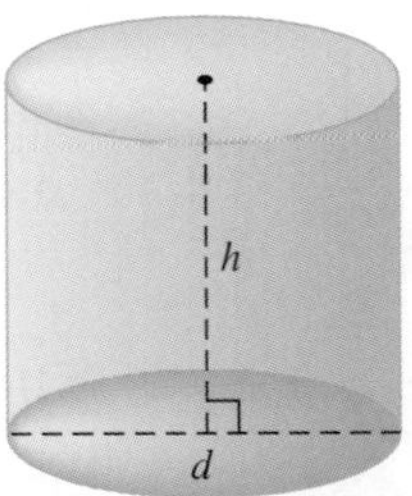

53. What is the ratio of the area of the circumscribed square to the area of the inscribed square?

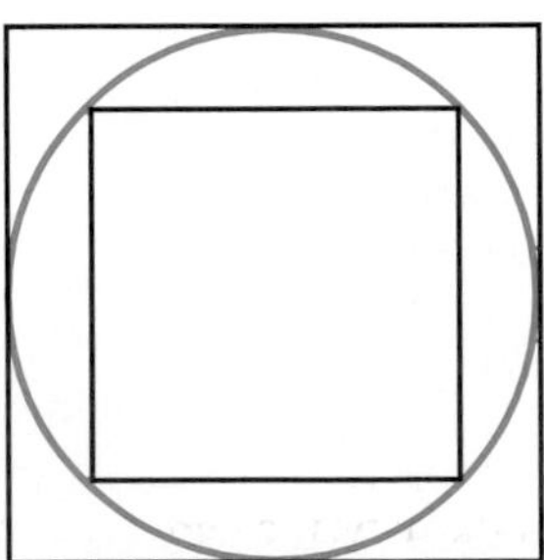

54. The diameter of the circle shown is 8 in. What is the perimeter of the inscribed square $ABCD$?

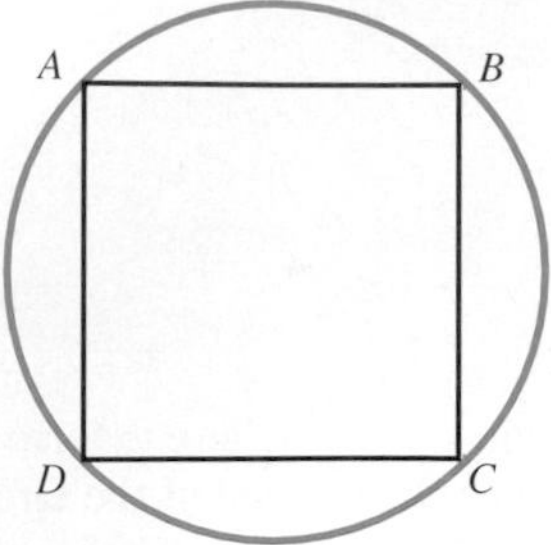

55. In the circle shown with center O, the radius is 6. $QTSR$ is an inscribed square. Find the value of $PQ^2 + PT^2 + PR^2 + PS^2$.

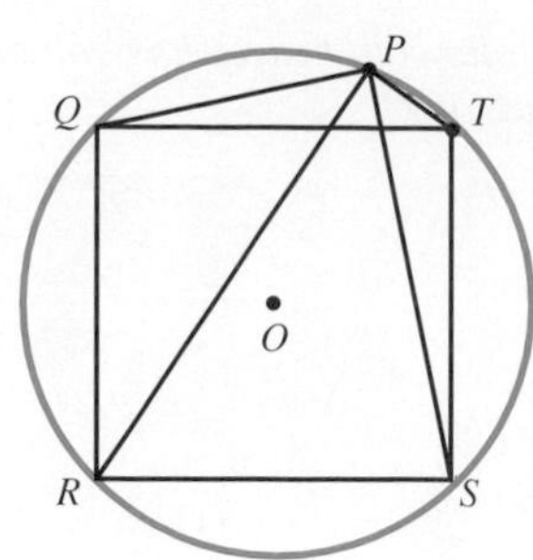

56. The square $JOSH$ is inscribed in a semicircle. What is the ratio of x to y?

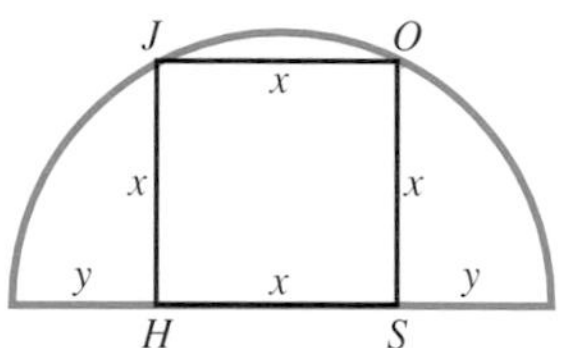

Answer to *For Group Discussion* on page 543: Euler's formula is $F + V - E = 2$.

9.6 Non-Euclidean Geometry, Topology, and Networks

At the beginning of this chapter is a list of the axioms and postulates from Euclid's famous *Elements.* Actually, Playfair's axiom on parallel lines was substituted for Euclid's own fifth postulate. However, to understand why the fifth caused trouble for so many mathematicians for so long, we must examine the original formulation.

John Playfair (1748–1819) wrote his *Elements of Geometry* in 1795. Playfair's Axiom is: Given a line k and a point P not on the line, there exists one and only one line m through P that is parallel to k. Playfair was a geologist, who fostered "uniformitarianism", the doctrine that geological processes long ago gave Earth its features, and processes today are the same kind as those in the past. He was the first to state that a river cuts its own valley.

Euclid's Fifth Postulate ("Parallel Postulate")

Euclid's fifth postulate states that if two lines (k and m in Figure 59) are such that a third line, n, intersects them so that the sum of the two interior angles (A and B) on one side of line n is less than two right angles, then the two lines, if extended far enough, will meet on the same side of n that has the sum of the interior angles less than two right angles.

Euclid's parallel postulate is quite different from the other nine postulates and axioms we listed. The others are simple statements that seem in complete agreement with our experience of the world around us. But the parallel postulate is long and wordy, and difficult to understand without a sketch.

The difference between the parallel postulate and the other axioms was noted by the Greeks, as well as later mathematicians. It was commonly believed

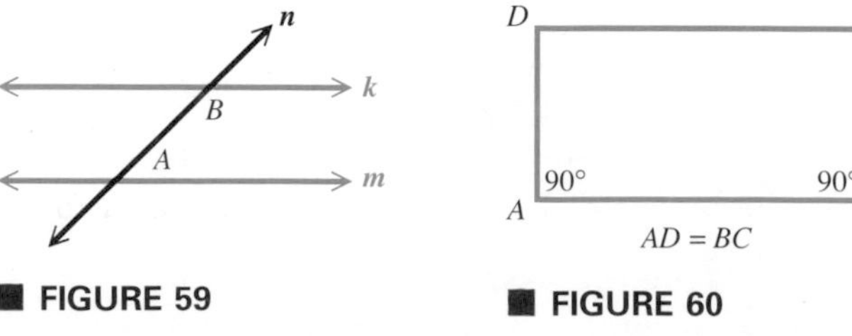

■ FIGURE 59

■ FIGURE 60

By the time of the Renaissance 17 centuries after Euclid, artists such as Leonardo da Vinci were studying the images of three-dimensional objects on two-dimensional surfaces. *Projective geometry* emerged, with ideas distinct from those of Euclid. In the nineteenth century even the hallowed axioms of Euclid were challenged, yielding *non-Euclidean geometries.* Modern geometry includes the important field of *topology,* in which figures can be stretched, twisted, and deformed (but not cut).

that this was not a postulate at all, but a theorem to be proved. For more than 2,000 years mathematicians tried repeatedly to prove it.

The most dedicated attempt came from an Italian Jesuit, Girolamo Saccheri (1667–1733). He attempted to prove the parallel postulate in an indirect way, so-called "reduction to absurdity." He would assume the postulate to be false and then show that the assumption leads to a contradiction of something true (an absurdity). Such a contradiction would thus prove the statement true.

Saccheri began with a quadrilateral, as in Figure 60. He assumed angles A and B to be right angles and sides AD and BC to be equal. His plan was as follows:

1. To assume that angles C and D are obtuse angles, and to show that this leads to a contradiction.
2. To assume that angles C and D are acute angles, and to show that this also leads to a contradiction.
3. Then if C and D can be neither acute nor obtuse angles, they must be right angles.
4. If C and D are both right angles, then it can be proved that the fifth postulate is true, and thus is a theorem rather than a postulate.

Saccheri had no trouble with part 1. However, he did not actually reach a contradiction in the second part, but produced some theorems so "repugnant" that he convinced himself he had vindicated Euclid. In fact, he published a book called in English *Euclid Freed of Every Flaw.*

Today we know that Saccheri was wrong—there is no contradiction to be found by assuming angles C and D acute angles. Thus, the fifth postulate is indeed an axiom, and not a theorem. It is *consistent* with Euclid's other axioms. Mathematicians had tried for 2,000 years to do the impossible.

The ten axioms of Euclid describe the world around us with remarkable accuracy. We now realize that the fifth postulate is necessary in Euclidean geometry to establish *flatness.* That is, the axioms of Euclid describe the geometry of *plane surfaces.* By changing the fifth postulate, we can describe the geometry of other surfaces. So, other geometric systems exist as much as Euclidean geometry exists, and they can even be demonstrated in our world. They are just not as familiar. A system of geometry in which the fifth postulate is changed is called a **non-Euclidean geometry.**

A song titled simply **Lobachevsky** appears on the comedy LP *Songs by Tom Lehrer* (Reprise RS-6216). The liner notes indicate that the song "is a description of one way to get ahead in mathematics (which happens to be the author's own academic specialty) or any other academic field." Find a copy of the album, and see what Lehrer suggests!

The Origins of Non-Euclidean Geometry

One non-Euclidean system was developed by three people working separately at about the same time. Early in the nineteenth century Carl Friedrich Gauss, one of the great mathematicians, worked out a consistent geometry replacing Euclid's fifth postulate. He never published his work, however, because he feared the ridicule of people who could not free themselves from habitual ways of thinking. Gauss first used the term "non-Euclidean." Nikolai Ivanovich Lobachevski (1793–1856) published a similar system in 1830 in the Russian language. At the same time, Janos Bolyai (1802–1860), a Hungarian army officer, worked out a similar system, which he published in 1832, not knowing about Lobachevski's work. Bolyai never recovered from the disappointment of not being the first, and did no further work in mathematics.

Lobachevski replaced Euclid's fifth postulate with:

> Angles C and D in the quadrilateral of Saccheri are acute angles.

This postulate of Lobachevski can be rephrased as follows:

> Through a point P off a line k (Figure 61), at least two different lines can be drawn parallel to k.

■ FIGURE 61

Compare this form of Lobachevski's postulate to the geometry of Euclid. How many lines can be drawn through P and parallel to k in Euclidean geometry? At first glance, the postulate of Lobachevski does not agree with what we know about the world around us. But this is only because we think of our immediate surroundings as being flat.

Many of the theorems of Euclidean geometry are valid for the geometry of Lobachevski, but many are not. For example, in Euclidean geometry, the sum of the measures of the angles in any triangle is 180°. In Lobachevskian geometry, the sum of the measures of the angles in any triangle is *less* than 180°. Also, triangles of different sizes can never have equal angles, so similar triangles do not exist.

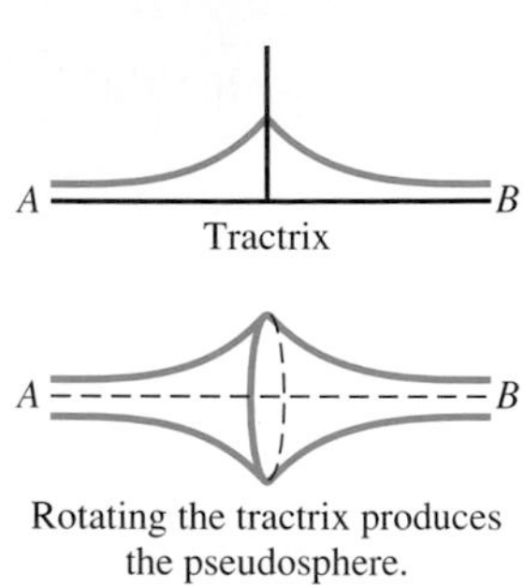

Rotating the tractrix produces the pseudosphere.

■ FIGURE 62

The geometry of Euclid can be represented on a plane. Since any portion of the earth that we are likely to see looks flat, Euclidean geometry is very useful for describing the everyday world around us. The non-Euclidean geometry of Lobachevski can be represented as a surface called a **pseudosphere.** This surface is formed by revolving a curve called a **tractrix** about the line AB in Figure 62.

For Further Thought

Artists must have a sound knowledge of geometry in order to paint realistically a world of solids in space. Our three-dimensional world must be represented in a convincing way on canvas or some other plane surface. For example, how should an artist paint a realistic view of railroad tracks going off to the horizon? In reality, the tracks are always at a constant distance apart, but they cannot be drawn that way except from overhead. The artist must make the tracks converge at a point. Only in this way will the scene look "real."

Beginning in the fifteenth century, artists led by Leone Battista Alberti, Leonardo da Vinci, and Albrecht Dürer began to study the problems of representing three dimensions in two. They found geometric methods of doing this. What artists initiated, mathematicians developed into a geometry different from that of Euclid—**projective geometry.**

Gerard Desargues (1591–1661), architect and engineer, was a technical advisor to the French government. He met Descartes at the siege of La Rochelle in 1628 and in the 1630s was a member of the Parisian group that included Descartes, Pascal, Fermat, and Mersenne. In 1636 and 1639 he published a treatise and proposals about the perspective section—Desargues had invented projective geometry. His works unfortunately were too difficult for the times; his terms were from botany and he did not use Cartesian symbolism. Desargues' geometric innovations were hidden for nearly 200 years. A manuscript by Desargues turned up in 1845, about 30 years after Jean-Victor Poncelet had rediscovered projective geometry.

For Group Discussion
Have a class member go to the board with a long straightedge, and draw a figure similar to the one shown here.

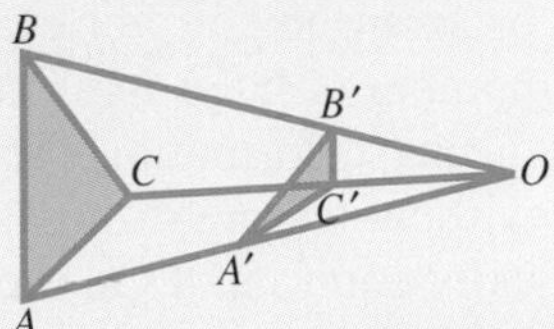

Then have the student, *very carefully,* follow these instructions.

1. Extend lines AC and $A'C'$ of the two triangles to meet in a point M.
2. Extend sides AB and $A'B'$ to meet in a point N.
3. Extend BC and $B'C'$ to meet in a point P.

Now, as a class, make a conjecture about what has happened. This exercise is an illustration of Desargues' theorem (stated in Exercise 12 of this section).

A second non-Euclidean system was developed by Georg Riemann (1826–1866). He pointed out the difference between a line that continues indefinitely and a line having infinite length. For example, a circle on the surface of a sphere continues indefinitely but does not have infinite length. Riemann developed the idea of geometry on a sphere and replaced Euclid's fifth postulate with:

Angles C and D of the Saccheri quadrilateral are obtuse angles.

In terms of parallel lines, Riemann's postulate becomes:

Through a point P off a line k, no line can be drawn that is parallel to k.

Riemannian geometry is important in navigation. "Lines" in this geometry are really *great circles,* or circles whose centers are at the center of the sphere. The shortest distance between two points on a sphere lies along an arc of a great circle. Great circle routes on a globe don't look at all like the shortest distance when the globe is flattened out to form a map, but this is part of the distortion that occurs when the earth is represented as a flat surface. See Figure 63. The sides of a triangle drawn on a sphere would be arcs of great circles. And, in Riemannian geometry, the sum of the measures of the angles in any triangle is *more* than 180°.

FIGURE 63

Georg Friedrich Bernhard Riemann (1826–1866) was a German mathematician. Though he lived a short time and published few papers, his work forms a basis for much modern mathematics. He made significant contributions to the theory of functions and the study of complex numbers as well as to geometry. Most calculus books today use the idea of a "Riemann sum" in defining the integral.

Riemann achieved a complete understanding of the non-Euclidean geometries of his day, expressing them on curved surfaces and showing how to extend them to higher dimensions. In 1915 Albert Einstein borrowed Riemannian geometry as the mathematical foundation of his general theory of relativity. The work of both Lobachevski and Riemann was slow to take hold in their day because both men were rather poor, neither was in the favor of the political powers in their regions, and the old established Euclidean ideas were generally regarded as above dispute.

Topology

This chapter began by suggesting that Euclidean geometry might seem rigid from a twentieth-century point of view. The plane and space figures studied in the Euclidean system are carefully distinguished by differences in size, shape, angularity, and so on. For a given figure such properties are permanent, and thus we can ask sensible questions about congruence and similarity. Suppose we studied "figures" made of rubber bands, as it were, "figures" that could be stretched, bent, or otherwise distorted without tearing or scattering. **Topology,** an important twentieth-century geometry, does just that.

Topological questions concern the basic structure of objects rather than size or arrangement. For example, a typical topological question has to do with the number of holes in an object, a basic structural property that does not change during deformation. You cannot deform a rubber ball to get a rubber band without tearing it—making a hole in it. Thus the two objects are not topologically equivalent. On the other hand, a doughnut and a coffee cup are topologically equivalent, since one could be stretched so as to form the other, without changing the basic structural property.

For Further Thought

Two examples of topological surfaces are the **Möbius strip** and the **Klein bottle.** The Möbius strip is a single-sided surface named after August Ferdinand Möbius (1790–1868), a pupil of Gauss.

To construct a Möbius strip, cut out a rectangular strip of paper, perhaps 3 cm by 25 cm. Paste together the two 3-cm ends after giving the paper a half-twist. To see how the strip now has only one side, mark an *x* on the strip and then mark another *x* on what appears to be the other "side." Begin at one of the *x*'s you have drawn, and trace a path along the strip. You will eventually come to the other *x* without crossing the edge of the strip.

A branch of chemistry called chemical topology studies the structures of chemical configurations. A recent advance in this area was the synthesis of the first molecular Möbius strip, which was formed by joining the ends of a double-stranded strip of carbon and oxygen atoms.

A mathematician confided
That a Möbius strip is one-sided.
And you'll get quite a laugh
If you cut one in half,
For it stays in one piece when divided.

Möbius strip

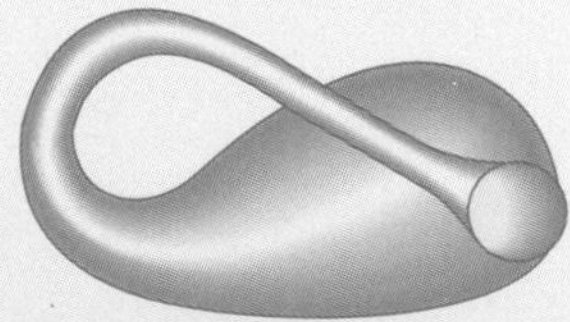

Klein bottle

Whereas a Möbius strip results from giving a paper *strip* a half-twist and then connecting it to itself, if we could do the same thing with a paper *tube* we would obtain a Klein bottle, named after Felix Klein (1849–1925). Klein produced important results in several areas, including non-Euclidean geometry and the early beginnings of group theory. (It is not possible to actually construct a Klein bottle.)

A mathematician named Klein
Thought the Möbius strip was divine.
Said he, "If you glue
The edges of two
You'll get a weird bottle like mine."

For Group Discussion

1. The Möbius strip has other interesting properties. With a pair of scissors, cut the strip lengthwise. Do you get two strips? Repeat the process with what you have obtained from the first cut. What happens?
2. Now construct another Möbius strip, and start cutting lengthwise about 1/3 of the way from one edge. What happens?
3. What would be the advantage of a conveyor belt with the configuration of a Möbius strip?

Topology in Medicine
Trypanosome is a parasite that causes sleeping sickness in humans. The genetic material for this parasite contains a large number of linked circles. If the drug ethidium bromide is introduced into the cells, these circles become hopelessly intertwined, and the parasite cannot reproduce. Thus, topology can be used to cure sleeping sickness!

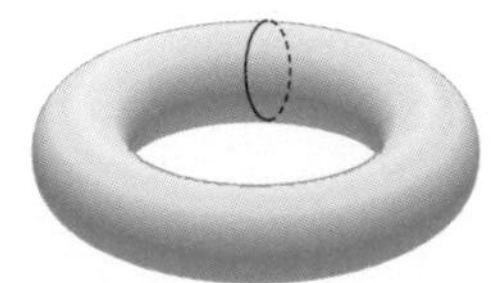

Torus One of the most useful figures in topology is the torus, a doughnut-like or tire-shaped surface. Its properties are different from those of a sphere, for example. Imagine a sphere covered with hair. You cannot comb the hairs in a completely smooth way; one fixed point remains, as you can find on your own head. In the same way, on the surface of Earth the winds are not a smooth system. There is a calm point somewhere. However, the hair on a torus could be combed smooth. The winds would be blowing smoothly on a planet shaped like a torus.

EXAMPLE 1 Decide if the figures in each pair are topologically equivalent.

(a) a football and a cereal box

If we assume that a football is made of a perfectly elastic substance like rubber or dough, it could be twisted or kneaded into the same shape as a cereal box. Thus, the two figures are topologically equivalent.

(b) a doughnut and an unzipped coat

A doughnut has one hole, while the coat has two (the sleeve openings). Thus, a doughnut could not be stretched and twisted into the shape of the coat without tearing another hole in it. Because of this, a doughnut and the coat are not topologically equivalent.

In topology, figures are classified according to their **genus**—that is, the number of cuts that can be made without cutting the figures into two pieces. The genus of an object is the number of holes in it. See Figure 64.

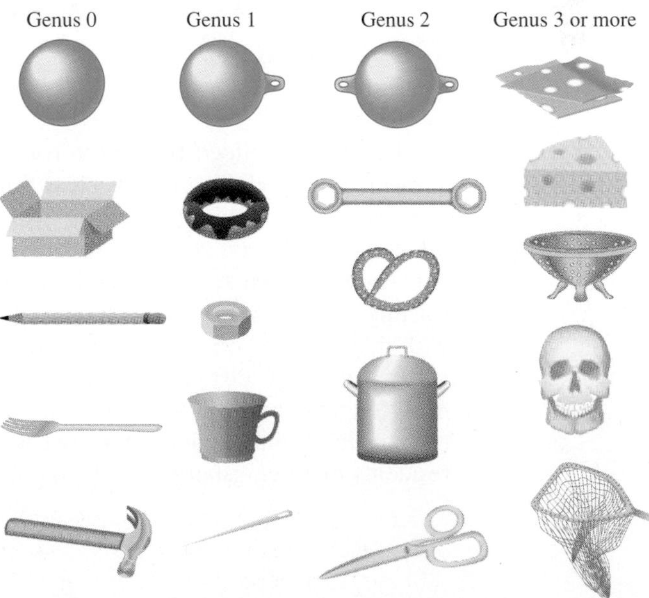

Genus of some common objects

FIGURE 64

Networks

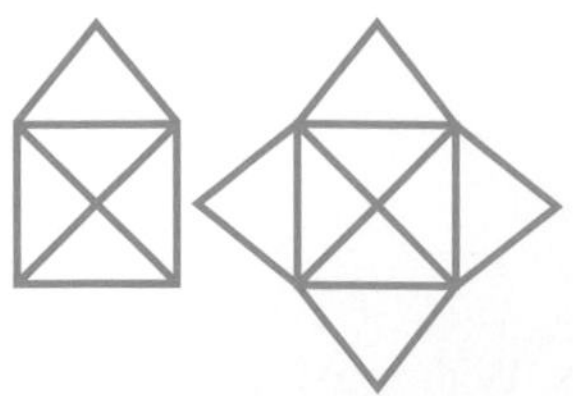

■ FIGURE 65

Another branch of modern geometry is graph theory. One topic of study in graph theory is networks.

A **network** is a diagram showing the various paths (or **arcs**) between points (called **vertices,** or **nodes**). A network can be thought of as a set of arcs and vertices. Figure 65 shows two examples of networks. The study of networks began formally with the so-called Koenigsburg Bridge problem as solved by Leonhard Euler (1707–1783). In Koenigsburg, Germany, the River Pregel flowed through the middle of town. There were two islands in the river. During Euler's lifetime, there were seven bridges connecting the islands and the two banks of the river.

The people of the town loved Sunday strolls among the various bridges. Gradually, a competition developed to see if anyone could find a route that crossed each of the seven bridges exactly once. The problem concerns what topologists today call the *traversability* of a network. No one could find a solution. The problem became so famous that in 1735 it reached Euler, who was then at the court of the Russian empress Catherine the Great. In trying to solve the problem, Euler began by drawing a network representing the system of bridges, as in Figure 66.

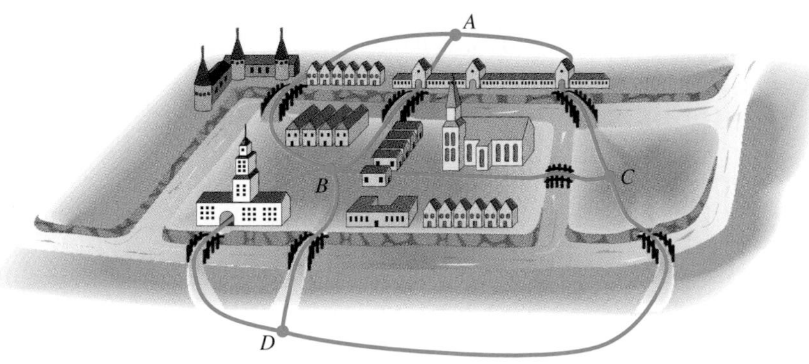

■ FIGURE 66

Euler first noticed that three routes meet at vertex A. Since 3 is an odd number, he called A an **odd vertex.** Since three routes meet at A, it must be a starting or an ending point for any traverse of the network. This must be true; otherwise, when you got to A on your second trip there would be no way to get out. An **even vertex,** one where an even number of routes meet, need not be a starting or an ending point. (Why is this?) Three paths also meet at C and D, with five paths meeting at B. Thus, B, C, and D are also odd vertices. An odd vertex must be a starting or an ending point of a traverse. Thus, all four vertices A, B, C, and D must be starting or ending points. Since a network can have only two starting or ending points (one of each), this network cannot be traversed. The residents of Koenigsburg were trying to do the impossible.

Euler's result can be summarized as follows.

Results on Vertices and Traversability

1. The number of odd vertices of any network is *even.* (That is, a network must have $2n$ odd vertices, where $n = 0, 1, 2, 3, \ldots$.)
2. A network with no odd vertices or exactly two odd vertices can be traversed. In the case of exactly two, start at one odd vertex and end at the other.
3. A network with more than two odd vertices cannot be traversed.

EXAMPLE 2 Decide whether the networks in Figure 67 and Figure 68 are traversable.

Since there are exactly two odd vertices (A and E) in Figure 67, it may be traversed. One way to traverse the network of the figure is to start at A, go through B to C, then back to A. (It is acceptable to go through a vertex as many times as needed.) Then go to E, to D, to C, and finally back to E. It is traversable.

In Figure 68, because vertices A, B, C, and D are all odd (five routes meet at each of them), this network is not traversable. (One of the authors of this text, while in high school, tried for hours to traverse it, not knowing he was attempting the impossible!)

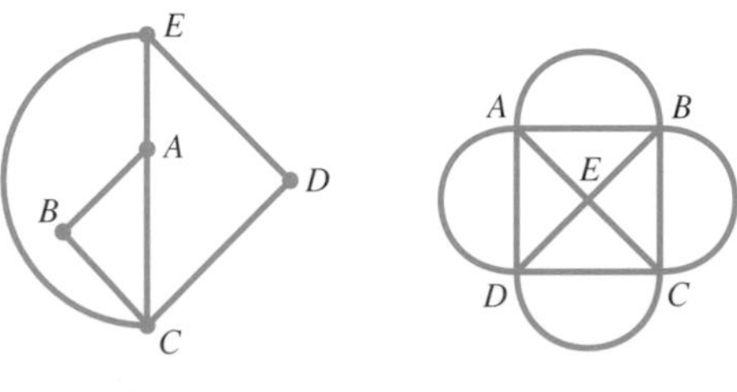

FIGURE 67 FIGURE 68

EXAMPLE 3 Figure 69 shows the floor plan of a house. Is it possible to travel through this house, going through each door exactly once?

Rooms A, C, and D have even numbers of doors, while rooms B and E have odd numbers of doors. If we think of the doors as the vertices of a graph, then the fact that we have exactly two odd vertices means that it is possible to travel through each door of the house exactly once. (Show how—start in room B or room E.)

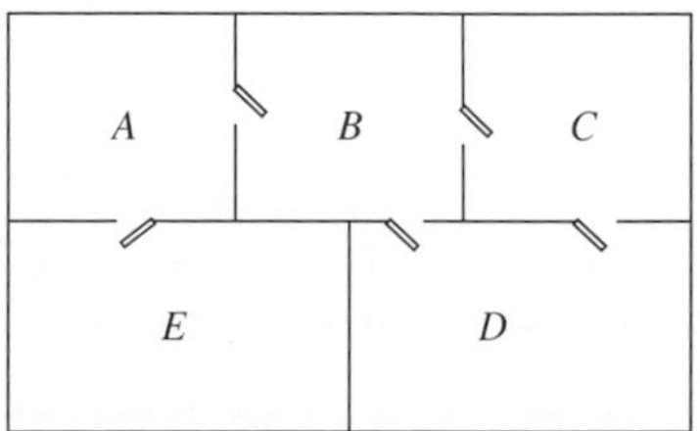

FIGURE 69

9.6 EXERCISES

The chart that follows characterizes certain properties of Euclidean and non-Euclidean geometries. Study it, and use it to respond to Exercises 1–10.

EUCLIDEAN	NON-EUCLIDEAN	
Dates back to about 300 B.C.	**Lobachevskian (about 1830)**	**Riemannian (about 1850)**
Lines have *infinite* length.		Lines have *finite* length.
Geometry on a plane	Geometry on a surface like a pseudosphere	Geometry on a sphere
Angles C and D of a Saccheri quadrilateral are *right* angles.	Angles C and D are *acute* angles.	Angles C and D are *obtuse* angles.
Given point P off line k, exactly *one* line can be drawn through P and parallel to k.	*More than one* line can be drawn through P and parallel to k.	*No* line can be drawn through P and parallel to k.
Typical triangle ABC	Typical triangle ABC	Typical triangle ABC
Two triangles can have the same size angles but different size sides (similarity as well as congruence).	Two triangles with the same size angles must have the same size sides (congruence only).	

1. In which geometry is the sum of the measures of the angles of a triangle equal to 180°?

2. In which geometry is the sum of the measures of the angles of a triangle greater than 180°?

3. In which geometry is the sum of the measures of the angles of a triangle less than 180°?

4. In a quadrilateral *ABCD* in Lobachevskian geometry, the sum of the measures of the angles must be ________________ 360°.
(less than/greater than)

5. In a quadrilateral *ABCD* in Riemannian geometry, the sum of the measures of the angles must be ________________ 360°.
(less than/greater than)

6. Suppose that *m* and *n* represent lines through *P* that are both parallel to *k*. In which geometry is this possible?

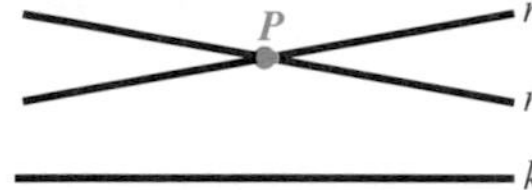

7. Suppose that *m* and *n* below *must* meet at a point. In which geometry is this possible?

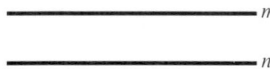

8. A globe representing the earth is a model for a surface in which geometry?

9. In which geometry is this statement possible? "Triangle *ABC* and triangle *DEF* are such that $\measuredangle A = \measuredangle D$, $\measuredangle B = \measuredangle E$, and $\measuredangle C = \measuredangle F$, and they have different areas."

10. Draw a figure (on a sheet of paper) as best you can showing the shape formed by the north pole *N* and two points *A* and *B* lying at the equator of a model of the earth.

11. Pappus, a Greek mathematician in Alexandria about 320 A.D., wrote a commentary on the geometry of the times. We will work out a theorem of his about a hexagon inscribed in two intersecting lines.

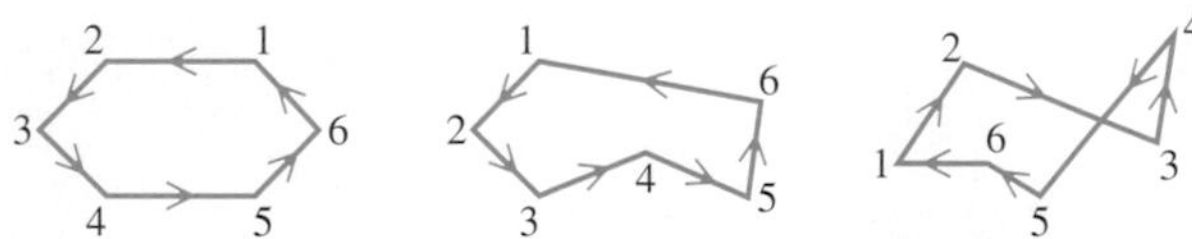

First we need to define an old word in a new way: a **hexagon** consists of any six lines in a plane, no three of which meet in the same point. As the figure shows, the vertices of several hexagons are labeled with numbers. Thus 1–2 represents a line segment joining vertices 1 and 2. Segments 1–2 and 4–5 are opposite sides of a hexagon, as are 2–3 and 5–6, and 3–4 and 1–6.

(a) Draw an angle less than 180°.

(b) Choose three points on one side of the angle. Label them 1, 5, 3 in that order, beginning with the point nearest the vertex.

(c) Choose three points on the other side of the angle. Label them 6, 2, 4 in that order, beginning with the point nearest the vertex.

(d) Draw line segments 1–6 and 3–4. Draw lines through the segments so they extend to meet in a point; call it *N*.
(e) Let lines through 1–2 and 4–5 meet in point *M*.
(f) Let lines through 2–3 and 5–6 meet in *P*.
(g) Draw a straight line through points *M*, *N*, and *P*.
(h) Write in your own words a theorem generalizing your result.

12. The following theorem comes from projective geometry:

Theorem of Desargues in a Plane Desargues' theorem states that in a plane, if two triangles are placed so that lines joining corresponding vertices meet in a point, then corresponding sides, when extended, will meet in three collinear points. (*Collinear* points are points lying on the same line.)

Draw a figure that illustrates this theorem.

In Exercises 13–20, each figure may be topologically equivalent to none or some of the objects labeled A–E. *List all topological equivalences (by letter) for each figure.*

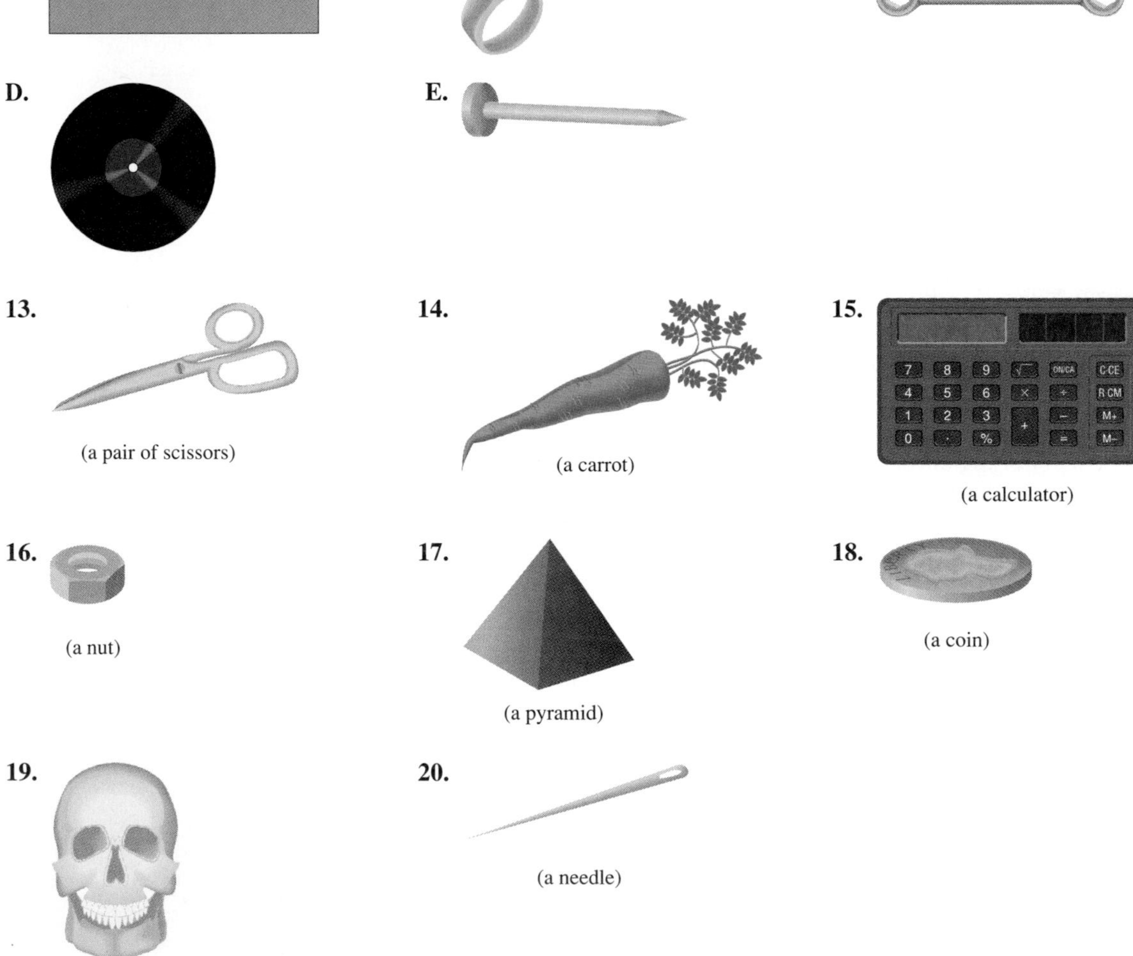

(a pair of scissors)

(a carrot)

(a calculator)

(a nut)

(a pyramid)

(a coin)

(a skull)

(a needle)

Someone once described a topologist as "a mathematician who doesn't know the difference between a doughnut and a coffee cup." This is due to the fact that both are of genus 1—they are topologically equivalent. Based on this interpretation, would a topologist know the difference between each of the following pairs of objects?

21. a spoon and a fork

22. a mixing bowl and a colander

23. a slice of American cheese and a slice of Swiss cheese

24. a penny loafer shoe and a sock

Give the genus of each of the following objects.

25. a compact disc

26. a phonograph record

27. a sheet of loose-leaf paper made for a three-ring binder

28. a sheet of loose-leaf paper made for a two-ring binder

29. a wedding band

30. a postage stamp

For each of the following networks, decide whether each lettered vertex is even *or* odd.

31.
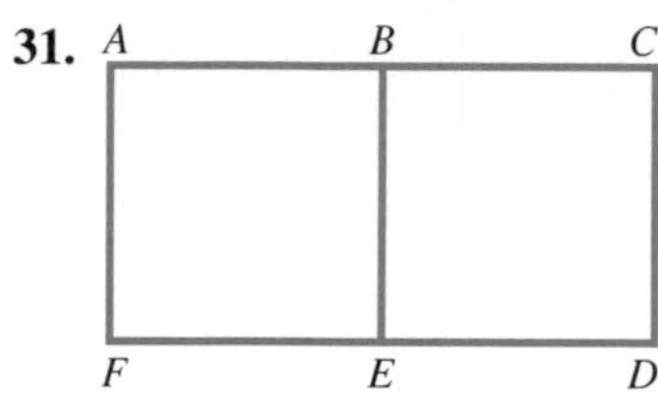

32.
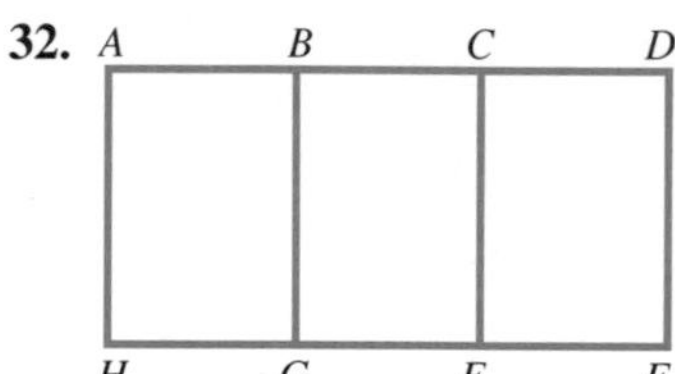

33.
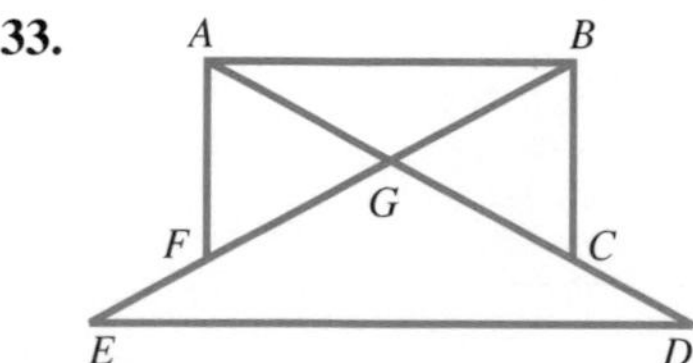

34.
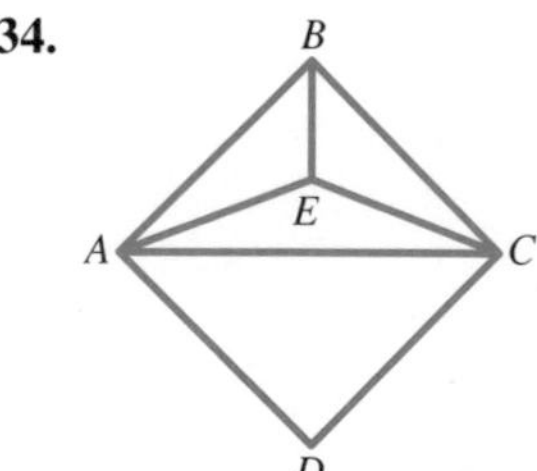

35.
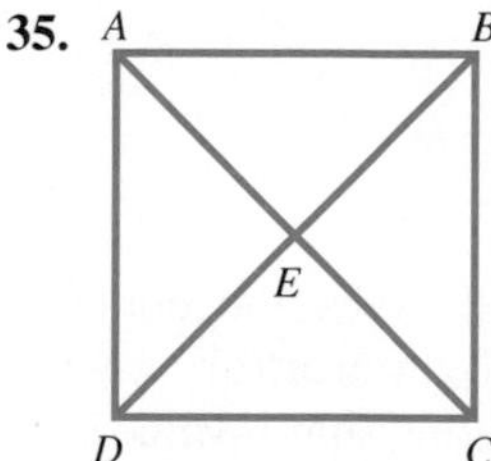

36.
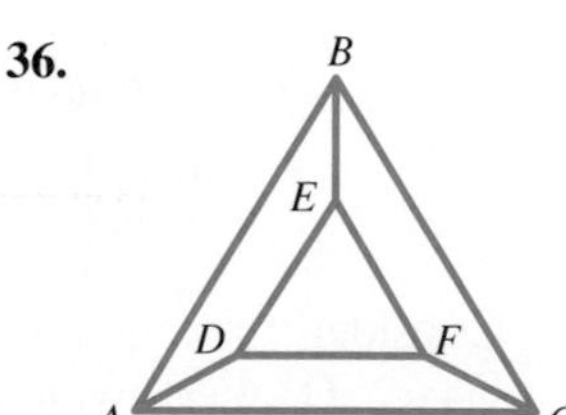

Decide whether or not each network is traversable. If a network is traversable, show how it can be traversed.

37.

38.

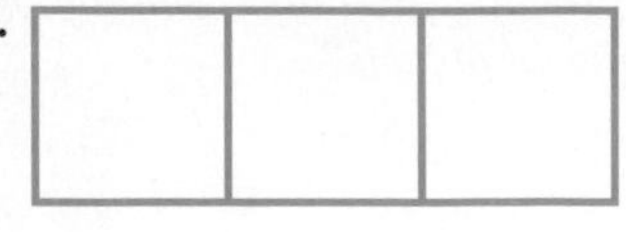

39.

40.

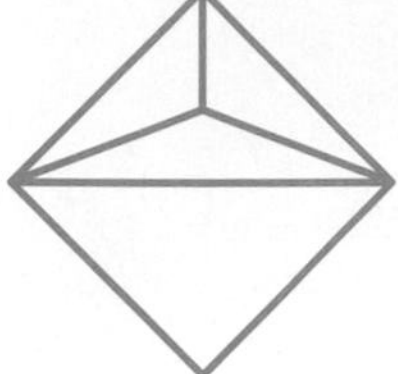

41.

42.

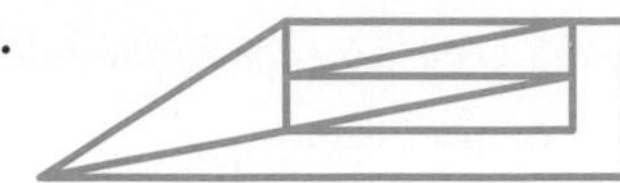

Is it possible to walk through each door of the following houses exactly once? If the answer is "yes", show how it can be done.

43.

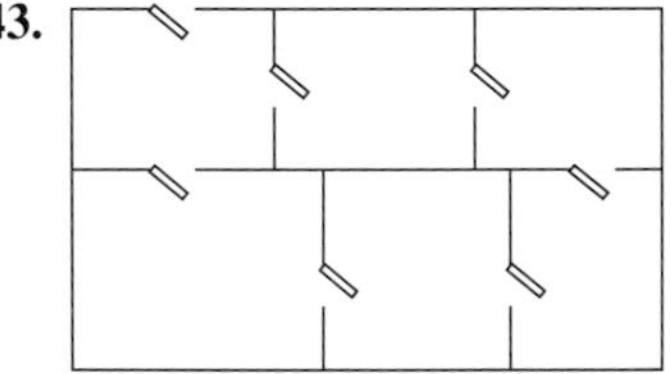

44.

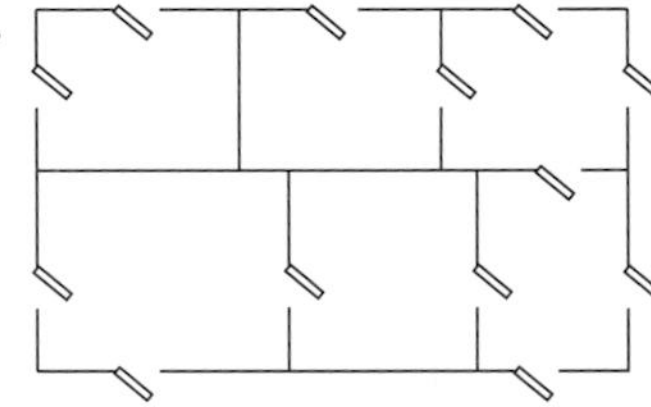

45.

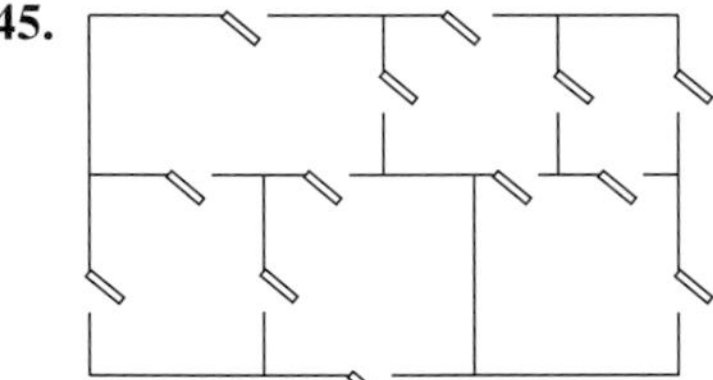

46.

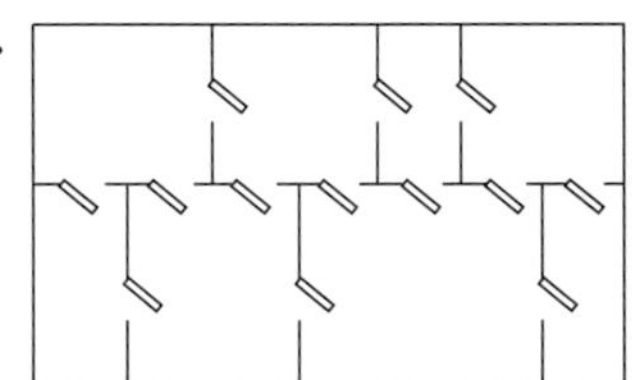

9.7 Chaos and Fractal Geometry

Does Chaos Rule the Cosmos?

—One of the ten great unanswered questions of Science, as found in the November 1992 issue of *Discover*

Consider the equation $y = kx(1 - x)$. Choosing $k = 2$ gives the equation $y = 2x(1 - x)$, which can be "iterated" by starting with an arbitrary x-value between 0 and 1, calculating the resulting y-value, substituting that y-value back in as x, calculating the resulting y-value, substituting that y-value back in as x, calculating another y-value, and so on. For example, a starting value of $x = .8$ produces the following sequence (which you can verify with a calculator):

$$.8, .32, .435, .492, .500, .500, .500, \quad \text{and so on.}$$

The sequence seems to begin randomly but quickly stabilizes at the value .500. A different initial x-value would produce another sequence which would also "converge" to .500. The value .500 can be called an *attractor* for the sequence generated by the equation $y = 2x(1 - x)$. The values of the sequence are "attracted" toward .500.

EXAMPLE 1 For the equation $y = kx(1 - x)$ with $k = 3$, begin with $x = .7$ and iterate with a calculator. What pattern emerges?

Using a Texas Instruments TI-82 calculator, we find the seventeenth through twentieth iterations give the sequence terms

$$\begin{gathered}.6354387337\\.6949690482\\.6359612107\\.6945436475.\end{gathered}$$

This sequence apparently converges in a manner different from the initial discussion, alternating between values near .636 and .695. Therefore, for $k = 3$, the sequence tends alternately toward two distinct attractors. ■

It happens that the equation of Example 1 exhibits the same behavior for any initial value of x between 0 and 1. You are asked to show this for several cases in the exercises.

EXAMPLE 2 In the equation of Example 1, change the multiplier k to 3.5, and find the forty-fourth through fifty-first terms. What pattern emerges?

Again, using a TI-82 calculator, and rounding to three decimal places, we get

$$.383,\ .827,\ .501,\ .875,\ .383,\ .827,\ .501,\ .875.$$

This sequence seems to stabilize around four alternating attractors, .383, .827, .501, and .875. ■

Notice that in our initial discussion, for $k = 2$, the sequence converged to *one* attractor. In Example 1, for $k = 3$, it converged to *two* attractors, and in Example 2, for $k = 3.5$, it converged to *four* attractors.

If k is increased further, it turns out that the number of attractors doubles over and over again, more and more often. In fact this doubling has occurred infinitely many times before k even gets as large as 4. When we look closely at groups of these doublings we find that they are always similar to earlier groups but on a smaller scale. This is called *self-similarity,* or *scaling,* an idea that is not new, but which has taken on new significance in recent years. Somewhere before k reaches 4, the resulting sequence becomes apparently totally random, with no attractors and no stability. This type of condition is one instance of what has come to be known in the scientific community as "chaos." This name came from an early paper by the mathematician James A. Yorke, of the University of Maryland at College Park.

The equation $y = kx(1 - x)$ does not look all that complicated, but the intricate behavior exhibited by it and similar equations has occupied some of the brightest minds (not to mention computers) in various fields—ecology, biology, physics, genetics, economics, mathematics—since about 1960. Such an equation might represent, for example, the population of some animal species where the value of k is determined by factors (such as food supply, or predators that prey on the species) that affect the increase or decrease of the population. Under certain conditions there is a long-run steady-state population (a single attractor). Under other conditions the population will eventually fluctuate between two alternating levels (two attractors), or four, or eight, and so on. But after a certain value of k, the long-term population becomes totally chaotic and unpredictable.

As long as k is small enough, there will be some number of attractors and the long-term behavior of the sequence (or population) is the same regardless of the initial x-value. But once k is large enough to cause chaos, the long-term behavior of the system will change drastically when the initial x-value is changed only slightly. For example, consider the following two sequences, both generated from $y = 4x(1 - x)$.

$$.600, .960, .154, .521, .998, .008, .032, \ldots$$
$$.610, .952, .183, .598, .962, .146, .499, \ldots$$

The fact that the two sequences wander apart from one another is partly due to roundoff errors along the way. But Yorke and others have shown that even "exact" calculations of the iterates would quickly produce divergent sequences just because of the slightly different initial values. This type of "sensitive dependence on initial conditions" was discovered (accidentally) back in the 1960s by Edward Lorenz when he was looking for an effective computerized model of weather patterns. He discerned the implication that any long-range weather predicting schemes might well be hopeless.

Patterns like those in the sequences above are more than just numerical oddities. Similar patterns apply to a great many phenomena in the physical, biological, and social sciences, many of them seemingly common natural systems that have been studied for hundreds of years. The measurement of a coastline, the description of the patterns in a branching tree, or a mountain range, or a cloud formation, or intergalactic cosmic dust, the prediction of weather patterns, the turbulent behavior of fluids of all kinds, the circulatory and neurological systems of the human body, fluctuations in populations and economic systems—these and many other phenomena remain mysteries, concealing their true nature somewhere beyond the reach of even our biggest and fastest computers.

Continuous phenomena are easily dealt with. A change in one quantity produces a predictable change in another. (For example, a little more pressure on the gas pedal produces a little more speed.) Mathematical functions that represent continuous events can be graphed by unbroken lines or curves, or perhaps smooth, gradually changing, surfaces. The governing equations for such phenomena are "linear," and extensive mathematical methods of solving them have been developed. On the other hand, erratic events associated with certain other equations are harder to describe or predict. The new science of chaos, made possible by modern computers, is opening up new ways to deal with such events.

One early attempt to deal with discontinuous processes in a new way, generally acknowledged as a forerunner of chaos theory, was that of the French mathematician René Thom, who, in the 1960s, applied the methods of topology.

To emphasize the feature of sudden change, Thom referred to events like a heartbeat, a buckling beam, a stock market crash, a riot, or a tornado, as *catastrophes.* He proved that all catastrophic events (in our four-dimensional space-time) are combinations of seven elementary catastrophes. (In higher dimensions the number quickly approaches infinity.)

Each of the seven elementary catastrophes has a characteristic topological shape. Two examples are shown in Figure 70. The figure at the left is called a *cusp.* The figure at the right is an *elliptic umbilicus* (a belly button with an oval cross-section). Thom's work became known as **catastrophe theory.**

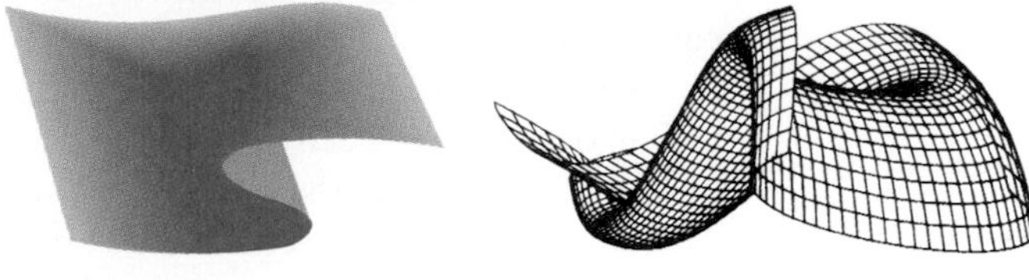

FIGURE 70

Computer graphics have been indispensable in the study of chaotic processes. The plotting of large numbers of points has revealed patterns that would otherwise have not been observed. (The underlying reasons for many of these patterns, however, have still not been explained.) The images shown in Figure 71 are created using chaotic processes.

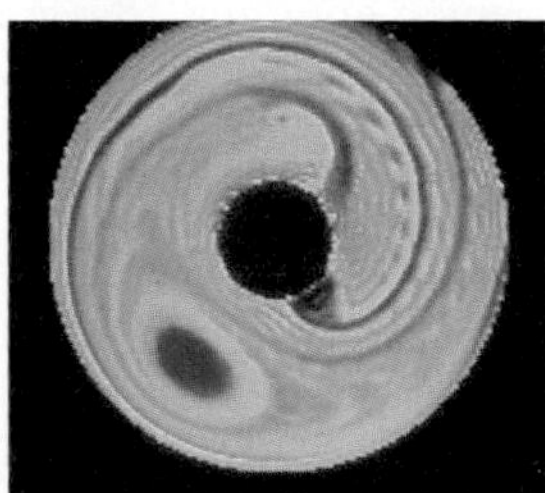
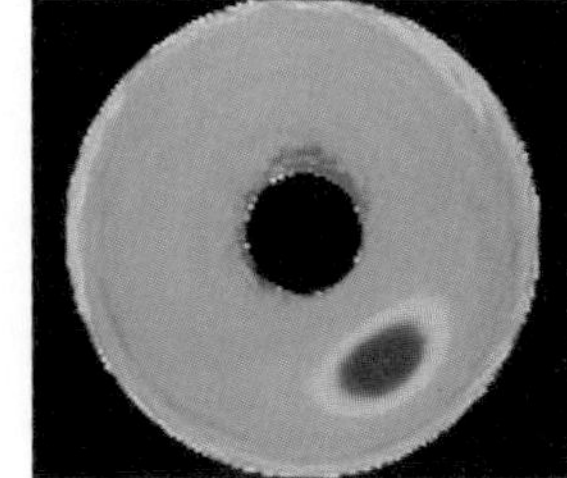

FIGURE 71

The surface of the earth, consisting of continents, mountains, oceans, valleys, and so on, has fractal dimension 2.2.

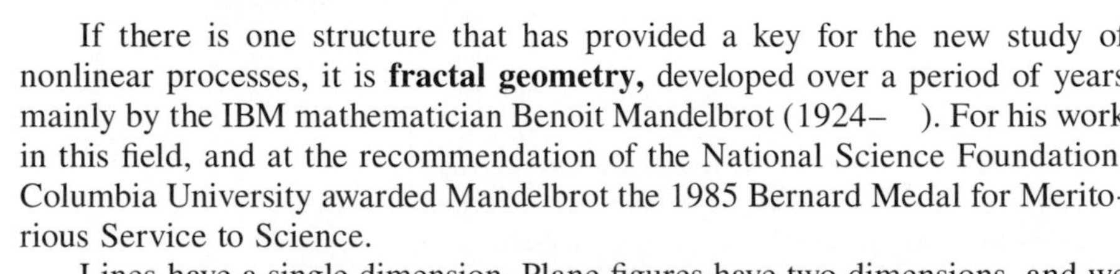

If there is one structure that has provided a key for the new study of nonlinear processes, it is **fractal geometry,** developed over a period of years mainly by the IBM mathematician Benoit Mandelbrot (1924–). For his work in this field, and at the recommendation of the National Science Foundation, Columbia University awarded Mandelbrot the 1985 Bernard Medal for Meritorious Service to Science.

Lines have a single dimension. Plane figures have two dimensions, and we live in a three-dimensional world. In a paper published in 1967, Mandelbrot investigated the idea of measuring the length of a coastline. He concluded that such a shape defies conventional Euclidean geometry and that rather than having a natural number dimension, it has a "fractional dimension." A coastline is an example of a *self-similar shape*—a shape that repeats itself over and over on different scales. From a distance, the bays and inlets cannot be individually observed, but as one moves closer they become more and more apparent. The

■ FIGURE 72

branching of a tree, from twig to limb to trunk, also exhibits a shape that repeats itself.

In the early twentieth century, the German mathematician H. von Koch investigated the so-called "Koch snowflake." It is shown in Figure 72. Starting with an equilateral triangle, each side then gives rise to another equilateral triangle. The process continues over and over, indefinitely, and a curve of infinite length is produced. The mathematics of Koch's era was not advanced enough to deal with such figures. However, using Mandelbrot's theory, it is shown that the Koch snowflake has dimension of about 1.26. This figure is obtained using a formula which involves logarithms. (Logarithms were introduced briefly in an earlier chapter.)

The theory of fractals is today being applied to many areas of science and technology. It has been used to analyze the symmetry of living forms, the turbulence of liquids, the branching of rivers, and price variation in economics. Hollywood has used fractals in the special effects found in some blockbuster movies. Figure 73 shows an example of a computer-generated fractal design.

An interesting account of the science of chaos is found in the popular 1987 book, *Chaos,* by James Gleick. Mandelbrot has published two books on fractals. They are *Fractals: Form, Chance, and Dimension* (1975), and *The Fractal Geometry of Nature* (1982).

Aside from providing a geometric structure for chaotic processes in nature, fractal geometry is viewed by many as a significant new art form. (To appreciate why, see the 1986 publication, *The Beauty of Fractals,* by H. O. Peitgen and P. H. Richter, which contains 184 figures, many in color.) Peitgen and others have also published *Fractals for the Classroom: Strategic Activities Volume One* (Springer-Verlag, 1991).

■ FIGURE 73

9.7 EXERCISES

Exercises 1–25 are taken from the November 1991 issue of Student Math Notes, *published by the National Council of Teachers of Mathematics. They were written by Tami Martin, School of Education, Boston University, and the authors wish to thank N.C.T.M. and Tami Martin for permission to reproduce this activity. Since the exercises should be done in numerical order, answers to all exercises (both even- and odd-numbered) appear in the answer section of this book.*

Most of the mathematical objects you have studied have dimensions that are whole numbers. For example, such solids as cubes and icosahedrons have dimension three. Squares, triangles, and many other planar figures are two-dimensional. Lines are one-dimensional, and points have dimension zero. Consider a square with side of length one. Gather several of these squares by cutting out or using patterning blocks.

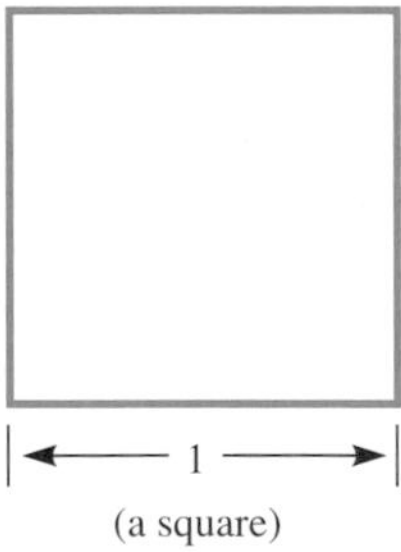

(a square)

1. What is the least number of these squares that can be put together edge to edge to form a larger square?

The size of a figure is calculated by counting the number of replicas *(small pieces) that make it up. Here, a replica is the original square with edges of length one.*

The original square is made up of one small square, so its size is one.

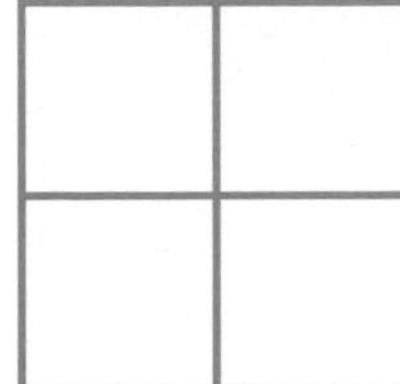

2. What is the size of the new square?

3. What is the length of each edge of the new square?

Similar figures have the same shape but are not necessarily the same size. The **scale factor** *between two similar figures can be found by calculating the ratio of corresponding edges:*

$$\frac{\text{new length}}{\text{old length}}.$$

4. What is the scale factor between the large square and the small square?

5. Find the ratio

$$\frac{\text{new size}}{\text{old size}}$$

for the two squares.

6. Form an even larger square that is three units long on each edge. Compare this square to the small square. What is the scale factor between the two squares? What is the ratio of new size to old size?

7. Form an even larger square that is four units long on each edge. Compare this square to the small square. What is the scale factor between the two squares? What is the ratio of the new size to the old size?

8. Complete the table for squares.

Scale Factor	2	3	4	5	6	10
Ratio of new size to old size						

9. How are the two rows in the table related?

Consider an equilateral triangle. The length of an edge of the triangle is one unit. The size of this triangle is one.

1

(an equilateral triangle)

10. What is the least number of equilateral triangles that can be put together edge to edge to form a similar larger triangle?

11. Complete the table for triangles.

Scale Factor	2	3	4	5	6	10
Ratio of new size to old size						

12. How does the relationship between the two rows in this table compare with the one you found in the table for squares?

One way to define the dimension, d, of a figure relates the scale factor, the new size, and the old size:

$$(\text{scale factor})^d = \frac{\text{new size}}{\text{old size}}.$$

Using a scale factor of two for squares or equilateral triangles, we can see that $2^d = 4/1$; *that is,* $2^d = 4$. *Since* $2^2 = 4$, *the dimension, d, must be two. This definition of dimension confirms what we already know—that squares and equilateral triangles are two-dimensional figures.*

13. Use this definition and your completed tables to confirm that the square and the equilateral triangle are two-dimensional figures for scale factors other than two.

Consider a cube, with edges of length one. Let the size of the cube be one.

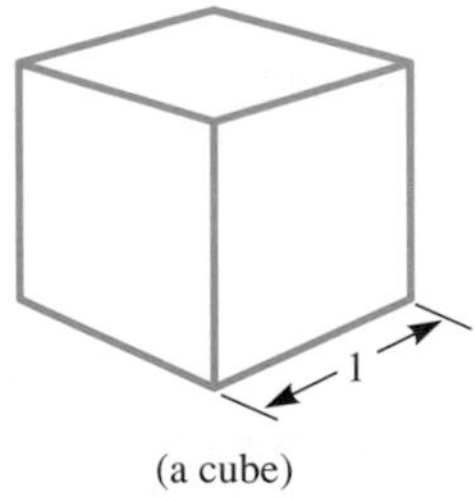

(a cube)

14. What is the least number of these cubes that can be put together face to face to form a larger cube?

15. What is the scale factor between these two cubes? What is the ratio of the new size to the old size for the two cubes?

16. Complete the table for cubes.

Scale Factor	2	3	4	5	6	10
Ratio of new size to old size						

17. How are the two rows in the table related?

18. Use the definition of dimension and a scale factor of two to verify that a cube is a three-dimensional object.

We have explored scale factors and sizes associated with two- and three-dimensional figures. Is it possible for mathematical objects to have fractional dimensions? Consider the following figure formed by replacing the middle third of a line segment of length one by an upside-down V, each of whose two sides are equal in length to the segment removed. The first four stages in the development of this figure are shown.

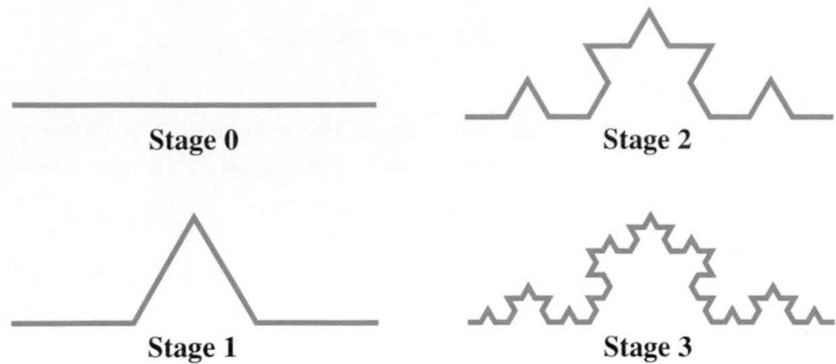

Finding the scale factor for this sequence of figures is difficult because the overall length of the figure remains the same while the number of pieces increases. To simplify the procedure, follow these steps.

Step 1: *Start with any stage (e.g., stage 1).*
Step 2: *Draw the next stage (e.g., stage 2) of the sequence and "blow it up" so that it contains an exact copy of the preceding stage (in this example, stage 1).*

Notice that stage 2 contains four copies, or replicas, of stage 1 and is three times as long as stage 1.

Length = 1, size = 1 (1 replica)

Stage 1

Length = 3, size = 4 (4 replicas)

Stage 2

19. The scale factor is equal to the ratio

$$\frac{\text{new length}}{\text{old length}}$$

between any two consecutive stages. The scale factor between stage 1 and stage 2 is _____.

20. The size can be determined by counting the number of replicas of stage 1 found in stage 2. Old size = 1, new size = _____.

Use the definition of dimension to compute the dimension, d, of the figure formed by this process: $3^d = 4/1$; *that is,* $3^d = 4$. *Since* $3^1 = 3$ *and* $3^2 = 9$, *for* $3^d = 4$ *the dimension of the figure must be greater than one but less than two:* $1 < d < 2$.

21. Use your calculator to estimate d. Remember that d is the exponent that makes 3^d equal 4. For example, since d must be between 1 and 2, try $d = 1.5$. But $3^{1.5} = 5.196\ldots$, which is greater than 4; thus d must be smaller than 1.5. Continue until you approximate d to three decimal places. (Use logarithms for an exact determination.)

The original figure was a one-dimensional line segment. By iteratively adding to the line segment, an object of dimension greater than one but less than two was generated. Objects with fractional dimension are known as **fractals.** *Fractals are infinitely self-similar objects* *formed by repeated additions to, or removals from, a figure. The object attained at the limit of the repeated procedure is the fractal.*

Next consider a two-dimensional object with sections removed iteratively. In each stage of the fractal's development, a triangle is removed from the center of each triangular region.

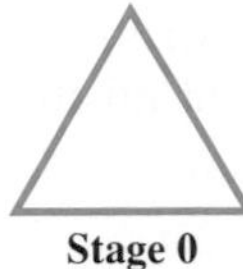
Stage 0

Stage 1

Stage 2

Stage 3

Use the process from the last example to help answer the following questions.

22. What is the scale factor of the fractal?

23. Old size = 1, new size = _____.

24. The dimension of the fractal is between what two whole number values?

25. Use the definition of dimension and your calculator to approximate the dimension of this fractal to three decimal places.

Use a calculator to determine the pattern of attractors for the equation $y = kx(1 - x)$ for the given value of k and the given initial value of x.

26. $k = 3.25, x = .7$

27. $k = 3.4, x = .8$

28. $k = 3.55, x = .7$

COLLABORATIVE INVESTIGATION
Generalizing the Angle Sum Concept

In this chapter we learn that the sum of the measures of the angles of a triangle is 180°. This fact can be extended to determine a formula for the sum of the measures of the angles of any convex polygon. To begin this investigation, divide up into groups of three or four students each. Spend the first few minutes preparing on a sheet of paper six figures as shown below.

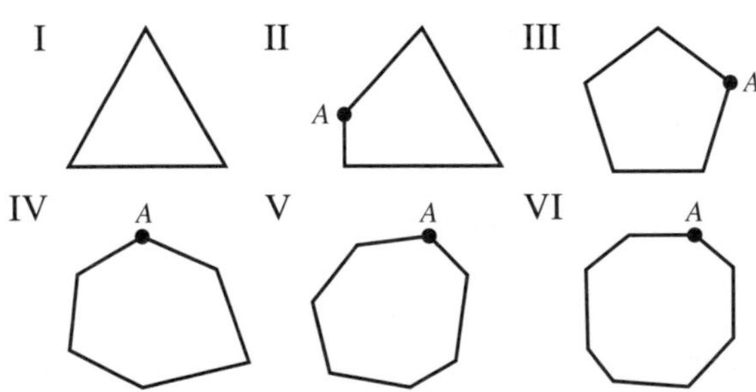

Now we define a diagonal from vertex A to be a segment from A to a non-adjacent vertex. The triangle in Figure I has no diagonals, but the polygons in Figures II through VI do have them.

Topics for Discussion

1. Choose someone in the group to draw all possible diagonals from vertex A in each figure.
2. Now complete the following table as a group.

Polygon	Number of Sides	Number of Triangles, t	Number of Degrees in Each Triangle	Sum of the Measures of All Angles of the Polygon, $t \cdot 180°$
I	_____	_____	_____	_____
II	_____	_____	_____	_____
III	_____	_____	_____	_____
IV	_____	_____	_____	_____
V	_____	_____	_____	_____
VI	_____	_____	_____	_____

3. Based on the table you completed, answer the following in order.
   (a) As suggested by the table, the number of triangles that a convex polygon can be divided into is _____ less than the number of sides.
   (b) Thus, if a polygon has s sides, it can be divided into _____ triangles.
   (c) From the table we see that the sum of the measures of all the angles of a polygon can be found from the expression $t \cdot 180°$. Thus, if a polygon has s sides, the sum of the measures of all the angles of the polygon is given by the expression (_____ − _____) · _____°.
4. Use your discovery from Exercise 3(c) to find the sum of the measures of all the angles of
   (i) a nonagon (ii) a decagon
   (iii) a 12-sided polygon.

Chapter 9 Test

1. Consider a 38° angle. Answer each of the following.
   (a) What is the measure of its complement?
   (b) What is the measure of its supplement?
   (c) Classify it as acute, obtuse, right, or straight.

Find the measure of each marked angle.

2.

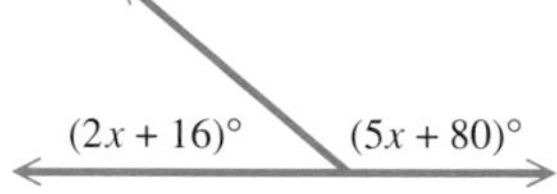

3.

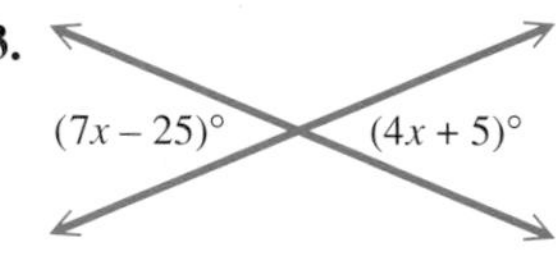

4.

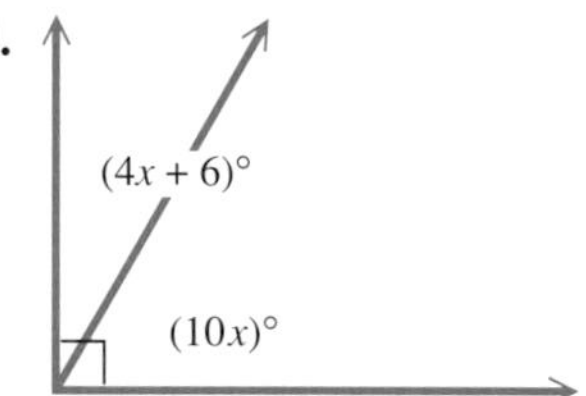

In Exercises 5 and 6, assume that lines m and n are parallel, and find the measure of each marked angle.

5.

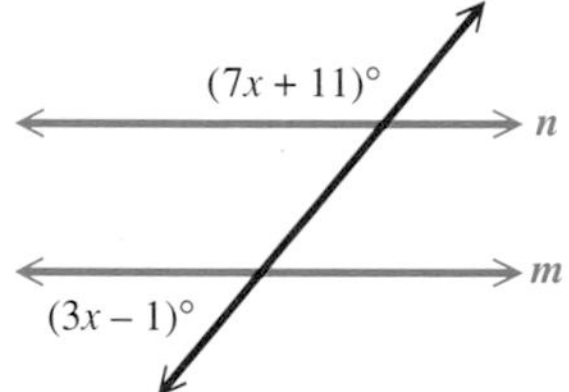

6.

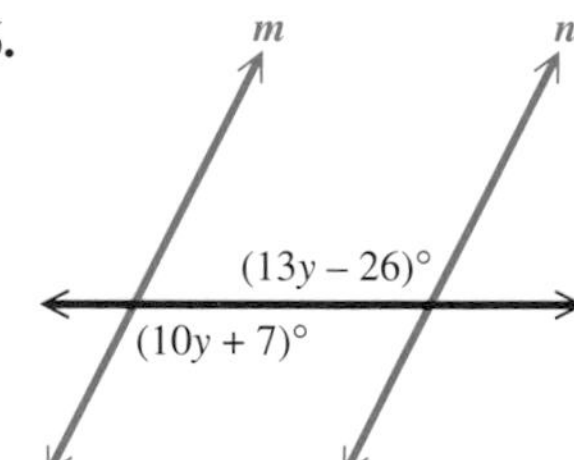

7. Explain why a rhombus must be a parallelogram, but a parallelogram might not be a rhombus.
8. Which one of the following statements is false?
   (a) A square is a rhombus.
   (b) The acute angles of a right triangle are complementary.
   (c) A triangle may have both a right angle and an obtuse angle.
   (d) A trapezoid may have non-parallel sides of the same length.

Identify each of the following curves as simple, closed, both, *or* neither.

9.

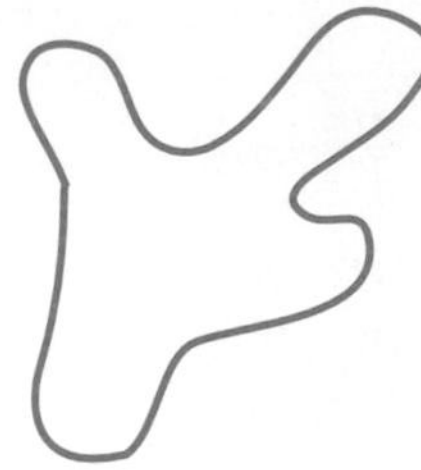

10.

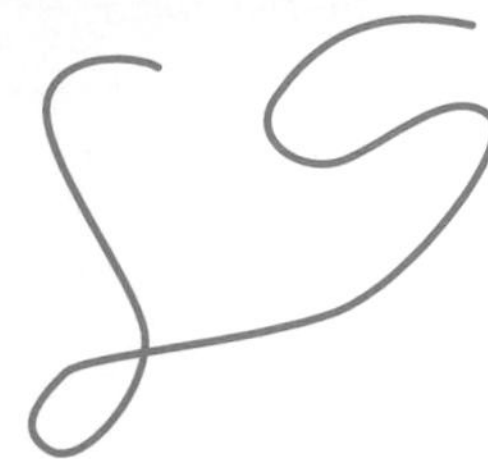

11. Find the measure of each angle in the triangle.

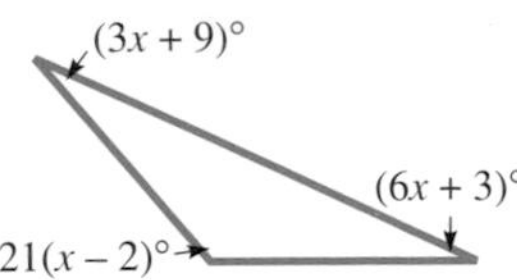

Find the area of each of the following figures.

12.

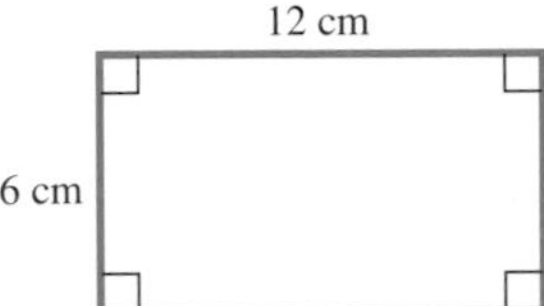

13.

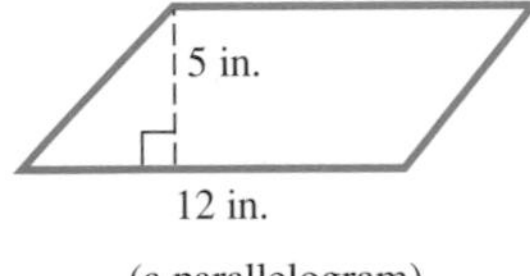

(a parallelogram)

14.

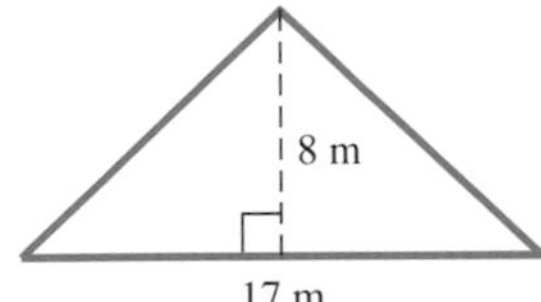

15.

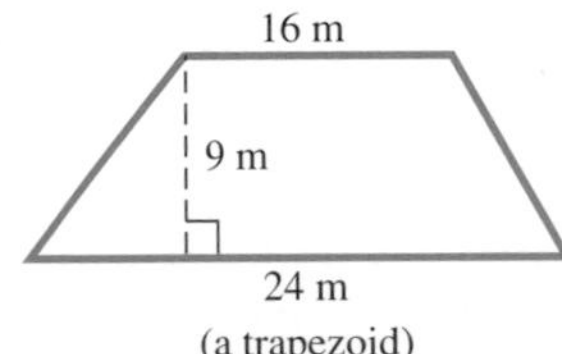

(a trapezoid)

---

16. If a circle has area 144π square inches, what is its circumference?

17. The length of a rectangle is 12 in. more than 3 times the width, and the perimeter is 72 in. Find the length and the width of the rectangle.

18. What is the area of the shaded portion of the figure? Use 3.14 as an approximation for π.

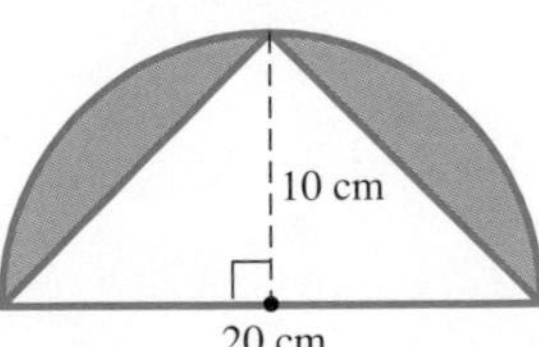

(a triangle within a semicircle)

19. Which congruence property for triangles can be used to show that $\Delta ABD \cong \Delta BAC$, given that $\measuredangle CAB = \measuredangle DBA$, and $DB = CA$? Give the additional information needed to apply the property.

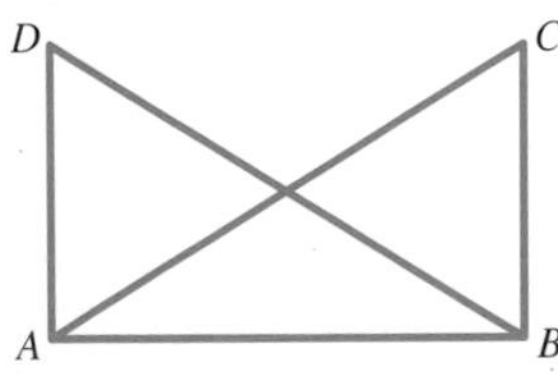

20. If a 30-ft pole casts a shadow 45 ft long, how tall is a pole whose shadow is 30 ft long at the same time?

21. What is the measure of a diagonal of a rectangle that has width 20 m and length 21 m?

Find **(a)** *the volume and* **(b)** *the surface area of each of the following space figures. When necessary, use 3.14 as an approximation for π.*

22.

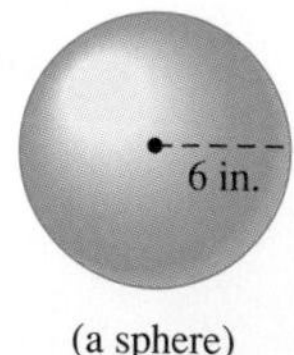

(a sphere)

23.

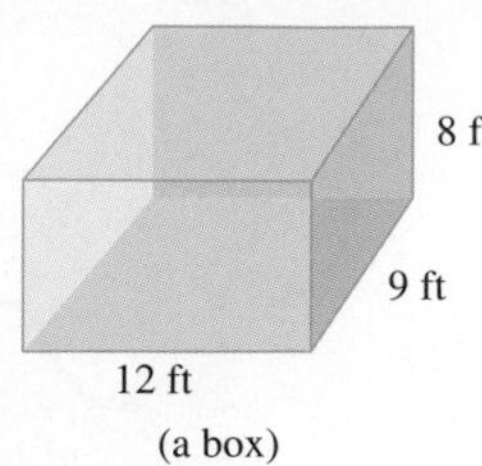

(a box)

24.

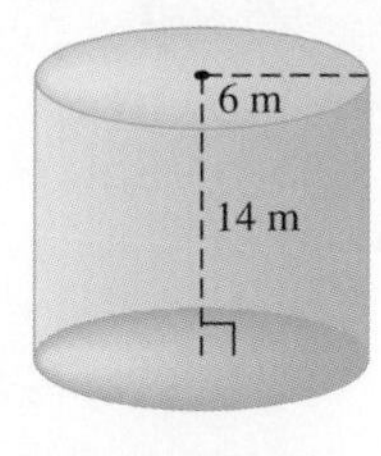

(a right circular cylinder)

25. List several main distinctions between Euclidean and non-Euclidean geometry.

Are the following pairs of objects topologically equivalent?

26. a page of a book and the cover of the same book

27. a pair of glasses with the lenses removed, and the Mona Lisa

28. Is it possible to traverse this network? If so, show how it can be done.

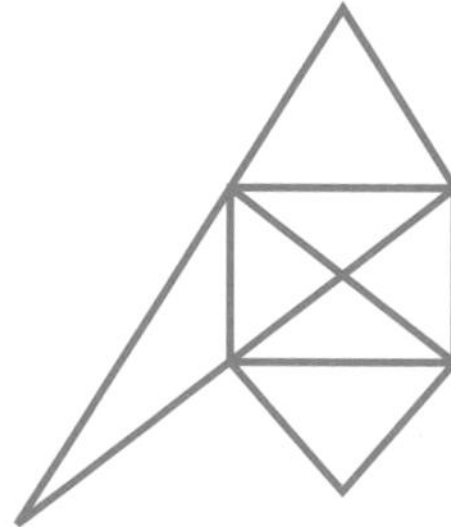

29. Use a calculator to determine the attractors for the sequence generated by the equation $y = 2.1x(1 - x)$, with initial value of $x = .6$.

30. Who is Benoit Mandelbrot?

10

Counting Methods

One of the most profound discoveries of our time occurred in 1953 when the double helical structure of DNA (deoxyribonucleic acid) was identified. DNA is the material of heredity. Since then, scientists have been trying to understand the complicated genetic code that determines who we are. An important application of this discovery has been the invention of DNA profiling. DNA profiling has been used in criminal cases, diagnosis of genetic diseases, and archaeology.

DNA is composed of four bases called adenine (A), guanine (G), thymine (T), and cytosine (C). The genetic code of an individual is made up of approximately six billion of these bases stored in a linear sequence. The nucleus of every cell contains a copy of this genetic code, tightly coiled in the shape of a double helix. It has been estimated that if all of the genetic material in an adult were fully extended, it would stretch the distance between Earth and the sun 50 times. Even the DNA found within a single cell has the potential to store 30 volumes of the Encyclopaedia Britannica three times over.

Any one of the four bases can occur anytime in the genetic code sequence. In a normal red blood cell, hemoglobin contains the genetic sequence

—ACTCCTG [A] GGAGGAGT—

whereas a person with sickle cell anemia will have exactly the same sequence except for a single base

—ACTCCTG [T] GGAGGAGT—

Thus, one error in a sequence of six billion letters causes this disease. This unique genetic sequence composed of the letters A, C, G, and T defines the makeup of an individual. In each location there are four choices for a letter. In two locations there are $4 \times 4 = 4^2 = 16$ choices. They are AA, AG, AT, AC, GA, GG, GT, GC, TA, TG, TT, TC, CA, CG, CT, and CC. Similarly, in three consecutive locations there are $4^3 = 64$ possibilities and in 6 billion locations there are $4^{6,000,000,000}$ different sequences possible. In reality 99% to 99.9% of this genetic code is identical for all humans and only .1% to 1% is unique to the individual. It is this small percentage that DNA profiling relies on. Counting methods like those above emphasize the uniqueness of humankind and of each individual. Small changes in the genetic code can have dramatic effects.

Source: Easteal, S., N. McLeod, and K. Reed, *DNA PROFILING Principles, Pitfalls and Potential,* Harwood Academic Publishers, 1991.

There are many situations where a *task* is to be performed and where it is useful to know how many different *results* are possible. Depending on the nature of the task, there are various ways to find the answer to the question, "How many possible results are there?" Commonly used strategies are the use of tables, charts and diagrams, observing patterns, drawing a sketch, and utilizing an appropriate equation or formula. And perhaps most importantly, remember that your own common sense can be the most helpful strategy of all. This chapter will present some approaches that have proved helpful over the years, but you should also feel free to apply your own reasoning skills to design your own methods. Counting problems will provide you the opportunity to utilize a great many of your mathematical skills.

10.1 Counting by Systematic Listing

Counting methods can be used to find the number of moves required to solve a Rubik's Cube. The scrambled cube must be modified so that each face is a solid color. Rubik's royalties from sales of the cube in Western countries made him Hungary's richest man.

The methods of counting presented in this section all involve coming up with an actual list of the possible results for a given task. This approach is only practical for fairly short lists. Other methods, developed in the remainder of the chapter, will enable us to often find "how many" without actually listing all the possibilities.

When listing possible results, it is extremely important to use a *systematic* approach. If we just start listing the possibilities as they happen to occur to us, we are very likely to miss some of them.

One-Part Tasks

The results for simple tasks consisting of one part can often be listed easily. For the task of tossing a single fair coin, for example, the list looks like this: *heads, tails.* There are two possible results. If the task is to roll a single fair die (a cube with faces numbered 1 through 6), the different results are 1, 2, 3, 4, 5, 6, a total of six possibilities.

EXAMPLE 1 Consider a club N with five members:

$$N = \{\text{Andy, Bill, Cathy, David, Evelyn}\},$$

or, as a shortcut,

$$N = \{A, B, C, D, E\}.$$

In how many ways can this group select a president (assuming all members are eligible)?

The task in this case is to select one of the five members as president. There are five possible results: A, B, C, D, and E. ■

Two-Part Tasks; Using Product Tables

TABLE 1

		Second Digit		
		1	2	3
First Digit	1	11	12	13
	2	21	22	23
	3	31	32	33

EXAMPLE 2 Determine the number of two-digit numbers that can be written using digits from the set {1, 2, 3}.

This task consists of two parts: choose a first digit and choose a second digit. The results for a two-part task can be pictured in a **product table** such as Table 1. From the table we obtain our list of possible results: 11, 12, 13, 21, 22, 23, 31, 32, 33. There are nine possibilities. ■

Bone dice were unearthed in the remains of a Roman garrison, Vindolanda, near the border between England and Scotland. Life on the Roman frontier was occupied with gaming as well as fighting. Some of the Roman dice were loaded in favor of 6 and 1.

Life on the American frontier was reflected in cattle brands that were devised to keep alive the memories of hardships, feuds, and romances. A rancher named Ellis from Paradise Valley in Arizona liked a pun now and then. He even designed his cattle brand in the shape of a pair of dice. You can guess that the pips were 6 and 1.

EXAMPLE 3 Determine the number of different possible results when two ordinary dice are rolled.

For clarity, assume the dice are easily distinguishable. Perhaps one is red and the other green. Then the task consists of two parts: roll the red die and roll the green die. Table 2 is a product table showing the thirty-six possible results. ◆

TABLE 2

		Green Die					
		1	**2**	**3**	**4**	**5**	**6**
Red Die	**1**	(1, 1)	(1, 2)	(1, 3)	(1, 4)	(1, 5)	(1, 6)
	2	(2, 1)	(2, 2)	(2, 3)	(2, 4)	(2, 5)	(2, 6)
	3	(3, 1)	(3, 2)	(3, 3)	(3, 4)	(3, 5)	(3, 6)
	4	(4, 1)	(4, 2)	(4, 3)	(4, 4)	(4, 5)	(4, 6)
	5	(5, 1)	(5, 2)	(5, 3)	(5, 4)	(5, 5)	(5, 6)
	6	(6, 1)	(6, 2)	(6, 3)	(6, 4)	(6, 5)	(6, 6)

You will want to refer back to Table 2 when various dice-rolling problems occur in the remainder of this chapter and the next.

EXAMPLE 4 Find the number of ways that club N of Example 1 can elect both a president and a secretary. Assume that all members are eligible, but that no one can hold both offices.

Again, the required task has two parts: determine the president and determine the secretary. Constructing Table 3 gives us the following list (where, for example, AB denotes president A and secretary B, while BA denotes president B and secretary A):

$$AB,\quad AC,\quad AD,\quad AE,\quad BA,\quad BC,\quad BD,\quad BE,\quad CA,\quad CB,$$
$$CD,\quad CE,\quad DA,\quad DB,\quad DC,\quad DE,\quad EA,\quad EB,\quad EC,\quad ED.$$

Notice that certain entries (down the main diagonal, from upper left to lower right) are omitted from the table, since the cases AA, BB, and so on would imply one person holding both offices. We see that there are twenty possibilities. ◆

TABLE 3

		Secretary				
		A	***B***	***C***	***D***	***E***
President	***A***		*AB*	*AC*	*AD*	*AE*
	B	*BA*		*BC*	*BD*	*BE*
	C	*CA*	*CB*		*CD*	*CE*
	D	*DA*	*DB*	*DC*		*DE*
	E	*EA*	*EB*	*EC*	*ED*	

EXAMPLE 5 Find the number of ways that club N can appoint a committee of two members to represent them at an association conference.

The required task again has two parts. In fact, we can refer to Table 3 again, but this time, the order of the two letters (people) in a given pair really makes no difference. For example, BD and DB are the same committee. (In Example 4, BD and DB were different results since the two people would be holding different offices.) In the case of committees, we eliminate not only the main diagonal entries, but also all entries below the main diagonal. Our resulting list contains ten possibilities:

$$AB, \quad AC, \quad AD, \quad AE, \quad BC, \quad BD, \quad BE, \quad CD, \quad CE, \quad DE.$$ ■

Tasks with Three or More Parts; Using Tree Diagrams

Problem Solving

When a task has more than two parts, it is not as easy to analyze it with a product table, since it would require a table of more than two dimensions, which is hard to construct on paper. Another helpful device is the **tree diagram,** which we use in the following examples.

EXAMPLE 6 Find the number of three-digit numbers that can be written using digits from the set $\{1, 2, 3\}$, assuming that **(a)** repeated digits are allowed, **(b)** repeated digits are not allowed.

(a) The task of constructing such a number has three parts: select the first digit, select the second digit, and select the third digit. As we move from left to right through the tree diagram in Figure 1, the tree branches at the first stage to all possibilities for the first digit. Then each first stage branch again branches, or splits, to the second stage, to all possibilities for the second digit. Finally, the third stage branching shows the third-digit possibilities. The list of possible results (27 of them) is shown at the right.

(b) For the case of non-repeating digits, we could construct a whole new tree diagram, as in Figure 2, or we could simply go down the list of numbers from the first tree diagram and strike out any that contain repeated digits. In either case we obtain only six possibilities. ■

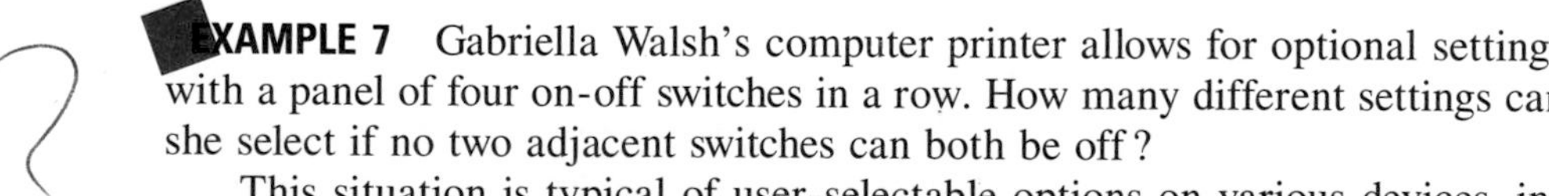

EXAMPLE 7 Gabriella Walsh's computer printer allows for optional settings with a panel of four on-off switches in a row. How many different settings can she select if no two adjacent switches can both be off?

This situation is typical of user-selectable options on various devices, including computer equipment, garage door openers, and other appliances. In Figure 3 we denote "on" and "off" with 1 and 0, respectively (a common practice). The number of possible settings is seen to be eight. Notice that each time on the tree diagram that a switch is indicated as off (0), the next switch can only be on (1). This is to satisfy the restriction that no two adjacent switches can both be off. ■

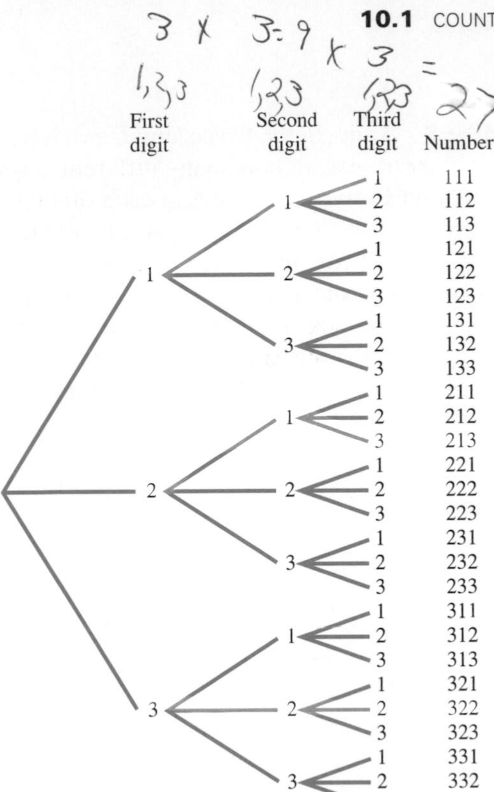

Tree diagram for three-digit numbers with digits from the set {1, 2, 3}

■ **FIGURE 1**

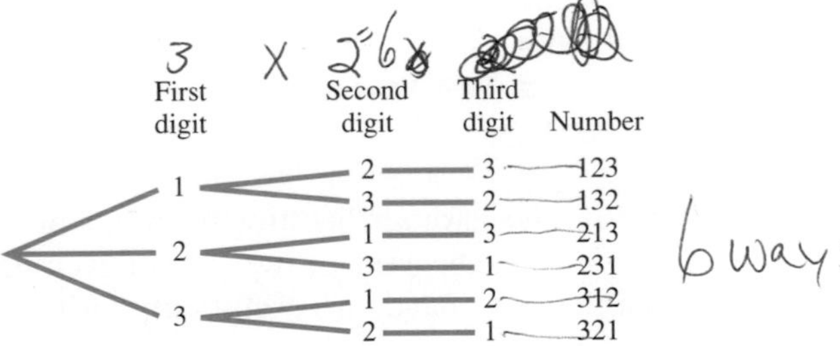

Tree diagram for non-repeating three-digit numbers with digits from the set {1, 2, 3}

■ **FIGURE 2**

First switch
Second switch
Third switch
Fourth switch
Switch settings
0101
0110
0111
1010
1011
1101
1110
1111

Tree diagram for printer settings

■ **FIGURE 3**

EXAMPLE 8 Andy, Betty, Clyde, and Dawn have tickets for four reserved seats in a row at a concert. In how many different ways can they seat themselves so that Andy and Betty will sit next to each other?

Here we have a four-part task: assign people to the first, second, third, and fourth seats. The tree diagram in Figure 4 again avoids repetitions, since no person can occupy more than one seat. Also, once *A* or *B* appears at any stage of the tree, the other one must occur at the next stage. (Why is this?) Notice that no splitting occurs from stage three to stage four since by that time there is only one person left unassigned. The right column in the figure shows the twelve possible seating arrangements. ■

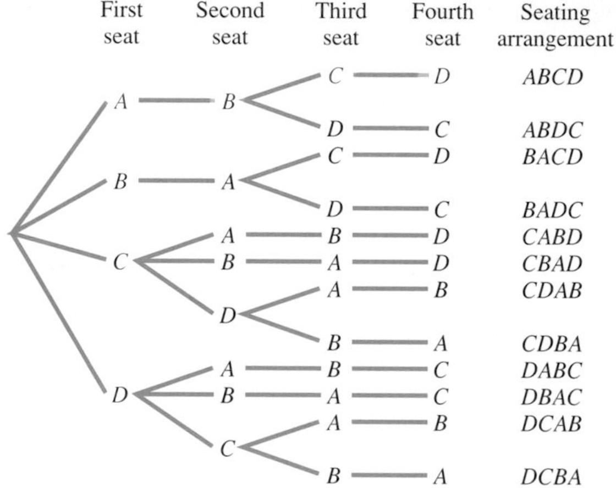

Tree diagram for concert seating

FIGURE 4

Although we have applied tree diagrams only to tasks with three or more parts, they can also be used for two-part or even simple, one-part tasks. Product tables, on the other hand, are useful only for two-part tasks.

Other Systematic Listing Methods

Product tables (for two-part tasks) and tree diagrams (for tasks of any number of parts) are useful methods for picturing and listing the possible results for given tasks. But you should feel free to create other approaches as well. To finish this section, we suggest some additional systematic ways to produce complete listings of possible results.

In Example 4, we used a product table (Table 3) to list all possible president-secretary pairs for the club $N = \{A, B, C, D, E\}$. We could also systematically construct the same list using a sort of alphabetical or left-to-right approach. First, consider the results where *A* is president. Any of the remaining members (*B*, *C*, *D*, or *E*) could then be secretary. That gives us the pairs *AB*, *AC*, *AD*, and *AE*. Next, assume *B* is president. The secretary could then be *A*, *C*, *D*,

The **"tree diagram"** on the map came from research on the feasibility of using motor-sailers (motor-driven ships with wind-sail auxiliary power) on the North Atlantic run. At the beginning of a run, weather forecasts and computer analysis are used to choose the best of the 45 million possible routes.

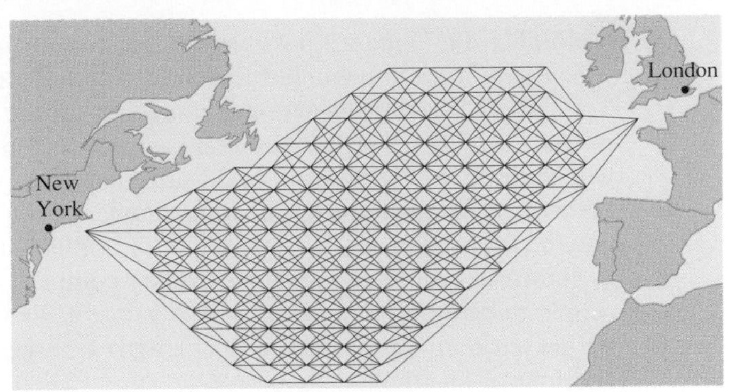

or *E*. We get the pairs *BA*, *BC*, *BD*, and *BE*. Continuing in order, we get the complete list just as in Example 4:

AB, *AC*, *AD*, *AE*, *BA*, *BC*, *BD*, *BE*, *CA*, *CB*,
CD, *CE*, *DA*, *DB*, *DC*, *DE*, *EA*, *EB*, *EC*, *ED*.

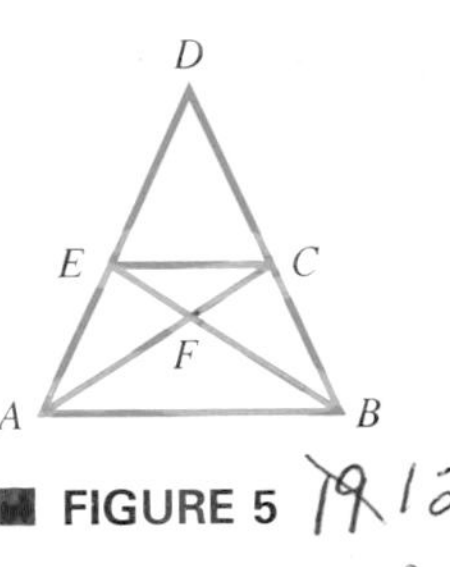

■ FIGURE 5

EXAMPLE 9 How many different triangles (of any size) are included in Figure 5?

One systematic approach is to label the points as shown, begin with *A*, and proceed in alphabetical order to write all three-letter combinations, then cross out the ones that are not triangles in the figure.

ABC, *ABD*, *ABE*, *ABF*, *ACD*, *ACE*, ~~*ACF*~~, ~~*ADE*~~, ~~*ADF*~~, *AEF*,
~~*BCD*~~, *BCE*, *BCF*, *BDE*, ~~*BDF*~~, ~~*BEF*~~, *CDE*, ~~*CDF*~~, *CEF*, ~~*DEF*~~

Finally, there are twelve different triangles in the figure. Why are *ACB* and *CBF* (and many others) not included in the list?

Another method might be first to identify the triangles consisting of a single region each: *DEC*, *ECF*, *AEF*, *BCF*, *ABF*. Then list those consisting of two regions each: *AEC*, *BEC*, *ABE*, *ABC*; and those with three regions each: *ACD*, *BED*. There are no triangles with four regions, but there is one with five: *ABD*. The total is again twelve. Can you think of other systematic ways of getting the same list? ■

Notice that in the first method shown in Example 9, the labeled points were considered in alphabetical order. In the second method, the single-region triangles were listed by using a top-to-bottom and left-to-right order. Using a definite system helps to ensure that we get a complete list.

Repetitive processes of certain types are called "iterations." In mathematical applications, a sequence of results is produced where each particular result is fed back through the process to produce the next result. An example of an iterative process is taking any whole number, multiplying its digits together to get a second whole number, then repeating the process, producing a chain of whole numbers that always eventually terminates with a single-digit number. Starting with 283, for example, we obtain

$$283 \to 48 \to 32 \to 6,$$

a chain of *length* 4, with *leading number* 283.

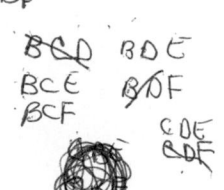

EXAMPLE 10 Find all possible number chains (of the type described above) which have a leading number less than 100 and terminate with 4.

The direct approach here might involve writing out the chain for each whole number from 0 through 99 and observing which ones terminate with 4. But it will be much quicker to work backwards and to employ a tree diagram. To begin the tree, write down the terminating number of the chain (4). Then, since 4 can only be factored as $1 \cdot 4$, $2 \cdot 2$, and $4 \cdot 1$, 4 must have come from 14, 22, or 41 as shown in Figure 6. Similar reasoning completes the tree diagram. Now the simplest chain that terminates with 4 is just 4. Reading from the tree diagram, we get ten chains altogether, one of length 1, three of length 2, two of length 3, and four of length 4.

4	$14 \to 4$	$27 \to 14 \to 4$	$39 \to 27 \to 14 \to 4$
	$22 \to 4$	$72 \to 14 \to 4$	$93 \to 27 \to 14 \to 4$
	$41 \to 4$		$89 \to 72 \to 14 \to 4$
			$98 \to 72 \to 14 \to 4$

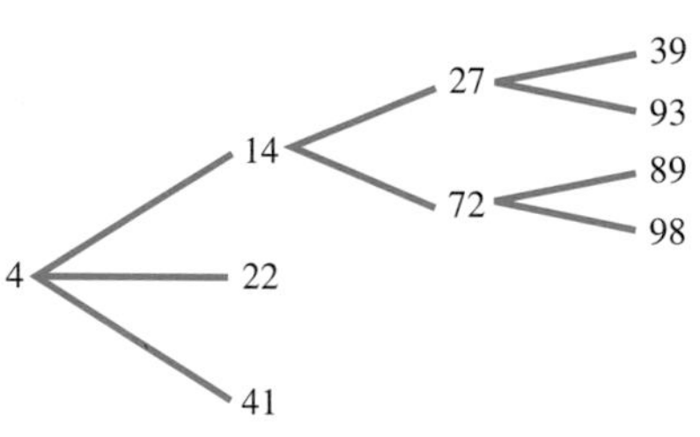

Tree diagram for number chains terminating at 4

FIGURE 6

10.1 EXERCISES

Refer to Examples 1 and 4, involving the club N = {Andy, Bill, Cathy, David, Evelyn}. *Assuming all members are eligible, but that no one can hold more than one office, list and count the different ways the club could elect the following groups of officers.*

1. a president and a treasurer
2. a president and a treasurer if the president must be a female
3. a president and a treasurer if the two officers must not be the same sex
4. a president, a secretary, and a treasurer, if the president and treasurer must be women
5. a president, a secretary, and a treasurer, if the president must be a man and the other two must be women
6. a president, a secretary, and a treasurer, if the secretary must be a woman and the other two must be men

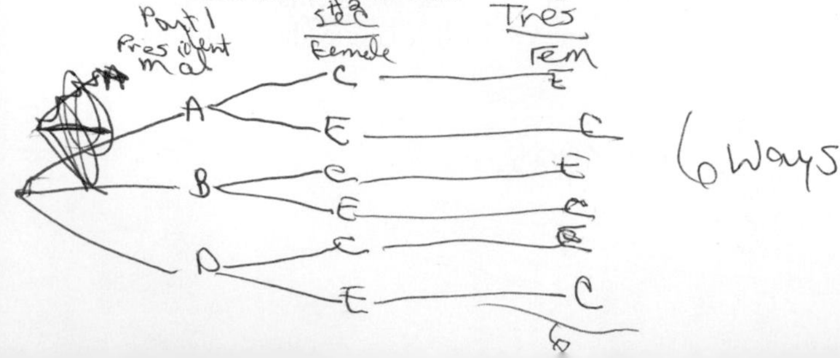

List and count the ways club N could appoint a committee of three members under the following conditions.

7. There are no restrictions.

8. The committee must include more men than women.

Refer to Table 2 (the product table for rolling two dice). Of the 36 possibilities, determine the number for which the sum (for both dice) is the following.

9. 2 **10.** 3 **11.** 4 **12.** 5 **13.** 6 **14.** 7

15. 8 **16.** 9 **17.** 10 **18.** 11 **19.** 12 **20.** odd

21. even **22.** from 6 through 8 inclusive

23. between 6 and 10 **24.** less than 5

25. Construct a product table showing all possible two-digit numbers using digits from the set {1, 2, 3, 4, 5, 6}.

Of the thirty-six numbers in the product table for Exercise 25, list the ones that belong to each of the following categories.

26. odd numbers **27.** numbers with repeating digits **28.** multiples of 6

29. prime numbers **30.** triangular numbers **31.** square numbers

32. Fibonacci numbers **33.** powers of 2

34. Construct a tree diagram showing all possible results when three fair coins are tossed. Then list the ways of getting the following results.
(a) at least two heads **(b)** more than two heads **(c)** no more than two heads
(d) fewer than two heads

35. Extend the tree diagram of Exercise 34 for four fair coins. Then list the ways of getting the following results.
(a) more than three tails **(b)** fewer than three tails **(c)** at least three tails
(d) no more than three tails

Determine the number of triangles (of any size) in each of the following figures.

36.

37.

38.

39.

Determine the number of squares (of any size) in each of the following figures.

40.

41.

42.

43.
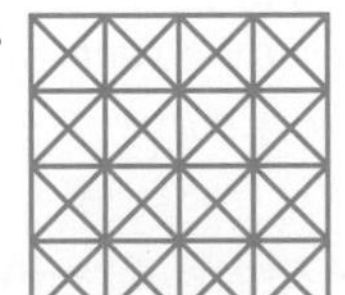

Consider only the smallest individual cubes and assume solid stacks (no gaps). Determine the number of cubes in each stack that are not visible.

44.

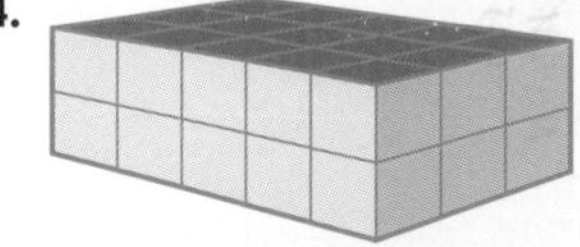

45.

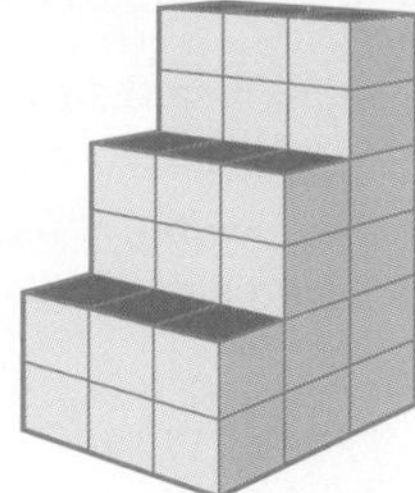

46.

47.

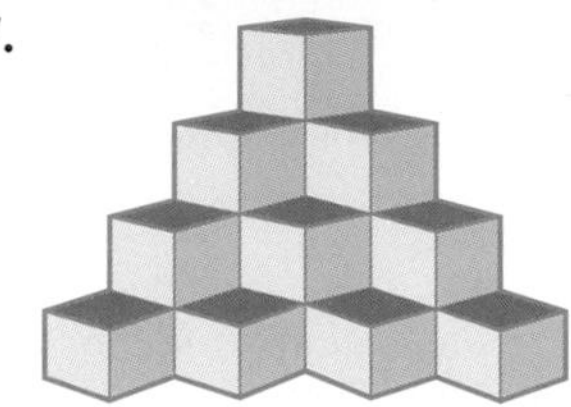

48. In the plane figure shown here, only movement that tends downward is allowed. Find the total number of paths from A to B.

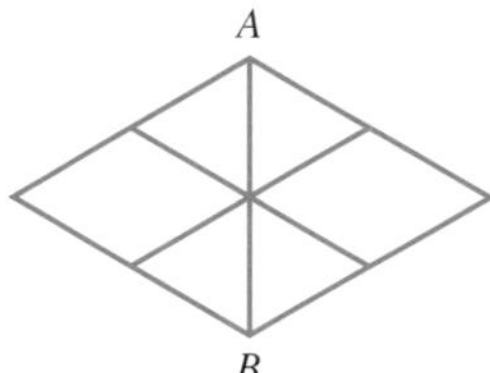

49. Find the number of paths from A to B in the figure shown here if the directions on various segments are restricted as shown.

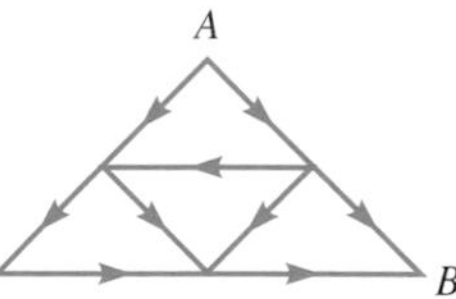

In each of Exercises 50–52, determine the number of different ways the given number can be written as the sum of two primes.

50. 30 **51.** 40 **52.** 95

53. A group of twelve strangers sat in a circle, and each one got acquainted only with the person to the left and the person to the right. Then all twelve people stood up and each one shook hands (once) with each of the others who was still a stranger. How many handshakes occurred?

54. Fifty people enter a single elimination chess tournament. (If you lose one game, you're out.) Assuming no ties occur, what is the number of games required to determine the tournament champion?

55. How many of the numbers from 10 through 100 have the sum of their digits equal to a perfect square?

56. How many three-digit numbers have the sum of their digits equal to 22?

57. How many integers between 100 and 400 contain the digit 2?

58. Jim Northington and friends are dining at the Bay Steamer Restaurant this evening, where a complete dinner consists of three items: (1) soup (clam chowder or minestrone) or salad (fresh spinach or shrimp), (2) sourdough rolls or bran muffin, and (3) entree (lasagna, lobster, or roast turkey). Jim selects his meal subject to the following restrictions. He cannot stomach more than one kind of seafood at a sitting. Also, whenever he tastes minestrone, he cannot resist having lasagna as well. And he cannot face the teasing he would receive from his companions if he were to order both spinach and bran. Use a tree diagram to determine the number of different choices Jim has.

For Exercises 59–61, refer to Example 7. How many different settings could Gabriella choose in each case?

59. No restrictions apply to adjacent switches.

60. No two adjacent switches can be off *and* no two adjacent switches can be on.

61. There are five switches rather than four, and no two adjacent switches can be on.

62. Determine the number of odd, non-repeating three-digit numbers that can be written using digits from the set {0, 1, 2, 3}.

63. A line segment joins the points (8, 12) and (53, 234) in the Cartesian plane. Including its endpoints, how many lattice points does this line segment contain? (A *lattice point* is a point with integer coordinates.)

64. In the pattern shown here, dots are one unit apart horizontally and vertically. If a segment can join any two dots, how many segments can be drawn with each of the following lengths?
(a) 1 **(b)** 2 **(c)** 3 **(d)** 4 **(e)** 5

65. Uniform length matchsticks are used to build a rectangular grid as shown here. If the grid is 15 matchsticks high and 28 matchsticks wide, how many matchsticks are used?

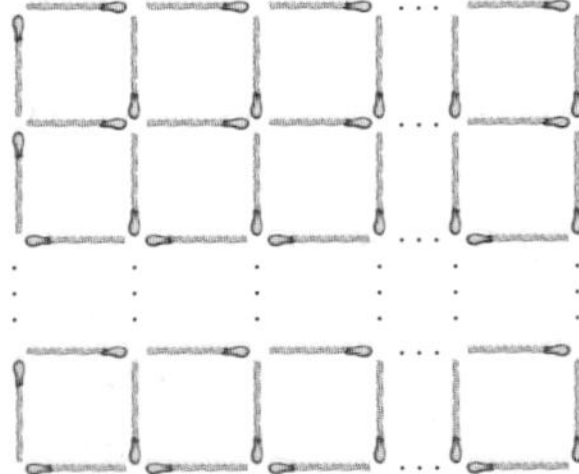

66. A square floor is to be tiled with square tiles as shown here with blue tiles on the main diagonals and red tiles everywhere else. (In all cases, both blue and red tiles must be used and the two diagonals must have a common blue tile at the center of the floor.)

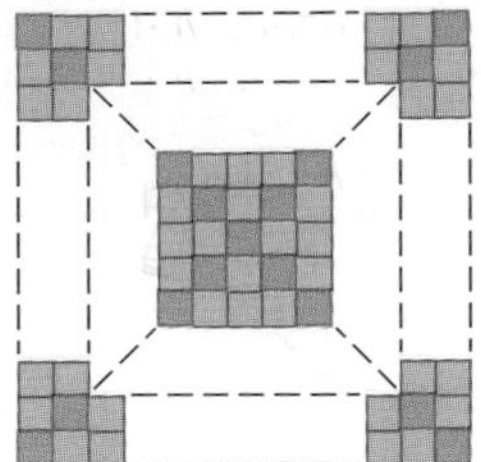

(a) If 81 blue tiles will be used, how many red tiles will be needed?
(b) For what numbers in place of 81 would this problem still be solvable?
(c) Find a formula expressing the number of red tiles required in general.

67. The text stated that number chains such as those of Example 10 always terminate with a single-digit number. This follows from the fact that adding the digits of a whole number with more than one digit always produces a sum smaller than the original number.
(a) Show that this last statement is true for any two-digit number. (*Hint:* Express the number as tu, where t is the tens digit and u is the units digit. Then the *value* of the number tu is $10t + u$.)
(b) Explain how part (a) implies that any two-digit number will eventually lead to a one-digit number.
(c) Show that adding the digits of any three-digit number will produce a sum smaller than the original number.

68. Kent Merrill and his son were among four father-and-son pairs who gathered to trade baseball cards. As each person arrived, he shook hands with anyone he had not known previously. Each person ended up making a different number of new acquaintances (0–6), except Kent and his son, who each met the same number of people. How many hands did Kent shake?

The Fundamental Counting Principle

In the previous section, we obtained complete lists of all possible results for various tasks. However, if the total number of possibilities is all we need to know, then an actual listing is usually unnecessary and is often very difficult or tedious to obtain, especially when the list is long. In this section, we develop ways to calculate "how many" using the *fundamental counting principle,* which is stated below.

The tree diagram in Figure 2 of the previous section can be used to illustrate the idea. It showed all possible non-repeating three-digit numbers formed from the set $\{1, 2, 3\}$. There were three stage-one branches, each of which split two ways to stage two. And each stage-two branch continued in just one way to stage three. Simply multiplying these three numbers gives the total number of possible results: $3 \cdot 2 \cdot 1 = 6$; and, the factors 3, 2, and 1 become evident without the tree diagram. We can reason that there are three possibilities for the first digit. After the first is chosen, two choices remain for the second digit. After the second is chosen, one choice remains for the third digit. This method of counting by products can be generalized as follows.

Fundamental Counting Principle

When a task consists of k separate parts, if the first part can be done in n_1 ways, the second part can be done in n_2 ways, and so on through the kth part, which can be done in n_k ways, then the total number of possible results for completing the task is given by the product

$$n_1 \cdot n_2 \cdot n_3 \cdot \ldots \cdot n_k.$$

All of the following examples utilize the fundamental counting principle. It may be helpful to set up a sequence of blanks for the various parts of the task, fill in the number of choices for each part as it is determined, and finally multiply the factors.

EXAMPLE 1 How many two-digit numbers are there in our (base ten) system of counting numbers? (While 80 is a two-digit number, 08 is not.)

The "task" is to select, or design, a two-digit number. This task consists of two parts: select the first digit and select the second digit. There are nine ways to choose the first digit (1 through 9), and then ten ways to choose the second digit (0 through 9). The total number of possibilities is $9 \cdot 10 = 90$.

In this example the second digit could have been chosen first, with ten choices possible. Then there are nine choices for the first digit. Again the total is $10 \cdot 9 = 90$.

Richard Dedekind (1831–1916) studied at the University of Göttingen, where he was Gauss' last student. His work was not recognized during his lifetime, but his treatment of the infinite and of what constitutes a real number are influential even today.

While on vacation in Switzerland, Dedekind met Georg Cantor. Dedekind was interested in Cantor's work on infinite sets. Perhaps because both were working in new and unusual fields of mathematics, such as number theory, and because neither received the professional attention he deserved during his lifetime, the two struck up a lasting friendship.

EXAMPLE 2 Find the number of two-digit numbers that do not contain repeated digits (for example, 66 is not allowed).

The basic task is again to select a two-digit number, and there are two parts: select the first digit and select the second digit. But a new restriction applies—no repetition of digits. There are nine choices for the first digit (1 through 9). Then nine choices remain for the second digit, since one nonzero digit has been used and cannot be repeated, but zero is now available. The total number is $9 \cdot 9 = 81$.

Starting with the second digit in Example 2 would have led to trouble. After observing that there are ten choices for the second digit, it would not be possible to decide on the number of choices for the first digit, since there is no way to know whether the second digit was zero or nonzero. To avoid this kind of ambiguity, it is usually best to start with any part of the task that has any special restrictions. In Example 2, the first digit is restricted in that it cannot be zero, so consider it first.

EXAMPLE 3 In how many ways can club N of the previous section elect a president and secretary if the secretary must be a man?

Since the special restriction applies to secretary, consider that office first. There are three choices: A, B, and D. Then four choices remain for president (the two men who were not chosen as secretary, together with the two women). The total number of ways is $3 \cdot 4 = 12$. ■

EXAMPLE 4 How many four-digit numbers are there in our system of counting numbers?

The task of selecting a four-digit number has four parts. There are no restrictions implied except that the first digit must not be zero. There are $9 \cdot 10 \cdot 10 \cdot 10 = 9{,}000$ (or $9 \cdot 10^3 = 9{,}000$) possible four-digit numbers. ■

EXAMPLE 5 In some states, auto license plates have contained three letters followed by three digits. How many different licenses are possible before a new scheme is necessary?

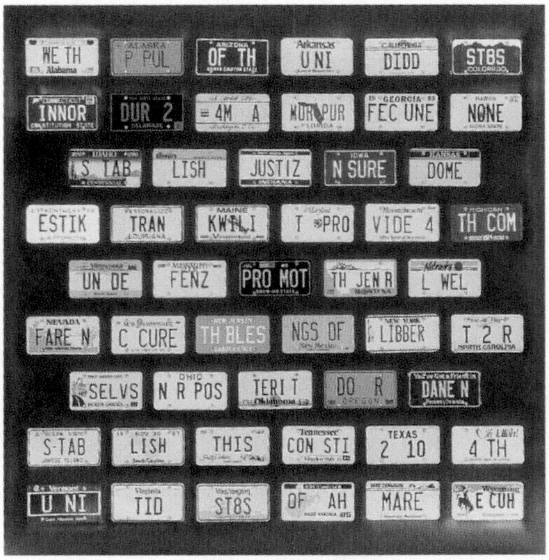

The basic task is to design a license number consisting of three letters followed by three digits. There are six component parts to this task. Since there are no restrictions on letters or digits to be used, the fundamental counting principle shows that there are

$$26^3 \cdot 10^3 = 17{,}576{,}000$$

possible licenses. (In practice a few of the possible sequences of letters are considered undesirable and are not used.) ■

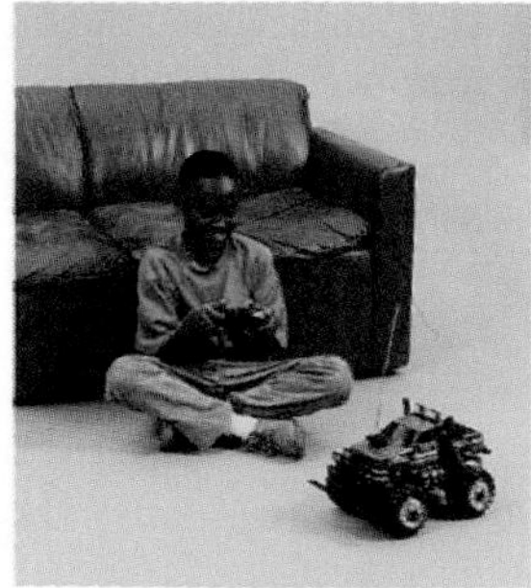

EXAMPLE 6 Mark decides to give his five remote control vehicles to his three younger brothers.

(a) In how many ways can he make the distribution?

This is a matter of distributing five objects among three recipients. Consider each of the five toys in turn. Notice that, for each one, there are three choices as to which brother will receive it. By the fundamental counting principle, the task can be completed in $3 \cdot 3 \cdot 3 \cdot 3 \cdot 3 = 3^5 = 243$ ways.

Notice that this problem is much harder if we try considering each brother in turn rather than considering the toys. In set theory we establish that the number of subsets of a set having n elements is 2^n. Therefore, the number of different *sets* of toys the first brother could be given is $2^5 = 32$ since there are 5 toys. But now the number of possibilities for the second brother cannot be easily found since the number of toys remaining is not known. Considering toys in turn was better since the choices for each successive toy did not depend on what was done with any previous toy.

(b) How many choices are there if the remote control jeep must go to brother Chris and the remote control Indy racer must go to either Chris or Scott?

In this case, two parts of the task are restricted. There is just one choice of brothers for the jeep, and just two choices for the Indy racer. The other three toys have three choices each. They can each go to any of the three brothers. The fundamental counting principle gives a total of $1 \cdot 2 \cdot 3 \cdot 3 \cdot 3 = 54$ choices.

EXAMPLE 7 Rework Example 8 of the previous section, this time using the fundamental counting principle.

Recall that Andy, Betty, Clyde, and Dawn (A, B, C, and D) are to seat themselves in a row so that A and B are side by side. Use the methods of this section and break the task into a series of three parts, each of which is simpler than the overall task.

The task is to seat four people (A, B, C, and D) in four adjacent seats (say 1, 2, 3, and 4) in such a way that A and B sit next to each other. One approach is to make three decisions as follows:

1	2	3	4
X	X	_	_
_	X	X	_
_	_	X	X

Seats available to A and B

1. Which pair of seats should A and B occupy? There are *three* choices ($1 - 2$, $2 - 3$, or $3 - 4$, as illustrated in the margin).
2. Which order should A and B take? There are *two* choices (A left of B, or B left of A).
3. Which order should C and D take? There are *two* choices (C left of D, or D left of C, not necessarily right next to each other).

Once the three decisions above are made, the seating order of the four people is decided. Using the fundamental counting principle, the total number of choices is $3 \cdot 2 \cdot 2 = 12$, the same result found in the previous section.

Factorials

This section began with a discussion of non-repeating three-digit numbers with digits from the set $\{1, 2, 3\}$. The number of possibilities was $3 \cdot 2 \cdot 1 = 6$, in keeping with the fundamental counting principle. That product can also be thought of as the total number of distinct *arrangements* of the three digits 1, 2, and 3. Similarly, the number of distinct arrangements of four objects, say A, B, C, and D, is, by the fundamental counting principle, $4 \cdot 3 \cdot 2 \cdot 1 = 24$. Since this type of product occurs so commonly in applications, we give it a special name and symbol as follows. For any counting number n, the product of *all* counting numbers from n down through 1 is called ***n* factorial,** and is denoted $n!$.

For Further Thought
A Classic Problem

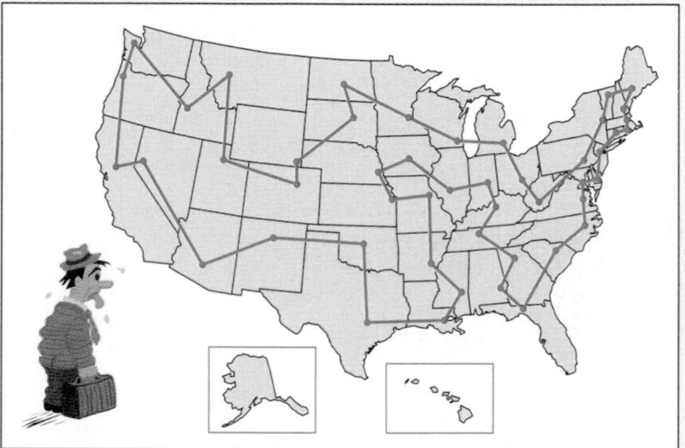

Suppose a salesperson wants to make exactly one visit to each capital city in the 48 contiguous states, starting and ending up at the same capital. What is the shortest route that would work?

The so-called "traveling salesman problem" (or TSP), in various versions, has baffled mathematicians for years. For the version stated above, there are 24!/3 possible routes to consider, and the huge size of this number is the reason the problem is difficult (even using the most powerful computers). In 1985, mathematician Shen Lin came up with the 10,628-mile route shown here, and even though he could not prove it was the shortest, he offered $100 to anyone who could find a shorter one.

In 1984, the Bell Laboratories mathematician, Narenda Karmarkar, devised a greatly improved algorithm to help solve the kinds of TSPs involving routing in the areas of transportation and communication. And in 1991, Donald L. Miller of Du Pont and Joseph R. Penky of Purdue University announced a solution to TSPs related to the efficiency of chemical and maintenance processes.

For Group Discussion

1. Within the accuracy limits of the drawing above, try to identify other routes that may be as good or better than the one shown.
2. Realistically, what factors other than *distance* might be important to the salesperson?
3. When a chemical company schedules the steps of a complex manufacturing process, what quantities might it want to minimize?

0! = 1
1! = 1
2! = 2
3! = 6
4! = 24
5! = 120
6! = 720
7! = 5,040
8! = 40,320
9! = 362,880
10! = 3,628,800

Short Table of Factorials
Their numerical value increases rapidly. The value of 100! is a number with 158 digits.

Factorial Formula

For any counting number n, the quantity n factorial is given by

$$n! = n(n-1)(n-2)\ldots 2 \cdot 1.$$

The first few factorial values are easily found by simple multiplication, but they rapidly become very large, as indicated in the margin. The use of a calculator is advised in most cases. (See the margin notes that follow.)

EXAMPLE 8 Evaluate the following expressions.

(a) $2! = 2 \cdot 1 = 2$

(b) $5! = 5 \cdot 4 \cdot 3 \cdot 2 \cdot 1 = 120$

(c) $6! = 6 \cdot 5 \cdot 4 \cdot 3 \cdot 2 \cdot 1 = 720$

(d) $6! - 3! = 720 - 6 = 714$

(e) $(6-3)! = 3! = 6$

(f) $\dfrac{6!}{3!} = \dfrac{6 \cdot 5 \cdot 4 \cdot \cancel{3} \cdot \cancel{2} \cdot \cancel{1}}{\cancel{3} \cdot \cancel{2} \cdot \cancel{1}} = 6 \cdot 5 \cdot 4 = 120$

(g) $\left(\dfrac{6}{3}\right)! = 2! = 2$

(h) $15! = 1.307674368 \times 10^{12}$

(i) $100! = 9.332621544 \times 10^{157}$

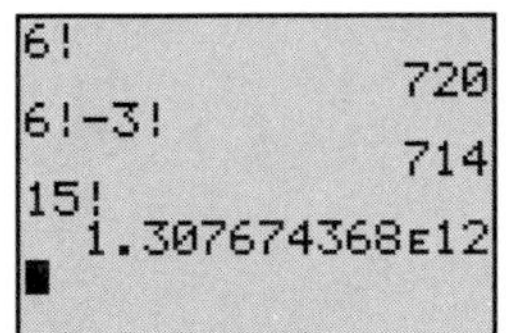

The results of Example 8(c), (d), and (h) are illustrated in this calculator screen.

Notice the difference between parts (d) and (e) and between parts (f) and (g) above. Parts (h) and (i) were found on a scientific calculator. Part (h) is exact, but part (i) is accurate only to the number of significant figures shown.

So that factorials will be defined for all whole numbers, including zero, it is common to define 0! as follows.

$$0! = 1$$

The definition 0! = 1 is illustrated here.

(We will see later that this special definition makes other results easier to state.)

Whenever we need to know the total number of ways to *arrange* a given number of distinct objects, we can use a factorial. The fundamental counting principle would do, but factorials provide a shortcut.

Arrangements of *n* Objects

The total number of different ways to arrange n distinct objects is $n!$.

EXAMPLE 9 Emily Barman has seven essays to include in her English 1A folder. In how many different orders can she arrange them?

The number of ways to arrange seven distinct objects is $7! = 5{,}040$.

EXAMPLE 10 Whenever Laura Stowe takes her fourteen daycare children to the park, they all want to be at the front of the line. How many different ways can they be arranged?

Fourteen children can be arranged in $14! = 87{,}178{,}291{,}200$ different ways.

10.2 EXERCISES

1. Explain the fundamental counting principle in your own words.

2. Describe how factorials can be used in counting problems.

For Exercises 3 and 4, n and m are counting numbers. Do the following: **(a)** *Say whether or not the given statement is true in general, and* **(b)** *explain your answer, using specific examples.*

3. $(n + m)! = n! + m!$

4. $(n \cdot m)! = n! \cdot m!$

Evaluate each expression without using a calculator.

5. $4!$

6. $7!$

7. $\frac{8!}{5!}$

8. $\frac{16!}{14!}$

9. $\frac{5!}{(5-2)!}$

10. $\frac{6!}{(6-4)!}$

11. $\frac{9!}{6!\,(6-3)!}$

12. $\frac{10!}{4!\,(10-4)!}$

13. $\frac{n!}{(n-r)!}$, where $n = 7$ and $r = 4$

14. $\frac{n!}{r!\,(n-r)!}$, where $n = 12$ and $r = 4$

Evaluate each expression using a calculator. (Some answers may not be exact.)

15. $11!$

16. $17!$

17. $\frac{12!}{7!}$

18. $\frac{15!}{9!}$

19. $\frac{13!}{(13-3)!}$

20. $\frac{16!}{(16-6)!}$

21. $\frac{20!}{10! \cdot 10!}$

22. $\frac{18!}{6! \cdot 12!}$

23. $\frac{n!}{(n-r)!}$, where $n = 23$ and $r = 10$

24. $\frac{n!}{r!\,(n-r)!}$, where $n = 28$ and $r = 15$

A panel containing three on-off switches in a row is to be set.

25. Assuming no restrictions on individual switches, use the fundamental counting principle to find the total number of possible panel settings.

26. Assuming no restrictions, construct a tree diagram to list all the possible panel settings of Exercise 25.

27. Now assume that no two adjacent switches can both be off. Explain why the fundamental counting principle does not apply.

28. Construct a tree diagram to list all possible panel settings under the restriction of Exercise 27.

29. Table 2 in the previous section shows that there are 36 possible outcomes when two fair dice are rolled. How many would there be if three fair dice were rolled?

30. How many five-digit numbers are there in our system of counting numbers?

Recall the club

$$N = \{\text{Andy, Bill, Cathy, David, Evelyn}\}.$$

In how many ways could they do each of the following?

31. line up all five members for a photograph

32. schedule one member to work in the office on each of five different days, assuming members may work more than one day

33. select a male and a female to decorate for a party

34. select two members, one to open their next meeting and another to close it, given that Bill will not be present

In the following exercises, counting numbers are to be formed using only digits from the set $\{0, 1, 2, 3, 4\}$. *Determine the number of different possibilities for each type of number described. (Note that* 201 *is a three-digit number while* 012 *is not.)*

35. three-digit numbers

36. even four-digit numbers

37. five-digit numbers with one pair of adjacent 0s and no other repeated digits

38. four-digit numbers beginning and ending with 3 and unlimited repetitions allowed

39. five-digit multiples of five that contain exactly two 3s and exactly one 0

40. six-digit numbers containing two 0s, two 2s, and two 4s

The Casa Loma Restaurant offers four choices in the soup and salad category (two soups and two salads), two choices in the bread category, and three choices in the entree category. Find the number of dinners available in each of the following cases.

41. One item is to be included from each of the three categories.

42. Only soup and entree are to be included.

Determine the number of possible ways to mark your answer sheet (with an answer for each question) for each of the following tests.

43. an eight-question true-or-false test

44. a thirty-question multiple-choice test with four answer choices for each question

---

Rogers Newman is making up his class schedule for next semester, which must include one class from each of the four categories shown here.

Category	Choices	Number of Choices
English	Medieval Literature Composition Modern Poetry	3
Mathematics	College Algebra Trigonometry	2
Computer Information Science	Introduction to Spreadsheets Advanced Word Processing C Programming BASIC Programming	4
Sociology	Social Problems Sociology of the Middle East Aging in America Minorities in America Women in American Culture	5

For each situation in Exercises 45–50, use the table above to determine the number of different sets of classes Rogers can take.

45. All classes shown are available.

46. He is not eligible for College Algebra or for BASIC Programming.

47. All sections of Minorities in America and Women in American Culture are filled already.

48. He does not have the prerequisites for Medieval Literature, Trigonometry, or C Programming.

49. Funding has been withdrawn for three of the computer courses and for two of the Sociology courses.

50. He must complete English Composition and Trigonometry next semester to fulfill his degree requirements.

---

51. Sean took two pairs of shoes, four pairs of pants, and six shirts on a trip. Assuming all items are compatible, how many different outfits can he wear?

52. A music equipment outlet stocks ten different guitars, four guitar cases, six amplifiers, and three effects processors, with all items mutually compatible and all suitable for beginners. How many different complete setups could Lionel choose to start his musical career?

53. Tadishi's ZIP code is 95841. How many ZIP codes, altogether, could be formed using all of those same five digits?

54. George Owen keeps four textbooks and three novels on his desk. In how many different ways can he arrange them in a row if

(a) the textbooks must be to the left of the novels?

(b) the novels must all be together?

(c) no two novels should be next to each other?

Andy, Betty, Clyde, Dawn, Evan, and Felicia have reserved six seats in a row at the theater, starting at an aisle seat. (Refer to Example 7 in this section.)

55. In how many ways can they arrange themselves? (*Hint:* Divide the task into the series of six parts shown below, performed in order.)

(a) If A is seated first, how many seats are available for him?

(b) Now, how many are available for B?

(c) Now, how many for C?

(d) Now, how many for D?

(e) Now, how many for E?

(f) Now, how many for F?

Now multiply together your six answers above.

56. In how many ways can they arrange themselves so that Andy and Betty will be next to each other?

1	2	3	4	5	6
X	X	_	_	_	_
_	X	X	_	_	_
_	_	X	X	_	_
_	_	_	X	X	_
_	_	_	_	X	X

Seats available to A and B

(*Hint:* First answer the following series of questions, assuming these parts are to be accomplished in order.)

(a) How many pairs of adjacent seats can A and B occupy?

(b) Now, given the two seats for A and B, in how many orders can they be seated?

(c) Now, how many seats are available for C?

(d) Now, how many for D?

(e) Now, how many for E?

(f) Now, how many for F?

57. In how many ways can they arrange themselves if the men and women are to alternate seats and a man must sit on the aisle? (*Hint:* First answer the following series of questions.)

(a) How many choices are there for the person to occupy the first seat, next to the aisle? (It must be a man.)

(b) Now, how many choices of people may occupy the second seat from the aisle? (It must be a woman.)

(c) Now, how many for the third seat? (one of the remaining men)

(d) Now, how many for the fourth seat? (a woman)

(e) Now, how many for the fifth seat? (a man)

(f) Now, how many for the sixth seat? (a woman)

58. In how many ways can they arrange themselves if the men and women are to alternate with either a man or a woman on the aisle? (*Hint:* First answer the following series of questions.)

(a) How many choices of people are there for the aisle seat?

(b) Now, how many are there for the second seat? (This person may not be of the same sex as the person on the aisle.)

(c) Now, how many choices are there for the third seat?

(d) Now, how many for the fourth seat?

(e) Now, how many for the fifth seat?

(f) Now, how many for the sixth seat?

10.3 Permutations and Combinations

Earlier we introduced factorials as a way of counting the number of *arrangements* of a given set of objects. For example, the members of the club $N = \{\text{Andy, Bill, Cathy, David, Evelyn}\}$ can arrange themselves in a row for a photograph in $5! = 120$ different ways. Using a factorial is generally more efficient than using the fundamental counting principle. We have also used previous methods, like tree diagrams and the fundamental counting principle, to answer questions like: How many ways can club N above elect a president, a secretary, and a treasurer if no one person can hold more than one office? This again is a matter of *arrangements.* The difference is that only three, rather than all five, of the members are involved in each case. A common way to rephrase the basic question here is as follows: *How many arrangements are there of five things taken three at a time?* The answer, by the fundamental counting principle, is $5 \cdot 4 \cdot 3 = 60$. The factors begin with 5 and proceed downward, just as in a factorial product, but do not go all the way to 1. (In this example the product stops when there are three factors.) We now generalize this idea.

Permutations

In the context of counting problems, arrangements are often called **permutations;** the number of permutations of n distinct things taken r at a time is written $P(n, r)$.* Since the number of objects being arranged cannot exceed the total number available, we assume for our purposes here that $r \leq n$. Applying the fundamental counting principle to arrangements of this type gives

$$P(n, r) = n(n - 1)(n - 2) \ldots [n - (r - 1)].$$

Simplification of the last factor gives the following formula.

Permutations Formula

The number of **permutations,** or *arrangements,* of n distinct things taken r at a time, where $r \leq n$, is given by

$$P(n, r) = n(n - 1)(n - 2) \ldots (n - r + 1).$$

The factors in this product begin at n and descend until the total number of factors is r. We now see that the number of ways in which club N can elect a president, a secretary, and a treasurer can be denoted $P(5, 3) = 5 \cdot 4 \cdot 3 = 60$.

EXAMPLE 1 Evaluate each permutation.

(a) $P(4, 2) = 4 \cdot 3 = 12$ (Begin at 4, descend until there are two factors.)

(b) $P(5, 2) = 5 \cdot 4 = 20$ (Begin at 5, use two factors.)

*Alternate notations are ${}_nP_r$ and P^n_r.

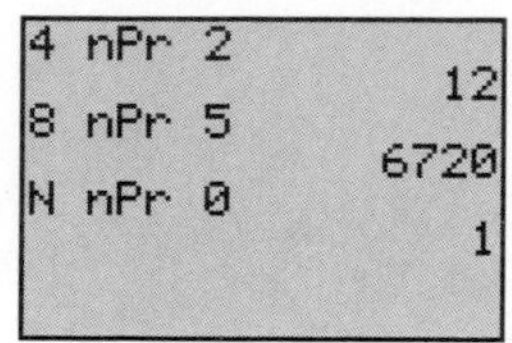

The results of Example 1(a) and (d) are supported here. For any whole number value of n, $P(n, 0) = 1$. See the final line of the display.

(c) $P(7, 3) = 7 \cdot 6 \cdot 5 = 210$ (Begin at 7, use three factors.)

(d) $P(8, 5) = 8 \cdot 7 \cdot 6 \cdot 5 \cdot 4 = 6{,}720$ (Begin at 8, use five factors.)

(e) $P(5, 5) = 5 \cdot 4 \cdot 3 \cdot 2 \cdot 1 = 120$ (Begin at 5, use five factors.)

Notice that $P(5, 5)$ is equal to 5!. It is true for all whole numbers n that $P(n, n) = n!$. (This is the number of possible arrangements of n distinct objects taken all n at a time.)

Permutations, in general, can also be related to factorials in the following way. Recall that

$$P(n, r) = n(n - 1)(n - 2) \ldots (n - r + 1).$$

Extending this product all the way down to 1 gives

$$n(n - 1)(n - 2) \ldots (n - r + 1)(n - r)(n - r - 1) \ldots 2 \cdot 1.$$

Then, dividing by exactly the same factors that were introduced into the product gives

$$\frac{n(n - 1)(n - 2) \ldots (n - r + 1)(n - r)(n - r - 1) \ldots 2 \cdot 1}{(n - r)(n - r - 1) \ldots 2 \cdot 1}.$$

This quotient is equal to $n!/(n - r)!$, and since it was obtained by *multiplying and dividing* $P(n, r)$ by the same quantity, it must be equal to $P(n, r)$. This formula can always be used to evaluate permutations.

Factorial Formula for Permutations

The number of **permutations,** or *arrangements,* of n distinct things taken r at a time, where $r \leq n$, can be calculated as

$$P(n, r) = \frac{n!}{(n - r)!}.$$

If n and r are very large numbers, a calculator with a factorial key and this formula will save a lot of work when finding permutations. See the margin note.

The formula above also shows that, for any whole number n,

$$P(n, 0) = \frac{n!}{(n - 0)!} = \frac{n!}{n!} = 1.$$

In other words, the number of arrangements of n things, taken 0 at a time, is 1. This is reasonable since there is exactly one way to arrange none of the n things.

EXAMPLE 2 Use the factorial formula to evaluate each permutation.

(a) $P(7, 3) = \frac{7!}{(7 - 3)!} = \frac{7!}{4!} = 210$

(b) $P(10, 4) = \frac{10!}{(10 - 4)!} = \frac{10!}{6!} = 5{,}040$

Change ringing, the English way of ringing church bells, combines mathematics and music. Bells are rung first in sequence, 1, 2, 3, ... Then the sequence is permuted ("changed"). On six bells, 720 different "changes" (different permutations of tone) can be rung: $P(6, 6) = 6!$.

Composers work out changes so that musically interesting and harmonious sequences occur regularly.

The church bells are swung by means of ropes attached to the wheels beside them. One ringer swings each bell, listening intently and watching the other ringers closely. If one ringer gets lost and stays lost, the rhythm of the ringing cannot be maintained; all the ringers have to stop.

A ringer can spend weeks just learning to keep a bell going and months learning to make the bell ring in exactly the right place. Errors of 1/4 second mean that two bells are ringing at the same time. Even errors of 1/10 second can be heard.

(c) $P(25, 0) = \dfrac{25!}{(25 - 0)!} = \dfrac{25!}{25!} = 1$

(d) $P(18, 12) = \dfrac{18!}{(18 - 12)!} = \dfrac{18!}{6!} = 8{,}892{,}185{,}702{,}400$

Concerning part (d), most calculators will not display this many digits, so you may obtain an answer such as 8.8921857×10^{12}. ■

Problem Solving

Permutations can be used any time we need to know the number of size-r arrangements that can be selected from a size-n set. The word *arrangement* implies an ordering, so we use permutations only in cases where the order of the items is important. Also, permutations apply only in cases where no repetition of items occurs. We summarize with the following guidelines.

Guidelines for Permutations

Permutations are applied only when

1. repetitions are not allowed, and
2. order is important.

Examples 3 and 4 involve cases that meet these guidelines.

EXAMPLE 3 How many non-repeating three-digit numbers can be written using digits from the set {3, 4, 5, 6, 7, 8}?

Repetitions are not allowed since the numbers are to be "non-repeating." (For example, 448 is not acceptable.) Also, order is important. (For example, 476 and 746 are *distinct* cases.) So we use permutations:

$$P(6, 3) = 6 \cdot 5 \cdot 4 = 120. \quad ■$$

The next example illustrates the common situation where a task involves multiple parts, and hence calls for the fundamental counting principle, but where the individual parts can be handled with permutations.

EXAMPLE 4 Suppose certain account numbers are to consist of two letters followed by four digits and then three more letters, where repetitions of letters or digits are not allowed *within* any of the three groups, but the last group of letters may contain one or both of those used in the first group. How many such accounts are possible?

The task of designing such a number consists of three parts.

1. Determine the first set of two letters.
2. Determine the set of four digits.
3. Determine the final set of three letters.

Each part requires an arrangement without repetitions, which is a permutation. Apply the fundamental counting principle and multiply together the results of the three parts.

$$\begin{aligned} P(26, 2) \cdot P(10, 4) \cdot P(26, 3) &= \underbrace{26 \cdot 25}_{\text{Part 1}} \cdot \underbrace{10 \cdot 9 \cdot 8 \cdot 7}_{\text{Part 2}} \cdot \underbrace{26 \cdot 25 \cdot 24}_{\text{Part 3}} \\ &= 650 \cdot 5{,}040 \cdot 15{,}600 \\ &= 51{,}105{,}600{,}000 \end{aligned}$$

Combinations

So far in this section, we have introduced permutations in order to evaluate the number of arrangements of n things taken r at a time, where repetitions are not allowed. The order of the items was important. Recall that club $N = \{$Andy, Bill, Cathy, David, Evelyn$\}$ could elect three officers in $P(5, 3) = 60$ different ways. With three-member committees, on the other hand, order is not important. The committees B, D, E and E, B, D are no different. The possible number of committees is not the number of arrangements of size 3. Rather, it is the number of *subsets* of size 3 (since the order of listing elements in a set makes no difference).

Subsets in this new context are called **combinations.** The number of combinations of n things taken r at a time (that is, the number of size r subsets, given a set of size n) is written $C(n, r)$.*

Here is a list of all the size-3 committees (subsets) of the club (set) $N = \{A, B, C, D, E\}$:

$$\{A, B, C\}, \quad \{A, B, D\}, \quad \{A, B, E\}, \quad \{A, C, D\}, \quad \{A, C, E\},$$
$$\{A, D, E\}, \quad \{B, C, D\}, \quad \{B, C, E\}, \quad \{B, D, E\}, \quad \{C, D, E\}.$$

There are ten subsets of 3, so ten is the number of three-member committees possible. Just as with permutations, repetitions are not allowed. For example, $\{E, E, B\}$ is not a valid three-member subset, just as EEB is not a valid three-member arrangement.

To see how to find the number of such subsets without listing them all, notice that each size-3 subset (combination) gives rise to six size-3 arrangements (permutations). For example, the single combination ADE yields these six permutations:

$$A, D, E \qquad D, A, E \qquad E, A, D \qquad A, E, D \qquad D, E, A \qquad E, D, A.$$

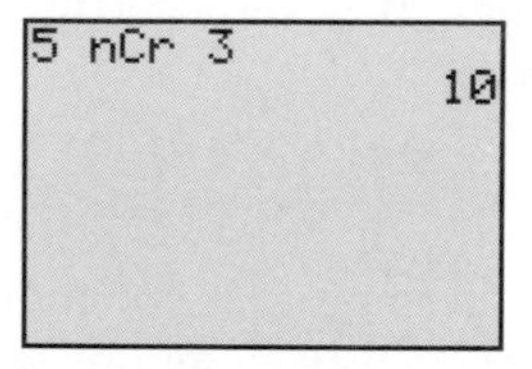

Calculators can perform operations involving combinations of n things taken r at a time. Compare this display to the result in the text.

Then there must be six times as many size-3 permutations as there are size-3 combinations, or in other words, one-sixth as many combinations as permutations. Therefore

$$C(5, 3) = \frac{P(5, 3)}{6} = \frac{5 \cdot 4 \cdot 3}{6} = 10.$$

The 6 appears in the denominator because there are six different ways to arrange a set of three things (since $3! = 3 \cdot 2 \cdot 1 = 6$). Generalizing from this example, r things can be arranged in $r!$ different ways, so we obtain the following formula.

*Alternate notations include ${}_nC_r$, C_r^n, and $\binom{n}{r}$.

Combinations Formula

The number of **combinations,** or *subsets,* of n distinct things taken r at a time, where $r \le n$, is given by

$$C(n, r) = \frac{P(n, r)}{r!} = \frac{n(n-1)(n-2)\ldots(n-r+1)}{r(r-1)(r-2)\ldots 2 \cdot 1}.$$

We saw earlier that permutations are expressible entirely in terms of factorials:

$$P(n, r) = \frac{n!}{(n-r)!}.$$

Using this formula, we obtain

$$C(n, r) = \frac{P(n, r)}{r!} = \frac{\frac{n!}{(n-r)!}}{r!} = \frac{n!}{r!\,(n-r)!}.$$

A B C D E F
Aaaaa aaaab aaaba. aaabb. aabaa. aabab.
G H I K L M
aabba aabbb abaaa. abaab. ababa. ababb.
N O P Q R S
abbaa. abbab. abbba. abbbb. baaaa. baaab.
T V W X Y Z
baaba. baabb. babaa. babab. babba. babbb.

"Biliteral cipher" (above) was invented by Francis Bacon early in the seventeenth century to code political secrets. This binary code, *a* and *b* in combinations of five, has 32 permutations. Bacon's "biformed alphabet" (bottom four rows) uses two type fonts to conceal a message in some straight text. The decoder deciphers a string of *a*'s and *b*'s, groups them by fives, then deciphers letters and words. This code was applied to Shakespeare's plays in efforts to prove Bacon the rightful author.

Factorial Formula for Combinations

The number of **combinations,** or *subsets,* of n distinct things taken r at a time, where $r \le n$, can be calculated as

$$C(n, r) = \frac{P(n, r)}{r!} = \frac{n!}{r!\,(n-r)!}.$$

With this result, combinations also can be computed using factorials. Using this formula together with the fact that $0! = 1$ gives, for any whole number n,

$$C(n, 0) = \frac{n!}{0!\,(n-0)!} = \frac{n!}{1 \cdot n!} = \frac{n!}{n!} = 1.$$

This means that there is exactly one combination of n things taken zero at a time. That is, a set of n objects has exactly one "empty" subset.

Problem Solving

The following guidelines stress that combinations have an important common property with permutations (repetitions are not allowed), as well as an important difference (order is *not* important with combinations). It is extremely important to realize this when solving problems.

Guidelines for Combinations

Combinations are applied only when

1. repetitions are not allowed, and
2. order is *not* important.

For Further Thought
Poker Hands

In 5-card poker, played with a standard 52-card deck, 2,598,960 different hands are possible. (See Example 6.) The desirability of the various hands depends upon their relative chance of occurrence, which, in turn, depends on the number of different ways they can occur, as shown in Table 4.

TABLE 4

Event E	Description of Event E	Number of Outcomes Favorable to E
Royal flush	Ace, king, queen, jack, and 10, all of the same suit	4
Straight flush	5 cards of consecutive denominations, all in the same suit (excluding royal flush)	36
Four of a kind	4 cards of the same denomination, plus 1 additional card	624
Full house	3 cards of one denomination, plus 2 cards of a second denomination	3,744
Flush	Any 5 cards all of the same suit (excluding royal flush and straight flush)	5,108
Straight	5 cards of consecutive denominations (not all the same suit)	10,200
Three of a kind	3 cards of one denomination, plus 2 cards of two additional denominations	54,912
Two pairs	2 cards of one denomination, plus 2 cards of a second denomination, plus 1 card of a third denomination	123,552
One pair	2 cards of one denomination, plus 3 additional cards of three different denominations	1,098,240
No pair	No two cards of the same denomination (and not all the same suit)	1,302,540
Total		2,598,960

For Group Discussion

As the table shows, a full house is a relatively rare occurrence. (Only four of a kind, straight flush, and royal flush are less likely.) To verify that there are 3,744 different full house hands possible, carry out the following steps.

1. Explain why there are $C(4, 3)$ different ways to select three aces from the deck.
2. Explain why there are $C(4, 2)$ different ways to select two 8s from the deck.
3. If "aces and 8s" (three aces and two 8s) is one *kind* of full house, show that there are $P(13, 2)$ different kinds of full house altogether.
4. Multiply the expressions from steps 1, 2, and 3 together. Explain why this product should give the total number of full house hands possible.

The set of 52 playing cards in the standard deck has four suits.

♠ spades
♥ hearts
♦ diamonds
♣ clubs

Ace is the unit card. Jacks, queens, and kings are "face cards." Each suit contains thirteen denominations: ace, 2, 3, . . . , 10, jack, queen, king. (In some games, ace rates above king, instead of counting as 1.)

EXAMPLE 5 Find the number of different subsets of size 2 in the set $\{a, b, c, d\}$. List them to check the answer.

A subset of size 2 must have two distinct elements, so repetitions are not allowed. And since the order in which the elements of a set are listed makes no difference, we see that order is not important. Use the combinations formula with $n = 4$ and $r = 2$.

$$C(4, 2) = \frac{P(4, 2)}{2!} = \frac{4 \cdot 3}{2 \cdot 1} = 6$$

or

$$C(4, 2) = \frac{4!}{2!\,(4-2)!} = \frac{4!}{2!\,2!} = 6$$

The six subsets of size 2 are $\{a, b\}$, $\{a, c\}$, $\{a, d\}$, $\{b, c\}$, $\{b, d\}$, $\{c, d\}$. ■

EXAMPLE 6 A common form of poker involves hands (sets) of five cards each, dealt from a standard deck consisting of 52 different cards (illustrated in the margin). How many different 5-card hands are possible?

A 5-card hand must contain five distinct cards, so repetitions are not allowed. Also, the order is not important since a given hand depends only on the cards it contains, and not on the order in which they were dealt or the order in which they are listed. Since order does not matter, use combinations:

$$C(52, 5) = \frac{P(52, 5)}{5!} = \frac{52 \cdot 51 \cdot 50 \cdot 49 \cdot 48}{5 \cdot 4 \cdot 3 \cdot 2 \cdot 1} = 2{,}598{,}960$$

or

$$C(52, 5) = \frac{52!}{5!\,(52-5)!} = \frac{52!}{5!\,47!} = 2{,}598{,}960.$$

In this case, we will refrain from listing them all. ■

EXAMPLE 7 Melvin wants to buy ten different books but can afford only four of them. In how many ways can he make his selection?

The four books selected must be distinct (repetitions are not allowed), and also the order of the four chosen has no bearing in this case, so we use combinations:

$$C(10, 4) = \frac{P(10, 4)}{4!} = \frac{10 \cdot 9 \cdot 8 \cdot 7}{4 \cdot 3 \cdot 2 \cdot 1} = 210 \text{ ways,}$$

or

$$C(10, 4) = \frac{10!}{4!\,(10-4)!} = \frac{10!}{4!\,6!} = 210 \text{ ways.}$$ ■

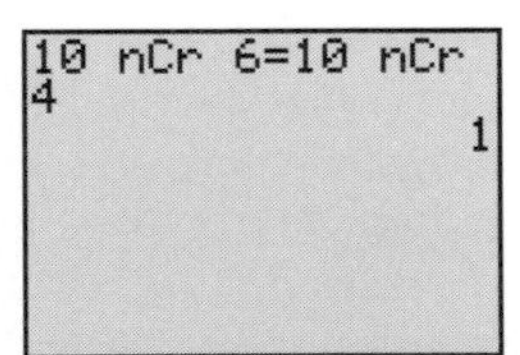

The "1" indicates that the statement

$C(10, 6) = C(10, 4)$

is true.

Notice that, according to our formula for combinations,

$$C(10, 6) = \frac{10!}{6!\,(10-6)!} = \frac{10!}{6!\,4!} = 210,$$

which is the same as $C(10, 4)$. In fact, Exercise 62 asks you to prove the fact that, in general, for all whole numbers n and r, with $r \leq n$,

$$C(n, r) = C(n, n - r).$$

See the margin note as well. It indicates that $C(10, 6)$ is equal to $C(10, 4)$.

EXAMPLE 8 How many different three-member committees could club N appoint so that exactly one woman is on the committee?

Recall that $N = \{\text{Andy, Bill, Cathy, David, Evelyn}\}$. Two members are women; three are men. Although the question mentioned only that the commit-

The illustration above is from the 1560s text ***Logistica,*** by the mathematician J. Buteo. Among other topics, the book discusses the number of possible throws of four dice and the number of arrangements of the cylinders of a combination lock. Note that "combination" is a misleading name for these locks since repetitions are allowed, and, also, order makes a difference.

tee must include exactly one woman, in order to complete the committee two men must be selected as well. Therefore the task of selecting the committee members consists of two parts:

1. Choose one woman.
2. Choose two men.

One woman can be chosen in $C(2, 1) = 2/1 = 2$ ways, and two men can be chosen in $C(3, 2) = (3 \cdot 2)/(2 \cdot 1) = 3$ ways. Using the fundamental counting principle gives $2 \cdot 3 = 6$ different committees. This number is small enough to check by listing the possibilities:

$$\{C, A, B\}, \{C, A, D\}, \{C, B, D\}, \{E, A, B\}, \{E, A, D\}, \{E, B, D\}.$$

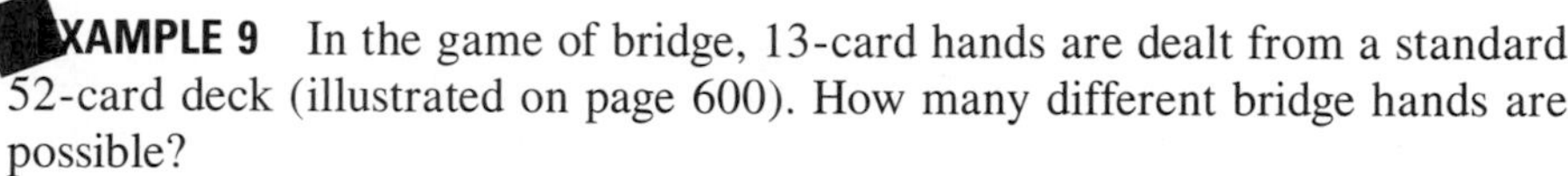

EXAMPLE 9 In the game of bridge, 13-card hands are dealt from a standard 52-card deck (illustrated on page 600). How many different bridge hands are possible?

As in most all card games, the order of the cards dealt to a hand is not important, so we use combinations and evaluate with a calculator:

$$C(52, 13) = \frac{52!}{13!\ 39!} = 635{,}013{,}559{,}600.$$

Problem Solving

As we have illustrated in this section, many counting problems involve selecting some of the items from a given set of items. The particular conditions of the problem will determine which specific technique to use. Since choosing the appropriate technique is critical, we offer the following guidelines.

Guidelines for Choosing a Counting Method

1. If selected items can be repeated, use the fundamental counting principle.
   Example: How many five-digit numbers are there?
   $$9 \cdot 10^4 = 90{,}000$$
2. If selected items cannot be repeated, and order is important, use permutations.
   Example: How many ways can three of seven people line up at a ticket counter?
   $$P(7, 3) = 7 \cdot 6 \cdot 5 = 210$$
3. If selected items cannot be repeated, and order is not important, use combinations.
   Example: How many ways can a committee of four be selected from a group of ten people?
   $$C(10, 4) = \frac{10 \cdot 9 \cdot 8 \cdot 7}{4 \cdot 3 \cdot 2 \cdot 1} = 210$$

If a task consists of multiple parts, consider one of the methods above for each of the individual parts, and then apply the fundamental counting principle to multiply all factors. Also, remember that not all counting problems are addressed by the above guidelines. Do not feel confined by any list of methods. Use your own insights. Since permutations and combinations are easily confused, we include the following table to emphasize both the similarities and the differences between them.

Permutations	**Combinations**
Number of ways of selecting r items out of n items	
Repetitions are not allowed.	
Order is important.	Order is not important.
Arrangements of n items taken r at a time	Subsets of n items taken r at a time
$P(n, r)$ $= n(n-1)(n-2)\ldots(n-r+1)$ $= \dfrac{n!}{(n-r)!}$	$C(n, r)$ $= \dfrac{n(n-1)(n-2)\ldots(n-r+1)}{r(r-1)(r-2)\ldots 2\cdot 1}$ $= \dfrac{n!}{r!\,(n-r)!}$

10.3 EXERCISES

Evaluate each of the following expressions.

1. $P(5, 2)$ **2.** $P(8, 3)$

3. $P(10, 0)$ **4.** $P(12, 5)$

5. $C(7, 3)$ **6.** $C(10, 6)$

7. $C(8, 4)$ **8.** $C(12, 8)$

Determine the number of permutations (arrangements) of each of the following.

9. 7 things taken 4 at a time

10. 8 things taken 5 at a time

11. 12 things taken 3 at a time

12. 41 things taken 2 at a time

Determine the number of combinations (subsets) of each of the following.

13. 5 things taken 4 at a time

14. 9 things taken 0 at a time

15. 11 things taken 7 at a time

16. 14 things taken 3 at a time

Use a calculator to evaluate each of these expressions.

17. $P(26, 8)$ **18.** $C(38, 12)$

In "Super Lotto," a California state lottery game, you select six distinct numbers from the counting numbers 1 through 51, hoping that your selection will match a random list selected by lottery officials.

19. How many different sets of six numbers can you select?

20. Diane Sypniewski always includes her age and her husband's age as two of the numbers in her Super Lotto selections. How many ways can she complete her list of six numbers?

21. Is it possible to evaluate $P(8, 12)$? Explain.

22. Is it possible to evaluate $C(6, 15)$? Explain.

23. Explain how permutations and combinations differ.

24. Explain how factorials are related to permutations.

25. How many different ways could first, second, and third place winners occur in a race with six runners competing?

26. John Young, a contractor, builds homes of eight different models and presently has five lots to build on. In how many different ways can he place homes on these lots? Assume five different models will be built.

27. How many different five-member committees could be formed from the 100 U.S. senators?

28. If any two points determine a line, how many lines are determined by seven points in a plane, no three of which are collinear?

29. Radio stations in the United States have call letters that begin with K or W (for west or east of the Mississippi River, respectively). Some have three call letters, such as WBZ in Boston, WLS in Chicago, and KGO in San Francisco. Assuming no repetition of letters, how many three-letter sets of call letters are possible?

30. Most stations that were licensed after 1927 have four call letters starting with K or W, such as WXYZ in Detroit or KRLD in Dallas. Assuming no repetitions, how many four-letter sets are possible?

31. Subject identification numbers in a certain scientific research project consist of three letters followed by three digits and then three more letters. Assume repetitions are not allowed within any of the three groups, but letters in the first group of three may occur also in the last group of three. How many distinct identification numbers are possible?

32. How many triangles are determined by twenty points in a plane, no three of which are collinear?

33. How many ways can a sample of five CD players be selected from a shipment of twenty-four players?

34. If the shipment of Exercise 33 contains six defective players, how many of the size-five samples would not include any of the defective ones?

35. In how many ways could twenty-five people be divided into five groups containing, respectively, three, four, five, six, and seven people?

36. Larry Sifford and seven of his friends were contemplating the drive back to Denver after a long day of skiing.
(a) If they brought two vehicles, how many choices do they have as to who will do the driving?
(b) Suppose that (due to an insurance limitation) one of the vehicles can only be driven by Larry, his wife, or their son, all of whom are part of the skiing group. Now how many choices of drivers are there?

37. Each team in an eight-team basketball league is scheduled to play each other team three times. How many games will be played altogether?

38. The Coyotes, a youth league baseball team, have seven pitchers, who only pitch, and twelve other players, all of whom can play any position other than pitcher. For Saturday's game, the coach has not yet determined which nine players to use nor what the batting order will be, except that the pitcher will bat last. How many different batting orders may occur?

39. A music class of eight girls and seven boys is having a recital. If each member is to perform once, how many ways can the program be arranged in each of the following cases?
(a) The girls must all perform first.
(b) A girl must perform first and a boy must perform last.
(c) Elisa and Doug will perform first and last, respectively.
(d) The entire program will alternate between the sexes.
(e) The first, eighth, and fifteenth performers must be girls.

40. Carole has eight errands to run today, five of them pleasant, but the other three unpleasant. How many ways can she plan her day in each of the following cases?
(a) She decides to put off the unpleasant errands to another day.
(b) She is determined to complete all eight errands today.
(c) She will work up her courage by starting with *at least* two pleasant errands and then completing the rest of the eight.
(d) She will begin and end the day with pleasant errands and will accomplish only six altogether.
(e) She will succeed in all eight by facing all three unpleasant errands first.

For Exercises 41–46, refer to the standard 52-card deck pictured in the text and notice that the deck contains four aces, twelve face cards, thirteen hearts (all red), thirteen diamonds (all red), thirteen spades (all black), and thirteen clubs (all black). Of the 2,598,960 *different five-card hands possible, decide how many would consist of the following cards.*

41. all diamonds

42. all black cards

43. all aces

44. four clubs and one non-club

45. two face cards and three non-face cards

46. two red cards, two clubs, and a spade

47. How many different three-number "combinations" are possible on a combination lock having 40 numbers on its dial? (*Hint:* "Combination" is a misleading name for these locks since repetitions are allowed and also order makes a difference.)

48. In a 7/39 lottery, you select seven distinct numbers from the set 1 through 39, where order makes no difference. How many different ways can you make your selection?

John Young (the contractor) is to build six homes on a block in a new subdivision. Overhead expenses have forced him to limit his line to two different models, standard and deluxe. (All standard model homes are the same and all deluxe model homes are the same.)

49. How many different choices does John have in positioning the six houses if he decides to build three standard and three deluxe models?

50. If John builds only two deluxe and four standards, how many different positionings can he use?

Because of his good work, John gets a contract to build homes on three additional blocks in the subdivision, with six homes on each block. He decides to build nine deluxe homes on these three blocks; two on the first block, three on the second, and four on the third. The remaining nine homes will be standard.

51. Altogether on the three-block stretch, how many different choices does John have for positioning the eighteen homes? (*Hint:* Consider the three blocks separately and use the fundamental counting principle.)

52. How many choices would he have if he built 2, 3, and 4 deluxe models on the three different blocks as before, but not necessarily on the first, second, and third blocks in that order?

53. **(a)** How many numbers can be formed using all six digits 2, 3, 4, 5, 6, and 7?
(b) Suppose all these numbers were arranged in increasing order: 234,567; 234,576; and so on. Which number would be 363rd in the list?

54. How many four-digit counting numbers are there whose digits are distinct and have a sum of 10, assuming the digit 0 is not used?

55. How many paths are possible from A to B if all motion must be to the right or downward?

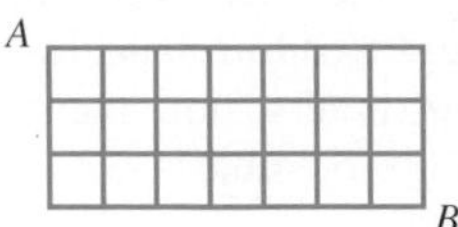

56. How many cards must be drawn (without replacement) from a standard deck of 52 to guarantee that at least two are from the same suit?

57. Judy Zahrndt wants to name her new baby so that her monogram (first, middle, and last initials) will be distinct letters in alphabetical order. How many different monograms could she select?

58. How many pairs of vertical angles are formed by eight distinct lines that all meet at a common point?

59. In how many ways can three men and six women be assigned to two groups so that neither group has more than six people?

60. Verify that $C(8, 3) = C(8, 5)$.

61. Verify that $C(12, 9) = C(12, 3)$.

62. Use the factorial formula for combinations to prove that in general, $C(n, r) = C(n, n - r)$.

10.4 Pascal's Triangle and the Binomial Theorem

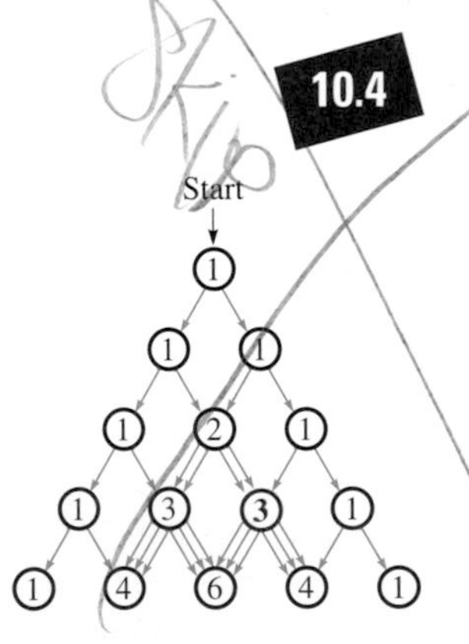

■ FIGURE 7

The triangular array in Figure 7 represents what we can call "random walks" that begin at START and proceed downward according to the following rule. At each circle (branch point), a coin is tossed. If it lands heads, we go downward to the left. If it lands tails, we go downward to the right. At each point, left and right are equally likely. In each circle we have recorded the number of different routes that could bring us to that point. For example, the colored 3 can be reached as the result of three different coin-tossing sequences: *htt, tht,* and *tth.*

Another way to generate the same pattern of numbers is to begin with 1s down both diagonals and then fill in the interior entries by adding the two numbers just above a given position (to the left and right). For example, the colored 28 in Table 5 is the result of adding 7 and 21 in the row above it.

By continuing to add pairs of numbers, we extend the array indefinitely downward, always beginning and ending each row with 1s. (The table shows just rows 0 through 10.) This unending "triangular" array of numbers is called **Pascal's triangle,** since Blaise Pascal wrote a treatise about it in 1653. There is evidence, though, that it was known as early as around 1100 and may have been

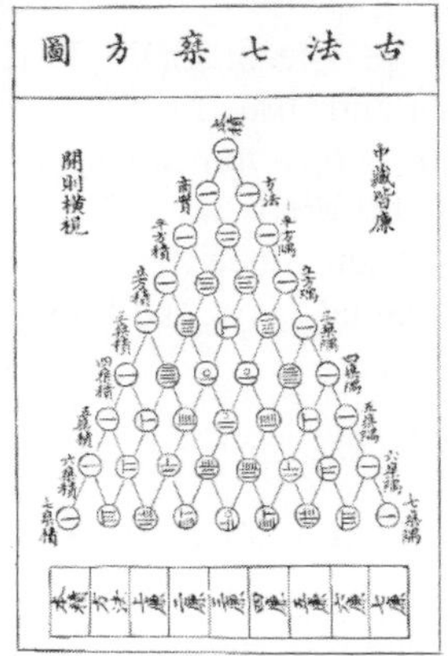

"Pascal's" triangle shown in the 1303 text, ***Szu-yuen Yu-chien*** (*The Precious Mirror of the Four Elements*) by the Chinese mathematician Chu Shih-chieh.

TABLE 5 Pascal's Triangle

Row Number		Row Sum
0	1	1
1	1 1	2
2	1 2 1	4
3	1 3 3 1	8
4	1 4 6 4 1	16
5	1 5 10 10 5 1	32
6	1 6 15 20 15 6 1	64
7	1 7 21 35 35 21 7 1	128
8	1 8 28 56 70 56 28 8 1	256
9	1 9 36 84 126 126 84 36 9 1	512
10	1 10 45 120 210 252 210 120 45 10 1	1,024

studied in China or India still earlier. At any rate, the "triangle" possesses many interesting properties. In counting applications, the most useful property is that, in general, entry number r in row number n is equal to $C(n, r)$—the number of *combinations* of n things taken r at a time. This correspondence is shown (through row 7) in Table 6.

TABLE 6

Row Number	
0	$C(0, 0)$
1	$C(1, 0)$ $C(1, 1)$
2	$C(2, 0)$ $C(2, 1)$ $C(2, 2)$
3	$C(3, 0)$ $C(3, 1)$ $C(3, 2)$ $C(3, 3)$
4	$C(4, 0)$ $C(4, 1)$ $C(4, 2)$ $C(4, 3)$ $C(4, 4)$
5	$C(5, 0)$ $C(5, 1)$ $C(5, 2)$ $C(5, 3)$ $C(5, 4)$ $C(5, 5)$
6	$C(6, 0)$ $C(6, 1)$ $C(6, 2)$ $C(6, 3)$ $C(6, 4)$ $C(6, 5)$ $C(6, 6)$
7	$C(7, 0)$ $C(7, 1)$ $C(7, 2)$ $C(7, 3)$ $C(7, 4)$ $C(7, 5)$ $C(7, 6)$ $C(7, 7)$
	and so on

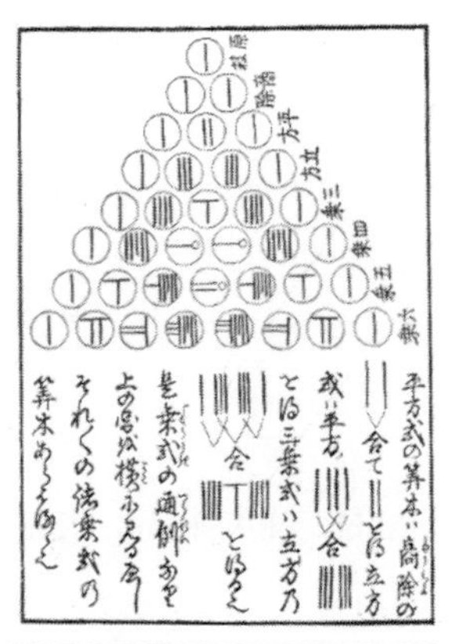

Japanese version of the triangle dates from the eighteenth century. The "stick numerals" evolved from bamboo counting pieces used on a ruled board. Possibly Omar Khayyam, twelfth-century Persian mathematician and poet, may also have divined its patterns in pursuit of algebraic solutions. (The triangle lists the coefficients of the binomial expansion.)

Having a copy of Pascal's triangle handy gives us another option for evaluating combinations. Any time we need to know the number of combinations of n things taken r at a time (that is, the number of subsets of size r in a set of size n), we can simply read entry number r of row number n. Keep in mind that the *first* row shown is *row number 0*. Also, the first entry of each row can be called entry number 0. This entry gives the number of subsets of size 0 (which is always 1 since there is only one empty set).

EXAMPLE 1 A group of ten people includes six women and four men. If five of these people are randomly selected to fill out a questionnaire, how many different samples of five people are possible?

Since this is simply a matter of selecting a subset of five from a set of ten (or combinations of ten things taken five at a time), we can read $C(10, 5)$ from row 10 of Pascal's triangle in Table 5. The answer is 252. ■

EXAMPLE 2 Among the 252 possible samples of five people in Example 1, how many of them would consist of exactly two women and three men?

Two women can be selected from six women in $C(6, 2)$ different ways, and three men can be selected from four men in $C(4, 3)$ different ways. These combination values can be read from Pascal's triangle. Then, since the task of obtaining two women and three men requires both individual parts, the fundamental counting principle tells us to multiply the two values:

$$C(6, 2) \cdot C(4, 3) = 15 \cdot 4 = 60.$$

EXAMPLE 3 If five fair coins are tossed, in how many different ways could exactly three heads be obtained?

There are various "ways" of obtaining exactly three heads because the three heads can occur on different subsets of the coins. For example, *hhtht* and *thhth* are just two of many possibilities. When such a possibility is written down, exactly three positions are occupied by an h, the other two by a t. Each distinct way of choosing three positions from a set of five positions gives a different possibility. (Once the three positions for h are determined, each of the other two positions automatically receives a t.) So our answer is just the number of size-three subsets of a size-five set, that is, the number of combinations of five things taken three at a time. We read this answer from row 5 of Pascal's triangle: $C(5, 3) = 10$.

Notice that row 5 of Pascal's triangle also provides answers to several other questions about tossing five fair coins. They are summarized in Table 7.

TABLE 7 Tossing Five Fair Coins

Number of Heads n	Ways of Obtaining Exactly n Heads	Listing
0	$C(5, 0) = 1$	ttttt
1	$C(5, 1) = 5$	htttt, thttt, tthtt, tttht, tttth
2	$C(5, 2) = 10$	hhttt, hthtt, httht, hltth, thhtt, ththt, thtth, tthht, ttһth, ttthh
3	$C(5, 3) = 10$	hhhtt, hhtht, hhtth, hthht, hthth, htthh, thhht, thhth, ththh, tthhh
4	$C(5, 4) = 5$	hhhht, hhhth, hhthh, hthhh, thhhh
5	$C(5, 5) = 1$	hhhhh

The Binomial Theorem

We will now look briefly at a totally different line of mathematical reasoning which also leads to Pascal's triangle, and applies it to a very important result in algebra. The "triangular" pattern of entries, it turns out, also occurs when "binomial" expressions are raised to various powers. (In algebra, any two-term expression, such as $x + y$, or $a + 2b$, or $w - 4$, is called a binomial.) The first few powers of the binomial $x + y$ are shown here, so that we can see a pattern.

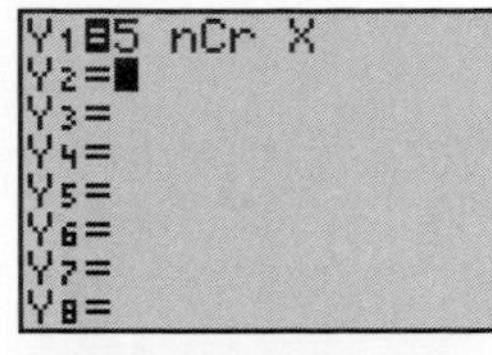

X	Y1	
0	1	
1	5	
2	10	
3	10	
4	5	
5	1	
6	0	

X=0

The table feature of the TI-82 calculator can be used to generate a row of Pascal's triangle. Why is there a "0" for Y_1 when $x = 6$?

$$
\begin{aligned}
(x + y)^0 &= 1 \\
(x + y)^1 &= x + y \\
(x + y)^2 &= x^2 + 2xy + y^2 \\
(x + y)^3 &= x^3 + 3x^2y + 3xy^2 + y^3 \\
(x + y)^4 &= x^4 + 4x^3y + 6x^2y^2 + 4xy^3 + y^4 \\
(x + y)^5 &= x^5 + 5x^4y + 10x^3y^2 + 10x^2y^3 + 5xy^4 + y^5
\end{aligned}
$$

The coefficients in any one of these expansions are just the entries of one of the rows of Pascal's triangle. The expansions can be verified by direct computation, using the distributive, associative, and commutative properties of algebra. For example,

$$
\begin{aligned}
(x + y)^3 &= (x + y) \cdot (x + y) \cdot (x + y) \\
&= (x + y) \cdot [(x + y) \cdot (x + y)] \\
&= (x + y) \cdot (x^2 + xy + yx + y^2) \\
&= (x + y) \cdot (x^2 + 2xy + y^2) \\
&= x^3 + 2x^2y + xy^2 + x^2y + 2xy^2 + y^3 \\
&= x^3 + 3x^2y + 3xy^2 + y^3.
\end{aligned}
$$

Since the coefficients of the binomial expansions give us the rows of Pascal's triangle, they are precisely the numbers we have been referring to as *combinations* in our study of counting. (In the study of algebra, they have traditionally been referred to as **binomial coefficients,** and have been denoted differently. For example, rather than $C(n, r)$, you may see the notation $\binom{n}{r}$.) As "binomial coefficients," they can still be thought of as answering the question "How many?" if we reason as follows. In the expansion for $(x + y)^4$, how many are there of the expression x^2y^2? By looking in the appropriate entry of row 4 of Pascal's triangle, we see that the answer is 6.

Generalizing the pattern shown by the six expansions above, we obtain the important result known as the **binomial theorem** (or sometimes known as the **general binomial expansion**).

Binomial Theorem

For any positive integer n,

$$
\begin{aligned}
(x + y)^n = C(n, 0) \cdot x^n &+ C(n, 1) \cdot x^{n-1}y + C(n, 2) \cdot x^{n-2}y^2 \\
&+ C(n, 3) \cdot x^{n-3}y^3 + \cdots \\
&+ C(n, n - 1) \cdot xy^{n-1} + C(n, n) \cdot y^n,
\end{aligned}
$$

or, using the factorial formula for combinations, and the fact that $C(n, 0) = C(n, n) = 1$,

$$
\begin{aligned}
(x + y)^n = x^n &+ \frac{n!}{(n - 1)!\,1!}x^{n-1}y + \frac{n!}{(n - 2)!\,2!}x^{n-2}y^2 \\
&+ \frac{n!}{(n - 3)!\,3!}x^{n-3}y^3 + \cdots \\
&+ \frac{n!}{1!\,(n - 1)!}xy^{n-1} + y^n.
\end{aligned}
$$

EXAMPLE 4 Write out the binomial expansion for $(x + y)^7$.

Reading the coefficients from row 7 of Pascal's triangle (Table 5), we obtain

$$(x + y)^7 = x^7 + 7x^6y + 21x^5y^2 + 35x^4y^3 + 35x^3y^4 + 21x^2y^5 + 7xy^6 + y^7. \blacksquare$$

Recall that "binomial" expressions can involve terms other than x and y. The binomial theorem still applies.

EXAMPLE 5 Write out the binomial expansion for $(2a + 5)^4$.

Initially, we get coefficients from Pascal's triangle (row 4), but after all the algebra is finished the final coefficients are different numbers:

$$\begin{aligned}(2a + 5)^4 &= (2a)^4 + 4(2a)^3 5 + 6(2a)^2 5^2 + 4(2a)5^3 + 5^4 \\ &= 2^4a^4 + 4(2^3a^3)5 + 6(2^2a^2)5^2 + 4(2a)5^3 + 5^4 \\ &= 16a^4 + 4 \cdot 8 \cdot a^3 \cdot 5 + 6 \cdot 4 \cdot a^2 \cdot 25 + 4 \cdot 2a \cdot 125 + 625 \\ &= 16a^4 + 160a^3 + 600a^2 + 1{,}000a + 625. \blacksquare\end{aligned}$$

10.4 EXERCISES

Read the following combination values directly from Pascal's triangle.

1. $C(4, 3)$ **2.** $C(5, 2)$ **3.** $C(6, 4)$ **4.** $C(7, 5)$

5. $C(8, 2)$ **6.** $C(9, 4)$ **7.** $C(9, 7)$ **8.** $C(10, 6)$

A committee of four Congressmen will be selected from a group of seven Democrats and three Republicans. Find the number of ways of obtaining each of the following.

9. exactly one Democrat **10.** exactly two Democrats

11. exactly three Democrats **12.** exactly four Democrats

Suppose eight fair coins are tossed. Find the number of ways of obtaining each of the following.

13. exactly three heads **14.** exactly four heads

15. exactly five heads **16.** exactly six heads

Kelly Melcher, searching for an Economics class, knows that it must be in one of nine classrooms. Since the professor does not allow people to enter after the class has begun, and there is very little time left, Kelly decides to try just four of the rooms at random.

17. How many different selections of four rooms are possible?

18. How many of the selections of Exercise 17 will fail to locate the class?

19. How many of the selections of Exercise 17 will succeed in locating the class?

20. What fraction of the possible selections will lead to "success"? (Give three decimal places.)

For a set of five elements, find the number of different subsets of each of the following sizes. (Use row 5 of Pascal's triangle to find the answers.)

21. 0 **22.** 1 **23.** 2 **24.** 3 **25.** 4 **26.** 5

27. How many subsets (of any size) are there for a set of five elements?

28. Find and explain the relationship between the row number and row sum in Pascal's triangle.

Over the years, many interesting patterns have been discovered in Pascal's triangle. Exercises 29 and 30 exhibit two such patterns.*

29. Name the next five numbers of the diagonal sequence indicated in the figure at the side.

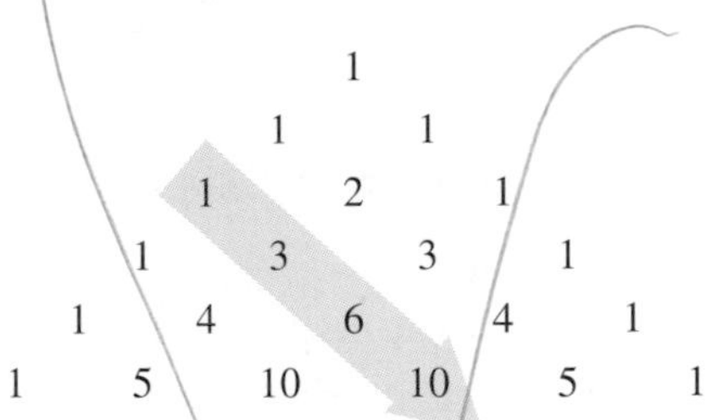

30. Complete the sequence of sums on the diagonals shown in the figure at the side. What pattern do these sums make? What is the name of this important sequence of numbers? The presence of this sequence in the triangle apparently was not recognized by Pascal.

Sums
1
1
2
3
5

1
1 1
1 2 1
1 3 3 1
1 4 6 4 1
1 5 10 10 5 1
1 6 15 20 15 6 1
1 7 21 35 35 21 7 1
1 8 28 56 70 56 28 8 1

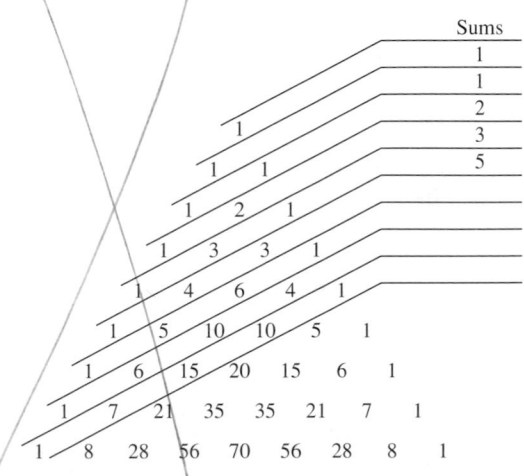

31. More than a century before Pascal's treatise on the "triangle" appeared, another work by the Italian mathematician Niccolo Tartaglia (1506–1559) came out and included the table of numbers shown here.

1	1	1	1	1	1
1	2	3	4	5	6
1	3	6	10	15	21
1	4	10	20	35	56
1	5	15	35	70	126
1	6	21	56	126	252
1	7	28	84	210	462
1	8	36	120	330	792

Explain the connection between Pascal's triangle and Tartaglia's "rectangle."

32. Construct another "triangle" by replacing every number in Pascal's triangle (rows **0** through **5**) by its remainder when divided by 2. What special property is shared by rows **2** and **4** of this new triangle?

*For example, see the article "Serendipitous Discovery of Pascal's Triangle" by Francis W. Stanley in *The Mathematics Teacher,* February 1975.

33. What is the next row that would have the same property as rows **2** and **4** in Exercise 32?

34. How many even numbers are there in row number **256** of Pascal's triangle? (Do Exercises 32 and 33 first.)

Write out the binomial expansion for each of the following powers.

35. $(x + y)^6$ **36.** $(x + y)^8$ **37.** $(z + 2)^3$

38. $(w + 3)^5$ **39.** $(2a + 5b)^4$ **40.** $(3d + 5f)^4$

41. $(b - h)^7$ (*Hint:* First change $b - h$ to $b + (-h)$.)

42. $(2n - 4m)^5$

43. How many terms appear in the binomial expansion for $(x + y)^n$?

44. Observe the pattern in this table and fill in the blanks to discover a formula for the rth term (the general term) of the binomial expansion for $(x + y)^n$.

Term Number	Coefficient	Variable Part	Term
1	1	x^n	$1\,x^n$
2	$\frac{n!}{(n-1)!\,1!}$	$x^{n-1}y$	$\frac{n!}{(n-1)!\,1!}x^{n-1}y$
3	$\frac{n!}{(n-2)!\,2!}$	$x^{n-2}y^2$	$\frac{n!}{(n-2)!\,2!}x^{n-2}y^2$
4	$\frac{n!}{(n-3)!\,3!}$	$x^{n-3}y^3$	$\frac{n!}{(n-3)!\,3!}x^{n-3}y^3$
⋮	⋮	⋮	⋮
r	$\frac{n!}{(n-r+1)!(r-1)!}$	______	______

Use the results of Exercise 44 to find the indicated term of each of the following expansions.

45. $(x + y)^{12}$; 5th term **46.** $(a + b)^{20}$; 16th term

10.5 Counting Problems Involving "Not" and "Or"

When counting the number of ways that an event can occur (or that a "task" can be done), it is sometimes easier to take an indirect approach. In this section we will discuss two indirect techniques: first, the *complements principle,* where we start by counting the ways an event would *not* occur, and then the *additive principle,* where we break an event into simpler component events, and count the number of ways one component *or* another would occur.

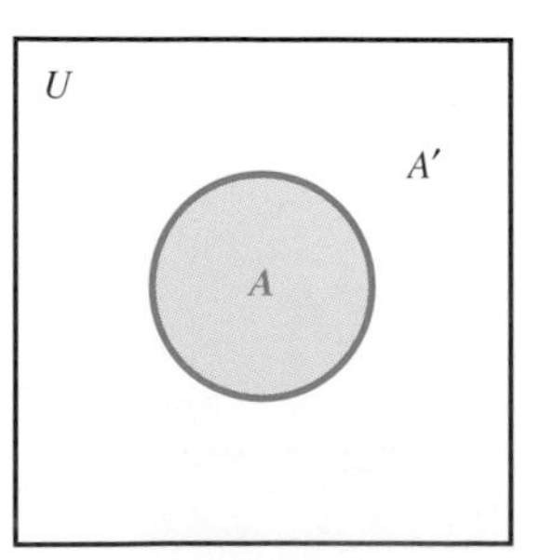

The complement of a set

■ **FIGURE 8**

Suppose U is the set of all possible results of some type and A is the set of all those results that satisfy a given condition. For any set S, the cardinal number of S (number of elements in S) is written $n(S)$, and the complement of S is written S'. Then Figure 8 suggests that

$$n(A) + n(A') = n(U), \qquad \text{or} \qquad n(A) = n(U) - n(A'),$$

as summarized here.

Complements Principle of Counting

If A is any set within the universal set U, then

$$n(A) = n(U) - n(A').$$

By this principle, the number of ways a certain condition can be satisfied is the total number of possible results minus the number of ways the condition would **not** be satisfied. (As we have seen earlier, in Chapters 2 and 3, the arithmetic operation of *subtraction* corresponds to the set operation of *complementation* and to the logical connective of *negation,* that is "not.")

EXAMPLE 1 For the set $S = \{a, b, c, d, e, f\}$, find the number of proper subsets.

Recall that a proper subset of S is any subset with fewer than all six elements. Several subsets of different sizes would satisfy this condition. However, it is easier to consider the one subset that is not proper, namely S itself. From set theory, we know that set S has a total of $2^6 = 64$ subsets. Thus, from the complements principle, the number of proper subsets is $64 - 1 = 63$. In words, the number of subsets that *are* proper is the total number of subsets minus the number of subsets that are *not* proper. ■

Consider the tossing of three fair coins. Since each coin will land either heads (h) or tails (t), the possible results can be listed as follows.

BCP

$$hhh, \quad hht, \quad hth, \quad thh, \quad htt, \quad tht, \quad tth, \quad ttt$$

(Even without the listing, we could have concluded that there would be eight possibilities. There are two possible outcomes for each coin, so the fundamental counting principle gives $2 \cdot 2 \cdot 2 = 2^3 = 8$.)

Now suppose we wanted the number of ways of obtaining *at least* one head. In this case, "at least one" means one or two or three. But rather than dealing with all three cases, we can note that "at least one" is the opposite (or complement) of "fewer than one" (which is zero). Since there is only one way to get zero heads (ttt), and there are a total of eight possibilities (as shown above), the complements principle gives the number of ways of getting at least one head: $8 - 1 = 7$. (The number of outcomes that include at least one head is the total number of outcomes minus the number of outcomes that do *not* include at least one head.) We find that indirect counting methods can often be applied to problems involving "at least," or "at most," or "less than," or "more than."

EXAMPLE 2 If four fair coins are tossed, in how many ways can at least one tail be obtained?

By the fundamental counting principle, $2^4 = 16$ different results are possible. Exactly one of these fails to satisfy the condition of "at least one tail" (namely, no tails, or $hhhh$). So our answer (from the complements principle) is $16 - 1 = 15$. ■

EXAMPLE 3 How many bridge hands (13 cards each) containing at least one heart are possible?

Example 9 of the third section of this chapter showed that the total number of bridge hands possible is 635,013,559,600. The ones that do not have at least one heart are the ones formed from the 39 cards in the deck that are not hearts. So the number of hands with no hearts is

$$C(39, 13) = \frac{39!}{13!\,26!} = 8{,}122{,}425{,}444.$$

By the complements principle, the number of hands with at least one heart is

$$635{,}013{,}559{,}600 - 8{,}122{,}425{,}444 = 626{,}891{,}134{,}156.$$

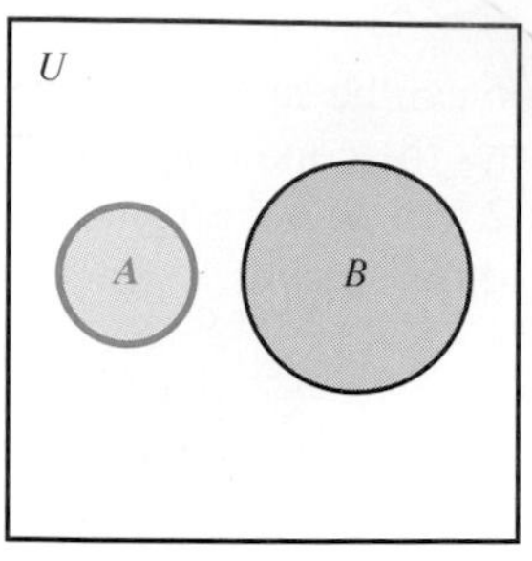

Disjoint sets

FIGURE 9

The complements principle is one way of counting indirectly. Another technique is related to Figure 9. If A and B are disjoint sets (have no elements in common), then writing the union of sets A and B as $A \cup B$ gives the principle stated here.

Special Additive Counting Principle

If A and B are disjoint sets, then

$$n(A \cup B) = n(A) + n(B)$$

(It is *special* because A and B are disjoint.) The principle states that if two conditions cannot both be satisfied together, then the number of ways that one **or** the other of them could be satisfied is found by adding the number of ways one of them could be satisfied to the number of ways the other could be satisfied. The idea is that we can sometimes analyze a set of possibilities by breaking it into a set of simpler component parts. (The arithmetic operation of *addition* corresponds to the set operation of *union* and to the logical connective of *disjunction,* that is "or.")

Results in Examples 3 and 4 are supported in this screen.

EXAMPLE 4 How many five-card hands consist of either all clubs or all red cards?

No hand that satisfies one of these conditions could also satisfy the other, so the two sets of possibilities (for all clubs and all red cards) are disjoint. Therefore, the special additive principle applies.

$$\begin{aligned} C(13, 5) + C(26, 5) &= \frac{13 \cdot 12 \cdot 11 \cdot 10 \cdot 9}{5 \cdot 4 \cdot 3 \cdot 2 \cdot 1} + \frac{26 \cdot 25 \cdot 24 \cdot 23 \cdot 22}{5 \cdot 4 \cdot 3 \cdot 2 \cdot 1} \\ &= 1{,}287 + 65{,}780 \\ &= 67{,}067 \end{aligned}$$

The special additive principle also extends to the union of three or more disjoint sets, as illustrated by the next example.

EXAMPLE 5 Terry White needs to take twelve more specific courses for a bachelors degree, including four in math, three in physics, three in computer science, and two in business. If five courses are randomly chosen from these twelve for next semester's program, how many of the possible selections would include at least two math courses?

Of all the information given here, what is important is that there are four math courses and eight other courses to choose from, and that five of them are being selected for next semester. If T denotes the set of selections that include at least two math courses, then we can write

$$T = A \cup B \cup C$$

where A = the set of selections with exactly two math courses,

B = the set of selections with exactly three math courses,

and C = the set of selections with exactly four math courses.

(In this case, *at least two* means exactly two **or** exactly three **or** exactly four.) The situation is illustrated in Figure 10.

By previous methods, we know that

$$n(A) = C(4, 2) \cdot C(8, 3) = 6 \cdot 56 = 336,$$
$$n(B) = C(4, 3) \cdot C(8, 2) = 4 \cdot 28 = 112,$$

and

$$n(C) = C(4, 4) \cdot C(8, 1) = 1 \cdot 8 = 8,$$

so that, by the additive principle,

$$n(T) = 336 + 112 + 8 = 456. \blacksquare$$

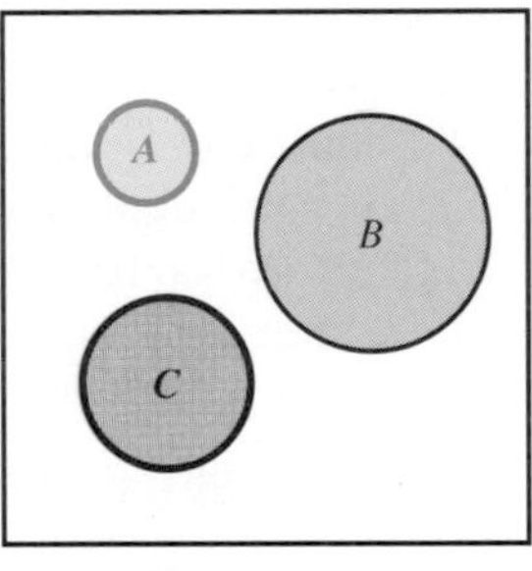

$T = A \cup B \cup C$

FIGURE 10

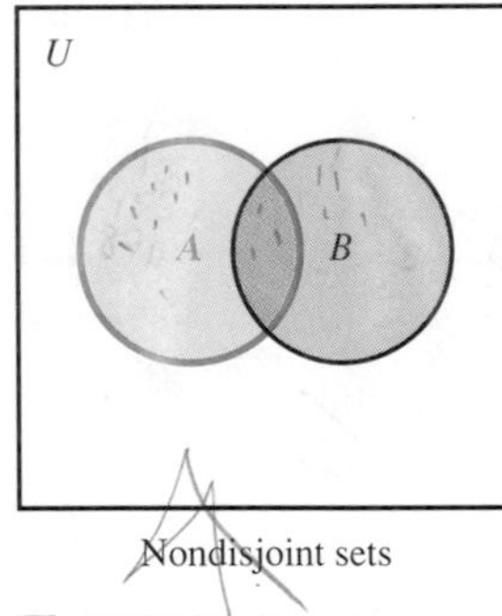

Nondisjoint sets

FIGURE 11

Figure 11 illustrates the case where two sets A and B are not disjoint, suggesting the general principle stated here.

General Additive Counting Principle

If A and B are any two sets, disjoint or not, then

$$n(A \cup B) = n(A) + n(B) - n(A \cap B)$$

(Recall that $A \cap B$ denotes the intersection of sets A and B.)

Problem Solving

By this principle, the number of elements in $A \cup B$ is found by adding the number of elements in A and the number in B. Since this process counts the overlapping region twice, once for A and once for B, it is necessary to subtract one occurrence of the overlapping region. This is often used in problem solving.

EXAMPLE 6 Table 8 categorizes a diplomatic delegation of 18 Congressional members as to political party and sex. If one of the members is chosen randomly to be spokesperson for the group, in how many ways could that person be a Democrat or a woman?

Using the general additive counting principle, we obtain

$$\begin{aligned} n(D \cup W) &= n(D) + n(W) - n(D \cap W) \\ &= 10 + 9 - 6 \\ &= 13. \end{aligned}$$

TABLE 8

	Men (M)	Women (W)	Totals
Republican (R)	5	3	8
Democrat (D)	4	6	10
Totals	9	9	18

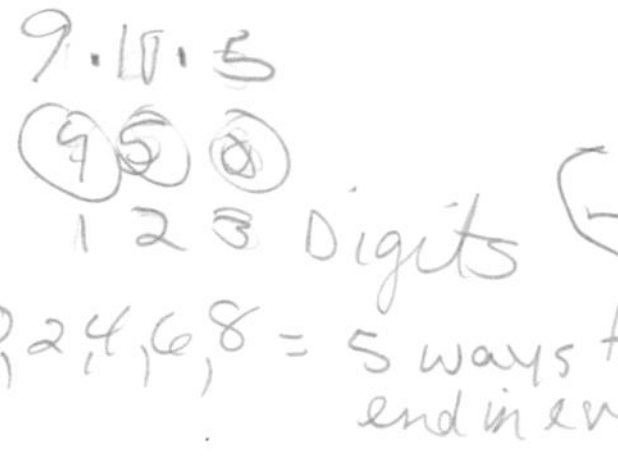

EXAMPLE 7 How many three-digit counting numbers are multiples of 2 or multiples of 5?

A multiple of 2 must end in an even digit (0, 2, 4, 6, or 8), so there are $9 \cdot 10 \cdot 5 = 450$ even three-digit counting numbers. A multiple of 5 must end in either 0 or 5, so there are $9 \cdot 10 \cdot 2 = 180$ of those. A multiple of both 2 and 5 is a multiple of 10 and must end in 0. There are $9 \cdot 10 \cdot 1 = 90$ of those, giving

$$450 + 180 - 90 = 540$$

possible three-digit numbers that are multiples of 2 or multiples of 5.

EXAMPLE 8 A single card is drawn from a standard 52-card deck.

(a) In how many ways could it be a heart or a king?

A single card can be both a heart and a king (the king of hearts), so use the general additive formula. There are thirteen hearts, four kings, and one card that is both a heart and a king:

$$13 + 4 - 1 = 16.$$

(b) In how many ways could the card be a club or a face card?

There are 13 clubs, 12 face cards, and 3 cards that are both clubs and face cards, giving

$$13 + 12 - 3 = 22.$$

EXAMPLE 9 How many subsets of a 25-element set have more than three elements?

It would be a real job to count directly all subsets of size 4, 5, 6, . . . , 25. It is much easier to count those with three or fewer elements:

There is	$C(25, 0) = 1$	size-0 subset.
There are	$C(25, 1) = 25$	size-1 subsets.
There are	$C(25, 2) = 300$	size-2 subsets.
There are	$C(25, 3) = 2{,}300$	size-3 subsets.

Since the total number of subsets is $2^{25} = 33{,}554{,}432$ (use a calculator), the number with more than three elements must be

$$33{,}554{,}432 - (1 + 25 + 300 + 2{,}300) = 33{,}554{,}432 - 2{,}626$$
$$= 33{,}551{,}806.$$

In Example 9, we used both the special additive formula (to get the number of subsets with no more than three elements) and the complements principle.

Problem Solving

As you work the exercises of this section, keep in mind that indirect methods may be best, and that you may also be able to use permutations, combinations, the fundamental counting principle, or listing procedures such as product tables or tree diagrams. Also, you may want to obtain combination values, when needed, from Pascal's triangle.

10.5 EXERCISES

1. Explain why the complements principle of counting is called an "indirect" method.
2. Explain the difference between the *special* and *general* additive principles of counting.

If you toss seven fair coins, in how many ways can you obtain each of the following results?

3. at least one head ("At least one" is the complement of "none.")
4. at least two heads ("At least two" is the complement of "zero or one.")
5. at least two tails
6. at least one of each (a head and a tail)

If you roll two fair dice (say red and green), in how many ways can you obtain each of the following? (Refer to Table 2 in the first section of this chapter.)

7. a 2 on the red die
8. a sum of at least 3
9. a 4 on at least one of the dice
10. a different number on each die

Among the 635,013,559,600 *possible bridge hands (*13 *cards each), how many contain the following cards? (The standard card deck was described in the third section of this chapter.)*

11. at least one card that is not a spade (complement of "all spades")
12. cards of more than one suit (complement of "all the same suit")
13. at least one face card (complement of "no face cards")
14. at least one diamond, but not all diamonds (complement of "no diamonds or all diamonds")

How many three-digit counting numbers meet the following requirements?

15. even or a multiple of 5
16. greater than 600 or a multiple of 10

If a given set has twelve elements, how many of its subsets have the given numbers of elements?

17. at most two elements
18. at least ten elements
19. more than two elements
20. from three through nine elements

Of a group of 50 *students,* 30 *enjoy music,* 15 *enjoy literature, and* 10 *enjoy both music and literature. How many of them enjoy the following?*

21. at least one of these two subjects (general additive principle) 35
22. neither of these two subjects (complement of "at least one")

If a single card is drawn from a standard 52-card deck, in how many ways could it be the following? (Use the general additive principle.)

23. a club or a jack 16
24. a face card or a black card
25. a diamond or a face card or a denomination greater than 10 (First note that this is the same as "A diamond or a denomination greater than 10" since every face card has denomination greater than 10. Do not consider ace to be greater than 10 in this case.) 22
26. a heart or a queen or a red card (First note that this is the same as "A queen or a red card." Why is this true?)

Table 4 in this chapter (For Further Thought) briefly described the various kinds of hands in five-card poker. A "royal flush" is a hand containing 10, *jack, queen, king, and ace all in the same suit. A "straight flush" is any five consecutive denominations in a common suit. The lowest would be ace,* 2, 3, 4, 5, *and the highest is* 9, 10, *jack, queen, king. (An ace high straight flush is not referred to as a straight flush since it has the special name "royal flush.") A "straight" contains five consecutive denominations but not all of the same suit, and a "flush" is any hand with all five cards the same suit, except a royal flush or a straight flush. A "three of a kind" hand contains exactly three of one denomination, and furthermore the remaining two must be of two* additional *denominations. (Why is this?)*

Verify each of the following. (Explain all steps of your argument.)

27. There are four ways to get a royal flush.
28. There are 36 ways to get a straight flush.
29. There are 5,108 ways to get a flush.
30. There are 10,200 ways to get a straight.
31. There are 624 ways to get four of a kind.
32. There are 54,912 ways to get three of a kind.

If three-digit numbers are formed using only digits from the set {0, 1, 2, 3, 4, 5, 6}, *how many will belong to the following categories?*

33. even numbers
34. multiples of 10
35. multiples of 100
36. multiples of 25
37. If license numbers consist of three letters followed by three digits, how many different licenses could be created having at least one letter or digit repeated? (*Hint:* Use the complements principle of counting.)
38. If two cards are drawn from a 52-card deck without replacement (that is, the first card is not replaced in the deck before the second card is drawn), in how many different ways is it possible to obtain a king on the first draw and a heart on the second? (*Hint:* Split this event into the two disjoint components "king of hearts and then another heart" and "non-heart king and then heart." Use the fundamental counting principle on each component, then apply the special additive principle.)
39. A committee of four faculty members will be selected from a department of twenty-five which includes professors Fontana and Spradley. In how many ways could the committee include at least one of these two professors?

Edward Roberts is planning a long-awaited driving tour, which will take him and his family on the southern route to the West Coast. Ed is interested in seeing the twelve national monuments listed here, but will have to settle for seeing just three of them since some family members are anxious to get to Disneyland.

New Mexico	Arizona	California
Gila Cliff Dwellings	Canyon de Chelly	Devils Postpile
Petroglyph	Organ Pipe Cactus	Joshua Tree
White Sands	Saguaro	Lava Beds
Aztec Ruins		Muir Woods
		Pinnacles

In how many ways could the three monuments chosen include the following?

40. sites in only one state
41. at least one site not in California
42. sites in fewer than all three states
43. sites in exactly two of the three states

44. In the figure here, a "segment" joins intersections and/or turning points. For example, the path from A straight up to C and then straight across to B consists of six segments. Altogether, how many paths are there from A to B that consist of exactly six segments?

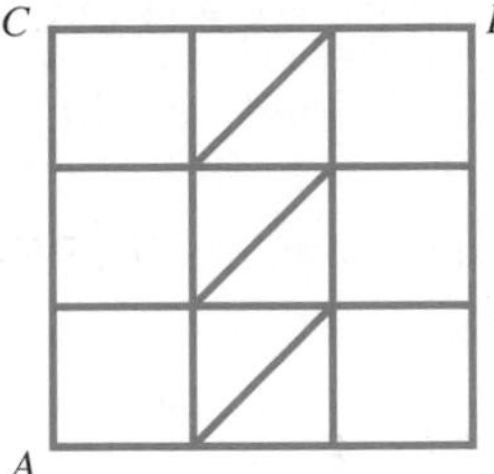

45. How many of the counting numbers 1 through 100 can be expressed as a sum of three or fewer powers of three (not necessarily distinct)? For example, 5 is such a number since $5 = 1 + 1 + 3$. (For one solution, see the October 1992 issue of *Mathematics Teacher,* page 497.)

46. Extend the general additive counting principle to three overlapping sets (as in the figure) to show that

$$n(A \cup B \cup C) = n(A) + n(B) + n(C) - n(A \cap B) - n(A \cap C) - n(B \cap C) + n(A \cap B \cap C).$$

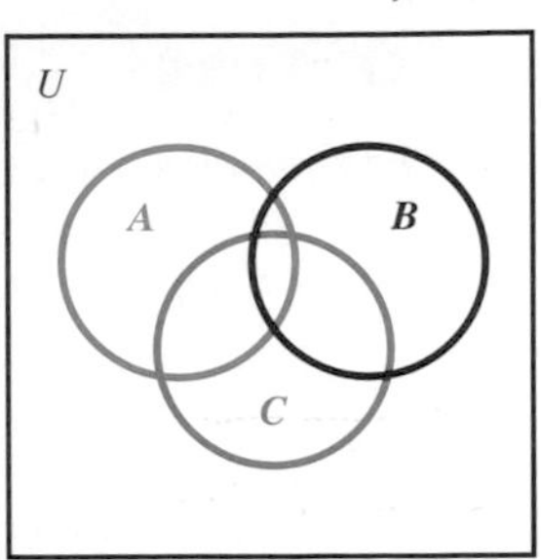

47. How many of the counting numbers 1 through 300 are *not* divisible by 2, 3, or 5? (*Hint:* Use the complements principle and the result of Exercise 46.)

48. How many three-digit counting numbers do not contain any of the digits 2, 5, 7, or 8?

49. Caralee Woods manages the shipping department of a firm that sells gold watches. If the watches can be shipped in packages of four, three, or one, in how many ways can a customer's order for fifteen watches be filled?

A Civil Air Patrol unit of fifteen members includes four officers. In how many ways can four of them be selected for a search and rescue mission in each of the following cases?

50. The search group must include at least one officer.

51. The search group must include at least two officers.

OLLABORATIVE INVESTIGATION
Approximating Factorials Using a Different Technique

In this chapter we learned one method of computing the exact value of a factorial using a product. Because factorials increase in magnitude so quickly, theoretical mathematics provides us a formula for approximating large factorials. Interestingly enough, the formula involves the irrational numbers π and e. It is known as **Stirling's formula,** and is as follows:

$$n! \approx \sqrt{2\pi n}\, n^n e^{-n}.$$

Like some other results in mathematics it is named for one person in spite of the fact that another person discovered it! According to David M. Burton, the mathematician Abraham De Moivre had just as much to do with discovering this formula as his friend, James Stirling.

As an example of an application of this formula, we observe that the exact value of 5! is 120. Using a TI-82 calculator, Stirling's formula gives the approximation of 5! as 118.019160, which is off by less than 2, an error of only 1.65%.

Topics for Discussion

Divide the class into several groups, and in each group be sure that there is someone who is proficient with a calculator. In each group, choose four values of n between 10 and 30. Then evaluate $n!$ using the factorial key of the calculator, and approximate $n!$ using Stirling's formula. Compare the two results in each case. How close are they? Discuss the results among the class.

Chapter 10 Test

If digits may be used from the set {0, 1, 2, 3, 4, 5, 6}, find the number of each of the following.

1. three-digit numbers
2. even three-digit numbers
3. three-digit numbers without repeated digits
4. three-digit multiples of five without repeated digits
5. Determine the number of triangles (of any size) in the figure shown here.

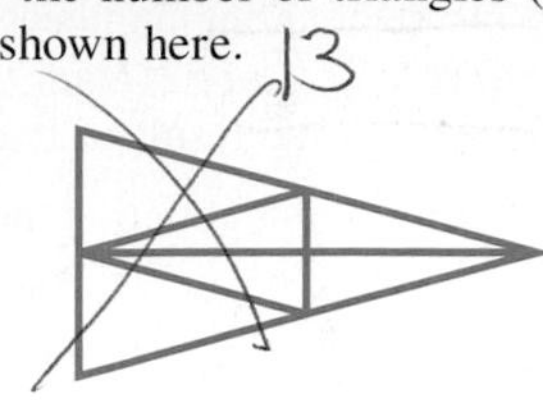

6. Construct a tree diagram showing all possible results when a fair coin is tossed four times, if the third toss must be different than the second.
7. How many non-repeating four-digit numbers have the sum of their digits equal to 30?
8. Using only digits from the set {0, 1, 2}, how many three-digit numbers can be written which have no repeated odd digits?

Evaluate the following expressions.

9. $5!$ 10. $\frac{8!}{5!}$ 11. $P(12, 4)$ 12. $C(7, 3)$

13. How many five-letter "words" without repeated letters are possible using the English alphabet? (Assume that any five letters make a "word.")
14. Using the Russian alphabet (which has 32 letters), and allowing repeated letters, how many five-letter "words" are possible?

If there are twelve players on a basketball team, find the number of choices the coach has in selecting each of the following.

15. four players to carry the team equipment
16. two players for guard positions and two for forward positions
17. five starters and five subs
18. a set of three or more of the players

Determine the number of possible settings for a row of four on-off switches under each of the following conditions.

19. There are no restrictions.
20. The first and fourth switches must be on.
21. The first and fourth switches must be set the same.
22. No two adjacent switches can be off.
23. No two adjacent switches can be set the same.
24. At least two switches must be on.

Four distinct letters are to be chosen from the set {A, B, C, D, E, F, G}. Determine the number of ways to obtain a subset that includes each of the following.

25. the letter D
26. both A and E
27. either A or E, but not both
28. equal numbers of vowels and consonants
29. more consonants than vowels
30. Write out and simplify the binomial expansion for $(x + 2)^5$.
31. If $C(n, r) = 495$ and $C(n, r + 1) = 220$, find the value of $C(n + 1, r + 1)$.
32. If you write down the second entry of each row of Pascal's triangle (starting with row 1), what sequence of numbers do you obtain?
33. Explain why there are $r!$ times as many permutations of n things taken r at a time as there are combinations of n things taken r at a time.

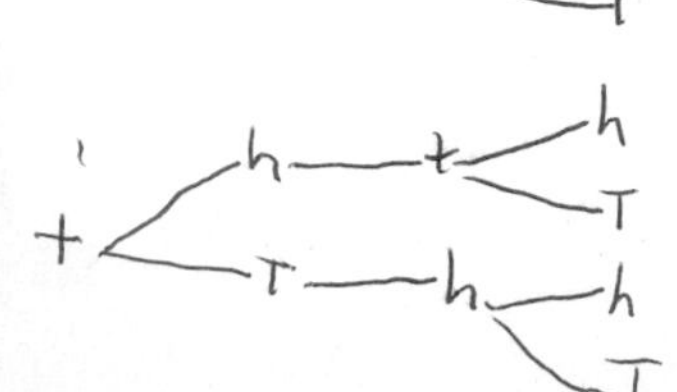

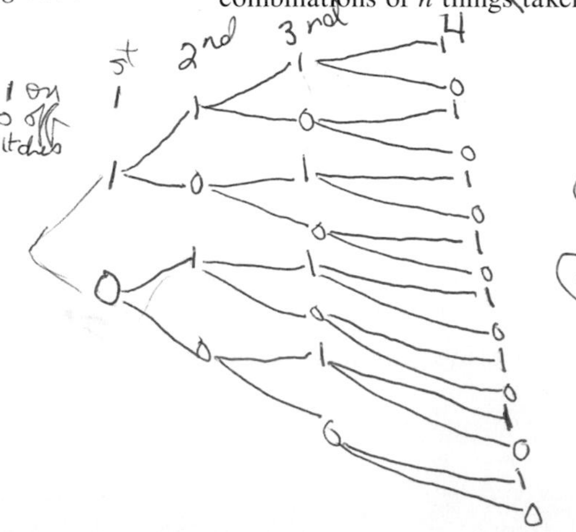

11

Probability

People are exposed to hazardous situations every day. Risk occurs in many different forms, whether it is from chemicals in the environment or behaviors that can produce detrimental consequences. Risk is the chance or probability that a harmful event will actually occur. The following list contains activities that carry an annual increased risk of death by 1 chance in a million.

Smoking 1.4 cigarettes
Spending 1 hour in a coal mine
Living 2 days in New York or Boston
Eating 40 teaspoons of peanut butter
Living 2 months with a cigarette smoker
Flying 1,000 miles in a jet
Traveling 300 miles in a car
Riding 10 miles on a bicycle

An enormous amount of effort in our society is spent on assessing risk. Ideally, people would like to live in a world with zero risk. However, almost every action or substance exposes a person to some risk. For example, in any one year the chance of being hit by a meteorite is approximately 1 in 1,700,000,000 while the chance of dying from smoking is about 1 in 170. Meteorites will likely harm no one in the United States whereas smoking is believed to be responsible for 420,000 deaths annually. Probability provides us with a measure of the likelihood that an event will occur and plays a central role in risk assessment. Knowledge about probability allows individuals to make informed decisions about their lives.

Sources: Rodricks, J., *Calculated Risk,* Cambridge University Press, 1992.
1995 Grolier Multimedia Encyclopedia.

9 Kittens = Find the Number of Choices in selecting each:

(A) 8 Kittens = $C(9,8) = 9$

(B) 2 kittens as gifts for Mary + 2 kittens as gifts for Bob
$C(9,2) \cdot C(7,2) = 756$

(C) a set of 2 or more kittens
$2^9 - [C(9,1) + C(9,0)] = 502$

"But is it probable," asked Pascal, "that probability gives assurance? Nothing gives certainty but truth; nothing gives rest but the sincere search for truth." When Pascal wrote that, he had gone to live at the Jansenist convent of Port-Royal after a carriage accident in 1654.

Pascal's notes on Christianity were collected after his death in the *Pensees* (thoughts). The above quotation is included. Another develops Pascal's "rule of the wager": If you bet God exists and live accordingly, you will have gained much even if God does not exist; if you bet the opposite and God does exist, you will have lost the reason for living right—hence everything.

If you go to a supermarket and select five pounds of peaches at 54¢ per pound, you can easily predict the amount you will be charged at the checkout counter: $5 \times \$.54 = \$2.70$. The amount charged for such purchases is a **deterministic phenomenon.** It can be predicted exactly on the basis of obtainable information, namely, in this case, number of pounds and cost per pound.

On the other hand, consider the problem faced by the produce manager of the market, who must order peaches to have on hand each day without knowing exactly how many pounds customers will buy during the day. The customer demand is an example of a **random phenomenon.** It fluctuates in such a way that its value on a given day cannot be predicted exactly with obtainable information.

The study of probability is concerned with such random phenomena. Even though we cannot be certain whether or not a given result will occur, we can often obtain a good measure of its *likelihood,* or **probability.** This chapter discusses various ways of finding and using probabilities.

Some mathematics of probability involved in games and gambling was studied in Italy as early as the fifteenth and sixteenth centuries, but a systematic mathematical theory of chance was not begun until 1654. In that year two French mathematicians, Pierre de Fermat (about 1601–1665) and Blaise Pascal (1623–1662), corresponded with each other regarding a problem posed by the Chevalier de Méré, a gambler and member of the aristocracy. *If the two players of a game are forced to quit before the game is finished, how should the pot be divided?* Pascal and Fermat solved the problem by developing basic methods of determining each player's chance, or probability, of winning.

An early contributor to the study of probability was the brilliant Dutch mathematician and scientist Christiaan Huygens (1629–1695), who, in 1657, wrote the first formal treatise on probability, which was based on the Pascal-Fermat correspondence. Swiss mathematician James Bernoulli (1654–1705) included a reprint of the Huygens paper and established other fundamental principles in his *Ars Conjectandi* (Art of Conjecture), which was published (after his death) in 1713.

Nearly all published work in probability theory until about 1800 was based on dice and other games of chance. The man usually credited with being the "father" of probability theory is the French mathematician Pierre Simon de Laplace (1749–1827); he was one of the first to apply probability to matters other than gambling.

Despite the work of Laplace and the urgings of a few important mathematicians in the nineteenth century, the general mathematical community failed to become interested in probability.

One event that eventually helped to bring probability theory into prominence was botanist Robert Brown's observation in 1828 of the irregular motion of pollen grains suspended in water. Such "Brownian motion", as it came to be called, was not understood until 1905 when Albert Einstein gave a complete explanation based on the treatment of molecular motion as a random phenomenon. (See the exercises for Section 6 of this chapter.)

The development of probability as a mathematical theory was mainly due to a line of remarkable scholars in Russia, including P. L. Chebyshev (1821–1894), his student A. A. Markov (1856–1922), and finally Andrei Nikolaevich Kolmogorov (born in 1903). Kolmogorov's *Foundations of the Theory of Probability* (1933) is an axiomatic treatment.

Probability and Odds

Probability

In the study of probability, we say that any observation, or measurement, of a random phenomenon is an **experiment.** The possible results of the experiment are called **outcomes,** and the set of all possible outcomes is called the **sample space.** Usually we are interested in some particular collection of the possible outcomes. Any such subset of the sample space is called an **event.** Outcomes that belong to the event are commonly referred to as "favorable outcomes", or "successes." Any time a success is observed, we say that the event has "occurred." The probability of an event, being a numerical measure of the event's likelihood, is determined in one of two ways, either *empirically* (experimentally) or *theoretically*. The following examples will clarify the difference between these two interpretations of probability.

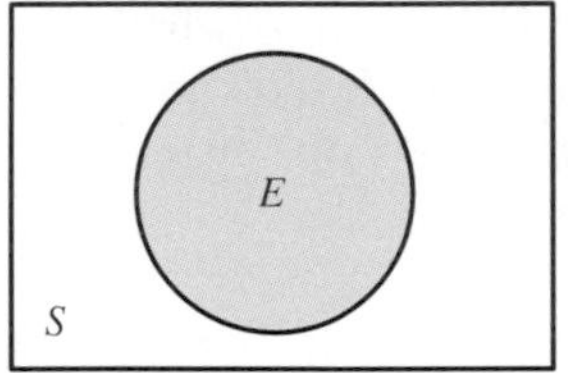

Every event is a subset of the sample space.

EXAMPLE 1 If a single coin is tossed, find the probability that it will land heads up.

There is no apparent reason for one side of a coin to land up any more often than the other (in the long run), so we would normally assume that heads and tails are equally likely. This assumption can be emphasized by saying that the coin is "fair." Now the experiment here is the tossing of a single fair coin, the sample space can be denoted $S = \{\text{h, t}\}$, and the event whose probability we seek is $E = \{\text{h}\}$. Since one of two possible outcomes is a head, the probability of heads is the quotient of 1 and 2.

$$\text{Probability (heads)} = \frac{1}{2}.$$

Symbolically, we can express this as

$$P(\text{h}) = \frac{1}{2} \quad \text{or} \quad P(E) = \frac{1}{2}.$$

EXAMPLE 2 If a styrofoam cup is tossed, find the probability that it will land on its top.

Intuitively, it seems that a cup will land on its side much more often than on its top or its bottom. But just how much more often is not clear. To get an idea, we performed the experiment of tossing such a cup 50 times. It landed on its side 44 times, on its top 5 times, and on its bottom just 1 time. By the frequency of "success" in this experiment, we concluded that

$$P(\text{top}) = \frac{5}{50} = \frac{1}{10}.$$

Notice that in Example 1, involving the tossing of a fair coin, the number of possible outcomes was obviously two, both were equally likely, and one of the outcomes was a head. No actual experiment was required. The desired probability was obtained *theoretically*. Theoretical probabilities apply to all kinds of games of chance (dice rolling, card games, roulette, lotteries, and so on), and also apparently to many phenomena in nature. Laplace, in his famous *Analytic Theory of Probability,* published in 1812, gave a formula that applies to any such theoretical probability, as long as the sample space S is finite and all outcomes are equally likely. (It is sometimes referred to as the *classical definition of probability.*)

Theoretical Probability Formula

If all outcomes in a sample space S are equally likely, and E is an event within that sample space, then the **theoretical probability** of event E is given by

$$P(E) = \frac{\text{number of favorable outcomes}}{\text{total number of outcomes}} = \frac{n(E)}{n(S)}.$$

On the other hand, Example 2 involved the tossing of a cup, where the likelihoods of the various outcomes were not intuitively clear. It took an actual experiment to arrive at a probability value of $1/10$. That value was found according to the *experimental,* or *empirical* probability formula.

Empirical Probability Formula

If E is an event that may happen when an experiment is performed, then the **empirical probability** of event E is given by

$$P(E) = \frac{\text{number of times event } E \text{ occurred}}{\text{number of times the experiment was performed}}.$$

Usually it is clear in applications which of the two probability formulas should be used. Here are two more examples.

EXAMPLE 3 Ina Trearty wants to have exactly two daughters. Assuming that boy and girl babies are equally likely, find her probability of success in each of the following cases.

(a) She has two children altogether.

The equal likelihood assumption here allows the use of theoretical probability. But how can we determine the number of favorable outcomes and the total number of possible outcomes? Based on counting methods, you may come up with the idea of using a tree diagram to enumerate the possibilities. This is shown in Figure 1. We obtain the sample space from the outcome column: $S = \{\text{gg, gb, bg, bb}\}$. The arrow shows the only outcome favorable to the event of exactly two daughters: $E = \{\text{gg}\}$. By the theoretical probability formula,

$$P(E) = \frac{n(E)}{n(S)} = \frac{1}{4}.$$

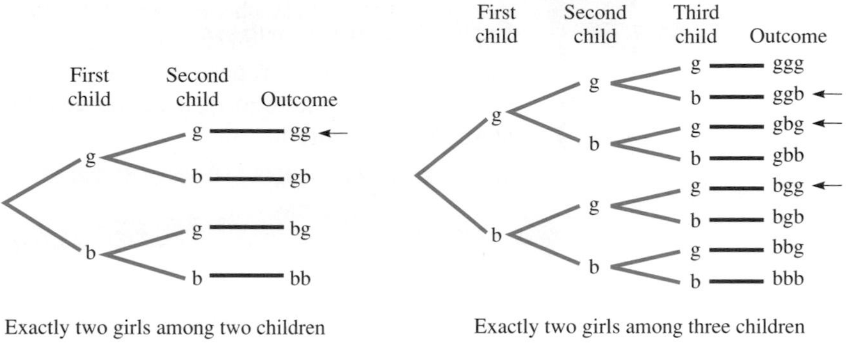

Exactly two girls among two children

FIGURE 1

Exactly two girls among three children

FIGURE 2

(b) She has three children altogether.

For three children altogether, we construct another tree diagram, as shown in Figure 2. In this case, $S = \{\text{ggg, ggb, gbg, gbb, bgg, bgb, bbg, bbb}\}$ and $E = \{\text{ggb, gbg, bgg}\}$, so $P(E) = 3/8$. ■

Although we assumed boy and girl babies to be equally likely in Example 3, male births typically occur a little more frequently. (At the same time, there are usually more females living at any given time, due to higher infant mortality rates among males and longer female life expectancy in general.)

EXAMPLE 4 In a recent year, births in the United States included 1.613 million males and 1.531 million females. If a person was selected randomly from the birth records of that year, what is the probability that the person would be a male?

Since male and female births are not exactly equally likely, and we have specific experimental data to go on here, we calculate the empirical probability.

$$P(\text{male}) = \frac{\text{number of males born that year}}{\text{total number of births that year}} = \frac{1{,}613{,}000}{1{,}613{,}000 + 1{,}531{,}000}$$
$$\approx .513$$ ■

Now think again about the cup of Example 2. If we tossed it 50 more times, we would have 100 total tosses upon which to base an empirical probability of the cup landing on its top. The new value would likely be (at least slightly) different from what we obtained before. It would still be an empirical probability, but would be "better" in the sense that it is based upon a larger set of outcomes. As we increase the number of tosses more and more, the resulting empirical probability values may approach some particular number. If so, that number can be defined as the theoretical probability of that particular cup landing on its top. This "limiting" value can only occur as the actual number of observed tosses approaches the total number of possible tosses of the cup. Since there are potentially an infinite number of possible tosses, we could never actually find the theoretical probability we want. But we can still assume such a number exists. And as the number of actual observed tosses increases, the resulting empirical probabilities should tend ever closer to the theoretical value. This very important principle is known as the **law of large numbers** (or sometimes as the "law of averages").

Law of Large Numbers

As an experiment is repeated more and more times, the proportion of outcomes favorable to any particular event will tend to come closer and closer to the theoretical probability of that event.

EXAMPLE 5 A fair coin was tossed 35 times, producing the following sequence of outcomes.

tthhh ttthh hthtt hhthh ttthh thttt hhthh

Calculate the ratio of heads to total tosses after the first toss, the second toss, and so on through all 35 tosses, and plot these ratios on a graph.

The first few ratios, to two decimal places, are .00, .00, .33, .50, .60, .50, ... The graph in Figure 3 shows how the fluctuations away from .50 become smaller as the number of tosses increases, and the ratios appear to approach .50 toward the right side of the graph, in keeping with the law of large numbers. ■

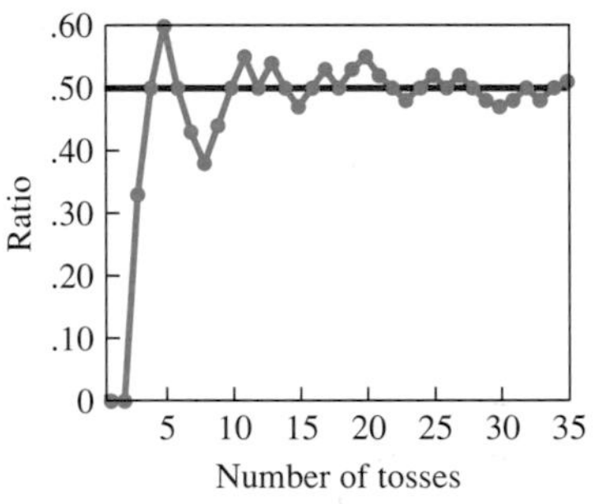

■ FIGURE 3

We see then that the law of large numbers provides an important connection between empirical and theoretical probabilities. Knowing an empirical probability value for an event allows us to *estimate* its theoretical probability. The more repetitions the estimate is based upon, the more reliable it is.

In a similar manner, knowing the theoretical probability of an event allows us to *predict* the fraction of times it will occur in a series of repeated experiments. (This is a function of deductive reasoning.) The prediction should be more accurate for larger numbers of repetitions. For example, if we toss a fair coin four times, and observe three heads, the proportion of heads is $3/4 = .75$, which is quite different than the expected proportion of .50, but it is not surprising for only four tosses. But if we toss the coin 40 times and observe 30 heads, the proportion of .75 is now far enough from .50 to cause us to question the "fairness" of the coin.

Probabilities, both empirical and theoretical, have been valuable tools in many areas of science. An important early example was the work of the Austrian monk Gregor Mendel, who used the idea of randomness to help establish the study of genetics. Mendel published his results in 1865, but they were largely ignored until 1900 when others rediscovered and recognized the importance of his work.

In an effort to understand the mechanism of character transmittal from one generation to the next in plants, Mendel counted the number of occurrences of various characteristics. He found that the flower color in certain pea plants obeyed this scheme:

Pure red crossed with pure white produces red.

Mendel theorized that red is "dominant" (symbolized with the capital letter R), while white is "recessive" (symbolized with the small letter r). The pure red parent carried only genes for red (R) and the pure white parent carried only genes for white (r). The offspring would receive one gene from each parent, hence one of the four combinations shown in the body of Table 1. Because every offspring receives one gene for red, that characteristic dominates and the offspring exhibits the color red.

TABLE 1 First to Second Generation

		Second Parent	
		r	**r**
First Parent	**R**	Rr	Rr
	R	Rr	Rr

TABLE 2 Second to Third Generation

		Second Parent	
		R	**r**
First Parent	**R**	RR	Rr
	r	rR	rr

Gregor Johann Mendel (1822–1884) came from a peasant family who managed to send him to school. By 1847 he had been ordained and was teaching at the Abbey of St. Thomas. He finished his education at the University of Vienna and returned to the abbey to teach mathematics and natural science. Mendel began to carry out experiments on plants in the abbey garden, notably pea plants, whose distinct traits (unit characters) he had puzzled over. In 1865 he published his results (the Czech stamp above commemorates the centennial). His work was not appreciated at the time even though he had laid the foundation of classical genetics. Mendel had established the basic laws of heredity: laws of unit characters, of dominance, and of segregation.

Now each of these second generation offspring, though exhibiting the color red, still carries one of each gene. So when two of them are crossed, each third generation offspring will receive one of the gene combinations shown in Table 2. Mendel theorized that each of these four possibilities would be equally likely, and produced experimental counts that were close enough to support this hypothesis. (In more recent years, some have accused Mendel, or his assistants, of fudging the experimental data, but his conclusions have not been disputed.)

EXAMPLE 6 Referring to Table 2, determine the probability that a third generation offspring will exhibit each of the following flower colors. Base the probabilities on the following sample space of equally likely outcomes: $S = \{RR, Rr, rR, rr\}$.

(a) red

Since red dominates white, any combination with at least one gene for red (R) will result in red flowers. Since three of the four possibilities meet this criterion, $P(\text{red}) = 3/4$.

(b) white

The combination rr is the only one with no gene for red, so $P(\text{white}) = 1/4$. ■

Due to the probabilistic laws of genetics, which began with Mendel's work, vast improvements have resulted in food supply technology, through hybrid animal and crop development. Also, the understanding and control of human diseases has been advanced by related studies. Some human hereditary diseases will be addressed in the exercises of this section.

Odds

Whereas probability compares the number of favorable outcomes to the total number of outcomes, **odds** compare the number of favorable outcomes to the number of unfavorable outcomes. Odds are commonly quoted, rather than probabilities, in horse racing, lotteries, and most other gambling situations. And the odds quoted are normally odds "against" rather than odds "in favor."

Odds

If all outcomes in a sample space are equally likely, a of them are favorable to the event E, and the remaining b outcomes are unfavorable to E, then the **odds in favor** of E are a to b, and the **odds against** E are b to a.

An odds ratio would normally be reduced, just as a fraction. For example, it is preferable to express odds of 12 to 1 rather than 48 to 4.

EXAMPLE 7 Jennifer has been promised one of six summer jobs, three of which would be at a nearby mountain resort. If she has equal chances for all six jobs, find the odds that she will land one at the resort.

Since three possibilities are favorable and three are not, the odds of working at the resort are 3 to 3, or 1 to 1. (The common factor of 3 has been divided out.) Odds of 1 to 1 are often termed "even odds," or a "50–50 chance." ■

EXAMPLE 8 Brad Rogers has purchased six tickets for an office raffle where the winner will receive a portable fax machine. If 51 tickets were sold altogether, and each has an equal chance of winning, what are the odds against Brad's winning the fax?

Brad has six chances to win and 45 chances to lose, so the odds against his winning are 45 to 6, or 15 to 2. ■

EXAMPLE 9 A friend invites you to play cards, and announces that your probability of winning the first hand is $7/11$. What are the odds in favor of your winning the first hand?

A probability of $7/11$ does not imply that there are necessarily 7 favorable outcomes and 11 total outcomes. (Why not?) But the *ratio* would be the same as this. Favorable to unfavorable outcomes will occur in the ratio 7 to $11 - 7 = 4$. So the odds in favor of your winning are 7 to 4. ■

TABLE 3 Number of Poker Hands in 5-Card Poker; Nothing Wild

Event E	Number of Outcomes Favorable to E
Royal flush	4
Straight flush	36
Four of a kind	624
Full house	3,744
Flush	5,108
Straight	10,200
Three of a kind	54,912
Two pairs	123,552
One pair	1,098,240
No pair	1,302,540
Total	2,598,960

The various kinds of hands possible in 5-card poker are summarized in Table 3. Since all of the 2,598,960 different hands are equally likely, the numbers from the table can be entered into the theoretical probability formula to answer questions like those in the following example.

EXAMPLE 10 Find the probability of being dealt each of the following hands in five-card poker. Use a calculator to obtain answers to eight decimal places.

As we have seen, 2,598,960 possible 5-card hands could be dealt from a standard deck. Since all these hands are equally likely, the theoretical probability formula applies. Use Table 3 to help with the following examples.

(a) a royal flush (the five highest cards of a single suit)
The table shows that there are 4 royal flushes, one for each suit, so that

$$P(\text{royal flush}) = \frac{4}{2{,}598{,}960} = \frac{1}{649{,}740} \approx .00000154.$$

(b) a full house (three of one denomination, two of another)
Again from the table,

$$P(\text{full house}) = \frac{3{,}744}{2{,}598{,}960} = \frac{6}{4{,}165} \approx .00144058.$$

(c) a spades flush (a hand containing 5 spades but not qualifying as a royal flush or even as a straight flush)
Of the 5,108 flush hands (shown in the table), 1/4 are spades (the other 3/4 being hearts, clubs, or diamonds), so

$$P(\text{spades flush}) = \frac{(1/4) \cdot 5{,}108}{2{,}598{,}960} = \frac{1{,}277}{2{,}598{,}960} \approx .00049135. \quad \blacksquare$$

Some of the ideas of set theory have been applied in this section, and others will appear as we proceed. We summarize up to this point as follows. For any probability experiment, the universal set is the set of all possible outcomes. In the present context, we call this the sample space. Any set within the universal set then becomes a subset of the sample space. We call such a subset an event. It is normally the probability, or likelihood, of a particular event that we are interested in. We will continue, in subsequent sections, to consider ways of evaluating and combining these probabilities and applying them in various cases.

11.1 EXERCISES

Suppose the spinner shown here is spun once. (Each number corresponds to a 120-degree sector, or one third of the circle.) Find **(a)** *the theoretical probability, and* **(b)** *the odds in favor, of each of the following events. Start by constructing the sample space for the experiment. For each exercise, list the favorable outcomes. Then form the probability fraction.*

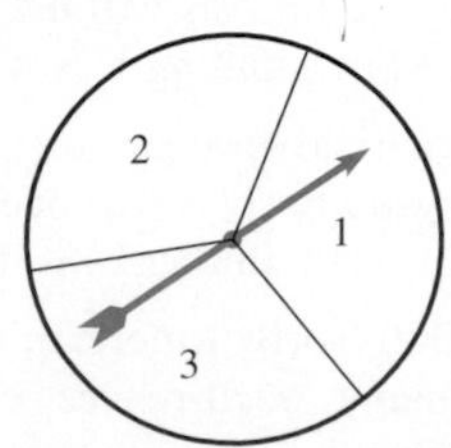

1. It will stop at 1.

2. It will stop at an odd number.

3. It will stop at an even number.

Suppose the same spinner is spun twice in succession to determine a two-digit number. Construct the sample space and refer to it for Exercises 4–7.

4. How many different two-digit numbers could be determined?

Find **(a)** *the probability, and* **(b)** *the odds in favor, of each of the following events.*

5. The resulting number will be even.

6. The resulting number will contain repeated digits.

7. The resulting number will be prime.

For the experiment of tossing two fair coins, say a dime and a nickel, find **(a)** *the probability, and* **(b)** *the odds in favor, of each of the following events. Start by writing out the sample space.*

8. The dime will land heads up.

9. The nickel will land heads up.

10. Two heads will occur.

11. Exactly one head will occur.

If three fair coins are to be tossed, find the probability of each of the following events. Start by writing out the sample space.

12. no heads

13. exactly one head

14. exactly two heads

15. three heads

The sample space for the rolling of two fair dice appeared in Table 2 of the first section in the previous chapter. Reproduce that table, but replace each of the 36 equally likely ordered pairs with its corresponding sum (for the two dice). Then find the probability of rolling each of the following sums.

16. 2	**17.** 3	**18.** 4	**19.** 5
20. 6	**21.** 7	**22.** 8	**23.** 9
24. 10	**25.** 11	**26.** 12	

In Exercises 27 and 28, compute answers to three decimal places.

27. In a hybrid corn research project, 200 seeds were planted, and 170 of them germinated. Find the empirical probability that any particular seed of this type will germinate.

28. In a certain state, 37,052 boys and 35,192 girls were born last year. Find the empirical probability that one of those births, chosen at random, would be **(a)** a boy; **(b)** a girl.

29. In Example 3, what would be Ina's probability of having exactly two daughters if she were to have four children altogether? (Use a tree diagram to construct the sample space.)

Mendel found no dominance in snapdragons (in contrast to peas) with respect to red and white flower color. When pure red and pure white parents are crossed (see Table 1), the resulting Rr *combination (one of each gene) produces second generation offspring with* pink *flowers. These second generation pinks, however, still carry one red and one white gene, so when they are crossed the third generation is still governed by Table 2.*

Find the following probabilities for third generation snapdragons.

30. P(red) **31.** P(pink) **32.** P(white)

Mendel also investigated various characteristics besides flower color. For example, round peas are dominant over recessive wrinkled peas. First, second, and third generations can again be analyzed using Tables 1 and 2, where R *represents round and* r *represents wrinkled.*

33. Explain why crossing pure round and pure wrinkled first generation parents will always produce round peas in the second generation offspring.

34. When second generation round pea plants (each of which carries both R and r genes) are crossed, find the probability that a third generation offspring will have
(a) round peas; **(b)** wrinkled peas.

Cystic fibrosis *is one of the most common inherited diseases in North America (including the United States), occurring in about 1 of every 2,000 Caucasian births and about 1 of every 250,000 non-Caucasian births. Even with modern treatment, victims usually die from lung damage by their early twenties. If we denote a cystic fibrosis gene with a* c *and a disease-free gene with a* C *(since the disease is recessive), then only a* cc *person will actually have the disease. Such persons would ordinarily die before parenting children, but a child can also inherit the disease from two* Cc *parents (who themselves are healthy, that is, have no symptoms but are "carriers" of the disease). This is like a pea plant inheriting white flowers from two red-flowered parents which both carry genes for white.*

35. Find the empirical probability (to four decimal places) that cystic fibrosis will occur in a randomly selected infant birth among U.S. Caucasians.

36. Find the empirical probability (to six decimal places) that cystic fibrosis will occur in a randomly selected infant birth among U.S. non-Caucasians.

37. Among 150,000 North American Caucasian births, about how many occurrences of cystic fibrosis would you expect?

Suppose that both partners in a marriage are cystic fibrosis carriers (a rare occurrence). Construct a chart similar to Table 2 and determine the probability of each of the following events.

38. Their first child will have the disease.

39. Their first child will be a carrier.

40. Their first child will neither have nor carry the disease.

Suppose a child is born to one cystic fibrosis carrier parent and one non-carrier parent. Find the probability of each of the following events.

41. The child will have cystic fibrosis.

42. The child will be a healthy cystic fibrosis carrier.

43. The child will neither have nor carry the disease.

Sickle-cell anemia *occurs in about 1 of every 500 black baby births and about 1 of every 160,000 non-black baby births. It is ordinarily fatal in early childhood. There is a test to identify carriers. Unlike cystic fibrosis, which is recessive, sickle-cell anemia is* **codominant.** *This means that inheriting two sickle-cell genes causes the disease, while inheriting just one sickle-cell gene causes a mild (non-fatal) version (which is called* **sickle-cell trait***). This is like a snapdragon plant manifesting pink flowers by inheriting one red gene and one white gene.*

Find the empirical probability of each of the following events (Exercises 44 and 45).

44. A randomly selected black baby will have sickle-cell anemia. (Give your answer to three decimal places.)

45. A randomly selected non-black baby will have sickle-cell anemia. (Give your answer to six decimal places.)

46. Among 80,000 births of black babies, about how many occurrences of sickle-cell anemia would you expect?

Find the theoretical probability of each of the following conditions in a child, both of whose parents have sickle-cell trait. (Exercises 47–49)

47. The child will have sickle-cell anemia.

48. The child will have sickle-cell trait.

49. The child will be healthy.

50. In the history of track and field, no woman has broken the 10-second barrier in the 100-meter run.
   (a) From the statement above, find the empirical probability that a woman runner will break the 10-second barrier next year.
   (b) Can you find the theoretical probability for the event of part (a)?
   (c) Is it possible that the event of part (a) will occur?

51. Is there any way a coin could fail to be "fair"? Explain.

52. On page 27 of their book *Descartes' Dream,* Philip Davis and Reuben Hersh ask the question, "Is probability real or is it just a cover-up for ignorance?" What do you think? Are some things truly random, or is everything potentially deterministic?

Three married couples arrange themselves randomly in six consecutive seats in a row. Find the probability of each event in Exercises 53–56. (Hint: In each case the denominator of the probability fraction will be 720.)

53. Each man will sit immediately to the left of his wife.

54. Each man will sit immediately to the left of a woman.

55. The women will be in three adjacent seats.

56. The women will be in three adjacent seats, as will the men.

57. If two distinct numbers are chosen randomly from the set $\{-2, -4/3, -1/2, 0, 1/2, 3/4, 3\}$, find the probability that they will be the slopes of two perpendicular lines.

58. At most horse-racing tracks, the "trifecta" is a particular race where you win if you correctly pick the "win," "place," and "show" horses (the first, second, and third place winners), in their proper order. If five horses of equal ability are entered in today's trifecta race, and you select an entry, what is the probability that you will be a winner?

59. If a dart hits the square target shown here at random, what is the probability that it will hit in a colored region? (*Hint:* Compare the area of the colored regions to the total area of the target.)

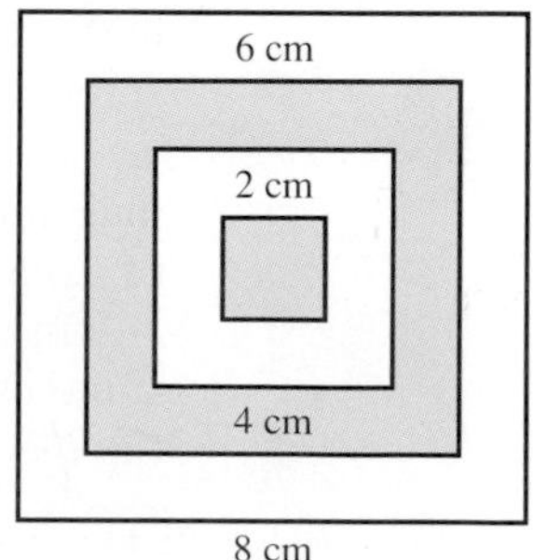

60. Suppose you plan to take four courses next term. If you select them randomly from a listing of ten courses, six of which are science courses, what is the probability that all four courses you select will be science courses?

Assuming that Marcy, Todd, and Jennifer are three of the 32 members of the class, and that three of the class members will be chosen randomly to deliver their reports during the next class meeting, find the probability (to six decimal places) of each of the following events (Exercises 61 and 62).

61. Marcy, Todd, and Jennifer are selected, in that order.

62. Marcy, Todd, and Jennifer are selected, in any order.

63. Theresa Piselli randomly selects two symphony performances to attend this season, choosing from a schedule of eight performances, two of which will feature works by Mozart. Find the probability that Theresa will select both of the Mozart programs.

64. When drawing cards without replacement from a standard 52-card deck, find the highest number of cards you could possibly draw and still get
(a) fewer than three black cards,
(b) fewer than six spades,
(c) fewer than four face cards,
(d) fewer than two kings.

In 5-card poker, find the probability of being dealt each of the following (Exercises 65–70). Give each answer to eight decimal places. (Refer to Table 3.)

65. a straight flush **66.** four of a kind

67. four aces **68.** one pair

69. a pair of 7s **70.** two pairs

71. If two distinct prime numbers are randomly selected from among the first eight prime numbers, what is the probability that their sum will be 24?

72. The digits 1, 2, 3, 4, and 5 are randomly arranged to form a five-digit number. Find the probability of each of the following events.
(a) The number is even.
(b) The first and last digits of the number are both even.

73. Two integers are randomly selected from the set $\{1, 2, 3, 4, 5, 6, 7, 8, 9\}$ and are added together. Find the probability that their sum is 11 if they are selected
(a) with replacement;
(b) without replacement.

*Numbers that are **palindromes** read the same forward and backward. For example, 30203 is a five-digit palindrome. If a single number is chosen randomly from each of the following sets, find the probability that it will be palindromic.*

74. the set of all two-digit numbers

75. the set of all three-digit numbers

11.2 Events Involving "Not" and "Or"

In the previous section, we set out and illustrated the basic ideas of probability. Remember that an empirical probability, based upon experimental observation, may be the best value available but is still only an approximation to the ("true") theoretical probability. For example, no human has ever been known to jump higher than eight feet vertically, so the empirical probability of such an event is zero. However, observing the rate at which high jump records have been broken, we suspect that the event is, in fact, possible and may one day occur. Hence it must have some non-zero theoretical probability, even though we have no way of assessing its exact value.

In the next three sections, we develop some properties of probability and some useful rules for evaluating probabilities. All of these results follow from the theoretical probability formula

$$P(E) = \frac{n(E)}{n(S)}.$$

Pierre Simon de Laplace (1749–1827) In 1773 Laplace began to solve the problem of why Jupiter's orbit seems to shrink and Saturn's orbit seems to expand. Eventually Laplace worked out the whole system. *Celestial Mechanics* resulted from almost a lifetime of work. In five volumes, it was published between 1799 and 1825 and gained for Laplace the reputation "Newton of France."

Laplace's work on probability was actually an adjunct to his celestial mechanics. He needed to demonstrate that probability is useful in interpreting scientific data. He also wrote a popular exposition of the system, which contains (in a footnote!) the "nebular hypothesis" that the sun and planets originated together in a cloud of matter, which then cooled and condensed into separate bodies.

Recall that this formula is only valid when all outcomes in the sample space S are equally likely. For the experiment of tossing two fair coins, we can write $S = \{\text{hh, ht, th, tt}\}$ and compute correctly that

$$P(\text{both heads}) = \frac{1}{4},$$

whereas if we define the sample space with non-equally likely outcomes as $S = \{\text{both heads, both tails, one of each}\}$, we are led to

$$P(\text{both heads}) = \frac{1}{3}, \quad \text{which is } \textit{wrong}.$$

(To convince yourself that $1/4$ is a better value than $1/3$, toss two fair coins 100 times or so to see what the empirical fraction seems to approach.)

Since any event E is a subset of the sample space S, we know that $0 \leq n(E) \leq n(S)$. Dividing all members of this inequality by $n(S)$ gives

$$\frac{0}{n(S)} \leq \frac{n(E)}{n(S)} \leq \frac{n(S)}{n(S)}, \qquad \text{or} \qquad 0 \leq P(E) \leq 1.$$

In words, the probability of any event is a number between 0 and 1, inclusive.

If event E is *impossible* (cannot happen), then $n(E)$ must be 0 (E is the empty set), so $P(E) = 0$. If event E is *certain* (cannot help but happen), then $n(E) = n(S)$, so $P(E) = n(E)/n(S) = n(S)/n(S) = 1$. These properties are summarized below.

Properties of Probability

Let E be an event from the sample space S. That is, E is a subset of S. Then

1. $\mathbf{0 \leq P(E) \leq 1}$ (The probability of an event is between 0 and 1, inclusive.)
2. $\mathbf{P(\emptyset) = 0}$ (The probability of an impossible event is 0.)
3. $\mathbf{P(S) = 1}$ (The probability of a certain event is 1.)

EXAMPLE 1 When a single fair die is rolled, find the probability of each of the following events.

(a) the number 2 is rolled

Since one of the six possibilities is a 2, $P(2) = 1/6$.

(b) a number other than 2 is rolled

There are 5 such numbers, 1, 3, 4, 5, and 6. $P(\text{a number other than } 2) = 5/6$.

(c) the number 7 is rolled

None of the possible outcomes is 7. $P(7) = 0/6 = 0$.

(d) a number less than 7 is rolled

All six of the possible outcomes are less than 7. $P(\text{a number less than } 7) = 6/6 = 1$. ■

Mary Somerville (1780–1872) is associated with Laplace because of her brilliant exposition of his *Celestial Mechanics.* She combined a deep understanding of science with the ability to communicate its concepts to the general public.

In her childhood Somerville was free to observe nature. She studied Euclid thoroughly and perfected her Latin so she could read Newton's *Principia.* In about 1816 she went to London and soon became part of its literary and scientific circles. She also corresponded with Laplace and other Continental scientists.

Somerville's book on Laplace's theories came out in 1831 with great acclaim. Then followed a panoramic book, *Connection of the Physical Sciences* (1834). A statement in one of its editions suggested that irregularities in the orbit of Uranus might indicate that a more remote planet, not yet seen, existed. This caught the eye of the scientist who worked out the calculations for Neptune's orbit.

Notice that parts (c) and (d) of Example 1 illustrate probability properties 2 and 3, respectively. Also notice that the events in parts (a) and (b) are *complements* of one another, and that their probabilities add up to 1. The same is true for any two complementary events; that is, $P(E) + P(E') = 1$. By rearranging terms, we can write this equation in two other equivalent forms, the most useful of which is stated in the following rule.

Complements Rule of Probability

The probability that an event E will occur is equal to one minus the probability that it will *not* occur.

$$P(E) = 1 - P(E')$$

If you have studied counting methods, you will notice that this rule is similar to the complements principle of counting (and indeed follows from it). The complements rule is illustrated in Figure 4.

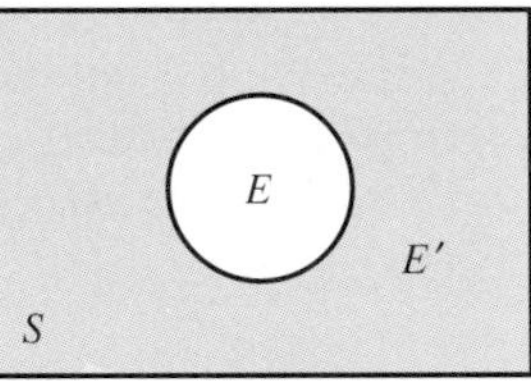

The logical connective "not" corresponds to "complement" in set theory.

$$P(\text{not } E) = P(S) - P(E) = 1 - P(E)$$

FIGURE 4

The following example illustrates how the complements rule allows probabilities to be found indirectly when this is easier.

EXAMPLE 2 When a single card is drawn from a standard 52-card deck, what is the probability of the event of drawing a card other than a king?

It is easier to count the kings than the non-kings. Let E be the event of *not* drawing a king. Then

$$P(E) = 1 - P(E') = 1 - \frac{n(E')}{n(S)} = 1 - \frac{4}{52} = \frac{48}{52} = \frac{12}{13}.$$

We will now consider some ways of combining events to obtain other events. First we look at composites involving the word "or." For event A and event B, we will want to find $P(A \text{ or } B)$. "A or B" is the event that at least one of the two individual events occurs.

■ FIGURE 5

EXAMPLE 3 If one number is selected randomly from the set {1, 2, 3, 4, 5, 6, 7, 8, 9, 10}, find the probability that it will be odd or a multiple of 3.

Take the sample space to be $S = \{1, 2, 3, 4, 5, 6, 7, 8, 9, 10\}$. Then, if we let A be the event that the selected number is odd and B be the event that it is a multiple of 3, we have

$$A = \{1, 3, 5, 7, 9\} \qquad \text{and} \qquad B = \{3, 6, 9\}.$$

Figure 5 shows the positioning of the 10 integers within the sample space, and we see that $A \cup B = \{1, 3, 5, 6, 7, 9\}$. The composite event *A or B* corresponds to the set $A \cup B$. So, by the theoretical probability formula,

$$P(A \text{ or } B) = \frac{6}{10} = \frac{3}{5}.$$ ■

In general we cannot just add the individual probabilities for a composite event of the form *A or B*. Doing so in Example 3 would have given us

$$\frac{5}{10} + \frac{3}{10} = \frac{8}{10} = \frac{4}{5}, \quad \text{which is } \textit{wrong}.$$

This would have counted the outcomes in the intersection region twice. The integers 3 and 9 are both odd *and* multiples of 3. The correct general rule is stated as follows. (Since the logical connectives *or* and *and* correspond to the set operations $\cup$ (union) and $\cap$ (intersection), respectively, we can state two versions of the rule.)

General Addition Rule of Probability

If A and B are any two events, then

$$P(A \text{ or } B) = P(A) + P(B) - P(A \text{ and } B),$$

or

$$P(A \cup B) = P(A) + P(B) - P(A \cap B).$$

The general addition rule is illustrated in Figure 6. This rule follows from the general additive counting principle:

$$n(A \cup B) = n(A) + n(B) - n(A \cap B).$$

For now, we will consider only events of the form "A and B" that are quite simple, as in Example 3. We will deal with more involved composites using "and" in the next section. Now we give another example using the general addition rule of probability.

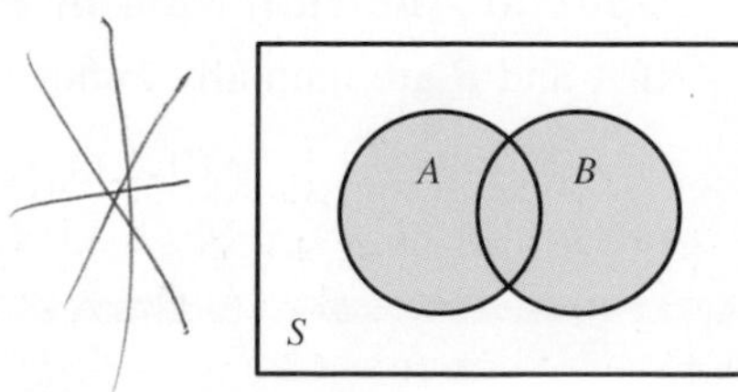

The logical connective "or" corresponds to "union" in set theory.

$P(A \text{ or } B) = P(A) + P(B) - P(A \text{ and } B)$

■ FIGURE 6

EXAMPLE 4 Of the 20 television programs to be aired this evening, Marc plans to watch one, which he will pick at random by throwing a dart at the TV schedule. If 8 of the programs are educational, 9 are interesting, and 5 are both educational and interesting, find the probability that the show he watches will have at least one of these attributes.

If E represents "educational," and I represents "interesting," then $P(E) = 8/20$, $P(I) = 9/20$, and $P(E \text{ and } I) = 5/20$. So

$$P(E \text{ or } I) = \frac{8}{20} + \frac{9}{20} - \frac{5}{20} = \frac{12}{20} = \frac{3}{5}.$$

In the following example, something special happens when we get to the "A and B" term of the addition rule.

EXAMPLE 5 Suppose a single card is to be drawn from a standard 52-card deck. Find the probability that it will be a spade or red.

There are 13 spades in the deck and 26 red cards. But the event "spade and red" cannot possibly occur since there are no red spades (the spades are all black). Therefore the third term in the general addition rule will be 0 (by probability property 2), and can be omitted. Hence

$$P(\text{spade or red}) = P(\text{spade}) + P(\text{red}) = \frac{13}{52} + \frac{26}{52} = \frac{39}{52} = \frac{3}{4}.$$

In Example 5, we saw that the event "spade and red" had probability zero since spade and red cannot possibly both occur for a single card drawn. In general, we say that two events that cannot both occur are called *mutually exclusive.* (As sets, such events are "disjoint.") There is no possible outcome for which both would occur.

Mutually Exclusive Events

For a particular performance of an experiment, any two events that cannot both occur are called **mutually exclusive.**

For events A and B that are mutually exclusive, the addition rule of probability takes the simpler form stated here.

Special Addition Rule of Probability

If A and B are mutually exclusive events for a given experiment, then

$$P(A \text{ or } B) = P(A) + P(B),$$

or

$$P(A \cup B) = P(A) + P(B).$$

The special addition rule is illustrated in Figure 7.

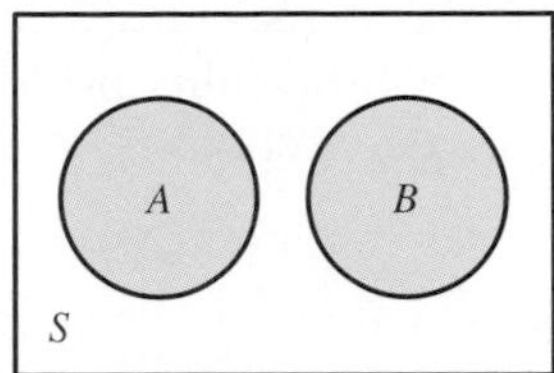

When A and B are mutually exclusive,

$P(A \text{ or } B) = P(A) + P(B)$.

■ **FIGURE 7**

We often encounter cases where composites of more than two events need to be considered. When each event involved is mutually exclusive of all the others, an extension of the special addition rule can be applied. We illustrate this in the following example.

x	$P(x)$
1	.05
2	.10
3	.20
4	.40
5	.10
6	.15

EXAMPLE 6 Ann figures there is only a slight chance she can finish her homework in just one hour tonight. In fact, if x represents the number of hours to be spent on homework, then she assigns probabilities to the various values of x as shown in the table at the side. (We are assuming that the actual time required will be rounded to the nearest hour and that it will not be less than 1 nor more than 6. The six time periods given are all mutually exclusive of one another since, for example, if she takes 3 hours, then she will not take 2 hours or 4 hours or any of the other choices.) Find the probability that she will finish in each of the following periods of time.

(a) fewer than 3 hours
"Fewer than 3" means 1 or 2. So $P(\text{fewer than } 3) = .05 + .10 = .15$.

(b) more than 2 hours

$$\begin{aligned} P(\text{more than } 2) &= P(3 \text{ or } 4 \text{ or } 5 \text{ or } 6) \\ &= P(3) + P(4) + P(5) + P(6) \\ &= .20 + .40 + .10 + .15 = .85 \end{aligned}$$

(c) more than 1 but no more than 5 hours

$$\begin{aligned} P(\text{more than 1 but no more than } 5) &= P(2) + P(3) + P(4) + P(5) \\ &= .10 + .20 + .40 + .10 = .80. \end{aligned}$$ ■

In Example 6, Ann figured her homework would take her one of the time intervals listed in the table. Therefore, one, two, three, four, five, and six hours were the only possible values for the amount of time to be spent on homework. The time spent on homework here is an example of a **random variable.** (It is "random" since we cannot predict which of its possible values will occur.) A table, like that of Example 6, which lists all possible values of a random variable, along with the probabilities that those values will occur, is called a **probability distribution** for that random variable. Since *all* possible values are listed, they make up the entire sample space, and so the listed probabilities must add up to 1 (by probability property 3). We will discuss probability distributions further in later sections.

11.2 EXERCISES

1. Patricia Quinlin has three office assistants. If A is the event that at least two of them are men and B is the event that at least two of them are women, are A and B mutually exclusive?

2. Vicki Valerin earned her college degree several years ago. Consider the following four events.

   Her alma mater is in the East.
   Her alma mater is a private college.
   Her alma mater is in the Northwest.
   Her alma mater is in the South.

   Are these events all mutually exclusive of one another?

3. Explain the difference between the "general" and "special" addition rules of probability, illustrating each one with an appropriate example.

For the experiment of rolling a single fair die, find the probability of each of the following events.

4. not less than 2
5. not prime
6. odd or less than 5
7. even or prime
8. odd or even
9. less than 3 or greater than 4

For the experiment of drawing a single card from a standard 52-card deck, find **(a)** *the probability, and* **(b)** *the odds in favor, of each of the following events.*

10. not an ace
11. king or queen
12. club or heart
13. spade or face card
14. not a heart, or a 7
15. neither a heart nor a 7

For the experiment of rolling an ordinary pair of dice, find the probability that the sum will be each of the following. (You may want to construct a table showing the sum for each of the 36 equally likely outcomes.)

16. 11 or 12
17. even or a multiple of 3
18. odd or greater than 9
19. less than 3 or greater than 9

20. Find the probability of getting a prime number in each of the following cases.
   (a) A number is chosen randomly from the set $\{1, 2, 3, 4, \ldots, 12\}$.
   (b) Two dice are rolled and the sum is observed.

21. Suppose, for a given experiment, A, B, C, and D are events, all mutually exclusive of one another, such that $A \cup B \cup C \cup D = S$ (the sample space). By extending the special addition rule of probability to this case, and utilizing probability property 3, what statement can you make?

If you are dealt a 5-card hand (this implies without replacement) from a standard 52-card deck, find the probability of getting each of the following. Refer to Table 3 of the previous section, and give answers to six decimal places.

22. a flush or three of a kind

23. a full house or a straight

24. a black flush or two pairs

25. nothing any better than two pairs

The table at the side gives golfer Jennifer Twomey's probabilities of scoring in various ranges on a par-70 course. In a given round, find the probability that her score will be each of the following (Exercises 26–30).

26. 95 or higher

27. par or above

28. in the 80s

29. not in the 90s

30. not in the 70s, 80s, or 90s

31. What are the odds of Jennifer's shooting below par?

x	$P(x)$
Below 60	.04
60–64	.06
65–69	.14
70–74	.30
75–79	.23
80–84	.09
85–89	.06
90–94	.04
95–99	.03
100 or above	.01

32. Anne Kelly randomly chooses a single ball from the urn shown here, and x represents the color of the ball chosen. Construct a complete probability distribution for the random variable x.

33. Let x denote the sum of two distinct numbers selected randomly from the set $\{1, 2, 3, 4, 5\}$. Construct the probability distribution for the random variable x.

34. Toss a pair of dice 50 times, keeping track of the number of times the sum is "less than 3 or greater than 9" (that is, 2, 10, 11, or 12).

(a) From your results, calculate an empirical probability for the event "less than 3 or greater than 9."

(b) By how much does your answer differ from the *theoretical* probability of Exercise 19?

For Exercises 35–38, let A be an event within the sample space S, and let $n(A) = a$ *and* $n(S) = s$.

35. Use the complements principle of counting to evaluate $n(A')$.

36. Use the theoretical probability formula to express $P(A)$ and $P(A')$.

37. Evaluate, and simplify, $P(A) + P(A')$.

38. What rule have you proved?

Suppose we want to form three-digit numbers using the set of digits $\{0, 1, 2, 3, 4, 5\}$. *For example,* 501 *and* 224 *are such numbers but* 035 *is not.*

39. How many such numbers are possible?

40. How many of these numbers are multiples of 5?

41. If one three-digit number is chosen at random from all those that can be made from the above set of digits, find the probability that the one chosen is not a multiple of 5.

42. An experiment consists of spinning both spinners shown here and multiplying the resulting numbers together. Find the probability that the resulting product will be even.

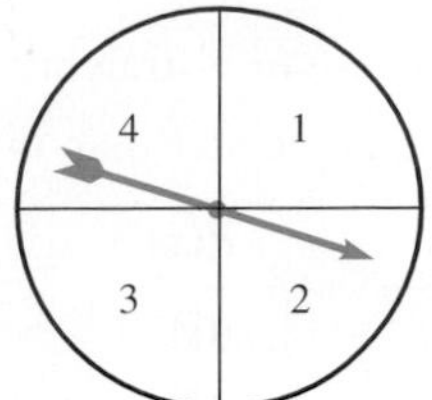

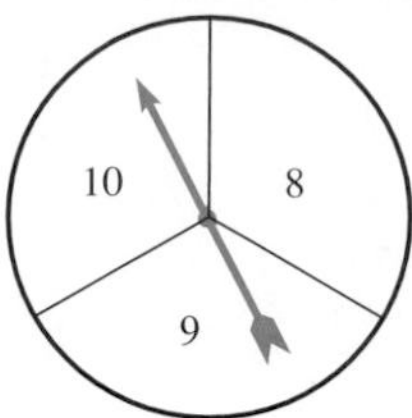

43. A bag contains fifty blue and fifty green marbles. Two marbles at a time are randomly selected. If they are both green, they are placed in box A; if both blue, in box B; if one is green and the other blue, in box C. After all marbles are drawn, what is the probability that the numbers of marbles in box A and box B are the same? (This problem is borrowed with permission from the December 1992 issue of *Mathematics Teacher,* page 736.)

11.3 Events Involving "And"

Several years ago, when Mattel Toys marketed the first talking Barbie doll to be released since the 1970s, they were not prepared for the controversy that resulted. Some of the Barbies were programmed to say "Math class is tough." The National Council of Teachers of Mathematics (NCTM), the American Association of University Women (AAUW), and numerous consumers voiced their complaints about the damage such a message could do to the self-confidence of children and to their attitudes toward school and mathematics.

Mattel subsequently agreed to erase the phrase from the microchip to be used in future doll production, and offered unhappy consumers the option to exchange the Barbie for one without the offensive message.

Incidentally, each Barbie was programmed to say four different statements, randomly selected from a pool of 270

continued on page 639

In the previous section, we developed rules for finding the probability of an event of the form *A or B*. Basically, we add the probabilities of event *A* and event *B*. But Example 3 in that section showed us that we must sometimes adjust that sum by subtracting the probability of the event *A and B*. The adjustment is necessary whenever *A* and *B* are not mutually exclusive.

We now consider how, in general, to find the probability of any event of the form *A and B*.

EXAMPLE 1 If one number is selected randomly from the set {1, 2, 3, 4, 5, 6, 7, 8, 9, 10}, find the probability that it will be odd and a multiple of 3.

As in Example 3 of the previous section,

$$S = \{1, 2, 3, 4, 5, 6, 7, 8, 9, 10\}, \quad A = \{1, 3, 5, 7, 9\}, \quad \text{and} \quad B = \{3, 6, 9\}.$$

The composite event *A and B* corresponds to the set $A \cap B = \{3, 9\}$. So, by the theoretical probability formula,

$$P(A \text{ and } B) = \frac{2}{10} = \frac{1}{5}.$$

The general formula for the probability of event *A and B* will require multiplying the individual probabilities of event *A* and event *B*. But, as Example 1 shows, this must be done carefully. Although $P(A) = 5/10$ and $P(B) = 3/10$, simply multiplying these two numbers would have given

$$\frac{5}{10} \cdot \frac{3}{10} = \frac{15}{100} = \frac{3}{20}, \quad \text{which is } wrong.$$

The correct procedure is to compute the probability of the second event on the assumption that the first event has happened (or is happening, or will happen—the timing is not important). This type of probability, computed assuming some special condition, is called a *conditional probability*.

prerecorded statements. Therefore, the probability of getting one that said "Math class is tough" was only

$$\frac{1 \cdot C(269, 3)}{C(270, 4)} \approx .015.$$

Some of the other messages included in the pool were "I love school, don't you?" "I'm studying to be a doctor," and "Let's study for the quiz."

Conditional Probability

The probability of event B, computed on the assumption that event A has happened, is called the **conditional probability of B, given A,** and is denoted $\mathbf{P(B|A)}$.

The assumed condition will normally reduce the sample space to a proper subset of what it was otherwise.

In Example 1 above, the assumption that the selected number is odd means that we then consider only the set $\{1, 3, 5, 7, 9\}$. Of these five elements, only two, namely 3 and 9, are multiples of 3. So, using a conditional probability as the second factor, we can write

$$P(A \text{ and } B) = P(A) \cdot P(B \mid A) = \frac{5}{10} \cdot \frac{2}{5} = \frac{1}{5}, \quad \text{which is correct.}$$

EXAMPLE 2 Given a family with two children, find the following.

(a) If it is known that at least one of the children is a girl, find the probability that both are girls. (Assume boys and girls are equally likely.)

The sample space for two children can be represented as $S = \{$gg, gb, bg, bb$\}$. In this case, the given condition that at least one child is a girl rules out the possibility bb, so the sample space is reduced to {gg, gb, bg}. The event "both girls" is satisfied by one of these three equally likely outcomes, so

$$P(\text{both girls} \mid \text{at least one girl}) = \frac{1}{3}.$$

(b) If the older child is a girl, find the probability that both children are girls.

Since the older child is a girl, the sample space reduces to {gg, gb}. (We assume the children are indicated in order of birth.) The event "both are girls" is satisfied by only one of the two elements in the reduced sample space, so

$$P(\text{both girls} \mid \text{older child a girl}) = \frac{1}{2}.$$

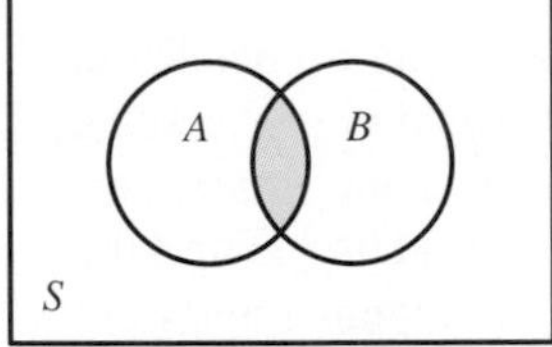

The logical connective "and" corresponds to "intersection" in set theory.

$P(A \text{ and } B) = P(A) \times P(B|A)$

FIGURE 8

Using a conditional probability, we can write the following *general multiplication rule of probability,* which is very similar to the fundamental counting principle. It is illustrated in Figure 8.

General Multiplication Rule of Probability

If A and B are any two events, then

$$P(A \text{ and } B) = P(A) \cdot P(B \mid A),$$

or

$$P(A \cap B) = P(A) \cdot P(B \mid A).$$

EXAMPLE 3 Each year, Roxanna Parker adds to her book collection a number of new publications that she believes will be of lasting value and interest. She has categorized each of her twenty 1997 acquisitions as hardcover or paperback and as fiction or nonfiction. The numbers of books in the various categories are shown in Table 4. If she randomly chooses one of these 20 books for this evening's reading, find the probability it will be each of the following types.

TABLE 4 1997 Books

	Fiction (*F*)	Nonfiction (*N*)	Totals
Hardcover (*H*)	3	5	8
Paperback (*P*)	8	4	12
Totals	11	9	20

(a) hardcover
Eight of the 20 books are hardcover, so $P(H) = 8/20 = 2/5$.

(b) fiction, given it is hardcover
The given condition that the book is hardcover reduces the sample space to eight books. Of those eight, just three are fiction. So $P(F \mid H) = 3/8$.

(c) hardcover and fiction
By the general multiplication rule of probability,

$$P(H \text{ and } F) = P(H) \cdot P(F \mid H) = \frac{2}{5} \cdot \frac{3}{8} = \frac{3}{20}.$$

In this case, part (c) is easier if we simply notice, directly from the table, that 3 of the 20 books are "hardcover and fiction." This verifies that the general multiplication rule of probability did give us the correct answer. ■

Who's out there? Carl Sagan has answered this by saying, "There must be other starfolk." Sagan, astronomer and exobiologist, has discussed the issue of life on other worlds in TV appearances and

continued on page 641

For Further Thought
The Birthday Problem

A classic problem (with a surprising result) involves the probability of a given group of people including at least one pair of people with the same birthday (the same day of the year, not necessarily the same year). This problem can be analyzed using the complements formula (Section 2) and the general multiplication rule of probability from this section. Suppose there are three people in the group. Then

P(at least one duplication of birthdays)

$= 1 - P$(no duplications) — Complements formula

$= 1 - P$(2nd is different than 1st and 3rd is different than 1st and 2nd)

$= 1 - \dfrac{364}{365} \cdot \dfrac{363}{365}$ — General multiplication rule

$\approx 1 - .992$

$= .008$

(To simplify the calculations, we have assumed 365 possible birth dates, ignoring February 29.)

in *The Cosmic Connection: An Extraterrestrial Perspective* (Dell paperback). Considering the great number of stars, some must have planets, and some of those planets must be conducive to life. In fact, some astronomers estimate the **odds against** life on Earth being the only life in the universe at one hundred billion billion to one. Also, the existence of intelligent life may not be unique to Earth. If intelligent beings are out there, some may have sent messages into space. Sagan urges us to maintain radio telescopes on a full-time basis for catching radio signals from outer space.

Not all experts expect eventual contact with extraterrestrial intelligence. For example, Freeman Dyson, a noted mathematical physicist and astronomer, says in his book, *Disturbing the Universe,* that after examining all the same evidence and arguments, he believes it is just as likely as not that there never was any other intelligent life out there. Dyson does agree, however, that we should keep looking.

By doing more calculations like the one above, we find that the smaller the group, the smaller the probability of a duplication; the larger the group, the larger the probability of a duplication. The table below shows the probability of at least one duplication for numbers of people through 50.

Number of People	Probability of at Least One Duplication	Number of People	Probability of at Least One Duplication
2	.003	27	.627
3	.008	28	.654
4	.016	29	.681
5	.027	30	.706
6	.040	31	.730
7	.056	32	.753
8	.074	33	.775
9	.095	34	.795
10	.117	35	.814
11	.141	36	.832
12	.167	37	.849
13	.194	38	.864
14	.223	39	.878
15	.253	40	.891
16	.284	41	.903
17	.315	42	.914
18	.347	43	.924
19	.379	44	.933
20	.411	45	.941
21	.444	46	.948
22	.476	47	.955
23	.507	48	.961
24	.538	49	.966
25	.569	50	.970
26	.598		

For Group Discussion

1. Based on the data shown in the table, what are the odds in favor of a duplication in a group of 30 people?
2. Estimate from the table the least number of people for which the probability of duplication is at least 1/2.
3. How small a group is required for the probability of a duplication to be *exactly* 0?
4. How large a group is required for the probability of a duplication to be *exactly* 1?

EXAMPLE 4 Suppose two cards are drawn successively, without replacement, from a standard 52-card deck. Find the probability that the first card is a queen and the second card is a face card.

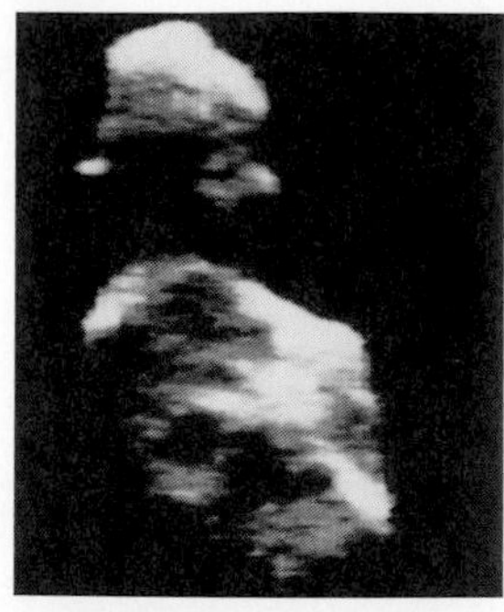

The NASA radar image above shows the two rocks that comprise the asteroid Toutatis, which flew within 2.2 million miles of Earth several years ago. A much smaller rock approached to within 106,000 miles barely two years earlier. In 1908, an asteroid actually entered the atmosphere, and disintegrated over Siberia, flattening nearly a thousand square miles of uninhabited forest.

Many craters on the planet's surface bear witness to impacts in earlier eras, including that of a 6- to 9-mile-wide asteroid on the Yucatan peninsula some 65 million years ago, which would likely have raised enough dust to darken the sky, chill the global climate, and thus destroy food plants and animals, including the dinosaurs.

A NASA report, based on empirical evidence, estimated a 1-in-10,000 chance that an asteroid measuring at least one-third of a mile wide will hit Earth "in our lifetime," destroying food crops and possibly ending civilization as we know it. An asteroid wider than 3 miles, colliding with Earth, could actually put the human species at risk of annihilation.

Let Q_1 denote queen on the first draw and let F_2 denote face card on the second draw. Then

$$P(Q_1 \text{ and } F_2) = P(Q_1) \cdot P(F_2 \mid Q_1) = \frac{4}{52} \cdot \frac{11}{51} = \frac{11}{663} \approx .017.$$

(Since the first card drawn was a queen, and is not replaced before the second card is drawn, the deck is then reduced to 51 cards, of which 11 are face cards. Therefore, $P(F_2 \mid Q_1) = 11/51$.)

In Example 4, since the drawing was done *without replacement*, our ability to evaluate the probability that the second card would be a face card depended upon knowing whether or not the first card was a face card. However, if we had drawn *with replacement*, returning the first card to the deck before drawing the second, the outcome on the second draw would not have depended upon what happened on the first draw. In that case, Q_1 and F_2 would be *independent* of one another. Independent events are defined generally as follows.

Independent Events

Two events A and B are called **independent** if knowledge about the occurrence of one of them has no effect on the probability of the other one, that is, if

$$P(B \mid A) = P(B).$$

The statement $P(A \mid B) = P(A)$ is equivalent to the one given in the definition. So, in practice, we can establish the independence of A and B by verifying either of the two equations.

EXAMPLE 5 A single card is drawn from a standard 52-card deck. Let B denote the event that the card is black, and let D denote the event that it is a diamond. Answer the following questions.

(a) Are events B and D independent?

For the unconditional probability of D, we get $P(D) = 13/52 = 1/4$. (Thirteen of the 52 cards are diamonds.) But for the conditional probability of D, given B, we have $P(D \mid B) = 0/26 = 0$. (None of the 26 black cards are diamonds.) Since the conditional probability $P(D \mid B)$ is different than the unconditional probability $P(D)$, B and D are not independent.

(b) Are events B and D mutually exclusive?

Mutually exclusive events, defined in the previous section, are events that cannot both occur for a given performance of an experiment. Since no card in the deck is both black and a diamond, B and D are mutually exclusive. (People sometimes get the idea that "mutually exclusive" and "independent" mean the same thing. But this example should convince you that this is not so.)

Whenever two events A and B are independent, conditional and unconditional probabilities will be no different. $P(B \mid A)$ will be the same as $P(B)$, so the condition can be ignored in calculating the probability. In this case, the general multiplication rule of probability reduces to the following special form.

Special Multiplication Rule of Probability

If A and B are independent events, then

$$P(A \textit{ and } B) = P(A) \cdot P(B),$$

or

$$P(A \cap B) = P(A) \cdot P(B).$$

EXAMPLE 6 If two cards are drawn *with replacement* from a standard 52-card deck, find the probability that the first card is a queen and the second card is a face card.

This is a repeat of Example 4, except that the drawing is done with replacement this time. Now, when the second card is drawn, all 52 cards are available, including all 12 face cards. Events Q_1 and F_2 are independent, and we use the special multiplication rule of probability.

$$P(Q_1 \textit{ and } F_2) = P(Q_1) \cdot P(F_2) = \frac{4}{52} \cdot \frac{12}{52} = \frac{3}{169} \approx .018,$$

a slightly different answer than in Example 4, where Q_1 and F_2 were not independent since the drawing was done without replacement. ■

The multiplication rules of probability, both the general and special forms, can be extended to cases where more than two events are involved.

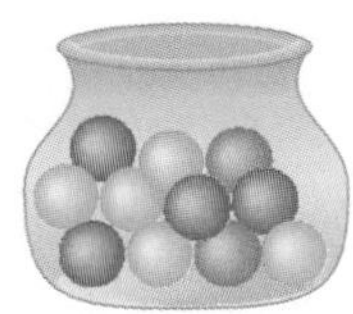

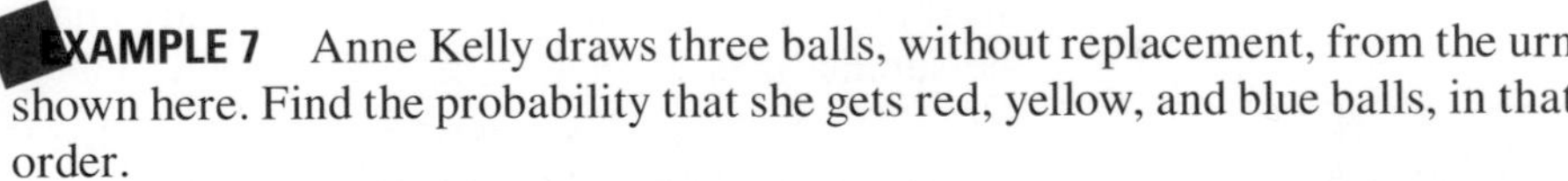

EXAMPLE 7 Anne Kelly draws three balls, without replacement, from the urn shown here. Find the probability that she gets red, yellow, and blue balls, in that order.

Using appropriate letters to denote the colors, and subscripts to indicate first, second, and third draws, the event we are after can be symbolized R_1 *and* Y_2 *and* B_3, so

$$\begin{aligned} P(R_1 \text{ and } Y_2 \text{ and } B_3) &= P(R_1) \cdot P(Y_2 \mid R_1) \cdot P(B_3 \mid R_1 \text{ and } Y_2) \\ &= \frac{4}{11} \cdot \frac{5}{10} \cdot \frac{2}{9} = \frac{4}{99} \approx .0404. \end{aligned}$$ ■

EXAMPLE 8 If five cards are drawn without replacement, find the probability that they are all hearts.

Each time a heart is drawn, the number of cards decreases by one and the number of hearts decreases by one. The probability of drawing only hearts is

$$\frac{13}{52} \cdot \frac{12}{51} \cdot \frac{11}{50} \cdot \frac{10}{49} \cdot \frac{9}{48} = \frac{33}{66{,}640} \approx .000495.$$

The probability found above can also be found by using the theoretical probability formula and combinations. The total possible number of 5-card hands, drawn without replacement, is $C(52, 5)$, and the number of those containing only hearts is $C(13, 5)$, so the required probability is

$$\frac{C(13, 5)}{C(52, 5)} = \frac{\dfrac{13!}{5!\,8!}}{\dfrac{52!}{5!\,47!}} \approx .000495. \blacksquare$$

Example 9 shows how some probability problems require the use of both addition and multiplication rules.

EXAMPLE 9 The local garage employs two mechanics, A and B, but you never know which mechanic will be working on your car. Your neighborhood consumer club has found that A does twice as many jobs as B, A does a good job three out of four times, and B does a good job only two out of five times. If you plan to take your car in for repairs, find the probability that a good job will be done.

Define the following events

A: work done by A

B: work done by B

G: good job done.

Since A does twice as many jobs as B, the (unconditional) probabilities of A and B are, respectively, $2/3$ and $1/3$. Since A does a good job three out of four times, the probability of a good job, given that A did the work, is $3/4$. And since B does well two out of five times, the probability of a good job, given that B did the work, is $2/5$. These last two probabilities are conditional. These four values can be summarized as $P(A) = 2/3$, $P(B) = 1/3$, $P(G \mid A) = 3/4$, and $P(G \mid B) = 2/5$. The event we are after, G, can occur in two mutually exclusive ways: A could do the work and do a good job $(A \cap G)$, or B could do the work and do a good job $(B \cap G)$. Therefore

$$\begin{aligned} P(G) &= P(A \cap G) + P(B \cap G) && \text{Special addition rule} \\ &= P(A) \cdot P(G \mid A) + P(B) \cdot P(G \mid B) && \text{General multiplication rule} \\ &= \frac{2}{3} \cdot \frac{3}{4} + \frac{1}{3} \cdot \frac{2}{5} \\ &= \frac{1}{2} + \frac{2}{15} = \frac{19}{30} \approx .633. \end{aligned}$$

The tree diagram in Figure 9 gives an alternate approach to finding this probability. Use the given information to draw the tree diagram, then find the probability of a good job by adding the probabilities from the indicated branches of the tree. ■

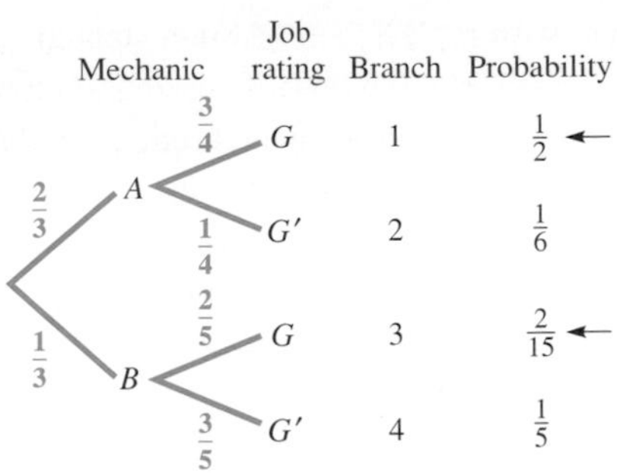

Garage mechanics experiment

FIGURE 9

EXAMPLE 10 Howard Shorthill is among five door prize winners at a Christmas party. The five winners are asked to choose, without looking, from a bag which, they are told, contains four candy canes and one $50 gift certificate. If Howard wants the gift certificate, can he improve his chance of getting it by drawing first among the five people?

We denote candy cane by C, gift certificate by G, and first draw, second draw, and so on by subscripts 1, 2, ... Then if Howard draws first, his probability of getting the gift certificate is

$$P(G_1) = \frac{1}{5}.$$

If he draws second, his probability of getting the gift certificate is

$$P(G_2) = P(C_1 \text{ and } G_2) = P(C_1) \cdot P(G_2 | C_1)$$
$$= \frac{4}{5} \cdot \frac{1}{4} = \frac{1}{5}, \quad \text{the same as before.}$$

For the third draw,

$$P(G_3) = P(C_1 \text{ and } C_2 \text{ and } G_3) = P(C_1) \cdot P(C_2 | C_1) \cdot P(G_3 | C_1 \text{ and } C_2)$$
$$= \frac{4}{5} \cdot \frac{3}{4} \cdot \frac{1}{3} = \frac{1}{5}, \quad \text{again the same value.}$$

Likewise, the probability of getting the gift certificate is $1/5$ when drawing fourth or when drawing fifth. Therefore, the order in which Howard draws does not affect his chances. ■

11.3 EXERCISES

For each of the following experiments, determine whether the two given events are independent.

1. A fair coin is tossed twice. The events are "head on the first" and "head on the second."
2. A pair of dice are rolled. The events are "even on the first" and "odd on the second."

3. Two cards are drawn, with replacement, from a standard 52-card deck. The events are "first card red" and "second card a club."

4. Two cards are drawn, without replacement, from a standard deck. The events are "first card an ace" and "second card a face card."

5. The answers are all guessed on a twenty-question multiple-choice test. The events are "first answer correct" and "last answer correct."

6. A committee of five is randomly selected from the 100 U.S. Senators. The events are "first member selected is a Republican" and "second member selected is a Republican." (Assume that there are both Republicans and non-Republicans in the Senate.)

One hundred college seniors attending a career fair at a major northeastern university were categorized according to sex and according to primary career motivation, as summarized here.

		Primary Career Motivation			
		Money	**Allowed to Be Creative**	**Sense of Giving to Society**	**Total**
Sex	**Male**	15	22	18	55
	Female	10	15	20	45
	Total	25	37	38	100

If one of these students is to be selected at random, find the probability that the student selected will satisfy each of the following conditions.

7. female

8. motivated primarily by creativity

9. not motivated primarily by money

10. male and motivated primarily by money

11. male, given that primary motivation is a sense of giving to society

12. motivated primarily by money or creativity, given that the student is a female

---

Suppose two cards are drawn with replacement *from a standard 52-card deck. Find the probability of each of the following events.*

13. a red card on the first draw, and a heart on the second

14. both red and neither one a heart

15. a 5 on the first draw, and an ace on the second

16. one 5 and one ace

Let two cards be dealt successively, without replacement, *from a standard 52-card deck. Find the probability of each of the following events (Exercises 17–27).*

17. spade second, given spade first

18. club second, given diamond first

19. ace second, given face card first

20. two face cards

21. spade first, then red card

22. a jack, then a queen

23. one jack and one queen

24. no face cards

25. The first card dealt is a jack and the second is a face card.

26. The first card dealt is a face card and the second is a jack.

27. The first card dealt is a club and the second is a king.

28. Given events A and B within the sample space S, the following sequence of steps establishes formulas that can be used to compute conditional probabilities. (The formula of step (b) involves probabilities, while that of step (d) involves cardinal numbers.) Justify each statement.

(a) $P(A \text{ and } B) = P(A) \cdot P(B \mid A)$

(b) Therefore, $P(B \mid A) = \dfrac{P(A \text{ and } B)}{P(A)}$.

(c) Therefore, $P(B \mid A) = \dfrac{n(A \text{ and } B)/n(S)}{n(A)/n(S)}$.

(d) Therefore, $P(B \mid A) = \dfrac{n(A \text{ and } B)}{n(A)}$.

Use the results of Exercise 28 to find each of the following probabilities when a single card is drawn from a standard 52-card deck.

29. $P(\text{queen} \mid \text{face card})$

30. $P(\text{face card} \mid \text{queen})$

31. $P(\text{red} \mid \text{diamond})$

32. $P(\text{diamond} \mid \text{red})$

33. If one number is chosen randomly from the integers 1 through 10, the probability of getting a number that is *odd and prime,* by the general multiplication rule, is

$$P(\text{odd}) \cdot P(\text{prime} \mid \text{odd}) = \frac{5}{10} \cdot \frac{3}{5} = \frac{3}{10}.$$

Compute, and compare, the product $P(\text{prime}) \cdot P(\text{odd} \mid \text{prime})$.

34. What does Exercise 33 imply, in general, about the probability of an event of the form *A and B*?

Sarah Devaney manages a discount supermarket which encourages volume shopping on the part of its customers. Sarah has discovered that, on any given weekday, 60 percent of the market's sales amount to more than \$50. That is, any given sale on such a day has a probability of .60 of being for more than \$50. Find the probability of each of the following events. (Give answers to two decimal places.)

35. The first two sales on Wednesday are both for more than \$50.

36. The first three sales on Wednesday are all for more than \$50.

37. None of the first three sales on Wednesday is for more than \$50.

38. Exactly one of the first three sales on Wednesday is for more than \$50.

One problem encountered by developers of the space shuttle program is air pollution in the area surrounding the launch site. A certain direction from the launch site is considered critical in terms of hydrogen chloride pollution from the exhaust cloud. It has been determined that weather conditions would cause emission cloud movement in the critical direction only 5% of the time.

Find probabilities for the following events (Exercises 39–42), assuming that probabilities for a particular launch in no way depend on the probabilities for other launches. (Give answers to two decimal places.)

39. A given launch will not result in cloud movement in the critical direction.

40. No cloud movement in the critical direction will occur during any of 5 launches.

41. Any 5 launches will result in at least one cloud movement in the critical direction.

42. Any 10 launches will result in at least one cloud movement in the critical direction.

43. One of the authors of this text has three sons and no daughters. Find the probability that a couple having three children will have all boys.

44. In Example 7, where Anne draws three balls without replacement, what would be her probability of getting one of each color, where the order does not matter?

Five men and three women are waiting to be interviewed for jobs. If they are all selected in random order, find the probability of each of the following events.

45. All the women will be interviewed first.

46. All the men will be interviewed first.

47. The first person interviewed will be a woman.

48. The second person interviewed will be a woman.

49. The last person interviewed will be a woman.

Let A and B be events, neither one certain and neither one impossible. For each statement in Exercises 50 and 51, **(a)** *determine whether it is true, and* **(b)** *explain why or why not.*

50. If A and B are mutually exclusive, then A and B are also independent.

51. If A and B are independent, then A and B are also mutually exclusive.

52. Roll a pair of dice until a sum of seven appears, keeping track of how many rolls it took. Repeat the process a total of 50 times, each time recording the number of rolls it took to get a sum of seven.

(a) Use your experimental data to compute an empirical probability (to two decimal places) that it would take at least three rolls to get a sum of seven.

(b) Find the theoretical probability (to two decimal places) that it would take at least three rolls to obtain a sum of seven.

53. Assuming boy and girl babies are equally likely, find the probability that it would take
(**a**) at least three births to obtain two girls,
(**b**) at least four births to obtain two girls,
(**c**) at least five births to obtain two girls.

54. Cards are drawn, without replacement, from an ordinary 52-card deck.
(**a**) How many must be drawn before the probability of obtaining at least one face card is greater than 1/2?
(**b**) How many must be drawn before the probability of obtaining at least one king is greater than 1/2?

Many everyday decisions, like who will drive to lunch, or who will pay for the coffee, are made by the toss of a (presumably fair) coin and using the criterion "heads, you will; tails, I will." This criterion is not quite fair, however, if the coin is biased (perhaps due to slightly irregular construction or wear). John von Neumann suggested a way to make perfectly fair decisions even with a possibly biased coin. If a coin, biased so that $P(\text{h}) = .5200$ *and* $P(\text{t}) = .4800$, *is tossed twice, find the following probabilities. (Give answers to four decimal places.)*

55. $P(\text{hh})$ **56.** $P(\text{ht})$ **57.** $P(\text{th})$ **58.** $P(\text{tt})$

59. Having completed Exercises 55–58, what do you think was von Neumann's scheme?

A certain brand of automatic garage door opener utilizes a transmitter control with six independent switches, each one set on or off. The receiver (wired to the door) must be set with the same pattern as the transmitter. Exercises 60–63 are based on ideas similar to those of the "birthday problem" in the "For Further Thought" feature in this section.*

60. How many different ways can the owner of one of these garage door openers set the switches?

61. If two residents in the same neighborhood each have one of this brand of opener, and both set the switches randomly, what is the probability to four decimal places that they will be opening each other's garage doors?

62. If five neighbors with the same type of opener set their switches independently, what is the probability of at least one pair of neighbors using the same settings? (Give your answer to four decimal places.)

63. What is the minimum number of neighbors who must use this brand of opener before the probability of at least one duplication of settings is greater than 1/2?

64. There are three cards, one that is green on both sides, one that is red on both sides, and one that is green on one side and red on the other. One of the three cards is selected randomly and laid on the table, with a red side up. What is the probability that it is also red on the other side?

In November, the rain in a certain valley tends to fall in storms of several days' duration. The unconditional probability of rain on any given day of the month is 0.500. *But the probability of rain on a day that follows a rainy day is* 0.700, *and the probability of rain on a day following a nonrainy day is* 0.400. *Find the probability of each of the following events. Give answers to three decimal places.*

65. rain on two consecutive days in November

66. rain on three consecutive days in November

67. rain on November 1 and 2, but not on November 3

68. rain on the first four days of November, given that October 31 was clear all day

69. rain throughout the first week of November, given that there was no rain on October 31

70. no rain in the first week of November, given that there was rain on October 31

In a certain four-engine vintage aircraft, now quite unreliable, each engine has a 10 percent chance of failure on any flight, as long as it is carrying its one-fourth share of the load. But if one engine fails, then the chance of failure increases to 20 percent for each of the other three engines. And if a second engine fails, each of the remaining two has a 30 percent chance of failure. Assuming that no two engines ever fail simultaneously, and that the aircraft can continue flying with as few as two operating engines, find each of the following for a given flight of this aircraft. (Give answers to four decimal places.)

71. the probability of no engine failures

72. the probability of exactly one engine failure

73. the probability of exactly two engine failures

74. the probability of a failed flight

* For more information, see "Matching Garage-Door Openers," by Bonnie H. Litwiller and David R. Duncan in the March 1992 issue of *Mathematics Teacher*, page 217.

In basketball "one-and-one" foul shooting is done as follows: if the player makes the first shot (1 point), he is given a second shot. If he misses the first shot, he is not given a second shot (see the tree diagram).

Karin Wagner, a basketball player, has a 60% foul shot record. (She makes 60% of her foul shots.) Find the probability that, on a given one-and-one foul shooting opportunity, Karin will score the following numbers of points.

First shot	Second shot	Branch	Total points
Point	Point	1	2
	No point	2	1
No point		3	0

75. no points **76.** one point **77.** two points

Binomial Probability

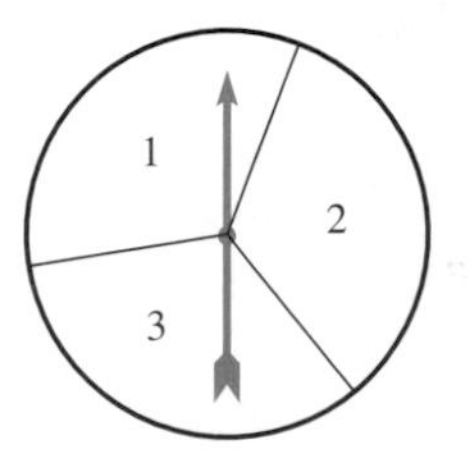

Suppose the spinner in the margin is spun twice, where we are interested in the number of times a 2 is obtained. (Assume each of the three regions contains a 120-degree arc, so that each is equally likely.) We can think of the outcome 2 as a "success," while outcomes 1 and 3 would be "failure." When the outcomes of an experiment are divided into just two categories, success and failure, the associated probabilities are called "binomial" (the prefix *bi* meaning *two*). Repeated performances of such an experiment, where the probability of success remains constant throughout all repetitions, are also known as repeated **Bernoulli trials** (after James Bernoulli).

If we let x denote the number of 2s obtained out of two spins, then x is a random variable, and we can construct its probability distribution. (Random variables and probability distributions were defined at the end of the second section of this chapter.) Apparently, x could possibly have values of 0, 1, or 2. Let's begin by listing the sample space (of equally likely outcomes) for the two spins:

$$S = \{(1, 1), (1, 2), (1, 3), (2, 1), (2, 2), (2, 3), (3, 1), (3, 2), (3, 3)\}.$$

James Bernoulli (1654–1705) is also known as Jacob or Jacques. He was charmed away from theology by the writings of Leibniz, became his pupil, and later headed the mathematics faculty at the University of Basel. His results in probability are contained in the *Art of Conjecture*, published in 1713, after his death. He also made many contributions to calculus and analytic geometry.

In each ordered pair here, the first and second entries give the outcomes on the first and second spins, respectively. There are nine outcomes, the number of 2s being 0 in four cases, 1 in four cases, and 2 in one case. From this information we construct the probability distribution in Table 5. In this particular case, we have an example of a **binomial probability distribution.** Notice that the sum of the probability column is 1, which agrees with property 3 of probability (the second section of this chapter), since all possible values of x have been listed.

In order to develop a general formula for binomial probabilities, we can consider another way to obtain the probability values in Table 5. The various spins of the spinner are independent of one another, and on each spin the probability of success (S) is $1/3$ and the probability of failure (F) is $2/3$. We will denote success on the first spin by S_1, failure on the second by F_2, and so on. Then, using the rules of probability, we have

$$\begin{aligned} P(x = 0) &= P(F_1 \text{ and } F_2) \\ &= P(F_1) \cdot P(F_2) \qquad \text{Special multiplication rule} \\ &= (2/3) \cdot (2/3) \\ &= 4/9, \end{aligned}$$

TABLE 5 Probability Distribution for Two Spins

x	$P(x)$
0	4/9
1	4/9
2	1/9
Sum = 9/9 = 1	

$$
\begin{aligned}
P(x = 1) &= P[(S_1 \text{ and } F_2) \text{ or } (F_1 \text{ and } S_2)] \\
&= P(S_1 \text{ and } F_2) + P(F_1 \text{ and } S_2) && \text{Special addition rule} \\
&= P(S_1) \cdot P(F_2) + P(F_1) \cdot P(S_2) && \text{Special multiplication rule} \\
&= (1/3) \cdot (2/3) + (2/3) \cdot (1/3) \\
&= 2/9 + 2/9 \\
&= 4/9,
\end{aligned}
$$

and

$$
\begin{aligned}
P(x = 2) &= P(S_1 \text{ and } S_2) \\
&= P(S_1) \cdot P(S_2) && \text{Special multiplication rule} \\
&= (1/3) \cdot (1/3) \\
&= 1/9.
\end{aligned}
$$

Notice the following pattern in the above calculations. There is only one way to get $x = 0$ (namely, F_1 and F_2). And there is only one way to get $x = 2$ (namely, S_1 and S_2). But there are two ways to get $x = 1$. One way is S_1 and F_2; the other is F_1 and S_2. There are two ways because the one success required can occur on the first spin or on the second spin. How many ways can exactly one success occur in two repeated trials? This question is equivalent to: How many size-one subsets are there of the set of two trials? The answer is $C(2, 1) = 2$. (The expression $C(2, 1)$ denotes "combinations of 2 things taken 1 at a time.") Each of the two ways to get exactly one success has a probability equal to $(1/3) \cdot (2/3)$, the probability of success times the probability of failure.

TABLE 6 Probability Distribution for Three Spins

x	$P(x)$
0	8/27
1	12/27
2	6/27
3	1/27
Sum = 27/27 = 1	

If the same spinner is spun three times rather than two, then x, the number of successes (2s) could have values of 0, 1, 2, or 3. Then the number of ways to get exactly 1 success is $C(3, 1) = 3$. They are S_1 and F_2 and F_3, F_1 and S_2 and F_3, and F_1 and F_2 and S_3. The probability of each of these three ways is $(1/3) \cdot (2/3) \cdot (2/3) = 4/27$. So $P(x = 1) = 3 \cdot (4/27) = 12/27 = 4/9$. The tree diagram in Figure 10 shows all possibilities for three spins, and Table 6 gives the associated probability distribution. (Notice in the tree diagram that the number of ways of getting two successes in three trials is also 3, in agreement

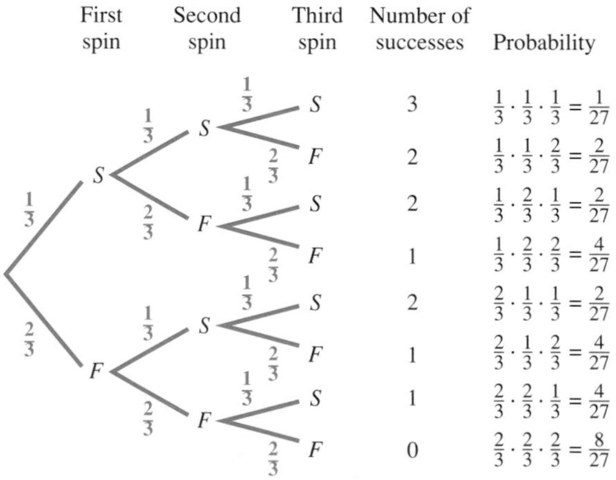

Tree diagram for three spins

FIGURE 10

with the fact that $C(3, 2) = 3$.) Notice also that, again, the sum of the $P(x)$ column is 1.

The pattern seen in these experiments can now be generalized to any binomial experiment. Let

n = the number of repeated trials,

p = the probability of success on any given trial,

$q = 1 - p$ = the probability of failure on any given trial,

and x = the number of successes that occur.

Note that p remains fixed throughout all n trials. This means that all trials are independent of one another. The random variable x (number of successes) can have any integer value from 0 through n. In general, x successes can be assigned among n repeated trials in $C(n, x)$ different ways, since this is the number of different subsets of x positions among a set of n positions. Also, regardless of which x of the trials result in successes, there will always be x successes and $n - x$ failures, so we multiply x number of ps and $n - x$ number of qs together.

All of this discussion is summarized in the following statement.

Binomial Probability Formula

When n independent repeated trials occur, where

p = probability of success and q = probability of failure

with p and q ($= 1 - p$) remaining constant throughout all n trials, the probability of exactly x successes is given by

$$P(x) = C(n, x)p^x q^{n-x}$$

Tables of binomial probability values are commonly available in statistics texts, for various values of p, often for n ranging up to about 20. Also, computer software packages for statistics will usually do these calculations for you automatically. For our purposes here, however, we will utilize the formula stated above. That way we can use any whole number value of n and any value of p from 0 through 1.

EXAMPLE 1 Find the probability of obtaining exactly three heads on five tosses of a fair coin.

Let heads be "success." Then this is a binomial experiment with $n = 5$, $p = 1/2$, $q = 1/2$, and $x = 3$. By the binomial probability formula,

$$P(3) = C(5, 3)\left(\frac{1}{2}\right)^3\left(\frac{1}{2}\right)^2 = 10 \cdot \frac{1}{8} \cdot \frac{1}{4} = \frac{5}{16}.$$

EXAMPLE 2 Find the probability of obtaining exactly two 5s on six rolls of a fair die.

Let 5 be "success." Then $n = 6$, $p = 1/6$, $q = 5/6$, and $x = 2$.

$$P(2) = C(6, 2)\left(\frac{1}{6}\right)^2\left(\frac{5}{6}\right)^4 = 15 \cdot \frac{1}{36} \cdot \frac{625}{1{,}296} = \frac{3{,}125}{15{,}552} \approx .201.$$

In the case of repeated independent trials, when an event involves more than one specific number of successes, we can employ the binomial probability formula along with the complements or addition rules.

EXAMPLE 3 A couple plans to have 5 children. Find the probability they will have more than 3 girls. (Assume girl and boy babies are equally likely.)

Let a girl be "success." Then $n = 5$, $p = q = 1/2$, and $x > 3$.

$$\begin{aligned} P(x > 3) &= P(x = 4 \text{ or } 5) \\ &= P(4) + P(5) \\ &= C(5, 4)\left(\frac{1}{2}\right)^4\left(\frac{1}{2}\right)^1 + C(5, 5)\left(\frac{1}{2}\right)^5\left(\frac{1}{2}\right)^0 \\ &= 5 \cdot \frac{1}{16} \cdot \frac{1}{2} + 1 \cdot \frac{1}{32} \cdot 1 \\ &= \frac{5}{32} + \frac{1}{32} = \frac{6}{32} = \frac{3}{16} = .1875. \end{aligned}$$

EXAMPLE 4 Rudy Preciado, a baseball player, has a well-established career batting average of .300. In a brief series with a rival team, Rudy will bat 10 times. Find the probability that he will get more than two hits in the series.

This "experiment" involves $n = 10$ repeated Bernoulli trials, with probability of success (a hit) given by $p = .3$ (which implies $q = 1 - .3 = .7$). Since, in this case, "more than 2" means "3 or 4 or 5 or 6 or 7 or 8 or 9 or 10" (which is eight different possibilities), it will be less work to apply the complements rule.

$$\begin{aligned} P(x > 2) &= 1 - P(x \leq 2) && \text{Complements rule} \\ &= 1 - P(x = 0 \text{ or } 1 \text{ or } 2) && \text{Only three different possibilities} \\ &= 1 - [P(0) + P(1) + P(2)] && \text{Special addition rule} \\ &= 1 - [C(10, 0)(.3)^0(.7)^{10} && \text{Binomial probability formula} \\ &\quad + C(10, 1)(.3)^1(.7)^9 + C(10, 2)(.3)^2(.7)^8] \\ &= 1 - [.0282 + .1211 + .2335] \\ &= 1 - .3828 \\ &= .6172. \end{aligned}$$

11.4 EXERCISES

For Exercises 1–24, give all numerical answers as common fractions reduced to lowest terms. For Exercises 25–59, give all numerical answers to three decimal places.

If three fair coins are tossed, find the probability of each of the following numbers of heads.

1. 0 **2.** 1 **3.** 2 **4.** 3

5. 1 or 2 **6.** at least 1 **7.** no more than 1 **8.** fewer than 3

9. Assuming boy and girl babies are equally likely, find the probability that a family with three children will have exactly two boys.

10. Pascal's triangle was shown in Table 5 of Section 4 in the previous chapter. Explain how the probabilities in Exercises 1–4 above relate to row 3 of the "triangle." (Recall that we referred to the topmost row of the triangle as "row number 0," and to the leftmost entry of each row as "entry number 0.")

11. Generalize the pattern of Exercise 10 to complete the following statement. If n fair coins are tossed, the probability of exactly x heads is the fraction whose numerator is entry number ______ of row number ______ in Pascal's triangle, and whose denominator is the sum of row number ______.

Use the pattern noted in Exercises 10 and 11 to find the probabilities of the following numbers of heads when seven fair coins are tossed.

12. 0 **13.** 1 **14.** 2 **15.** 3 **16.** 4 **17.** 5 **18.** 6 **19.** 7

A fair die is rolled three times. A 4 is considered "success," while all other outcomes are "failures." Find the probability of each of the following numbers of successes.

20. 0 **21.** 1 **22.** 2 **23.** 3

24. Exercises 10 and 11 established a way of using Pascal's triangle rather than the binomial probability formula to find probabilities of different numbers of successes in coin tossing experiments. Explain why the same process would not work for Exercises 20–23.

In the remaining exercises of this section, give all numerical answers to three decimal places.

For n repeated independent trials, with constant probability of success p for all trials, find the probability of exactly x successes in each of the following cases.

25. $n = 5, \quad p = 1/3, \quad x = 4$

26. $n = 10, \quad p = .7, \quad x = 5$

27. $n = 20, \quad p = 1/8, \quad x = 2$

28. $n = 30, \quad p = .6, \quad x = 22$

For Exercises 29–31, refer to Example 4 in the text.

29. Does Rudy's probability of a hit really remain constant at exactly .300 through all ten times at bat? Explain your reasoning.

30. If Rudy's batting average is exactly .300 going into the series, and that value is based on exactly 1,200 career hits out of 4,000 previous times at bat, what is the greatest his average could possibly be (to three decimal places) when he goes up to bat the tenth time of the series? What is the least his average could possibly be when he goes up to bat the tenth time of the series?

31. Do you think the use of the binomial probability formula was justified in Example 4, even though p is not strictly constant? Explain your reasoning.

Robin Strang is taking a ten-question multiple-choice test for which each question has three answer choices, only one of which is correct. Robin decides on answers by rolling a fair die and marking the first answer choice if the die shows 1 or 2, the second if it shows 3 or 4, and the third if it shows 5 or 6. Find the probability of each of the following events.

32. exactly four correct answers

33. exactly seven correct answers

34. fewer than three correct answers

35. at least seven correct answers

It is known that a certain prescription drug produces undesirable side effects in 30 percent of all patients who use it. Among a random sample of eight patients using the drug, find the probability of each of the following events.

36. None have undesirable side effects.

37. Exactly one has undesirable side effects.

38. Exactly two have undesirable side effects.

39. More than two have undesirable side effects.

In a certain state, it has been shown that only 50 percent of the high school graduates who are capable of college work actually enroll in colleges. Find the probability that, among nine capable high school graduates in this state, each of the following numbers will enroll in college (Exercises 40–43).

40. exactly 4 **41.** from 4 through 6

42. none **43.** all 9

44. At a large midwestern university, 40 percent of all students have their own personal computers. If five students at that university are selected at random, find the probability that exactly three of them have their own computers.

45. If it is known that 65 percent of all orange trees will survive a hard frost, then what is the probability that at least half of a group of six trees will survive such a frost?

46. An extensive survey revealed that, during a certain Presidential election campaign, 44 percent of the political columns in a certain group of major newspapers were favorable to the incumbent President. If a sample of fifteen of these columns is selected at random, what is the probability that exactly six of them will be favorable?

In the case of n independent repeated Bernoulli trials, the formula developed in this section gives the probability of exactly x successes. Sometimes, however, we are interested not in the event that exactly x successes will occur in n trials, but rather the event that the first success will occur on the xth trial. For example, consider the probability that, in a series of coin tosses, the first success (head) will occur on the fourth toss. This implies a failure first, then a second failure, then a third failure, and finally a success. Symbolically, the event is F_1 *and* F_2 *and* F_3 *and* S_4. *The probability of this sequence of outcomes is* $q \cdot q \cdot q \cdot p$, or $q^3 \cdot p$. *In general, if the probability of success stays constant at p (which implies a probability of failure of* $q = 1 - p$*), then the probability that the first success will occur on the xth trial can be computed as follows.*

$$P(F_1 \text{ and } F_2 \text{ and } \ldots \text{ and } F_{x-1} \text{ and } S_x) = q^{x-1} \cdot p.$$

47. Explain why, in the formula above, there is no combination factor, such as the $C(n, x)$ in the binomial probability formula.

48. Assuming male and female babies are equally likely, find the probability that a family's fourth child will be their first daughter.

49. If the probability of getting caught when you exceed the speed limit on a certain stretch of highway is .35, find the probability that the first time you will get caught is the fifth time that you exceed the speed limit.

50. If 20 percent of all workers in a certain industry are union members, and workers in this industry are selected successively at random, find the probability that the first union member to occur will be on the sixth selection.

51. If a certain type of rocket always has a three percent chance of an aborted launching, find the probability that the first launch to be aborted is the 25th launch.

Harvey is standing on the corner tossing a coin. He decides he will toss it 10 times, each time walking 1 block north if it lands heads up, and 1 block south if it lands tails up. In each of the following exercises, find the probability that he will end up in the indicated location. (In each case, ask how many successes, say heads, would be required and use the binomial formula. Some ending positions may not be possible with 10 tosses.) The random process involved here illustrates what we call a **random walk.** *It is a simplified model of Brownian motion, mentioned in the chapter introduction. Further applications of the idea of a random walk are found in the section on simulation at the end of this chapter.*

52. 10 blocks north of his corner

53. 6 blocks north of his corner

54. 6 blocks south of his corner

55. 5 blocks south of his corner

56. 2 blocks north of his corner

57. at least 2 blocks north of his corner

58. at least 2 blocks from his corner

59. on his corner

Expected Value

x	$P(x)$
1	.05
2	.10
3	.20
4	.40
5	.10
6	.15

The probability distribution at the side, from Example 6 in the second section of this chapter, shows the probabilities assigned by Ann to the various lengths of time her homework may take on a given night. If Ann's friend Fred asks her how many hours her studies will take, what would be her best guess? Six different time values are possible, with some more likely than others. One thing

Ann could do is to calculate a "weighted average" by multiplying each possible time value by its probability and then adding the six products.

$$1(.05) + 2(.10) + 3(.20) + 4(.40) + 5(.10) + 6(.15)$$
$$= .05 + .20 + .60 + 1.60 + .50 + .90 = 3.85.$$

Thus 3.85 hours is the **expected value** (or the *mathematical expectation*) of the quantity of time to be spent. A general definition of expected value follows.

Expected Value

If a random variable x can have any of the values $x_1, x_2, x_3, \ldots, x_n$, and the corresponding probabilities of these values occurring are $P(x_1), P(x_2), P(x_3), \ldots, P(x_n)$, then the expected value of x is

$$x_1 \cdot P(x_1) + x_2 \cdot P(x_2) + x_3 \cdot P(x_3) + \cdots + x_n \cdot P(x_n).$$

EXAMPLE 1 Find the expected number of boys for a three-child family (that is, the expected value of the number of boys).

The sample space for this experiment is

$$S = \{\text{ggg, ggb, gbg, gbb, bgg, bgb, bbg, bbb}\}.$$

The probability distribution is shown in the table below, along with the products and their sum, which gives the expected value.

Number of boys, x	Probability, $P(x)$	Product, $x \cdot P(x)$
0	$\frac{1}{8}$	0
1	$\frac{3}{8}$	$\frac{3}{8}$
2	$\frac{3}{8}$	$\frac{6}{8}$
3	$\frac{1}{8}$	$\frac{3}{8}$
	Expected value:	$\frac{12}{8} = \frac{3}{2}$

The expected number of boys is 3/2, or 1.5. This result should agree with your intuition. Since boys and girls are equally likely, "half" the children are expected to be boys. ■

Notice that the expected value for the number of boys in the family is itself an impossible value. So, as in this example, the expected value itself may never occur. Many times the expected value actually cannot occur; it is only a kind of long run average of the various values that could occur. (See a discussion of statistics for a more detailed coverage of averages.) If we recorded the number of boys in many different three-child families, then by the law of large numbers the greater the number of families observed, the closer the observed average should be to the expected value.

The idea of expected value can be important in analyzing games, as illustrated in Examples 2 through 4.

EXAMPLE 2 A player pays \$1 for the privilege of rolling a single fair die once. If he rolls a 6, he receives a "payoff" of \$5 (*net winnings* would be \$4, which is payoff minus cost to play). If he rolls anything other than 6, he receives nothing back (*net winnings:* payoff minus cost is $\$0 - \$1 = -\$1$). Find the expected net winnings of this game.

List the given information in a table.

Number Rolled	Payoff	Net Winnings	Probability	Product
6	\$5	\$4	$\frac{1}{6}$	$\$\frac{4}{6}$
1–5	\$0	$-\$1$	$\frac{5}{6}$	$-\$\frac{5}{6}$
			Expected value:	$-\$\frac{1}{6}$

The expected net winnings are

$$-\$\frac{1}{6} \approx -17¢.$$

The player would not lose 17¢ on any single play. But, if he plays this game repeatedly, then, in the long run, he should expect to lose about 17¢ per play *on the average.* If he played 100 times, he would win sometimes and lose sometimes but should come out in the end about $100 \cdot (17¢) = \$17$ in the hole. ■

A game in which the expected net winnings are zero is called a **fair game.** The game in Example 2 has negative expected net winnings, so it is unfair against the player. A game with positive expected net winnings is unfair in favor of the player.

EXAMPLE 3 In one state lottery, a player buys a card for \$1. The player marks any three digits on the card. Any three digits are acceptable, including (for example) 008 or 051. A player selecting the winning number receives a payoff of \$600. Find the expected net winnings for the player.

The probability of selecting all three digits correctly is

$$\frac{1}{10} \cdot \frac{1}{10} \cdot \frac{1}{10} = \frac{1}{1{,}000}, \qquad \text{and incorrectly is} \qquad 1 - \frac{1}{1{,}000} = \frac{999}{1{,}000}.$$

So the expected net winnings are

$$(\$599)\frac{1}{1{,}000} + (-\$1)\frac{999}{1{,}000} = -\$.40.$$

The expected loss is 40¢ per play. ■

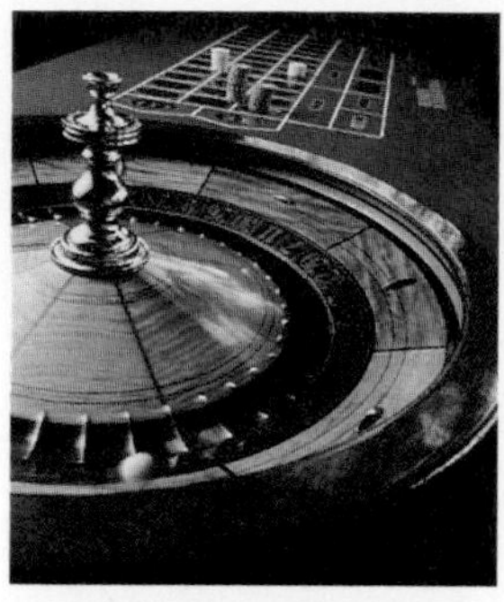

Roulette ("little wheel") was invented in France in the seventeenth or early eighteenth century. It has been a featured game of chance in the gambling casino of Monte Carlo.

You are looking straight down into a typical roulette wheel, the disk inside a shallow bowl. The disk is divided into red and black alternating compartments, numbered 1 to 36 (but not in that order). There is a compartment also for 0 (and for 00 in the United States). In roulette, the wheel is set in motion, and an ivory ball is thrown into the bowl opposite to the direction of the wheel. When the wheel stops, the ball comes to rest in one of the compartments—the number and color determine who wins.

The players bet against the banker (person in charge of the pool of money) by placing money or equivalent chips in spaces on the roulette table corresponding to the wheel's colors or numbers. Bets can be made on one number or several, on odd or even, on red or black, or on combinations. The banker pays off according to the odds against the particular bet(s). For example, the classic payoff for a winning single number is $36 for each $1 bet.

Games in a gambling casino are usually set up so that they are (at least slightly) unfair in favor of the house.

EXAMPLE 4 One simple type of *roulette* is played with an ivory ball and a wheel set in motion. The wheel contains thirty-eight compartments. Eighteen of the compartments are black, eighteen red, one is labeled "zero," and one is labeled "double zero." (These last two are neither black nor red.) In this case, assume the player places $1 on either red or black. If the player picks the correct color of the compartment in which the ball finally lands, the payoff is $2; otherwise the payoff is zero. Find the expected net winnings.

By the expected value formula, expected net winnings are

$$(\$1)\frac{18}{38} + (-\$1)\frac{20}{38} = -\$\frac{1}{19}.$$

The expected loss here is $1/19, or about 5.3¢, per play. ■

Expected values are often useful when decisions must be made in the midst of uncertain conditions, in business and other areas.

EXAMPLE 5 Patricia Gainey, a developer, wants to bid on a parcel of real estate. In light of uncertain city zoning ordinances, she figures that there is a .40 probability that she could make a profit of $150,000 on the land, and a .60 probability that her profit would be $80,000. What is Patricia's expected profit if she bids on the parcel?

The variable profit has two possible values. By the expected value formula, the expected profit is

$$(\$150{,}000) \cdot (.40) + (\$80{,}000) \cdot (.60) = \$108{,}000. \quad ■$$

In deciding whether to go ahead with a bid in Example 5, Patricia would most likely compare the expected profit value arrived at there with expected profits associated with other available investment options.

For Further Thought
Expected Value of Games of Chance

Slot machines are a popular game for those who want to lose their money with very little mental effort. We cannot calculate an expected value applicable to all slot machines since payoffs vary from machine to machine. But we can calculate the "typical expected value."

A player operates a slot machine by pulling a handle after inserting a coin. Three reels inside the machine then rotate, and come to rest in some random order. Assume that the three reels show the pictures listed in Table 7. For example, of the 20 pictures on the first reel, 2 are cherries, 5 are oranges, 5 are plums, 2 are bells, 2 are melons, 3 are bars, and 1 is the number 7.

For Further Thought
Expected Value of Games of Chance (Continued)

TABLE 7

	Reels		
Pictures	**1**	**2**	**3**
Cherries	2	5	4
Oranges	5	4	5
Plums	5	3	3
Bells	2	4	4
Melons	2	1	2
Bars	3	2	1
7s	1	1	1
Totals	20	20	20

A picture of cherries on the first reel, but not on the second, leads to a payoff of 3 coins (*net* winnings: 2 coins); a picture of cherries on the first two reels, but not the third, leads to a payoff of 5 coins (*net* winnings: 4 coins). All other possible payoffs are as shown in Table 8.

TABLE 8

Winning Combinations	Number of Ways	Probability	Number of Coins Received	Net Winnings (in coins)	Probability Times Winnings
1 cherry (on first reel)	$2 \times 15 \times 20 = 600$	600/8,000	3	2	1,200/8,000
2 cherries (on first two reels)	$2 \times 5 \times 16 = 160$	160/8,000	5	4	640/8,000
3 cherries	$2 \times 5 \times 4 = 40$	40/8,000	10	9	360/8,000
3 oranges	$5 \times 4 \times 5 = 100$	100/8,000	10	9	900/8,000
3 plums	$5 \times 3 \times 3 = 45$	45/8,000	14	13	585/8,000
3 bells	$2 \times 4 \times 4 = 32$	32/8,000	18	17	544/8,000
3 melons (jackpot)	$2 \times 1 \times 2 = 4$	4/8,000	100	99	396/8,000
3 bars (jackpot)	$3 \times 2 \times 1 = 6$	6/8,000	200	199	1,194/8,000
3 7s (jackpot)	$1 \times 1 \times 1 = 1$	1/8,000	500	499	499/8,000
Totals	988				6,318/8,000

Since, according to Table 8, there are 2 ways of getting cherries on the first reel, 15 ways of *not* getting cherries on the second reel, and 20 ways of getting anything on the third reel, we have a total of $2 \times 15 \times 20 = 600$ ways

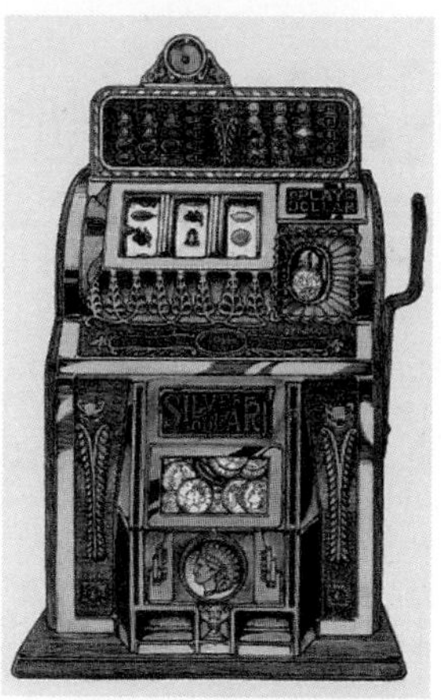

The first Silver Dollar Slot Machine was fashioned in 1929 by the Fey Manufacturing Company, San Francisco, inventors of the 3-reel, automatic payout machine (1895).

of getting a net payoff of 2. Since there are 20 pictures per reel, there are a total of $20 \times 20 \times 20 = 8{,}000$ possible combinations. Hence, the probability of receiving a net payoff of 2 coins is 600/8,000. Table 8 takes into account all *winning* combinations, with the necessary products for computing expectation added in the last column. However, since a *nonwinning* combination can occur in $8{,}000 - 988 = 7{,}012$ ways (with winnings of -1 coin), the product $(-1) \cdot 7{,}012/8{,}000$ must also be included. Hence, the expected value of this particular slot machine is

$$\frac{6{,}318}{8{,}000} + (-1) \cdot \frac{7{,}012}{8{,}000} = -.087 \text{ coin.}$$

On a machine costing one dollar per play, the expected *loss* (per play) is

$$(.087)(1 \text{ dollar}) = 8.7 \text{ cents.}$$

Actual slot machines in Nevada and Atlantic City vary in expected loss per dollar of play from about 3 cents to about 50 cents, with the better payoffs usually coming in the larger establishments. But you still lose.

Table 9 comes from an article by Andrew Sterrett in *The Mathematics Teacher* (March 1967), in which he discusses rules for various games of chance and calculates their expected values. He uses expected values to find expected times it would take to lose $1,000 if you played continually at the rate of $1 per play and one play per minute.

TABLE 9

Game	Expected Value	Days	Hours	Minutes
Roulette (with one 0)	−$.027	25	16	40
Roulette (with 0 and 00)	−$.053	13	4	40
Chuck-a-luck	−$.079	8	19	46
Keno (one number)	−$.200	3	11	20
Numbers	−$.300	2	7	33
Football pool (4 winners)	−$.375	1	20	27
Football pool (10 winners)	−$.658	1	1	19

For Group Discussion

1. Explain why the entries of the "Net winnings" column of Table 8 are all one fewer than the corresponding entries of the "Number of coins received" column.
2. Verify all values in the "Number of ways" column of Table 8. (Refer to Table 7.)
3. In order to make your money last as long as possible in a casino, which game should you play?

EXAMPLE 6 David Shea, an importer, is considering the purchase of a shipment of foreign dolls. Drawing on his considerable experience, Shea calculates that the probabilities of being able to resell the shipment for \$10,000, \$9,000, \$8,000, or \$7,000 are .18, .26, .41, and .15, respectively. In order to have an expected profit on the shipment of at least \$2,000, how much can Shea afford to pay for the purchase?

The expected revenue (or income) from resales can be found as shown in the following table.

Income, x	**Probability,** $P(x)$	**Product,** $x \cdot P(x)$
\$10,000	.18	\$1,800
\$9,000	.26	\$2,340
\$8,000	.41	\$3,280
\$7,000	.15	\$1,050
		Expected revenue: \$8,470

Now, in general, we have the relationship

$$\text{profit} = \text{revenue} - \text{cost}.$$

Therefore, in terms of expectations,

$$\text{expected profit} = \text{expected revenue} - \text{cost}.$$

So $\$2{,}000 = \$8{,}470 - \text{cost}$ or $\text{cost} = \$8{,}470 - \$2{,}000 = \$6{,}470.$

Shea can pay up to \$6,470 and still maintain an expected profit of at least \$2,000. ◆

EXERCISES

1. A couple who are planning their future say, "We expect to have 2.5 daughters." Explain what this statement means.

A certain game consists of rolling a single fair die and pays off as follows: \$3 for a 6, \$2 for a 5, \$1 for a 4, and no payoff otherwise.

2. Find the expected winnings for this game.
3. What is a fair price to pay to play this game?

For Exercises 4 and 5, consider a game consisting of rolling a single fair die, with payoffs as follows. If an even number of spots turns up, you receive that many dollars. But if an odd number of spots turns up, you must pay that many dollars.

4. Find the expected net winnings of this game.
5. Is this game fair, or unfair against the player, or unfair in favor of the player?
6. A certain game involves tossing 3 fair coins, and it pays 10¢ for 3 heads, 5¢ for 2 heads, and 3¢ for 1 head. Is 5¢ a fair price to pay to play this game? (That is, does the 5¢ cost to play make the game fair?)
7. In a form of roulette slightly different from that in Example 4, a more generous management supplies a wheel having only thirty-seven compartments, with eighteen red, eighteen black, and one zero. Find the expected net winnings if you bet on red in this game.

If two cards are drawn from a standard 52-card deck, find the expected number of spades in each of the following cases (Exercises 8 and 9).

8. The drawing is done with replacement.
9. The drawing is done without replacement.

10. In a certain mathematics class, the probabilities have been empirically determined for various numbers of absences on any given day. These values are shown in the table below. Find the expected number of absences on a given day. (Give the answer to two decimal places.)

Number absent	0	1	2	3	4
Probability	.12	.27	.30	.18	.13

11. An insurance company will insure a $100,000 home for its total value for an annual premium of $300. If the probability of total loss for such a home in a given year is .002, and you assume that either total loss or no loss will occur, what is the company's expected annual gain (or profit) on each such policy?

A college foundation raises funds by selling raffle tickets for a new car worth $25,000.

12. If 500 tickets are sold for $100 each, determine
- **(a)** the expected *net* winnings of a person buying one of the tickets,
- **(b)** the total profit for the foundation, assuming they had to purchase the car,
- **(c)** the total profit for the foundation, assuming the car was donated.

13. If 750 tickets are sold for $100 each, determine
- **(a)** the expected *net* winnings of a person buying one of the tickets,
- **(b)** the total profit for the foundation, assuming they had to purchase the car,
- **(c)** the total profit for the foundation, assuming the car was donated.

Five thousand raffle tickets are sold. One first prize of $1,000, two second prizes of $500 each, and five third prizes of $100 each are to be awarded, with all winners selected randomly.

14. If you purchased one ticket, what are your expected winnings?

15. If you purchased two tickets, what are your expected winnings?

16. If the tickets were sold for $1 each, how much profit goes to the raffle sponsor?

17. An amusement park, considering adding some new attractions, conducted a study over several typical days and found that, of 10,000 families entering the park, 830 brought just one child (defined as younger than age twelve), 2,370 brought two children, 4,980 brought three children, 1,210 brought four children, 260 brought five children, 180 brought six children, and 170 brought no children at all. Find the expected number of children per family attending this park. (Round your answer to the nearest tenth.)

18. Five cards are numbered 1 through 5. Two of these cards are chosen randomly (without replacement), and the numbers on them are added. Find the expected value of this sum.

19. Most members of the Mathematical Association of America in a recent year paid for membership dues and regular association publications according to the table below, where M indicates the *American Mathematical Monthly*, G indicates *Mathematics Magazine*, and J indicates the *College Mathematics Journal*. The third row of the table shows how many members in a survey of 1,000 fell into each category. For those members surveyed, find the expected amount paid (to the nearest cent).

Membership category	M	G	J	M + G	M + J	G + J	M + G + J
Amount paid	$108	$90	$94	$126	$130	$112	$148
Number in category	97	144	106	51	195	186	221

20. In a certain California city, projections for the next year are that there is a 30% chance that electronics jobs will increase by 1,200, a 50% chance that they will increase by 500, and a 20% chance that they will *decrease* by 800. What is the expected change in the number of electronics jobs in that city in the next year?

21. In one version of the game *keno*, the house has a pot containing 80 balls, numbered 1 through 80. A player buys a ticket for $1 and marks one number on it (from 1 to 80). The house then selects 20 of the 80 numbers at random. If the number selected by the player is among the 20 selected by the management, the player is paid $3.20. Find the expected net winnings for this game.

A game show contestant is offered the option of receiving a set of new kitchen appliances worth \$1,800, or accepting a chance to win either a luxury vacation worth \$4,000 or a boat worth \$5,000. The contestant's probabilities of winning the vacation or the boat are .25 and .15, respectively.

22. If the contestant were to turn down the appliances and go for one of the other prizes, what would be the expected winnings?

23. Purely in terms of monetary value, what is the contestant's wiser choice?

*The table below illustrates how a salesman for Levi Strauss & Co. rates his accounts by considering the existing volume of each account plus potential additional volume.**

1	2	3	4	5	6	7
Account Number	**Existing Volume**	**Potential Additional Volume**	**Probability of Getting Additional Volume**	**Expected Value of Additional Volume**	**Existing Volume plus Expected Value of Additional Volume**	**Classification**
1	\$15,000	\$ 10,000	.25	\$2,500	\$17,500	
2	40,000	0	—	—	40,000	
3	20,000	10,000	.20	2,000		
4	50,000	10,000	.10	1,000		
5	5,000	50,000	.50			
6	0	100,000	.60			
7	30,000	20,000	.80			

Use the table to work Exercises 24–27.

24. Compute the missing expected values in column 5.

25. Compute the missing amounts in column 6.

26. In column 7, classify each account according to this scheme: Class A if the column 6 value is \$55,000 or more; Class B if the column 6 value is at least \$45,000 but less than \$55,000; Class C if the column 6 value is less than \$45,000.

27. Considering all seven of this salesman's accounts, compute the total additional volume he can "expect" to get.

28. Recall that in the game keno of Exercise 21, the house randomly selects 20 numbers from the counting numbers 1–80. In the variation called 6-spot keno, the player pays 60¢ for his ticket and marks 6 numbers of his choice. If the 20 numbers selected by the house contain at least 3 of those chosen by the player, he gets a payoff according to this scheme.

3 of the player's numbers among the 20	\$ .35
4 of the player's numbers among the 20	2.00
5 of the player's numbers among the 20	60.00
6 of the player's numbers among the 20	1,250.00

Find the player's expected net winnings in this game.

* This information was provided by James McDonald of Levi Strauss & Co., San Francisco.

11.6 Simulation

An important area within probability theory is the process called **simulation.** It is possible to study a complicated, or unclear, phenomenon by *simulating,* or imitating, it with a simpler phenomenon involving the same basic probabilities. For example, recall from the first section of this chapter Mendel's discovery that when two Rr pea plants (red-flowered but carrying both red and white genes) are crossed, the offspring will have red flowers if an R gene is received from either parent, or from both. This is because red is dominant and white is recessive. Table 2, reproduced in the margin, shows that three of the four equally likely possibilities result in red-flowered offspring.

		Second Parent	
		R	**r**
First Parent	**R**	RR	Rr
	r	rR	rr

Now suppose we want to know (or at least approximate) the probability that three offspring in a row will have red flowers. It is much easier (and quicker) to toss coins than to cross pea plants. And the equally likely outcomes, heads and tails, can be used to simulate the transfer of the equally likely genes, R and r. If we toss two coins, say a nickel and a penny, then we can interpret the results as follows.

hh $\Rightarrow$ RR $\Rightarrow$ red gene from first parent, red gene from second parent
$\Rightarrow$ red flowers

ht $\Rightarrow$ Rr $\Rightarrow$ red gene from first parent, white gene from second parent
$\Rightarrow$ red flowers

th $\Rightarrow$ rR $\Rightarrow$ white gene from first parent, red gene from second parent
$\Rightarrow$ red flowers

tt $\Rightarrow$ rr $\Rightarrow$ white gene from first parent, white gene from second parent
$\Rightarrow$ white flowers

Although nothing is certain for a few tosses, the law of large numbers indicates that larger and larger numbers of tosses should become better and better indicators of general trends in the genetic process.

Simulation methods, also called "Monte Carlo" methods (see the exercises), have been successfully used in many areas of scientific study throughout most of this century. Most practical applications require huge numbers of random digits, so computers are used to produce them. A computer, however, cannot toss coins. It must use an algorithmic process, programmed into the computer, which is called a "random number generator." It

continued on page 664

EXAMPLE 1 Toss two coins fifty times and use the results to approximate the probability that the crossing of Rr pea plants will produce three successive red-flowered offspring.

We actually tossed two coins 50 times and got the following sequence.

th, hh, th, tt, th, hh, ht, th, ht, th, hh, hh,
tt, th, hh, ht, ht, ht, ht, th, hh, hh, hh, tt,
ht, tt, hh, ht, ht, hh, tt, tt, tt, th, tt, tt, hh,
ht, ht, ht, hh, tt, th, hh, tt, hh, ht, tt, tt, tt

By the color interpretation described above, this gives the following sequence of flower colors in the offspring.

red–red–red–white–red–red–red–red–red–red–red–red–white–
red–red–red–red–red–red–red–red–red–red–white–red–white–
red–red–red–red–white–white–white–red–white–white–red–red–
red–red–red–white–red–red–white–red–red–white–white–white

is very difficult to avoid all nonrandom patterns in the results, so the digits produced are called "pseudorandom" numbers. They must pass a battery of tests of randomness before being "approved for use."

Recently, computer scientists and physicists have been encountering unexpected difficulties with even the most sophisticated *random-number generators*. As reported in the December 19, 1992, issue of *Science News* (page 422), it seems that a random number generator can pass all the tests, and work just fine for some simulation applications, but then produce faulty answers when used with a different simulation. Therefore, random number generators apparently cannot be approved for all uses in advance, but must be carefully checked along with each new simulation application proposed.

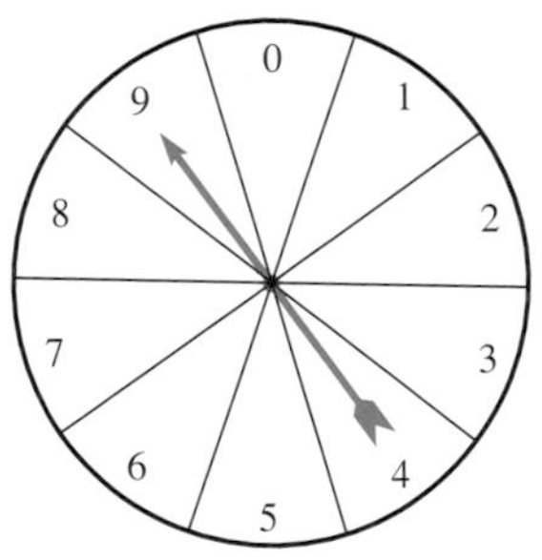

■ **FIGURE 11**

We now have an experimental list of 48 sets of three successive plants, the 1st, 2nd, and 3rd entries, then the 2nd, 3rd, and 4th entries, and so on. Do you see why there are 48 in all? Now we just count up the number of these sets of three that are "red-red-red." Since there are 20 of those, our empirical probability of three successive red offspring, obtained through simulation, is $20/48 = 5/12$, or about .417. By applying the special multiplication rule (since all outcomes here are independent of one another), we find that the theoretical value is $(3/4)^3 = 27/64$, or about .422, so our approximation obtained by simulation is very close. ■

In human births boys and girls are (essentially) equally likely. Therefore, an individual birth can be simulated by tossing a fair coin, letting a head correspond to a girl and a tail to a boy.

EXAMPLE 2 A sequence of 40 actual coin tosses produced the results below.

bbggb, gbbbg, gbgbb, bggbg, bbbbg, gbbgg, gbbgg, bgbbg

(For every head we have written g, for girl; for every tail, b, for boy.) Refer to this sequence to answer the following questions.

(a) How many pairs of two successive births are represented by the above sequence?

Beginning with the 1st–2nd pair and ending with the 39th–40th pair, there are 39 pairs.

(b) How many of those pairs consist of both boys?

By observing the sequence, we count 11 pairs consisting of both boys.

(c) Find the empirical probability, based on this simulation, that two successive births will both be boys. Give your answer to three decimal places.

Utilizing parts (a) and (b), we have $11/39 \approx .282$. ■

Another way to simulate births is to generate a random sequence of digits, perhaps interpreting even digits as girls and odd digits as boys. The digits might be generated by spinning the spinner in Figure 11. It turns out that many kinds of phenomena can be simulated using random digits, so we can save lots of effort by using the spinner to obtain a table of random digits, like in Table 10, and then use that table to carry out any simulation experiment that is needed. Notice that the 250 random digits in Table 10 have been grouped conveniently so that we can easily follow down a column or across a row.

EXAMPLE 3 A couple plans to have five children. Use random number simulation to estimate the probability they will have more than three boys.

Let each sequence of five digits, as they appear in Table 10, represent a family with five children, and (arbitrarily) associate odd digits with boys, even digits with girls. (Recall that 0 is even.) Verify that, of the fifty families simulated, ten of them have more than 3 boys (4 boys or 5 boys). Therefore the estimation (empirical value) is

$$P(\text{more than 3 boys}) = 10/50 = .20.$$ ■

TABLE 10

51592	73219
77876	55707
36500	48007
40571	65191
04822	06772
53033	94928
92080	15709
01587	39922
36006	96365
63698	14655
17297	65587
22841	76905
91979	12369
96480	54219
74949	89329
76896	90060
47588	06975
45521	05050
02472	69774
55184	78351
40177	11464
84861	84086
86937	51497
20931	12307
22454	68009

Recall that the theoretical value for the probability estimated in Example 3, above, was obtained in Example 3 of Section 4 in this chapter. It was .1875. So our estimate above was fairly close. In light of the law of large numbers, a larger sampling of random digits (more than 50 simulated families) would most likely yield a closer approximation. Very extensive tables of random digits are readily available in statistical research publications. Computers can also be programmed to generate sequences of "pseudo-random" digits, which serve the same purposes. In most simulation experiments, much larger samples than we are using here are necessary to obtain reliable results. But the examples here adequately illustrate the procedure.

EXAMPLE 4 Use random number simulation to estimate the probability that two cards drawn from a standard deck with replacement will both be of the same suit.

Use this correspondence: 0 and 1 mean clubs, 2 and 3 mean diamonds, 4 and 5 mean hearts, 6 and 7 mean spades, 8 and 9 are disregarded. Now read down the columns of Table 10. Suppose we (arbitrarily) use the first digit of each five-digit group. The first time from top to bottom gives the sequence

$$5-7-3-4-0-5-0-3-6-1-2-7-7-4-4-0-5-4-2-2.$$

(Five 8s and 9s were omitted.) Starting again at the top, we obtain

$$7-5-4-6-0-1-3-1-6-7-1-5-0-0-6-7-1-5-1-6.$$

(Again, there happened to be five 8s and 9s.) This 40-digit sequence of digits yields the sequence of suits shown next.

hearts–spades–diamonds–hearts–clubs–hearts–clubs–diamonds–spades–clubs–diamonds–spades–spades–hearts–hearts–clubs–hearts–hearts–diamonds–diamonds–spades–hearts–hearts–spades–clubs–clubs–diamonds–clubs–spades–spades–clubs–hearts–clubs–clubs–spades–spades–clubs– hearts–clubs–spades

Verify that, of the 39 successive pairs of suits (hearts–spades, spades–diamonds, diamonds–hearts, etc.), 9 of them are pairs of the same suit. This makes the estimated probability $9/39 \approx .23$. (For comparison, the theoretical value is .25.)

11.6 EXERCISES

1. Explain why, in Example 1, fifty tosses of the coins produced only 48 sets of three successive offspring.
2. Use the sequence of flower colors of Example 1 to approximate the probability that *four* successive offspring will all have red flowers.
3. Should the probability of two successive girl births be any different from that of two successive boy births?
4. Simulate 40 births by tossing coins yourself, and obtain an empirical probability for two successive girls.

5. Use Table 10 to simulate fifty families with three children. Let 0–4 correspond to boys and 5–9 to girls, and use the middle grouping of three digits (159, 787, 650, and so on). Estimate the probability of exactly two boys in a family of three children. Compare the estimation with the theoretical probability of exactly two boys in a family of three children, which is $3/8 = .375$.

Exercises 75–77 of Section 3 involved one-and-one foul shooting in basketball. Karin, who had a 60% foul-shooting record, had probabilities of scoring 0, 1, *or* 2 *points of* .40, .24, *and* .36, *respectively. Use* Table 10 *(with digits 0–5 representing hit and 6–9 representing miss) to simulate* 50 *one-and-one shooting opportunities for Karin. Begin at the top of the first column (5, 7, 3, etc., to the bottom), then move to the second column (1, 7, 6, etc.), going until 50 one-and-one opportunities are obtained. (Notice that some "opportunities" involve one shot while others involve two shots.) Keep a tally of the numbers of times* 0, 1, *and* 2 *points are scored. From the tally, find the empirical probability that, on a given opportunity, Karin will score as follows. (Round your answers to two decimal places.)*

Number of Points	Tally
0	
1	
2	

6. no points **7.** 1 point **8.** 2 points

Exercises 52–59 of Section 4 illustrated a simple version of the idea of a "random walk." Atomic particles released in nuclear fission also move in a random fashion. During World War II, John von Neumann and Stanislaw Ulam used simulation with random numbers to study particle motion in nuclear reactions. Von Neumann coined the name "Monte Carlo" for the methods used, and since then the terms "Monte Carlo methods" and "simulation methods" have often been used with very little distinction.

The figure suggests a model for random motion in two dimensions. Assume that a particle moves in a series of 1-unit "jumps," each one in a random direction, any one of 12 *equally likely possibilities. One way to choose directions is to roll a fair die and toss a fair coin. The die determines one of the directions 1–6, coupled with heads on the coin. Tails on the coin reverses the direction of the die, so that the die coupled with tails gives directions 7–12. Thus* 3*h (meaning* 3 *with the die and heads with the coin) gives direction* 3; 3*t gives direction* 9 *(opposite to* 3*); and so on.*

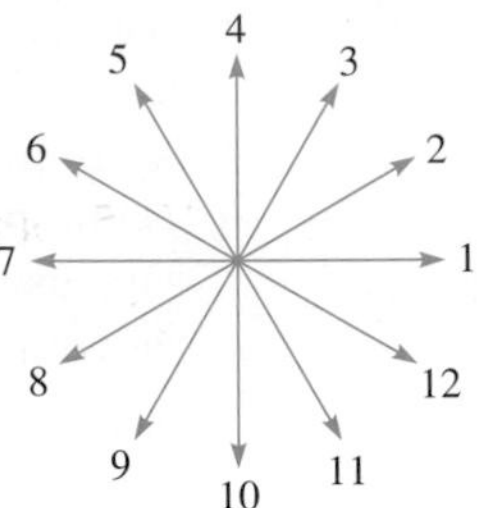

9. Simulate the motion described above with 10 rolls of a die (and tosses of a coin). Draw the 10-jump path you get. Make your drawing accurate enough so you can estimate (by measuring) how far from its starting point the particle ends up.

10. Repeat the experiment of Exercise 9 four more times. Measure distance from start to finish for each of the 5 "random trips." Add these 5 distances and divide the sum by 5, to arrive at an "expected net distance" for such a trip.

Consider another two-dimensional random walk governed by the following conditions.

1. *Start out from a given street corner, and travel one block north. At each intersection:*
2. *Turn left with probability 1/6.*
3. *Go straight with probability 2/6 (= 1/3).*
4. *Turn right with probability 3/6 (= 1/2).*

(Never turn around.)

11. Explain how a fair die could be used to simulate this random walk.

12. Use Table 10 to simulate this random walk. For every 1 encountered in the table, turn left and proceed for another block. For every 2 or 3, go straight and proceed for another block. For every 4, 5, or 6, turn right and proceed for another block. Disregard all other digits. (Do you see how this scheme satisfies the probabilities given above?) This time begin at the upper right corner of the table, running down the column 9, 7, 7, and so on, to the bottom. Then start at the top of the next column to the left, 1, 0, 0, and so on, to the bottom. When these two columns of digits are used up, stop the "walk." Describe, in terms of distance and direction, where you have ended up relative to your starting point.

COLLABORATIVE INVESTIGATION
Approximating π Using Probability

(The information in this investigation was obtained from *Burton's History of Mathematics: An Introduction,* Third Edition, by David M. Burton, published by Wm. C. Brown, 1995, page 441.)

The following problem was posed by Georges Louis Leclerc, Compte de Buffon (1707–1788) in his *Histoire Naturelle* in 1777.

> A large plane area is ruled with equidistant parallel lines, the distance between two consecutive of the series being a. A thin needle of length $l < a$ is tossed randomly onto the plane. What is the probability that the needle will intersect one of these lines?

The answer to this problem is found using integral calculus, and the probability p is shown to be $p = \frac{2l}{\pi a}$. Because this probability involves π, an experiment can be set up so that one can find an approximate value for π. This observation was first made by the mathematician Pierre Simon Laplace, and was performed experimentally by Johann Wolf, a professor of astronomy at Bern, in 1853. In this investigation, we will duplicate this experiment.

Topics for Discussion

Divide the class into groups of 3 or 4 students each. Each group should prepare on a sheet of blank paper a series of parallel lines 45 millimeters apart. Obtain a thin needle-like object, such as a matchstick, which measures 36 millimeters in length. Have each group assign one student the job of dropping the needle onto the piece of paper, another the job of determining whether the needle "hits" or "misses" a line, and another the job of keeping tally. Drop the needle 100 times, and record the number of hits.

When the entire class has completed the experiment, find a class total for the number of attempts and the number of hits. Compute the quotient

$$\frac{\text{hits}}{\text{attempts}}$$

and call it q. If we use this value for the theoretical probability of the intersections and then solve

$$\frac{72}{45\pi} = q,$$

we find the experimental approximation of π.

(*Note:* Wolf tossed his needle 5,000 times and it hit a line 2,532 times, leading to an experimental value of π equal to 3.1596, which is off by less than .02 from the actual value of π.)

Chapter 11 Test

1. Explain the difference between *empirical* and *theoretical* probability.

2. State the *law of large numbers,* and use coin tossing to illustrate it.

A single card is chosen at random from a standard 52-card deck. Find the odds against its being each of the following.

3. a heart

4. a red queen

5. a king or a black face card

The chart at the right represents genetic transmission of cystic fibrosis. C *denotes a normal gene while* c *denotes a cystic fibrosis gene. (Normal is dominant.) Both parents in this case are* Cc, *which means that they inherited one of each gene, and are therefore carriers but do not have the disease.*

		Second Parent	
		C	c
First Parent	C		Cc
	c		

6. Complete the chart, showing all four equally likely gene combinations.

7. Find the probability that a child of these parents will also be a carrier without the disease.

8. What are the odds against a child of these parents actually having cystic fibrosis?

The manager of a pizza parlor (which operates seven days a week) allows each of three employees to select one day off next week. Assuming the selection is done randomly, find the probability of each of the following events.

9. All three select different days.

10. All three select the same day.

11. Exactly two of them select the same day.

Two distinct numbers are randomly selected from the set {1, 2, 3, 4, 5}. Find the probability of each of the following events.

12. Both numbers are even.

13. Both numbers are prime.

14. The sum of the two numbers is odd.

15. The product of the two numbers is odd.

A three-member committee is selected randomly from a group consisting of three men and two women.

16. Let x denote the number of men on the committee, and complete the probability distribution table.

x	$P(x)$
0	0
1	
2	
3	

17. Find the probability that the committee members are not all men.

18. Find the expected number of men on the committee.

A pair of dice are rolled. Find the following.

19. the odds against "doubles" (the same number on both dice)

20. the probability of a sum greater than 2

21. the odds against a sum of "7 or 11"

22. the probability of a sum that is even and less than 5

Jeff Cole has an .82 chance of making par on each hole of golf that he plays. Today he plans to play just three holes. Find the probability of each of the following events. Round answers to three decimal places.

23. He makes par on all three holes.

24. He makes par on exactly two of the three holes.

25. He makes par on at least one of the three holes.

26. He makes par on the first and third holes but not on the second.

Two cards are drawn, without replacement, from a standard 52-card deck. Find the probability of each of the following events.

27. Both cards are red.

28. Both cards are the same color.

29. The second card is a queen, given that the first card is an ace.

30. The first card is a face card and the second is black.

Refer to the sequence of 40 coin tosses in Example 2 of Section 6.

31. How many triples of 3 successive births are listed?

32. How many of those triples consist of all girls?

33. What is the empirical probability that 3 successive births will be all girls?

12

Statistics

Some of the results from one study of the relationship between smoking and lung cancer are shown in the table. The first row lists the average number of cigarettes smoked per day and the second row contains the corresponding mortality rate due to lung cancer in deaths per 100,000 people.

Cigarettes/Day	0	5	15	30	45
Deaths/100,000	30	132	256	447	606

Statistics is used to analyze and interpret these numerical data. If the points are plotted in the *xy*-coordinate plane, it appears that the data are roughly linear. Using a statistical technique called *regression*, a researcher can find a line that gives the best fit for the data. Both the data points and the "best fit" line are shown in the graph. The equation of this line is $y = 12.6x + 55.2$.

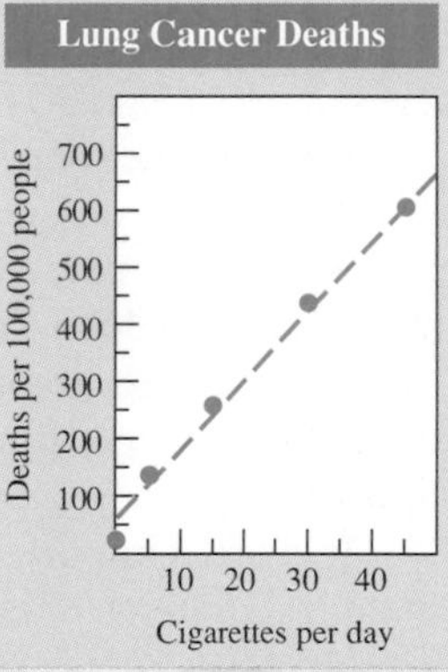

What do these results mean? Since the relationship is linear, each additional cigarette per day increases the risk of lung cancer death by the same amount. The slope of the regression line is 12.6. This implies that in a group of 100,000 people, if each member of the group smokes one additional cigarette per day, it will result in approximately 12.6 additional lung cancer deaths.

Source: Kahn, H. and C. Sempos, *Statistical Methods in Epidemiology,* Oxford University Press, 1989.

Various statistical options on the TI-82.

Most scientific and graphing calculators will perform all computations we describe in this chapter, including sorting of data (once they are entered). So your purpose in studying this material should be to understand the meanings of the various computed quantities. It is recommended that the actual computations, especially for sizable collections of data, be done on a calculator (or a computer with suitable statistical software).

Governments collect and analyze an amazing quantity of "statistics"; the census, for example, is a vast project of gathering data. The census is not a new idea; two thousand years ago Mary and Joseph traveled to Bethlehem to be counted in a census. Long before, the Egyptians had recorded numerical information that is still being studied. For a long time, in fact, "statistics" referred to information about the government. The word itself comes from the Latin *statisticus,* meaning "of the state." The term was easily transferred during the nineteenth century to numerical information of other kinds and then to methods for analyzing that information.

The development of current statistical theories, methods, tests, and experimental designs is due to the efforts of many people. For example, John Gaunt analyzed the weekly Bills of Mortality that recorded deaths in London in the first half of the seventeenth century. He published his *Observations* in 1662, noting that male births were more numerous than female, but eventually the numbers of both sexes came to be about equal. From these beginnings developed insurance companies, founded on the predictability of deaths as shown in mortality tables. From the field of biology, Sir Francis Galton and Karl Pearson in the nineteenth century made important contributions to statistical theory. William S. Gossett, a student of Pearson, produced some of the first results concerning small samples. He felt obliged to publish under the name "Student"; his basic findings came to be known as Student's *t*-test—an important statistical tool.

In the twentieth century, statistical theory and methodology have been highly developed and have become crucial in nearly all governmental, academic, social, and business fields. One surge of development occurred in the 1920s as a result of Sir Ronald Fisher's study of agricultural methods. Then, in the 1960s, business and the social sciences became highly oriented toward statistical methods of data analysis. Gertrude Mary Cox was a major writer and leader throughout the middle part of the century in furthering the recognition and usage of statistics. And John Tukey has written extensively in the area of practical data analysis through graphical and numerical methods. Tukey's approach has become increasingly popular in recent years and is known as **exploratory data analysis.**

It is often important in statistics to distinguish between a **population,** which includes *all* items of interest, and a **sample,** which includes *some* (but ordinarily not all) of the items in the population. For example, to predict the outcome of an approaching presidential election, we may be interested in a population of many millions of voter preferences (those of all potential voters in the country).

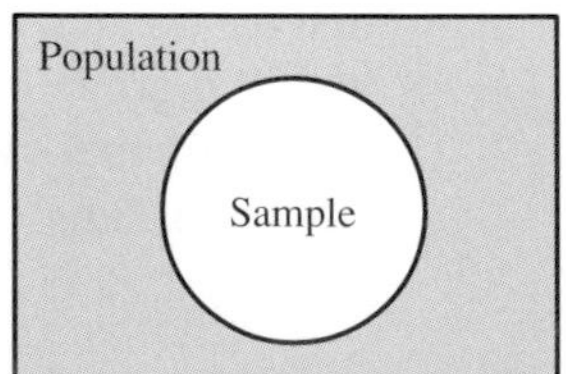

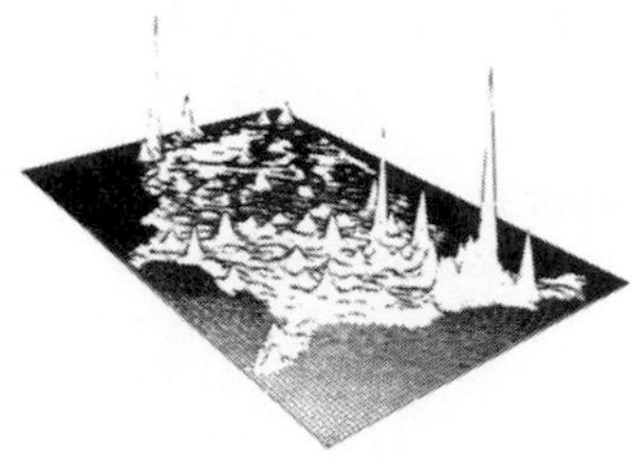

Population Densities There are many different kinds of graphs. For example, the computer-generated graph above shows U.S. population densities by peaks. The higher the peak, the denser the population in that area. The map shows the densest population to be in the New York City area. The blank areas in the West indicate fewer than two people per square mile.

As a practical matter, however, even national polling organizations with considerable resources will obtain only a relatively small sample, say 2,000, of those preferences. In general, a sample can be any subset of a population.

The study of statistics can be divided into two main areas. The first, **descriptive statistics,** has to do with collecting, organizing, summarizing, and presenting data (information). As we shall see, the main tools of descriptive statistics are tables of numbers, various kinds of graphs, and various calculated quantities, such as averages. These tools can be applied to both samples and populations. Whether a particular collection of data is a sample or a population is a relative matter. For example, if you are interested in how your class compares academically to other students in the school, then you might take the grade point averages of all those in your class as a sample within the population of all GPAs for the entire school. To see how your school compares to others across the country, you might take all GPAs for your school as a sample within the population of GPAs for all students in the country. In one case, your school's GPAs comprise the population; in the other case, they are the sample. Although the descriptive tools apply to both, we shall generally think of the data collections in the early sections of the chapter as samples.

Later in the chapter, we deal with some aspects of the second main area of statistics, namely **inferential statistics,** which has to do with drawing inferences or conclusions (making conjectures) about populations based on information from samples. It is in this area that we can best see the relationship between probability and statistics.

A population of 10,000

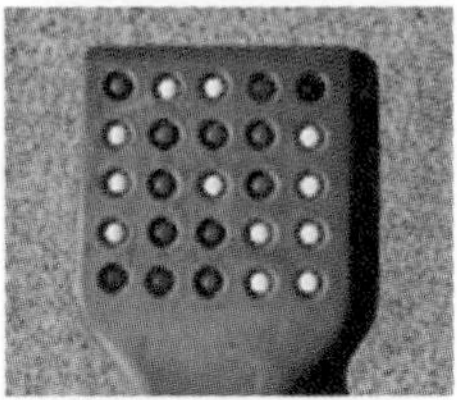

A random sample of 25

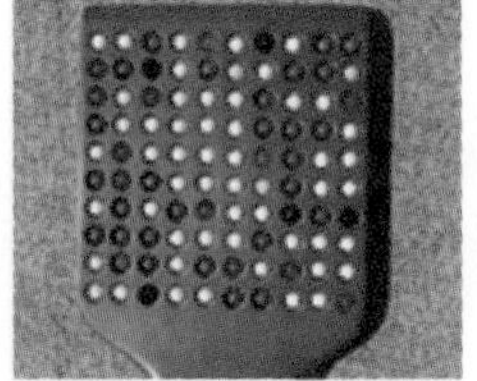

A random sample of 100

The Lansford Publishing Company, of San Jose, California, has produced a sampling demonstrator which consists of a large bowl containing 10,000 beads of various colors, along with several paddles for easily drawing samples of various sizes. Beneath the bowl, at the side, we show two random samples from the bowl, the first of which consists of 25 beads, 9 of which are green. From this sample we can "infer" that the bowl (the population) must contain about $9/25$, or 36%, green beads, that is, about 3,600. This is a matter of *inductive reasoning.* From a particular sample we have made a conjecture about the population in general. If we increase the size of our sample, then the proportion of green beads in the larger sample will most likely give a better estimate of the proportion in the population. The second sample at the side contains 100 beads, 28 of which are green. So the new estimate would be about $28/100$, or 28%, that is, 2,800 green beads in the bowl.

In fact, the bowl is known to contain 30%, or 3,000 green beads. So our larger sample estimate of 2,800 was considerably more accurate than the smaller sample estimate of 3,600. Knowing the true proportion of green beads in the population, we can turn the situation around, using *deductive reasoning* to predict the proportion of green beads in a particular sample. Our prediction for a sample of 100 beads would be 30% of 100, or 30. The sample, being random, may or may not end up containing exactly 30 green beads.

To summarize this discussion: If we know what a population is like, then probability theory enables us to conclude what is likely to happen in a sample (deductive reasoning); if we know what a sample is like, then inferential statistics enables us to draw inferences about the population (inductive reasoning).

12.1 Frequency Distributions and Graphs

Information that has been collected but not yet organized or processed is called **raw data.** It is often **quantitative** (or **numerical**), but can also be **qualitative** (or **nonnumerical**), as illustrated in Table 1.

TABLE 1 Examples of Raw Data

The number of siblings in ten different families (quantitative data):
3, 1, 2, 1, 5, 4, 3, 3, 8, 2

The makes of six different automobiles (qualitative data):
Toyota, Ford, Nissan, Toyota, Chevrolet, Honda

Quantitative data are generally more useful when they are **sorted,** or arranged in numerical order. In sorted form, the first list in Table 1 appears as

$$1, 1, 2, 2, 3, 3, 3, 4, 5, 8.$$

When a data set includes many repeated items, it can be organized into a **frequency distribution,** which lists the distinct data values (x) along with their frequencies (f). The frequency designates the number of times the corresponding item occurred in the data set. It is also helpful to show the **relative frequency** of each distinct item. This is the fraction, or percentage, of the data set represented by the item. If n denotes the total number of items, and a given item, x, occurred f times, then the relative frequency of x is f/n. Example 1 illustrates these ideas.

EXAMPLE 1 The 25 members of a psychology class were polled as to the number of siblings in their individual families. Construct a frequency distribution and a relative frequency distribution for their responses, which are shown here.

$$2, 3, 1, 3, 3, 5, 2, 3, 3, 1, 1, 4, 2, 4, 2, 5, 4, 3, 6, 5, 1, 6, 2, 2, 2$$

The data here range from a low of 1 to a high of 6. The frequencies (obtained by inspection) and relative frequencies are shown in Table 2. ◆

TABLE 2 Frequency and Relative Frequency Distributions for Numbers of Siblings

Number x	Frequency f	Relative Frequency f/n
1	4	$4/25 = 16\%$
2	7	$7/25 = 28\%$
3	6	$6/25 = 24\%$
4	3	$3/25 = 12\%$
5	3	$3/25 = 12\%$
6	2	$2/25 = 8\%$

The numerical data of Table 2 can be more easily interpreted with the aid of a graph. A common technique is to construct a **histogram.** A series of

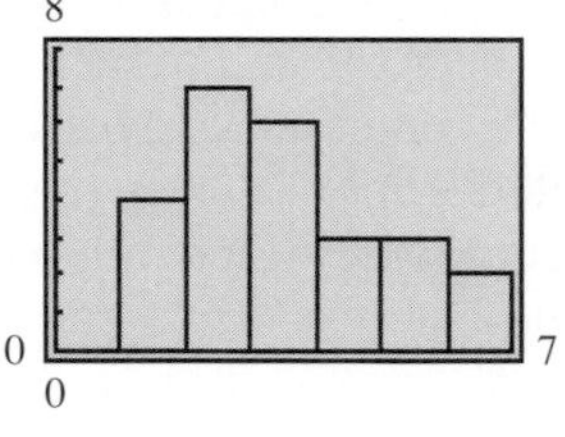

This histogram was generated using the data in Table 2. Compare with Figure 1.

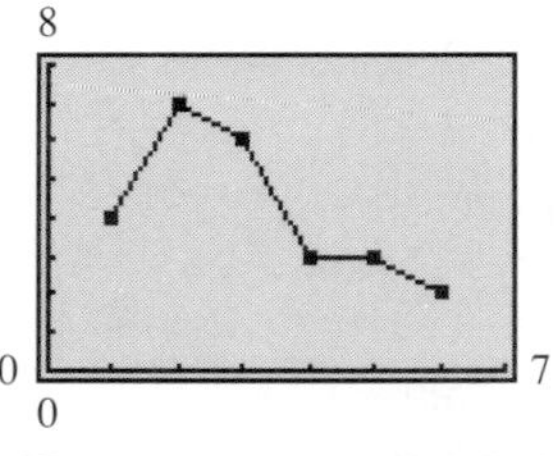

This line graph resembles the frequency polygon in Figure 2. It was generated using the data in Table 2.

rectangles, whose lengths represent the frequencies, are placed next to one another as shown in Figure 1. On each axis, horizontal and vertical, a label and the numerical scale should be shown.

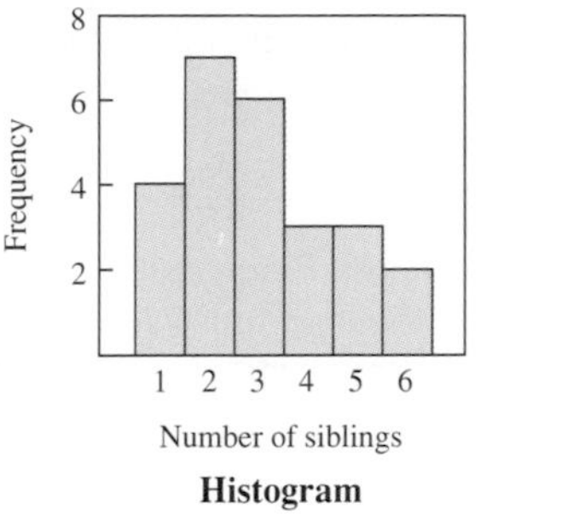

Histogram

FIGURE 1

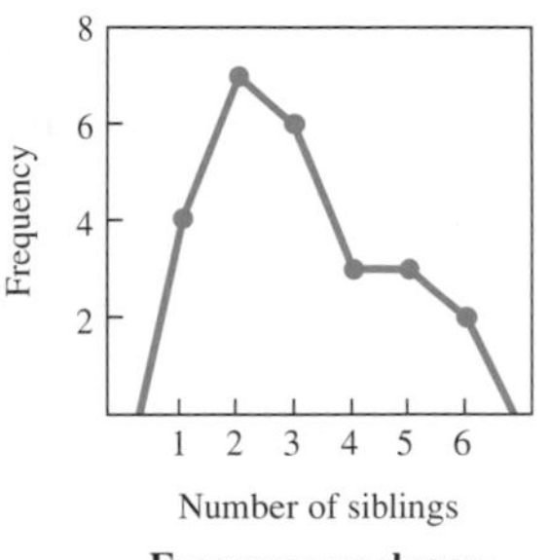

Frequency polygon

FIGURE 2

The information shown in the histogram of Figure 1 can also be conveyed by a **frequency polygon,** as is done in Figure 2. Simply plot a single point at the appropriate height for each frequency, connect the points with a series of connected line segments, and complete the polygon with segments that trail down to the axis beyond 1 and 6.

Grouped Frequency Distributions

Data sets containing large numbers of items are often arranged into groups, or *classes.* All data items are assigned to their appropriate classes, and then a **grouped frequency distribution** can be set up and a graph displayed. Although there are no fixed rules for establishing the classes, most statisticians agree on a few general guidelines, such as the following.

Guidelines for the Classes of a Grouped Frequency Distribution

1. Make sure each data item will fit into one, and only one, class.
2. Try to make all classes the same width.
3. Make sure the classes do not overlap.
4. Use from 5 to 12 classes. (Too few or too many classes can obscure the tendencies in the data.)

EXAMPLE 2 Forty students, selected randomly in the school cafeteria on a Monday morning, were asked to estimate the number of hours they had spent studying in the past week (including both in-class and out-of-class time). Their responses are recorded here.

18	60	72	58	20	15	12	26	16	29
26	41	45	25	32	24	22	55	30	31
55	39	29	44	29	14	40	31	45	62
36	52	47	38	36	23	33	44	17	24

Tabulate a grouped frequency distribution and a grouped relative frequency distribution and construct a histogram and a frequency polygon for the given data.

Scanning the data, we see that they range from a low of 12 to a high of 72. If we let our classes run from 10 through 19, from 20 through 29, and so on up to 70 through 79, all guidelines above will be met and the common class width will be a convenient 10 units. We go through the data set, tallying each item into the appropriate class. The tally totals produce class frequencies, which in turn produce relative frequencies, as shown in Table 3. The histogram is displayed in Figure 3.

TABLE 3 Grouped Frequency and Relative Frequency Distributions for Weekly Study Times

Class Limits	Tally	Frequency f	Relative Frequency f/n
10–19	𝍸 𝍷	6	$6/40 = 15.0\%$
20–29	𝍸 𝍸 𝍷	11	$11/40 = 27.5\%$
30–39	𝍸 𝍷𝍷𝍷𝍷	9	$9/40 = 22.5\%$
40–49	𝍸 𝍷𝍷	7	$7/40 = 17.5\%$
50–59	𝍷𝍷𝍷𝍷	4	$4/40 = 10.0\%$
60–69	𝍷𝍷	2	$2/40 = 5.0\%$
70–79	𝍷	1	$1/40 = 2.5\%$

Total: $n = 40$

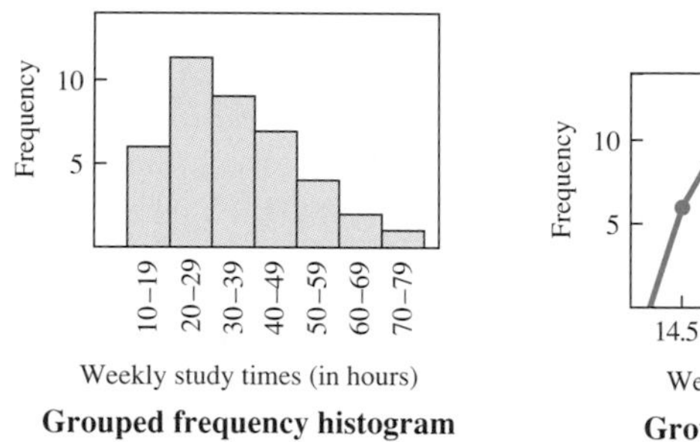

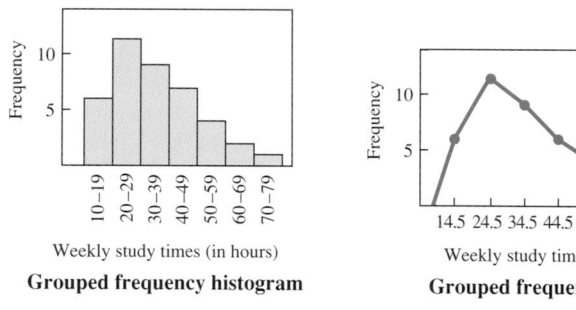

Grouped frequency histogram

FIGURE 3

Frequency
10
5
14.5 24.5 34.5 44.5 54.5 64.5 74.5
Weekly study times (in hours)

Grouped frequency polygon

FIGURE 4

In Table 3 (and Figure 3) the numbers 10, 20, 30, and so on are called the **lower class limits.** They are the smallest possible data values within the respective classes. The numbers 19, 29, 39, and so on are called the **upper class limits.** The common **class width** for the distribution is the difference of any two successive lower class limits (such as 30–20), or of any two successive upper class limits (such as 59–49). The class width for this distribution is 10.

To construct a frequency polygon, notice that, in a *grouped* frequency distribution, the data items in a given class are generally not all the same. We can obtain the "middle" value, or **class mark,** by adding the lower and upper class limits and dividing this sum by 2. We locate all the class marks along the horizontal axis and plot points above the class marks. The heights of the plotted points represent the class frequencies. The resulting points are connected just as for an ordinary (non-grouped) frequency distribution. The result, in this case, is shown in Figure 4. ■

For Further Thought

When fair coins are tossed, the results on particular tosses cannot be reliably predicted. As more and more coins are tossed, however, the proportions of heads and tails become more predictable. This is a consequence of the "law of large numbers." For example, if five coins are tossed, then the resulting number of heads, denoted x, is a "random variable," whose possible values are 0, 1, 2, 3, 4, and 5. If the five coins are tossed repeatedly, say 64 separate times, then the binomial probability formula can be used to get **expected frequencies** (or **theoretical frequencies**), as shown in the table below. The first two columns of the table comprise an **expected frequency distribution** for 64 tosses of five fair coins. In an actual experiment, we could obtain **observed frequencies** (or **empirical frequencies**), which would most likely differ somewhat from the expected frequencies. But 64 repetitions of the experiment should be enough to provide fair consistency between expected and observed values.

Number of Heads x	Expected Frequency e	Observed Frequency o
0	2	
1	10	
2	20	
3	20	
4	10	
5	2	

For Group Discussion

Have members of your group toss five coins a total of 64 times, keeping a record of the results.

1. Enter your experimental results in the third column of the table above, producing an **observed frequency distribution.**
2. Construct two histograms, one from the expected frequency distribution and one from your observed frequency distribution.
3. Compare the two histograms, and explain why they are different.

Stem-and-Leaf Displays

TABLE 4 Grouped Frequency Distribution for Weekly Study Times

Class Limits	Frequency
10–19	6
20–29	11
30–39	9
40–49	7
50–59	4
60–69	2
70–79	1

Notice that in Table 3 the tally marks themselves give a good impression of how the data are distributed. (It is clear, for example, that the highest frequency occurs in the class 20–29 and that the items are concentrated more heavily in the lower classes generally.) In fact the tally marks are almost like a histogram turned on its side. Nevertheless, once the tallying is done, the tally marks are usually dropped and the grouped frequency distribution is presented as in Table 4.

The pictorial advantage of the tally marks is now lost. Furthermore, we cannot tell, from the grouped frequency distribution itself (or from the tally marks either, for that matter), what any of the original items were. We only know, for example, that there were seven items in the class 40–49. We do not know specifically what any of them were.

To avoid these hindrances, we can employ one of the tools of exploratory data analysis, namely the **stem-and-leaf display,** which is illustrated in Example 3.

EXAMPLE 3 Present the study times data of Example 2 in a stem-and-leaf display.

Going back to Example 2 for the original raw data, we arrange the numbers in Table 5. The tens digits, to the left of the vertical line, are the "stems," while the corresponding ones digits are the "leaves." We have entered all items from the first row of the original data, from left to right, then the items from the second row, and so on through the fourth row. ■

TABLE 5 Stem-and-Leaf Display for Weekly Study Times

Stem	Leaves
1	8 5 2 6 4 7
2	0 6 9 6 5 4 2 9 9 3 4
3	2 0 1 9 1 6 8 6 3
4	1 5 4 0 5 7 4
5	8 5 5 2
6	0 2
7	2

Notice that the stem-and-leaf display of Example 3 conveys at a glance the same pictorial impressions that a histogram would convey without the need for constructing the drawing. And it also preserves the exact data values.

Bar Graphs and Pie Charts

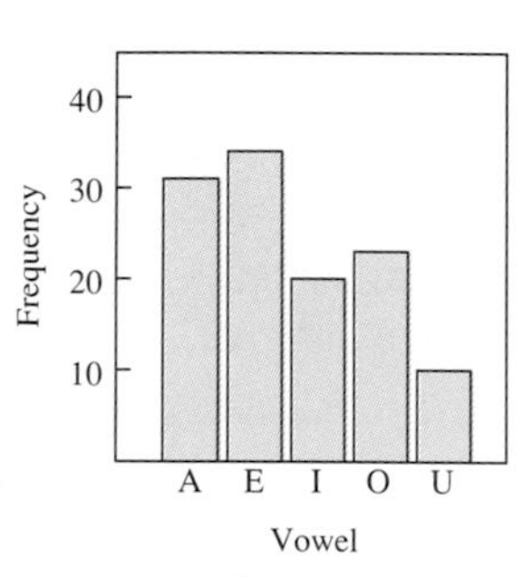

■ **FIGURE 5**

A frequency distribution of nonnumerical observations can be presented in the form of a **bar graph,** which is similar to a histogram except that the rectangles (bars) are usually not touching one another and are sometimes arranged horizontally rather than vertically. The bar graph of Figure 5 shows the frequencies of occurrence of the vowels A, E, I, O, and U in this paragraph.

A graphical alternative to the bar graph is the so-called **circle graph,** or **pie chart,** which uses a circle to represent the total of all the categories and divides the circle into sectors, or wedges (like pieces of pie), whose sizes show the relative magnitudes of the categories. The angle around the entire circle measures 360°. Then, for example, a category representing 20% of the whole should correspond to a sector whose central angle is 20% of 360°, that is, $.20(360°) = 72°$.

EXAMPLE 4 Suzanne Sitlington found that, during her first semester of college, her expenses fell into categories as shown in Table 6. Present this information in a circle graph.

The central angle of the sector for food is $.30(360°) = 108°$. Rent is $.25(360°) = 90°$. Calculate the other four angles similarly. Then draw a circle and measure the appropriate angles with a protractor. The completed circle graph appears in Figure 6. ■

TABLE 6 Student Expenses

Expenses	Percent of Total
Food	30%
Rent	25%
Entertainment	15%
Clothing	10%
Books	10%
Other	10%

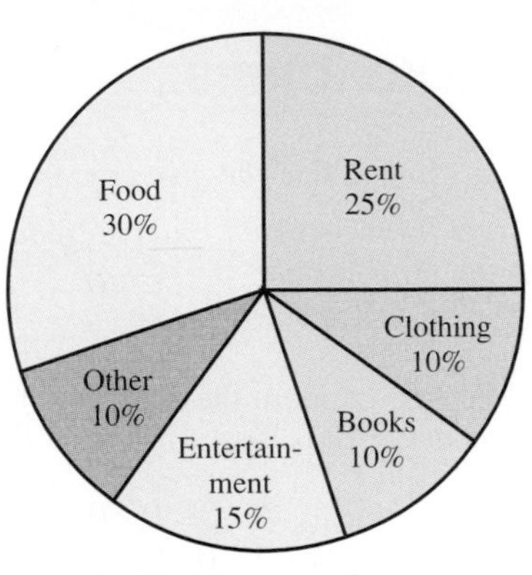

FIGURE 6

A circle graph shows, at a glance, the relative magnitudes of various categories. If we are interested in demonstrating how a quantity *changes,* say with respect to time, we can use another type of illustration, usually called a **line graph.** We connect a series of line segments which rise or fall with time, according to the magnitude of the quantity being illustrated. To compare the patterns of change for two or more quantities, we can even plot multiple line graphs together. (A line graph looks somewhat like a frequency polygon, but the quantities graphed are not necessarily frequencies.)

EXAMPLE 5 Suzanne, from Example 4, wanted to keep track of her major expenses, food and rent, over the course of four years of college (eight semesters), in order to see how each one's budget percentage changed with time and also how the two compared with each other. Use the data she collected (Table 7) to show this information in a line graph, and state any significant conclusions that are apparent from the graph.

A comparison line graph for the given data (Figure 7) shows that the food percentage stayed fairly constant over the four years (at close to 30%), while the rent percentage, starting several points below food, rose steadily, surpassing food after the fourth semester and finishing significantly higher than food.

TABLE 7 Food and Rent Expense Percentages

Semester	Food	Rent
First	30%	25%
Second	31	26
Third	30	28
Fourth	29	29
Fifth	28	34
Sixth	31	34
Seventh	30	37
Eighth	29	38

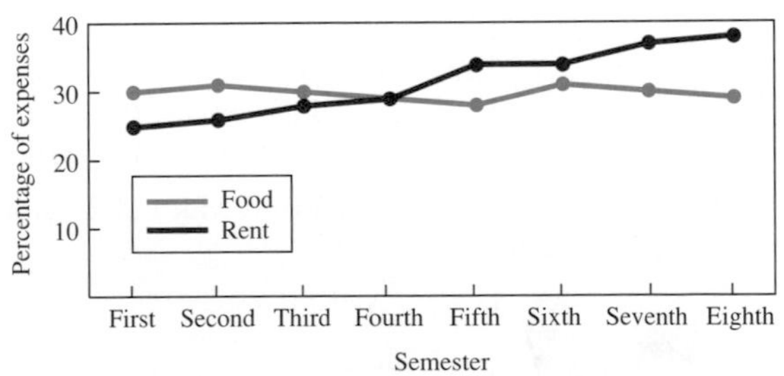

Comparison line graph

FIGURE 7

12.1 EXERCISES

In Exercises 1 and 2, use the given data to do the following:

(a) *Construct frequency and relative frequency distributions, in a table similar to Table 2.*
(b) *Construct a histogram.*
(c) *Construct a frequency polygon.*

1. The following data are the numbers of questions omitted on a math exam by the 30 members of the class.

1 0 3 0 0 2 1 2 2 0 0 5 1 1 3
4 2 0 2 0 1 0 1 2 3 3 4 0 1 0

2. The following data are quiz scores for the members of an economics class.

8 5 6 10 4 7 2 7 6 3 1 7 4 9
5 9 2 6 5 4 6 6 8 4 10 8 9 7

In each of Exercises 3–6, use the given data to do the following:

(a) *Construct grouped frequency and relative frequency distributions, in a table similar to Table 3. (In each case, follow the suggested guidelines for class limits and class width.)*
(b) *Construct a histogram.*
(c) *Construct a frequency polygon.*

3. The heights (in inches) of the 54 starting players in a baseball tournament were as follows.

53 51 65 62 61 55 59 52 62
64 48 54 64 57 51 67 60 49
49 59 54 52 53 60 58 60 64
52 56 56 58 66 59 62 50 58
60 63 64 52 60 58 63 53 56
58 61 55 50 65 56 61 55 54

Use five classes with a uniform class width of 5 inches, and use a lower limit of 45 inches for the first class.

4. The following raw data represent the monthly account balances (to the nearest dollar) for a sample of 50 brand new charge card users.

138 78 175 46 79 118 90 163 88 107
126 154 85 60 42 54 62 128 114 73
129 130 81 67 119 116 145 105 96 71
100 145 117 60 125 130 94 88 136 112
118 84 74 62 81 110 108 71 85 165

Use seven classes with a uniform width of 20 dollars, where the lower limit of the first class is 40 dollars.

5. The following data represent the daily high temperatures (in degrees Fahrenheit) for the month of June in a southwestern U.S. city.

79	84	88	96	102	104	99	97	92	94
85	92	100	99	101	104	110	108	106	106
90	82	74	72	83	107	111	102	97	94

Use nine classes with a uniform width of 5 degrees, where the lower limit of the first class is 70 degrees.

6. The following data represent IQ scores of a group of 50 tenth graders.

113	109	118	92	130	112	114	117	122	115
127	107	108	113	124	112	111	106	116	118
121	107	118	118	110	124	115	103	100	114
104	124	116	123	104	135	121	126	116	111
96	134	98	129	102	103	107	113	117	112

Use nine classes with a uniform width of 5, where the lower limit of the first class is 91.

In each of Exercises 7–10, construct a stem-and-leaf display for the given data. In each case, treat the ones digits as the leaves. For any single-digit data, use a stem of 0.

7. On a certain date approaching midseason, the teams in the National Basketball Association had won the following numbers of games.

20	29	11	26	11	12	7	26	18
19	14	13	22	9	25	11	10	15
10	22	23	31	8	24	15	24	15

8. The students in a calculus class were asked how many college units they had accumulated to date. Their responses were as follows.

12	4	13	12	21	22	15	17	33	24
32	42	26	11	53	62	42	25	13	8
54	18	21	14	19	17	38	17	20	10

9. Following are the daily round-trip distances to school (in miles) for 30 randomly chosen students at a community college in California.

12	30	10	11	18	26	34	18	8	12
26	14	5	22	4	25	9	10	6	21
18	18	9	16	44	23	4	13	36	8

10. The following data represent net yards gained per game by National Football League running backs who played during a given week of the season.

28	19	36	73	37	88	67	33	54	123	79
12	39	45	22	58	7	73	30	43	24	36
51	43	33	55	40	29	112	60	94	86	62
42	29	18	25	41	3	49	102	16	32	46

Bar graphs (as well as other kinds of graphs) are often drawn in special ways in order to catch attention or to emphasize certain information. The example shown here pertains to the changing nature of U.S. cities throughout this century. (The picket fence is a common symbol of the suburban "American dream.")

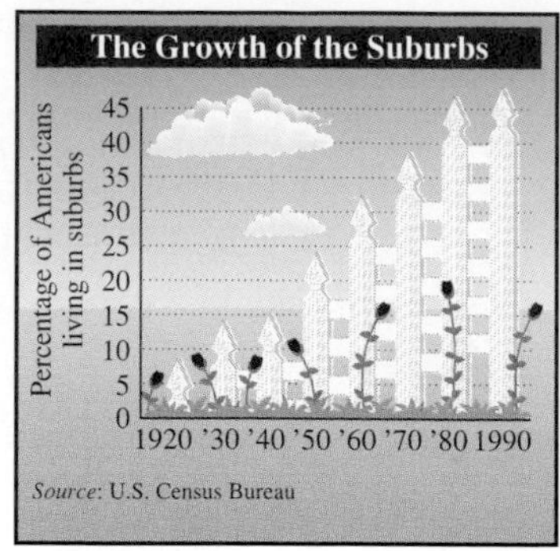

Refer to the graph for Exercises 11–13.

11. About what percentage of Americans lived in suburbs in the following years?
(a) 1920 **(b)** 1950 **(c)** 1990

12. In which decade was there an increase of about 10%?

13. In which decade was the percentage closest to 50%?

Example 5 used a line graph to compare two quantities. Bar graphs are often used for comparison purposes as well.

The bar graphs here show "balance of trade" data for the three years leading up to the North American Free Trade Agreement (NAFTA) of 1992 (ratified in 1993). Refer to these bar graphs for Exercises 14–21. Note that a country's "trade deficit" is the amount by which imports exceed exports, while a "trade surplus" is the amount by which exports exceed imports.

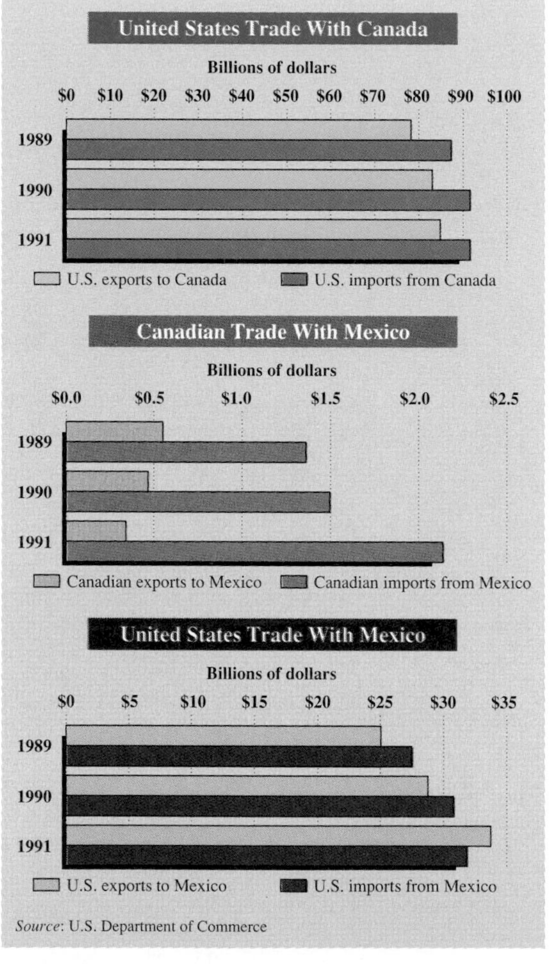

14. What was the dollar amount of Canadian imports from Mexico in 1990?

15. What was the approximate value of U.S. exports to Canada in 1991?

16. What trade category declined over the 1989–1991 period?

17. Over the three-year period, which country was able to steadily increase its trade surplus over which other country?

18. Over the three-year period, which country was able to steadily decrease (but not eliminate) its trade deficit with which other country?

19. Over the three-year period, which country changed a deficit to a surplus (with which other country)?

20. In 1990, what was the approximate U.S. net combined trade deficit with both Canada and Mexico?

21. (a) If the period 1991–1998 saw a 50% increase in U.S. exports to Mexico, what would be the approximate value of those exports in 1998?
(b) Find out the most recent data on U.S. exports to Mexico, and discuss how close the scenario in part (a) is to reality.

The two circle graphs here show revenue sources for U.S. cities in 1980 and 1990. Refer to them for Exercises 22 and 23.

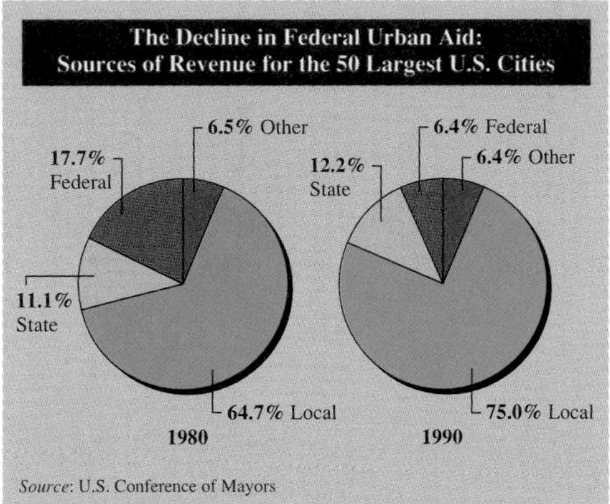

22. How much less was the federal support percentage in 1990 than in 1980?

23. How much more was the local support percentage in 1990 than in 1980?

24. A survey by the Bureau of Labor Statistics asked American workers how they were trained for their jobs. The percentages who responded in various categories are shown in the table below. Use the information in the table to draw a circle graph.

Principal Source of Training	Approximate Percentage of Workers
Trained in school	33%
Informal on-the-job training	25
Formal training from employers	12
Trained in military, or correspondence or other courses	10
No particular training, or could not identify any	20

25. Bureau of Labor Statistics data for a recent year showed that the average annual earnings of American workers corresponded to educational level as shown in the table below. Draw a bar graph that shows this information.

Educational Level	Average Annual Earnings
Less than 4 years of high school	$19,168
High school graduate	24,308
Four years of college	38,620

Dick Stratton, wishing to retire at age 60, is studying the comparison line graph here, which shows (under certain assumptions) how the net worth of his retirement savings (initially $200,000 at age 60) will change as he gets older, and as he withdraws living expenses from savings. Refer to the graph for Exercises 26–29.

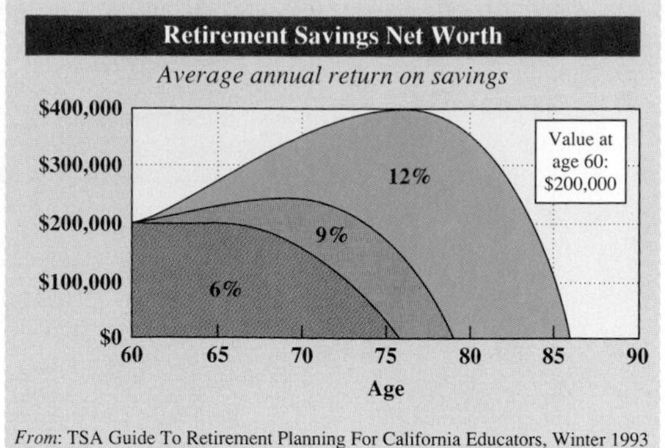

From: TSA Guide To Retirement Planning For California Educators, Winter 1993

26. Assuming Dick can maintain an average annual return of 9%, how old will he be when his money runs out?

27. If he could earn an average of 12% annually, what maximum net worth would Dick achieve? At about what age would the maximum occur?

28. Suppose Dick reaches age 70, in good health, and the average annual return has proved to be 6%.

(a) About how much longer can he expect his money to last?

(b) What options might he consider, in order to extend that time?

29. At age 77, about how many times more will Dick's net worth be if he averages a 12% return than if he averages a 9% return?

Stem-and-leaf displays can be modified in various ways in order to obtain a reasonable number of stems. For example, the following data, representing the measured masses (in grams) of thirty mineral samples in a geology lab, are shown in a ***double-stem*** *display in Table 8.*

60.7	41.4	50.6	39.5	46.4	58.1	49.7	38.8	61.6	55.2
47.3	52.7	62.4	59.0	44.9	35.6	36.2	40.6	56.9	42.6
34.7	48.3	55.8	54.2	33.8	51.3	50.1	57.0	42.8	43.7

TABLE 8 Stem-and-Leaf Display for Mineral Sample Masses

(30–34)	3	4.7	3.8				
(35–39)	3	9.5	8.8	5.6	6.2		
(40–44)	4	1.4	4.9	0.6	2.6	2.8	3.7
(45–49)	4	6.4	9.7	7.3	8.3		
(50–54)	5	0.6	2.7	4.2	1.3	0.1	
(55–59)	5	8.1	5.2	9.0	6.9	5.8	7.0
(60–64)	6	0.7	1.6	2.4			

30. Describe how the stem-and-leaf display of Table 8 was constructed.

31. Explain why Table 8 is called a "double stem" display.

32. In general, how many stems (total) are appropriate for a stem-and-leaf display? Explain your reasoning.

33. According to the *1995 Information Please Almanac,* the highest temperatures (in degrees Fahrenheit) ever recorded in the 50 states were as follows.

112	100	127	120	134	118	105	110	109	113
100	118	117	116	118	121	114	114	105	109
107	112	114	115	118	117	118	122	106	110
116	108	110	121	113	120	119	111	104	111
120	113	120	117	105	110	118	112	114	114

Present these data in a double stem display.

34. The table here shows commonly accepted percentages of occurrence for the various letters in English language usage. (Code breakers have carefully analyzed these percentages as an aid in deciphering secret codes.) For example, notice that E is the most commonly occurring letter, followed by T, A, O, N, and so on. The letters Q and Z occur least often. Referring to Figure 5 in the text, would you say that the relative frequencies of occurrence of the vowels in the associated paragraph were typical or unusual? Explain your reasoning.

Letter	Percent	Letter	Percent
E	13	L	3 1/2
T	9	C, M, U	3
A, O	8	F, P, Y	2
N	7	W, G, B	1 1/2
I, R	6 1/2	V	1
S, H	6	K, X, J	1/2
D	4	Q, Z	1/5

The percentages shown in Exercise 34 are based on a very large sampling of English language text. Since they are based upon experiment, they are "empirical" rather than "theoretical." By converting each percent in that table to a decimal fraction, you can produce an ***empirical probability distribution.*** *For example, if a single letter is randomly selected from a randomly selected passage of text, the probability that it will be an* E *is* .13. *The probability that a randomly selected letter would be a vowel (*A, E, I, O, *or* U*) is* .385 (.08 + .13 + .065 + .08 + .03).

35. Rewrite the distribution shown in Exercise 34 as an empirical probability distribution. Give values to three decimal places. Note that the 26 probabilities in this distribution—one for each letter of the alphabet—should add up to 1 (except for, perhaps, a slight round-off error).

36. From your distribution of Exercise 35, construct an empirical probability distribution just for the vowels A, E, I, O, and U. (*Hint:* Divide each vowel's probability, from Exercise 35, by .385 to obtain a distribution whose five values add up to 1.) Give values to three decimal places.

37. Construct an appropriately labeled bar chart from your distribution of Exercise 36.

38. Based on the occurrences of vowels in the paragraph represented by Figure 5, construct a probability distribution for the vowels. The frequencies are: A–31, E–34, I–20, O–23, U–10. Give probabilities to three decimal places.

39. Is the probability distribution of Exercise 38 theoretical or empirical? Is it different from the distribution of Exercise 36? Which one is more accurate? Explain your reasoning.

40. Convert the grouped frequency distribution of Table 3 to an empirical probability distribution, using the same classes and giving probability values to three decimal places.

41. Recall that the distribution of Exercise 40 was based on weekly study times for a sample of 40 students. Suppose one of those students was chosen randomly. Using your distribution, find the probability that the study time in the past week for the student selected would have been in each of the following ranges.

(a) 30–39 hours **(b)** 40–59 hours **(c)** fewer than 30 hours
(d) at least 50 hours

The 40 members of a recreation class were asked to name their favorite sport. The table below shows the numbers who responded in various ways.

Sport	Number of Class Members
Sailing	9
Hang gliding	5
Bungee jumping	7
Sky diving	3
Canoeing	12
Rafting	4

Use this information in Exercises 42–45.

42. If a member of this class is selected at random, what is the probability that the favorite sport of the person selected is bungee jumping?

43. Based on the data above, construct a probability distribution, giving probabilities to three decimal places.

44. **(a)** Is the distribution of Exercise 43 theoretical or is it empirical?
(b) Explain your answer to part (a).

45. Explain why a frequency polygon trails down to the axis at both ends while a line graph ordinarily does not.

12.2 Measures of Central Tendency

Video Recyclers, a local business that sells "previously viewed" movies, had the following daily sales over a one-week period:

$$\$305, \quad \$285, \quad \$240, \quad \$376, \quad \$198, \quad \$264.$$

It would be desirable to have a single number to serve as a kind of representative value for this whole set of numbers—that is, some value around which all the numbers in the set tend to cluster, a kind of "middle" number or a **measure of central tendency.** Three such measures are discussed in this section: the mean, the median, and the mode. (The "weighted mean," also discussed in this section, is a special application of the mean.)

The most common measure of central tendency is the **mean** (more properly called the **arithmetic mean**). The mean of a sample is denoted $\overline{x}$ (read "x bar"), while the mean of a complete population is denoted μ (the lower case Greek letter *mu*). Inferential statistics often involves both sample means and the population mean in the same discussion, but for our purposes here, data sets are considered to be samples, so we will use $\overline{x}$. The mean of a set of data items is found by adding up all the items and then dividing the sum by the number of items. (The mean is what most people associate with the word "average.") Since adding up, or summing, a list of items is a common procedure in statistics, we will make use of the common symbol for "summation," which is Σ (the capital

Greek letter *sigma*). Thus the sum of n items, say $x_1, x_2, \ldots, x_n$, can be denoted $\Sigma x = x_1 + x_2 + \cdots + x_n$, and the mean is found as follows.

> **Mean**
>
> The **mean** of n data items, $x_1, x_2, \ldots, x_n$, is given by the formula
>
> $$\bar{x} = \frac{\Sigma x}{n}.$$

Now use this formula to find the central tendency of the daily sales figures for Video Recyclers:

$$\text{Mean} = \bar{x} = \frac{\Sigma x}{n} = \frac{305 + 285 + 240 + 376 + 198 + 264}{6} = \frac{1{,}668}{6}$$
$$= 278.$$

The mean value (the "average daily sales") for the week is \$278.

Many calculators find the mean automatically when a set of data items are entered. To recognize these calculators, look for a key marked $\boxed{\bar{x}}$, or perhaps $\boxed{\mu}$. The student is encouraged to make full use of the capabilities of the calculator, but also to understand the formulas involved in the calculations.

EXAMPLE 1 Last year's annual sales for eight different flower shops were

\$374,910	\$321,872	\$242,943	\$351,147
\$382,740	\$412,111	\$334,089	\$262,900.

The sum is \$2,682,712. Now find the mean.

$$\bar{x} = \frac{\Sigma x}{n} = \frac{2{,}682{,}712}{8} = 335{,}339$$

The mean annual sales amount is \$335,339. ■

Grade-point averages are very familiar to college students. For example, the table below shows the units and grades earned by one student last term.

Course	Grade	Units
Mathematics	A	3
History	C	3
Chemistry	B	5
Art	B	2
PE	A	1

In one common system of finding a grade-point average, an A grade is assigned 4 points, with 3 points for B, 2 for C, and 1 for D. Find the grade-point average by multiplying the number of units for a course and the number assigned to each grade, and then adding these products. Then divide this sum by the total number of units. Work as follows.

Course	Grade	Grade Points	Units	(Grade Points) × (Units)
Mathematics	A	4	3	12
History	C	2	3	6
Chemistry	B	3	5	15
Art	B	3	2	6
PE	A	4	1	4
		Totals:	14	43

$$\text{Grade-point average} = \frac{43}{14} = 3.07 \quad \text{(rounded)}$$

This student earned a grade-point average of 3.07.

The calculation of a grade-point average is an example of a *weighted mean,* since the grade points for each course grade must be weighted according to the number of units of the course. (For example, five units of A is better than 2 units of A.) The number of units is called the *weighting factor*.

The discussion of grade-point averages above suggests the following general definition.

Weighted Mean

The **weighted mean** of n numbers, $x_1, x_2, \ldots, x_n$, that are weighted by the respective factors $f_1, f_2, \ldots, f_n$ is given by the formula

$$\overline{w} = \frac{\Sigma x \cdot f}{\Sigma f}.$$

In words, the weighted mean of a group of (weighted) items is the sum of all products of items times weighting factors, divided by the sum of all weighting factors.

The weighted mean formula is commonly used to find the mean for a frequency distribution. In this case, the weighting factors are the frequencies.

For Further Thought

In baseball statistics, a player's "batting average" gives the average number of hits per time at bat. For example, a player who has gotten 84 hits in 250 times at bat has a batting average of $84/250 = .336$. This "average" can be interpreted as the empirical probability of that player's getting a hit the next time at bat.

The following are actual comparisons of hits and at-bats for two major league players in the 1989 and 1990 seasons. The numbers illustrate a puzzling statistical occurrence known as **Simpson's paradox.** (This information was reported by Richard J. Friedlander on page 845 of the November 1992 issue of the *MAA Journal.*)

	Dave Justice			Andy Van Slyke		
	Hits	**At-bats**	**Batting Average**	**Hits**	**At-bats**	**Batting Average**
1989	12	51	_____	113	476	_____
1990	124	439	_____	140	493	_____
Combined (1989–90)	_____	_____	_____	_____	_____	_____

For Group Discussion

1. Fill in the ten blanks in the table, giving batting averages to three decimal places.
2. Which player had a better average in 1989?
3. Which player had a better average in 1990?
4. Which player had a better average in 1989 and 1990 combined?
5. Did the results above surprise you? How can it be that one player's batting average leads another's for each of two years, and yet trails the other's for the combined years?

EXAMPLE 2 Find the mean salary for a small company which pays annual salaries to its employees as shown in the following frequency distribution.

Salary x	Number of Employees f
\$12,000	8
\$16,000	11
\$18,500	14
\$21,000	9
\$34,000	2
\$50,000	1

According to the weighted mean formula, we can set up the work as follows.

Salary x	Number of Employees f	Salary × Number $x \cdot f$
\$12,000	8	\$ 96,000
\$16,000	11	\$176,000
\$18,500	14	\$259,000
\$21,000	9	\$189,000
\$34,000	2	\$ 68,000
\$50,000	1	\$ 50,000
	Totals: 45	\$838,000

$$\text{Mean salary} = \frac{\$838{,}000}{45} = \$18{,}622 \quad \text{(rounded)}$$ ■

For some data sets the mean can be a misleading indicator of average. Consider Shady Sam, who runs a small business that employs five workers at the following annual salaries.

$11,500, $11,950, $12,800, $14,750, $15,000

The employees, knowing that Sam accrues vast profits to himself, decide to go on strike and demand a raise. To get public support, they go on television and tell about their miserable salaries, pointing out that the mean salary in the company is only

$$\bar{x} = \frac{\$11{,}500 + \$11{,}950 + \$12{,}800 + \$14{,}750 + \$15{,}000}{5}$$

$$= \frac{\$66{,}000}{5} = \$13{,}200.$$

The local television station schedules an interview with Sam to investigate. In preparation, Sam calculates the mean salary of *all* workers (including his own salary of $195,000):

$$\bar{x} = \frac{\$11{,}500 + \$11{,}950 + \$12{,}800 + \$14{,}750 + \$15{,}000 + \$195{,}000}{6}$$

$$= \frac{\$261{,}000}{6} = \$43{,}500.$$

When the T.V. crew arrives, Sam calmly assures them that there is no reason for his employees to complain since the company pays a generous mean salary of $43,500.

The employees, of course, would argue that when Sam included his own salary in the calculation, it caused the mean to be a misleading indicator of average. This was so because Sam's salary is not typical. It lies a good distance away from the general grouping of the items (salaries). An extreme value like this is referred to as an *outlier*. Since a single outlier can have a significant effect on the value of the mean, we say that the mean is "highly sensitive to extreme values."

Another measure of central tendency, which is not so sensitive to extreme values, is the **median.** This is the value that divides a group of numbers into two parts, with half the numbers below the median and half above it.

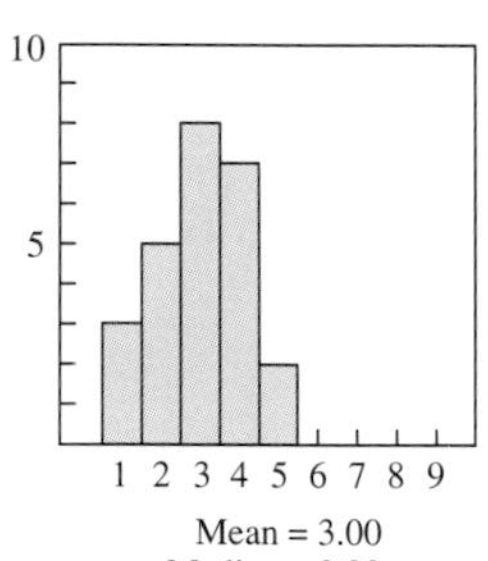

Mean = 3.00
Median = 3.00

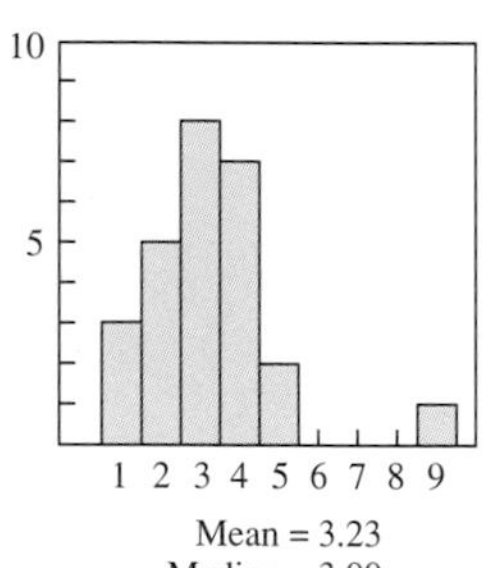

Mean = 3.23
Median = 3.00

The introduction of a single "outlier" above increased the mean by 8 percent but left the median unaffected.

Outliers should usually be considered as *possible* errors in the data.

Median

To find the **median** of a group of items:

1. Arrange the items in numerical order (from smallest to largest).
2. If the number of items is *odd,* the median is the middle item in the list.
3. If the number of items is *even,* the median is the mean of the two middle items.

For Shady Sam's business, all salaries (including Sam's), arranged in numerical order, are

$11,500, $11,950, $12,800, $14,750, $15,000, $195,000,

so

$$\text{median} = \frac{\$12{,}800 + \$14{,}750}{2} = \frac{\$27{,}550}{2} = \$13{,}775.$$

This figure is a representative average, based on all six salaries, that the employees would probably agree is reasonable.

EXAMPLE 3 Find the median of each list of numbers.

(a) 6, 7, 12, 13, 18, 23, 24

This list is already in numerical order. The number of values in the list, 7, is odd, so the median is the middle value, or 13.

(b) 17, 15, 9, 13, 21, 32, 41, 7, 12

First, place the numbers in numerical order from smallest to largest.

7, 9, 12, 13, 15, 17, 21, 32, 41

The middle number can now be picked out; the median is 15.

(c) 147, 159, 132, 181, 174, 253

First write the numbers in numerical order.

132, 147, 159, 174, 181, 253

Since the list contains an even number of items, namely 6, there is no single middle item. Find the median by taking the mean of the two middle items, 159 and 174.

$$\frac{159 + 174}{2} = \frac{333}{2} = 166.5.$$

In the case of a frequency distribution, locating the middle item (the median) is a bit different. First find the total number of items in the set by adding the frequencies ($n = \Sigma f$). Then the median is the item whose *position* is given by the following formula.

Position of the Median in a Frequency Distribution

$$\text{Position of median} = \frac{n + 1}{2} = \frac{\Sigma f + 1}{2}$$

It is important to notice that this formula gives only the *position,* and not the actual value, of the median. The next example shows how the formula is used to find the median.

EXAMPLE 4 Find the medians for the following distributions.

(a)

Value	Frequency	Cumulative Frequency
1	1	1
2	3	4
3	2	6
4	4	10
5	8	18
6	2	20
	Total: 20	

Adding the frequencies shows that there are 20 items total, so

$$\text{position of median} = \frac{20 + 1}{2} = \frac{21}{2} = 10.5.$$

The median, then, is the average of the tenth and eleventh items. To find these items, make use of *cumulative frequencies,* which tell, for each different value, how many items have that value or a lower value. Since the value 4 has a cumulative frequency of 10 $(1 + 3 + 2 + 4 = 10)$, that is, 10 items have a value of 4 or less, and 5 has a cumulative frequency of 18, the tenth item is 4 and the eleventh item is 5, making the median $(4 + 5)/2 = 9/2 = 4.5$.

(b)

Value	Frequency	Cumulative Frequency
2	5	5
4	8	13
6	10	23
8	6	29
10	6	35
	Total: 35	

Here there are 35 items total, so

$$\text{position of median} = \frac{35 + 1}{2} = \frac{36}{2} = 18.$$

From the cumulative frequency column, the fourteenth through the twenty-third items are all 6s. This means the eighteenth item is a 6, so the median is 6.

The third important measure of central tendency is the **mode.** If ten students earned scores on a business law examination of

74, 81, 39, 74, 82, 80, 100, 92, 74, 85,

then we notice that more students earned the score 74 than any other score. This fact makes 74 the mode of this list.

A note in *The Wall Street Journal* claimed that during a televised presentation of the Academy Awards, the average winner took 1 minute and 39 seconds to thank 7.8 friends, relatives, and loyal supporters for "making it all possible." Would these numbers refer to means, medians, or modes?

Mode

The **mode** of a data set is the value that occurs most often.

EXAMPLE 5 Find the mode for each set of data.

(a) 51, 32, 49, 49, 74, 81, 92
The number 49 occurs more often than any other. Therefore, 49 is the mode. Note that it is not necessary to place the numbers in numerical order when looking for the mode.

(b) 482, 485, 483, 485, 487, 487, 489
Both 485 and 487 occur twice. This list is said to have *two* modes, or to be *bimodal.* (See the discussion below.)

(c) 10,708, 11,519, 10,972, 17,546, 13,905, 12,182
No number here occurs more than once. This list has no mode.

(d)

Value	Frequency
19	1
20	3
22	8 ← Greatest frequency
25	7
26	4
28	2

The frequency distribution shows that the most frequently occurring value (and thus the mode) is 22. ■

That we have included the mode (along with the mean and the median) as a measure of *central tendency* is traditional, probably because many important kinds of data sets do have their most frequently occurring values "centrally" located. However, there is no reason why the mode cannot be one of the smallest values in the set or one of the largest. In such a case, the mode really is not a good measure of "central tendency."

When the data items being studied are non-numeric, the mode may be the only measure of central tendency that can be used. For example, the bar graph of Figure 5 in Section 1 showed frequencies of occurrence of vowels in a sample paragraph. Since the vowels A, E, I, O, and U are not numbers, they cannot be added, so their mean does not exist. Furthermore, the vowels cannot be arranged in any meaningful numerical order, so their median does not exist either. The mode, however, does exist. As the bar graph shows, the mode is the letter E.

Sometimes, a distribution will contain two distinct items that both occur more often than any other items such as in Example 5(b) above. Such a distribution is called **bimodal** (literally, "two modes"). This term is commonly applied even when the two modes do not have exactly the same frequency. When a distribution has three or more different items sharing the highest frequency of occurrence, that information is not often useful. (Too many modes tend to obscure the significance of these "most frequent" items.) We will say that such a distribution has *no* mode.

As shown in Section 1, data are sometimes presented in a stem-and-leaf display in order to give a graphical impression of their distribution. We can also calculate measures of central tendency from a stem-and-leaf display. The median and mode are more easily identified when the "leaves" are **ranked** (arranged in numerical order) on their "stems." In Table 9, we have rearranged the leaves of Table 5 in Section 1 (which showed the weekly study times from Example 2 of that section).

TABLE 9 Stem-and-Leaf Display for Weekly Study Times, with Leaves Ranked

1	2	4	5	6	7	8					
2	0	2	3	4	4	5	6	6	9	9	9
3	0	1	1	2	3	6	6	8	9		
4	0	1	4	4	5	5	7				
5	2	5	5	8							
6	0	2									
7	2										

EXAMPLE 6 Find the mean, median, and mode for the data of Table 9.

If you have entered the data into a calculator with statistical capabilities, the mean will be automatically computed for you. Otherwise, add all items (reading from the stem-and-leaf display) and divide by $n = 40$.

$$\bar{x} = \frac{12 + 14 + 15 + \cdots + 60 + 62 + 72}{40} = \frac{1{,}395}{40} = 34.875$$

In this case, $n = 40$ (an even number), so the median is the average of the twentieth and twenty-first items, in order. Counting leaves, we see that these will be the third and fourth items on the stem 3. So

$$\text{median} = \frac{31 + 32}{2} = 31.5.$$

By inspection, we see that 29 occurred three times and no other value occurred that often, so

$$\text{mode} = 29.$$

The most useful way to analyze a data set often depends on whether the distribution is **symmetric** or **non-symmetric.** In a "symmetric" distribution, as we move out from the central point, the pattern of frequencies is the same (or nearly so) to the left and to the right. In a "non-symmetric" distribution, the patterns to the left and right are different. Figure 8 shows several types of symmetric distributions, while Figure 9 shows some non-symmetric distributions. A non-symmetric distribution with a tail extending out to the left is called **skewed to the left.** If the tail extends out to the right, the distribution is **skewed to the right.** Notice that a bimodal distribution may be either symmetric or non-symmetric.

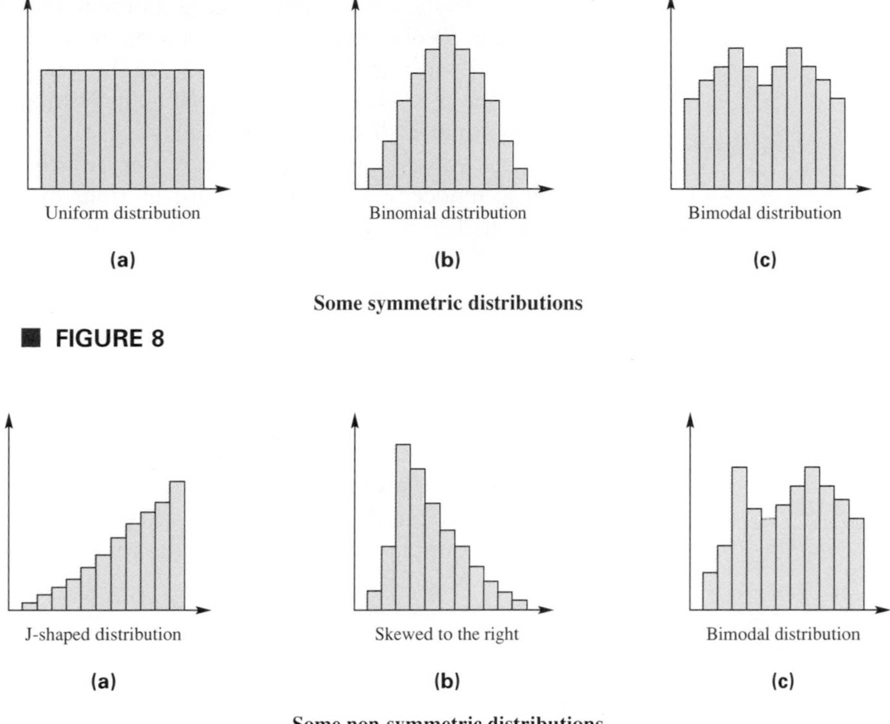

Some symmetric distributions

FIGURE 8

Some non-symmetric distributions

FIGURE 9

We summarize here the measures discussed in this section.

Common Measures of Central Tendency

The **mean** of a set of numbers is found by adding all the values in the set and dividing by the number of values.

The **median** is a kind of "middle" number. To find the median, first arrange the values in numerical order. For an *odd* number of values, the median is the middle value in the list. For an *even* number of values, the median is the mean of the two middle values.

The **mode** is the value that occurs most often. Some sets of numbers have two most frequently occurring values and are **bimodal.** Other sets have no mode at all (if no value occurs more often than the others or if more than two values occur most often).

We conclude this section with a summary of the relative advantages and disadvantages of the mean, the median, and the mode as measures of central tendency.

For distributions of numeric data, the mean and median will always exist, while the mode may not exist. On the other hand, for non-numeric data, it may be that none of the three measures exists. Or it may be that only the mode exists.

Generally, even a single change in the data may cause the mean to change, while the median and mode may not be affected at all. Therefore, the mean is the most "sensitive" of the measures. In the case of a symmetric distribution, the mean, the median, and the mode (if a single mode exists) will all be equal. In a non-symmetric distribution, the mean is often unduly affected by relatively few extreme values, and therefore may not be a good representative measure of central tendency. For example, distributions of salaries, or of family incomes, often include a few values that are much higher than the bulk of the items. In such cases, the median is a more useful measure. Any time there are many more items on one side of the mean than on the other, extreme non-symmetry is indicated and the median is most likely a better choice than the mean.

The mode is the only one of the three measures covered here that must always be equal to one of the data items of the distribution. In fact, more of the data items are equal to the mode than to any other number. A fashion shop planning to stock only one hat size for next season would want to know the mode (the most common) of all hat sizes among their potential customers. Likewise, a designer of family automobiles would be interested in the most common family size. In examples like these, designing for the mean or the median might not be right for anyone.

As we will see in the next section, measures of central tendency do not completely characterize data sets. We will also want to consider the degree of spread in the data.

EXERCISES

For each of the following lists of data, calculate **(a)** *the mean,* **(b)** *the median, and* **(c)** *the mode or modes (if any). Round mean values to the nearest tenth.*

1. 3, 7, 12, 16, 23

2. 21, 25, 32, 48, 53, 62

3. 128, 230, 196, 224, 196, 233

4. 26, 31, 46, 31, 26, 29, 31

5. 3.1, 4.5, 6.2, 7.1, 4.5, 3.8, 6.2, 6.3

6. 14,320, 16,950, 17,330, 15,470

7. .78, .93, .66, .94, .87, .62, .74, .81

8. .53, .03, .28, .18, .39, .28, .14, .22, .04

9. 12.3, 45.6, 78.9, 1.2, 34.5, 67.8, 90.1

10. .3, .8, .4, .3, .7, .9, .2, .1, .5, .9, .6

11. 128, 131, 136, 125, 132, 128, 125, 127

12. 8.97, 5.64, 2.31, 1.02, 4.35, 7.68

The tables here show population figures for the world's twelve most populous countries (1994) and the twelve most populous cities (midyear 1995 estimates). "City" here means a population cluster, which may consist of a city or a city with adjoining cities or clusters if separated by less than one mile. (Source of data: 1995 Information Please Almanac, *pages 133 and 130)*

Most Populous Countries

Rank	Country	Population (in millions)
1	China	1,192.0
2	India	911.6
3	United States	260.8
4	Indonesia	199.7
5	Brazil	155.3
6	Russia	147.8
7	Pakistan	126.4
8	Japan	125.0
9	Bangladesh	116.6
10	Nigeria	98.1
11	Mexico	91.8
12	Germany	81.2

Most Populous Cities

Rank	City	Population (in thousands)
1	Tokyo-Yokohama, Japan	28,447
2	Mexico City, Mexico	23,913
3	São Paulo, Brazil	21,539
4	Seoul, South Korea	19,065
5	New York, United States	14,638
6	Osaka-Kobe-Kyoto, Japan	14,060
7	Bombay, India	13,532
8	Calcutta, India	12,885
9	Rio de Janeiro, Brazil	12,788
10	Buenos Aires, Argentina	12,232
11	Manila, Philippines	11,342
12	Moscow, Russia	10,769

Find **(a)** *the mean population, and* **(b)** *the median population for each of the following.*

13. the 5 most populous countries

14. the 12 most populous countries

15. the 5 most populous cities

16. the 12 most populous cities

Amid concerns about global warming and rain forest destruction, the World Resources Institute reported the five countries with the greatest average yearly forest loss for a recent five-year period, as well as the five greatest emitters of chlorofluorocarbons in a one-year period. The estimates are shown here.

Country	Average Yearly Forest Loss (in square miles)
Brazil	9,700
Colombia	3,400
Indonesia	2,400
Mexico	2,375
Ivory Coast	2,000

Country	Chlorofluorocarbon Emissions (in metric tons)
United States	130,000,000
Japan	95,000,000
C.I.S.	67,000,000
Germany	34,000,000
United Kingdom	25,000,000

17. Find the mean value of yearly forest loss per country for the five countries listed. Give your answer to the nearest 100 square miles.

18. Find the mean value of chlorofluorocarbon emissions per country for the five countries listed. Give your answer to the nearest one million metric tons.

While doing an experiment, a physics student recorded the following sequence of elapsed times (in seconds) in a lab notebook: 2.16, 22.2, 2.96, 2.20, 2.73, 2.28, 2.39.

19. Find the mean.

20. Find the median.

The student, when reviewing the calculations later, decided that the entry 22.2 should have been recorded as 2.22, and made that change in the listing.

21. Find the mean for the new list.

22. Find the median for the new list.

23. Which measure, the mean or the median, was affected more by correcting the error?

24. In general, which measure, mean or median, is affected less by the presence of an extreme value in the data?

Luis Rebonne earned the following scores on his six math tests last semester.

88, 95, 91, 32, 90, 95

25. Find the mean, the median, and the mode for Luis's scores.

26. Which of the three averages probably is the best indicator of Luis's ability?

27. If Luis's instructor gives him a chance to replace his score of 32 by taking a "make-up" exam, what must he score on the make-up to get an overall average (mean) of 93?

Exercises 28–30 give frequency distributions for sets of data values. For each set find the **(a)** *mean (to the nearest hundredth),* **(b)** *median, and* **(c)** *mode or modes (if any).*

28.

Value	Frequency
2	5
4	1
6	8
8	4

29.

Value	Frequency
25	4
28	7
30	10
31	3
32	15

30.

Value	Frequency
603	13
597	8
589	9
598	12
601	6
592	4

31. A company has five employees with a salary of $17,500, eight with a salary of $18,000, four with a salary of $20,300, two with a salary of $24,500, seven with a salary of $26,900, and one with a salary of $115,500. Find the mean salary for the employees (to the nearest hundred dollars).

Find the grade point average for each of the following students. Assume A = 4, B = 3, C = 2, D = 1, F = 0. *Round to the nearest hundredth.*

32.

Units	Grade
4	C
7	B
3	A
3	F

33.

Units	Grade
2	A
6	B
5	C

The table below gives the land area and population of the eleven founding members of the Commonwealth of Independent States. (All were former republics of the Soviet Union.) Use the information from the table for Exercises 34–37.

Name	Area (square miles)	Population
Armenia	11,506	3,373,000
Azerbaijan	33,436	7,222,000
Belarus	80,155	10,480,000
Kazakhstan	1,049,156	16,992,000
Kyrgystan	76,641	4,409,000
Moldova	13,012	4,460,000
Russia	6,592,850	151,436,000
Tajikistan	55,251	5,252,000
Turkmenistan	188,456	3,631,000
Ukraine	233,090	53,125,000
Uzbekistan	172,742	20,453,000

34. Find the mean area (to the nearest square mile) for these 11 states.

35. Discuss the meaningfulness of the average calculated in Exercise 34.

36. Find the mean population (to the nearest 1,000) for the 11 states.

37. Discuss the meaningfulness of the average calculated in Exercise 36.

The table below lists the ten most common languages (other than English) spoken at home by Americans. The numbers in the right column indicate the change (increase or decrease) from 1980 to 1990 in the number of Americans speaking the different languages. (Source of data: 1995 Information Please Almanac, *page 835) Use the table for Exercises 38–41. (The rankings are given for information only. They will not enter into your calculations.)*

Most Common Non-English Languages Among Americans

Rank	Language	Change in Number of People
1	Spanish	5,789,839
2	French	129,901
3	German	−59,644
4	Italian	−324,631
5	Chinese	617,476
6	Tagalog	391,289
7	Polish	−102,667
8	Korean	350,766
9	Vietnamese	303,801
10	Portuguese	68,759

38. Find the mean change per language for all ten languages.

39. Find the mean change per language for the seven categories with increases.

40. Find the mean change per language for the three categories with decreases.

41. Use your answers for Exercises 39 and 40 to find the weighted mean for the change per language for all ten categories. Compare with the value calculated in Exercise 38.

42. The table below shows the "value to the winner" for each of the three so-called "triple crown" horse races in 1994. Find the average value to winner for the three races.

Race	Value to Winner
Belmont Stakes	$392,280
Kentucky Derby	$628,800
Preakness Stakes	$447,720

The table at the right shows the medal standings for the 1994 Winter Olympics in Lillehammer, Norway. Use the given information for Exercises 43–46.

Medal Standings for the 1994 Winter Olympics

Nation	Gold	Silver	Bronze	Total
Norway	10	11	5	26
Germany	9	7	8	24
Russia	11	8	4	23
Italy	7	5	8	20
United States	6	5	2	13
Canada	3	6	4	13
Switzerland	3	4	2	9
Austria	2	3	4	9
South Korea	4	1	1	6
Finland	0	1	5	6

Calculate each of the following for all nations shown. For all calculations of mean, round answers to one decimal place.

43. the mean number of gold medals

44. the median number of silver medals

45. the mode, or modes, for the number of bronze medals

46. each of the following for the total number of medals
(a) mean **(b)** median **(c)** mode or modes

In each of Exercises 47 and 48, use the given stem-and-leaf display to identify **(a)** *the mean,* **(b)** *the median, and* **(c)** *the mode (if any) for the data represented.*

47. The display here represents prices (to the nearest dollar) charged by 23 different auto repair shops for a new alternator (installed). Give answers to the nearest cent.

9	7
10	2 4
10	5 7 9
11	1 3 4 4
11	5 5 8 8 9
12	0 4 4
12	5 7 7 9
13	8

48. The display here represents scores achieved on a 100-point biology midterm exam by the 34 members of the class.

4	7
5	1 3 6
6	2 5 5 6 7 8 8
7	0 4 5 6 7 7 8 8 8 8 9
8	0 1 1 3 4 5 5
9	0 0 0 1 6

49. Yolanda Tubalinal's Business professor lost his grade book, which contained Yolanda's five test scores for the course. A summary of the scores (each of which was an integer from 0 to 100) indicates that the mean was 88, the median was 87, and the mode was 92. (The data set was not bimodal.) What is the lowest possible number among the missing scores?

50. Explain what an "outlier" is and how it affects measures of central tendency.

51. A food processing company that packages individual cups of instant soup wishes to find out the best number of cups to include in a package. In a survey of 22 consumers, they found that five prefer a package of 1, five prefer a package of 2, three prefer a package of 3, six prefer a package of 4, and three prefer a package of 6.
(a) Calculate the mean, median, and mode values for preferred package size.
(b) Which measure in part (a) should the food processing company use?
(c) Explain your answer to part (b).

52. The following are scores earned by 15 college students on a 20-point math quiz.

0, 1, 3, 14, 14, 15, 16, 16, 17, 17, 18, 18, 18, 19, 20

(a) Calculate the mean, median, and mode values for these scores.
(b) Which measure in part (a) is most representative of the data?
(c) Explain your answer to part (b).

In each of Exercises 53–56, begin a list of the given numbers, in order, starting with the smallest one. Continue the list only until the median of the listed numbers is a multiple of 4. *Stop at that point and find* **(a)** *the number of numbers listed, and* **(b)** *the mean of the listed numbers (to two decimal places).*

53. counting numbers

54. prime numbers

55. Fibonacci numbers

56. triangular numbers

57. Seven consecutive whole numbers add up to 147. What is the result when their mean is subtracted from their median?

58. If the mean, median, and mode are all equal for the set $\{70, 110, 80, 60, x\}$, find the value of x.

59. Vince Straub wants to include a fifth number, n, along with the numbers 2, 5, 8, and 9 so that the mean and median of the five numbers will be equal. How many choices does Vince have for the number n, and what are those choices?

60. Refer to the salary data of Example 2. Explain what is wrong with simply calculating the mean salary as follows.

$$\begin{aligned}\bar{x} &= \frac{\Sigma x}{n}\\ &= \frac{\$12{,}000 + \$16{,}000 + \$18{,}500 + \$21{,}000 + \$34{,}000 + \$50{,}000}{6}\\ &= \$25{,}250\end{aligned}$$

For Exercises 61–63, refer to the grouped frequency distribution shown here.

Class Limits	Frequency f
21–25	5
26–30	3
31–35	8
36–40	12
41–45	21
46–50	38
51–55	35
56–60	20

61. Can you identify any specific data items that occurred in this sample?

62. Can you compute the actual mean for this sample?

63. Describe how you might approximate the mean for this sample. Justify your procedure.

12.3 Measures of Dispersion

	A	B
	5	1
	6	2
	7	7
	8	12
	9	13
Mean	7	7
Median	7	7

The mean is a good indicator of the central tendency of a set of data values, but it does not give the whole story about the data. To see why, compare distribution A with distribution B in the table at the side.

Both distributions of numbers have the same mean (and the same median also), but beyond that, they are quite different. In the first, 7 is a fairly typical value; but in the second, most of the values differ quite a bit from 7. What is needed here is some measure of the **dispersion,** or *spread,* of the data. Two of the most common measures of dispersion, the *range* and the *standard deviation,* are discussed in this section.

The **range** of a data set is a straightforward measure of dispersion defined as follows.

Range

For any set of data, the **range** of the set is given by

Range = (highest value in the set) − (lowest value in the set).

For a short list of data, calculation of the range is simple. For a more extensive list, it is more difficult to be sure you have accurately identified the highest and lowest values.

Hand-held calculators with statistical functions make this easy. Once the data are entered, the calculator may actually show the range (among other things), or will at least sort the data and identify the minimum and maximum items. (The associated symbols may be something like MIN Σ and MAX Σ, or min$X$ and max$X$.) Given these two values, a simple subtraction produces the range.

In distribution A, the highest value is 9 and the lowest is 5. Thus

$$\text{Range} = \text{highest} - \text{lowest} = 9 - 5 = 4.$$

In distribution B,

$$\text{Range} = 13 - 1 = 12.$$

Dive	Mark	Myrna
1	28	27
2	22	27
3	21	28
4	26	6
5	18	27
Mean	23	23
Median	22	27
Range	10	22

The range can be misleading if it is interpreted unwisely. For example, suppose three judges for a diving contest assign points to Mark and Myrna on five different dives, as shown in the table.

The ranges for the divers make it tempting to conclude that Mark is a more consistent diver than Myrna. However, a closer check indicates that Myrna is actually more consistent, with the exception of one very poor score. That score, 6, is an outlier which, if not actually recorded in error, must surely be due to some special circumstance. (Notice that the outlier does not seriously affect Myrna's median score, which is more typical of her overall performance than is her mean score.)

One of the most useful measures of dispersion, the **standard deviation,** is based on *deviations from the mean* of the data values. To find how much each value deviates from the mean, first find the mean, and then subtract the mean from each data value.

EXAMPLE 1 Find the deviations from the mean for the data values

32, 41, 47, 53, 57.

Add these values and divide by the total number of values, 5. This process shows that the mean is 46. To find the deviations from the mean, subtract 46 from each data value.

Data value	32	41	47	53	57
Deviation	**−14**	**−5**	**1**	**7**	**11**

$32 - 46 = -14$ $\qquad$ $57 - 46 = 11$

(To check your work, add the deviations. The sum of deviations for a set of data is always 0.)

It is perhaps tempting now to find a measure of dispersion by finding the mean of the deviations. However, this number always turns out to be 0 no matter how much dispersion is shown by the data; this is because the positive deviations just cancel out the negative ones. This problem of positive and negative numbers canceling each other can be avoided by *squaring* each deviation. (The square of a negative number is positive.) The following chart shows the squares of the deviations for the data above.

Data value	32	41	47	53	57
Deviation from mean	−14	−5	1	7	11
Square of deviation	**196**	**25**	**1**	**49**	**121**

$(-14) \cdot (-14) = 196$ $\qquad$ $11 \cdot 11 = 121$

An average of the squared deviations could now be found by dividing their sum by the number of data values n (5 in this case), which we would do if our data values comprised a population. However, since we are considering the data to be a sample, we divide by $n - 1$ instead.* The average that results is itself a measure of dispersion, called the *variance,* but a more common measure is obtained by taking the square root of the variance. This gives, in effect, a kind of average of the deviations from the mean, and is called the sample **standard deviation.** It is denoted by the letter s. (The standard deviation of a population is denoted σ, the lowercase Greek letter sigma.)

Continuing our calculations from the chart above, we obtain

$$s = \sqrt{\frac{196 + 25 + 1 + 49 + 121}{4}} = \sqrt{\frac{392}{4}} = \sqrt{98} \approx 9.90.$$

Most calculators, even basic types, will find square roots, such as $\sqrt{98}$, to as many digits as you need. The necessary key usually looks something like $\boxed{\sqrt{x}}$. In this book, we will normally give from two to four significant figures for such calculations.

The algorithm (process) described above for finding the sample standard deviation can be summarized as follows.

*Although the reasons cannot be explained at this level, dividing by $n - 1$ rather than n produces a sample measure that is more accurate for purposes of inference. In most cases the results using the two different divisors differ only slightly.

Calculation of Standard Deviation

Let a sample of n numbers $x_1, x_2, \ldots, x_n$ have mean $\bar{x}$. Then the **sample standard deviation,** s, of the numbers is given by

$$s = \sqrt{\frac{\Sigma(x - \bar{x})^2}{n - 1}}.$$

The individual steps involved in this calculation are as follows.

1. Calculate $\bar{x}$, the mean of the numbers.
2. Find the deviations from the mean.
3. Square each deviation.
4. Sum the squared deviations.
5. Divide the sum in step 4 by $n - 1$.
6. Take the square root of the quotient in step 5.

The description above helps show why standard deviation measures the amount of spread in a data set. But for actual calculation purposes, we recommend the use of a scientific calculator, or a statistical calculator, that will do all the detailed steps automatically. To emphasize the advantage of this, we illustrate both methods in the following example.

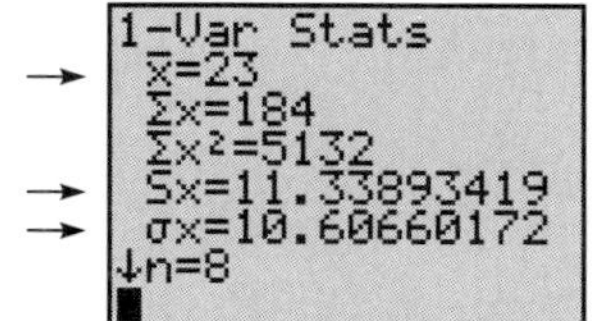

The arrows point to the mean and the sample and population standard deviations. See Example 2.

EXAMPLE 2 Find the standard deviation of the sample

$$7, 9, 18, 22, 27, 29, 32, 40$$

by using **(a)** the step-by-step process, and **(b)** the statistical functions of a calculator.

(a) Carry out the six steps summarized above.

Step 1: Find the mean of the values.

$$\frac{7 + 9 + 18 + 22 + 27 + 29 + 32 + 40}{8} = 23$$

Step 2: Find the deviations from the mean.

Data values	7	9	18	22	27	29	32	40
Deviations	−16	−14	−5	−1	4	6	9	17

Step 3: Square each deviation.

Squares of deviations: 256 196 25 1 16 36 81 289

Step 4: Sum the squared deviations.

$$256 + 196 + 25 + 1 + 16 + 36 + 81 + 289 = 900$$

Step 5: Divide by $n - 1 = 8 - 1 = 7$.

$$900/7 \approx 128.57$$

Step 6: Take the square root.

$$\sqrt{128.57} \approx 11.3$$

(b) Enter the eight data values. (The key for entering data may look something like this: [Σ+]. Find out which key it is on your calculator.) Then press the key for standard deviation. (This one may look like

[STDEV] or [SD] or [s_{n-1}] or [σ_{n-1}].

If your calculator also has a key that looks like [σ_n], it is probably for *population* standard deviation, which involves dividing by n rather than by $n - 1$, as mentioned earlier.) The result should again be 11.3. (If you *mistakenly* used the population standard deviation key, the result is 10.6 instead.) ◆

For data given in the form of a frequency distribution, some calculators allow entry of both values and frequencies, or each value can be entered separately the number of times indicated by its frequency. Then press the standard deviation key.

The following example is included only to strengthen your understanding of frequency distributions and standard deviation, not as a practical algorithm for calculating.

EXAMPLE 3 Find the sample standard deviation for the frequency distribution shown below.

Value	Frequency
2	5
3	8
4	10
5	2

Complete the calculation by extending the table as shown here. To find the numbers in the "Deviation" column, first find the mean, and then subtract the mean from the numbers in the "Value" column.

Value	Frequency	Value Times Frequency	Deviation	Squared Deviation	Squared Deviation Times Frequency
2	5	10	−1.36	1.8496	9.2480
3	8	24	−.36	.1296	1.0368
4	10	40	.64	.4096	4.0960
5	2	10	1.64	2.6896	5.3792
Sums	25	84			19.76

$$\bar{x} = \frac{84}{25} = 3.36$$

$$s = \sqrt{\frac{19.76}{24}} \approx \sqrt{.8233} \approx .91$$ ◆

Central tendency and dispersion (or "spread tendency") are different and independent properties of a set of data. Sometimes one is more critical, sometimes the other. For example, in selecting which of two baskets of a dozen

Good average, poor consistency

Good consistency, poor average

In this case, a good average (greater central tendency) is more desirable than good consistency (lesser dispersion).

tomatoes to purchase (at equal costs), you would choose the one with the higher average weight (per tomato) rather than the one with the more consistent weight (smaller standard deviation). This maximizes the total weight of your purchase. (See the illustration at the side.)

On the other hand, a marksman is more concerned (to a certain point) with consistency than with average placement. The shooter of the bottom target at the side requires only a minor adjustment of his sights; that of the top target will require considerably more effort. (In general, consistent errors can be dealt with and corrected more easily than random errors.)

Standard deviation (or any other measure of dispersion) indicates the degree to which the numbers in a data set are "spread out." The larger the measure, the more spread. Beyond that, it is difficult to describe exactly what standard deviation measures. There is one useful result, however, named for the Russian mathematician Pafnuty Lvovich Chebyshev (1821–1894), which applies to all data sets.

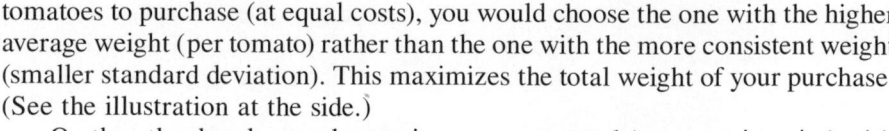

Chebyshev's Theorem

For any set of numbers, regardless of how they are distributed, the fraction of them that lie within k standard deviations of their mean (where $k > 1$) is *at least*

$$1 - \frac{1}{k^2}.$$

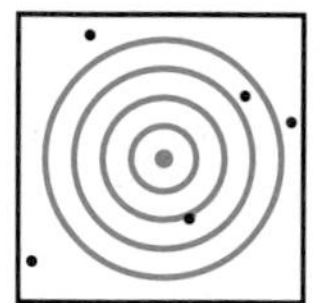
Good average, poor consistency

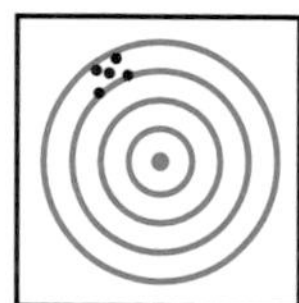
Good consistency, poor average

In this case, good consistency (lesser dispersion) is more desirable than a good average (central tendency).

Be sure to notice the words *at least* in the theorem. In certain distributions the fraction of items within k standard deviations of the mean may be more than $1 - 1/k^2$, but in no case will it ever be less. The theorem is meaningful for any value of k greater than 1 (integer or noninteger).

EXAMPLE 4 What is the minimum percentage of the items in a data set which lie within 3 standard deviations of the mean?

With $k = 3$, we calculate

$$1 - \frac{1}{3^2} = 1 - \frac{1}{9} = \frac{8}{9} \approx .889 = 88.9\%.$$

The minimum percentage is 88.9%.

Recall that inferential statistics involves drawing conclusions about populations based on data from samples. (See the chapter introduction.) Finding an "empirical" probability, as discussed earlier, is really a matter of drawing a statistical inference. Example 2 of the first section in the probability chapter concluded that, since a styrofoam cup landed on its top five times out of 50 tosses, $P(\text{top}) = 5/50 = 1/10$. That probability was an inference about the population of *all* possible tosses of that cup, based on a particular sample of 50 of those tosses.

We often use inferential techniques to compare two (or more) populations by comparing samples that come from those populations.

Pafnuty Lvovich Chebyshev (1821–1894) was a Russian mathematician known mainly for his work on the theory of prime numbers. He also worked with statistics. Chebyshev and the French mathematician and statistician **Jules Bienaymé** (1796–1878) developed, independently of one another, an important inequality of probability that is now known as the Bienaymé-Chebyshev inequality.

Founder of the St. Petersburg Mathematical School, Chebyshev also taught at the University of St. Petersburg. His *Theory of Congruences* was used as a text in Russia for many years.

EXAMPLE 5 Two companies, A and B, sell 12-ounce jars of instant coffee. Five jars of each were randomly selected from markets, and the contents were carefully weighed, with the following results.

A: 12.02, 12.08, 11.99, 11.96, 11.99

B: 12.40, 12.21, 12.36, 12.22, 12.27

Find **(a)** which company provides more coffee in their jars and **(b)** which company fills its jars more consistently.

From the given data, we calculate the mean and standard deviation values for both samples.

Sample A	Sample B
$\bar{x}_A = 12.008$	$\bar{x}_B = 12.292$
$s_A = .0455$	$s_B = .0847$

(a) Since $\bar{x}_B$ is greater than $\bar{x}_A$, we *infer* that Company B most likely provides more coffee (greater mean).

(b) Since s_A is less than s_B, we *infer* that Company A seems more consistent (smaller standard deviation). ■

Concerning Example 5 above, we point out some important reservations. The conclusions drawn are quite tentative because the samples were small. We could place more confidence in our inferences if we used larger samples, for then it would be more likely that the samples were accurate representations of their respective populations. In a more detailed study of inferential statistics, you would learn techniques best suited to small samples as well as those best suited to larger samples. You would also learn to state your inferences more precisely so that the degree of uncertainty is conveyed along with the basic conclusion.

12.3 EXERCISES

1. If your calculator finds both kinds of standard deviation, the sample standard deviation and the population standard deviation, which of the two will be a larger number for a given set of data? (*Hint:* Recall the difference between how the two standard deviations are calculated.)

2. If your calculator finds only one kind of standard deviation, explain how you would determine whether it is sample or population standard deviation (assuming your calculator manual is not available).

Find **(a)** *the range, and* **(b)** *the standard deviation for each sample (Exercises 3–12). Round fractional answers to the nearest hundredth.*

3. 2, 5, 6, 8, 9, 11, 15, 19

4. 8, 5, 12, 8, 9, 15, 21, 16, 3

5. 25, 34, 22, 41, 30, 27, 31

6. 67, 83, 55, 68, 77, 63, 84, 72, 65

7. 318, 326, 331, 308, 316, 322, 310, 319, 324, 330

8. 5.7, 8.3, 7.4, 6.6, 7.4, 6.8, 7.1, 8.0, 8.5, 7.9, 7.1, 7.4, 6.9, 8.2

9. 84.53, 84.60, 84.58, 84.48, 84.72, 85.62, 85.03, 85.10, 84.96

10. 206.3, 210.4, 209.3, 211.1, 210.8, 213.5, 212.6, 210.5, 211.0, 214.2

11.

Value	Frequency
9	3
7	4
5	7
3	5
1	2

12.

Value	Frequency
14	8
16	12
18	15
20	14
22	10
24	6
26	3

Use Chebyshev's theorem for Exercises 13–28.

Find the least possible fraction of the numbers in a data set lying within the given number of standard deviations of the mean. Give answers as standard fractions reduced to lowest terms.

13. 2 **14.** 4 **15.** 7/2 **16.** 11/4

Find the least possible percentage (to the nearest tenth of a percent) of the items in a distribution lying within the given number of standard deviations of the mean.

17. 3 **18.** 5 **19.** 5/3 **20.** 5/2

In a certain distribution of numbers, the mean is 70 *and the standard deviation is* 8. *At least what fraction of the numbers are between the following pairs of numbers? Give answers as standard fractions reduced to lowest terms.*

21. 54 and 86 **22.** 46 and 94 **23.** 38 and 102 **24.** 30 and 110

In the same distribution (mean 70 *and standard deviation* 8*), find the largest fraction of the numbers that could meet the following requirements. Give answers as standard fractions reduced to lowest terms.*

25. less than 54 or more than 86

26. less than 50 or more than 90

27. less than 42 or more than 98

28. less than 52 or more than 88

Katherine Steinbacher owns a minor league baseball team. Each time the team wins a game, Katherine pays the nine starting players, the manager and two coaches bonuses, which are certain percentages of their regular salaries. The amounts paid are listed here.

$80, $105, $120, $175, $185, $190, $205, $210, $215, $300, $320, $325

Use the distribution of bonuses above for Exercises 29–34.

29. Find the mean of the distribution.

30. Find the standard deviation of the distribution.

31. How many of the bonus amounts are within one standard deviation of the mean?

32. How many of the bonus amounts are within two standard deviations of the mean?

33. What does Chebyshev's theorem say about the number of the amounts that are within two standard deviations of the mean?

34. Explain any discrepancy between your answers for Exercises 32 and 33.

William Tobey manages the service department of a trucking company. Each truck in the fleet utilizes an electronic engine control module, which must be replaced when it fails. Long-lasting modules are desirable, of course, but a preventive replacement program can also avoid costly breakdowns on the highway. For this purpose it is desirable that the modules be fairly consistent in their lifetimes; that is, they should all last about the same number of miles before failure, so that the timing of preventive replacements can be done accurately. William tested a sample of 20 Brand A modules, and they lasted 43,560 *high-*

way miles on the average (mean), with a standard deviation of 2,116 *miles. The listing below shows how long each of another sample of 20 Brand B modules lasted.*

50,660, 41,300, 45,680, 48,840, 47,300,
51,220, 49,100, 48,660, 47,790, 47,210,
50,050, 49,920, 47,420, 45,880, 50,110,
49,910, 47,930, 48,800, 46,690, 49,040

Use the data above for Exercises 35–37.

35. According to the sampling that was done, which brand of module has the longer average life (in highway miles)?

36. Which brand of module apparently has a more consistent (or uniform) length of life (in highway miles)?

37. If Brands A and B are the only modules available, which one should William purchase for his maintenance program? Explain your reasoning.

Utilize the following sample for Exercises 38–43.

13, 14, 16, 18, 20, 22, 25

38. Compute the mean and standard deviation for the sample (each to the nearest hundredth).

39. Now add 5 to each item of the sample above and compute the mean and standard deviation for the new sample.

40. Go back to the original sample. This time subtract 10 from each item, and compute the mean and standard deviation of the new sample.

41. Based on your answers for Exercises 38–40, make conjectures about what happens to the mean and standard deviation when all items of the sample have the same constant k added or subtracted.

42. Go back to the original sample again. This time multiply each item by 3, and compute the mean and standard deviation of the new sample.

43. Based on your answers for Exercises 38 and 42, make conjectures about what happens to the mean and standard deviation when all items of the sample are multiplied by the same constant k.

44. In Section 2 we showed that the mean, as a measure of central tendency, is highly sensitive to extreme values. Which measure of dispersion, covered in this section, would be more sensitive to extreme values? Illustrate your answer with one or more examples.

45. The Quaker Oats Company conducted a survey to determine whether or not a proposed premium to be included in boxes of their cereal was appealing enough to generate new sales. Four cities were used as test markets, where the cereal was distributed with the premium, and four cities as control markets, where the cereal was distributed without the premium. The eight cities were chosen on the basis of their similarity in terms of population, per capita income, and total cereal purchase volume. The results follow.

		Percent Change in Average Market Share per Month
Test Cities	1	+18
	2	+15
	3	+7
	4	+10
Control Cities	1	+1
	2	−8
	3	−5
	4	0

(a) Find the mean of the change in market share for the four test cities.
(b) Find the mean of the change in market share for the four control cities.
(c) Find the standard deviation of the change in market share for the test cities.
(d) Find the standard deviation of the change in market share for the control cities.
(e) Find the difference between the means of (a) and (b). This difference represents the estimate of the percent change in sales due to the premium.
(f) The two standard deviations from (c) and (d) were used to calculate an "error" of ± 7.95 for the estimate in (e). With this amount of error, what are the smallest and largest estimates of the increase in sales?

On the basis of the interval estimate of part (f) the company decided to mass produce the premium and distribute it nationally.

In a skewed distribution, the mean will be farther out toward the tail than the median, as shown in the sketch.

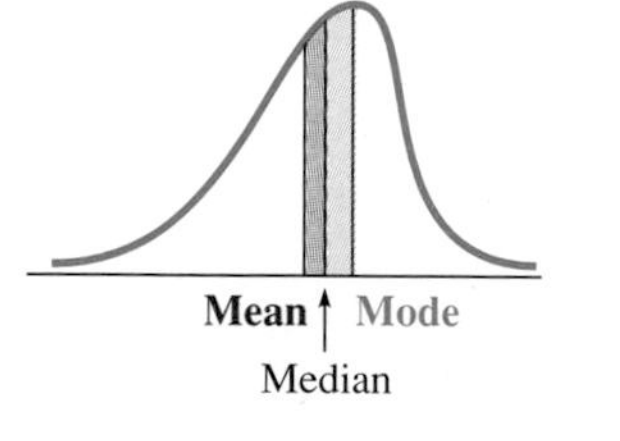

Skewed to the left

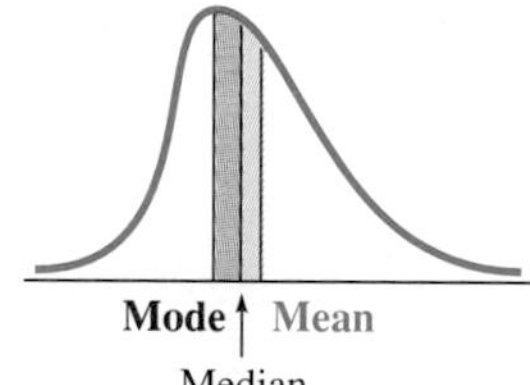

Skewed to the right

A common way of measuring the degree of skewness, which involves both central tendency and dispersion, is with the **skewness coefficient,** *calculated as follows.*

$$\frac{3 \times (\text{mean} - \text{median})}{\text{standard deviation}}$$

46. Under what conditions would the skewness coefficient be each of the following?
(a) positive **(b)** negative

47. Explain why the mean of a skewed distribution is always farther out toward the tail than the median.

48. Suppose that the mean length of patient stay (in days) in U.S. hospitals is 2.7 days, with a standard deviation of 7.1 days. Make a sketch of how this distribution may look.

For Exercises 49–51, refer to the grouped frequency distribution shown here. (Also refer to Exercises 61–63 in the previous section.)

Class Limits	Frequency f
21–25	5
26–30	3
31–35	8
36–40	12
41–45	21
46–50	38
51–55	35
56–60	20

49. Can you identify any specific data items that occurred in this sample?

50. Can you compute the actual standard deviation for this sample?

51. Describe how you might approximate the standard deviation for this sample. Justify your procedure.

52. Suppose the frequency distribution of Example 3 involved 50 or 100 (or even more) distinct data values, rather than just four. Explain why the procedure of that example would then be very inefficient.

53. A "J-shaped" distribution can be skewed to the right as well as to the left.
(a) In a J-shaped distribution skewed to the right, which data item would be the mode, the largest or the smallest item?
(b) In a J-shaped distribution skewed to the left, which data item would be the mode, the largest or the smallest item?
(c) Explain why the mode is a weak measure of central tendency for a J-shaped distribution.

Measures of Position

Measures of central tendency and measures of dispersion give us an effective way of characterizing an overall set of data. Central tendency has to do with where, along a number scale, the overall data set is centered. Dispersion has to do with how much the data set is spread out from the center point. And Chebyshev's theorem, stated in the previous section, tells us in a general sense what portions of the data set may be dispersed different amounts from the center point.

In many cases, we are especially interested in certain individual items in the data set, rather than in the set as a whole. We still need a way of measuring how an item fits into the collection, how it compares to other items in the collection, or even how it compares to another item in another collection. There are several common ways of creating such measures. Since they measure an item's position within the data set, they are usually called **measures of position.**

The *z*-score

Each individual item in a sample can be assigned a **z-score,** which is defined as follows.

The *z*-score

If x is a data item in a sample with mean $\overline{x}$ and standard deviation s, then the **z-score** of x is given by

$$z = \frac{x - \overline{x}}{s}.$$

Since $x - \overline{x}$ gives the amount by which x differs (or deviates) from the mean $\overline{x}$, $(x - \overline{x})/s$ gives the number of standard deviations by which x differs from $\overline{x}$. Notice that z will be positive if x is greater than $\overline{x}$ but negative if x is less than $\overline{x}$. Chebyshev's theorem assures us that, in any distribution whatsoever, at least 89% (roughly) of the items will lie within three standard deviations of the mean. That is, at least 89% of the items will have z-scores between -3 and 3. In fact, many common distributions, especially symmetric ones, have considerably more than 89% of their items within three standard deviations of the mean (as we will see in the next section). Hence, a z-score greater than 3 or less than -3 is rare.

EXAMPLE 1 Two friends, Reza and Weihua, who take different history classes, received their midterm exams back on the same day. Reza's score was 86 while Weihua's was only 78. Which student did relatively better, given the class data shown here?

	Reza	Weihua
Class Mean	73	69
Class Standard Deviation	8	5

Calculate as follows.

$$\text{Reza:} \qquad z = \frac{86 - 73}{8} = 1.625$$

$$\text{Weihua:} \quad z = \frac{78 - 69}{5} = 1.8.$$

Since Weihua's z-score is higher, she was positioned relatively higher within her class than Reza was within his class.

Percentiles

When you take the Scholastic Aptitude Test (SAT), or any other standardized test taken by large numbers of students, your raw score is usually converted to a **percentile** score, which is defined as follows.

Percentile

If approximately n percent of the items in a distribution are less than the number x, then x is the ***n*th percentile** of the distribution.

For example, if you scored at the eighty-third percentile on the SAT, it means that you outscored approximately 83% of all those who took the test. (It does *not* mean that you got 83% of the answers correct.) Since the percentile score positions an item within the data set, it is another "measure of position." The following example approximates percentiles for a fairly small collection of data.

EXAMPLE 2 The following are the numbers of dinner customers served by a restaurant on 40 consecutive days. (The numbers have been ranked lowest to highest.)

46	51	52	55	56	56	58	59	59	59
61	61	62	62	63	63	64	64	64	65
66	66	66	67	67	67	68	68	69	69
70	70	71	71	72	75	79	79	83	88

For this data set, find each of the following.

(a) The thirty-fifth percentile

The thirty-fifth percentile can be taken as the item below which 35 percent of the items are ranked. Since 35 percent of 40 = (.35)(40) = 14, we take the 15th item, or 63, as the thirty-fifth percentile.

(b) The ninety-fifth percentile

Since 95 percent of 40 = (.95)(40) = 38, the ninety-fifth percentile can be taken as the thirty-ninth item, or 83. ■

Technically, percentiles were originally conceived as a set of 99 values (not necessarily data items) along the scale that would divide the data set into 100 equal parts. They were computed only for very large data sets. With smaller data sets, as in Example 2 above, dividing the data into 100 parts would necessarily leave many of those parts empty. However, the modern techniques of exploratory data analysis seek to apply the percentile concept to even small data sets. Thus we use approximation techniques as in Example 2.

Another option is to divide the data into a lesser number of equal (or nearly equal) parts, as shown next.

Deciles and Quartiles

Basically, **deciles** are the nine values (denoted $D_1, D_2, \ldots, D_9$) along the scale that divide a data set into ten (approximately) equal parts, and **quartiles** are the three values (Q_1, Q_2, and Q_3) that divide a data set into four (approximately) equal parts. Since deciles and quartiles serve to position particular items within portions of a distribution, they are also "measures of position."

A decile is always equivalent to a certain percentile. So it can be approximated just as in Example 2.

EXAMPLE 3 Find the fourth decile for the dinner customer data of Example 2.

Refer to the ranked data table. The fourth decile is the fortieth percentile, and 40% of 40 = (.40)(40) = 16. We take the fourth decile to be the seventeenth item, or 64. ■

Although the three quartiles are also equivalent to corresponding percentiles, notice that the second quartile, Q_2, is also equivalent to the median, a measure of central tendency introduced in Section 2. A common convention for computing quartiles goes back to the way we computed the median.

Finding Quartiles

For any set of data (ranked in order from smallest to largest):

The **second quartile,** Q_2, is just the median, the middle item when the number of items is odd, or the mean of the two middle items when the number of items is even.

The **first quartile,** Q_1, is the median of all items below Q_2.

The **third quartile,** Q_3, is the median of all items above Q_2.

EXAMPLE 4 Find the three quartiles for the data of Example 2.

Refer to the ranked data. The two middle data items are 65 and 66, so

$$Q_2 = \frac{65 + 66}{2} = 65.5.$$

The first 20 items (an even number) are all below Q_2, and the two middle items in that set are 59 and 61.

$$Q_1 = \frac{59 + 61}{2} = 60.$$

The last 20 items are above Q_2.

$$Q_3 = \frac{69 + 70}{2} = 69.5.$$

The Box Plot

The final topic of this section is another tool from exploratory data analysis. It involves the median (a measure of central tendency), the range (a measure of dispersion), and the first and third quartiles (measures of position), all incorporated into a simple visual display. It is called a **box plot** (or a **box-and-whisker plot**).

Box Plot

For a given set of data, a **box plot** (or **box-and-whisker plot**) consists of a rectangular box positioned above a numerical scale, extending from Q_1 to Q_3, with the value of Q_2 (the median) indicated within the box, and with "whiskers" (line segments) extending to the left and right from the box out to the minimum and maximum data items.

EXAMPLE 5 Construct a box plot for the weekly study times data of Example 2 in Section 1.

In order to determine the quartiles and the minimum and maximum values more easily, we use the stem-and-leaf display (with ranked leaves), given in Table 9 of Section 2:

Stem	Leaves
1	2 4 5 6 7 8
2	0 2 3 4 4 5 6 6 9 9 9
3	0 1 1 2 3 6 6 8 9
4	0 1 4 4 5 5 7
5	2 5 5 8
6	0 2
7	2

The median (determined earlier in Example 6 of Section 2) is

$$\frac{31 + 32}{2} = 31.5.$$

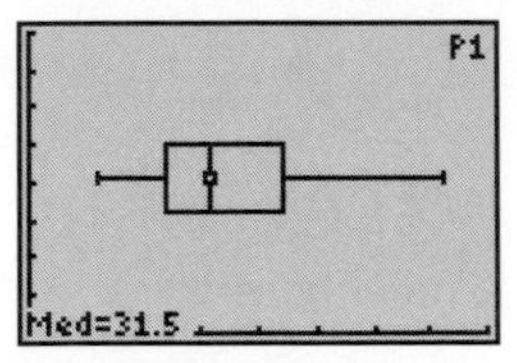

This box plot corresponds to the results of Example 5. It indicates the median in the display at the bottom. The TRACE function of the TI-82 will locate the minimum, maximum, and quartile values as well.

From the stem-and-leaf display,

$$Q_1 = \frac{24 + 24}{2} = 24 \quad \text{and} \quad Q_3 = \frac{44 + 45}{2} = 44.5.$$

The minimum and maximum items are 12 and 72. The box plot is shown in Figure 10. ■

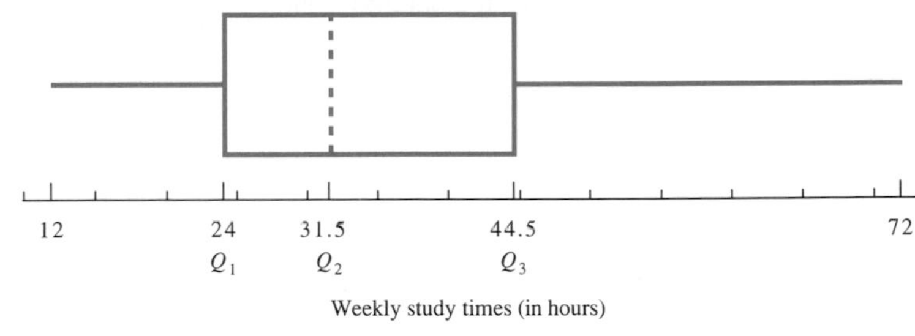

FIGURE 10

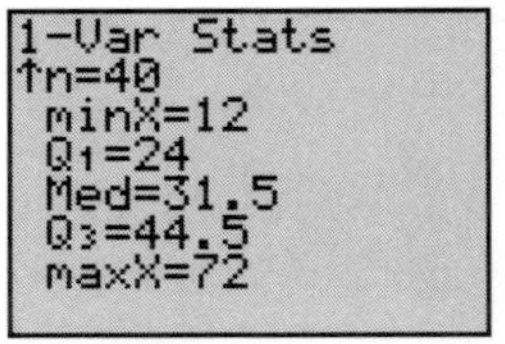

This screen supports the results of Example 5.

Notice in Figure 10 that the box plot clearly conveys the following:

1. central tendency (the location of the median);
2. the location of the middle half of the data (the extent of the box);
3. dispersion (the range is the extent of the whiskers);
4. skewness (the non-symmetry of both the box and the whiskers).

Some calculators, such as the TI-82, will automatically compute the necessary values and display a box plot for a data set. (See the margin note.)

EXERCISES

For each of Exercises 1–4, make use of z-scores.

1. In a calculus class, Emily Francke scored 5 on a quiz for which the class mean and standard deviation were 4.6 and 2.1, respectively. Charles Gomez scored 6 on another quiz for which the class mean and standard deviation were 4.9 and 2.3, respectively. Relatively speaking, which student did better?
2. In Saturday's track meet, Gregory Syftestad, a high jumper, jumped 6 feet 3 inches. Conference high jump marks for the past season had a mean of 6 feet even and a standard deviation of 3.5 inches. Alexander Szabo, Gregory's teammate, achieved 18 feet 4 inches in the long jump. In that event the conference's season average (mean) and standard deviation were 16 feet 6 inches and 1 foot 10 inches, respectively. Relative to this past season in this conference, which athlete had a better performance on Saturday?

3. The lifetimes of Brand A tires are distributed with mean 45,000 miles and standard deviation 4,500 miles, while Brand B tires last for only 38,000 miles on the average (mean) with standard deviation 2,080 miles. Jutta's Brand A tires lasted 37,000 miles and Arvind's Brand B tires lasted 35,000 miles. Relatively speaking, within their own brands, which driver got the better wear?

4. In a certain lake, the trout average 12 inches in length with a standard deviation of 2.75 inches. The bass average 4 pounds in weight with a standard deviation of .8 pound. If Imelda caught an 18-inch trout and Timothy caught a 6-pound bass, then relatively speaking, which catch was the better trophy?

Refer to the dinner customers data of Example 2. Approximate each of the following. (Use the methods illustrated in this section.)

5. the fifteenth percentile

6. the seventy-fifth percentile

7. the third decile

8. the eighth decile

For Exercises 9–20, refer to the top 20 scorers data for the 1993–1994 season of the National Basketball Association.

Leading Scorers	Games Played	Total Points	Average Points per Game
David Robinson, San Antonio	80	2,383	29.8
Shaquille O'Neal, Orlando	81	2,377	29.4
Hakeem Olajuwon, Houston	80	2,184	27.3
Dominique Wilkins, L.A. Clippers	74	1,923	26.0
Karl Malone, Utah	82	2,039	24.9
Patrick Ewing, N.Y. Knicks	79	1,939	24.5
Mitch Richmond, Sacramento	78	1,823	23.4
Scottie Pippen, Chicago	72	1,587	22.0
Charles Barkley, Phoenix	65	1,402	21.6
Alonzo Mourning, Charlotte	60	1,287	21.5
Glen Rice, Miami	81	1,708	21.1
Latrell Sprewell, Golden State	82	1,720	21.0
Danny Manning, Atlanta	68	1,403	20.6
Joe Dumars, Detroit	69	1,410	20.4
Derrick Coleman, New Jersey	77	1,559	20.3
Ron Harper, L.A. Clippers	75	1,508	20.1
Clifford Robinson, Portland	82	1,647	20.1
Kevin Johnson, Phoenix	67	1,340	20.0
Reggie Miller, Indiana	79	1,574	19.9
Jim Jackson, Dallas	82	1,576	19.2

Compute z-scores for Exercises 9–12.

9. Scottie Pippen's games played

10. Mitch Richmond's total points

11. Shaquille O'Neal's average points per game

12. David Robinson's total points

Determine who occupied each of the following positions.

13. the eighty-fifth percentile in average points per game

14. the forty-fifth percentile in games played

15. the eighth decile in total points scored

16. the ninth decile in average points per game

Determine who was relatively highest in the groupings of Exercises 17 and 18.

17. Charles Barkley in total points, Danny Manning in average points per game, or Joe Dumars in games played

18. Reggie Miller in total points, Kevin Johnson in games played, or Ron Harper in average points per game

19. Construct a stem-and-leaf plot with double stems for the games played data.

20. Construct a box plot for the average points per game data.

21. What does your box plot of Exercise 20 indicate about each of the following?
(a) the central tendency of these data
(b) the dispersion of these data
(c) the skewness of these data

22. The text stated that, for *any* distribution of data, at least 89% of the items will be within three standard deviations of the mean. Why couldn't we just move some items farther out from the mean to obtain a new distribution that would violate this condition?

23. Describe the basic difference between a measure of central tendency and a measure of position.

This chapter has introduced three major characteristics: central tendency, dispersion, and position, and has developed various ways of measuring them in numerical data. In each of Exercises 24–28, a new measure is described. Explain in each case which of the three characteristics you think it would measure and why.

24. **Midrange** $= \dfrac{\text{minimum item} + \text{maximum item}}{2}$

25. **Midquartile** $= \dfrac{Q_1 + Q_3}{2}$

26. **Coefficient of Variation** $= \dfrac{s}{\bar{x}} \cdot 100$

27. **Interquartile range** $= Q_3 - Q_1$

28. **Semi-interquartile range** $= \dfrac{Q_3 - Q_1}{2}$

29. The "skewness coefficient" was defined in the exercises of the previous section to be

$$\frac{3 \times (\bar{x} - Q_2)}{s}.$$

Is this a measure of individual data items or of the overall distribution?

30. In a national standardized test, Jennifer scored at the ninety-second percentile. If 67,500 individuals took the test, about how many scored higher than Jennifer did?

31. Let the three quartiles (from smallest to largest) for a large population of scores be denoted Q_1, Q_2, and Q_3.
(a) Is it necessarily true that $Q_2 - Q_1 = Q_3 - Q_2$?
(b) Explain your answer to part (a).

In Exercises 32–35, answer yes *or* no *and explain your answer. (Consult the exercises above for definitions of some of these measures.)*

32. Is the midquartile necessarily the same as the median?

33. Is the midquartile necessarily the same as the midrange?

34. Is the interquartile range necessarily half the range?

35. Is the semi-interquartile range necessarily half the interquartile range?

36. Omer and Alessandro participated in the standardization process for a new state-wide chemistry test. Within the large group participating, their raw scores and corresponding z-scores were as shown here.

	Raw score	z-score
Omer	60	.69
Alessandro	72	1.67

Find the overall mean and standard deviation of the distribution of scores. (Give answers to two decimal places.)

*Since the National Football League began keeping official statistics in 1932, the passing effectiveness of quarterbacks has been rated by several different methods. The current system, adopted in 1973, is based on four performance components considered most important: completions, touchdowns, yards gained, and interceptions, as percentages of the number of passes attempted. The actual computation of the ratings seems to be poorly understood by fans and media sportscasters alike, but can be accomplished with the following formula.**

$$\text{Rating} = \frac{\left(250 \times \frac{C}{A}\right) + \left(1{,}000 \times \frac{T}{A}\right) + \left(12.5 \times \frac{Y}{A}\right) + 6.25 - \left(1{,}250 \times \frac{I}{A}\right)}{3},$$

$$\text{where} \quad \begin{aligned} A &= \text{attempted passes} \\ C &= \text{completed passes} \\ T &= \text{touchdown passes} \\ Y &= \text{yards gained} \\ I &= \text{interceptions} \end{aligned}$$

In addition to the weighting factors (coefficients) appearing in the formula, the four category ratios are limited to the following maximums.

$$.775 \text{ for } \frac{C}{A}, \qquad .11875 \text{ for } \frac{T}{A}, \qquad 12.5 \text{ for } \frac{Y}{A}, \qquad .095 \text{ for } \frac{I}{A}$$

These limitations are intended to prevent any one component of performance from having an undue effect on the overall rating. They are not often invoked, but in special cases can have a significant effect. (See Exercises 48–51.)

The formula above rates all passers against the same performance standard and is applied, for example, after a single game, an entire season, or a career.

* Joe Farabee of American River College (a long-time friend and colleague of author Heeren) derived the formula after sifting through a variety of descriptions of the rating system in sports columns and NFL publications.

Based on the 1994 regular-season statistics shown here for Steve Young of the San Francisco 49ers and Troy Aikman of the Dallas Cowboys, compute each of the following (to one decimal place).

	Young	Aikman
Attempts	461	361
Completions	324	233
Yards	3,969	2,676
TDs	35	13
Interceptions	10	12

37. Young's season rating

38. Aikman's season rating

The 1994 regular season ratings are shown here (ranked by conference) for the fifteen leading passers in the National Conference (N.F.C.) and the fourteen in the American Conference (A.F.C.). Find the requested measures in Exercises 39–42.

N.F.C. Passer	Rating Points	A.F.C. Passer	Rating Points
Young, San Francisco	112.8	Marino, Miami	89.2
Favre, Green Bay	90.7	Elway, Denver	85.7
Everett, New Orleans	84.9	Kelly, Buffalo	84.6
Aikman, Dallas	84.9	Montana, Kansas City	83.6
George, Atlanta	83.3	Humphries, San Diego	81.6
Erickson, Tampa Bay	82.5	Hostetler, Raiders	81.0
Moon, Minnesota	79.9	O'Donnell, Pittsburgh	78.9
Walsh, Chicago	77.9	Esiason, New York Jets	77.3
Cunningham, Philadelphia	74.4	Blake, Cincinnati	76.9
Miller, Rams	73.6	Bledsoe, New England	73.6
Brown, New York Giants	72.5	Testaverde, Cleveland	70.7
Schroeder, Arizona	68.4	Mirer, Seattle	70.2
Mitchell, Detroit	62.0	Klingler, Cincinnati	65.7
Beuerlein, Arizona	61.6	Tolliver, Houston	62.6
Shuler, Washington	59.6		

39. the three quartiles for the N.F.C. ratings

40. the three quartiles for the A.F.C. ratings

41. the three quartiles for all ratings (the N.F.C. and A.F.C. figures combined)

42. the ninetieth percentile for the combined figures

43. Construct box plots for both conferences, one above the other in the same drawing.

Refer to the box plots of Exercise 43 for Exercises 44–47. In each case, identify the conference for which the given measure is greater, and estimate about how much greater it is.

44. the median

45. the midquartile

46. the range

47. the interquartile range

Sid Luckman of the Chicago Bears, the passing champion of the 1943 season, had the following pertinent statistics in that year.

202 attempts, 110 completions, 28 touchdowns, 2,194 yards, 12 interceptions

Use these numbers in the rating formula for the following exercises.

48. Compute Luckman's 1943 rating, being careful to replace each ratio in the formula with its allowed maximum if necessary.

49. Compute Luckman's rating assuming no restrictions on the ratios.

50. Which component of Luckman's passing game that year was apparently very unusual, even among high caliber passers?

51. Compare the two ratings of Exercises 48 and 49 with Steve Young's all-time record season rating of 112.8 (achieved in 1994).

12.5 The Normal Distribution

A random variable that can take on only certain fixed values is called a **discrete random variable.** For example, the number of heads in 5 tosses of a coin is discrete since its only possible values are 0, 1, 2, 3, 4, and 5. A variable whose values are not restricted in this way is a **continuous random variable.** For example, the diameter of camellia blossoms would be a continuous variable, spread over a scale perhaps from 5 to 25 centimeters. The values would not be restricted to whole numbers, or even to tenths, or hundredths, etc. Although in practice we may not measure the diameters to more accuracy than, say, tenths of centimeters, they could theoretically take on any of the infinitely many values from 5 to 25. In terms of set theory, we can say that a discrete random variable can take on only a countable number of values, whereas a continuous random variable can take on an uncountable number of values.

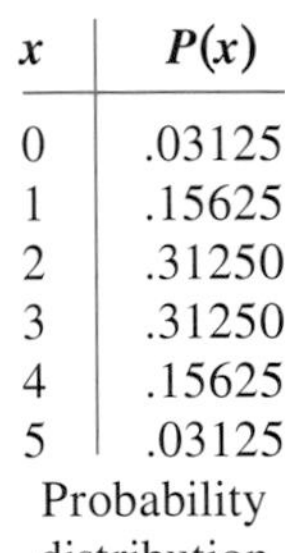

x	$P(x)$
0	.03125
1	.15625
2	.31250
3	.31250
4	.15625
5	.03125

Probability distribution

In contrast to most distributions discussed earlier in this chapter, which were empirical (based on observation), the distributions covered in this section and the next are theoretical (based on theoretical probabilities). A knowledge of theoretical distributions enables us to identify when actual observations are inconsistent with stated assumptions, which is the key to inferential statistics.

The theoretical probability distribution for the discrete random variable "number of heads" when 5 fair coins are tossed is shown here, and Figure 11 shows the corresponding histogram. The probability values can be found using the binomial probability formula or using Pascal's triangle.

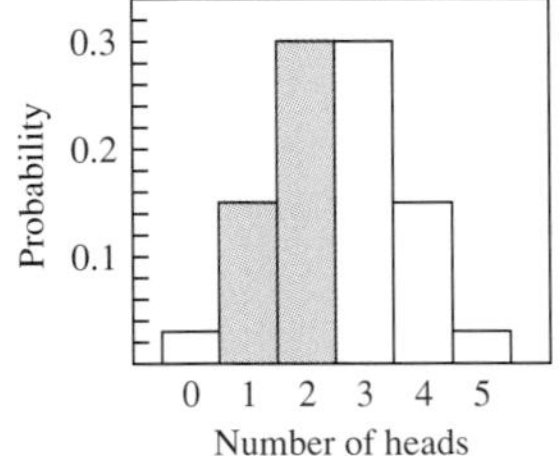

FIGURE 11

Since each rectangle in Figure 11 is 1 unit wide, the *area* of the rectangle is also equal to the probability of the corresponding number of heads. The area, and thus the probability, for the event "1 head or 2 heads" is shaded in the figure. The graph consists of 6 distinct rectangles since "number of heads" is a *discrete* variable with 6 possible values. The sum of the 6 rectangular areas is exactly 1 square unit.

In contrast to the discrete "number of heads" distribution above, a probability distribution for camellia blossom diameter cannot be tabulated or graphed in quite the same way, since this variable is *continuous.* The graph would be smeared out into a "continuous" bell-shaped curve (rather than a set of rectangles) as shown in Figure 12. The vertical scale on the graph in this case shows what we call "probability density," the probability per unit along the horizontal axis.

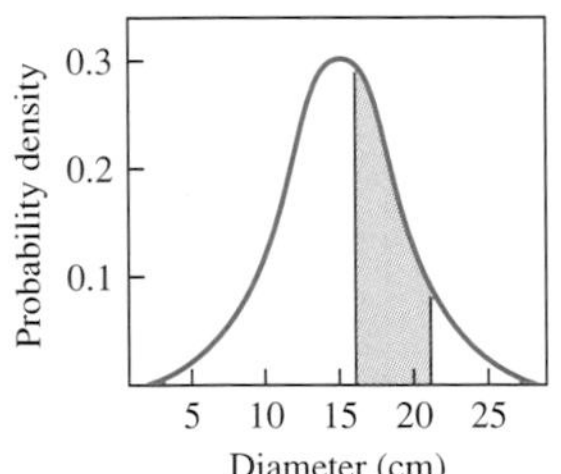

FIGURE 12

This particular curve is highest at a diameter value of 15 cm, its center point, and drops off rapidly and equally toward a zero level in both directions. A symmetric, bell-shaped curve of this type is called a **normal curve.** Any random variable whose graph has this characteristic shape is said to have a **normal distribution.** A great many continuous random variables have this type of normal distribution, and even the distributions of discrete variables can often be approximated closely by normal curves. Thus, normal curves are well worth studying. (Of course, not all distributions are normal or approximately so—for example, income distributions usually are not.)

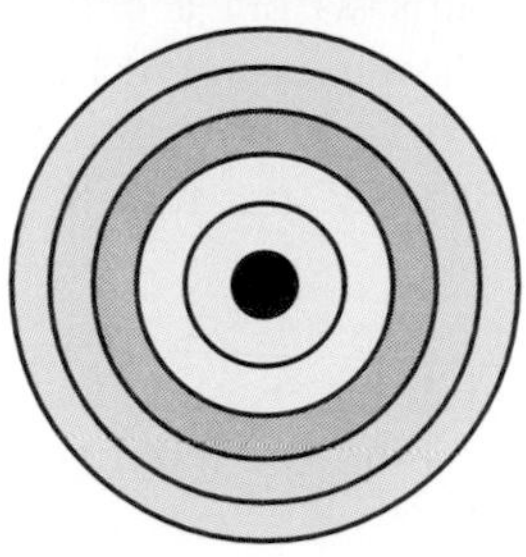

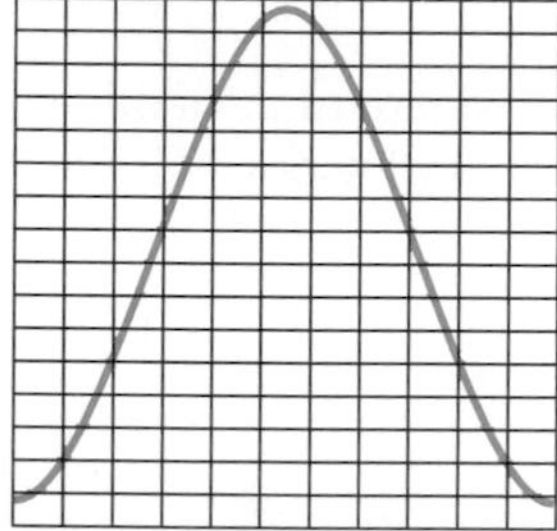

A normal distribution occurs in darts if the player, always aiming at the bull's-eye, tosses a fairly large number of times, and the aim on each toss is affected by independent random errors.

On a normal curve, if the quantity shown on the horizontal axis is the number of standard deviations from the mean, rather than values of the random variable itself, then we call the curve the **standard normal curve.** In that case, the horizontal dimension measures numbers of standard deviations, with the value being 0 at the mean of the distribution. The number of standard deviations, associated with a point on the horizontal axis, is the "standard score" for that point. It is the same as the z-score defined in the previous section.

The area under the curve along a certain interval is numerically equal to the probability that the random variable will have a value in the corresponding interval. The total area under the standard normal curve is exactly 1 square unit.

The area of the shaded region in Figure 12 is equal to the probability of a randomly chosen blossom having a diameter in the interval from exactly 16.4 cm to exactly 21.2 cm. To find that area, we first convert 16.4 and 21.2 to standard deviation units. With this approach, we can deal with any normal curve by comparing it to the standard normal curve. A single table of areas for the standard normal curve then provides probabilities for all normal curves, as will be shown later in this section.

The normal curve was first developed by Abraham De Moivre (1667–1754), but his work went unnoticed for many years. It was independently redeveloped by Pierre Laplace and Carl Friedrich Gauss (1777–1855). Gauss found so many uses for this curve that it is sometimes called the *Gaussian curve.* Several properties of the normal curve are summarized below and are illustrated in Figure 13.

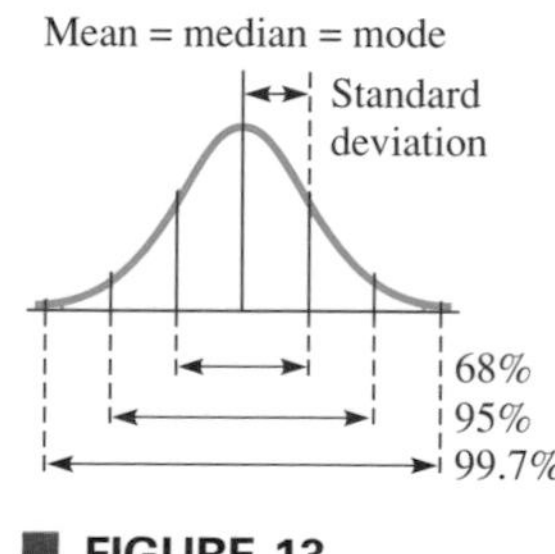

FIGURE 13

Properties of the Normal Curve

The graph of a normal curve is bell-shaped and symmetric about a vertical line through its center.

The mean, median, and mode of a normal curve are all equal and occur at the center of the distribution.

Empirical Rule About 68% of all data values of a normal curve lie within 1 standard deviation of the mean (in both directions), about 95% within 2 standard deviations, and about 99.7% within 3 standard deviations.

Close but Never Touching
When a curve approaches closer and closer to a line, without ever actually meeting it (as a normal curve approaches the horizontal axis), the line is called an **asymptote,** and the curve is said to approach the line **asymptotically.**

The empirical rule indicates that a very small percentage of the items in a normal distribution will lie more than 3 standard deviations from the mean (approximately 0.3%, divided equally between the upper and lower tails of the distribution). As we move away from the center, the curve falls rapidly, and then more and more gradually, toward the axis. But it *never* actually reaches the axis. No matter how far out we go, there is always a chance (though very small) of an item occurring even farther out. Theoretically then, the range of a true normal distribution is infinite.

It is important to realize that the percentage of items within a certain interval is equivalent to the probability that a randomly chosen item will lie in that interval. Thus one result of the empirical rule is that, if we choose a single number at random from a normal distribution, the probability that it will lie within 1 standard deviation of the mean is about .68.

EXAMPLE 1 Suppose 300 chemistry students take a midterm exam and that the distribution of their scores can be treated as normal. Find the number of scores falling into each of the following intervals.

(a) Within 1 standard deviation of the mean

By the empirical rule, 68% of all scores lie within 1 standard deviation of the mean. Since there is a total of 300 scores, the number of scores within 1 standard deviation is

$$(68\%)(300) = (.68)(300) = 204.$$

(b) Within 2 standard deviations of the mean

A total of 95% of all scores lie within 2 standard deviations of the mean. This would be

$$(.95)(300) = 285.$$ ■

Most questions we need to answer about normal distributions involve regions other than those within 1, 2, or 3 standard deviations of the mean. For example, we might need the percentage of items within 1 1/2 or 2 1/5 standard deviations of the mean, or perhaps the area under the curve from .8 to 1.3 standard deviations above the mean. In such cases, we need more than the empirical rule. The traditional approach is to refer to a table of area values, such as Table 10 which appears on page 721. Computer software packages designed for statistical uses will usually produce the required probability values on command, and several of the advanced hand-held calculators also have this capability. Those tools are recommended. As an optional approach, we will illustrate the use of Table 10 here.

The table gives the fraction of all scores in a normal distribution that lie between the mean and z standard deviations from the mean. *Because of the symmetry of the normal curve, the table can be used for values above the mean or below the mean.* All of the items in the table can be thought of as corresponding to the area under the curve. The total area is arranged to be 1.000 square unit, with .500 square units on each side of the mean. The table shows that at 3.27 standard deviations from the mean, essentially all of the area is accounted for. Whatever remains beyond is so small that it does not appear in the first three decimal places.

The column under A gives the proportion of the area under the entire curve that is between $z = 0$ and a positive value of z.

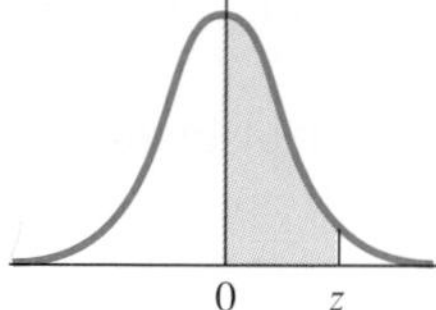

Because the curve is symmetric about the 0-value, the area between $z = 0$ and a *negative* value of z can be found by using the corresponding positive value of z.

TABLE 10 Areas Under the Standard Normal Curve

z	*A*	*z*	*A*	*z*	*A*	*z*	*A*	*z*	*A*	*z*	*A*
.00	.000	.56	.212	1.12	.369	1.68	.454	2.24	.488	2.80	.497
.01	.004	.57	.216	1.13	.371	1.69	.455	2.25	.488	2.81	.498
.02	.008	.58	.219	1.14	.373	1.70	.455	2.26	.488	2.82	.498
.03	.012	.59	.222	1.15	.375	1.71	.456	2.27	.488	2.83	.498
.04	.016	.60	.226	1.16	.377	1.72	.457	2.28	.489	2.84	.498
.05	.020	.61	.229	1.17	.379	1.73	.458	2.29	.489	2.85	.498
.06	.024	.62	.232	1.18	.381	1.74	.459	2.30	.489	2.86	.498
.07	.028	.63	.236	1.19	.383	1.75	.460	2.31	.490	2.87	.498
.08	.032	.64	.239	1.20	.385	1.76	.461	2.32	.490	2.88	.498
.09	.036	.65	.242	1.21	.387	1.77	.462	2.33	.490	2.89	.498
.10	.040	.66	.245	1.22	.389	1.78	.463	2.34	.490	2.90	.498
.11	.044	.67	.249	1.23	.391	1.79	.463	2.35	.491	2.91	.498
.12	.048	.68	.252	1.24	.393	1.80	.464	2.36	.491	2.92	.498
.13	.052	.69	.255	1.25	.394	1.81	.465	2.37	.491	2.93	.498
.14	.056	.70	.258	1.26	.396	1.82	.466	2.38	.491	2.94	.498
.15	.060	.71	.261	1.27	.398	1.83	.466	2.39	.492	2.95	.498
.16	.064	.72	.264	1.28	.400	1.84	.467	2.40	.492	2.96	.499
.17	.068	.73	.267	1.29	.402	1.85	.468	2.41	.492	2.97	.499
.18	.071	.74	.270	1.30	.403	1.86	.469	2.42	.492	2.98	.499
.19	.075	.75	.273	1.31	.405	1.87	.469	2.43	.493	2.99	.499
.20	.079	.76	.276	1.32	.407	1.88	.470	2.44	.493	3.00	.499
.21	.083	.77	.279	1.33	.408	1.89	.471	2.45	.493	3.01	.499
.22	.087	.78	.282	1.34	.410	1.90	.471	2.46	.493	3.02	.499
.23	.091	.79	.285	1.35	.412	1.91	.472	2.47	.493	3.03	.499
.24	.095	.80	.288	1.36	.413	1.92	.473	2.48	.493	3.04	.499
.25	.099	.81	.291	1.37	.415	1.93	.473	2.49	.494	3.05	.499
.26	.103	.82	.294	1.38	.416	1.94	.474	2.50	.494	3.06	.499
.27	.106	.83	.297	1.39	.418	1.95	.474	2.51	.494	3.07	.499
.28	.110	.84	.300	1.40	.419	1.96	.475	2.52	.494	3.08	.499
.29	.114	.85	.302	1.41	.421	1.97	.476	2.53	.494	3.09	.499
.30	.118	.86	.305	1.42	.422	1.98	.476	2.54	.495	3.10	.499
.31	.122	.87	.308	1.43	.424	1.99	.477	2.55	.495	3.11	.499
.32	.126	.88	.311	1.44	.425	2.00	.477	2.56	.495	3.12	.499
.33	.129	.89	.313	1.45	.427	2.01	.478	2.57	.495	3.13	.499
.34	.133	.90	.316	1.46	.428	2.02	.478	2.58	.495	3.14	.499
.35	.137	.91	.319	1.47	.429	2.03	.479	2.59	.495	3.15	.499
.36	.141	.92	.321	1.48	.431	2.04	.479	2.60	.495	3.16	.499
.37	.144	.93	.324	1.49	.432	2.05	.480	2.61	.496	3.17	.499
.38	.148	.94	.326	1.50	.433	2.06	.480	2.62	.496	3.18	.499
.39	.152	.95	.329	1.51	.435	2.07	.481	2.63	.496	3.19	.499
.40	.155	.96	.332	1.52	.436	2.08	.481	2.64	.496	3.20	.499
.41	.159	.97	.334	1.53	.437	2.09	.482	2.65	.496	3.21	.499
.42	.163	.98	.337	1.54	.438	2.10	.482	2.66	.496	3.22	.499
.43	.166	.99	.339	1.55	.439	2.11	.483	2.67	.496	3.23	.499
.44	.170	1.00	.341	1.56	.441	2.12	.483	2.68	.496	3.24	.499
.45	.174	1.01	.344	1.57	.442	2.13	.483	2.69	.496	3.25	.499
.46	.177	1.02	.346	1.58	.443	2.14	.484	2.70	.497	3.26	.499
.47	.181	1.03	.349	1.59	.444	2.15	.484	2.71	.497	3.27	.500
.48	.184	1.04	.351	1.60	.445	2.16	.485	2.72	.497	3.28	.500
.49	.188	1.05	.353	1.61	.446	2.17	.485	2.73	.497	3.29	.500
.50	.192	1.06	.355	1.62	.447	2.18	.485	2.74	.497	3.30	.500
.51	.195	1.07	.358	1.63	.449	2.19	.486	2.75	.497	3.31	.500
.52	.199	1.08	.360	1.64	.450	2.20	.486	2.76	.497	3.32	.500
.53	.202	1.09	.362	1.65	.451	2.21	.487	2.77	.497	3.33	.500
.54	.205	1.10	.364	1.66	.452	2.22	.487	2.78	.497		
.55	.209	1.11	.367	1.67	.453	2.23	.487	2.79	.497		

Carl Friedrich Gauss (1777–1855) was one of the greatest mathematical thinkers of history. In his *Disquisitiones arithmeticae,* published in 1798, he pulled together work by predecessors and enriched and blended it with his own into a unified whole. The book is regarded by many as the true beginning of the theory of numbers.

Of his many contributions to science, the statistical method of least squares is the most widely used today in astronomy, biology, geodesy, physics, and the social sciences. Gauss took special pride in his contributions to developing the method. Despite an aversion to teaching, he taught an annual course in the method for the last twenty years of his life.

It has been said that Gauss was the last person to have mastered all of the mathematics known in his day. Eric Temple Bell, in *Mathematics Queen and Servant of Science,* said that just number theory, one of the four major divisions of higher mathematics, was beyond the complete mastery of any two men, and that the other three divisions, geometry, algebra, and analysis (calculus) were even more extensive. Considering that Bell said this in 1951, the expanse of mathematics today, nearly five decades later, is virtually inconceivable.

EXAMPLE 2 Use the normal curve table to find the percent of all scores that lie between the mean and the following values.

(a) One standard deviation above the mean

Here $z = 1.00$ (the number of standard deviations, written as a decimal to the nearest hundredth). Refer to Table 10. Find 1.00 in the z column. The table entry is .341, so 34.1% of all values lie between the mean and one standard deviation above the mean.

Another way of looking at this is to say that the area in color in Figure 14 represents 34.1% of the total area under the normal curve.

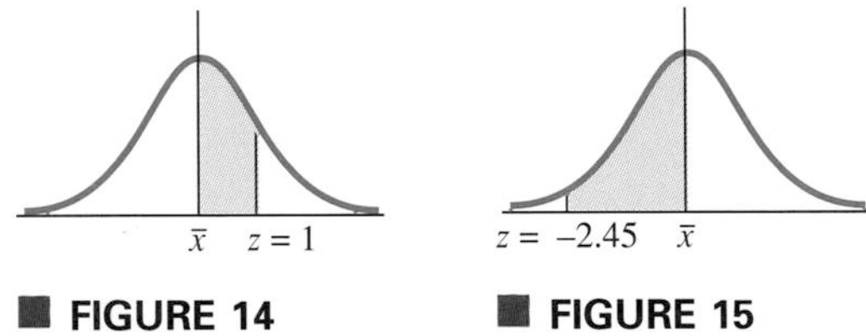

FIGURE 14 **FIGURE 15**

(b) 2.45 standard deviations below the mean

Even though we go *below* the mean here (to the left), the normal curve table still works since the normal curve is symmetrical about its mean. Find 2.45 in the z column. A total of .493, or 49.3%, of all values lie between the mean and 2.45 standard deviations below the mean. This region is shaded in Figure 15.

EXAMPLE 3 The lengths of long-distance phone calls placed in a certain city are distributed normally with mean 6 minutes and standard deviation 2 minutes. If 1 call is randomly selected from phone company records, what is the probability that it will have lasted more than 10 minutes?

Here 10 minutes is two standard deviations more than the mean. The probability of such a call is equal to the shaded region in Figure 16.

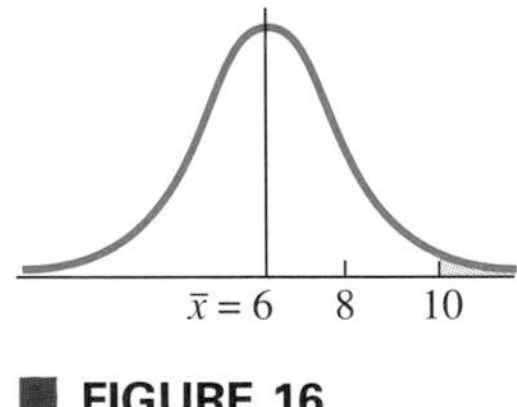

FIGURE 16

From Table 10, the area between the mean and two standard deviations above is .477 ($z = 2.00$). The total area to the right of the mean is .500. Find the area from $z = 2.00$ to the right by subtracting: $.500 - .477 = .023$. So the probability of a call exceeding 10 minutes is .023.

EXAMPLE 4 Find the total areas indicated in the regions in color in each of Figures 17 and 18.

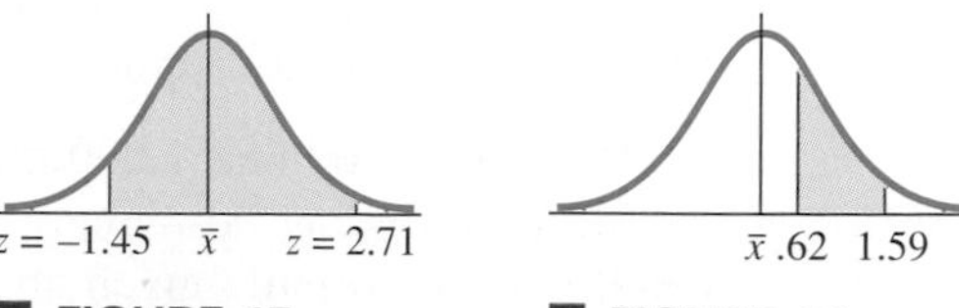

FIGURE 17 **FIGURE 18**

For Figure 17, find the area from 1.45 standard deviations below the mean to 2.71 standard deviations above the mean. From Table 10, $z = 1.45$ leads to an area of .427, while $z = 2.71$ leads to .497. The total area is the sum of these, or $.427 + .497 = .924$.

To find the indicated area in Figure 18, refer again to Table 10. From the table, $z = .62$ leads to an area of .232, while $z = 1.59$ gives .444. To get the area between these two values of z, subtract the areas: $.444 - .232 = .212$.

Examples 2, 3, and 4 emphasize the *equivalence* of three quantities, as follows.

Meaning of Normal Curve Areas

In a standard normal curve, the following three quantities are equivalent.

1. **Percentage** (of total items that lie in an interval)
2. **Probability** (of a randomly chosen item lying in an interval)
3. **Area** (under the normal curve along an interval)

Which quantity we think of depends upon how a particular question is formulated. They are all evaluated by using A-values from Table 10.

In general, when we use Table 10, z is the z-score of a particular data item x. Recall from the previous section that the z-score is related to the mean $\bar{x}$ and the standard deviation s of the distribution by the formula

$$z = \frac{x - \bar{x}}{s}.$$

For example, suppose a normal curve has mean 220 and standard deviation 12. To find the number of standard deviations that a given value is from the mean, use the formula above. For the value 247,

$$z = \frac{247 - 220}{12} = \frac{27}{12} = 2.25,$$

so 247 is 2.25 standard deviations above the mean. For 204,

$$z = \frac{204 - 220}{12} = \frac{-16}{12} \approx -1.33,$$

so 204 is 1.33 standard deviations *below* the mean. (We know it is below rather than above the mean since the z-score is negative rather than positive.)

A Basic Consumer (t)issue
The following describes an actual event, but the names are withheld.

It all started when a reporter on consumer issues for a Midwest TV station received a complaint that rolls of Brand X toilet paper manufactured by Company Y did not have the number of sheets claimed on the wrapper. Brand X is supposed to have 375 sheets, but three rolls of it were found by reporters to have 360, 361, and 363. Company Y executives were shocked. They said that the **odds against** six rolls having fewer than 375 sheets each are 1 billion to 1. Company executives counted sheets and found that several rolls of Brand X actually had 380 sheets each (machines count the sheets only in 10s). TV reporters made an independent count, and their results agreed with Company Y.

What happened the first time?

Well, the reporters hadn't actually counted sheets, but had measured rolls and divided the length of a roll by the length of one sheet. Small variations in length can be expected, which add up over a roll, giving false results. The wrong quantitative method, so to speak, was applied the first time.

This story perhaps points up the distinction in probability and statistics between **discrete** values and **continuous** values.

EXAMPLE 5 In one area, the average motorist drives about 1,200 miles per month, with standard deviation 150 miles. Assume that the number of miles is closely approximated by a normal curve, and find the percent of all motorists driving the following distances.

(a) Between 1,200 and 1,600 miles per month

Start by finding how many standard deviations 1,600 miles is above the mean. Use the formula given above.

$$z = \frac{1{,}600 - 1{,}200}{150} = \frac{400}{150} \approx 2.67$$

From Table 10, .496, or 49.6%, of all motorists drive between 1,200 and 1,600 miles per month.

(b) Between 1,000 and 1,500 miles per month

As shown in Figure 19, values of z must be found for both 1,000 and 1,500.

$$\text{For } 1{,}000\text{:} \quad z = \frac{1{,}000 - 1{,}200}{150} = \frac{-200}{150} \approx -1.33.$$

$$\text{For } 1{,}500\text{:} \quad z = \frac{1{,}500 - 1{,}200}{150} = \frac{300}{150} = 2.00.$$

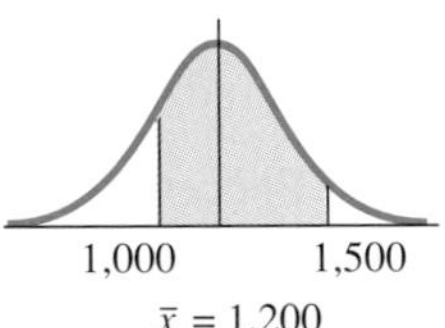

FIGURE 19

From Table 10, $z = -1.33$ leads to an area of .408, while $z = 2.00$ gives .477. This means a total of $.408 + .477 = .885$, or 88.5%, of all motorists drive between 1,000 and 1,500 miles per month. ■

The examples above have given a data value and then required that z be found. The next example gives z and asks for the data value.

EXAMPLE 6 A particular normal distribution has mean $\bar{x} = 81.7$ and standard deviation $s = 5.21$. What data value from the distribution would correspond to $z = -1.35$?

Start with the formula

$$z = \frac{x - \bar{x}}{s}.$$

Here $z = -1.35$, $\overline{x} = 81.7$, $s = 5.21$, and x is unknown. Substitute the given values into the formula to get

$$-1.35 = \frac{x - 81.7}{5.21}.$$

Solve this equation for x by first multiplying both sides by 5.21.

$$-1.35(5.21) = \frac{x - 81.7}{5.21}(5.21)$$

$$-7.0335 = x - 81.7$$

Add 81.7 to each side to get $74.6665 = x$.

Rounding to the nearest tenth, the required data value is 74.7. ■

12.5 EXERCISES

Identify each of the following variable quantities as discrete *or* continuous.

1. the number of heads in 30 tossed coins

2. the number of babies born in one day at a certain hospital

3. the average weight of babies born in a week

4. the heights of seedling fir trees at six weeks of age

5. the time as shown on a digital watch

6. the time as shown on a watch with a sweep hand

Suppose 100 geology students measure the mass of an ore sample. Due to human error and limitations in the reliability of the balance, not all the readings are equal. The results are found to closely approximate a normal curve, with mean 37 g and standard deviation 1 g.

Use the symmetry of the normal curve and the empirical rule to estimate the number of students reporting readings in the following ranges.

7. more than 37 g **8.** more than 36 g

9. between 36 and 38 g **10.** between 36 and 39 g

On standard IQ tests, the mean is 100, with a standard deviation of 15. The results come very close to fitting a normal curve. Suppose an IQ test is given to a very large group of people. Find the percent of people whose IQ scores fall into the following categories.

11. less than 100 **12.** greater than 115

13. between 70 and 130 **14.** more than 145

Find the percent of area under a normal curve between the mean and the given number of standard deviations from the mean. (Note that positive indicates above the mean, while negative indicates below the mean.)

15. 2.50 **16.** .81 **17.** -1.71 **18.** -2.04

Find the percent of the total area under the normal curve between the given values of z.

19. $z = 1.41$ and $z = 2.83$

20. $z = -1.74$ and $z = -1.02$

21. $z = -3.11$ and $z = 1.44$

22. $z = -1.98$ and $z = 1.98$

Find a value of z such that the following conditions are met.

23. 5% of the total area is to the right of z.

24. 1% of the total area is to the left of z.

25. 15% of the total area is to the left of z.

26. 25% of the total area is to the right of z.

The Alva light bulb has an average life of 500 hr, with a standard deviation of 100 hr. The length of life of the bulb can be closely approximated by a normal curve. An amusement park buys and installs 10,000 such bulbs. Find the total number that can be expected to last the following amounts of time.

27. at least 500 hr

28. between 500 and 650 hr

29. between 650 and 780 hr

30. between 290 and 540 hr

31. less than 740 hr

32. less than 410 hr

The chickens at Ben and Ann Rice's farm have a mean weight of 1,850 g with a standard deviation of 150 g. The weights of the chickens are closely approximated by a normal curve. Find the percent of all chickens having the following weights.

33. more than 1,700 g

34. less than 1,800 g

35. between 1,750 and 1,900 g

36. between 1,600 and 2,000 g

A box of oatmeal must contain 16 oz. The machine that fills the oatmeal boxes is set so that, on the average, a box contains 16.5 oz. The boxes filled by the machine have weights that can be closely approximated by a normal curve. What fraction of the boxes filled by the machine are underweight if the standard deviation is as follows?

37. .5 oz **38.** .3 oz

39. .2 oz **40.** .1 oz

41. In nutrition, the recommended daily allowance of vitamins is a number set by the government to guide an individual's daily vitamin intake. Actually, vitamin needs vary drastically from person to person, but the needs are closely approximated by a normal curve. To calculate the recommended daily allowance, the government first finds the average need for vitamins among people in the population and the standard deviation. The **recommended daily allowance** is then defined as the mean plus 2.5 times the standard deviation. What fraction of the population will receive adequate amounts of vitamins under this plan?

Find the recommended daily allowance for each vitamin if the mean need and standard deviation are as follows. (See Exercise 41.)

42. mean need = 1,800 units;
standard deviation = 140 units

43. mean need = 159 units;
standard deviation = 12 units

Assume the following distributions are all normal, and use the areas under the normal curve given in Table 10 to find the appropriate areas.

44. A machine that fills quart milk cartons is set up to average 32.2 oz per carton, with a standard deviation of 1.2 oz. What is the probability that a filled carton will contain less than 32 oz of milk?

45. The mean clotting time of blood is 7.45 sec, with a standard deviation of 3.6 sec. What is the probability that an individual's blood-clotting time will be less than 7 sec or greater than 8 sec?

46. The average size of the fish caught in Lake Amotan is 12.3 in., with a standard deviation of 4.1 in. Find the probability that a fish caught there will be longer than 18 in.

47. To be graded extra large, an egg must weigh at least 2.2 oz. If the average weight for an egg is 1.5 oz, with a standard deviation of .4 oz, how many of five dozen eggs would you expect to grade extra large?

Kimberly Workman teaches a course in marketing. She uses the following system for assigning grades to her students.

Grade	Score in Class
A	Greater than $\bar{x} + (3/2)s$
B	$\bar{x} + (1/2)s$ to $\bar{x} + (3/2)s$
C	$\bar{x} - (1/2)s$ to $\bar{x} + (1/2)s$
D	$\bar{x} - (3/2)s$ to $\bar{x} - (1/2)s$
F	Below $\bar{x} - (3/2)s$

What percent of the students receive the following grades?

48. A **49.** B **50.** C

51. Do you think this system would be more likely to be fair in a large freshman class in psychology or in a graduate seminar of five students? Why?

A teacher gives a test to a large group of students. The results are closely approximated by a normal curve. The mean is 74 with a standard deviation of 6. The teacher wishes to give As to the top 8% of the students and Fs to the bottom 8%. A grade of B is given to the next 15%, with Ds given similarly. All other students get Cs. Find the bottom cutoff (rounded to the nearest whole number) for the following grades. (Hint: read Table 10 backwards.)

52. A **53.** B **54.** C **55.** D

A normal distribution has mean 94.2 and standard deviation 7.68. Follow the method of Example 6 and find data values corresponding to the following values of z. Round to the nearest tenth.

56. $z = .72$

57. $z = 1.44$

58. $z = -2.39$

59. $z = -3.87$

60. What percentage of the items lie within 1.25 standard deviations of the mean
(a) in any distribution (by Chebyshev's theorem)?
(b) in a normal distribution (by Table 10)?

61. Explain the difference between the answers to parts (a) and (b) in Exercise 60.

EXTENSION

How to Lie with Statistics

The statement that there are "lies, damned lies, and statistics" is attributed to Benjamin Disraeli, Queen Victoria's prime minister. Other people have said even stronger things about statistics. This often intense distrust of statistics has come about because of a belief that "you can prove anything with numbers." It must be admitted that there is often a conscious or unconscious distortion in many published statistics. The classic book on distortion in statistics is *How to Lie with Statistics,* by Darrell Huff. This section quotes some common methods of distortion that Huff gives in his book.

> **The Sample with the Built-In Bias**
> A house-to-house survey purporting to study magazine readership was once made in which a key question was: What magazines does your household read? When the results were tabulated and analyzed it appeared that a great many people loved *Harper's* and not very many read *True Story*. Now there were publishers' figures around at the time that showed very clearly that *True Story* had more millions of circulation than *Harper's* had hundreds of thousands. Perhaps we asked the wrong kind of people, the designers of the survey said to themselves. But no, the questions had been asked in all sorts of neighborhoods all around the country. The only reasonable conclusion then was that a good many of the respondents, as people are called when they answer such questions, had not told the truth. About all the survey had uncovered was snobbery.
>
> In the end it was found that if you wanted to know what certain people read it was no use asking them. You could learn a good deal more by going to their houses and saying you wanted to buy old magazines and what could be had? Then all you had to do was count the *Yale Reviews* and the *Love Romances*. Even that dubious device, of course, does not tell you what people read, only what they have been exposed to.
>
> Similarly, the next time you learn from your reading that the average American (you hear a good deal about him these days, most of it faintly improbable) brushes his teeth 1.02 times a day—a figure pulled out of the air, but it may be as good as anyone else's—ask yourself a question. How can anyone have found out such a thing? Is a woman who has read in countless advertisements that non-brushers are social offenders going to confess to a stranger that she does not brush her teeth regularly? The statistic may have meaning to one who wants to know only what people say about tooth-brushing but it does not tell a great deal about the frequency with which bristle is applied to incisor.

The Well-Chosen Average

A common trick is to use a different kind of average each time, the word "average" having a very loose meaning. It is a trick commonly used, sometimes in innocence but often in guilt, by people wishing to influence public opinion or sell advertising space. When you are told that something is an average you still don't know very much about it unless you can find out which of the common kinds of average it is—mean, median or mode.

Try your skepticism on some items from "A letter from the Publisher" in *Time* magazine. Of new subscribers it said, "Their median age is 34 years and their average family income is $7,270 a year." An earlier survey of "old TIMErs" had found that their "median age was 41 years. . . . Average income was $9,535. . . ." The natural question is why, when median is given for ages both times, the kind of average for incomes is carefully unspecified. Could it be that the mean was used instead because it is bigger, thus seeming to dangle a richer readership before advertisers?

Be careful when reading charts or graphs—often there is no numerical scale, or no units are given. This makes the chart pretty much meaningless. Huff has a good example of this:

The Little Figures That Are Not There

Before me are wrappers from two boxes of Grape-Nuts Flakes. They are slightly different editions, as indicated by their testimonials: one cites Two-Gun Pete and the other says "If you want to be like Hoppy . . . you've got to eat like Hoppy!" Both offer charts to show ("Scientists *proved* it's true!") that these flakes "start giving you energy in 2 minutes!" In one case the chart hidden in these forests of exclamation points has numbers up the side; in the other case the numbers have been omitted. This is just as well, since there is no hint of what the numbers mean. Both show a steeply climbing brown line ("energy release"), but one has it starting one minute after eating Grape-Nuts Flakes, the other 2 minutes later. One line climbs about twice as fast as the other, suggesting that even the draftsman didn't think these graphs meant anything.

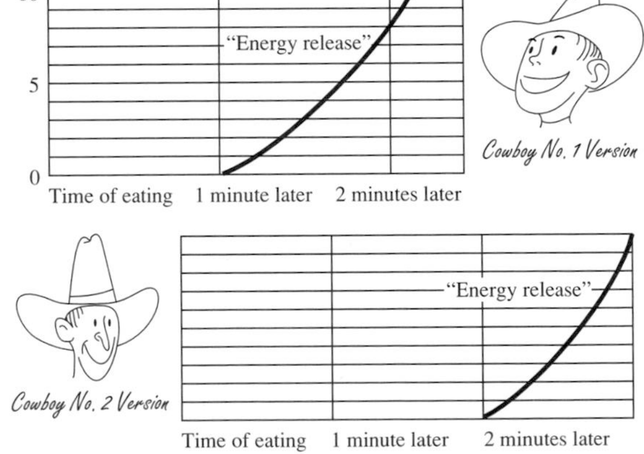

The Gee-Whiz Graph

About the simplest kind of statistical picture, or graph, is the line variety. It is very useful for showing trends, something practically everybody is interested in showing or knowing about or spotting or deploring or forecasting. We'll let our graph show how national income increased ten percent in a year.

Stand Up and Be Counted

According to an Associated Press release in 1988 (*The Sacramento Union,* September 28), the U.S. Census Bureau anticipated that the 1990 census would cost up to $3 billion (the cost in 1980 was $1.1 billion). Even so, there would be significant problems finding the necessary workers to conduct the census and getting response from the populace. Census Director John G. Keane cited a deteriorating climate in the nation for taking censuses and surveys, due to Americans' attitude toward being counted.

Negative attitudes toward statistics and polls are understandable when we are aware of some of the distorted methods used in surveys. Hopefully, as more people learn what statistics are and what they are not, what they can do and what they cannot do, their abuses can be curbed and their legitimate uses can be strengthened.

Begin with paper ruled into squares. Name the months along the bottom. Indicate billions of dollars up the side. Plot your points and draw your line, and your graph will look like this:

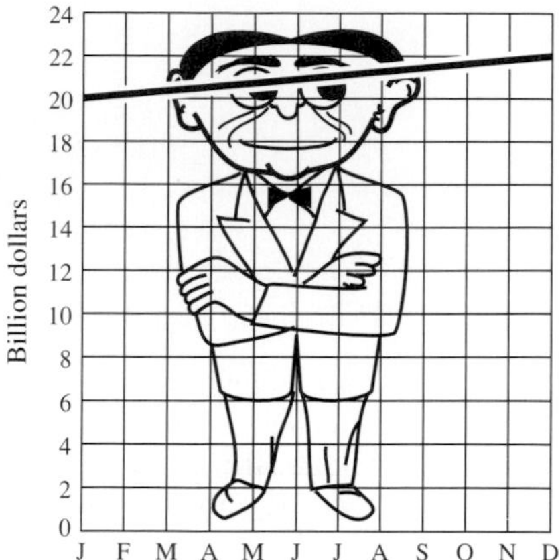

Now that's clear enough. It shows what happened during the year and it shows it month by month. He who runs may see and understand, because the whole graph is in proportion and there is a zero line at the bottom for comparison. Your ten percent looks like ten percent—an upward trend that is substantial but perhaps not overwhelming.

That is very well if all you want to do is convey information. But suppose you wish to win an argument, shock a reader, move him into action, sell him something. For that, this chart lacks schmaltz. Chop off the bottom.

Now that's more like it. (You've saved paper too, something to point out if any carping person objects to your misleading graphics.) The figures are the same and so is the curve. It is the same graph. Nothing has been falsified—except the impression that it gives. But what the hasty reader sees now is a national income line that has climbed halfway up the paper in twelve months, all because most of the chart isn't there any more. Like the missing parts of speech in sentences that you met in grammar classes, it is "understood." Of course, the eye doesn't "understand" what isn't there, and a small rise has become, visually, a big one.

Now that you have practiced to deceive, why stop with truncating? You have a further trick available that's worth a dozen of that. It will make your modest rise of ten percent look livelier than one hundred percent is entitled to look. Simply change the proportion between the side and the bottom. There's no rule against it, and it does give your graph a prettier shape. All you have to do is let each mark up the side stand for only one-tenth as many dollars as before.

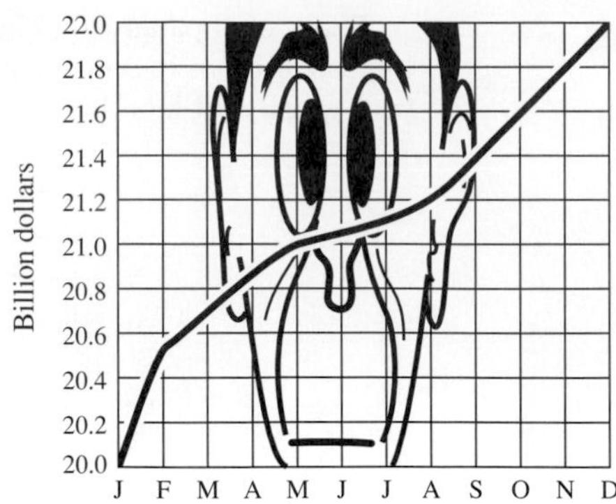

That *is* impressive, isn't it? Anyone looking at it can just feel prosperity throbbing in the arteries of the country. It is a subtler equivalent of editing "National income rose ten percent" into ". . . climbed a whopping ten percent." It is vastly more effective, however, because it contains no adjectives or adverbs to spoil the illusion of objectivity. There's nothing anyone can pin on you.

Suppose Diana makes twice as much money as Mike. One way to show this is with a graph using silver dollars to represent the income of each. If we used one silver dollar for Mike and two for Diana, we would be fine. But it is more common to use proportional dollars. It is common to find a dollar of one size used for Mike, with one twice as wide for Diana. This is wrong—the larger dollar actually has four times the area, giving the impression that Diana earns four times as much as Mike. Huff gives another example of this:

The One-Dimensional Picture
Newsweek once showed how "U.S. Old Folks Grow Older" by means of a chart on which appeared two male figures, one representing the 68.2-year life expectancy of today, the other the 34-year life expectancy of 1879–1889. It was the same old story. One figure was twice as tall as the other and so would have had eight times the bulk or weight. This picture sensationalized facts in order to make a better story. It would be called a form of yellow journalism.

Cause and Effect

Many people often assume that just because two things changed together one caused the other. The classic example of this is the fact that teachers' salaries and liquor consumption increased together over the last few decades. Neither of these caused the other; rather, both were caused by the same underlying growth in national prosperity. Another case of faulty cause and effect reasoning is in the claim that going to college raises your income. See Huff's comments on page 648.

Reams of pages of figures have been collected to show the value in dollars of a college education, and stacks of pamphlets have been published to bring these figures—and conclusions more or less based on them—to the attention of potential students. I am not quarreling with the intention. I am in favor of education myself, particularly if it includes a course in elementary statistics. Now these figures have pretty conclusively demonstrated that people who have gone to college make more money than people who have not. The exceptions are numerous, of course, but the tendency is strong and clear.

The only thing wrong is that along with the figures and facts goes a totally unwarranted conclusion. . . . It says that these figures show that if *you* (your son, your daughter) attend college you will probably earn more money than if you decide to spend the next four years in some other manner. This unwarranted conclusion has for its basis the equally unwarranted assumption that since college-trained folks make more money, they make it because they went to college. Actually we don't know but that these are the people who would have made more money even if they had not gone to college. There are a couple of things that indicate rather strongly that this is so. Colleges get a disproportionate number of two groups of people: the bright and the rich. The bright might show good earning power without college knowledge. And as for the rich ones . . . well, money breeds money in several obvious ways. Few children of rich parents are found in low-income brackets whether they go to college or not.

Extrapolation

This refers to predicting the future, based only on what has happened in the past. However, it is very rare for the future to be just like the past—something will be different. One of the best examples of extrapolating incorrectly is given by Mark Twain:

> In the space of one hundred and seventy-six years the Lower Mississippi has shortened itself two hundred and forty-two miles. That is an average of a trifle over one mile and a third per year. Therefore, any calm person, who is not blind or idiotic, can see that in the Old Oölitic Silurian Period, just a million years ago next November, the Lower Mississippi River was upward of one million three hundred thousand miles long, and stuck out over the Gulf of Mexico like a fishing rod. And by the same token any person can see that seven hundred and forty-two years from now the Lower Mississippi will be only a mile and three-quarters long, and Cairo and New Orleans will have joined their streets together, and be plodding comfortably along under a single mayor and a mutual board of aldermen. There is something fascinating about science. One gets such wholesale returns of conjecture out of such a trifling investment of fact.

Extension Exercises

The Norwegian stamp at the side features two graphs.

1. What does the solid line represent?
2. Has it increased much?
3. What can you tell about the dashed line?
4. Does the graph represent a long period of time or a brief period of time?

The illustration at the side shows the decline in the value of the British pound for a certain period.

$2.40

5. Calculate the percent of decrease in the value by using the formula

Percent of decrease

$$= \frac{\text{old value} - \text{new value}}{\text{old value}}.$$

$1.72

6. We estimate that the smaller banknote shown has about 50% less area than the larger one. Do you think this is close enough?

Several advertising claims are given below. Decide what further information you might need before deciding to accept the claim.

7. 98% of all Toyotas ever sold in the United States are still on the road.

8. Sir Walter Raleigh pipe tobacco is 44% fresher.

9. Eight of 10 dentists responding to a survey preferred Trident Sugarless Gum.

10. A Volvo has 2/3 the turning radius of a Continental.

11. A Ford LTD is as quiet as a glider.

12. *Wall Streel Journal* circulation has increased, as shown by the graph below.

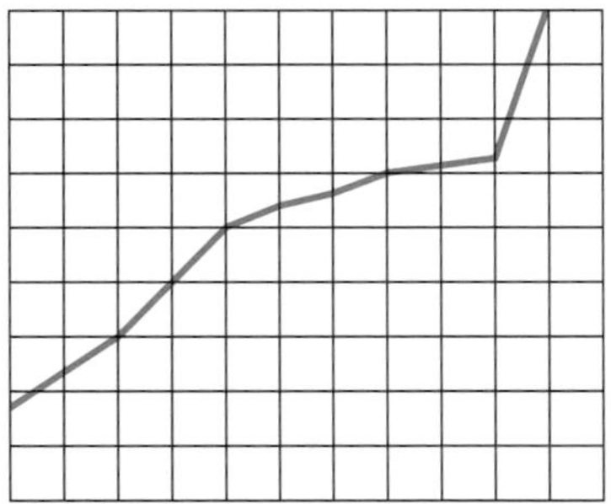

Exercises 13–16 come from Huff's book. Decide how these exercises describe possibly misleading uses of numbers.

13. Each of the following maps shows what portion of our national income is now being taken and spent by the federal government. The map on the left does this

The Darkening Shadow
(Western Style)

(Eastern Style)

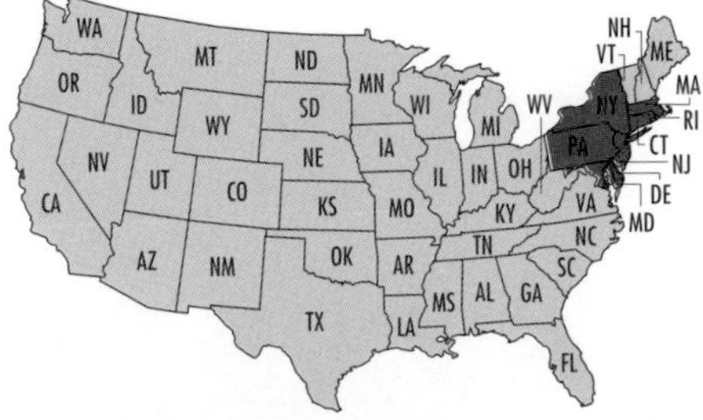

To show we aren't cheating, we added MD, DE, and RI for good measure

by shading the areas of most of the states *west* of the Mississippi to indicate that federal spending has become equal to the total incomes of the people of those states. The map on the right does this for states *east* of the Mississippi.

14. Long ago, when Johns Hopkins University had just begun to admit women students, someone not particularly enamored of coeducation reported a real shocker: Thirty three and one-third percent of the women at Hopkins had married faculty members!

15. The death rate in the Navy during the Spanish-American War was nine per thousand. For civilians in New York City during the same period it was sixteen per thousand. Navy recruiters later used these figures to show that it was safer to be in the Navy than out of it.

16. If you should look up the latest available figures on influenza and pneumonia, you might come to the strange conclusion that these ailments are practically confined to three southern states, which account for about eighty percent of the reported cases.

When a sample is selected from a population, it is important to decide whether the sample is reasonably representative of the entire population. For example, a sample of the general population that is only 20% women should make us suspicious. The same is true of a questionnaire asking, "Do you like to answer questionnaires?"

In each exercise, pick the choice that gives the most representative sample. (Sampling techniques are a major branch of inferential statistics.*)*

17. A factory has 10% management employees, 30% clerical employees, and 60% assembly-line workers. A sample of 50 is chosen to discuss parking.
(a) 4 management, 21 clerical, 25 assembly-line
(b) 6 management, 15 clerical, 29 assembly-line
(c) 8 management, 9 clerical, 33 assembly-line

18. A college has 35% freshmen, 28% sophomores, 21% juniors, and 16% seniors. A sample of 80 is chosen to discuss methods of electing student officers.
(a) 22 freshmen, 22 sophomores, 24 juniors, 12 seniors
(b) 24 freshmen, 20 sophomores, 22 juniors, 14 seniors
(c) 28 freshmen, 23 sophomores, 16 juniors, 13 seniors

19. A computer company has plants in Boca Raton, Jacksonville, and Tampa. The plant in Boca Raton produces 42% of all the company's output, with 27% and 31% coming from Jacksonville and Tampa, respectively. A sample of 120 parts is chosen for quality testing.
(a) 38 from Boca Raton, 39 from Jacksonville, 43 from Tampa
(b) 43 from Boca Raton, 37 from Jacksonville, 40 from Tampa
(c) 50 from Boca Raton, 31 from Jacksonville, 39 from Tampa

20. At one resort, 56% of all guests come from the Northeast, 29% from the Midwest, and 15% from Texas. A sample of 75 guests is chosen to discuss the dinner menu.
(a) 41 from the Northeast, 21 from the Midwest, 13 from Texas
(b) 45 from the Northeast, 18 from the Midwest, 12 from Texas
(c) 47 from the Northeast, 20 from the Midwest, 8 from Texas

Use the information supplied in the following problems to solve for the given variables.

21. An insurance agency has 7 managers, 25 agents, and 18 clerical employees. A sample of 10 is chosen.

(a) Let m be the number of managers in the sample, a the number of agents, and c the number of clerical employees. Find m, a, and c, if

$$c = m + 2$$
$$a = 2c.$$

(b) To check that your answer is reasonable, calculate the numbers of the office staff that *should* be in the sample, if all groups are represented proportionately.

22. A small college has 12 deans, 24 full professors, 39 associate professors, and 45 assistant professors. A sample of 20 employees is chosen to discuss the graduation speaker.

(a) Let d be the number of deans in the sample, f the number of full professors, a the number of associate professors, and s the number of assistant professors. Find d, f, a, and s if

$$f = 2d$$
$$a = f + d + 1$$
$$a = s.$$

(*Hint:* there are 2 deans.)

(b) To check that your answer is reasonable, calculate the number of each type of employee that *should* be in the sample, if all groups are represented proportionately.

12.6 Regression and Correlation

TABLE 11 Age Versus Income

Resident	Age	Annual Income
A	19	2,150
B	23	2,550
C	27	3,250
D	31	3,150
E	36	4,250
F	40	4,200
G	44	4,350
H	49	5,000
I	52	4,950
J	54	5,650

One very important branch of inferential statistics, called **regression analysis,** is used to compare quantities or variables, to discover relationships that exist between them, and to formulate those relationships in useful ways. For example, suppose a sociologist gathers data on a few (say ten) of the residents of a small village in a remote country in order to get an idea of how annual income relates to age in that village. The data are shown here (in Table 11).

The first step in analyzing these data is to graph the results, as shown in the **scatter diagram** of Figure 20. (Modern graphing calculators will plot scatter diagrams.)

Once a scatter diagram has been produced, we can draw a curve that best fits the pattern exhibited by the sample data points. This curve can have any one of many characteristic shapes, depending on how the quantities involved are related. We would like to infer a graph for the entire population of points for the related quantities, but we only have available a sample of those points, so the best-fitting curve for the sample points is called an **estimated regression curve.** If, as in the example above, the points in the scatter diagram seem to lie

Francis Galton (1822–1911) learned to read at age three, was interested in mathematics and machines, but was an indifferent mathematics student at Trinity College, Cambridge. After several years as a traveling gentleman of leisure, he became interested in researching methods of predicting weather. It was during this research on weather that Galton developed early intuitive notions of correlation and regression and posed the problem of multiple regression.

Galton's key statistical work is *Natural Inheritance*. In it, he set forth his ideas on regression and correlation. Examining the relative heights of fathers and sons led him to seek a unit-free measure of association. He discovered the correlation coefficient while pondering Alphonse Bertillon's scheme for classifying criminals by physical characteristics. It was a major contribution to statistical method.

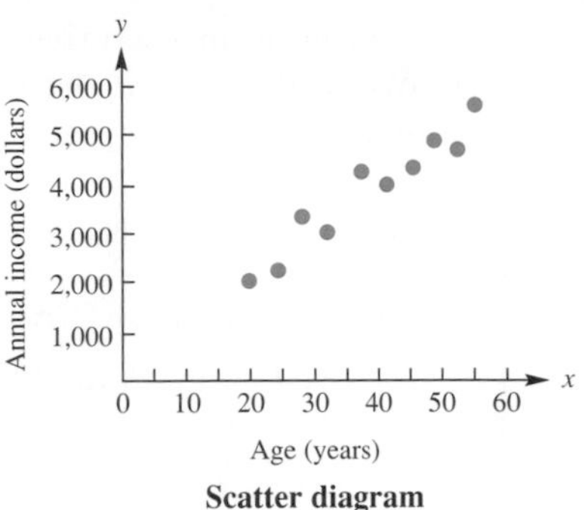

Scatter diagram

FIGURE 20

approximately along a straight line, the relation is assumed to be linear, and the line that best fits the data points is called the **estimated regression line.**

If we let x denote age and y denote income in the data of Table 11, and assume that the best-fitting curve is a line, then the equation of that line will take the form

$$y = ax + b,$$

where a is the slope of the line and b is the y-coordinate of the y-intercept (the y-value at which the line, if extended, would intersect the y-axis).

To completely identify the estimated regression line, we must find the values of the "regression coefficients" a and b, which requires some calculation. In Figure 21, a tentative line has been drawn through the scatter diagram.

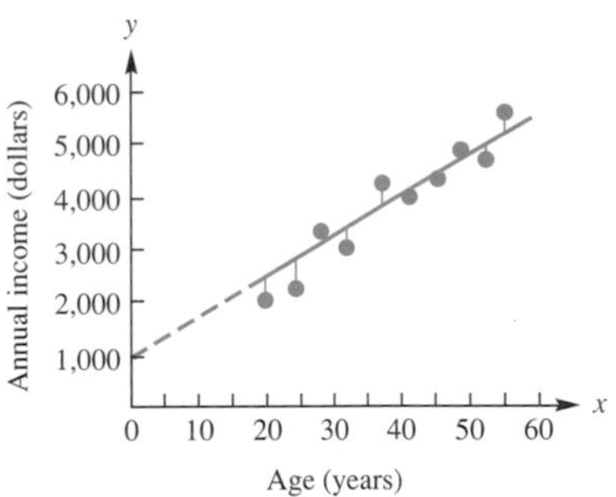

A tentative estimated regression line

FIGURE 21

For each x-value in the data set, the corresponding y-value usually differs from the value it would have if the data point were exactly on the line. These differences are shown in the figure by vertical segments. Choosing another line would make some of these differences greater and some lesser. The most common procedure is to choose the line where the sum of the squares of all these differences is minimized. This is called the **method of least squares,** and the resulting line is called the **least squares line.**

In the equation of the least squares line, the variable y' can be used to distinguish the *predicted* values (which would give points on the least squares line) from the *observed* values y (those occurring in the data set).

The least squares criterion mentioned above leads to specific values of a and b. We shall not give the details, which involve differential calculus, but the results are given here. (Σ—the Greek letter *sigma*—represents summation just as in earlier sections.)

Regression Coefficient Formulas

The **least squares line** $y' = ax + b$ that provides the best fit to the data points $(x_1, y_1), (x_2, y_2), \ldots, (x_n, y_n)$ has

$$a = \frac{n(\Sigma xy) - (\Sigma x)(\Sigma y)}{n(\Sigma x^2) - (\Sigma x)^2}$$

and

$$b = \frac{\Sigma y - a(\Sigma x)}{n}.$$

EXAMPLE 1 Find the equation of the least squares line for the age and income data given earlier. Graph the line.

Start with the two columns on the left in the following chart (these just repeat the original data). Then find the products $x \cdot y$, and the squares x^2.

	x	y	$x \cdot y$	x^2
	19	2,150	40,850	361
	23	2,550	58,650	529
	27	3,250	87,750	729
	31	3,150	97,650	961
	36	4,250	153,000	1,296
	40	4,200	168,000	1,600
	44	4,350	191,400	1,936
	49	5,000	245,000	2,401
	52	4,950	257,400	2,704
	54	5,650	305,100	2,916
Sums:	375	39,500	1,604,800	15,433

From the chart, $\Sigma x = 375$, $\Sigma y = 39{,}500$, $\Sigma xy = 1{,}604{,}800$, and $\Sigma x^2 = 15{,}433$. Also, $n = 10$ since there are 10 pairs of values. Now find a with the formula given above.

$$a = \frac{10(1{,}604{,}800) - 375(39{,}500)}{10(15{,}433) - (375)^2}$$

$$= \frac{1{,}235{,}500}{13{,}705} \approx 90.15$$

Finally, use this value of a to find b.

$$b = \frac{39{,}500 - 90.15(375)}{10} \approx 569.4$$

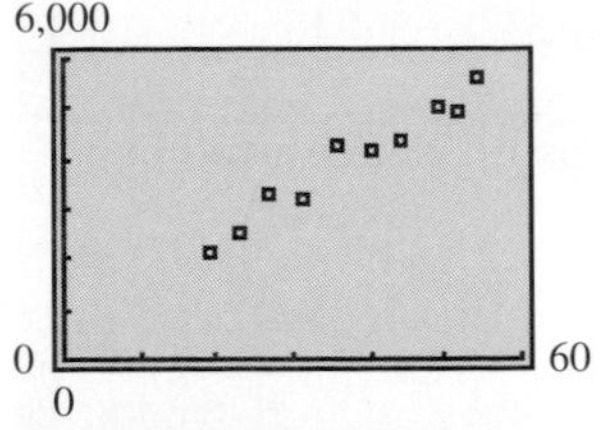

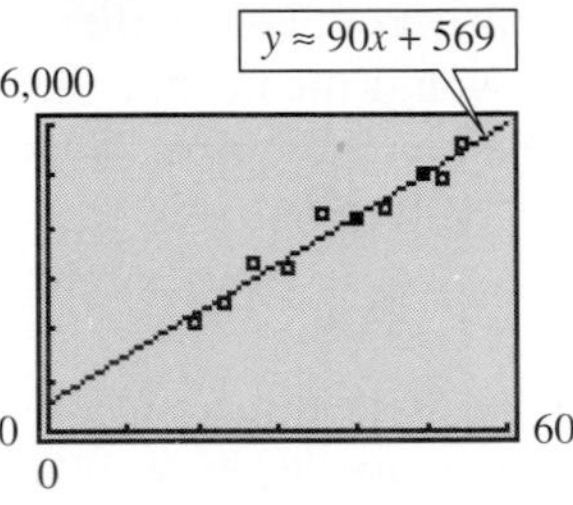

The information in Figure 22 and the accompanying discussion is supported in these screens.

The equation of the least squares line (with coefficients rounded to whole numbers) is

$$y' = 90x + 569.$$

Letting $x = 20$ in this equation gives $y' = 2{,}369$, and $x = 50$ implies $y' = 5{,}069$. The two points (20, 2,369) and (50, 5,069) are used to graph the regression line in Figure 22. Notice that the intercept coordinates (0, 569) also fit the extended line.

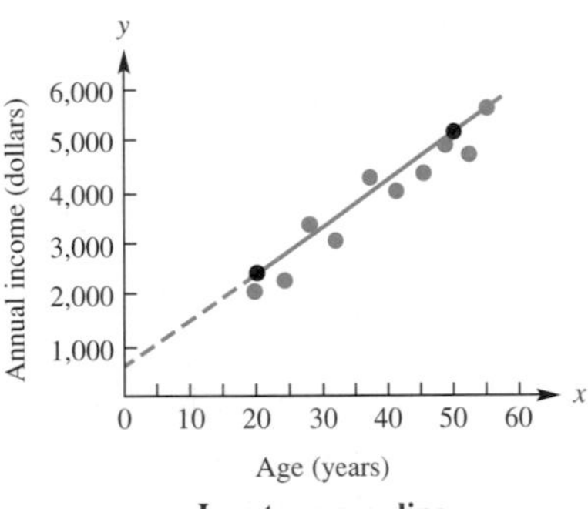

FIGURE 22

A computer, or a scientific, statistical, or graphing calculator, is recommended for finding regression coefficients. (See the margin notes.) That way, tedious calculations, like in Example 1, can be avoided and the regression line produced automatically.

EXAMPLE 2 Use the result of Example 1 to predict the income of a village resident who is 35 years old.

Use the equation $y' = 90x + 569$ and replace x with 35.

$$y' = 90(35) + 569 = 3{,}719$$

Based on the data given earlier, a 35-year-old will make about \$3,719 per year.

Correlation

Once an equation for the line of best fit (the least squares line) has been found, it is reasonable to ask, "Just how good is this line for predictive purposes?" If the points already observed fit the line quite closely, then future pairs of scores can be expected to do so. If the points are widely scattered about even the "best-fitting" line, then predictions are not likely to be accurate. In general, the closer the *sample* data points lie to the least squares line, the more likely it is that the entire *population* of (x, y) points really do form a line, that is, that x and y really are related linearly. Also, the better the fit, the more confidence we can have that our least squares line (based on the sample) is a good estimator of the true population line.

One common measure of the strength of the linear relationship in the sample is called the **sample correlation coefficient,** denoted r. It is calculated from the sample data according to the following formula.

Sample Correlation Coefficient Formula

In linear regression, the strength of the linear relationship is measured by the correlation coefficient

$$r = \frac{n(\Sigma xy) - (\Sigma x)(\Sigma y)}{\sqrt{n(\Sigma x^2) - (\Sigma x)^2} \cdot \sqrt{n(\Sigma y^2) - (\Sigma y)^2}}.$$

The value of r is always between -1 and 1, or perhaps equal to -1 or 1. Values of exactly 1 or -1 indicate that the least squares line goes exactly through all the data points. If r is close to 1 or -1, but not exactly equal, then the line comes "close" to fitting through all the data points, and we say that the linear correlation between x and y is "strong." If r is equal, or nearly equal, to 0, there is no linear correlation, or the correlation is weak. This would mean that the points are a totally disordered conglomeration. Or it may be that the points form an ordered pattern but one which is not linear. (More is said about this case following Example 3.) If r is neither close to 0 nor close to 1 or -1, we might describe the linear correlation as "moderate."

A positive value of r indicates a regression line with positive slope and we say the linear relationship between x and y is direct; as x increases, y also increases. A negative r-value means the line has negative slope, so there is an inverse relationship between x and y; as x increases, y decreases.

The slope a and y-intercept b of the regression equation, along with the correlation coefficient r, are given. Compare with Examples 1 and 3.

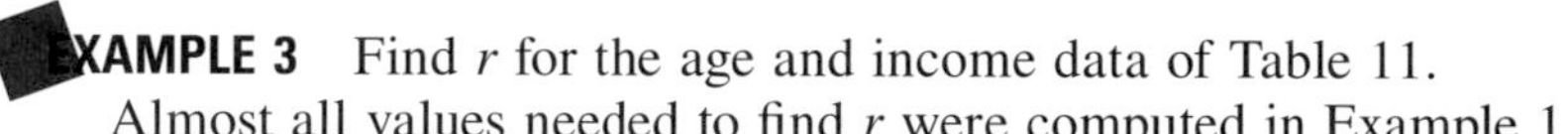

EXAMPLE 3 Find r for the age and income data of Table 11.

Almost all values needed to find r were computed in Example 1.

$$n = 10 \qquad \Sigma x = 375$$
$$\Sigma y = 39{,}500 \qquad \Sigma xy = 1{,}604{,}800$$
$$\Sigma x^2 = 15{,}433$$

The only missing value is Σy^2. Squaring each y in the original data and adding the squares gives

$$\Sigma y^2 = 167{,}660{,}000.$$

Now use the formula to find r.

$$r = .98 \text{ (to two decimal places)}$$

This value of r very close to 1 shows that age and income in this village are highly correlated. The fact that r is positive indicates that the linear relationship is direct; as age increases, income also increases. ◆

As mentioned earlier, least squares curves are not always least squares lines. If a scatter diagram reveals that the quantities represented by the data may not be linearly related, then a best-fitting curve of some other type can be sought. For example, for any set of paired data, the TI-82 graphing calculator will compute best-fit curves of the types shown in Table 12 (all of which are examined in an earlier chapter of this text). Notice that an inverse square law, a common kind of relationship in nature, would be included in power functions (with $b = -2$).

TABLE 12 Curve-Fitting Options on the TI-82

Type of Curve	Equation
Linear	$y = ax + b$
Quadratic	$y = ax^2 + bx + c$
Cubic	$y = ax^3 + bx^2 + cx + d$
Quartic	$y = ax^4 + bx^3 + cx^2 + dx + e$
Logarithmic	$y = a + b \ln x$
Exponential	$y = a \cdot b^x$
Power	$y = a \cdot x^b$

The calculator provides correlation coefficients also for most types of regression curves, so it is easy to try the various types of curves to see which type will fit the sample data most closely. Computation of the various coefficients is more involved for the non-linear curves, and we do not show those formulas. For Example 4, we use the TI-82.

EXAMPLE 4 Find a reasonable curve to fit the following data, which relates the time employees have been working to the time they take to assemble a certain part.

Time on job, x (in weeks)	1	2	4	7	9	10	1	3	6	13	15	8
Time for task, y (in minutes)	8	6	5	3	2	2	10	4	2	1	1	2

The scatter diagram is shown in Figure 23. The relationship is probably not linear. We tried four possibilities, as shown in Table 13.

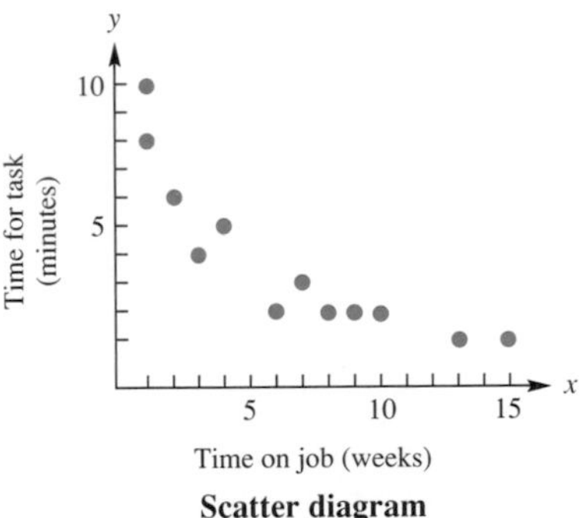

FIGURE 23

TABLE 13 Time on Job Versus Time for a Task

Curve Type	Correlation Coefficient, r	a	b
Linear	$-.857$	$-.536$	7.360
Logarithmic	$-.963$	8.484	-2.974
Exponential	$-.954$	8.297	.855
Power	$-.958$	9.970	$-.775$

The negative r-values obtained here verify the inverse nature of the relationship (which was clear from the scatter diagram, also). All four of the curve types provide a fairly good correlation (r close to -1). But the best fit is apparently the logarithmic curve ($r = -.963$). Specifically the best of all logarithmic curves is

$$y = 8.484 - 2.974 \ln x.$$

The graph of this curve is superimposed upon the scatter diagram in Figure 24. 

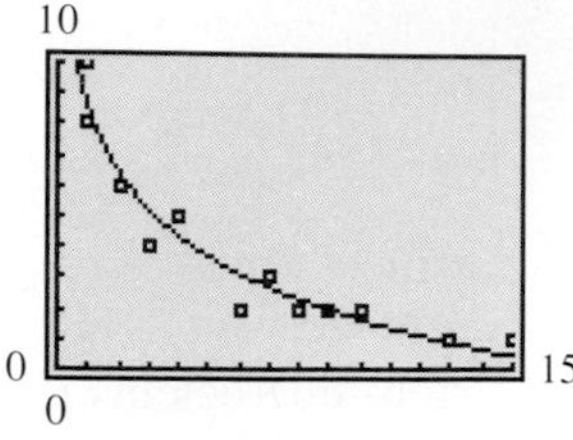

This is a calculator-generated screen corresponding to Figure 24. It uses the data of Table 13 and the corresponding logarithmic curve.

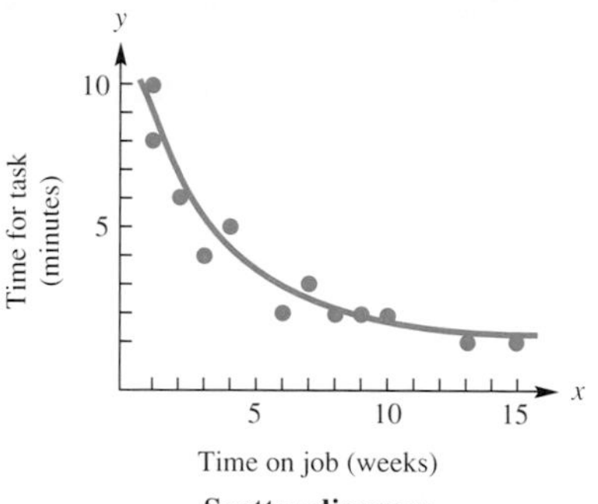

FIGURE 24

12.6 EXERCISES

1. In a study to determine the linear relationship between the length (in decimeters) of an ear of corn (y) and the amount (in tons per acre) of fertilizer used (x), the following data were collected.

$$n = 10 \qquad \Sigma xy = 75$$
$$\Sigma x = 30 \qquad \Sigma x^2 = 100$$
$$\Sigma y = 24 \qquad \Sigma y^2 = 80$$

(a) Find an equation for the least squares line.
(b) Find the coefficient of correlation.
(c) If 3 tons per acre of fertilizer are used, what length (in decimeters) would the equation in (a) predict for an ear of corn?

2. In an experiment to determine the linear relationship between temperatures on the Celsius scale (y) and on the Fahrenheit scale (x), a student got the following results.

$$n = 5 \qquad \Sigma xy = 28{,}050$$
$$\Sigma x = 376 \qquad \Sigma x^2 = 62{,}522$$
$$\Sigma y = 120 \qquad \Sigma y^2 = 13{,}450$$

(a) Find an equation for the least squares line.
(b) Find the reading on the Celsius scale that corresponds to a reading of 120° Fahrenheit, using the equation of part (a).
(c) Find the coefficient of correlation.

3. A sample of 10 adult men gave the following data on their heights and weights.

Height (inches) (x)	62	62	63	65	66	67	68	68	70	72
Weight (pounds) (y)	120	140	130	150	142	130	135	175	149	168

(a) Find the equation of the least squares line.
(b) Using the results of (a), predict the weight of a man whose height is 60 inches.
(c) What would be the predicted weight of a man whose height is 70 inches?
(d) Compute the coefficient of correlation.

4. The table below gives reading ability scores and IQs for a group of 10 individuals.

Reading (x)	83	76	75	85	74	90	75	78	95	80
IQ (y)	120	104	98	115	87	127	90	110	134	119

(a) Plot a scatter diagram with reading on the horizontal axis.
(b) Find the equation of a regression line.
(c) Use the regression line equation to estimate the IQ of a person with a reading score of 65.

5. Sales, in thousands of dollars, of a certain company are shown here.

Year (x)	0	1	2	3	4	5
Sales (y)	48	59	66	75	80	90

Find the equation of the least squares line. Find the coefficient of correlation.

The data for the remaining exercises were adapted from the 1995 Information Please Almanac.

In each case, obtain the least squares regression line and the linear correlation coefficient, and make a statement describing the linear correlation (for example, as strong, moderate, *or* weak*).*

6. Largest U.S. Cities

Rank	City	Population (x)	Mayor's Salary (y)
1	New York	7,311,966	$130,000
2	Los Angeles	3,489,779	123,778
3	Chicago	2,768,483	115,000
4	Houston	1,690,180	133,000
5	Philadelphia	1,552,572	110,000
6	San Diego	1,148,851	65,300
7	Dallas	1,022,497	2,600
8	Phoenix	1,012,230	37,500
9	Detroit	1,012,110	117,000
10	San Antonio	966,437	3,000

7. World's Most Populous Cities

Rank	City	Population, x (in thousands)	Area, y (in square miles)
1	Tokyo-Yokohama, Japan	28,447	1,089
2	Mexico City, Mexico	23,913	522
3	São Paulo, Brazil	21,539	451
4	Seoul, South Korea	19,065	342
5	New York, United States	14,638	1,274
6	Osaka-Kobe-Kyoto, Japan	14,060	495
7	Bombay, India	13,532	95
8	Calcutta, India	12,885	209
9	Rio de Janeiro, Brazil	12,788	260
10	Buenos Aires, Argentina	12,232	535

8.

Animal	Average Gestation or Incubation Period, x (days)	Record Life Span, y (years)
Cat	63	26
Dog	63	24
Duck	28	15
Elephant	624	71
Goat	151	17
Guinea pig	68	6
Hippopotamus	240	49
Horse	336	50
Lion	108	29
Parakeet	18	12
Pig	115	22
Rabbit	31	15
Sheep	151	16

9. In this case, relate **(a)** period of revolution (y) to mean distance from sun (x), and **(b)** period of rotation (y) to mean distance from sun (x).

Planet	Period of Revolution (days)	Period of Rotation (hours)	Mean Distance From Sun (millions of miles)
Mercury	88	1,416	36
Venus	225	5,832	67
Earth	365	24	93
Mars	687	25	142
Jupiter	4,329	10	484
Saturn	10,753	11	887
Uranus	30,660	17	1,784
Neptune	60,225	16	2,796
Pluto	90,520	153	3,666

COLLABORATIVE INVESTIGATION

Combining Sets of Data

Divide your class into two separate groups, one consisting of the women and the other consisting of the men. Each group is to select a recorder to write the group's results. As a group, carry out the following tasks. (You may want to devise a way to allow members of the group to provide personal data anonymously.)

1. Record the number of members (n) in your group.
2. Collect shoe sizes (x) and heights in inches (y) for all members of the group.
3. Compute the mean, median, and mode(s) if any for each of the two sets of data.
4. Compute the standard deviation of each of the two sets of data.
5. Construct a box plot for each of the two sets of data.
6. Plot a scatter diagram for the x-y data collected.
7. Find the equation of the least squares regression line ($y' = ax + b$).
8. Evaluate the correlation coefficient (r).
9. Evaluate the strength of the linear relationship between shoe size and height for your group.

Now re-combine your two groups into one. Discuss and carry out the following tasks.

1. If possible, compute the mean of the heights for the combined group, using only the means for the two individual groups and the number of members in each of the two groups. If this is not possible, explain why and describe how you *could* find the combined mean. Obtain the combined mean.
2. Do the same as in item 1 above for the median of the heights for the combined group.
3. Do the same for the mode of the heights for the combined group.
4. Fill in the table below, and discuss any apparent relationships among the computed statistics.

	Number of Members	**Mean**	**Median**	**Mode**
Women				
Men				
Combined				

Chapter 12 Test

The bar graphs here show trends in several economic indicators over the period 1989–1994. Refer to these graphs for Exercises 1–4.

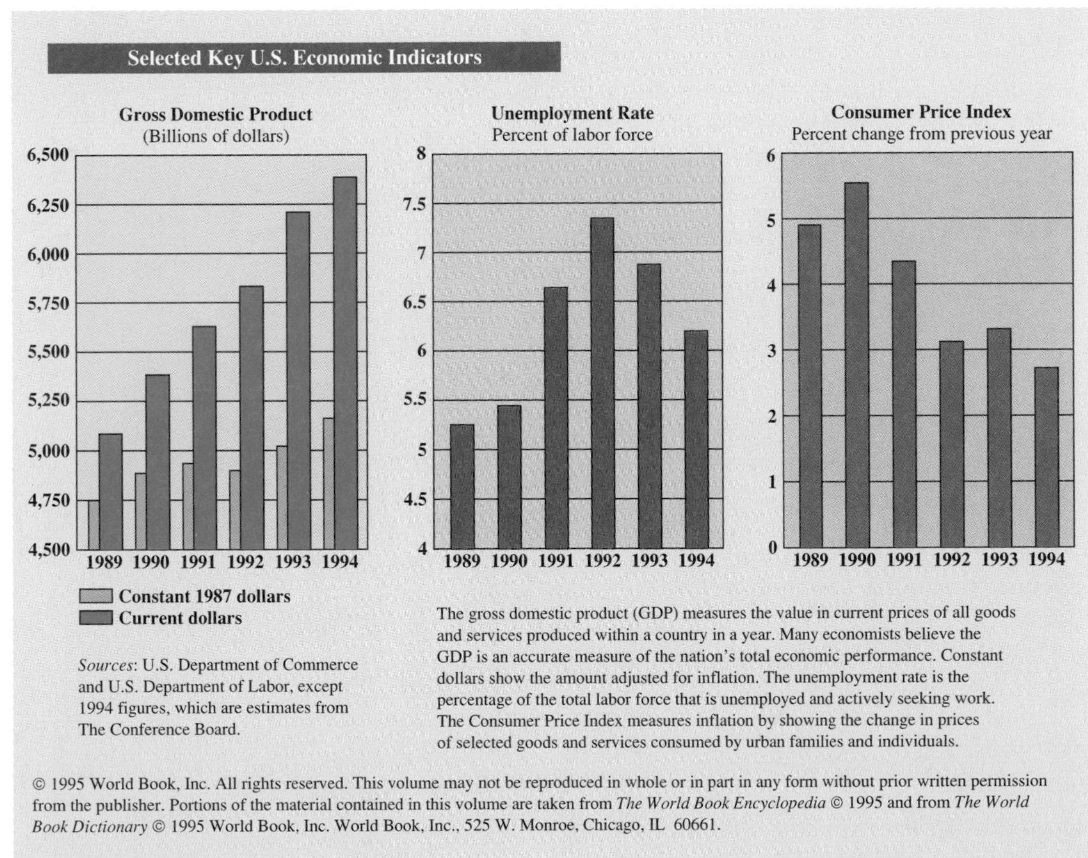

Selected Key U.S. Economic Indicators
Gross Domestic Product
(Billions of dollars)
6,500
6,250
6,000
5,750
5,500
5,250
5,000
4,750
4,500
1989 1990 1991 1992 1993 1994
Constant 1987 dollars
Current dollars
Unemployment Rate
Percent of labor force
8
7.5
7
6.5
6
5.5
5
4.5
4
1989 1990 1991 1992 1993 1994
Consumer Price Index
Percent change from previous year
6
5
4
3
2
1
0
1989 1990 1991 1992 1993 1994
Sources: U.S. Department of Commerce and U.S. Department of Labor, except 1994 figures, which are estimates from The Conference Board.
The gross domestic product (GDP) measures the value in current prices of all goods and services produced within a country in a year. Many economists believe the GDP is an accurate measure of the nation's total economic performance. Constant dollars show the amount adjusted for inflation. The unemployment rate is the percentage of the total labor force that is unemployed and actively seeking work. The Consumer Price Index measures inflation by showing the change in prices of selected goods and services consumed by urban families and individuals.

1. About what was the gross domestic product (current dollars) in 1992?
2. Over the six-year period, about what was the lowest percent change in consumer price index, and what year did it occur?

3. About what was the highest unemployment rate, and what year did it occur?

4. Describe the apparent trends in both constant dollar and current dollar gross domestic product. Describe how these two quantities seem to relate to one another over the years, and explain why this is so.

Nannette Williams, a sales representative for a publishing company, recorded the following numbers of client contacts for the twenty-two days that she was on the road in the month of March. Use the given data for Exercises 5–7.

12	8	15	11	20	18	14	22	13	26	17
19	16	25	19	10	7	18	24	15	30	24

5. Construct grouped frequency and relative frequency distributions. Use five uniform classes of width 5 where the first class has a lower limit of 6. (Round relative frequencies to two decimal places.)

6. From your frequency distribution of Exercise 5, construct **(a)** a histogram and **(b)** a frequency polygon. Use appropriate scales and labels.

7. For the data above, how many uniform classes would be required if the first class had limits 7–9?

In Exercises 8–11, find the indicated measures for the following frequency distribution.

Value	4	6	8	10	12	14
Frequency	2	5	10	8	4	1

8. the mean

9. the median

10. the mode

11. the range

12. The following data are exam scores achieved by the students in a physics class. Arrange the data into a stem-and-leaf display with leaves ranked.

79	43	65	84	77	70	52	61	80	75	68	48	55	78	71
38	45	64	67	73	67	50	67	91	84	33	49	61	79	72

Use the stem-and-leaf display shown here for Exercises 13–18.

Stem	Leaves
4	3 3 4
4	6 7 9 9
5	0 1 1 2 3 3 3 4
5	5 6 7 8 8 9
6	1 2 2 3
6	5 7
7	2 4
7	8
8	0
8	7

Compute the measures required in Exercises 13–17.

13. the median

14. the mode(s), if any

15. the range

16. the third decile

17. the fifteenth percentile

18. Construct a box plot for the above data, showing values for the five important quantities on the numerical scale.

A certain training institute gives a standardized test to large numbers of applicants nationwide. The resulting scores form a normal distribution with mean 75 and standard deviation 10. Find the percent of all applicants with scores as follows. (Use the empirical rule.)

19. between 55 and 95

20. greater than 105 or less than 45

21. less than 65

22. between 85 and 95

In a certain young forest, the heights of the fir trees are normally distributed with mean 10.6 meters and standard deviation 3.1 meters. If a single tree is selected randomly, find the probability (to the nearest thousandth) that its height will fall in each of the following intervals.

23. less than 8.5 meters

24. between 14.3 and 16.2 meters

The 1994 Major League Baseball season effectively ended on August 11 when the Players Association went on strike. The table below gives statistics through that date for each of the three divisions of the two leagues. In each case, $n =$ number of teams in the division, $\bar{x} =$ average (mean) number of games won, and $s =$ standard deviation of number of games won. Refer to the table for Exercises 25–27.

American League			National League		
Eastern Division	**Central Division**	**Western Division**	**Eastern Division**	**Central Division**	**Western Division**
$n = 5$	$n = 5$	$n = 4$	$n = 5$	$n = 5$	$n = 4$
$\bar{x} = 59$	$\bar{x} = 60.0$	$\bar{x} = 49.8$	$\bar{x} = 60.4$	$\bar{x} = 57.4$	$\bar{x} = 53.3$
$s = 7.3$	$s = 7$	$s = 2.2$	$s = 10$	$s = 8$	$s = 4.6$

25. Overall, who had the lowest winning average, the Eastern teams, the Central teams, or the Western teams?

26. Overall, where were division teams the most "consistent" in number of games won, Eastern, Central, or Western?

27. Find (to the nearest tenth) the average number of games won for all 28 teams.

Carry out the following for the paired data values shown here. (In Exercises 29–31, give all calculated values to two decimal places.)

x	1	4	6	7
y	9	7	8	1

28. Plot a scatter diagram.

29. Find the equation for the least squares regression line.

30. Use your equation from Exercise 29 to predict y when $x = 3$.

31. Find the sample correlation coefficient for the given data.

32. Evaluate the strength of the linear relationship for the above data. How confident are you in your evaluation? Explain.

33. Relate the concepts of *inferential statistics* and *inductive reasoning*.

13

Consumer Mathematics

It is easy to put off saving for the future. Retirement may seem distant and there are often many reasons for not starting a savings plan. Does waiting to contribute to a retirement fund have a dramatic effect on the amount of money that an individual has at age 65? Consider two retirement plans. In Plan A a person begins saving at age 20 and deposits $2,000 in a retirement fund on each birthday from age 21 to 30. After age 30 no more contributions are made. In Plan B a person waits until age 30 to start saving and then deposits $2,000 in the retirement fund on every birthday between the ages of 31 and 65. If the retirement fund pays 12% interest, which plan will have the largest accumulation of money? The table lists the amount of money (in dollars) in the fund for each plan at various ages.

Age	Plan A	Plan B
20	0	0
25	12,706	0
30	35,097	0
35	61,864	12,706
40	109,007	35,097
45	192,108	74,559
50	338,560	144,105
55	596,659	266,668
60	1,051,517	482,665
65	1,853,133	863,327

Plan A results in the accumulation of $1,853,133, whereas Plan B produces $863,327. Early contributions can make a substantial difference. In this chapter you will learn about compound interest. Compound interest over time is the reason for the growth of a relatively small amount of money into over a million dollars.

As participants in the American (and international) economy, most all of us at times wish to obtain something for which we either don't presently have sufficient money or don't want to commit our own money. In such a case, a common choice is to borrow money (that is "take out a *loan*") for the purchase, thus putting ourselves into debt. Sometimes, on the other hand, we will have more money than we need, for certain periods, in which case we may loan our money to someone else. This can involve depositing funds into a simple savings account or into one of many other kinds of "investments."

In the situations mentioned above, we borrow (or lend) some money, for which we expect to pay (or receive) a fee. The mathematics involved in relating that fee to the amount borrowed or lent (and evaluating certain other related costs) is what we mean here by "consumer mathematics."

Wise consumers, in deciding on a purchase, not only carefully compare the quality and cost of the goods, but also carefully compare the features, including cost, of various sources of financing. Accepting a financing offer from a retailer selling you a car, furniture, or other item may seem an easy choice, but can often be very costly compared to other options.

The first section of this chapter covers the basic mathematical concepts involved in borrowing and lending, and relates the same ideas to the subject of economic inflation. Sections 2, 3, and 4 deal specifically with borrowing situations, and Section 5 deals with investing (mainly in stocks). By studying this material, you will gain some useful tools for managing your money, for comparing alternative sources of borrowed money, and for considering potential investments.

13.1 Interest and Inflation

When you borrow money from a bank, credit union, or other lender, the basic fee you pay for the privilege of using the money until it is repaid is called **interest.** Likewise, when you deposit money in a bank (or other savings institution), the bank pays interest to you for the privilege of using your money.

The amount of a loan or a deposit is called the **principal.** The amount of interest is usually given as a percent of the principal. This percent is called the **rate of interest** (or the **interest rate,** or simply the **rate**). The rate of interest is always assumed to be an annual rate unless otherwise stated.

Interest calculated only on principal is called **simple interest.** Interest calculated on principal plus any previously earned interest is called **compound interest.** Generally, simple interest is used for loans of one year or less, while compound interest is used for longer loans. Banks always pay compound interest on money deposited.

Simple Interest

The calculation of simple interest was illustrated in an earlier chapter. The applicable formula is restated here.

Simple Interest

If P = principal, r = annual interest rate, and t = time (in years), then the **simple interest** I is given by

$$I = Prt.$$

EXAMPLE 1 Find the simple interest if \$6,350 is borrowed for 7 months at 5%.

Here $P = \$6{,}350$, $r = 5\% = .05$, and $t = 7/12$ (since 7 months is 7/12 of a year). The interest is

$$I = Prt = (\$6{,}350)(.05)\left(\frac{7}{12}\right) = \$185.21,$$

rounded to the nearest cent. 

Throughout this chapter, a scientific calculator will be essential.

Note that the simple interest formula is best applied with the aid of a calculator. As we proceed through the chapter, increasingly complex formulas will be developed and a financial or scientific calculator will be essential.

Future Value and Present Value

In Example 1 above, at the end of 7 months the borrower would have to repay

$$\text{Principal} + \text{interest} = \$6{,}350 + \$185.21 = \$6{,}535.21.$$

The total amount repaid is sometimes called the **maturity value** (or simply the **value**) of the loan. We will generally refer to it as the **future value,** or **future amount,** since when a loan is being set up, repayment will be occurring in the future. We will use A to denote future value. In this sense the original principal, denoted P, can also be thought of as **present value.** The future value depends on the principal (present value) and the interest as follows.

$$A = P + I = P + Prt = P(1 + rt).$$

We summarize this relationship in the following formulas.

Future and Present Values for Simple Interest

If a principal P is borrowed at simple interest for t years at an annual interest rate of r, then the **future value** of the loan, denoted A, is given by

$$A = P(1 + rt).$$

Likewise, the **present value,** P, is given by

$$P = \frac{A}{1 + rt}.$$

King Hammurabi tried to hold interest rates at 20 percent for both silver and gold, but moneylenders ignored his decrees.

EXAMPLE 2 Alex Ortega took out a simple interest loan for \$210 to purchase textbooks and school supplies. If the annual interest rate is 9% and he must repay the loan after 5 months, find the future value (the maturity value) of the loan.

Applying the future value formula for simple interest,

$$A = P(1 + rt) = \$210\left[1 + (.09)\left(\frac{5}{12}\right)\right] = \$217.88.$$

At the end of 5 months, Alex will need to repay \$217.88.

EXAMPLE 3 Suppose that Alex (Example 2) is granted a 5-month *deferral* of the \$210 payment rather than a loan. That is, instead of incurring interest for 5 months, he will have to pay just the \$210 at the end of that period. If he has extra money right now, and if he can earn 6% simple interest on savings, what lump sum must he deposit now in order to have the \$210 available after 5 months?

We apply the present value formula for simple interest.

$$P = \frac{A}{1 + rt} = \frac{\$210}{1 + (.06)(5/12)} = \$204.88.$$

A deposit of \$204.88 now, growing at 6% simple interest, will grow to \$210 over a 5-month period.

Compound Interest

As mentioned earlier, interest paid on principal plus interest is called compound interest. To see how this works, suppose \$1,000 is deposited in a savings account paying a 4% interest rate. During the first year that the money is on deposit, the account would earn interest of

$$I = Prt = (\$1{,}000)(.04)(1) = \$40.$$

This amount of interest would be *credited* (added) to the account, so that at the beginning of the second year the account would contain the amount

$$\$1{,}000 + \$40 = \$1{,}040.$$

During the second year, the account would earn interest of

$$(\$1{,}040)(.04)(1) = \$41.60.$$

At the beginning of the third year, the account would contain the amount

$$\$1{,}040 + \$41.60 = \$1{,}081.60.$$

The interest earned during the second year was more than the interest earned during the first year. The interest earned during the first year was added to the principal and then used in calculating the amount of interest paid during the second year. This is the way compounding of interest works. (As we shall see below, interest is commonly calculated and credited to the account more often than yearly.)

Problem Solving

As we develop a future value formula for compound interest, watch for the use of problem-solving strategies. Exercise 30 at the end of this section will ask you to identify them.

Generalizing from the illustration of compound interest above, we now develop a formula that will give the amount in the account at any future time, that is the **future value** of the account. Realistically, earned interest can be credited to an account at time intervals other than one year (usually more often). For example, it can be done semiannually, quarterly, monthly, or daily. (Daily compounding is quite common.) This time interval is called the **compounding period** (or simply the **period**). Start with the following definitions.

$$P = \text{original principal deposited}$$
$$r = \text{annual interest rate}$$
$$m = \text{number of periods per year}$$
$$n = \text{total number of periods}$$

During each individual compounding period, interest is earned according to the simple interest formula, and as interest is added, the beginning principal increases from one period to the next. During the first period, the interest earned is given by

$$\begin{aligned} \text{Interest} &= (\text{Principal})(\text{rate})(\text{time}) \\ &= P(r)\left(\frac{1}{m}\right) && \text{One period} = \frac{1}{m}\text{ year} \\ &= P\left(\frac{r}{m}\right). && \text{Rewrite } (r)\left(\frac{1}{m}\right) \text{ as } \left(\frac{r}{m}\right). \end{aligned}$$

At the end of the first period, the account then contains

$$\begin{aligned} \text{Ending amount} &= \text{beginning amount} + \text{interest} \\ &= P + P\left(\frac{r}{m}\right) \\ &= P\left(1 + \frac{r}{m}\right). && \text{Factor } P \text{ from both terms.} \end{aligned}$$

Now during the second period, the interest earned is given by

$$\begin{aligned} \text{Interest} &= [\text{Principal}](\text{rate})(\text{time}) \\ &= \left[P\left(1 + \frac{r}{m}\right)\right](r)\left(\frac{1}{m}\right) \\ &= P\left(1 + \frac{r}{m}\right)\left(\frac{r}{m}\right), \end{aligned}$$

so that the account ends the second period containing

$$\begin{aligned} \text{Ending amount} &= \text{beginning amount} + \text{interest} \\ &= P\left(1 + \frac{r}{m}\right) + P\left(1 + \frac{r}{m}\right)\left(\frac{r}{m}\right) \\ &= P\left(1 + \frac{r}{m}\right)\left[1 + \frac{r}{m}\right] && \text{Factor } P\left(1 + \frac{r}{m}\right) \text{ from both terms.} \\ &= P\left(1 + \frac{r}{m}\right)^2. \end{aligned}$$

Consider one more period, namely the third. The interest earned is

$$\begin{aligned} \text{Interest} &= [\text{Principal}](\text{rate})(\text{time}) \\ &= \left[P\left(1 + \frac{r}{m}\right)^2\right](r)\left(\frac{1}{m}\right) \\ &= P\left(1 + \frac{r}{m}\right)^2\left(\frac{r}{m}\right), \end{aligned}$$

The following item, broadcast on CBS Radio News and reproduced here with permission, shows how one community plans to use the power of compound interest to help it plan for the future. (News broadcast © CBS Inc. 1977. All rights reserved. Originally broadcast over the CBS Radio Network as part of the Hourly News by Charles Osgood on March 2, 1977.)

"Eli Hooker, of Union City, Michigan, doesn't want his community to go broke when the tricentennial rolls around in 2076. Hooker says the local Bicentennial Committee last year didn't have enough money. He was chairman. So, to plan ahead, he has collected twenty-five dollars apiece from 42 residents of Union City, squirreling it away in a local bank. At seven-percent interest, compounded, the fund would have a million dollars by 2076.

"Hooker hopes that will be enough for the parades and fireworks a tricentennial celebration will require. But Eli Hooker isn't being an entirely selfless patriot about all this: he hopes one of the big events on July 4, 2076, will be the unveiling of a monument to the 42 people of Union City who contributed 99 years in advance."

Unfortunately, that million dollars won't be worth as much then as we might think. If the community decides to hire people to parade around in historical costumes, it might have paid them $4 per hour in 1976. The going wage in 2076, assuming 7% annual inflation for a hundred years, would be well over $3,000 per hour.

so the account ends the third period containing

$$\begin{aligned}\text{Ending amount} &= \text{beginning amount} + \text{interest}\\ &= P\left(1+\frac{r}{m}\right)^2 + P\left(1+\frac{r}{m}\right)^2\left(\frac{r}{m}\right)\\ &= P\left(1+\frac{r}{m}\right)^2\left[1+\frac{r}{m}\right]\\ &= P\left(1+\frac{r}{m}\right)^3.\end{aligned}$$

Factor $P\left(1+\frac{r}{m}\right)^2$ from both terms.

In order to clearly see patterns in the equations above, we summarize our results in Table 1. Observing the patterns occurring in the first three rows of the table, we can apply inductive reasoning to arrive at corresponding expressions for period number n.

TABLE 1 Compound Amount

Period Number	Beginning Amount	Interest Earned During Period	Ending Amount
1	P	$P\left(\frac{r}{m}\right)$	$P\left(1+\frac{r}{m}\right)$
2	$P\left(1+\frac{r}{m}\right)$	$P\left(1+\frac{r}{m}\right)\left(\frac{r}{m}\right)$	$P\left(1+\frac{r}{m}\right)^2$
3	$P\left(1+\frac{r}{m}\right)^2$	$P\left(1+\frac{r}{m}\right)^2\left(\frac{r}{m}\right)$	$P\left(1+\frac{r}{m}\right)^3$
⋮	⋮	⋮	⋮
n	$P\left(1+\frac{r}{m}\right)^{n-1}$	$P\left(1+\frac{r}{m}\right)^{n-1}\left(\frac{r}{m}\right)$	$P\left(1+\frac{r}{m}\right)^n$

The right-hand column of the table shows that, after algebraic simplification, the ending amount for a given period is just the previous ending amount multiplied by one more factor of $\left(1+\frac{r}{m}\right)$. The future value formula for compound interest is taken from the lower right entry of the table, and a slight rearrangement gives the corresponding present value formula.

Future and Present Values for Compound Interest

If P dollars are deposited at an annual interest rate of r, compounded m times per year, and the money is left on deposit for a total of n periods, then the **future value,** A (the final amount on deposit), is given by

$$A = P\left(1+\frac{r}{m}\right)^n,$$

and the **present value,** P, is given by

$$P = \frac{A}{\left(1+\frac{r}{m}\right)^n}.$$

Use this pattern to find the proper table entry.

$n \backslash i \rightarrow$ 1 1/2%
↓ ↓
20 → **1.3469**

The formulas above can generally be applied using a calculator with a y^x key. Alternatively, we can use Table 2, which provides values of the quantity $\left(1 + \frac{r}{m}\right)^n$ for a limited number of interest rates and compounding frequencies. (The table will suffice for some but not all examples and exercises presented here. And it will sometimes not provide the accuracy required.) To use the table, first identify the *rate per period:* $i = r/m$. Then determine the *number of periods, n*. For example, if the compounding is to occur for 5 years, then $n = 5m$. The i-value determines a column in the table and the n-value determines a row. The table entry where that row and that column intersect gives the value of $\left(1 + \frac{r}{m}\right)^n$, which is the compound amount of \$1, at interest i per period, for n periods. To complete your calculation of future or present value, multiply or divide the appropriate amount by this table value. (See Example 4(a).) To illustrate, interest of 6% compounded quarterly for 5 years gives $i = 6\%/4 = 1\ 1/2\%$, $n = (5)(4) = 20$ (quarters). The column for 1 1/2% and the row for 20 periods intersect at the value 1.3469.

EXAMPLE 4 Find the future value (final amount on deposit) and the amount of interest earned for the following deposits.

(a) \$18,950 at 6% compounded quarterly for 5 years

We have $P = \$18{,}950$, $r = 6\% = .06$, $m = 4$. And over a time of 5 years, we obtain $n = 5m = (5)(4) = 20$.

Using the future value formula and a calculator, we get

$$A = \$18{,}950\left(1 + \frac{.06}{4}\right)^{20} = \$25{,}522.90.$$

To find the interest earned, subtract the amount originally deposited (present value) from the future value:

$$\begin{aligned} \text{Interest earned} &= \text{future value} - \text{present value} \\ &= \$25{,}522.90 - \$18{,}950 \\ &= \$6{,}572.90. \end{aligned}$$

For comparison, recall that the Table 2 value obtained earlier for $\left(1 + \frac{r}{m}\right)^n$ was 1.3469. So use of the table produces

$$A \approx \$18{,}950(1.3469) = \$25{,}523.76.$$

(Note the lack of accuracy in this value. It is off by 86 cents.)

(b) \$6,429 at 4.2% compounded monthly for 30 months

Here $P = \$6{,}429$, $r = 4.2\% = .042$, $m = 12$, and $n = 30$, so

$$A = \$6{,}429\left(1 + \frac{.042}{12}\right)^{30} = \$7{,}139.45.$$

Interest earned = future value − present value = \$7,139.45 − \$6,429 = \$710.45. (Note that in this case, Table 2 is not applicable since $i = .042/12 = .0035$ is not included in the table, and also $n = 30$ is beyond the scope of the table.) ■

TABLE 2 Compound Amount of $1 (for n periods at interest $i = r/m$ per period)

$n \backslash i$	1/4%	1/2%	1%	1 1/4%	1 1/2%	2%	2 1/2%	3%	3 1/2%	4%	4 1/2%	5%	6%
1	1.0025	1.0050	1.0100	1.0125	1.0150	1.0200	1.0250	1.0300	1.0350	1.0400	1.0450	1.0500	1.0600
2	1.0050	1.0100	1.0201	1.0252	1.0302	1.0404	1.0506	1.0609	1.0712	1.0816	1.0920	1.1025	1.1236
3	1.0075	1.0151	1.0303	1.0380	1.0457	1.0612	1.0769	1.0927	1.1087	1.1249	1.1412	1.1576	1.1910
4	1.0100	1.0202	1.0406	1.0509	1.0614	1.0824	1.1038	1.1255	1.1475	1.1699	1.1925	1.2155	1.2625
5	1.0126	1.0253	1.0510	1.0641	1.0773	1.1041	1.1314	1.1593	1.1877	1.2167	1.2462	1.2763	1.3382
6	1.0151	1.0304	1.0615	1.0774	1.0934	1.1262	1.1597	1.1941	1.2293	1.2653	1.3023	1.3401	1.4185
7	1.0176	1.0355	1.0721	1.0909	1.1098	1.1487	1.1887	1.2299	1.2723	1.3159	1.3609	1.4071	1.5036
8	1.0202	1.0407	1.0829	1.1045	1.1265	1.1717	1.2184	1.2668	1.3168	1.3686	1.4221	1.4775	1.5938
9	1.0227	1.0459	1.0937	1.1183	1.1434	1.1951	1.2489	1.3048	1.3629	1.4233	1.4861	1.5513	1.6895
10	1.0253	1.0511	1.1046	1.1323	1.1605	1.2190	1.2801	1.3439	1.4106	1.4802	1.5530	1.6289	1.7908
11	1.0278	1.0564	1.1157	1.1464	1.1779	1.2434	1.3121	1.3842	1.4600	1.5395	1.6229	1.7103	1.8983
12	1.0304	1.0617	1.1268	1.1608	1.1956	1.2682	1.3449	1.4258	1.5111	1.6010	1.6959	1.7959	2.0122
13	1.0330	1.0670	1.1381	1.1753	1.2136	1.2936	1.3785	1.4685	1.5640	1.6651	1.7722	1.8856	2.1329
14	1.0356	1.0723	1.1495	1.1900	1.2318	1.3195	1.4130	1.5126	1.6187	1.7317	1.8519	1.9799	2.2609
15	1.0382	1.0777	1.1610	1.2048	1.2502	1.3459	1.4483	1.5580	1.6753	1.8009	1.9353	2.0789	2.3966
16	1.0408	1.0831	1.1726	1.2199	1.2690	1.3728	1.4845	1.6047	1.7340	1.8730	2.0224	2.1829	2.5404
17	1.0434	1.0885	1.1843	1.2351	1.2880	1.4002	1.5216	1.6528	1.7947	1.9479	2.1134	2.2920	2.6928
18	1.0460	1.0939	1.1961	1.2506	1.3073	1.4282	1.5597	1.7024	1.8575	2.0258	2.2085	2.4066	2.8543
19	1.0486	1.0994	1.2081	1.2662	1.3270	1.4568	1.5987	1.7535	1.9225	2.1068	2.3079	2.5270	3.0256
20	1.0512	1.1049	1.2202	1.2820	1.3469	1.4859	1.6386	1.8061	1.9898	2.1911	2.4117	2.6533	3.2071
21	1.0538	1.1104	1.2324	1.2981	1.3671	1.5157	1.6796	1.8603	2.0594	2.2788	2.5202	2.7860	3.3996
22	1.0565	1.1160	1.2447	1.3143	1.3876	1.5460	1.7216	1.9161	2.1315	2.3699	2.6337	2.9253	3.6035
23	1.0591	1.1216	1.2572	1.3307	1.4084	1.5769	1.7646	1.9736	2.2061	2.4647	2.7522	3.0715	3.8197
24	1.0618	1.1272	1.2697	1.3474	1.4295	1.6084	1.8087	2.0328	2.2833	2.5633	2.8760	3.2251	4.0489
25	1.0644	1.1328	1.2824	1.3642	1.4509	1.6406	1.8539	2.0938	2.3632	2.6658	3.0054	3.3864	4.2919

EXAMPLE 5 The Cebelinskis' daughter will need $7,000 in 5 years to help pay for her college education. What lump sum, deposited today at 8% compounded quarterly, will produce the necessary amount?

This question requires that we find the present value P based on the following.

Future value:	$A = \$7{,}000$
Annual rate:	$r = 8\% = .08$
Periods per year:	$m = 4$
Total number of periods:	$n = (5)(4) = 20$

The present value formula gives

$$P = \frac{A}{\left(1 + \frac{r}{m}\right)^n} = \frac{\$7{,}000}{\left(1 + \frac{.08}{4}\right)^{20}} = \$4{,}710.80.$$

Assuming interest of 8% compounded quarterly can be maintained, a deposit of $4,710.80 today will grow to $7,000 five years in the future. ◆

The next example will involve the use of logarithms. For help with those steps, see the discussion of logarithms in an earlier chapter (or see your calculator manual).

EXAMPLE 6 In a savings account paying 5% interest, compounded daily, when will the amount in the account be twice the original principal?

In the future value formula, the future value must equal two times the present value, so we substitute $2P$ for A.

$$2P = P\left(1 + \frac{r}{m}\right)^n$$

$$2 = \left(1 + \frac{r}{m}\right)^n \qquad \text{Divide both sides by } P.$$

$$2 = \left(1 + \frac{.05}{365}\right)^n \qquad \text{Insert the values of } r \text{ and } m.$$

$$\log 2 = \log\left(1 + \frac{.05}{365}\right)^n \qquad \text{Take log of both sides.}$$

$$\log 2 = n \log\left(1 + \frac{.05}{365}\right) \qquad \text{Use a property of logarithms.}$$

Now solving for n gives

$$n = \frac{\log 2}{\log\left(1 + \frac{.05}{365}\right)} = 5{,}060 \quad \text{(rounding to the nearest whole number).}$$

Since n denotes the number of periods, which is days in this case, the required amount of time is 5,060 days, or 13 years, 315 days (ignoring leap years).

Effective Annual Yield

Banks, credit unions, and other savings institutions often give two quantities when advertising the rates they will pay. The first, the actual annualized interest rate, is the **nominal rate** (the "named" or "stated" rate). The second quantity is the equivalent rate that would produce the same final amount, or future value, at the end of one year if the interest being paid were simple rather than compound. This is called the "effective rate," or more commonly the **effective annual yield.** Since the interest is normally compounded multiple times per year, the yield will usually be somewhat higher than the nominal rate.

EXAMPLE 7 What is the effective annual yield of an account paying a nominal rate of 6.50%, compounded quarterly?

From the given data, $r = .065$ and $m = 4$. Suppose we deposited $P = \$1$ and left it for 1 year ($n = 4$). Then the compound future value formula gives

$$A = \left(1 + \frac{.065}{4}\right)^4 \approx 1.06660.$$

The initial deposit of \$1, after 1 year, has grown to \$1.06660. As usual, interest earned = future value − present value = 1.0666 − 1 = .0666 = 6.66%. A nominal rate of 6.50% results in an effective annual yield of 6.66%.

Generalizing the procedure of Example 7 gives the following formula.

Effective Annual Yield

A nominal interest rate of r, compounded m times per year, is equivalent to an **effective annual yield** of

$$Y = \left(1 + \frac{r}{m}\right)^m - 1.$$

Do you think you could use the above formula to find the nominal rate if you knew the effective annual yield and the number of compounding periods per year? (See Exercises 51 and 52.)

Consumers can make practical use of the effective annual yield concept when shopping for loans or savings opportunities. A borrower should seek the lowest yield available, while a depositor should look for the highest.

EXAMPLE 8 Curtis Johnson wants to deposit \$2,800 into a savings account and has narrowed his choices to the three institutions represented here. Which is the best choice?

Institution	Rate on Deposits of \$1,000 to \$5,000
Friendly Credit Union	5.08% annual rate, compounded monthly
Premier Savings	5.16% annual yield
Neighborhood Bank	4.93%, compounded daily

Compare the effective annual yields for the three institutions:

$$\text{For Friendly, } Y = \left(1 + \frac{.0508}{12}\right)^{12} - 1 = .0520 = 5.20\%.$$

For Premier, $Y = 5.16\%$.

$$\text{For Neighborhood, } Y = \left(1 + \frac{.0493}{365}\right)^{365} - 1 = .0505 = 5.05\%.$$

The best of the three yields, 5.20%, is offered by Friendly Credit Union.

Inflation

The compounding of increases, as reflected in the future value of compound interest, also occurs in situations other than the earning or paying of interest. An important example is **inflation,** the periodic increase in the cost of living. In periods of rapid inflation, costs go up very quickly. Necessities, such as food, often experience the greatest increases, but most all other items go up in cost also.

Inflation in an economy is usually expressed as a monthly or annual rate of price increases, estimated by government agencies in a systematic way. In the United States, the Bureau of Labor Statistics publishes **consumer price index (CPI)** figures, which reflect the prices of certain items purchased by large numbers of people. The items include such things as food, housing, automobiles,

Consumer Price Index (CPI-U) 1960 to 1995 [1982–84 = 100]

Year	CPI-U	Pct. Change in CPI-U
1960	29.6	1.7
1961	29.9	1.0
1962	30.2	1.0
1963	30.6	1.3
1964	31.0	1.3
1965	31.5	1.6
1966	32.4	2.9
1967	33.4	3.1
1968	34.8	4.2
1969	36.7	5.5
1970	38.8	5.7
1971	40.5	4.4
1972	41.8	3.2
1973	44.4	6.2
1974	49.3	11.0
1975	53.8	9.1
1976	56.9	5.8
1977	60.6	6.5
1978	65.2	7.6
1979	72.6	11.3
1980	82.4	13.5
1981	90.9	10.3
1982	96.5	6.2
1983	99.6	3.2
1984	103.9	4.3
1985	107.6	3.6
1986	109.6	1.9
1987	113.6	3.6
1988	118.3	4.1
1989	124.0	4.8
1990	130.7	5.4
1991	136.2	4.2
1992	140.3	3.0
1993	144.5	3.0
1994	148.2	2.6
1995	152.4	2.8

Source: Bureau of Labor Statistics

fuel, and clothing. Items such as yachts or expensive jewelry are not included. For example, the primary index, representing "all urban consumers," went up 2.6% from December of 1993 to December of 1994, so that the **annual inflation rate** for that year was 2.6%. Since 1960, U.S. annual inflation rates have varied from a low of around 1% to a high of around 13.5%.

Your own personal experience of inflation, or that of your family, may or may not closely reflect the published index figures. You may not be hit as hard if you do not normally buy many of the items included in the government survey. But if you rely heavily on gasoline for work travel, or on certain high-priced foods because of medical restrictions, you may feel an increase greater than the average. Also, residents of rural areas often experience a lower inflation rate than those of urban areas. In spite of these variations, the consumer price index does give a useful measure of general price increases.

Although CPI figures are reported on a monthly basis, annual figures give adequate information for most consumers. Furthermore, prices do not actually make upward jumps at just certain points in time (like quarterly) as account values do under interest compounding. Prices fluctuate on a continual basis throughout the year. Thus it is more appropriate for inflationary estimates to use the formula for "continuous compounding" (which was introduced in an earlier chapter). This formula is based on the assumption that compounding occurs not just monthly, or daily, but *continuously*. As interest (or price increase) just begins to occur, it is immediately "credited to the account" and begins to earn more interest.

Future Value for Continuous Compounding

If an initial deposit of P dollars earns continuously compounded interest at an annual rate r for a period of t years, then the **future value,** A, is given by

$$A = Pe^{rt}.$$

EXAMPLE 9 Suppose you earn a salary of $24,000 per year. About what salary would you need 20 years from now to maintain your purchasing power in case the inflation rate were to persist at each of the following levels?

(a) 3% (approximately the 1995 level)

In this case we can use the continuous compounding future value formula with $P = \$24{,}000$, $r = .03$, and $t = 20$. Using the e^x key on a calculator, we find

$$A = Pe^{rt} = (\$24{,}000)e^{(.03)(20)} = \$43{,}730.85.$$

The required salary 20 years from now would be about $44,000.

(b) 13% (approximately the 1980 level)

For this level of inflation, we would have

$$A = Pe^{rt} = (\$24{,}000)e^{(.13)(20)} = \$323{,}129.71.$$

The required salary 20 years from now would be about $323,000.

Inflation causes prices to climb rapidly over a period of time, with some prices rising much more quickly than others.

(c) 1% (approximately the 1962 level)
For this level, we obtain

$$A = Pe^{rt} = (\$24{,}000)e^{(.01)(20)} = \$29{,}313.67.$$

The required salary 20 years from now would be about $29,000. ◆

Notice that in Example 9, we stated the required salaries as rounded numbers, $44,000, $323,000, and $29,000. In the first place, the given inflation rates were rounded. Also, even the more precise CPI figures published by the government are only an attempt to reflect cost of living on a very average basis. Finally, the assumption that any given rate will continue for 20 years is itself a speculative estimation. For all these reasons (and more), you should regard all inflationary projections into the future as only approximations.

Especially when working with quantities like inflation, where continual fluctuations and inexactness prevail, we often develop rough "rules of thumb" for obtaining quick estimates. One example is the estimation of the **years to double,** which is the number of years it takes for the general level of prices to double for a given annual rate of inflation. One way to estimate the years to double is to use Table 2 with n representing number of years and i representing annual inflation rate.

EXAMPLE 10 Given an annual inflation rate of 3%, what will be the years to double?

Read down the 3% column of Table 2 and locate the first entry that is 2 or greater. Now read left across this row to find the value of n, the number of years. This value is the approximate years to double. (It is actually the smallest whole number of years that will at least double prices.) In this example, the answer is 24 years. ◆

The estimation method of Example 10 has the drawback that Table 2 is limited in scope. Furthermore, the table really assumes a single jump in prices once a year, whereas inflation more realistically involves a continual increase throughout the year. An estimation method that avoids these shortcomings can be obtained from the future value formula for continuous compounding. Start with the formula.

$$\begin{aligned}
A &= Pe^{rt} \\
2P &= Pe^{rt} && \text{Prices are to double.} \\
2 &= e^{rt} && \text{Divide both sides by } P. \\
\ln 2 &= rt && \text{Take the natural log of both sides.} \\
t &= \frac{\ln 2}{r} && \text{Solve for } t. \\
t &= \frac{100 \ln 2}{100r} && \text{Multiply top and bottom by 100.} \\
\text{years to double} &\approx \frac{70}{\text{annual inflation rate}} && 100 \ln 2 \approx 70
\end{aligned}$$

(Since r is the inflation rate as a *decimal,* $100r$ is the inflation rate as a *percent.*)

The result above is usually called the **rule of 70.** The result should be rounded *up* to the next whole number of years. The rule of 70 would thus give the same estimate we got in Example 10 for a 3% rate of inflation, since 70/3 = 23 1/3 and 23 1/3 rounds up to 24.

EXAMPLE 11 Use the rule of 70 to estimate the years to double for an annual inflation rate of 7%.

By the formula,

$$\text{years to double} \approx \frac{70}{7} = 10.$$

With a 7% inflation rate, prices would double in about 10 years.

13.1 EXERCISES

Remember that answers obtained using Table 2 may lack accuracy and thus not agree exactly with answers given in the back of the text. Also, unless otherwise known, assume 12 months per year, 30 days per month, and 365 days per year whenever such equivalences are needed.

Find the simple interest owed for each of the following loans.

1. $1,200 at 10% for 1 year
2. $8,000 at 8% for 1 year
3. $450 at 9% for 9 months
4. $5,000 at 7.5% for 4 months
5. $2,675 at 9.2% for $2\frac{1}{2}$ years
6. $1,580 at 6.75% for 32 months

Find the future value of each of the following deposits if the account pays **(a)** *simple interest, and* **(b)** *interest compounded annually.*

7. $500 at 4% for 3 years
8. $1,000 at 5% for 6 years
9. $2,500 at 7% for 4 years
10. $5,000 at 6% for 2 years

Solve the following interest-related problems.

11. Ezra Mason was late on his property tax payment to the county. He owed $7,500 and paid the tax 4 months late. The county charges a penalty of 10% simple interest. Find the amount of the penalty.
12. Charles Paisley bought a new supply of police uniforms. He paid $625 for the uniforms and agreed to pay for them in 9 months at 13% simple interest. Find the amount of interest that he will owe.
13. Elanna Williams opened a security service on March 1. To pay for office furniture and guard dogs, Williams borrowed $12,500 at the bank and agreed to pay the loan back in 7 months at 12% simple interest. Find the *total amount* she must repay.
14. David Fontana is owed $180 by the Internal Revenue Service for overpayment of last year's taxes. The I.R.S. will repay the amount at 8% simple interest. Find the *total amount* Fontana will receive if the interest is paid for 9 months.

Find the missing final amount (future value) and/or interest earned.

	Principal	Rate	Compounded	Time	Final Amount	Compound Interest
15.	$ 975	8%	quarterly	4 years	$1,338.47	______
16.	$1,150	7%	semiannually	6 years	$1,737.73	______
17.	$ 480	6%	semiannually	9 years	______	$337.17
18.	$2,370	10%	quarterly	5 years	______	______
19.	$7,500	$5\frac{1}{2}$%	annually	25 years	______	______
20.	$3,450	12%	semiannually	10 years	______	______

For each of the following deposits, find the future value *(final amount on deposit) when compounding occurs* **(a)** *annually,* **(b)** *semiannually, and* **(c)** *quarterly.*

	Principal	Rate	Time
21.	\$1,000	8%	3 years
23.	\$8,000	5%	5 years

	Principal	Rate	Time
22.	\$ 4,000	6%	6 years
24.	\$15,000	8%	6 years

Occasionally a bank may actually pay interest compounded continuously. For each of the following deposits, find the interest earned *if interest is compounded* **(a)** *semiannually,* **(b)** *quarterly,* **(c)** *monthly,* **(d)** *daily, and* **(e)** *continuously. (Recall that interest = future value − present value.)*

	Principal	Rate	Time
25.	\$ 500	7.3%	4 years
26.	\$1,550	8.5%	33 months (assume 1,003 days)

27. Describe the effect of interest being compounded more and more often. In particular, how good is continuous compounding?

Work the following interest-related problems.

28. Chris Siragusa takes out an 11% simple interest loan today which will be repaid 15 months from now with a payoff amount of \$796.25. What amount is Chris borrowing?

29. What amount can you borrow today if it must be repaid in 5 months with simple interest at 10% and you know that at that time you will be able to repay no more than \$1,500?

30. In the development of the future value formula for compound interest in the text at least four specific problem solving strategies were employed. Identify (name) as many of them as you can and describe their use in this case.

Find the present value for each of the following future amounts. (Table 2 may be applied in some cases, but may limit the accuracy of answers.)

31. \$1,000 (6% compounded annually for 5 years)

32. \$14,000 (4% compounded quarterly for 3 years)

33. \$9,860 (8% compounded semiannually for 10 years)

34. \$15,080 (5% compounded monthly for 4 years)

Ron and Diane want to establish an account that will help their 21-year-old son retire when he reaches age 61. For each of the following interest rates find the lump sum they must deposit today so that \$500,000 will be available for retirement.

35. 8% compounded quarterly

36. 12% compounded quarterly

37. 8% compounded daily

38. 12% compounded daily

Suppose a savings and loan pays a nominal rate of 7% on savings deposits. Find the effective annual yield *if interest is compounded as stated in Exercises 39–45. (Give answers to the nearest thousandth of a percent.)*

39. annually

40. semiannually

41. quarterly

42. monthly

43. daily

44. 1,000 times per year

45. 10,000 times per year

46. Judging from Exercises 39–45, what do you suppose is the effective annual yield if a nominal rate of 7% is compounded continuously? Explain your reasoning.

On July 21, 1995, Schools Federal Credit Union in Sacramento posted the following savings rates for certificates of deposit (CDs) in several categories:

		Balance \$10,000	Balance \$50,000
12-month	**Rate**	6.00	6.10
	Yield	6.14	6.24
30-month	**Rate**	6.50	6.60
	Yield	6.66	6.77

47. If you put \$40,000 into a 12-month CD, how much would you have in 1 year?

48. How often does compounding occur in these accounts?

Solve the following interest-related problems.

49. How long would it take to double your money in an account paying 9% compounded semiannually? (Give your answer in years plus days, ignoring leap years.)

50. After what time period would the interest earned equal the original principal if the account pays 9% compounded daily? (Give your answer in years plus days, ignoring leap years.)

51. At Greenway Savings interest is compounded monthly and the effective annual yield is 5.85%. What is the nominal rate?

52. In Fair Oaks, one bank pays a nominal rate of 6.80% compounded daily on deposits. A second bank produces the same yield as the first but only compounds interest quarterly.

(a) What nominal rate does the second bank pay?

(b) Which bank should Dave choose if he has $3,000 to deposit for 7 months? (Assume that the second bank pays no interest on funds deposited for less than an entire quarter.)

(c) Which bank should Carole choose if she has $5,000 to deposit for 1 year?

Use Table 2 to estimate the number of years it would take for the general level of prices to double at the following annual inflation rates.

53. 3% **54.** 4%

55. 5% **56.** 6%

Use the rule of 70 to estimate the years to double for the following annual inflation rates.

57. 1% **58.** 2%

59. 8% **60.** 9%

Use the rule of 70 to estimate the annual inflation rate (to the nearest tenth of a percent) that would cause the general level of prices to double in the following time periods.

61. 6 years **62.** 15 years

Work the following inflation-related problems.

63. Derive a rule for estimating the "years to triple," that is, the number of years it would take for the general level of prices to triple for a given annual inflation rate.

64. What would you call your rule of Exercise 63? Explain your reasoning.

The 1996 prices of several items are given below. Find the estimated future costs required to fill the blanks in the chart. (Give a number of significant figures consistent with the 1996 cost figures provided.)

Item	1996 Cost	2001 Cost 2% Inflation	2016 Cost 2% Inflation	2001 Cost 10% Inflation	2016 Cost 10% Inflation
65. House	$95,000	______	______	______	______
66. Meal Deal	$3.99	______	______	______	______
67. Gallon of gasoline	$1.34	______	______	______	______
68. Small car	$11,500	______	______	______	______

Work the following interest-related problems.

69. Mr. and Mrs. Ghardellini are close to retirement. They are selling a piece of land to a developer. The couple will be in a lower tax bracket in 4 years, and they wish to defer receipt of the money until then. Find the lump sum that the developer can deposit today, at 10% compounded quarterly, so that enough will be available to pay $200,000 to the couple in 4 years.

70. Radiology, Inc. will need $38,000 in 2 1/2 years when a new laser machine becomes available. What lump sum can they invest today, at 9% compounded semiannually, to have the required amount for the machine?

EXTENSION

Annuities

Some important concepts in the mathematics of finance depend upon being able to sum the terms of a **geometric sequence.** In a geometric sequence each term, after the first, is generated by multiplying the previous term by the **common ratio,** a number remaining constant throughout the sequence. For example, a geometric sequence with first term 3 and common ratio 2 starts out as follows:

$$3, 6, 12, 24, 48, 96, \ldots .$$

In general, if the first term is denoted a, the common ratio is denoted r, and there are n terms altogether, then the complete sequence is

$$a, ar, ar^2, ar^3, \ldots, ar^{n-1}.$$

(Verify by inductive reasoning that the nth term really is ar^{n-1}.)

The sum of all n terms of the geometric sequence above can be written

$$S = a + ar + ar^2 + \cdots + ar^{n-2} + ar^{n-1}.$$

Multiply both sides of this equation by r:

$$Sr = ar + ar^2 + ar^3 + \cdots + ar^{n-1} + ar^n.$$

Now position these two equations, one below the other, and subtract one from the other, as follows.

$$\begin{array}{rcccccc} S = a & + ar & + ar^2 & + \cdots + ar^{n-2} & & + ar^{n-1} \\ Sr = ar & + ar^2 & + ar^3 & + \cdots + ar^{n-1} & & + ar^n \\ \hline S - Sr = (a - ar) & + (ar - ar^2) & + (ar^2 - ar^3) & + \cdots + (ar^{n-2} & - ar^{n-1}) & + (ar^{n-1} - ar^n) \end{array}$$

We can now rearrange and regroup the terms on the right to obtain the following.

$$S - Sr = a + (ar - ar) + (ar^2 - ar^2) + \cdots + (ar^{n-1} - ar^{n-1}) - ar^n$$

Notice that the right side of this result consists mostly of pairs of identical terms, one positive and one negative, which cancel each other out, so we get

$$S - Sr = a - ar^n.$$

Factor both sides:

$$S(1 - r) = a(1 - r^n).$$

Solve for the sum S:

$$S = \frac{a(1 - r^n)}{1 - r}.$$

The Sum of a Geometric Sequence

If a geometric sequence has first term a and common ratio r, and has n terms altogether, then the sum of all n terms is given by

$$S = \frac{a(1 - r^n)}{1 - r}.$$

Now let's see how the summation formula above can be applied to systematic savings or investment plans. Consider the following situation.

> When I deposit money into a certain account, I can earn interest at an annual rate of r, compounded m times per year. Assume I will make a regular deposit of R (in dollars) *at the end* of each compounding period. What will be the total value of my account after n periods?

This type of account, which receives regular periodic deposits *at the end* of each compounding period, is called an **ordinary annuity** account. The total term (time period) over which deposits are made, consisting of the n compounding periods, is called the **accumulation period** of the annuity.

As each deposit R goes into the account, it begins to earn interest and grows to some amount according to the future value formula for compound interest. In the table below we list the n deposits, starting with the last one, which is made at the very end of the nth compounding period. Since the last deposit (deposit number n) has no time to accumulate interest, its final value is just R.

Value of an Annuity Account

Deposit Number	Future Value of Deposit
n	R
$n - 1$	$R\left(1 + \frac{r}{m}\right)$
$n - 2$	$R\left(1 + \frac{r}{m}\right)^2$
$\vdots$	$\vdots$
3	$R\left(1 + \frac{r}{m}\right)^{n-3}$
2	$R\left(1 + \frac{r}{m}\right)^{n-2}$
1	$R\left(1 + \frac{r}{m}\right)^{n-1}$

(Why is the exponent in the future value of the first deposit $n - 1$ rather than n?)

The future value of the account must be the sum of the future values of all the individual deposits, which is the sum of the entries of the right hand column of the table. But those entries comprise a geometric sequence. Its first term is R, the common ratio is $\left(1 + \frac{r}{m}\right)$, and the number of terms is n. So, in this case, the sum of the sequence (the **future value of the annuity**) is given by

$$\text{Sum} = \frac{R\left[1 - \left(1 + \frac{r}{m}\right)^n\right]}{1 - \left(1 + \frac{r}{m}\right)}.$$

A little algebra produces the following formula.

Future Value of an Ordinary Annuity

If money in an annuity account earns interest at an annual rate r, compounded m times per year, and regular deposits of amount R are made into this account at the end of each compounding period, then the final value (or **future value**) of the account at the end of n periods is given by

$$V = \frac{R\left[\left(1 + \frac{r}{m}\right)^n - 1\right]}{\frac{r}{m}}.$$

EXAMPLE Kevin Coleman sets up an ordinary annuity account with his credit union. Starting a month from today he will deposit \$100 per month, and deposits will earn 7.2% per year, compounded monthly. Find **(a)** the value of Kevin's account 10 years from today, **(b)** the total of all his deposits over the 10 years, and **(c)** the total interest earned.

(a) In this case, $R = \$100$, $r = 7.2\% = .072$, $m = 12$, and $n = 10 \times 12 = 120$. The formula for future value of an ordinary annuity gives

$$V = \frac{\$100\left[\left(1 + \frac{.072}{12}\right)^{120} - 1\right]}{\frac{.072}{12}} = \$17{,}500.30.$$

(b) Over 10 years, monthly deposits of \$100 will total

$$\$100 \times 10 \times 12 = \$12{,}000.$$

(c)
$$\begin{aligned}\text{Interest earned} &= \text{Future value} - \text{Total deposits}\\ &= \$17{,}500.30 - \$12{,}000\\ &= \$5{,}500.30.\end{aligned}$$

The future value of Kevin's account, at the end of 10 years, will be \$17,500.30. This value results from a total of \$12,000 in deposits which, over those 10 years, earn an additional \$5,500.30 in interest.

Extension Exercises

For each of the following ordinary annuity accounts, find **(a)** *the future value of the account (at the end of the accumulation period),* **(b)** *the total of all deposits, and* **(c)** *the total interest earned.*

	Regular Deposit	Compounded	Annual Interest Rate	Accumulation Period
1.	\$1,000	yearly	8.5%	10 years
2.	\$2,000	yearly	10.0%	20 years
3.	\$50	monthly	6.0%	5 years
4.	\$75	monthly	7.2%	10 years
5.	\$20	weekly	5.2%	3 years
6.	\$30	weekly	7.8%	5 years

An annuity account into which regular deposits are made at the beginning *of each compounding period, rather than at the end, is called an* **annuity due.** *(It is usually assumed that a deposit is made at the end of compounding period number n so that the future value of an annuity due is the accumulation of n + 1 deposits rather than just n.)*

7. If the interest rate, regular deposit amount, compounding frequency, and accumulation period are all the same in both cases, would the future value of an annuity due be less or more than that of an ordinary annuity? Explain.

8. Modify the derivation of the formula for future value of an ordinary annuity (presented in the text) to show that the **future value of an annuity due** is given by

$$V = \frac{R\left[\left(1 + \frac{r}{m}\right)^{n+1} - 1\right]}{\frac{r}{m}}.$$

Use the formula of Exercise 8 to find the future value of each annuity due given below.

	Regular Deposit	Compounded	Annual Interest Rate	Accumulation Period
9.	\$50	monthly	6.0%	5 years
10.	\$75	monthly	7.2%	10 years
11.	\$20	weekly	5.2%	3 years
12.	\$30	weekly	7.8%	5 years

13. Do you think it is practical for a financial institution (or insurance company) to guarantee to pay you a set interest rate for years into the future on your annuity account? Why or why not?

Interview some of your friends or acquaintances, teachers, or other employees of schools, hospitals, or other non-profit organizations and ask them about "tax deferred annuities" (TDAs) or "tax sheltered annuities" (TSAs). Explain the following features of these accounts.

14. their basic purpose

15. their tax advantages

16. their interest rates (as compared to those of regular savings accounts)

17. penalties for excessive deposits or for early withdrawals

18. the "annuitization period" (as opposed to the accumulation period)

13.2 Consumer Credit

When you borrow your neighbor's lawnmower, you (they hope) return the whole thing all at once when you are finished with it. However, when *money* is borrowed (in any sizable amount), timely repayment is much more likely if the debt is repaid over a period of time in a series of smaller payments.

Borrowing to finance purchases, and repaying with periodic payments is called **installment buying.** In this section we discuss the two general types

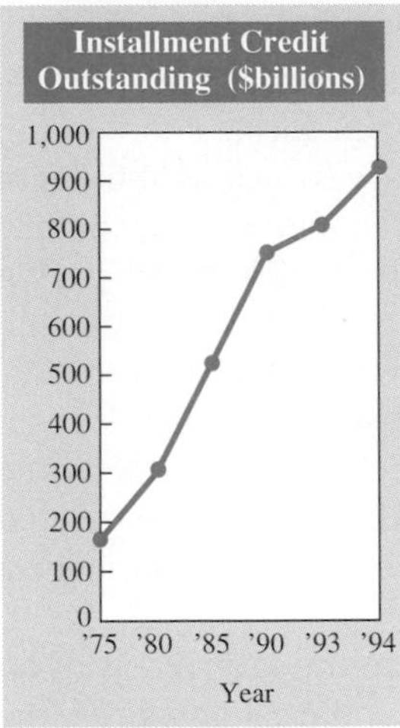

The above line graph, based on Federal Reserve data, shows a steady historical growth of consumer borrowing. (Beware of the distorted horizontal scale here.)

of consumer installment credit. **Closed-end** credit involves borrowing a set amount up front and paying a series of equal payments until the loan is paid off. Furniture, appliances, and cars are commonly financed through closed-end credit (sometimes called **fixed installment loans**).

With **open-end** credit, there is no fixed number of payments—the consumer continues paying until no balance is owed. With open-end credit, additional credit is often extended before the initial amount is paid off. Examples of open-end credit include most department store charge accounts and bank charge cards such as MasterCard and VISA.

Closed-End Credit

Loans set up under closed-end credit are often based on **add-on interest.** This means that if an amount P is borrowed, the annual interest rate is to be r, and payments will extend over t years, then the required interest comes from the simple interest formula: $I = Prt$. You simply "add on" this amount of interest to the principal borrowed to arrive at the total debt (or amount to be repaid).

$$\begin{aligned}\text{Amount to be repaid} &= \text{Amount borrowed} + \text{Interest due}\\ &= P + Prt\end{aligned}$$

This total debt is then equally divided among the payments (usually monthly) to be made over the t years.

EXAMPLE 1 The Hoangs are a newlywed couple. They buy \$2,700 worth of furniture and appliances to furnish their first apartment. They pay \$500 down and agree to pay the balance at an 8% add-on rate for 2 years. Find the total amount they owe and their monthly payment.

They finance a total of \$2,700 − \$500 = \$2,200. The interest charge is

$$I = Prt$$
$$I = \$2{,}200 \times .08 \times 2 = \$352.$$

The total amount owed is

$$\$2{,}200 + \$352 = \$2{,}552,$$

with a monthly payment of

$$\frac{\$2{,}552}{24} = \$106.33.$$ ■

Notice that the repayment amount in Example 1 was exactly the same ($P + Prt$) as the final amount A in a savings account paying a simple interest rate r for t years. (See the previous section.) But in the case of savings the bank keeps all of your money for the entire time period. In Example 1, the Hoangs did not keep the full principal amount for the full time period. They repaid it in 24 monthly installments. The 8% add-on rate turns out to be equivalent to a much higher "true annual interest" rate, as we shall see in the next section.

Another type of fixed installment loan, the **real estate mortgage,** usually involves much larger principal amounts and longer repayment periods than are involved in the consumer loans covered in this section. We will discuss mortgages in Section 4.

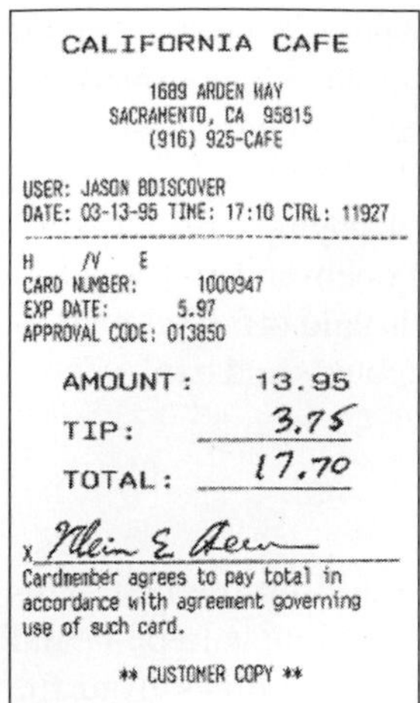

CALIFORNIA CAFE

1689 ARDEN WAY
SACRAMENTO, CA 95815
(916) 925-CAFE

USER: JASON BDISCOVER
DATE: 03-13-95 TIME: 17:10 CTRL: 11927

CARD NUMBER: 1000947
EXP DATE: 5.97
APPROVAL CODE: 013850

AMOUNT: 13.95
TIP: 3.75
TOTAL: 17.70

Cardmember agrees to pay total in accordance with agreement governing use of such card.

** CUSTOMER COPY **

■ FIGURE 1

Open-End Credit

With a typical department store account, or bank card, a credit limit is established initially and the consumer can make many purchases during a month (up to the credit limit). The required monthly payment can vary from a set minimum (which may depend on the account balance) up to the full balance. Some people carry a balance on their charge accounts from month to month, never paying them off completely.

Historically, multiple-copy *sales slip* (or *charge slip*) forms have been provided to merchants by the bank card companies. The customer signs this form and receives a copy of it at the time of making a purchase. Recently, some merchants have begun to generate generic sales slips on their own computerized registers, which print the type of card (perhaps coded) and account number on the slip. An example is shown in Figure 1.

At the end of each billing period, the customer commonly receives an **itemized billing,** a statement giving a list of purchases and cash advances, the total balance owed, the minimum payment required, and perhaps other account information. Normally a portion of the billing statement is to be removed and returned with the payment. The example of an itemized Discover® Card statement shown in Figure 2 reflects an additional feature, the Cashback Bonus® Award. Many banks today will pay you back some small percentage of your purchases as an incentive to choose their card.

account number	6011 0009 4700 6171
new balance	$ 505.82
payment due date	January 19, 1995
minimum payment due	$ 15.00
amount enclosed	$ 505.82

3 MISSION AV
CAR CA

DISCOVER CARD SERVICES
PO BOX 860210
PASADENA CA 91186-0210

1000947 8200000000015

Closing Date: December 25, 1994 page 1 of 2

Cashback Bonus®	this period	to date
qualified purchases	$ 522.08	$ 13,984.03
Cashback Bonus award earned	5.22	124.84
Cashback Bonus anniversary date	April 25	

Account Summary

account number 1000347
credit limit $ 7,000.00
credit available $ 6,494.00

previous balance		$ 533.74
payments and credits	−	550.00
purchases and miscellaneous charges	+	522.08
cash advances	+	0.00
balance transfers	+	0.00
FINANCE CHARGES	+	0.00
new balance	=	$ 505.82

To avoid additional finance charges, pay your entire new balance by January 19, 1995.

Transactions

payments and credits

DEC.	19	PAYMENT - THANK YOU	$-550.00

purchases

NOV.	24	HOMETOWN BUFFET #112 N HIGHLANDS CA	$ 22.60
NOV.	27	BROOKFIELD'S REST #2 SACRAMENTO CA	18.51
DEC.	3	MONTEREY PLAZA HOTEL MONTEREY CA	36.00
DEC.	5	THE MONTEREY HOTEL MONTEREY CA	256.17
DEC.	13	PRICE CLUB #414 N HIGHLANDS CA	112.81
DEC.	14	LUMBERJACK BUILDING MA SACRAMENTO CA	3.69
DEC.	18	CHEF LIU CARMICHAEL CA	36.00
DEC.	19	TOWER BOOKS 314 SACRAMENTO CA	16.30

If your income stops - how will you pay this bill? CreditSafe(R) Plus could help pay your Discover(R) Card bill when you can't. Sign up for CreditSafe Plus today! CreditSafe Plus, important protection for your Discover Card account. To enroll or for information call 1-800-453-0280.

Take advantage of travel savings with the enclosed ValueFinders(R) Coupons!

Please see following page for additional information.

■ FIGURE 2

Credit card use seems to still be accelerating. The American Bankers Association reported that consumers charged more than $165 billion on their bank cards in 1987. According to the report "The 1995 Credit Card Issuers' Guide," published by Faulkner & Gray Inc. of Chicago, IL, just the four main card groups, VISA, MasterCard, American Express, and Discover, accounted for over $474 billion in charges in 1993.

Since the tax deduction for credit card interest was completely phased out in 1991, more and more people are paying their bills in full each month. But not all cards allow the complete avoidance of interest charges. (For example, in some cases interest is charged from the date of purchase.) Be sure to consider this and the other features of any card *carefully* before you select it for your own use.

Any charges beyond the cash price of an item are called **finance charges.** Finance charges may include interest, annual fee, credit insurance coverage, a time payment differential, or carrying charges.

A few open-end lenders still use one of the simpler methods of calculating finance charges, such as the **unpaid balance method** which is illustrated in Example 2. Typically, you would begin with the unpaid account balance at the *end of the previous month* (or *billing period*) and apply to it the current monthly interest rate (or the daily rate multiplied by the number of days in the current billing period). Any purchases or returns in the current period do not affect the finance charge calculation.

EXAMPLE 2 The table below shows Anna Bell O'Malley's VISA account activity for a 2-month period. If the bank charges interest of 1.6% per month on the unpaid balance, and there are no other finance charges, find the missing quantities in the table, and the total finance charges for the 2 months.

Month	Unpaid Balance at Beginning of Month	Finance Charge	Purchases During Month	Returns	Payment	Unpaid Balance at End of Month
July	$179.82	______	$101.58	$ 0	$30	______
August	______	______	$ 19.84	$42.95	$85	______

The July finance charge is 1.6% of $179.82, which is $2.88. Then, adding any finance charge and purchases and subtracting any returns and payments, we arrive at the unpaid balance at the end of July (and the beginning of August):

$$\$179.82 + \$2.88 + \$101.58 - \$30 = \$254.28.$$

A similar calculation gives an August finance charge of

$$1.6\% \text{ of } \$254.28 = .016 \times \$254.28 = \$4.07,$$

and an unpaid balance at the end of August of

$$\$254.28 + \$4.07 + \$19.84 - \$42.95 - \$85 = \$150.24.$$

The total finance charges for the 2 months were

$$\$2.88 + \$4.07 = \$6.95.$$

Most open-end credit plans now calculate finance charges by the **average daily balance method.** From the bank's point of view, this seems fairer since it considers balances on all days of the billing period and thus comes closer to charging consumers for the credit they actually utilize. The average daily balance method is illustrated in Example 3.

EXAMPLE 3 The activity in Connie Rabara's MasterCard account for one billing period is shown below. If the previous balance (on March 3) was $209.46, and the bank charges 1.5% per month on the average daily balance, find **(a)** the average daily balance for the next billing (April 3) and **(b)** the finance charge to appear on the April 3 billing.

March 3	Billing date	
March 12	Payment	$50.00
March 17	Clothes	$28.46
March 20	Mail order	$31.22
April 1	Auto parts	$59.10

Learning the following rhyme will help you in problems like the one found in Example 3.

Thirty days hath September,
April, June, and November.
All the rest have thirty-one,
Save February which has
twenty-eight days clear,
And twenty-nine in each leap
year.

(a) First make a table that shows the beginning date of the billing period and the dates of all transactions in the billing period. Along with each of these, compute the balance due on that date.

Date	Balance Due
March 3	\$209.46
March 12	\$209.46 − \$50 = \$159.46
March 17	\$159.46 + \$28.46 = \$187.92
March 20	\$187.92 + \$31.22 = \$219.14
April 1	\$219.14 + \$59.10 = \$278.24

Next, tabulate the balance due figures, along with the number of days until the balance changed. Multiply each balance amount by the number of days. The sum of these products gives the "sum of the daily balances."

Date	Balance Due	Number of Days Until Balance Changed	(Balance Due) × (Number of Days)
March 3	\$209.46	9	\$1,885.14
March 12	\$159.46	5	\$ 797.30
March 17	\$187.92	3	\$ 563.76
March 20	\$219.14	12	\$2,629.68
April 1	\$278.24	2	\$ 556.48
		Totals: 31	\$6,432.36

$$\text{Average daily balance} = \frac{\text{Sum of daily balances}}{\text{Days in billing period}} = \frac{\$6{,}432.36}{31} = \$207.50.$$

Rabara will pay a finance charge based on the average daily balance of \$207.50.

(b) The finance charge for the April 3 billing will be

$$1.5\% \text{ of } \$207.50 = .015 \times \$207.50 = \$3.11.$$ ■

Bank card and other open-end credit accounts normally require some minimum payment each month, which depends in some way on your balance. But if you need to delay payment, you can carry most of your balance from month to month, and pay the resulting finance charges. Whatever method is used for calculating finance charges, you can avoid finance charges altogether in most cases by paying your entire new balance by the due date each month. This is a wise practice in fact, since these types of credit accounts, although convenient, are relatively expensive. Purchases charged to your account during the month will not be billed to you until your next billing date. And even after that, most lenders allow a grace period (usually about 25 days) between the billing date and the payment due date. That means that, with careful timing, you can purchase an item and not have to actually pay for it for up to nearly 2 months (without paying interest).

Although the open-end credit examples in this section have mainly addressed the actual interest charges, there are other features of these accounts that can be at least as important as interest. For example:

1. Is an annual fee charged? If so, how much is it?
2. Is a special "introductory" rate offered? If so, how long will it last?
3. Are there other incentives, such as rebates, credits toward certain purchases, or return of interest charges for long-time use?

You will be asked to consider some of these factors in the exercises that follow.

13.2 EXERCISES

Suppose you buy appliances costing $1,750 at a store charging 11% add-on interest, and you make a $500 down payment.

1. Find the total amount you will be financing.
2. Find the total interest if you will pay off the loan over a 2-year period.
3. Find the total amount owed.
4. The amount from Exercise 3 is to be repaid in 24 monthly installments. Find the monthly payment.

Suppose you want to buy a new car that costs $12,000. You have no cash—only your old car, which is worth $3,000 as a trade-in.

5. How much do you need to finance to buy the new car?
6. The dealer says the interest rate is 9% add-on for 4 years. Find the total interest.
7. Find the total amount owed.
8. Find the monthly payment.

In the following exercises, find the total interest and the monthly payment using the add-on method of calculating interest. Round each answer to the nearest cent.

	Amount of Loan	Length of Loan	Interest Rate
9.	$3,700	3 years	9%
10.	$1,746	24 months	10%
11.	$ 400	1 year	13%
12.	$2,896	18 months	9.2%

	Amount of Loan	Length of Loan	Interest Rate
13.	$ 158	10 months	13%
14.	$2,100	24 months	12%
15.	$2,700	36 months	11%
16.	$ 798	9 months	10.3%

Work the following problems. Give monetary answers to the nearest cent unless directed otherwise.

17. The Giordanos buy $7,000 worth of furniture for their new home. They pay $2,000 down. The store charges 10% add-on interest. The Giordanos will pay off the furniture in 30 monthly payments (2 1/2 years). Find the monthly payment.
18. Find the monthly payment required to pay off an auto loan of $6,780 over 3 years if the add-on interest rate is 11.3%.
19. The total purchase price of a new home entertainment system is $15,480. If the down payment is $3,500 and the balance is to be financed over 48 months at 12% add-on interest, what is the monthly payment?
20. What are the monthly payments Crystal Clymo pays on a loan of $790 for a period of 10 months if 9% add-on interest is charged?

21. Jeff Klinger has misplaced the sales contract for his car and cannot remember the amount he originally financed. He does know that the add-on interest rate was 8.8% and the loan required a total of 36 monthly payments of $196.62 each. How much did Jeff borrow (to the nearest dollar)?

22. Kelly Ng is making monthly payments of $318.35 to pay off a 2 1/2 year loan for $7,500. What is her add-on interest rate (to the nearest tenth of a percent)?

23. How long (in years) will it take Matt Byerly to pay off a $10,000 loan with monthly payments of $245.83 if the add-on interest rate is 9.5%?

24. How many monthly payments must Jawann make on a $7,000 loan if he pays $354 a month and the add-on interest rate is 10.69%?

Find the finance charge on each of the following open-end charge accounts. Assume interest is calculated on the unpaid balance of the account. Round to the nearest cent.

	Unpaid Balance	Monthly Interest Rate
25.	$244.88	1.8%
26.	$500.19	1.6%
27.	$167.99	1.78%
28.	$877.11	1.49%

Complete each of the following tables, showing the unpaid balance at the end of each month. Assume a monthly interest rate of 1.5% on the unpaid balance.

29.

Month	Unpaid Balance at Beginning of Month	Finance Charge	Purchases During Month	Returns	Payment	Unpaid Balance at End of Month
February	$297.11	______	$86.14	0	$50	______
March	______	______	109.83	$15.75	60	______
April	______	______	39.74	0	72	______
May	______	______	56.29	18.09	50	______

30.

Month	Unpaid Balance at Beginning of Month	Finance Charge	Purchases During Month	Returns	Payment	Unpaid Balance at End of Month
October	$554.19	______	$128.72	$23.15	$125	______
November	______	______	291.64	0	170	______
December	______	______	147.11	17.15	150	______
January	______	______	27.84	139.82	200	______

31.

Month	Unpaid Balance at Beginning of Month	Finance Charge	Purchases During Month	Returns	Payment	Unpaid Balance at End of Month
August	$822.91	______	$155.01	$38.11	$100	______
September	______	______	208.75	0	75	______
October	______	______	56.30	0	90	______
November	______	______	190.00	83.57	150	______

32.

Month	Unpaid Balance at Beginning of Month	Finance Charge	Purchases During Month	Returns	Payment	Unpaid Balance at End of Month
March	$1,522.83	______	$308.13	$74.88	$250	______
April	______	______	488.35	0	350	______
May	______	______	134.99	18.12	175	______
June	______	______	157.72	0	190	______

Find the finance charge for each of the following charge accounts. Assume interest is calculated on the average daily balance of the account.

	Average Daily Balance	Monthly Interest Rate
33.	$ 138.20	1 1/2%
34.	$ 350.49	1 3/4%
35.	$ 746.11	1.667%
36.	$1,490.74	1.805%

Find the average daily balance for each of the following credit card accounts. Assume one month between billing dates (using the proper number of days in the month). Then find the finance charge, if interest is 1.5% per month on the average daily balance.

37. Previous balance: $906.14

May 9	Billing date	
May 17	Payment	$200
May 30	Dinner	$ 46.11

38. Previous balance: $228.95

January 27	Billing date	
February 9	Cheese	$11.08
February 13	Returns	$26.54
February 20	Payment	$29
February 25	Repairs	$71.19

39. Previous balance: $312.78

June 11	Billing date	
June 15	Returns	$106.45
June 20	Jewelry	$115.73
June 24	Car rental	$ 74.19
July 3	Payment	$115

40. Previous balance: $714.58

August 17	Billing date	
August 21	Mail order	$ 14.92
August 23	Returns	$ 25.41
August 27	Beverages	$ 31.82
August 31	Payment	$108
September 9	Returns	$ 71.14
September 11	Plane ticket	$110
September 14	Cash advance	$100

Assume no purchases or returns are made in Exercises 41 and 42.

41. At the beginning of a 31-day billing period, Eleanora Garza has an unpaid balance of $720 on her credit card. Three days before the end of the billing period, she pays $600. Find her finance charge at 1.5% per month using the following methods.
(a) unpaid balance method
(b) average daily balance method

42. Jack Dominick's VISA bill dated April 14 shows an unpaid balance of $1,070. Five days before May 14, the end of the billing period, Dominick makes a payment of $900. Find his finance charge at 1.6% per month using the following methods.
(a) unpaid balance method
(b) average daily balance method

One version of the "90 Days Same as Cash" promotion is offered by a "major purchase card," which establishes an account charging 1.8583% interest per month on the account balance. Interest charges are added to the balance each month, becoming part of the balance on which interest is computed the next month. If you pay off the original purchase charge within 3 months, all interest charges are cancelled. Otherwise you are liable for all the interest. Suppose you purchase $2,400 worth of carpeting under this plan.

43. Find the interest charge added to the account balance at the end of
(a) the first month, **(b)** the second month, **(c)** the third month.

44. Suppose you pay off the account 1 day late (3 months plus 1 day). What total interest amount must you pay? (For purpose of calculating interest, ignore the extra day.)

45. Treating the 3 months as 1/4 year, find the equivalent simple interest rate for this purchase (to the nearest tenth of a percent).

Carole's bank card account charges 1.5% per month on the average daily balance as well as the following special fees:

Cash advance fee:	*2% (not less than $2 nor more than $10)*
Late payment fee:	*$15*
Over-the-credit-limit fee:	*$15*

In the month of June, Carole's average daily balance was $1,846. She was on vacation during the month and did not get her account payment in on time, which resulted in a late payment and also resulted in charges accumulating to a sum above her credit limit. She also used her card for six $100 cash advances while on vacation. Find the following based on account transactions in that month.

46. interest charges to the account

47. special fees charged to the account

Write out your responses to each of the following.

48. Is it possible to use a bank credit card for your purchases without paying anything for credit? If so, explain how.

49. Obtain applications or descriptive brochures for several different bank card programs, compare their features (including those in fine print), and explain which deal would be best for you, and why.

50. Research and explain the difference, if any, between a "credit" card and a "debit" card.

51. Many charge card offers include the option of purchasing credit insurance coverage, which would make your monthly payments if you became disabled and could not work and/or would pay off the account balance if you died. Find out the details on at least one such offer, and discuss why you would or would not accept it.

52. Make a list of "special incentives" offered by bank cards you are familiar with, and briefly describe the pros and cons of each one.

53. Early in 1995, one bank offered a card with a "low introductory rate" of 8.9%, good through the end of the year. And furthermore, you could receive back a percentage (up to 100%!) of all interest you pay, as shown in the table.

Use your card for:	2 years	5 years	10 years	15 years	20 years
Get back:	10%	25%	50%	75%	100%

(As soon as you take a refund, the time clock starts over.) Since you can eventually claim all your interest payments back, is this card a good deal? Explain why or why not.

54. Either recall a car-buying experience you have had, or visit a new-car dealer and interview a sales person. Write a description of the procedure involved in purchasing a car on credit.

Daniel and Nora Onishi are considering two bank card offers that are the same in all respects except that Bank A charges no annual fee and charges monthly interest of 1.6% on the unpaid balance while Bank B charges a $40 annual fee and monthly interest of 1.3% on the unpaid balance. From their records, the Onishis have found that the unpaid balance they tend to carry from month to month is quite consistent and averages $900.

55. Estimate their total yearly cost to use the card if they choose the card from
(a) Bank A **(b)** Bank B.

56. Which card is their better choice?

57. If Bank B raised its annual fee to $50 (and continued to charge 1.3% per month interest), then how high a monthly interest rate (to the nearest tenth of a percent) could Bank A charge and still offer the Onishis a more economical card?

58. Under the initial charges (1.6% per month for Bank A and 1.3% per month plus $40 per year for Bank B), what is the least average monthly unpaid balance (to the nearest cent) that would make Bank B's card a better deal for the Onishis?

13.3 Truth in Lending

The 1960s were a decade of heightened awareness of environmental and consumer issues. Congress enacted a number of measures designed to safeguard the health, safety, and well-being of American consumers. One of these was the Consumer Credit Protection Act, which was passed in 1968, took effect in 1969, and has commonly been known as the **Truth in Lending Act.**

In this section we discuss two major issues addressed in the law:

1. How can I tell the true annual interest rate a lender is charging?
2. How much of the finance charge am I entitled to save if I decide to pay off a loan sooner than originally scheduled?

To receive detailed information on Truth in Lending, you may write to:

Division of Consumer and Community Affairs
Board of Governors of the Federal Reserve System
Washington, DC 20551

Annual Percentage Rate (APR)

Question 1 above arose because lenders were computing and describing the interest they charged in several different ways. For example, how does 1.5% per month at Sears compare to 9% per year add-on interest at a furniture store? Truth in Lending standardized the so-called "true annual interest rate", or **annual percentage rate,** commonly denoted **APR.** All sellers (car dealers, stores, banks, insurance agents, credit card companies, and the like) must disclose the APR when you ask, and the contract must state the APR whether or not you ask. This enables a borrower to more easily compare the true costs of different loans.

Theoretically, you as a borrower should not need to calculate APR yourself, but it is possible to verify the value stated by the lender if you wish. Since the formulas for finding APR are quite involved, it is easiest to use a table provided by the Federal Reserve Bank. We show an abbreviated version in Table 3. It will identify APR values to the nearest half percent from 10.0% to 16.0%, which should do in most cases. (You should be able to obtain a more complete table from your local bank.) Our table is designed to apply to loans requiring *monthly* payments and extending over the most common lengths for consumer loans from 6 to 60 months. Apart from home mortgages, which are discussed in the next section, these conditions characterize most closed-end consumer loans.

Table 3 relates the following three quantities.

APR = true annual interest rate (shown across the top)

n = total number of scheduled monthly payments (shown down the left side)

h = finance charge per \$100 of amount financed (shown in the body of the table)

Later in this section, we will give the formula connected with this table. As you will see, though, it expresses h in terms of n and APR and is *not* easily solved to express APR in terms of n and h. Hence we have provided the table for determining APR values.

The use of Table 3 is illustrated in the following example, where we find the true annual interest rate (APR) for Example 1 of the previous section.

TABLE 3 Annual Percentage Rate (APR) for Monthly Payment Loans

Number of Monthly Payments (n)	Annual Percentage Rate (APR)												
	10.0%	10.5%	11.0%	11.5%	12.0%	12.5%	13.0%	13.5%	14.0%	14.5%	15.0%	15.5%	16.0%
	(Finance charge per $100 of amount financed) (h)												
6	$2.94	$3.08	$3.23	$3.38	$3.53	$3.68	$3.83	$3.97	$4.12	$4.27	$4.42	$4.57	$4.72
12	5.50	5.78	6.06	6.34	6.62	6.90	7.18	7.46	7.74	8.03	8.31	8.59	8.88
18	8.10	8.52	8.93	9.35	9.77	10.19	10.61	11.03	11.45	11.87	12.29	12.72	13.14
24	10.75	11.30	11.86	12.42	12.98	13.54	14.10	14.66	15.23	15.80	16.37	16.94	17.51
30	13.43	14.13	14.83	15.54	16.24	16.95	17.66	18.38	19.10	19.81	20.54	21.26	21.99
36	16.16	17.01	17.86	18.71	19.57	20.43	21.30	22.17	23.04	23.92	24.80	25.68	26.57
48	21.74	22.90	24.06	25.23	26.40	27.58	28.77	29.97	31.17	32.37	33.59	34.81	36.03
60	27.48	28.96	30.45	31.96	33.47	34.99	36.52	38.06	39.61	41.17	42.74	44.32	45.91

EXAMPLE 1 Recall that the Hoangs paid $500 down on a $2,700 purchase and agreed to pay the balance at an 8% add-on rate for 2 years. Find the APR for their loan.

As shown in the previous section, the total amount financed was

$$\text{Purchase price} - \text{down payment} = \$2{,}700 - \$500 = \$2{,}200,$$

and the finance charge (interest) was

$$I = Prt = \$2{,}200 \times .08 \times 2 = \$352.$$

Use the table for this example as follows.

14.5%
↓ ↑
24 → 15.80

Next find the finance charge per $100 of the amount financed. To do this, divide the finance charge by the amount financed, then multiply by $100.

$$\begin{pmatrix}\text{Finance charge per} \\ \$100 \text{ financed}\end{pmatrix} = \frac{\text{Finance charge}}{\text{Amount financed}} \times \$100$$

$$= \frac{\$352}{\$2{,}200} \times \$100 = \$16.$$

This amount, $16, represents h, the finance charge per $100 of the amount financed. Since the loan was to be paid over 24 months, look down to the "24 monthly payments" row of the table ($n = 24$). Then look across the table for the h-value closest to $16.00, which is $15.80. From that point, read up the column to find the APR, 14.5% (to the nearest half percent). In this case, an 8% add-on rate is equivalent to an APR of 14.5%. ■

EXAMPLE 2 After a down payment on your new car, you still owe $7,454. You agree to repay the balance in 48 monthly payments of $200 each. What is the APR on your loan?

First find the finance charge.

$$\begin{aligned}\text{Finance charge} &= \text{Total payments} - \text{Amount financed} \\ &= 48 \times \$200 - \$7{,}454 \\ &= \$2{,}146\end{aligned}$$

Now find the finance charge per $100 financed as in Example 1.

$$\begin{pmatrix}\text{Finance charge per}\\ \text{\$100 financed}\end{pmatrix} = \frac{\text{Finance charge}}{\text{Amount financed}} \times \$100$$

$$= \frac{\$2{,}146}{\$7{,}454} \times \$100 = \$28.79$$

Find the "48 payments" row of the table, read across to find the number closest to 28.79, which is 28.77. From there read up to find the APR, which is 13.0%. ■

Unearned Interest

Question 2 at the beginning of the section arises when a borrower decides, for one reason or another, to pay off a closed-end loan earlier than originally scheduled. In such a case, it turns out that the lender has not loaned as much money for as long as planned and so has not really "earned" the full finance charge originally disclosed.

If a loan is paid off early, the amount by which the original finance charge is reduced is called the **unearned interest.** We will discuss two common methods of calculating unearned interest, the **actuarial method** and the **rule of 78.** The Truth-in-Lending Act requires that the method for calculating this refund (or reduction) of finance charge be disclosed at the time the loan is initiated. Whichever method is used, the borrower may not, in fact, save all the unearned interest, since the lender is entitled to impose an **early payment penalty** to recover certain costs. (See Exercises 35, 36, and 40.) A lender's intention to impose such a penalty in case of early payment must also be disclosed at initiation of the loan.

We will now state and illustrate both methods of computing unearned interest.

Unearned Interest—Actuarial Method

For a closed-end loan requiring *monthly* payments, which is paid off earlier than originally scheduled, let

R = regular monthly payment

k = remaining number of scheduled payments (*after* current payment)

h = finance charge per $100, corresponding to a loan with the same APR and k monthly payments.

Then the **unearned interest, *u*,** is given by

$$u = kR\left(\frac{h}{\$100 + h}\right).$$

Do you see why this formula should give the unearned interest for a loan paid off early? (Exercises 37–39 will address this question.)

EXAMPLE 3 Suppose you get an unexpected pay raise and want to pay off your car loan of Example 2 at the end of 3 years rather than paying for 4 years as originally agreed. Find **(a)** the unearned interest (the amount you will save by retiring the loan early), and **(b)** the "payoff amount" (the amount required to pay off the loan at the end of 3 years).

(a) From Example 2, recall that $R = \$200$ and APR $= 13.0\%$. The current payment is payment number 36, so $k = 48 - 36 = 12$. Use Table 3, with 12 payments and APR 13.0%, to obtain $h = \$7.18$. Finally, by the actuarial method formula,

$$u = 12 \times \$200\left(\frac{\$7.18}{\$100 + \$7.18}\right) = \$160.78.$$

By the actuarial method, you will save \$160.78 in interest by terminating the loan early.

(b) The payoff amount can be determined as follows.

$$\begin{aligned}\text{Payoff amount} &= \begin{pmatrix}\text{Payment}\\ \text{number 36}\end{pmatrix} + \begin{pmatrix}\text{Sum of scheduled payments}\\ \text{after payment number 36}\end{pmatrix} - \begin{pmatrix}\text{Unearned}\\ \text{interest}\end{pmatrix}\\ &= \$200 + 12 \times \$200 - \$160.78\\ &= \$2{,}439.22.\end{aligned}$$

The required payment to terminate the loan at the end of 36 months is \$2,439.22.

The actuarial method, since it is based on the APR value, is probably the better method. However, some lenders still use the second method, the **rule of 78,** which is based on an approximating assumption. The rule of 78 works as follows.

Unearned Interest—Rule of 78

For a closed-end loan requiring *monthly* payments, which is paid off earlier than originally scheduled, let

F = original finance charge

n = number of payments originally scheduled

k = remaining number of scheduled payments (*after* current payment).

Then the unearned interest, u, is given by

$$u = \frac{k(k + 1)}{n(n + 1)} \times F.$$

(You can justify the above formula, and see how the method got its name, by carrying out the steps of the For Further Thought feature at the end of this section.)

EXAMPLE 4 Again assume that the loan in Example 2 is paid off at the time of the thirty-sixth monthly payment. This time use the rule of 78 to find **(a)** the unearned interest, and **(b)** the payoff amount.

(a) From Example 2, the original finance charge is $F = \$2{,}146$. Also, $n = 48$ and $k = 12$. By the rule of 78,

$$u = \frac{12(12 + 1)}{48(48 + 1)} \times \$2{,}146 = \$142.34.$$

By the rule of 78, you will save \$142.34 in interest, which is \$18.44 *less* than your savings by the actuarial method.

(b) In this case,

$$\begin{aligned}\begin{matrix}\text{Payoff}\\ \text{amount}\end{matrix} &= \begin{pmatrix}\text{Payment}\\ \text{number 36}\end{pmatrix} + \begin{pmatrix}\text{Sum of scheduled payments}\\ \text{after payment number 36}\end{pmatrix} - \begin{pmatrix}\text{Unearned}\\ \text{interest}\end{pmatrix}\\ &= \$200 + 12 \times \$200 - \$142.34\\ &= \$2{,}457.66.\end{aligned}$$

The payoff amount is \$2,457.66, which is \$18.44 *more* than the payoff amount calculated by the actuarial method in Example 3.

When the rule of 78 was first introduced into financial law (by the Indiana legislature in 1935), loans were ordinarily written for 1 year or less, interest rates were relatively low, and loan amounts were less than they tend to be today. For these reasons the rule of 78 was acceptably accurate then. Today however, with very accurate tables and/or calculators readily available, the rule of 78 is used much less often than it formerly was.*

Suppose you want to compute unearned interest accurately (so you don't trust the rule of 78), but the APR value, or the number of scheduled payments, or the number of remaining payments (or at least one of the three) is not included in Table 3. Then what? Actually, in the actuarial method, you can evaluate h (the finance charge per \$100 financed) using the same formula that we used to generate Table 3 in the first place.

Finance Charge per $100 Financed

If a closed-end loan requires n equal monthly payments and APR denotes the true annual interest rate for the loan (as a decimal), then $\boldsymbol{h}$, the **finance charge per \$100 financed,** is given by

$$h = \frac{n \times \dfrac{\text{APR}}{12} \times \$100}{1 - \left(1 + \dfrac{\text{APR}}{12}\right)^{-n}} - \$100.$$

EXAMPLE 5 Sergei borrowed \$4,000 to pay for music equipment for his band. His loan contract states an APR of 12.9% and stipulates 28 monthly payments of \$166.20 each. Sergei decides to pay the loan in full at the time of his nineteenth payment. Find **(a)** the unearned interest, and **(b)** the payoff amount.

*For an excellent discussion of the shortcomings of the rule of 78, see Alonzo F. Johnson's article, "The Rule of 78: A Rule that Outlived Its Useful Life," in *Mathematics Teacher,* September 1988.

(a) First find h from the finance charge formula just given. (Remember to use the remaining number of payments, 28–19, as the value of n.)

$$h = \frac{9 \times \frac{.129}{12} \times \$100}{1 - \left(1 + \frac{.129}{12}\right)^{-9}} - \$100 = \$5.45.$$

Next use the actuarial formula for unearned interest.

Regular monthly payment: $R = \$166.20$
Remaining number of payments: $k = 28 - 19 = 9$
Finance charge per \$100: $h = \$5.45$

$$u = 9 \times \$166.20 \times \frac{\$5.45}{\$100 + \$5.45} = \$77.31.$$

The amount of interest Sergei will save is \$77.31.

(b)
$$\begin{aligned}\text{Payoff amount} &= \begin{pmatrix}\text{Payment}\\ \text{number 19}\end{pmatrix} + \begin{pmatrix}\text{Sum of scheduled payments}\\ \text{after payment number 19}\end{pmatrix} - \begin{pmatrix}\text{Unearned}\\ \text{interest}\end{pmatrix}\\ &= \$166.20 + 9 \times \$166.20 - \$77.31\\ &= \$1{,}584.69.\end{aligned}$$

To pay off the loan, Sergei must pay \$1,584.69. ■

For Further Thought

To see how the rule of 78 (for unearned interest) got its name, consider a 12-month loan. First compute the sum of the natural numbers 1 through 12:

$$1 + 2 + 3 + 4 + 5 + 6 + 7 + 8 + 9 + 10 + 11 + 12 = 78.$$

Imagine the total finance charge being divided into 78 equal parts and the loan principal into 12 equal parts. *Assume* that, for the various months of the loan, the fraction of the loan principal used by the borrower and hence the fraction of the finance charge owed, would be as shown here:

Month	Fraction of Loan Principal Used by Borrower	Fraction of Finance Charge Owed
1	12/12	12/78
2	11/12	11/78
3	10/12	10/78
4	9/12	9/78
5	8/12	8/78
6	7/12	7/78
7	6/12	6/78
8	5/12	5/78
9	4/12	4/78
10	3/12	3/78
11	2/12	2/78
12	1/12	1/78
		78/78 = 1

Now suppose the loan is paid off early, say with four payments remaining. The portion (fraction) of the finance charge that should be saved can be seen by starting from the bottom of the right column of the table:

$$\frac{1}{78} + \frac{2}{78} + \frac{3}{78} + \frac{4}{78} = \frac{1 + 2 + 3 + 4}{78} = \frac{10}{78}.$$

For Group Discussion

Divide into three (or more) groups. The groups are to repeat the development above for the following loan situations. (Additional groups choose other numbers.)

(a) 10 payments total; paid in full with 3 payments remaining
(b) 18 payments total; paid in full with 5 payments remaining
(c) 24 payments total; paid in full with 8 payments remaining

1. As above, obtain for your situation the portion of the finance charge that should be saved.
2. Determine how the numerator and denominator of your fraction relate to the remaining number of payments k and the total number of payments n, respectively.
3. Have you proved the formula for the rule of 78? (*Hint:* Write the fraction of the finance charge to be saved as N/D.) Provide justification for each statement below.

$$N = 1 + 2 + \cdots + (k - 1) + k$$

Also

$$N = k + (k - 1) + \cdots + 2 + 1$$
$$2N = (k + 1) + (k + 1) + \cdots + (k + 1) + (k + 1)$$
$$2N = k(k + 1)$$
$$N = \frac{k(k + 1)}{2}$$

Likewise,

$$D = \frac{n(n + 1)}{2}$$

Hence

$$\frac{N}{D} = \frac{\frac{k(k + 1)}{2}}{\frac{n(n + 1)}{2}} = \frac{k(k + 1)}{n(n + 1)}.$$

Bring your groups back together and combine your efforts, if necessary, to justify each statement above.

4. Explain why the name "rule of 78" is not generally an appropriate name for this method of computing unearned interest.

13.3 EXERCISES

Find the APR (true annual interest rate), to the nearest half percent, for each of the following.

	Amount Financed	Finance Charge	Number of Payments
1.	$1,200	$ 46	6
2.	$4,800	$ 630	18
3.	$5,000	$ 850	24
4.	$4,900	$1,120	48

Find the monthly payment for each of the following. Round to the nearest cent.

	Purchase Price	Down Payment	Finance Charge	Number of Payments
5.	$2,000	$ 500	$ 100	24
6.	$4,280	$ 280	$1,200	36
7.	$3,950	0	$ 950	48
8.	$9,000	$3,000	$1,500	60

Find the APR (true annual interest rate), to the nearest half percent, for each of the following.

	Purchase Price	Down Payment	Add-on Interest Rate	Number of Payments
9.	$ 4,190	$ 390	8%	18
10.	$ 3,250	$ 750	9%	48
11.	$ 7,480	$2,200	6%	12
12.	$12,800	$4,500	7%	36

For each of the following loans, find **(a)** *the finance charge, and* **(b)** *the APR.*

13. Frank Barry financed a $540 television with 18 monthly payments of $33.40 each.

14. Pat Peterson bought a horse trailer for $4,090. She paid $1,150 down and paid the remainder at $112.50 per month for 2 1/2 years.

15. Mike Karelius still owed $720 on his new refrigerator after the down payment. He agreed to pay monthly payments for a year at 6% add-on interest.

16. Phil Givant paid off a $12,000 car loan over 4 years with monthly payments of $304.35 each.

Each of the following loans was paid in full before its due date. **(a)** *Obtain the value of h from Table 3. Then* **(b)** *use the actuarial method to find the amount of unearned interest, and* **(c)** *find the payoff amount.*

	Regular Payment	APR	Remaining Number of Scheduled Payments After Payoff
17.	$285.20	10.5%	6
18.	$410.00	15.0%	18
19.	$595.80	12.0%	12
20.	$314.50	11.5%	24

Each of the following loans was paid off early. Find the unearned interest by **(a)** *the actuarial method, and* **(b)** *the rule of 78.*

	Amount Financed	Regular Monthly Payment	Total Number of Payments Scheduled	Remaining Number of Scheduled Payments After Payoff
21.	\$ 1,650	\$101.39	18	6
22.	\$12,460	\$319.03	48	12
23.	\$ 9,850	\$209.28	60	12
24.	\$ 6,730	\$230.02	36	18

Each of the following loans was paid in full before its due date. **(a)** *Obtain the value of h from the appropriate formula. Then* **(b)** *use the actuarial method to find the amount of unearned interest, and* **(c)** *find the payoff amount.*

	Regular Payment	APR	Remaining Number of Scheduled Payments After Payoff
25.	\$148	9.1%	3
26.	\$430	10.25%	5

Wei-Jen Luan needs to borrow \$3,000 to supplement her down payment on a classic automobile. She can either borrow the \$3,000 from a finance company at 7% add-on interest for 4 years or take a credit union loan that will require 48 monthly payments of \$78.25 each. Use this information for the following exercises.

27. Find the APR (to the nearest half percent) for each loan and decide which one is Wei-Jen's better choice.

28. Suppose Wei-Jen takes the credit union loan. At the time of her fortieth payment she decides that this loan is a nuisance and decides to pay it off. If the credit union uses the rule of 78 for computing unearned interest, how much will she save by paying in full now?

29. What would Wei-Jen save in interest if she paid in full at the time of the fortieth payment and the credit union used the actuarial method for computing unearned interest?

30. Under the conditions of Exercise 29, what amount must Wei-Jen come up with to pay off her loan?

To convert an add-on interest rate to its corresponding APR, some people recommend using the formula

$$\text{APR} = \frac{2n}{n+1} \times r,$$

where r is the add-on rate and n is the total number of payments.

31. Apply the given formula to calculate the APR (to the nearest half percent) for the loan of Example 1 ($r = .08$, $n = 24$).

32. Compare your APR value in Exercise 31 to the value of Example 1. What do you conclude?

Exercises 33 and 34 also refer to Example 1 in the text.

33. Describe why, in the example, the APR and the add-on rate differ. Which one is more legitimate? Why?

34. In the example, the Hoangs paid off a \$2,200 loan over 2 years (24 monthly payments), and their 8% add-on rate turned out to be equivalent to 14.5% APR. Compute the APR equivalent to 8% add-on given that they were paying over the following time periods.
(a) 3 years **(b)** 4 years **(c)** 5 years

A certain retailer's credit contract designates the rule of 78 for computing unearned interest, and also imposes a "prepayment penalty." In case of any payoff earlier than the due date, they will charge an additional 10% of the original finance charge. Find the least value of k (remaining payments after payoff) that would result in any net savings in each of the following cases.

35. 24 payments originally scheduled

36. 36 payments originally scheduled

The actuarial method of computing unearned interest assumes that, throughout the life of the loan, the borrower is paying interest at the rate given by APR for money actually being used by the borrower. When contemplating complete payoff along with the current payment, think of the k future payments as applying to a separate loan with the same APR, h being the finance charge per \$100 of that *loan. Refer to the following formula.*

$$u = kR\left(\frac{h}{\$100 + h}\right)$$

37. Describe in words the quantity represented by $\frac{h}{\$100 + h}$.

38. Describe in words the quantity represented by kR.

39. Explain why the product of the two quantities above represents unearned interest.

Write out your responses to the following exercises.

40. Give reasons why a lender may be justified in imposing a prepayment penalty.

41. Discuss reasons that a borrower may want to pay off a loan early.

42. Find out what federal agency you can contact if you have questions about compliance with the Truth in Lending Act. (Any bank, or retailer's credit department, should be able to help you with this.)

13.4 Buying a House

Heating a house is another cost that may get you involved with banks and interest rates after you finally get a roof over your head. The roof you see above does more than keep the rain off. It holds glass tube solar collectors, part of the solar heating system in the building.

For many decades, home ownership has been considered a centerpiece of the "American dream." And since homes and condominiums have generally kept pace over the years with the rate of inflation, they have also been sound investments. In recent years, however, as economic and social factors affecting home ownership have become more complicated, it is increasingly important that individuals consider a number of questions when contemplating the purchase of a home. In this section we will point out some of these questions and will focus on the major ones whose answers require some mathematical calculations. For most people, a home represents the largest purchase of their lifetime, and it is certainly worth careful consideration.

In the process of deciding whether to buy a home, you will probably want to do some house-shopping to see what is available and what the prevailing prices are. As you begin to learn more toward the decision to buy, your "shopping" can be extended to include lenders. Banks, savings and loan companies, credit unions, and others are potential sources of your loan, and comparing a number of these can be worth your while.

A loan for a substantial amount, extending over a lengthy time interval (perhaps up to 30 or even 40 years), for the purpose of buying a home or other property or real estate, and for which the property is pledged as security for the loan, is called a **mortgage.** (In some areas, a mortgage may also be called a **deed of trust,** or a **security deed.**) The time until final payoff is called the **term** of the mortgage. The portion of the purchase price of the home which the buyer pays initially is called the **down payment.** And basically, the **principal amount of the mortgage** (the amount borrowed) is found by subtracting the down payment from the purchase price.

Even if you are fortunate enough to have the cash to pay outright for your home, or to pay a very substantial down payment, you still may be better off to make as small a down payment as possible, take a sizable mortgage, and invest your extra funds in other ways. This decision basically involves how favorable a mortgage you can negotiate and what alternative investment opportunities are available.

We will now discuss the common types of mortgages. We group them into two categories, **fixed-rate mortgages** and **adjustable-rate mortgages.**

The 1986 Tax Reform Act provides another advantage to homeowners and also encourages buyers to choose minimum down payments and maximum mortgages. Basically, interest on the amount originally borrowed for the home purchase is still tax-deductible whereas most other interest expense is not. A homeowner who has decreased his home debt over the years (through principal payments) can take out a so-called *home equity loan* for other purposes and still deduct the interest payments (on the amount up to the original home mortgage). The rates on home equity loans (second mortgages) typically run a couple of percentage points higher than on first mortgages.

Before you seek the advantages of borrowing against your home, remember that indebtedness has its drawbacks too. You should carefully consider the costs and risks of such an action.

Fixed-Rate Mortgages

With a **fixed-rate mortgage,** the interest rate will remain constant throughout the term, and the initial principal balance, together with interest due on the loan, is repaid to the lender through regular (constant) periodic (we assume monthly) payments. This is called **amortizing** the loan. (Other items, in addition to principal and interest, which may affect the monthly payment amount, will be discussed later.)

The regular monthly payment needed to amortize a loan depends on the amount financed, the term of the loan, and the interest rate. It can be found by using the following formula.

Regular Monthly Payment

The **regular monthly payment** required to repay a loan of P dollars, together with interest at an annual rate r, over a term of t years, is given by

$$R = \frac{P\left(\frac{r}{12}\right)\left(1 + \frac{r}{12}\right)^{12t}}{\left(1 + \frac{r}{12}\right)^{12t} - 1}.$$

EXAMPLE 1 Find the monthly payment necessary to amortize a \$75,000 mortgage at 7.5% annual interest for 15 years.

Using the formula above, we find

$$R = \frac{\$75{,}000\left(\frac{.075}{12}\right)\left(1 + \frac{.075}{12}\right)^{(12)(15)}}{\left(1 + \frac{.075}{12}\right)^{(12)(15)} - 1} = \$695.26.$$

With a programmable calculator, you can store the formula above and minimize the work of entering the needed numbers. Another approach is to use a tool such as Table 4, which gives payment values (per \$1,000 principal) for typical ranges of mortgage terms and interest rates. The entries in the table are given to five decimal places so that accuracy to the nearest cent can be obtained for normal mortgage amounts. Example 2 illustrates the use of Table 4.

TABLE 4 Monthly Payments to Repay Principal and Interest on a $1,000 Mortgage

Annual Rate (r)	Term of Mortgage (Years) (t)					
	5	10	15	20	25	30
6.0%	$19.33280	$11.10205	$8.43857	$7.16431	$6.44301	$5.99551
6.5%	19.56615	11.35480	8.71107	7.45573	6.75207	6.32068
7.0%	19.80120	11.61085	8.98828	7.75299	7.06779	6.65302
7.5%	20.03795	11.87018	9.27012	8.05593	7.38991	6.99215
8.0%	20.27639	12.13276	9.55652	8.36440	7.71816	7.33765
8.5%	20.51653	12.39857	9.84740	8.67823	8.05227	7.68913
9.0%	20.75836	12.66758	10.14267	8.99726	8.39196	8.04623
9.5%	21.00186	12.93976	10.44225	9.32131	8.73697	8.40854
10.0%	21.24704	13.21507	10.74605	9.65022	9.08701	8.77572
10.5%	21.49390	13.49350	11.05399	9.98380	9.44182	9.14739
11.0%	21.74242	13.77500	11.36597	10.32188	9.80113	9.52323
11.5%	21.99261	14.05954	11.68190	10.66430	10.16469	9.90291
12.0%	22.24445	14.34709	12.00168	11.01086	10.53224	10.28613
12.5%	22.49794	14.63762	12.32522	11.36141	10.90354	10.67258
13.0%	22.75307	14.93107	12.65242	11.71576	11.27835	11.06200
13.5%	23.00985	15.22743	12.98319	12.07375	11.65645	11.45412
14.0%	23.26825	15.52664	13.31741	12.43521	12.03761	11.84872
14.5%	23.52828	15.82868	13.65501	12.79998	12.42163	12.24556
15.0%	23.78993	16.13350	13.99587	13.16790	12.80831	12.64444
15.5%	24.05319	16.44105	14.33990	13.53881	13.19745	13.04517
16.0%	24.31806	16.75131	14.68701	13.91256	13.58889	13.44757

EXAMPLE 2 Find the monthly payment necessary to amortize a $98,000 mortgage at 9% for 25 years.

In Table 4, read down to the 9% row and across to the column for 25 years, to find the entry 8.39196. Since this is the monthly payment amount needed to amortize a loan of $1,000, and our loan is for $98,000, our required monthly payment is

$$98 \times \$8.39196 = \$822.41.$$

So that the borrower pays interest only on the money actually owed in a month, interest on real-estate loans is computed on the decreasing balance of the loan. Each equal monthly payment is first applied toward interest for the previous month. The remainder of the payment is then applied toward reduction of the principal amount owed. Payments in the early years of a real-estate loan are mostly interest; only a small amount goes toward reducing the principal. The amount of interest decreases each month, so that larger and larger amounts of the payment will apply to the principal. During the last years of the loan, most of the monthly payment is applied toward the principal. (See Table 5 on page 786.)

Many lenders supply an **amortization schedule** (or **repayment schedule**) showing the allotment of payments for interest and principal, and the principal balance, for each month over the entire life of the loan. Tables showing these breakdowns are available. However, the following steps show how the computations work. (Notice that Step 1 applies the simple interest formula $I = Prt$.)

Step 1: $\begin{matrix}\text{Interest for}\\ \text{the month}\end{matrix} = \begin{pmatrix}\text{Old balance}\\ \text{of principal}\end{pmatrix}\begin{pmatrix}\text{Annual}\\ \text{interest rate}\end{pmatrix}\left(\frac{1}{12}\text{ year}\right)$

Step 2: $\begin{matrix}\text{Payment on}\\ \text{principal}\end{matrix} = \begin{pmatrix}\text{Monthly}\\ \text{payment}\end{pmatrix} - \begin{pmatrix}\text{Interest for}\\ \text{the month}\end{pmatrix}$

Step 3: $\begin{matrix}\text{New balance}\\ \text{of principal}\end{matrix} = \begin{pmatrix}\text{Old balance}\\ \text{of principal}\end{pmatrix} - \begin{pmatrix}\text{Payment on}\\ \text{principal}\end{pmatrix}$

This sequence of steps is done at the end of each month. The new balance obtained in Step 3 becomes the Step 1 old balance for the next month.

EXAMPLE 3 The Petersons paid for many years on a \$60,000 mortgage with a term of 30 years and a 4.5% interest rate. Prepare an amortization schedule for the first two months of their mortgage.

Since Table 4 does not include interest rates as low as 4.5%, we use the regular monthly payment formula.

$$R = \frac{\$60{,}000\left(\frac{.045}{12}\right)\left(1 + \frac{.045}{12}\right)^{(12)(30)}}{\left(1 + \frac{.045}{12}\right)^{(12)(30)} - 1} = \$304.01.$$

Now apply Steps 1 through 3.

Step 1: Interest for the month = (\$60,000)(.045)(1/12) = \$225

Step 2: Payment on principal = \$304.01 − \$225 = \$79.01

Step 3: New balance of principal = \$60,000 − \$79.01 = \$59,920.99

Now starting with an old balance of \$59,920.99, repeat the same steps for the second month.

Interest for the month = (\$59,920.99)(.045)(1/12) = \$224.70

Payment on principal = \$304.01 − \$224.70 = \$79.31

New balance of principal = \$59,920.99 − \$79.31 = \$59,841.68.

These calculations are summarized here.

Amortization Schedule

Payment Number	Interest Payment	Principal Payment	Balance of Principal
Initially	→		\$60,000.00
1	\$225.00	\$79.01	\$59,920.99
2	\$224.70	\$79.31	\$59,841.68

Homes and condominiums are sometimes available for a very low down payment from the inventory of HUD (the Department of Housing and Urban Development). For current information on these opportunities, consult your local real estate companies.

TABLE 5 Amortization Schedules for a $60,000, 30-year Mortgage

4.5% Interest Monthly Payment: $304.01				14.5% Interest Monthly Payment: $734.73			
Payment Number	**Interest Payment**	**Principal Payment**	**Balance of Principal**	**Payment Number**	**Interest Payment**	**Principal Payment**	**Balance of Principal**
Initially →			60,000.00	Initially →			60,000.00
1	225.00	79.01	59,920.99	1	725.00	9.73	59,990.27
2	224.70	79.31	59,841.68	2	724.88	9.85	59,980.42
3	224.41	79.60	59,762.08	3	724.76	9.97	59,970.45
12	221.68	82.33	59,032.06	12	723.63	11.11	59,875.11
60	205.47	98.54	54,694.75	60	714.96	19.77	59,149.53
175	152.47	151.54	40,506.65	175	656.05	78.69	54,214.82
176	151.90	152.11	40,354.54	176	655.10	79.64	54,135.18
236	113.60	190.41	30,102.65	236	571.02	163.72	47,092.76
237	112.88	191.13	29,911.52	237	569.04	165.70	46,927.07
240	110.73	193.28	29,333.83	240	562.96	171.78	46,417.87
303	59.33	244.68	15,575.66	303	368.65	366.09	30,142.47
304	58.41	245.60	15,330.05	304	364.22	370.51	29,771.96
359	2.27	301.74	302.88	359	17.44	717.29	725.96
360	1.14	302.88	0.00	360	8.77	725.96	0.00
Totals:	49,444.03	60,000.00		Totals:	204,504.88	60,000.00	

Prevailing mortgage interest rates have varied considerably over the years. Table 5 shows portions of the amortization schedule for the Petersons' loan of Example 3, and also shows what the corresponding values would have been had their interest rate been 14.5% rather than 4.5%. A computer easily calculates and prints the interest, principal, and balance figures for all 360 payments, but in the table we show the breakdowns only for selected months (payment numbers). In particular, notice how much interest is involved in this home mortgage. At the (low) 4.5% rate, $49,444.03 in interest was paid along with the $60,000 principal. At a rate of 14.5%, the interest alone would total the huge sum of $204,504.88.

In addition to principal and interest on their mortgage, homeowners must also pay at least two other sizable periodic expenses: **property taxes** (usually imposed by the city or county) and **homeowners insurance** (covering potential damage from fire, unusual weather, and so forth). Often these items are paid out of a **reserve account** (also called an **escrow** or an **impound** account) kept by the lender. In this case the borrower must pay enough each month, along with amortization costs, so that the reserve account will be sufficient to pay taxes and insurance premiums when they come due. (In some cases, the homeowner pays taxes directly to the taxing authority and insurance premiums directly to an insurance company, totally separate from mortgage payments to the lender. But in any case, taxes and insurance must be considered as part of the expense of owning a home.)

EXAMPLE 4 Mr. and Mrs. Carlton have acquired an \$85,000 mortgage at 8.5% for 20 years. The property taxes (\$840 per year) and homeowners insurance (\$480 per year) on their new home will be paid from a lender's impound account. Find the monthly payment they must make to the lender.

The monthly amortization costs of the loan (principal and interest) are found using the Table 4 entry 8.67823.

$$\text{Principal and interest payment} = 85 \times \$8.67823 = \$737.65$$

Monthly payments for taxes and insurance will be one twelfth of the annual amounts.

$$\text{Tax and insurance payment} = \frac{\$840 + \$480}{12} = \$110$$

Total monthly payment required:

$$\text{Principal and Interest} + \text{Taxes and Insurance}$$
$$= \$737.65 + \$110 = \$847.65.$$

Adjustable-Rate Mortgages

In recent years, the lending industry has introduced many variations on the basic fixed-rate mortgage. The purpose is partly to cut down on the risk to lenders, and partly to make home loans accessible to more people. When you see a surprisingly low interest rate advertised, most likely it is for an **adjustable-rate mortgage (ARM),** also known as a **variable-rate mortgage (VRM).** These loans generally start out with a lower rate than similar fixed-rate loans, but the rate changes periodically, reflecting changes in prevailing rates. To understand how ARMs work, you must know about the following features: *indexes, margins, discounts, caps, negative amortization,* and *convertibility*.

Your ARM interest rate may change every 1, 3, or 5 years (occasionally more frequently). The frequency of change in rate is called the **adjustment period.** For example, an ARM with an adjustment period of 1 year is called a 1-year ARM. When the rate changes, your payment normally changes also. These adjustments are caused by fluctuations in an **index** upon which your rate is based. A variety of indexes are used, including the 1-, 3-, and 5-year U.S. Treasury security rates. Also used are national and regional "cost of funds" indexes. There are subtle differences in the various indexes, and you should be familiar with them before signing up for a particular ARM.

To determine your interest rate, the lender will add to the applicable index a few percentage points called the **margin.** The index and the margin are both important in determining the cost of the loan.

EXAMPLE 5 Suppose you pay \$20,000 down on an \$80,000 house and take out a 1-year ARM for a 30-year term. The lender uses the 1-year Treasury index (presently at 8%) and a 3% margin.

(a) Find your monthly payment for the first year.
First find the amount of your mortgage:

$$\begin{aligned}\text{Mortgage amount} &= \text{Cost of house} - \text{Down payment}\\ &= \$80{,}000 - \$20{,}000\\ &= \$60{,}000.\end{aligned}$$

Your first year interest rate will be

$$\text{ARM interest rate} = \text{Index rate} + \text{Margin} = 8\% + 3\% = 11\%.$$

Now from Table 4 (using 11% over 30 years) we obtain 9.52323, so

$$\text{First year monthly payment} = 60(\$9.52323) = \$571.39.$$

(b) Suppose that, after a year, the 1-year Treasury index has increased to 9.7%. Find your monthly payment for the second year.

During the first year, a small amount of the mortgage principal has been paid, so in effect we will now have a new "mortgage amount." (Also, the term will now be 1 year less than the original term.) The amortization schedule for the first year (not shown here) yields a loan balance, after the twelfth monthly payment, of \$59,729.98.

Also, for the second year,

$$\text{ARM interest rate} = \text{Index rate} + \text{Margin} = 9.7\% + 3\% = 12.7\%.$$

Since 12.7% is not included in Table 4, we resort to the regular monthly payment formula, using the new mortgage balance and also using 29 years for the remaining term.

$$\begin{aligned}\text{Second year monthly payment} &= \frac{\$59{,}729.98\left(\frac{.127}{12}\right)\left(1 + \frac{.127}{12}\right)^{348}}{\left(1 + \frac{.127}{12}\right)^{348} - 1}\\ &= \$648.78.\end{aligned}$$

The first ARM interest rate adjustment has caused the monthly payment to rise to \$648.78, which is an increase of \$77.39 over the initial monthly payment.

Sometimes initial rates offered on an ARM are less than the sum of the index and the margin. Such a **discount** is often the result of the seller (a new home builder, for example) paying the lender an amount in order to reduce the buyer's initial interest rate. This arrangement is called a "seller buydown." Discounts result in lower initial monthly payments, but are often combined with higher initial fees, or even an increase in the price of the house. Also, when the discount expires, you may experience sizable increases in your monthly payments.

EXAMPLE 6 In Example 5, suppose that a seller buydown discounts your initial (first year) rate by 2%. Find your first year and second year monthly payments.

The 8% index rate is discounted to 6%. Adding the 3% margin yields a net first year rate of 9% (rather than the 11% of Example 5), so the Table 4 entry is 8.04623, and

$$\text{First year monthly payment} = 60 \times \$8.04623 = \$482.77.$$

The amortization schedule shows a balance at the end of the first year of $59,590.13. The discount now expires, and the index has increased to 9.7%, so for the second year,

$$\text{ARM interest rate} = \text{Index rate} + \text{Margin} = 9.7\% + 3\% = 12.7\%.$$

Using the monthly payment formula, with $r = 12.7\%$ and $t = 29$,

$$\text{Second-year monthly payment} = \frac{\$59{,}590.13\left(\frac{.127}{12}\right)\left(1 + \frac{.127}{12}\right)^{348}}{\left(1 + \frac{.127}{12}\right)^{348} - 1}$$

$$= \$647.26.$$

In this case, the initial monthly payment of $482.77 looks considerably better than the $571.39 of Example 5, but at the start of year two, monthly payments jump by $164.49. Such an adjustment may cause you to experience a case of "payment shock." ■

To reduce the risk of payment shock, be sure the ARM you obtain includes **cap** features. An **interest rate cap** limits the amount your interest rate can increase. A **periodic cap** limits how much the rate can increase from one adjustment period to the next (typically about 1% per 6 months or 2% per year). An **overall cap** limits how much the rate can increase over the life of the loan (typically about 5% total). Most all ARMs must have overall rate caps by law, and many also have periodic caps. Some ARMs also have **payment caps,** which limit how much the *payment* can increase at each adjustment. However, since such a cap limits the payment but not the interest rate, it is possible that your payment may not cover even the interest owed. In that case you could find yourself owing more principal at the end of an adjustment period than you did at the beginning. This situation is called **negative amortization.** To make up the lost interest, the lender will increase your payments more in the future, or may extend the term of the loan.

Another way to reduce risk is with a **convertibility** feature, which allows you, at certain points in time, to elect to convert the mortgage to a fixed-rate mortgage. This would make sense if you decide you can't handle the uncertainty of periodic adjustments in your payments. If you elect to convert, your fixed rate will be determined by prevailing rates at the time of conversion. If you think you may want to completely, or partially, pay off the principal of the loan ahead of schedule, try to negotiate an ARM with no **prepayment penalty.** This means that you will be able to make early principal payments without paying any special fees, and your early payment(s) will go totally toward reducing the principal. Caps, convertibility, and prepayments will be addressed further in some of the exercises at the end of this section.

Life Insurance in Force (in millions of dollars)

	1945	1993
Ordinary	101,550	6,428,434
Group	22,172	4,456,338
Industrial	27,675	20,451
Credit	365	199,518

Government-backed mortgages, including FHA (Federal Housing Administration) and VA (Veterans Administration) loans, carry a government guarantee to protect the lender in case the borrower fails to repay the loan. Those who do not qualify for these loans, and obtain a conventional loan instead, usually are required to buy private mortgage insurance (PMI) as part of the loan package.

According to the chart above (data source: American Council of Life Insurance), life insurance coverage of consumer credit has grown far faster (percentage-wise) over the past 50 years than any other life insurance category.

Closing Costs

Besides the down payment, and the periodic costs of home ownership, including principal and interest payments, property taxes, and insurance, there are some significant one-time expenses which apply to both fixed-rate and adjustable-rate mortgages and are paid when the mortgage is originally set up. Lumped together, these charges are called **closing costs** (or **settlement charges**). The **closing,** or **settlement,** occurs when all details of the transaction have been determined and the final contracts are signed. It is the formal process by which the ownership title is transferred from seller to buyer. Within 3 days after your loan application, the lender is required to deliver or mail to you certain materials which include a "good faith estimate" of your closing costs. That is the time to ask questions, clarify all proposed costs, and, if you wish, shop around for alternate providers of settlement services. At the time of the closing, it will likely be too late to avoid paying for services that have already been provided. We will illustrate typical closing costs in the following example.

EXAMPLE 7 For a $58,000 mortgage, the borrower was charged the following closing costs.

Loan origination fee (1% of mortgage amount)	______
Broker loan fee	$1,455
Lender document and underwriting fees	375
Lender tax and wire fees	205
Fee to title company	200
Title insurance fee	302
Title reconveyance fee	65
Document recording fees	35

Compute the total closing costs for this mortgage.

"Loan origination fees" are commonly referred to as **points.** Each "point" amounts to 1% of the mortgage amount. By imposing points, the lender can effectively raise the interest rate without raising monthly payments (since points are normally paid at closing rather than over the life of the loan). In this case, "one point" translates to $580. Adding this to the other amounts listed gives total closing costs of $3,217. ■

Refinancing means initiating a new home mortgage and paying off the old one. In times of relatively low rates, lenders often encourage homeowners to refinance (presumably at a lower rate than the present loan carries). Here is a rule of thumb to help you make that decision.

You will probably not recover the closing costs of a new mortgage if either of the following conditions applies.

1. The reduction in interest rate is less than 2%.
2. You are likely to sell the house within the next 5 years.

Not all mortgage loans will involve the same closing costs. You should seek to know just what services are being provided and whether their costs are reasonable. There is considerable potential for abuse in the area of closing costs. Therefore, in 1974 Congress passed the **Real Estate Settlement Procedures Act (RESPA)** to standardize Truth-in-Lending practices in the area of consumer mortgages. The lender is required to provide you with clarifying information, including the APR for your loan. The APR, the cost of your loan as a yearly rate, will probably be higher than the stated interest rate since it must reflect all points and other fees paid directly for credit as well as the actual interest.

EXERCISES

Find the monthly payment needed to amortize principal and interest for each of the following fixed-rate mortgages. You can use either the regular monthly payment formula or Table 4, as appropriate.

	Loan Amount	Interest Rate	Term
1.	\$50,000	13.0%	15 years
2.	\$30,000	11.0%	20 years
3.	\$48,500	9.8%	25 years
4.	\$85,000	7.5%	14 years
5.	\$27,750	15.5%	30 years
6.	\$95,450	8.9%	30 years
7.	\$132,500	7.6%	20 years
8.	\$105,000	5.5%	10 years

Complete the first one or two months (as required) of the following amortization schedules for fixed-rate mortgages.

9. Mortgage: \$43,500
Interest rate: 10.0%
Term of loan: 30 years

Amortization Schedule

Payment Number	Total Payment	Interest Payment	Principal Payment	Balance of Principal
1	(a) ______	(b) ______	(c) ______	(d) ______

10. Mortgage: \$51,000
Interest rate: 9.5%
Term of loan: 20 years

Amortization Schedule

Payment Number	Total Payment	Interest Payment	Principal Payment	Balance of Principal
1	(a) ______	(b) ______	(c) ______	(d) ______

11. Mortgage: \$58,500
Interest rate: 8.5%
Term of loan: 25 years

Amortization Schedule

Payment Number	Total Payment	Interest Payment	Principal Payment	Balance of Principal
1	(a) ______	(b) ______	(c) ______	(d) ______
2	(e) ______	(f) ______	(g) ______	(h) ______

12. Mortgage: \$64,000
Interest rate: 11.0%
Term of loan: 15 years

Amortization Schedule

Payment Number	Total Payment	Interest Payment	Principal Payment	Balance of Principal
1	(a) ______	(b) ______	(c) ______	(d) ______
2	(e) ______	(f) ______	(g) ______	(h) ______

13. Mortgage: $28,000
Interest rate: 8.2%
Term of loan: 5 years

Amortization Schedule

Payment Number	Total Payment	Interest Payment	Principal Payment	Balance of Principal
1	(a) ______	(b) ______	(c) ______	(d) ______
2	(e) ______	(f) ______	(g) ______	(h) ______

14. Mortgage: $82,000
Interest rate: 12.5%
Term of loan: 22 years

Amortization Schedule

Payment Number	Total Payment	Interest Payment	Principal Payment	Balance of Principal
1	(a) ______	(b) ______	(c) ______	(d) ______
2	(e) ______	(f) ______	(g) ______	(h) ______

Find the total monthly payment, including taxes and insurance, on each of the following mortgages.

	Mortgage	Interest Rate	Term of Loan	Annual Taxes	Annual Insurance
15.	$48,700	11%	30 years	$ 509	$176
16.	$34,500	12%	20 years	$ 660	$105
17.	$59,200	9%	25 years	$ 775	$287
18.	$32,200	10%	10 years	$ 172	$165
19.	$69,890	8.8%	25 years	$1,080.19	$423.74
20.	$53,760	11.3%	15 years	$ 840.74	$209.77

Suppose $70,000 is owed on a house after the down payment is made. The monthly payment for principal and interest at 10.5% for 30 years is $70 \times \$9.14739 = \$640.32$.

21. How many monthly payments will be made?

22. What is the total amount that will be paid for principal and interest?

23. The total interest charged is the total amount paid minus the amount financed. What is the total interest?

24. Which is more—the amount financed or the total interest paid? By how much?

You may remember seeing home mortgage interest rates fluctuate widely in a period of not too many years. The following exercises show the effect of changing rates. Refer to Table 5, which compared the amortization of a $60,000, 30-year mortgage for rates of 4.5% and 14.5%. Give values of each of the following for **(a)** *a 4.5% rate, and* **(b)** *a 14.5% rate.*

25. monthly payments

26. percentage of first monthly payment that goes toward principal

27. balance of principal after 1 year

28. balance of principal after 20 years

29. the first monthly payment that includes more toward principal than toward interest

30. amount of interest included in final payment

Suppose a $60,000 mortgage is to be amortized at 7.5% interest. Find the total amount of interest that would be paid for each of the following terms.

31. 10 years **32.** 20 years **33.** 30 years **34.** 40 years

For each of the following adjustable-rate mortgages, find **(a)** *the initial monthly payment,* **(b)** *the monthly payment for the second adjustment period, and* **(c)** *the change in monthly payment at the first adjustment. (The "adjusted balance" is the principal balance at the time of the first rate adjustment. Assume no caps apply.)*

	Beginning Balance	Term	Initial Index Rate	Margin	Adjustment Period	Adjusted Index Rate	Adjusted Balance
35.	$75,000	20 years	6.5%	2.5%	1 year	8.0%	$73,595.52
36.	$44,500	30 years	7.2%	2.75%	3 years	6.6%	$43,669.14

Jeff Walenski has a 1-year ARM for $50,000 over a 20-year term. The margin is 2% and the index rate starts out at 7.5% and increases to 10.0% at the first adjustment. The balance of principal at the end of the first year is $49,119.48. The ARM includes a periodic rate cap of 2% per adjustment period. (Use this information for Exercises 37–40.)

37. Find **(a)** the interest owed and **(b)** the monthly payment due for the first month of the first year.

38. Find **(a)** the interest owed and **(b)** the monthly payment due for the first month of the second year.

39. What is the monthly payment adjustment at the end of the first year?

40. If the index rate has dropped slightly at the end of the second year, will the third year monthly payments necessarily drop? Why or why not?

For Exercises 41–44, refer to the following table of closing costs for the purchase of a $95,000 house requiring a 20% down payment.

Title insurance premium	$200
Document recording fee	25
Loan fee (two points)	______
Appraisal fee	225
Prorated property taxes	545
Prorated fire insurance premium	190

Find each of the following.

41. the mortgage amount

42. the loan fee

43. the total closing costs

44. the total amount of cash required of the buyer at closing (including down payment)

On the basis of material in this section, or on your own research, give brief written responses to each of the following.

45. Suppose your ARM allows conversion to a fixed rate loan at each of the first five adjustment dates. Describe circumstances under which you would want to convert.

46. Describe each of the following types of mortgages.

Graduated payment
Balloon payment
Growing equity
Shared equity
Partnership

47. Should a home buyer always pay the smallest down payment that will be accepted? Explain.

48. Should a borrower always choose the shortest term available in order to minimize the total interest expense? Explain.

49. Under what conditions would an ARM probably be a better choice than a fixed-rate mortgage?

50. Why are second-year payments (slightly) less in Example 6 than in Example 5 even though the interest rate is exactly 12.7% in both cases?

51. Do you think that the discount in Example 6 actually makes the overall cost of the mortgage less? Explain.

52. Discuss the term "payment shock" mentioned at the end of Example 6.

53. Find out what is meant by the following terms and describe some of the features of each.

FHA-backed mortgage
VA-backed mortgage
Conventional mortgage

13.5 Investing in the Stock Market

Buying stock in a corporation makes you a part owner of the corporation. In return for the money you invest in stocks, you share in any profits the company makes. Your share of the company profits is called a **dividend.** With any luck, the company will do well and prosper. If this happens, many other people will want your stock, and they will be willing to pay you a good price for your shares so that you can sell at a profit.

On the other hand, if the company does not do well, then fewer people will want its stock, and the price will fall. The price of a share of stock is determined by the law of supply and demand at institutions called **stock exchanges.** The largest stock exchange is the New York Stock Exchange (NYSE), located on Wall Street in New York City.

The public does not go directly to the exchange to buy and sell stock. Instead, members of the public buy their stock through **stockbrokers,** people who do have access to the exchange. Stockbrokers charge a fee for buying or selling stock, as discussed below.

Today a large proportion of stock is owned by pension funds. Pension funds collect money from a group of people, such as certain government workers or the employees of a particular company. This money is then invested by the people in charge of the fund. When an employee retires, money is taken from the fund to pay monthly retirement checks to the employee. Employees obviously want to be certain they will have retirement checks when they retire. For this reason, the people running the pension fund must be very careful and conservative in their investments.

In this book we are interested mainly in following the prices of stocks rather than in methods for deciding which stocks to buy. You can find the current stock prices in many daily newspapers or from one of the computer on-line services that provide this information. The table on the next page shows only a small portion of the activities on the New York Stock Exchange for a recent day of trading.

	52 Weeks Hi	52 Weeks Lo	Stock	Sym	Div	Yld %	PE	Vol 100s	Hi	Lo	Close	Net Chg
s	37¼	21⅝	AEGON NV	AEG		...	14	88	36⅛	35¾	35⅞	− ¼
	5	**3⅜**	♣**Aeroflex**	**ARX**		**...**	**7**	**403**	**4⅝**	**4⅜**	**4⅜**	**− ¼**
n	27⅞	25	AetnaMIPS pfA		2.37	8.7	...	261	27½	27¼	27⅜	...
	54⅞	42¼	AetnaLife	AET	2.76	4.4	12	3539	64⅝	62⅝	63	−1¼
	14¾	8¼	♣AgnicoEgl	AEM	.10	.7	...	959	14	13⅝	13½	− ½
	18¼	14½	AgreeRlty	ADC	1.80	10.8	...	101	17	16⅝	16⅝	− ⅜
	23¾	15¼	Ahmanson	AHM	.88	3.9	15	3697	22½	22⅛	22½	+ ¼
	26⅞	24½	Ahmanson pfD		2.40	9.2	...	68	26⅛	26	26⅛	− ⅛
	26⅛	22⅛	Ahmanson pfC		2.10	8.1	...	46	25⅞	25¾	25⅞	− ⅛
	52⅝	39¾	Ahmanson pfD		3.00	5.9	...	81	51½	51	51	− ¼
	36⅜	25⅝	Ahold	AHO	.79e	2.2	...	63	35⅝	34⅞	35¼	...
	3	7/16	vjAileen	AEE		...	...	182	7/16	7/16	7/16	...
	59⅝	42½	AirProduct	APD	1.04f	1.8	20	1531	58⅛	57½	57⅞	− ⅜
	30⅞	18	AirbornFrght	ABF	.30	1.4	14	283	21½	21⅛	21½	...
	29⅞	19⅝	♣Airgas	ARG		...	29	694	28½	28	28¼	− ⅛
	17	10⅞	Airlease	FLY	2.00	13.0	11	78	15⅝	15⅜	15⅜	− ⅛
	31⅜	23⅞	AirTouch	ATI		...	cc	5584	30⅝	30⅛	30½	+ ⅜
	34⅜	21½	AK Steel	AKS		...	2	1154	32⅞	32⅛	32½	+ ⅛
n	34¼	24⅛	AK Steel pf		2.15	6.5	...	30	33¼	32¾	33	...
	25¾	22	AlaPwr pfA		1.90	7.5	...	24	25¼	25⅛	25¼	+ ¼
	23½	18⅜	AlaPwr pfC		1.60	7.0	...	11	22¾	22¾	22¾	...
	24½	18	AlaPwr pfB		1.70	7.2	...	10	23½	23	23½	− ¼
	25½	21⅞	AlaPwr pfH		1.90	7.5	...	66	25½	25¼	25⅜	+ ¼
▲	19⅝	13⅛	AlaskaAir	ALK	.05j	...	21	2547	19⅞	18¾	19⅝	+ ¾
▲	25¼	16⅛	♣AlbanyInt	AIN	.40f	1.6	27	1142	25⅞	25¼	25¾	+ ½
	17⅝	12¼	Albemarle	ALB	.20	1.3	19	422	15¾	15⅝	15¾	...
	32½	20¾	♣AlbertoCl	ACV	.32	1.1	17	91	30⅝	30⅛	30¼	− ⅛
	28	19	♣AlbertoCl A	ACVA	.32	1.2	14	479	25¾	25⅝	25⅝	− ⅛
	32½	26	Albertsons	ABS	.52	1.7	18	3036	30¼	29¾	30	− ¼
▲	35⅞	22½	♣Alcan	AL	.30	.8	18	27831	36⅜	33⅞	35⅞	...
	24½	15⅜	AlcatelAlsthom	ALA	.61e	3.2	...	2239	19⅝	19⅛	19¼	− ⅛
	81⅝	53	AlcoStd	ASN	1.04	1.3	54	647	79⅞	79¼	79⅞	+ ⅝
	48	25	♣AlexBrown	AB	.70	1.5	11	882	46⅜	45½	45⅞	− ¼
	26⅜	17⅝	Alex&Alex	AAL	.10	.4	dd	653	25¼	25	25¼	...
	58⅝	48⅞	Alexanders	ALX		...	cc	12	57	56⅝	56⅝	− ⅛
	13¼	11⅛	AllAmTerm	AAT	1.08a	8.4	...	406	12⅞	12¾	12⅞	...
	163	141	AlleghanyCp	Y	stk	...	10	132	163	161½	161½	−1½
	23	18	AllegLud	ALS	.48	2.2	54	532	22⅛	21⅞	22⅛	+ ⅛
	25⅛	19¾	AllegPwr	AYP	1.64	6.9	13	1707	24	23⅝	23⅞	− ⅛
▲	31⅛	15¾	AllenGp	ALN	.20	.6	27	3713	32⅜	31	31⅝	+ ¾
	30⅞	23⅝	Allergan	AGN	.48f	1.8	16	1656	27	26⅜	27	+ ⅜
n	20	15½	AllncAll	AMO	1.03e	5.4	...	156	19⅝	19¼	19¼	...
	22⅝	15⅜	AllncCapMgt	AC	1.64	8.7	11	504	19⅛	18½	18⅞	− ⅛
	10⅛	4⅜	AllncEntn	CDS		...	23	604	9⅜	9	9	− ¼
	10¼	8⅜	AllncGlblEnv	AEF		...	...	125	9¾	9⅝	9¾	+ ⅛
	12⅞	8⅞	AllianceWrld II	AWF	1.42a	12.5	...	1307	11½	11⅛	11⅜	...
	14	9	AllianceWrld	AWG	1.29	11.0	...	367	11⅞	11¾	11¾	− ⅛
	43⅜	27¼	AlliantTech	ATK		...	dd	89	43	42⅝	42⅝	− ½
	29⅞	22¼	♣AlldIrishBk	AIB	1.34e	4.6	...	131	29⅛	28⅞	28⅞	− ¼
	28⅛	25⅝	AlldIrishBk pf		2.97	10.8	...	27	27⅝	27½	27⅝	+ ⅜
▲	20½	12	AlliedPdts	ADP	.03e	.1	12	250	20⅞	20⅝	20⅝	− ⅛
	46⅝	30⅝	AlliedSgnl	ALD	.78	1.7	16	7308	46⅝	45½	45⅝	− ⅞
▲	22⅞	14⅝	♣AllmrPC	APY	.16	.7	10	1298	23¼	22¾	23	+ ⅛
	10½	8¾	AllmrST	ALM	.84a	8.3	...	51	10¼	10⅛	10⅛	...
	9¾	8¼	AllmonTr	GSO	.73e	8.1	...	196	9⅛	9	9	− ⅛
▲	33¼	22⅝	♣Allstate	ALL	.78	2.4	11	26964	33⅜	32⅛	32⅝	− ⅝
n	44¾	35¾	Allstate exch	PME	.56p	...	...	826	44½	44¼	44½	− ¼
	31⅜	23¼	ALLTEL	AT	.96	3.7	18	2151	25¾	25¼	25¾	...
	7½	4¾	Allwaste	ALW		...	15	1386	5⅝	5¼	5⅜	+ ⅛
▲	35½	23¾	Alumax	AMX		...	10	2298	36¼	35¼	36⅛	+ ⅝
s▲	56½	36⅞	♣Alcoa	AA	.90	1.6	12	18677	57¼	56⅛	56¾	+ ⅞
	24⅝	17	♣Alza	AZA		...	32	3896	23¾	23⅜	23⅝	+ ⅛
	7⅝	4½	AmaxGold	AU		...	dd	2105	6⅛	6	6	− ⅛
n	56¼	42¾	AmaxGold pfB		3.75	7.6	...	19	49½	49¼	49½	+ ⅜
	41½	30¼	Ambac	ABK	.54	1.3	10	651	40¾	40½	40⅝	+ ⅛
	23⅜	16¾	♣AmcastInd	AIZ	.52	2.6	10	113	19⅞	19⅝	19⅞	+ ¼
	53⅛	43¾	AmerHess	AHC	.60	1.3	cc	1676	48½	47¾	48	− ⅛
	22¾	15¼	Amerco pfA		2.13	9.6	...	135	22⅜	22⅛	22⅛	− ¼
n	16⅜	6⅜	AmWestAir B	AWA		...	...	1968	13⅜	13⅛	13⅜	+ ⅛
n	7⅜	1½	AmWestAir wt			...	...	369	5⅝	5¼	5½	+ ⅛
	8⅞	8	AmAdjTerm96	BDJ	.39a	4.5	...	175	8¾	8¾	8¾	...
	8¾	7½	AmAdjTerm97	CDJ	.48	5.6	...	407	8 11/16	8⅝	8⅝	...
	8⅝	7⅜	AmAdjTerm98	DDJ	.51	6.1	...	604	8⅜	8⅜	8⅜	...
	8⅝	7⅛	AmAdjTerm99	EDJ	.54	6.5	...	614	8¼	8¼	8¼	...
	10⅜	8⅞	AmAnnuity	AAG	.06e	.6	10	36	9¾	9⅝	9¾	+ ⅛
	3½	**1½**	**AmBknote**	**ABN**		**...**	**dd**	**336**	**2¼**	**2⅛**	**2¼**	**+ ⅛**
	42⅛	31⅞	AmBrand	AMB	2.00	5.1	11	6603	39⅜	38⅝	39¼	− ⅛
s	20½	12¾	AmBusnPdts	ABP	.56	2.9	14	133	19⅜	19	19⅛	− ¼
	19⅝	16	AmCapBdFd	ACB	1.54	7.9	...	59	19½	19⅜	19⅜	+ ⅛
	21	18	AmCapCvSec	ACS	1.44e	7.0	...	3	20⅝	20⅝	20⅝	...
	7½	6¼	AmCapIncTr	ACD	.65a	9.1	...	187	7¼	7⅛	7⅛	...
	12¼	6½	AmEagle	FLI	.12	1.1	11	231	11½	11⅜	11⅜	− ⅛
	35¾	29⅛	AmElecPwr	AEP	2.40	6.9	13	1202	34⅞	34½	34⅝	− ¼
	37	25⅞	AmExpress	AXP	.90	2.5	13	7484	36½	36	36¼	− ⅛
	50⅛	38¼	AmExpressDECS	AXD	2.30	4.6	...	812	49⅞	49¼	49⅞	+ ½
	27⅝	21⅝	AmFnl	AFG	1.00	3.8	cc	945	26⅞	26⅜	26⅝	+ ⅛
	14¾	9	SavanFood	SFI	.10m	.9	48	242	11⅛	10⅞	11	...
s	27⅞	19⅞	Sbarro	SBA	.76	3.2	15	1580	24¼	23⅝	24⅛	+ ⅜
s	23⅜	20½	SCANA	SCG	1.44	6.5	14	1930	22¼	22	22⅛	− ⅛
	13⅜	7¼	Schawk A	SGK	.26	3.1	16	15	8½	8⅜	8⅜	− ⅛
	50¼	33¼	Scherer	SHR		...	25	618	47	45⅞	46	− ⅝
s	45½	30½	ScheringPl	SGP	1.16	2.6	17	11708	44⅛	43⅛	44	+ ¾
	66⅝	50	♣Schlumbgr	SLB	1.50f	2.4	27	1923	63⅜	62¾	63¼	+ ¼
	13¾	9⅜	SchroderAsian	SHF		...	...	504	12⅛	12	12	...
s	49⅛	16⅞	♣SchwabC	SCH	.24	.5	29	2756	47	46	46	− ⅞
s	24⅞	16¾	SciAtlanta	SFA	.06	.3	27	2790	22⅞	22⅜	22½	− ½
	19¾	14	Scotsman	SCT	.10	.5	12	316	18⅝	18⅜	18⅜	− ⅛
s	51⅝	25⅜	ScottPaper	SPP	.40	.8	21	4894	49¼	47¾	47¾	−1
	7⅜	3¾	♣ScotLiqdGld	SGD	.10e	2.2	5	77	4⅝	4½	4½	...
	33½	26¾	ScrippsEW	SSP	.52f	1.6	19	291	32⅝	32⅛	32⅛	− ⅜
s	20½	14⅜	ScudAsiaFd	SAF	5.08e	31.3	...	171	16½	16⅛	16¼	− ⅜
	11⅛	8⅝	ScudNewEur	NEF		...	...	691	11⅛	10⅞	11	− ⅛
x	14	10⅜	ScudrWldInco	SWI	1.36a	10.6	...	124	12⅞	12¾	12⅞	+ ⅛
	17½	11¾	SeaCont A	SCRA	.77	4.5	13	863	17	16¼	17	+ ⅛
	17⅞	15⅝	SeaCont pfC		2.10	12.0	...	12	17¾	17½	17½	...
	47¼	40½	SeaCont pfE		4.00	8.8	...	41	45½	45¼	45½	+ ⅛
▲	45¼	21½	SeagateTch	SEG		...	13	15141	45⅜	44½	44⅞	− ⅛
	35¾	25½	Seagram	VO	.60	1.7	3	7559	34⅞	34⅜	34¾	+ ¼
	20	16¼	SeagullEngy	SGO		...	dd	2138	18	17⅝	17⅞	+ ⅜
	49⅝	29	SealedAir	SEE		...	24	404	49⅝	48⅞	48⅞	− ⅝
▲x	**30⅜**	**21½**	♣**Sears**	**S**	**1.60a**	**5.0**	**7**	**27588**	**32⅜**	**30⅜**	**32¼**	**+2⅛**
	27⅜	24⅜	♣Sears pf		2.22	8.5	...	98	26⅜	26¼	26¼	− ⅛
	18¼	14½	SecCapIndl	SCN	.94	5.7	28	57	16⅜	16¼	16⅜	+ ⅛
n	25⅞	24	SecCapInd pfA			...	...	220	24⅞	24⅝	24⅞	...
	18⅞	15½	SecCapPac	PTR	1.15	6.5	24	452	17⅞	17¾	17¾	...
	25⅜	20	SecCapPac pfA		1.75	7.7	...	51	22⅝	22½	22⅝	...
n	25¼	24	SecCapPac pfB		.23p	...	...	51	25⅛	24⅞	24⅞	− ⅛
	26¼	19¼	SecurityCT	SRC	.48	1.9	9	485	25¾	25	25	− ¾
	37	18⅞	Seitel	SEI		...	24	951	30½	30	30¼	+ ⅛
▲	14	10⅝	SeligQual	SQF	.94	6.7	...	144	14¼	13⅞	14⅛	+ ¼
	12⅜	9¾	SeligSelct	SEL	.84	6.9	...	87	12⅜	12¼	12¼	− ⅛
	9¼	7⅞	SrHighIncII	SAL	.88e	9.9	...	507	9⅛	8⅞	8⅞	− ⅛
	9¼	7⅝	SrHighInc	ARK	.88e	9.6	...	606	9⅛	9	9⅛	+ ⅛
	9⅜	7¾	SrStratInco	SSN	.95e	10.9	...	174	8⅞	8¾	8¾	− ⅛
	38¼	21	♣SensorElec	SRM	.22	.8	20	20639	27	26⅜	26¾	+ ⅜
	31¾	17⅝	SequaA	SQAA	.60	2.0	dd	42	30⅛	29½	29¾	− ½
	35⅛	21	SequaB	SQAB	.50	1.5	dd	6	33¾	33¾	33¾	− ¼
▲	32¾	24⅛	♣ServiceCp	SRV	.44	1.3	21	3133	33⅜	32¾	33⅜	+ ¾
	7⅜	**4⅛**	**SvcMdse**	**SME**		**...**	**13**	**7243**	**7¼**	**6⅝**	**6¾**	**− ⅜**
	27¼	21½	SvcMastr	SVM	.96	3.5	15	493	27¼	26¾	27⅛	+ ¼
n	14½	6¾	ShandongPwr	SH	.21p	...	...	440	7⅝	6¾	6⅞	− ¼
	37⅝	23⅜	ShangPetro	SHI	1.47e	4.4	...	344	33⅞	33¼	33¼	− ¼
	17⅞	12	ShawInd	SHX	.30	1.8	20	3954	16¾	16¼	16⅜	− ½
	32½	16⅜	ShawmutNtl	SNC	.88	2.8	19	2546	31½	31¼	31½	...
n	28	25	ShawmutNtl dep pfX		1.10e	4.1	...	56	27⅛	26⅞	27	− ⅛
	27⅜	24⅜	ShawmutNtl pf		2.33	8.9	...	231	26⅛	25⅞	26⅛	+ ⅛
	10½	2	ShawmutNtl wt			...	...	11	9¾	9⅝	9⅝	− ¼
	13¾	7½	ShelbyWill	SY	.28	2.2	74	106	12½	12⅛	12½	− ⅛
	75⅝	63¾	ShellTrans	SC	3.21e	4.4	16	29	73	72½	72⅞	+ ¼
	38	29¾	♣SherwinWill	SHW	.64	1.8	16	3180	36⅛	35½	35⅞	− ⅛
	8¾	5⅛	Sherwood	SHD		...	9	54	8½	8¼	8⅜	+ ⅛
	15⅞	9⅛	Shoneys	SHN		...	10	505	12⅞	12¾	12¾	...
	13½	8⅝	ShopkoStr	SKO	.44	3.5	11	1267	13⅛	12½	12¾	− ½
	19	11½	Showboat	SBO	.10	.5	20	181	18¾	18⅜	18⅜	− ⅛
	24⅜	17¾	ShurgardStor	SHU	1.84	7.7	...	671	24	23⅜	23¾	+ ¼
	33⅝	22⅛	♣SierraHlth	SIE		...	14	533	24⅝	23½	24½	+ ⅞
	21⅞	18	SierraPac	SRP	1.12	5.2	12	484	21⅝	21¼	21½	− ⅛
	8⅛	4	SignlAprl	SIA		...	dd	11	5½	5½	5½	...
s▲	25½	13½	SignetBk	SBK	.68	2.7	11	3236	25⅝	24¾	25	− ½
▲	45⅜	21¼	SiliconGrph	SGI		...	35	6698	45⅝	44⅛	44⅝	+ ¼
	27⅞	22½	♣SimonPrpty	SPG	1.97	7.9	27	2320	25¼	24½	24⅞	− ¼
s	17⅜	12⅛	SingaporeFd	SGF	2.89e	17.8	...	128	16½	16¼	16¼	− ¼
	34⅜	22	♣Singer	SEW	.20	.7	14	346	27	26	26¾	− ⅛
n	11⅜	8½	SinterMetal A	SNM		...	...	38	11⅜	11⅛	11¼	...
	12½	7	SitheEngys	SYT		...	19	315	9⅛	9⅛	9⅛	− ⅛
	13½	9¼	♣SizelerProp	SIZ	1.12	11.1	16	300	10⅛	10	10⅛	...
	7⅛	5	SizzlerInt	SZ	.16	2.7	24	261	6⅛	5⅞	5⅞	− ⅛
	22⅛	16⅜	Skyline	SKY	.48	2.6	14	273	18⅝	18⅜	18½	...
	17¼	13⅝	SmartFinal	SMF	.20	1.2	19	359	16⅝	16½	16⅝	+ ⅛
	29⅞	19⅛	SmithAO	AOS	.60	2.2	10	397	28⅛	27½	27½	− ⅝
	26⅛	21¾	SmithResdntl	SRW	1.90	7.9	...	14	24	23⅞	24	...
	4⅞	11/16	vjSmithCorona	SCO		...	...	1708	¾	11/16	¾	+ 1/16
	19⅜	11	♣SmithInt	SII		...	16	1131	16⅜	16⅛	16¼	...
	27⅝	19⅛	SmithFood	SFD	.60	3.1	11	580	19⅝	19¼	19⅜	...
▲	46⅛	30⅝	♣SmithklBeech	SBH	1.30e	2.8	...	237	46⅝	46	46⅝	+ ¾
	45⅜	27⅝	♣SmithklBeech eq	SBE	1.30e	2.9	...	3222	45⅜	44⅛	45¼	+1
	25	20¾	Smucker A	SJMA	.52f	2.3	18	59	22¼	21¾	22¼	+ ¼
	22¾	18¾	Smucker B	SJMB	.52f	2.6	16	30	19¾	19⅝	19⅝	...
	40⅞	29	SnapOn	SNA	1.08	2.7	17	524	40¾	40⅝	40¾	...
	19⅛	11⅝	♣SnyderOil	SNY	.26	2.2	dd	4120	12	11⅝	12	+ ⅜
	26	18⅞	♣SnyderOil pfA		1.50	7.7	...	3441	19⅜	19	19⅜	+ ¼
	25¼	10⅝	♣SofmorDk	SDG		...	dd	2631	21	20¼	20⅝	− ⅛

Source: The Wall Street Journal, July 14, 1995.

On a typical day of trading at the **New York Stock Exchange,** the scene seems to be one of confusion and chaos. Most of the people assembled here, however, are involved in an enterprise that requires a high degree of sophistication and expertise.

According to David A. Vise, in the *1988 World Book Year Book*, it all began in 1792 when a small group of merchants and auctioneers first met in the Wall Street area to trade stocks. Today, about 50 million Americans invest in stocks, either directly or through mutual funds. If we count investments held by pension funds, banks, and other institutions, well over 100 million Americans are represented in the stock market. The combined value of the stocks listed on the NYSE (more than 1,500 of them) is over $2 trillion.

To read the listings in the table, first find the line corresponding to the company whose stock interests you. Next to the company name (which may be abbreviated) is the company's unique ticker symbol (a code consisting of from one to several capital letters). For example, the symbol for Albertsons is ABS.

$$32\frac{1}{2} \quad 26 \quad \text{Albertsons} \quad \text{ABS} \quad .52 \quad 1.7 \quad 18 \quad 3036 \quad 30\frac{1}{4} \quad 29\frac{3}{4} \quad 30 \quad -\frac{1}{4}$$

The headings at the tops of the columns indicate the meanings of the numbers across the row.

The numbers 32 1/2 and 26 in front of the company's name show that 32 1/2 (or $32.50) is the highest price that the stock has reached in the last 52 weeks, while 26 (or $26.00) is the lowest. (Over the years, it has been common to give stock prices to the nearest eighth of a dollar.) After the symbol of the company is .52, showing that the company is currently paying a dividend of 52¢ per share of stock. As mentioned above, a dividend is money paid by a company to owners of its stock. Dividends go up when the company is doing well and go down when business is bad. Then comes 1.7, the **current yield** on the company's stock, in percent. The dividend of 52¢ per share represents 1.7% of the current purchase price of the stock.

The 18 after the 1.7 is the price-earnings ratio, the purchase price per share divided by the earnings per share. After 18 is 3036, the number of shares sold that day, in hundreds. On the day in question, a total of

$$3036 \times 100 = 303{,}600$$

shares of Albertsons stock were sold. After 3036 comes a list of the prices reached by the stock during the day. The first number, 30 1/4, shows the highest price reached by the stock during the day, while 29 3/4 was the lowest. The next number, 30, says that the stock closed for the day at $30.00 per share. This last price is −1/4, or $.25 per share lower, than the last price of the previous day.

EXAMPLE 1 Use the stock table to find the required amounts below.

(a) The highest price for the last 52 weeks for Aetna Life & Casualty (AET)
Find the correct line from the stock table.

$$64\frac{7}{8} \quad 42\frac{1}{4} \quad \text{AetnaLife} \quad \text{AET} \quad 2.76 \quad 4.4 \quad 12 \quad 3539 \quad 64\frac{3}{8} \quad 62\frac{5}{8} \quad 63 \quad -1\frac{1}{4}$$

The highest price for this stock for the last 52 weeks is 64 7/8, or $64.875, the left-most number.

(b) The dividend for Sherwin Williams (SHW)

The dividend is given just after the company symbol. From the table, find the dividend for Sherwin Williams to be $.64, or 64¢. (Notice the special symbol ♣ preceding the company name in the listing. This means that Sherwin Williams is one of a number of companies whose annual reports can be obtained free by readers of the *Wall Street Journal*.) ■

Wall Street Games
Beginning in November 1988, about 15,000 students in some 1,000 community colleges, four-year colleges, and graduate programs competed in the first AT&T Collegiate Investment Challenge, a promotional program for the educational board game *Wall Street Games.* (Maybe you were one of the competitors.)

Starting with $500,000 in play money, a computerized network with the latest stock prices, and a toll-free telephone number to place orders, the students played the market for a 4-month period. First and second finishers won $25,000 and $10,000 respectively (in *real* money). Other prizes were a trip to New York, a week in the Bahamas, watches, and athletic shoes.

EXAMPLE 2 Find the cost for 100 shares of Sears stock, at the high price for the day.

Find the total cost of a stock purchase by multiplying the price per share times the number of shares purchased. From the stock table, the high price for the day for Sears was $32 3/8 per share. First change $32 3/8 to the decimal $32.375, then multiply.

$$\text{Total cost} = 100 \times \$32.375 = \$3,237.50$$

One hundred shares of this stock would cost $3,237.50, plus any broker's fees.

You may have noticed that the Sears listing in the table appeared in boldface print. This feature distinguishes a stock whose price changed by 5% or more (as long as the previous day's closing price was at least $2). Various other special features occur throughout the listings and are explained where the listings are published. Here are a few additional examples.

1. An underlined row in the listings indicates a stock with an especially high volume of trading for the day relative to its usual volume. In the case of the NYSE listings, the 40 largest volume percentage leaders are indicated in this way.
2. A "pf" between the company name and symbol indicates a "preferred" stock, a special category of stock issued by some companies in addition to their ordinary, or "common" stock. A company must pay dividends to holders of its preferred stock before it can pay dividends to holders of its common stock. Additional advantages of preferred stocks include attractive yields, stable dividend income, and tax advantages for corporate investors.
3. The symbol ▲ at the left indicates a new 52-week high, while the symbol ▼ indicates a new 52-week low.

Finding Commissions

To buy or sell a stock, it is necessary to use a broker. The broker has representatives at the exchange who will execute a buyer's order. The broker will charge a **commission** for executing an order. Commission rates formerly were set by stock exchange rules and did not vary from broker to broker. Since 1974, however, commissions are competitive and vary considerably among brokers.

In recent years, **discount brokers** have become popular. These brokers merely buy and sell stock, and offer little in the way of additional services, such as research or suggestions on stocks to buy or sell.

The rates charged depend on whether the order is for a **round lot** of shares (multiples of 100) or an **odd lot** (any portion of an order for fewer than 100 shares). Odd lot orders often involve an **odd-lot differential:** an extra charge of $1/8, or 12.5¢, per share.

Typical expenses in buying and selling stock through a discount broker are as follows. (There is often a minimum commission charge.)

Buying Stock	***Selling Stock***
Broker's commission of 2.5% of purchase price plus any odd-lot differential	Broker's commission of 2.5% of selling price plus any odd-lot differential plus SEC fee (see below)

The SEC fee is set by the Securities and Exchange Commission (the SEC), a federal agency that regulates stock markets. The fee goes toward funding the activities of the SEC and is assessed on stock sales only (not purchases). It is approximately 1¢ per $300 in value (or fraction of $300). For example, to approximate the fee for a sale of $1,600, first divide $1,600 by $300:

$$\frac{\$1{,}600}{\$300} \approx 5.3.$$

Since the fee is 1¢ per $300, or *fraction of $300,* round 5.3 *up* to 6. The fee is then approximately 6 × 1¢ = 6¢.

EXAMPLE 3 Ranieka Brooks bought 75 shares of stock, paying $26 3/4 per share. Find her total cost for the purchase.

The basic cost of the stock is found by multiplying the price per share and the number of shares:

$$\begin{aligned}\text{Price per share} \times \text{number of shares} &= \$26\ 3/4 \times 75\\ &= \$26.75 \times 75 = \$2{,}006.25.\end{aligned}$$

The broker's commission is 2.5% of this amount, or

$$\begin{aligned}\text{Broker's commission} &= 2.5\% \times \$2{,}006.25\\ &= .025 \times \$2{,}006.25 = \$50.16.\end{aligned}$$

Since Brooks bought 75 shares (and not a multiple of 100 shares) she must pay an odd-lot differential of $1/8, or 12.5¢, per share. For 75 shares, this odd-lot differential amounts to

$$75 \text{ shares} \times \$1/8 \text{ per share} = \$9.38.$$

Find the total cost of the shares of stock as follows.

$2,006.25	Basic cost
50.16	Broker's commission
9.38	Odd-lot differential
$2,065.79	

EXAMPLE 4 Find the amount received by a person selling 700 shares of a stock at $63 5/8 per share.

The basic price of the stock is

$$700 \times \$63\ 5/8 = \$44{,}537.50.$$

Find the SEC fee as described above, First,

$$\frac{\$44{,}537.50}{\$300} = 148.458.$$

Round 148.458 up to 149. The SEC fee is then

$$149 \times 1¢ = \$1.49.$$

The broker's commission is 2.5% of the basic price, or

$$.025 \times \$44,537.50 = \$1,113.44.$$

Find the seller's proceeds as follows.

\$44,537.50	Basic price
− 1,113.44	Broker's fee
− 1.49	SEC fee
\$43,422.57	

There is no odd-lot differential since the number of shares is a multiple of 100. ■

Stock prices have historically followed a general upward trend, but with lots of variation within that general trend. A time of increasing prices is called a *bull* market while a time of decreasing prices is a *bear* market.

A strong bull market that began in August, 1982 (with the average more than tripling over 5 years) suddenly came to a crashing halt on October 19, 1987 (the day now known as *Black Monday*). The Dow Jones Industrial Average dropped 508 points on that single day, losing 22.6% of its value. Even the crash of October 28, 1929, which precipitated the Great Depression, was only a 12.8% drop. The 1987 crash was followed by a shaky period, but soon another period of gain was established and investors began to regain their confidence in the market. By January 24, 1989, some 15 months after Black Monday, the Dow experienced its first closing mark that exceeded the closing price immediately preceding Black Monday. Figure 3 shows a graphical depiction of the effect of Black Monday within a context of about 15 years.

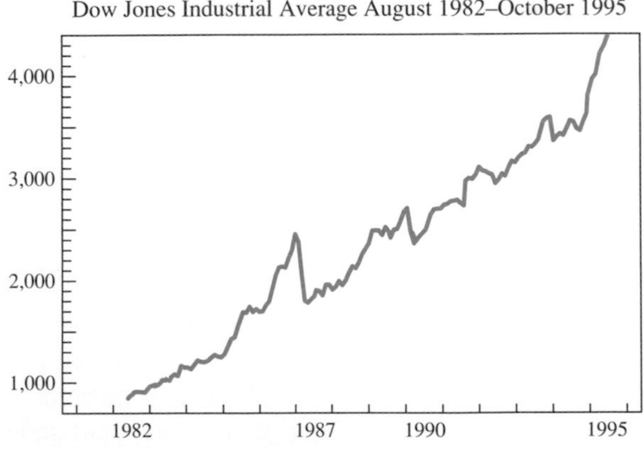

■ **FIGURE 3**

The reasons for the 1987 collapse have been debated, and new policies and regulations have been adopted in an attempt to curb such drastic occurrences in the future. However, market fluctuations apparently depend upon many factors, not the least of which is the prevailing mood of investors, which in turn stem from a variety of causes, some known and some unknown.

Free copies of any of the following investment brochures can be obtained by writing to:
Consumer Brochures
Investment Company Institute
P.O. Box 27850
Washington, DC 20038–7850

What Is a Mutual Fund?
Eight Basics of Bond Fund Investing
Money Market Mutual Funds
Reading the Mutual Fund Prospectus
Planning for College? (on employing mutual funds)
Discipline (on dollar-cost averaging)
Investing: Start Now!
Time to Celebrate Life (on retirement planning)
Retirement Plan Distributions
A Close Look at Closed-End Funds
On Track With Unit Investment Trusts

Mutual Funds

Trying to decide which stocks to buy is not easy. And after buying a stock, there is always the question of whether and when to sell. Most of us have known of people who bought a stock for no special reason, paid no attention to it for years, and then discovered it had done very well for them. In general, however, that scenario is the exception rather than the rule. When dealing in the stock of a particular company, it is wise to stay informed about the progress of that company in order to avoid serious losses that may be predictable. Therefore, owning stocks can take a lot of time.

To avoid these problems, many people buy **mutual funds.** In a mutual fund the money committed by small investors is pooled, to be used by the managers of the fund to buy stocks. In theory, the full-time professional manager of the fund can carefully watch the market and buy and sell at just the right time.

In practice, however, there is no proof that the managers of mutual funds are any smarter than anybody else. The records of mutual funds are not necessarily better than random selection of stocks. However, mutual funds do offer certain advantages:

1. Stocks may be purchased on a monthly payment plan.
2. Dividends can be given to you in cash or in the form of additional shares (called "reinvestment of dividends").
3. Diversification can be achieved with modest investment amounts. (A mutual fund normally has no more than four or five percent of its money in a single stock.)
4. Many funds sell shares directly with no commission charge. (However, this advantage may be countered by higher "load factors," or management fees.)

There are presently several thousand sizable mutual funds, whose data are reported daily by the National Association of Securities Dealers (NASD) in extensive tables similar to the stock table illustrated earlier. The listings report fund investment objectives, short and long term returns, and daily changes in fund values, rankings, and so forth. Some funds invest in stocks only, some contain bonds only, and others are a blended mix of stocks and bonds.

For Further Thought
Summary of Investments

The following summary of various types of investments is based on average cases and is intended only as a general comparison of investment opportunities. There are numerous exceptions to the characteristics shown. For example, though the chart shows no selling fees for mutual funds, this is really only true for the so-called "no-load" funds. The "loaded" funds do charge "early redemption fees" or other kinds of sales charges. That is not to say that a no-load fund is necessarily better. For example, the absence of sales fees may be offset by higher "management fees."

These various costs, as well as other important factors, will be clearly specified in a fund's prospectus and should be studied carefully before a purchase decision is made. Some expertise is not entirely unnecessary for mutual fund investing, just less so than with certain other options, like direct stock purchases, raw land, or collectibles.

To use the chart, read across the columns to find the general characteristics of the investment. Liquidity refers to the ability to get cash from the investment quickly.

Investment	Protection of Principal	Protection Against Inflation	Rate of Return (%)	Certainty of Return	Selling Fees	Liquidity	Long-term Growth	Requires Work From You	Expert Knowledge Required
Cash	excellent	none	0			excellent	no	no	no
Ordinary life insurance	good	poor	2–4	excellent	high	excellent	no	no	no
Savings bonds	excellent	poor	4–5	excellent	none	excellent	no	no	no
Bank savings	excellent	poor	2–4	excellent	none	excellent	no	no	no
Credit union	excellent	poor	3–4	excellent	none	excellent	no	no	no
Corporation bonds	good	poor	6–8	excellent	low	good	no	some	some
Tax-free municipal bonds	excellent	poor	3–6	excellent	medium	good	no	no	some
Corporation stock	good	good	4–12	good	medium	good	yes	some	some
Preferred stock	good	poor	5–8	excellent	medium	good	no	some	some
Mutual funds	good	good	4–12	good	none	good	yes	no	no
Your own home	good	good	4–10	good	medium	poor	yes	yes	no
Mortgages on the homes of others	fair	poor	5–8	fair	high	poor	no	yes	some
Raw land	fair	good	?	fair	medium	poor	yes	no	yes
Rental properties	good	good	4–8	fair	medium	poor	yes	yes	some
Stamps, coins, antiques	fair	good	0–10	fair	high	poor	yes	yes	yes

For Group Discussion

List **(a)** some investment types you would recommend considering and **(b)** some you would recommend avoiding for a friend whose main investment objective is as follows.

1. Avoid losing money.
2. Avoid losing purchasing power.
3. Avoid having to learn about investments.
4. Avoid having to work on investments.
5. Be sure to realize a gain.
6. Be able to "cash in" at any time.
7. Get as high a return as possible.
8. Be able to "cash in" without paying fees.
9. Realize quick profits.
10. Realize growth over the long term.

EXERCISES

Find each of the following from the stock table in the text. Give money answers in dollars.

1. high for the day for Alcan Aluminum (AL)
2. low for the day for Airborne Freight (ABF)
3. closing price for Singer Company (SEW)
4. change from the previous day for Sizzler International (SZ)
5. 52-week high for Agnico-Eagle Mines (AEM)
6. 52-week low for Sears (S) (*not* Sears preferred)
7. dividend for Aetna Life & Casualty (AET)
8. dividend for Snap-on Incorporated (SNA)
9. low for the day for Alex Brown (AB)
10. high for the day for Sherwin Williams (SHW)
11. sales for the day for Airtouch Communications (ATI)
12. sales for the day for Scott Paper (SPP)
13. yield for American Brands (AMB)
14. yield for Snyder Oil Corporation (SNY) (*not* the preferred stock)

Find the cost for each of the following stock purchases. Ignore any broker's fees.

Stock	Symbol	Number of Shares	Transaction Price
15. Alcoa	AA	500	close
16. Allstate	ALL	300	low
17. Showboat	SBO	400	low
18. Seagram	VO	600	low

Find the cost for each of the following stock purchases. Include typical discount broker's fees as described in the text, and any odd-lot differentials (charged only on the odd-lot portion of the order).

Stock	Symbol	Number of Shares	Transaction Price
19. Alaska Air Group	ALK	40	low
20. Shoneys	SHN	50	high
21. Sierra Pacific	SRP	425	close
22. Albemarle Corporation	ALB	635	low

Find the amount received by the sellers of the following stocks. Use the sales expenses described in the text. (Note that each of Exercises 27–30 involves an odd-lot differential on a portion of the sale.)

23. 500 shares at 32 7/8
24. 700 shares at 51 1/4
25. 1,400 shares at 8 7/8
26. 800 shares at 41 3/8
27. 420 shares at 50 1/4
28. 660 shares at 62 1/8
29. 950 shares at 29 3/4
30. 1,220 shares at 18 5/8

Refer to the stock table in the text to work the following problems.

31. Ricardo Espinoza bought 70 shares of Albertsons (ABS) and 120 shares of Sears (S), both at the low price for the day. Find the total cost of this purchase including broker's commissions and any odd-lot differentials.

32. Betty Harrison purchased 20 shares of Alco Standard (ASN) and 40 shares of American Eagle (FLI), both at the closing price for the day. Find her total cost, including broker's commissions and any odd-lot differentials.

33. Margaret Decker bought 200 shares of Sears (preferred) at the low for the day and sold 140 shares of American Brands (AMB) at the high for the day. Considering all applicable expenses, what *net* amount did she take in or pay out for these combined transactions?

34. Mehta Farak sold 125 shares of Allstate (ALL) and bought 90 shares of Alcoa (AA), both at the low price of the day. Including all applicable expenses, what *net* amount did Mehta take in or pay out for these transactions?

Write a short essay in response to each of the following. Some research may be required.

35. Describe the concept of "dollar-cost averaging," and relate it to advantage number 1 of mutual funds as listed in the text.

36. Discuss advantage number 3 of mutual funds as listed in the text.

OLLABORATIVE INVESTIGATION

To Buy or to Rent?

Divide your class into groups of at least four students each. Each group is to first read the following.

Mark and Emily, a young couple with a child, live in a rented apartment. After taxes, Mark earns $24,000 per year and Emily's part-time job brings in $6,000 per year. In the foreseeable future, their earnings will probably just keep pace with inflation. They are presently operating on the following monthly budget.

Rent	$700
Food	400
Daycare	350
Clothing	280
Utilities	80
Entertainment	200
Savings	300
Other	190

They have accumulated $19,000 in savings.

Having found a house they would like to buy, and having consulted several lenders, they have discovered that, at best, purchasing the house would involve costs as shown here.

Price of house	$85,000
Down payment required (20%)	______
Closing costs (required at closing):	
Loan fee (1 point)	______
Appraisal fee	200
Title insurance premium	250
Document & recording fees	45
Prorated property taxes	430
Prorated fire insurance premium	190
Other	150
Total:	______
Fixed-rate 30-year mortgage at 9.5%	
Annual taxes:	$900
Annual fire insurance:	$360

In order to own their own home, Mark and Emily are willing to cut their clothing and entertainment expenditures by 50% each, and could decrease their savings allotments by 30%. (Some savings will still be necessary to provide for additional furnishings they would need, and for expenses of an expanding family in the future.) They also figure that the new and yard would require $70 monthly for maintenance and that utilities will be twice what they have been in the apartment. Other than that, their present budget allotments would remain the same.

Now divide your group into two "subgroups." The first subgroup is to answer these questions:

1. What amount would the home purchase cost Mark and Emily in immediate expenditures?
2. Do they have enough cash on hand?

The second subgroup is to answer these questions:

3. What amount would the house cost Mark and Emily on an ongoing monthly basis?
4. What amount will their monthly budget allow toward this cost?
5. Can they meet the monthly expenses of owning the house?

Within your group, decide whether Mark and Emily can afford to purchase the house. Select a representative to report your findings to the class.

Compare the evaluations of the various groups, and try to resolve any discrepancies.

Chapter 13 Test

Find all monetary answers to the nearest cent. Use tables from the chapter if necessary.

Find the future value of each of the following deposits.

1. $100 for 5 years at 6% simple interest
2. $50 for 2 years at 8% compounded quarterly
3. Find the effective annual yield to the nearest hundredth of a percent for an account paying 6% compounded monthly.
4. Use the rule of 70 to estimate the years to double at an inflation rate of 12%.
5. What amount deposited today in an account paying 10% compounded semiannually would grow to $100,000 in 10 years?
6. Darcy Randal's MasterCard statement shows an average daily balance of $680. Find the interest due if the rate is 1.6% per month.

Greg McRill buys a turquoise necklace for his wife on their anniversary. He pays $4,000 for the necklace with $1,000 down. The dealer charges add-on interest of 8% per year. McRill agrees to make payments for 24 months.

7. Find the total amount of interest he will pay.
8. Find the monthly payment.
9. Find the APR value for this loan (to the nearest half percent).
10. Find the unearned interest if he repays the loan in full with six payments remaining. Use the most accurate method available.
11. Newark Hardware wants to include financing terms in their advertising. If the price of a floor waxer is $150 and the finance charge with no down payment is $7 over a 6-month period (six equal monthly payments), find the true annual percentage rate (APR).
12. Explain what a mutual fund is and what its advantages are.

Find the monthly payment required for each of the following home loans.

13. The mortgage is for $50,000 at a fixed rate of 12% for 20 years. Amortize principal and interest only.
14. The purchase price of the home is $88,000. The down payment is 20%. Interest is at 10.5% fixed for a term of 30 years. Annual taxes are $900 and annual insurance is $600.
15. If the lender in Exercise 14 charges two "points," how much does that add to the cost of the loan?
16. Explain in general what "closing costs" are. Are they different from "settlement charges"?
17. To buy your home you obtain a 1-year ARM with 2.25% margin and a 2% periodic rate cap. The index starts at 7.85% but has increased to 10.05% by the first adjustment date. What interest rate will you pay during the second year?

Refer to the stock table in Section 5 of this chapter to find the following.

18. the 52-week high for Airlease (FLY)
19. the sales for the day for Allied Products (ADP)
20. the cost to buy 1,500 shares of Sherwood (SHD) at the low price of the day, including a typical discount broker's commission as described in the chapter

14

Matrices and Their Applications

Although cryptography has been used by the military for centuries, only recently has the public found a need for it. Because of the large amount of electronic communication today, businesses, banks, and private individuals use cryptography to keep information secret. The process of disguising plain text into a cipher is called encryption. When a message is received, the cipher must be decoded into the original plain text using decryption.

Codes that encrypt a letter of the alphabet in the same way each time are usually easy to break. Suppose a simple code changes letters into numbers using the following technique: A = 01, B = 02, C = 03, . . . , Z = 26. After a few messages a person may be able to decipher the code because letters in the English language do not occur with equal likelihood. The most frequently used letter in the alphabet is E, which occurs approximately 13% of the time, whereas the least frequent is Z, which occurs only .1% of the time. One can quickly determine which number is associated with E and which one is associated with Z. According to one sample, the percentage *P* of occurrence for the most frequent letters in English text are as shown in the table.

Letter	*P*	Letter	*P*	Letter	*P*
E	13.0%	T	9.3%	N	7.8%
R	7.7%	O	7.4%	I	7.4%
A	7.3%	S	6.3%	D	4.4%

Matrices can be used to create sophisticated encryption methods that code blocks of letters rather than a single letter at one time. As a result, the letter E would be coded differently in the words THE and EAT. To decipher the message, the inverse of the matrix is required.

Sources: Schneier, B., *Applied Cryptography: Protocols, Algorithms, and Source Code in C,* John Wiley & Sons, Inc., 1994.

Sinkov, A., *Elementary Cryptanalysis: A Mathematical Approach,* Random House, 1968.

This chapter is concerned with the mathematical object called a *matrix,* which can be defined as follows.

Matrix

A **matrix** is a rectangular array of numbers. (The plural of *matrix* is *matrices.*) The numbers in the array are called the **entries,** or *elements,* of the matrix. Each horizontal line of entries is called a **row,** while each vertical line of entries is called a **column.**

Sometimes a matrix is simply a table, a convenient way of summarizing numerical data. Other times, it may reveal numerical patterns or show special relationships between two quantities. Examples already seen elsewhere in this book include magic squares, number base tables, addition and multiplication tables, probability and frequency distributions, compound interest tables, amortization schedules, and stock market listings. As you will see in this chapter, when a rectangular array is thought of as a matrix, particular symbols, notation, and manipulation techniques are utilized.

As an example, suppose you run the snack bar at the campus cinema and want to keep track of the revenue from your various snacks. For the Friday night showings of *Casablanca* with Humphrey Bogart, you list the dollar amounts as in Table 1.

TABLE 1 Friday Night Snack Bar Revenues

	Candy	Popcorn	Soft drinks
7 P.M. show	\$0	\$12	\$10
9 P.M. show	\$5	\$8	\$9

Following common mathematical practice, you can convey the information of this table in a matrix, enclosing the entries in brackets and denoting the matrix with an appropriate capital letter (in this case F, for Friday) as follows.

$$F = \begin{array}{l} \textbf{7 P.M. show} \\ \textbf{9 P.M. show} \end{array} \overset{\begin{array}{ccc} \text{Candy} & \text{Popcorn} & \text{Drinks} \\ \downarrow & \downarrow & \downarrow \end{array}}{\begin{bmatrix} 0 & 12 & 10 \\ 5 & 8 & 9 \end{bmatrix}}$$

For example, the specific entry in row 1, column 3, namely 10, indicates that the soft drink revenue at the 7 P.M. show was \$10.

Our main interest in this chapter will be to develop some algebraic operations on matrices, which make them useful in practical applications. We will see matrix F again in Section 1, where matrices will be added, subtracted, and multiplied by numbers. Section 2 involves multiplying matrices by other matrices. In Sections 3 and 4, we will discover how matrices can be used to solve systems of linear equations. Solving such systems was an important motivation for the English mathematicians who, in the 1850s, originated the development of matrix theory. These men were Arthur Cayley (1821–1895) and James Sylvester (1814–1897). (Linear systems were solved by the *elimination method* in an earlier chapter.) Section 5 of the chapter will introduce some basic ideas from *game theory,* a practical modern branch of mathematics.

Throughout the chapter we will discover various properties of matrices as a mathematical system. Some of these will probably be what you would expect, while others may surprise you.

14.1 Basic Operations on Matrices

Recall from the definition above that a matrix is a rectangular array of numbers. The number of rows and columns in a matrix determines its *size,* or its **order.** For example, we say that matrix F above is a 2×3 (read "2 by 3") matrix, or that it has order 2×3. (Notice that, when specifying the order of a matrix, we give the number of rows first and then the number of columns.)

If a matrix has the same number of rows as columns, it is called a **square matrix.** A matrix with only one row is called a **row matrix.** A matrix with only one column is called a **column matrix.** A matrix in which all entries are zero is called a **zero matrix.**

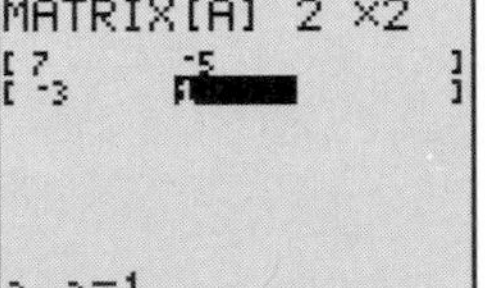

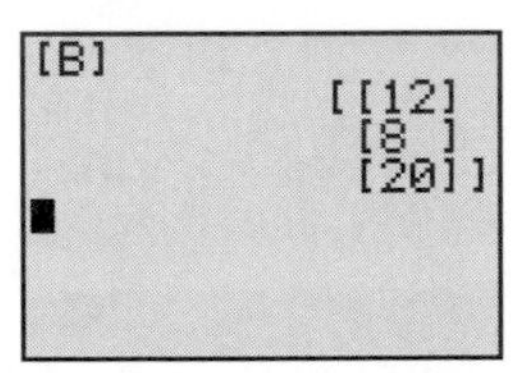

The figures above show how matrices A and B of Example 1 appear on a graphing calculator screen.

EXAMPLE 1 Give the order of each of the following matrices, and identify any that are square matrices, row matrices, column matrices, or zero matrices.

$$A = \begin{bmatrix} 7 & -5 \\ -3 & 1 \end{bmatrix} \quad B = \begin{bmatrix} 12 \\ 8 \\ 20 \end{bmatrix} \quad C = \begin{bmatrix} 1 & 4 \\ 9 & 5 \\ 6 & 2 \end{bmatrix} \quad D = \begin{bmatrix} -5 & 0 & 0 \\ 2 & 3 & 1 \\ 9 & 4 & 0 \end{bmatrix}$$

$$E = \begin{bmatrix} 10 & 50 & 30 \end{bmatrix} \quad F = \begin{bmatrix} 0 & 0 & 0 \\ 0 & 0 & 0 \end{bmatrix} \quad G = \begin{bmatrix} 6 \end{bmatrix}$$

The orders of the seven given matrices are as follows.

Matrix	A	B	C	D	E	F	G
Order	2×2	3×1	3×2	3×3	1×3	2×3	1×1

A, D, and G are square matrices. E and G are row matrices. B and G are column matrices. F is a zero matrix. ◆

Notice that matrix G (in Example 1) is both a row matrix and a column matrix. Notice also that, even though G has only a single entry, G is a matrix and not a number. Pure numbers are commonly distinguished from matrices by referring to the numbers as *scalars.* For example, 6 is a scalar, whereas [6] is a matrix. This distinction is sometimes important, so the brackets must always be included around the entries of a matrix.

Equal Matrices

Two matrices are said to be **equal** if they have the same order and if all corresponding entries are equal.

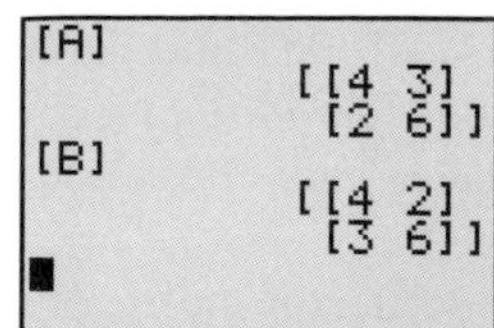

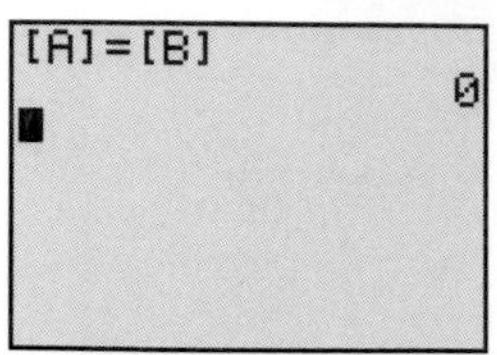

The fact that matrices A and B are not equal is illustrated by the 0 returned by the calculator in the bottom screen. Had they been equal, a 1 would have been returned.

EXAMPLE 2 In each case, determine whether matrices A and B are equal.

(a) $A = \begin{bmatrix} 4 & 3 \\ 2 & 6 \end{bmatrix} \quad B = \begin{bmatrix} 4 & 2 \\ 3 & 6 \end{bmatrix}$

Even though A and B have the same order (2×2) and include the same numbers as entries, not all *corresponding* entries are equal. For example, the row 2, column 1 entry of matrix A is 2, while the corresponding entry of matrix B is 3. Matrices A and B are not equal.

(b) $A = \begin{bmatrix} 1 & 2 & 3 \\ 6 & 5 & 1 \end{bmatrix} \quad B = \begin{bmatrix} 1 & 2 \\ 6 & 5 \end{bmatrix}$

Matrix A has order 2×3, while B has order 2×2. Since they do not have the same order, matrices A and B are not equal.

(c) $A = [12 \quad -2 \quad 4] \quad B = [12 \quad -2 \quad 4]$

A and B have the same order (1×3) and all corresponding entries are equal, so matrices A and B are equal.

In the remainder of this section, we introduce three operations on matrices: addition, subtraction, and scalar multiplication. In each case the result is another matrix. Addition (and subtraction) can only be done when the two matrices to be added (or subtracted) have the same order. If a matrix operation can be done, we say the result is **computable.** However, if any condition (such as different orders) makes the operation impossible, we say the result is **not computable.**

[A]
[[2 -1 3 2]
[1 0 4 2]]
[B]
[[5 6 -2 5]
[3 -1 4 7]]

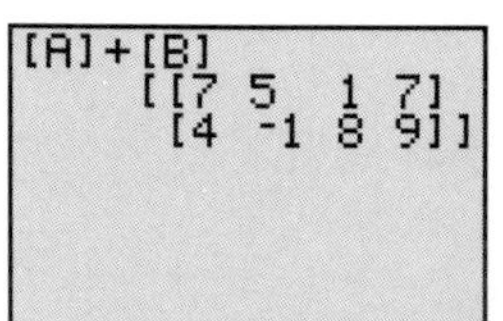

The result of Example 3 is supported in these two screens.

Addition of Matrices

The **sum** of two matrices of the same order is found by *adding* corresponding entries. If two matrices have different orders, their sum is not computable.

EXAMPLE 3 Let $A = \begin{bmatrix} 2 & -1 & 3 & 2 \\ 1 & 0 & 4 & 2 \end{bmatrix}$ and $B = \begin{bmatrix} 5 & 6 & -2 & 5 \\ 3 & -1 & 4 & 7 \end{bmatrix}$. Find $A + B$.

Add the corresponding entries.

$$\begin{aligned} A + B &= \begin{bmatrix} 2 & -1 & 3 & 2 \\ 1 & 0 & 4 & 2 \end{bmatrix} + \begin{bmatrix} 5 & 6 & -2 & 5 \\ 3 & -1 & 4 & 7 \end{bmatrix} \\ &= \begin{bmatrix} 2+5 & -1+6 & 3+(-2) & 2+5 \\ 1+3 & 0+(-1) & 4+4 & 2+7 \end{bmatrix} \\ &= \begin{bmatrix} 7 & 5 & 1 & 7 \\ 4 & -1 & 8 & 9 \end{bmatrix} \end{aligned}$$

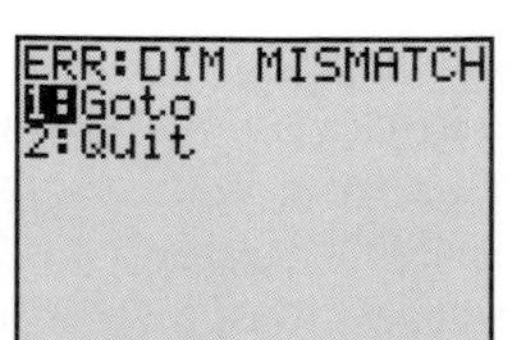

This screen, indicating a dimension (or order) mismatch, was the result of attempting to add matrices A and B as defined in Example 4.

EXAMPLE 4 Let $A = \begin{bmatrix} 2 & 1 \\ 3 & 4 \end{bmatrix}$ and let $B = \begin{bmatrix} 4 & -1 & 0 \\ 2 & 5 & 0 \end{bmatrix}$. Find $A + B$.

Since A is 2×2 and B is 2×3, the matrices have different orders. Their sum is not computable.

EXAMPLE 5 Matrix F from the chapter introduction showed snack bar revenues for the Friday night movie showings. The corresponding Saturday night revenues are shown here.

$$S = \begin{array}{l} \text{7 P.M. show} \\ \text{9 P.M. show} \end{array} \begin{bmatrix} 4 & 15 & 11 \\ 4 & 10 & 10 \end{bmatrix}$$

(Columns: Candy, Popcorn, Drinks)

Derive a matrix that shows the various revenues for the weekend (Friday and Saturday combined).

The sum of matrices F and S will show the required quantities, since it adds corresponding entries in all six matrix positions.

$$\begin{aligned} W &= F + S \\ &= \begin{bmatrix} 0 & 12 & 10 \\ 5 & 8 & 9 \end{bmatrix} + \begin{bmatrix} 4 & 15 & 11 \\ 4 & 10 & 10 \end{bmatrix} \\ &= \begin{bmatrix} 0+4 & 12+15 & 10+11 \\ 5+4 & 8+10 & 9+10 \end{bmatrix} \\ &= \begin{array}{l} \text{7 P.M. show} \\ \text{9 P.M. show} \end{array} \begin{bmatrix} 4 & 27 & 21 \\ 9 & 18 & 19 \end{bmatrix} \end{aligned}$$

(Columns: Candy, Popcorn, Drinks)

Specifically, total weekend popcorn revenues from the 9 P.M. show was $18. (Other similar amounts can also be read from matrix W.)

Subtraction of Matrices

The **difference** of two matrices of the same order is found by *subtracting* corresponding entries. If two matrices have different orders, their difference is not computable.

```
[A]
     [[4  -3]
      [-1 5 ]]
[B]
     [[-2 4]
      [1  5]]
```

```
[A]-[B]
     [[6  -7]
      [-2 0 ]]
```

The result of Example 6 is supported in these two screens.

EXAMPLE 6 Find the indicated difference.

$$\begin{bmatrix} 4 & -3 \\ -1 & 5 \end{bmatrix} - \begin{bmatrix} -2 & 4 \\ 1 & 5 \end{bmatrix} = \begin{bmatrix} 4-(-2) & -3-4 \\ -1-1 & 5-5 \end{bmatrix} = \begin{bmatrix} 6 & -7 \\ -2 & 0 \end{bmatrix}$$

EXAMPLE 7 Refer to the revenue matrices of Example 5. Find a matrix that shows, for each of the six revenue categories, how much more (or less) was taken in on Saturday night than on Friday night. Call this matrix E (for *E*xcess revenue).

Compute the following difference.

$$\begin{aligned} E &= S - F \\ &= \begin{bmatrix} 4 & 15 & 11 \\ 4 & 10 & 10 \end{bmatrix} - \begin{bmatrix} 0 & 12 & 10 \\ 5 & 8 & 9 \end{bmatrix} \\ &= \begin{bmatrix} 4-0 & 15-12 & 11-10 \\ 4-5 & 10-8 & 10-9 \end{bmatrix} \end{aligned}$$

$$= \begin{array}{l} \text{7 P.M. show} \\ \text{9 P.M. show} \end{array} \overset{\begin{array}{ccc} \text{Candy} & \text{Popcorn} & \text{Drinks} \end{array}}{\begin{bmatrix} 4 & 3 & 1 \\ -1 & 2 & 1 \end{bmatrix}}$$

This result shows, for example, that candy sales at the 9 P.M. show were \$1 *less* on Saturday than on Friday. (Revenues in all other categories were *greater* on Saturday.) 

As mentioned earlier, a real number is called a *scalar* in distinction to a matrix. Multiplication of a matrix by a scalar is defined as follows.

Scalar Multiplication of Matrices

The **product of a scalar and a matrix** is found by *multiplying each entry* of the matrix by the scalar.

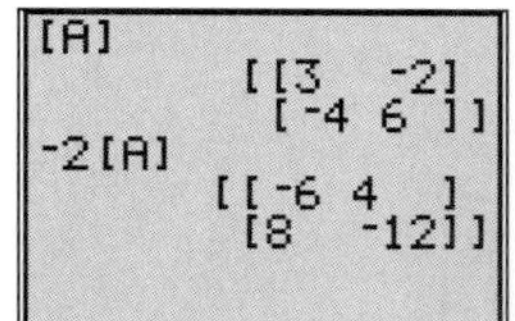

This screen supports the result of Example 8.

EXAMPLE 8 Let $A = \begin{bmatrix} 3 & -2 \\ -4 & 6 \end{bmatrix}$ and let b represent the scalar -2, or $b = -2$. Find bA, that is, $-2A$.

Multiply each element of A by -2.

$$-2A = -2\begin{bmatrix} 3 & -2 \\ -4 & 6 \end{bmatrix} = \begin{bmatrix} -2(3) & -2(-2) \\ -2(-4) & -2(6) \end{bmatrix} = \begin{bmatrix} -6 & 4 \\ 8 & -12 \end{bmatrix}$$

EXAMPLE 9 Refer to Example 5. Suppose the weekend revenues, shown in matrix W, are to be equally divided by two campus clubs sponsoring the sales. Find a matrix showing the amounts each club should get from each of the six revenue categories.

Since there are two clubs, the required amounts will be shown by the following product.

$$\frac{1}{2}W = \frac{1}{2}\begin{bmatrix} 4 & 27 & 21 \\ 9 & 18 & 19 \end{bmatrix} = \begin{array}{l} \text{7 P.M. show} \\ \text{9 P.M. show} \end{array} \overset{\begin{array}{ccc} \text{Candy} & \text{Popcorn} & \text{Drinks} \end{array}}{\begin{bmatrix} 2 & 13.5 & 10.5 \\ 4.5 & 9 & 9.5 \end{bmatrix}}$$

For example, each club gets \$13.50 from the popcorn sales at the 7 P.M. shows.

One of our objectives throughout this chapter will be to observe what we can about the way matrices behave as mathematical entities. We want to notice some properties possessed by matrices and by the operations that apply to them. For example, is addition of matrices commutative? Is it associative? What about subtraction of matrices? What about multiplication?

Problem Solving

Notice that addition, subtraction, and (scalar) multiplication *of matrices* have been defined in this section in terms of addition, subtraction, and multiplication *of real numbers.* Therefore, we should expect to be able to deduce certain properties of these operations on matrices by considering what we already know about the corresponding operations on real numbers. This idea of extending our previous knowledge to new or different situations is a key concept in mathematics (as well as other areas). As you study the following example, try to observe specific problem-solving techniques at work.

EXAMPLE 10 Consider the set of all matrices of the same order. Is addition of matrices, on this set, a *commutative* operation?

The question here pertains to all matrices of "the same order." Let us start by narrowing our attention to a particular order, say 2×3. And let us consider two particular 2×3 matrices, say

$$A = \begin{bmatrix} 1 & 3 & 5 \\ 2 & 4 & 6 \end{bmatrix} \quad \text{and} \quad B = \begin{bmatrix} 2 & 4 & 6 \\ 8 & 10 & 12 \end{bmatrix}.$$

Check to see whether $A + B = B + A$:

$$A + B = \begin{bmatrix} 1+2 & 3+4 & 5+6 \\ 2+8 & 4+10 & 6+12 \end{bmatrix} = \begin{bmatrix} 3 & 7 & 11 \\ 10 & 14 & 18 \end{bmatrix}$$

and

$$B + A = \begin{bmatrix} 2+1 & 4+3 & 6+5 \\ 8+2 & 10+4 & 12+6 \end{bmatrix} = \begin{bmatrix} 3 & 7 & 11 \\ 10 & 14 & 18 \end{bmatrix}.$$

By inspection, these two results are, indeed, equal. On the basis of this result, you may be led to conclude (inductively) that the answer to the original question is "yes." But we really have not verified the result except for the particular matrices A and B. Let us extend our attention to *all* 2×3 matrices by considering matrices

$$C = \begin{bmatrix} a & c & e \\ b & d & f \end{bmatrix} \quad \text{and} \quad D = \begin{bmatrix} g & i & k \\ h & j & l \end{bmatrix}$$

for *any* real numbers $a, b, c, \ldots, l$. Now

$$C + D = \begin{bmatrix} a+g & c+i & e+k \\ b+h & d+j & f+l \end{bmatrix}$$

and

$$D + C = \begin{bmatrix} g+a & i+c & k+e \\ h+b & j+d & l+f \end{bmatrix}.$$

We can now use our previous knowledge that addition is, in fact, commutative on the set of real numbers to conclude that $a + g = g + a, c + i = i + c$, etc., and that therefore $C + D = D + C$. This conclusion is based on *deductive* reasoning, which is totally reliable (unlike inductive reasoning).

The desired result has still only been established for 2×3 matrices (by considering C and D). Our final step is to observe that our deduction with C and D would still be valid for two equal order matrices of *any* order. (Why is this so?) Therefore, addition of matrices, on *any* set of equal order matrices, is indeed commutative. ■

14.1 EXERCISES

Perform the indicated matrix operations. If a required result is not computable, state why not.

1. $\begin{bmatrix} 3 & 2 \\ 5 & 1 \end{bmatrix} + \begin{bmatrix} 8 & -1 \\ 4 & 3 \end{bmatrix}$

2. $\begin{bmatrix} -6 & 9 \\ 2 & 4 \end{bmatrix} + \begin{bmatrix} -2 & 5 \\ -1 & 3 \end{bmatrix}$

3. $\begin{bmatrix} 2 & 8 & -1 \\ 4 & 0 & 3 \end{bmatrix} - \begin{bmatrix} -1 & 5 & 2 \\ 0 & 4 & 3 \end{bmatrix}$

4. $\begin{bmatrix} -1 & 2 & 4 \\ -2 & 3 & 0 \end{bmatrix} - \begin{bmatrix} 2 & -1 & 4 \\ -5 & 6 & 9 \end{bmatrix}$

5. $\begin{bmatrix} -2 & 3 & 1 \\ 4 & 0 & -2 \\ 5 & -1 & 6 \end{bmatrix} + \begin{bmatrix} -1 & 1 & 1 \\ 4 & 2 & 3 \\ -1 & 4 & 7 \end{bmatrix}$

6. $\begin{bmatrix} 3 & 4 & -1 \\ 7 & -8 & 2 \\ 9 & -1 & 3 \end{bmatrix} - \begin{bmatrix} -4 & 2 & -1 \\ 3 & -1 & 8 \\ 4 & 9 & 0 \end{bmatrix}$

7. $\begin{bmatrix} -4 & 3 & 2 \\ -8 & 0 & 1 \\ 4 & 2 & 1 \end{bmatrix} - \begin{bmatrix} 3 & -4 \\ 2 & -3 \\ 8 & 9 \end{bmatrix}$

8. $\begin{bmatrix} -1 & 6 & 2 \\ 3 & 4 & 7 \\ 9 & 8 & 2 \end{bmatrix} + \begin{bmatrix} -4 & 9 & 2 & 0 \\ -3 & 1 & 4 & 0 \\ 8 & 9 & 7 & 0 \end{bmatrix}$

9. $\begin{bmatrix} 3 \\ 4 \\ 2 \end{bmatrix} + \begin{bmatrix} -1 \\ 9 \\ 8 \end{bmatrix} - \begin{bmatrix} 2 \\ 1 \\ 3 \end{bmatrix}$

10. $\begin{bmatrix} 7 \\ 8 \\ 9 \end{bmatrix} + \begin{bmatrix} 1 \\ 9 \\ 3 \end{bmatrix} - \begin{bmatrix} -4 \\ 3 \\ 8 \end{bmatrix}$

11. $\begin{bmatrix} -2 & 4 \\ 0 & 9 \end{bmatrix} + \begin{bmatrix} -8 & 1 \\ 3 & 6 \end{bmatrix} - \begin{bmatrix} 4 & 7 \\ -2 & 5 \end{bmatrix}$

12. $\begin{bmatrix} -3 & 2 \\ 4 & 8 \end{bmatrix} + \begin{bmatrix} -9 & -1 \\ 6 & 3 \end{bmatrix} + \begin{bmatrix} -2 & 5 \\ 0 & 8 \end{bmatrix}$

Let $A = \begin{bmatrix} -4 & 0 \\ 3 & -4 \end{bmatrix}$ *and let* $B = \begin{bmatrix} 2 & -1 \\ -4 & 0 \end{bmatrix}$. *Compute the following.*

13. $A + B$

14. $A - B$

15. $2A$

16. $3B$

17. $-4B$

18. $-5A$

19. $2A + 3B$

20. $-7A + 4B$

21. $(-1)A + A$

A dietician prepares a diet specifying the allowable amounts of four main food groups: group I, meats; group II, fruits and vegetables; group III, breads and starches; and group IV, milk products. Amounts are given in appropriate units (1 ounce for meat, 1/2 cup for fruits and vegetables, 1 slice for bread, and 8 ounces for milk).

22. The numbers of units for breakfast for the four food groups respectively are 2, 1, 2, and 1; for lunch, 3, 2, 2, and 1; and for dinner, 4, 3, 2, and 1. Write a 3×4 matrix to display this information.

23. The amounts of fat, carbohydrates, and protein in the food groups respectively are as follows:

Fat: 5, 0, 0, 10 Carbohydrates: 0, 10, 15, 12 Protein: 7, 1, 2, 8.

Display this information in a 4×3 matrix.

24. Assume there are 8 calories per unit of fat, 4 calories per unit of carbohydrate, and 5 calories per unit of protein. Show these data in a 3×1 matrix.

Don and Connie run a business (D&C Creations) in California building and selling gold rush era tables and chairs. They maintain factories in Sloughhouse (F1) and Nevada City (F2) and ship their products to warehouses in Sacramento (W1), San Francisco (W2), and Reno (W3). The table here shows shipping costs per item from each factory to each warehouse.

Route	Cost Per Item
F1 to W1	$12
F1 to W2	20
F1 to W3	25
F2 to W1	16
F2 to W2	25
F2 to W3	30

25. Construct a 2×3 matrix, C, showing the data for shipping cost per item.

26. On April 1, the freight company that *D&C* utilizes adjusted per item shipping costs as shown here. Construct a 2×3 matrix, A, showing the shipping cost adjustments.

Route	Adjustment
F1 to W1	$3
F1 to W2	−2
F1 to W3	2
F2 to W1	4
F2 to W2	2
F2 to W3	−2

27. Show an appropriate matrix operation whose result gives *D&C*'s shipping costs per item after April 1.

28. Each month, Don and Connie send to each of their warehouses 10 model-A tables, 12 model-B tables, 5 model-C tables, 15 model-A chairs, 20 model-B chairs, and 8 model-C chairs. Exhibit this information in a 2×3 matrix (with tables represented in the top row).

29. The matrix here shows stock on hand at the *D&C* Reno warehouse as of September 1.

$$R = \begin{matrix} & \begin{matrix} A & B & C \end{matrix} \\ \begin{matrix} \text{Tables} \\ \text{Chairs} \end{matrix} & \begin{bmatrix} 45 & 35 & 20 \\ 65 & 40 & 35 \end{bmatrix} \end{matrix}$$

If no stock is sent out of the warehouse during September, how much stock will be on hand October 1? Use the result of Exercise 28 and matrix methods.

30. In Exercise 29, how many model-B chairs were on hand October 1 at the Reno warehouse?

31. In Exercise 29, how many model-C tables were on hand October 1 at the Reno warehouse?

32. The number of each model held on September 1 at the Sacramento and San Francisco warehouses of *D&C* are:

$$S = \begin{bmatrix} 22 & 25 & 38 \\ 31 & 34 & 35 \end{bmatrix} \quad \text{and} \quad F = \begin{bmatrix} 30 & 32 & 28 \\ 43 & 47 & 30 \end{bmatrix}.$$

Find the total inventory in all three warehouses on September 1.

33. Suppose the Sacramento warehouse of *D&C* shipped the following numbers of items during September:

$$K = \begin{bmatrix} 5 & 10 & 8 \\ 11 & 14 & 15 \end{bmatrix}.$$

Find the stock on hand October 1, taking into account the numbers of items received and shipped during the month.

Matrix K shows the weights of four men and four women at the beginning of a diet designed to produce weight loss. Matrix M shows the weights after the diet.

$$K = \begin{matrix} \text{Men} \\ \text{Women} \end{matrix} \begin{bmatrix} 160 & 158 & 172 & 193 \\ 132 & 143 & 119 & 157 \end{bmatrix} \qquad M = \begin{matrix} \text{Men} \\ \text{Women} \end{matrix} \begin{bmatrix} 154 & 148 & 163 & 178 \\ 132 & 154 & 112 & 136 \end{bmatrix}$$

34. Show a matrix operation whose result gives the weight losses of all eight people on the diet.

35. In Exercise 34, how much weight did the third man lose?

36. In Exercise 34, what result did the second woman experience?

Find all real values of the variables that will make each of the following equalities true. In case the given equality is impossible, state why.

37. $\begin{bmatrix} x & y & z \\ 2 & 5 & 1 \end{bmatrix} = \begin{bmatrix} 7 & 2 \\ 4 & 5 \\ 2 & 1 \end{bmatrix}$

38. $\begin{bmatrix} 3 & 4 \\ z & x \end{bmatrix} = [3z \quad 4x]$

39. $\begin{bmatrix} 5 & y \\ z & 2 \end{bmatrix} = \begin{bmatrix} 5 & z \\ y & 2 \end{bmatrix}$

40. $[x \quad y \quad z] = [2x \quad 3y \quad 4z]$

41. $\begin{bmatrix} 6 & y \\ k & m \end{bmatrix} = \begin{bmatrix} 6 & 3 \\ 8 & 1 \end{bmatrix}$

42. $\begin{bmatrix} 5 & 7 \\ 1 & 8 \end{bmatrix} = \begin{bmatrix} a & b \\ 1 & c \end{bmatrix}$

43. $\begin{bmatrix} x+2 & 5 & 9 \\ 1 & 3 & y \end{bmatrix} = \begin{bmatrix} 2x+1 & z-3 & 9 \\ 1 & 3 & 8 \end{bmatrix}$

44. $\begin{bmatrix} 2 & m+5 & a \\ 1 & n-3 & -1 \end{bmatrix} = \begin{bmatrix} 2 & 2m+6 & 2a \\ 1 & 4n-12 & -1 \end{bmatrix}$

45. $\begin{bmatrix} -2 & 6 \\ x & 2 \end{bmatrix} + \begin{bmatrix} x & y \\ 3 & 5 \end{bmatrix} = \begin{bmatrix} 8 & 2 \\ 7 & 7 \end{bmatrix}$

46. $\begin{bmatrix} 5 & m & 3 \\ 2 & y & 5 \end{bmatrix} - \begin{bmatrix} 3 & 4 & a \\ 1 & 6 & a \end{bmatrix} = \begin{bmatrix} 2 & 5 & 2 \\ 1 & 4 & 1 \end{bmatrix}$

The dates in the calendar display for February 1998 form a 4×7 *matrix, as shown here.*

February 1998

Sun	Mon	Tue	Wed	Thu	Fri	Sat
1	2	3	4	5	6	7
8	9	10	11	12	13	14
15	16	17	18	19	20	21
22	23	24	25	26	27	28

47. Will the month of March on a calendar ever form a matrix? Why or why not?

48. What is the next year following 1998 in which February will again form a matrix? (*Hint:* You *may* want to consider "leap years" or "perpetual calendars." You can look up these concepts in this text or elsewhere.)

Example 10 in this section established the commutative property for addition of matrices. Use similar reasoning in Exercises 49–56 to determine whether each statement is true *or* false *for matrices A, B, and C of the same order and for scalars a and b.*

49. $A - B = B - A$

50. $A + (B + C) = (A + B) + C$ (associative property for addition)

51. $A + B$ is a matrix. (closure property)

52. If 0 is a matrix of the same order as A, containing only zero entries, then $0 + A = A$ and $A + 0 = A$. (identity property for addition)

53. If $-A = (-1)A$, then $A + (-A) = 0$ and $-A + A = 0$. (inverse property for addition)

54. $A - B = A + (-B)$, where $-B = (-1)B$.

55. $a(B + C) = aB + aC$

56. $(a + b)C = aC + bC$

57. Does the set of all 2×2 matrices and the operation of addition form a *mathematical system?* Explain why or why not. (*Hint:* Look up the definition of "mathematical system.")

58. Does the system of Exercise 57 form a *group?* Explain why or why not. (*Hint:* Look up the definition of a mathematical "group.")

14.2 Multiplication of Matrices

D&C Creations is a business that manufactures and sells unfinished tables and chairs. (See Exercises 25–33 of Section 1.) Table 2 shows the units of materials required for each table, while Table 3 shows the per unit cost of materials.

TABLE 2 Materials Required for a Table

		Materials	
		Wood	**Hardware**
Furniture	**Table**	8	5

TABLE 3 Cost of Materials

		Cost Per Unit
Materials	**Wood**	\$2
	Hardware	\$3

From the tabulated data we see that each table manufactured requires 8 units of wood and 5 units of hardware, and wood costs \$2 per unit while hardware costs \$3 per unit.

The cost of materials for each table will then be

$$8 \cdot (\$2) + 5 \cdot (\$3) = \$16 + \$15 = \$31.$$

(Make sure you understand this calculation.) To see how matrix notation can be of use here, first exhibit the material requirements for a table in the **row** matrix M:

$$M = \begin{array}{c} \\ \text{Table} \end{array} \begin{array}{c} \begin{array}{cc} \text{Wood} & \text{Hardware} \end{array} \\ \begin{bmatrix} 8 & 5 \end{bmatrix} \end{array}$$

Matrix M relates the furniture item (table—on the left) to its materials (wood and hardware—at the top). Also exhibit the data on cost per unit (for materials) in the **column** matrix C:

$$C = \begin{array}{c} \text{Wood} \\ \text{Hardware} \end{array} \begin{array}{c} \text{Cost per unit} \\ \begin{bmatrix} 2 \\ 3 \end{bmatrix} \end{array}$$

Matrix C relates materials (on the left) to cost (at the top).

Now define *row-column multiplication* as follows.

Row-Column Multiplication

If a row matrix and a column matrix have equal numbers of entries, then their **row-column product** is the *scalar* obtained by multiplying all of their corresponding entries (first with first, second with second, and so on) and adding the resulting products. Always position the row matrix first (on the left) and the column matrix second (on the right). If the row matrix and column matrix have different numbers of entries, their row-column product is not computable.

We can now see that the cost of materials for each table ($31 as calculated above) is the row-column product of M and C.

$$\begin{aligned} MC &= \text{Table} \begin{array}{c} \begin{array}{cc} \text{Wood} & \text{Hardware} \end{array} \\ \begin{bmatrix} 8 & 5 \end{bmatrix} \end{array} \cdot \begin{array}{c} \text{Wood} \\ \text{Hardware} \end{array} \begin{array}{c} \text{Cost per unit} \\ \begin{bmatrix} 2 \\ 3 \end{bmatrix} \end{array} \\ &= 8 \cdot 2 + 5 \cdot 3 \\ &= 31. \end{aligned}$$

Notice that matrix M relates furniture (table) to materials and matrix C relates materials to cost, so the product MC relates furniture (table) to cost. These relationships are illustrated in Figure 1.

The next section (in For Further Thought) refers to Cramer's rule, a method of solving linear systems using determinants. Actually, the basic properties of determinants were recognized by Leibniz long before Cramer published the theory. It was a century after Cramer's publication that three British mathematicians (to be profiled in subsequent pages) combined to formulate the more general theory of matrices. James Sylvester gave the name "matrix" to any rectangular array from which a determinant could be formed. About this time William Hamilton used arrays to perform linear "transformations." He had already conceived of systems without the commutative law for multiplication. (Matrix multiplication is not commutative!) At last, Arthur Cayley published the essential theory of matrices, influenced both by the form of matrices and by the brief notation of a system of linear equations in terms of coefficients and constants.

Sylvester's "matrix" was apt at the time when determinants had supremacy. The word had the general meaning of something in which something else developed or formed. In fact, "matrix" goes back to the Latin *mater* (for "mother") and a related form meaning "womb." Now that matrices are independent of determinants, the term is not so meaningful. A kind of role reversal has taken place, as it were: matrix theory has put determinants in the shadow.

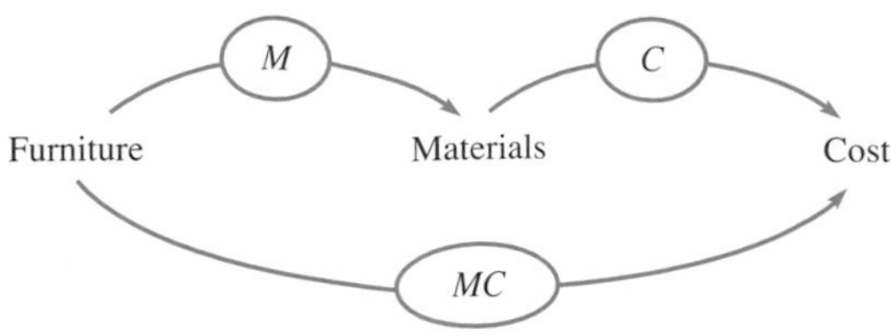

FIGURE 1
Quantities Related by Matrices M and C and their Row-Column Product MC

The row-column multiplication of M and C is reiterated below, without the labels, and with arrows emphasizing which entries are multiplied together.

$$MC = [8 \quad 5] \cdot \begin{bmatrix} 2 \\ 3 \end{bmatrix} = 8 \cdot 2 + 5 \cdot 3 = 31$$

Recall that, for a row-column product to be computable, the following conditions are required.

1. The first (left) matrix must be a row matrix, and the second (right) matrix must be a column matrix.
2. The two matrices must have the same number of entries.

Row-column multiplication is important because it is the basis for defining matrix multiplication in general. To show the usefulness of general matrix multiplication, we continue our illustration with *D&C* Creations.

It is found that each *chair* made by *D&C* requires 6 units of wood and 4 units of hardware. Representing this information in the row matrix $N = [6 \quad 4]$, we can now discover the material cost of a chair by computing another row-column product.

$$NC = [6 \quad 4] \cdot \begin{bmatrix} 2 \\ 3 \end{bmatrix} = 6 \cdot 2 + 4 \cdot 3 = 24$$

Note that N relates a chair to its materials, and C relates those materials to their costs. Thus NC relates a chair to its cost ($24).

Now observe that the known relationships between furniture and materials (for *both tables and chairs*) can be combined into a single matrix, say R, which shows materials needed for both types of furniture.

$$R = \begin{matrix} & \textbf{Wood} & \textbf{Hardware} \\ \textbf{Table} & 8 & 5 \\ \textbf{Chair} & 6 & 4 \end{matrix}$$

We can now exhibit both row-column products MC and NC within a single matrix product, RC, as follows.

$$RC = \begin{matrix} & \textbf{Wood} & \textbf{Hardware} \\ \textbf{Table} & 8 & 5 \\ \textbf{Chair} & 6 & 4 \end{matrix} \cdot \begin{matrix} & \textbf{Cost per unit} \\ \textbf{Wood} & 2 \\ \textbf{Hardware} & 3 \end{matrix}$$

The computation of this matrix product involves two row-column products: row 1 of R times the column of C, and row 2 of R times the column of C. The actual calculation appears as follows.

$$RC = \begin{array}{l} \\ \textbf{Table} \\ \textbf{Chair} \end{array} \begin{array}{c} \textbf{Cost} \\ \begin{bmatrix} 8 \cdot 2 + 5 \cdot 3 \\ 6 \cdot 2 + 4 \cdot 3 \end{bmatrix} \end{array}$$

$$= \begin{array}{l} \\ \textbf{Table} \\ \textbf{Chair} \end{array} \begin{array}{c} \textbf{Cost} \\ \begin{bmatrix} 31 \\ 24 \end{bmatrix} \end{array}$$

This result shows that the materials (wood and hardware) cost \$31 for each table and \$24 for each chair.

Without labels, the above matrix product appears as follows.

$$RC = \begin{bmatrix} 8 & 5 \\ 6 & 4 \end{bmatrix} \cdot \begin{bmatrix} 2 \\ 3 \end{bmatrix} = \begin{bmatrix} 31 \\ 24 \end{bmatrix}$$

In this product, the 31 (a scalar) in row 1, column 1 is simply the row-column product of row 1 from R with column 1 from C. The 24 in row 2, column 1 is the row-column product of row 2 from R with column 1 from C. These row-column products are computable only because the rows of R have the same number of entries as the column of C. In other words, R has as many columns as C has rows (2 in both cases).

In general, the following diagram shows when the matrix product AB is computable and what the order of the product will be.

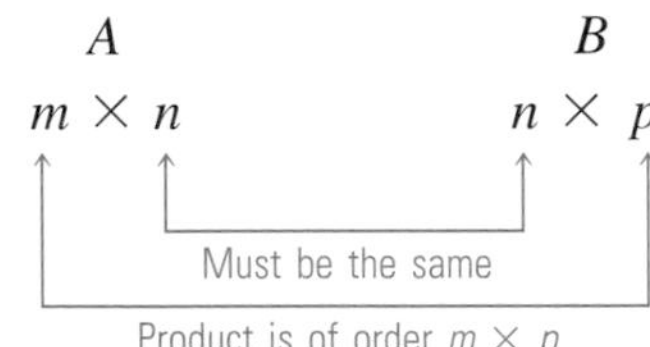

In summary, the product of two matrices can be defined as follows.

Multiplication of Two Matrices

If A is an $m \times n$ matrix and B is an $n \times p$ matrix, then the **product of the two matrices,** AB, is an $m \times p$ matrix where each particular entry, say the entry in row i, column j, is the row-column product of row i in matrix A with column j in matrix B. (If the number of columns in A is different from the number of rows in B, the product AB is not computable.)

EXAMPLE 1 Find each matrix product AB if it exists.

(a) $A = \begin{bmatrix} 2 & -1 & 3 & 1 \\ 0 & 1 & 0 & 2 \end{bmatrix} \qquad B = \begin{bmatrix} -1 & 2 & 3 \\ 0 & 0 & -1 \\ 1 & 2 & 0 \\ 0 & -1 & 0 \end{bmatrix}$

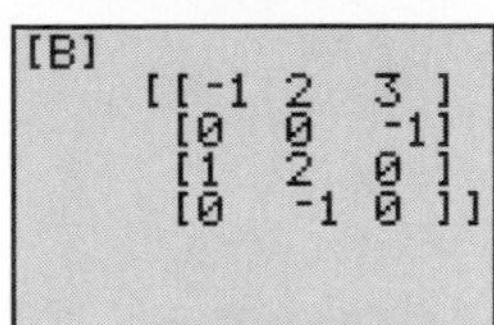

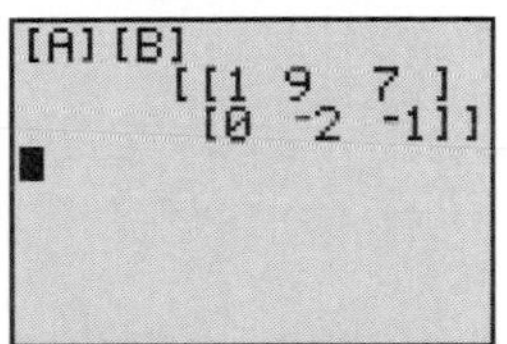

[A] [B]
[[1 9 7]
[0 -2 -1]]

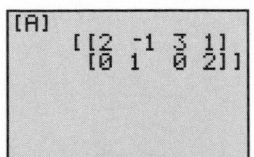

These screens illustrate how the product AB in Example 1(a) is found with a graphing calculator. Attempting to find BA would lead to an error message, however.

By the definition above, since A is of order 2×4 and B is of order 4×3, the product exists and will be of order 2×3.

$$AB = \begin{bmatrix} \square & \square & \square \\ \square & \square & \square \end{bmatrix}$$

Each entry of AB is the appropriate row-column product. For example, the row 1, column 2 entry is found as shown in Figure 2.

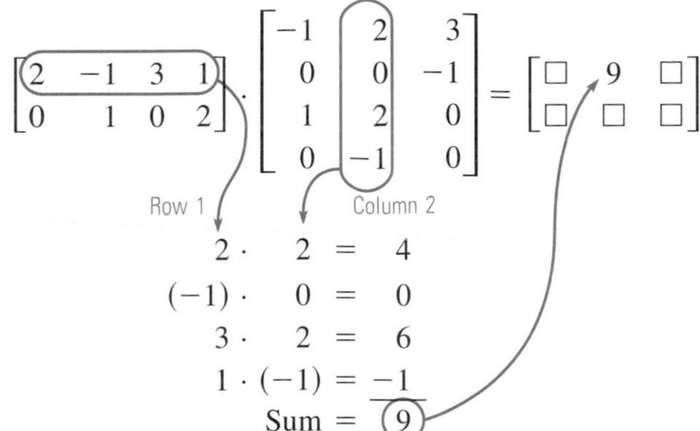

■ **FIGURE 2**
Finding the Row 1, Column 2 Entry of a Matrix Product

Compute the other five entries similarly to find that

$$AB = \begin{bmatrix} 1 & 9 & 7 \\ 0 & -2 & -1 \end{bmatrix}.$$

(b) $A = \begin{bmatrix} -2 & 1 \\ 3 & 4 \end{bmatrix}$ $\quad B = \begin{bmatrix} -1 & 2 \\ 0 & 1 \\ 4 & 0 \end{bmatrix}$

Since A is of order 2×2 and B is of order 3×2 (that is, the number of columns of A is different from the number of rows of B), the product AB is not computable. ■

The following example establishes an important fact about matrix multiplication.

EXAMPLE 2 Decide in each case whether or not $AB = BA$.

(a) $A = \begin{bmatrix} 6 & -2 \\ 1 & 4 \\ 0 & 3 \end{bmatrix}$ $\quad B = \begin{bmatrix} 2 & -1 & 0 & 4 \\ 1 & 3 & 2 & -2 \end{bmatrix}$

In this case, AB is computable but BA is not. (Why not?) Therefore AB and BA cannot be equal. (There is no need to compute AB.)

```
[A]
        [[4 2]
         [6 2]]
[B]
        [[5 1]
         [3 4]]
```

```
[A][B]
      [[26 12]
       [36 14]]
[B][A]
      [[26 12]
       [36 14]]
```

The results of Example 2(b) are illustrated on these screens.

(b) $A = \begin{bmatrix} 4 & 2 \\ 6 & 2 \end{bmatrix} \qquad B = \begin{bmatrix} 5 & 1 \\ 3 & 4 \end{bmatrix}$

Here, AB and BA are both computable.

$$AB = \begin{bmatrix} 4 & 2 \\ 6 & 2 \end{bmatrix} \cdot \begin{bmatrix} 5 & 1 \\ 3 & 4 \end{bmatrix} = \begin{bmatrix} 26 & 12 \\ 36 & 14 \end{bmatrix}$$

$$BA = \begin{bmatrix} 5 & 1 \\ 3 & 4 \end{bmatrix} \cdot \begin{bmatrix} 4 & 2 \\ 6 & 2 \end{bmatrix} = \begin{bmatrix} 26 & 12 \\ 36 & 14 \end{bmatrix}$$

In this case, AB and BA are equal.

(c) $A = \begin{bmatrix} -2 & 3 \\ 2 & -1 \end{bmatrix} \qquad B = \begin{bmatrix} -1 & -3 \\ -2 & 2 \end{bmatrix}$

Again, AB and BA are both computable.

$$AB = \begin{bmatrix} -2 & 3 \\ 2 & -1 \end{bmatrix} \cdot \begin{bmatrix} -1 & -3 \\ -2 & 2 \end{bmatrix} = \begin{bmatrix} -4 & 12 \\ 0 & -8 \end{bmatrix}$$

$$BA = \begin{bmatrix} -1 & -3 \\ -2 & 2 \end{bmatrix} \cdot \begin{bmatrix} -2 & 3 \\ 2 & -1 \end{bmatrix} = \begin{bmatrix} -4 & 0 \\ 8 & -8 \end{bmatrix}$$

In this case, AB and BA are *not* equal! ◆

```
[A]
      [[-2 3 ]
       [2  -1]]
[B]
      [[-1 -3]
       [-2 2 ]]
```

```
[A][B]
      [[-4 12]
       [0  -8]]
[B][A]
      [[-4 0 ]
       [8  -8]]
```

The results of Example 2(c) are illustrated on these screens.

Suppose we had asked the following question: Given two matrices A and B such that AB and BA are both computable, are AB and BA *sometimes* equal, *always* equal, or *never* equal? In light of parts (b) and (c) of Example 2, AB and BA are *sometimes* equal. Generally, the condition "sometimes" is established by exhibiting one case where the statement is true and another case where it is false. The conditions "always" and "never" take more than examples for their proofs. (See Example 5.)

Part (c) of Example 2 shows that AB and BA are not always equal. Therefore *matrix multiplication is not commutative*. This fact delayed the acceptance of matrix theory during its early development since mathematicians tended to doubt the usefulness of a non-commutative multiplication operation. The significance of this result is emphasized in the following statement.

Matrix Multiplication Is *Not* Commutative

Given two matrices A and B, even when the products AB and BA are both computable, AB does not always equal BA. (In fact, AB and BA are equal only in rare cases.)

EXAMPLE 3 Recall that *D&C* Creations makes unfinished tables and chairs. The matrices

$$R = \begin{array}{c} \\ \textbf{Table} \\ \textbf{Chair} \end{array} \begin{array}{c} \begin{array}{cc} \textbf{Wood} & \textbf{Hardware} \end{array} \\ \begin{bmatrix} 8 & 5 \\ 6 & 4 \end{bmatrix} \end{array} \quad \text{and} \quad C = \begin{array}{c} \textbf{Wood} \\ \textbf{Hardware} \end{array} \begin{array}{c} \textbf{Cost per unit} \\ \begin{bmatrix} 2 \\ 3 \end{bmatrix} \end{array}$$

relate furniture to materials and materials to cost, respectively.

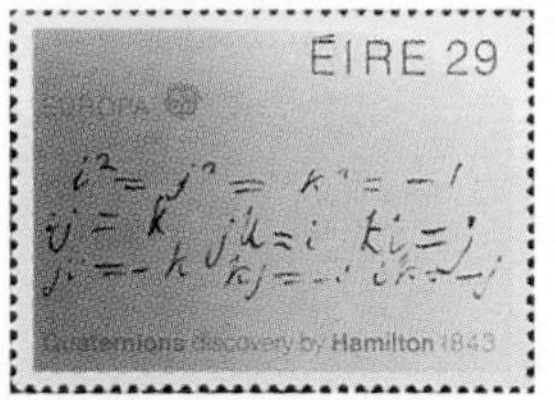

William Rowan Hamilton (1805–1865) was a prodigy as a linguist and mathematician. His work involved quaternions. (A **quaternion** is a kind of hypercomplex number since it represents a force acting in three dimensions rather than in the plane. Quaternions involve the symbols *j* and *k* as well as *i*, the imaginary number discussed elsewhere.) Hamilton could add and subtract quaternions, but he could not find any product that did not violate the commutative law, $AB = BA$.

In the year 1843, after fifteen years of searching for a way to multiply quaternions, it is said that Hamilton was strolling along the Royal Canal in Dublin when he realized suddenly that a mathematical system could be consistent without obeying the commutative law. (As the text indicates, multiplication of matrices is *not* commutative.) This happened at Brougham Bridge, and he carved on it the famous formula

$$i^2 = j^2 = k^2 = ijk = -1.$$

This insight freed algebra and paved the way for Cayley.

Don and Connie have observed that most potential customers prefer to buy furniture that is already finished, so they have begun to finish their tables and chairs. They find that each table requires 3 units of paint, each chair requires 2 units of paint, and paint costs \$1 per unit.

(a) Expand the original matrices R and C to reflect this new information.

The expanded matrices are as follows.

$$R = \begin{array}{c} \text{Table} \\ \text{Chair} \end{array} \overset{\begin{array}{ccc} \text{Wood} & \text{Hardware} & \text{Paint} \\ \downarrow & \downarrow & \downarrow \end{array}}{\begin{bmatrix} 8 & 5 & 3 \\ 6 & 4 & 2 \end{bmatrix}} \quad \text{and} \quad C = \begin{array}{c} \text{Wood} \\ \text{Hardware} \\ \text{Paint} \end{array} \overset{\text{Cost per unit}}{\begin{bmatrix} 2 \\ 3 \\ 1 \end{bmatrix}}$$

Notice that, in their new versions, R still relates furniture (on the left) to materials (at the top) and C still relates materials (at the left) to cost (at the top).

(b) Compute the new product RC, if possible, and explain its meaning.

The new product RC is still computable. A 2×3 matrix times a 3×1 matrix will produce a 2×1 matrix.

$$RC = \begin{array}{c} \text{Table} \\ \text{Chair} \end{array} \overset{\begin{array}{ccc} \text{Wood} & \text{Hardware} & \text{Paint} \\ \downarrow & \downarrow & \downarrow \end{array}}{\begin{bmatrix} 8 & 5 & 3 \\ 6 & 4 & 2 \end{bmatrix}} \cdot \begin{array}{c} \text{Wood} \\ \text{Hardware} \\ \text{Paint} \end{array} \overset{\text{Cost per unit}}{\begin{bmatrix} 2 \\ 3 \\ 1 \end{bmatrix}} = \begin{array}{c} \text{Table} \\ \text{Chair} \end{array} \overset{\text{Cost per unit}}{\begin{bmatrix} 34 \\ 26 \end{bmatrix}}$$

The result shows that (including paint) the materials will now cost \$34 for each table and \$26 for each chair. ■

EXAMPLE 4 Let $M = \begin{bmatrix} -8 & 6 \\ 2 & 1 \end{bmatrix}$ and $I = \begin{bmatrix} 1 & 0 \\ 0 & 1 \end{bmatrix}$. Find MI and IM.

Use the definition of matrix multiplication to find MI.

$$MI = \begin{bmatrix} -8 & 6 \\ 2 & 1 \end{bmatrix} \cdot \begin{bmatrix} 1 & 0 \\ 0 & 1 \end{bmatrix}$$

$$= \begin{bmatrix} -8 \cdot 1 + 6 \cdot 0 & -8 \cdot 0 + 6 \cdot 1 \\ 2 \cdot 1 + 1 \cdot 0 & 2 \cdot 0 + 1 \cdot 1 \end{bmatrix} = \begin{bmatrix} -8 & 6 \\ 2 & 1 \end{bmatrix} = M$$

Here $MI = M$. Multiplying in the same way will show that $IM = M$. ■

Matrix I of Example 4 preserves the identity of matrix M for multiplication and is called the **identity matrix.**

Identity matrices of any square order can be found. For example, the 3×3 identity matrix is displayed here.

$$\begin{bmatrix} 1 & 0 & 0 \\ 0 & 1 & 0 \\ 0 & 0 & 1 \end{bmatrix} \qquad 3 \times 3 \text{ identity matrix}$$

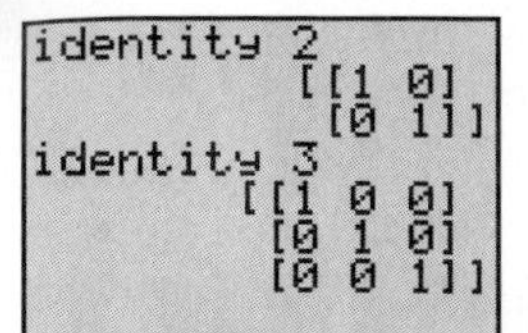

The identity matrices of orders 2 and 3 are shown in this screen.

Notice the similarity between I, the identity for matrix multiplication, and 1, the identity for multiplication of real numbers. Just as $a \cdot 1 = a$ and $1 \cdot a = a$ for every real number a, it is also true that $MI = M$ and $IM = M$ for any square matrix M and identity matrix I of the same order.

EXAMPLE 5 Prove that the equalities $MI = M$ and $IM = M$ are both *always* true for 2×2 matrices M.

Let $M = \begin{bmatrix} a & c \\ b & d \end{bmatrix}$ for any real numbers a, b, c, and d. Then

$$MI = \begin{bmatrix} a & c \\ b & d \end{bmatrix} \cdot \begin{bmatrix} 1 & 0 \\ 0 & 1 \end{bmatrix} = \begin{bmatrix} a \cdot 1 + c \cdot 0 & a \cdot 0 + c \cdot 1 \\ b \cdot 1 + d \cdot 0 & b \cdot 0 + d \cdot 1 \end{bmatrix} = \begin{bmatrix} a & c \\ b & d \end{bmatrix} = M$$

and

$$IM = \begin{bmatrix} 1 & 0 \\ 0 & 1 \end{bmatrix} \cdot \begin{bmatrix} a & c \\ b & d \end{bmatrix} = \begin{bmatrix} 1 \cdot a + 0 \cdot b & 1 \cdot c + 0 \cdot d \\ 0 \cdot a + 1 \cdot b & 0 \cdot c + 1 \cdot d \end{bmatrix} = \begin{bmatrix} a & c \\ b & d \end{bmatrix} = M.$$

We complete this section with an application of matrices to communication networks. The ideas apply to a great variety of situations, including computer networks, telephone systems, genetic evolution, spread of rumors or diseases, and many other kinds of communication among countries, individuals, businesses, or other social and economic groups.

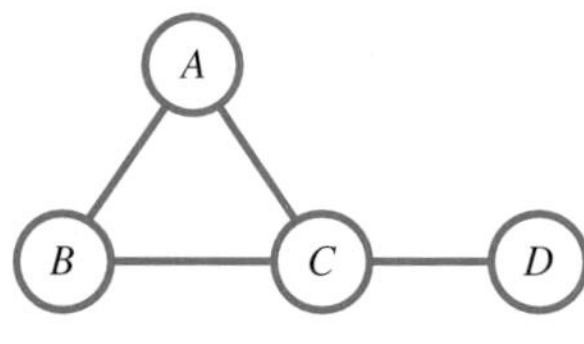

FIGURE 3
A Small Computer Network

Suppose four computers in a small office are connected as shown in Figure 3. Computer A is connected directly to computers B and C, C is connected to A, B, and D, and so on. All direct connections can be shown in the **communication matrix** M, where an entry of 1 indicates a direct connection while a 0 indicates no direct connection. (The main diagonal of the matrix, from upper left to lower right, contains all 0s since a computer is not considered to be directly connected to itself.)

$$M = \begin{matrix} & \begin{matrix} A & B & C & D \end{matrix} \\ \begin{matrix} A \\ B \\ C \\ D \end{matrix} & \begin{bmatrix} 0 & 1 & 1 & 0 \\ 1 & 0 & 1 & 0 \\ 1 & 1 & 0 & 1 \\ 0 & 0 & 1 & 0 \end{bmatrix} \end{matrix}$$

Define the column matrix K (of all 1s) as follows.

$$K = \begin{bmatrix} 1 \\ 1 \\ 1 \\ 1 \end{bmatrix}$$

Then compute the product MK.

$$MK = \begin{bmatrix} 0 & 1 & 1 & 0 \\ 1 & 0 & 1 & 0 \\ 1 & 1 & 0 & 1 \\ 0 & 0 & 1 & 0 \end{bmatrix} \cdot \begin{bmatrix} 1 \\ 1 \\ 1 \\ 1 \end{bmatrix} = \begin{bmatrix} 2 \\ 2 \\ 3 \\ 1 \end{bmatrix}$$

Verify, by comparing Figure 3, that *MK* gives, for each computer in the network, the number of other computers with which it has a direct connection.

Notice that *A* and *D* are not directly connected. To communicate with *D*, *A* must go through one or more intermediaries. For example, *A* could send data to *D* through the single intermediary *C* or by way of the two intermediaries *B* and *C*.

The following example utilizes the matrix M^2. (Exponents on matrices work just as on real numbers.)

EXAMPLE 6 For matrix *M* above, find M^2 and explain its meaning.

$$M^2 = MM = \begin{bmatrix} 0 & 1 & 1 & 0 \\ 1 & 0 & 1 & 0 \\ 1 & 1 & 0 & 1 \\ 0 & 0 & 1 & 0 \end{bmatrix} \cdot \begin{bmatrix} 0 & 1 & 1 & 0 \\ 1 & 0 & 1 & 0 \\ 1 & 1 & 0 & 1 \\ 0 & 0 & 1 & 0 \end{bmatrix} = \begin{bmatrix} 2 & 1 & 1 & 1 \\ 1 & 2 & 1 & 1 \\ 1 & 1 & 3 & 0 \\ 1 & 1 & 0 & 1 \end{bmatrix}$$

Each entry in M^2 indicates the number of different ways one computer can send data to another through a single intermediary. For example, the 1 in row 3, column 2 shows that *C* can get to *B* through a single intermediary in just 1 way. (Figure 3 shows that the one way is through *A*.) ◆

Practical applications of matrices often involve orders much greater than what we have illustrated here, and can require tremendous amounts of calculation. (A 30 × 20 matrix times a 20 × 40 matrix requires 24,000 individual products of entries, plus 1,200 sums of 20 terms each. And the original entries may involve several decimal places.)

Fortunately, modern computers, and even handheld calculators, can do matrix operations directly.

[A]
[[2.8 8.1 7.7]
[3.6 4.6 6.4]
[4.7 8.2 5.5]
[1.9 9.3 8.9]
[7.3 4.6 7.8]
[3.5 2.5 4.2]]

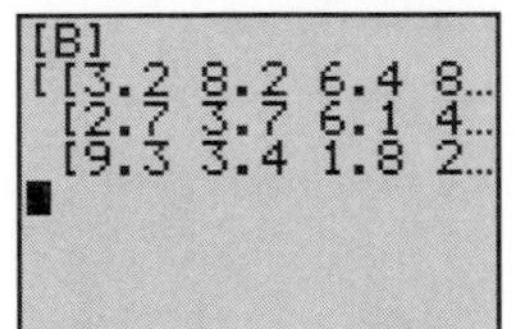

[A][B]
[[102.44 79.11 ...
[83.46 68.3 ...
[88.33 87.58 ...
[113.96 80.25 ...
[108.32 103.4 ...
[57.01 52.23 ...

The computation in Example 7 is easily accomplished with a graphing calculator having matrix capability. Notice that the second and third screens are incomplete. The user must scroll to the right in order to read the hidden entries (indicated by the ellipsis . . .).

EXAMPLE 7 Compute the following matrix product.

$$\begin{bmatrix} 2.8 & 8.1 & 7.7 \\ 3.6 & 4.6 & 6.4 \\ 4.7 & 8.2 & 5.5 \\ 1.9 & 9.3 & 8.9 \\ 7.3 & 4.6 & 7.8 \\ 3.5 & 2.5 & 4.2 \end{bmatrix} \cdot \begin{bmatrix} 3.2 & 8.2 & 6.4 & 8.6 & 2.9 \\ 2.7 & 3.7 & 6.1 & 4.8 & 8.6 \\ 9.3 & 3.4 & 1.8 & 2.5 & 4.7 \end{bmatrix}$$

The 90 required products (2.8 times 3.2, and so on) could be done individually, but with a calculator we just enter the two matrices *A* and *B* and then enter matrix *A* times matrix *B*. (Most of the work is in keying in the entries.) The resulting product is

$$\begin{bmatrix} 102.44 & 79.11 & 81.19 & 82.21 & 113.97 \\ 83.46 & 68.30 & 62.62 & 69.04 & 80.08 \\ 88.33 & 87.58 & 90.00 & 93.53 & 110.00 \\ 113.96 & 80.25 & 84.91 & 83.23 & 127.32 \\ 108.32 & 103.40 & 88.82 & 104.36 & 97.39 \\ 57.01 & 52.23 & 45.21 & 52.60 & 51.39 \end{bmatrix}.$$ ◆

The ability to accomplish all the tedious calculations on a machine helps make matrix theory of practical use. For example, if *D&C* Creations expanded to a product line of, say, 30 items utilizing 20 materials (rather than 2 or 3 of each), the matrices *R* and *C* (relating products to materials and materials to cost) would become quite large, but the relationship between products and material costs would still be exhibited in the single matrix product *RC*.

14.2 EXERCISES

Compute each of the following products, if possible. If a product is not computable, state why not.

1. $\begin{bmatrix} -3 & 2 \\ 4 & 1 \end{bmatrix} \cdot \begin{bmatrix} -2 & 0 \\ 1 & 3 \end{bmatrix}$

2. $\begin{bmatrix} -5 & 2 \\ 1 & 4 \end{bmatrix} \cdot \begin{bmatrix} -1 & 4 \\ 3 & 0 \end{bmatrix}$

3. $\begin{bmatrix} 0 & -2 \\ 5 & 1 \end{bmatrix} \cdot \begin{bmatrix} -3 & 6 \\ 1 & 4 \end{bmatrix}$

4. $\begin{bmatrix} -3 & 5 \\ 1 & 0 \end{bmatrix} \cdot \begin{bmatrix} 8 & -2 \\ 1 & 7 \end{bmatrix}$

5. $\begin{bmatrix} 1 & 3 \\ 4 & 1 \end{bmatrix} \cdot \begin{bmatrix} 2 & 1 & 0 \\ 5 & 2 & 3 \end{bmatrix}$

6. $\begin{bmatrix} -5 & 6 \\ 3 & 4 \end{bmatrix} \cdot \begin{bmatrix} -3 & 5 & 0 \\ 0 & 1 & 5 \end{bmatrix}$

7. $\begin{bmatrix} 3 & 2 \\ 5 & 1 \\ 0 & 4 \end{bmatrix} \cdot \begin{bmatrix} 2 & -1 \\ 1 & 3 \end{bmatrix}$

8. $\begin{bmatrix} -1 & 0 \\ 4 & 1 \\ 2 & 0 \end{bmatrix} \cdot \begin{bmatrix} -5 & 0 \\ 4 & 2 \end{bmatrix}$

9. $\begin{bmatrix} -5 & 1 \\ 6 & 2 \end{bmatrix} \cdot \begin{bmatrix} -3 & 2 \\ 4 & 1 \\ 5 & 8 \end{bmatrix}$

10. $\begin{bmatrix} -7 & 3 \\ 9 & 2 \end{bmatrix} \cdot \begin{bmatrix} 4 & -1 \\ 2 & -5 \\ 3 & 6 \end{bmatrix}$

11. $\begin{bmatrix} -5 & 1 & 3 \\ 2 & 0 & 4 \\ 3 & 0 & 2 \end{bmatrix} \cdot \begin{bmatrix} -2 & 0 & 0 \\ 4 & 1 & 0 \\ 0 & 1 & 0 \end{bmatrix}$

12. $\begin{bmatrix} -3 & 4 & 0 \\ 2 & -1 & 5 \\ 0 & 4 & 1 \end{bmatrix} \cdot \begin{bmatrix} 0 & 4 & 0 \\ 2 & 0 & 3 \\ 0 & 5 & 1 \end{bmatrix}$

13. Harry's Donuts, a small neighborhood bakery, sells four main items: sweet rolls, bread, cakes, and pies. Matrix *A* below shows the number of units of the main ingredients needed for these items.

$A =$

	Eggs	Flour	Sugar	Shortening	Milk
Sweet rolls	1	4	1/4	1/4	1
Bread	0	3	0	1/4	0
Cakes	4	3	2	1	1
Pies	0	1	0	1/3	0

The cost, in cents per egg or per cup, of each ingredient when purchased in large lots or in small lots is given by matrix *B*.

$B =$

	PURCHASE OPTION	
	Large lot	Small lot
Eggs	5	5
Flour	8	10
Sugar	10	12
Shortening	12	15
Milk	5	6

Use matrix multiplication to find a matrix representing the comparative costs per item for the two purchase options (buying in large lots or in small lots).

Suppose a day's orders at Harry's (see Exercise 13) consist of 20 dozen sweet rolls, 200 loaves of bread, 50 cakes, and 60 pies.

14. Write this day's orders as a 1×4 matrix *D*, and use matrix multiplication to write as a matrix the amount of each ingredient needed to fill the day's orders.

15. Use matrix multiplication to find a matrix representing the costs to fill the day's orders under the two purchase options.

Refer to the expanded matrices of Example 3, part (a). Assume that Don and Connie have now expanded their product line to include desks, each requiring 10 units of wood, 7 units of hardware, and 4 units of paint.

16. Set up a matrix R relating furniture to materials for the complete line of *D&C* Creations.

17. Set up and compute an appropriate matrix product that will relate furniture to cost of materials.

18. Don and Connie have found that the *labor* required for assembling and finishing is four hours per table, two hours per chair, and five hours per desk, and that their labor cost is $9 per hour. Write a column matrix relating furniture to labor cost.

19. Combine your answers from Exercises 17 and 18 to obtain a 3×2 matrix relating furniture (table, chair, desk) to cost of resources (materials, labor).

20. Write a column matrix relating furniture to selling price. Assume the selling price of each furniture item must cover its own material and labor costs, plus $10 in overhead expenses, plus a profit ($20 per table, $10 per chair, and $25 per desk).

In Ohio, high school football teams are ranked by the Harbin Football Team Rating System. A team is granted one level 1 point for each team it defeats, and is also granted one level 2 point for each time that a team it defeats beats another team. Teams A, B, C, and D all played one another. The results are shown in matrix H, where an entry of 1 indicates a win. For example, A beat C only, B beat both A and D, and so on.

$$H = \begin{array}{c} \\ A \\ B \\ C \\ D \end{array} \begin{array}{c} \begin{matrix} A & B & C & D \end{matrix} \\ \begin{bmatrix} 0 & 0 & 1 & 0 \\ 1 & 0 & 0 & 1 \\ 0 & 1 & 0 & 1 \\ 1 & 0 & 0 & 0 \end{bmatrix} \end{array}$$

21. Complete the table below for these four teams.

Team	Level 1 Points	Level 2 Points	Total Points
A	____	____	____
B	____	____	____
C	____	____	____
D	____	____	____

22. Let K be a 4×1 matrix with every entry a 1.
(a) Compute the product HK.
(b) What does HK represent?

23. **(a)** Compute H^2K.
(b) What does H^2K represent?

24. **(a)** Compute $HK + H^2K$.
(b) What does $HK + H^2K$ represent?

25. Is it true that $(H + H^2)K = HK + H^2K$? If so, verify it by direct computation.

26. Is it true that $(I + H)HK = HK + H^2K$? If so, verify it by direct computation. (In this case, I is the 4×4 identity matrix.)

Consider the matrices $A = \begin{bmatrix} 1 & 2 \\ 3 & 2 \end{bmatrix}$, $B = \begin{bmatrix} -1 & -3 \\ 2 & 2 \end{bmatrix}$, $C = \begin{bmatrix} 1 & 2 \\ -1 & -3 \end{bmatrix}$, and $I = \begin{bmatrix} 1 & 0 \\ 0 & 1 \end{bmatrix}$.

The following statements illustrate (but do not prove) several algebraic properties of the system of 2×2 matrices. In each case, verify the given statement by direct computation.

27. AB is computable and is of order 2×2. (closure property)

28. $AI = IA = A$. (identity property)

29. $(AB)C = A(BC)$. (associative property)

30. $(A + B)C = AC + BC$. (the right distributive property for addition)

31. $(A - B)C = AC - BC$. (the right distributive property for subtraction)

32. Explain why the verifications in Exercises 27–31 do not actually prove the corresponding properties for the system of 2×2 matrices.

In Exercises 33–36, determine whether $AB = BA$.

33. $A = \begin{bmatrix} 8 & 6 \\ 3 & 5 \end{bmatrix}$ $B = \begin{bmatrix} 10 & 14 \\ 7 & 3 \end{bmatrix}$

34. $A = \begin{bmatrix} -5/3 & -2/3 \\ -2 & -1 \end{bmatrix}$ $B = \begin{bmatrix} -3 & 2 \\ 6 & -5 \end{bmatrix}$

35. $A = \begin{bmatrix} 2 & 7 \\ -3 & 4 \end{bmatrix}$ $B = \begin{bmatrix} 6 & -4 \\ 1 & 2 \end{bmatrix}$

36. $A = \begin{bmatrix} 1 & 0 \\ 1 & 1 \end{bmatrix}$ $B = \begin{bmatrix} 1 & 1 \\ 0 & 1 \end{bmatrix}$

37. Find an original example of 2×2 matrices A and B such that $AB = BA$.

38. Let $M = \begin{bmatrix} 1 & 1 \\ 1 & 1 \end{bmatrix}$.
(a) Compute M^2.
(b) Compute M^3.
(c) Compute M^4.
(d) Without actually computing it, what is M^{10}?

39. (a) Prove that $SI = S$ for every 3×3 matrix S.

(b) Does part (a) establish that $I = \begin{bmatrix} 1 & 0 & 0 \\ 0 & 1 & 0 \\ 0 & 0 & 1 \end{bmatrix}$ is the identity for multiplication of 3×3 matrices? Explain why or why not.

40. In Example 6, how do you interpret the entry 3 in row 3, column 3 of the matrix M^2?

41. Refer to Example 6. Compute the matrix M^3 and interpret the meaning of its entries.

Assume that a fifth computer is introduced into the network referred to in Example 6. It is connected as shown here.

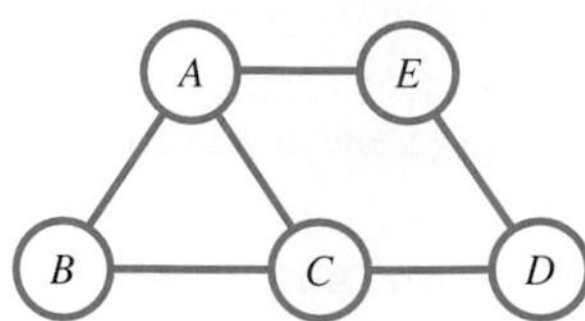

42. Write the communication matrix M for this new network.

43. Compute M^2.

44. According to M^2, how many ways can computer E send data to C through a single intermediary?

45. Without computing M^3, how many channels of communication would it show from B to E? What are those channels?

46. How many channels of communication from B to E would be shown in the matrix M^4?

The real numbers 2/5 and 5/2 are "multiplicative inverses" of each other because $(2/5) \cdot (5/2) = 1$, which is the identity for multiplication of real numbers. Similarly, two matrices A and B are multiplicative inverses of each other if $AB = I$ and $BA = I$, where I is the identity (of the same order in both cases) for multiplication of matrices. (In Section 4, we will address matrix inverses in more detail.)

47. In verifying that A and B are inverses, is it really necessary to compute both AB and BA? Why or why not?

48. What can you say about the orders of A and B if they are to be multiplicative inverses of each other?

Use the definition of matrix inverses (preceding Exercise 47) to determine whether each of the following pairs are inverses of each other.

49. $\begin{bmatrix} 1 & 0 \\ 0 & -1 \end{bmatrix}$ and $\begin{bmatrix} 1 & 0 \\ 0 & -1 \end{bmatrix}$

50. $\begin{bmatrix} 0 & 1 \\ 1 & 0 \end{bmatrix}$ and $\begin{bmatrix} 0 & 1 \\ 1 & 0 \end{bmatrix}$

51. $\begin{bmatrix} 2 & 1 \\ 4 & 3 \end{bmatrix}$ and $\begin{bmatrix} 3/2 & -1/2 \\ -2 & 1 \end{bmatrix}$

52. $\begin{bmatrix} -3 & 2 \\ 6 & -5 \end{bmatrix}$ and $\begin{bmatrix} -5/3 & -2/3 \\ -2 & -1 \end{bmatrix}$

53. $\begin{bmatrix} 3 & 1 & 4 \\ 1 & 0 & 1 \end{bmatrix}$ and $\begin{bmatrix} -1 & 1 \\ 0 & 1 \\ 1 & -1 \end{bmatrix}$

54. $\begin{bmatrix} -2 & 1/3 & 1 \\ 1/4 & 3 & 2 \end{bmatrix}$ and $\begin{bmatrix} 1 & 3 \\ 1/2 & 3/2 \end{bmatrix}$

55. $\begin{bmatrix} 5 & 6 \\ -10 & -13 \end{bmatrix}$ and $\begin{bmatrix} 13/5 & 6/5 \\ -2 & 1 \end{bmatrix}$

56. $\begin{bmatrix} 1 & 1 \\ 1 & -1 \end{bmatrix}$ and $\begin{bmatrix} 1/2 & 1/2 \\ 1/2 & 1/2 \end{bmatrix}$

57. $\begin{bmatrix} 1 & 3 & 3 \\ 1 & 4 & 3 \\ 1 & 3 & 4 \end{bmatrix}$ and $\begin{bmatrix} 7 & -3 & -3 \\ -1 & 1 & 0 \\ -1 & 0 & 1 \end{bmatrix}$

58. $\begin{bmatrix} -1 & 0 & 2 \\ 3 & 1 & 0 \\ 0 & 2 & -3 \end{bmatrix}$ and $\begin{bmatrix} -1/5 & 4/15 & -2/15 \\ 3/5 & 1/5 & 2/5 \\ 2/5 & 2/15 & -1/15 \end{bmatrix}$

59. Explain in your own words what a *row-column product* is, and how it is related to multiplication of matrices in general.

60. Explain why, in Example 7, the matrix product involves 90 products of entries.

Use a calculator to compute the following matrix products.

61. $\begin{bmatrix} 23.8 & 10.6 & 12.8 & 21.9 \\ 15.8 & 24.5 & 11.6 & 14.8 \\ 13.8 & 10.6 & 13.3 & 14.7 \\ 16.4 & 12.8 & 11.1 & 15.9 \\ 21.3 & 12.3 & 13.2 & 32.1 \end{bmatrix} \cdot \begin{bmatrix} 8.23 & 7.45 & 8.25 & 6.27 & 9.91 & 7.29 \\ 2.29 & 6.52 & 3.35 & 5.94 & 6.37 & 6.73 \\ 2.58 & 9.88 & 1.28 & 7.53 & 1.23 & 3.28 \\ 8.45 & 5.27 & 6.37 & 4.38 & 4.39 & 3.46 \end{bmatrix}$

62. $\begin{bmatrix} 4.56 & 9.87 & 2.91 \\ 6.37 & 2.81 & 6.81 \\ 5.28 & 4.79 & 4.21 \\ 5.28 & 6.97 & 1.38 \\ 3.84 & 5.28 & 3.91 \\ 4.68 & 2.38 & 1.54 \\ 4.44 & 5.55 & 6.66 \\ 9.63 & 8.52 & 7.41 \end{bmatrix} \cdot \begin{bmatrix} 8.23 & 4.19 & 2.28 & 9.28 & 8.27 \\ 1.32 & 6.58 & 7.17 & 8.23 & 9.82 \\ 3.37 & 8.26 & 1.49 & 6.54 & 9.87 \end{bmatrix}$

14.3 Matrix Row Operations and Systems of Equations

Linear systems of equations were solved in an earlier chapter by the elimination method. That method can be streamlined into a systematic procedure based on matrices. Matrix methods are particularly suitable for computer solutions of large systems of equations having many unknowns.

To begin, consider a system of three equations and three unknowns such as

$$\begin{aligned} a_1x + b_1y + c_1z &= d_1 \\ a_2x + b_2y + c_2z &= d_2 \\ a_3x + b_3y + c_3z &= d_3. \end{aligned}$$

This system can be written in an abbreviated form as the following 3×4 matrix.

$$\begin{bmatrix} a_1 & b_1 & c_1 & d_1 \\ a_2 & b_2 & c_2 & d_2 \\ a_3 & b_3 & c_3 & d_3 \end{bmatrix}$$

(Recall that each number in a matrix is called an *element,* or *entry*.) The last column of this matrix consists of the constants from the right in the original system of equations. By using a vertical line to set that column apart from the others (which consist of the coefficients of the variables), we obtain the following **augmented matrix.** (The square coefficient matrix is "augmented" with the column of constants.)

$$\left[\begin{array}{ccc|c} a_1 & b_1 & c_1 & d_1 \\ a_2 & b_2 & c_2 & d_2 \\ a_3 & b_3 & c_3 & d_3 \end{array}\right]$$

Each row of this augmented matrix represents one of the equations in the original system. By making a series of legitimate transformations in the matrix, we can eventually obtain a matrix that will produce an equivalent simpler system, one whose solution will be apparent. (Two systems of equations are *equivalent* if they have identical solutions.)

The trick is to use only matrix transformations such that the system corresponding to each successive matrix is equivalent to the previous system. The allowed transformations are the same as those used in the elimination method. In matrix language, they are called **matrix row operations.** They are listed here.

NAMES MATH EDIT
5↑identity
6:randM(
7:augment(
8:rowSwap(
9:row+(
0:*row(
A:*row+(

Options 8, 9, 0, and A in this menu allow the user to perform the matrix row operations described in the text.

Matrix Row Operations

For any nonzero real number k and any augmented matrix of a system of linear equations, the following **matrix row operations** will produce the matrix of an equivalent system:

1. **interchanging any two rows of a matrix,**
2. **multiplying the elements of any row of a matrix by the same nonzero number k, and**
3. **adding a common multiple of the elements of one row to the corresponding elements of another row.**

Our objective is to use row operations to transform the augmented matrix above into the form

$$\left[\begin{array}{ccc|c} 1 & 0 & 0 & a \\ 0 & 1 & 0 & b \\ 0 & 0 & 1 & c \end{array}\right]$$

since the corresponding system will then be

$$\begin{cases} 1 \cdot x + 0 \cdot y + 0 \cdot z = a \\ 0 \cdot x + 1 \cdot y + 0 \cdot z = b \\ 0 \cdot x + 0 \cdot y + 1 \cdot z = c \end{cases} \quad \text{or} \quad \begin{cases} x = a \\ y = b. \\ z = c \end{cases}$$

The solution set is then $\{(a, b, c)\}$.

In effect, we use row operations to produce the elements of the identity

$$\begin{array}{ccc} 1 & 0 & 0 \\ 0 & 1 & 0 \\ 0 & 0 & 1 \end{array}$$

to the left of the vertical line, and then read the solution (the variable values) from the right side of the vertical line. Any number of variables may occur in the system to be solved. For example, if we were solving a system of two equations with two unknowns, we would want to obtain a matrix of the form

$$\left[\begin{array}{cc|c} 1 & 0 & a \\ 0 & 1 & b \end{array}\right].$$

This procedure is called the **Gauss-Jordan method.** It is illustrated in the following examples.

EXAMPLE 1 Solve the linear system

$$\begin{aligned} 3x - 4y &= 1 \\ 5x + 2y &= 19. \end{aligned}$$

The equations should all be in the same form, with the variable terms in the same order on the left, and the constant term on the right. Begin by writing the augmented matrix.

$$\left[\begin{array}{cc|c} 3 & -4 & 1 \\ 5 & 2 & 19 \end{array}\right]$$

The goal is to transform this augmented matrix into the form

$$\left[\begin{array}{cc|c} 1 & 0 & k \\ 0 & 1 & j \end{array}\right]$$

for real numbers k and j. Once the augmented matrix is in this form, the matrix can be rewritten as a linear system to get

$$\begin{aligned} x &= k \\ y &= j. \end{aligned}$$

The necessary transformations are performed as follows. It is best to work in columns beginning in each column with the element that is to become a 1. In the augmented matrix

$$\left[\begin{array}{cc|c} 3 & -4 & 1 \\ 5 & 2 & 19 \end{array}\right]$$

there is a 3 in the first row, first column position. Use row operation 2, multiplying each entry in the first row by $1/3$ to get a 1 in this position. (This step is abbreviated as $(1/3)$R1.)

$$\left[\begin{array}{cc|c} \mathbf{1} & \mathbf{-4/3} & \mathbf{1/3} \\ 5 & 2 & 19 \end{array}\right] \quad \frac{1}{3}\text{R1}$$

Get a 0 in the second row, first column by multiplying each element of the first row by -5 and adding the result to the corresponding element in the second row, using row operation 3.

$$\left[\begin{array}{cc|c} 1 & -4/3 & 1/3 \\ \mathbf{0} & \mathbf{26/3} & \mathbf{52/3} \end{array}\right] \quad -5\text{R1} + \text{R2}$$

Get a 1 in the second row, second column by multiplying each element of the second row by $3/26$, using row operation 2.

$$\left[\begin{array}{cc|c} 1 & -4/3 & 1/3 \\ \mathbf{0} & \mathbf{1} & \mathbf{2} \end{array}\right] \quad \frac{3}{26}\text{R2}$$

Finally, get a 0 in the first row, second column by multiplying each element of the second row by $4/3$ and adding the result to the corresponding element in the first row, using row operation 3.

$$\left[\begin{array}{cc|c} \mathbf{1} & \mathbf{0} & \mathbf{3} \\ 0 & 1 & 2 \end{array}\right] \quad \frac{4}{3}\text{R2} + \text{R1}$$

This last matrix corresponds to the system

$$\begin{aligned} x &= 3 \\ y &= 2, \end{aligned}$$

which has the solution set $\{(3, 2)\}$. This solution could have been read directly from the third column of the final matrix. ■

For Further Thought

The **determinant** of the 2×2 matrix $\begin{bmatrix} a & c \\ b & d \end{bmatrix}$ is symbolized with vertical "fences" rather than brackets:

$$\begin{vmatrix} a & c \\ b & d \end{vmatrix}.$$

This determinant stands for a real number: by definition, the product of the two numbers on the diagonal from NW to SE minus the product of the two numbers on the diagonal from SW to NE. (See Exercises 33–36 in the next section.) For example,

$$\begin{vmatrix} 4 & 1 \\ 6 & -2 \end{vmatrix} = 4(-2) - 6(1) = -14.$$

In 1750, Gabriel Cramer (1704–1752) published the theory of determinants. It provided a way of solving linear systems which predated more general matrix methods by a century. Cramer's rule is still used today. The system

$$3x + y = 5$$
$$2x - y = 10$$

is solved by first forming three determinants as follows.

$D = \begin{vmatrix} 3 & 1 \\ 2 & -1 \end{vmatrix}$ (the determinant of the coefficient matrix)

$D_x = \begin{vmatrix} 5 & 1 \\ 10 & -1 \end{vmatrix}$ (the same as D but with the x-coefficients replaced by the constants)

$D_y = \begin{vmatrix} 3 & 5 \\ 2 & 10 \end{vmatrix}$ (the same as D but with the y-coefficients replaced by the constants)

Now $D = 3(-1) - 2(1) = -5$, $D_x = 5(-1) - 10(1) = -15$, and $D_y = 3(10) - 2(5) = 20$. The solution values are then

$$x = \frac{D_x}{D} = \frac{-15}{-5} = 3 \quad \text{and} \quad y = \frac{D_y}{D} = \frac{20}{-5} = -4.$$

In other words, the solution set of the system of equations is $\{(3, -4)\}$.

For Group Discussion

1. Apply Cramer's rule to the following system.

$$2x - 4y = 14$$
$$5x + 3y = -30$$

2. Apply Cramer's rule to the following system.

$$3x - 4y = 6$$
$$-12x + 16y = 10$$

3. Why do you think Cramer's rule failed to produce a solution for the system in problem 2 above? Could another method be used to solve the system?

Grassmann Discovered at Last In 1844, the year after Hamilton overcame his difficulties with quaternions, a wider system than Hamilton's appeared. This was the theory of extension by Hermann Grassmann (1809–1877). (Here quaternions are a special case.) Grassmann had a leaning toward philosophy, which tended to distort his meaning for most readers. His work was ignored. Even his revision in 1862 was ignored. Grassmann eventually gave up on mathematics.

Grassmann was ahead of his time and was not appreciated until the early twentieth century, when his work was applied to the general theory of relativity.

EXAMPLE 2 At the Evergreen Ranch, 6 goats and 5 sheep sell for \$305, while 2 goats and 9 sheep cost \$285. Find the cost of a goat and the cost of a sheep.

To solve this problem, first write a system of equations. Then use the Gauss-Jordan method to solve the system. Begin by identifying the variables.

$$x = \text{cost of a goat} \quad \text{and} \quad y = \text{cost of a sheep}$$

From the given information, 6 goats and 5 sheep cost a total of \$305, so that

$$6x + 5y = 305.$$

Also, 2 goats and 9 sheep cost a total of \$285, so that

$$2x + 9y = 285.$$

We must now find the common solution for these two equations, that is, the solution of the linear system

$$\begin{aligned} 6x + 5y &= 305 \\ 2x + 9y &= 285. \end{aligned}$$

Write the augmented matrix and then apply the appropriate row operations.

$$\left[\begin{array}{cc|c} 6 & 5 & 305 \\ 2 & 9 & 285 \end{array}\right]$$

$$\left[\begin{array}{cc|c} 1 & 5/6 & 305/6 \\ 2 & 9 & 285 \end{array}\right] \quad \frac{1}{6}\text{R1}$$

$$\left[\begin{array}{cc|c} 1 & 5/6 & 305/6 \\ 0 & 22/3 & 550/3 \end{array}\right] \quad -2\text{R1} + \text{R2}$$

$$\left[\begin{array}{cc|c} 1 & 5/6 & 305/6 \\ 0 & 1 & 25 \end{array}\right] \quad \frac{3}{22}\text{R2}$$

$$\left[\begin{array}{cc|c} 1 & 0 & 30 \\ 0 & 1 & 25 \end{array}\right] \quad -\frac{5}{6}\text{R2} + \text{R1}$$

From the rightmost column, read $x = 30$, $y = 25$. Each goat costs \$30, while each sheep costs \$25.

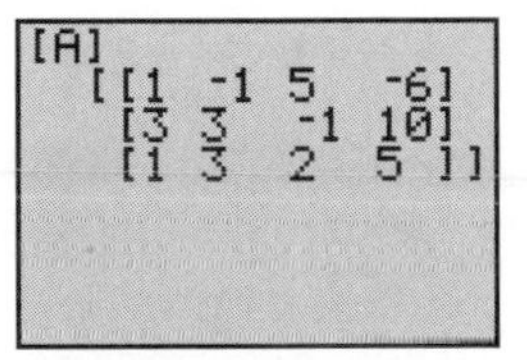

This is the augmented matrix for the original system in Example 3.

EXAMPLE 3 Use the Gauss-Jordan method to solve the system

$$\begin{aligned} x - y + 5z &= -6 \\ 3x + 3y - z &= 10 \\ x + 3y + 2z &= 5. \end{aligned}$$

Since the system is in proper form, begin by writing the augmented matrix of the linear system.

$$\left[\begin{array}{ccc|c} 1 & -1 & 5 & -6 \\ 3 & 3 & -1 & 10 \\ 1 & 3 & 2 & 5 \end{array}\right]$$

The final matrix is to be of the form

$$\left[\begin{array}{ccc|c} 1 & 0 & 0 & m \\ 0 & 1 & 0 & n \\ 0 & 0 & 1 & p \end{array}\right],$$

where m, n, and p are real numbers. This final form of the matrix gives the system $x = m$, $y = n$, and $z = p$, so the solution set is $\{(m, n, p)\}$.

The original augmented matrix already has a 1 in the first row, first column. Get a 0 in the second row, first column by multiplying each element in the first row by -3 and adding the result to the corresponding element in the second row, using row operation 3.

$$\left[\begin{array}{rrr|r} 1 & -1 & 5 & -6 \\ 0 & 6 & -16 & 28 \\ 1 & 3 & 2 & 5 \end{array}\right] \quad -3R1 + R2$$

Now, to change the last element in the first column to a 0, use row operation 3 and multiply each element of the first row by -1, and add the result to the corresponding element of the third row.

$$\left[\begin{array}{rrr|r} 1 & -1 & 5 & -6 \\ 0 & 6 & -16 & 28 \\ 0 & 4 & -3 & 11 \end{array}\right] \quad -1R1 + R3$$

The same procedure is used to transform the second and third columns. For both of these columns perform the additional step of getting a 1 in the appropriate position of each column. Do this by multiplying the elements of the row by the reciprocal of the number in that position. (As row operations are performed, any fractions appearing in various positions can be reduced to lowest terms.)

$$\left[\begin{array}{rrr|r} 1 & -1 & 5 & -6 \\ 0 & 1 & -8/3 & 14/3 \\ 0 & 4 & -3 & 11 \end{array}\right] \quad \frac{1}{6}R2$$

$$\left[\begin{array}{rrr|r} 1 & 0 & 7/3 & -4/3 \\ 0 & 1 & -8/3 & 14/3 \\ 0 & 4 & -3 & 11 \end{array}\right] \quad R2 + R1$$

$$\left[\begin{array}{rrr|r} 1 & 0 & 7/3 & -4/3 \\ 0 & 1 & -8/3 & 14/3 \\ 0 & 0 & 23/3 & -23/3 \end{array}\right] \quad -4R2 + R3$$

$$\left[\begin{array}{rrr|r} 1 & 0 & 7/3 & -4/3 \\ 0 & 1 & -8/3 & 14/3 \\ 0 & 0 & 1 & -1 \end{array}\right] \quad \frac{3}{23}R3$$

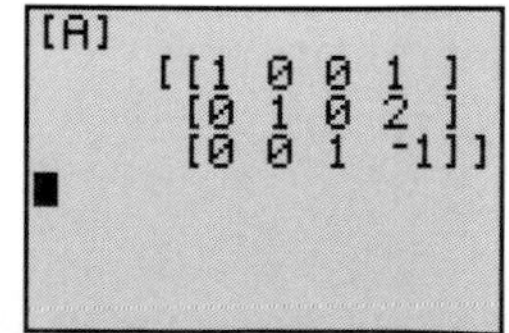

This is the augmented matrix for the final system in Example 3.

$$\left[\begin{array}{rrr|r} 1 & 0 & 0 & 1 \\ 0 & 1 & -8/3 & 14/3 \\ 0 & 0 & 1 & -1 \end{array}\right] \quad -\frac{7}{3}R3 + R1$$

$$\left[\begin{array}{rrr|r} 1 & 0 & 0 & 1 \\ 0 & 1 & 0 & 2 \\ 0 & 0 & 1 & -1 \end{array}\right] \quad \frac{8}{3}R3 + R2$$

The linear system associated with this final matrix is

$$\begin{aligned} x &= 1 \\ y &= 2 \\ z &= -1, \end{aligned}$$

and the solution set is $\{(1, 2, -1)\}$. ■

14.3 EXERCISES

Let $A = \begin{bmatrix} -4 & 0 & 1 \\ 3 & -4 & 2 \\ -1 & 2 & 1 \end{bmatrix}$ *and* $B = \begin{bmatrix} 2 & -1 & 3 \\ -4 & 0 & 1 \\ 1 & 3 & -3 \end{bmatrix}$.

Apply the row operations indicated in each case, and write out the resulting matrix. (We work here with a square matrix rather than an augmented matrix just to practice row operations.)

1. Interchange rows 1 and 3 of A.

2. Interchange rows 2 and 3 of B.

3. Multiply each element of row 2 of A by -3.

4. Double each element in row 1 of A.

5. Multiply each element of row 2 of B by -4, and add the results to row 1.

6. Multiply each element of row 3 of A by 5, and add the results to row 2.

7. Multiply each element of row 1 of A by 3, and add the results to row 3.

8. Multiply each element of row 3 of B by -1, and add the results to row 2.

9. Form matrix $A + B$. Multiply each element of row 2 of $A + B$ by -3, and add the results to row 1.

10. Form matrix $B - A$. Multiply each element of row 1 of $B - A$ by 5, and add the results to row 3.

State which row operation must be applied to matrix A or B from above to get the following matrices.

11. to A; $\begin{bmatrix} 3 & -4 & 2 \\ -4 & 0 & 1 \\ -1 & 2 & 1 \end{bmatrix}$

12. to B; $\begin{bmatrix} 3 & 2 & 0 \\ -4 & 0 & 1 \\ 1 & 3 & -3 \end{bmatrix}$

13. to A; $\begin{bmatrix} -4 & 0 & 1 \\ 6 & -8 & 4 \\ -1 & 2 & 1 \end{bmatrix}$

14. to A; $\begin{bmatrix} -4 & 0 & 1 \\ 3 & -4 & 2 \\ 1/2 & -1 & -1/2 \end{bmatrix}$

15. to B; $\begin{bmatrix} 2 & -1 & 3 \\ 0 & -2 & 7 \\ 1 & 3 & -3 \end{bmatrix}$

16. to B; $\begin{bmatrix} -1 & -10 & 12 \\ -4 & 0 & 1 \\ 1 & 3 & -3 \end{bmatrix}$

Find the sequence of row operations that will change each of the following matrices into the identity matrix, if possible.

17. $\begin{bmatrix} 1 & 2 \\ 0 & -1 \end{bmatrix}$

18. $\begin{bmatrix} 2 & 1 \\ 1 & -1 \end{bmatrix}$

19. $\begin{bmatrix} -2 & 2 \\ 4 & 1 \end{bmatrix}$

20. $\begin{bmatrix} 0 & -1 \\ -2 & 0 \end{bmatrix}$

21. $\begin{bmatrix} 6 & 3 & 0 \\ 0 & 4 & 0 \\ 0 & 0 & 2 \end{bmatrix}$

22. $\begin{bmatrix} -2 & 2 & 0 \\ 0 & 1 & 0 \\ 1 & 1 & 1 \end{bmatrix}$

Write the augmented matrix for each system. Do not try to solve.

23. $\begin{aligned} 2x + 3y &= 11 \\ x + 2y &= 8 \end{aligned}$

24. $\begin{aligned} 3x + 5y &= -13 \\ 2x + 3y &= -9 \end{aligned}$

25. $\begin{aligned} x + 5y &= 6 \\ 3x - 4y &= 1 \end{aligned}$

26. $2x + 7y = 1$
$5x + y = -15$

27. $2x + y + z = 3$
$3x - 4y + 2z = -7$
$x + y + z = 2$

28. $4x - 2y + 3z = 4$
$3x + 5y + z = 7$
$5x - y + 4z = 7$

29. $x + y = 6$
$2y + z = 2$
$z = 2$

30. $x = 6$
$y + 2z = 2$
$x - 3z = 2$

Use matrix row operations to solve the following systems of equations.

31. $x + y = 5$
$x - y = -1$

32. $x + 2y = 5$
$2x + y = -2$

33. $x + y = -3$
$2x - 5y = -6$

34. $3x - 2y = 4$
$3x + y = -2$

35. $2x - 3y = 10$
$2x + 2y = 5$

36. $6x + y = 5$
$5x + y = 3$

37. $2x - 5y = 10$
$3x + y = 15$

38. $4x - y = 3$
$-2x + 3y = 1$

39. $2x - 3y = 2$
$4x - 6y = 1$

40. $x + 2y = 1$
$2x + 4y = 3$

41. $x + y = -1$
$y + z = 4$
$x + z = 1$

42. $x - z = -3$
$y + z = 9$
$x + z = 7$

43. $x + y - z = 6$
$2x - y + z = -9$
$x - 2y + 3z = 1$

44. $x + 3y - 6z = 7$
$2x - y + z = 1$
$x + 2y + 2z = -1$

45. $-x + y = -1$
$y - z = 6$
$x + z = -1$

46. $x + y = 1$
$2x - z = 0$
$y + 2z = -2$

47. $3x + 2y - w = 0$
$2x + z + 2w = 5$
$x + 2y - z = -2$
$2x - y + z + w = 2$

48. $x + 3y - 2z - w = 9$
$4x + y + z + 2w = 2$
$-3x - y + z - w = -5$
$x - y - 3z - 2w = 2$

Use the Gauss-Jordan method to solve each problem.

49. Linda Ramirez is a building contractor. If she hires 7 day laborers and 2 concrete finishers, her payroll for the day is \$692, while 1 day laborer and 5 concrete finishers cost \$476. Find the daily wage charge for each type of worker.

50. A biologist wants to grow two types of algae, types *A* and *B*. She has available 15 gal of nutrient I and 26 gal of nutrient II. A vat of algae *A* needs 2 gal of nutrient I and 3 gal of nutrient II, while a vat of algae *B* needs 1 gal of I and 2 gal of II. How many vats of each type of algae should the biologist grow in order to use all the nutrients?

51. To make his portrait bust of Millard Fillmore, Harry bought 2 lb of dark clay and 3 lb of light clay, paying \$13 for the clay. He later needed one more pound of dark clay and 2 lb of light clay, costing \$7 altogether. How much did he pay per pound for each type of clay?

52. The perimeter of a triangle is 21 cm. If two sides are of equal length, and the third side is 3 cm longer than one of the equal sides, find the lengths of the sides.

53. The secretary of the local consumer group bought some decals at 8¢ each and some bumper stickers at 10¢ each to give to the members. He spent a total of \$15.52. If he bought a total of 170 items, how many of each kind did he buy?

54. The Matrix and the Patrix, a local musical group, is coming to play at a school festival. The reporter for the school newspaper doesn't know how many guitarists and how many members of the rhythm section there are in the group. Janet, one of the guitarists, says that, not counting herself, there are three times as many members of the rhythm group as guitarists. Steve, a member of the rhythm section, says that, not counting himself, the number of members of the rhythm section is one less than twice the number of guitarists. How many of each are there in the group?

14.4 Matrix Inverses; Input-Output Models

Two matrices A and B are said to be **inverses** of each other if $AB = I$ and $BA = I$, where I is the appropriate identity matrix (the same one in both cases). For example, the matrices

$$\begin{bmatrix} 3 & 5 \\ 1 & 2 \end{bmatrix} \quad \text{and} \quad \begin{bmatrix} 2 & -5 \\ -1 & 3 \end{bmatrix}$$

are inverses of each other, since

$$\begin{bmatrix} 3 & 5 \\ 1 & 2 \end{bmatrix} \cdot \begin{bmatrix} 2 & -5 \\ -1 & 3 \end{bmatrix} = \begin{bmatrix} 1 & 0 \\ 0 & 1 \end{bmatrix}$$

and

$$\begin{bmatrix} 2 & -5 \\ -1 & 3 \end{bmatrix} \cdot \begin{bmatrix} 3 & 5 \\ 1 & 2 \end{bmatrix} = \begin{bmatrix} 1 & 0 \\ 0 & 1 \end{bmatrix}.$$

The inverse of matrix A is written A^{-1} (read "A inverse").

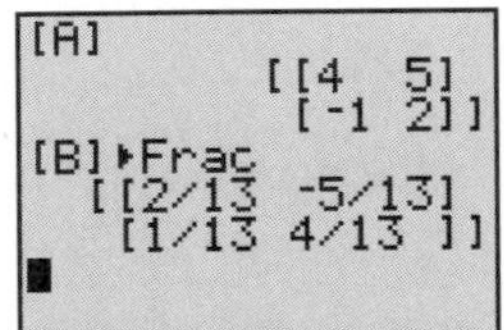

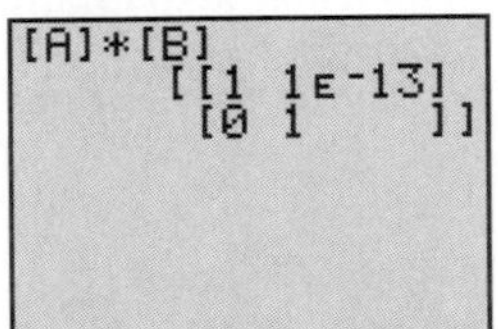

These screens support the result of Example 1. Notice that the entry in row 1, column 2 of the product is 1×10^{-13}, which is as close as the calculator can get to 0 for this computation. *One must always be aware of the limitations of technology!*

Matrix Inverses

Matrices A and A^{-1} are inverses of each other if

$$AA^{-1} = I \quad \textbf{and} \quad A^{-1}A = I.$$

EXAMPLE 1 Determine whether or not

$$\begin{bmatrix} 4 & 5 \\ -1 & 2 \end{bmatrix} \quad \text{and} \quad \begin{bmatrix} 2/13 & -5/13 \\ 1/13 & 4/13 \end{bmatrix}$$

are inverses of each other.

Find the two products.

$$\begin{bmatrix} 4 & 5 \\ -1 & 2 \end{bmatrix} \cdot \begin{bmatrix} 2/13 & -5/13 \\ 1/13 & 4/13 \end{bmatrix} = \begin{bmatrix} 8/13 + 5/13 & -20/13 + 20/13 \\ -2/13 + 2/13 & 5/13 + 8/13 \end{bmatrix} = \begin{bmatrix} 1 & 0 \\ 0 & 1 \end{bmatrix}$$

and

$$\begin{bmatrix} 2/13 & -5/13 \\ 1/13 & 4/13 \end{bmatrix} \cdot \begin{bmatrix} 4 & 5 \\ -1 & 2 \end{bmatrix} = \begin{bmatrix} 8/13 + 5/13 & 10/13 - 10/13 \\ 4/13 - 4/13 & 5/13 + 8/13 \end{bmatrix} = \begin{bmatrix} 1 & 0 \\ 0 & 1 \end{bmatrix}.$$

Since both products are equal to I (the 2×2 identity matrix in this case), the two given matrices *are* inverses of each other. ■

Only square matrices can possibly have inverses. (Why is this so?) Furthermore, not all square matrices have inverses. (See Example 4.) As we develop a way to find inverses, we will at first consider only square matrices of order 2×2.

First, by the definition of matrix inverse, if A is a 2×2 matrix and A^{-1} is its inverse, then $A^{-1}A = I$. Therefore if we can find the matrix multiplier that changes A into I, then we have found A^{-1}.

To begin our search for such a multiplier, we observe that applying any of the three row operations of the previous section to a 2×2 matrix has the same effect as multiplying that matrix by another (appropriate) 2×2 matrix. For example, let $M = \begin{bmatrix} 2 & 4 \\ 3 & 5 \end{bmatrix}$ and multiply M by the 2×2 multiplier $\begin{bmatrix} 0 & 1 \\ 1 & 0 \end{bmatrix}$:

$$\begin{bmatrix} 0 & 1 \\ 1 & 0 \end{bmatrix} \cdot \begin{bmatrix} 2 & 4 \\ 3 & 5 \end{bmatrix} = \begin{bmatrix} 0 \cdot 2 + 1 \cdot 3 & 0 \cdot 4 + 1 \cdot 5 \\ 1 \cdot 2 + 0 \cdot 3 & 1 \cdot 4 + 0 \cdot 5 \end{bmatrix} = \begin{bmatrix} 3 & 5 \\ 2 & 4 \end{bmatrix}.$$

This particular multiplier has the same effect as row operation 1 (interchanging the two rows). In Exercises 27–30, you will see that the other two row operations are also equivalent to multiplying by (appropriate) matrix multipliers.

James Joseph Sylvester (1814–1897) was working as an actuary in the early 1850s in London when Cayley (see facing page) was practicing law there. They met, and while not collaborators as such, they worked side-by-side on the algebra of invariants. Sylvester thought up "matrix" and a number of other terms such as "invariant" and "syzygy" (look that up!). Sylvester's early career was hampered by English law, which required a religious test (Sylvester was Jewish). Thus he did not get his degree from Cambridge until 1872. Sylvester, in contrast to Cayley, taught at several institutions, including a brief stay at the University of Virginia (1841). He was at Johns Hopkins from 1877 to 1883, where his influence helped to establish mathematical research at the university level. He founded the *American Journal of Mathematics.*

Sylvester was also a poet, and he published his *Laws of Verse* in 1870, applying principles of "phonetic syzygy."

Since a product of two (or more) 2×2 matrices is another 2×2 matrix, a sequence of row operations, applied in succession, has the net effect of multiplying by some single matrix. Now we know that

$$A^{-1}A = I \qquad \text{and also} \qquad A^{-1}I = A^{-1},$$

so that the multiplier, A^{-1}, that changes A to I will also change I to A^{-1}. This means that any sequence of row operations that changes A to I will also change I to A^{-1}. This fact enables us to use row operations to find A^{-1}. Form a new augmented matrix of the form $[A|I]$ and apply row operations to change A (on the left) to I. The I (on the right) will automatically become A^{-1}, as illustrated here.

$[A|I]$

Apply row
operations
to obtain

$[I|A^{-1}]$

EXAMPLE 2 Find the inverse of the matrix $A = \begin{bmatrix} 2 & -5 \\ -1 & 3 \end{bmatrix}$.

Form the augmented matrix $[A|I]$ and then change A to I on the left, using row operations as in Section 3.

$$[A|I]$$

$$\left[\begin{array}{cc|cc} 2 & -5 & 1 & 0 \\ -1 & 3 & 0 & 1 \end{array}\right]$$

Arthur Cayley (1821–1895) showed such an early affinity for mathematics that he went to Cambridge instead of entering his father's business. Even while Cayley was practicing law (1848–1862), he wrote numerous theoretical papers on mathematics. During this time he and James Sylvester were founding the theory of invariants, out of which grew Cayley's *Memoir on the Theory of Matrices* (1857). Invariants are algebraic expressions that remain virtually unaltered during certain "transformations," such as flips, rotations, and stretches.

In 1863 Cayley became Sadlerian Professor of Pure Mathematics at Cambridge. Except for a lecture series at Johns Hopkins in 1881, his life centered around Cambridge. His work, in contrast, extended to nearly every branch of mathematics. He helped to open the horizons of algebra to include systems that worked even where the commutative law didn't hold. He also helped to free geometry from its three-dimensional Euclidean limitations. Cayley theorized about geometry in n dimensions as early as 1843.

First obtain a 1 in row 1, column 1:

$$\left[\begin{array}{cc|cc} 1 & -5/2 & 1/2 & 0 \\ -1 & 3 & 0 & 1 \end{array}\right] \quad \tfrac{1}{2}\text{R1}$$

Next obtain a 0 in row 2, column 1:

$$\left[\begin{array}{cc|cc} 1 & -5/2 & 1/2 & 0 \\ 0 & 1/2 & 1/2 & 1 \end{array}\right] \quad \text{R1} + \text{R2}$$

Next obtain a 1 in row 2, column 2:

$$\left[\begin{array}{cc|cc} 1 & -5/2 & 1/2 & 0 \\ 0 & 1 & 1 & 2 \end{array}\right] \quad 2\text{R2}$$

Finally obtain a 0 in row 1, column 2:

$$\left[\begin{array}{cc|cc} 1 & 0 & 3 & 5 \\ 0 & 1 & 1 & 2 \end{array}\right] \quad \tfrac{5}{2}\text{R2} + \text{R1}$$

$$[I \mid A^{-1}]$$

This result shows that $A^{-1} = \begin{bmatrix} 3 & 5 \\ 1 & 2 \end{bmatrix}$. You can verify this by computing AA^{-1} and $A^{-1}A$. Both of these products should equal I, the 2×2 identity matrix. ■

The procedure above generalizes to a square matrix of any order (3×3, 4×4, and so on), as summarized here.

Finding the Inverse of a Matrix

Given any square matrix A, its inverse A^{-1} (if there is an inverse) is found by performing the following steps:

1. Form the augmented matrix $[A \mid I]$.
2. Use row operations to obtain a 1 in row 1, column 1.
3. Obtain 0s in all other column 1 positions.
4. Obtain a 1 in row 2, column 2.
5. Obtain 0s in all other column 2 positions.
6. Continue until A is changed into I.
7. The right half of the final result is now A^{-1}.

If any one of these steps is impossible, then matrix A has no inverse.

EXAMPLE 3 Find the inverse of $B = \begin{bmatrix} 1 & 3 & 3 \\ 1 & 4 & 3 \\ 1 & 3 & 4 \end{bmatrix}$.

Perform the steps listed above.

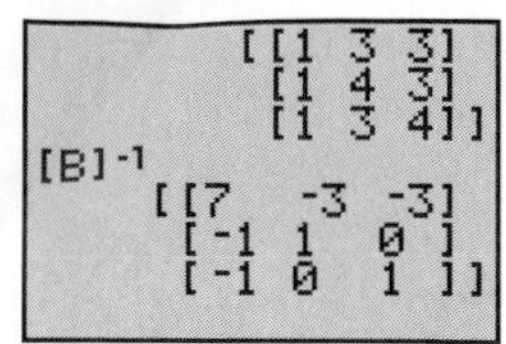

Matrix B of Example 3 is defined at the top of this screen. Its inverse, B^{-1}, is displayed at the bottom.

Step 1:

$$\left[\begin{array}{ccc|ccc} 1 & 3 & 3 & 1 & 0 & 0 \\ 1 & 4 & 3 & 0 & 1 & 0 \\ 1 & 3 & 4 & 0 & 0 & 1 \end{array}\right]$$

Step 2: There is already a 1 in row 1, column 1.

Step 3:

$$\left[\begin{array}{ccc|ccc} 1 & 3 & 3 & 1 & 0 & 0 \\ 0 & 1 & 0 & -1 & 1 & 0 \\ 0 & 0 & 1 & -1 & 0 & 1 \end{array}\right] \quad \begin{array}{l} \\ -\text{R1} + \text{R2} \\ -\text{R1} + \text{R3} \end{array}$$

Step 4: There is already a 1 in row 2, column 2.

Step 5:

$$\left[\begin{array}{ccc|ccc} 1 & 0 & 3 & 4 & -3 & 0 \\ 0 & 1 & 0 & -1 & 1 & 0 \\ 0 & 0 & 1 & -1 & 0 & 1 \end{array}\right] \quad \begin{array}{l} -3\text{R2} + \text{R1} \\ \\ \\ \end{array}$$

Step 6:

$$\left[\begin{array}{ccc|ccc} 1 & 0 & 0 & 7 & -3 & -3 \\ 0 & 1 & 0 & -1 & 1 & 0 \\ 0 & 0 & 1 & -1 & 0 & 1 \end{array}\right] \quad \begin{array}{l} -3\text{R3} + \text{R1} \\ \\ \\ \end{array}$$

Step 7:

$$B^{-1} = \begin{bmatrix} 7 & -3 & -3 \\ -1 & 1 & 0 \\ -1 & 0 & 1 \end{bmatrix}$$

To check, show that $BB^{-1} = I$ and $B^{-1}B = I$. ■

EXAMPLE 4 Let $C = \begin{bmatrix} 2 & 1 \\ 6 & 3 \end{bmatrix}$. Find C^{-1}.

Proceed as before.

$$\left[\begin{array}{cc|cc} 2 & 1 & 1 & 0 \\ 6 & 3 & 0 & 1 \end{array}\right]$$

$$\left[\begin{array}{cc|cc} 1 & 1/2 & 1/2 & 0 \\ 6 & 3 & 0 & 1 \end{array}\right] \quad \frac{1}{2}\text{R1}$$

$$\left[\begin{array}{cc|cc} 1 & 1/2 & 1/2 & 0 \\ 0 & 0 & -3 & 1 \end{array}\right] \quad -6\text{R1} + \text{R2}$$

We cannot go further. There is no way to obtain a 1 in row 2, column 2 without losing the 0 just obtained in row 2, column 1. Thus, matrix C has no inverse. ■

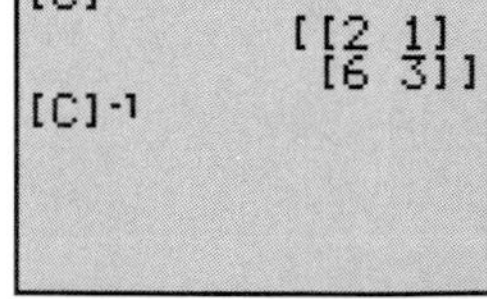

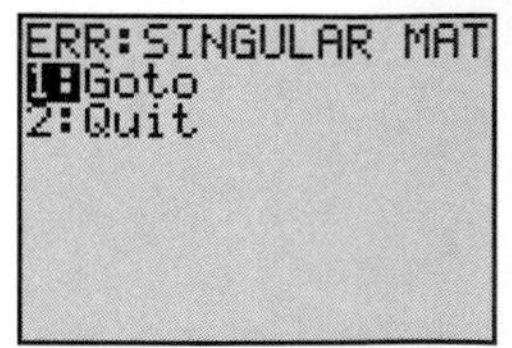

Matrix C, as defined in Example 4, has no inverse. Such a matrix is called a *singular matrix,* as seen on the bottom screen.

The matrix operations introduced in this chapter (addition, subtraction, multiplication by a scalar, and multiplication of two matrices) form the basis of *matrix algebra,* which is similar (but not identical) to real number algebra. Matrix algebra has many applications.

The ability to perform algebraic operations on matrices enables us to formulate the matrix solution of systems of equations in a simpler way. For example, in Example 1 of Section 3, the system

$$\begin{aligned} 3x - 4y &= 1 \\ 5x + 2y &= 19 \end{aligned}$$

was solved using row operations. Notice now that if we arrange the x and y coefficients, the variables of the system, and the constants of the system into the three separate matrices

$$A = \begin{bmatrix} 3 & -4 \\ 5 & 2 \end{bmatrix}, \qquad X = \begin{bmatrix} x \\ y \end{bmatrix}, \qquad \text{and} \qquad B = \begin{bmatrix} 1 \\ 19 \end{bmatrix},$$

the system can be written as the single matrix equation

$$AX = B.$$

Verify that the statement

$$\begin{bmatrix} 3 & -4 \\ 5 & 2 \end{bmatrix} \cdot \begin{bmatrix} x \\ y \end{bmatrix} = \begin{bmatrix} 1 \\ 19 \end{bmatrix}$$

is equivalent to the original system of equations.

Now we solve the matrix equation $AX = B$ using matrix algebra as follows.

$$\begin{aligned} AX &= B \\ A^{-1}(AX) &= A^{-1}B && \text{Multiply both sides by } A^{-1}. \\ (A^{-1}A)X &= A^{-1}B && \text{Use the associative property on the left.} \\ IX &= A^{-1}B && A^{-1}A = I \text{ by the definition of inverse.} \\ X &= A^{-1}B && IX = X \text{ by the definition of identity.} \end{aligned}$$

When multiplying both sides of a matrix equation by another matrix (as was done above), be sure to multiply in the same order on both sides of the equation, since matrix multiplication is not commutative.

The matrix X, which now contains the variable values, that is, the solution of the system, is simply the product $A^{-1}B$. We summarize this result as follows.

Solving a Linear System Using Matrix Algebra

A system of linear equations represented by the matrix equation

$$AX = B$$

can be solved by finding A^{-1} and computing the product $A^{-1}B$. If A^{-1} does not exist, the system does not have a unique solution.

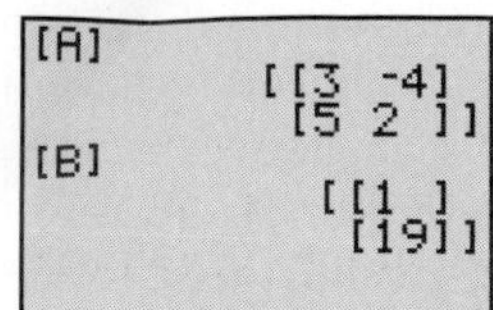

A graphing calculator with matrix capability can easily accomplish the system-solving procedure described in Example 5.

EXAMPLE 5 Solve the system

$$\begin{aligned} 3x - 4y &= 1 \\ 5x + 2y &= 19 \end{aligned}$$

using matrix algebra.

First find A^{-1} as outlined in this section.

$$\left[\begin{array}{cc|cc} 3 & -4 & 1 & 0 \\ 5 & 2 & 0 & 1 \end{array}\right]$$

$$\left[\begin{array}{cc|cc} 1 & -4/3 & 1/3 & 0 \\ 5 & 2 & 0 & 1 \end{array}\right] \quad \frac{1}{3}\text{R1}$$

$$\left[\begin{array}{cc|cc} 1 & -4/3 & 1/3 & 0 \\ 0 & 26/3 & -5/3 & 1 \end{array}\right] \quad -5\text{R1} + \text{R2}$$

$$\left[\begin{array}{cc|cc} 1 & -4/3 & 1/3 & 0 \\ 0 & 1 & -5/26 & 3/26 \end{array}\right] \quad \frac{3}{26}\text{R2}$$

$$\left[\begin{array}{cc|cc} 1 & 0 & 1/13 & 2/13 \\ 0 & 1 & -5/26 & 3/26 \end{array}\right] \quad \frac{4}{3}\text{R2} + \text{R1}$$

Now $$X = A^{-1}B = \begin{bmatrix} 1/13 & 2/13 \\ -5/26 & 3/26 \end{bmatrix} \cdot \begin{bmatrix} 1 \\ 19 \end{bmatrix} = \begin{bmatrix} 3 \\ 2 \end{bmatrix}$$

so the solution set of the system is $\{(3, 2)\}$. ◆

It may have occurred to you that this matrix algebra method of solving a linear system is really no better than the row operation method of the last section. In fact, finding A^{-1} here required the same sequence of row operations that we used before. In addition, we then had to compute the product $A^{-1}B$.

The real advantage of this last method is that modern calculators (and computers) can easily compute matrix inverses as well as products. So when using a calculator, finding $A^{-1}B$ simply requires entering matrices A and B from the system of equations and then calculating an inverse and a product on the machine. (And these calculations are no more work for large systems than for small ones—except that there are more numbers to enter.)

EXAMPLE 6 Use matrix algebra to solve this system.

$$\begin{aligned} 3x - 5y + 2z &= 8 \\ x + 3y - z &= 5 \\ x - 11y + 4z &= -2 \end{aligned}$$

Start with the coefficient matrix

$$A = \begin{bmatrix} 3 & -5 & 2 \\ 1 & 3 & -1 \\ 1 & -11 & 4 \end{bmatrix}.$$

When we try to calculate A^{-1}, we find that A does not have an inverse. (Verify this.) Therefore the given system does not have a unique solution. ◆

Input-Output Models

The economist W. W. Leontief, winner of a Nobel Prize in 1973, developed a method for analyzing an economy which involves matrices. The method is generally known as a **Leontief model,** or an **input-output model.** Suppose that an economy produces n commodities. The production of each uses some (perhaps all) of the other commodities in the economy. Thus, the production of oil requires oil, as well as steel, electricity, and so on. The amounts of all commodities used in the production of one unit of each commodity can be written as an $n \times n$ matrix, called the *input-output matrix,* or the **technological matrix** of the system. Denote the technological matrix A.

Let each of the following be an $n \times 1$ matrix showing the appropriate amounts of all commodities in the economy.

P = Gross Production (total of all commodities produced)

U = Internal Use (amounts of commodities used up in the production process itself)

D = Demand for net production (amounts of commodities needed above and beyond internal use)

The gross production must be sufficient to supply internal use plus the demand for net production:

$$P = U + D.$$

Consider an economy with only two basic products, wheat and oil. Assume that the production of 1 metric ton of wheat requires 1/4 metric ton of wheat and 1/3 metric ton of oil. The production of 1 metric ton of oil requires 1/12 metric ton of wheat and 1/9 metric ton of oil. This leads to the following technological matrix.

$$A = \text{INPUT}\quad \begin{matrix} & \text{OUTPUT} \\ & \begin{matrix}\text{Wheat} & \text{Oil}\end{matrix} \\ \begin{matrix}\text{Wheat} \\ \text{Oil}\end{matrix} & \begin{bmatrix} 1/4 & 1/12 \\ 1/3 & 1/9 \end{bmatrix}, \end{matrix} \qquad \text{or simply} \qquad A = \begin{bmatrix} 1/4 & 1/12 \\ 1/3 & 1/9 \end{bmatrix}$$

(Notice that input appears on the left of the matrix and output on the top.)

We now claim that the internal use matrix is given by

$$U = AP.$$

To illustrate, suppose the gross production was

$$P = \begin{matrix} & \text{Gross production} \\ \begin{matrix}\text{Wheat} \\ \text{Oil}\end{matrix} & \begin{bmatrix} 864 \\ 4{,}320 \end{bmatrix}. \end{matrix}$$

Then

$$\begin{pmatrix}\text{Wheat used}\\ \text{internally}\end{pmatrix} = \begin{pmatrix}\text{Wheat}\\ \text{used}\\ \text{per unit}\\ \text{of wheat}\\ \text{produced}\end{pmatrix} \cdot \begin{pmatrix}\text{Units of}\\ \text{wheat}\\ \text{produced}\end{pmatrix} + \begin{pmatrix}\text{Wheat}\\ \text{used}\\ \text{per unit}\\ \text{of oil}\\ \text{produced}\end{pmatrix} \cdot \begin{pmatrix}\text{Units}\\ \text{of oil}\\ \text{produced}\end{pmatrix}$$

$$= \frac{1}{4} \cdot 864 + \frac{1}{12} \cdot 4{,}320$$

$$= 576$$

and

$$\begin{pmatrix}\text{Oil used}\\ \text{internally}\end{pmatrix} = \begin{pmatrix}\text{Oil used}\\ \text{per unit}\\ \text{of wheat}\\ \text{produced}\end{pmatrix} \cdot \begin{pmatrix}\text{Units of}\\ \text{wheat}\\ \text{produced}\end{pmatrix} + \begin{pmatrix}\text{Oil used}\\ \text{per unit}\\ \text{of oil}\\ \text{produced}\end{pmatrix} \cdot \begin{pmatrix}\text{Units}\\ \text{of oil}\\ \text{produced}\end{pmatrix}$$

$$= \frac{1}{3} \cdot 864 + \frac{1}{9} \cdot 4{,}320$$

$$= 768$$

so

$$U = \begin{array}{l} \\ \text{Wheat} \\ \text{Oil} \end{array} \begin{array}{c} \text{Internal use} \\ \begin{bmatrix} 576 \\ 768 \end{bmatrix} \end{array}.$$

But this is just what we obtain from the product AP:

$$AP = \begin{bmatrix} 1/4 & 1/12 \\ 1/3 & 1/9 \end{bmatrix} \cdot \begin{bmatrix} 864 \\ 4{,}320 \end{bmatrix} = \begin{bmatrix} 576 \\ 768 \end{bmatrix}.$$

Since $U = AP$, the equation

$$P = U + D$$

now becomes

$$P = AP + D \qquad \text{or} \qquad P - AP = D.$$

This equation, however, gives D in terms of P. The usual situation is that D is known and we need to find P. So we use matrix algebra to solve for P.

$$P - AP = D$$

$$IP - AP = D \qquad \text{Replace } P \text{ with } IP\text{, where } I \text{ is the identity matrix with the same order as } A.$$

$$(I - A)P = D \qquad \text{Distributive property}$$

$$(I - A)^{-1}(I - A)P = (I - A)^{-1}D \qquad \text{Multiply both sides by } (I - A)^{-1}.$$

$$IP = (I - A)^{-1}D \qquad \text{Replace } (I - A)^{-1}(I - A) \text{ with } I.$$

$$P = (I - A)^{-1}D \qquad \text{Replace } IP \text{ with } P.$$

We now have a method of finding the gross production, P, required to meet a demand for net production, D.

```
[A]▸Frac
    [[1/4 1/12]
     [1/3 1/9 ]]
identity 2-[A]▸F
rac
  [[3/4  -1/12]
   [-1/3 8/9  ]]
```

```
(identity 2-[A])
-1▸Frac
 [[32/23 3/23 ]
  [12/23 27/23]]
```

```
[D]
         [[500 ]
          [1000]]
(identity 2-[A])
-1*[D]
  [[826.0869565]
   [1434.782609]]
```

As you read through Example 7, follow these screens from top to bottom to see how a graphing calculator accomplishes the computation.

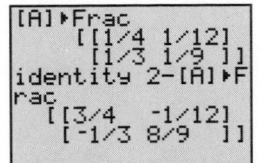

EXAMPLE 7 Find the gross production required in the wheat/oil economy above if the demand for net production is

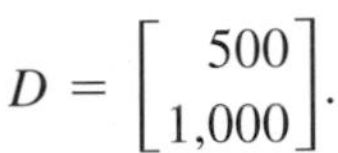

$$D = \begin{bmatrix} 500 \\ 1{,}000 \end{bmatrix}.$$

First compute $I - A$:

$$I - A = \begin{bmatrix} 1 & 0 \\ 0 & 1 \end{bmatrix} - \begin{bmatrix} 1/4 & 1/12 \\ 1/3 & 1/9 \end{bmatrix} = \begin{bmatrix} 3/4 & -1/12 \\ -1/3 & 8/9 \end{bmatrix}.$$

Obtain the inverse of $I - A$ (using the procedure described earlier, or a calculator).

$$(I - A)^{-1} = \begin{bmatrix} 32/23 & 3/23 \\ 12/23 & 27/23 \end{bmatrix}$$

Now

$$P = (I - A)^{-1} \cdot D = \begin{bmatrix} 32/23 & 3/23 \\ 12/23 & 27/23 \end{bmatrix} \cdot \begin{bmatrix} 500 \\ 1{,}000 \end{bmatrix} = \begin{bmatrix} 826 \\ 1{,}435 \end{bmatrix}.$$

(Final amounts have been rounded to whole numbers.) A demand for net production of 500 metric tons of wheat and 1,000 metric tons of oil requires a gross production of 826 metric tons of wheat and 1,435 metric tons of oil.

14.4 EXERCISES

Find the inverse of each matrix that has an inverse.

1. $\begin{bmatrix} 1 & 0 \\ 0 & -1 \end{bmatrix}$ **2.** $\begin{bmatrix} 0 & 1 \\ 1 & 0 \end{bmatrix}$ **3.** $\begin{bmatrix} 2 & 1 \\ 4 & 3 \end{bmatrix}$ **4.** $\begin{bmatrix} -3 & 2 \\ 6 & -5 \end{bmatrix}$

5. $\begin{bmatrix} 1 & 1 \\ 1 & 1 \end{bmatrix}$ **6.** $\begin{bmatrix} -1 & -1 \\ 1 & 1 \end{bmatrix}$ **7.** $\begin{bmatrix} 1 & 2 & 0 \\ 0 & 1 & 0 \\ 1 & -1 & 1 \end{bmatrix}$ **8.** $\begin{bmatrix} 0 & 1 & 0 \\ 0 & 0 & -2 \\ 1 & -1 & 0 \end{bmatrix}$

9. $\begin{bmatrix} 4 & 3 & 3 \\ -1 & 0 & -1 \\ -4 & -4 & -3 \end{bmatrix}$ **10.** $\begin{bmatrix} 1 & 0 & 2 \\ -1 & 0 & -2 \\ 1 & 1 & 1 \end{bmatrix}$

Use matrix algebra to find the unique solution of each system. If the system has no unique solution, state why.

11. $\begin{aligned} 4x + y &= 5 \\ 2x + y &= 3 \end{aligned}$

12. $\begin{aligned} 3x - 7y &= 31 \\ 2x - 4y &= 18 \end{aligned}$

13. $\begin{aligned} 5x - y &= 6 \\ -10x + 2y &= 15 \end{aligned}$

14. $\begin{aligned} 5x - y &= 14 \\ x + 8y &= 11 \end{aligned}$

15. $\begin{aligned} x + y - z &= 6 \\ 2x - y + z &= -9 \\ x - 2y + 3z &= 1 \end{aligned}$

16. $\begin{aligned} 4x + 2y - 3z &= 6 \\ x - 4y + z &= -4 \\ -x + 2z &= 2 \end{aligned}$

17. $\begin{aligned} 2x - y + 3z &= 0 \\ x + 2y - z &= 5 \\ 2y + z &= 1 \end{aligned}$

18. $\begin{aligned} x + y &= 1 \\ 2x - z &= 0 \\ y + 2z &= -2 \end{aligned}$

19. $\begin{aligned} 2x - 3y + z &= 4 \\ x + y - z &= 1 \\ -5y + 3z &= 2 \end{aligned}$

20. $\begin{aligned} x + 3y &= 7 \\ 2x - y - 5z &= 12 \\ 7y + 5z &= 2 \end{aligned}$

In each of Exercises 1 and 2, the given matrix was its own inverse. In each of the following cases, find the missing elements to make the given matrix its own inverse.

21. $\begin{bmatrix} 3 & \square \\ 2 & \square \end{bmatrix}$ **22.** $\begin{bmatrix} \square & 8 \\ \square & 5 \end{bmatrix}$ **23.** $\begin{bmatrix} 2 & \square \\ \square & -2 \end{bmatrix}$ **24.** $\begin{bmatrix} \square & 5 \\ -7 & \square \end{bmatrix}$

Assuming that a and b are real numbers, determine whether each of the following statements is sometimes true, always true, *or* never true. *If sometimes true, give the conditions under which it is true.*

25. $\begin{bmatrix} a & 0 \\ 0 & a \end{bmatrix}$ and $\begin{bmatrix} b & 0 \\ 0 & b \end{bmatrix}$ are inverses of each other.

26. $\begin{bmatrix} 0 & a \\ a & 0 \end{bmatrix}$ and $\begin{bmatrix} 0 & b \\ b & 0 \end{bmatrix}$ are inverses of each other.

27. Recall that "row operation 2" is to multiply a given row by some nonzero constant. Show that the following product has the effect (on the second matrix) of multiplying its first row by k.

$$\begin{bmatrix} k & 0 \\ 0 & 1 \end{bmatrix} \cdot \begin{bmatrix} 2 & 4 \\ 3 & 5 \end{bmatrix}$$

28. What 2×2 multiplier, on the left, will change

$$\begin{bmatrix} 2 & 4 \\ 3 & 5 \end{bmatrix} \quad \text{to} \quad \begin{bmatrix} 2 & 4 \\ 3k & 5k \end{bmatrix}$$

(that is, multiply the second row by the constant k)?

29. "Row operation 3" is to add some multiple of one row to another row. Show that the following product has the effect (on the second matrix) of adding twice its first row elements to its corresponding second row elements.

$$\begin{bmatrix} 1 & 0 \\ 2 & 1 \end{bmatrix} \cdot \begin{bmatrix} 2 & 4 \\ 3 & 5 \end{bmatrix}$$

30. What 2×2 multiplier, on the left, will change

$$\begin{bmatrix} 2 & 4 \\ 3 & 5 \end{bmatrix} \quad \text{to} \quad \begin{bmatrix} 11 & 19 \\ 3 & 5 \end{bmatrix}$$

(that is, add three times the second row to the first row)?

The matrix in Example 4 had no inverse. Here are several other 2×2 *matrices that do not have inverses:*

$$\begin{bmatrix} 5 & 1 \\ 10 & 2 \end{bmatrix}, \quad \begin{bmatrix} 6 & -3 \\ -8 & 4 \end{bmatrix}, \quad \begin{bmatrix} 17 & 22 \\ 34 & 44 \end{bmatrix}.$$

31. Describe the characteristic shared by the above matrices that causes them to not have inverses.

32. Find two additional 2×2 matrices, neither of which has an inverse.

The ***determinant*** *of a* 2×2 *matrix is denoted by vertical bars (similar to absolute value signs), so that, for example, the determinant of* $\begin{bmatrix} 2 & 6 \\ 5 & 8 \end{bmatrix}$ *is denoted* $\begin{vmatrix} 2 & 6 \\ 5 & 8 \end{vmatrix}$. *The numerical* ***value*** *of a determinant is defined as follows:*

$$\begin{vmatrix} a & c \\ b & d \end{vmatrix} = ad - bc.$$

Evaluate each determinant.

33. $\begin{vmatrix} 7 & 4 \\ 5 & 3 \end{vmatrix}$ **34.** $\begin{vmatrix} 2 & 8 \\ -3 & -6 \end{vmatrix}$ **35.** $\begin{vmatrix} -4 & 8 \\ 5 & -10 \end{vmatrix}$ **36.** $\begin{vmatrix} 3 & -2 \\ 6 & -4 \end{vmatrix}$

37. Extend the concept of a determinant to 3×3 matrices, and show how to evaluate a 3×3 determinant. (You may want to consult an algebra book.)

38. Give an example of a 3×3 determinant with the value 0.

A matrix is called ***singular*** *if its determinant has the value 0.*

39. Explain why a system of linear equations has no unique solution if its coefficient matrix is singular.

40. If A is a singular square matrix, what can you say about A^{-1}?

Use a calculator to find the inverse of each matrix. Give all elements to three decimal places.

41. $\begin{bmatrix} 1.2 & 3.7 \\ -4.8 & 5.5 \end{bmatrix}$ **42.** $\begin{bmatrix} 7/6 & 2/11 \\ 5/13 & 1/9 \end{bmatrix}$

43. $\begin{bmatrix} 14.8 & 18.1 & 11.6 \\ 8.2 & 12.6 & 9.5 \\ 15.1 & 13.8 & 7.4 \end{bmatrix}$ **44.** $\begin{bmatrix} 125 & 391 & 416 \\ 277 & 143 & 209 \\ 610 & 157 & 504 \end{bmatrix}$

45. In the real number system, how many numbers do not have multiplicative inverses? What are they?

46. In the system of 2×2 matrices, how many matrices do not have multiplicative inverses? What are they?

47. Assume the same basic wheat/oil economy described in the text. Find the gross production required for a net production of 750 metric tons of wheat and 1,000 metric tons of oil. Round final amounts to whole numbers.

48. Find the gross production required in Exercise 47 if production of 1 metric ton of wheat consumes 1/3 metric ton of oil (and no wheat), while the production of 1 metric ton of oil consumes 1/5 metric ton of wheat (and no oil). Round final amounts to whole numbers.

49. Consider an economy depending on three basic commodities: agriculture, manufacturing, and transportation. Suppose 1/4 unit of manufacturing and 1/2 unit of transportation are required to produce 1 unit of agriculture, 1/2 unit of agriculture and 1/4 unit of transportation are required to produce 1 unit of manufacturing, and 1/4 unit of agriculture and 1/4 unit of manufacturing are required to produce 1 unit of transportation. How many units of each commodity should be produced to satisfy a demand for 1,000 units of each commodity? Round final amounts to whole numbers.

50. A simple economy depends on three commodities: oil, corn, and coffee. Production of one unit of oil requires .1 unit of oil, .2 unit of corn, and no units of coffee. To produce one unit of corn requires .2 unit of oil, .1 unit of corn, and .05 unit of coffee. To produce one unit of coffee requires .1 unit of oil, .05 unit of corn, and .1 unit of coffee. Find the gross production required to give a net production of 1,000 units each of oil, corn, and coffee. Round final amounts to whole numbers.

EXTENSION

Cryptography

Cryptography, or **code theory,** is one of the areas being continually affected by advancing mathematical theory. "Secret codes," of various forms, have been used throughout history, for purposes that range from international intrigue to childhood pastimes. Today, with more and more information being stored and exchanged within computerized networks, businesses and other interests depend upon advanced cryptographic systems for protection of their data.

A common type of code is constructed by replacing each letter of the alphabet with another letter or symbol, such as

A = 3 C = # E = +
B = $ D = 9 F = *

and so on. A code like this is very easy for an expert to break. This is because mere substitution of one symbol for another does not change the relative frequency of occurrence of the symbols. Thus, in the substitution of symbols for letters in the English language, + would occur most frequently in any reasonably long message simply because + represents E, and E is the most common letter used in English. (We say "reasonably long message" since in some short messages, such as "summer sessions seldom stimulate", the letter E might not be the most common.) Code breakers have spent a lot of time deciding on the frequency of occurrence of letters. The sample referred to on the opening page of this chapter yields the following percentages for all letters of the alphabet.

Letter	Percent	Letter	Percent	Letter	Percent
E	13.0	D	4.4	G, W	1.6
T	9.3	H, L	3.5	V	1.3
N	7.8	C	3.0	B	.9
R	7.7	F	2.8	X	.5
I, O	7.4	P, U	2.7	K, Q	.3
A	7.3	M	2.5	J	.2
S	6.3	Y	1.9	Z	.1

Governments and computer users need to develop more sophisticated methods of coding and decoding messages. Several of the methods depend on matrix theory. To see how, assign numbers to letters of the alphabet. For simplicity, let A correspond to 1, B to 2, C to 3, and so on. Ignore punctuation and let 27 correspond to a blank space.

A = 1	G = 7	M = 13	S = 19	Y = 25
B = 2	H = 8	N = 14	T = 20	Z = 26
C = 3	I = 9	O = 15	U = 21	space = 27
D = 4	J = 10	P = 16	V = 22	
E = 5	K = 11	Q = 17	W = 23	
F = 6	L = 12	R = 18	X = 24	

Break the message into groups of three letters (and spaces) each. For example, the message SOME CODES ARE EASILY BROKEN becomes

SOM E_C ODE S_A RE_ EAS ILY _BR OKE N_ _

where ___ represents a blank. Each group of letters is then converted to a 3×1 matrix. For example, the letters S, O, and M convert to 19, 15, and 13, respectively, so that

$$\begin{bmatrix} \text{S} \\ \text{O} \\ \text{M} \end{bmatrix} \quad \text{corresponds to} \quad \begin{bmatrix} 19 \\ 15 \\ 13 \end{bmatrix}.$$

Our message corresponds to the following sequence of matrices.

$$\begin{bmatrix} 19 \\ 15 \\ 13 \end{bmatrix} \begin{bmatrix} 5 \\ 27 \\ 3 \end{bmatrix} \begin{bmatrix} 15 \\ 4 \\ 5 \end{bmatrix} \begin{bmatrix} 19 \\ 27 \\ 1 \end{bmatrix} \begin{bmatrix} 18 \\ 5 \\ 27 \end{bmatrix} \begin{bmatrix} 5 \\ 1 \\ 19 \end{bmatrix} \begin{bmatrix} 9 \\ 12 \\ 25 \end{bmatrix} \begin{bmatrix} 27 \\ 2 \\ 18 \end{bmatrix} \begin{bmatrix} 15 \\ 11 \\ 5 \end{bmatrix} \begin{bmatrix} 14 \\ 27 \\ 27 \end{bmatrix}$$

Now choose any 3×3 "multiplier" matrix that has an inverse. Let us choose matrix M as follows.

$$M = \begin{bmatrix} 1 & 3 & 3 \\ 1 & 4 & 3 \\ 1 & 3 & 4 \end{bmatrix}$$

Finding the product of M and each of the preceding matrices gives

$$\begin{bmatrix} 103 \\ 118 \\ 116 \end{bmatrix} \begin{bmatrix} 95 \\ 122 \\ 98 \end{bmatrix} \begin{bmatrix} 42 \\ 46 \\ 47 \end{bmatrix}$$

and so on. (See Exercise 5.) The process of translating the message into this series of matrices is called "encoding." In effect, it disguises the original sequence of matrices. The entries of these product matrices can then be transmitted as the message. The recipient of the message divides the numbers into groups of 3 and converts each group of 3 into a matrix. After multiplying these matrices by the matrix M^{-1}, the message can be read.

A code of this type is simple to use but hard to decode for anyone not knowing the multiplication matrix M. For further information on codes, see *The Codebreakers* by David Kahn. Further information on the substitution of symbols for letters is given in the Sherlock Holmes story "The Dancing Men" and in Edgar Allan Poe's "The Gold Bug" (see the marginal note at the left).

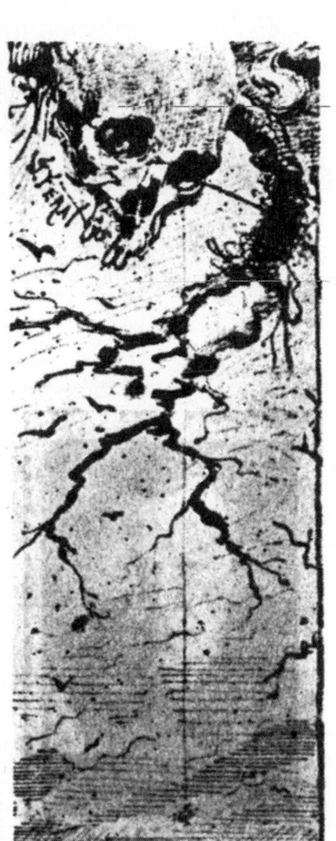

"Shoot from the left eye of the death's head. . . ." In Poe's tale *The Gold-Bug* (1843), the image of a skull on an old piece of parchment starts Legrand, the central character, on a treasure hunt. In the course of it he must decode a message by comparing the frequency of its given symbols with the frequency of letters in the English language. The illustration shows the gold bug (instead of a bullet) dropped from the left eye socket of the skull.

Extension Exercises

1. The message

THE THREE TENORS

is to be encoded by using the matrix method described in the text. The message is broken into groups of two letters (and spaces). Find the sequence of matrices derived from the given message.

2. Use the multiplier matrix $M = \begin{bmatrix} -1 & 2 \\ 2 & -5 \end{bmatrix}$ to encode the sequence of matrices of Exercise 1, and list the resulting sequence of 2×1 matrices.

3. A message that was encoded with the multiplier matrix of Exercise 2 was received as the following sequence of numbers.

$$39, -98, -19, 34, -7, 10, 3, -11, 36,$$
$$-99, 21, -62, 7, -18, 33, -87, 41, -109$$

What multiplier matrix should be used for decoding this message?

4. Find the original message (of Exercise 3).

5. Finish encoding the message SOME CODES ARE EASILY BROKEN given in the text.

6. The person receiving the message of Exercise 5 (SOME CODES ARE EASILY BROKEN) also received the following encoded message (already converted to matrix form).

$$\begin{bmatrix} 99 \\ 119 \\ 107 \end{bmatrix} \begin{bmatrix} 116 \\ 134 \\ 135 \end{bmatrix} \begin{bmatrix} 84 \\ 85 \\ 102 \end{bmatrix} \begin{bmatrix} 98 \\ 125 \\ 102 \end{bmatrix} \begin{bmatrix} 45 \\ 51 \\ 51 \end{bmatrix} \begin{bmatrix} 81 \\ 84 \\ 102 \end{bmatrix} \begin{bmatrix} 153 \\ 173 \\ 180 \end{bmatrix}$$

Decode this second message, assuming it was encoded using the multiplier matrix given in the text.

7. Suppose you are a special agent. The encoding multiplier matrix for today is $M = \begin{bmatrix} 2 & 1 & 2 \\ 3 & 4 & 3 \\ 3 & 1 & 2 \end{bmatrix}$. You need to send the message

DESTROY FILES IMMEDIATELY

to another agent. What sequence of numbers would you transmit?

8. After your transmission of Exercise 7, you receive the response

45, 90, 51, 83, 172, 88, 51, 89, 55, 88, 177, 108, 63, 107, 88.

What is the message?

9. Could the encoding matrix (the multiplier) used in the type of cryptography discussed here be of order 4×4? Explain why or why not.

10. What are the restrictions (if any) on the type of matrix used as the encoding multiplier?

11. Explain why the matrix type code discussed here would be difficult to break.

12. The following message is written in a code in which the frequency of the symbols is the main key to the solution.

)?−−8))6*+8506*3×6;4?*7*&×*−6.48()6)985)?(8+2:;48)81&?(;46*3)6*;
48&(+8(*598+.8()8=8(5*−8−5(81?098;4&+)&15*50:)6)6*;?6;6&*0?−7

Find the frequency of each symbol. By comparing high frequency symbols with the high frequency letters in English (see the text above), try to decipher the message. (*Hint:* look for repeated two-symbol combinations and double letters for added clues. Try to identify vowels first.)

14.5 Matrices and Game Theory*

In this section, we discuss a large and still growing branch of mathematics called **game theory.** While the name may suggest something frivolous, the applications certainly are not. Game theory was developed in the 1940s as a means of analyzing competitive situations in business, warfare, and social settings.

A **two-player game,** the most common case, involves two opponents. These opponents can be two persons, two teams of persons, two countries, and so on. One player may even be a nonpersonal entity. For example, imagine that you are a California citrus farmer in the winter playing a strategic game against nature. You can protect your crop against freezing by burning smudge pots at night at a cost of $1,000. If you burn smudge pots, you will be able to sell your crop for a net profit of $14,000, provided that the freeze does develop. If you do nothing, you will either lose $3,000 in planting costs if the freeze comes, or will gain $8,000 profit if no freeze occurs.

To analyze this situation we set up a **payoff matrix,** which shows the farmer's possible actions, or **strategies,** on the left and nature's strategies (the "States of Nature") on the top. The entries of the matrix are the amounts of profit (or loss) for the various combinations of strategy choices.

		STATES OF NATURE	
		Freeze	No freeze
FARMER'S	Use smudge pots	$14,000	$7,000
STRATEGIES:	Don't use them	−$3,000	$8,000

Economic Behavior In 1944 a new direction in economics was set by the *Theory of Games and Economic Behavior,* containing the famous minimax theorem. It was written by mathematician John von Neumann (see page 853) and economist Oskar Morgenstern. The book's impact relates to the question about which mathematics is proper to economics. Beginning with Antoine Cournot in the 1830s, economists looked to physics for analogies in thermodynamics or mechanics since physics had been so successfully "mathematized." The game-theoretic approach was a break with the past, showing how mathematics other than the calculus could give models of economic behavior.

Several criteria may be used in the decision making. If you are an optimist, you might choose the **maximax** criterion, which selects the action that *maximizes* the *maximum* gain. In our example, if you use smudge pots, your maximum gain would be $14,000, while if you do not use smudge pots, your maximum gain would be $8,000. Using the maximax criterion, you would choose the action that gives the larger of the two maximums and use smudge pots.

On the other hand, you might be a pessimist. In that case, the **maximin** criterion, which *maximizes* the *minimum* payoff, would suit you just fine. The minimum payoffs under the two actions are $7,000, with smudge pots, and −$3,000 if nothing is done. The larger of these occurs if smudge pots are used, so that this is the appropriate action for this criterion, too.

If you are neither an optimist nor a pessimist, and if you can assign some probability p to the "freeze" state of nature, then you can compromise by choosing the action with the greatest expected payoff. Suppose you believe (the weather forecaster has convinced you) that the probability of a freeze is .4. The probability of no freeze would then be .6, with the expected payoffs as follows.

$$\text{If smudge pots are used:} \quad 14{,}000(.4) + 7{,}000(.6) = \$9{,}800$$
$$\text{If no pots are used:} \quad -3{,}000(.4) + 8{,}000(.6) = \$3{,}600$$

* The material in this section depends somewhat on the ideas of probability and expected value, which were discussed in an earlier chapter.

Con(traption) Game This stamp from Hungary pictures Turk, an automaton built in 1769 that ostensibly played chess—it was actually operated from within by a concealed chess player. Turk was exhibited in the U.S. in the 1830s by Johann Maelzel, inventor of the metronome (and a friend of Beethoven). Edgar Allan Poe's essay on Turk deduces the person inside. Poe observed *Schlumberger,* a European chess player, just before and after Turk's chess game, but never during. Schlumberger is psyched out by Thomas Gavin in his novel *King-kill* (Random House, 1977): Schlumberger hides within Turk to avoid the ordeal of playing chess as his own man. (The title refers to the origin of "checkmate" in the Arabic *shah mat*—"the king is left unable to escape.")

Turk eventually wound up in the Chinese Museum, Philadelphia. It was wrecked by a fire there in 1854. Fire also destroyed Ajeeb, another chess automaton, this time at Coney Island in 1929. The U.S. Chess Master Harry Nelson Pillsbury operated Ajeeb at the turn of the century.

This criterion, choosing the maximum of the expected payoffs, also indicates that smudge pots should be used. Note that, as your subjective beliefs about the probability of a freeze change, so would the expected payoffs. To assign probabilities to such unique events requires a subjective or personal assessment of the event. It works out here that pots should be used as long as you feel the probability of a freeze is greater than 1/18.

In the example discussed here, all three criteria led to the same action. However, if the cost of using the smudge pots were much larger, or the profit when they were used in the event of a freeze were much smaller, the situation would look very different. Also, for the payoff matrix to be helpful the effect of all important factors must somehow be included. At the least, the use of a payoff matrix does clarify the problem and permit an orderly approach to considering the alternatives.

Notice that in a two-player game player A's strategies (1, 2, and so on) are shown at the left of the payoff matrix, and player B's strategies (I, II, and so on) are shown at the top of the payoff matrix. Each row of the game matrix (the payoff matrix) represents one of A's possible strategies, while each column represents one of B's possible strategies. Each entry in the matrix is the payoff that results for that particular choice of strategies. We assume that if the payoff is a positive number, A receives that amount from B; if it is negative, A pays B. Under this assumption, one player must lose whatever the other wins. This condition defines a **zero-sum game.** Not all games are zero-sum games. The stock market, for example, is *not*. Stocks can go up or down according to outside forces. (In effect, money enters this system from sources other than investors.) Therefore, it is possible that all investors can make money or that they can all lose.

The remainder of this section will be restricted. We discuss two-player, zero-sum games only.

EXAMPLE 1 Assume the payoffs here are dollar amounts. What are the results of each possible combination of strategy choices?

$$\begin{array}{cc} & B \\ & \begin{array}{cc} \text{I} & \text{II} \end{array} \\ A \begin{array}{c} 1 \\ 2 \end{array} & \begin{bmatrix} 2 & -1 \\ -3 & 4 \end{bmatrix} \end{array}$$

From the payoff matrix, we observe the following.

If A applies strategy 1 and B applies strategy I, then A receives \$2 from B.

If A applies strategy 1 and B applies strategy II, then A pays B \$1.

If A applies strategy 2 and B applies strategy I, then A pays B \$3.

If A applies strategy 2 and B applies strategy II, then A receives \$4 from B.

An $m \times n$ (matrix) game is one in which player A has m possible strategies (rows) and player B has n strategies (columns). For example, a 3×3 game is shown here.

$$\begin{array}{cc} & B \\ & \begin{array}{ccc} \text{I} & \text{II} & \text{III} \end{array} \\ A \begin{array}{c} 1 \\ 2 \\ 3 \end{array} & \begin{bmatrix} -3 & -6 & 10 \\ 3 & 0 & 1 \\ 5 & -4 & -8 \end{bmatrix} \end{array}$$

From B's viewpoint, strategy II is better than strategy I no matter which strategy A selects, since a gain of \$6 is better than a gain of \$3; breaking even is better than paying \$3 to A; and a gain of \$4 is better than a loss of \$5. Therefore, B should never select strategy I. In a situation like this, strategy II is said to *dominate* strategy I (or strategy I is *dominated* by II). Column II dominates column I since *every* column II payoff is *less than* the corresponding column I payoff.

From A's viewpoint, no strategy (row) in the original matrix above dominates another, since for a row to dominate another, its payoffs must all be greater than the corresponding payoffs in the other row. (Verify that this condition does not occur in the payoff matrix above.)

It is sometimes possible to reduce the order of an $m \times n$ game by eliminating dominated strategies. (No player would ever choose to employ a dominated strategy anyway.)

EXAMPLE 2 Reduce the 3×3 game matrix above by eliminating any dominated strategies.

As observed above, A sees no dominant strategies initially but B would eliminate strategy I, so we arrive at the following reduced matrix.

$$\begin{array}{cc} & B \\ & \begin{array}{cc} \text{II} & \text{III} \end{array} \\ A \begin{array}{c} 1 \\ 2 \\ 3 \end{array} & \begin{bmatrix} -6 & 10 \\ 0 & 1 \\ -4 & -8 \end{bmatrix} \end{array}$$

But now that strategy I for B has been removed, it turns out that strategy 2 for A dominates strategy 3. Therefore, we can finally reduce the game to a 2×2 matrix as follows.

$$\begin{array}{cc} & B \\ & \begin{array}{cc} \text{II} & \text{III} \end{array} \\ A \begin{array}{c} 1 \\ 2 \end{array} & \begin{bmatrix} -6 & 10 \\ 0 & 1 \end{bmatrix} \end{array}$$ ◆

Which strategies should be chosen by A and B? The goal of game theory is to find *optimum* strategies, those which are most profitable (or least costly) for a player. The resulting payoff, assuming both players select their respective optimum strategies, is called the ***value*** of the game. We usually find optimum strategies using the ***minimax*** criterion, in which we *minimize* the *maximum* possible loss. For example, a game for players A and B is shown here.

$$\begin{array}{cc} & B \\ & \begin{array}{ccc} \text{I} & \text{II} & \text{III} \end{array} \\ A \begin{array}{c} 1 \\ 2 \\ 3 \end{array} & \begin{bmatrix} -7 & 8 & -1 \\ 4 & 7 & 0 \\ -9 & 3 & -6 \end{bmatrix} \end{array}$$

In this game, player A's greatest possible losses for his or her strategy choices are \$7, \$0, and \$9, respectively. To minimize the maximum possible loss, player A should select strategy 2. In the same way, B's greatest possible losses are \$4, \$8, and \$0, so that B can minimize possible losses by choosing strategy III. Thus, the payoff will be \$0 (strategy 2 for A and III for B), which is the value of the game. A game whose value is \$0 is called a **fair game.**

The pair of strategies (2, III) determined by the minimax criterion is called the **saddle point** of the game. The saddle point is the pair of strategies for which the entry in the payoff matrix is both the *minimum* in its row and *maximum* in its column. This fact makes the saddle point easy to determine, if it exists. For example, in the matrix below, by the definition above, the saddle point is (4, I) and the value of the game is \$3. (Whenever there are two or more saddle points, they must all yield equal values.)

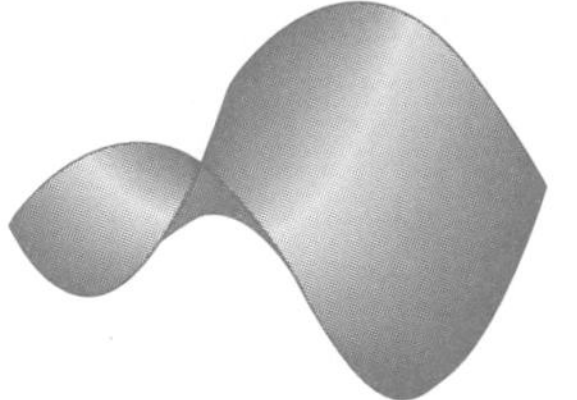

The name **saddle point** comes from a saddle. The seat of the saddle is the maximum from one direction and the minimum from another direction.

$$\begin{array}{cc} & B \\ & \begin{array}{cc} \text{I} & \text{II} \end{array} \\ A \begin{array}{c} 1 \\ 2 \\ 3 \\ 4 \end{array} & \begin{bmatrix} 2 & 2 \\ 0 & 4 \\ 1 & 6 \\ 3 & 7 \end{bmatrix} \end{array}$$

Unfortunately, not every $m \times n$ game has a saddle point. When there is no saddle point, another approach must be used. Both players will have to *mix* their strategies. If this mixing were to be done in some systematic fashion, the competitor would soon guess it and play accordingly. For this reason, it is best to mix strategies according to previously determined probabilities. For example, if a player has only two strategies and has decided to select them with equal probability, the random choice can be made by tossing a coin. Heads could represent one strategy; tails the other. Although this would result in the two strategies being used about equally over the long run, on a particular play no player could predetermine which strategy would be used. Some other device, such as a spinner, a die, or random numbers, is necessary for more than two strategies, or when the predetermined probability is different from 1/2.

To see how to decide on the probability with which a particular strategy is selected, return to the example about the smudge pots. Suppose you must decide several times each season whether or not to use smudge pots, and you don't really know the probability of a freeze. We need to find the fraction of the time (the probability) that you should use smudge pots. There is no saddle point for this game. Thus, let x represent the probability that you will choose strategy 1, with $1 - x$ the probability that you will choose strategy 2.

$$\begin{array}{cc} & \text{NATURE} \\ & \begin{array}{cc} \text{I} & \text{II} \end{array} \\ \text{YOU} \begin{array}{c} 1 \\ 2 \end{array} & \begin{bmatrix} \$14{,}000 & \$7{,}000 \\ -\$3{,}000 & \$8{,}000 \end{bmatrix} \end{array}$$

John von Neumann (1903–1957) achieved much in a life cut short by cancer. Stories told about him made him a living legend. In 1933 he joined the Institute for Advanced Study, newly founded, and was one of its original six mathematicians. By then he had already written important papers on set theory, quantum mechanics, and the minimax theorem. After 1940 he turned to the applications of mathematics, working with Stanislaw Ulam during World War II and extending game theory with Morgenstern. Von Neumann was intensely interested in the theory and use of computers and automata. His interest in artificial intelligence is expressed in his lecture notes published as *The Computer and the Brain* (Yale paperback).

For Nature's strategy I (a freeze) your expected profit is

$$E_{\text{I}} = 14{,}000x - 3{,}000(1 - x).$$

For Nature's strategy II (no freeze) your expected profit is

$$E_{\text{II}} = 7{,}000x + 8{,}000(1 - x).$$

You will maximize your profit in the long run if you choose a value of x for which $E_{\text{I}} = E_{\text{II}}$. (This fact was proved in 1928 by John von Neumann. We will use it here without proof.) To make $E_{\text{I}} = E_{\text{II}}$, solve as follows.

$$\begin{aligned} E_{\text{I}} &= E_{\text{II}} \\ 14{,}000x - 3{,}000(1 - x) &= 7{,}000x + 8{,}000(1 - x) \\ 17{,}000x - 3{,}000 &= -1{,}000x + 8{,}000 \\ 18{,}000x &= 11{,}000 \\ x &= \frac{11}{18} \end{aligned}$$

Thus, the optimum strategy here is to use strategy 1 (use smudge pots) 11/18 of the time, and strategy 2 (don't use them) a total of

$$1 - x = 1 - \frac{11}{18} = \frac{7}{18}$$

of the time.

EXAMPLE 3 Find the farmer's expected gain from the smudge pot strategy outlined above.

We can substitute 11/18 for x in either E_{I} or E_{II}. Using E_{I} gives

$$14{,}000\left(\frac{11}{18}\right) - 3{,}000\left(\frac{7}{18}\right) = \frac{133{,}000}{18} = \$7{,}388.89.$$

This result, \$7,388.89, is the value of the game. The value comes out the same no matter which player's optimum strategy and expected gain are considered. ■

14.5 EXERCISES

In the following games, assume that player A chooses rows and player B chooses columns. Eliminate all dominated strategies from each of the following games.

1. $\begin{bmatrix} 1 & 4 \\ 4 & -1 \\ 3 & 5 \\ -4 & 0 \end{bmatrix}$ **2.** $\begin{bmatrix} 2 & 3 & 1 & -5 \\ -1 & 5 & 4 & 1 \\ 1 & 0 & 2 & -3 \end{bmatrix}$ **3.** $\begin{bmatrix} 8 & 12 & -7 \\ -2 & 1 & 4 \end{bmatrix}$ **4.** $\begin{bmatrix} 6 & 2 \\ -1 & 10 \\ 3 & 5 \end{bmatrix}$

Where it exists, find the saddle point and the value of the game for each of the following.

5. $\begin{bmatrix} -6 & 2 \\ -1 & -10 \\ 3 & 5 \end{bmatrix}$ **6.** $\begin{bmatrix} 7 & 8 \\ -2 & 15 \end{bmatrix}$ **7.** $\begin{bmatrix} 3 & -4 & 1 \\ 5 & 3 & 2 \end{bmatrix}$

8. $\begin{bmatrix} 2 & 3 & 1 \\ -1 & 4 & -7 \\ 5 & 2 & 0 \\ 8 & -4 & -1 \end{bmatrix}$ **9.** $\begin{bmatrix} 1 & 4 & -3 & 1 & -1 \\ 2 & 5 & 0 & 4 & 10 \\ 1 & -3 & -2 & 5 & 2 \end{bmatrix}$ **10.** $\begin{bmatrix} -4 & 2 & -3 & -7 \\ 4 & 3 & 5 & 9 \end{bmatrix}$

Solve each of the following problems.

11. Hillsdale College has sold out all tickets for a jazz concert to be held in the stadium. If it rains, the show will have to be moved into the gym, which has a much smaller seating capacity. The dean must decide in advance whether to set up the seats and the stage in the gym or in the stadium, or both, just in case. The matrix below shows the net profit in each case.

		NATURE	
		Rain	No rain
ACTIONS	Set up stadium	−$1,550	$1,500
	Set up gym	$1,000	$1,000
	Set up both	$750	$1,400

(a) What action should the dean take if he is an optimist?
(b) If he is a pessimist?
(c) If the weather forecaster predicts rain with a probability of .6, what action should the dean take to maximize expected profit?

12. A community is considering an anti-smoking campaign. The city manager has been asked to recommend one of three possible actions: a campaign for everyone in the community over age 10, a campaign for youths only, or no campaign at all. The two states of nature are a true cause-effect relationship between smoking and cancer or no such relationship. The cost to the community (including loss of life and productivity) in each case is shown below.*

	Cause-effect relationship	No such relationship
Campaign for all	$100,000	$800,000
Campaign for youth	$2,820,000	$20,000
No campaign	$3,100,100	$0

* This problem is based on an article by B. G. Greenburg in the September 1969 issue of the *Journal of the American Statistical Association.*

(a) What action should the manager recommend if she is an optimist?
(b) If she is a pessimist?
(c) If the Director of Public Health estimates that the probability of a true cause-effect relationship is .8, which action should be recommended to minimize expected costs?

13. The research department of Allied Manufacturing has developed a new process that they feel will result in an improved product. Management must decide whether or not to market the new product. The new product may be better than the old, or it may not be better. If the new product is better, and the company decides to market it, sales should increase by \$50,000. If it is not better and they replace the old product with the new product on the market, they will lose \$25,000 to competitors. If they decide not to market the new product, they will lose \$40,000 if it is better and research costs of \$10,000 if it is not.
(a) Prepare a payoff matrix.
(b) If management believes that the probability is .4 that the new product is better, find the expected profits under each possible course of action, and find the best action.

14. Suppose Allied Manufacturing (of Exercise 13) decides to put their new product on the market with a big advertising campaign. At the same time, they find out that their major competitor, Bates Manufacturing, has also decided on a big advertising campaign for its version of the product. The payoff matrix below shows the increased sales in millions for Allied, which also represent decreases for Bates.

BATES

$$\text{ALLIED}\quad \begin{array}{c} \text{TV} \\ \text{Radio} \end{array} \overset{\begin{array}{cc} \text{TV} & \text{Radio} \end{array}}{\begin{bmatrix} 1.0 & -.7 \\ -.5 & .5 \end{bmatrix}}$$

Find the optimum strategy for Allied, and find the value of the game.

In each of the following games, find the optimum strategy for each player, and find the value of the game. Identify any fair games.

15. $\begin{bmatrix} 5 & 1 \\ 3 & 4 \end{bmatrix}$ **16.** $\begin{bmatrix} -4 & 5 \\ 3 & -4 \end{bmatrix}$ **17.** $\begin{bmatrix} -2 & 4 \\ 0 & 5 \end{bmatrix}$ **18.** $\begin{bmatrix} 0 & -4 \\ 4 & 0 \end{bmatrix}$

19. In the game of matching coins, two players each flip a coin. If both coins match (both show heads or both show tails), player *A* wins \$1. If there is no match, player *B* wins \$1, as in the matrix below. Find the optimum strategies for each player, and find the value of the game.

$$\begin{bmatrix} 1 & -1 \\ -1 & 1 \end{bmatrix}$$

20. Explain the meaning of a "zero-sum" game. Give an example of a zero-sum game and an example of a non–zero-sum game.

21. Explain these terms: *maximax criterion, maximin criterion,* and *minimax criterion.*

22. Explain why, in a game matrix, a dominant row has entries that are *greater* than those of the dominated row, whereas a dominant column has entries that are *less* than those of the dominated column.

23. Explain what is meant by a "mixed" strategy.

Refer to the original payoff matrix in the text for the citrus farmer. Assume the farmer is neither an optimist nor a pessimist. Each of Exercises 24–26 gives a probability of having a freeze. In each case find **(a)** *the expected payoff for strategy 1 (use smudge pots) and* **(b)** *the expected payoff for strategy 2 (don't use them).*

24. 1/17 **25.** 1/18 **26.** 1/19

27. Discuss how your answers to Exercises 24–26 relate to the claim in the text that "pots should be used as long as you feel the probability of a freeze is greater than 1/18."

28. Refer to Example 3 in the text. Use E_{II} to find the farmer's expected gain for Nature's strategy II. Is the value the same as was obtained in the example using E_I?

In Example 3 it was stated that the value of a game (with no saddle point) comes out the same no matter which player's optimum strategy and expected gain are considered. Work the following exercises to consider the smudge pot question from Nature's viewpoint. Assume the original payoff matrix and suppose that Nature will randomly impose a freeze a certain fraction of the time, given by x.

29. Derive an expression for E_1, the expected payoff in case the farmer chooses to use smudge pots.

30. Derive an expression for E_2, the expected payoff in case the farmer chooses not to use smudge pots.

31. Equate E_1 and E_2 from Exercises 29 and 30 to find Nature's optimum strategy.

32. Find the value (expected payoff) of the game when Nature employs the strategy of Exercise 31. Is the value the same as that arrived at in Example 3?

Another citrus farmer facing the smudge pot decision has calculated the following payoff matrix. Refer to it for Exercises 33–37.

	Freeze	No freeze
Use smudge pots	\$8,000	\$7,000
Don't use them	−\$5,000	\$10,000

33. Which strategy would an optimist choose?

34. Which strategy would a pessimist choose?

35. If the farmer is neither an optimist nor a pessimist but is able to reasonably assess the probability of a freeze, how large must that probability be for the farmer's preferred strategy to be to use smudge pots?

36. The payoff matrix above still has no saddle point, so the farmer's best plan is probably to use a mixed strategy. Use the procedure described just before Example 3 to decide what fraction of the time the farmer should use smudge pots.

37. By using smudge pots the fraction of the time computed in Exercise 36, what expected gain does the farmer have? (Compare with Example 3.)

38. For the citrus farmer problem, devise a new 2×2 payoff matrix for which an optimist and a pessimist would both choose not to use smudge pots. Describe conditions that might give rise to the payoffs in your matrix.

COLLABORATIVE INVESTIGATION
Algebraic Structure of the System of 2 × 2 Matrices

In this activity, the algebraic structure of the system of 2 × 2 matrices will be compared with that of the system of real numbers.

1. First, as an entire class, review the following description of the real number system to clarify the meanings of the various properties.

The set of real numbers R, along with the operations of addition $(+)$ and multiplication $(\cdot)$ and the relation of equality $(=)$, is a **mathematical system.** Specifically, R is an **abelian (commutative) group** under the operation $+$ because the following five properties are satisfied. For all a, b, and c in R:

(1) $a + b$ is in R. (closure property of $+$)
(2) $a + b = b + a$. (commutative property of $+$)
(3) $(a + b) + c = a + (b + c)$. (associative property of $+$)
(4) there is an element 0 in R such that $a + 0 = a$. (identity property of $+$)
(5) there exists an element $-a$ in R such that $a + (-a) = 0$. (inverse property of $+$)

Furthermore, the system R is an **associative ring** because the following three additional properties are satisfied. For all a, b, and c in R:

(6) $a \cdot b$ is in R. (closure property of $\cdot$)
(7) $a \cdot (b \cdot c) = (a \cdot b) \cdot c$. (associative property of $\cdot$)
(8) $a \cdot (b + c) = a \cdot b + a \cdot c$ and $(b + c) \cdot a = b \cdot a + c \cdot a$. (the left and right distributive properties)

In addition, R is a **commutative ring with unit element** because of the following two properties. For all a and b in R:

(9) $a \cdot b = b \cdot a$. (commutative property of $\cdot$)
(10) there is an element 1 in R such that $a \cdot 1 = 1 \cdot a = a$. (identity property of $\cdot$)

Finally, R is a **field** because of one more property. For all a in R (other than 0):

(11) there exists an element a^{-1} in R such that $a \cdot a^{-1} = 1$. (inverse property of $\cdot$)

In summary, the real numbers comprise a field because

A. all real numbers are an abelian group under $+$, and
B. all nonzero real numbers are an abelian group under $\cdot$.

2. Next, divide into groups of three or four students each. Each small group is to consider the system of all 2 × 2 matrices with real number entries, together with the operations of $+$ and $\cdot$ and the relation of $=$. Investigate the eleven properties listed above, in turn, for this new system. In each case, try examples, or make use of corresponding properties of the real numbers to prove the results, or use results established in this chapter (including appropriate exercises). Decide, within your group, on answers to the following questions.

(1) Do the 2 × 2 matrices form a mathematical field? If not, which ones of the eleven properties fail to hold?
(2) If the system of matrices is not a field, is it at least a commutative ring with unit element?
(3) If not, is it at least an associative ring?
(4) If not, is it at least a commutative group under addition?

3. Once the individual groups have decided on their opinions, come back together and share your reasoning. Try to come up with a firm conclusion as a class.

Chapter 14 Test

Let $A = \begin{bmatrix} 2 & -1 & 3 \\ 2 & 0 & 1 \\ 4 & 1 & 3 \end{bmatrix}$ *and* $B = \begin{bmatrix} -3 & 0 & 1 \\ 0 & 2 & -1 \\ -1 & 0 & 3 \end{bmatrix}$. *Perform the indicated matrix operations.*

1. $A + B$ **2.** $A - B$ **3.** AB **4.** BA **5.** $2A$ **6.** $-3B + A$

Apply the following row operations to matrix A from above.

7. Interchange rows 2 and 3.

8. Multiply each element of row 2 of A by -3, and add the results to row 1.

Find each of the following products, when possible. If not possible, state why not.

9. $\begin{bmatrix} -2 & 1 \\ 3 & 4 \end{bmatrix} \cdot \begin{bmatrix} 2 \\ 1 \end{bmatrix}$

10. $\begin{bmatrix} 3 & 4 & 1 \\ 2 & 0 & 8 \end{bmatrix} \cdot \begin{bmatrix} 1 & 0 & 2 \\ 0 & 0 & 1 \\ 3 & 0 & 2 \end{bmatrix}$

11. $[1 \quad 6 \quad 1] \cdot \begin{bmatrix} 3 \\ 2 \\ 1 \end{bmatrix}$

12. $\begin{bmatrix} 8 & 9 & 7 \\ 2 & 1 & 3 \end{bmatrix} \cdot \begin{bmatrix} 2 & 3 \\ 4 & 2 \end{bmatrix}$

Find the inverse of each of the following matrices, where possible.

13. $\begin{bmatrix} 2 & 1 \\ 5 & 3 \end{bmatrix}$

14. $\begin{bmatrix} 1 & 0 & 1 \\ 2 & 3 & 1 \\ 1 & 3 & 2 \end{bmatrix}$

Set up the augmented matrix for each of the following systems of equations. Do not solve the systems.

15. $\begin{aligned} 9x - 4y &= 20 \\ 5x + 10y &= 12 \end{aligned}$

16. $\begin{aligned} 2x + y - 6z &= 6 \\ 5x + 12z &= 15 \\ 6y + 3z &= 8 \end{aligned}$

Solve each system of equations by the Gauss-Jordan method (Exercises 17 and 18).

17. $\begin{aligned} 5x + 6y &= 15 \\ 3x + 7y &= 9 \end{aligned}$

18. $\begin{aligned} 3x + 4y + z &= 10 \\ x - 3y - 5z &= -3 \\ 2x + 7y &= 1 \end{aligned}$

19. If the system of equations

$$\begin{aligned} x + y &= 1 \\ 2x - y &= -7 \end{aligned}$$

is written as $AX = B$, what exactly is represented by each of the following?
(a) A **(b)** X **(c)** B

20. Given the matrix equation $AX = B$, express X in terms of A and B.

21. Use matrix algebra to solve the system of equations of Exercise 19.

22. Explain the conditions under which the matrix product AB is computable.

The economy of a small country involves bananas and coal. Production of 1 unit of bananas consumes .3 unit of bananas and .1 unit of coal. Production of 1 unit of coal consumes .2 unit of bananas and .4 unit of coal.

23. Set up the technological matrix for this economy.

24. Find the gross production required to meet a demand for net production of 400 units of bananas and 300 units of coal.

25. Remove any dominated strategies from the following game.

$$\begin{bmatrix} 6 & 8 & -3 & 4 \\ 2 & 0 & 2 & -1 \end{bmatrix}$$

26. Identify any saddle points in the following game.

$$\begin{bmatrix} -3 & 1 & 4 & 2 \\ 8 & 9 & 6 & 4 \\ 9 & 10 & 15 & 19 \end{bmatrix}$$

27. Find the optimum strategy for player A and the value of the following game.

$$\begin{array}{cc} & \quad B \\ & \begin{array}{cc} \text{I} & \text{II} \end{array} \\ A \begin{array}{c} 1 \\ 2 \end{array} & \begin{bmatrix} -3 & -4 \\ -4 & 5 \end{bmatrix} \end{array}$$

28. Find two different pairs of values for the missing entries in the matrix

$$A = \begin{bmatrix} \square & -3 \\ 5 & \square \end{bmatrix}$$

that will cause A to be its own inverse.

Arturo Hogan makes silver jewelry inlaid with turquoise. Each ring requires 1.6 units of silver and .7 unit of turquoise, each bracelet requires 5.8 units of silver and 7.5 units of turquoise, and each necklace requires 26.3 units of silver and 18.8 units of turquoise. Arturo pays \$5.50 per unit for silver and \$2.10 per unit for turquoise.

29. Set up a matrix M relating jewelry items (ring, bracelet, necklace) to materials (silver, turquoise), and another matrix C relating materials to cost (per unit).

30. Perform an appropriate matrix operation to obtain a matrix relating jewelry items to cost (per item) for materials.

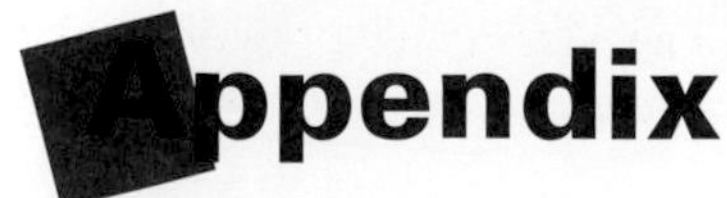

Appendix

The Metric System

Joseph Louis Lagrange (1736–1813) was born in Turin, Italy, and became a professor at age 19. In 1776 he came to Berlin at the request of Frederick the Great to take the position Euler left. A decade later Lagrange settled permanently in Paris. Napoleon was among many who admired and honored him.

Lagrange's greatest work was in the theory and application of calculus. He carried forward Euler's work of putting calculus on firm algebraic ground in his theory of functions. His *Analytic Mechanics* (1788) applied calculus to the motion of objects. Lagrange's contributions to algebra had great influence on Galois and hence the theory of groups. He also wrote on number theory; he proved, for example, that every integer is the sum of at most four squares. His study of the moon led to methods for finding longitude.

The metric system of weights and measures is based on decimals and powers of ten. For this reason, calculations and changes of units in the metric system are much easier to do than in the English system that we now use. The metric system was developed by a committee of the French Academy just after the French Revolution of 1789. The president of the committee was the mathematician Joseph Louis Lagrange.

The advantages of the metric system can be seen when compared to our English system. In the English system, one inch is one-twelfth of a foot, while one foot is one-third of a yard. One mile is equivalent to 5,280 feet, or 1,760 yards. Obviously, there is no consistency in subdivisions. In the metric system, prefixes are used to indicate multiplications or divisions by powers of ten. For example, the basic unit of length in the metric system is the *meter* (which is a little longer than one yard). To indicate one thousand meters, attach the prefix "kilo-" to get kilometer. To indicate one one-hundredth of a meter, use the prefix "centi-" to obtain centimeter. These prefixes are examples of those used for lengths, weights, and volumes in the metric system. A complete list of the prefixes of the metric system is shown in the table below, with the most commonly used prefixes appearing in heavy type.

Metric Prefixes

Prefix	Multiple	Prefix	Multiple
exa	1,000,000,000,000,000,000	deci	.1
peta	1,000,000,000,000,000	**centi**	.01
tera	1,000,000,000,000	**milli**	.001
giga	1,000,000,000	micro	.000001
mega	1,000,000	nano	.000000001
kilo	1,000	pico	.000000000001
hecto	100	femto	.000000000000001
deka	10	atto	.000000000000000001

Length, Area, and Volume

Lagrange urged the committee devising the metric system to find some natural measure for length from which weight and volume measures could be derived. It was decided that one **meter** would be the basic unit of length, with a meter being defined as one ten-millionth of the distance from the equator to the North Pole. According to Paul G. Hewitt, in *Conceptual Physics,* 7th edition (HarperCollins, 1993):

> This distance was thought at the time to be close to 10,000 kilometers. One ten-millionth of this, the meter, was carefully determined and marked off by means of scratches on a bar of platinum-iridium alloy. This bar is kept at the International Bureau of Weights and Measures in France. The standard meter in France has since been calibrated in terms of the wavelength of light—it is 1,650,763.73 times the wavelength of orange light emitted by the atoms of the gas krypton-86. The meter is now defined as being the length of the path traveled by light in vacuum during a time interval of 1/299,792,458 of a second.

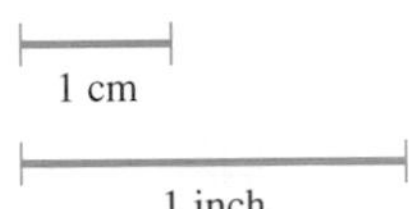

■ **FIGURE 1**

To obtain measures longer than one meter, Greek prefixes were added. For measures smaller than a meter, Latin prefixes were used. A meter is a little longer than a yard (about 39.37 inches). Shorter lengths are commonly measured in micrometers, centimeters, and millimeters. A **centimeter** is one one-hundredth of a meter and is about 2/5 of an inch. (See Figure 1.) Centimeter is symbolized cm.

Because the metric system is based on decimals and powers of ten, conversions within the system are very easy. They simply involve multiplying and dividing by powers of ten. For example, to convert 2.5 m to centimeters, multiply 2.5 by 100 (since 100 cm = 1 m) to obtain 250 cm. On the other hand, to convert 18.6 cm to meters, divide by 100 to obtain .186 m. Other conversions are made in the same manner, using the meanings of the prefixes. Can you see why 42 m is equal to 42,000 millimeters (mm)?

How far away is each village in meters? In miles?

Long distances are usually measured in kilometers. A **kilometer** is 1,000 meters. (According to a popular dictionary, the word *kilometer* may be pronounced with the accent on either the first or the second syllable. Scientists usually stress the second syllable.) A kilometer (symbolized km) is equal to about .6 mile. Figure 2 shows the ratio of 1 kilometer to 1 mile.

Conversions from meters to kilometers, and vice versa, are made by multiplying or dividing by 1,000 as necessary. For example, 37 kilometers equals 37,000 meters, while 583 meters equals .583 km.

The area of a figure can be measured in square metric units. Figure 3 shows a square that is 1 cm on each side; thus, it is a **square centimeter (cm^2).** One square meter (m^2) is the area of a square with sides one meter long. How many cm^2 are in one m^2?

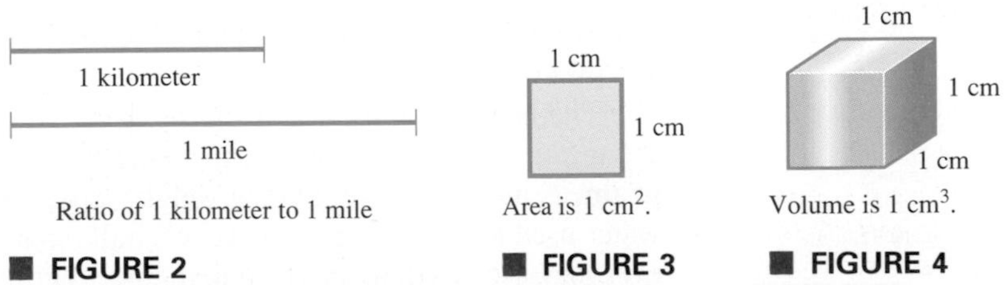

Ratio of 1 kilometer to 1 mile

■ **FIGURE 2**

Area is 1 cm^2.

■ **FIGURE 3**

Volume is 1 cm^3.

■ **FIGURE 4**

The volume of a three-dimensional figure is measured in cubic units. If, for example, the dimensions are given in centimeters, the volume may be determined by the appropriate formula from geometry, and it will be in **cubic centimeters (cm^3).** See Figure 4 for a sketch of a box whose volume is one cm^3.

A Comparison of Distances

Length in Meters	Approximate Related Distances
10^{19}	Distance to the North Star
10^{12}	Distance of Saturn from the sun
10^{11}	Distance of Venus from the sun
10^{9}	Diameter of the sun
10^{8}	Diameter of Jupiter
10^{7}	Diameter of Earth; distance from Washington, D.C. to Tokyo
10^{6}	Distance from Chicago to Wichita, Kansas
10^{5}	Average distance across Lake Michigan
10^{4}	Average width of the Grand Canyon
10^{3}	Length of the Golden Gate Bridge
10^{2}	Length of a football field
10^{1}	Average height of a two-story house
10^{0}	Width of a door
10^{-1}	Width of your hand
10^{-2}	Diameter of a piece of chalk
10^{-3}	Thickness of a dime
10^{-4}	Thickness of a piece of paper
10^{-5}	Diameter of a red blood cell
10^{-7}	Thickness of a soap bubble
10^{-8}	Average distance between molecules of air in a room
10^{-9}	Diameter of a molecule of oil
10^{-14}	Diameter of an atomic nucleus
10^{-15}	Diameter of a proton

Volume and Weight

In the metric system, one **liter** is the quantity assigned to the volume of a box that is 10 cm on a side. (See Figure 5.) A liter is a little more than a quart. (See Figure 6.) Notice the advantage of this definition over the equivalent one in the English system—using a ruler marked in cm, a volume of 1 liter (symbolized 1 L) can be constructed. On the other hand, given a ruler marked in inches, it would be difficult to construct a volume of 1 quart.

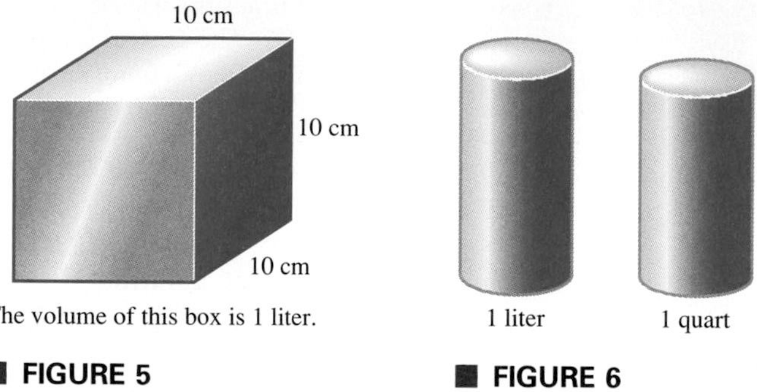

■ FIGURE 5

■ FIGURE 6

The prefixes mentioned earlier are used throughout the metric system, so one **milliliter (ml)** is one one-thousandth of a liter, one **centiliter (cl)** is one one-hundredth of a liter, one **kiloliter (kl)** is 1,000 liters, and so on. Milliliters are used extensively in science and medicine. Many beverages are now sold by milliliters and by liters. For example, 750 ml is a common size for wine bottles, and many soft drinks are now sold in 1- and 2-liter bottles.

Because of the way a liter is defined as the volume of a box 10 cm on a side,

$$1 \text{ L} = 10 \text{ cm} \times 10 \text{ cm} \times 10 \text{ cm} = 1{,}000 \text{ cm}^3$$

or $\frac{1}{1{,}000} \text{ L} = 1 \text{ cm}^3$.

Since $\frac{1}{1{,}000} \text{ L} = 1 \text{ ml}$, we have the following relationship.

$$\mathbf{1 \text{ ml} = 1 \text{ cm}^3}$$

For example, the volume of a box which is 8 cm by 6 cm by 5 cm may be given as 240 cm^3 or as 240 ml.

The box in Figure 4 is 1 cm by 1 cm by 1 cm. The volume of this box is 1 cm^3, or 1 ml. By definition, the weight of the water that fills such a box is **1 gram.** A nickel five-cent piece weighs close to 5 grams, or 5 g. The volume of water used to define a gram is very small, so a gram is a very small weight. For everyday use, a **kilogram,** or one thousand grams, is more practical. A kilogram is about 2.2 pounds. A common abbreviation for kilogram is the word **kilo.**

Anders Celsius and his scale for measuring temperature are honored on this Swedish stamp. The original scale had the freezing point of water at 100° and the boiling point at 0°, but biologist Carl von Linne inverted the scale, giving us the familiar Celsius scale of today.

Extremely small weights can be measured with **milligrams (mg)** and **centigrams (cg).** These measures are so small that they, like centiliters and milliliters, are used mainly in science and medicine.

Temperature

In the metric system temperature is measured in **degrees Celsius.** On the Celsius temperature scale, water freezes at 0° and boils at 100°. These two numbers are easier to remember than the corresponding numbers on the Fahrenheit scale, 32° and 212°.

The thermometer in Figure 7 shows some typical temperatures in both Fahrenheit and Celsius. For example, room temperature is 22°C, or 70°F, and body temperature is about 37°C, or 99°F.

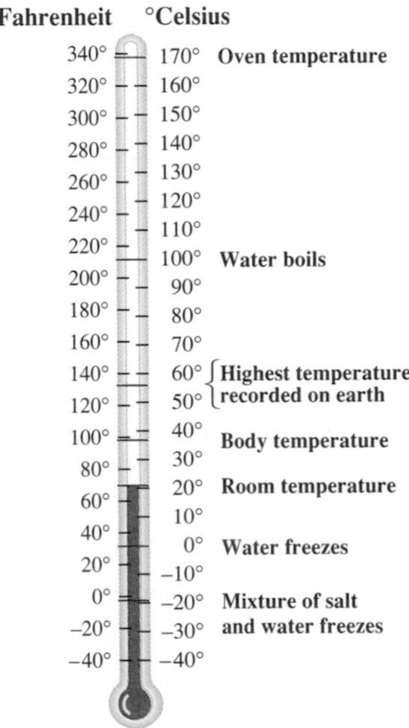

FIGURE 7

Switching a country to the metric system requires a tremendous educational campaign. The stamps shown here were issued as part of one such campaign. They show the various metric units and items measured in those units.

The formulas given below can be used to convert between Celsius and Fahrenheit temperatures.

Celsius-Fahrenheit Conversion Formulas

To convert a reading from Fahrenheit to Celsius, use the formula

$$C = \frac{5(F - 32)}{9}.$$

To convert from Celsius to Fahrenheit, use the formula

$$F = \frac{9}{5}C + 32.$$

Metric Conversions

Due to legislation enacted by Congress, the metric system is used in the United States, and an ultimate goal is for the two systems to be in use, side-by-side, with public acceptance of both systems. Many industries now use the metric system. In particular, industries that export a great many goods are using the metric system, since this is compatible with most of the countries with which they trade.

Some scientific calculators are programmed to do conversions between the English and metric systems. Approximate conversions can be made with the aid of the tables below.

Metric to English

To Convert From	To	Multiply By
meters	yards	1.0936
meters	feet	3.2808
meters	inches	39.37
kilometers	miles	.6214
grams	pounds	.0022
kilograms	pounds	2.20
liters	quarts	1.0567
liters	gallons	.2642

English to Metric

To Convert From	To	Multiply By
yards	meters	.9144
feet	meters	.3048
inches	meters	.0254
miles	kilometers	1.609
pounds	grams	454
pounds	kilograms	.454
quarts	liters	.9464
gallons	liters	3.785

APPENDIX EXERCISES

Perform the following conversions by multiplying or dividing by the appropriate power of 10.

1. 8 m to millimeters
2. 14.76 m to centimeters
3. 8,500 cm to meters
4. 250 mm to meters
5. 68.9 cm to millimeters
6. 3.25 cm to millimeters
7. 59.8 mm to centimeters
8. 3.542 mm to centimeters
9. 5.3 km to meters
10. 9.24 km to meters
11. 27,500 m to kilometers
12. 14,592 m to kilometers

Use a metric ruler to perform the following measurements, first in centimeters, then in millimeters.

13. |————————|
14. |————————————|
15. |————————————————————|

16. Based on your measurement of the line segment in Exercise 13, one inch is about how many centimeters? how many millimeters?

Perform the following conversions by multiplying or dividing by the appropriate power of 10.

17. 6 L to centiliters
18. 4.1 L to milliliters
19. 8.7 L to milliliters
20. 12.5 L to centiliters
21. 925 cl to liters
22. 412 ml to liters
23. 8,974 ml to liters
24. 5,639 cl to liters
25. 8,000 g to kilograms
26. 25,000 g to kilograms
27. 5.2 kg to grams
28. 12.42 kg to grams
29. 4.2 g to milligrams
30. 3.89 g to centigrams
31. 598 mg to grams
32. 7,634 cg to grams

Use the formulas given in the text to perform the following conversions. If necessary, round to the nearest degree.

33. 86°F to Celsius

34. 536°F to Celsius

35. −114°F to Celsius

36. −40°F to Celsius

37. 10°C to Fahrenheit

38. 25°C to Fahrenheit

39. −40°C to Fahrenheit

40. −15°C to Fahrenheit

Solve each of the following problems. Refer to geometry formulas as necessary.

41. One nickel weighs 5 g. How many nickels are in 1 kg of nickels?

42. Sea water contains about 3.5 g salt per 1,000 ml of water. How many grams of salt would be in one liter of sea water?

43. Helium weighs about .0002 g per milliliter. How much would one liter of helium weigh?

44. About 1,500 g sugar can be dissolved in a liter of warm water. How much sugar could be dissolved in one milliliter of warm water?

45. Northside Foundry needed seven metal strips, each 67 cm long. Find the total cost of the strips, if they sell for $8.74 per meter.

46. Uptown Dressmakers bought fifteen pieces of lace, each 384 mm long. The lace sold for $54.20 per meter. Find the cost of the fifteen pieces.

47. Imported marble for desk tops costs $174.20 per square meter. Find the cost of a piece of marble 128 cm by 174 cm.

48. A special photographic paper sells for $63.79 per square meter. Find the cost to buy 80 pieces of the paper, each 9 cm by 14 cm.

49. An importer received some special coffee beans in a box 82 cm by 1.1 m by 1.2 m. Give the volume of the box, both in cubic centimeters and cubic meters.

50. A fabric center receives bolts of woolen cloth in crates 1.5 m by 74 cm by 97 cm. Find the volume of a crate, both in cubic centimeters and cubic meters.

51. A medicine is sold in small bottles holding 800 ml each. How many of these bottles can be filled from a vat holding 160 L of the medicine?

52. How many 2-liter bottles of soda pop would be needed for a wedding reception if 80 people are expected, and each drinks 400 ml of soda?

Perform the following conversions. Use a calculator and/or the table in the text as necessary.

53. 982 yd to meters

54. 12.2 km to miles

55. 125 mi to kilometers

56. 1,000 mi to kilometers

57. 1,816 g to pounds

58. 1.42 lb to grams

59. 47.2 lb to grams

60. 7.68 kg to pounds

61. 28.6 L to quarts

62. 59.4 L to quarts

63. 28.2 gal to liters

64. 16 qt to liters

Metric measures are very common in medicine. Since we convert among metric measures by moving the decimal point, errors in locating the decimal point in medical doses are not unknown. Decide whether the following doses of medicine seem reasonable *or* unreasonable.

65. Take 2 kg of aspirin three times a day.

66. Take 4 L of Kaopectate every evening at bedtime.

67. Take 25 ml of cough syrup daily.

68. Soak your feet in 6 L of hot water.

69. Inject 1/2 L of insulin every morning.

70. Apply 40 g of salve to a cut on your finger.

Select the most reasonable choice for each of the following.

71. length of an adult cow
(a) 1 m **(b)** 3 m **(c)** 5 m

72. length of a Cadillac
(a) 1 m **(b)** 3 m **(c)** 5 m

73. distance from Seattle to Miami
(a) 500 km **(b)** 5,000 km **(c)** 50,000 km

74. length across an average nose
(a) 3 cm **(b)** 30 cm **(c)** 300 cm

75. distance across this page
(a) 1.93 mm **(b)** 19.3 mm **(c)** 193 mm

76. weight of this book
(a) 1 kg **(b)** 10 kg **(c)** 1,000 kg

77. weight of a large automobile
(a) 1,300 kg **(b)** 130 kg **(c)** 13 kg

78. volume of a 12-ounce bottle of beverage
(a) 35 ml **(b)** 355 ml **(c)** 3,550 ml

79. height of a person
(a) 180 cm **(b)** 1,800 cm **(c)** 18 cm

80. diameter of the earth
(a) 130 km **(b)** 1,300 km **(c)** 13,000 km

81. length of a long freight train
(a) 8 m **(b)** 80 m **(c)** 800 m

82. volume of a grapefruit
(a) 1 L **(b)** 4 L **(c)** 8 L

83. the length of a pair of Levis
(a) 70 cm **(b)** 700 cm **(c)** 7 cm

84. a person's weight
(a) 700 kg **(b)** 7 kg **(c)** 70 kg

85. diagonal measure of the picture tube of a table model TV set
(a) 5 cm **(b)** 50 cm **(c)** 500 cm

86. width of a standard bedroom door
(a) 1 m **(b)** 3 m **(c)** 5 m

87. thickness of a standard audiotape cassette
(a) .9 mm **(b)** 9 mm **(c)** 90 mm

88. length of an edge of a record album
(a) 300 mm **(b)** 30 mm **(c)** 3,000 mm

89. the temperature at the surface of a frozen lake
(a) 0°C **(b)** 10°C **(c)** 32°C

90. the temperature in the middle of Death Valley on a July afternoon
(a) 25°C **(b)** 40°C **(c)** 65°C

91. surface temperature of desert sand on a hot summer day
(a) 30°C **(b)** 60°C **(c)** 90°C

92. boiling water
(a) 100°C **(b)** 120°C **(c)** 150°C

93. air temperature on a day when you need a sweater
(a) 30°C **(b)** 20°C **(c)** 10°C

94. air temperature when you go swimming
(a) 30°C **(b)** 15°C **(c)** 10°C

95. temperature when baking a cake
(a) 120°C **(b)** 170°C **(c)** 300°C

96. temperature of bath water
(a) 35°C **(b)** 50°C **(c)** 65°C

97. temperature in a sauna
(a) 25°C **(b)** 60°C **(c)** 90°C

98. temperature in a freezer
(a) 32°C **(b)** 10°C **(c)** −5°C

Answers to Selected Exercises

In this section we provide the answers to most of the odd-numbered section, extension, and appendix exercises in the preceding chapters (not including those requiring a conceptual written response). (We also provide even-numbered exercise answers to the Chapter Tests.)

To further assist you with your study of mathematics, you may want to obtain a copy of the *Study Guide and Solution Manual* that goes with this book (ISBN 0-673-98371-4). This Guide contains worked-out solutions to all of the exercises (except for the Chapter Tests) whose answers are given in this section of the text. It also contains brief chapter summaries that review key points, provide extra examples, and enumerate major topic objectives. Your college bookstore either has this *Guide* or can order it for you.

CHAPTER 1 Approach to Problem Solving

1.1 Exercises (Page 7)

1. deductive **3.** inductive **5.** deductive **7.** deductive **9.** inductive **13.** 38 **15.** 1 **17.** 11/13 **19.** 36 **21.** 1 **23.** There are many possibilities. One such list is 10, 20, 30, 40, 50, ... **25.** $(12{,}345 \times 9) + 6 = 111{,}111$ **27.** $15{,}873 \times 35 = 555{,}555$ **29.** $11{,}111 \times 11{,}111 = 123{,}454{,}321$ **31.** $2 + 4 + 8 + 16 + 32 = 64 - 2$ **33.** $3 + 9 + 27 + 81 + 243 = \frac{3(243 - 1)}{2}$ **35.** $\frac{1}{1 \cdot 2} + \frac{1}{2 \cdot 3} + \frac{1}{3 \cdot 4} + \frac{1}{4 \cdot 5} + \frac{1}{5 \cdot 6} = \frac{5}{6}$ **37.** 11,325 **39.** 125,250 **41.** 7,875 **43.** 2,550 **45.** 0 2 10 30 60 90 102 90 60 (To find any term, choose the term directly above it and add to it the two preceding terms. If there are fewer than two terms, add as many as there are.) **47.** E (One, Two, Three, and so on.) **49.** **(f)** The final result will always be half of the number added in Step (b). **51.** 111,111,111; 222,222,222; 333,333,333; 72 **53.** A cycle will form that has its sixth number equal to the first number chosen. For example, choosing 5 and 6 gives 5, 6, 1.4, .4, 1, 5, 6 as the first seven terms.

1.2 Exercises (Page 17)

1. 38 **3.** 126 **5.** 660 **7.** 2,410 **9.** **(a)** 6, 9, 14, 21 **(b)** Add 9 to 21 to get 30. **(c)** $5^2 + 5 = 30$. The results agree. **11.** $(1{,}234 \times 8) + 4 = 9{,}876$ **13.** $1{,}010{,}101 \times 1{,}010{,}101 = 1{,}020{,}304{,}030{,}201$ **15.** $1 + 2 + 3 + 4 + 5 + 4 + 3 + 2 + 1 = 5^2$ **17.** $4^2 + 4 = 5^2 - 5$ **19.** $16 + 17 + 18 + 19 + 20 = 21 + 22 + 23 + 24$ **21.** 5,050 **23.** 138,075 **25.** 625 **27.** 22,801 **29.** $S = n(n + 1)$ **33.** row 1: 28, 36; row 2: 36, 49, 64; row 3: 35, 51, 70, 92;

row 4: 28, 45, 66, 91, 120; row 5: 18, 34, 55, 81, 112, 148; row 6: 8, 21, 40, 65, 96, 133, 176
35. $8(1) + 1 = 9 = 3^2$; $8(3) + 1 = 25 = 5^2$; $8(6) + 1 = 49 = 7^2$; $8(10) + 1 = 81 = 9^2$
37. The pattern is 1, 0, 0, 1, 0, 0, ... **39.** For $n = 5$, the pattern is 1, 0, 2, 2, 0, 1, 0, 2, 2, 0, ... For $n = 6$, the pattern is 1, 0, 3, 4, 3, 0, 1, 0, 3, 4, 3, 0, ... **41.** The difference is 4. 10 and 16 are in column 4. **43.** The difference is 2. 5 and 6 are in column 2. **45.** **(a)** a triangular number **(b)** a triangular number
47. 289 **49.** 92 **51.** 148 **53.** $N_n = \frac{n(7n - 5)}{2}$
55. a triangular number **57.** a perfect cube

1.3 Exercises (Page 27)

1. 1 **3.** \$98 **5.** 17 days **7.** 14 **9.** 16 **11.** 585, 666, 747, 828, 909 (The largest two-digit palindromic number is 99, so the sum must be 9 + 9 = 18.) **13.** It is a palindrome, since it reads the same backward as forward. **15.** 1949 **17.** B, since either Bill or Bob must be the tallest person. **19.** Eve has \$5 and Adam has \$7. **21.** 170 pounds **23.** 14 (The correct problem is

$$\begin{array}{r} 435 \\ 826 \\ +\ 147 \\ \hline 1{,}408.) \end{array}$$

25.

6	1	8
7	5	3
2	9	4

27. Start the two timers together. When the 5-minute timer runs out, start boiling the egg. When the 9-minute timer runs out, turn it over and time another 9 minutes. When it is finished, the egg will have boiled (9 − 5) + 9 = 13 minutes. **29.** 25 pitches (The visiting team's pitcher retires 24 consecutive batters through the first eight innings, using only one pitch per batter. His team does not score either. Going into the bottom of the ninth tied 0–0, the first batter for the home team hits his first pitch for a home run. The pitcher threw 25 pitches and loses the game by a score of 1–0.) **31.** 7 1/2 minutes (Boil them all at the same time.) **33.** The two children row across. One stays on the opposite bank and the other returns. One soldier rows across, and the child on the opposite bank then rows back. The two children row across. One stays and the other returns. Now another soldier rows across. This process continues until all the soldiers are across. **35.** For three weighings, first balance four against four. Of the lighter four, balance two against the other two. Finally, of the lighter two, balance them one against the other. To find the bad coin in two weighings, divide the eight coins into groups of 3, 3, 2. Weigh the groups of three against each other on the scale. If the groups weigh the same, the fake is in the two left out and can be found in one additional weighing. If the two groups of three do not weigh the same, pick the lighter group. Choose any two of the coins and weigh them. If one of these is lighter, it is the fake; if they weigh the same, then the third coin is the fake. **37.** P represents 9, A represents 1, and B represents 0.

39.

41. 28

43. Only once—after you have subtracted 10 from 100, you are no longer subtracting from 100.

45.

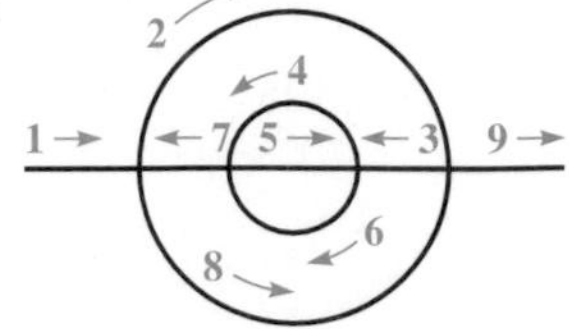

47. none **49.** 9 years old

51. (a), since perfect squares must have a units digit of 0, 1, 4, 9, 6, or 5. $(23{,}784{,}855{,}888{,}784 = (4{,}876{,}972)^2)$
53. 6

	X	X
X		X
X	X	

One of several possibilities

55. The products will always differ by 1.

57. 10 **59.** They must be February and March. This can occur only in a non-leap year. The next time it will happen is in 1998. **61.** "Madam, I'm Adam."

1.4 Exercises (Page 35)

1. 33.2 **3.** 6.82 **5.** 2.457 **7.** 639 **9.** 54.76 **11.** 753.571 **13.** 40.84101
15. 1.043353839 **17.** 6.283185307 **19.** $95^2 = 9{,}025$ **21.** negative **23.** 1 **25.** the same as
27. an error message; Division by 0 is not allowed. **29.** negative **31.** If the last digit is 6, it truncates. If it is 7, it rounds off. **33.** The result is the three-digit number you started with. Dividing by 7, then 13, and then 11 is the same as dividing by 1,001. A three-digit number *abc* multiplied by 1,001 gives *abcabc,* so we have just reversed the process. **35.** LOOSE **37.** ShOES **39.** Answers will vary. **41.** 11 **43.** 7 **45.** (b)
47. (b) **49.** (c) **51.** 360 miles **53.** 45 miles per gallon **55.** between 1989 and 1990; 262.3 thousand
57. 456.75 thousand in defense; 365.4 thousand in construction **59.** (d) **61.** Inkjet **63.** 38.5%
65. $8.38 billion **67.** pie chart **69.** St. Louis **71.** cold front

Chapter 1 Test (Page 49)

1. inductive **2.** deductive **3.** 1/9 **4.** $65{,}359{,}477{,}124{,}183 \times 68 = 4{,}444{,}444{,}444{,}444{,}444$
5. 351 **6.** 500,500 **7.** 65; $65 = 1 + 7 + 13 + 19 + 25$ **8.** 1, 8, 21, 40, 65, 96, 133, 176; The pattern is 1, 0, 1, 0, 1, 0, 1, 0, ... **10.** 8 **11.** 11 **12.** 9 **13.** 6 **14.** Neither is correct, since $3^3 = 27$. **15.** 0 **16.** The sum of the digits is always 9. **17.** 4.171330723 **18.** 3.375 **19.** (b)
20. **(a)** $249,000 **(b)** true

CHAPTER 2 Sets

2.1 Exercises (Page 56)

1. {1, 2, 3, 4} **3.** {0, 1, 2, 3, 4, 5, 6} **5.** {6, 7, 8, 9, 10, 11, 12, 13, 14}
7. {−15, −13, −11, −9, −7, −5, −3, −1} **9.** {2, 4, 8, 16, 32, 64, 128, 256}
11. {1, 1/3, 1/9, 1/27, 1/81, 1/243} **13.** {0, 2, 4, 6, 8, 10, 12, 14} **15.** {21, 22, 23, . . .}
17. {Democrat, Republican} **19.** {4, 8, 12, . . .} **21.** {1, 1/2, 1/3, 1/4, . . .}
In Exercises 23–27, there may be other acceptable descriptions.
23. $\{x \mid x$ is a rational number$\}$ **25.** $\{x \mid x$ is a movie released this year$\}$ **27.** $\{x \mid x$ is an odd counting number less than 100$\}$ **29.** finite **31.** finite **33.** infinite **35.** infinite **37.** 7 **39.** 500 **41.** 26
43. 19 **45.** 28 **47.** well defined **49.** not well defined **51.** not well defined **53.** $\in$
55. $\notin$ **57.** $\in$ **59.** $\notin$ **61.** false **63.** true **65.** true **67.** true **69.** false **71.** true
73. true **75.** true **77.** false **79.** false **81.** true **85.** {2} and {3, 4} (Other examples are possible.) **87.** $\{a, b\}$ and $\{a, c\}$ (Other examples are possible.) **89.** **(a)** $\{r\}$, $\{g, s\}$, $\{c, s\}$, $\{v, r\}$, $\{g, r\}$, $\{c, r\}$, $\{s, r\}$ **(b)** $\{v, g, r\}$, $\{v, c, r\}$, $\{v, s, r\}$, $\{g, c, r\}$, $\{g, s, r\}$, $\{c, s, r\}$

2.2 Exercises (Page 64)

1. $\not\subseteq$ **3.** $\subseteq$ **5.** $\subseteq$ **7.** $\not\subseteq$ **9.** both **11.** $\subseteq$ **13.** both **15.** neither **17.** true
19. false **21.** true **23.** true **25.** true **27.** true **29.** true **31.** false **33.** false
35. true **37.** false **39.** 8; 7 **41.** 64; 63 **43.** 32; 31 **45.** {2, 3, 5, 7, 9, 10} **47.** {2}

49. {1, 2, 3, 4, 5, 6, 7, 8, 9, 10} **51.** {High cost, Entertaining, Fixed schedule, Current films, Low cost, Flexible schedule, Older films} **53.** {High cost, Fixed schedule, Current films} **55.** {Low cost, Flexible schedule, Older films} **57.** ∅ **59.** *ABCD, ABCE, ABDE, ACDE, BCDE* **61.** *AB, AC, AD, AE, BC, BD, BE, CD, CE, DE* **63.** one way (none present) **65.** They are the same: $32 = 2^5$. The number of ways that people from a group of five can gather is the same as the number of subsets there are of a set of five elements. **67.** 15

2.3 Exercises (Page 76)

1. {a, c} **3.** {a, b, c, d, e, f} **5.** {a, b, c, d, e, f, g} **7.** {b, d, f} **9.** {d, f} **11.** ∅ **13.** {a, b, c, e, g} **15.** {a, c, e, g} **17.** {a} **19.** {e, g} **21.** {d, f} **23.** {e, g}

In Exercises 25–29, there may be other acceptable descriptions.

25. the set of all elements that either are in *A*, or are not in *B* and not in *C* **27.** the set of all elements that are in *C* but not in *B*, or are in *A* **29.** the set of all elements that are in *A* but not in *C*, or in *B* but not in *C* **31.** $U = \{e, h, c, l, b\}$ **33.** $\{l, b\}$ **35.** $\{e, h, c, l, b\}$ **37.** the set of all tax returns showing business income or filed in 1996 **39.** the set of all tax returns filed in 1996 without itemized deductions **41.** the set of all tax returns with itemized deductions or showing business income, but not selected for audit **43.** always true **45.** always true **47.** not always true **49.** always true **51.** always true **53.** **(a)** {1, 3, 5, 2} **(b)** {1, 2, 3, 5} **(c)** For any sets *X* and *Y*, $X \cup Y = Y \cup X$. **55.** **(a)** {1, 3, 5, 2, 4} **(b)** {1, 3, 5, 2, 4} **(c)** For any sets *X*, *Y*, and *Z*, $X \cup (Y \cup Z) = (X \cup Y) \cup Z$. **57.** **(a)** {4} **(b)** {4} **(c)** For any sets *X* and *Y*, $(X \cup Y)' = X' \cap Y'$. **59.** **(a)** {1, 3, 5} **(b)** For any set *X*, $X \cup \emptyset = X$. **61.** true **63.** false **65.** false **67.** true **69.** true **71.** $A \times B = \{(2, 4), (2, 9), (8, 4), (8, 9), (12, 4), (12, 9)\}$; $B \times A = \{(4, 2), (4, 8), (4, 12), (9, 2), (9, 8), (9, 12)\}$ **73.** $A \times B = \{$(d, p), (d, i), (d, g), (o, p), (o, i), (o, g), (g, p), (g, i), (g, g)$\}$; $B \times A = \{$(p, d), (p, o), (p, g), (i, d), (i, o), (i, g), (g, d), (g, o), (g, g)$\}$ **75.** $n(A \times B) = 6$; $n(B \times A) = 6$ **77.** $n(A \times B) = 210$; $n(B \times A) = 210$ **79.** 3

81.

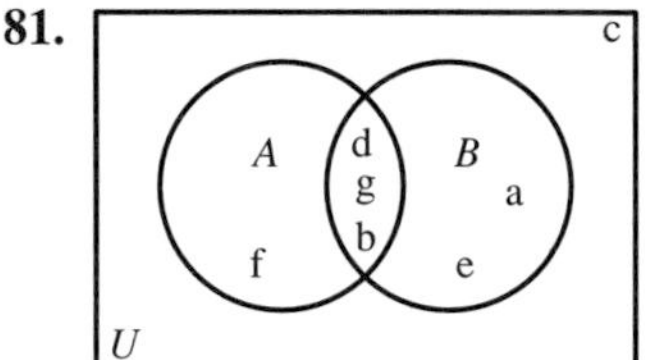

83.

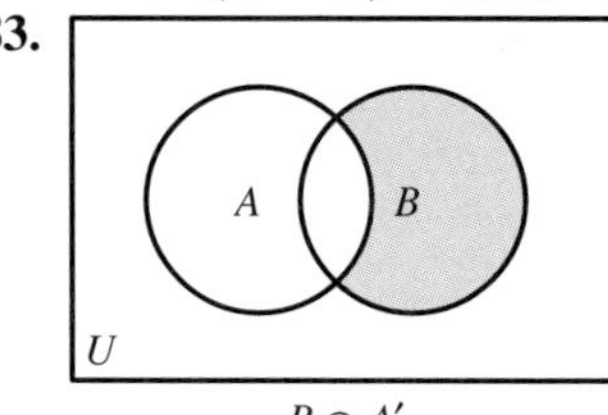

$B \cap A'$

85.

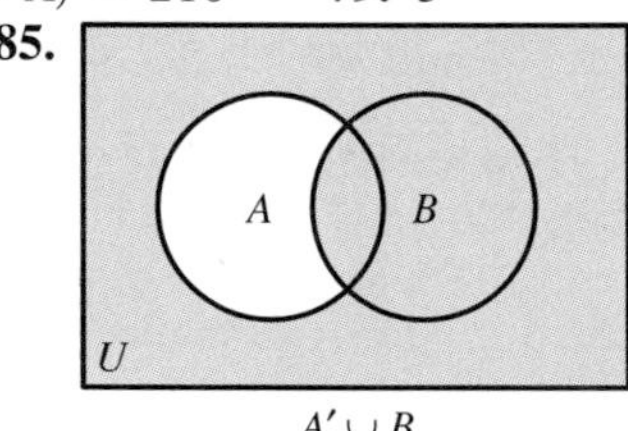

$A' \cup B$

87.

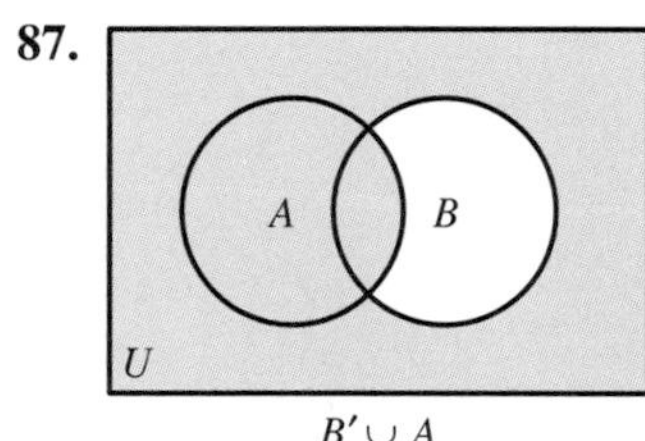

$B' \cup A$

89.

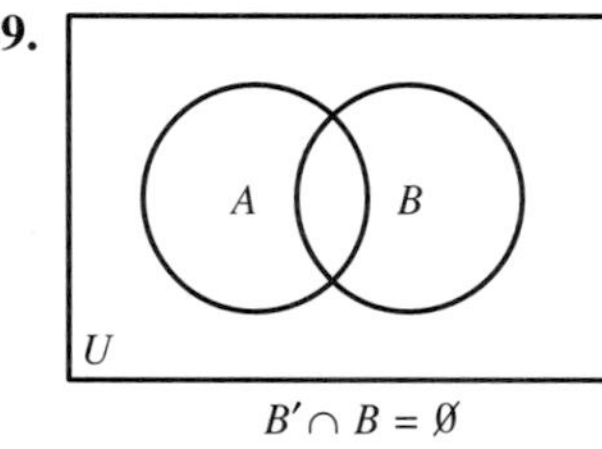

$B' \cap B = \emptyset$

91.

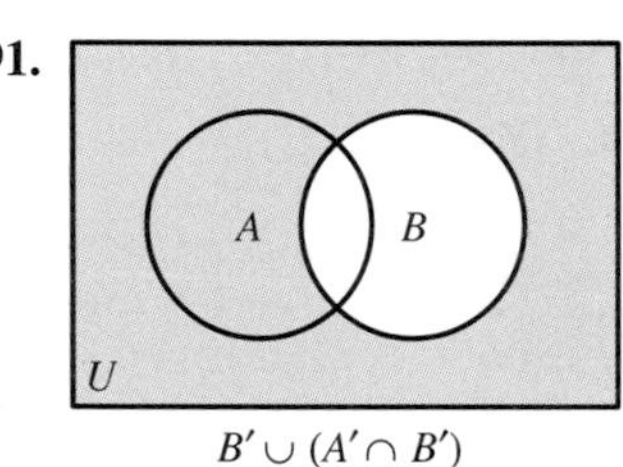

$B' \cup (A' \cap B')$

93.

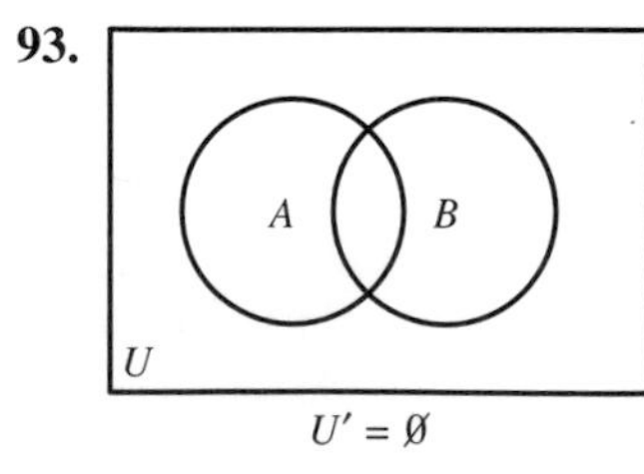

$U' = \emptyset$

95.

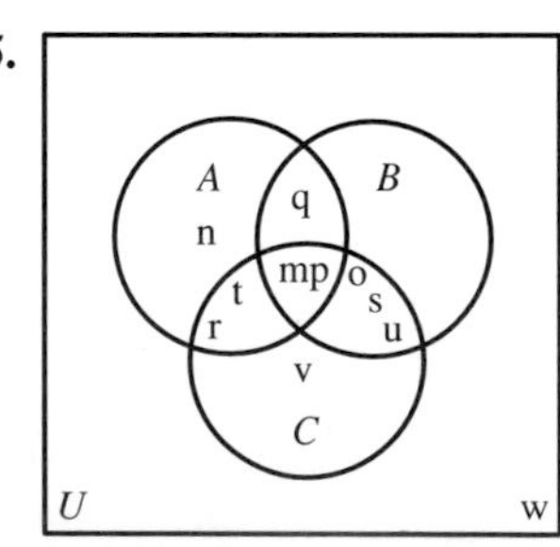

97.

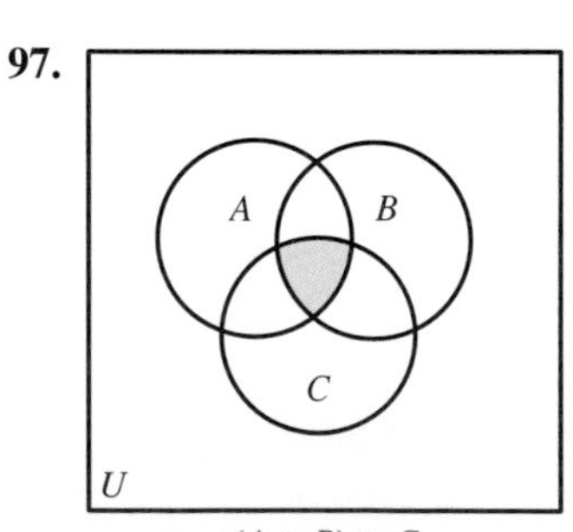

$(A \cap B) \cap C$

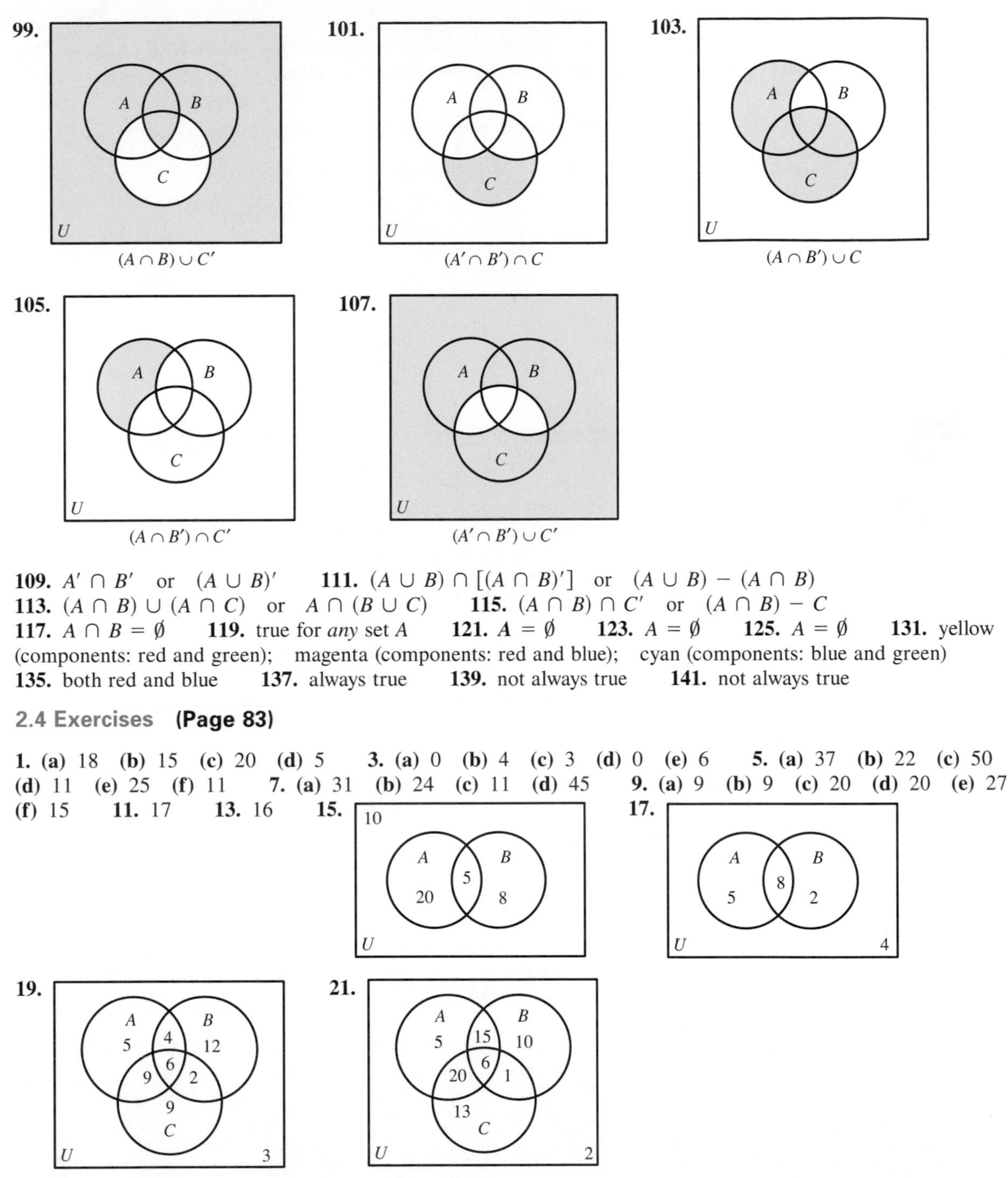

99. $(A \cap B) \cup C'$ **101.** $(A' \cap B') \cap C$ **103.** $(A \cap B') \cup C$

105. $(A \cap B') \cap C'$ **107.** $(A' \cap B') \cup C'$

109. $A' \cap B'$ or $(A \cup B)'$ **111.** $(A \cup B) \cap [(A \cap B)']$ or $(A \cup B) - (A \cap B)$ **113.** $(A \cap B) \cup (A \cap C)$ or $A \cap (B \cup C)$ **115.** $(A \cap B) \cap C'$ or $(A \cap B) - C$ **117.** $A \cap B = \emptyset$ **119.** true for *any* set A **121.** $A = \emptyset$ **123.** $A = \emptyset$ **125.** $A = \emptyset$ **131.** yellow (components: red and green); magenta (components: red and blue); cyan (components: blue and green) **135.** both red and blue **137.** always true **139.** not always true **141.** not always true

2.4 Exercises (Page 83)

1. **(a)** 18 **(b)** 15 **(c)** 20 **(d)** 5 **3.** **(a)** 0 **(b)** 4 **(c)** 3 **(d)** 0 **(e)** 6 **5.** **(a)** 37 **(b)** 22 **(c)** 50 **(d)** 11 **(e)** 25 **(f)** 11 **7.** **(a)** 31 **(b)** 24 **(c)** 11 **(d)** 45 **9.** **(a)** 9 **(b)** 9 **(c)** 20 **(d)** 20 **(e)** 27 **(f)** 15 **11.** 17 **13.** 16 **15.** (see diagram) **17.** (see diagram) **19.** (see diagram) **21.** (see diagram)

25. 4, 8, 16 **27.** 1, 2, 3, 4, 5, 6, 7, 8, 9, 10, 11, 12, 13, 14, 15 (all except 16) **29.** 5, 8, 13 **31.** **(a)** none **(b)** 52 **(c)** 44

2.5 Exercises (Page 93)

1. (Other correspondences are possible.)
{I, II, III}
$\updownarrow$ $\updownarrow$ $\updownarrow$
{x, y, z}

3. (Other correspondences are possible.)
{a, d, i, t, o, n}
$\updownarrow$ $\updownarrow$ $\updownarrow$ $\updownarrow$ $\updownarrow$ $\updownarrow$
{a, n, s, w, e, r}

5. 11 **7.** 0 **9.** $\aleph_0$

11. $\aleph_0$ **13.** $\aleph_0$ **15.** 12 **17.** $\aleph_0$ **19.** both **21.** equivalent **23.** equivalent

25. {2, 4, 6, 8, . . . , $2n$, . . .}
$\updownarrow$ $\updownarrow$ $\updownarrow$ $\updownarrow$ $\updownarrow$
{1, 2, 3, 4, . . . , n, . . .}

27. {1,000,000, 2,000,000, 3,000,000, . . . , 1,000,000n, . . .}
$\updownarrow$ $\updownarrow$ $\updownarrow$ $\updownarrow$
{ 1, 2, 3, . . . , n, . . .}

29. {2, 4, 8, 16, 32, . . . , 2^n, . . .}
$\updownarrow$ $\updownarrow$ $\updownarrow$ $\updownarrow$ $\updownarrow$ $\updownarrow$
{1, 2, 3, 4, 5, . . . , n, . . .}

31. Not always true. For example, let A = the set of counting numbers, B = the set of real numbers.
33. Not always true. For example, A could be the set of all subsets of the set of reals. Then $n(A)$ would be an infinite number *greater* than c.
35. **(a)** Rays emanating from point P will establish a geometric pairing of the points on the semicircle with the points on the line.

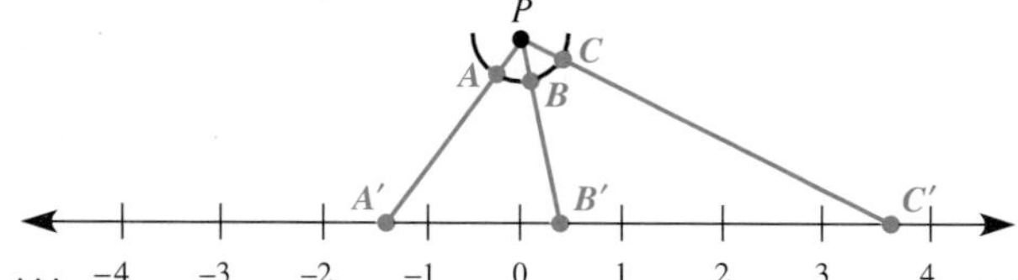

(b) The set of real numbers is infinite (having been placed in a one-to-one correspondence with a proper subset of itself).

37. {3, 6, 9, 12, . . . , $3n$, . . .}
$\updownarrow$ $\updownarrow$ $\updownarrow$ $\updownarrow$ $\updownarrow$
{6, 9, 12, 15, . . . , $3n + 3$, . . .}

39. {3/4, 3/8, 3/12, 3/16, . . . , $3/(4n)$, . . .}
$\updownarrow$ $\updownarrow$ $\updownarrow$ $\updownarrow$ $\updownarrow$
{3/8, 3/12, 3/16, 3/20, . . . , $3/(4n + 4)$, . . .}

41. {1/9, 1/18, 1/27, . . . , $1/(9n)$, . . .}
$\updownarrow$ $\updownarrow$ $\updownarrow$ $\updownarrow$
{1/18, 1/27, 1/36, . . . , $1/(9n + 9)$, . . .}

Chapter 2 Test (Page 96)

1. {a, b, c, d, e} **2.** {a, b, d} **3.** {c, f, g, h} **4.** {a, c} **5.** true **6.** false **7.** true **8.** true **9.** false **10.** true **11.** true **12.** true **13.** 8 **14.** 15

Answers may vary for Exercises 15–18.

15. the set of odd integers between -4 and 10 **16.** the set of months of the year **17.** $\{x \mid x$ is a negative integer$\}$ **18.** $\{x \mid x$ is a multiple of 8 between 20 and 90$\}$ **19.** $\subseteq$ **20.** neither

21.

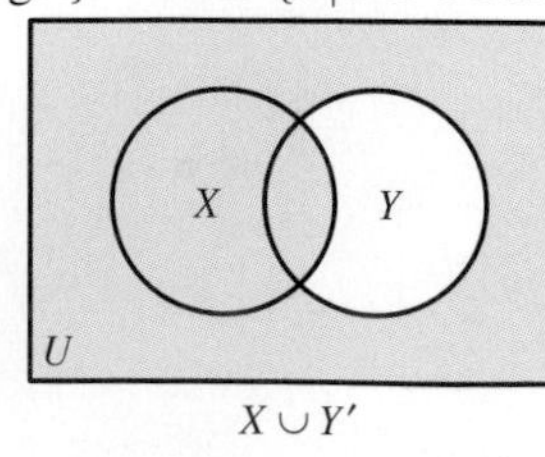

$X \cup Y'$

22.

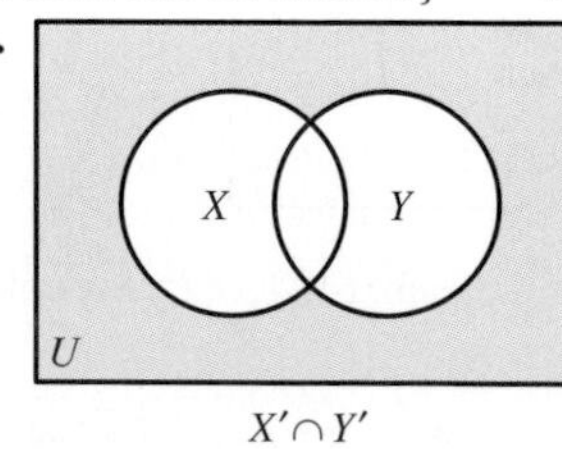

$X' \cap Y'$

23.

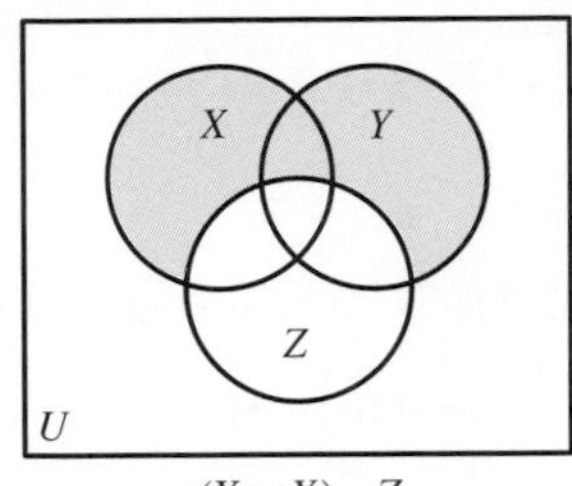

$(X \cup Y) - Z$

24.

X Y Z U

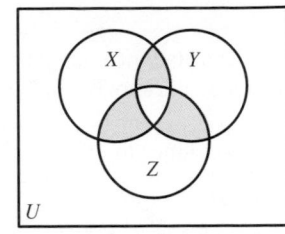

$[(X \cap Y) \cup (Y \cap Z) \cup (X \cap Z)] - (X \cap Y \cap Z)$

25. {Electric razor} **26.** {Adding machine, Barometer, Pendulum clock, Thermometer} **27.** {Electric razor} **30. (a)** 16 **(b)** 32 **(c)** 33 **(d)** 45 **(e)** 14 **(f)** 26

CHAPTER 3 Logic

3.1 Exercises (Page 103)

1. statement **3.** not a statement **5.** statement **7.** statement **9.** not a statement **11.** not a statement **13.** statement **15.** not compound **17.** compound **19.** compound **21.** compound **23.** The flowers are not to be watered. **25.** Some rain fell in southern California today. **27.** At least one student present will not get another chance. **29.** No people have all the luck. **31.** Someone does not like Sara Lee. **33.** $x \leq 3$ **35.** $p < 4$ **39.** She does not have blue eyes. **41.** She has blue eyes and he is 43 years old. **43.** She does not have blue eyes or he is 43 years old. **45.** She does not have blue eyes or he is not 43 years old. **47.** It is not the case that she does not have blue eyes and he is 43 years old. **49.** $p \wedge \sim q$ **51.** $\sim p \vee q$ **53.** $\sim(p \vee q)$ or, equivalently, $\sim p \wedge \sim q$ **57.** true **59.** true **61.** true **63.** false **65.** false **67.** C **69.** B **71.** A, B **73.** A, B

3.2 Exercises (Page 115)

1. It must be true. **3.** F **7.** T **9.** T **11.** F **13.** T **15.** T **17.** T **19.** a disjunction **21.** T **23.** T **25.** T **27.** T **29.** F **31.** T **33.** T **35.** T **37.** 4 **39.** 16 **41.** 128 **43.** 6

In Exercises 45–59 and later in this chapter, we give the truth values found in the order in which they appear in the final column of the truth table, when the truth table is constructed in the format described in the section.

45. FFTF **47.** FTTT **49.** TTTT **51.** FFFT **53.** TFFF **55.** FFFFTFFF **57.** FTFTTTTT **59.** TTTTTTTTTTTTFTTT **61.** You can't pay me now and you can't pay me later. **63.** It is not summer or there is snow. **65.** I did not say yes or she did not say no. **67.** $5 - 1 \neq 4$ or $9 + 12 = 7$ **69.** Neither Dasher nor Dancer will lead Santa's sleigh next Christmas. **71.** T **73.** T **75.** FTTF **77.** F **79.** T

3.3 Exercises (Page 122)

1. true **3.** true **5.** false **7.** true **11.** If it's in *USA Today,* then you can believe it. **13.** If the person is Kathi Callahan, then her area code is 708. **15.** If you're a soldier, then you maintain your weapon. **17.** If it's a koala, then it doesn't live in Mississippi. **19.** If it's an alligator, then it cannot live in these waters. **21.** T **23.** F **25.** T **27.** If I do not major in mathematics, then I pass my psychology course. **29.** If I study in the library, then I major in mathematics and I pass my psychology course. **31.** If I do not pass my psychology course, then I do not major in mathematics or I study in the library. **33.** $s \rightarrow d$ **35.** $\sim d \rightarrow \sim s$ **37.** $d \vee (c \rightarrow s)$ **39.** $\sim s \rightarrow d$ **41.** T **43.** F **45.** T **47.** F **49.** T **51.** T **55.** TTTF **57.** TTFT **59.** TTTT; tautology **61.** TFTF **63.** TTTTTTFT **65.** TTTTTTTTTTTTTTFT **67.** You give your plants tender, loving care and they do not flourish. **69.** She doesn't and he will not. **71.** The person is a resident of Boise and is not a resident of Idaho. **73.** You do not give your plants tender, loving care or they flourish. **75.** She does or he will. **77.** The person is not a resident of Boise or is a resident of Idaho. **79.** equivalent **81.** equivalent **83.** equivalent **85.** not equivalent

Extension Exercises (Page 126)

1. $p \wedge (r \vee q)$ **3.** $q \vee [p \wedge (q \vee \sim p)]$; The statement simplifies to q.
5. $(\sim p \vee q) \vee (\sim p \vee \sim q)$; The statement simplifies to T. **7.**

9.

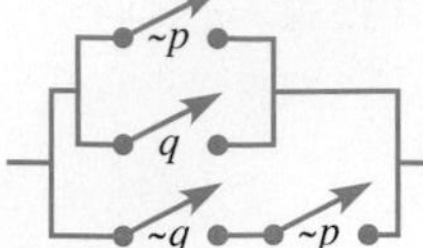

11. The statement simplifies to $(\sim p \wedge r) \wedge \sim q$.

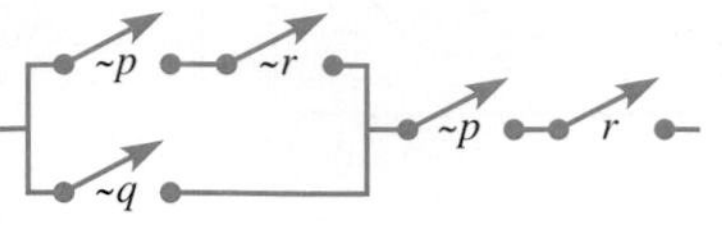

13. The statement simplifies to T.

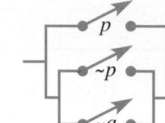

15. \$262.80

3.4 Exercises (Page 132)

Wording may vary in the answers to Exercises 1–9.
1. (a) If I follow, then you lead. **(b)** If you do not lead, then I will not follow. **(c)** If I do not follow, then you do not lead. **3. (a)** If I were rich, then I would have a nickel for each time that happened. **(b)** If I did not have a nickel for each time that happened, then I would not be rich. **(c)** If I were not rich, then I would not have a nickel for each time that happened. **5. (a)** If it contains calcium, then it's milk. **(b)** If it's not milk, then it does not contain calcium. **(c)** If it does not contain calcium, then it's not milk. **7. (a)** If it gathers no moss, then it is a rolling stone. **(b)** If it is not a rolling stone, then it gathers moss. **(c)** If it gathers moss, then it is not a rolling stone. **9. (a)** If there's fire, then there's smoke. **(b)** If there's no smoke, then there's no fire. **(c)** If there's no fire, then there's no smoke. **11. (a)** $\sim q \rightarrow p$ **(b)** $\sim p \rightarrow q$ **(c)** $q \rightarrow \sim p$ **13. (a)** $\sim q \rightarrow \sim p$ **(b)** $p \rightarrow q$ **(c)** $q \rightarrow p$ **15. (a)** $(q \vee r) \rightarrow p$ **(b)** $\sim p \rightarrow (\sim q \wedge \sim r)$ **(c)** $(\sim q \wedge \sim r) \rightarrow \sim p$
19. If I finish studying, then I'll go to the party. **21.** If it is Wednesday, then yesterday was Tuesday. **23.** If a number is a whole number, then it is an integer. **25.** If you are in Fort Lauderdale, then you are in Florida. **27.** If one is elected, then one is an environmentalist. **29.** If the principal hires more teachers, then the school board approves. **31.** If a number is an integer, then it is rational. **33.** If pigs fly, then Newt will be a liberal. **35.** If the figure is a parallelogram, then it is a four-sided figure with opposite sides parallel. **37.** If the figure is a square, then it is a rectangle with two adjacent sides equal. **39.** If an integer has a units digit of 0 or 5, then it is divisible by 5. **41.** (d) **45.** true **47.** true **49.** false **51.** contrary **53.** consistent **55.** contrary **57.** For example: That man is Arnold Parker; That man sells books.

3.5 Exercises (Page 136)

1. valid **3.** invalid **5.** invalid **7.** valid **9.** invalid **11.** invalid **13.** invalid **15.** One possible conclusion is "All expensive things make you feel good." **17.** invalid **19.** valid **21.** invalid **23.** valid **25.** invalid **27.** invalid **29.** valid **33.** Fred, Juanita, Evita, Butch, Li

3.6 Exercises (Page 146)

1. valid by reasoning by transitivity **3.** valid by modus ponens **5.** fallacy by fallacy of the converse **7.** valid by modus tollens **9.** fallacy by fallacy of the inverse **11.** valid by disjunctive syllogism **13.** valid **15.** invalid **17.** invalid **19.** invalid **21.** valid **23.** valid
25. Every time something squeaks, I use WD-40.
Every time I use WD-40, I go to the hardware store.
———
Every time something squeaks, I go to the hardware store.

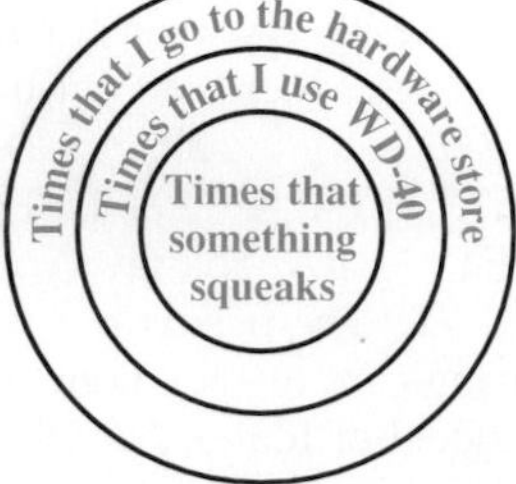

27. valid **29.** invalid **31.** invalid **33.** valid **35.** valid

Answers in Exercises 39–45 may be replaced by their contrapositives.

39. If he is your son, then he cannot do logic. **41.** If the person is a teetotaler, then the person is not a pawnbroker. **43.** If it is an opium-eater, then it has no self-command. **45.** If it is written on blue paper, then it is filed. **47. (a)** $r \rightarrow p$ **(b)** $\sim r \rightarrow \sim q$ **(c)** $s \rightarrow \sim p$ **(d)** Your sons are not fit to serve on a jury. **49. (a)** $s \rightarrow r$ **(b)** $p \rightarrow q$ **(c)** $q \rightarrow \sim r$ **(d)** Guinea pigs don't appreciate Beethoven. **51. (a)** $p \rightarrow q$ **(b)** $\sim u \rightarrow \sim s$ **(c)** $t \rightarrow \sim r$ **(d)** $q \rightarrow s$ **(e)** $v \rightarrow p$ **(f)** $\sim r \rightarrow \sim u$ **(g)** Opium-eaters do not wear white kid gloves.

Chapter 3 Test (Page 150)

1. $5 + 3 \neq 9$ **2.** There is a good boy who does not deserve favour. **3.** All people here can play this game. **4.** It comes to that and I am here. **5.** My mind is not made up or you can change it. **6.** $\sim p \rightarrow q$ **7.** $p \vee \sim q$ **8.** $p \rightarrow \sim q$ **9.** It is not broken and you can fix it. **10.** It is broken if and only if you can't fix it. **11.** F **12.** F **13.** T **14.** T **15.** For a conditional statement to be false, the antecedent must be true and the consequent must be false. For a conjunction to be true, both component statements must be true.

In Exercises 17–18, we give the truth values found in the order in which they appear in the final column of the truth table, when the truth table is constructed in the format described in the chapter.

17. TFFF **18.** TTTT (tautology) **19.** true **20.** false

Wording may vary in the answers to Exercises 21–25.

21. If it is a rational number, then it is a real number. **22.** If a polygon is a rectangle, then it is a quadrilateral. **23.** If a number is divisible by 6, then it is divisible by 2. **24.** If she cries, then she is hurt. **25. (a)** If the graph helps me understand it, then a picture paints a thousand words. **(b)** If a picture doesn't paint a thousand words, then the graph won't help me understand it. **(c)** If the graph doesn't help me understand it, then a picture doesn't paint a thousand words. **26. (a)** $(q \wedge r) \rightarrow \sim p$ **(b)** $p \rightarrow (\sim q \vee \sim r)$ **(c)** $(\sim q \vee \sim r) \rightarrow p$ **27.** valid **28. (a)** B **(b)** F **(c)** C **(d)** D **29.** valid **30.** invalid

CHAPTER 4 Numeration and Mathematical Systems

4.1 Exercises (Page 161)

1. 2,412 **3.** 3,005,231 **5.** 𓍢𓍢𓍢𓍢𓎆𓎆𓏺𓏺𓏺𓏺𓏺𓏺𓏺 **7.** 𓆐𓆐𓆐𓆼𓆼𓆼𓆼𓆼𓆼𓎆𓎆𓎆𓎆𓎆𓎆𓎆𓎆𓎆

9. 𓂭𓂭𓂭𓂭𓆼𓆼𓆼𓆼𓆼𓆼𓍢𓍢𓍢𓍢𓍢 **11.** 𓂭𓂭𓂭𓂭𓆼𓆼𓆼𓆼𓆼𓍢𓍢𓍢𓍢𓍢𓍢𓎆𓎆𓎆𓎆𓎆 **13.** 𓂭𓂭𓂭𓂭𓂭𓆼𓆼𓆼𓆼𓍢𓍢𓍢𓍢 **15.** 246

17. 4,902 **19.** 六
十
三 **21.** 二
千
四
百
十
六

23. 二
百
零
二 to 二
百
二
十 **25.** 九
百
六
十 to 一
千
二
百
七
十
九 **27.** 392 **29.** 6,168 **31.** 22 **33.** 1,263

35. 57 **37.** 1,116 **39.** 533,000 shekels **45.** 99,999 **47.** 3,124 **49.** $10^d - 1$ **51.** $7^d - 1$

Exercises (Page 170)

1. $(3 \times 10^1) + (7 \times 10^0)$ **3.** $(2 \times 10^3) + (8 \times 10^2) + (1 \times 10^1) + (5 \times 10^0)$
5. $(3 \times 10^3) + (6 \times 10^2) + (2 \times 10^1) + (8 \times 10^0)$
7. $(1 \times 10^7) + (3 \times 10^6) + (6 \times 10^5) + (0 \times 10^4) + (6 \times 10^3) + (0 \times 10^2) + (9 \times 10^1) + (0 \times 10^0)$
9. 73 **11.** 5,072 **13.** 50,602,003 **15.** 89 **17.** 32 **19.** 109 **21.** 722 **23.** 6
25. 207 **27.** 23 **29.** 4,536 **31.** **33.**

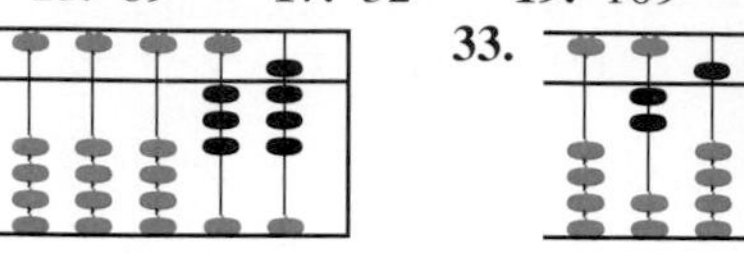

35. 1,764 **37.** 28,084 **39.** 3,035,154 **41.** 496 **43.** 217,204 **45.** 410 **47.** 26,598

4.3 Exercises (Page 179)

1. 1, 2, 3, 4, 5, 6, 10, 11, 12, 13, 14, 15, 16, 20, 21, 22, 23, 24, 25, 26 **3.** 1, 2, 3, 4, 5, 6, 7, 8, 10, 11, 12, 13, 14, 15, 16, 17, 18, 20, 21, 22 **5.** 13_{five}; 20_{five} **7.** $\text{B6E}_{\text{sixteen}}$; $\text{B70}_{\text{sixteen}}$ **9.** 3 **11.** 11
13. smallest: $1{,}000_{\text{three}} = 27$; largest: $2{,}222_{\text{three}} = 80$ **15.** 14 **17.** 11 **19.** 956 **21.** 881
23. 28,854 **25.** 139 **27.** 5,601 **29.** 321_{five} **31.** $10{,}011_{\text{two}}$ **33.** 93_{sixteen} **35.** $2{,}131{,}101_{\text{five}}$
37. $1{,}001{,}001{,}010_{\text{two}}$ **39.** $102{,}112{,}101_{\text{three}}$ **41.** $111{,}134_{\text{six}}$ **43.** 32_{seven} **45.** $11{,}651_{\text{seven}}$
47. $11{,}110{,}111_{\text{two}}$ **49.** 467_{eight} **51.** $11{,}011{,}100_{\text{two}}$ **53.** $\text{2D}_{\text{sixteen}}$ **55.** 37_{eight} **57.** 1,427
59. 1011000 **61.** 1110010 **63.** CHUCK
65. 1001111111001011011001100101110000111011101110011 **67.** 27 and 63 **69.** pennies, nickels, and quarters **73.** 7 **75.** no **77.** no **79.** no **81.** no **83.** no

4.4 Exercises (Page 191)

1. 5 **3.** 6 **5.** row 2: 0, 6, 10; row 3: 9, 0, 9; row 4: 0, 4, 0, 8, 0; row 5: 1, 9, 2, 7; row 6: 6, 0, 0; row 7: 4, 11, 6, 8, 3, 5; row 8: 8, 4, 0, 0; row 9: 6, 3, 9, 3, 9, 6, 3; row 10: 6, 4, 0, 10, 8, 6, 4; row 11: 10, 9, 8, 7, 6, 5, 4, 3, 2 **7.** yes **9.** row 1: 0; row 2: 0; row 3: 0, 1, 2; row 4: 0, 1, 2 **11.** yes
13. yes (0 is its own inverse, 1 and 4 are inverses of each other, and 2 and 3 are inverses of each other.)
15. yes **17.** yes (1 is the identity element.) **19.** 3 **21.** 4 **25.** 0700 **27.** 0000 **29.** false
31. true **33.** 3 **35.** 3 **37.** 1 **39.** 10 **43.** 5 **45.** 4
47. **(a)** row 1: 0; row 2: 2, 3, 4, 5, 6, 0, 1; row 3: 3, 4, 5, 6, 0, 1, 2; row 4: 4, 5, 6, 0, 1, 2, 3; row 5: 5, 6, 0, 1, 2, 3, 4; row 6: 6, 0, 1, 2, 3, 4, 5 **(b)** All four properties are satisfied. **(c)** 0 is its own inverse, 1 and 6 are inverses of each other, 2 and 5 are inverses of each other, and 3 and 4 are inverses of each other.
49. **(a)** 1 **(b)** All properties are satisfied. **(c)** 1 is its own inverse, and 2 is its own inverse.
51. **(a)** row 2: 1, 3, 7; row 3: 3, 0; row 4: 3, 2, 1; row 5: 6, 7, 4; row 7: 3, 1, 6, 4; row 8: 6, 5, 2 **(b)** no inverse property **(c)** 1 is its own inverse, as is 8; 2 and 5 are inverses of each other, as are 4 and 7; 3 has no inverse, and 6 has no inverse. **53.** {3, 10, 17, 24, 31, 38, . . .} **55.** identity **57.** 100,000
59. 62 **61.** **(a)** 365 **(b)** Friday **63.** Chicago: July 23 and 29; New Orleans: July 5 and August 16; San Francisco: August 9 **65.** $-i$ **67.** i **69.** Sunday **71.** Friday **73.** Jan., Oct. **75.** June
77. incorrect **79.** 0 **81.** 9

4.5 Exercises (Page 201)

1. all properties; 1 is the identity element; 1 is its own inverse, as is 4; 2 and 3 are inverses.
3. commutative, associative, and identity properties; 1 is the identity element; 2, 3, and 4 have no inverses.
5. all properties; 1 is the identity element; 1, 3, 5, and 7 are their own inverses. **7.** closure and commutative properties **9.** all properties; A is the identity element; J and U are inverses; A and T are their own inverses. **11.** a **13.** a **15.** row b: d; row c: d, b; row d: b, c **17.** associative, commutative, identity (U), closure **19.** Answers may vary. One possibility is as follows.

	a	***b***	***c***	***d***
a	*a*	*b*	*c*	*d*
b	*b*	*a*	*d*	*c*
c	*c*	*d*	*a*	*b*
d	*d*	*c*	*b*	*a*

21. no **23.** **(a)** true **(b)** true **(c)** true **(d)** true **25.** $a + b + c = 1$ or $a = 0$
27. **(a)** $a = 0$ **(b)** $a = 0$ **29.** Each side simplifies to e. **31.** Each side simplifies to d.
33.

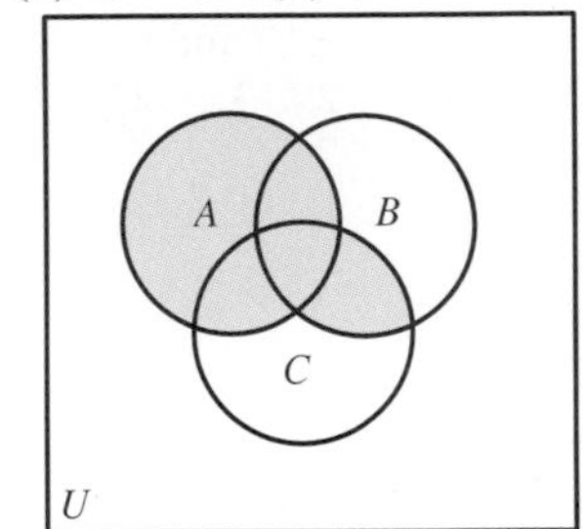

$A \cup (B \cap C) = (A \cup B) \cap (A \cup C)$

35. TTTTTFFF for each final column

4.6 Exercises (Page 210)

3. yes **5.** yes **7.** no: inverse **9.** no: closure **11.** no: inverse **13.** no: identity, inverse **15.** yes **17.** yes **21.** S **23.** N **25.** M **27.** M **29.** N **31.** R **33.** T **35.** yes **37.** yes **39.** all **41.** yes **43.** yes **45.** no **47.** yes **49.** no **51.** yes

Chapter 4 Test (Page 213)

1. ancient Egyptian; 2,536 **2.** 8,364
3. $(6 \times 10^4) + (0 \times 10^3) + (9 \times 10^2) + (2 \times 10^1) + (3 \times 10^0)$ **4.** 114 **5.** 38 **6.** 43,020
7. $111{,}010_{\text{two}}$ **8.** $24{,}341_{\text{five}}$ **9.** 256_{eight} **10.** $101{,}101{,}010{,}010_{\text{two}}$ **11.** less repetition of symbols
12. fewer different symbols to learn **13.** fewer digits in numerals **14.** 8 **15.** 9 **16.** 0 **17.** 6
18. 1 **19.** 0

20.

+	0	1	2	3	4	5
0	0	1	2	3	4	5
1	1	2	3	4	5	0
2	2	3	4	5	0	1
3	3	4	5	0	1	2
4	4	5	0	1	2	3
5	5	0	1	2	3	4

21.

×	0	1	2	3	4	5
0	0	0	0	0	0	0
1	0	1	2	3	4	5
2	0	2	4	0	2	4
3	0	3	0	3	0	3
4	0	4	2	0	4	2
5	0	5	4	3	2	1

22. {5, 17, 29, 41, 53, 65, . . .} **23.** {7, 15, 23, 31, 39, 47, . . .} **24.** 75 **26.** **(a)** yes **(b)** o
27. **(a)** yes **28.** **(a)** no **29.** **(a)** no **30.** **(a)** no

CHAPTER 5 Number Theory

5.1 Exercises (Page 220)

1. true **3.** false **5.** false **7.** true **9.** false **11.** 1, 2, 3, 4, 6, 12 **13.** 1, 2, 4, 5, 10, 20 **15.** 1, 2, 4, 13, 26, 52 **17.** 1, 2, 3, 4, 5, 6, 8, 10, 12, 15, 20, 24, 30, 40, 60, 120 **19.** **(a)** no **(b)** yes **(c)** no **(d)** yes **(e)** no **(f)** no **(g)** yes **(h)** no **(i)** no **21.** **(a)** yes **(b)** yes **(c)** yes **(d)** no **(e)** yes **(f)** no **(g)** no **(h)** no **(i)** yes **23.** **(a)** yes **(b)** yes **(c)** no **(d)** yes **(e)** yes **(f)** no **(g)** yes **(h)** yes **(i)** no **25.** **(a)** no **(b)** no **(c)** no **(d)** yes **(e)** no **(f)** no **(g)** no **(h)** no **(i)** no **27.** **(a)** no **(b)** yes **(c)** no **(d)** no **(e)** no **(f)** no **(g)** yes **(h)** no **(i)** no
29. 101, 103, 107, 109, 113, 127, 131, 137, 139, 149, 151, 157, 163, 167, 173, 179, 181, 191, 193, 197, 199
33. The last four digits must form a number divisible by 16. The number 456,882,320 is divisible by 16, since 2,320 is divisible by 16: $2{,}320 = 16 \cdot 145$. **35.** $2^2 \cdot 3 \cdot 5^2$ **37.** $5^2 \cdot 17$ **39.** $3 \cdot 5 \cdot 59$
41. $3^2 \cdot 5^2 \cdot 7$ **43.** yes **45.** no **47.** yes **49.** no **51.** 840 **53.** 2,520 **57.** 0, 2, 4, 6, 8 **59.** 0, 4, 8 **61.** 0, 6 **63.** 6 **65.** 9 **67.** 12 **69.** 27 **71.** leap year **73.** not a leap year **75.** leap year **77.** not a leap year

5.2 Exercises (Page 229)

1. true **3.** true **5.** true **7.** true **9.** true **11.** The sum of the proper divisors is 496: $1 + 2 + 4 + 8 + 16 + 31 + 62 + 124 + 248 = 496$. **13.** 8,191 is prime; 33,550,336

5. $1 + 1/2 + 1/3 + 1/6 = 2$ **17.** abundant **19.** deficient **21.** 12, 18, 20, 24 **23.** $1 + 3 + 5 + 7 + 9 + 15 + 21 + 27 + 35 + 45 + 63 + 105 + 135 + 189 + 315 = 975$, and $975 > 945$, so 945 is abundant. **25.** $1 + 2 + 4 + 8 + 16 + 32 + 37 + 74 + 148 + 296 + 592 = 1{,}210$ and $1 + 2 + 5 + 10 + 11 + 22 + 55 + 110 + 121 + 242 + 605 = 1{,}184$. **27.** $3 + 11$ **29.** $3 + 23$ **31.** Let $a = 5$ and $b = 3$; $11 = 5 + 2 \cdot 3$ **33.** 71 and 73 **35.** 137 and 139 **37.** **(a)** $5 = 9 - 4 = 3^2 - 2^2$ **(b)** $11 = 36 - 25 = 6^2 - 5^2$ **39.** $5^2 + 2 = 27 = 3^3$ **41.** 41 is prime. **43.** 83 is prime. **47.** 1,231 **49.** 971

5.3 Exercises (Page 236)

1. true **3.** true **5.** false **7.** true **9.** true **11.** 10 **13.** 120 **15.** 7 **17.** 12 **19.** 10 **21.** 6 **23.** 12 **25.** 12 **27.** 70 **31.** 120 **33.** 672 **35.** 840 **37.** 180 **39.** 1,260 **41.** 1,680 **43.** **(a)** $p^b q^c r^c$ **(b)** $p^a q^a r^b$ **45.** 30 **47.** 15 **49.** 2,880 **51.** **(a)** 6 **(b)** 36 **53.** **(a)** 18 **(b)** 216 **55.** p and q are relatively prime. **59.** 144th **61.** 48 **63.** \$600; 25 books

5.4 Exercises (Page 243)

1. 2,584 **3.** 46,368 **5.** $(1 + \sqrt{5})/2$ **7.** $1 + 1 + 2 + 3 + 5 + 8 = 21 - 1$; Each expression is equal to 20. **9.** $1 + 2 + 5 + 13 + 34 + 89 = 144$; Each expression is equal to 144. **11.** $13^2 - 5^2 = 144$; Each expression is equal to 144. **13.** $1 - 2 + 5 - 13 + 34 - 89 = -8^2$; Each expression is equal to -64. **15.** (There are other ways to do this.) **(a)** $37 = 34 + 3$ **(b)** $40 = 34 + 5 + 1$ **(c)** $52 = 34 + 13 + 5$ **17.** **(a)** The greatest common factor of 10 and 4 is 2, and the greatest common factor of $F_{10} = 55$ and $F_4 = 3$ is $F_2 = 1$. **(b)** The greatest common factor of 12 and 6 is 6, and the greatest common factor of $F_{12} = 144$ and $F_6 = 8$ is $F_6 = 8$. **(c)** The greatest common factor of 14 and 6 is 2, and the greatest common factor of $F_{14} = 377$ and $F_6 = 8$ is $F_2 = 1$. **19.** **(a)** $5 \cdot 34 - 13^2 = 1$ **(b)** $13^2 - 3 \cdot 55 = 4$ **(c)** $2 \cdot 89 - 13^2 = 9$ **(d)** The difference will be 25, since we are obtaining the squares of the terms of the Fibonacci sequence. $13^2 - 1 \cdot 144 = 25 = 5^2$. **21.** 199 **23.** Each sum is two less than a Lucas number. **25.** **(a)** $8 \cdot 18 = 144$; Each expression is equal to 144. **(b)** $8 + 21 = 29$; Each expression is equal to 29. **(c)** $8 + 18 = 2 \cdot 13$; Each expression is equal to 26. **27.** 3, 4, 5 **29.** 16, 30, 34 **31.** The sums are 1, 1, 2, 3, 5, 8, 13. They are terms of the Fibonacci sequence. **33.** $(1 + \sqrt{5})/2 \approx 1.618033989$ and $(1 - \sqrt{5})/2 \approx -.618033989$. The decimal places have the same digits. **35.** 377 **37.** 17,711

Extension Exercises (Page 248)

1.

2	7	6
9	5	1
4	3	8

3.

11	10	4	23	17
18	12	6	5	24
25	19	13	7	1
2	21	20	14	8
9	3	22	16	15

5.

15	16	22	3	9
8	14	20	21	2
1	7	13	19	25
24	5	6	12	18
17	23	4	10	11

7.

24	9	12
3	15	27
18	21	6

Magic sum is 45.

9.

$\frac{17}{2}$	12	$\frac{1}{2}$	4	$\frac{15}{2}$
$\frac{23}{2}$	$\frac{5}{2}$	$\frac{7}{2}$	7	8
2	3	$\frac{13}{2}$	10	11
5	6	$\frac{19}{2}$	$\frac{21}{2}$	$\frac{3}{2}$
$\frac{11}{2}$	9	$\frac{25}{2}$	1	$\frac{9}{2}$

Magic sum is $32\frac{1}{2}$.

11. 479 **13.** 467 **15.** 269 **17.** **(a)** 73 **(b)** 70 **(c)** 74 **(d)** 69
19. **(a)** 7 **(b)** 22 **(c)** 5 **(d)** 4 **(e)** 15 **(f)** 19 **(g)** 6 **(h)** 23 **21.**

30	39	48	1	10	19	28
38	47	7	9	18	27	29
46	6	8	17	26	35	37
5	14	16	25	34	36	45
13	15	24	33	42	44	4
21	23	32	41	43	3	12
22	31	40	49	2	11	20

23. Each sum is equal to 34. **25.** Each sum is equal to 68. **27.** Each sum is equal to 9,248.
29. Each sum is equal to 748. **31.**

16	2	3	13
5	11	10	8
9	7	6	12
4	14	15	1

The second and third columns are interchanged.

33.

18	20	10
8	16	24
22	12	14

35.

39	48	57	10	19	28	37
47	56	16	18	27	36	38
55	15	17	26	35	44	46
14	23	25	34	43	45	54
22	24	33	42	51	53	13
30	32	41	50	52	12	21
31	40	49	58	11	20	29

Magic sum is 238.

37. 260 **39.** $52 + 45 + 16 + 17 + 54 + 43 + 10 + 23 = 260$ **41.**

5	13	21	9	17
6	19	2	15	23
12	25	8	16	4
18	1	14	22	10
24	7	20	3	11

Chapter 5 Test (Page 252)

1. false **2.** true **3.** true **4.** true **5.** true **6** **(a)** yes **(b)** yes **(c)** no **(d)** yes **(e)** yes **(f)** no **(g)** yes **(h)** yes **(i)** no **7.** **(a)** composite **(b)** neither **(c)** prime **8.** $2^5 \cdot 3^2 \cdot 5$ **10.** **(a)** deficient **(b)** perfect **(c)** abundant **11.** (c) **12.** 41 and 43 **13.** 90 **14.** 360 **15.** Monday **16.** 46,368 **17.** $89 - (8 + 13 + 21 + 34) = 13$; Each expression is equal to 13. **18.** **(a)** 1, 5, 6, 11, 17, 28, 45, 73 **(b)** The process will yield 19 for any term chosen. **19.** (a)

CHAPTER 6 The Real Number System

6.1 Exercises (Page 261)

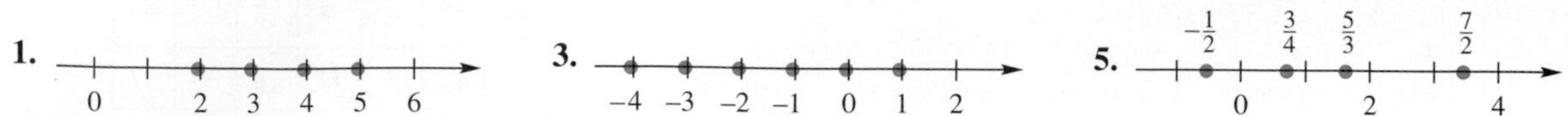

7. **(a)** 3, 7 **(b)** 0, 3, 7 **(c)** −9, 0, 3, 7 **(d)** −9, −5/4, −3/5, 0, 3, 5.9, 7 **(e)** $-\sqrt{7}$, $\sqrt{5}$ **(f)** All are real numbers. **9.** true **11.** true **13.** false **15.** true **17.** true **19.** true **21.** **(a)** −5 **(b)** 5 **23.** **(a)** 6 **(b)** 6 **27.** 3; 3; 3; −3 **29.** {1, 2, 3, 4, 5, 6} **31.** {12, 14, 16, 18, . . .} **33.** ∅ **35.** {3, −3} **37.** {0, 5, 10, 15, . . .}

Answers may vary in Exercises 39–43.

39. 1/2, 5/8, 1 3/4 **41.** −3 1/2, −2/3, 3/7 **43.** $\sqrt{5}$, $\sqrt{2}$, $-\sqrt{3}$ **45.** false **47.** true **49.** false **51.** true **53.** true **55.** false **57.** true **59.** true **61.** 3 **63.** −7 **65.** 3 **67.** −4 **69.** (a), (b), (c), (e) **71.** (c), (e) **73.** (c), (e) **75.** (a), (b), (c), (e) **77.** (d), (e) **79.** 1989, 1992, and 1993 **81.** 1989, 1990, 1991, and 1992 **83.** $302.7 million

6.2 Exercises (Page 271)

1. true **3.** false **5.** true **7.** −20 **9.** −4 **11.** −11 **13.** 9 **15.** 20 **17.** 4 **19.** 24 **21.** −1,296 **23.** 6 **25.** −6 **27.** 0 **29.** −6 **31.** −4 **33.** 27 **35.** 45 **37.** −2 **39.** (a), (b), (c) **41.** commutative property of addition **43.** associative property of addition **45.** inverse property of addition **47.** identity property of multiplication **49.** identity property of addition **51.** distributive property **53.** closure property of addition **55.** **(a)** −2, 2 **(b)** commutative **(c)** yes; Choose $a = b$. For example, $a = b = 2$: $2 - 2 = 2 - 2$. **57.** **(a)** messing up your room **(b)** spending money **(c)** decreasing the volume on your portable radio **59.** identity **61.** no **65.** −81 **67.** 81 **69.** −81 **71.** −81 **73.** −396 million dollars **75.** **(a)** 50,395 feet **(b)** 37,486 feet **(c)** 1,345 feet **(d)** 8,274 feet **77.** −184 meters **79.** 112°F **81.** −1 yard **83.** 14,776 feet **85.** 469 B.C. **87.** 14 feet **89.** **(a)** 74 **(b)** −22 **(c)** −42 **(d)** 14 **(e)** 7

6.3 Exercises (Page 284)

1. true **3.** true **5.** 1/3 **7.** −3/7

Answers will vary in Exercises 9 and 11. We give three of infinitely many possible answers.

9. 6/16, 9/24, 12/32 **11.** −10/14, −15/21, −20/28 **13.** **(a)** 1/3 **(b)** 1/4 **(c)** 2/5 **(d)** 1/3 **15.** the dots in the intersection of the triangle and the rectangle as a part of the dots in the entire figure **17.** **(a)** Carlton **(b)** Dykler **(c)** Dykler **(d)** Bishop **(e)** Crowe and Marshall; 1/2 **19.** 1/2 **21.** 43/48 **23.** −5/24 **25.** 23/56 **27.** 27/20 **29.** 5/12 **31.** 1/9 **33.** 3/2 **35.** 3/2 **37.** 81 1/2 dollars **39.** **(a)** 4 1/2 cups **(b)** $\frac{7}{8}$ cup **41.** 13/3 **43.** 29/10 **45.** 6 3/4 **47.** 6 1/8 **49.** −17 7/8 **51.** 3 **53.** 3/7 **55.** −103/89 **57.** $2 + \cfrac{1}{6 + \cfrac{1}{2}}$ **59.** $4 + \cfrac{1}{1 + \cfrac{1}{2 + \cfrac{1}{1 + \cfrac{1}{2}}}}$

61. 25/9 **63.** 5/8 **65.** 19/30 **67.** −3/4 **69.** 14/19 **71.** 13/29 **73.** 5/2 **75.** It gives the rational number halfway between the two integers. **77.** .75 **79.** .1875 **81.** $.\overline{27}$ **83.** $.\overline{285714}$ **85.** 2/5 **87.** 17/20 **89.** 467/500 **91.** 8/9 **93.** 6/11 **95.** 13/30 **97.** 2 **99.** repeating **101.** terminating **103.** terminating **105.** **(a)** $.\overline{3}$ or .333 . . . **(b)** $.\overline{6}$ or .666 . . . **(c)** $.\overline{9}$ or .999 . . .

6.4 Exercises (Page 296)

1. rational **3.** irrational **5.** rational **7.** rational **9.** irrational **11.** rational **13.** irrational **15.** **(a)** $.\overline{8}$ **(b)** irrational, rational

The number of digits shown will vary among calculator models in Exercises 17–23.
17. 6.244997998 **19.** 3.885871846 **21.** 29.73213749 **23.** 1.060660172 **25.** $5\sqrt{2}$; 7.071067812 **27.** $5\sqrt{3}$; 8.660254038 **29.** $12\sqrt{2}$; 16.97056275 **31.** $5\sqrt{6}/6$; 2.041241452 **33.** $\sqrt{7}/2$; 1.322875656 **35.** $\sqrt{21}/3$; 1.527525232 **37.** $3\sqrt{17}$ **39.** $4\sqrt{7}$ **41.** $10\sqrt{2}$ **43.** $3\sqrt{3}$ **45.** 392,000 square miles **47.** $8\sqrt{5}$ feet; 17.9 feet **49.** 2.5 seconds **51.** The result is 3.1415929, which agrees with the first seven digits in the decimal form of π. **53. (a)** 3.1415926 **(b)** 3.141592653589 **(c)** 3.14159265358979 **55.** They are only rational *approximations* of π. **57. (a)** $2\sqrt[3]{2}$ **(b)** $3\sqrt[3]{2}$ **(c)** $2\sqrt[3]{3}$ **(d)** $5\sqrt[3]{2}$ **59.** 5.8 feet; 69 inches

6.5 Exercises (Page 307)

1. true **3.** false **5.** false **7.** false **9.** false **11.** 11.315 **13.** −4.215 **15.** .8224 **17.** 47.5 **19.** 31.6 **21.** gasoline in 1986 **23.** false **25. (a)** $11,042.49 **(b)** $8,997.26 **(c)** $3,901.35 **27. (a)** .031 **(b)** .035 **29.** 297 **31. (a)** 78.4 **(b)** 78.41 **33. (a)** .1 **(b)** .08 **35. (a)** 12.7 **(b)** 12.69 **37.** 42% **39.** 36.5% **41.** .8% **43.** 210% **45.** 20% **47.** 1% **49.** 37 1/2% **51.** 150% **55. (a)** 5 **(b)** 24 **(c)** 8 **(d)** .5 or 1/2 **(e)** 600 **57.** No, it is $49.50. **59. (a)** 33 1/3% **(b)** 25% **(c)** 40% **(d)** 33 1/3% **61.** 124.8 **63.** 2.94 **65.** 150% **67.** 600 **69.** 1.4% **71.** (a) **73.** (c) **75.** 16.3% **77.** $533 **79.** 284% **81.** 16.1% **83.** $4.50 **85.** $.75 **87. (a)** 3; 2; 5 **(b)** 5; 5

Extension Exercises (Page 314)

1. $12i$ **3.** $-15i$ **5.** $i\sqrt{3}$ **7.** $5i\sqrt{3}$ **9.** −5 **11.** −18 **13.** −40 **15.** $\sqrt{2}$ **17.** $3i$ **19.** 6 **23.** 1 **25.** −1 **27.** $-i$ **29.** i

Chapter 6 Test (Page 315)

1. (a) 10 **(b)** 0, 10 **(c)** −8, 0, 10 **(d)** −8, −4/3, −.6, 0, 3.9, 10 **(e)** $-\sqrt{6}$, $\sqrt{2}$ **(f)** All are real numbers. **3.** {1, 2, 3} **4.** Answers may vary. Three examples are 1/2, 3/4, 5/6. **5.** true **6. (a)** 1 **(b)** 3 **(c)** 10 **7.** $7,506 million **8. (a)** E **(b)** A **(c)** B **(d)** D **(e)** F **(f)** C **9.** 177°F **10.** $.2 million; −$4.4 million; $4.1 million **11. (a)** Hickman **(b)** Camp and Levinson **(c)** Hickman **(d)** Cooper and Cornett; 2/5 **12.** 11/16 **13.** 57/160 **14.** −2/5 **15.** 3/2 **16.** 4/25 **17.** .45 **18.** $.41\overline{6}$ **19.** 18/25 **20.** 58/99 **21. (a)** irrational **(b)** rational **(c)** rational **(d)** rational **(e)** irrational **22. (a)** 12.247448714 **(b)** $5\sqrt{6}$ **23. (a)** 4.913538149 **(b)** $13\sqrt{7}/7$ **24. (a)** −45.254834 **(b)** $-32\sqrt{2}$ **26.** 13.81 **27.** −.315 **28.** 38.7 **29.** −24.3 **30. (a)** 9.04 **(b)** 9.045 **31.** 16.65 **32.** 101.5 **33.** (c) **34.** 5.45% **35. (a)** 26 2/3% **(b)** 66 2/3%

CHAPTER 7 The Basic Concepts of Algebra

7.1 Exercises (Page 326)

1. yes **3.** no **5.** yes **7.** (b) **9.** {−1} **11.** {3} **13.** {−7} **15.** {0} **17.** $\left\{-\frac{5}{3}\right\}$ **19.** $\left\{-\frac{1}{2}\right\}$ **21.** {2} **23.** {−2} **25.** {7} **27.** {−5} **31.** 2 (that is, 10^2) **33.** {4} **35.** {0} **37.** {0} **39.** {2,000} **41.** {25} **43.** {40} **45.** identity; contradiction **47.** contradiction; ∅ **49.** conditional; {0} **51.** identity; {all real numbers} **53.** 2,820 million dollars; 1996 (when $x = 7$) **55. (a)** 230 billion dollars **(b)** 275 billion dollars **(c)** 300 billion dollars **(d)** 335 billion dollars **57.** L should not appear on both sides of the equation. **59.** $r = \frac{d}{t}$ **61.** $b = \frac{A}{h}$ **63.** $a = P - b - c$ **65.** $h = \frac{2A}{b}$ **67.** $h = \frac{S - 2\pi r^2}{2\pi r}$ or $h = \frac{S}{2\pi r} - r$ **69.** $F = \frac{9}{5}C + 32$ **71.** $H = \frac{A - 2LW}{2W + 2L}$

.2 Exercises (Page 338)

3. $x - 13$ **5.** $7 + x$ **7.** $8(x + 12)$ **9.** $\frac{x}{6}$ **11.** $\frac{x}{12}$ **13.** $\frac{6}{7}x$ **15.** (d) **17.** 47 Democrats, 53 Republicans **19.** 1,037 **21.** highest grade: 97; lowest grade: 63 **23.** sorting mail: 1 1/2 hours; selling stamps: 3 hours; supervising: 2 hours **25.** 2 **27.** Labrador retrievers: 120,879; Rottweilers: 95,445 **29.** The leading state, North Carolina, employs 105,633 people. Second is Kentucky with 61,648 people. **31.** 35 milliliters **33.** $350 **35.** $14.15 **37.** 160 gallons **39.** 3 1/3 gallons **41.** 4 liters **43.** 25 milliliters **45.** (d) **47.** $5,000 at 3%; $1,000 at 5% **49.** $8,500 at 4%; $3,500 at 5% **51.** $40,000 at 3%; $110,000 at 4% **53.** **(a)** 174 **(b)** 106 **(c)** 123 **(d)** 73 **(e)** 84 **55.** **(a)** $40,740,000 **(b)** $58,200,000 **(c)** $110,580,000 **57.** 13 quarters, 23 half-dollars **59.** 17 pennies, 17 dimes, 10 quarters **61.** 11 two-cent pieces, 33 three-cent pieces **63.** 305 students, 105 nonstudents **67.** 3.718 hours **69.** 1.715 hours **71.** 7.51 meters per second **73.** 10.08 meters per second **75.** 530 miles **79.** 10 hours **81.** 1 3/4 hours **83.** 10:00 A.M. **85.** 18 miles

7.3 Exercises (Page 353)

1. (c) **3.** $\frac{4}{3}$ **5.** $\frac{4}{3}$ **7.** $\frac{15}{2}$ **9.** $\frac{1}{5}$ **11.** $\frac{24}{5}$ **15.** true **17.** false **19.** true **21.** $\{35\}$ **23.** $\{27\}$ **25.** $\{-1\}$ **27.** $\{10\}$ **29.** $13\frac{1}{3}$ spears **31.** 6.875 ounces **33.** $706.20 **35.** $23.40 **37.** 4 feet **39.** 45 heads **41.** $338 **43.** 12,500 fish **45.** yes: 2,520; no: 1,260; undecided: 420 **47.** 30-count size **49.** 31-ounce size **51.** 32-ounce size **53.** 4 **55.** 1 **57.** **(a)**

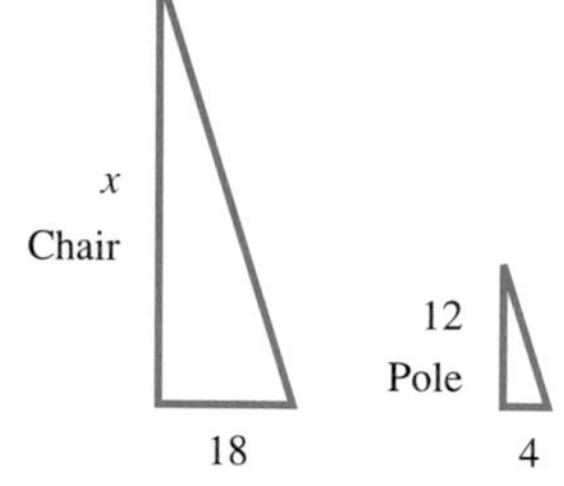

(b) 54 feet **59.** $294 **61.** $331 **63.** $4,850 **65.** 9 **67.** 125 **69.** $\frac{4}{9}$ **71.** **(a)** increases **(b)** decreases **73.** $40.32 **75.** 20 miles per hour **77.** about 302 pounds **79.** 100 pounds per square inch **81.** 20 pounds per square foot **83.** 144 feet

7.4 Exercises (Page 364)

1. $(-\infty, 3)$ **3.** $(-\infty, 8]$ **5.** $(8, \infty)$ **7.** $[2, \infty)$ **9.** $[-3, 4]$ **11.** $(2, 9]$ **13.** $\{x \mid x < 4\}$ **15.** $\{x \mid x \geq 1.5\}$ **17.** $\{x \mid -3 \leq x < 10\}$ **19.** $\{x \mid -4 \leq x \leq -2\}$ **21.** **(a)** A **(b)** D **(c)** C **(d)** B

23. $(2, \infty)$

0 2

25. $(-\infty, -3]$

−3 0

27. $[5, \infty)$

0 5

29. $[7, \infty)$

0 7

31. $(-3, \infty)$

−3 0

33. $(-\infty, 23/5)$

0 $\frac{23}{5}$

35. $(-\infty, -11]$

−11 −6 0

37. $(2, 11)$

0 2 11

39. $[-14/3, -4/3]$

$-\frac{14}{3}$ $-\frac{4}{3}$ 0

41. $[-19/2, 35/2]$

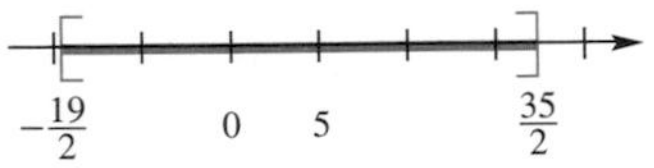

43. April, May, June, July

45. January, February, March, August, September, October, November, December **47.** 2 miles **49.** at least 80 **51.** 50 miles **53.** 26 tapes

7.5 Exercises (Page 375)

1. 625 **3.** $\frac{25}{9}$ **5.** -32 **7.** -8 **9.** -81 **13.** (c) **15.** $\frac{1}{49}$ **17.** $-\frac{1}{49}$ **19.** -128 **21.** $\frac{16}{5}$ **23.** 125 **25.** $\frac{25}{16}$ **27.** $\frac{9}{20}$ **29.** 1 **31.** 1 **33.** 0 **35.** reciprocal; additive inverse **37.** (d) **39.** x^{16} **41.** 5 **43.** $\frac{1}{27}$ **45.** $\frac{1}{81}$ **47.** $\frac{1}{t^7}$ **49.** $9x^2$ **51.** $\frac{1}{a^5}$ **53.** x^{11} **55.** r^6 **57.** $-\frac{56}{k^2}$ **59.** $\frac{1}{z^4}$ **61.** $-\frac{3}{r^7}$ **63.** $\frac{27}{a^{18}}$ **65.** $\frac{x^5}{y^2}$ **67.** (d) **69.** 2.3×10^2 **71.** 2×10^{-2} **73.** 6,500 **75.** .0152 **77.** 6×10^5 **79.** 2×10^5 **81.** 2×10^5 **83.** 635,000,000,000 **85.** 1.15×10^6 **87.** 1.5×10^2; .00023; .006 **89.** .000002 **91.** 4.2×10^{42} **93.** about 15,300 seconds **95.** 5.87×10^{12} miles **97.** 4.6×10^7 meters **99.** **(a)** 1.75×10^6 **(b)** about \$$9.63 \times 10^{10}$

7.6 Exercises (Page 385)

1. $x^2 - x + 3$ **3.** $9y^2 - 4y + 4$ **5.** $6m^4 - 2m^3 - 7m^2 - 4m$ **7.** $-2x^2 - 13x + 11$ **9.** $x^2 - 5x - 24$ **11.** $28r^2 + r - 2$ **13.** $12x^5 + 8x^4 - 20x^3 + 4x^2$ **15.** $4m^2 - 9$ **17.** $16m^2 + 16mn + 4n^2$ **19.** $25r^2 + 30rt^2 + 9t^4$ **21.** $-2z^3 + 7z^2 - 11z + 4$ **23.** $m^2 + mn - 2n^2 - 2km + 5kn - 3k^2$ **25.** $a^2 - 2ab + b^2 + 4ac - 4bc + 4c^2$ **27.** (a) **31.** $2m^2(4m^2 + 3m - 6)$ **33.** $4k^2m^3(1 + 2k^2 - 3m)$ **35.** $2(a + b)(1 + 2m)$ **37.** $(m - 1)(2m^2 - 7m + 7)$ **39.** $(2s + 3)(3t - 5)$ **41.** $(t^3 + s^2)(r - p)$ **43.** $(8a + 5b)(2a - 3b)$ **45.** $(5z - 2x)(4z - 9x)$ **47.** $(x - 5)(x + 3)$ **49.** $(y + 7)(y - 5)$ **51.** $6(a - 10)(a + 2)$ **53.** $3m(m + 1)(m + 3)$ **55.** $(3k - 2p)(2k + 3p)$ **57.** $(5a + 3b)(a - 2b)$ **59.** $x^2(3 - x)^2$ **61.** $2a^2(4a - b)(3a + 2b)$ **65.** $(3m - 2)^2$ **67.** $2(4a - 3b)^2$ **69.** $(2xy + 7)^2$ **71.** $(x + 6)(x - 6)$ **73.** $(y + w)(y - w)$ **75.** $(3a + 4)(3a - 4)$ **77.** $(5s^2 + 3t)(5s^2 - 3t)$ **79.** $(p^2 + 25)(p + 5)(p - 5)$ **81.** $(2 - a)(4 + 2a + a^2)$ **83.** $(5x - 3)(25x^2 + 15x + 9)$ **85.** $(3y^3 + 5z^2)(9y^6 - 15y^3z^2 + 25z^4)$ **87.** $(x + y)(x - 5)$ **89.** $(m - 2n)(p^4 + q)$ **91.** $(2z + 7)^2$ **93.** $(10x + 7y)(100x^2 - 70xy + 49y^2)$ **95.** $(5m^2 - 6)(25m^4 + 30m^2 + 36)$ **97.** $(6m - 7n)(2m + 5n)$ **99.** $4^2 + 2^2 = 20$ and $(4 + 2)(4 + 2) = 36$; $20 \neq 36$

7.7 Exercises (Page 393)

1. $\{-3, 9\}$ **3.** $\left\{\frac{7}{2}, -\frac{1}{5}\right\}$ **5.** $\{-3, 4\}$ **7.** $\{-7, -2\}$ **9.** $\left\{-\frac{1}{2}, \frac{1}{6}\right\}$ **11.** $\{-2, 4\}$ **13.** $\left\{-\frac{5}{2}, 3\right\}$ **15.** $\{\pm 8\}$ **17.** $\{\pm\sqrt{7}\}$ **19.** $\{\pm 2\sqrt{6}\}$ **21.** $\emptyset$ **23.** $\{4 \pm \sqrt{3}\}$ **25.** $\left\{\frac{5 \pm \sqrt{13}}{2}\right\}$ **27.** $\left\{\frac{2 \pm \sqrt{3}}{2}\right\}$ **29.** $\left\{\frac{1 \pm \sqrt{3}}{2}\right\}$ **31.** $\left\{\frac{1 \pm \sqrt{5}}{2}\right\}$ **33.** $\left\{\frac{-1 \pm \sqrt{2}}{2}\right\}$ **35.** $\left\{\frac{1 \pm \sqrt{29}}{2}\right\}$ **37.** $\emptyset$ **43.** 0; (c) **45.** 121; (a) **47.** 360; (b) **49.** 5.2 seconds **51.** Find s when $t = 0$. **53.** 3 minutes **55.** 3.4 seconds **57.** 2.3, 5.3, 5.8 **59.** 412.3 feet **61.** eastbound ship: 80 miles; southbound ship: 150 miles **63.** 5, 12, and 13 centimeters **65.** length: 2 centimeters; width: 1.5 centimeters **67.** 1 foot **69.** length: 26 meters; width: 16 meters **71.** 18 inches **73.** 5 or 14

Chapter 7 Test (Page 398)

1. {2} **2.** {4} **3.** identity; {all real numbers} **4.** $v = \dfrac{S + 16t^2}{t}$
5. Hawaii: 4,021 square miles; Maui: 728 square miles; Kauai: 551 square miles
6. 5 liters **7.** 2.2 hours **8.** 8 slices for \$2.19 **9.** 2,300 miles **10.** 200 amps
11. $(-\infty, 4]$ **12.** $(-2, 6]$

13. (c) **14.** at least 82 **15.** $\dfrac{16}{9}$ **16.** -64 **17.** $\dfrac{64}{27}$ **18.** 0 **19.** $\dfrac{216}{p^4}$ **20.** $\dfrac{1}{m^{14}}$
21. 3×10^{-4} **22.** 400,000,000,000,000 **23.** $\$1.2 \times 10^{10}$ **24.** $4k^2 + 6k + 10$
25. $15x^2 - 14x - 8$ **26.** $16x^4 - 9$ **27.** $3x^3 + 20x^2 + 23x - 36$
28. One example is $t^5 + 2t^4 + 3t^3 - 4t^2 + 5t + 6$. **29.** $(2p - 3q)(p - q)$ **30.** $(10x + 7y)(10x - 7y)$
31. $(3y - 5x)(9y^2 + 15yx + 25x^2)$ **32.** $(4 - m)(x + y)$ **33.** $\left\{-\dfrac{3}{2}, \dfrac{1}{3}\right\}$ **34.** $\left\{\dfrac{1 \pm \sqrt{29}}{2}\right\}$
35. .87 second

CHAPTER 8 Functions, Graphs, and Systems of Equations and Inequalities

8.1 Exercises (Page 406)

1. origin **3.** y **5.** 2 **7.** **(a)** I **(b)** III **(c)** II **(d)** IV **(e)** none
9. **(a)** I or III **(b)** II or IV **(c)** II or IV **(d)** I or III **11–20.**

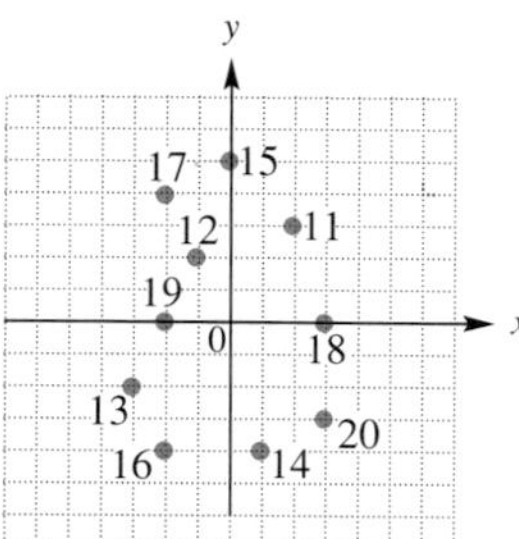

21. **(a)** $\sqrt{34}$ **(b)** $\left(\dfrac{1}{2}, \dfrac{5}{2}\right)$ **23.** **(a)** $\sqrt{61}$ **(b)** $\left(\dfrac{1}{2}, 1\right)$ **25.** **(a)** $\sqrt{146}$ **(b)** $\left(-\dfrac{1}{2}, \dfrac{3}{2}\right)$
27. B **29.** D **31.** $x^2 + y^2 = 36$ **33.** $(x + 1)^2 + (y - 3)^2 = 16$
35. $x^2 + (y - 4)^2 = 3$ **37.** center: (0, 0); radius: r **39.** center: $(-2, 3)$; radius: 2
41. center: $(-5, 7)$; radius: 9 **43.** center: (2, 4); radius: 4 **45.**

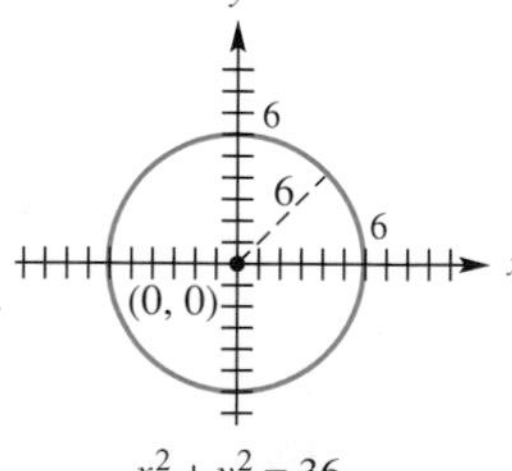

47.

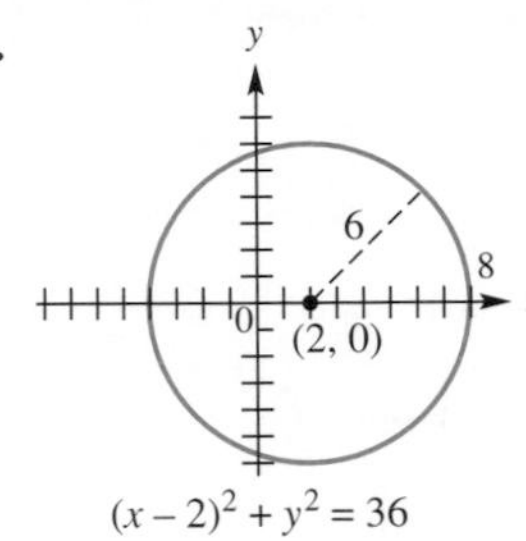

49.

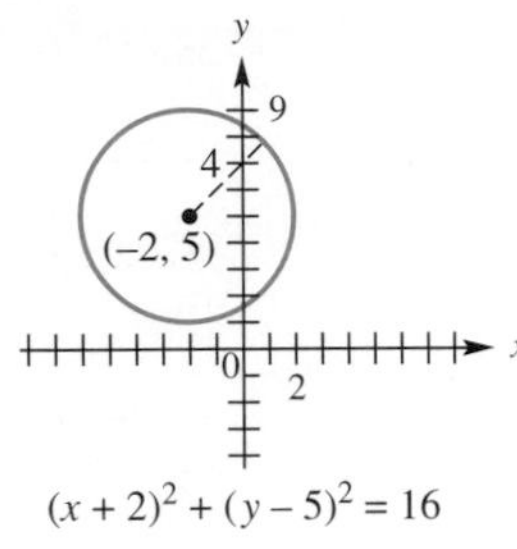

51.

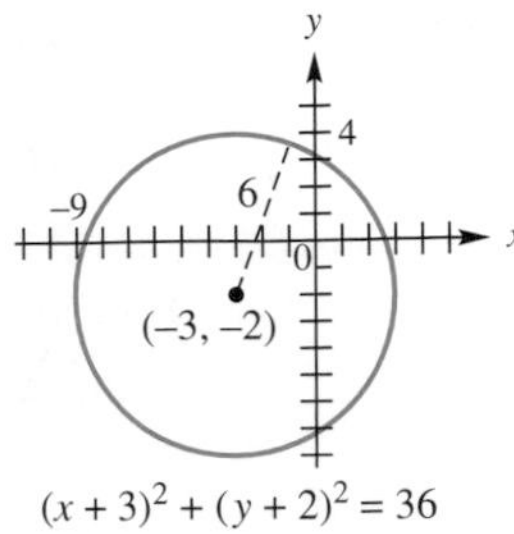

57. $(-3, 4)$ is a solution for each of the following equations: $(x-1)^2 + (y-4)^2 = 16$; $(x+6)^2 + y^2 = 25$; $(x-5)^2 + (y+2)^2 = 100$ **63.** (b)

8.2 Exercises (Page 415)

1. $(0, 5), \left(\frac{5}{2}, 0\right), (1, 3), (2, 1)$ **3.** $(0, -4), (4, 0), (2, -2), (3, -1)$ **5.** $(0, 4), (5, 0), \left(3, \frac{8}{5}\right), \left(\frac{5}{2}, 2\right)$

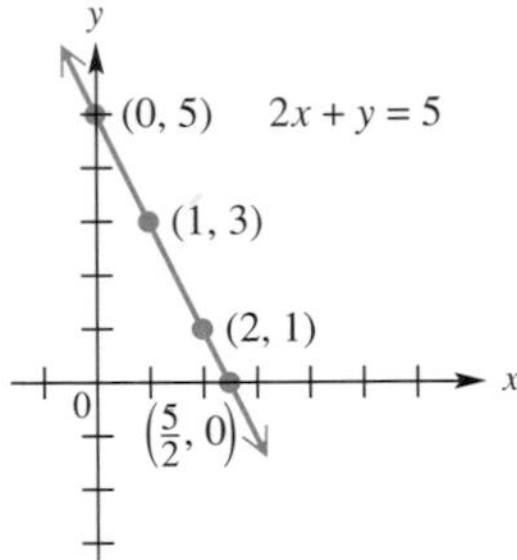

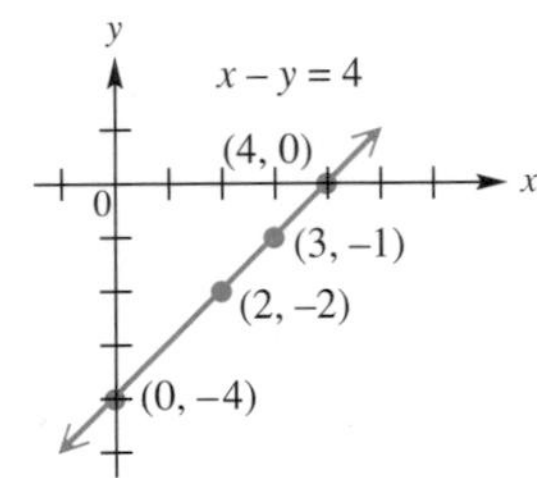
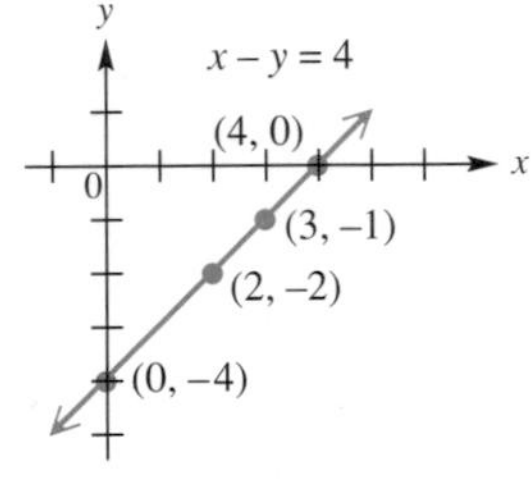

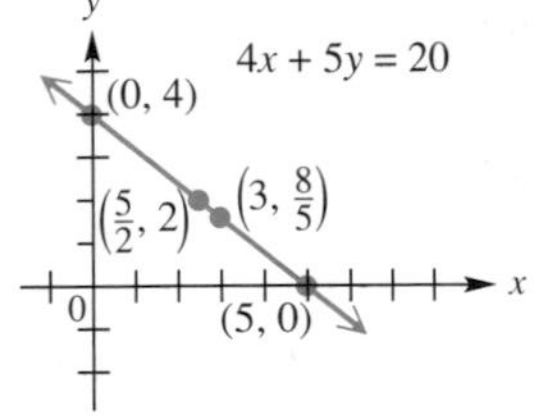

7.

x	y
0	4
8/3	0
2	1
4	−2

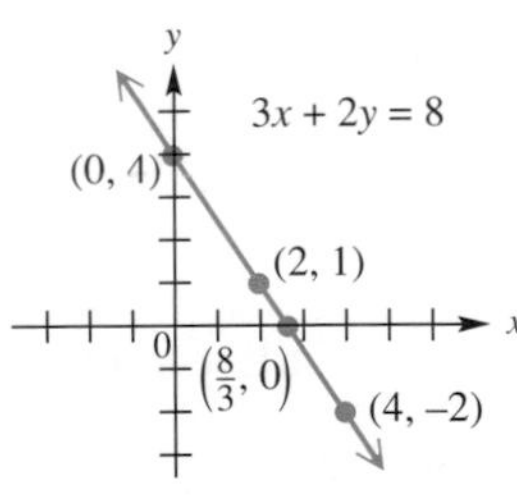

11. (a) **13.** $(4, 0)$; $(0, 6)$

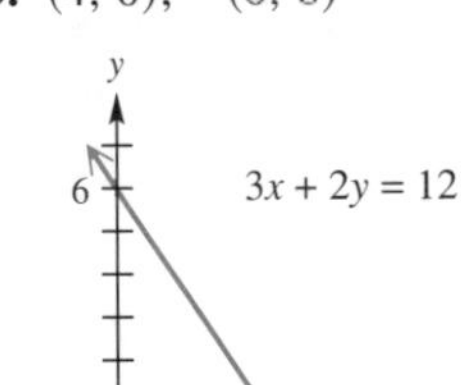

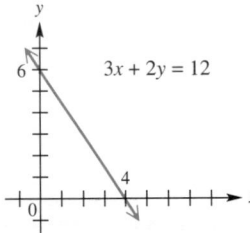

15. $(2, 0)$; $\left(0, \frac{5}{3}\right)$ **17.** $\left(\frac{5}{2}, 0\right)$; $(0, -5)$ **19.** $(2, 0)$; $\left(0, -\frac{2}{3}\right)$

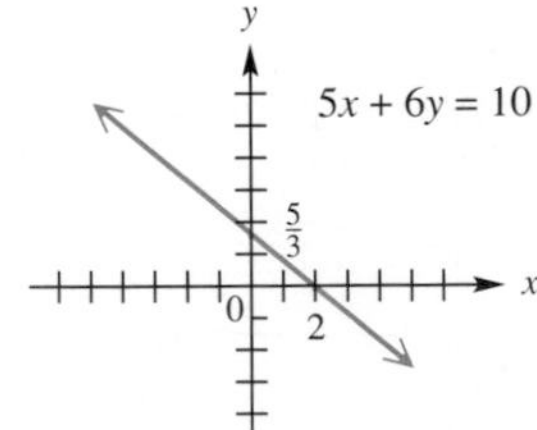

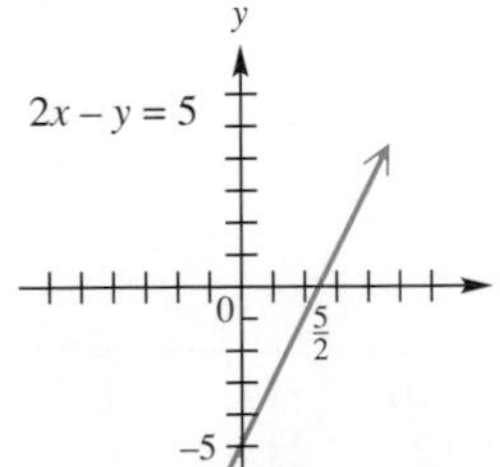

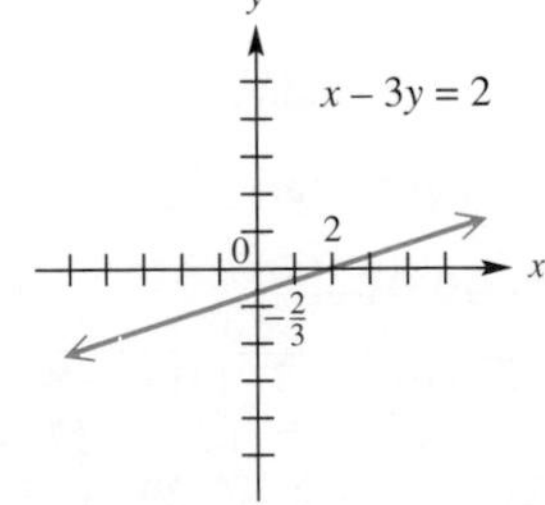

21. (0, 0); (0, 0)

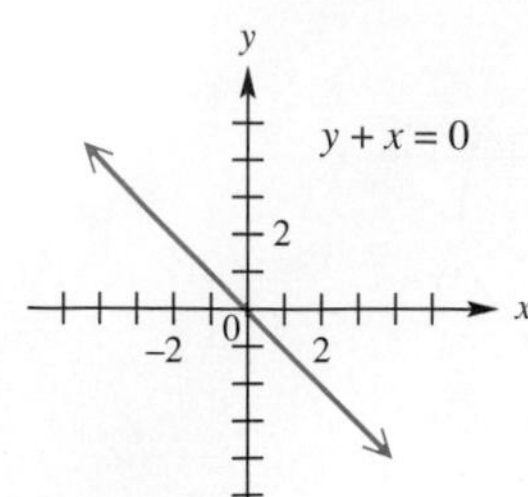

23. (0, 0); (0, 0)

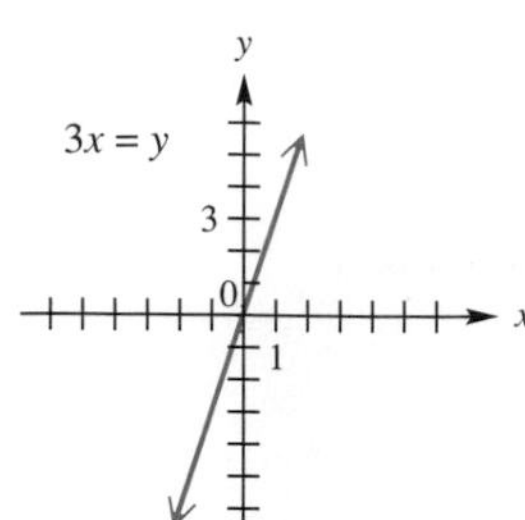

25. (2, 0); none

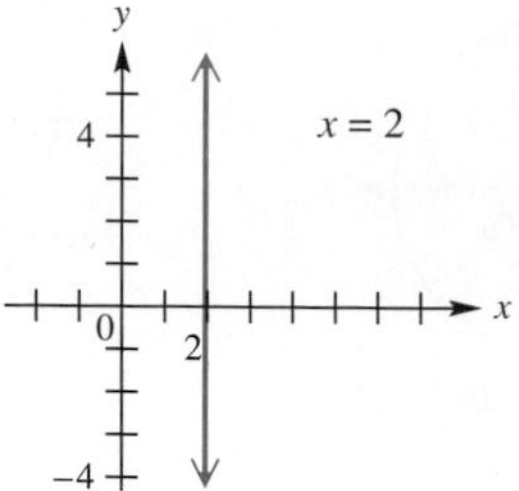

27. none; (0, 4)

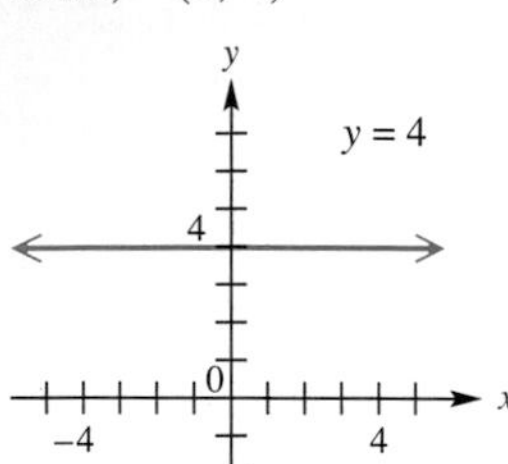

29. (c) **31.** (a) **33.** (d)

35. (b) **37.** (a), (b), (c), (d), (f) **39.** 8 **41.** $\frac{5}{6}$ **43.** 0 **45.** positive

47. negative **49.** zero **51.**

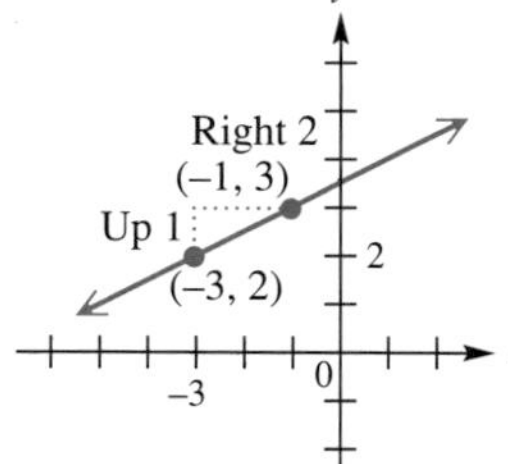

53.

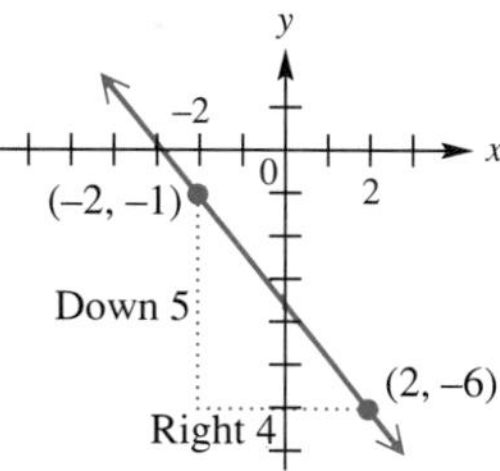

55.

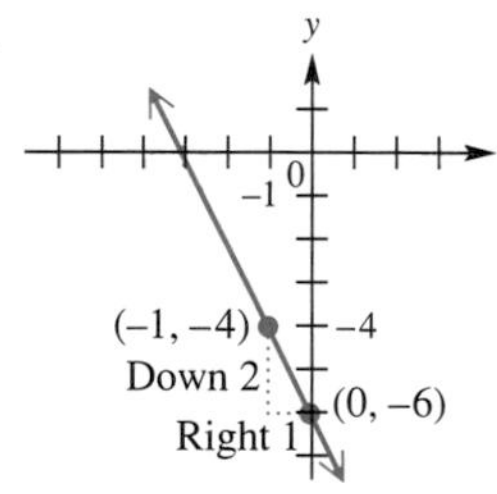

57.

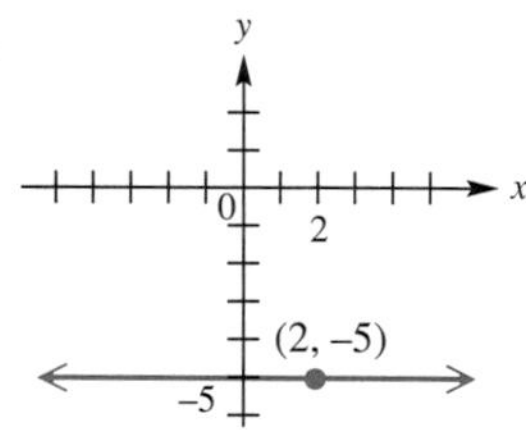

59. parallel **61.** perpendicular

63. neither parallel nor perpendicular **65.** $\frac{7}{10}$ **67. (a)** \$.92 **(b)** It means an *increase* in price.

69. 6.25 billion minutes per year

8.3 Exercises (Page 423)

1. (a) **3.** (c) **5.** (b) **7.** (d) **9.** $3x + 4y = 10$ **11.** $2x + y = 18$
13. $x - 2y = -13$ **15.** $y = 12$ **17.** $4x - y = 12$ **21.** $x = 9$ **23.** $x = .5$
25. $y = 8$ **27.** $2x - y = 2$ **29.** $x + 2y = 8$ **31.** $y = 5$ **33.** $x = 7$ **35.** $y = 5x + 15$
37. $y = -\frac{2}{3}x + \frac{4}{5}$ **39.** $y = \frac{2}{5}x + 5$ **41. (a)** $y = -x + 12$ **(b)** -1 **(c)** (0, 12)
43. (a) $y = -\frac{5}{2}x + 10$ **(b)** $-\frac{5}{2}$ **(c)** (0, 10) **45. (a)** $y = \frac{2}{3}x - \frac{10}{3}$ **(b)** $\frac{2}{3}$ **(c)** $\left(0, -\frac{10}{3}\right)$

47.

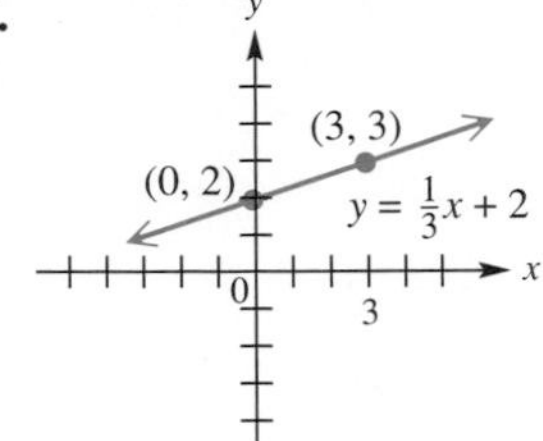

49.

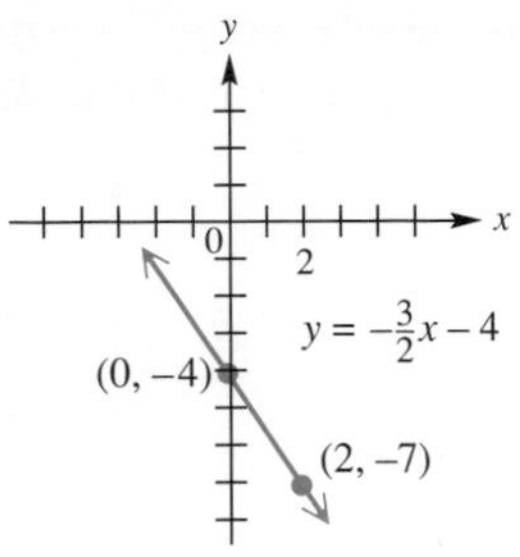

51.

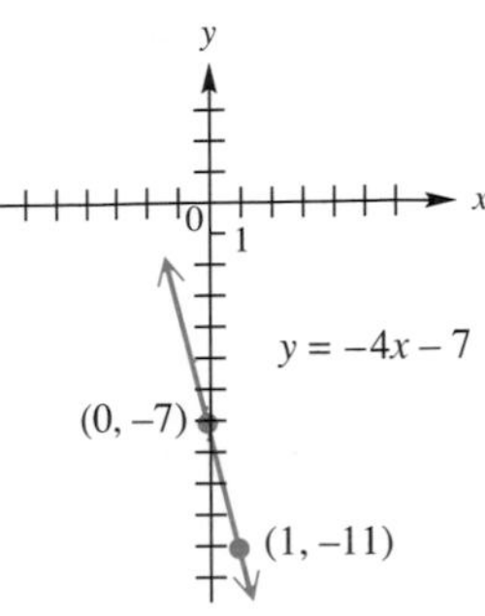

53. $3x - y = 19$ **55.** $x - 2y = 2$ **57.** $x + 2y = 18$ **59.** $y = 7$ **61.** $y = -22.125x + 220$
63. $y = -\dfrac{791}{6}x + 29{,}750$

8.4 Exercises (Page 433)

3. function; domain: {5, 3, 4, 7}; range: {1, 2, 9, 3} **5.** not a function; domain: {2, 0}; range: {4, 2, 6} **7.** not a function; domain: {1, 2, 3, 5}; range: {10, 15, 19, −27} **9.** function; domain: $(-\infty, \infty)$; range: $(-\infty, 4]$ **11.** not a function; domain: $[-4, 4]$; range: $[-3, 3]$ **13.** function; domain: $(-\infty, \infty)$ **15.** not a function; domain: $[0, \infty)$ **17.** not a function; domain: $(-\infty, \infty)$ **19.** function; domain: $[0, \infty)$ **21.** function; domain: $(-\infty, 0) \cup (0, \infty)$ **23.** function (also a linear function); domain: $(-\infty, \infty)$ **25.** function; domain: $[-\frac{1}{2}, \infty)$
27. function; domain: $(-\infty, -3) \cup (-3, 3) \cup (3, \infty)$ **29.** 5 **31.** 2 **33.** −1 **35.** −13
37. domain and range: $(-\infty, \infty)$ **39.** domain and range: $(-\infty, \infty)$

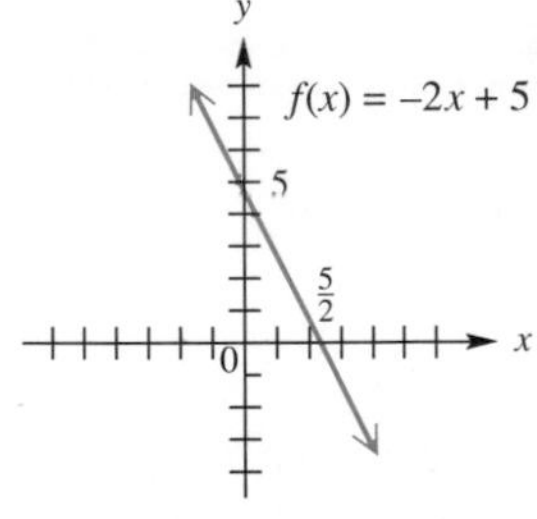

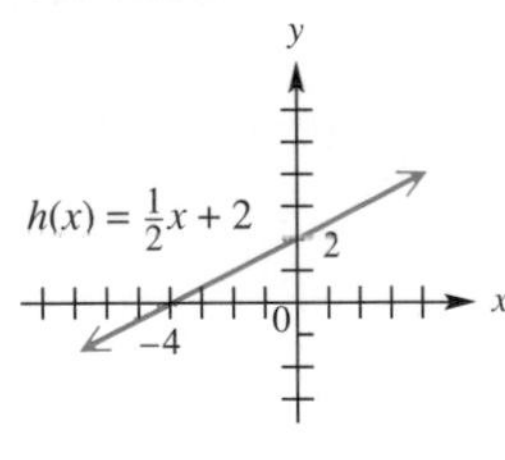

41. domain and range: $(-\infty, \infty)$ **43.** domain: $(-\infty, \infty)$; range: {5}

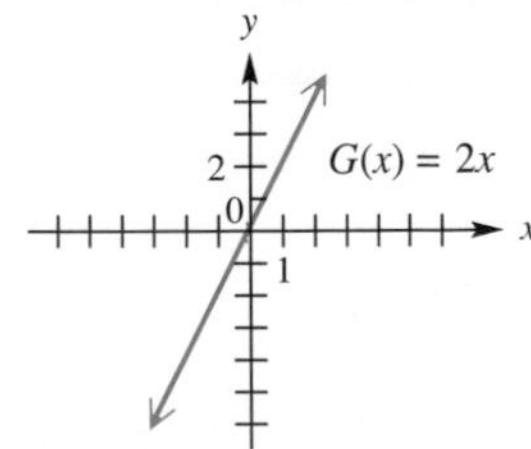

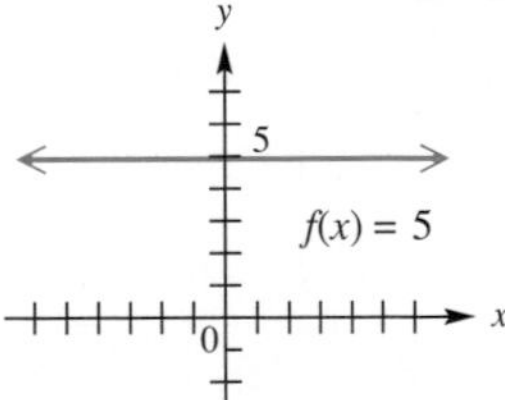

45. line; −2; $-2x + 4$; −2; 3; −2 **47.** (d)
49. **(a)** \$0; \$.50; \$1.00; \$1.50 **(b)** \$.50x **(c)**

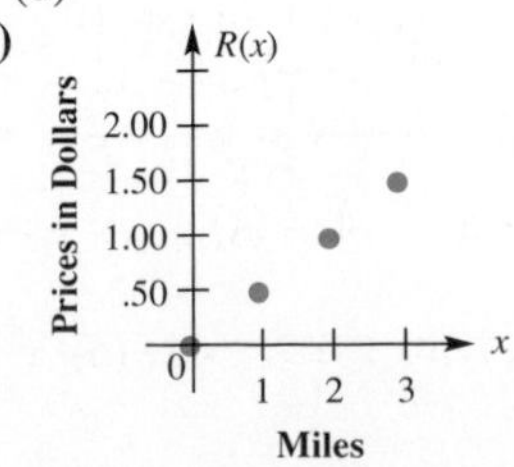

51. $f(3) = 7$

53. $f(x) = -3x + 5$ **55.** **(a)** 29,562 **(b)** 29,316 **(c)** 28,947
(d) In 1986 (when $x = 2$), the number of post offices was approximately 29,439.
57. **(a)** $C(x) = .02x + 200$ **(b)** $R(x) = .04x$ **(c)** 10,000 **(d)**

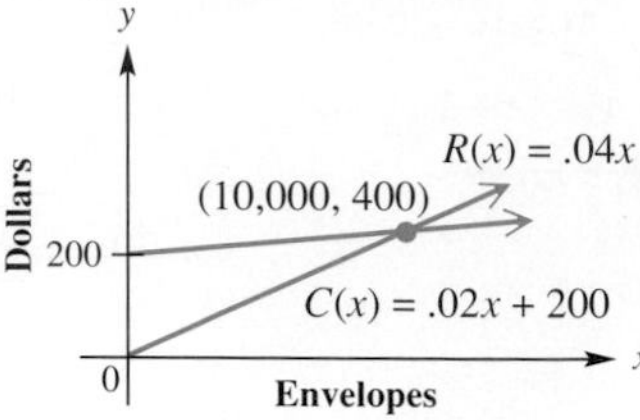

For $x < 10{,}000$, a loss
For $x > 10{,}000$, a profit

59. **(a)** $C(x) = 3.00x + 2{,}300$ **(b)** $R(x) = 5.50x$ **(c)** 920 **(d)**

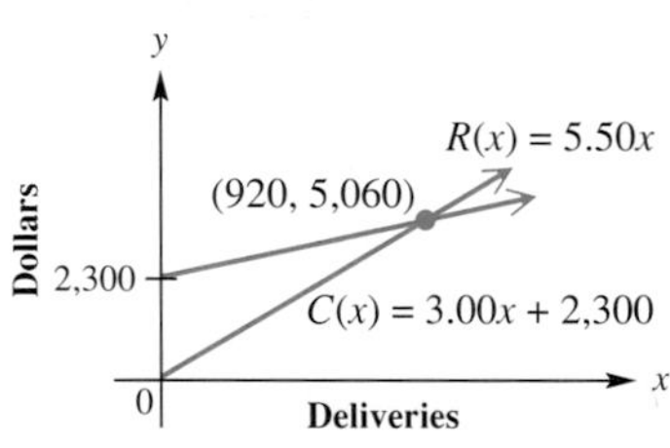

For $x < 920$, a loss
For $x > 920$, a profit

8.5 Exercises (Page 443)

3. (0, 0) **5.** (0, 4) **7.** (1, 0) **9.** (−3, −4) **11.** downward; narrower
13. upward; wider **17.** **(a)** I **(b)** IV **(c)** II **(d)** III **19.** H **21.** B **23.** D **25.** F

27.

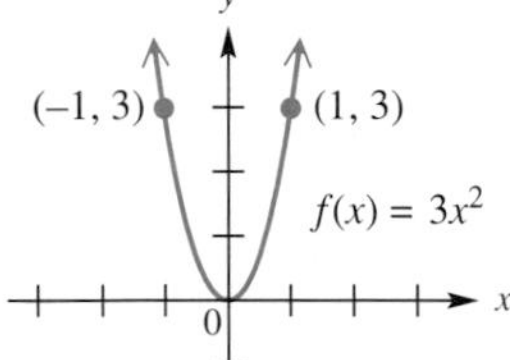

29.

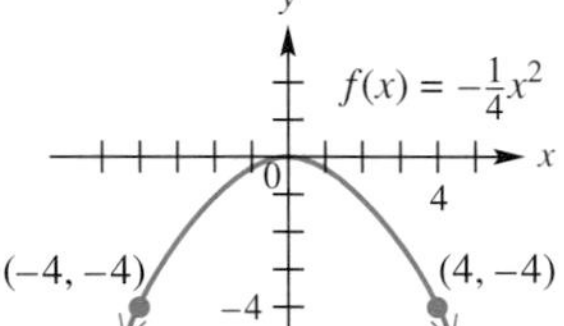

31.

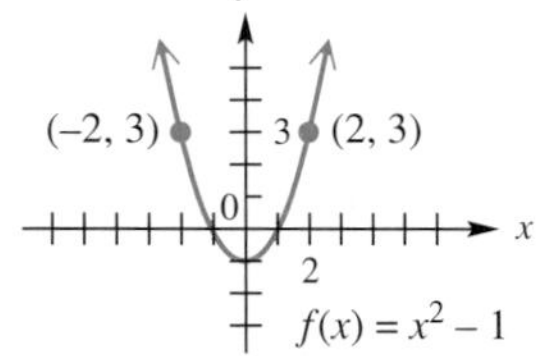

33.

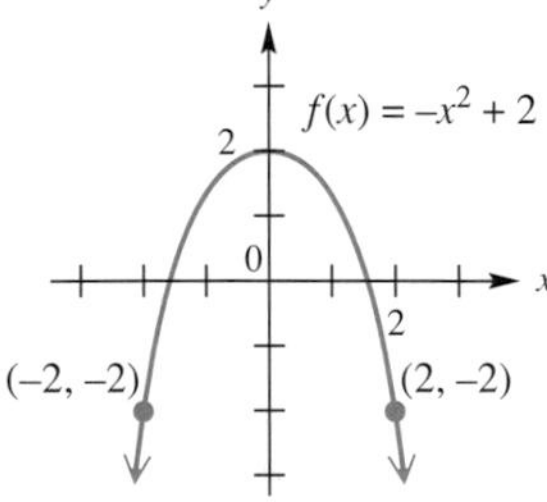

35.

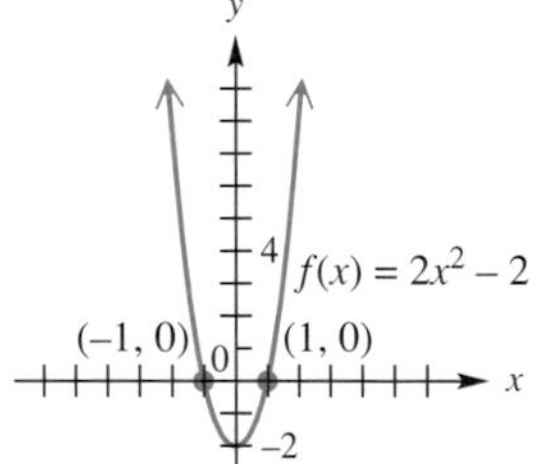

37.

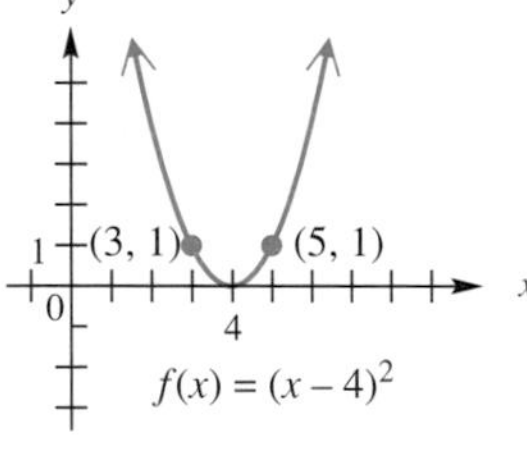

39.

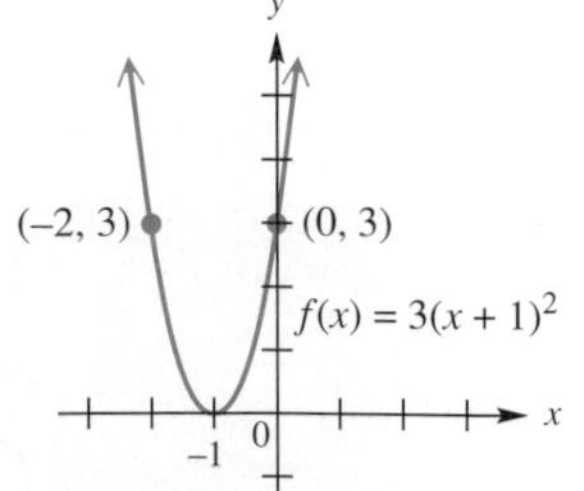

41.

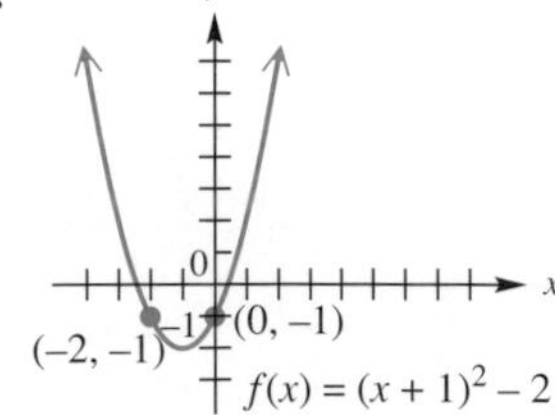

43.

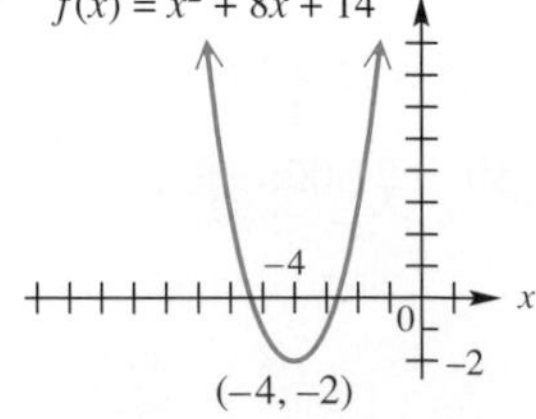

45.

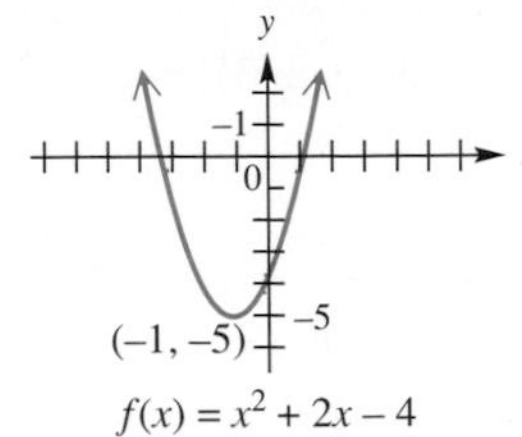

47.

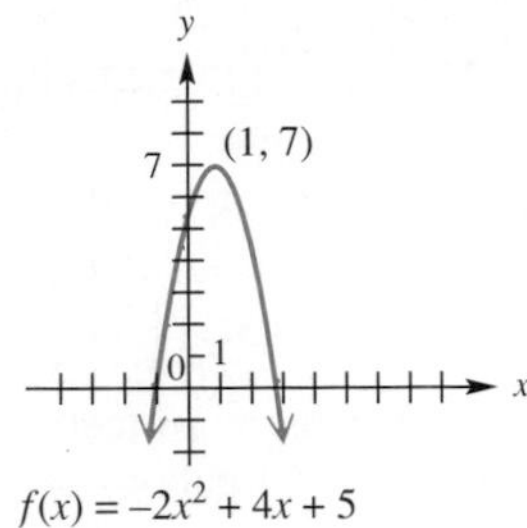

49. The parabola opens downward, indicating $a < 0$. **51.** Since $x = 2.1 \approx 2$, the year corresponding to 2, which is 1990, is the year of maximum number of jobs. **53.** 25 meters **55.** in 1995; 68 million barrels daily **57.** 16 feet; 2 seconds **59.** 4.1 seconds; 81.6 meters

8.6 Exercises (Page 455)

1. rises; falls **3.** does not **5.** rises; falls **7.** does not **9.** 2.56425419972 **11.** 1.25056505582 **13.** 7.41309466896 **15.** .0000210965628481 **17.**

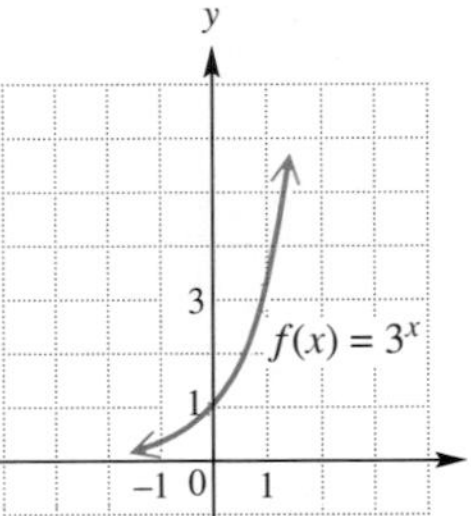

19.

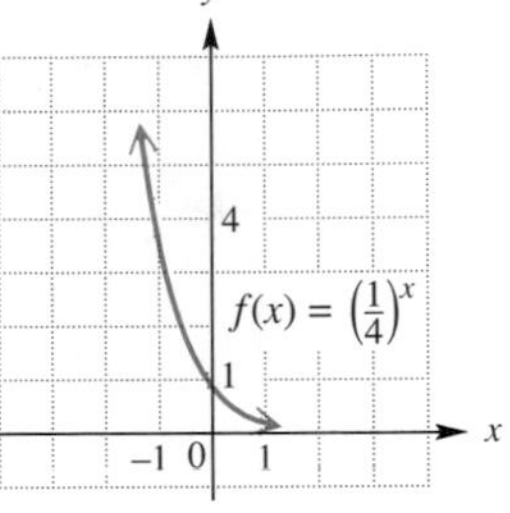

21. 20.0855369232 **23.** .018315638889 **25.** $2 = \log_4 16$

27. $-3 = \log_{2/3}\left(\frac{27}{8}\right)$ **29.** $2^5 = 32$ **31.** $3^1 = 3$ **33.** 1.38629436112 **35.** -1.0498221245

37.

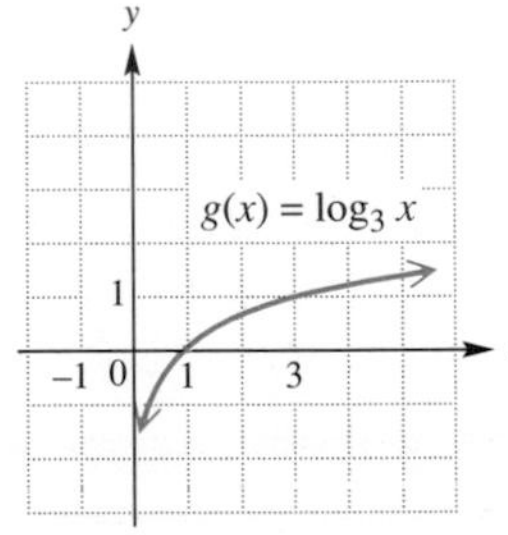

39.

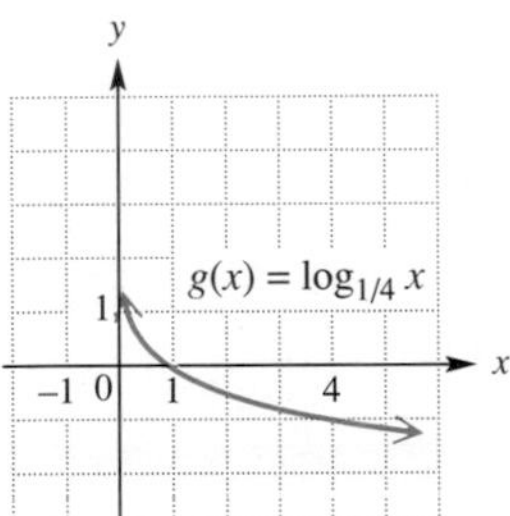

41. $7,686.32 **43.** $76,855.95

45. **(a)** $26,281.78 **(b)** $32,100.64 **(c)** $41,218.03 **47.** **(a)** .5°C **(b)** .35°C **49.** **(a)** 1.6°C **(b)** .5°C **51.** **(a)** 172.3 million **(b)** 310.6 million **(c)** It will be twice as large. **53.** **(a)** 100 grams **(b)** 31.25 grams **(c)** .30 gram (to the nearest hundredth) **55.** 6.67 years after it was purchased **57.** about $360,000 **59.** In 1991, the average salary was about $800,000. **61.** **(a)** 3.10 grams **(b)** 1.19 grams **(c)** 29 years

8.7 Exercises (Page 471)

3. yes **5.** no **7.** $\{(2, 2)\}$

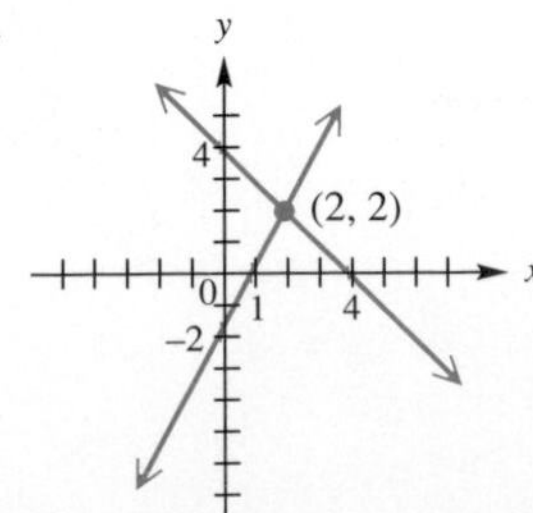

9. $\{(3, -1)\}$ **11.** $\{(2, -3)\}$ **13.** $\left\{\left(\frac{3}{2}, -\frac{3}{2}\right)\right\}$

15. $\left\{\left(\frac{6-2y}{7}, y\right)\right\}$ **17.** $\{(2, -4)\}$ **19.** $\emptyset$ **21.** $\{(1, 2)\}$ **23.** $\left\{\left(\frac{22}{9}, \frac{22}{3}\right)\right\}$ **25.** $\{(2, 3)\}$
27. $\{(5, 4)\}$ **29.** $\left\{\left(-5, -\frac{10}{3}\right)\right\}$ **31.** $\{(2, 6)\}$ **35.** $\{(1, 4, -3)\}$ **37.** $\{(0, 2, -5)\}$
39. $\left\{\left(-\frac{7}{3}, \frac{22}{3}, 7\right)\right\}$ **41.** $\{(4, 5, 3)\}$ **43.** $\{(2, 2, 2)\}$ **45.** $\left\{\left(\frac{8}{3}, \frac{2}{3}, 3\right)\right\}$
47. **(a)** 1978 and 1982 **(b)** just less than 500,000 **49.** length: 78 feet; width: 36 feet
51. length: 12 feet; width: 5 feet **53.** 6 gallons of 25%; 14 gallons of 35% **55.** 3 liters of pure acid; 24 liters of 10% **57.** 14 kilograms of nuts; 16 kilograms of cereal **59.** 76 general admission; 108 with student identification **61.** 28 dimes; 66 quarters **63.** \$1,000 at 2%; \$2,000 at 4%
65. CGA monitor: \$400; VGA monitor: \$500 **67.** 6 units of yarn: 2 units of thread **69.** dark clay: \$5 per kilogram; light clay: \$4 per kilogram **71.** freight train: 50 kilometers per hour; express train: 80 kilometers per hour **73.** boat: 21 miles per hour; current: 3 miles per hour **75.** $x + y + z = 180$; angle measures: 70°, 30°, 80° **77.** first: 20°; second: 70°; third: 90° **79.** shortest: 12 centimeters; middle: 25 centimeters; longest: 33 centimeters **81.** A: 180 cases; B: 60 cases; C: 80 cases **83.** 10 pounds of jelly beans; 2 pounds of chocolate eggs; 3 pounds of marshmallow chicks

8.8 Exercises (Page 479)

1.
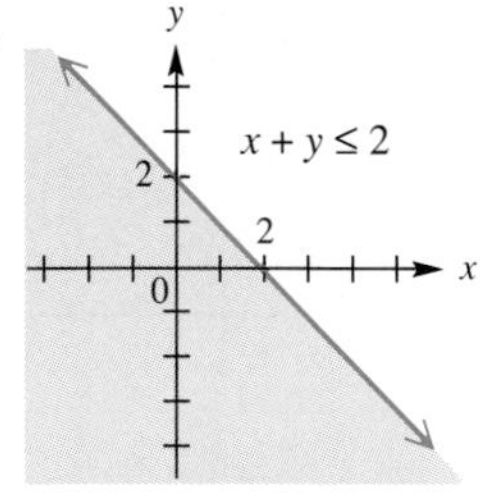

3.
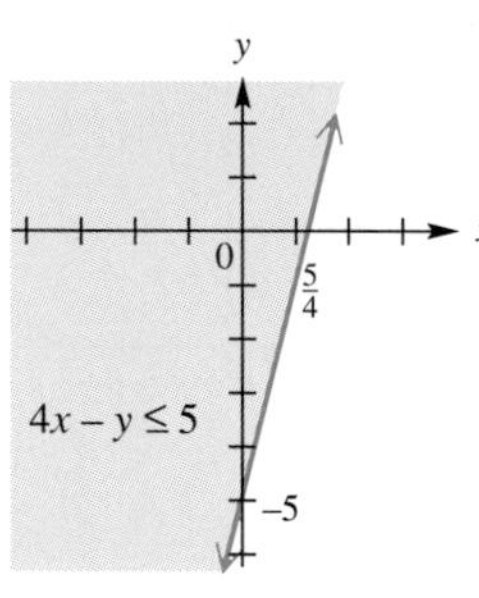

5.
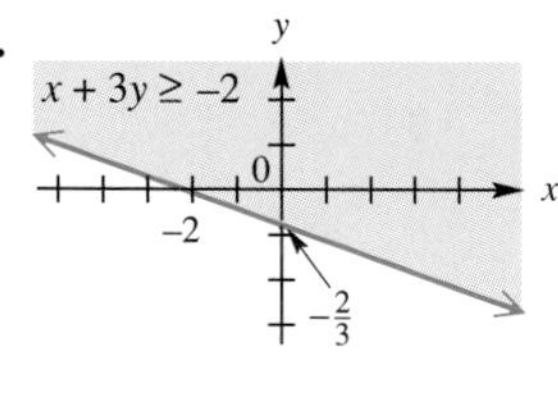

7.
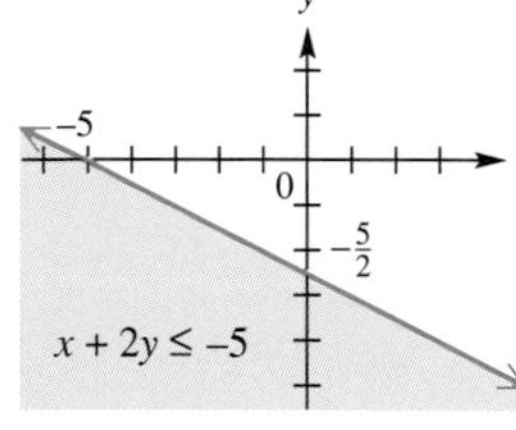

9.
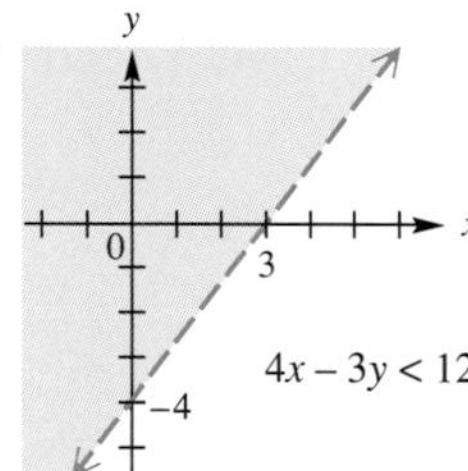

11.
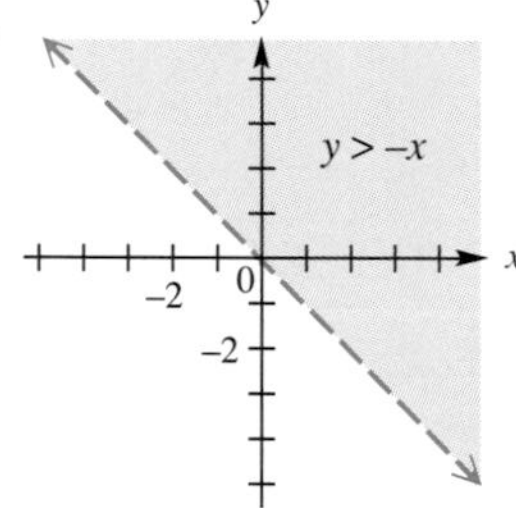

13.
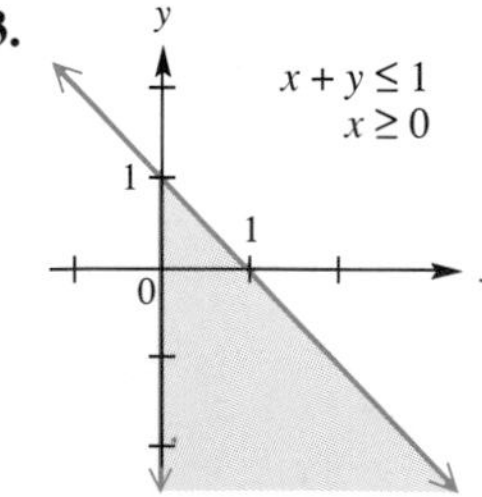

15.
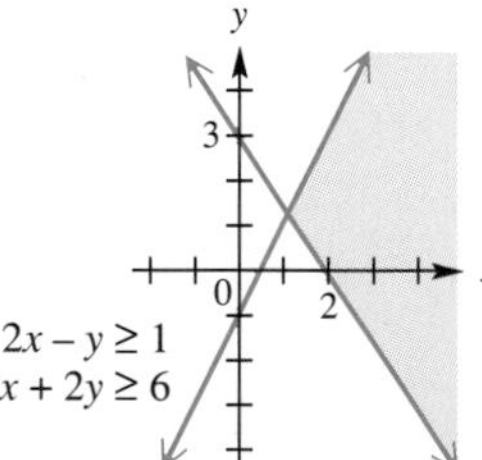

17.
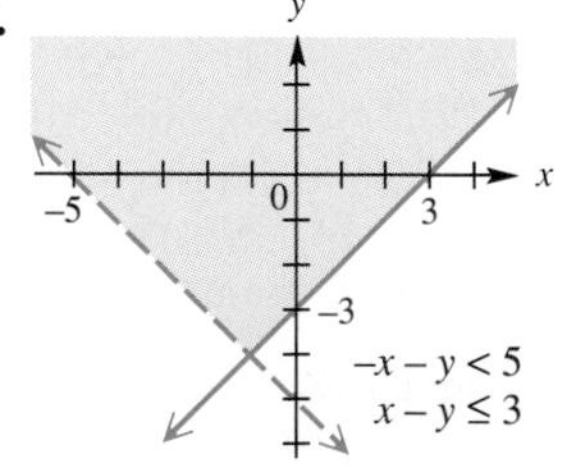

Extension Exercises (Page 481)

1. maximum of 65 at (5, 10); minimum of 8 at (1, 1) **3.** $\left(\frac{6}{5}, \frac{6}{5}\right)$; $\frac{42}{5}$ **5.** $\left(\frac{17}{3}, 5\right)$; $\frac{49}{3}$
7. Use 3 3/4 servings of A and 1 7/8 servings of B, for a minimum cost of \$1.69.
9. Manufacture 1,600 type 1 and 0 type 2, for a maximum revenue of \$160.
11. Make 3 batches of cakes and 6 batches of cookies, for a maximum profit of \$210.
13. Ship no medical kits and 4,000 containers of water.

Chapter 8 Test (Page 484)

1. $\sqrt{41}$ **2.** $(x + 1)^2 + (y - 2)^2 = 9$

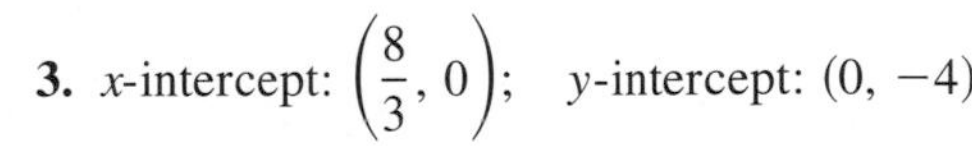
3. x-intercept: $\left(\frac{8}{3}, 0\right)$; y-intercept: $(0, -4)$

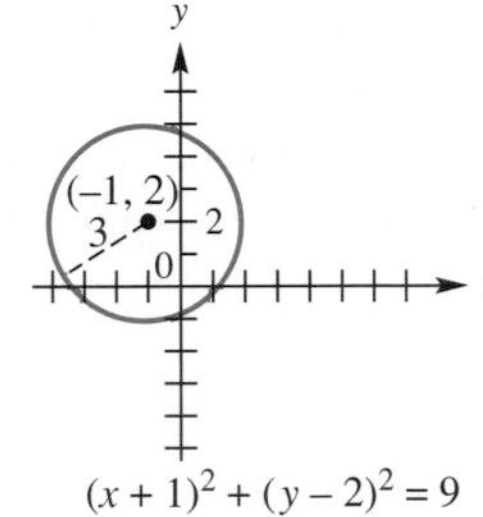

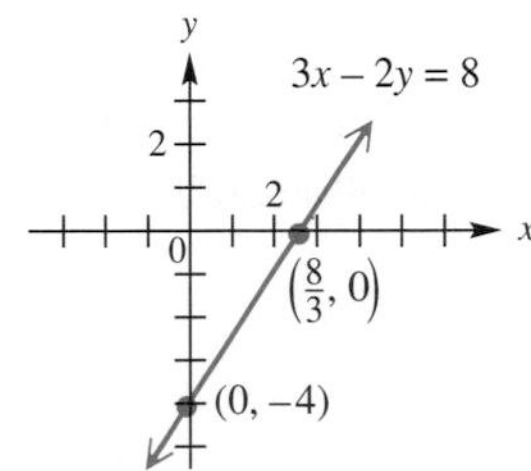

4. $\frac{2}{7}$ **5. (a)** $y = -\frac{2}{5}x + \frac{13}{5}$ **(b)** $y = -\frac{1}{2}x - \frac{3}{2}$ **(c)** $y = -\frac{1}{2}x + 2$ **6.** (b) **7.** 11,989
8. (a) It is the slope of the line. **(b)** It is the annual increase in the number of cases served.
9. (a) 1989 and 1990 **(b)** 1991 and 1992 **(c)** 1991 **10. (a)** $(-\infty, \infty)$ **(b)** 22 **11.** $\left(-\infty, \frac{9}{4}\right]$
12. 500 calculators; \$30,000 **13.** axis: $x = -3$; vertex: $(-3, 4)$; domain: $(-\infty, \infty)$; range: $(-\infty, 4]$

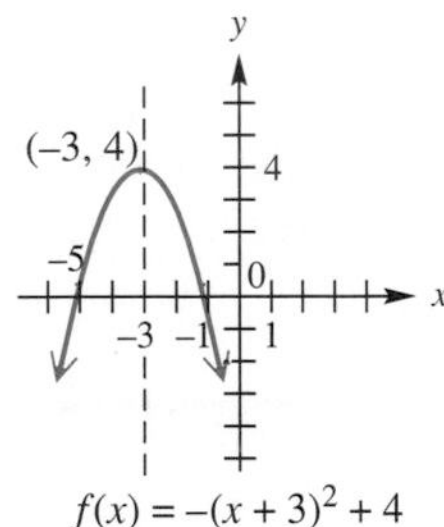

14. 80 feet by 160 feet **15. (a)** 2116.31264888 **(b)** .157237166314 **(c)** 3.15955035878 **16.** (a)
17. (a) \$13,521.90 **(b)** \$13,529.96 **18. (a)** $A(20) \approx 3{,}727$ million, which is off by 27 million.
(b) 5,342 million **(c)** 6,395 million **19.** $\{(4, -2)\}$ **20.** $\{(2, 0, -1)\}$ **21.** $\{(1, 2z + 3, z)\}$
22. 75 to A; 125 to B **23.** \$20,000 at 10%; \$50,000 at 6%; \$70,000 at 5%
24. (a) **25.**

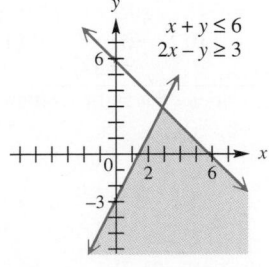

y
$x + y \leq 6$
$2x - y \geq 3$
6
0
2
6
x
−3

CHAPTER 9 Geometry

9.1 Exercises (Page 496)

1. true **3.** false **5.** true **7.** false **9.** true **11.** true **13.** false

15. **(a)** $\overline{AB}$ **(b)** A B **17.** **(a)** $\overleftrightarrow{CB}$ **(b)** A B C

19. **(a)** $\overrightarrow{BC}$ **(b)** B C D **21.** **(a)** $\overrightarrow{BA}$ **(b)** A B

23. **(a)** $\overline{CA}$ **(b)** A B C **25.** F **27.** D **29.** B **31.** E

There may be other correct forms of the answers in Exercises 33–39.
33. $\overline{MO}$ **35.** $\overline{MO}$ **37.** $\emptyset$ **39.** $\overline{OP}$ **41.** 62° **43.** 1° **45.** $(90 - x)°$ **47.** 48° **49.** 154° **51.** $(180 - y)°$ **53.** $\measuredangle CBD$ and $\measuredangle ABE$; $\measuredangle CBE$ and $\measuredangle DBA$ **55.** **(a)** 52° **(b)** 128° **57.** 107° and 73° **59.** 75° and 75° **61.** 139° and 139° **63.** 65° and 115° **65.** 35° and 55° **67.** 49° and 49° **69.** 48° and 132° **71.** 25° **73.** 30° **77.** Measures are given in numerical order, starting with angle 1: 55°, 65°, 60°, 65°, 60°, 120°, 60°, 60°, 55°, 55°.

9.2 Exercises (Page 505)

1. false **3.** false **5.** true **7.** true **9.** true **13.** both **15.** closed **17.** closed **19.** neither **21.** convex **23.** convex **25.** not convex **27.** right, scalene **29.** acute, equilateral **31.** right, scalene **33.** right, isosceles **35.** obtuse, scalene **37.** acute, isosceles **43.** $A = 50°$; $B = 70°$; $C = 60°$ **45.** $A = B = C = 60°$ **47.** $A = B = 52°$; $C = 76°$ **49.** 165° **51.** 170° **53.** **(e)** It reinforces the concept that the sum of the measures of the angles of a triangle is 180°, since the pencil has gone through one-half of a complete rotation.

9.3 Exercises (Page 516)

1. true **3.** true **5.** false **7.** 12 cm^2 **9.** 5 cm^2 **11.** 8 $in.^2$ **13.** 4.5 cm^2 **15.** 418 mm^2 **17.** 8 cm^2 **19.** 3.14 cm^2 **21.** 1,017.36 m^2 **23.** 4 m **25.** 300 ft, 400 ft, 500 ft **27.** 46 ft **29.** 23,800.10 ft^2 **31.** perimeter **33.** 12 in.; 12π in.; 36π $in.^2$ **35.** 5 ft; 10π ft; 25π ft^2 **37.** 6 cm; 12 cm; 36π cm^2 **39.** 10 in.; 20 in.; 20π in. **41.** 14.5 **43.** 7 **45.** 5.1 **47.** 5 **49.** 5 **51.** 2.4 **53.** **(a)** 20 cm^2 **(b)** 80 cm^2 **(c)** 180 cm^2 **(d)** 320 cm^2 **(e)** 4 **(f)** 3; 9 **(g)** 4; 16 **(h)** n^2 **55.** \$800 **57.** n^2 **59.** 80 **61.** 76.26 **63.** 132 ft^2 **65.** 5,376 cm^2 **67.** 145.34 m^2 **69.** 14-in. pizza **71.** 14-in. pizza **73.** 26 in. **75.** 625 ft^2 **77.** 648 $in.^2$ **79.** $\frac{(4 - \pi)r^2}{4}$

9.4 Exercises (Page 530)

1. SSS; $AB = AB$, since AB is equal to itself. **3.** SAS; $DB = DB$, since DB is equal to itself. **5.** ASA; $AC = AC$, since AC is equal to itself. **7.** 67°; 67° **9.** 6 in. **13.** $\measuredangle A$ and $\measuredangle P$; $\measuredangle C$ and $\measuredangle R$; $\measuredangle B$ and $\measuredangle Q$; $\overline{AC}$ and $\overline{PR}$; $\overline{CB}$ and $\overline{RQ}$; $\overline{AB}$ and $\overline{PQ}$ **15.** $\measuredangle H$ and $\measuredangle F$; $\measuredangle K$ and $\measuredangle E$; $\measuredangle HGK$ and $\measuredangle FGE$; $\overline{HK}$ and $\overline{FE}$; $\overline{GK}$ and $\overline{GE}$; $\overline{HG}$ and $\overline{FG}$ **17.** $\measuredangle P = 78°$; $\measuredangle M = 46°$; $\measuredangle A = \measuredangle N = 56°$ **19.** $\measuredangle T = 74°$; $\measuredangle Y = 28°$; $\measuredangle Z = \measuredangle W = 78°$ **21.** $\measuredangle T = 20°$; $\measuredangle V = 64°$; $\measuredangle R = \measuredangle U = 96°$ **23.** $a = 20$; $b = 15$ **25.** $a = 6$; $b = 15/2$ **27.** $x = 6$ **29.** $x = 110$ **31.** $c = 111\ 1/9$ **33.** 30 m **35.** 500 m, 700 m **37.** 112.5 ft **39.** 506 2/3 ft **41.** $c = 17$ **43.** $a = 13$ **45.** $c = 50$ m **47.** yes **49.** The sum of the squares of the two shorter sides of a right triangle is equal to the square of the longest side. **51.** (3, 4, 5) **53.** (7, 24, 25) **55.** (12, 16, 20) **59.** (3, 4, 5) **61.** (7, 24, 25) **65.** (4, 3, 5) **67.** (8, 15, 17) **71.** 24 m **73.** 16 ft **75.** 4.55 ft **77.** 19 ft, 3 in. **79.** 28 ft, 10 in. **81.** **(a)** $\frac{1}{2}(a + b)(a + b)$ **(b)** PWX: $\frac{1}{2}ab$; PZY: $\frac{1}{2}ab$; PXY: $\frac{1}{2}c^2$ **(c)** $\frac{1}{2}(a + b)(a + b) = \frac{1}{2}ab + \frac{1}{2}ab + \frac{1}{2}c^2$. When simplified, this gives $a^2 + b^2 = c^2$. **83.** $24 + 4\sqrt{6}$ **85.** 10 **87.** $256 + 64\sqrt{3}$ **89.** 5 in.

Extension Exercises (Page 540)

1. $\sin\theta = 3/5$, $\cos\theta = 4/5$, $\tan\theta = 3/4$ **3.** $\sin\theta = 4/5$, $\cos\theta = 3/5$, $\tan\theta = 4/3$ **5.** $\sin\theta = 5/13$, $\cos\theta = 12/13$, $\tan\theta = 5/12$ **7. (a)** complements **(b)** $90° - \theta$ **(c)** Each is equal to a/c. **(d)** $\sin\theta = \cos(90° - \theta)$ **9.** $\sin 30° = 1/2$, $\cos 30° = \sqrt{3}/2$, $\tan 30° = 1/\sqrt{3}$ or $\sqrt{3}/3$, $\sin 60° = \sqrt{3}/2$, $\cos 60° = 1/2$, $\tan 60° = \sqrt{3}$ **11.** 36.9° **13.** 46.4° **15.** 9.0 ft **17.** 11.1 ft **19.** 25.4 ft **21.** 67° **23.** 33.2 m

9.5 Exercises (Page 548)

1. true **3.** true **5.** false **7. (a)** 3 3/4 m^3 **(b)** 14 3/4 m^2 **9. (a)** 96 $in.^3$ **(b)** 130.4 $in.^2$ **11. (a)** 267,946.67 ft^3 **(b)** 20,096 ft^2 **13. (a)** 549.5 cm^3 **(b)** 376.8 cm^2 **15. (a)** 65.94 m^3 **(b)** 71.74 m^2 **17.** 168 $in.^3$ **19.** 1,969.10 cm^3 **21.** 427.29 cm^3 **23.** 508.68 cm^3 **25.** 2,415,766.67 m^3 **27.** .523 m^3 **29.** 12 in.; 288π $in.^3$; 144π $in.^2$ **31.** 5 ft; $\frac{500}{3}\pi$ ft^3; 100π ft^2 **33.** 2 cm; 4 cm; 16π cm^2 **35.** 1 m; 2 m; $\frac{4}{3}\pi$ m^3 **37.** volume **39.** $\sqrt[3]{2}x$ **41.** \$8,100 **43.** \$37,500 **45.** 2.5 **47.** 6 **49.** 210 $in.^3$ **51.** $\frac{62{,}500}{3}\pi$ $in.^3$ **53.** 2 to 1 **55.** 288

9.6 Exercises (Page 560)

1. Euclidean **3.** Lobachevskian **5.** greater than **7.** Riemannian **9.** Euclidean **11. (a)–(g)**

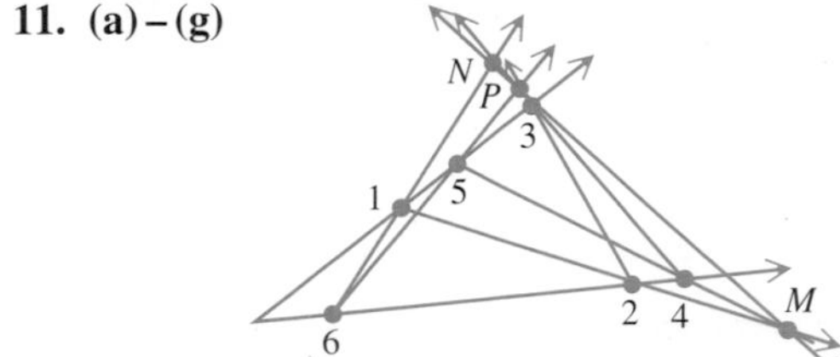

(h) Suppose that a hexagon is inscribed in an angle. Let each pair of opposite sides be extended so as to intersect. Then the three points of intersection thus obtained will lie in a straight line. **13.** C **15.** A, E **17.** A, E **19.** none of them **21.** no **23.** yes **25.** 1 **27.** 3 **29.** 1 **31.** *A*, *C*, *D*, and *F* are even; *B* and E are odd. **33.** *A*, *B*, *C*, and *F* are odd; *D*, *E*, and *G* are even. **35.** *A*, *B*, *C*, and *D* are odd; *E* is even. **37.** traversable **39.** not traversable **41.** traversable **43.** yes

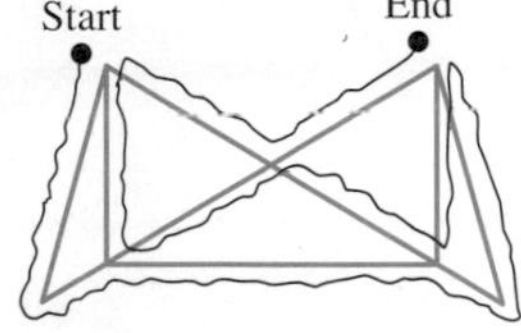

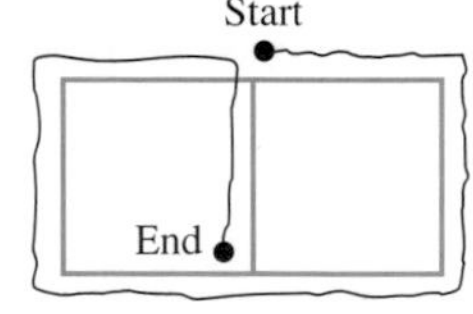

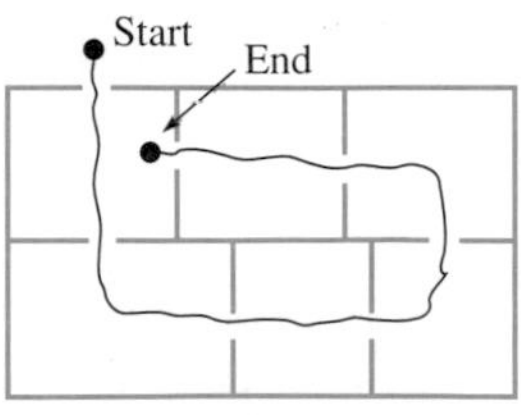

45. no

9.7 Exercises (Page 568)

1. 4 **2.** 4 **3.** 2 **4.** 2/1 or 2 **5.** 4/1 or 4 **6.** 3/1 or 3; 9/1 or 9 **7.** 4/1 or 4; 16/1 or 16 **8.** 4, 9, 16, 25, 36, 100 **9.** Each ratio in the bottom row is the square of the scale factor in the top row. **10.** 4 **11.** 4, 9, 16, 25, 36, 100 **12.** Each ratio in the bottom row is again the square of the scale factor in the top row. **13.** Answers will vary. Some examples are: $3^d = 9$, thus $d = 2$; $5^d = 25$, thus $d = 2$; $4^d = 16$, thus $d = 2$. **14.** 8 **15.** 2/1 or 2; 8/1 or 8 **16.** 8, 27, 64, 125, 216, 1,000 **17.** Each ratio in the bottom row is the cube of the scale factor in the top row. **18.** Since $2^3 = 8$, the value of d in $2^d = 8$ must be 3. **19.** 3/1 or 3 **20.** 4 **21.** 1.262 or ln 4/ln 3 **22.** 2/1 or 2 **23.** 3 **24.** It is between 1 and 2. **25.** 1.585 or ln 3/ln 2 **27.** .842, .452, .842, .452, . . . The two attractors are .842 and .452.

Chapter 9 Test (Page 572)

1. **(a)** 52° **(b)** 142° **(c)** acute **2.** 40°, 140° **3.** 45°, 45° **4.** 30°, 60° **5.** 130°, 50° **6.** 117°, 117° **8.** (c) **9.** simple, closed **10.** neither **11.** 30°, 45°, 105° **12.** 72 cm^2 **13.** 60 $in.^2$ **14.** 68 m^2 **15.** 180 m^2 **16.** 24π in. **17.** length: 30 in.; width: 6 in. **18.** 57 cm^2 **19.** SAS; $AB = AB$, since AB is equal to itself. **20.** 20 ft **21.** 29 m **22.** **(a)** 904.32 $in.^3$ **(b)** 452.16 $in.^2$ **23.** **(a)** 864 ft^3 **(b)** 552 ft^2 **24.** **(a)** 1,582.56 m^3 **(b)** 753.60 m^2 **26.** yes **27.** no **28.** yes **29.** The only attractor is .524.

End
Start

30. He is the mathematician who developed fractal geometry.

CHAPTER 10 Counting Methods

10.1 Exercises (Page 582)

1. *AB, AC, AD, AE, BA, BC, BD, BE, CA, CB, CD, CE, DA, DB, DC, DE, EA, EB, EC, ED*; 20 ways **3.** *AC, AE, BC, BE, CA, CB, CD, DC, DE, EA, EB, ED*; 12 ways **5.** *ACE, AEC, BCE, BEC, DCE, DEC*; 6 ways **7.** *ABC, ABD, ABE, ACD, ACE, ADE, BCD, BCE, BDE, CDE*; 10 ways **9.** 1 **11.** 3 **13.** 5 **15.** 5 **17.** 3 **19.** 1 **21.** 18 **23.** 15 **25.**

	1	**2**	**3**	**4**	**5**	**6**
1	11	12	13	14	15	16
2	21	22	23	24	25	26
3	31	32	33	34	35	36
4	41	42	43	44	45	46
5	51	52	53	54	55	56
6	61	62	63	64	65	66

27. 11, 22, 33, 44, 55, 66

29. 11, 13, 23, 31, 41, 43, 53, 61 **31.** 16, 25, 36, 64 **33.** 16, 32, 64 **35.**

First coin | Second coin | Third coin | Fourth coin | Result

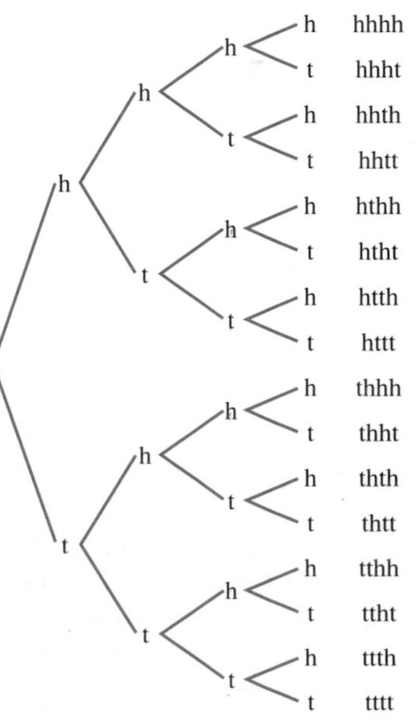

(a) tttt **(b)** hhhh, hhht, hhth, hhtt, hthh, htht, htth, thhh, thht, thth, tthh **(c)** httt, thtt, ttht, ttth, tttt **(d)** hhhh, hhht, hhth, hhtt, hthh, htht, htth, httt, thhh, thht, thth, thtt, tthh, ttht, ttth **37.** 16 **39.** 44 **41.** 17 **43.** 72 **45.** 12 **47.** 10 **49.** 6 **51.** 3 **53.** 54 **55.** 18 **57.** 138 **59.** 16 **61.** 13 **63.** 4 **65.** 883

10.2 Exercises (Page 591)

3. (a) no **5.** 24 **7.** 336 **9.** 20 **11.** 84 **13.** 840 **15.** 39,916,800 **17.** 95,040 **19.** 1,716 **21.** 184,756 **23.** $4.151586701 \times 10^{12}$ **25.** $2 \cdot 2 \cdot 2 = 8$ **29.** 216 **31.** $5! = 120$ **33.** $3 \cdot 2 = 6$ **35.** $4 \cdot 5^2 = 100$ **37.** $3 \cdot 4 \cdot 3 \cdot 2 = 72$ **39.** $6 \cdot 3^2 \cdot 1 = 54$ **41.** $4 \cdot 2 \cdot 3 = 24$ **43.** $2^8 = 256$ **45.** $3 \cdot 2 \cdot 4 \cdot 5 = 120$ **47.** $3 \cdot 2 \cdot 4 \cdot 3 = 72$ **49.** $3 \cdot 2 \cdot 1 \cdot 3 = 18$ **51.** $2 \cdot 4 \cdot 6 = 48$ **53.** $5! = 120$ **55. (a)** 6 **(b)** 5 **(c)** 4 **(d)** 3 **(e)** 2 **(f)** 1; $6 \cdot 5 \cdot 4 \cdot 3 \cdot 2 \cdot 1 = 720$ **57. (a)** 3 **(b)** 3 **(c)** 2 **(d)** 2 **(e)** 1 **(f)** 1; $3 \cdot 3 \cdot 2 \cdot 2 \cdot 1 \cdot 1 = 36$

10.3 Exercises (Page 602)

1. 20 **3.** 1 **5.** 35 **7.** 70 **9.** 840 **11.** 1,320 **13.** 5 **15.** 330 **17.** 62,990,928,000 **19.** $C(51, 6) = 18{,}009{,}460$ **25.** $P(6, 3) = 120$ **27.** $C(100, 5) = 75{,}287{,}520$ **29.** $2 \cdot P(25, 2) = 1{,}200$ **31.** $P(26, 3) \cdot P(10, 3) \cdot P(26, 3) = 175{,}219{,}200{,}000$ **33.** $C(24, 5) = 42{,}504$ **35.** $C(25, 3) \cdot C(22, 4) \cdot C(18, 5) \cdot C(13, 6) = 2.47365374256 \times 10^{14}$ **37.** $\frac{8 \cdot 21}{2} = 84$ **39. (a)** $8! \cdot 7! = 203{,}212{,}800$ **(b)** $8 \cdot 7 \cdot 13! = 348{,}713{,}164{,}800$ **(c)** $13! = 6{,}227{,}020{,}800$ **(d)** $8! \cdot 7! = 203{,}212{,}800$ **(e)** $8 \cdot 7 \cdot 6 \cdot 12! = 160{,}944{,}537{,}600$ **41.** $C(13, 5) = 1{,}287$ **43.** 0 (impossible) **45.** $C(12, 2) \cdot C(40, 3) = 652{,}080$ **47.** $40^3 = 64{,}000$ **49.** $C(6, 3) = 20$ **51.** $C(6, 2) \cdot C(6, 3) \cdot C(6, 4) = 4{,}500$ **53. (a)** $6! = 720$ **(b)** 523,647 **55.** $C(10, 3) = 120$ **57.** $C(24, 2) = 300$ **59.** $C(9, 3) + C(9, 4) = 210$ **61.** Each is equal to 220.

10.4 Exercises (Page 608)

1. 4 **3.** 15 **5.** 28 **7.** 36 **9.** $C(7, 1) \cdot C(3, 3) = 7$ **11.** $C(7, 3) \cdot C(3, 1) = 105$ **13.** $C(8, 3) = 56$ **15.** $C(8, 5) = 56$ **17.** $C(9, 4) = 126$ **19.** 56 **21.** 1 **23.** 10 **25.** 5 **27.** 32 **29.** . . . , 15, 21, 28, 36, 45, . . . ; (These are the triangular numbers.) **31.** The rows of Tartaglia's rectangle correspond to the diagonals of Pascal's triangle. **33.** row 8 **35.** $x^6 + 6x^5y + 15x^4y^2 + 20x^3y^3 + 15x^2y^4 + 6xy^5 + y^6$ **37.** $z^3 + 6z^2 + 12z + 8$ **39.** $16a^4 + 160a^3b + 600a^2b^2 + 1{,}000ab^3 + 625b^4$ **41.** $b^7 - 7b^6h + 21b^5h^2 - 35b^4h^3 + 35b^3h^4 - 21b^2h^5 + 7bh^6 - h^7$ **43.** $n + 1$ **45.** $n = 12, r = 5: \frac{12!}{8!\,4!}x^8y^4 = 495x^8y^4$

10.5 Exercises (Page 615)

3. $2^7 - 1 = 127$ **5.** 120 **7.** 6 **9.** $6 + 6 - 1 = 11$ **11.** $635{,}013{,}559{,}600 - 1 = 635{,}013{,}559{,}599$ **13.** $635{,}013{,}559{,}600 - C(40, 13) = 622{,}980{,}336{,}720$ **15.** $9 \cdot 10 \cdot 6 = 540$ **17.** $C(12, 0) + C(12, 1) + C(12, 2) = 79$ **19.** $2^{12} - 79 = 4{,}017$ **21.** $30 + 15 - 10 = 35$ **23.** $13 + 4 - 1 = 16$ **25.** $13 + 12 - 3 = 22$ **33.** $6 \cdot 7 \cdot 4 = 168$ **35.** $6 \cdot 1 \cdot 1 = 6$ **37.** $26^3 \cdot 10^3 - P(26, 3) \cdot P(10, 3) = 6{,}344{,}000$ **39.** $C(25, 4) - C(23, 4) = 3{,}795$ **41.** $C(12, 3) - C(5, 3) = 210$ **43.** $C(12, 3) - (15 + 60) = 145$ **45.** 40 **47.** $300 - (150 + 100 + 60 - 50 - 30 - 20 + 10) = 80$ **49.** 15 **51.** $C(15, 4) - [C(11, 4) + C(11, 3) \cdot C(4, 1)] = 375$

Chapter 10 Test (Page 618)

1. $6 \cdot 7 \cdot 7 = 294$ **2.** $6 \cdot 7 \cdot 4 = 168$ **3.** $6 \cdot 6 \cdot 5 = 180$ **4.** $6 \cdot 5 \cdot 1 = 30$ end in 0; $5 \cdot 5 \cdot 1 = 25$ end in 5; $30 + 25 = 55$ **5.** 13

6. First toss — Second toss — Third toss — Fourth toss

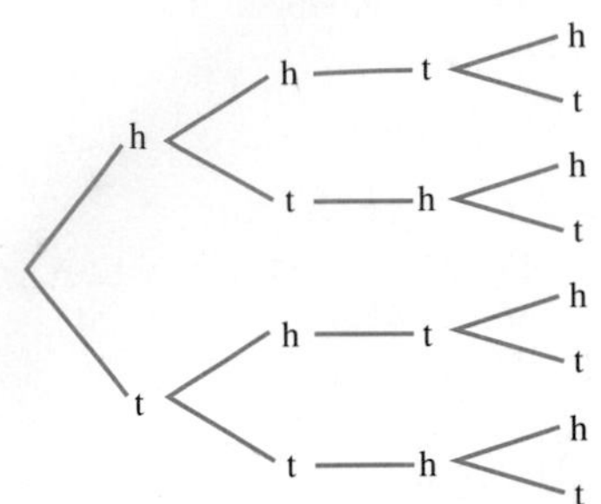

7. $4! = 24$ **8.** 12 **9.** 120

10. 336 **11.** 11,880 **12.** 35 **13.** $P(26, 5) = 7{,}893{,}600$ **14.** $32^5 = 33{,}554{,}432$ **15.** $C(12, 4) = 495$ **16.** $C(12, 2) \cdot C(10, 2) = 2{,}970$ **17.** $C(12, 5) \cdot C(7, 5) = 16{,}632$ **18.** $2^{12} - [C(12, 0) + C(12, 1) + C(12, 2)] = 4{,}017$ **19.** $2^4 = 16$ **20.** $2^2 = 4$ **21.** $2 \cdot 2^2 = 8$ **22.** 8 **23.** 2 **24.** $16 - (1 + 4) = 11$ **25.** $C(6, 3) = 20$ **26.** $C(5, 2) = 10$ **27.** $2 \cdot C(5, 3) = 20$ **28.** $C(5, 2) = 10$ **29.** $C(5, 4) + C(2, 1) \cdot C(5, 3) = 25$ **30.** $x^5 + 10x^4 + 40x^3 + 80x^2 + 80x + 32$ **31.** $495 + 220 = 715$ **32.** the counting numbers

CHAPTER 11 Probability

11.1 Exercises (Page 627)

1. **(a)** 1/3 **(b)** 1 to 2 **3.** **(a)** 1/3 **(b)** 1 to 2 **5.** **(a)** 1/3 **(b)** 1 to 2 **7.** **(a)** 4/9 **(b)** 4 to 5 **9.** **(a)** 1/2 **(b)** 1 to 1 **11.** **(a)** 1/2 **(b)** 1 to 1 **13.** 3/8 **15.** 1/8 **17.** $2/36 = 1/18$ **19.** $4/36 = 1/9$ **21.** $6/36 = 1/6$ **23.** $4/36 = 1/9$ **25.** $2/36 = 1/18$ **27.** $170/200 = .850$ **29.** $6/16 = 3/8$ **31.** $2/4 = 1/2$ **35.** $1/2{,}000 = .0005$ **37.** 75 **39.** $2/4 = 1/2$ **41.** 0 **43.** $2/4 = 1/2$ **45.** $1/160{,}000 \approx .000006$ **47.** 1/4 **49.** 1/4 **53.** $3 \cdot 1 \cdot 2 \cdot 1 \cdot 1 \cdot 1 = 6$; $6/720 = 1/120 \approx .0083$ **55.** $4 \cdot 3! \cdot 3! = 144$; $144/720 = 1/5 = .2$ **57.** $2/C(7, 2) = 2/21 \approx .095$ **59.** $(6^2 - 4^2 + 2^2)/8^2 = 3/8 = .375$ **61.** $1/P(32, 3) \approx .000034$ **63.** $1/C(8, 2) = 1/28 \approx .036$ **65.** $36/2{,}598{,}960 \approx .00001385$ **67.** $48/2{,}598{,}960 \approx .00001847$ **69.** $84{,}480/2{,}598{,}960 \approx .03250531$ **71.** $3/28 \approx .107$ **73.** **(a)** $8/9^2 = 8/81 \approx .099$ **(b)** $4/C(9, 2) = 1/9 \approx .111$ **75.** $(9 \cdot 10)/(9 \cdot 10^2) = 1/10$

11.2 Exercises (Page 636)

1. yes **5.** 1/2 **7.** 5/6 **9.** 2/3 **11.** **(a)** 2/13 **(b)** 2 to 11 **13.** **(a)** 11/26 **(b)** 11 to 15 **15.** **(a)** 9/13 **(b)** 9 to 4 **17.** 2/3 **19.** 7/36 **21.** $P(A) + P(B) + P(C) + P(D) = 1$ **23.** .005365 **25.** .971285 **27.** .76 **29.** .93 **31.** 6 to 19 **33.**

x	$P(x)$
3	.1
4	.1
5	.2
6	.2
7	.2
8	.1
9	.1

35. $n(A') = s - a$ **37.** $P(A) + P(A') = 1$ **39.** 180 **41.** 2/3 **43.** 1

11.3 Exercises (Page 645)

1. independent **3.** independent **5.** independent **7.** $45/100 = 9/20$ **9.** $75/100 = 3/4$ **11.** $18/38 = 9/19$ **13.** $(26/52) \cdot (13/52) = 1/8$ **15.** $(4/52) \cdot (4/52) = 1/169$ **17.** $12/51 = 4/17$ **19.** 4/51 **21.** $(13/52) \cdot (26/51) = 13/102$ **23.** $2 \cdot (4/663) = 8/663$

25. $(4/52) \cdot (11/51) = 11/663$ **27.** $(1/52) \cdot (3/51) + (12/52) \cdot (4/51) = 1/52$ **29.** $1/3$ **31.** 1
33. 3/10 (the same) **35.** .36 **37.** .06 **39.** .95 **41.** .23 **43.** 1/8 **45.** 1/56 **47.** 3/8
49. 3/8 **51. (a)** not true **53. (a)** 3/4 **(b)** 1/2 **(c)** 5/16 **55.** .2704 **57.** .2496
61. $1/64 \approx .0156$ **63.** 10 **65.** .350 **67.** .105 **69.** .047 **71.** $(.90)^4 = .6561$
73. $C(4, 2) \cdot (.10) \cdot (.20) \cdot (.70)^2 = .0588$ **75.** .40 **77.** .36

11.4 Exercises (Page 652)

1. 1/8 **3.** 3/8 **5.** 3/4 **7.** 1/2 **9.** 3/8 **11.** x; n; n **13.** 7/128 **15.** 35/128
17. 21/128 **19.** 1/128 **21.** 25/72 **23.** 1/216 **25.** .041 **27.** .268 **33.** .016 **35.** .020
37. .198 **39.** .448 **41.** .656 **43.** .002 **45.** .883 **49.** .062 **51.** .014
53. $45/1{,}024 \approx .044$ **55.** 0 **57.** $(210 + 120 + 45 + 10 + 1)/1{,}024 = 193/512 \approx .377$
59. $252/1{,}024 = 63/256 \approx .246$

11.5 Exercises (Page 660)

3. \$1 **5.** unfair in favor of the player **7.** $-\$1/37 \approx -2.7¢$ **9.** 1/2 **11.** \$100
13. (a) −\$66.67 **(b)** \$50,000 **(c)** \$75,000 **15.** \$1 **17.** 2.8 **19.** \$118.72 **21.** −20¢
23. Accept the appliances (since \$1,800 > \$1,750). **25.** column 6: 22,000, 51,000, 30,000, 60,000, 46,000 **27.** \$106,500

11.6 Exercises (Page 665)

3. no **5.** 18/50 = .36 (This is quite close to .375, the theoretical value.) **7.** 13/50 = .26
9. Answers will vary.

Chapter 11 Test (Page 667)

3. 3 to 1 **4.** 25 to 1 **5.** 11 to 2 **6.**

	C	c
C	CC	Cc
c	cC	cc

7. 1/2 **8.** 3 to 1

9. $(7/7) \cdot (6/7) \cdot (5/7) = 30/49$ **10.** $(7/7) \cdot (1/7) \cdot (1/7) = 1/49$ **11.** $1 - (30/49 + 1/49) = 18/49$
12. $C(2, 2)/C(5, 2) = 1/10$ **13.** $C(3, 2)/C(5, 2) = 3/10$ **14.** 6/10 = 3/5 **15.** 3/10
16.

x	$P(x)$
0	0
1	3/10
2	6/10
3	1/10

17. 9/10 **18.** 18/10 = 9/5 **19.** 5 to 1 **20.** $1 - 1/36 = 35/36$ **21.** 7 to 2

22. 4/36 = 1/9 **23.** $(.82)^3 \approx .551$ **24.** $C(3, 2)(.82)^2(.18) \approx .363$ **25.** $1 - (.18)^3 \approx .994$
26. $(.82)(.18)(.82) \approx .121$ **27.** 25/102 **28.** 25/51 **29.** 4/51 **30.** 3/26 **31.** 38
32. 1 **33.** .026

CHAPTER 12 Statistics

12.1 Exercises (Page 678)

1. (a)

x	f	f/n
0	10	$10/30 \approx 33\%$
1	7	$7/30 \approx 23\%$
2	6	$6/30 = 20\%$
3	4	$4/30 \approx 13\%$
4	2	$2/30 \approx 7\%$
5	1	$1/30 \approx 3\%$

(b)

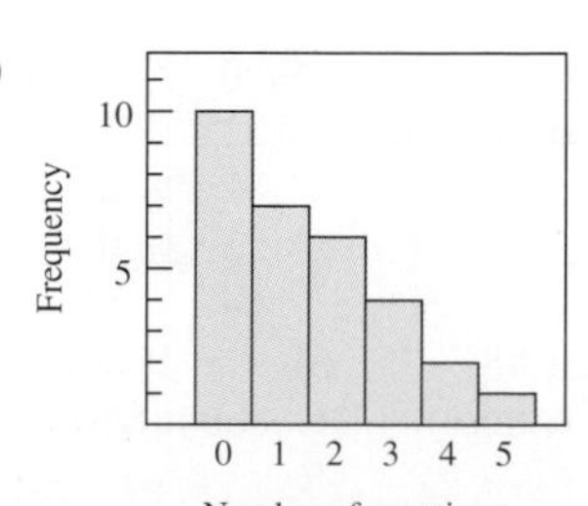

(c)

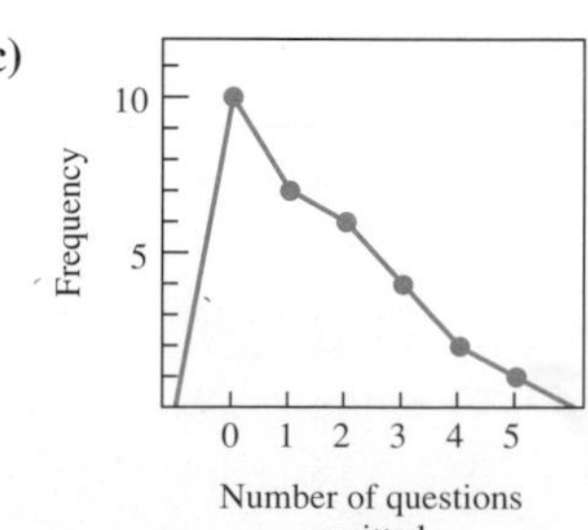

3. (a)

Class Limits	Tally	Frequency f	Relative Frequency f/n
45–49	𝍷𝍷𝍷	3	$3/54 \approx 5.6\%$
50–54	𝍸 𝍸 𝍷𝍷𝍷𝍷	14	$14/54 \approx 25.9\%$
55–59	𝍸 𝍸 𝍸 𝍷	16	$16/54 \approx 29.6\%$
60–64	𝍸 𝍸 𝍸 𝍷𝍷	17	$17/54 \approx 31.5\%$
65–69	𝍷𝍷𝍷𝍷	4	$4/54 \approx 7.4\%$
	Total:	$n = 54$	

(b)

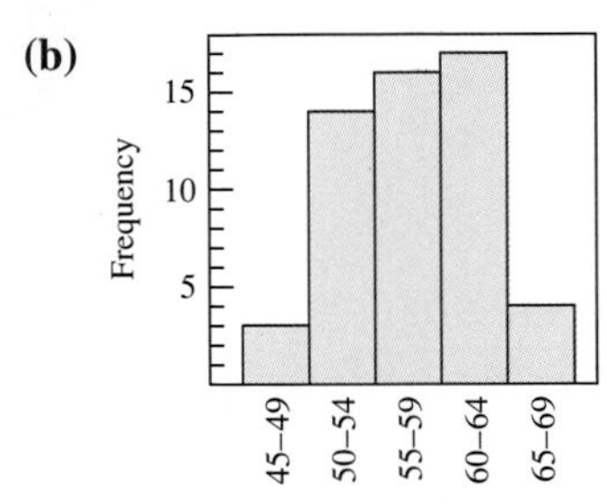

(c)

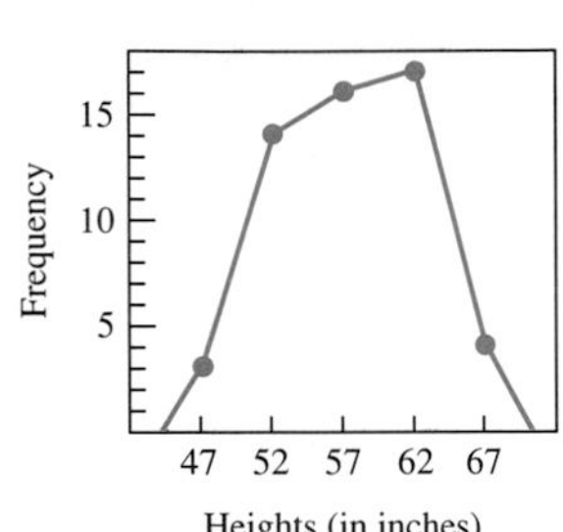

5. (a)

Class Limits	Tally	Frequency f	Relative Frequency f/n
70–74	𝍷𝍷	2	$2/30 \approx 6.7\%$
75–79	𝍷	1	$1/30 \approx 3.3\%$
80–84	𝍷𝍷𝍷	3	$3/30 = 10.0\%$
85–89	𝍷𝍷	2	$2/30 \approx 6.7\%$
90–94	𝍸	5	$5/30 \approx 16.7\%$
95–99	𝍸	5	$5/30 \approx 16.7\%$
100–104	𝍸 𝍷	6	$6/30 = 20.0\%$
105–109	𝍷𝍷𝍷𝍷	4	$4/30 \approx 13.3\%$
110–114	𝍷𝍷	2	$2/30 \approx 6.7\%$
	Total:	$n = 30$	

(b)

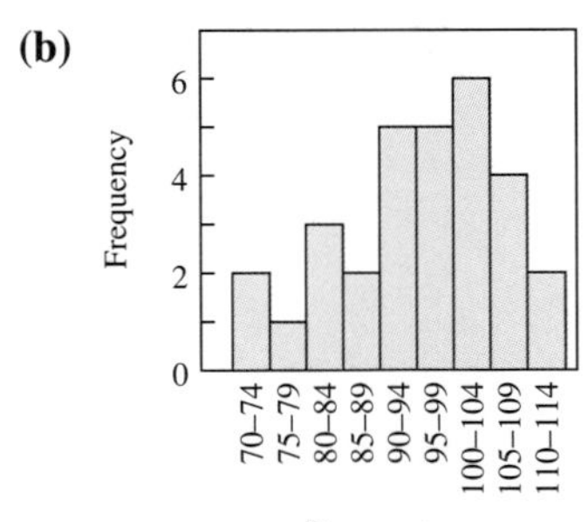

(c)

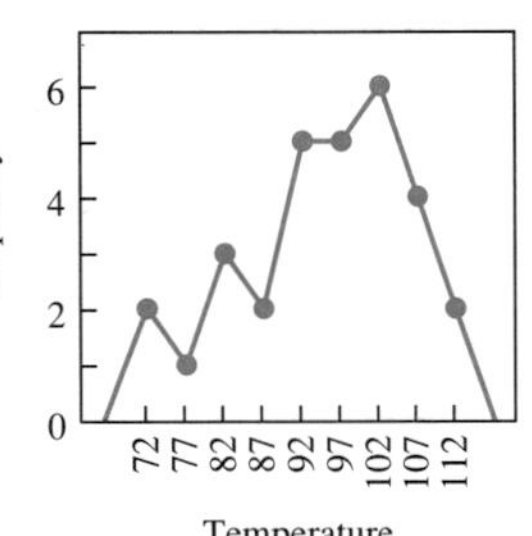

7.

Stem	Leaves
0	7 9 8
1	1 1 2 8 9 4 3 1 0 5 0 5 5
2	0 9 6 6 2 5 2 3 4 4
3	1

9.

Stem	Leaves
0	8 5 4 9 6 9 4 8
1	2 0 1 8 8 2 4 0 8 8 6 3
2	6 6 2 5 1 3
3	0 4 6
4	4

11. (a) about 8% **(b)** about 24% **(c)** about 48% **13.** the 1980s **15.** about $85 billion
17. Mexico over Canada **19.** U.S. with Mexico **21.** about $50 billion **23.** 10.3%

25.

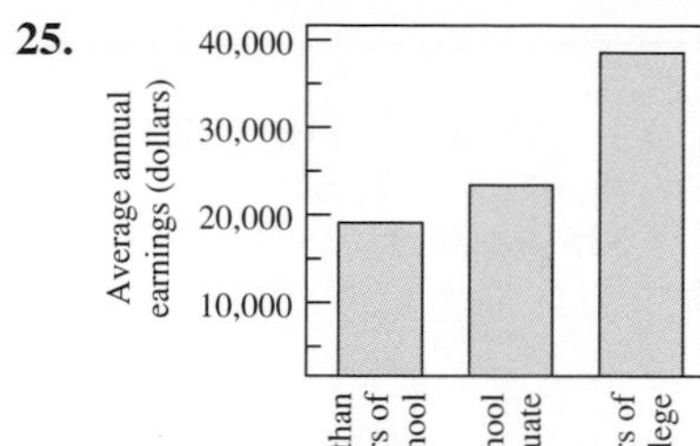

27. about \$400,000; about 76 years **29.** about four times as much

33.

(100–104)	10	0 0 4
(105–109)	10	5 9 5 9 7 6 8 5
(110–114)	11	2 0 3 4 4 2 4 0 0 3 1 1 3 0 2 4 4
(115–119)	11	8 8 7 6 8 5 8 7 8 6 9 7 8
(120–124)	12	0 1 2 1 0 0 0
(125–129)	12	7
(130–134)	13	4

35.

Letter	Probability
E	.13
T	.09
A, O	.08
N	.07
I, R	.065
S, H	.06
D	.04
L	.035
C, M, U	.03
F, P, Y	.02
W, G, B	.015
V	.01
K, X, J	.005
Q, Z	.002

37.

Probability
.5
.4
.3
.2
.1
A E I O U
Vowel

41. **(a)** .225 **(b)** .275 **(c)** .425 **(d)** .175

43.

Sport	Probability
Sailing	.225
Hang gliding	.125
Bungee jumping	.175
Sky diving	.075
Canoeing	.300
Rafting	.100

12.2 Exercises (Page 694)

1. **(a)** 12.2 **(b)** 12 **(c)** none **3.** **(a)** 201.2 **(b)** 210 **(c)** 196 **5.** **(a)** 5.2 **(b)** 5.35 **(c)** 4.5 and 6.2 **7.** **(a)** .8 **(b)** .795 **(c)** none **9.** **(a)** 47.2 **(b)** 45.6 **(c)** none **11.** **(a)** 129 **(b)** 128 **(c)** 125 and 128 **13.** **(a)** 543.9 million **(b)** 260.8 million **15.** **(a)** 21,520,000 **(b)** 21,539,000 **17.** 4,000 square miles **19.** 5.27 sec **21.** 2.42 sec **23.** the mean **25.** mean = 81.8; median = 90.5; mode = 95 **27.** 99 **29.** **(a)** 29.97 **(b)** 30 **(c)** 32 **31.** \$24,600 **33.** 2.77 **39.** 1,093,119 **41.** $[(1{,}093{,}119 \times 7) + (-162{,}314 \times 3)]/(7 + 3) = 716{,}489$. This is the same as Exercise 38. **43.** 5.5 **45.** 4 **47.** **(a)** \$116.30 **(b)** \$115.00 **(c)** none **49.** 83 **51.** **(a)** mean = 3; median = 3; mode = 4 **(b)** mode **53.** **(a)** 7 **(b)** 4 **55.** **(a)** 8 **(b)** 6.75 **57.** 0 **59.** three choices: 1, 6, 16 **61.** no

12.3 Exercises (Page 705)

1. the sample standard deviation **3. (a)** 17 **(b)** 5.53 **5. (a)** 19 **(b)** 6.27 **7. (a)** 23 **(b)** 7.75 **9. (a)** 1.14 **(b)** .37 **11. (a)** 8 **(b)** 2.41 **13.** 3/4 **15.** 45/49 **17.** 88.9% **19.** 64% **21.** 3/4 **23.** 15/16 **25.** 1/4 **27.** 4/49 **29.** \$202.50 **31.** six **33.** at least nine **35.** Brand B ($\bar{x}_B = 48{,}176 > 43{,}560$) **39.** 23.29; 4.35 **45. (a)** 12.5 **(b)** -3.0 **(c)** 4.9 **(d)** 4.2 **(e)** 15.5 **(f)** 7.55 and 23.45 **49.** no **53. (a)** smallest **(b)** largest

12.4 Exercises (Page 713)

1. Charles (since $z = .48 > .19$) **3.** Arvind (since $z = -1.44 > -1.78$) **5.** 58 **7.** 62 **9.** $-.55$ **11.** 2.07 **13.** Hakeem Olajuwon **15.** Karl Malone **17.** Danny Manning (since $z = -.63$ is greater than either $-.97$ or -1.00)

19.

	Stem	Leaves
(60–64)	6	0
(65–69)	6	5 8 9 7
(70–74)	7	4 2
(75–79)	7	9 8 7 5 9
(80–84)	8	0 1 0 2 1 2 2 2

21. (a) The median is 21.3. **(b)** The range is 10.6. **(c)** The data are skewed to the right. **29.** the overall distribution **31. (a)** no **37.** 112.8 **39.** $Q_1 = 68.4$; $Q_2 = 77.9$; $Q_3 = 84.9$ **41.** $Q_1 = 70.45$; $Q_2 = 77.9$; $Q_3 = 84.1$

43.

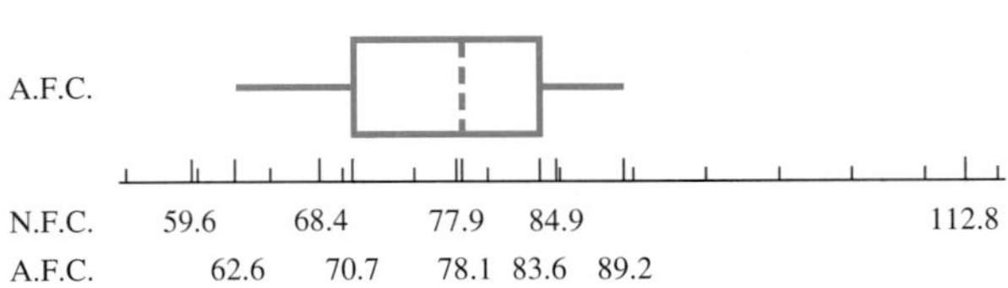

45. A.F.C.; about 1/2 unit **47.** N.F.C.; about 4 units **49.** 114.2

12.5 Exercises (Page 725)

1. discrete **3.** continuous **5.** discrete **7.** 50 **9.** 68 **11.** 50% **13.** 95% **15.** 49.4% **17.** 45.6% **19.** 7.7% **21.** 92.4% **23.** 1.64 **25.** -1.03 or -1.04 **27.** 5,000 **29.** 640 **31.** 9,920 **33.** 84.1% **35.** 37.8% **37.** .159 or 15.9% **39.** .006 or .6% **41.** .994 or 99.4% **43.** 189 units **45.** .888 **47.** about 2 eggs **49.** 24.1% **53.** 78 **55.** 66 **57.** 105.3 **59.** 64.5

Extension Exercises (Page 731)

1. We have no way of telling. **3.** We can tell only that it rises and then falls. **5.** 28% **7.** How long have Toyotas been sold in the United States? How do other makes compare? **9.** The dentists preferred Trident Sugarless Gum to what? Which and how many dentists were surveyed? What percentage responded? **11.** Just how quiet *is* a glider, really? **13.** The maps convey their impressions in terms of *area* distribution, whereas personal income distribution may be quite different. The map on the left probably implies too high a level of government spending, while that on the right implies too low a level. **15.** By the time the figures were used, circumstances may have changed greatly. (The Navy was much larger.) Also, New York City was most likely not typical of the nation as a whole. **17.** (b) **19.** (c) **21. (a)** $m = 1$, $a = 6$, $c = 3$ **(b)** There should be 1.4 managers, 5 agents, and 3.6 clerical employees.

12.6 Exercises (Page 740)

1. (a) $y' = .3x + 1.5$ **(b)** $r = .20$ **(c)** 2.4 decimeters **3. (a)** $y' = 3.35x - 78.4$ **(b)** 123 lb **(c)** 156 lb **(d)** $r = .66$ **5.** $y' = 8.06x + 49.52$; $r = .996$ **7.** $y' = .0276x + 50.1$; $r = .413$; The linear correlation is moderate. **9. (a)** $y' = 24.02x - 4{,}582$; $r = .989$; The linear correlation is strong. **(b)** $y' = -.4972x + 1{,}384$; $r = -.347$; The linear correlation is moderate at best.

Chapter 12 Test (Page 743)

1. \$5,840 billion **2.** 2.7%; 1994 **3.** 7.3%; 1992

5.

Class Limits	Frequency f	Relative Frequency f/n
6–10	3	$3/22 \approx .14$
11–15	6	$6/22 \approx .27$
16–20	7	$7/22 \approx .32$
21–25	4	$4/22 \approx .18$
26–30	2	$2/22 \approx .09$

6. **(a)**

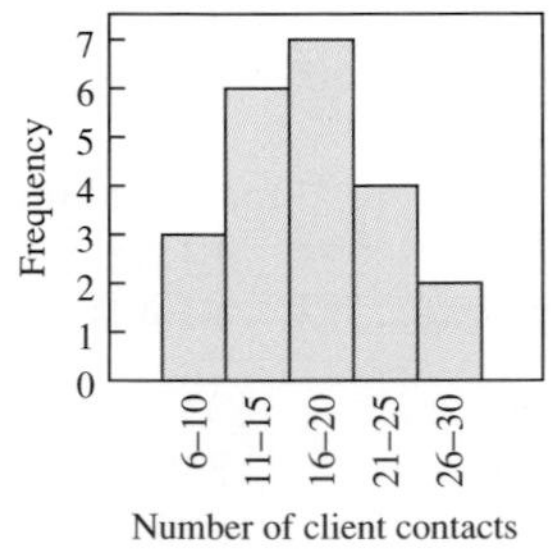

(b)

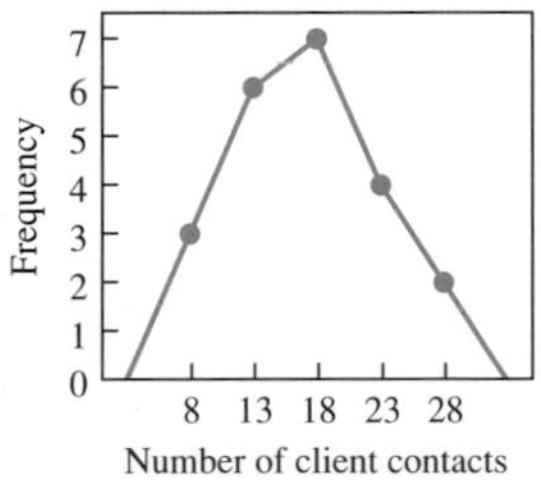

7. 8 **8.** 8.67 **9.** 8 **10.** 8 **11.** 10 **12.**

3	3 8
4	3 5 8 9
5	0 2 5
6	1 1 4 5 7 7 7 8
7	0 1 2 3 5 7 8 9 9
8	0 4 4
9	1

13. 55.5

14. 53 **15.** 44 **16.** 51 **17.** 47 **18.**

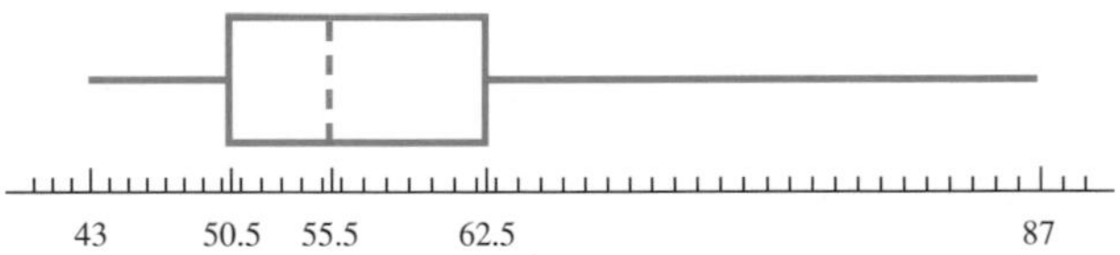

19. about 95% **20.** about .3% **21.** about 16% **22.** about 13.5% **23.** .248 **24.** .082
25. Western **26.** Western **27.** 57.0 **28.**

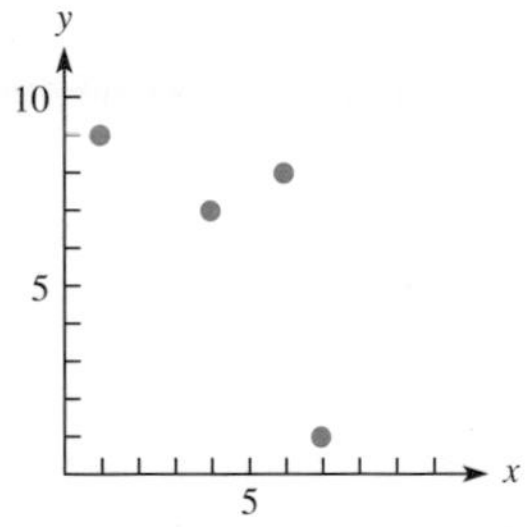

29. $y' = -.98x + 10.64$

30. 7.70 **31.** −.72

CHAPTER 13 Consumer Mathematics

13.1 Exercises (Page 758)

1. \$120 **3.** \$30.38 **5.** \$615.25 **7.** **(a)** \$560 **(b)** \$562.43 **9.** **(a)** \$3,200 **(b)** \$3,276.99
11. \$250 **13.** \$13,375 **15.** \$363.47 **17.** \$817.17 **19.** \$28,600.44; \$21,100.44
21. **(a)** \$1,259.71 **(b)** \$1,265.32 **(c)** \$1,268.24 **23.** **(a)** \$10,210.25 **(b)** \$10,240.68 **(c)** \$10,256.30
25. **(a)** \$166.08 **(b)** \$167.79 **(c)** \$168.96 **(d)** \$169.53 **(e)** \$169.55 **29.** \$1,440 **31.** \$747.26
33. \$4,499.98 **35.** \$21,035.00 **37.** \$20,388.25 **39.** 7.000% **41.** 7.186% **43.** 7.250%
45. 7.251% **47.** \$42,456 **49.** 7 years, 319 days **51.** 5.70% **53.** 24 years

55. 15 years **57.** 70 years **59.** 9 years **61.** 11.7%

63. $\text{years to triple} = \dfrac{100 \ln 3}{\text{annual inflation rate}} \approx \dfrac{110}{\text{annual inflation rate}}$

65. \$105,000; \$142,000; \$157,000; \$702,000 **67.** \$1.48; \$2.00; \$2.21; \$9.90 **69.** \$134,724.99

Extension Exercises (Page 763)

1. **(a)** \$14,835.10 **(b)** \$10,000 **(c)** \$4,835.10 **3.** **(a)** \$3,488.50 **(b)** \$3,000 **(c)** \$488.50 **5.** **(a)** \$3,374.70 **(b)** \$3,120 **(c)** \$254.70 **9.** \$3,555.94 **11.** \$3,398.08

13.2 Exercises (Page 769)

1. \$1,250 **3.** \$1,525 **5.** \$9,000 **7.** \$12,240 **9.** \$999; \$130.53 **11.** \$52; \$37.67 **13.** \$17.12; \$17.51 **15.** \$891; \$99.75 **17.** \$208.33 **19.** \$369.38 **21.** \$5,600 **23.** 5 years **25.** \$4.41 **27.** \$2.99

29.

Month	Unpaid Balance at Beginning of Month	Finance Charge	Purchases During Month	Returns	Payment	Unpaid Balance at End of Month
February	\$297.11	\$4.46	\$86.14	\$0	\$50	\$337.71
March	337.71	5.07	109.83	15.75	60	376.86
April	376.86	5.65	39.74	0	72	350.25
May	350.25	5.25	56.29	18.09	50	343.70

31.

Month	Unpaid Balance at Beginning of Month	Finance Charge	Purchases During Month	Returns	Payment	Unpaid Balance at End of Month
August	\$822.91	\$12.34	\$155.01	\$38.11	\$100	\$852.15
September	852.15	12.78	208.75	0	75	998.68
October	998.68	14.98	56.30	0	90	979.96
November	979.96	14.70	190.00	83.57	150	951.09

33. \$2.07 **35.** \$12.44 **37.** \$772.63; \$11.59 **39.** \$312.91; \$4.69 **41.** **(a)** \$10.80 **(b)** \$9.93 **43.** **(a)** \$44.60 **(b)** \$45.43 **(c)** \$46.27 **45.** 22.7% **47.** \$42 **55.** **(a)** \$172.80 **(b)** \$180.40 **57.** 1.7%

13.3 Exercises (Page 780)

1. 13.0% **3.** 15.5% **5.** \$66.67 **7.** \$102.08 **9.** 14.5% **11.** 11.0% **13.** **(a)** \$61.20 **(b)** 14.0% **15.** **(a)** \$43.20 **(b)** 11.0% **17.** **(a)** \$3.08 **(b)** \$51.13 **(c)** \$1,945.27 **19.** **(a)** \$6.62 **(b)** \$443.92 **(c)** \$7,301.48 **21.** **(a)** \$22.44 **(b)** \$21.49 **23.** **(a)** \$130.92 **(b)** \$115.37 **25.** **(a)** \$1.52 **(b)** \$6.65 **(c)** \$585.35 **27.** finance company APR: 12.5%; credit union APR: 11.5%; choose credit union **29.** \$26.15 **31.** 15.5% **35.** 8

13.4 Exercises (Page 791)

1. \$632.62 **3.** \$433.90 **5.** \$362.00 **7.** \$1,075.53 **9.** **(a)** \$381.74 **(b)** \$362.50 **(c)** \$19.24 **(d)** \$43,480.76 **11.** **(a)** \$471.06 **(b)** \$414.38 **(c)** \$56.68 **(d)** \$58,443.32 **(e)** \$471.06 **(f)** \$413.97 **(g)** \$57.09 **(h)** \$58,386.23 **13.** **(a)** \$570.42 **(b)** \$191.33 **(c)** \$379.09 **(d)** \$27,620.91 **(e)** \$570.42 **(f)** \$188.74 **(g)** \$381.68 **(h)** \$27,239.23 **15.** \$520.86 **17.** \$585.30 **19.** \$702.30 **21.** 360 **23.** \$160,515.20 **25.** **(a)** \$304.01 **(b)** \$734.73 **27.** **(a)** \$59,032.06 **(b)** \$59,875.11 **29.** **(a)** payment 176 **(b)** payment 304 **31.** \$25,465.20 **33.** \$91,030.80 **35.** **(a)** \$674.79 **(b)** \$746.36 **(c)** an increase of \$71.57 **37.** **(a)** \$395.83 **(b)** \$466.07 **39.** \$65.02 **41.** \$76,000 **43.** \$2,705

13.5 Exercises (Page 802)

1. \$36.625 **3.** \$26.75 **5.** \$14.75 **7.** \$2.76 **9.** \$45.50 **11.** 558,400 shares **13.** 5.1% **15.** \$28,375 **17.** \$7,350 **19.** \$773.75 **21.** \$9,369.06 **23.** \$16,026.01 **25.** \$12,113.95 **27.** \$20,574.16 **29.** \$27,548.74 **31.** \$5,881.94 **33.** \$11.75 net paid out

Chapter 13 Test (Page 804)

1. \$130 **2.** \$58.58 **3.** 6.17% **4.** 6 years **5.** \$37,688.95 **6.** \$10.88 **7.** \$480 **8.** \$145 **9.** 14.5% **10.** \$35.63 **11.** 16.0% **13.** \$550.54 **14.** \$768.98 **15.** \$1,408 **17.** 12.1% **18.** \$17 **19.** 25,000 shares **20.** \$12,684.38

CHAPTER 14 Matrices and Their Applications

14.1 Exercises (Page 812)

1. $\begin{bmatrix} 11 & 1 \\ 9 & 4 \end{bmatrix}$ **3.** $\begin{bmatrix} 3 & 3 & -3 \\ 4 & -4 & 0 \end{bmatrix}$ **5.** $\begin{bmatrix} -3 & 4 & 2 \\ 8 & 2 & 1 \\ 4 & 3 & 13 \end{bmatrix}$

7. It is not computable because the matrices do not have the same order. **9.** $\begin{bmatrix} 0 \\ 12 \\ 7 \end{bmatrix}$ **11.** $\begin{bmatrix} -14 & -2 \\ 5 & 10 \end{bmatrix}$

13. $A + B = \begin{bmatrix} -2 & -1 \\ -1 & -4 \end{bmatrix}$ **15.** $2A = \begin{bmatrix} -8 & 0 \\ 6 & -8 \end{bmatrix}$ **17.** $-4B = \begin{bmatrix} -8 & 4 \\ 16 & 0 \end{bmatrix}$

19. $2A + 3B = \begin{bmatrix} -2 & -3 \\ -6 & -8 \end{bmatrix}$ **21.** $(-1)A + A = \begin{bmatrix} 0 & 0 \\ 0 & 0 \end{bmatrix}$

23.

	Fat	Carbohydrates	Protein
I	5	0	7
II	0	10	1
III	0	15	2
IV	10	12	8

25. $C =$

	W1	W2	W3
F1	12	20	25
F2	16	25	30

27. $C + A = \begin{bmatrix} 15 & 18 & 27 \\ 20 & 27 & 28 \end{bmatrix}$ **29.** $\begin{bmatrix} 45 & 35 & 20 \\ 65 & 40 & 35 \end{bmatrix} + \begin{bmatrix} 10 & 12 & 5 \\ 15 & 20 & 8 \end{bmatrix} = \begin{bmatrix} 55 & 47 & 25 \\ 80 & 60 & 43 \end{bmatrix}$

31. 25 model-*C* tables **33.** $\begin{bmatrix} 22 & 25 & 38 \\ 31 & 34 & 35 \end{bmatrix} + \begin{bmatrix} 10 & 12 & 5 \\ 15 & 20 & 8 \end{bmatrix} - \begin{bmatrix} 5 & 10 & 8 \\ 11 & 14 & 15 \end{bmatrix} = \begin{bmatrix} 27 & 27 & 35 \\ 35 & 40 & 28 \end{bmatrix}$

35. 9 pounds **37.** The given equality is impossible because the two matrices have different orders. **39.** $\{(y, z) \mid y = z\}$ **41.** $k = 8, m = 1, y = 3$ **43.** $x = 1, y = 8, z = 8$

45. The given equality is impossible. In the first row, first column of the matrix $\begin{bmatrix} x & y \\ 3 & 5 \end{bmatrix}$, x must be equal to 10. But in the second row, first column of the matrix $\begin{bmatrix} -2 & 6 \\ x & 2 \end{bmatrix}$, x must be equal to 4.

47. No. On a calendar, there are seven columns. In order for the days of a given month to form a matrix, the number of days in that month must be divisible by 7. March has 31 days, which is not divisible by 7; therefore the month of March on a calendar will never form a matrix. **49.** false **51.** true **53.** true **55.** true

14.2 Exercises (Page 824)

1. $\begin{bmatrix} 8 & 6 \\ -7 & 3 \end{bmatrix}$ **3.** $\begin{bmatrix} -2 & -8 \\ -14 & 34 \end{bmatrix}$ **5.** $\begin{bmatrix} 17 & 7 & 9 \\ 13 & 6 & 3 \end{bmatrix}$ **7.** $\begin{bmatrix} 8 & 3 \\ 11 & -2 \\ 4 & 12 \end{bmatrix}$ **9.** It is not computable because the number of columns in the first matrix is not equal to the number of rows in the second matrix.

11. $\begin{bmatrix} 14 & 4 & 0 \\ -4 & 4 & 0 \\ -6 & 2 & 0 \end{bmatrix}$

13. $AB =$

	PURCHASE OPTION Large lot	Small lot
Sweet rolls	47.5	57.75
Bread	27	33.75
Cakes	81	95
Pies	12	15

15. $(DA)B =$

	Cost (large lot)	Cost (small lot)
Day's orders	11,120	13,555

17. $RC =$

	Cost per unit
Table	34
Chair	26
Desk	45

19.

	Materials	Labor
Table	34	36
Chair	26	18
Desk	45	45

21.

Team	Level 1 pts	Level 2 pts	Total pts
A	1	2	3
B	2	2	4
C	2	3	5
D	1	1	2

23. **(a)** $H^2K = \begin{bmatrix} 2 \\ 2 \\ 3 \\ 1 \end{bmatrix}$ **(b)** H^2K represents the number of Level 2 points earned by each team.

25. $(H + H^2)K = \begin{bmatrix} 3 \\ 4 \\ 5 \\ 2 \end{bmatrix}$; Yes, $(H + H^2)K = HK + H^2K$. **27.** $AB = \begin{bmatrix} 3 & 1 \\ 1 & -5 \end{bmatrix}$ **29.** $(AB)C = \begin{bmatrix} 2 & 3 \\ 6 & 17 \end{bmatrix}$; $A(BC) = \begin{bmatrix} 2 & 3 \\ 6 & 17 \end{bmatrix}$

31. $(A - B)C = \begin{bmatrix} -3 & -11 \\ 1 & 2 \end{bmatrix}$; $AC - BC = \begin{bmatrix} -3 & -11 \\ 1 & 2 \end{bmatrix}$

33. $AB = \begin{bmatrix} 122 & 130 \\ 65 & 57 \end{bmatrix}$; $BA = \begin{bmatrix} 122 & 130 \\ 65 & 57 \end{bmatrix}$; Yes, $AB = BA$.

35. $AB = \begin{bmatrix} 19 & 6 \\ -14 & 20 \end{bmatrix}$; $BA = \begin{bmatrix} 24 & 26 \\ -4 & 15 \end{bmatrix}$; No, AB is not equal to BA.

39. **(a)** Let $S = \begin{bmatrix} a & b & c \\ d & e & f \\ g & h & j \end{bmatrix}$; $I = \begin{bmatrix} 1 & 0 & 0 \\ 0 & 1 & 0 \\ 0 & 0 & 1 \end{bmatrix}$.

Then $SI = \begin{bmatrix} a \cdot 1 + b \cdot 0 + c \cdot 0 & a \cdot 0 + b \cdot 1 + c \cdot 0 & a \cdot 0 + b \cdot 0 + c \cdot 1 \\ d \cdot 1 + e \cdot 0 + f \cdot 0 & d \cdot 0 + e \cdot 1 + f \cdot 0 & d \cdot 0 + e \cdot 0 + f \cdot 1 \\ g \cdot 1 + h \cdot 0 + j \cdot 0 & g \cdot 0 + h \cdot 1 + j \cdot 0 & g \cdot 0 + h \cdot 0 + j \cdot 1 \end{bmatrix} = \begin{bmatrix} a & b & c \\ d & e & f \\ g & h & j \end{bmatrix} = S.$

43. $M^2 = \begin{bmatrix} 3 & 1 & 1 & 2 & 0 \\ 1 & 2 & 1 & 1 & 1 \\ 1 & 1 & 3 & 0 & 2 \\ 2 & 1 & 0 & 2 & 0 \\ 0 & 1 & 2 & 0 & 2 \end{bmatrix}$ **45.** 2 ways: B to C to A to E and B to C to D to E

49. $\begin{bmatrix} 1 & 0 \\ 0 & -1 \end{bmatrix} \cdot \begin{bmatrix} 1 & 0 \\ 0 & -1 \end{bmatrix} = \begin{bmatrix} 1 & 0 \\ 0 & 1 \end{bmatrix}$; Yes, they are inverses.

51. $\begin{bmatrix} 2 & 1 \\ 4 & 3 \end{bmatrix} \cdot \begin{bmatrix} 3/2 & -1/2 \\ -2 & 1 \end{bmatrix} = \begin{bmatrix} 1 & 0 \\ 0 & 1 \end{bmatrix}$; $\begin{bmatrix} 3/2 & -1/2 \\ -2 & 1 \end{bmatrix} \cdot \begin{bmatrix} 2 & 1 \\ 4 & 3 \end{bmatrix} = \begin{bmatrix} 1 & 0 \\ 0 & 1 \end{bmatrix}$; Yes, they are inverses.

53. No, they are not inverses. **55.** $\begin{bmatrix} 5 & 6 \\ -10 & -13 \end{bmatrix} \cdot \begin{bmatrix} 13/5 & 6/5 \\ -2 & 1 \end{bmatrix} = \begin{bmatrix} 1 & 12 \\ 0 & -25 \end{bmatrix}$; No, they are not inverses.

57. $\begin{bmatrix} 1 & 3 & 3 \\ 1 & 4 & 3 \\ 1 & 3 & 4 \end{bmatrix} \cdot \begin{bmatrix} 7 & -3 & -3 \\ -1 & 1 & 0 \\ -1 & 0 & 1 \end{bmatrix} = \begin{bmatrix} 1 & 0 & 0 \\ 0 & 1 & 0 \\ 0 & 0 & 1 \end{bmatrix}$; $\begin{bmatrix} 7 & -3 & -3 \\ -1 & 1 & 0 \\ -1 & 0 & 1 \end{bmatrix} \cdot \begin{bmatrix} 1 & 3 & 3 \\ 1 & 4 & 3 \\ 1 & 3 & 4 \end{bmatrix} = \begin{bmatrix} 1 & 0 & 0 \\ 0 & 1 & 0 \\ 0 & 0 & 1 \end{bmatrix}$;
Yes, they are inverses.

61. Product $= \begin{bmatrix} 438.227 & 488.299 & 387.747 & 404.496 & 415.265 & 362.598 \\ 341.127 & 470.054 & 321.549 & 396.768 & 391.883 & 369.323 \\ 296.377 & 380.795 & 260.023 & 314.025 & 285.172 & 266.426 \\ 327.277 & 399.097 & 293.671 & 332.085 & 327.514 & 297.122 \\ 508.767 & 538.464 & 438.303 & 446.607 & 446.589 & 392.418 \end{bmatrix}$

14.3 Exercises (Page 833)

1. $\begin{bmatrix} -1 & 2 & 1 \\ 3 & -4 & 2 \\ -4 & 0 & 1 \end{bmatrix}$ **3.** $\begin{bmatrix} -4 & 0 & 1 \\ -9 & 12 & -6 \\ -1 & 2 & 1 \end{bmatrix}$ **5.** $\begin{bmatrix} 18 & -1 & -1 \\ -4 & 0 & 1 \\ 1 & 3 & -3 \end{bmatrix}$ **7.** $\begin{bmatrix} -4 & 0 & 1 \\ 3 & -4 & 2 \\ -13 & 2 & 4 \end{bmatrix}$

9. $\begin{bmatrix} 1 & 11 & -5 \\ -1 & -4 & 3 \\ 0 & 5 & -2 \end{bmatrix}$ **11.** Interchange rows 1 and 2. **13.** Double each element of row 2. **15.** Multiply each element of row 1 by 2, and add the results to row 2. **17.** Multiply row 2 by -1. Multiply row 2 by -2 and add to row 1. **19.** Multiply row 1 by $-1/2$. Multiply row 1 by -4 and add to row 2. Multiply row 2 by $1/5$. Multiply row 2 by 1 and add to row 1. **21.** Multiply row 1 by $1/6$. Multiply row 2 by $1/4$. Multiply row 2 by $-1/2$ and add to row 1. Multiply row 3 by $1/2$. **23.** $\left[\begin{array}{cc|c} 2 & 3 & 11 \\ 1 & 2 & 8 \end{array}\right]$ **25.** $\left[\begin{array}{cc|c} 1 & 5 & 6 \\ 3 & -4 & 1 \end{array}\right]$

27. $\left[\begin{array}{ccc|c} 2 & 1 & 1 & 3 \\ 3 & -4 & 2 & -7 \\ 1 & 1 & 1 & 2 \end{array}\right]$ **29.** $\left[\begin{array}{ccc|c} 1 & 1 & 0 & 6 \\ 0 & 2 & 1 & 2 \\ 0 & 0 & 1 & 2 \end{array}\right]$ **31.** $\{(2, 3)\}$ **33.** $\{(-3, 0)\}$ **35.** $\{(7/2, -1)\}$

37. $\{(5, 0)\}$ **39.** $\emptyset$ **41.** $\{(-2, 1, 3)\}$ **43.** $\{(-1, 23, 16)\}$ **45.** $\{(3, 2, -4)\}$ **47.** $\{(-1, 2, 5, 1)\}$
49. A laborer charges \$76 per day and a concrete finisher charges \$80 per day. **51.** Dark clay costs \$5 per lb, light clay costs \$1 per lb. **53.** 74 decals and 96 bumper stickers were purchased.

14.4 Exercises (Page 843)

1. $\begin{bmatrix} 1 & 0 \\ 0 & -1 \end{bmatrix}$ **3.** $\begin{bmatrix} 3/2 & -1/2 \\ -2 & 1 \end{bmatrix}$ **5.** no inverse **7.** $\begin{bmatrix} 1 & -2 & 0 \\ 0 & 1 & 0 \\ -1 & 3 & 1 \end{bmatrix}$ **9.** $\begin{bmatrix} 4 & 3 & 3 \\ -1 & 0 & -1 \\ -4 & -4 & -3 \end{bmatrix}$

11. $(1, 1)$ **13.** There is no unique solution because the coefficient matrix has no inverse. **15.** $(-1, 23, 16)$

17. $(2, 1, -1)$ **19.** There is no unique solution because the coefficient matrix has no inverse. **21.** $\begin{bmatrix} 3 & -4 \\ 2 & -3 \end{bmatrix}$

23. The missing elements can be any two numbers with a product of -3. **25.** sometimes true (true whenever the numbers a and b are reciprocals of each other)

27. $\begin{bmatrix} k & 0 \\ 0 & 1 \end{bmatrix} \cdot \begin{bmatrix} 2 & 4 \\ 3 & 5 \end{bmatrix} = \begin{bmatrix} 2k + 0\cdot3 & 4k + 0\cdot5 \\ 0\cdot2 + 1\cdot3 & 0\cdot4 + 1\cdot5 \end{bmatrix} = \begin{bmatrix} 2k & 4k \\ 3 & 5 \end{bmatrix}$ (The first row is multiplied by k.)

29. $\begin{bmatrix} 1 & 0 \\ 2 & 1 \end{bmatrix} \cdot \begin{bmatrix} 2 & 4 \\ 3 & 5 \end{bmatrix} = \begin{bmatrix} 1\cdot2 + 0\cdot3 & 1\cdot4 + 0\cdot5 \\ 2\cdot2 + 1\cdot3 & 2\cdot4 + 1\cdot5 \end{bmatrix} = \begin{bmatrix} 2 & 4 \\ 2\cdot2 + 3 & 2\cdot4 + 5 \end{bmatrix}$ **33.** 1 **35.** 0

41. $\begin{bmatrix} .226 & -.152 \\ .197 & .049 \end{bmatrix}$ **43.** $\begin{bmatrix} -.871 & .602 & .594 \\ 1.905 & -1.511 & -1.047 \\ -1.774 & 1.590 & .876 \end{bmatrix}$ **45.** one; 0 **47.** 1,174 metric tons of wheat and 1,565 metric tons of oil **49.** 3,077 units of agriculture; 2,564 units of manufacturing; 3,179 units of transportation

Extension Exercises (Page 847)

1. $\begin{bmatrix} 20 \\ 8 \end{bmatrix} \begin{bmatrix} 5 \\ 27 \end{bmatrix} \begin{bmatrix} 20 \\ 8 \end{bmatrix} \begin{bmatrix} 18 \\ 5 \end{bmatrix} \begin{bmatrix} 5 \\ 27 \end{bmatrix} \begin{bmatrix} 20 \\ 5 \end{bmatrix} \begin{bmatrix} 14 \\ 15 \end{bmatrix} \begin{bmatrix} 18 \\ 19 \end{bmatrix}$ **3.** $\begin{bmatrix} -5 & -2 \\ -2 & -1 \end{bmatrix}$

5. $\begin{bmatrix} 103 \\ 118 \\ 116 \end{bmatrix} \begin{bmatrix} 95 \\ 122 \\ 98 \end{bmatrix} \begin{bmatrix} 42 \\ 46 \\ 47 \end{bmatrix} \begin{bmatrix} 103 \\ 130 \\ 104 \end{bmatrix} \begin{bmatrix} 114 \\ 119 \\ 141 \end{bmatrix} \begin{bmatrix} 65 \\ 66 \\ 84 \end{bmatrix} \begin{bmatrix} 120 \\ 132 \\ 145 \end{bmatrix} \begin{bmatrix} 87 \\ 89 \\ 105 \end{bmatrix} \begin{bmatrix} 63 \\ 74 \\ 68 \end{bmatrix} \begin{bmatrix} 176 \\ 203 \\ 203 \end{bmatrix}$

7. 51, 89, 55, 88, 177, 108, 89, 201, 114, 40, 90, 49, 83, 192, 102, 49, 106, 62, 19, 51, 23, 69, 116, 89, 131, 264, 156

14.5 Exercises (Page 854)

1. $\begin{bmatrix} 4 & -1 \\ 3 & 5 \end{bmatrix}$ **3.** $\begin{bmatrix} 8 & -7 \\ -2 & 4 \end{bmatrix}$ **5.** (3, I); 3 **7.** (2, III); 2 **9.** (2, III); 0 **11.** **(a)** Set up stadium **(b)** Set up gym **(c)** Set up both **13.** **(a)**

	Better	**Not**
Market New	50,000	−25,000
Don't	−40,000	−10,000

(b) \$5,000 and −\$22,000; Market new **15.** 1: 1/5; 2: 4/5; I: 3/5; II: 2/5; value = 17/5
17. The saddle point gives a value of 0. The game is fair. The first player should select strategy (row) 2, while the second player should select strategy (column) I. **19.** Both A and B should choose either strategy with a probability of 1/2. The value of the game is 0, a fair game. **25.** **(a)** \$7,388.89 **(b)** \$7,388.89
29. $7{,}000x + 7{,}000$ **31.** Nature should impose a freeze 1/18 of the time. **33.** strategy 2 **35.** greater than 3/16 **37.** \$7,187.50

Chapter 14 Test (Page 858)

1. $\begin{bmatrix} -1 & -1 & 4 \\ 2 & 2 & 0 \\ 3 & 1 & 6 \end{bmatrix}$ **2.** $\begin{bmatrix} 5 & -1 & 2 \\ 2 & -2 & 2 \\ 5 & 1 & 0 \end{bmatrix}$ **3.** $\begin{bmatrix} -9 & -2 & 12 \\ -7 & 0 & 5 \\ -15 & 2 & 12 \end{bmatrix}$ **4.** $\begin{bmatrix} -2 & 4 & -6 \\ 0 & -1 & -1 \\ 10 & 4 & 6 \end{bmatrix}$ **5.** $\begin{bmatrix} 4 & -2 & 6 \\ 4 & 0 & 2 \\ 8 & 2 & 6 \end{bmatrix}$

6. $\begin{bmatrix} 11 & -1 & 0 \\ 2 & -6 & 4 \\ 7 & 1 & -6 \end{bmatrix}$ **7.** $\begin{bmatrix} 2 & -1 & 3 \\ 4 & 1 & 3 \\ 2 & 0 & 1 \end{bmatrix}$ **8.** $\begin{bmatrix} -4 & -1 & 0 \\ 2 & 0 & 1 \\ 4 & 1 & 3 \end{bmatrix}$ **9.** $\begin{bmatrix} -3 \\ 10 \end{bmatrix}$ **10.** $\begin{bmatrix} 6 & 0 & 12 \\ 26 & 0 & 20 \end{bmatrix}$

11. [16] **12.** There is no product, since the first matrix has a different number of columns than the second has rows. **13.** $\begin{bmatrix} 3 & -1 \\ -5 & 2 \end{bmatrix}$ **14.** $\begin{bmatrix} 1/2 & 1/2 & -1/2 \\ -1/2 & 1/6 & 1/6 \\ 1/2 & -1/2 & 1/2 \end{bmatrix}$ **15.** $\left[\begin{array}{cc|c} 9 & -4 & 20 \\ 5 & 10 & 12 \end{array}\right]$ **16.** $\left[\begin{array}{ccc|c} 2 & 1 & -6 & 6 \\ 5 & 0 & 12 & 15 \\ 0 & 6 & 3 & 8 \end{array}\right]$

17. {(3, 0)} **18.** {(4, −1, 2)} **19.** **(a)** $\begin{bmatrix} 1 & 1 \\ 2 & -1 \end{bmatrix}$ **(b)** $\begin{bmatrix} x \\ y \end{bmatrix}$ **(c)** $\begin{bmatrix} 1 \\ -7 \end{bmatrix}$ **20.** $X = A^{-1}B$ **21.** {(−2, 3)}

23. $A = \begin{array}{l} \\ \text{Bananas} \\ \text{Coal} \end{array} \begin{array}{c} \begin{array}{cc} \text{Bananas} & \text{Coal} \end{array} \\ \begin{bmatrix} .3 & .2 \\ .1 & .4 \end{bmatrix} \end{array}$ **24.** 750 units of bananas, 625 units of coal **25.** $\begin{bmatrix} 8 & -3 \\ 0 & 2 \end{bmatrix}$

26. 9 **27.** Select strategy one 9/10 of the time and strategy two 1/10 of the time; value is −31/10.

28. 4 and −4, or −4 and 4

29. $M = \begin{array}{l} \\ \text{Ring} \\ \text{Bracelet} \\ \text{Necklace} \end{array} \begin{array}{c} \begin{array}{cc} \text{Silver} & \text{Turquoise} \end{array} \\ \begin{bmatrix} 1.6 & .7 \\ 5.8 & 7.5 \\ 26.3 & 18.8 \end{bmatrix} \end{array}; \quad C = \begin{array}{l} \\ \text{Silver} \\ \text{Turquoise} \end{array} \begin{array}{c} \text{Cost per unit} \\ \begin{bmatrix} 5.50 \\ 2.10 \end{bmatrix} \end{array}$

30. $MC = \begin{bmatrix} 1.6 & .7 \\ 5.8 & 7.5 \\ 26.3 & 18.8 \end{bmatrix} \cdot \begin{bmatrix} 5.50 \\ 2.10 \end{bmatrix} = \begin{array}{l} \\ \text{Ring} \\ \text{Bracelet} \\ \text{Necklace} \end{array} \begin{array}{c} \text{Cost per item} \\ \begin{bmatrix} 10.27 \\ 47.65 \\ 184.13 \end{bmatrix} \end{array}$

Appendix The Metric System

Appendix Exercises (Page 864)

1. 8,000 mm **3.** 85 m **5.** 689 mm **7.** 5.98 cm **9.** 5,300 m **11.** 27.5 km **13.** 2.54 cm; 25.4 mm **15.** 5 cm; 50 mm **17.** 600 cl **19.** 8,700 ml **21.** 9.25 L **23.** 8.974 L **25.** 8 kg **27.** 5,200 g **29.** 4,200 mg **31.** .598 g **33.** 30°C **35.** −81°C **37.** 50°F **39.** −40°F **41.** 200 nickels **43.** .2 g **45.** $40.99 **47.** $387.98 **49.** 1,082,400 cm^3; 1.0824 m^3 **51.** 200 bottles **53.** 897.9 m **55.** 201.1 km **57.** 3.995 lb **59.** 21,428.8 g **61.** 30.22 qt **63.** 106.7 L **65.** unreasonable **67.** reasonable **69.** unreasonable **71.** (b) **73.** (b) **75.** (c) **77.** (a) **79.** (a) **81.** (c) **83.** (a) **85.** (b) **87.** (b) **89.** (a) **91.** (b) **93.** (c) **95.** (b) **97.** (b)

Acknowledgments

Literary Permissions

Chapter 1

21 From *How to Solve It: A New Aspect of Mathematical Method* by George Polya. Copyright 1945, © 1973 renewed by Princeton University Press. Excerpt reprinted with permission of Princeton University Press.

34 (Figure 6a) Graph, "The U.S. Economy is Growing" from *The New York Times,* February 27, 1994. Copyright © 1994 by the New York Times Company. Reprinted by permission.

34 (Figure 6c) Graph, "Dow Jones Last 10 Days" from *New Orleans Times-Picayune,* May 30, 1992. Reprinted by permission.

Chapter 3

107 (Adaptation) From "Raymond Smullyan" by Ira Mothner, *Mathematical People.* Copyright © 1985 by Birkhauser Boston. Reprinted by permission.

145 (Excerpt) From "Education Helps Control Alcoholism" by Sue Rusche, *The Sacramento Union,* July 9, 1988. Copyright © 1988 by The Sacramento Union. Reprinted by permission.

Chapter 4

194–195 (Table) "Months with a Friday the 13th" from "An Aid to the Superstitious" by G. L. Ritter, S. R. Lowry, H. B. Woodruff, and T. L. Isenhour, *The Mathematics Teacher,* May 1977. Copyright © 1977 The National Council of Teachers of Mathematics, Inc. Reprinted by permission.

Chapter 5

219–220 This program was written by Charles W. Gantner and is provided courtesy of Texas Instruments.

224 Six Quotations (pp. 134, 144, 169, 177, 189, 191) from *The Penguin Dictionary of Curious and Interesting Numbers* by David Wells (Penguin Books, 1986) copyright © David Wells, 1986.

228 Except from "A Taylor-made Plug for Wiles' Proof" from *The College Mathematics Journal,* March 1995, pp. 100–105. Reprinted by permission of the Mathematical Association of America.

231 This program was written by Tim Tilton and is provided courtesy of Texas Instruments.

Chapter 9

556–557 (Figure) From "Möbius Molecule: Synthesis with a Twist" by David Walba, *Science News,* Vol. 122, July 17, 1982, p. 36. Reprinted by permission of the author.

568–571 (Exercises 1–25) Pages 1–4 from *National Council of Teachers of Mathematics Student Math Notes,* November 1991. Reprinted by permission of the National Council of Teachers of Mathematics.

Chapter 10

581 (Map) From "Ride with the Wind" by Linda Garmon, *Science News,* Vol. 119, June 27, 1981. Reprinted by permission of Wind Ship Development Corporation.

Chapter 11

630 (Exercise 74) Problem #6 from *Mathematics Teacher,* January 1989, p. 38. Reprinted by permission.

630 (Exercise 75) Problem #2 from *Mathematics Teacher,* November 1989, p. 626. Reprinted by permission.

638 (Exercise 43) Problem #8 from *Mathematics Teacher,* December 1992, p. 736. Reprinted by permission.

Chapter 12

680 (Exercises 11–13) Adapted graph "The Growth of the Suburbs and the Black Migration to the Cities" from *The 1993 World Book Year Book.* Copyright © 1993 World Book, Inc. By permission of the publisher.
680 (Exercises 14–21) Adapted graph "U.S. Trade with Canada and Mexico" from *The 1993 World Book Year Book.* Copyright © 1993 World Book, Inc. By permission of the publisher.
681 (Exercises 22–23) Graph "The Decline in Federal Urban Aid . . ." from *The 1993 World Book Year Book.* Copyright © 1993 World Book, Inc. By permission of the publisher.
682 (Exercises 26–29) Graph "Retirement Savings Net Worth" from *TSA Guide to Retirement Planning for California Educators,* Winter 1993. Reprinted by permission of John L. Kattman.
727–734 From *How to Lie with Statistics* by Darrell Huff, illustrated by Irving Geis. Copyright 1954 and renewed © 1982 by Darrell Huff and Irving Geis. Reprinted by permission of W. W. Norton & Company, Inc.
743–744 (Exercises 1–4) Adapted graph "Selected U.S. Economic Indicators" from *The 1993 World Book Year Book.* Copyright © 1993 World Book, Inc. By permission of the publisher.

Chapter 13

750 Originally broadcast over the CBS Radio Network as part of the Hourly News by Charles Osgood on March 2, 1977. Copyright © 1977 CBS, Inc. All Rights Reserved. Reprinted by permission.
795 (Table) From New York Stock Exchange, *The Wall Street Journal,* July 14, 1995. Copyright © 1995 Dow Jones and Company, Inc. All Rights Reserved Worldwide. Reprinted by permission of the Wall Street Journal.

Photo Credits

Unless otherwise acknowledged, all photographs are the property of Scott, Foresman and Company. Page abbreviations are as follows: (T)top, (C)center, (B)bottom, (L)left, (R)right.

Front End Sheets

Heliostat, Peter Menzel/Stock Boston
DNA sequences, Hank Morgan/SS/Photo Researchers
Sunbathers, Craig Newbauer/Peter Arnold, Inc.
Computerized rotary motor, Greenlar/Image Works
Single desktop computer, Tim Davis/Photo Researchers
Central computer, Andy Bernhaut/Photo Researchers
Bar code scanner, Charles Gupton/Stock Boston
Money globe, Superstock, Inc.

Back End Sheets

Deforestation, Gregory G. Dimijian/Photo Researchers
Maxwell Smart, Photofest
Retired couple, Will & Deni McIntyre/Photo Researchers
Cat scan, Simon Fraser/SPL/Photo Researchers
Genetic engineer, Geoff Tompkinson/SPL/Photo Researchers
Bicyclist & female smokers, Superstock, Inc.

Chapter 1

13 Corbis-Bettmann Archive
21 Courtesy, Stanford University News Service
31(T) Courtesy, Texas Instruments Incorporated
41 Library of Congress

Chapter 2

53 Baveria-Verlag
57 Corbis-UPI/Bettmann
84 Hans Reinhard/Bruce Coleman Inc.
88 Superstock, Inc.

Chapter 3

99 Corbis-Bettmann Archive
101 Erich Lessing/Art Resource

109, 112 Library of Congress
117 Photofest
133 Corbis-Bettmann Archive
145 Eveready Battery Company, Inc.

Chapter 4

153 John Buitenkant/Photo Researchers
154(B) Bridgeman/Art Resource
154(T) Courtesy, Prindle, Weber & Schmidt, Inc.
155 Egyptian Expedition/Rogers Fund/Metropolitan Museum of Art
156 Owen Franken/Stock Boston
158 Trans. #5001(3) American Museum of Natural History
160 British Museum
165 Trinity College, Cambridge
167(B) Courtesy, IBM Corporation
174 Corbis-Bettmann Archive
179 Registered trademarks of Kellogg Company
187 Tom Wurt/Stock Boston
205 Courtesy Professor Noether

Chapter 5

218 H. C. Williams
228 Courtesy, Princeton University
232 Giraudon/Art Resource
240 © The Walt Disney Company
241(L) Roy Morsch/Stock Market
242(L) Art Resource
243 James Randkley/Tony Stone Images
247(L) Corbis-Bettmann Archive
247(R) Benjamin Franklin Collection/Sterling Memorial Library/Yale University
248 Stadtmuseum, Aachen/Superstock, Inc.
249 Metropolitan Museum of Art/Harris Brisbane Dick Fund, 1943

Chapter 6

262 Times-Picayune
282 Focus on Sports, Inc.
301 © 1987 Universal Press Syndicate. Reprinted with permission. All rights reserved.
302 John Hornsby
305 Carl Purcell/Photo Researchers

Chapter 7

347 David R. Frazier Photolibrary
350 Photo Collection/California Institute of Technology Archives Archives
392 Nigel Chalcraft/Check Six

Chapter 8

406 David R. Frazier Photolibrary
413 William Johnson/Stock Boston
439 National Optical Astronomy Observatories
480 Stanford University News Service

Chapter 9

488 From Eve's History
502 Alinari/Art Resource
504 Corbis-Bettmann Archive
516 Tony Arruza/Tony Stone Images
522 George Arthur Plimpton Collection, Columbia University
525 Peter French/DRK Photo
529 Gay Bumgarner/Tony Stone Images
533 Mount Rushmore National Park, U.S. Dept. of the Interior, National Park Service/Photo: Paul Harsled
543(TR) From *On Growth and Form* by Arch Wentworth Thompson. Cambridge University Press. Abridged edition edited by John Tyler Bonner
543(TL) 2072 from *On Growth and Form* by Arch Wentworth Thompson. Cambridge University Press. Abridged edition edited by John Tyler Bonner.
548 Superstock, Inc.
552 Corbis-Hulton/UPI Bettmann
554 Superstock, Inc.
556 Corbis-Bettmann Archive
567(TL & TR) Courtesy of Philip P. Marcus
567(BL) NASA
568 Stephen Johnson/Tony Stone Images

Chapter 10

577 British Museum
587 National Museum of American Art, Washington, D.C./Art Resource
596 Washington National Cathedral

Chapter 11

632 Stock Montage, Inc./Historical Pictures Service
640 Kobal Collection/Superstock, Inc.
642 JPL/NASA
649 Museum fur Volkerlunde und Schweizerishches
657 Superstock, Inc.
659 From David G. Christensens' *Slot Machines, A Pictorial History 1889–1972*

Chapter 12

671(T) Drawn by SYMVU, Harvard University Software Program, Laboratory for Computer Graphics and Spatial Analysis, Harvard Graduate School of Design
705 Sovfoto
714 AP/Wide World
735 Library of Congress

Chapter 13

748 Oriental Institute Museum, The Univ. of Chicago
757 Brown Brothers
766(R) Courtesy NOVUS Services, Inc.
767 Wells Fargo Bank
782 Cameramann International, Ltd.
796 David R. Frazier Photolibrary

Chapter 14

836 Stock Montage, Inc.
837 Granger Collection
853 Granger Collection

Appendix

861 Cameramann International, Ltd.

dex

Pronunciation Guide

a	**a**ct, b**a**t	j	**j**ust, fu**dg**e	Œ	as in German sch**ö**n or in French f**eu**
ā	c**a**pe, w**ay**	k	**k**eep, to**k**en	R	rolled r as in French **r**ouge or in German **r**ot
â	d**a**re, M**a**ry	KH	as in Scottish lo**ch** or in German i**ch**	sh	**sh**oe, fi**sh**
ä	**a**lms, c**a**lm	N	as in French bo**n** or u**n**	th	**th**in, pa**th**
ch	**ch**ild, bea**ch**	o	**o**x, w**a**sp	u	**u**p, l**o**ve
e	s**e**t, m**e**rry	ō	**o**ver, n**o**	û	**u**rge, b**u**rn
ē	**e**qual, b**ee**	o͝o	b**oo**k, p**oo**r	y	**y**es, on**i**on
@	like a in **a**lone or e in syst**e**m	oo	**oo**ze, f**oo**l	z	**z**eal, la**z**y
g	**g**ive, be**g**	ô	**ou**ght, r**aw**	ZH	trea**s**ure, mira**g**e
i	**i**f, b**i**g	oi	**oi**l, j**oy**		
ī	**i**ce, b**i**te	ou	**ou**t, c**ow**		

A

D

E

F

P

Q

R

T

Rain forest deforestation, an example of which is shown here, is one factor contributing to the increase of atmospheric carbon dioxide. See page 400 to see how functions can be used to model the buildup of carbon dioxide in the atmosphere and the corresponding threat of global warming.

Geometry is used by researchers developing a new imaging technique that will produce dynamic images of physiological processes, unlike the CAT scanner shown here which produces only static images. See page 487 for more information.

Counting methods can be used to determine the number of possible DNA sequences there are for a given species. Such information is important for genetic engineers such as the one shown here investigating recombinant protein. See page 575.

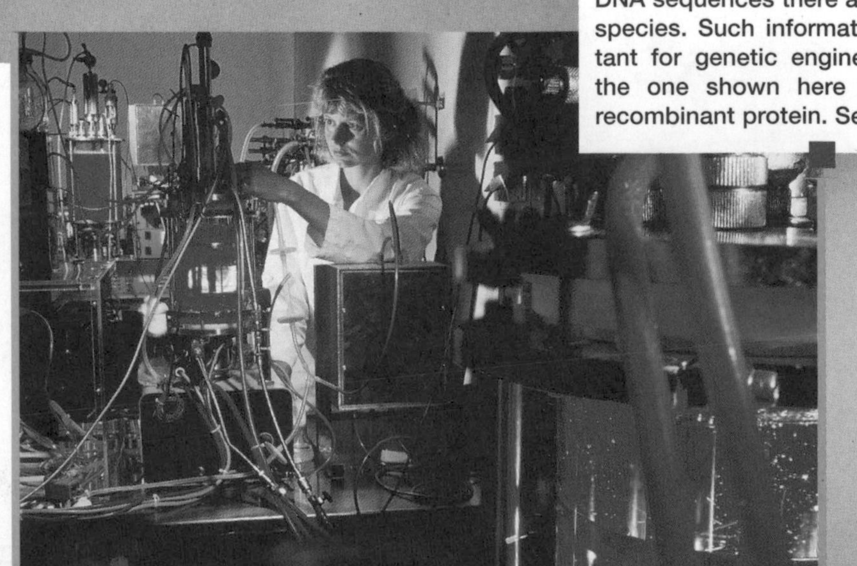